ISBN 978-0-265-98915-9
PIBN 10919782

DEPARTMENT OF THE INTERIOR
ALBERT B. FALL, Secretary

UNITED STATES GEOLOGICAL SURVEY
GEORGE OTIS SMITH, Director

Water-Supply Paper 471

SURFACE WATER SUPPLY OF THE UNITED STATES
1918

PART I. NORTH ATLANTIC SLOPE DRAINAGE BASINS

NATHAN C. GROVER, Chief Hydraulic Engineer
C. H. PIERCE, C. C. COVERT, and G. C. STEVENS, District Engineers

Prepared in cooperation with the States of
MAINE, VERMONT, MASSACHUSETTS, and NEW YORK

WASHINGTON
GOVERNMENT PRINTING OFFICE
1921

CONTENTS.

ILLUSTRATIONS.

SURFACE WATER SUPPLY OF THE NORTH ATLANTIC SLOPE DRAINAGE BASINS, 1918.

AUTHORIZATION AND SCOPE OF WORK.

This volume is one of a series of 14 reports presenting results of measurements of flow made on streams in the United States during the year ending September 30, 1918.

The data presented in these reports were collected by the United States Geological Survey under the following authority contained in the organic law (20 Stat. L., p. 394):

Provided, That this officer [the Director] shall have the direction of the Geological Survey and the classification of public lands and examination of the geological structure, mineral resources, and products of the national domain.

The work was begun in 1888 in connection with special studies relating to irrigation in the arid West. Since the fiscal year ending June 30, 1895, successive sundry civil bills passed by Congress have carried the following item and appropriations:

For gaging the streams and determining the water supply of the United States, and for the investigation of underground currents and artesian wells, and for the preparation of reports upon the best methods of utilizing the water resources.

Annual appropriations for the fiscal years ending June 30, 1895–1919.

1895	$12,500.00
1896	20,000.00
1897 to 1900, inclusive	50,000.00
1901 to 1902, inclusive	100,000.00
1903 to 1906, inclusive	200,000.00
1907	150,000.00
1908 to 1910, inclusive	100,000.00
1911 to 1917, inclusive	150,000.00
1918	175,000.00
1919	148,244.10

In the execution of the work many private and State organizations have cooperated, either by furnishing data or by assisting in collecting data. Acknowledgments for cooperation of the first kind are made in connection with the description of each station affected; cooperation of the second kind is acknowledged on page 11.

Measurements of stream flow have been made at about 4,510 points in the United States and also at many points in Alaska and the Hawaiian Islands. In July, 1918, 1,180 gaging stations were being maintained by the Survey and the cooperating organizations. Many miscellaneous discharge measurements are made at other points. In connection with this work data were also collected in regard to precipitation, evaporation, storage reservoirs, river profiles, and water power in many sections of the country and will be made available in water-supply papers from time to time. Information in regard to publications relating to water resources is presented in the appendix to this report.

DEFINITION OF TERMS.

The volume of water flowing in a stream—the "run-off" or "discharge"—is expressed in various terms, each of which has become associated with a certain class of work. These terms may be divided into two groups—(1) those that represent a rate of flow, as second-feet, gallons per minute, miners' inches, and discharge in second-feet per square mile, and (2) those that represent the actual quantity of water, as run-off in depth in inches, acre-feet, and millions of cubic feet. The principal terms used in this series of reports are second-feet, second-feet per square mile, run-off in inches, and acre-feet. They may be defined as follows:

"Second-feet" is an abbreviation for "cubic feet per second." A second-foot is the rate of discharge of water flowing in a channel of rectangular cross section 1 foot wide and 1 foot deep at an average velocity of 1 foot per second. It is generally used as a fundamental unit from which others are computed.

"Second-feet per square mile" is the average number of cubic feet of water flowing per second from each square mile of area drained, on the assumption that the run-off is distributed uniformly both as regards time and area.

"Run-off (depth in inches)" is the depth to which an area would be covered if all the water flowing from it in a given period were uniformly distributed on the surface. It is used for comparing run-off with rainfall, which is usually expressed in depth in inches.

An "acre-foot," equivalent to 43,560 cubic feet, is the quantity required to cover an acre to the depth of 1 foot. The term is commonly used in connection with storage for irrigation.

The following terms not in common use are here defined:

"Stage-discharge relation;" an abbreviation for the term "relation of gage height to discharge."

"Control;" a term used to designate the section or sections of the stream below the gage which determine the stage-discharge relation at the gage. It should be noted that the control may not be the same section or sections at all stages.

The "point of zero flow" for a gaging station is that point on the gage—the gage height—to which the surface of the river would fall if were no flow.

EXPLANATION OF DATA.

The data presented in this report cover the year beginning October 1, 1917, and ending September 30, 1918. At the beginning of January in most parts of the United States much of the precipitation in the preceding three months is stored as ground water, in the form of snow or ice, or in ponds, lakes, and swamps, and this stored water passes off in the streams during the spring break-up. At the end of September, on the other hand, the only stored water available for run-off is possibly a small quantity in the ground; therefore the run-off for the year beginning October 1 is practically all derived from precipitation within that year.

The base data collected at gaging stations consist of records of stage, measurements of discharge, and general information used to supplement the gage heights and discharge measurements in determining the daily flow. The records of stage are obtained either from direct readings on a staff gage or from a water-stage recorder that gives a continuous record of the fluctuations. Measurements of discharge are made with a current meter. (See Pls. I, II.) The general methods are outlined in standard textbooks on the measurement of river discharge.

From the discharge measurements rating tables are prepared that give the discharge for any stage, and these rating tables, when applied to the gage heights, give the discharge from which the daily, monthly, and yearly mean discharge is determined.

The data presented for each gaging station in the area covered by this report comprise a description of the station, a table giving results of discharge measurements, a table showing the daily discharge of the stream, and a table of monthly and yearly discharge and run-off.

If the base data are insufficient to determine the daily discharge, tables giving daily gage heights and results of discharge measurements are published.

The description of the station gives, in addition to statements regarding location and equipment, information in regard to any conditions that may affect the constancy of the stage-discharge relation, covering such subjects as the occurrence of ice, the use of the stream for log driving, shifting of control, and the cause and effect of backwater; it gives also information as to diversions that decrease the flow at the gage, artificial regulation, maximum and minimum recorded stages, and the accuracy of the records.

The table of daily discharge gives, in general, the discharge in second-feet corresponding to the mean of the gage heights read each day. At stations on streams subject to sudden or rapid diurnal fluctuations the discharge obtained from the rating table and the mean daily gage height may not be the true mean discharge for the day.

If such stations are equipped with water-stage recorders the mean daily discharge may be obtained by averaging discharge at regular intervals during the day or by using the discharge integrator, an instrument operating on the principle of the planimeter and containing as an essential element the rating curve of the station.

In the table of monthly discharge the column headed "Maximum" gives the mean flow for the day when the mean gage height was highest. As the gage height is the mean for the day it does not indicate correctly the stage when the water surface was at crest height and the corresponding discharge was consequently larger than given in the maximum column. Likewise, in the column headed "Minimum" the quantity given is the mean flow for the day when the mean gage height was lowest. The column headed "Mean" is the average flow in cubic feet for each second during the month. On this average flow computations recorded in the remaining columns, which are defined on page 8, are based.

ACCURACY OF FIELD DATA AND COMPUTED RESULTS.

The accuracy of stream-flow data depends primarily (1) on the permanence of the stage-discharge relation and (2) on the accuracy of observation of stage, measurements of flow, and interpretation of records.

A paragraph in the description of the station or footnotes added to the tables gives information regarding (1) the permanence of the stage-discharge relation, (2) precision with which the discharge rating curve is defined, (3) refinement of gage readings, (4) frequency of gage readings, and (5) methods of applying daily gage heights to the rating table to obtain the daily discharge.[1]

For the rating tables "well defined" indicates, in general, that the rating is probably accurate within 5 per cent; "fairly well defined," within 10 per cent; "poorly defined," within 15 to 25 per cent. These notes are very general and are based on the plotting of the individual measurements with reference to the mean rating curve.

The monthly means for any station may represent with high accuracy the quantity of water flowing past the gage, but the figures showing discharge per square mile and depth of run-off in inches may be subject to gross errors caused by the inclusion of large noncontributing districts in the measured drainage area, by lack of information concerning water diverted for irrigation or other use, or by inability to interpret the effect of artificial regulation of the flow of the river above the station. "Second-feet per square mile" and "Run-off (depth in inches)" are therefore not computed if such errors appear probable. The computations are also omitted for

[1] For a more detailed discussion of the accuracy of stream-flow data see Grover, N. C., and Hoyt, J. C., [Accu]racy of stream-flow data: U. S. Geol. Survey Water-Supply Paper 400, pp. 53-59, 1916.

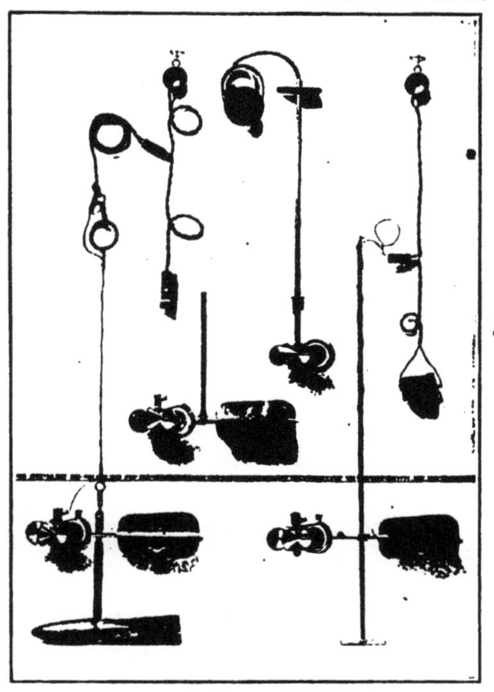

A. PRICE CURRENT METERS.

B. TYPICAL GAGING STATION.

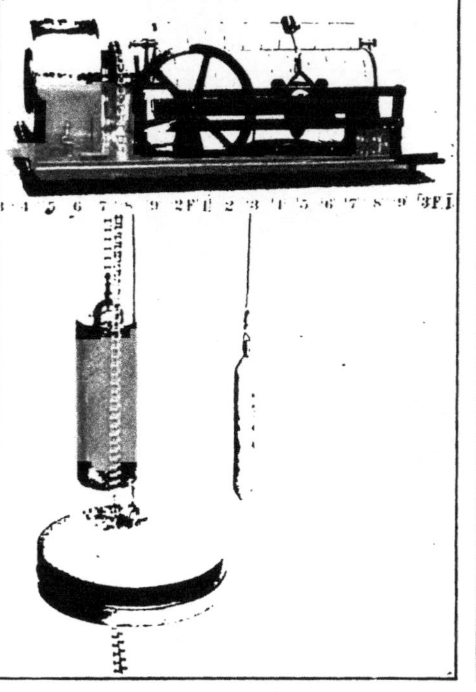

A. STEVENS CONTINUOUS.

B. GURLEY PRINTING.

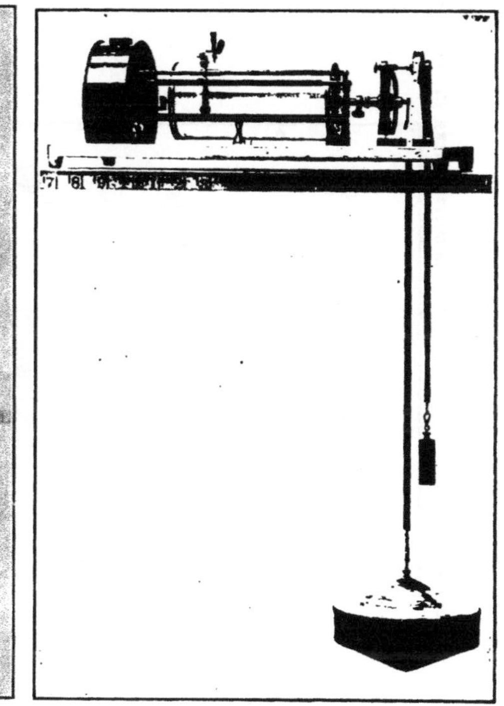

C. FRIEZ.

WATER-STAGE RECORDERS.

stations on streams draining areas in which the annual rainfall is less than 20 inches. All figures representing "second-feet per square mile" and "run-off (depth in inches)" previously published by the Survey should be used with caution because of possible inherent but unknown sources of error.

The table of monthly discharge gives only a general idea of the flow at the station and should not be used for other than preliminary estimates; the tables of daily discharge allow more detailed studies of the variation in flow. It should be borne in mind, however, that the observations in each succeeding year may be expected to throw new light on data previously published.

COOPERATION.

The hydrometric work in Maine was carried on in cooperation with the public utilities commission, Benjamin F. Cleaves, chairman, and Paul L. Bean, chief engineer.

In Vermont the work was carried on in cooperation with the State, Horace F. Graham, governor, and Herbert M. McIntosh, State engineer.

The work in New Hampshire was done in cooperation with the commission on water conservation and water power, George B. Leighton, commissioner.

The work in Massachusetts was carried on in cooperation with the commission on waterways and public lands, John N. Cole, chairman.

Financial assistance has been rendered by the New England Power Co., the Turners Falls Power & Electric Co., the Connecticut Valley Lumber Co., the Holyoke Water Power Co., the International Paper Co., the Connecticut Power Co., the Eastern Connecticut Power Co., Profile Falls Power Co., and the W. H. McElwain Co.

Work in the State of New York has been conducted under cooperative agreements with the State engineer and surveyor and, since July 1, 1911, with the division of waters of the State conservation commission.

The water-stage recorder on Hudson River at Spier Falls, N. Y., was inspected by an employee of the Adirondack Electric Power Corporation, Glens Falls, N. Y.

The station on Rappahannock River near Fredericksburg, Va., was maintained in cooperation with the Spottsylvania Power Co.

DIVISION OF WORK.

The data for stations in New England were collected and prepared for publication under the direction of C. H. Pierce, district engineer. The work in Maine was under the immediate supervision of A. F. McAlary, assistant engineer of the public utilities commission, who was assisted by H. A. Lancaster. The other assistants in New England were O. W. Hartwell, H. W. Fear, M. R. Stackpole, J. W. Moulton, A. N. Weeks, and Hope Hearn.

Data for stations in New York were collected and prepared for publication under the direction of C. C. Covert, district engineer, who was assisted by O. W. Hartwell, E. D. Burchard, A. H. Davison, W. A. James, and Helen Kimmey.

For stations in New Jersey, Maryland, and Virginia, the data were collected and prepared for publication under the direction of G. C. Stevens, district engineer, who was assisted by H. J. Jackson, B. L. Hopkins, M. I. Walters, and J. W. Moulton.

GAGING-STATION RECORDS.

ST. JOHN RIVER BASIN.

ST. JOHN RIVER AT VAN BUREN, MAINE.

LOCATION.—At new international bridge at Van Buren, Aroostook County, about 14 miles above Grand Falls.

DRAINAGE AREA.—8,270 square miles.

RECORDS AVAILABLE.—May 4, 1908, to September 30, 1918.

GAGE.—Gage used since May 6, 1912, painted vertically on second pier from Van Buren end of bridge; zero of gage, 407.69 feet above sea level. From 1908 to 1911 stage was read on a vertical rod attached to pier of sawdust carrier of Hammond's mill, about 700 feet below international bridge, but as published, readings are reduced to datum of bridge gage. Gage read by W. H. Scott.

DISCHARGE MEASUREMENTS.—Made from international bridge.

CHANNEL AND CONTROL.—Control practically permanent. Banks high, rocky, cleared, and not subject to overflow except in very high freshets.

EXTREMES OF DISCHARGE.—Maximum stage recorded during year, 24.5 feet at 8.10 a. m. May 2 (discharge, 104,000 second-feet); minimum stage recorded, 1.45 feet at 6.30 a. m. October 1 (discharge, 1,820 second-feet). Discharge estimated at 1,520 second-feet several times in February and March (stage-discharge relation affected by ice).

ICE.—Stage-discharge relation seriously affected by ice, usually from December to March; estimates based on gage heights at Grand Falls and rating curve derived from measurements at Van Buren.

REGULATION.—The little storage above for log driving probably does not materially affect the flow.

ACCURACY.—Stage-discharge relation practically permanent except when affected by ice. Rating curve well defined. Gage read to tenths twice daily. Daily discharge ascertained by applying mean daily gage height to rating table. Records good.

COOPERATION.—Winter-gage heights at Grand Falls furnished by H. S. Ferguson, consulting engineer.

No discharge measurements were made at this station during the year ending September 30, 1918.

Daily discharge, in second-feet, of St. John River at Van Buren, Maine, for the year ending Sept. 30, 1918.

Day.	Oct.	Nov.	Dec.	Jan.	Feb.	Mar.	Apr.	May.	June.	July.	Aug.	Sept.
1	1,820	47,000	6,820	2,700	1,880	1,640	2,580	87,500	22,900	17,200	24,400	2,880
2	2,700	52,500	6,570	2,700	1,880	1,640	3,580	104,000	21,800	15,700	25,500	2,880
3	3,060	46,000	6,840	2,580	1,690	1,640	4,080	102,000	20,800	15,700	21,200	2,880
4	3,640	38,200	7,110	2,580	1,690	1,640	4,990	94,000	19,500	15,100	16,900	3,060
5	4,700	32,300	6,700	2,460	1,690	1,640	5,840	81,000	18,200	13,600	14,200	3,250
6	5,140	27,800	6,190	2,460	1,690	1,640	7,400	69,000	16,900	12,200	12,500	3,440
7	6,760	24,700	6,070	2,240	1,640	1,520	10,100	58,600	16,000	11,100	11,600	4,920
8	8,980	22,600	5,400	2,240	1,640	1,520	12,900	57,500	15,700	10,300	11,100	6,050
9	9,760	20,800	5,090	2,360	1,640	1,520	14,100	63,000	15,700	12,200	11,400	6,760
10	9,500	19,500	4,990	2,360	1,640	1,520	14,500	66,000	15,400	26,200	10,800	7,240
11	8,470	17,900	4.800	2,360	1,560	1,520	14,900	67,800	15,400	46,500	10,000	6,050
12	7,970	17,200	4,600	2,360	1,560	1,520	15,300	67,800	13,900	46,500	8,470	5,360
13	8,220	15,700	4,330	2,360	1,600	1,520	17,400	61,900	13,600	41,000	7,480	4,480
14	8,470	13,900	4,240	2,240	1,560	1,520	20,500	59,200	14,500	35,400	7,240	4,480
15	7,970	12,800	4,240	2,300	1,690	1,560	23,700	61,900	17,900	31,100	6,760	5,360
16	8,720	12,200	4,160	2,240	1,640	1,520	25,600	59,200	18,800	31,900	6,050	7,480
17	9,500	11,900	4,240	2,140	1,640	1,520	35,200	53,500	16,600	32,300	5,590	11,100
18	9,240	12,200	3,990	2,140	1,640	1,520	42,500	48,000	13,900	31,500	5,820	10,300
19	10,000	12,200	3,900	2,140	1,520	1,520	37,700	43,500	12,500	29,800	5,820	10,000
20	10,000	11,600	3,900	2,140	1,520	1,640	34,500	41,000	11,100	27,000	5,820	11,400
21	9,500	11,000	3,990	2,140	1,520	1,780	32,300	39,600	10,000	24,000	5,360	13,600
22	8,470	10,200	4,080	2,140	1,520	1,780	31,900	38,600	9,240	20,800	4,920	15,100
23	7,970	8,790	3,900	2,080	1,520	1,640	34,500	38,600	9,760	19,500	4,700	18,800
24	7,970	8,960	3,740	2,030	1,520	1,520	40,600	37,200	17,200	18,200	4,050	20,800
25	7,720	8,150	3,500	2,030	1,640	1,520	48,000	35,000	20,400	16,900	3,840	18,500
26	9,500	6,320	3,580	2,030	1,640	1,520	50,000	32,800	32,300	15,700	3,840	15,700
27	10,800	4,800	3,580	2,030	1,690	1,560	49,500	29,000	27,000	15,100	3,440	13,600
28	14,500	4,510	3,280	1,930	1,640	1,690	50,000	26,600	22,200	16,000	3,440	13,600
29	15,400	4,990	2,840	1,930	....	1,930	53,000	26,200	19,500	14,800	3,250	16,900
30	15,700	5,840	2,840	1,930	....	2,030	63,600	27,000	17,200	13,300	3,250	21,800
31	24,400	....	2,700	1,930	....	2,300	....	25,100	....	14,800	3,060	....

NOTE.—Stage-discharge relation affected by ice Nov. 23 to Apr. 17; discharge for this period determined from gage heights at Grand Falls and rating curve derived from measurements at Van Buren.

Monthly discharge of St. John River at Van Buren, Maine, for the year ending Sept. 30, 1918.

[Drainage area, 8,270 square miles.]

Month.	Discharge in second-feet.				Run-off (depth in inches on drainage area).
	Maximum.	Minimum.	Mean.	Per square mile.	
October	24,400	1,820	8,920	1.08	1.24
November	52,500	4,510	18,100	2.19	2.44
December	7,110	2,700	4,570	.553	.64
January	2,700	1,930	2,240	.271	.31
February	1,880	1,520	1,630	.197	.21
March	2,300	1,520	1,630	.197	.23
April	63,600	2,580	26,700	3.23	3.60
May	104,000	25,100	54,900	6.64	7.66
June	32,300	9,240	17,500	2.11	2.35
July	46,500	10,300	22,300	2.70	3.11
August	25,500	3,060	8,770	1.06	1.22
September	21,800	2,880	9,590	1.16	1.29
The year	104,000	1,520	14,800	1.79	24.30

MACHIAS RIVER BASIN.

MACHIAS RIVER AT WHITNEYVILLE, MAINE.

LOCATION.—At a wooden highway bridge in Whitneyville, Washington County, 200 feet below a storage dam and 4 miles above Machias.

DRAINAGE AREA.—465 square miles.

RECORDS AVAILABLE.—October 17, 1903, to September 30, 1918.

GAGE.—Chain installed on the wooden highway bridge October 10, 1911; prior to October 3, 1905, chain gage on the Washington County Railroad bridge, three-fourths of a mile downstream; October 3, 1905, to October 9, 1911, staff gage on highway bridge at datum of present chain gage. Gage read by I. S. Albee.

DISCHARGE MEASUREMENTS.—Made from railroad bridge or by wading.

CHANNEL AND CONTROL.—Practically permanent.

EXTREMES OF DISCHARGE.—Maximum stage recorded during year, 10.0 feet at 3 p. m. April 22 and 3.30 p. m. April 23 (discharge, 5,900 second-feet); minimum stage recorded 3.25 feet on August 3, 4, 5, 6, and 7 (discharge, 160 second-feet).

ICE.—River usually remains open at the gage but ice farther downstream occasionally affects the stage-discharge relation.

REGULATION.—Opening and closing of gates in storage dam immediately above station each day during low stages of the river cause considerable fluctuation; some log driving every year and jams of short duration occasionally occur.

ACCURACY.—Stage-discharge relation practically permanent except when affected by ice. Rating curve well defined between 100 and 4,000 second-feet. Gage read to tenths once daily, except from December 15 to March 30, when it was read three times a week. Daily discharge ascertained by applying mean daily gage height to rating table and making corrections for effect of ice during the winter. Records fair.

Discharge measurements of Machias River at Whitneyville, Maine, during the year ending Sept. 30, 1918.

Date.	Made by—	Gage height.	Discharge.	Date.	Made by—	Gage height.	Discharge.
		Feet.	*Sec.-ft.*			*Feet.*	*Sec.-ft.*
Jan. 5	A. F. McAlary	a 4.30	308	Mar. 16	A. F. McAlary	a 4.80	474
Feb. 16	.....do	a 5.1	538	Aug. 11	H. A. Lancaster	4.22	640

a Stage-discharge relation affected by ice.

Daily discharge, in second-feet, of Machias River at Whitneyville, Maine, for the year ending Sept. 30, 1918.

Day.	Oct.	Nov.	Dec.	Jan.	Feb.	Mar.	Apr.	May.	June.	July.	Aug.	Sept.
1	770	1,380	598	360	270	860	920	4,150	1,380	490	244	178
2	980	1,380	544	360	270	800	1,250	4,800	1,380	387	200	221
3	860	1,240	540	360	270	800	1,860	3,750	1,380	387	160	221
4	711	1,240	520	340	270	800	2,200	3,150	1,240	387	160	314
5	654	1,100	490	310	270	800	2,500	2,950	1,240	387	160	362
6	711	1,100	460	310	270	740	2,660	2,570	1,100	438	160	412
7	860	1,100	460	310	270	680	2,750	2,030	1,100	860	160	412
8	860	980	440	310	270	660	2,950	1,860	1,240	1,700	200	412
9	980	980	440	340	270	640	2,950	1,540	1,240	1,540	200	412
10	980	980	410	360	270	580	3,150	1,540	1,240	1,380	490	412
11	1,240	1,240	410	360	270	520	2,950	1,540	1,100	1,380	654	412
12	1,700	1,100	410	360	270	490	2,950	1,860	1,100	1,100	682	412
13	2,200	1,100	410	360	270	490	2,750	2,030	1,240	980	740	362
14	1,860	980	410	360	270	460	2,750	2,390	1,240	860	740	362
15	1,540	860	410	360	390	460	2,750	2,570	1,240	770	740	362
16	1,540	770	410	360	540	470	2,750	2,950	1,240	711	740	362
17	1,540	711	410	360	520	490	2,950	3,150	1,240	711	740	362
18	1,540	711	410	360	490	520	2,950	3,350	1,100	711	626	362
19	1,700	711	390	360	490	520	2,950	3,350	1,100	711	571	464
20	1,860	711	360	360	490	540	2,950	3,150	1,100	711	517	682
21	1,860	711	360	360	490	580	2,950	2,750	1,100	654	464	1,380
22	1,540	711	360	310	490	580	5,900	2,570	1,100	598	412	1,860
23	1,540	860	360	290	490	580	5,900	2,390	1,860	544	412	1,940
24	1,540	1,100	360	270	520	600	5,600	2,210	1,700	544	412	1,940
25	2,030	1,100	360	270	580	640	5,240	2,030	1,540	544	362	1,460
26	1,860	1,240	360	270	640	640	4,360	1,860	1,240	544	314	1,240
27	1,540	1,380	360	270	740	660	3,150	1,700	1,100	544	267	2,120
28	1,540	1,380	360	270	860	680	3,150	1,700	860	490	221	3,950
29	1,540	1,100	360	270	......	720	3,550	1,540	711	490	178	3,150
30	1,540	770	360	270	......	740	3,750	1,540	598	438	178	2,750
31	1,540	......	360	270	......	800	......	1,380	......	338	178	......

NOTE.—Stage-discharge relation affected by ice Dec. 3 to Apr. 5; discharge for this period computed from gage heights corrected for effect of ice by means of three discharge measurements, observer's notes, and weather records.

Monthly discharge of Machias River at Whitneyville, Maine, for the year ending Sept. 30, 1918.

[Drainage area, 465 square miles.]

Month.	Discharge in second-feet.				Run-off (depth in inches on drainage area).
	Maximum.	Minimum.	Mean.	Per square mile.	
October	2,210	654	1,390	2.99	3.45
November	1,380	711	1,020	2.19	2.44
December	598	360	416	.895	1.03
January	360	270	325	.699	.81
February	860	270	411	.884	.92
March	860	460	630	1.35	1.56
April	5,900	920	3,180	6.84	7.63
May	4,800	1,380	2,460	5.29	6.10
June	1,860	598	1,200	2.58	2.88
July	1,700	338	720	1.55	1.79
August	740	160	396	.854	.98
September	3,950	178	976	2.10	2.34
The year	5,900	160	1,090	2.34	31.93

UNION RIVER BASIN.

WEST BRANCH OF UNION RIVER AT AMHERST, MAINE.

LOCATION.—At highway bridge three-fourths of a mile west of Amherst post office, Hancock County, on road to Bangor, 1 mile below highway bridge at old tannery dam.

DRAINAGE AREA.—140 square miles.

RECORDS AVAILABLE.—July 25, 1909, to September 30, 1918.

GAGE.—Chain, installed June 2, 1910, at same datum as old vertical gage nailed to log abutment; read by Mrs. Emma Sumner.

DISCHARGE MEASUREMENTS.—Made from downstream side of the bridge.

CHANNEL AND CONTROL.—Gravel; unlikely to change except in unusual flood.

EXTREMES OF DISCHARGE.—Maximum stage recorded during year, 10.9 feet at 9 a. m. and 4 p. m. April 24 (discharge, 1,440 second-feet); minimum stage recorded, 5.2 feet at 8 a. m. and 4 p. m. October 5 (discharge, 16 second-feet); minimum discharge estimated as 12 second-feet February 9 and 10, but stage-discharge relation was affected by ice at the time.

ICE.—Surface ice forms to a considerable thickness and anchor ice is found at the measuring section; stage-discharge relation seriously affected.

REGULATION.—Regimen of stream only slightly affected by operation of the few log-driving dams above the station.

ACCURACY.—Stage-discharge relation practically permanent except as affected by backwater from ice and occasional log jams. Rating curve well defined below 1,100 second-feet. Gage read to half-tenths twice daily, except from December 1 to March 30, when it was read three times a week. Daily discharge ascertained by applying mean daily gage height to rating table and making corrections for effect of ice during the winter. Records fair.

Discharge measurements of West Branch of Union River at Amherst, Maine, during the year ending Sept. 30, 1918.

Date.	Made by—	Gage height.	Discharge.	Date.	Made by—	Gage height.	Discharge.
		Feet.	*Sec.-ft.*			*Feet.*	*Sec.-ft.*
Dec. 20	A. F. McAlary	[a]9.25	201	June 15	H. A. Lancaster	5.74	76
20	.....do	[a]7.80	68	Sept. 5	.....do	5.47	36.2
Mar. 22	H. A. Lancaster	[a]9.11	179	5	.....do	5.47	35.9
June 15	.....do	5.74	68				

[a] Stage-discharge relation affected by ice.

Daily discharge, in second-feet, of West Branch of Union River at Amherst, Maine, for the year ending Sept. 30, 1918.

Day.	Oct.	Nov.	Dec.	Jan.	Feb.	Mar.	Apr.	May.	June.	July.	Aug.	Sept.
1	62	392	240	100	68	160	420	1,280	190	217	94	27
2	39	438	240	94	68	210	480	1,200	190	226	87	24
3	39	461	240	94	50	200	540	1,000	114	236	87	29
4	68	461	240	94	34	190	560	930	107	304	87	29
5	16	438	240	94	39	200	640	930	107	304	74	37
6	50	415	240	115	44	210	680	800	62	245	74	63
7	68	438	240	135	34	200	740	508	80	325	62	107
8	74	438	230	135	24	190	780	304	80	284	62	144
9	80	415	230	135	12	180	832	199	80	484	114	94
10	159	392	230	135	12	175	832	182	62	438	217	39
11	174	392	230	130	29	175	860	438	62	369	208	29
12	217	347	200	130	50	175	900	438	62	304	190	39
13	255	369	200	130	74	190	800	347	62	264	174	50
14	304	347	200	135	88	210	864	392	68	144	129	87
15	347	347	210	135	74	210	930	392	68	159	68	56
16	347	325	210	135	80	210	1,040	264	50	264	34	39
17	325	304	210	135	88	175	1,040	284	44	508	50	44
18	461	284	200	130	100	145	1,040	264	62	461	80	44
19	532	245	200	120	100	145	965	199	74	392	62	122
20	347	208	200	120	106	145	897	159	68	325	56	347
21	347	174	190	120	100	135	930	159	50	304	62	532
22	392	144	180	130	88	175	1,320	159	74	255	62	556
23	392	245	190	135	88	190	1,400	174	107	245	56	392
24	392	284	160	130	74	210	1,440	152	208	208	50	325
25	532	245	135	115	62	230	1,400	182	166	190	50	325
26	410	240	130	106	74	240	1,360	129	114	174	44	580
27	182	230	120	94	88	240	1,160	136	56	144	39	930
28	190	220	115	80	115	260	1,120	122	144	129	34	864
29	208	210	115	74	....	300	1,000	94	107	114	34	656
30	245	200	115	68	....	350	1,200	80	50	107	29	580
31	415	....	105	68	....	350	....	80	....	107	29	....

NOTE.—Stage-discharge relation affected by ice Nov. 26 to Apr. 8 and Apr. 11-12. Discharge for these periods computed from gage heights corrected for effect of ice by means of three discharge measurements, observer's notes, and weather records.

Monthly discharge of West Branch of Union River at Amherst, Maine, for the year ending Sept. 30, 1918.

[Drainage area, 140 square miles].

Month.	Discharge in second-feet.				Run-off (depth in inches on drainage area).
	Maximum.	Minimum.	Mean.	Per square mile.	
October	532	16	247	1.76	2.03
November	461	144	322	2.30	2.57
December	240	105	193	1.38	1.59
January	135	68	115	.821	.95
February	115	12	66.5	.475	.49
March	350	135	206	1.47	1.70
April	1,440	420	939	6.71	7.49
May	1,280	80	386	2.76	3.18
June	208	44	92.3	.659	.74
July	508	107	265	1.89	2.18
August	217	29	80.6	.576	.66
September	930	24	240	1.71	1.91
The year	1,440	12	263	1.88	25.49

PENOBSCOT RIVER BASIN.

WEST BRANCH OF PENOBSCOT RIVER AT MILLINOCKET, MAINE.

Location.—At Quakish Lake dam and Millinocket mill of Great Northern Paper Co., at Millinocket, Penobscot County.

Drainage area.—1,880 square miles.

Records available.—January 11, 1901, to September 30, 1918.

Gages.—Water-stage recorder at Quakish Lake dam and gages in fore bay and tail-race at mill.

Channel and control.—Crest of concrete dam.

Discharge.—Flow computed by considering the flow over the dam, the flow through the wheels, and the water used through log sluices and filters. The wheels were rated at Holyoke, Mass., before being placed in position, and were tested later by numerous tube-float and current-meter measurements. Ratings for four new wheels installed in 1917 are based on acceptance test on one unit after installation, the discharge at various gate openings being measured by the use of Pitot tubes. When the flow of the river is less than 3,000 second-feet, all the water generally flows through the wheels of the mill.

Ice.—Determination of discharge not seriously affected by ice; Ferguson Pond, just above entrance to canal, eliminates effect from anchor ice.

Regulation.—Dams at outlets of North Twin and Ripogenus lakes store water on a surface of about 73 square miles, with a capacity of about 41.5 billion cubic feet. Except during the time (usually in August) when excess water has to be supplied for log driving on the river below Millinocket and for a short time during the spring freshet, run-off is regulated by storage. Determination corrected for storage.

Cooperation.—Records furnished by engineers of Great Northern Paper Co.

Monthly discharge of West Branch of Penobscot River at Millinocket, Maine, for the year ending Sept. 30, 1918.

[Drainage area, 1,880 square miles].

Month.	Discharge in second-feet.			Run-off (depth in inches on drainage area).
	Observed.	Corrected for storage.		
	Mean.	Mean.	Per square mile.	
October	2,920	3,140	1.67	1.92
November	3,450	3,610	1.92	2.14
December	2,900	1,520	.809	.93
January	2,780	627	.334	.39
February	3,460	300	.160	.17
March	3,940	206	.110	.13
April	3,380	8,180	4.35	4.85
May	2,970	8,190	4.36	5.03
June	2,940	2,510	1.34	1.50
July	4,800	5,480	2.91	3.36
August	3,080	2,170	1.15	1.33
September	2,820	2,400	1.28	1.43
The year	3,290	3,210	1.71	23.18

WEST BRANCH OF PENOBSCOT RIVER NEAR MEDWAY, MAINE.

Location.—Just above Nichatou Rapids, half a mile above mouth of East Branch of Penobscot River and town of Medway, Penobscot County, and 2 miles below East Millinocket.

Drainage area.—2,100 square miles.

Records available.—February 20, 1916, to September 30, 1918.

Gages.—Chain on left bank; read by A. T. Reed; Gurley 7-day water-stage recorder on left bank installed August 4, 1916.

Discharge measurements.—Made from cable.

Channel and control.—Bed fairly smooth at measuring section; covered with rocks and boulders above and below gage. Channel divides a few hundred feet below gage, but practically entire flow passes to left of Nichatou Island. Control formed by Nichatou Island and head of Nichatou Rapids; somewhat shifting.

Extremes of discharge.—Maximum stage during year from water-stage recorder, 7.11 feet at 1 p. m. July 16 (discharge, 11,500 second-feet); minimum stage during year from water-stage recorder, 2.09 feet at 10 a. m. September 2 (discharge, from extension of rating curve, about 1,140 second-feet).

1916–1918: Maximum stage recorded, 9.88 feet at 1 p. m. June 18, 1917 (discharge, from extension of rating curve, about 20,000 second-feet); minimum stage recorded, 1.45 feet at 9.45 a. m. January 7, 1917 (discharge, 585 second-feet).

Ice.—Ice forms along both banks, but the main channel remains open; stage-discharge relation not seriously affected.

Regulation.—Flow at ordinary stages completely regulated by dams and storage reservoirs above station.

Accuracy.—Stage-discharge relation shifted slightly at time of high water in June, 1917. Rating curve used previous to June, 1917, well defined below 12,000 second-feet; curve used subsequent to that date well defined between 2,000 and 12,000 second-feet. Daily discharge ascertained by discharge integrator. Records fair.

Discharge measurements of West Branch of Penobscot River near Medway, Maine, during the year ending Sept. 30, 1918.

Date.	Made by—	Gage height.	Discharge.	Date.	Made by—	Gage height.	Discharge.
		Feet.	*Sec.-ft.*			*Feet.*	*Sec.-ft.*
May 25	Clark and Lancaster....	4.38	3,970	May 26	Clark and Lancaster....	4.33	3,880
26	.....do.................	3.48	2,490	July 16	H. A. Lancaster........	7.14	11,500

Daily discharge, in second-feet, of West Branch of Penobscot River near Medway, Maine, for the year ending Sept. 30, 1918.

Day.	Oct.	Nov.	Dec.	Jan.	Feb.	Mar.	Apr.	May.	June.	July.	Aug.	Sept.
1	2,400	4,500	3,000	4,400	4,150	5,700	5,600	4,000	4,000	3,200	3,600	4,000
2	2,700	4,150	2,750	4,350	4,100	5,600	5,800	4,450	3,650	3,200	3,550	2,000
3	2,750	4,100	3,100	4,650	3,450	4,800	5,800	4,350	3,750	3,150	3,500	2,400
4	2,700	3,600	3,000	4,550	3,750	4,800	5,900	4,400	4,200	2,950	3,500	2,600
5	2,900	4,000	3,200	4,200	3,700	5,100	5,900	3,700	4,200	3,250	3,500	2,700
6	3,200	3,700	3,250	2,900	3,900	5,400	6,000	3,700	3,550	2,950	3,300	2,750
7	2,450	3,800	3,250	3,900	4,445	5,200	4,500	4,200	3,250	2,600	3,300	2,800
8	2,600	5,800	3,200	3,700	4,250	5,000	5,000	4,300	3,250	4,200	3,300	2,700
9	3,300	9,700	3,050	3,700	5,100	5,000	4,900	4,150	2,650	4,500	3,500	3,000
10	2,850	8,800	3,300	3,700	4,700	4,200	4,400	4,000	3,000	4,300	3,500	3,100
11	2,750	7,300	3,350	3,900	5,400	5,300	4,250	4,000	3,000	4,000	2,950	2,850
12	3,050	4,550	3,500	3,550	5,400	5,000	4,200	3,400	3,100	3,550	3,750	2,900
13	2,900	4,150	3,350	3,250	5,000	4,800	4,050	3,650	3,000	3,500	3,750	2,900
14	2,700	3,750	3,550	3,750	4,900	4,900	3,600	4,000	3,100	5,400	4,050	3,200
15	2,800	3,400	3,550	3,620	4,800	4,650	3,950	3,950	3,100	10,000	4,450	2,900
16	3,250	3,340	3,550	3,580	4,900	4,700	4,450	3,950	2,650	11,100	4,250	3,400
17	3,150	3,350	3,500	3,400	4,750	....	4,500	3,700	2,900	9,400	4,250	3,400
18	2,950	2,750	3,300	3,450	4,750	....	4,400	3,400	3,150	8,500	3,650	3,200
19	2,870	3,800	3,350	3,400	5,600	....	4,300	3,000	3,050	8,500	4,150	3,400
20	3,400	4,850	3,350	2,800	5,300	....	4,200	3,500	3,100	8,400	4,050	3,500
21	2,950	3,900	3,350	3,050	4,600	....	3,650	4,200	3,150	7,900	3,100	3,700
22	3,350	3,550	3,200	3,150	5,200	....	3,900	4,000	3,600	8,200	3,100	3,250
23	3,850	3,410	3,000	3,200	5,800	....	4,150	4,150	3,550	7,300	2,950	3,400
24	3,750	3,500	3,000	3,350	3,880	4,600	3,600	4,250	4,350	6,700	2,950	3,300
25	3,500	2,750	2,600	3,450	5,600	5,200	3,650	3,800	4,200	5,500	2,550	3,250
26	3,900	3,500	3,300	3,350	6,100	5,300	4,100	3,800	3,950	4,100	3,300	3,100
27	4,150	3,600	3,100	2,850	6,000	5,200	3,900	3,900	3,800	3,550	2,800	3,300
28	3,500	3,450	3,400	3,200	5,800	5,400	3,400	4,200	3,700	3,300	3,050	3,500
29	3,900	3,300	3,400	3,350	....	5,400	3,350	4,100	3,700	3,300	3,300	3,400
30	3,900	2,750	3,000	3,800	....	6,000	3,550	4,200	3,050	3,450	3,400	3,250
31	4,400	....	4,150	3,900	....	5,300	....	4,350	....	3,450	3,400	....

NOTE.—Average discharge Mar. 17-23 estimated at 5,000 second-feet by comparison with records at West Enfield and observer's once-daily gage readings.

Monthly discharge of West Branch of Penobscot River near Medway, Maine, for the year ending Sept. 30, 1918.

[Drainage area, 2,100 square miles.]

Month.	Discharge in second-feet.				Run-off (depth in inches on drainage area).
	Maximum.	Minimum.	Mean.	Per square mile.	
October	4,400	2,400	3,190	1.52	1.75
November	9,700	2,750	4,240	2.02	2.25
December	4,150	2,600	3,260	1.55	1.79
January	4,650	2,800	3,590	1.71	1.97
February	6,100	3,450	4,830	2.30	2.40
March	6,000	....	5,080	2.42	2.79
April	6,000	3,350	4,430	2.11	2.35
May	4,450	3,000	3,960	1.89	2.18
June	4,350	2,650	3,420	1.63	1.82
July	11,100	2,600	5,270	2.51	2.89
August	4,450	2,550	3,480	1.66	1.91
September	4,000	2,000	3,110	1.48	1.65
The year	11,100	2,000	3,980	1.90	25.75

NOTE.—The monthly discharge in second-feet per square mile and the run-off depth in inches do not represent the natural run-off from the basin because of storage. (See "Regulation.")

PENOBSCOT RIVER AT WEST ENFIELD, MAINE.

LOCATION.—At steel highway bridge 1,000 feet below mouth of Piscataquis River and 3 miles west of Enfield railroad station, Penobscot County.

DRAINAGE AREA.—6,600 square miles.

RECORDS AVAILABLE.—January 1, 1902, to September 30, 1918.

GAGES.—Friez water-stage recorder on left bank, downstream side of left bridge abutment, used since December 11, 1912, standard chain gage on upstream side of bridge used prior to that date; gages set to same datum.

DISCHARGE MEASUREMENTS.—Made from bridge.

CHANNEL AND CONTROL.—Channel at gage broken by four bridge piers; straight above and below the gage. Banks high, rocky, and not subject to overflow. Control is at Passadumkeag Rips, about 5 miles below the gage; a wing dam at this point is overflowed at about gage height 5.5 feet.

EXTREMES OF DISCHARGE.—Maximum stage during year, from water-stage recorder, 11.2 feet at 8 p. m. May 2 (discharge, 40,700 second-feet); minimum stage during year from water-stage recorder, 2.30 feet at 11 p. m. October 1 (discharge, 3,840 second-feet).

ICE.—Stage-discharge relation usually affected by ice from December to April; discharge ascertained by comparison with records at Sunkhaze Rips collected by Thomas W. Clark.

REGULATION.—Flow since 1900 largely controlled by storage, principally in the lakes tributary to the West Branch. Results not corrected for storage.

ACCURACY.—Stage-discharge relation practically permanent except as affected by ice and occasionally by logs. Rating curve well defined. Operation of water-stage recorder satisfactory throughout the year. Daily discharge ordinarily ascertained by applying to rating table average gage height taken from recorder sheets and corrections for effect of ice during the winter; at times of serious fluctuation in stage the daily discharge is ascertained by using the average of 12 two-hour periods. Records excellent.

COOPERATION.—Gage-height record and several discharge measurements furnished by Thomas W. Clark, hydraulic engineer, Oldtown, Maine. Discharge measurements also made by students of the University of Maine, under the direction of Prof. H. S. Boardman.

Discharge measurements of Penobscot River at West Enfield, Maine, during the year ending Sept. 30, 1918.

Date.	Made by—	Gage height.	Discharge.	Date.	Made by—	Gage height.	Discharge.
		Feet.	*Sec.-ft.*			*Feet.*	*Sec.-ft.*
Oct. 7	H. A. Lancaster	3.15	5,950	Feb. 7	McAlary and Lancaster.	[a] 5.84	4,870
19	University of Maine students	4.94	9,990	Aug. 27	T. W. Clark	[b] 3.22	5,440

[a] Stage-discharge relation affected by ice.
[b] Stage-discharge relation affected by log jam.

Daily discharge, in second-feet, of Penobscot River at West Enfield, Maine, for the year ending Sept. 30, 1918.

Day.	Oct.	Nov.	Dec.	Jan.	Feb.	Mar.	Apr.	May.	June.	July.	Aug.	Sept.
1	4,170	34,100	7,040	4,900	5,100	8,000	13,000	31,300	9,120	7,570	9,870	6,010
2	4,390	28,000	6,910	5,000	5,200	8,300	17,200	38,600	8,680	7,960	8,970	5,650
3	4,960	24,200	6,800	5,200	5,300	8,300	21,100	37,100	8,970	7,570	8,680	4,730
4	5,070	21,500	7,400	5,200	4,800	7,200	25,300	32,400	8,970	7,170	8,680	5,070
5	5,190	19,500	7,600	5,300	4,700	7,400	24,700	28,800	8,540	6,780	8,120	4,960
6	5,420	18,000	7,300	5,300	4,600	7,900	26,000	25,000	7,960	6,650	7,960	4,960
7	5,650	17,000	6,400	5,100	4,900	8,000	26,000	23,000	7,710	6,650	7,960	4,840
8	5,190	16,600	6,200	5,300	5,000	7,700	28,300	23,000	8,680	11,700	7,710	4,730
9	5,420	18,600	5,800	5,400	5,100	7,400	29,900	20,800	9,120	30,000	7,570	4,840
10	5,770	19,700	5,800	5,500	5,400	7,400	29,400	18,800	7,960	33,500	7,710	5,070
11	5,770	17,600	6,200	5,300	5,200	6,700	29,400	17,800	7,570	27,000	7,300	5,190
12	6,770	15,200	6,300	5,300	5,200	7,300	27,000	17,400	7,570	22,500	6,600	4,960
13	8,540	12,300	6,100	5,300	5,600	7,400	25,000	16,800	7,840	21,100	6,800	4,960
14	10,200	10,800	6,000	4,900	5,600	7,300	24,000	16,800	8,120	20,800	7,000	6,010
15	10,500	10,200	6,000	5,000	5,800	7,300	24,700	19,100	8,400	25,500	7,600	6,910
16	11,500	9,720	6,100	5,200	6,000	7,200	28,800	19,100	8,120	31,600	8,500	7,170
17	12,100	9.420	5,900	5,200	6,200	6,900	31,800	17,000	7,710	31,000	8,100	7,170
18	11,600	8,970	5,800	5,200	5,900	6,400	31,800	14,600	7,440	28,000	8,100	6,780
19	10,800	8,260	5,900	5,200	5,900	7,200	30,200	14,100	7,040	25,700	7,200	7,440
20	11,300	9,720	6,000	5,200	6,700	7,300	28,300	12,600	6,910	22,500	6,300	9,800
21	13.700	9,720	6,300	4,800	6,700	7,400	26,500	12,100	7,710	20,400	5,700	14,800
22	13.700	8,680	6,200	4,600	5,900	8,000	27,800	11,500	8,260	19,700	5,600	17,800
23	13,700	8,260	6.000	4,700	6.400	8,300	34,400	11,500	11.300	18,800	5.200	15,800
24	12,700	8,540	5,500	5,100	6,900	8,400	36,800	10,700	17,400	17,400	4,900	14,600
25	13,500	9,120	5.200	5,000	5,500	8,400	37,100	10,800	14,400	15,800	5,300	13,400
26	19,000	8,100	5,800	5,200	6,800	9,100	34,700	10,700	12,000	13,700	5,300	12,500
27	18.800	8.400	5.200	5,200	7,700	9,100	31,600	10,000	10,500	12,000	5,400	14,300
28	17,200	6,800	5,300	4,700	7,800	9,400	28,600	9,270	9,720	10,700	5,400	22,200
29	16,200	6,900	5,300	4,500	....	9,700	26,500	9,570	9,270	9,720	5,500	20,400
30	16,600	7,440	5,100	4,600	....	10,500	26,800	9,720	8,680	9,870	5,650	18,000
31	24,700	....	5,000	4,900	....	11,500	....	9,120	....	10,300	5,530	....

NOTE.—Stage-discharge relation affected by ice Nov. 26–29 and Dec. 3 to Apr. 8; discharge for this period computed from gage heights corrected for effect of ice by means of one discharge measurement at West Enfield and numerous discharge measurements and other data at Sunk Haze. Stage-discharge relation affected by log jams Aug. 12–29; determinations of discharge for this period based on observed gage heights corrected for effect of logs by means of one discharge measurement at West Enfield and data at Sunk Haze.

Monthly discharge of Penobscot River at West Enfield, Maine, for the year ending Sept. 30, 1918.

[Drainage area, 6,600 square miles.]

Month	Discharge in second-feet.				Run-off (depth in inches on drainage area).
	Maximum.	Minimum.	Mean.	Per square mile.	
October	24,700	4,170	10,600	1.60	1.85
November	34,100	6,800	13,700	2.07	2.31
December	7,600	5,000	6,080	.922	1.06
January	5,500	4,500	5,070	.769	.89
February	7,800	4,600	5,730	.874	.91
March	11,500	6,400	8,010	1.21	1.40
April	37,100	13,000	27,800	4.22	4.71
May	38,600	9,120	18,000	2.72	3.14
June	17,400	6,910	9,060	1.37	1.53
July	33,500	6,650	17,400	2.63	3.03
August	9,870	4,900	6,960	1.06	1.22
September	22,200	4,730	9,370	1.42	1.58
The year	38,600	4,170	11,500	1.74	23.63

EAST BRANCH OF PENOBSCOT RIVER AT GRINDSTONE, MAINE.

LOCATION.—At Bangor & Aroostook Railroad bridge half a mile south of railroad station at Grindstone, Penobscot County, one-eighth mile above Grindstone Falls, and 8 miles above confluence with West Branch at Medway.

DRAINAGE AREA.—1,100 square miles; includes 270 square miles of Chamberlain Lake drainage.

RECORDS AVAILABLE.—October 23, 1902, to September 30, 1918.

GAGE.—Chain attached to railroad bridge; read by R. D. Porter.

DISCHARGE MEASUREMENTS.—Made from railroad bridge.

CHANNEL AND CONTROL.—Practically permanent; stream confined by abutments of bridge and broken by one pier at ordinary stages; velocity of current medium at moderate and high stages but sluggish at low water.

EXTREMES OF DISCHARGE.—Maximum stage recorded during year, 10.7 feet at 4 p. m. July 9 (discharge, 12,900 second-feet); minimum stage recorded, 4.4 feet at 7 a. m. October 1 (discharge, 290 second-feet). Minimum discharge estimated as 210 second-feet from February 10–17, when stage-discharge relation was affected by ice.

ICE.—Ice forms to a considerable thickness at the gage and down to the head of Grindstone Falls, and although the falls usually remain open during the greater part of the winter, the stage-discharge relation is somewhat affected.

REGULATION.—Several dams maintained at outlets of a number of lakes and ponds near source of river are regulated for log driving; during the summer and fall gates are generally left open. The basin of the East Branch since about 1840 includes about 270 square miles of territory draining into Chamberlain Lake that formerly drained into the St. John River basin, the diversion being made through what is known as the Telos canal. Results not corrected for storage and diversions.

ACCURACY.—Stage-discharge relation occasionally affected by backwater from log jams at station and at Grindstone Falls immediately below, and by ice during winter. Rating curve well defined between 300 and 9,000 second-feet. Gage read to half-tenths once daily (except Sundays), except from November 27 to March 30, when it was read three times a week. Daily discharge ascertained by applying mean daily gage height to rating table and making corrections for effect of ice during the winter. Record fair for moderate and high stages but uncertain for low stages.

Discharge measurements of East Branch of Penobscot River at Grindstone, Maine, during the year ending Sept. 30, 1918.

Date.	Made by—	Gage height.	Discharge.	Date.	Made by—	Gage height.	Discharge.
		Feet.	*Sec.-ft.*			*Feet.*	*Sec.-ft.*
Dec. 17	A. F. McAlary	a 5.15	406	May 1	H. A. Lancaster	8.04	5,170
Jan. 28	.....do	a 5.21	289	18	.....do	6.88	3,080
Mar. 1	.....do	a 5.80	583	Aug. 1	.....do	6.61	2,460
27	H. A. Lancaster	a 5.65	554	Sept. 3	.....do	5.44	996

a Stage-discharge relation affected by ice.

Daily discharge, in second-feet, of East Branch of Penobscot River at Grindstone, Maine, for the year ending Sept. 30, 1918.

Day.	Oct.	Nov.	Dec.	Jan.	Feb.	Mar.	Apr.	May.	June.	July.	Aug.	Sept.
1	290	7,800	560	250	250	580	1,150	5,340	1,530	2,250	2,550	1,000
2	415	6,380	560	250	250	600	2,100	8,100	1,650	2,250	2,550	1,050
3	390	5,590	560	250	250	540	2,700	6,110	1,810	2,250	2,550	950
4	365	4,800	560	250	250	470	4,240	5,100	1,670	2,250	2,200	770
5	365	3,860	560	250	250	470	4,440	4,600	1,160	2,100	1,950	620
6	390	3,490	520	250	250	470	4,240	4,240	1,530	1,950	2,100	580
7	450	3,490	520	270	250	470	4,600	5,340	1,280	2,200	1,950	560
8	470	3,160	500	290	250	470	5,100	4,870	1,400	3,320	1,810	600
9	470	2,850	500	290	250	470	5,100	4,650	1,450	12,600	1,950	620
10	500	2,550	500	320	210	440	4,650	3,860	1,530	9,000	1,810	590
11	470	2,400	470	320	210	420	4,240	3,760	1,280	7,500	1,600	560
12	500	2,320	420	240	210	420	4,240	3,400	1,400	7,800	1,400	500
13	730	2,020	420	340	210	420	4,440	3,160	2,020	9,000	1,280	560
14	1,000	1,740	420	320	210	420	4,600	4,050	1,950	8,000	1,280	815
15	1,340	1,600	420	340	210	420	4,870	5,340	1,810	6,930	1,400	750
16	1,600	1,600	420	340	210	420	5,850	5,100	1,810	9,300	1,400	620
17	1,600	1,460	420	360	210	420	5,850	3,160	1,810	6,380	1,280	620
18	1,340	1,460	420	390	230	420	5,590	3,000	1,160	5,100	1,150	620
19	1,340	1,460	420	420	230	440	5,340	2,800	1,160	4,650	1,050	815
20	1,340	1,340	420	390	230	440	4,870	2,550	2,250	3,670	950	815
21	1,500	1,220	420	360	230	470	4,600	1,950	2,400	4,000	950	1,460
22	1,400	1,220	420	360	250	470	4,440	1,950	2,700	4,240	860	1,450
23	1,280	1,100	390	390	270	500	5,100	2,250	5,000	4,050	860	1,340
24	1,160	1,000	360	360	290	500	7,210	1,950	5,850	3,860	860	1,280
25	2,850	950	360	340	320	540	6,380	1,950	2,700	3,490	950	1,160
26	4,240	815	360	320	390	560	6,110	1,950	2,850	3,160	1,050	1,050
27	3,160	820	340	290	470	560	5,340	1,950	2,400	3,000	1,050	1,100
28	3,000	700	320	290	530	560	4,600	1,950	2,400	2,900	950	1,670
29	2,850	620	290	270	......	600	3,860	1,950	2,250	2,850	860	1,500
30	3,160	560	270	270	......	780	4,240	1,950	2,250	2,550	950	1,400
31	8,400	......	250	270	......	940	......	1,810	......	3,000	950	......

NOTE.—Stage-discharge relation affected by ice from Dec. 27 to Apr. 3; discharge for this period computed from gage heights corrected for effect of ice by means of four discharge measurements, observer's notes, and weather records. Discharge estimated for Sundays (gage not read).

Monthly discharge of East Branch of Penobscot River at Grindstone, Maine, for the year ending Sept. 30, 1918.

[Drainage area, 1,100 square miles.]

Month.	Discharge in second-feet.				Run-off (depth in inches on drainage area).
	Maximum.	Minimum.	Mean.	Per square mile.	
October	8,400	290	1,560	1.42	1.64
November	7,800	560	2,350	2.14	2.39
December	560	250	431	.392	.45
January	420	250	315	.286	.33
February	530	210	263	.239	.25
March	940	420	507	.461	.53
April	7,210	1,150	4,670	4.25	4.74
May	8,100	1,810	3,550	3.23	3.72
June	5,850	1,160	2,080	1.89	2.11
July	12,600	1,950	4,700	4.27	4.92
August	2,550	860	1,440	1.31	1.51
September	1,670	500	914	.831	.93
The year	12,600	210	1,900	1.73	23.52

MATTAWAMKEAG RIVER AT MATTAWAMKEAG, MAINE.

LOCATION.—At Maine Central Railroad bridge at village of Mattawamkeag, Penobscot County, half a mile above mouth of river.

DRAINAGE AREA.—1,500 square miles.

RECORDS AVAILABLE.—August 26, 1902, to September 30, 1918.

GAGE.—Chain fastened to railroad bridge; read by W. T. Mincher.

DISCHARGE MEASUREMENTS.—Made from the bridge; low-water measurements made by wading at a point about a mile above station.

CHANNEL AND CONTROL.—Practically permanent; channel at bridge broken by two piers.

EXTREMES OF DISCHARGE.—Maximum stage recorded during year, 9.9 feet at 5 p. m. April 26 (discharge, 12,400 second-feet); minimum stage recorded, 3.90 feet at 7 a. m. October 1 (discharge, 560 second-feet). Minimum discharge estimated as 340 second-feet on February 7 when stage-discharge relation was affected by ice.

ICE.—Stage-discharge relation usually affected by ice for several months each winter.

REGULATION.—Dams are maintained at outlets of several large lakes and ponds but the stored water is used only for log driving.

ACCURACY.—Stage-discharge relation occasionally affected by backwater from log jams and, during winter, by ice. Rating curve well defined below 15,000 second-feet. Gage read to tenths twice daily, except from December 16 to March 28, when it was read twice a week. Daily discharge ascertained by applying mean daily gage height to rating table and making corrections for effect of ice during the winter. Records good.

Discharge measurements of Mattawamkeag River at Mattawamkeag, Maine, during the year ending Sept. 30, 1918.

Date.	Made by—	Gage height.	Discharge.	Date.	Made by—	Gage height.	Discharge.
		Feet.	*Sec.-ft.*			*Feet.*	*Sec.-ft.*
Jan. 7	A. F. McAlary	[a] 6.30	657	May 16	H. A. Lancaster	6.63	4,270
Feb. 8	.....do	[a] 5.76	466	June 22	.....do	4.87	1,420
Mar. 5	.....do	[a] 6.6	1,010	July 30	.....do	5.07	1,690
30	H. A. Lancaster	[a] 6.7	1,250	Sept. 7	.....do	3.94	575
Apr. 10	.....do	[b] 8.44	7,300				

[a] Stage-discharge relation affected by ice.
[b] Stage-discharge relation possibly affected by high stage of Penobscot River.

Daily discharge, in second-feet, of Mattawamkeag River at Mattawamkeag, Maine, for the year ending Sept. 30, 1918.

Day.	Oct.	Nov.	Dec.	Jan.	Feb.	Mar.	Apr.	May.	June.	July.	Aug.	Sept.
1	590	6,140	1,000	500	560	940	1,500	9,690	1,230	1,400	1,510	590
2	730	6,850	1,000	540	560	1,000	1,950	10,200	1,340	1,230	1,450	590
3	850	7,100	1,000	540	560	1,050	2,500	10,500	1,570	1,280	1,340	590
4	940	6,610	940	540	540	1,050	3,000	10,200	1,570	1,180	1,180	590
5	895	6,140	900	540	470	1,000	3,600	9,690	1,510	985	1,080	590
6	895	5,240	900	600	390	1,050	4,400	8,900	1,450	940	1,080	590
7	1,080	4,810	900	660	340	1,050	4,600	8,120	1,570	1,130	1,180	620
8	1,230	4,600	900	620	470	1,080	4,800	7,600	1,820	2,090	1,130	590
9	1,180	4,400	1,150	620	500	1,080	6,100	7,100	1,820	4,200	1,080	620
10	1,230	4,000	940	620	470	1,150	7,400	6,370	1,820	5,460	1,080	730
11	1,400	3,610	900	620	440	1,100	8,100	6,370	1,820	5,910	1,080	730
12	1,690	3,230	840	620	420	1,100	8,640	6,140	1,690	5,460	985	655
13	2,380	2,870	800	620	360	1,050	8,640	5,460	1,950	5,680	1,080	690
14	3,230	2,530	840	620	420	1,050	8,380	4,600	2,230	5,910	1,080	850
15	4,000	2,380	800	620	540	1,000	8,380	4,400	2,380	6,370	1,080	940
16	4,400	2,230	700	620	620	1,000	8,900	4,000	2,380	6,850	1,080	1,280
17	4,600	2,090	640	600	740	1,000	9,690	4,000	1,570	6,370	985	1,280
18	4,400	2,090	620	600	740	940	9,960	4,000	1,570	6,140	895	1,280
19	4,000	1,950	620	600	740	940	9,690	3,610	1,570	5,910	850	1,820
20	4,000	1,950	620	600	740	1,000	9,420	3,040	1,510	5,460	730	2,380
21	4,810	1,820	620	600	780	1,000	8,640	2,700	1,510	4,810	655	4,200
22	5,020	1,820	620	600	810	1,000	8,640	2,530	1,510	4,600	590	5,460
23	4,810	1,690	620	600	810	940	9,690	2,380	1,570	4,200	620	5,910
24	4,600	1,510	620	600	810	940	10,800	2,230	2,090	3,800	655	5,910
25	4,400	1,280	620	600	810	1,000	11,900	1,950	2,230	3,040	655	5,460
26	4,600	1,080	620	560	840	1,000	12,400	1,570	2,090	2,700	655	5,020
27	4,810	995	600	560	840	1,050	11,900	1,510	1,820	2,380	620	4,810
28	4,810	940	560	560	900	1,150	11,300	1,340	1,820	2,090	590	6,370
29	4,600	900	560	560	......	1,250	10,500	1,400	1,690	1,820	590	7,350
30	4,400	940	540	560	......	1,250	9,960	1,400	1,690	1,690	590	7,600
31	5,240	......	500	560	......	1,250	......	1,280	......	1,690	590	......

NOTE.—Stage-discharge relation affected by ice Nov. 28 to Apr. 11; discharge for this period computed from gage heights corrected for effect of ice by means of five discharge measurements, observer's notes, and weather records.

Monthly discharge of Mattawamkeag River at Mattawamkeag, Maine, for the year ending Sept. 30, 1918.

[Drainage area, 1,500 square miles.]

Month.	Discharge in second-feet.				Run-off (depth in inches on drainage area).
	Maximum.	Minimum.	Mean.	Per square mile.	
October	5,240	590	3,090	2.06	2.38
November	7,100	900	3,130	2.09	2.33
December	1,150	500	758	.505	.58
January	620	540	589	.393	.45
February	900	340	615	.410	.43
March	1,250	940	1,050	.700	.81
April	12,400	1,500	7,840	5.23	5.84
May	10,500	1,280	4,980	3.32	3.83
June	2,380	1,230	1,750	1.17	1.30
July	6,850	940	3,640	2.43	2.80
August	1,510	590	925	.617	.71
September	7,600	590	2,540	1.69	1.89
The year	12,400	340	2,580	1.72	23.35

PISCATAQUIS RIVER NEAR FOXCROFT, MAINE.

Location.—At highway bridge known as Lows Bridge, halfway between Guilford and Foxcroft, Piscataquis County, three-fourths of a mile above mouth of Black Stream and 3 miles below Mill Stream.

Drainage area.—286 square miles.

Records available.—August 17, 1902, to September 30, 1918.

Gage.—Staff attached to left abutment of bridge; read by A. F. D. Harlow.

Discharge measurements.—At medium and high stages made from bridge; at low stages made by wading either above or below the bridge.

Channel and control.—Practically permanent; banks are high and are overflowed only during extreme floods.

Extremes of discharge.—Maximum open-water stage recorded during year, 7.8 feet at 7.30 a. m. October 31 (discharge, 5,310 second feet; a stage of 8.6 feet was recorded at 5 p. m. April 3, but the water was probably held back by an ice jam); minimum stage recorded, 1.9 feet several times during August and September (discharge, 51 second-feet). Minimum discharge estimated as 17 second-feet several times during January, when stage-discharge relation was affected by ice.

Ice.—Stage-discharge relation affected by ice during some winters.

Regulation.—The stream is used to develop power at several manufacturing plants above the station; distribution of flow somewhat affected by operation of wheels.

Accuracy.—Stage-discharge relation occasionally affected by backwater from log jams and by ice during winter. Rating curve well defined between 20 and 4,000 second-feet. Gage read to tenths twice daily. Daily discharge ascertained by applying mean daily gage height to rating table and making corrections for effect of ice during the winter. Some uncertainty exists in regard to accuracy of gage heights and the effect of diurnal fluctuation. Records fair.

Discharge measurements of Piscataquis River near Foxcroft, Maine, during the year ending Sept. 30, 1918.

Date.	Made by—	Gage height.	Discharge.	Date.	Made by—	Gage height.	Discharge.
		Feet.	*Sec.-ft.*			*Feet.*	*Sec.-ft.*
Jan. 14	A. F. McAlary	a 4.27	180	July 31	H. A. Lancaster	2.94	341
Feb. 18	.....do	a 4.38	202	Sept. 22	.....do	3.64	792
Mar. 26	H. A. Lancaster	a 4.56	251	23	.....do	3.02	404

a Stage-discharge relation affected by ice.

Daily discharge, in second-feet, of Piscataquis River near Foxcroft, Maine, for the year ending Sept. 30, 1918.

Day.	Oct.	Nov.	Dec.	Jan.	Feb.	Mar.	Apr.	May.	June.	July.	Aug.	Sept.
1	175	2,430	380	80	58	560	640	3,200	355	305	355	51
2	175	1,700	380	80	46	560	1,150	4,110	305	260	355	72
3	175	1,240	380	90	24	380	2,400	3,200	440	222	240	110
4	175	1,020	380	110	19	380	2,300	2,100	380	222	260	145
5	175	925	380	100	24	280	2,200	1,240	330	222	260	145
6	260	800	380	90	24	175	2,300	1,240	260	190	240	120
7	470	800	380	36	24	200	2,400	1,020	90	305	355	160
8	470	680	380	28	24	100	2,200	720	222	840	280	100
9	470	680	380	80	24	58	2,210	500	380	2,540	240	110
10	410	640	300	100	24	120	2,000	500	440	2,000	240	110
11	280	570	240	100	51	100	1,800	500	440	1,420	222	90
12	280	640	240	64	19	72	1,700	440	440	1,420	222	120
13	470	680	200	17	46	72	1,420	440	330	1,330	222	110
14	500	640	200	31	58	90	1,420	380	330	1,240	190	160
15	640	605	200	22	31	100	1,700	355	330	1,420	190	160
16	570	605	200	24	28	100	1,700	470	380	1,330	190	175
17	585	605	200	24	19	64	2,210	720	355	2,100	132	175
18	410	380	200	19	160	90	2,210	680	260	1,510	64	190
19	410	380	200	22	46	110	1,800	410	260	1,330	110	205
20	410	380	200	90	51	110	1,800	680	190	1,020	190	440
21	640	440	145	72	200	145	1,700	680	145	840	160	380
22	470	440	160	72	72	145	2,540	640	145	500	132	640
23	440	440	64	110	110	145	2,980	640	2,760	500	132	470
24	305	500	46	145	31	260	2,980	535	760	500	90	355
25	760	570	80	72	120	360	2,540	470	680	500	64	355
26	1,150	570	80	58	330	260	2,000	440	680	440	80	500
27	925	640	58	17	145	260	1,800	440	605	305	80	2,320
28	1,060	640	64	28	145	260	1,420	440	570	355	72	1,600
29	970	640	72	28	......	300	1,510	380	260	380	120	720
30	1,060	500	72	40	......	330	2,100	260	305	500	100	535
31	4,830	......	90	58	......	500	......	500	......	355	51	......

NOTE.—Stage-discharge relation affected by ice Dec. 10 to Apr. 8; discharge for this period computed from gage heights corrected for effect of ice by means of three discharge measurements, observer's notes, and weather records.

Monthly discharge of Piscataquis River near Foxcroft, Maine, for the year ending Sept. 30, 1918.

[Drainage area, 286 square miles.]

Month.	Discharge in second-feet.				Run-off (depth in inches on drainage area).
	Maximum.	Minimum.	Mean.	Per square mile.	
October	4,830	175	647	2.26	2.61
November	2,430	380	726	2.54	2.83
December	380	46	217	.759	.88
January	145	17	61.6	.214	.25
February	330	19	69.8	.244	.25
March	560	64	216	.755	.87
April	2,980	640	1,970	6.89	7.69
May	4,110	260	914	3.19	3.68
June	2,760	90	448	1.57	1.75
July	2,540	190	852	2.98	3.44
August	355	51	182	.636	.73
September	2,320	51	377	1.32	1.47
The year	4,830	17	557	1.95	26.45

PASSADUMKEAG RIVER AT LOWELL, MAINE.

LOCATION.—About 400 feet below dam and highway bridge at Lowell, Penobscot County, and 10 miles above mouth of river.

DRAINAGE AREA.—301 square miles.

RECORDS AVAILABLE.—October 1, 1915, to September 30, 1918.

GAGES.—Chain and staff gages on left bank; from October 1, 1915, to October 1, 1917, chain and staff gages on right bank half a mile below the highway bridge; read by F. A. Lord. Staff above dam for supplementary use during winter.

DISCHARGE MEASUREMENTS.—Made from cable near gage.

CHANNEL AND CONTROL.—Channel rough and somewhat irregular; control about 100 feet below gage; practically permanent.

EXTREMES OF DISCHARGE.—Maximum stage recorded during year, 3.30 feet several times during April and May (discharge, 1,490 second-feet); minimum stage recorded, 1.40 feet at 8 a. m. August 30 (discharge, 127 second-feet).

1916-1918: Maximum stage recorded, 5.8 feet at 9.30 a. m. April 26, 1917 (discharge, 2,460 second-feet); minimum stage recorded, 1.40 feet at 8 a. m. August 30, 1918 (discharge, 127 second-feet).

ICE.—Stage-discharge relation usually affected by ice from December to April.

REGULATION.—Distribution of flow somewhat affected by use of storage reservoirs above station. A small dam and mill 400 feet above the gage cause fluctuations in stage for a short time each day when mill is in operation.

ACCURACY.—Stage-discharge relation practically permanent, except when affected by backwater due to logs on control or to ice. Gage read to half-tenths once daily. Rating curve well defined between 90 and 2,000 second-feet. Daily discharge ascertained by applying gage height to rating table and making corrections for effect of ice during the winter. Records fair.

COOPERATION.—Discharge measurements made by engineers employed by T. W. Clark, hydraulic engineer, Oldtown, Maine.

Discharge measurements of Passadumkeag River at Lowell, Maine, during the year ending Sept. 30, 1918.

Date.	Made by—	Gage height.	Discharge.	Date.	Made by—	Gage height.	Discharge.
		Feet.	*Sec.-ft.*			*Feet.*	*Sec.-ft.*
Oct. 6	Pressey and Lancaster.	1.67	191	Mar. 12	H. A. Lancaster	a 1.84	226
24	H. A. Lancaster	2.18	481	Apr. 3	....do	2.56	758
Nov. 2	Clark and Lancaster	2.52	749	4	....do	2.70	843
28	H. A. Lancaster	2.15	436	Sept. 18	....do	1.14	94
Jan. 30	....do	a 1.77	182	18	....do	1.17	110
30	....do	a 1.77	180				

a Stage-discharge relation affected by ice.

Daily discharge, in second-feet, of Passadumkeag River at Lowell, Maine, for the year ending Sept. 30, 1918.

Day.	Oct.	Nov.	Dec.	Jan.	Feb.	Mar.	Apr.	May.	June.	July.	Aug.	Sept.
1	178	712	382	190	180	300	382	1,490	712	478	275	138
2	163	712	382	180	180	300	669	1,490	712	444	275	138
3	178	712	382	180	180	270	588	1,440	669	478	252	138
4	178	628	382	180	180	270	845	1,380	628	382	252	138
5	178	628	353	180	180	250	845	1,330	669	382	252	138
6	194	628	353	180	180	230	845	1,220	628	353	231	138
7	212	550	353	180	180	230	845	1,220	669	353	252	150
8	275	550	380	180	180	230	890	1,220	800	480	231	150
9	275	550	353	180	180	230	935	1,070	760	550	275	178
10	300	588	350	180	180	230	980	1,020	760	710	252	212
11	326	478	350	180	180	230	1,120	980	710	890	300	212
12	382	478	350	180	180	230	1,070	1,020	670	840	326	212
13	478	478	350	180	180	230	1,070	1,020	630	800	275	212
14	669	444	330	180	180	230	935	980	630	756	252	231
15	669	444	326	180	180	230	935	980	630	756	252	275
16	669	444	330	180	180	230	980	980	590	756	252	275
17	712	412	326	180	180	210	1,070	1,020	550	800	231	275
18	669	353	300	180	190	212	1,070	935	510	756	252	275
19	588	300	300	180	210	212	513	935	480	756	231	252
20	628	353	300	180	210	212	513	890	440	712	231	353
21	669	353	275	180	230	231	1,070	756	410	669	231	382
22	513	353	275	180	230	231	1,170	800	380	669	212	628
23	513	353	275	180	230	231	1,330	712	440	588	252	712
24	478	382	275	180	250	252	1,440	800	510	513	252	760
25	513	444	252	180	252	252	1,490	800	510	478	178	760
26	588	478	230	180	270	252	1,440	756	510	382	194	800
27	628	513	230	180	270	252	1,380	756	510	326	194	940
28	669	444	210	180	300	252	1,330	756	513	330	194	980
29	628	478	210	180	......	275	1,070	756	478	326	178	980
30	628	444	210	180	......	300	1,380	712	478	300	127	980
31	669	......	210	180	......	326	......	712	......	300	138	......

NOTE.—Stage-discharge relation affected by ice Dec. 8, 10–14, 16; Dec. 26 to Feb. 24; and Feb. 26 to Mar. 17. Discharge for these periods computed from gage heights corrected for effect of ice by means of three discharge measurements and gage heights at dam. Corrections made for operation of gates July 8, 28; and for log jams June 8–27, July 8–13, and Sept. 24–30.

Monthly discharge of Passadumkeag River at Lowell, Maine, for the year ending Sept. 30, 1918.

[Drainage area, 301 square miles.]

Month.	Discharge in second-feet.				Run-off (depth in inches on drainage area).
	Maximum.	Minimum.	Mean.	Per square mile.	
October	712	163	465	1.54	1.78
November	712	300	489	1.62	1.81
December	382	210	309	1.03	1.19
January	190	180	180	.598	.69
February	300	180	204	.678	.71
March	326	210	246	.817	.94
April	1,490	382	1,010	3.36	3.75
May	1,490	712	998	3.32	3.83
June	800	380	586	1.95	2.18
July	890	300	558	1.85	2.13
August	326	127	235	.781	.90
September	980	138	400	1.33	1.48
The year	1,490	127	474	1.57	21.39

KENDUSKEAG STREAM NEAR BANGOR, MAINE.

LOCATION.—At highway bridge at Sixmile Falls, 6 miles northwest of Bangor, Penobscot County, and 7 miles below mouth of Black Stream.

DRAINAGE AREA.—191 square miles. See "Diversions."

RECORDS AVAILABLE.—September 15, 1908, to September 30, 1918.

GAGE.—Chain attached to bridge; read by Fred Cort.

DISCHARGE MEASUREMENTS.—Made from the bridge.

CHANNEL AND CONTROL.—Practically permanent; channel broken by one pier at the bridge.

EXTREMES OF DISCHARGE.—Maximum stage recorded during year, 8.7 feet at 7.35 a. m. April 4 (discharge, 4,370 second-feet); minimum stage recorded, 1.7 feet several times in June and September (discharge, 29 second-feet).

ICE.—Stage-discharge relation seriously affected by ice for several months.

DIVERSIONS.—An artificial cut was made for log driving through a low divide between Souadabscook Stream and Black Stream, which enters the Kenduskeag about 7 miles above the gaging station. During high stages of the Souadabscook part of its waters finds its way through the artificial cut into the Kenduskeag; at low stages of the Souadabscook all the flow continues down its own channel; Black Stream probably sends its waters only to the Kenduskeag.

ACCURACY.—Stage-discharge relation probably permanent except when affected by ice. Rating curve well defined below 3,600 second-feet. Gage read to tenths twice daily during open-water period; three times a week from December 25 to March 26. Daily discharge ascertained by applying mean daily gage height to rating table and making corrections for effect of ice during the winter. Records good for ordinary stages.

Discharge measurements of Kenduskeag Stream near Bangor, Maine, during the year ending Sept. 30, 1918.

Date.	Made by—	Gage height.	Discharge.	Date.	Made by—	Gage height.	Discharge.
		Feet.	*Sec.-ft.*			*Feet.*	*Sec.-ft.*
Dec. 24	A. F. McAlary	[a] 2.80	69	Apr. 1	A. F. McAlary	[a] 7.35	1,760
Jan. 26	do	[a] 2.98	59	July 5	H. A. Lancaster	1.75	32.7
Feb. 25	do	[a] 4.47	210				

[a] Stage-discharge relation affected by ice.

Daily discharge, in second-feet, of Kenduskeag Stream near Bangor, Maine, for the year ending Sept. 30, 1918.

Day.	Oct.	Nov.	Dec.	Jan.	Feb.	Mar.	Apr.	May.	June.	July.	Aug.	Sept.
1	76	1,750	311	60	60	860	1,800	790	48	42	123	29
2	84	1,480	265	60	60	760	3,000	1,240	48	37	99	33
3	76	1,360	206	60	60	680	4,000	1,060	37	37	84	37
4	84	890	206	60	54	620	4,370	790	37	37	68	29
5	68	790	181	60	48	540	3,930	538	37	37	61	29
6	84	615	170	60	48	380	2,950	538	29	48	76	33
7	91	538	170	60	48	430	2,460	463	29	76	61	29
8	91	500	150	60	54	380	2,370	343	29	181	76	29
9	107	576	140	60	60	360	2,050	375	29	392	68	33
10	115	392	125	60	60	360	1,120	327	29	740	123	37
11	150	296	115	54	60	340	1,540	280	29	1,180	159	42
12	194	234	100	54	60	330	1,540	343	33	1,480	170	37
13	265	206	100	48	60	310	1,500	265	37	1,610	170	76
14	392	206	100	60	68	330	1,200	250	29	1,970	181	140
15	359	194	100	68	68	330	1,060	234	37	2,950	206	206
16	250	206	90	68	76	340	1,000	181	29	2,550	181	296
17	206	181	90	68	90	330	1,000	206	33	2,050	159	427
18	170	181	90	76	100	330	945	170	33	1,000	150	500
19	159	234	90	84	100	340	740	132	29	790	115	538
20	463	296	90	90	115	360	655	115	33	615	84	840
21	655	392	84	90	130	360	538	107	37	538	54	1,480
22	538	375	100	90	140	360	890	107	42	538	61	1,610
23	463	427	76	84	160	360	1,480	91	68	615	76	1,610
24	392	463	68	76	180	330	1,420	91	280	538	91	1,610
25	1,060	538	68	68	210	330	1,180	76	234	392	91	1,480
26	1,750	463	68	60	440	380	840	68	150	206	76	1,480
27	1,360	538	68	60	760	410	655	61	99	194	61	1,480
28	1,000	463	68	60	820	460	538	61	68	206	54	1,610
29	1,120	410	68	60	......	800	500	76	61	181	42	1,180
30	1,120	343	60	60	......	1,200	538	61	48	159	37	840
31	1,680	......	60	60	......	1,400	......	61	......	140	37	......

NOTE.—Stage-discharge relation affected by ice Dec. 6 to Apr. 3; discharge for this period computed from gage heights corrected for effect of ice by means of four discharge measurements, observer's notes, and weather records.

Monthly discharge of Kenduskeag Stream near Bangor, Maine, for the year ending Sept. 30, 1918.

[Drainage area, 191 square miles.]

Month.	Discharge in second-feet.				Run-off (depth in inches on drainage area).
	Maximum.	Minimum.	Mean.	Per square mile.	
October	1,750	68	472	2.47	2.85
November	1,750	181	518	2.71	3.02
December	311	60	119	.623	.72
January	90	48	65.7	.344	.40
February	820	48	150	.780	.81
March	1,400	310	487	2.55	2.94
April	4,370	500	1,590	8.34	9.31
May	1,240	61	306	1.60	1.84
June	280	29	58.7	.307	.34
July	2,950	37	694	3.63	4.18
August	206	37	99.8	.523	.60
September	1,610	29	593	3.10	3.46
The year	4,370	29	429	2.25	30.47

KENNEBEC RIVER BASIN.

MOOSEHEAD LAKE AT EAST OUTLET, MAINE.

LOCATION.—At wharf at east outlet of lake, 8 miles from Kineo, Piscataquis County.

DRAINAGE AREA.—1,240 square miles.

RECORDS AVAILABLE.—April 1, 1895, to September 30, 1918.

GAGE.—Staff at end of boat landing; two datums have been used at east outlet; the first (or original datum) is 1,011.30 feet above mean sea level and about 10 feet below sills of outlet gates; gage is read to this datum; the second, to which all gage readings published to and including 1911 have been referred, is 10 feet higher; that is, the zero is at the sill of the gates; as it is believed that low water may go below the sill of the gates (zero of second datum), gage heights since 1912 are published as read—that is, to original datum.

REGULATION.—The lake is regulated to a capacity of 23,735 million cubic feet. The dam at the east outlet is controlled by 39 gates, the sills of the gates being at elevations varying from 8.0 feet to 11.4 feet. At extreme low stages the flow from the lake is controlled not by the gates but by a bar above the dam at a gage height of about 9 feet. The records show only fluctuations in the level of the lake and are used in the studies of regulation of the lake and in computing the natural flow of the Kennebec at The Forks.

COOPERATION.—Record furnished by Hollingsworth & Whitney Co.

Daily gage height, in feet, of Moosehead Lake at east outlet, Maine, for the year ending Sept. 30, 1918.

Day.	Oct.	Nov.	Dec.	Jan.	Feb.	Mar.	Apr.	May.	June.	July.	Aug.	Sept.
1	15.6	15.55			13.05		11.4	15.5		16.75		
2				14.6		11.95					16.7	15.0
3	15.5		16.0				11.5		17.2	16.65		
4				14.5	12.95	11.9						14.9
5	15.4	16.0	15.95						17.2	16.6	16.55	
6					12.8	11.85	11.8	16.4				
7		16.1	15.7	14.3							16.5	14.8
8	15.2				12.75	11.75	12.05	16.6	17.2	16.6	16.4	
9		16.1		14.2								14.9
10	15.15		15.8				12.45	16.65	17.05	16.8		
11					12.65	11.7						14.5
12	15.15	16.1	15.75	14.0	12.5		12.6		17.05	17.0	16.4	
13						11.65		16.9				14.4
14		16.2	15.7	13.9					17.0		16.25	
15	15.0				12.45	11.6	13.0	17.1		17.0		
16	15.0	16.25		13.75								
17			15.5					17.2	16.9	17.1	16.0	14.2
18				13.8	12.3	11.6						14.1
19	14.9	16.25	15.4				13.7		16.8	17.1	15.95	
20					12.2	11.6		17.3				
21				13.6							15.8	
22	14.9		15.25			11.55	14.1	17.3	16.7	17.0		
23		16.2		13.55	12.0						15.75	14.3
24	14.9		15.1				14.4		16.8	17.0		
25				13.45	12.0	11.5		17.3				
26	14.9	16.2	15.0				14.7		16.9	16.9	15.5	14.5
27					12.0	11.5		17.3				
28		16.15	14.9	13.25					16.9		15.4	
29	15.0						15.1			16.8		
30		16.1		13.2		11.4					15.3	14.6
31	15.3		14.7					17.3		16.7		

KENNEBEC RIVER AT THE FORKS, MAINE.

Location.—At wooden highway bridge, 2,000 feet above mouth of Dead River, at The Forks, Somerset County.

Drainage area.—1,570 square miles.

Records available.—September 28, 1901, to September 30, 1918.

Gages.—Chain on bridge, a vertical staff on timber retaining wall on left bank, 75 feet above bridge, and a Gurley 7-day water-stage recorder on left abutment, recorder set to read the same as chain gage at low water, but gives lower readings than chain gage at high water; used during summer months only. Chain gage read by S. C. Durgin.

Discharge measurements.—Made from the bridge.

Channel and control.—Channel at bridge is subject to slight changes in section; control is occasionally affected by backwater from Dead River.

Extremes of discharge.—Maximum stage recorded during year, from water-stage recorder, 6.19 feet at 10 a. m. May 2 (discharge, 9,670 second-feet); minimum stage recorded, 1.10 feet on August 15, 16, and 17 (discharge, 580 second-feet).

Ice.—Stage-discharge relation seriously affected by ice for several months.

Regulation.—Flow regulated by storage in Moosehead Lake. During May, June, July, and August the operation of Indian Pond for log driving causes a large diurnal fluctuation. Records of monthly discharge have been reduced to natural flow by adding or subtracting the amount of water stored in or released from Moosehead Lake.

Accuracy.—Stage-discharge relation occasionally affected by backwater from Dead River and by ice during the winter. Rating curve fairly well defined, a table of relation being used to convert discharge rating for chain gage to a corresponding rating for water-stage recorder. Water-stage recorder in operation October 1-12 and April 25 to September 30; chain gage read to half-tenths once daily. Daily discharge when water-stage recorder was in operation determined by use of discharge integrator. When water-stage recorder was not in operation, discharge ascertained by applying daily gage height to rating table and making corrections for effect of ice during the winter. Records fair for period when water-stage recorder was in operation and poor during remainder of year.

Discharge measurements of Kennebec River at The Forks, Maine, during the year ending Sept. 30, 1918.

Date.	Made by—	Gage height.	Discharge.	Date.	Made by—	Gage height.	Discharge.
		Feet.	*Sec.-ft.*			*Feet.*	*Sec.-ft.*
Jan 23	A. F. McAlary	[a]3.80	2,390	Apr. 25	A. F. McAlary	[b]3.20	2,100
Feb. 12	.....do	[a]4.30	2,440	Sept. 27	H. A. Lancaster	1.48	842
Mar. 19	.....do	2.33	1,580				

[a] Stage-discharge relation affected by ice.
[b] Gage height affected by backwater from Dead River.

Daily discharge, in second-feet, of Kennebec River at The Forks, Maine, for the year ending Sept. 30, 1918.

Day.	Oct.	Nov.	Dec.	Jan.	Feb.	Mar.	Apr.	May.	June.	July.	Aug.	Sept.
1	3,000	3,170	3,320	2,900	2,300	1,500	1,100	3,300	3,650	3,100	2,900	2,900
2	2,900	2,330	3,320	2,600	2,300	1,500	3,300	4,350	3,700	3,200	2,800	2,800
3	3,000	2,330	3,320	2,600	2,300	1,250	2,900	3,850	3,400	3,200	2,500	2,650
4	3,100	1,510	3,320	2,600	2,300	1,250	2,400	3,500	2,700	3,250	3,300	2,650
5	3,000	1,300	3,320	2,600	2,300	1,250	2,100	1,700	3,350	3,300	3,500	2,650
6	3,000	1,100	3,170	2,600	2,300	1,250	1,960	1,800	3,250	2,550	3,200	2,650
7	3,000	1,010	3,100	2,600	2,600	1,250	1,960	1,850	2,950	2,900	3,100	2,650
8	3,300	1,300	3,100	2,600	2,600	1,300	1,960	3,700	2,700	3,200	3,250	2,600
9	2,600	1,960	3,000	2,500	2,600	1,300	1,960	3,400	2,850	3,650	3,050	2,500
10	2,500	1,960	3,000	2,500	2,500	1,300	1,960	3,100	3,400	3,250	3,000	2,450
11	2,500	1,960	2,900	2,500	2,500	1,400	1,960	3,400	3,400	2,600	2,700	2,400
12	2,100	1,960	2,900	2,500	2,500	1,400	1,740	1,550	2,950	3,400	2,900	3,000
13	2,600	1,960	2,900	2,500	1,950	1,450	1,740	1,400	3,000	2,700	2,750	2,850
14	2,600	1,960	2,900	2,500	1,900	1,500	1,510	3,700	3,000	3,500	2,650	2,800
15	2,460	1,850	2,900	2,600	1,850	1,500	1,960	3,400	3,000	4,200	2,500	2,800
16	2,460	1,850	2,900	2,600	1,850	1,550	2,740	3,300	2,850	4,350	2,700	2,800
17	2,200	1,850	2,900	2,500	1,800	1,550	3,320	3,400	2,950	3,800	3,750	2,800
18	1,960	1,850	3,000	2,500	1,700	1,550	3,320	4,600	2,800	3,400	2,950	2,750
19	1,510	1,850	3,300	2,500	1,700	1,550	2,740	3,550	3,000	3,550	2,700	2,600
20	1,510	1,850	3,200	2,500	1,600	1,550	2,460	4,800	2,850	3,650	2,650	2,600
21	1,510	1,850	3,200	2,300	1,500	1,550	2,200	3,050	3,000	3,800	2,550	2,500
22	1,620	2,200	3,000	2,300	1,400	1,550	2,200	5,000	2,900	3,200	2,500	1,380
23	1,510	2,330	2,900	2,400	1,400	1,500	2,460	3,800	1,500	3,200	2,550	1,080
24	1,740	2,460	2,900	2,400	1,450	1,500	3,320	4,050	1,000	3,300	2,900	900
25	2,080	2,330	2,900	2,400	1,500	1,500	3,300	3,300	850	3,200	3,050	800
26	1,960	2,460	2,900	2,600	1,550	1,500	2,100	3,300	750	3,000	3,000	750
27	1,960	2,460	2,700	2,600	1,550	1,500	2,000	3,100	3,000	3,000	2,950	2,100
28	1,960	2,330	2,700	2,600	1,550	1,500	2,000	2,800	3,000	2,300	2,900	2,350
29	1,850	3,170	3,000	2,500	....	1,500	1,800	3,200	3,000	3,650	2,800	2,200
30	1,740	3,640	3,000	2,500	....	1,250	2,400	1,000	3,060	3,050	2,850	1,700
31	3,320	....	3,000	2,300	....	1,250	....	3,000	....	3,200	2,900	....

NOTE.—Stage-discharge relation affected by ice Dec. 7 to Mar. 2, Mar. 7–13, and Apr. 2–5; discharge for these periods computed from gage heights corrected for effect of ice by means of two discharge measurements, records of discharge from Moosehead Lake, and weather records.

Monthly discharge of Kennebec River at The Forks, Maine, for the year ending Sept. 30, 1918.

[Drainage area, 1,570 square miles.]

Month.	Discharge in second-feet.			Corrected run-off (depth in inches on drainage area).
	Observed.	Corrected for storage.		
	Mean.	Mean.	Per square mile.	
October	2,340	1,920	1.22	1.41
November	2,070	3,060	1.95	2.18
December	3,030	1,360	.866	1.00
January	2,520	630	.401	.46
February	1,980	550	.350	.36
March	1,430	730	.465	.54
April	2,300	6,930	4.42	4.93
May	3,200	5,720	3.64	4.20
June	2,790	2,170	1.38	1.54
July	3,280	3,160	2.01	2.32
August	2,900	1,100	.701	.81
September	2,320	1,580	1.01	1.13
The year	2,520	2,410	1.54	20.88

KENNEBEC RIVER AT WATERVILLE, MAINE.

Location.—At dam and mill of Hollingsworth & Whitney Co. at Waterville, Kennebec County, 2 miles above Sebasticook River and 3½ miles above Messalonskee Stream.

Drainage area.—4,270 square miles.

Records available.—March 22, 1892, to Sept. 30, 1918.

Gages.—Rod gages in pond above dam and in tailrace of mill. A water-stage recorder is used to obtain a record of height of water in tailrace and head on the wheels.

Determination of discharge.—Daily discharge values are the sums of the discharge through several wheels, through the logway, and over the spillway, as computed from one set of observations per day on several gages. When flow is less than about 3,500 second-feet all the water is used through the wheels.

Ice.—Stage-discharge relation not as a rule affected by ice; in most years winter flow passes through wheels of mill.

Regulation.—Numerous power plants and much storage above station; results not corrected for storage.

Accuracy.—Daily discharge as given is the sum of the discharge through several wheels and over the spillway, as determined from one set of observations per day on several gages. Owing to the possibility of changes in stage and uncertainties of ratings of the wheels, and the spillway, the determinations may differ appreciably from the true mean daily discharge. Therefore the records as published can be considered only fair. Errors in determinations for individual days are probably compensatory, and may be largely eliminated in the computed mean discharge for a month or a year.

Cooperation.—Records furnished by Hollingsworth & Whitney Co.

Daily discharge, in second-feet, of Kennebec River at Waterville, Maine, for the year ending Sept. 30, 1917.

Day.	Oct.	Nov.	Dec.	Jan.	Feb.	Mar.	Apr.	May.	June.	July.	Aug.	Sept.
1	2,790	3,820	23,500	4,320	4,020	3,840	14,600	17,800	17,400	14,500	34,000	8,160
2	5,000	3,820	17,200	4,180	4,230	3,850	14,400	20,300	15,500	14,300	19,100	7,470
3	3,920	3,970	7,740	4,150	4,930	4,110	11,700	22,200	13,200	13,300	12,200	7,690
4	3,830	3,820	6,680	3,940	2,390	1,360	10,600	23,600	15,200	9,790	11,600	6,970
5	2,860	3,640	9,610	3,970	4,320	4,800	11,800	20,200	15,200	11,700	10,300	6,540
6	2,970	4,690	9,190	4,730	4,000	3,850	12,900	18,800	11,000	13,300	9,230	5,800
7	2,740	3,330	9,630	4,400	4,130	3,870	40,900	16,600	10,500	12,900	5,440	5,810
8	2,700	3,280	9,150	4,360	3,820	3,850	40,000	16,700	10,600	11,500	3,890	6,320
9	4,440	3,230	7,720	4,610	3,870	3,880	37,500	15,000	14,000	12,000	5,640	4,960
10	4,700	3,550	6,350	4,510	3,930	4,060	28,900	11,400	11,100	12,000	6,430	6,890
11	4,760	3,930	7,920	4,450	2,420	1,190	19,100	11,400	14,800	10,800	10,700	6,620
12	3,950	2,340	7,240	4,250	4,340	4,790	17,900	14,500	61,000	11,800	12,500	4,640
13	3,940	4,030	4,700	3,920	3,840	4,290	11,500	16,100	76,500	12,000	10,700	5,230
14	3,460	3,770	4,870	1,320	3,900	3,900	12,800	12,400	53,800	12,000	9,340	5,000
15	3,190	3,540	3,540	5,000	4,920	4,050	14,500	16,600	45,700	9,440	7,360	4,810
16	4,670	3,300	3,540	6,510	4,120	3,850	14,400	12,900	42,000	12,400	7,410	3,510
17	3,760	2,940	100	5,430	4,400	3,950	15,100	13,000	41,000	11,900	7,660	5,060
18	3,380	3,620	4,390	5,090	2,040	1,580	13,500	11,100	88,500	11,900	8,160	4,580
19	3,670	2,280	5,140	6,060	4,880	4,950	14,900	18,200	78,800	10,600	5,980	4,790
20	5,320	3,580	4,480	5,290	3,700	4,220	18,300	9,780	48,600	10,500	8,090	5,470
21	11,100	3,670	4,670	3,660	4,390	3,960	20,200	12,900	44,600	10,700	8,130	5,480
22	6,650	3,010	4,660	5,370	4,340	3,950	23,500	12,900	41,000	9,230	9,850	5,270
23	6,810	3,260	5,720	4,140	4,000	4,400	27,500	16,400	37,400	10,700	9,350	4,380
24	5,080	3,910	6,710	4,020	4,050	4,420	30,000	14,600	29,300	10,300	8,570	5,460
25	4,010	7,500	6,300	3,640	916	1,870	27,200	19,400	27,400	10,000	15,000	4,800
26	4,330	4,240	7,130	4,500	4,840	4,700	20,500	18,700	23,700	4,520	13,000	4,800
27	3,710	4,820	5,720	4,820	4,800	5,620	20,500	16,300	17,000	4,030	3,950	4,860
28	3,260	4,290	4,950	4,060	3,890	12,200	19,000	15,800	13,500	3,980	19,700	4,740
29	3,000	4,680	5,710	4,580	....	25,900	18,200	13,700	11,900	3,900	8,460	4,590
30	3,930	4,010	4,610	4,180	....	23,600	12,500	14,100	11,700	4,020	8,100	3,250
31	3,910	....	2,990	[illegible],080	....	18,400	....	17,800	....	12,100	8,470	....

Monthly discharge of Kennebec River at Waterville, Maine, for the year ending Sept. 30, 1917.

[Drainage area, 4,270 square miles.]

Month.	Discharge in second-feet.				Run-off (depth in inches on drainage area).
	Maximum.	Minimum.	Mean.	Per square mile.	
October	11,100	2,700	4,250	0.996	1.15
November	7,500	2,280	3,800	.890	.99
December	23,500	100	6,830	1.60	1.84
January	6,510	1,320	4,440	1.04	1.20
February	4,930	916	3,910	.916	.95
March	25,900	1,190	5,910	1.38	1.59
April	40,900	10,600	19,800	4.64	5.18
May	23,600	9,780	15,800	3.70	4.27
June	88,500	10,500	31,400	7.35	8.20
July	14,500	3,900	10,400	2.44	2.81
August	34,000	3,890	10,300	2.41	2.78
September	8,160	3,250	5,470	1.28	1.43
The year	88,500	100	10,200	2.39	32.39

NOTE.—The monthly discharge in second-feet per square mile and the run-off in depth in inches do not represent the natural flow from the basin because of artificial storage. The yearly discharge and run-off doubtless represent more nearly the natural flow, for probably little stored water is held over from year to year.

Daily discharge, in second-feet, of Kennebec River at Waterville, Maine, for the year ending Sept. 30, 1918.

Day.	Oct.	Nov.	Dec.	Jan.	Feb.	Mar.	Apr.	May.	June.	July.	Aug.	Sept.
1	4,760	20,200	3,980	3,930	3,400	3,920	12,100	24,400	7,250	5,980	4,450	2,380
2	4,410	14,200	3,390	3,890	3,300	3,950	12,700	33,900	6,730	4,490	4,570	4,320
3	4,870	11,500	5,300	3,830	2,340	3,330	52,900	26,600	7,340	4,440	4,560	4,190
4	4,260	9,480	4,830	3,780	2,950	4,380	32,100	21,100	5,660	3,910	4,460	4,310
5	4,240	7,060	4,590	3,880	3,220	3,900	28,200	19,100	5,050	5,150	5,130	3,820
6	3,970	5,690	4,710	3,770	2,880	3,930	23,100	17,400	4,130	4,430	4,640	3,900
7	3,570	5,230	4,710	3,870	2,940	3,860	18,900	9,820	4,780	3,320	4,680	3,900
8	5,030	4,980	4,150	3,980	2,900	3,860	21,600	16,400	4,740	5,220	4,690	2,850
9	4,590	4,260	3,160	3,880	3,029	3,890	20,900	16,400	4,600	4,740	4,680	3,890
10	4,190	5,020	4,750	3,880	443	2,590	20,900	12,700	6,040	7,660	4,980	3,500
11	4,050	3,440	4,130	3,980	2,970	3,900	18,200	12,800	4,750	10,900	5,200	3,130
12	3,980	5,680	3,880	3,880	3,360	3,610	16,100	11,600	4,730	7,320	5,220	3,870
13	4,040	4,260	3,880	2,760	3,490	3,590	17,600	11,800	4,620	7,660	4,590	3,910
14	2,820	4,840	3,860	3,980	3,670	3,160	11,500	10,200	4,690	10,100	4,500	3,910
15	5,340	4,340	3,860	3,860	3,670	3,860	13,300	20,500	4,800	13,100	4,600	3,240
16	5,090	4,340	2,130	3,880	3,780	3,910	14,700	16,900	4,060	3,880	4,470	4,130
17	5,100	4,340	3,960	3,860	2,200	1,840	17,900	12,800	5,180	12,100	4,510	3,870
18	4,830	3,040	3,830	3,860	3,620	3,830	18,600	12,200	4,660	11,100	3,600	3,880
19	4,590	4,610	3,830	2,180	3,870	3,810	20,400	13,600	4,040	8,410	5,140	3,830
20	4,600	4,030	3,930	1,760	3,150	3,810	15,200	7,280	4,330	8,410	4,410	4,080
21	3,820	3,860	3,930	3,100	3,830	3,830	11,800	9,900	4,430	7,320	4,290	6,630
22	4,190	3,860	4,000	3,100	3,670	3,900	14,300	8,770	4,240	8,350	3,430	7,250
23	3,910	3,880	2,580	4,520	3,660	4,230	17,400	10,100	5,670	6,970	3,640	6,150
24	3,860	4,210	3,840	3,780	2,630	5,280	21,100	8,820	11,100	6,050	3,360	5,540
25	3,940	3,480	2,970	3,650	3,930	5,550	22,000	8,820	9,070	6,280	2,430	5,000
26	10,100	4,540	4,110	3,680	3,770	5,380	19,200	7,970	6,700	5,090	4,660	4,630
27	8,770	3,860	3,890	2,060	3,860	6,030	15,700	8,990	4,400	4,410	4,560	8,210
28	6,130	3,860	3,890	3,100	3,880	6,450	13,000	4,460	4,000	3,020	4,240	21,900
29	8,720	2,000	3,520	3,640	....	6,110	16,200	4,590	5,060	4,620	4,060	12,300
30	8,020	4,430	2,410	3,350	....	7,880	14,800	4,920	4,410	4,540	4,550	9,730
31	17,100	....	3,920	3,340	....	5,760	....	8,990	....	4,530	4,300	....

Monthly discharge of Kennebec River at Waterville, Maine, for the year ending Sept. 30, 1918.

[Drainage area, 4,270 square miles.]

Month.	Discharge in second-feet.				Run-off (depth in inches on drainage area).
	Maximum.	Minimum.	Mean.	Per square mile.	
October	17,100	2,820	5,370	1.26	1.45
November	20,200	2,000	5,620	1.32	1.47
December	5,300	2,130	3,870	.906	1.04
January	4,520	1,760	3,550	.831	.96
February	3,930	443	3,230	.756	.79
March	7,880	1,840	4,300	1.01	1.16
April	52,900	11,500	19,100	4.47	4.99
May	33,900	4,460	13,300	3.11	3.58
June	11,100	4,000	5,370	1.26	1.41
July	13,100	3,020	6,560	1.54	1.78
August	5,220	2,430	4,400	1.03	1.19
September	21,900	2,380	5,410	1.27	1.42
The year	52,900	443	6,680	1.56	21.24

NOTE.—The monthly discharge in second-feet per square mile and the run-off in depth in inches do not represent the natural flow from the basin because of artificial storage. The yearly discharge and run-off doubtless represent more nearly the natural flow, for comparatively little stored water is held over from year to year.

DEAD RIVER AT THE FORKS, MAINE.

LOCATION.—One-eighth mile above farmhouse of Jeremiah Durgin, 1¼ miles west of The Forks, Somerset County.

DRAINAGE AREA.—878 square miles.

RECORDS AVAILABLE.—September 29, 1901, to August 15, 1907; and March 16, 1910, to September 30, 1918.

GAGE.—Staff bolted to large boulder on left bank; read by H. J. Farley.

DISCHARGE MEASUREMENTS.—Made from cable 700 feet above gage.

CHANNEL AND CONTROL.—Stream bed rough; control practically permanent.

EXTREMES OF DISCHARGE.—Maximum stage recorded during year, 5.4 feet at 8.30 a. m. May 30 (discharge, 11,300 second-feet); minimum stage recorded, 0.2 foot on September 12, 13, and 17 (water held back by logging dams, exact discharge not determined).

ICE.—Stage-discharge relation seriously affected by ice.

REGULATION.—A number of dams on lakes above; used for log driving during May and June.

ACCURACY.—Stage-discharge relation practically permanent except when ice is present. Rating curve well defined above 400 second-feet. Gage read to half-tenths twice daily except from December 30 to April 1, when it was read three times a week. Some uncertainty in regard to accuracy of gage heights. Daily discharge ascertained by applying mean daily gage height to rating table, and making corrections for effect of ice during the winter. Records fair.

Discharge measurements of Dead River at The Forks, Maine, during the year ending Sept. 30, 1918.

Date.	Made by—	Gage height.	Discharge.	Date.	Made by—	Gage height.	Discharge.
		Feet.	*Sec.-ft.*			*Feet.*	*Sec.-ft.*
Jan. 3	A. F. McAlary	[a]2.30	308	Sept. 27	H. A. Lancaster	2.42	2,620
Feb. 12	.....do	[a]1.70	278	28	.....do	2.92	3,560
Mar. 19	.....do	[a]2.48	431				

[a] Stage-discharge relation affected by ice.

Daily discharge, in second-feet, of Dead River at The Forks, Maine, for the year ending Sept. 30, 1918.

Day.	Oct.	Nov.	Dec.	Jan.	Feb.	Mar.	Apr.	May.	June.	July.	Aug.	Sept.
1	965	6,140	510	320	280	1,300	6,800	6,140	965	1,030	50	462
2	665	5,530	610	320	280	1,250	7,130	6,140	840	780	50	462
3	370	6,790	610	320	280	1,250	6,460	5,830	720	560	50	462
4	370	2,750	610	320	280	1,150	5,530	6,140	560	325	75	462
5	415	2,290	510	320	280	1,100	4,970	6,140	462	240	50	415
6	510	1,780	510	320	280	960	4,220	5,530	370	160	50	370
7	840	1,700	510	370	280	900	3,990	5,240	257	200	50	415
8	720	1,540	500	400	280	840	3,990	5,530	200	370	50	415
9	720	1,390	500	400	280	720	3,990	5,830	160	720	75	415
10	665	1,240	500	400	280	600	4,220	5,240	160	1,030	100	370
11	610	1,240	500	400	280	560	3,550	6,140	160	1,240	50	....
12	462	1,170	320	400	280	460	2,750	5,530	160	1,390	50	....
13	720	1,170	320	400	280	420	2,120	5,240	224	1,540	50	....
14	965	1,390	320	400	280	370	1,780	4,460	308	1,320	130	....
15	840	1,240	240	400	280	320	2,030	3,990	397	1,100	240	....
16	965	1,100	240	400	280	320	2,750	3,770	510	965	224	....
17	1,100	1,100	240	400	320	320	4,220	3,550	415	902	160	....
18	840	965	240	400	370	370	4,970	3,140	415	840	160	....
19	840	965	240	400	460	430	4,970	2,290	343	720	100	....
20	720	965	240	400	560	720	4,710	2,200	325	720	100	240
21	720	902	320	400	600	840	3,770	1,940	325	610	100	840
22	665	840	320	400	720	960	3,990	1,700	462	610	100	1,700
23	610	720	320	400	840	1,050	5,530	1,390	780	510	100	1,620
24	560	720	320	400	900	1,050	6,140	1,390	2,030	370	90	902
25	1,100	610	320	400	1,050	1,100	6,460	1,170	1,700	240	50	665
26	2,750	610	320	400	1,150	1,300	6,790	1,100	1,540	160	462	500
27	2,380	610	320	280	1,300	1,550	5,830	1,100	1,540	160	415	1,780
28	2,200	560	320	280	1,300	1,950	6,140	1,240	1,540	160	462	3,340
29	2,200	560	320	280	....	2,300	3,990	1,100	1,540	100	370	3,140
30	3,990	415	320	280	....	2,800	6,790	4,710	1,460	100	415	2,560
31	6,790	....	320	280	....	4,500	....	965	....	75	370	....

NOTE.—Stage-discharge relation affected by ice from Dec. 8 to Apr. 1; discharge for this period computed from gage heights corrected for effect of ice by means of three discharge measurements, observer's reports, and weather records. Discharge estimated as averaging 75 second-feet Sept. 11-19; water held back by logging dams. (Some uncertainty in regard to accuracy of gage heights during this period.)

Monthly discharge of Dead River at The Forks, Maine, for the year ending Sept. 30, 1918.

[Drainage area, 878 square miles.]

Month.	Discharge in second-feet.				Run-off (depth in inches on drainage area).
	Maximum.	Minimum.	Mean.	Per square mile.	
October	6,790	370	1,230	1.41	1.63
November	6,790	415	1,630	1.86	2.08
December	610	240	380	.433	.50
January	400	280	364	.415	.48
February	1,300	280	502	.572	.60
March	4,500	320	1,090	1.24	1.43
April	7,130	1,780	4,690	5.34	5.96
May	6,140	965	3,740	4.26	4.91
June	2,030	160	696	.793	.88
July	1,540	75	621	.707	.82
August	462	50	155	.177	.20
September	3,340	....	742	.845	.94
The year	7,130	....	1,320	1.50	20.43

SEBASTICOOK RIVER AT PITTSFIELD, MAINE.

LOCATION.—At steel highway bridge just above Maine Central Railroad bridge in Pittsfield, Somerset County.

DRAINAGE AREA.—320 square miles.

RECORDS AVAILABLE.—July 27, 1908, to September 30, 1918.

Gage.—Chain attached to highway bridge; read by C. D. Morrill.

Discharge measurements.—Made from the highway bridge.

Channel and control.—Practically permanent; banks high and rocky and not subject to overflow.

Extremes of discharge.—Maximum stage recorded during year, 5.72 feet at 2.35 p. m. April 8 (discharge, 2,840 second-feet); minimum stage recorded, 2.38 feet at 3.10 p. m. February 23 (discharge, 69 second-feet).

Ice.—Stage-discharge relation not seriously affected by ice, as the rapid fall and the proximity of the power plant immediately above station tend to keep river open.

Regulation.—About 800 feet upstream from the station is the dam of the American Woolen Co. (Pioneer mills) and the Smith Textile Co.; and about half a mile farther upstream is the dam of the American Woolen Co.'s Waverly mill; the storage of water at these dams causes diurnal fluctuation at the gage.

Accuracy.—Stage-discharge relation has apparently changed slightly at times. Rating curve well defined between 70 and 4,000 second-feet. Gage read to half-tenths twice daily from October 1 to February 1, and to hundredths from February 2 to September 30. Owing to lack of exact information in regard to the stage at night when the mills are shut down, determinations of mean daily discharge are not published.

The following discharge measurement was made by A. F. McAlary:

November 30, 1917: Gage height, 3.64 feet; discharge, 551 second-feet.

Twice-daily discharge, in second-feet, of Sebasticook River at Pittsfield, Maine, for the year ending Sept. 30, 1918.

Day.	Oct.		Nov.		Dec.		Jan.		Feb.		Mar.	
	A. M.	P. M.	A. M.	P. M.	A. M.	P. M.	A. M.	P. M.	A. M.	P. M.	A. M	P. M.
1	331	376	1,320	1,320	376	424	376	331	......	......	376	376
2	331	376	1,320	1,380	450	450	376	376	331	158	310	154
3	331	354	1,320	1,320	560	560	424	331	400	218	154	145
4	331	331	1,320	1,320	475	502	376	331	475	376	340	372
5	331	289	1,210	1,210	424	400	400	376	475	376	331	331
6	250	250	1,160	1,210	424	376	250	250	657	331	372	354
7	180	197	1,050	1,380	376	400	376	354	376	542	376	386
8	250	331	1,320	1,210	376	376	376	376	340	336	414	400
9	331	354	1,160	1,210	331	354	376	376	475	197	386	164
10	331	376	815	475	400	400	376	376	200	145	174	180
11	331	354	657	590	400	450	400	376	542	376	434	400
12	270	376	475	590	450	400	376	180	297	400	424	376
13	310	232	502	530	424	450	148	......	297	372	386	400
14	232	214	475	530	502	475	......	......	289	376	376	340
15	250	310	475	502	214	214	......	......	331	336	376	344
16	289	354	475	475	214	214	......	......	434	154	400	142
17	331	354	475	354	400	400	......	......	170	145	142	142
18	289	354	331	331	400	400	......	......	367	354	340	331
19	310	376	424	475	376	354	......	......	376	344	367	340
20	331	310	424	424	354	354	......	......	354	331	354	331
21	180	180	331	400	354	331	......	......	386	354	340	331
22	197	331	331	376	354	376	......	......	424	354	354	331
32	289	376	331	331	214	214	......	......	354	69	354	104
24	310	376	376	400	331	354	......	......	197	148	133	133
25	310	331	424	475	180	180	......	......	400	331	340	331
26	289	354	530	590	331	354	......	......	367	354	367	340
27	310	331	475	502	354	376	......	......	331	331	405	390
28	331	400	424	475	400	376	......	......	367	331	400	386
29	475	475	......	354	400	376	......	......	......	......	390	386
30	475	475	......	502	214	250	......	......	......	......	424	310
31	815	1,160	......	......	657	530	......	......	......	......	465	578

Twice-daily discharge, in second-feet, of Sebasticook River at Pittsfield, Maine, for the year ending Sept. 30, 1918—Continued.

Day.	Apr.		May.		June.		July.		Aug.		Sept.	
	A. M.	P. M.	A. M.	P. M.	A. M.	P. M.	A. M.	P. M.	A. M.	P. M.	A. M.	P. M.
1	1,000	1,050	1,490	1,550	475	250	414	400	465	424	118	118
2	1,470	1,550	1,910	1,910	343	343	424	414	450	414	118	164
3	2,160	2,680	2,010	1,910	486	465	450	424	424	197	354	354
4	2,810	2,780	1,850	1,610	475	450	289	281	214	214	400	367
5	2,810	2,810	1,670	1,550	450	400	424	414	450	424	400	376
6	2,740	2,550	1,610	1,490	414	354	424	258	450	414	386	376
7	2,680	2,680	1,160	1,050	450	390	281	289	434	414	354	133
8	2,810	2,840	717	774	344	232	465	480	450	400	104	104
9	2,740	2,740	952	952	281	289	530	492	450	414	354	331
10	2,680	2,740	887	815	439	400	542	530	424	148	400	376
11	2,620	2,550	883	624	439	376	590	560	197	190	414	376
12	2,420	2,480	644	644	434	400	624	624	386	400	400	354
13	2,220	2,100	765	732	414	400	657	530	450	424	414	367
14	2,030	2,030	774	757	424	376	560	590	424	400	400	145
15	2,060	2,060	694	694	376	250	1,250	1,210	414	386	96	96
16	2,100	2,030	644	603	164	174	1,280	1,210	424	414	354	331
17	1,970	1,970	590	578	424	424	1,160	1,130	424	187	331	376
18	2,030	1,970	560	376	400	376	1,100	1,160	180	190	376	367
19	1,910	1,850	354	354	376	367	1,100	1,120	424	414	386	354
20	1,670	1,550	542	530	400	376	1,050	924	424	400	400	386
21	1,490	1,490	530	502	376	367	860	815	400	376	439	530
22	1,670	1,670	530	486	376	154	952	815	414	386	492	450
23	1,910	1,890	519	480	250	232	765	694	400	376	475	530
24	2,010	2,030	502	475	450	450	732	694	395	148	530	486
25	2,100	1,970	486	289	424	434	657	624	180	180	519	475
26	1,890	1,730	270	270	465	450	644	578	180	180	502	475
27	1,670	1,470	486	480	475	465	590	376	232	124	694	774
28	1,320	1,380	475	475	444	424	400	400	118	164	732	560
29	1,430	1,380	492	475	439	262	560	530	164	124	548	502
30	1,320	1,300	270	270	258	250	502	450	118	118	603	590
31	......	......	519	465	......	......	475	434	112	104	......	......

NOTE.—Times of gage height readings varied from 6 to 10 a. m. and from noon to 6 p. m. One or more of the mills above the gage were in operation 24 hours a day, except Sundays, during greater part of the time from October, 1916, to September, 1918.

ANDROSCOGGIN RIVER BASIN.

ANDROSCOGGIN RIVER AT ERROL DAM, N. H.

LOCATION.—At Errol dam, 1 mile above Errol, Coos County.

DRAINAGE AREA.—1,095 square miles.

RECORDS AVAILABLE.—January 1, 1905, to September 30, 1918.

GAGE.—Movable rod gage; readings taken daily from sill of deep gate No. 6; elevation of zero of gage or sill of gate, 1,231.3 feet above mean sea level.

DISCHARGE.—Computed from discharge through 14 gates in the dam by means of coefficients determined from a few discharge measurements.[1]

ICE.—Stage-discharge relation little affected by ice.

REGULATION.—Errol dam regulates the storage of Umbagog Lake, the lower of the Rangeley series of lakes, comprising the principal storage of Androscoggin River and amounting to nearly 20 billion cubic feet, and also a recently developed storage site on Magalloway River created by the Aziscohos dam, which amounts to about 9.6 billion cubic feet, thus making the total storage about 29.6 billion cubic feet. Errol dam is about 5 miles below outlet of Umbagog Lake and about 3. 5 miles below mouth of Magalloway River, thus making this stream one of the feeders of Umbagog Lake. Results not corrected for storage.

COOPERATION.—Records obtained and computations of daily discharge made under direction of Walter H. Sawyer, agent for Union Water Power Co., Lewiston, Maine.

[1] See U. S. Geol. Survey Water-Supply Paper 321, p. 61.

Daily discharge, in second-feet, of Androscoggin River at Errol dam, N. H., for the year ending Sept. 30, 1918.

Day.	Oct.	Nov.	Dec.	Jan.	Feb.	Mar.	Apr.	May.	June.	July.	Aug.	Sept.
1	1,930	808	2,270	2,050	2,000	1,970	1,950	2,500	1,120	1,540	2,140	1,760
2	1,900	1,490	2,360	2,100	2,030	2,140	2,070	2,470	1,180	1,630	2,190	1,820
3	1,940	1,660	2,410	2,100	1,980	2,240	2,160	2,350	1,520	1,650	2,190	1,930
4	1,920	1,610	2,990	2,050	1,910	2,160	2,200	2,120	1,530	1,820	2,180	1,930
5	1,720	1,560	2,310	2,000	1,910	2,110	2,230	1,960	1,520	1,940	2,180	1,920
6	1,690	1,560	2,200	1,830	1,850	2,130	2,030	1,940	1,480	1,870	2,190	1,950
7	1,870	1,630	2,110	1,920	1,880	2,150	1,940	1,940	1,340	1,650	2,160	2,010
8	1,890	1,520	2,100	1,970	1,910	2,180	1,940	1,870	1,290	1,600	1,980	1,950
9	1,900	1,560	2,300	1,970	1,980	2,200	1,960	1,770	1,460	1,630	1,460	1,950
10	1,890	1,540	2,220	1,980	2,000	2,290	1,960	1,170	1,560	1,810	1,030	1,780
11	1,920	1,520	2,310	1,970	1,990	2,480	1,940	830	1,680	1,920	1,390	1,870
12	2,010	1,580	2,270	1,900	1,990	2,340	1,940	1,530	1,760	1,830	1,760	1,870
13	1,900	1,940	2,120	1,880	1,940	2,240	1,740	1,690	1,590	1,650	1,910	1,950
14	1,980	2,160	2,020	1,900	1,960	2,200	1,450	894	1,400	1,230	2,090	1,540
15	1,710	2,150	2,060	1,810	1,960	2,070	1,340	896	1,590	1,180	2,160	1,900
16	1,930	2,150	2,100	1,950	2,000	2,070	1,410	900	1,680	1,370	2,140	2,020
17	1,990	2,070	2,130	2,030	2,050	2,070	1,690	896	1,660	1,580	2,140	1,980
18	1,860	2,130	2,070	1,840	2,160	2,010	1,900	818	1,770	1,630	2,140	1,540
19	1,790	2,360	2,030	2,080	2,240	1,940	2,010	1,230	1,810	1,720	2,140	1,420
20	1,750	2,280	1,980	1,940	2,400	2,010	2,050	1,560	1,790	1,840	2,050	1,400
21	1,800	2,200	1,810	2,160	2,430	2,010	1,990	1,530	1,790	1,820	2,080	(a)
22	1,800	2,120	1,730	2,140	2,430	1,970	2,070	1,500	1,320	1,770	2,130	835
23	1,770	2,030	2,210	2,120	2,190	1,950	2,080	1,500	1,100	1,800	2,130	1,350
24	1,800	2,130	1,980	2,140	2,020	2,010	2,130	990	1,120	1,950	2,130	485
25	1,600	2,290	2,020	2,160	1,900	2,060	2,170	909	1,140	2,140	2,120	622
26	1,790	2,260	2,000	2,180	1,800	2,060	2,170	1,630	1,130	2,170	2,000	329
27	1,890	2,210	2,040	2,180	1,790	2,060	2,180	1,600	1,260	2,180	1,790	55
28	2,010	2,560	2,050	2,090	1,910	2,090	2,180	1,500	1,540	2,180	1,760	197
29	1,990	2,370	2,000	2,020	....	2,090	2,280	1,480	1,540	2,130	1,760	374
30	972	2,370	2,060	1,980	....	2,090	2,370	1,080	1,720	2,060	1,770	915
31	(a)	....	2,040	1,980	....	1,980	....	1,080	....	2,050	1,770	....

a Mills shut down; water held back by dams.

Monthly discharge of Androscoggin River at Errol dam, N. H., for the year ending Sept. 30, 1918.

[Drainage area, 1,095 square miles.]

Month.	Discharge in second-feet.				Run-off (depth in inches on drainage area).
	Maximum.	Minimum.	Mean.	Per square mile.	
October	2,010	(a)	1,770	1.62	1.87
November	2,560	803	1,930	1.76	1.96
December	2,990	1,730	2,140	1.95	2.25
January	2,180	1,810	2,010	1.83	2.11
February	2,430	1,790	2,020	1.84	1.92
March	2,480	1,950	2,110	1.93	2.22
April	2,370	1,340	1,990	1.82	2.03
May	2,500	818	1,490	1.36	1.57
June	1,810	1,100	1,480	1.35	1.51
July	2,180	1,180	1,790	1.63	1.88
August	2,190	1,030	1,970	1.80	2.08
September	2,020	(a)	1,390	1.27	1.42
The year	2,990	(a)	1,840	1.68	22.82

a Mills shut down; water held back by dams.

NOTE.—The monthly discharge in second-feet per square mile and the run-off in depth in inches do not represent the natural run-off from the basin because of storage. (See "Regulation.")

ANDROSCOGGIN RIVER AT BERLIN, N. H.

Location.—At the upper or sawmill dam of the Berlin Mills Co. at Berlin, Coos County.

Drainage area.—1,350 square miles.

Records available.—October 1, 1913, to September 30, 1918.

Gages.—Fixed gages are maintained in the river above the forebay racks and in the tailrace immediately below the outlet of the wheels; these gages are referred to the same datum, and the differences in the readings give the head on the wheels; a gage is also attached to each wheel gate, from which the wheel-gate opening can be ascertained.

Determination of discharge.—Discharge computed from curves prepared from Holyoke tests of the wheel runners, using the head and gate openings as ascertained from the gages. Quantity of water wasted over the dam is computed by the Francis formula for discharge over weirs.

Ice. Stage-discharge relation not affected by ice.

Regulation.—Under the agreement between the power users on Androscoggin River, the flow at Berlin, N. H., is maintained at a minimum of 1,550 second-feet and at such a point above 1,550 second-feet as is consistent with the constant maintenance of that quantity. Final regulation of the river is made at Pontocook dam, N. H., above which is a pond containing about a day's supply; the primary regulation is made at Errol, N. H., about 30 miles above Berlin.

Cooperation.—Gages are under the direction of George P. Abbott, of the Berlin Mills Co., and discharge record is furnished for publication by Walter H. Sawyer, agent for Union Water Power Co., Lewiston, Maine.

Daily discharge, in second-feet, of Androscoggin River at Berlin, N. H., for the year ending Sept. 30, 1918.

Day.	Oct.	Nov.	Dec.	Jan.	Feb.	Mar.	Apr.	May.	June.	July.	Aug.	Sept.
1	2,000	4,000	2,300	2,300	2,000	2,000	2,400	3,700	1,950	1,900	1,950	1,650
2	1,800	3,000	2,300	2,200	2,000	2,100	3,200	3,900	2,000	1,900	1,900	1,650
3	1,700	3,000	2,300	2,300	2,000	2,100	3,800	3,500	2,000	1,900	1,900	1,640
4	1,800	3,000	2,400	2,300	2,000	2,100	3,500	3,500	1,950	1,900	1,950	1,650
5	1,900	3,200	2,700	2,300	2,100	2,100	3,300	3,200	2,000	1,950	1,900	1,620
6	1,800	2,700	2,600	2,200	1,800	1,900	2,600	3,000	1,900	1,950	1,950	1,620
7	2,000	2,400	2,500	2,200	1,800	2,000	2,400	2,900	1,950	1,900	1,900	1,620
8	2,000	2,400	2,400	2,200	1,800	2,100	3,000	2,900	2,000	1,900	1,950	1,650
9	2,100	2,400	2,300	2,200	1,900	2,000	2,900	2,900	2,200	1,900	2,400	1,650
10	2,100	2,300	2,300	2,100	1,900	2,000	2,900	2,200	1,950	1,900	2,200	1,640
11	2,100	2,300	2,400	2,300	2,000	2,100	2,900	2,200	1,950	1,900	2,000	1,630
12	2,100	2,000	2,400	2,300	2,100	2,100	2,600	2,300	1,950	1,900	1,950	1,650
13	1,800	1,900	2,500	2,300	1,900	2,100	2,600	2,300	1,850	1,950	1,900	1,650
14	1,800	2,100	2,300	2,100	1,900	2,200	2,800	2,700	1,850	2,100	1,900	1,650
15	1,900	2,100	2,200	2,000	2,000	2,100	2,600	2,700	1,850	2,000	1,900	1,620
16	2,100	2,200	2,200	2,100	1,800	2,000	2,600	2,000	1,950	2,000	1,950	1,600
17	2,100	2,300	2,200	2,200	1,800	2,000	2,800	2,000	1,950	1,990	1,950	1,650
18	2,100	2,300	2,400	2,100	1,900	2,100	2,900	1,900	2,000	1,990	1,950	1,650
19	2,100	2,300	2,400	2,100	2,100	2,000	2,600	1,900	1,950	1,990	1,950	1,570
20	2,100	2,400	2,400	2,100	2,300	2,000	2,600	2,000	1,950	1,950	1,900	1,750
21	2,100	2,600	2,300	2,100	2,400	2,100	2,800	1,900	1,950	2,000	1,850	2,000
22	1,800	2,500	2,300	2,100	2,400	2,200	2,900	2,000	1,950	1,800	1,860	1,900
23	1,800	2,400	2,200	2,300	2,100	2,100	3,100	2,000	1,950	1,900	1,900	1,650
24	1,900	2,300	2,200	2,300	2,100	2,000	3,300	1,900	1,950	1,900	1,920	1,600
25	2,200	2,200	(a)	2,200	2,300	2,100	3,200	1,900	1,950	1,950	1,900	1,650
26	2,100	2,200	2,200	2,300	2,100	2,200	3,000	2,000	1,950	1,900	1,700	1,680
27	2,100	2,200	2,100	2,300	2,100	2,200	2,900	2,000	1,950	1,900	1,620	1,850
28	2,100	2,300	2,300	2,300	2,000	2,200	3,000	1,950	1,950	1,900	1,620	1,650
29	2,400	2,300	2,200	2,400	....	2,300	3,300	1,950	1,950	1,950	1,650	1,600
30	3,600	2,300	2,200	2,200	....	2,300	3,200	1,900	2,000	1,900	1,650	1,600
31	6,300	....	2,200	2,000	....	2,300	....	1,950	....	1,900	1,650	....

a Mills shut down; water held back by dams.

Monthly discharge of Androscoggin River at Berlin, N. H., for the year ending Sept. 30, 1918.

[Drainage area, 1,350 square miles.]

Month.	Discharge in second-feet.				Run-off (depth in inches on drainage area).
	Maximum.	Minimum.	Mean.	Per square mile.	
October	6,300	1,700	2,190	1.62	1.87
November	4,000	1,900	2,450	1.81	2.02
December	2,700	(a)	2,220	1.67	1.92
January	2,400	2,000	2,210	1.64	1.89
February	2,400	1,800	2,020	1.50	1.56
March	2,300	1,900	2,100	1.56	1.80
April	3,800	2,400	2,920	2.16	2.41
May	3,900	1,900	2,420	1.79	2.06
June	2,200	1,850	1,960	1.45	1.62
July	2,100	1,800	1,930	1.43	1.65
August	2,400	1,620	1,890	1.40	1.61
September	2,000	1,570	1,670	1.24	1.38
The year	6,300	(a)	2,170	1.61	21.79

a Mills shut down; water held back by dams.

NOTE.—The monthly discharge in second-feet per square mile and the run-off depth in inches do not represent the natural run-off from the basin because of storage. (See "Regulation.")

ANDROSCOGGIN RIVER AT RUMFORD, MAINE.

LOCATION.—At two dams of Rumford Falls Power Co. at Rumford.

DRAINAGE AREA.—2,090 square miles.

RECORDS AVAILABLE.—May 18, 1892, to September 30, 1918.

GAGES.—One in pond above each dam and in tailraces of power station and mills.

DISCHARGE.—Computed from discharge over the dam by use of the Francis weir formula with modified coefficient, and the quantities passing through the various wheels of the power station and mills, which have been carefully rated.

ICE.—Stage-discharge relation little affected by ice.

REGULATION.—Storage in Rangeley system of lakes at headwaters of Androscoggin River aggregates about 29.6 billion cubic feet. The stored water is regulated in the interests of the water-power users above and below. Results not corrected for storage.

COOPERATION.—Records obtained and computations made by Mr. Charles A. Mixer, engineer, Rumford Falls Power Co.

Daily discharge, in second-feet, of Androscoggin River at Rumford, Maine, for the year ending Sept. 30, 1918.

Day.	Oct.	Nov.	Dec.	Jan.	Feb.	Mar.	Apr.	May.	June.	July.	Aug.	Sept.
1	2,440	7,650	2,690	2,330	2,360	2,950	7,290	11,180	2,560	2,600	2,470	1,490
2	2,360	4,660	2,450	2,370	2,280	2,840	10,570	9,730	2,200	2,560	2,510	1,600
3	2,360	4,010	2,710	2,430	2,390	2,640	12,430	7,410	2,550	2,530	2,440	2,040
4	2,480	3,080	2,740	2,440	2,290	2,830	9,910	6,600	2,480	1,910	1,920	2,230
5	2,530	3,320	2,900	2,430	2,210	2,830	7,360	5,130	2,540	2,430	2,450	2,100
6	2,640	3,290	2,860	2,160	1,990	2,850	6,330	4,890	2,460	2,560	2,540	2,060
7	1,920	3,050	2,670	2,370	1,880	2,740	5,640	5,350	2,560	1,990	2,580	2,060
8	2,610	2,780	2,500	2,360	1,870	2,750	6,510	5,290	2,920	2,770	2,600	1,810
9	2,600	2,560	1,930	2,330	1,880	2,770	6,300	4,800	2,320	2,740	3,730	2,070
10	2,550	2,650	2,550	2,390	1,970	2,500	6,110	4,210	2,530	2,800	4,350	2,110
11	2,540	2,000	2,520	2,330	2,060	2,710	5,580	5,310	2,670	2,770	2,960	2,050
12	2,510	2,670	2,500	2,380	2,290	2,750	5,180	4,050	2,500	2,790	2,730	2,020
13	2,640	2,620	2,640	2,430	2,400	2,770	4,720	3,960	2,560	2,810	2,550	2,200
14	2,640	2,630	2,530	2,310	2,400	2,760	4,160	6,020	2,510	3,320	2,550	2,590
15	2,490	2,730	2,480	2,320	2,350	2,770	4,990	5,670	2,560	3,360	2,570	1,720
16	2,640	2,920	2,710	2,320	2,480	2,730	6,540	4,570	1,950	2,850	2,580	2,100
17	2,560	2,970	2,530	2,310	2,670	2,300	6,730	3,820	2,390	2,750	2,550	2,130
18	2,540	2,470	2,620	2,630	2,590	2,700	7,060	3,640	2,520	2,850	2,110	2,240
19	2,560	3,030	2,650	2,570	2,510	2,720	5,600	2,830	2,490	2,690	2,500	2,480
20	2,560	2,850	2,690	2,320	2,660	2,740	4,600	3,140	2,420	2,650	2,520	2,530
21	1,970	2,960	2,760	2,250	2,990	2,830	4,360	3,100	2,460	1,790	2,500	4,180
22	2,490	3,000	2,610	2,240	3,030	3,500	5,880	3,970	3,500	2,480	2,330	3,700
23	2,490	2,960	2,540	2,460	3,140	3,970	6,650	2,910	5,920	2,330	2,340	2,700
24	2,350	2,870	2,180	2,530	3,250	3,430	7,290	2,630	3,440	2,000	2,410	2,560
25	4,730	2,110	2,000	2,520	2,830	3,790	6,410	2,440	2,990	2,170	1,830	2,650
26	3,860	2,100	2,450	2,540	2,860	3,920	5,380	1,950	2,720	2,120	2,290	3,290
27	3,080	2,020	2,420	2,460	3,220	3,690	5,370	2,500	2,650	2,190	2,260	11,240
28	3,040	2,350	2,390	2,290	3,210	3,620	5,300	2,680	2,610	2,100	2,050	6,750
29	3,780	2,640	2,370	2,460	....	3,820	6,210	2,630	2,930	2,360	2,120	3,830
30	8,320	2,780	2,540	2,520	....	4,390	9,280	2,580	1,990	2,530	2,090	3,130
31	15,210	....	2,280	2,400	....	5,280	....	2,470	....	2,540	2,070	....

Monthly discharge of Androscoggin River at Rumford, Maine, for the year ending Sept. 30, 1918.

[Drainage area, 2,090 square miles.]

Month.	Discharge in second-feet.				Run-off (depth in inches on drainage area).
	Maximum.	Minimum.	Mean.	Per square mile.	
October	15,210	1,920	3,270	1.56	1.80
November	7,650	2,000	2,990	1.43	1.60
December	2,900	1,930	2,530	1.21	1.40
January	2,630	2,160	2,390	1.14	1.31
February	3,250	1,870	2,500	1.20	1.25
March	5,280	2,300	3,130	1.50	1.73
April	12,430	4,160	6,520	3.12	3.48
May	11,180	1,950	4,430	2.12	2.44
June	5,920	1,950	2,700	1.29	1.44
July	3,360	1,790	2,530	1.21	1.40
August	4,350	1,830	2,500	1.20	1.38
September	11,240	1,490	2,860	1.37	1.53
The year	15,210	1,490	3,200	1.53	20.76

NOTE.—The monthly discharge in second-feet per square mile and the run-off depth in inches do not represent the natural run-off from the basin because of storage. (See "Regulation.") The indicated minimum discharge usually occurs on Sundays when water is held back by dams.

MAGALLOWAY RIVER AT AZISCOHOS DAM, MAINE.

LOCATION.—At Aziscohos dam, Oxford County, 15 miles above mouth.

DRAINAGE AREA.—215 square miles.

RECORDS AVAILABLE.—January 1, 1912, to September 30, 1918.

GAGE.—Vertical staff in two sections, the lower attached to one of the concrete buttresses of the dam and the upper on the concrete gate tower.

DETERMINATION OF DISCHARGE.—Discharge determined from readings of gate openings. Gates have been rated by current-meter measurements at a station about a mile below the dam.

REGULATION.—The storage of about 9,593 million cubic feet is completely regulated, and the discharge corresponds to requirements of water users below. The operation of the gates is planned to maintain as nearly as possible a constant flow at Berlin, N. H. Results not corrected for storage.

COOPERATION.—Discharge computed and furnished for publication by Walter H. Sawyer, agent Union Water Power Co., Lewiston, Maine.

Monthly discharge of Magalloway River at Aziscohos dam, Maine, for the year ending Sept. 30, 1918.

[Drainage area, 215 square miles.]

Month.	Discharge in second-feet.				Run-off (depth in inches on drainage area).
	Maximum.	Minimum.	Mean.	Per square mile.	
October	1,720	90	596	2.77	3.19
November	1,560	92	349	1.62	1.81
December	2,200	1,490	1,790	8.33	9.60
January	2,050	1,440	1,680	7.81	9.00
February	1,680	46	757	3.52	3.66
March	619	49	124	.577	.67
April	77	58	69	.321	.36
May	1,030	79	180	.837	.96
June	1,240	88	535	2.49	2.78
July	167	147	153	.712	.82
August	1,100	161	272	1.27	1.46
September	259	154	177	.823	.92
The year	2,200	46	558	2.60	35.23

NOTE.—The monthly discharge in second-feet per square mile and the run-off in depth in inches do not represent the natural run-off from the basin because of storage. (See Regulation.)

LITTLE ANDROSCOGGIN RIVER NEAR SOUTH PARIS, MAINE.

LOCATION.—At left end of old dam at Bisco Falls, 200 feet below highway bridge and 5½ miles above South Paris, Oxford County.

DRAINAGE AREA.—75 square miles.

RECORDS AVAILABLE.—September 14, 1913, to September 30, 1918.

GAGE.—Chain on left bank installed April 16, 1914; original gage, a vertical staff, was destroyed by ice March 2, 1914; from March 18 to April 9, 1914, a chain gage on a footbridge was used; all gages referred to same datum and at practically the same place. Gage read by G. A. Jackson.

DISCHARGE MEASUREMENTS.—Made from highway bridge or by wading.

CHANNEL AND CONTROL.—At low and medium stages water flows through opening at left of old stone dam; opening was enlarged by high water of April 9, 1914; water flows over dam at gage height 5.30 feet.

EXTREMES OF DISCHARGE.—Maximum stage recorded during year, 8.3 feet at 5 p. m. September 26 (discharge, 1,970 second-feet); minimum stage recorded, 1.16 feet at 8 p. m. August 4 (discharge, 8 second-feet).

1914–1918: Maximum stage recorded, 9.3 feet at 7 a. m. July 9, 1915 (discharge, 2,970 second-feet); minimum stage recorded, 0.7 foot at 6 p. m. August 16 (discharge, 1 second-foot).

ICE.—Control remains open throughout the winter; stage-discharge relation not affected by ice.

REGULATION.—Storage at Snows Falls, 1½ miles above the station, and at West Paris, 4 miles above, has some effect on regimen of stream.

ACCURACY.—Stage-discharge relation changed at the time of high water April 9, 1914; otherwise practically permanent. Rating curve well defined below 700 second-feet and fairly well defined between 700 and 1,800 second-feet. Gage read to tenths once daily. Daily discharges ascertained by applying daily gage height to rating table. Records good except for times of sudden changes in stage, when the number of gage readings is insufficient to determine accurately the mean daily flow.

No discharge measurements were made during the year.

Daily discharge, in second-feet, of Little Androscoggin River near South Paris, Maine, for the year ending Sept. 30, 1918.

Day.	Oct.	Nov.	Dec.	Jan.	Feb.	Mar.	Apr.	May.	June.	July.	Aug.	Sept.
1	37	219	54	24	26	132	558	650	100	54	14	11
2	30	140	50	24	24	132	1,080	458	100	47	13	13
3	26	124	50	24	24	108	1,080	325	92	47	11	13
4	26	112	47	30	24	112	760	303	92	40	8	29
5	29	92	54	30	24	108	442	259	76	34	47	24
6	76	84	50	34	24	100	458	249	68	34	47	24
7	54	68	40	29	24	92	442	259	100	116	47	29
8	68	64	34	24	24	92	442	219	92	124	54	24
9	100	64	47	32	24	96	411	219	92	140	372	24
10	47	54	54	26	24	76	372	199	100	124	325	18
11	54	47	50	26	30	76	325	239	100	116	189	20
12	47	58	40	29	30	76	303	219	92	124	124	20
13	61	54	40	24	30	72	325	169	92	140	124	34
14	92	47	47	32	30	68	348	270	76	149	314	34
15	80	54	47	34	30	68	348	249	68	140	458	29
16	61	47	34	40	30	72	325	219	34	124	281	24
17	54	54	34	37	30	61	336	219	34	76	124	24
18	47	40	37	34	30	68	325	199	40	76	100	18
19	54	54	37	32	30	72	360	199	34	47	84	384
20	47	47	34	29	34	100	360	189	24	47	68	270
21	34	47	34	29	24	104	384	124	24	40	68	270
22	50	47	29	32	26	159	426	100	535	47	54	219
23	50	54	34	34	24	169	372	100	585	40	47	199
24	47	54	37	32	29	179	325	76	303	29	34	219
25	179	54	32	24	26	189	259	76	219	24	24	270
26	108	47	34	26	92	219	249	84	140	29	29	1,970
27	76	54	34	26	159	259	239	92	108	24	24	760
28	124	50	24	24	149	259	219	92	76	29	18	512
29	124	47	24	24	......	303	303	92	68	24	14	336
30	140	47	24	24	......	325	426	100	47	24	13	303
31	426	......	24	26	......	411	......	100	......	20	11	......

NOTE.—Discharge estimated Oct. 2, Dec. 30 to Jan. 5, and Feb. 3–19; consideration being given to temperature and rainfall data.

Monthly discharge of Little Androscoggin River near South Paris, Maine, for the year ending Sept. 30, 1918.

[Drainage area, 75 square miles.]

Month.	Discharge in second-feet.				Run-off (depth in inches on drainage area).
	Maximum.	Minimum.	Mean.	Per square mile.	
October	426	26	79.0	1.05	1.21
November	219	40	67.5	.900	1.00
December	54	24	39.0	.520	.60
January	40	24	29.2	.389	.45
February	159	24	38.4	.512	.53
March	411	61	141	1.88	2.17
April	1,080	219	420	5.60	6.25
May	650	76	205	2.73	3.15
June	585	24	120	1.60	1.78
July	149	20	68.7	.916	1.06
August	458	8	101	1.35	1.56
September	1,970	11	204	2.72	3.04
The year	1,970	8	126	1.68	22.80

PRESUMPSCOT RIVER BASIN.

PRESUMPSCOT RIVER AT OUTLET OF SEBAGO LAKE, MAINE.

Location.—At outlet dam at Sebago Lake and hydroelectric plant at Eel Weir Falls, 1 mile below lake outlet.

Drainage area.—436 square miles.

Records available.—January 1, 1887, to September 30, 1918. All data from 1887 to 1911 recomputed and published in the second annual report of Maine State Water Storage Commission.

Gages.—On bulkhead of gatehouse at outlet dam, and in fore bay and tailrace of power plant.

Discharge.—Prior to March, 1904, discharge was determined from records of opening of gates in dam; since March, 1904, flow from lake has been recorded by three Allen meters, one on each of three pairs of 30-inch Hercules wheels; wheels and recording meters checked by current-meter measurements, brake tests of wheels, and electrical readings of the generator output. Water wasted at regulating gates is measured from records of gate openings and coefficients determined from current-meter measurements.

Ice.—Stage-discharge relation not affected by ice.

Regulation.—Sebago Lake (area, 46 square miles) is under complete regulation. Results not corrected for storage.

Cooperation.—Record in cubic feet per minute furnished by S. D. Warren Co.; record in second-feet computed by engineers of United States Geological Survey.

Daily discharge, in second-feet, of Presumpscot River at outlet of Sebago Lake, Maine, for the year ending Sept. 30, 1918.

Day.	Oct.	Nov.	Dec.	Jan.	Feb.	Mar.	Apr.	May.	June.	July.	Aug.	Sept.
1	765	705	813	807	807	654	542	445	539	502	764	230
2	788	773	273	818	804	633	528	438	170	584	704	263
3	803	808	820	803	235	135	524	470	366	590	679	689
4	803	212	745	797	472	676	490	445	558	186	252	672
5	878	743	742	817	816	668	533	237	575	526	678	619
6	798	780	783	299	820	707	569	508	601	652	746	622
7	278	817	787	788	919	699	187	537	628	187	741	650
8	790	783	723	780	901	707	558	435	547	675	715	262
9	798	747	237	801	918	708	568	444	212	619	574	629
10	800	770	773	805	811	236	572	594	587	644	534	627
11	790	235	742	783	494	722	547	507	570	693	128	647
12	805	787	818	769	490	718	585	205	600	699	598	689
13	778	740	830	323	830	715	504	528	498	594	589	641
14	208	760	808	760	818	735	172	514	651	199	692	622
15	792	752	825	799	806	728	497	563	575	565	661	277
16	803	778	372	796	792	709	474	548	199	611	716	592
17	777	782	825	804	258	249	502	591	504	664	634	613
18	773	238	813	511	505	771	542	545	559	505	172	617
19	795	797	825	412	789	760	501	192	600	683	692	577
20	733	730	818	373	794	757	598	477	679	569	707	548
21	198	748	822	402	808	693	248	546	626	133	707	421
22	805	798	822	541	785	639	422	555	488	643	801	148
23	830	705	337	730	777	597	458	571	65	677	753	598
24	787	668	752	801	216	190	496	564	412	689	708	607
25	710	282	240	805	741	687	533	484	518	661	257	570
26	803	788	733	803	722	595	591	221	582	682	747	566
27	777	825	822	239	676	688	533	504	535	583	730	409
28	192	785	835	522	633	613	149	560	588	258	737	335
29	770	648	822	803	......	551	628	588	555	624	748	169
30	770	762	288	811	......	536	536	473	242	642	774	604
31	720	......	805	816	......	138	......	528	......	651	600	......

Monthly discharge of Presumpscot River at outlet of Sebago Lake, Maine, for the year ending Sept. 30, 1918.

[Drainage area, 436 square miles.]

Month.	Discharge in second-feet.				Run-off (depth in inches on drainage area).
	Maximum.	Minimum.	Mean.	Per square mile.	
October	878	192	713	1.64	1.89
November	825	212	691	1.58	1.76
December	835	240	701	1.61	1.86
January	818	239	681	1.56	1.80
February	919	216	676	1.55	1.61
March	771	135	597	1.37	1.58
April	628	149	486	1.11	1.24
May	594	192	478	1.10	1.27
June	679	65	494	1.13	1.26
July	699	133	555	1.27	1.46
August	774	128	630	1.44	1.66
September	689	148	517	1.19	1.33
The year	919	65	602	1.38	18.72

NOTE.—The monthly discharge does not represent the natural flow from the basin because of artificial storage. The yearly discharge and run-off probably represent more nearly the natural flow, because comparatively little stored water is held over from year to year.

SACO RIVER BASIN.

SACO RIVER AT CORNISH, MAINE.

LOCATION.—At highway bridge at Cornish, York County, half a mile below mouth of Ossipee River.

DRAINAGE AREA.—1,300 square miles.

RECORDS AVAILABLE.—June 4, 1916, to September 30, 1918.

GAGE.—Chain attached to bridge; read by S. J. Elliott and A. H. Guimont.

DISCHARGE MEASUREMENTS.—Made from bridge.

CHANNEL AND CONTROL.—Channel covered with sand and boulders; broken by one pier at bridge.

EXTREMES OF DISCHARGE.—Maximum stage recorded during year, 5.6 feet at 3 p. m. April 7 (discharge, 7,560 second-feet); minimum stage recorded, 0.74 foot at 9.30 a. m. September 15 (discharge, 644 second-feet). Minimum discharge estimated as 350 second-feet several times in January and February; stage-discharge relation affected by ice at the time.

1916–1918: Maximum stage recorded, 9.4 feet at 6.30 a. m. June 18, 1917 (approximate discharge, from extension of rating curve, 17,400 second-feet); minimum open-water stage recorded, 0.8 foot several times in August and September, 1917 (discharge, 635 second-feet).

ICE.—Ice forms to considerable thickness; stage relation seriously affected during most winters.

REGULATION.—Distribution of flow probably not seriously affected by power developments above the gage.

ACCURACY.—Stage-discharge relation has apparently shifted since station was first established; present rating curve fairly well defined between 1,000 and 7,000 second-feet. Gage read to half-tenths twice daily, except from December 14 to March 27, when it was read three times a week. Daily discharge ascertained by applying daily gage height to rating table and making corrections for effect of ice during the winter. Records fair.

Discharge measurements of Saco River at Cornish, Maine, during the year ending Sept. 30, 1918.

Date.	Made by—	Gage height.	Discharge.	Date.	Made by—	Gage height.	Discharge.
		Feet.	*Sec.-ft.*			*Feet.*	*Sec.-ft.*
Jan. 11	A. F. McAlary..........	a 2.40	851	Apr. 12	H. A. Lancaster........	5.11	6,440
Feb. 15	.....do..................	a 2.65	691	May 9	.....do..................	4.26	4,850
Mar. 14	.....do..................	a 3.43	1,360				

a Stage-discharge relation affected by ice.

Daily discharge, in second-feet, of Saco River at Cornish, Maine, for the year ending Sept. 30, 1918.

Day.	Oct.	Nov.	Dec.	Jan.	Feb.	Mar.	Apr.	May.	June.	July.	Aug.	Sept.
1	845	1,830	960	700	440	960	3,690	5,280	1,530	1,730	1,180	1,020
2	810	1,630	1,000	700	440	960	5,460	5,640	1,730	1,830	1,140	915
3	845	2,040	920	700	440	960	6,420	5,640	1,530	1,630	1,100	880
4	915	3,210	920	700	500	1,000	6,860	5,830	1,530	1,530	880	915
5	880	3,690	880	680	540	1,100	7,090	5,600	1,260	1,440	1,100	1,060
6	880	3,370	880	440	560	1,200	7,320	5,400	1,180	1,530	985	1,060
7	845	3,530	880	500	600	1,200	7,560	5,200	1,350	1,350	985	950
8	1,020	3,210	840	600	620	1,200	7,320	5,000	1,180	1,440	985	810
9	1,020	2,770	840	740	640	1,250	7,090	4,800	1,180	1,630	1,530	985
10	1,020	2,380	800	800	640	1,200	6,860	4,560	1,260	1,530	1,630	985
11	1,020	2,040	800	860	660	1,200	6,640	4,560	1,440	1,530	1,830	1,020
12	1,100	2,040	800	620	660	1,200	6,640	4,380	1,350	1,630	1,730	1,020
13	1,140	1,730	800	380	540	1,250	6,220	3,860	1,440	1,730	1,630	1,020
14	1,060	1,730	740	560	680	1,350	5,830	3,690	1,440	1,930	1,530	845
15	1,180	1,630	700	680	700	1,350	5,830	3,530	1,530	2,040	1,530	680
16	1,140	1,530	700	840	600	1,350	5,460	3,690	1,400	2,040	1,530	1,060
17	1,100	1,440	680	800	350	1,350	5,460	3,530	1,300	2,260	1,350	1,060
18	1,180	1,440	680	800	500	1,450	5,460	3,210	1,250	2,380	1,350	1,060
19	1,140	1,440	680	640	560	1,550	5,460	3,370	1,250	2,040	1,260	1,180
20	1,140	1,530	680	380	540	1,650	5,460	2,910	1,250	2,040	1,180	1,260
21	1,260	1,630	700	560	660	1,750	5,460	2,630	1,350	2,150	1,100	1,830
22	1,180	1,630	740	680	660	1,850	5,460	2,500	1,500	1,930	1,140	1,930
23	1,260	1,440	780	800	600	2,000	5,640	2,380	2,150	1,530	1,020	2,040
24	1,100	1,440	780	740	600	2,100	5,830	2,260	2,630	1,440	985	2,040
25	1,260	1,140	700	620	740	2,200	5,830	1,930	2,630	1,350	1,020	2,040
26	1,440	1,100	700	350	840	2,300	5,830	1,730	2,630	1,180	1,020	2,380
27	1,440	1,000	700	350	960	2,500	5,460	1,730	2,500	1,180	1,020	4,920
28	1,350	960	700	520	960	2,600	5,460	1,730	2,260	1,100	1,020	4,740
29	1,830	960	700	740	....	2,700	5,100	1,730	2,150	1,260	985	4,920
30	1,530	920	700	920	....	2,900	5,100	1,530	1,930	1,260	915	5,100
31	2,040	....	700	800	....	3,100	....	1,730	....	1,180	1,060	....

NOTE.—Stage-discharge relation affected by ice Nov. 27 to Mar. 30; discharge for this period computed from gage heights corrected for effect of ice by means of three discharge measurements, observer's notes, weather records, and comparative records of power plant at Hiram, plus records of Ossipee. Discharge estimated May 5–9 and June 16–22 by comparative hydrograph.

Monthly discharge of Saco River at Cornish, Maine, for the year ending Sept. 30, 1918.

[Drainage area, 1,390 square miles.]

Month.	Discharge in second-feet.				Run-off (depth in inches on drainage area).
	Maximum.	Minimum.	Mean.	Per square mile.	
October	2,040	810	1,160	0.892	1.03
November	3,690	920	1,880	1.45	1.62
December	1,000	680	777	.598	.69
January	920	350	652	.502	.58
February	960	350	615	.473	.49
March	3,100	960	1,640	1.26	1.45
April	7,560	3,690	5,980	4.60	5.13
May	5,830	1,530	3,600	2.77	3.19
June	2,630	1,180	1,640	1.26	1.41
July	2,380	1,100	1,640	1.26	1.45
August	1,830	880	1,220	.938	1.08
	5,100	680	1,720	1.32	1.47
r	7,560	350	1,880	1.45	19.59

OSSIPEE RIVER AT CORNISH, MAINE.

Location.—At highway bridge in Cornish, York County, 1¼ miles above confluence with Saco River.

Drainage area.—448 square miles.

Records available.—July 5, 1916, to September 30, 1918.

Gage.—Chain attached to bridge; read by O. W. Adams.

Discharge measurements.—Made from bridge.

Channel and control.—Bed covered with sand and gravel; possibly somewhat shifting; broken by one pier at bridge.

Extremes of discharge.—Maximum stage recorded during year, 4.15 feet at 4 p. m. April 4 (discharge, 2,610 second-feet); minimum stage recorded, 0.90 foot at 6 p. m. September 14 (discharge, 320 second-feet). Minimum discharge estimated as 240 second-feet several times during January and February; stage-discharge relation affected by ice at the time.

1916–1918: Maximum stage recorded, 7.25 feet at 6 a. m. June 18, 1917 (approximate discharge, from extension of rating curve, 6,480 second-feet); minimum open-water stage recorded, 0.90 foot at 6 p. m. September 14, 1918 (discharge, 320 second-feet).

Ice.—Ice forms to considerable thickness; stage-discharge relation seriously affected during most winters.

Regulation.—Flow regulated by dams at Kezar Falls and at outlet of Great Ossipee Lake.

Accuracy.—Stage-discharge relation practically permanent except when affected by ice. Rating curve well defined between 350 and 2,400 second-feet. Gage read to half-tenths once a day except from January 1 to February 25, when it was read three or four times a week. Daily discharge, ascertained by applying gage height to rating table and making corrections for effect of ice during the winter. Records fair.

Discharge measurements of Ossipee River at Cornish, Maine, during the year ending Sept. 30, 1918.

Date.	Made by—	Gage height.	Discharge.	Date.	Made by—	Gage height.	Discharge.
		Feet.	*Sec.-ft.*			*Feet.*	*Sec.-ft.*
Jan. 10	A. F. McAlary	[a]1.61	220	Apr. 11	H. A. Lancaster	3.66	2,150
Feb. 15	do	[a]2.23	232	12	do	3.49	1,990
Mar. 13	do	[a]2.97	406	May 9	do	2.50	1,160

[a] Stage-discharge relation affected by ice.

Daily discharge, in second-feet, of Ossipee River at Cornish, Maine, for the year ending Sept. 30, 1918.

Day.	Oct.	Nov.	Dec.	Jan.	Feb.	Mar.	Apr.	May.	June.	July.	Aug.	Sept.
1	390	520	300	290	260	360	1,320	1,820	500	500	360	375
2	360	550	310	290	270	360	1,820	1,910	480	500	350	375
3	375	575	320	290	250	360	2,460	1,730	480	480	340	375
4	360	600	320	290	250	390	2,560	1,560	460	440	360	390
5	350	600	300	300	250	420	2,560	1,500	440	420	360	375
6	390	600	310	290	250	420	2,270	1,400	390	420	350	360
7	375	575	300	290	250	420	2,270	1,400	375	420	340	330
8	375	500	310	290	250	420	2,180	1,320	390	440	350	330
9	375	480	320	270	250	420	2,180	1,160	390	440	960	350
10	360	480	320	250	250	420	2,180	1,000	420	460	850	350
11	360	480	320	250	250	420	2,090	1,000	460	460	815	360
12	375	460	320	250	240	400	2,000	1,000	440	480	660	350
13	390	440	310	250	240	400	2,000	1,000	460	500	600	340
14	405	420	310	250	240	390	1,910	1,080	460	480	550	320
15	405	405	310	240	240	340	1,820	1,000	460	460	550	330
16	405	405	320	270	240	340	1,640	1,000	420	460	500	340
17	405	390	310	260	240	360	1,640	1,000	420	460	420	340
18	390	375	310	260	240	390	1,730	920	375	500	390	340
19	390	350	300	250	260	390	1,730	850	390	500	390	500
20	405	350	290	250	270	400	1,640	750	390	460	375	525
21	405	350	290	250	260	560	1,640	720	390	600	360	815
22	405	360	290	260	250	660	2,000	690	460	405	360	780
23	405	375	290	250	240	720	2,000	690	720	390	340	600
24	420	390	300	250	260	840	2,000	550	720	375	330	525
25	500	390	300	250	270	1,000	1,910	500	750	360	350	550
26	420	405	300	250	290	1,150	1,730	480	690	360	360	815
27	405	400	300	250	310	1,260	1,640	550	690	340	360	1,240
28	410	380	290	250	310	1,300	1,480	550	630	350	360	1,730
29	440	340	290	250	......	1,400	1,400	550	600	375	360	1,730
30	480	310	290	250	......	1,320	1,400	550	525	375	375	1,560
31	520	......	290	250	......	1,320	......	500	......	390	375	......

NOTE.—Stage-discharge relation affected by ice from Nov. 27 to Mar. 28; discharge for this period computed from gage heights corrected for effect of ice by means of three discharge measurements, observer's notes, and weather records. Discharge estimated Oct. 28 to Nov. 1, Mar. 31, and May 5.

Monthly discharge of Ossipee River at Cornish, Maine, for the year ending Sept. 30, 1918.

[Drainage area, 448 square miles.]

Month.	Discharge in second-feet.				Run-off (depth in inches on drainage area).
	Maximum.	Minimum.	Mean.	Per square mile.	
October	520	350	402	.897	1.03
November	600	310	442	.987	1.10
December	320	290	305	.681	.79
January	300	240	263	.587	.68
February	310	240	256	.571	.59
March	1,400	340	624	1.39	1.60
April	2,560	1,320	1,910	4.26	4.75
May	1,910	480	989	2.21	2.55
June	750	375	492	1.10	1.23
July	600	340	439	.980	1.13
August	960	330	445	.993	1.14
September	1,730	320	590	1.32	1.47
The year	2,560	240	596	1.33	18.06

MERRIMACK RIVER BASIN.

PEMIGEWASSET RIVER AT PLYMOUTH, N. H.

LOCATION.—At two-span highway bridge in Plymouth, Grafton County, three-fourths of a mile below mouth of Bakers River.

DRAINAGE AREA.—615 square miles.

RECORDS AVAILABLE.—January 1, 1886, to September 30, 1918.

GAGES.—Vertical staff gage in three sections; two lower sections about 40 feet above the bridge; upper section on bridge abutment; used since July 1, 1907. Chain gage on upstream side of bridge used from September 4, 1903, to June 30, 1907. The datum of the staff is 1.11 feet higher than that of the chain gage.

DISCHARGE MEASUREMENTS.—Made from upstream side of bridge at ordinary and high stages. At extremely low stages measurements made by wading.

CHANNEL AND CONTROL.—Right channel is rocky and practically permanent; left channel covered with fine gravel which shifts occasionally. Control section for low stages is gravel bed of river and has changed somewhat at various times. At high stages the banks are overflowed below the bridge and the control is somewhat indefinite.

EXTREMES OF DISCHARGE.—Maximum open-water stage recorded, 1912–1918: 15.42 feet at 7 a. m. March 28, 1913 (approximate discharge, from extension of rating curve, 18,700 second-feet); a gage height of 18.17 feet was recorded at 4 p. m. February 25, 1915, but stage-discharge relation was probably affected by ice at the time: Minimum stage recorded, 0.64 foot at 7 a. m. September 20, 1913 (discharge, 71 second-feet); an estimated discharge of 60 second-feet occurred September 21, 1913.

ICE.—River freezes over and stage-discharge relation is usually affected by ice from December to March.

REGULATION.—There are several small ponds on Bakers River and other tributaries, but practically no storage regulation. At very low stages the paper mill at Livermore Falls is obliged to shut down several times daily, and at these times the ponding of water affects the distribution of flow at Plymouth.

ACCURACY.—Stage-discharge relation practically permanent from April, 1912, to September, 1918, except when affected by ice. Rating curve well defined below 15,000 second-feet. Gage read to half inches twice daily, except Sundays. Daily discharge ascertained by applying mean daily gage height to rating table, and making corrections for effect of ice during the winter. Sunday discharge estimated by hydrograph comparisons with records at other gaging stations. Records good.

Records from October 1, 1911, to December 31, 1913, previously published have been revised by means of additional discharge measurements. Estimates for high stages prior to October 1, 1911, which have been published in various water-supply papers of the Geological Survey, are probably too high.

COOPERATION.—Gage-height records furnished by proprietors of locks and canals on Merrimack River, Arthur T. Safford, engineer.

Discharge measurements of Pemigewasset River at Plymouth, N. H., during 1912–1918.

Date.	Made by—	Gage height.	Discharge.	Date.	Made by—	Gage height.	Discharge.
1912.		*Feet.*	*Sec.-ft.*	1914.		*Feet.*	*Sec.-ft.*
Jan. 27	Coffin and Moore	[a]1.90	349	Oct. 7	Reported by A. T. Safford	−0.08	149
29	R. J. Coffin	[a]1.80	355				
29	Adams and Coffin	[a]1.81	374	1915.			
Feb. 3	.....do	[a]1.74	343	Aug. 28	Pierce and Thweatt	1.96	1,690
12	.....do	[a]1.58	280	Nov. 24	Hardin Thweatt	1.55	728
18	C. R. Adams	[a]1.60	291				
28	Adams and Coffin	[a]1.90	290	1916.			
Mar. 6	.....do	[a]1.82	293	Apr. 17	Hardin Thweatt	3.90	3,440
12	Smead and Moore	[a]1.70	304	18	.....do	5.06	5,000
Apr. 19	C. R. Adams	5.90	6,160	May 18	Thweatt and Mansur	7.68	8,280
				19	.....do	5.38	5,290
1913.				June 20	Pierce and Thweatt	5.13	4,920
Aug. 20	Reported by A. T. Safford	.104	200	1918.			
				May 17	Pierce and Weeks	2.50	1,700
				Nov. 18	H. W. Fear	2.88	2,180

[a] Stage-discharge relation affected by ice.

NOTE.—Six discharge measurements made in March and April, 1919, were used in determining the rating curve for high stages.

Daily discharge, in second-feet, of Pemigewasset River at Plymouth, N. H., for the years ending Sept. 30, 1912–1918.

Day.	Oct.	Nov.	Dec.	Jan.	Feb.	Mar.	Apr.	May.	June.	July.	Aug.	Sept.
1911–12.												
1	960	1,030	1,400	700	370	270	5,200	2,450	4,720	266	265	335
2	870	997	1,100	660	410	270	5,800	2,200	4,000	255	290	353
3	900	870	870	660	343	280	5,450	2,050	3,180	247	390	406
4	1,130	810	750	600	340	250	2,450	1,950	2,160	242	700	390
5	3,170	760	660	540	330	250	1,500	1,880	1,600	232	422	406
6	1,850	720	900	620	330	293	3,350	1,900	1,260	222	390	422
7	1,240	997	997	580	310	290	6,100	3,340	1,650	222	314	353
8	1,080	1,570	870	540	320	290	11,600	3,340	1,170	222	278	500
9	932	1,170	780	520	310	310	5,190	2,770	900	232	266	314
10	780	997	870	700	290	300	3,650	2,610	700	222	266	320
11	780	965	997	620	270	290	3,090	2,820	728	212	5,460	302
12	690	1,030	1,320	520	260	304	3,110	3,230	700	212	3,760	290
13	600	1,170	3,170	470	290	350	3,450	3,920	630	208	1,450	290
14	540	1,240	2,300	460	290	560	3,370	6,830	585	215	728	296
15	480	1,100	1,570	450	290	640	3,310	3,290	545	222	482	300
16	425	1,030	1,320	520	290	1,150	6,210	2,300	555	282	450	326
17	375	900	1,100	410	280	1,800	10,300	4,280	565	282	377	585
18	690	997	840	390	291	2,500	8,270	3,700	700	222	300	482
19	4,500	1,650	870	390	300	2,300	6,570	2,900	545	242	353	406
20	2,550	1,170	810	3,600	310	2,700	5,510	2,100	500	227	365	422
21	1,660	1,030	870	1,000	310	2,900	5,160	2,870	450	250	326	1,750
22	2,160	965	932	620	310	2,200	4,820	4,230	422	341	290	1,000
23	2,670	810	4,060	600	300	1,800	10,200	3,290	375	482	302	606
24	2,420	720	4,820	520	290	1,600	6,930	2,610	329	302	326	545
25	1,480	780	2,420	490	290	1,300	4,130	2,610	314	290	400	466
26	1,170	765	1,570	400	290	1,150	3,860	1,930	353	266	302	397
27	1,200	750	1,320	349	310	960	4,600	1,600	314	266	422	365
28	997	780	1,060	355	290	920	3,950	1,450	302	260	466	353
29	900	2,550	900	365	270	1,200	3,290	967	290	255	390	400
30	810	1,850	800	350	....	6,100	2,660	3,020	278	242	326	606
31	810	....	780	400	....	5,500	....	4,660	....	242	314	....

Daily discharge, in second-feet, of Pemigewasset River at Plymouth, N. H., for the years ending Sept. 30, 1912–1918—Continued.

Day.	Oct.	Nov.	Dec.	Jan.	Feb.	Mar.	Apr.	May.	June.	July.	Aug.	Sept.
1912–13.												
1	728	967	600	2,710	1,280	420	10,500	1,900	2,300	314	278	106
2	482	2,660	565	1,800	1,600	450	4,500	1,650	1,700	326	266	186
3	436	1,500	2,100	1,650	1,350	430	2,970	1,360	1,260	314	269	194
4	422	1,010	1,750	5,560	1,150	415	2,560	1,250	1,130	302	266	190
5	408	756	1,600	3,550	1,000	400	2,820	1,220	931	290	290	190
6	385	728	1,800	2,660	850	365	2,600	1,170	728	300	266	186
7	365	652	2,770	1,950	700	370	2,300	1,050	652	326	255	145
8	365	9,550	1,600	1,600	730	365	2,200	981	600	302	250	106
9	377	5,460	1,050	1,400	640	360	1,450	895	545	290	242	91
10	358	3,000	931	1,600	590	355	1,220	786	525	278	240	91
11	365	2,100	895	1,350	525	350	1,400	700	482	565	242	108
12	408	1,700	786	2,300	500	540	2,200	652	450	466	232	79
13	1,100	1,400	756	1,700	480	680	3,500	585	436	400	222	74
14	625	1,400	652	1,450	500	620	3,230	565	422	365	217	90
15	490	2,100	600	1,150	525	4,050	3,700	545	400	358	194	178
16	482	1,700	565	1,050	515	12,100	3,760	482	390	314	186	202
17	482	1,350	565	1,250	500	9,200	3,020	652	422	302	180	74
18	436	1,130	555	1,700	465	6,100	2,400	750	545	290	186	91
19	450	967	676	3,500	435	5,200	2,300	728	525	290	194	128
20	600	826	1,800	2,500	420	6,600	2,100	630	500	285	186	71
21	466	700	1,560	1,600	450	14,100	1,650	525	450	278	128	60
22	390	652	1,450	3,300	590	16,500	1,360	545	425	272	113	208
23	605	676	1,500	2,050	1,130	5,680	1,220	2,000	408	266	113	5,030
24	7,050	900	1,130	1,950	480	4,280	2,050	3,650	390	266	140	1,220
25	9,610	786	1,010	2,050	480	5,030	2,450	3,000	390	260	194	545
26	5,240	714	756	1,800	465	14,600	3,290	2,450	365	255	212	390
27	3,500	676	585	1,600	435	6,770	2,900	1,840	341	248	204	358
28	2,450	652	605	1,350	420	18,700	2,710	1,320	326	242	208	330
29	1,600	652	650	1,150	....	7,440	2,510	4,820	300	314	198	302
30	1,180	680	728	850	....	4,500	2,820	4,180	290	302	194	242
31	1,010	....	1,840	1,050	....	3,240	....	3,070	....	290	150	....
1913–14.												
1	242	981	1,340	525	1,180	620	2,300	5,080	525	296	290	525
2	255	700	525	525	1,180	7,000	5,780	3,550	482	392	278	365
3	2,400	630	525	500	1,050	12,400	4,400	4,000	450	450	266	353
4	1,260	605	605	490	950	9,890	2,610	4,820	436	385	186	314
5	750	565	585	490	985	7,100	2,000	4,820	3,230	350	242	296
6	545	545	555	475	835	6,330	1,650	4,820	1,220	408	222	280
7	482	482	550	450	770	5,560	1,560	5,680	750	377	222	266
8	408	525	1,600	440	740	4,660	1,600	4,080	565	422	232	266
9	341	600	1,450	420	715	3,760	5,460	4,600	525	390	227	266
10	326	11,100	1,220	420	600	2,970	5,130	5,250	482	365	222	266
11	314	4,280	1,130	415	530	2,300	3,020	4,500	450	353	212	266
12	400	2,300	756	415	510	2,200	3,500	3,450	65	350	242	255
13	700	1,500	786	400	460	2,160	4,230	2,820	341	565	232	220
14	605	1,220	750	385	460	2,000	2,820	2,610	300	545	232	186
15	565	1,050	756	440	450	1,000	2,610	2,200	290	390	255	266
16	500	900	786	480	440	756	2,400	2,000	326	341	235	242
17	482	728	630	460	430	728	2,400	1,800	302	326	222	232
18	450	700	565	450	430	931	2,970	1,600	302	290	212	222
19	440	652	535	440	440	895	4,000	1,600	290	300	212	222
20	436	1,360	525	430	450	700	14,100	1,900	290	302	232	200
21	7,670	1,800	500	415	435	676	18,400	1,900	300	290	290	186
22	2,710	1,260	500	415	420	600	9,220	1,800	302	278	278	186
23	1,260	1,000	530	400	400	565	6,210	1,600	290	266	270	186
24	770	859	600	390	420	500	4,280	1,500	290	266	266	204
25	826	786	530	490	420	482	3,970	1,560	278	255	255	186
26	2,000	682	575	1,020	400	482	3,750	1,320	302	255	232	186
27	3,450	585	510	1,120	390	652	3,550	1,130	290	255	232	200
28	2,400	545	480	1,440	375	5,780	5,300	1,050	300	242	222	222
29	1,400	545	530	1,460	....	3,500	5,780	826	266	186	314	232
30	1,260	600	555	1,300	....	2,710	7,540	700	302	290	750	222
31	1,320	....	545	1,200	....	2,450	....	600	....	341	1,220	....

Daily discharge, in second-feet, of Pemigewasset River at Plymouth, N. H., for the years ending Sept. 30, 1912–1918—Continued.

Day.	Oct.	Nov.	Dec.	Jan.	Feb.	Mar.	Apr.	May.	June.	July.	Aug.	Sept.
1914–15.												
1	194	186	341	265	600	8,610	810	4,280	615	474	1,000	700
2	212	186	341	295	620	9,500	700	3,500	570	3,020	1,090	555
3	194	242	630	410	670	8,110	585	2,400	500	5,070	1,800	540
4	190	212	1,010	315	700	6,430	600	1,950	474	3,000	1,200	535
5	186	242	605	225	680	5,620	786	1,630	450	2,300	1,640	500
6	186	232	540	260	760	5,330	721	1,420	450	2,480	1,680	443
7	186	212	482	220	940	4,800	770	1,110	450	1,560	1,290	429
8	186	215	365	900	1,350	3,750	1,340	1,310	418	913	1,150	458
9	186	222	500	590	1,200	2,630	2,520	1,480	408	12,900	1,050	458
10	194	255	450	500	920	2,100	2,800	1,510	422	4,820	1,130	450
11	190	222	408	450	760	1,980	5,500	1,320	408	3,009	877	436
12	186	222	408	480	740	1,680	10,800	985	474	2,070	742	432
13	186	186	332	450	700	1,430	6,560	913	474	1,450	770	429
14	186	186	290	430	670	1,250	3,970	1,080	474	1,170	1,220	415
15	222	200	450	400	640	1,110	3,210	896	436	1,110	1,000	383
16	186	290	482	400	1,400	1,260	3,220	850	535	985	1,090	371
17	186	1,900	500	380	4,000	1,200	3,210	834	600	949	985	383
18	186	585	466	380	2,300	770	3,000	850	1,750	950	1,150	390
19	186	500	194	940	1,700	1,000	2,870	834	1,050	1,080	985	394
20	266	326	190	7,300	1,400	949	2,850	896	750	1,050	742	397
21	278	408	186	3,900	1,250	1,000	2,920	778	676	1,090	565	408
22	266	380	320	2,450	1,100	1,050	2,200	742	615	1,080	1,000	2,050
23	255	365	300	1,500	1,050	1,070	1,580	700	482	1,220	5,950	728
24	255	365	310	1,700	900	913	1,400	664	458	1,220	8,500	566
25	220	314	260	1,500	10,300	877	2,000	560	450	1,150	1,760	540
26	186	290	340	1,400	16,200	842	3,860	530	443	1,430	2,080	450
27	186	302	340	1,300	10,100	770	3,550	1,010	400	1,400	1,450	540
28	186	545	350	1,200	8,800	815	2,770	965	422	1,240	931	640
29	186	450	186	1,050	....	859	2,370	859	405	1,110	800	586
30	186	365	360	700	....	810	2,160	700	390	1,260	742	500
31	186	....	186	640	....	756	....	652	....	1,050	770	....
1915–16.												
1	450	585	1,220	1,330	3,750	2,050	10,900	4,230	2,070	1,070	535	359
2	458	555	958	1,550	4,970	1,550	6,100	4,180	1,360	1,080	450	341
3	466	525	896	1,500	3,860	1,350	4,660	4,340	1,150	6,670	429	325
4	474	535	826	1,400	2,850	1,200	3,650	2,920	1,700	4,230	397	326
5	515	500	670	1,200	1,950	1,150	2,730	3,060	3,550	3,400	450	320
6	859	458	585	1,300	1,250	1,100	2,800	2,630	2,420	2,300	440	335
7	688	470	595	2,500	1,100	1,050	2,870	2,700	2,590	1,750	429	341
8	565	500	515	1,300	850	1,000	2,320	2,770	1,980	1,170	429	341
9	615	482	490	1,100	800	980	2,180	2,780	2,120	1,080	1,820	314
10	565	466	482	900	870	1,100	1,890	2,350	4,230	967	2,350	302
11	525	466	474	800	800	1,000	2,020	1,770	6,330	688	1,200	290
12	500	458	462	760	700	900	2,560	1,980	3,500	676	949	278
13	482	443	450	740	700	840	2,820	1,450	3,290	1,080	700	272
14	450	470	466	720	660	940	3,230	1,130	2,820	810	565	255
15	458	535	466	700	720	840	2,370	967	2,100	700	490	266
16	742	585	490	660	740	800	2,900	1,080	2,010	615	474	3,020
17	615	605	490	600	820	860	3,650	2,720	2,400	565	422	1,120
18	565	515	515	560	800	840	4,720	11,200	6,330	540	429	545
19	545	482	700	500	720	780	4,620	5,300	4,870	575	390	605
20	525	1,560	2,160	480	660	720	3,360	3,360	4,870	515	374	525
21	490	2,180	1,260	470	600	720	3,060	2,550	3,460	535	359	482
22	450	1,060	985	700	620	700	3,780	2,200	2,450	595	353	450
23	450	859	913	2,300	640	640	6,100	2,230	1,750	1,800	335	515
24	432	721	786	2,970	760	640	7,490	2,070	1,260	1,750	443	1,120
25	422	682	770	2,510	780	620	4,820	2,120	1,130	1,470	422	931
26	429	652	2,300	2,200	1,450	700	4,740	1,960	1,820	742	408	700
27	520	630	5,670	2,370	4,050	860	3,800	1,770	1,380	615	436	535
28	585	590	2,860	4,130	3,850	1,880	3,260	1,350	1,200	1,110	462	515
29	565	570	2,250	6,270	2,550	2,900	3,230	1,090	1,380	676	436	458
30	575	1,200	1,920	4,820	....	4,400	3,550	1,380	1,130	565	415	2,240
31	580	....	1,560	3,890	....	7,700	....	3,020	....	466	397	....

Daily discharge, in second-feet, of Pemigewasset River at Plymouth, N. H., for the years ending Sept. 30, 1912–1918—Continued.

Day.	Oct.	Nov.	Dec.	Jan.	Feb.	Mar.	Apr.	May.	June.	July.	Aug.	Sept.
1916-17.												
1	1,320	450	5,850	400	475	820	5,400	3,550	3,050	2,060	436	680
2	913	580	2,900	500	490	640	6,110	3,110	3,110	1,630	371	700
3	700	585	1,700	415	450	540	5,130	2,920	3,170	1,290	365	565
4	664	535	1,440	400	450	528	3,790	2,630	3,700	895	320	482
5	620	490	1,400	390	490	508	3,150	2,660	2,870	981	415	422
6	565	468	2,020	510	440	540	3,300	2,650	2,480	810	359	397
7	450	466	1,720	640	460	500	4,790	2,630	2,200	714	266	397
8	460	436	1,280	690	450	485	3,550	2,480	1,880	670	266	397
9	422	429	1,110	680	475	500	2,610	2,770	4,900	630	222	380
10	422	429	1,300	585	465	500	1,980	2,770	3,550	575	358	365
11	468	500	1,050	550	320	480	1,680	2,630	5,250	545	415	347
12	390	450	850	440	350	520	1,650	3,170	11,600	615	365	347
13	384	422	826	425	440	500	1,430	3,050	8,140	700	341	284
14	976	466	688	490	420	475	1,500	2,920	4,610	652	341	341
15	640	415	640	840	420	500	1,530	4,180	3,980	590	353	275
16	535	436	600	1,600	400	480	1,560	2,670	2,800	555	359	235
17	490	408	550	1,250	400	465	1,500	2,500	4,820	545	474	247
18	466	408	590	1,060	420	525	2,540	2,800	13,500	482	714	314
19	450	432	470	850	400	470	3,310	2,730	5,650	482	540	218
20	1,490	458	450	700	420	430	5,780	3,550	3,610	520	450	218
21	1,520	500	425	640	430	490	6,980	5,080	3,980	490	408	212
22	1,130	486	425	650	400	520	8,240	3,600	2,610	400	595	212
23	770	380	700	665	420	700	9,550	2,950	2,000	408	422	222
24	664	5,560	600	620	400	920	7,500	4,210	1,820	422	422	240
25	605	3,340	500	610	380	2,050	5,130	3,270	2,350	482	1,070	314
26	600	1,700	450	570	400	3,210	4,850	2,710	1,770	422	640	212
27	545	1,130	415	540	420	4,090	3,470	2,480	1,460	415	525	272
28	500	1,200	405	490	590	9,280	3,000	2,320	1,290	390	408	284
29	470	940	390	520	....	9,860	2,800	2,540	1,090	380	422	255
30	458	2,420	350	550	....	6,570	4,340	2,660	3,110	384	470	235
31	474	....	300	540	....	4,690	....	2,820	....	450	895	....
1917-18.												
1	346	3,690	415	350	280	1,700	6,430	4,970	700	605	320	306
2	272	2,280	430	335	250	1,300	7,160	4,620	840	595	266	390
3	332	1,680	395	250	300	1,100	9,500	3,100	742	620	366	353
4	332	1,400	440	235	180	980	4,720	2,610	605	450	290	341
5	332	1,130	408	250	290	900	3,110	2,240	525	490	272	302
6	700	994	360	300	266	900	2,500	1,900	500	466	341	284
7	590	895	340	250	220	850	3,050	2,680	510	515	365	320
8	443	786	315	200	260	770	3,600	2,630	1,240	575	353	275
9	458	714	290	280	235	730	3,350	2,120	640	565	1,600	290
10	450	688	325	250	180	715	3,940	1,560	700	525	2,550	344
11	401	664	375	260	195	730	3,000	3,810	1,170	595	1,580	240
12	347	640	295	235	260	625	2,770	2,730	931	545	826	326
13	415	610	360	225	290	600	2,200	1,680	1,700	664	652	341
14	700	575	425	300	300	670	2,480	4,120	1,130	1,130	545	326
15	500	565	400	300	350	640	2,820	3,180	859	877	565	365
16	490	555	350	350	500	640	4,280	2,110	700	652	585	443
17	555	525	305	300	550	650	4,500	1,700	595	525	443	377
18	458	525	456	260	500	750	4,950	1,430	585	555	415	408
19	394	525	456	300	500	920	3,000	1,260	500	482	394	1,180
20	415	474	360	300	509	1,250	2,300	1,110	474	458	365	859
21	700	458	415	310	470	1,750	2,300	1,050	466	390	365	2,980
22	436	555	420	350	600	2,700	4,660	913	615	390	338	2,300
23	429	535	420	320	700	4,030	4,610	770	1,700	390	365	1,280
24	408	490	415	300	750	3,680	4,340	700	2,300	390	314	1,090
25	395	450	365	335	770	3,520	2,920	640	1,700	390	320	2,350
26	1,130	480	395	275	759	3,480	2,370	615	1,090	401	326	1,890
27	714	490	375	250	1,000	3,210	2,250	676	786	308	341	10,000
28	1,130	350	355	225	1,800	2,680	2,240	1,030	700	320	341	4,290
29	1,310	380	355	250	....	2,850	2,870	670	640	347	278	2,920
30	1,400	400	335	275	....	3,450	4,550	676	615	365	266	1,680
31	11,200	....	320	275	....	4,180	....	786	....	338	278	....

NOTE.—Stage discharge relation affected by ice Dec. 30, 1911, to Apr. 8, 1912; Jan. 7 to Mar. 21, 1913; Dec. 19, 1913, to Mar. 2, 1914; Dec. 22, 1914, to Feb. 28, 1915; Jan. 2-22 and Feb. 4 to Mar. 31, 1916; Dec. 15, 1916, to Mar. 25, 1917; Nov. 26, 1917, to Mar. 21, 1918; discharge for these periods determined from gage heights corrected for effect of ice. Discharge on Sundays (gage not read) estimated by hydrograph comparison with records of flow of other rivers.

Monthly discharge of Pemigewasset River at Plymouth, N. H., for the years ending Sept. 30, 1912–1918.

[Drainage area, 615 square miles].

Month.	Discharge in second-feet.				Run-off (depth in inches on drainage area).
	Maximum.	Minimum.	Mean.	Per square mile.	
1911-12.					
October	4,560	375	1,320	2.15	2.48
November	2,550	720	1,070	1.74	1.94
December	4,820	660	1,390	2.26	2.61
January	3,600	349	626	1.02	1.18
February	410	260	306	.498	.54
March	6,100	250	1,320	2.15	2.48
April	11,600	1,500	5,100	8.29	9.25
May	6,330	967	2,860	4.65	5.36
June	4,720	278	1,030	1.67	1.86
July	482	208	251	.408	.47
August	5,460	255	683	1.11	1.28
September	1,750	290	466	.758	.85
The year	11,600	208	1,370	2.23	30.30
1912-13.					
October	9,610	353	1,380	2.24	2.58
November	9,550	630	1,600	2.60	2.90
December	2,770	555	1,110	1.80	2.08
January	5,560	850	1,970	3.20	3.69
February	1,600	420	685	1.11	1.16
March	18,700	350	4,850	7.89	9.10
April	10,500	1,220	2,790	4.54	5.06
May	4,820	482	1,480	2.41	2.78
June	2,300	290	621	1.01	1.13
July	565	242	312	.507	.58
August	290	113	210	.341	.39
September	5,030	60	375	.610	.68
The year	18,700	60	1,460	2.37	32.13
1913-14.					
October	7,670	242	1,190	1.93	2.22
November	11,100	482	1,330	2.16	2.41
December	1,600	480	710	1.15	1.33
January	1,460	385	603	.980	1.13
February	1,180	375	602	.979	1.02
March	12,400	482	2,980	4.85	5.59
April	18,400	1,560	4,750	7.72	8.61
May	5,680	600	2,750	4.47	5.15
June	3,280	266	495	.805	.90
July	565	186	336	.546	.63
August	1,220	186	290	.472	.54
September	525	186	251	.408	.46
The year	18,400	186	1,360	2.21	29.99
1914-15.					
October	278	186	203	.330	.38
November	1,900	186	354	.576	.64
December	1,010	186	391	.636	.73
January	7,300	220	1,060	1.72	1.98
February	16,200	600	2,590	4.21	4.38
March	9,500	756	2,560	4.16	4.80
April	10,800	600	2,720	4.42	4.93
May	4,280	530	1,230	2.00	2.31
June	1,750	390	548	.891	.99
July	12,900	474	2,050	3.33	3.84
August	5,950	565	1,360	2.21	2.55
September	2,050	371	535	.870	.97
The year	16,200	186	1,290	2.10	28.50

Monthly discharge of Pemigewasset River at Plymouth, N. H., for the years ending Sept. 30, 1912–1918—Continued.

Month.	Discharge in second-feet.				Run-off (depth in inches on drainage area).
	Maximum.	Minimum.	Mean.	Per square mile.	
1915–16.					
October	859	422	534	0.868	1.00
November	2,180	443	678	1.10	1.23
December	5,670	450	1,130	1.84	2.12
January	6,270	470	1,720	2.80	3.23
February	4,970	600	1,550	2.52	2.72
March	7,700	620	1,380	2.24	2.58
April	10,900	1,890	3,870	6.29	7.02
May	11,200	967	2,730	4.44	5.12
June	6,330	1,130	2,620	4.26	4.75
July	6,670	466	1,310	2.13	2.46
August	2,350	335	587	.954	1.10
September	3,020	255	615	1.00	1.12
The year	11,200	255	1,560	2.54	34.45
1916–17.					
October	1,520	384	656	1.07	1.23
November	5,560	380	895	1.46	1.63
December	5,850	300	1,040	1.69	1.95
January	1,600	390	639	1.04	1.20
February	590	320	431	.701	.73
March	9,860	430	1,700	2.76	3.18
April	9,550	1,430	3,940	6.41	7.15
May	5,080	2,320	2,980	4.85	5.59
June	13,500	1,090	3,880	6.31	7.04
July	2,060	380	662	1.08	1.24
August	1,070	222	452	.735	.85
September	700	212	334	.543	.61
The year	13,500	212	1,470	2.39	32.40
1917–18.					
October	11,200	245	922	1.50	1.73
November	3,680	350	815	1.33	1.48
December	450	290	375	.610	.70
January	350	200	277	.450	.52
February	1,800	150	462	.751	.78
March	4,180	600	1,700	2.76	3.18
April	9,500	2,200	3,760	6.11	6.82
May	4,970	615	1,940	3.15	3.63
June	2,300	466	875	1.42	1.58
July	1,130	308	513	.834	.96
August	2,550	266	534	.868	1.00
September	10,000	240	1,290	2.10	2.34
The year	11,200	150	1,120	1.82	24.72

Days of deficiency in discharge of Pemigewasset River at Plymouth, N. H., during the years ending Sept. 30, 1912–1918.

Discharge in second-feet per square mile.	Discharge in second feet.	Theoretical horsepower per foot of fall.	Days of deficiency in discharge.						
			1911–12.	1912–13.	1913–14.	1914–15.	1915–16.	1916–17.	1917–18.
0.1	62	....	....	1	....	....	....	....	....
.15	93	....	....	9	....	....	....	....	....
.2	123	....	....	14	....	....	....	....	....
.3	185	21.0	....	21	....	....	....	....	2
.4	246	28.0	19	46	39	47	....	10	15
.5	308	35.0	73	78	87	62	6	19	55
.6	369	41.9	111	100	105	78	18	37	107
.7	430	48.9	136	122	132	108	32	96	139
.8	492	55.9	147	145	165	135	77	148	164
.9	554	62.9	159	162	193	150	100	182	183
1.0	615	69.9	170	177	213	163	124	198	208
1.1	677	76.9	179	194	220	175	141	214	223
1.2	738	83.9	191	206	228	186	160	225	238
1.3	800	90.9	201	213	241	201	170	226	248
1.4	861	97.8	208	216	245	212	186	232	253
1.5	923	105	222	219	247	222	191	236	260
1.6	984	112	230	224	250	228	198	239	262
1.75	1,080	123	243	232	257	251	208	242	266
1.9	1,170	133	250	240	260	263	221	246	275
2.05	1,260	143	259	248	267	274	229	248	279
2.25	1,390	158	264	256	276	282	240	258	283
2.5	1,540	175	269	264	282	296	245	262	286
2.75	1,700	193	278	277	290	302	249	267	292
3.0	1,850	210	281	289	298	307	258	272	299
3.5	2,160	245	289	300	299	315	273	278	308
4.0	2,460	280	300	311	310	322	292	283	314
5.0	3,080	350	314	327	320	334	318	313	333
7.0	4,310	489	342	341	336	346	343	338	351
10.0	6,150	699	358	352	354	353	358	355	360
15.0	9,230	1,050	363	357	360	359	364	360	362
20.0	12,300	1,400	366	361	362	363	366	364	365
25.0	15,400	1,750	366	363	364	364	....	365	....
30.0	18,500	2,100	....	364	365	365	....	....	....
35.0	21,500	2,440	....	365	....	....	....	....	....

NOTE.—The above table gives the theoretical horsepower per foot of fall that may be developed at different rates of discharge and shows the number of days on which the discharge and corresponding horsepower were respectively less than the amounts given in the columns for discharge and horsepower. In using this table allowance should be made for the various losses, the principal ones being the wheel loss, which may be as large as 20 per cent, and the head loss, which may be as large as 5 per cent.

MERRIMACK RIVER AT FRANKLIN JUNCTION, N. H.

LOCATION.—At covered wooden bridge of Boston & Maine Railroad 1 mile below confluence of Pemigewasset and Winnepesaukee rivers, at Franklin Junction, Merrimack County.

DRAINAGE AREA.—1,460 square miles.

RECORDS AVAILABLE.—July 8, 1903, to September 30, 1918.

GAGE.—Standard chain gage fastened to floor of bridge on upstream side over the west channel; read by F. R. Roers. A gage painted on the downstream right-hand side of the center pier gives results considerably in error for low stages.

DISCHARGE MEASUREMENTS.—Made from upstream side of bridge.

CHANNEL AND CONTROL.—Coarse gravel and boulders; fairly permanent.

EXTREMES OF DISCHARGE.—Maximum stage recorded during year, 13.0 feet at 7 a. m. October 31 (discharge, 18,000 second-feet); minimum stage recorded, 4.0 feet at 6 a. m. August 26, 6. a. m. August 31, and 6 a. m. September 13 (discharge, 1,030 second-feet).

1903–1918: Maximum stage recorded, 19.5 feet at 5 p. m. April 21, 1914 (discharge by extension of rating curve, 32,300 second-feet); minumum stage recorded 3.30 feet October 4, 1903 (discharge by extension of rating curve, 250 second-feet).

ICE.—Stage-discharge relation usually affected by ice during the winter.

REGULATION.—Flow affected by storage in Winnepesaukee, Squam, and New Found lakes, and by the operation of mills above the station.

ACCURACY.—Stage-discharge relation subject to slight changes. Rating curve fairly well defined below 10,000 second-feet. Gage read to half-tenths once or twice daily, except on Sundays and numerous other days with no readings. Gage not read from January 24 to February 26. Readings of doubtful accuracy. Daily discharge ascertained by applying mean gage height to rating table. Records poor.

COOPERATION.—Gage heights furnished by the proprietors of locks and canals on Merrimack River.

Discharge measurements of Merrimack River at Franklin Junction, N. H., during the year ending Sept. 30, 1918.

[Made by M. R. Stackpole.]

Date.	Gage height.	Discharge.	Date.	Gage height.	Discharge.
	Feet.	*Sec.-ft.*		*Feet.*	*Sec.-ft.*
Dec. 20	[a] 4.62	1,260	Feb. 26	[a] 5.93	1,360
Jan. 21	[a] 5.65	983	Mar. 25	6.07	3,570

[a] Stage-discharge relation affected by ice.

Daily discharge, in second-feet, of Merrimack River at Franklin Junction, N. H., for the year ending Sept. 30, 1918.

Day.	Oct.	Nov.	Dec.	Mar.	Apr.	May.	June.	July.	Aug.	Sept.
1	1,440	6,200	1,530	4,300	6,000	6,000	1,620	1,820	1,220	1,100
2	1,440	3,790	1,440	3,600	13,600	5,800	1,650	1,530	1,260	1,150
3	1,440	3,120	1,350	3,000	15,500	5,600	1,720	1,530	1,220	1,170
4	1,440	2,800	1,300	2,800	8,510	4,660	1,620	1,400	1,200	1,220
5	1,440	2,540	1,260	2,040	7,250	4,500	1,620	1,350	1,170	1,220
6	1,400	2,260	1,260	2,040	6,830	4,448	1,480	1,350	1,170	1,170
7	1,600	2,160	1,260	1,930	6,300	4,480	1,480	1,300	1,170	1,170
8	1,600	2,040	....	2,040	6,000	4,130	1,530	1,350	1,170	1,200
9	1,530	1,930	....	1,930	5,800	3,450	1,600	1,530	1,620	1,260
10	1,620	1,820	....	2,000	6,410	2,970	1,720	1,440	4,130	1,260
11	1,620	1,750	....	2,040	6,200	3,790	1,820	1,400	3,400	1,170
12	1,530	1,720	....	2,040	6,000	3,300	2,040	1,350	2,820	1,220
13	1,550	1,620	....	1,930	5,800	2,820	2,540	1,300	1,620	1,080
14	1,550	1,530	....	2,040	5,300	4,300	2,160	1,700	1,600	1,170
15	1,620	1,440	....	2,040	4,840	5,800	1,820	1,930	1,530	1,300
16	1,620	1,620	....	1,930	5,200	3,790	1,700	1,720	1,350	1,400
17	1,720	1,440	....	1,950	5,600	3,120	1,620	1,620	1,300	1,350
18	1,720	1,450	....	1,930	5,200	2,680	1,620	1,620	1,300	1,300
19	1,620	1,440	....	2,160	5,200	2,400	1,530	1,440	1,350	1,440
20	1,530	1,530	....	1,820	5,800	2,280	1,440	1,350	1,260	2,280
21	1,800	1,530	....	1,820	6,200	2,040	1,440	1,300	1,300	2,820
22	1,530	1,530	....	2,040	6,410	2,040	1,620	1,260	1,260	2,700
23	1,530	1,530	....	2,280	6,000	1,820	2,800	1,260	1,260	2,680
24	1,440	1,530	....	4,400	6,620	1,820	3,620	1,170	1,260	3,960
25	1,530	1,500	....	3,620	5,800	1,720	3,120	1,260	1,200	3,450
26	2,820	1,480	....	3,450	5,020	1,600	2,040	1,260	1,170	3,790
27	2,280	1,440	....	3,450	4,480	1,530	2,040	1,250	1,260	14,000
28	2,150	1,480	....	3,120	4,100	1,930	1,930	1,250	1,260	8,720
29	2,040	1,500	....	1,930	4,480	1,930	1,820	1,260	1,260	5,800
30	2,680	1,480	....	5,200	5,200	1,820	1,800	1,260	1,170	3,450
31	17,900	....	....	5,500	....	1,820	....	1,260	1,080	....

NOTE.—Discharge on Sundays and other days gage was not read estimated by comparison with records obtained at several other stations.

Monthly discharge of Merrimack River at Franklin Junction, N. H., for the year ending Sept. 30, 1918.

[Drainage area, 1,460 square miles.]

Month.	Discharge in second-feet.				Run-off (depth in inches on drainage area).
	Maximum.	Minimum.	Mean.	Per square mile.	
October	17,900	1,400	2,220	1.52	1.75
November	6,200	1,440	1,970	1.35	1.51
December			1,100	.753	.87
January			930	.637	.73
February			1,230	.842	.88
March	5,500	1,820	2,660	1.82	2.10
April	15,500	4,100	6,390	4.38	4.89
May	6,000	1,530	3,240	2.22	2.56
June	3,620	1,440	1,890	1.29	1.44
July	1,820	1,170	1,410	.966	1.11
August	4,130	1,080	1,490	1.02	1.18
September	14,000	1,080	2,570	1.76	1.96
The year	17,900		2,260	1.55	20.98

NOTE.—Mean monthly discharge for December, January, and February estimated at 1.7 times discharge of Pemigewasset River at Plymouth plus discharge from Lake Winnepesaukee at Lakeport.

MERRIMACK RIVER AT LAWRENCE, MASS.

LOCATION.—At dam of Essex Co., in Lawrence, Essex County.

DRAINAGE AREA.—Total of Merrimack River basin above Lawrence, 4,663 square miles; net drainage area, exclusive of diverted parts of Nashua and Sudbury River and Lake Cochituate basins, 4,452 square miles.

RECORDS AVAILABLE.—January 1, 1880, to September 30, 1918.

COMPUTATIONS OF DISCHARGE.—Accurate record is kept of the flow over the dam and through the various wheels and gates. This flow includes the water wasted into the Merrimack from the Nashua, Sudbury, and Cochituate drainage basins. Estimates of the quantity wasted from these basins is furnished by the Metropolitan Water and Sewerage Board of Boston and subtracted from the quantity measured at Lawrence to obtain the net flow from the net drainage area of 4,452 square miles.

DIVERSIONS.—Practically the entire flow of the South Branch of Nashua River, Sudbury River, and Lake Cochituate is diverted for use by the Metropolitan water district of Boston.

REGULATION.—Flow regulated to some extent by storage in Lake Winnepesaukee. The low-water flow of the stream is affected by operation of various power plants above Lawrence.

STORAGE.—There are several reservoirs in the basin. It is estimated that the water surface is about 3.5 per cent of the entire drainage area.

COOPERATION.—The entire record has been furnished by R. A. Hale, principal assistant engineer of the Essex Co.; rearranged in form for climatic year by engineers of the Geological Survey.

Daily discharge, in second-feet, of Merrimack River at Lawrence, Mass., for the year ending Sept. 30, 1918.

Day.	Oct.	Nov.	Dec.	Jan.	Feb.	Mar.	Apr.	May.	June.	July.	Aug.	Sept.
1	2,316	14,902	2,499	985	2,491	7,823	18,380	10,832	3,128	4,063	2,085	278
2	2,268	10,312	1,516	2,442	1,768	6,967	20,716	13,487	2,239	3,554	2,033	211
3	2,305	6,772	4,405	2,754	629	6,298	25,296	14,325	4,778	2,931	1,241	2,551
4	2,256	4,754	3,658	2,613	2,446	7,253	26,928	12,562	3,488	688	173	2,521
5	2,276	5,753	3,364	1,572	2,620	6,609	22,954	9,671	3,590	3,641	2,155	2,246
6	1,304	4,868	3,457	281	2,558	6,179	17,944	9,560	3,497	2,337	2,179	2,108
7	302	4,481	3,843	2,348	2,516	6,170	14,461	8,388	3,570	688	2,021	1,116
8	2,457	4,243	2,616	2,390	2,344	6,230	13,955	7,826	2,163	3,967	2,035	31
9	3,220	3,873	786	2,246	1,786	4,741	13,366	7,868	536	3,512	2,139	1,812
10	3,266	2,433	3,719	1,973	603	4,266	13,694	7,425	4,018	2,591	1,551	2,011
11	3,432	802	3,175	1,950	2,561	5,817	14,112	6,032	3,420	2,390	2,060	1,977
12	1,739	4,521	2,854	1,532	2,767	5,068	13,222	6,366	3,397	2,667	5,154	2,031
13	1,492	3,973	2,631	539	2,666	5,034	11,698	8,096	3,414	2,165	3,987	2,075
14	271	3,737	2,558	3,095	2,818	5,001	10,861	6,733	3,674	570	3,601	1,324
15	2,946	2,821	2,012	2,853	3,208	5,309	12,609	7,821	3,011	2,959	3,274	373
16	3,202	3,507	688	2,766	2,917	3,929	14,495	9,418	2,363	3,200	2,650	2,090
17	3,192	2,572	2,710	2,896	1,558	3,816	15,489	7,974	4,598	2,986	2,150	2,105
18	2,548	589	2,704	1,651	4,630	7,284	15,261	6,056	3,363	3,453	546	2,251
19	2,852	3,727	2,664	1,476	4,083	8,141	14,825	5,286	3,048	3,883	2,208	2,595
20	2,028	3,326	2,833	1,249	5,972	8,277	12,572	6,299	2,869	2,477	2,362	3,077
21	449	2,887	3,279	3,089	6,518	9,327	11,581	4,747	3,015	599	2,560	2,581
22	2,910	3,018	2,557	2,767	6,279	11,684	13,143	4,908	2,196	2,379	2,392	4,172
23	3,052	3,659	779	2,625	6,714	13,984	16,153	4,281	895	2,774	2,230	7,605
24	2,774	2,669	2,545	2,542	5,139	15,576	16,998	4,366	6,427	2,745	1,202	5,966
25	3,212	1,056	1,160	2,508	6,364	17,505	15,294	2,964	6,469	2,516	405	5,031
26	3,901	4,487	4,196	1,683	6,855	16,937	13,742	2,576	5,666	2,514	1,634	4,402
27	3,413	3,689	3,561	587	7,431	16,463	11,233	5,221	4,976	1,481	1,921	8,402
28	3,616	2,989	2,835	2,616	7,779	15,185	9,615	4,188	4,410	304	1,940	18,165
29	5,112	692	2,029	2,662	....	14,231	8,925	3,876	2,812	2,446	1,982	13,546
30	4,722	3,098	540	2,435	....	14,003	8,684	1,388	780	2,213	1,974	10,180
31	7,362	....	2,663	2,409	....	15,455	....	5,141	....	2,169	1,298	....

NOTE.—Table shows the actual flow at Lawrence; not corrected for water wasted by the Metropolitan Water and Sewerage Board.

Weekly discharge, in second-feet, of Merrimack River at Lawrence, Mass., for the year ending Sept. 30, 1918.

[Weeks arranged in order of dryness.]

Week ending Sunday—	Measured at Lawrence (total drainage area 4,663 square miles).	Wasting into Merrimack River from diverted drainage basins (211 square miles.)	From net drainage area of 4,452 square miles.	Per square mile of net drainage area.
Sept. 8	1,541	6	1,535	0.345
Sept. 1	1,575	7	1,568	.352
Sept. 15	1,658	26	1,632	.367
Aug. 4	1,766	16	1,750	.393
Jan. 13	1,854	44	1,810	.407
Oct. 7, 1917	1,861	12	1,849	.415
Jan. 6	1,901	36	1,865	.419
Aug. 25	1,908	8	1,900	.427
Aug. 11	2,020	12	2,008	.451
July 28	2,102	11	2,091	.470
Feb. 10	2,125	17	2,108	.473
Feb. 3	2,144	20	2,124	.477
Jan. 27	2,257	24	2,233	.502
Oct. 14, 1917	2,268	16	2,252	.506
Jan. 20	2,284	59	2,225	.500
Dec. 30, 1917	2,409	80	2,329	.523
Oct. 21, 1917	2,460	20	2,440	.548
Dec. 23, 1917	2,504	91	2,413	.542
Dec. 16, 1917	2,520	56	2,464	.553
July 14	2,552	19	2,533	.569
July 7	2,557	22	2,535	.569
Feb. 17	2,642	84	2,558	.575

Weekly discharge, in second-feet, of Merrimack River at Lawrence, Mass., for the year ending Sept. 30, 1918—Continued.

Week ending Sunday—	Measured at Lawrence (total drainage area 4,663 square miles).	Wasting into Merrimack River from diverted drainage basin (211 square miles).	From net drainage area of 4,452 square miles.	Per square mile of net drainage area.
Sept. 22	2,696	68	2,628	0.590
Dec. 2, 1917	2,710	64	2,646	.594
July 21	2,787	21	2,766	.621
June 23	2,855	31	2,824	.634
Nov. 25, 1917	2,906	63	2,843	.639
Aug. 18	3,052	9	3,043	.684
June 9	3,089	12	3,077	.691
Nov. 18, 1917	3,103	65	3,038	.682
Dec. 9, 1917	3,161	80	3,081	.692
Oct. 28, 1917	3,268	109	3,159	.710
June 16	3,328	11	3,317	.745
June 2	3,597	15	3,582	.805
Nov. 11, 1917	3,779	108	3,671	.825
May 26	4,305	13	4,292	.964
June 30	4,506	38	4,468	1.004
Mar. 17	4,853	242	4,611	1.036
Feb. 24	5,619	267	5,352	1.202
Mar. 10	5,921	236	5,685	1.277
Mar. 3	7,074	369	6,705	1.506
May 19	7,340	55	7,285	1.636
May 12	7,638	78	7,560	1.698
Nov. 4, 1917	7,705	135	7,570	1.700
Sept. 29	9,017	134	8,883	1.995
Mar. 24	10,610	197	10,413	2.339
May 5	11,069	172	10,897	2.448
Apr. 14	12,986	78	12,908	2.899
Apr. 28	13,740	173	13,567	3.047
Apr. 21	13,976	116	13,860	3.113
Mar. 31	15,683	130	15,553	3.493
Apr. 7	20,954	99	20,855	4.684

Monthly discharge of Merrimack River at Lawrence, Mass., for the year ending Sept. 30, 1918.

Month.	Mean discharge in second-feet.				Run-off.		Rainfall in inches.
	Measured at Lawrence (total drainage area, 4,663 square miles).	Wasting into Merrimack from diverted drainage basins (211 square miles).	From net drainage area of 4,452 square miles.	Per square mile of net drainage area.	Depth in inches on drainage area.	Per cent of rainfall.	
October	2,780	49	2,731	0.613	0.707	12.6	5.60
November	4,007	82	3,925	.882	.984	91.1	1.08
December	2,608	77	2,531	.569	.656	23.4	2.80
January	2,114	38	2,076	.466	.537	18.6	2.88
February	3,786	142	3,644	.819	.853	29.5	2.89
March	9,050	220	8,830	1.983	2.286	103.9	2.20
April	14,973	117	14,856	3.337	3.724	126.7	2.94
May	6,925	67	6,858	1.540	1.776	82.2	2.16
June	3,394	22	3,372	.757	.845	22.3	3.79
July	2,478	18	2,460	.553	.638	19.8	3.23
August	2,101	9	2,092	.470	.542	19.1	2.84
September	3,828	58	3,770	.847	.945	12.3	7.70
The year	4,837	75	4,762	1.070	14.493	36.1	40.12

[illegible].—The monthly discharge in second-feet per square mile and the run-off in depth in inches, shown [illegible]able, do not represent the natural flow from the basin because of artificial storage.

SMITH RIVER NEAR BRISTOL, N. H.

Location.—At highway bridge in South Alexandria, 3 miles from Bristol, Grafton County.

Drainage area.—78.5 square miles (measured on Walker map).

Records available.—May 11 to September 30, 1918.

Gage.—Vertical staff attached to downstream side of left abutment of highway bridge; read by George Perry and Archie Flanders.

Discharge measurements.—Made from downstream side of highway bridge or by wading.

Channel and control.—Channel rough and covered with boulders; control ledge rock and boulders 130 feet below gage.

Extremes of discharge.—Maximum stage recorded during period May 11 to September 30, 2.08 feet at 6 p. m. May 14 (discharge, 311 second-feet); minimum stage recorded during period, 0.70 foot at various times during July, August, and September (discharge, 11 second-feet).

Ice.—Ice forms to a considerable thickness during winter; stage-discharge relation affected.

Regulation.—The operation of the few small mills above the gage does not greatly affect the distribution of flow. Several small lakes in the basin; but little if any storage regulation.

Accuracy.—Stage-discharge relation probably permanent except when affected by ice. Rating curve well defined between 10 and 600 second-feet. Gage read to hundredths twice daily. Daily discharge ascertained by applying mean daily gage height to rating table. Records good.

Discharge measurements of Smith River near Bristol, N. H., during the year ending Sept. 30, 1918.

Date.	Made by—	Gage height.	Discharge.
		Feet.	*Sec.-ft.*
May 13	A. N. Weeks	1.30	[a] 106
19	C. H. Pierce	1.23	85
July 28	....do	.72	12.8

[a] Results uncertain; measurement not used in developing rating curve.

Note.—Several additional discharge measurements obtained subsequent to Sept. 30 were used in determining the rating curve.

Daily discharge, in second-feet, of Smith River near Bristol, N.H., for the year ending Sept. 30, 1918.

Day.	May.	June.	July.	Aug.	Sept.	Day.	May.	June.	July.	Aug.	Sept.
1		52	32	11	20	16	167	46	24	29	33
2		49	33	13	22	17	129	42	23	31	32
3		39	22	13	23	18	108	35	21	27	31
4		32	24	11	25	19	92	34	20	23	33
5		26	23	11	18	20	82	35	18	20	35
6		28	22	12	11	21	84	28	18	14	43
7		43	24	11	12	22	84	92	16	13	67
8		46	26	18	13	23	67	86	14	13	62
9		38	26	22	14	24	52	56	14	11	58
10		38	26	33	13	25	52	49	14	11	50
11	150	46	25	65	15	26	58	46	13	11	242
12	116	52	24	52	20	27	72	41	11	11	262
13	100	82	25	52	28	28	62	39	11	13	268
14	282	69	26	55	28	29	46	37	13	13	248
15	265	52	26	26	28	30	46	33	15	14	248
						31	50		14	20	

Note.—Daily discharge Sept. 21-25 estimated by comparison with records at gaging stations in near-by drainage basins.

Monthly discharge of Smith River near Bristol, N. H., for the year ending Sept. 30, 1918.

[Drainage area, 78.5 square miles.]

Month.	Discharge in second-feet.				Run-off (depth in inches on drainage area).
	Maximum.	Minimum.	Mean.	Per square mile.	
May 11-13	282	46	103	1.31	1.02
June	92	26	46.4	.591	.66
July	33	11	20.7	.264	.30
August	65	11	21.9	.279	.32
September	268	11	66.7	.850	.95

CONTOOCOOK RIVER NEAR ELMWOOD, N. H.

LOCATION.—At covered highway bridge on county road between Hancock and Greenfield, Hillsboro County, half a mile below mouth of Kimball Brook and 1½ miles south of Elmwood railroad station.

DRAINAGE AREA.—168 square miles (measured on topographic maps).

RECORDS AVAILABLE.—September 20, 1917, to September 30, 1918.

GAGE.—Chain on upstream side of bridge; read by Mrs. G. M. Elliott.

DISCHARGE MEASUREMENTS.—Made from bridge or by wading.

CHANNEL AND CONTROL.—Stream bed is covered with boulders and gravel. Control at low stages is rock ledge about 50 feet below gage and is well defined; at high stages control is probably at a storage dam about 3 miles downstream.

EXTREMES OF DISCHARGE.—Maximum stage recorded during year, 7.33 feet at 1 p. m. April 3 (discharge, 1,790 second-feet); a stage of 7.50 feet occurred at 1 p. m. March 23, but stage-discharge relation was affected by ice at the time; minimum stage recorded, 1.48 feet at 6.15 a. m. August 23 (discharge, 19 second-feet).

ICE.—River is usually covered with ice for several months during the winter.

REGULATION.—Considerable storage has been developed in Nubanusit Lake and other reservoirs on the main river and tributaries. Water power is used at various places on the river above the station; the first dam above the gage is at North Peterboro, 4 miles upstream.

ACCURACY.—Stage-discharge relation probably permanent, except when affected by ice. Rating curve fairly well defined between 50 and 1,200 second-feet. Gage read twice daily to hundredths, except from December 11 to April 4, when it was read once daily. Daily discharge ascertained by applying mean daily gage height to rating table. Records fair.

Discharge measurements of Contoocook River near Elmwood, N. H., during the years ending Sept. 30, 1917-18.

Date.	Made by—	Gage height.	Discharge.	Date.	Made by—	Gage height.	Discharge.
1917.		Feet.	Sec.-ft.	1918.		Feet.	Sec.-ft.
Sept. 7	M. R. Stackpole	2.58	130	Feb. 2	M. R. Stackpole	[a]3.42	120
20	....do	2.16	74	Mar. 9	H. W. Fear	[a]4.61	388
Dec. 10	....do	[a]2.63	104	Apr. 5	....do	5.57	1,020
				8	....do	4.64	674
				Aug. 21	J. W. Moulton	2.38	101

[a] Stage-discharge relation affected by ice.

Daily discharge, in second-feet, of Contoocook River near Elmwood, N. H., for period of Sept. 20, 1917, to Sept. 30, 1918.

Day.	Sept.	Oct.	Nov.	Dec.	Jan.	Feb.	Mar.	Apr.	May.	June.	July.	Aug.	Sept.
1		49	800	104	58	118	660	1,110	530	182	104	73	68
2		58	437	126	73	118	594	1,420	594	111	104	78	45
3		84	292	104	84	78	498	1,780	498	126	78	63	26
4		97	224	104	90	45	467	1,370	353	118	49	58	30
5		97	224	104	90	41	437	990	268	111	49	73	45
6		111	246	111	37	26	408	765	292	111	68	54	68
7		58	224	104	41	26	408	627	303	126	63	54	54
8		68	162	84	49	26	437	695	257	118	63	37	41
9		104	172	68	134	26	380	660	224	90	104	68	49
10		111	152	78	152	49	328	800	234	111	97	68	73
11		97	104	131	134	78	303	627	292	126	73	58	68
12		84	172	118	134	111	280	594	213	126	84	63	45
13		126	182	73	97	152	280	530	246	118	73	58	78
14		84	172	68	90	152	303	627	467	143	49	64	73
15		97	172	118	104	152	353	910	353	104	49	58	54
16		126	172	73	118	192	303	835	292	84	84	54	37
17		118	172	68	134	213	303	730	268	97	68	49	68
18		111	97	90	143	234	353	765	224	97	97	54	73
19		111	90	104	134	234	437	660	152	90	111	58	111
20	78	126	118	118	104	437	498	530	192	90	90	62	172
21	84	90	111	118	97	467	765	467	202	84	73	68	353
22	84	73	111	126	90	627	1,030	870	224	303	84	63	192
23	73	78	162	90	104	660	1,190	910	246	498	90	45	162
24	49	118	224	78	104	594	1,150	765	213	380	90	68	118
25	68	530	152	73	118	562	1,110	594	172	303	78	37	104
26	84	353	224	68	118	594	1,110	467	143	162	84	26	380
27	84	224	172	84	111	730	910	353	192	126	68	58	1,460
28	84	257	134	78	104	730	660	353	213	111	45	63	594
29	104	202	104	97	111		695	353	280	118	54	63	328
30	68	303	104	104	118		800	353	152	78	73	63	234
31		1,110		37	118		870		172		73	63	

NOTE.—Stage-discharge relation affected by ice from Nov. 30 to Apr. 2; daily discharge determined from gage heights corrected for effect of ice by means of three discharge measurements and weather records. Gage not read Apr. 1-2 and Aug. 13-21; discharge estimated.

Monthly discharge of Contoocook River near Elmwood, N. H., for the year ending Sept. 30, 1918.

[Drainage area, 168 square miles.]

Month.	Discharge in second-feet.				Run-off (depth in inches on drainage area).
	Maximum.	Minimum.	Mean.	Per square mile.	
October	1,110	49	170	1.01	1.16
November	800	90	196	1.17	1.30
December	134	37	93.7	.558	.64
January	152	37	103	.613	.71
February	730	26	267	1.59	1.66
March	1,190	280	591	3.52	4.06
April	1,780	353	750	4.46	4.98
May	594	143	273	1.62	1.87
June	498	78	148	.881	.98
July	111	45	76.4	.455	.52
August	78	26	58.9	.351	.40
September	1,460	26	173	1.03	1.15
The year	1,780	26	241	1.43	19.43

BLACKWATER RIVER NEAR CONTOOCOOK, N. H.

Location.—At covered highway bridge in town of Webster, 150 feet north of Webster-Hopkinton town line, 1.1 miles from Tyler flag station, Boston & Maine Railroad, and 3½ miles from Contoocook, Merrimack County, N. H.

Drainage area.—131 square miles (measured on Walker maps).

Records available.—May 16 to September 30, 1918.

Gage.—Chain on downstream side of bridge; read by H. F. Corliss.

Discharge measurements.—Made from bridge or by wading.

Channel and control.—Channel deep at and above the gage. Control is at site of old dam about 100 feet below the gage; probably permanent.

Extremes of stage.—Maximum stage recorded May 16 to September 30, 1918, 7.55 feet at 6.55 p. m. September, 28; minimum stage recorded, 2.10 feet at 8.15 a. m. August 7.

Ice.—River usually freezes over during the winter.

Regulation.—A small amount of storage has been developed in Pleasant Pond (New London). Several small mills above the gage, but distribution of flow not seriously affected.

Accuracy.—Stage-discharge relation probably permanent. Rating curve well defined below 1,600 second-feet. Gage read twice daily to hundredths. Daily discharge ascertained by applying mean daily gage height to rating table. Results good.

Discharge measurements of Blackwater River near Contoocook, N. H., during the year ending Sept. 30, 1918.

Date.	Made by—	Gage height.	Discharge.
		Feet.	*Sec.-ft.*
May 16	A. N. Weeks	4.00	333
20	C. H. Pierce	3.19	161
June 6	O. W. Hartwell	2.59	75

Note.—Several discharge measurements obtained subsequent to Sept. 30, 1918, were used in determining the rating curve.

Daily gage height, in feet, of Blackwater River near Contoocook, N. H., for the year ending Sept. 30, 1918.

Day.	May.	June.	July.	Aug.	Sept.	Day.	May.	June.	July.	Aug.	Sept.
1		105	69	40	44	16	311	86	73	78	63
2		97	66	41	43	17	250	73	66	69	56
3		85	63	40	40	18	210	69	65	61	52
4		79	63	37	40	19	173	69	62	53	67
5		75	62	37	39	20	164	62	58	48	94
6		72	59	32	38	21	147	63	54	48	164
7		70	64	34	37	22	139	102	49	44	260
8		73	65	43	36	23	131	173	46	43	260
9		75	65	120	37	24	118	210	48	41	173
10		73	69	192	35	25	109	173	45	48	139
11		81	68	250	33	26	102	147	46	45	192
12		92	66	192	32	27	94	117	48	40	719
13		106	63	147	37	28	102	94	45	37	1,020
14		118	68	115	46	29	114	81	43	40	955
15		109	69	88	54	30	117	75	40	41	547
						31	108		40	49	

Monthly discharge of Blackwater River near Contoocook for the year ending Sept. 30, 1918.

[Drainage area, 131 square miles.]

Month.	Discharge in second-feet.				Run-off (depth in inches on drainage area).
	Maximum.	Minimum.	Mean.	Per square mile.	
May 16-31	311	94	149	1.14	0.68
June	210	62	96.8	.739	.82
July	69	40	58.3	.445	.51
August	250	32	70.4	.537	.62
September	1,020	32	178	1.36	1.52

SUNCOOK RIVER AT NORTH CHICHESTER, N. H.

Location.—About 100 feet below highway bridge and 500 feet from Chichester depot, North Chichester, Merrimack County, 2½ miles above mouth of Little Suncook River.

Drainage area.—157 square miles (measured on plane-table sheets).

Records available.—May 21 to September 30, 1918.

Gage.—Vertical staff attached to tree on left bank; Sanborn water-stage recorder temporarily installed at same place.

Discharge measurements.—Made from bridge or by wading.

Channel and control.—Stream bed covered with gravel and other alluvial deposits. Low-water control at head of rapids about 150 feet below gage; at high water the control is probably formed by crest of an old dam near Epsom.

Extremes of discharge.—Maximum stage May 21 to September 30, 1918, from water-stage recorder, 5.0 feet at 12 noon September 27 (discharge, 800 second-feet); minimum stage, from water-stage recorder, 1.2 feet several times in July and September (discharge, 16 second-feet).

Ice.—River is covered with ice for several months during the winter.

Regulations.—Storage has been developed at several points above Pittsfield. The operation of mills at Pittsfield causes a large variation in discharge during days when the mills are in operation.

Accuracy.—Stage-discharge relation probably permanent except when affected by ice. Rating curve fairly well defined between 20 and 800 second-feet. Staff gage read twice daily to half-tenths and used for comparison with water-stage recorder. Daily discharge ascertained by applying mean daily gage height to rating table from water-stage recorder. Records good.

Discharge measurements of Suncook River at North Chichester, N. H., during the year ending Sept. 30, 1918.

Date.	Made by—	Gage height.	Discharge.
		Feet.	*Sec.-ft.*
May 21	A. N. Weeks	2.40	195
22	C. H. Pierce	1.80	70
June 6	O. W. Hartwell	1.30	21.4

Note.—Several discharge measurements obtained subsequent to Sept. 30 were used in determining the discharge rating curve.

Daily discharge, in second-feet, of Suncook River at North Chichester, N. H. for the year ending Sept. 30, 1918.

Day.	May.	June.	July.	Aug.	Sept.	Day.	May.	June.	July.	Aug.	Sept.
1	....	28	94	103	28	16	....	28	103	94	46
2	....	32	103	103	17	17	....	121	103	52	57
3	....	78	103	41	70	18	....	85	121	24	78
4	....	85	28	20	52	19	....	94	85	85	70
5	....	85	94	78	52	20	....	78	57	85	70
6	....	78	57	85	64	21	112	78	85	85	180
7	....	103	82	103	46	22	94	64	150	94	344
8	....	46	103	112	14	23	103	94	112	57	191
9	....	28	94	78	85	24	85	180	103	36	130
10	....	94	85	130	57	25	57	103	94	24	112
11	....	85	70	130	57	26	52	112	70	94	170
12	....	112	78	140	57	27	112	85	28	103	685
13	....	94	28	41	57	28	94	85	14	94	488
14	....	112	20	78	36	29	94	52	64	94	296
15	....	52	103	41	17	30	28	28	103	85	213
						31	85	....	103	41	....

NOTE.—Water-stage recorder not in operation May 21 and May 31 to June 5; daily discharge computed from twice-daily readings of staff gage.

Monthly discharge of Suncook River at North Chichester, N. H., for the year ending Sept. 30, 1918.

[Drainage area, 157 square miles.]

Month.	Discharge in second-feet.				Run-off (depth in inches on drainage area).
	Maximum.	Minimum.	Mean.	Per square mile.	
May 21-31	112	28	83.3	0.530	0.21
June	180	28	79.9	.509	.57
July	150	14	80.2	.511	.59
August	140	20	78.4	.499	.58
September	685	14	128	.815	.91

SOUHEGAN RIVER AT MERRIMACK, N. H.

LOCATION.—At head of Atherton Falls, 7 miles below mouth of Beaver Brook and 1½ miles above confluence of Souhegan and Merrimack rivers at Merrimack, Hillsboro County.

DRAINAGE AREA.—168 square miles.

RECORDS AVAILABLE.—July 13, 1909, to September 30, 1918.

GAGES.—Gurley printing water-stage recorder on left bank about 350 feet above the falls; used since October 15, 1913. A vertical staff was used from July 13, 1909, to April 11, 1911, when it was washed out. From April 12, 1911, to October 14, 1913, a chain gage attached to a tree on left bank 350 feet above the falls was used.

DISCHARGE MEASUREMENTS.—Made by wading below the falls at low stages or from cable at high stages.

CHANNEL AND CONTROL.—The channel opposite the gage is a pool in which velocity is very low. The control of this pool is a rock ledge at the head of Atherton Falls and is permanent.

ICE.—Ice forms on control for short periods in the winter, slightly affecting stage-discharge relation.

EXTREMES OF DISCHARGE.—Maximum stage, from water-stage recorder, 5.92 feet at 8 p. m. March 26 (discharge, 1,830 second-feet); minimum stage, from water-stage recorder, 2.03 feet at 6 p. m. August 16 (discharge, 25 second-feet).

1909–1918: Maximum stage recorded, 9.6 feet on August 5, 1915 (discharge from extension of rating curve, about 4,930 second-feet); minimum stage recorded, 1.90 feet at 8 a. m. September 8, 1909 (discharge, 15 second-feet).

REGULATION.—Flow affected by the operation of the mills at Milford, about 8 miles above.

ACCURACY.—Stage-discharge relation permanent except when affected by ice for short periods. Rating curve well defined below 2,000 second-feet. Operation of water-stage recorder satisfactory except for periods noted in footnote to daily discharge table. Daily discharge ascertained by applying mean of 24 hourly gage heights to rating table. Records good for periods when water-stage recorder was in operation.

Discharge measurements of Souhegan River at Merrimack, N. H., during the year ending Sept. 30, 1918.

Date.	Made by—	Gage height.	Dis-charge.
		Feet.	*Sec.-ft.*
Jan. 16	M. R. Stackpole	a 2.80	99
Feb. 11	H. W. Fear	a 2.55	71

a Stage-discharge relation affected by ice.

Daily discharge, in second-feet, of Souhegan River at Merrimack, N. H., for the year ending Sept. 30, 1918.

Day.	Oct.	Nov.	Dec.	Jan.	Feb.	Mar.	Apr.	May.	June.	July.	Aug.	Sept.
1	32	506	72	60	98	700	1,140	303	200	60	42	45
2	34	307	102	70	96	600	1,330	570	200	52	42	40
3	40	232	104	72	80	510	1,500	510	160	52	37	35
4	37	175	114	82	80	480	1,070	371	115	52	39	35
5	39	152	130	82	82	450	830	299	110	52	35	45
6	42	162	128	74	82	420	638	260	105	52	36	55
7	42	142	118	82	78	410	545	246	105	50	34	60
8	36	138	112	78	78	400	515	225	105	46	36	60
9	40	120	106	86	82	390	488	201	105	42	39	50
10	46	116	82	90	78	320	496	185	110	60	43	35
11	46	114	68	94	70	310	442	180	120	75	64	40
12	51	90	90	98	80	310	393	165	130	70	44	40
13	52	90	80	95	86	310	393	162	130	65	49	45
14	46	104	78	90	100	340	398	175	130	60	62	44
15	48	100	88	95	120	420	748	210	120	55	64	50
16	70	108	90	100	145	420	830	188	110	46	38	50
17	92	96	84	105	170	406	665	162	90	70	33	45
18	52	92	92	110	200	406	540	135	80	84	45	60
19	51	74	98	105	240	460	474	106	130	90	50	110
20	57	74	96	100	420	535	380	108	64	90	50	300
21	49	92	102	95	580	665	371	118	45	85	60	380
22	34	86	104	95	640	950	692	122	300	80	60	300
23	58	84	106	90	700	1,230	950	120	480	70	60	210
24	62	142	96	90	620	1,330	665	118	400	65	55	150
25	315	228	90	90	600	1,260	560	110	200	60	50	110
26	331	182	92	88	700	1,300	434	105	160	55	45	400
27	207	125	100	88	740	1,010	380	105	140	50	35	1,500
28	170	92	96	88	740	775	327	102	125	40	40	640
29	225	96	90	86	......	802	299	140	110	35	50	360
30	198	92	90	92	......	860	303	180	70	33	50	250
31	610	......	74	98	......	980	......	200	......	32	50	......

NOTE.—Stage-discharge relation affected by ice Jan. 12 to Feb. 12. Discharge estimated Feb. 13 to Mar. 15, May 23 to June 17, June 22 to July 28, and Aug. 17 to Sept. 30 from observer's readings and comparative hydrographs of Ashuelot, Contoocook, and Pemigewasset rivers.

Monthly discharge of Souhegan River at Merrimack, N. H., for the year ending Sept. 30, 1918.

[Drainage area, 168 square miles.]

Month.	Discharge in second-feet.				Run-off (depth in inches on drainage area).
	Maximum.	Minimum.	Mean.	Per square mile.	
October	610	32	104	.619	0.71
November	506	74	140	.833	.93
December	130	72	95.9	.571	.66
January	110	60	89.3	.532	.61
February	740	70	•278	1.65	1.73
March	1,330	310	637	3.79	4.37
April	1,500	299	627	3.73	4.16
May	570	102	199	1.18	1.36
June	480	45	148	.881	.98
July	90	32	59.0	.351	.40
August	64	33	46.4	.276	.32
September	1,500	35	185	1.10	1.23
The year	1,500	32	216	1.29	17.45

SOUTH BRANCH OF NASHUA RIVER BASIN (WACHUSETT DRAINAGE BASIN) NEAR CLINTON, WORCESTER COUNTY, MASS.

LOCATION.—At Wachusett dam near Clinton.

DRAINAGE AREA.—119 square miles 1896 to 1907; 118.19 square miles 1908–1913, 108.84 square miles 1914–1918.

RECORDS AVAILABLE.—July, 1896, to September, 1918.

REGULATION.—Flow affected by storage in Wachusett reservoir and other ponds. Beginning with 1897, the determinations of discharge have been corrected for gain or loss in the reservoir and ponds, so that the record shows approximately the natural flow of the stream.

The yield per square mile is the yield of the drainage area including the water surfaces. For the years 1897 to 1902, inclusive, the water surface amounted to 2.2 per cent of the total area; 1903, 2.4 per cent; 1904, 3.6 per cent; 1905, 4.1 per cent; 1906, 5.1 per cent; 1907, 6.0 per cent; 1908 and subsequent years, 7.0 per cent.

COOPERATION.—Record furnished by the Metropolitan Water and Sewerage Board of Boston; rearranged in form of climatic year by engineers of the Geological Survey.

Yield and rainfall in South Branch of Nashua River basin (Wachusett drainage area near Clinton, Mass., for year ending Sept. 30, 1918.

[Drainage area, 108.84 square miles.]

Month.	Total yield (million gallons).	Yield per square mile.		Run-off.		Rainfall (inches).
		Million gallons per day.	Second-feet.	Depth on drainage area (inches).	Per cent of rainfall.	
October	1,871.8	0.555	0.858	0.99	16.4	6.03
November	1,021.3	.313	.484	.54	43.1	1.25
December	1,312.4	.389	.602	.69	29.9	2.31
January	1,634.3	.484	.749	.86	29.0	2.97
February	6,166.6	2.024	3.131	3.26	76.6	4.25
March	8,727.4	2.590	4.008	4.61	206.0	2.24
April	5,249.0	1.608	2.487	2.78	80.1	3.47
May	2,271.6	.673	1.042	1.20	112.8	1.07
June	1,707.2	.523	.809	.90	19.8	4.57
July	943.6	.280	.433	.50	17.9	2.80
August	536.4	.159	.246	.28	9.9	2.82
September	1,968.6	.603	.933	1.04	14.5	7.18
The year	33,410.2	.841	1.302	17.65	43.1	40.96

Summary of yield and rainfall in South Branch of Nashua River basin (Wachusett drainage area) near Clinton, Mass., for years ending Sept. 30, 1897–1918.

[Drainage area, 108.84 square miles.]

Month.	Total yield (million gallons).	Yield per square mile.		Run-off.		Rainfall (inches).
		Million gallons per day.	Second-feet.	Depth on drainage area (inches).	Per cent of rainfall.	
October	37,317.6	0.502	0.777	0.90	23.6	3.82
November	51,750.0	.720	1.114	1.24	34.3	3.62
December	81,514.9	1.098	1.700	1.96	52.0	3.77
January	87,401.9	1.178	1.824	2.10	57.9	3.63
February	95,523.5	1.413	2.186	2.28	60.0	3.80
March	189,288.3	2.550	3.946	4.55	112.9	4.03
April	151,902.1	2.115	3.272	3.65	98.9	3.69
May	87,522.3	1.179	1.825	2.10	63.8	3.29
June	55,854.4	.778	1.205	1.34	35.6	3.76
July	31,822.5	.429	.664	.76	18.8	4.04
August	30,889.9	.416	.644	.74	17.9	4.14
September	23,615.7	.329	.509	.57	15.9	3.59
The year	924,403.1	1.057	1.635	22.19	49.1	45.18

SUDBURY RIVER AND LAKE COCHITUATE BASINS NEAR FRAMINGHAM AND COCHITUATE, MIDDLESEX COUNTY, MASS.

DRAINAGE AREA.—Area of Sudbury basin from 1875 to 1878, inclusive, was 77.8 square miles; 1879–80, 78.2 square miles; 1881–1916, 75.2 square miles. Area of Cochituate basin from 1863 to 1909, inclusive, was 18.87 square miles; 1910, 17.8 square miles; 1911 to 1918, 17.58 square miles.

RECORDS AVAILABLE.—Of Sudbury River, January, 1875, to September, 1918; of Lake Cochituate, January, 1863, to September, 1918. Sudbury River and Lake Cochituate have been studied by the engineers of the city of Boston, the State Board of Health of Massachusetts, and the Metropolitan Water and Sewerage Board; records of rainfall have been kept in the Sudbury basin since 1875 and in the Cochituate basin since 1852, but the Cochituate basin records are considered of doubtful accuracy previous to 1872.

REGULATION.—The greater part of the flow from these basins is controlled by storage reservoirs constructed by the city of Boston and the Metropolitan Water and Sewerage Board. Lake Cochituate, which drains into Sudbury River a short distance below Framingham, is controlled as a storage reservoir by the Metropolitan Waterworks. In the Sudbury River basin the water surfaces exposed to evaporation have been increased from time to time by the construction of additional storage reservoirs. From 1875 to 1878, inclusive, the water surface amounted to 1.9 per cent of the total area; from 1879 to 1884, to 3 per cent; 1885 to 1893, to 3.4 per cent; 1894 to 1897, to 3.9 per cent; 1898 and subsequent years, 6.5 per cent.

DETERMINATION OF DISCHARGE.—In determining the run-off of the Sudbury and Cochituate drainage areas the water diverted for the municipal supply of Framingham, Natick, and Westboro, which discharge their sewerage outside the basins, is taken into consideration; the results, however, are probably less accurate since the sewerage diversion works were constructed. Water from the Wachusetts drainage area also passes into the reservoirs in the Sudbury basin and must be measured to determine the yield of the Sudbury basin; the small errors unadvoidable in the measurement of large quantities of water decrease the accuracy of the determination of the Sudbury water supply during months of low yield for years subsequent to 1897.

COOPERATION.—Record furnished by the Metropolitan Water and Sewerage Board of Boston: rearranged in form of climatic year by engineers of the Geological Survey.

Yield and rainfall in Sudbury River basin near Framingham, Mass., for year ending Sept. 30, 1918.

[Drainage area, 75.2 square miles.]

Month.	Total yield (million gallons).	Yield per square mile.		Run-off.		Rainfall (inches).
		Million gallons per day.	Second-feet.	Depth on drainage area (inches).	Per cent of rainfall.	
October	1,123.8	0.482	0.746	0.860	15.2	5.65
November	989.1	.438	.678	.757	57.6	1.31
December	886.7	.380	.589	.678	24.2	2.81
January	635.5	.273	.422	.486	14.0	3.47
February	3,808.3	1.809	2.798	2.914	81.3	3.58
March	5,091.3	2.187	3.384	3.896	156.2	2.50
April	3,306.2	1.466	2.267	2.530	57.1	4.43
May	1,490.7	.639	.989	1.141	98.8	1.16
June	417.1	.185	.286	.319	8.7	3.65
July	224.3	.096	.149	.171	4.2	4.07
August	−125.8	−.054	−.083	−.096	−6.0	1.61
September	1,437.9	.637	.986	1.100	12.8	8.60
The year	19,285.1	.702	1.086	14.756	34.5	42.84

Summary of yield and rainfall in Sudbury River basin near Framingham, Mass., for the years ending Sept. 30, 1876–1918.

[Drainage area, 75.2 square miles.]

Month.	Total yield (million gallons).	Yield per square mile.		Run-off.		Rainfall (inches).
		Million gallons per day.	Second-feet.	Depth on drainage area (inches).	Per cent of rainfall.	
October	41,361.7	0.412	0.638	0.74	19.3	3.82
November	70,586.2	.728	1.126	1.26	34.4	3.66
December	94,755.1	.945	1.462	1.69	44.3	3.81
January	118,068.9	1.178	1.823	2.10	51.5	4.08
February	151,709.5	1.660	2.568	2.67	64.8	4.12
March	271,950.1	2.713	4.198	4.84	112.5	4.30
April	189,208.9	1.951	3.019	3.37	95.5	3.53
May	106,338.0	1.060	1.640	1.89	58.0	3.26
June	46,735.5	.482	.746	.83	27.8	2.99
July	17,588.6	.175	.271	.31	8.5	3.64
August	23,291.0	.232	.359	.41	10.6	3.87
September	21,599.7	.223	.345	.38	11.3	3.37
The year	1,153,193.2	.976	1.510	20.49	46.1	44.45

Yield and rainfall in Lake Cochituate basin near Cochituate, Mass., for year ending Sept. 30, 1918.

[Drainage area, 17.58 square miles.]

Month.	Total yield (million gallons).	Yield per square mile.		Run-off.		Rainfall (inches).
		Million gallons per day.	Second-feet.	Depth on drainage area (inches).	Per cent of rainfall.	
October	361.9	0.664	1.027	1.18	18.6	6.33
November	280.2	.531	.822	.92	71.9	1.28
December	363.0	.666	1.030	1.19	44.1	2.70
January	276.0	.506	.783	.90	27.6	3.26
February	874.1	1.776	2.748	2.86	75.3	3.80
March	1,023.6	1.878	2.906	3.35	148.2	2.26
April	700.5	1.328	2.054	2.29	49.7	4.61
May	333.4	.612	.947	1.09	99.1	1.10
June	109.9	.208	.322	.36	10.8	3.34
July	88.3	.162	.251	.29	8.0	3.64
August	−17.5	−.032	−.050	−.06	−4.3	1.41
September	425.9	.808	1.250	1.40	16.3	8.58
The year	4,819.3	.759	1.174	15.77	37.2	42.31

Summary of yield and rainfall in Lake Cochituate basin near Cochituate, Mass., for the years ending Sept. 30, 1864–1918.

[Drainage area, 17.58 square miles.]

Month.	Total yield (million gallons).	Yield per square mile.		Run-off.		Rainfall (inches).
		Million gallons per day.	Second-feet.	Depth on drainage area (inches).	Per cent of rainfall.	
October	15,573.6	0.519	0.803	0.93	22.9	4.06
November	21,263.5	.733	1.134	1.26	32.6	3.86
December	26,825.8	.895	1.385	1.60	44.7	3.58
January	32,552.6	1.086	1.682	1.94	50.3	3.86
February	41,150.2	1.507	2.332	2.45	62.5	3.92
March	64,116.7	2.139	3.309	3.82	89.5	4.27
April	47,959.6	1.653	2.558	2.85	81.8	3.48
May	28,883.9	.966	1.495	1.72	48.6	3.54
June	13,507.9	.466	.721	.80	26.3	3.04
July	7,823.9	.261	.404	.47	12.6	3.72
August	11,140.1	.372	.576	.66	16.2	4.07
September	11,305.9	.390	.603	.67	18.8	3.57
The year	322,108.7	.912	1.411	19.17	42.6	44.97

THAMES RIVER BASIN.

QUINEBAUG RIVER AT JEWETT CITY, CONN.

LOCATION.—About 1,000 feet below railroad bridge and 570 feet below mouth of canal from Slater Mills (Pachaug River), Jewett City, town of Griswold, New London County.

DRAINAGE AREA.—712 square miles (measured on topographic maps).

RECORDS AVAILABLE.—July 17 to September 30, 1918.

GAGES.—Gurley 7-day graph water-stage recorder on left bank, referred to gage datum by a hook gage inside the well; an inclined staff gage is used for auxiliary readings. Recorder inspected by A. B. Ambot.

DISCHARGE MEASUREMENTS.—Made from cable.

CHANNEL AND CONTROL.—Bed of gravel and alluvial deposits. Control for low stages is fairly well defined riffle a few hundred feet below the gages; at high stages the control is at head of rapids 2½ miles below the gage.

EXTREMES OF DISCHARGE.—Maximum stage July 17 to September 30, from water-stage recorder, 9.42 feet at 3 p. m. September 27 (discharge, 3,430 second-feet); minimum stage July 17 to September 30, from water-stage recorder, 4.22 feet at midnight July 28 (water held back by dams) (discharge, from extension of rating curve, 104 second-feet).

ICE.—Probably little, if any, effect from ice during the winter.

REGULATION.—The flow of Pachaug River, which drains 59.7 square miles and enters Quinebaug River through the canal 570 feet above the gage, is under almost complete regulation. Numerous small reservoirs and power plants on the main river and tributaries above the station also affect the distribution of flow. The operation of mills at Jewett City causes a large variation in discharge.

ACCURACY.—Stage-discharge relation probably permanent. Rating curve well defined between 200 and 6,000 second-feet. Operation of water-stage recorder satisfactory except for short period as stated in footnote to daily-discharge table. Daily discharge ascertained by use of discharge integrator. Records good.

The following discharge measurement was made by H. W. Fear:

Sept. 21, 1918: Gage height, 7.61 feet; discharge, 1,800 second-feet.[1]

[1] Ten discharge measurements made subsequent to Sept. 30 were used in determining the discharge rating curve.

Daily discharge, in second-feet, of Quinebaug River at Jewett City, Conn., for the year ending Sept. 30, 1918.

Day.	July.	Aug.	Sept.	Day.	July.	Aug.	Sept.	Day.	July.	Aug.	Sept.
1	......	880	195	11	......	500	355	21	245	490	1,800
2	......	780	200	12	......	1,060	350	22	445	490	1,580
3	......	540	370	13	......	850	465	23	485	465	1,500
4	......	405	390	14	......	780	510	24	530	370	1,180
5	......	620	365	15	......	680	495	25	500	145	950
6	......	620	395	16	......	740	550	26	490	370	940
7	......	620	280	17	510	550	600	27	345	375	2,750
8	......	540	175	18	510	305	700	18	130	355	2,700
9	......	660	380	19	520	530	1,400	29	430	365	2,050
10	......	560	375	20	390	510	1,360	30	445	355	1,700
								31	600	200	......

NOTE.—Water-stage recorder not in operation Sept. 15-18; discharge estimated.

Monthly discharge of Quinebaug River at Jewett City, Conn., for the year ending Sept. 30, 1918.

[Drainage area, 712 square miles.]

Month.	Discharge in second-feet.				Run-off (depth in inches on drainage area).
	Maximum.	Minimum.	Mean.	Per square mile.	
July 17-31	600	130	438	0.615	0.34
August	1,060	145	537	.754	.87
September	2,750	175	902	1.27	1.42

CONNECTICUT RIVER BASIN.

CONNECTICUT RIVER AT FIRST LAKE, NEAR PITTSBURG, N. H.

LOCATION.—At the outlet of First Lake, 6 miles northeast of Pittsburg, Coos County.

DRAINAGE AREA.—81.4 square miles (from surveys by engineers of the Connecticut Valley Lumber Co.).

RECORDS AVAILABLE.—April 1, 1917, to September 30, 1918.

GAGES.—Gurley 7-day water-stage recorder on right bank about one-fourth mile below the outlet dam; installed in July, 1918; inclined staff gage at same site installed in November, 1917, and used in determining sluice-gate ratings; scales on gate frames indicate amount of sluice-gate openings; staff gage in lake above dam.

DISCHARGE MEASUREMENT.—Made from log bridge half a mile below gage, by wading, or from cable 200 feet above gage.

CHANNEL AND CONTROL.—Bed rough; rock bottom. Channel at cable section has been improved by removal of rocks and ledges. Control for river gage is rock ledge that extends completely across the stream; about 3 feet of fall immediately below ledge.

COMPUTATION OF DISCHARGE.—Beginning July 28, 1918, discharge determined from water-stage recorder. Previous to installation of water-stage recorder discharge through three sluice gates, 6 feet, 8 feet, and 20 feet in width, determined from gate ratings based on current-meter measurements and comparative readings of river gage, or from daily readings of river gage when gates remained at same opening for 24 hours. Discharge through one water wheel, used when slasher was in operation determined from figures of water-wheel efficiency and power output.

ICE.—Practically no effect from ice on the control section for river gage; formation of ice in the sluice-gate openings materially changes conditions at gates.

REGULATION.—About 4.1 billion cubic feet of storage has been developed in lakes and ponds above the gage; records of monthly discharge have been corrected for effect of storage in First Lake but not for effects of storage in lakes tributary to First Lake.

ACCURACY.—Stage-discharge relation for river gage practically permanent. Rating curve for river gage well defined below 800 second-feet. Operation of water-stage recorder satisfactory from its installation July 28, 1918. Rating curves for middle and upper leaves of 6-foot and 8-foot gates fairly well defined for periods used. Rating curves for lower sections of gates and for conditions of weir discharge somewhat uncertain. Daily discharge for January, February, March, and July to September 30, 1918, ascertained by applying gage height at river gage to rating table; daily discharge for other periods ascertained by applying records of gate openings to rating table and giving due consideration to times of opening and closing gates and changes in gate settings. Records good for periods when river gage was used and fair for periods when records of gate openings were used.

Daily gage height, in feet, of First Lake near Pittsburg, N. H., for the year ending Sept. 30, 1918.

Day.	Oct.	Nov.	Dec.	Jan.	Feb.	Mar.	Apr.	May.	June.	July.	Aug.	Sept.
1	20.0	16.5	18.9	11.9	6.1	3.8	3.4	13.1	20.0	17.9	19.3	12.5
2	19.7	16.8	18.7	11.7	5.9	3.9	3.8	14.0	19.9	17.8	19.2	12.2
3	19.5	16.9	18.5	11.5	5.8	3.8	4.3	14.6	19.9	17.7	19.1	11.9
4	19.3	17.5	18.2	11.3	5.6	3.8	4.7	15.0	19.8	17.6	19.0	11.5
5	19.3	17.9	18.0	11.0	5.5	3.8	5.0	15.4	19.6	17.6	18.7	11.3
6	19.2	18.1	18.0	10.8	5.4	3.8	5.2	15.9	19.5	17.5	18.5	10.9
7	19.1	18.1	17.9	10.5	5.3	3.8	5.5	16.4	19.4	17.4	18.3	10.6
8	18.9	18.4	17.7	10.4	5.2	3.8	5.7	17.0	19.4	17.3	18.0	10.1
9	18.7	18.6	17.5	10.2	5.1	3.8	6.0	17.6	19.4	17.3	17.8	9.9
10	18.5	18.9	17.4	10.0	5.0	3.8	6.1	18.0	19.4	17.3	17.8	9.5
11	18.3	19.2	17.2	9.7	4.9	3.8	6.2	18.3	19.4	17.4	17.4	9.1
12	18.0	19.4	17.0	9.5	4.9	3.8	6.3	19.0	19.4	17.4	17.2	8.9
13	17.4	19.5	16.8	9.4	4.7	3.8	6.5	19.2	19.3	17.2	16.8	8.4
14	17.2	19.7	16.6	9.2	4.6	3.8	6.6	19.6	19.2	17.3	16.4	8.3
15	17.3	19.9	16.4	9.0	4.5	3.8	6.8	20.1	19.1	17.4	16.1	7.9
16	17.2	20.0	16.2	8.9	4.4	3.7	7.1	20.5	19.0	18.0	15.7	7.7
17	17.0	20.1	16.0	8.6	4.3	3.8	7.4	20.5	18.9	18.1	15.4	7.5
18	16.8	20.2	15.8	8.4	4.2	3.7	7.6	20.6	19.0	18.3	15.3	7.2
19	16.5	20.3	15.5	8.2	4.2	3.7	7.8	20.6	19.0	18.4	15.2	7.1
20	16.3	20.4	15.2	8.0	4.1	3.7	8.0	20.5	18.9	18.8	15.0	7.0
21	16.3	20.4	14.9	7.8	4.1	3.5	8.1	20.5	18.8	18.9	14.8	7.0
22	16.1	20.2	14.7	7.6	4.0	3.4	8.3	20.4	18.6	19.0	14.7	7.3
23	15.8	20.1	14.5	7.4	4.0	3.4	8.7	20.3	18.5	19.2	14.5	7.8
24	15.6	19.9	14.2	7.3	3.9	3.4	9.2	20.2	18.5	19.2	14.4	8.2
25	15.4	19.8	13.9	7.1	3.9	3.4	9.7	20.1	18.4	19.2	14.2	8.5
26	15.2	19.8	13.6	6.9	3.9	3.4	10.0	20.0	18.4	19.3	13.9	8.7
27	14.9	19.6	13.4	6.7	3.8	3.4	10.3	20.1	18.4	19.4	13.7	9.2
28	14.6	19.5	13.1	6.5	3.8	3.4	10.6	20.3	18.2	19.3	13.5	9.8
29	14.5	19.3	12.8	6.4	....	3.5	11.0	20.4	18.0	19.1	13.2	10.2
30	14.4	19.0	12.5	6.3	....	3.4	12.0	20.3	18.0	19.1	12.9	10.8
31	15.4	....	12.2	6.2	....	3.5	....	20.2	....	19.3	12.7	....

Discharge measurements of Connecticut River at First Lake, near Pittsburg, N. H., during the year ending Sept. 30, 1918.

Date.	Made by—	Gage height.	Discharge.	Date.	Made by—	Gage height.	Discharge.
		Feet.	*Sec.-ft.*			*Feet.*	*Sec.-ft.*
Nov. 3[a]	C. H. Pierce	1.72	37.2	Nov. 7[a]	M. R. Stackpole	2.66	332
3[a]	M. R. Stackpole	1.72	39.6	7[a]	.....do	2.66	328
4[a]	.....do	2.07	99	8[a]	.....do	1.86	58
4[c]	C. H. Pierce	2.07	111	8[a]	J. P. Locke	1.86	64
4[a]	.....do	2.33	208	9[a]	M. R. Stackpole	2.20	151
4[a]	M. R. Stackpole	2.33	184	9[a]	.....do	2.20	148
5[a]	.....do	1.96	66	Apr. 29[b]	.....do	1.53	12.3
5[a]	.....do	1.96	75	29[b]	.....do	1.53	13.3
6[a]	.....do	2.20	140	29[c]	.....do	1.53	27.9
6[a]	.....do	2.20	145	May 10[d]	.....do	2.71	374
7[a]	.....do	2.50	253	18[d]	.....do	[e] 5.3	433
7[a]	.....do	2.50	267				

[a] Measurement made about half a mile below gage; practically no inflow between gage and measuring section. Section rough and conditions unsuitable for current-meter measurements.
[b] Measurement made by wading 300±feet above gage.
[c] Measurement made about half a mile below gage; considerable inflow between gage and measuring section; results of measurement not corrected for inflow. Section rough and conditions unsuitable for current-meter measurements.
[d] Measurement made about half a mile below gage; results of measurement corrected for inflow between gage and measuring section. Section rough and conditions unsuitable for current-meter measurements.
[e] Stage-discharge relation affected by log jam on control.

NOTE.—Measurements made at cable section except as noted. Twenty-three discharge measurements made subsequent to September 30 were used in determining the discharge rating curve.

Daily discharge, in second-feet, of Connecticut River at First Lake, near Pittsburg, N. H., for the year ending Sept. 30, 1918.

Day.	Oct.	Nov.	Dec.	Jan.	Feb.	Mar.	Apr.	May.	June.	July.	Aug.	Sept.
1	151	92	410	356	186	84	33	13	255	203	330	419
2	90	17	407	376	182	84	7	13	17	205	363	398
3	104	38	392	350	179	82	7	14	233	183	345	371
4	202	47	377	348	175	82	8	15	231	186	376	404
5	164	20	204	407	171	79	8	15	238	183	450	450
6	169	33	191	387	164	82	8	56	342	90	419	451
7	269	105	311	325	169	82	8	15	356	190	435	432
8	331	36	297	303	194	82	8	19	260	63	503	409
9	328	37	285	281	181	82	26	72	51	171	427	406
10	328	23	280	294	182	82	8	173	342	185	505	415
11	389	24	270	303	175	79	8	16	376	15	511	419
12	385	24	286	303	167	79	8	193	375	207	547	417
13	444	25	407	281	157	79	9	181	369	279	547	457
14	426	26	360	298	150	79	9	202	287	332	543	434
15	431	26	350	332	139	79	9	162	95	15	452	402
16	431	27	345	345	132	77	9	279	296	149	447	375
17	460	27	349	360	125	80	9	279	267	292	193	347
18	586	27	520	330	119	79	10	269	196	291	231	331
19	551	28	494	303	113	78	10	269	351	308	240	315
20	519	104	423	290	110	75	31	350	349	443	245	310
21	532	353	365	273	107	57	10	279	321	373	245	120
22	500	348	319	260	104	52	10	270	335	411	245	10
23	469	243	441	252	98	52	10	216	259	416	240	10
24	460	295	503	240	92	53	10	259	267	364	240	10
25	535	259	486	232	89	53	11	264	249	308	236	10
26	507	246	392	224	87	54	35	264	179	369	233	10
27	476	313	520	216	84	54	11	279	374	432	280	11
28	383	392	309	205	84	54	11	270	358	411	333	11
29	307	414	302	201	......	55	11	292	305	358	382	11
30	120	334	313	197	......	54	38	287	196	303	423	11
31	87	......	440	194	......	55	......	274	......	209	443	......

Monthly discharge of Connecticut River at First Lake, near Pittsburg, N. H., for the year ending Sept. 30, 1918.

[Drainage area, 81.4 square miles.]

Months.	Observed discharge (second-feet).			Gain or lost in storage in First Lake (millions of cubic feet).	Discharge corrected for storage (second-feet).		Run off (depth in inches on drainage area).
	Maximum.	Minimum.	Mean.		Mean.	Per square mile.	
October	586	87	359	− 555.8	151	1.85	2.13
November	414	17	133	+ 421.5	296	3.64	4.06
December	520	191	366	− 772.3	78	.958	1.10
January	407	194	292	− 615.0	62	.762	.88
February	194	84	140	− 215.9	51	.627	.65
March	84	52	70.9	− 29.1	60	.737	.85
April	38	7	13.0	+ 838.6	337	4.14	4.62
May	292	13	176	+ 934.2	525	6.45	7.44
June	376	17	268	− 266.6	165	2.03	2.26
July	443	15	256	+ 156.6	314	3.86	4.45
August	547	193	368	− 754.7	86	1.06	1.18
September	457	10	272	− 201.7	194	2.38	2.66
The year	586	7	228	− 1,060.2	193	2.38	32.26

NOTE.—Not corrected for effect of storage in Second Lake.

CONNECTICUT RIVER AT ORFORD, N. H.

LOCATION.—At covered highway bridge between Orford, N. H., and Fairlee, Vt., 10 miles downstream (by river) from mouth of Waits River.

DRAINAGE AREA.—3,100 square miles.

RECORDS AVAILABLE.—August 6, 1900, to September 30, 1918.

GAGES.—Inclined staff on left bank 25 feet below bridge; chain attached to upstream side of bridge is also used at certain stages.

DISCHARGE MEASUREMENTS.—Open-water measurements made from cable.

CHANNEL AND CONTROL.—Channel wide and deep, with gravelly bottom; control for high stages is at the dam at Wilder, 20 miles below station.

EXTREMES OF DISCHARGE.—Maximum stage recorded during year, 21.5 feet at 7 a. m. April 3 (discharge, 29,300 second-feet); minimum stage recorded, 3.08 feet at 6 p. m. August 30 (discharge, 920 second-feet).

1900–1918: Maximum stage recorded, 33.4 feet at 12 noon March 28, 1913 (discharge, by extension of rating curve, about 57,300 second-feet); minimum 24-hour discharge, 288 second-feet, September 28, 1908.

ICE.—Stage-discharge relation seriously affected by ice, usually from December to March; ice cover usually remains in place throughout the winter.

REGULATION.—About 4,100 million cubic feet of storage has been developed at First and Second Connecticut lakes and tributary streams above Pittsburg. There are several power plants above the station, but the operation of these mills does not seriously affect the distribution of flow.

ACCURACY.—Stage-discharge relation affected at times by use of flashboards at Wilder dam and, during the winter, by ice. Several rating curves were used during the year, depending upon the condition of flashboards. Gage read to tenths twice daily. Daily discharge ascertained by applying mean daily gage height to rating table. Records good.

Discharge measurements of Connecticut River at Orford, N. H., during the year ending Sept. 30, 1918.

Date.		Made by—	Gage height.	Discharge.	Date.		Made by—	Gage height.	Discharge.
			Feet.	*Sec.-ft.*				*Feet.*	*Sec.-ft.*
Oct.	9	M. R. Stackpole	8.46	5,380	Apr.	6	H. W. Fear	16.85	19,700
	9	.....do	8.38	5,460		15	M. R. Stackpole	11.71	10,400
Nov.	1	.....do	19.72	25,400		15	.....do	11.82	10,900
	10	.....do	7.89	5,030	May	23	.....do	7.72	5,230
Dec.	3	H. W. Fear	[a] 7.00	2,650	June	14	H. W. Fear	7.04	4,420
Jan.	3	M. R. Stackpole	[a] 5.48	1,460		14	.....do	7.21	4,780
	23	.....do	[a] 5.90	1,540	July	21[b]	C. H. Pierce	6.22	2,340
Feb.	14	.....do	[a] 5.70	1,290		22[c]	H. W. Fear	5.86	2,310
Mar.	8	.....do	[a] 7.90	2,820	Aug.	22[c]	J. W. Moulton	3.86	1,390
	21	.....do	[a] 8.52	3,360	Sept.	2[c]	.....do	4.46	1,570

[a] Stage-discharge relation affected by ice.
[b] 5 feet of flashboards on dam at Wilder; mill not running (Sunday).
[c] 5 feet of flashboards on dam at Wilder; mill in operation.

Daily discharge, in second-feet, of Connecticut River at Orford, N. H., for the year ending Sept. 30, 1918.

Day.	Oct.	Nov.	Dec.	Jan.	Feb.	Mar.	Apr.	May.	June.	July.	Aug.	Sept.
1	2,350	24,300	2,070	1,720	1,250	4,800	20,300	16,700	4,840	2,850	1,770	1,260
2	2,840	23,100	2,510	1,590	1,200	4,580	24,700	20,500	5,460	2,950	2,430	1,670
3	3,020	18,000	2,670	1,590	1,200	4,360	28,900	20,500	5,080	2,850	2,220	2,310
4	3,020	11,900	2,670	1,410	1,200	3,700	26,000	16,400	2,850	2,470	1,920	1,840
5	3,500	8,800	2,510	1,410	1,250	3,300	22,700	12,600	2,380	2,110	1,820	1,780
6	4,920	7,340	2,510	1,530	1,350	3,200	20,300	10,400	2,030	1,950	1,670	1,620
7	6,480	6,370	2,350	1,470	1,250	3,020	16,500	9,770	2,110	2,030	1,620	1,520
8	6,000	5,850	2,070	1,590	1,250	2,750	15,000	9,920	3,050	2,110	1,970	1,520
9	5,280	5,460	2,070	1,470	1,250	2,590	15,700	10,100	5,460	2,500	3,630	1,720
10	5,040	4,840	2,070	1,530	1,150	2,590	17,000	8,600	5,330	2,710	8,520	2,310
11	4,800	4,720	1,980	1,590	1,150	2,430	15,000	9,770	3,710	2,860	7,120	2,700
12	4,140	4,240	1,790	1,590	1,150	2,350	12,800	11,300	2,650	3,100	4,940	2,380
13	4,140	4,020	1,590	1,590	1,150	2,210	11,200	10,700	3,710	3,100	3,910	1,670
14	4,580	3,800	1,530	1,590	1,200	2,210	10,600	13,300	4,500	3,100	3,140	1,350
15	4,360	3,400	1,590	1,470	1,250	2,210	10,600	17,800	5,330	3,540	3,050	1,520
16	4,360	3,200	1,720	1,350	1,350	2,000	12,600	17,400	5,080	4,430	3,820	1,960
17	4,800	3,100	1,790	1,530	1,590	2,070	15,000	14,200	4,380	4,330	2,620	2,540
18	5,280	3,000	1,860	1,530	1,720	2,210	16,900	10,100	3,710	3,630	2,240	2,540
19	4,580	3,000	1,860	1,530	2,000	2,280	16,700	8,020	3,270	3,100	1,960	2,700
20	4,250	3,000	2,000	1,650	2,000	2,590	13,600	6,900	3,050	2,710	1,520	3,620
21	4,580	2,910	2,140	1,650	2,280	3,400	10,800	6,360	2,650	2,430	1,370	6,220
22	4,800	3,000	2,140	1,590	2,590	5,280	11,600	5,840	2,650	2,360	1,210	10,100
23	4,580	3,000	2,280	1,470	2,750	7,320	14,100	5,080	3,600	2,030	1,160	10,200
24	4,030	2,910	2,140	1,410	2,930	8,040	15,000	4,840	4,610	1,720	1,160	8,600
25	4,250	2,730	2,210	1,410	2,840	8,530	15,400	4,260	5,700	1,620	1,210	7,600
26	5,280	2,550	2,140	1,410	2,840	9,180	14,000	2,850	5,700	1,580	1,060	9,330
27	5,520	2,460	1,860	1,410	3,920	9,050	11,900	2,650	4,720	1,520	1,210	14,800
29	5,640	2,190	1,860	1,470	4,590	8,160	10,200	4,610	3,930	1,430	1,010	17,800
29	5,760	2,020	1,860	1,410	....	8,160	10,600	4,840	3,050	1,430	1,010	16,300
30	6,960	2,020	1,720	1,410	....	9,440	12,600	4,960	2,650	1,430	910	12,100
31	21,800	....	1,720	1,300	....	12,500	....	4,380	....	1,520	1,060	....

NOTE.—Stage-discharge relation affected by ice from Nov. 24 to Mar 31; daily discharge determined from gage heights corrected for affect of ice by means of six discharge measurements, observer's notes, and weather records.

Monthly discharge of Connecticut River at Orford, N. H., for the year ending Sept. 30, 1918.

[Drainage area, 3,100 square miles.]

Month.	Observed discharge (second-feet).			Gain or loss in storage at First Connecticut Lake (millions of cubic feet).	Discharge corrected for storage (second-feet).		Run-off (depth in inches on drainage area).
	Maximum.	Minimum.	Mean.		Mean.	Per square mile.	
October	21,800	2,350	5,190	− 555.8	4,980	1.61	1.86
November	24,300	2,020	5,910	+ 421.5	6,070	1.96	2.19
December	2,670	1,530	2,040	− 772.3	1,750	.565	.65
January	1,720	1,300	1,510	− 615.0	1,280	.413	.48
February	4,590	1,150	1,840	− 215.9	1,750	.565	.59
March	12,500	2,000	4,730	− 29.1	4,720	1.52	1.75
April	28,900	10,200	15,600	+ 838.6	15,900	5.13	5.72
May	20,500	2,650	9,860	+ 984.2	10,200	3.29	3.79
June	5,700	2,030	3,910	− 266.6	3,810	1.23	1.37
July	4,430	1,430	2,500	+ 156.6	2,440	.787	.91
August	8,520	910	2,380	− 754.7	2,100	.677	.78
September	17,800	1,260	5,120	− 201.7	5,040	1.63	1.82
The year	28,900	910	5,050	−1,060.2	5,020	1.62	21.91

CONNECTICUT RIVER AT SUNDERLAND, MASS.

Location.—At five-span steel highway bridge at Sunderland, Franklin County, on road leading to South Deerfield, 18 miles in a direct line and 24 miles by river above dam at Holyoke. Deerfield River enters from west about 8 miles above station.

Drainage area.—8,000 square miles.

Records available.—March 31, 1904, to September 30, 1918.

Gages.—Chain on downstream side of bridge read by V. Lawer. Sanborn water-stage recorder installed September 3, 1916.

Discharge measurements.—Made from highway bridge.

Channel and control.—Channel deep; bottom of coarse gravel and alluvial deposits. Control at low stages not well defined, but practically permanent. At high stages the control is at the crest of the dam at Holyoke.

Extremes of discharge.—Maximum stage recorded during year, 21.6 feet at 6 p. m. April 3 (discharge, 70,200 second-feet); minimum stage recorded, 0.6 foot at 6 a. m. August 26 (discharge, 700 second-feet).

1904–1918: Maximum stage recorded, 30.7 feet during the night of March 28, 1913, determined by leveling from flood marks (discharge, computed from extension of rating curve,[1] about 108,000 second-feet); minimum stage recorded, 0.6 foot September 28, 1914, and August 26, 1918 (discharge, 700 second-feet).

Ice.—The river usually freezes over early in the winter, but the ice is likely to break up at times of sudden rises in stage and at those times it occasionally forms ice jams at Northampton, 10 miles below the station, causing several feet of backwater at the gage.

Regulation.—Distribution of flow affected by operation of power plants at Turners Falls, and by regulation of Deerfield River. (See Deerfield River at Charlemont, Mass.) The effect of the regulation is shown by low water at the gage on Sundays and Mondays. Storage in Somerset reservoir and First Connecticut Lake has little effect on the monthly discharge as measured at Sunderland.

[1] Taken from revised rating curve and supersedes figures published in previous reports.

Accuracy.—Stage-discharge relation practically permanent except when affected by ice. Rating curve (fig. 1) used in revision of records is well defined between 1,000 and 75,000 second-feet. Chain gage read to half-tenths twice daily; gage heights from water-stage recorder used for stages below 10.0 feet (24,700 second-feet). Daily discharge ascertained by applying gage height to rating table and making correction for effect of ice during winter. Records previously published have been revised by means of a more accurately determined rating curve making use of all discharge measurements. Records good.

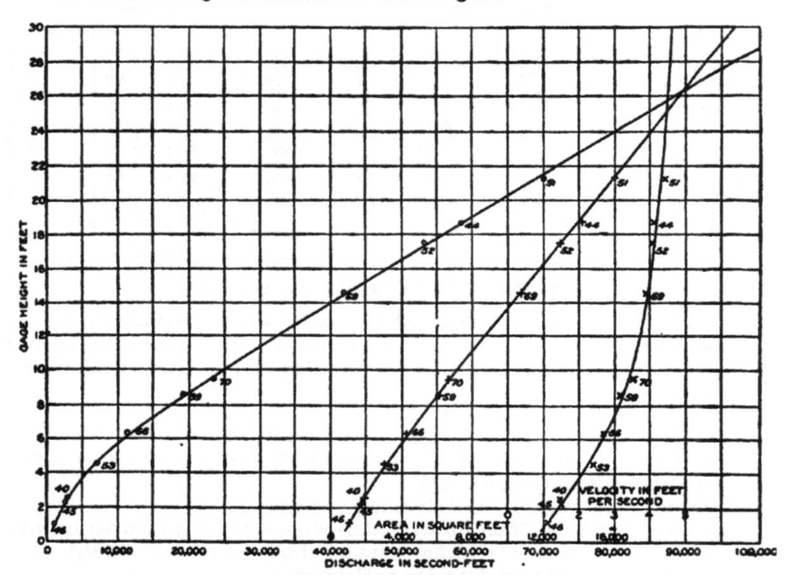

Figure 1.—Rating curves for Connecticut River at Sunderland, Mass. Measurements 40–70 were made during period 1913–1919. Measurements made when stage-discharge relation was affected by ice not shown on diagram.

Discharge measurements of Connecticut River at Sunderland, Mass., during 1913–1918.

Date.	Made by—	Gage height.	Discharge.	Date.	Made by—	Gage height.	Discharge.
1913.		*Feet.*	*Sec.-ft.*	1916.		*Feet.*	*Sec.-ft.*
Aug. 10	C. H. Pierce	2.54	2,940	Jan. 22	Pierce and Barnes	[a] 9.03	[c] 8,500
				Feb. 1	R. S. Barnes	[a] 16.94	46,800
1914.				4	.....do	[a] 15.88	33,500
Jan. 17	R. S. Barnes	[a] 4.20	4,700	Mar. 24	Hardin Thweatt	[a] 8.27	8,490
Mar. 5	Pierce and Barnes	[a] 13.42	26,400	31	.....do	[a] 19.05	50,900
Apr. 30	.....do	18.69	58,400	Dec. 7	A. H. Davison	8.60	19,300
Aug. 20	C. H. Pierce	2.22	2,530				
Nov. 2	R. S. Barnes	1.10	1,180	1917.			
Dec. 22	.....do	[a] 3.60	2,760	Jan. 3	A. H. Davison	[a] 5.92	6,490
				Feb. 1	.....do	[a] 6.36	6,700
1915.				Mar. 3	.....do	[a] 8.44	10,600
Jan. 9	R. S. Barnes	[a] 5.88	5,780				
Feb. 7	.....do	[a] 6.45	7,800	1918.			
24	.....do	[a] 7.15	9,040	Jan. 9	M. R. Stackpole	[a] 5.27	4,450
27	.....do	21.27	[b] 70,000	Feb. 11	.....do	[a] 3.53	1,680
28	.....do	17.50	[b] 53,200	Mar. 17	.....do	[a] 7.40	6,330
Sept. 25	Hardin Thweatt	4.48	7,060	June 12	H. W. Fear	6.31	11,300

[a] Stage-discharge relation affected by ice.
[b] Measurement recomputed since publication in Water-Supply Papers 401 and 415.
[c] Partly estimated.

Note.—Two discharge measurements obtained in April, 1919, were used in determining the rating curve.

Daily discharge, in second-feet, of Connecticut River at Sunderland, Mass., for the years ending Sept. 30, 1904–1918.

Day.	Oct.	Nov.	Dec.	Jan.	Feb.	Mar.	Apr.	May.	June.	July.	Aug.	Sept.
1904.												
1							34,200	62,100	12,000	5,410	4,400	3,850
2							42,000	54,900	9,240	6,950	3,500	3,500
3							43,100	48,600	9,240	6,950	3,330	3,020
4							36,100	44,700	8,960	5,830	3,670	2,720
5							31,200	40,800	8,160	5,200	3,850	2,720
6							31,500	36,100	7,660	4,990	3,670	3,330
7							34,200	33,100	9,240	4,790	3,500	3,330
8							38,800	29,600	11,000	4,790	3,330	5,200
9							47,000	26,600	15,700	4,590	2,450	5,200
10							56,900	25,500	17,400	4,400	3,170	4,790
11							58,100	25,800	13,300	4,030	3,670	4,400
12							54,100	28,100	10,100	3,670	3,500	4,030
13							48,600	28,500	8,420	3,170	2,720	3,670
14							42,300	26,200	7,910	3,330	2,580	3,500
15							35,700	22,100	6,950	3,170	2,450	13,000
16							30,000	22,500	6,270	3,170	1,960	19,900
17							26,200	25,100	5,830	3,170	3,020	14,700
18							24,700	28,100	5,410	3,670	3,020	13,300
19							25,500	32,700	4,400	3,330	3,170	12,000
20							25,100	42,300	3,850	3,500	3,670	9,520
21							23,200	44,300	4,030	3,330	8,420	7,660
22							21,700	40,000	4,210	3,020	7,910	6,950
23							22,100	33,800	3,850	2,720	8,160	8,160
24							23,200	27,400	3,670	2,080	7,910	9,240
25							26,200	22,500	3,670	2,720	8,160	8,420
26							32,700	20,300	3,670	3,020	7,180	7,910
27							36,900	19,500	3,670	2,720	6,720	11,000
28							47,800	17,800	3,500	2,870	5,830	12,000
29							68,500	15,300	3,670	3,020	4,790	12,000
30							69,300	14,000	4,030	3,170	4,210	13,600
31						37,700		13,300		3,670	4,030	
1904–5.												
1	18,400	9,520	6,960	2,300	2,100	2,000	93,300	22,100	8,690	12,000	27,000	7,180
2	17,800	8,690	6,960	2,600	2,000	2,000	79,200	22,500	7,660	9,520	21,400	10,100
3	18,400	8,420	8,420	2,700	2,000	2,000	63,700	21,400	6,950	7,420	17,400	16,400
4	18,100	7,910	9,240	2,700	2,000	2,000	53,300	20,300	6,490	13,600	15,700	33,800
5	16,000	7,660	7,660	2,700	1,900	2,000	45,500	19,500	6,270	22,100	12,600	42,300
6	14,000	7,420	7,180	2,700	2,000	2,100	45,500	20,300	6,950	21,700	8,960	35,000
7	11,300	6,720	7,420	2,700	2,000	2,200	49,400	21,000	7,180	17,000	7,910	27,700
8	9,810	6,950	7,180	2,600	2,100	2,200	43,900	21,000	8,960	12,600	7,660	22,500
9	8,690	6,490	7,180	2,900	2,100	2,300	37,700	20,600	9,810	9,520	6,950	18,100
10	8,160	6,270	6,720	3,000	2,200	2,300	33,800	19,500	8,960	7,660	6,270	14,700
11	8,420	5,830	5,200	3,200	2,000	2,400	38,400	18,800	7,910	7,180	6,720	12,000
12	9,240	5,620	4,800	3,200	2,000	2,400	49,000	17,400	6,720	6,270	8,420	12,600
13	9,810	5,200	4,400	3,300	2,200	2,600	47,800	16,700	6,720	5,830	8,420	17,400
14	9,520	5,200	4,000	3,300	2,300	2,600	42,300	16,000	7,420	5,620	7,910	16,700
15	9,520	5,830	3,700	3,200	2,300	2,600	38,400	14,700	8,420	5,200	7,420	13,300
16	9,240	5,830	3,300	3,000	2,300	2,600	34,600	15,000	8,960	4,790	7,660	11,000
17	8,160	5,830	3,200	3,000	2,300	2,600	30,800	16,700	8,960	4,590	12,600	9,520
18	8,160	5,200	2,700	2,900	2,300	2,600	27,400	16,700	7,660	4,990	16,000	9,810
19	7,420	4,790	2,900	2,700	2,000	2,900	24,000	16,000	6,720	4,990	14,000	39,200
20	7,180	5,200	2,700	2,700	2,200	4,000	21,000	14,700	6,950	5,410	11,300	39,200
21	7,420	5,200	2,600	2,600	2,100	5,200	19,900	13,600	7,910	5,830	8,960	34,200
22	25,500	6,030	2,400	2,300	2,100	5,600	22,500	12,600	15,000	5,830	7,910	33,100
23	22,500	7,420	2,400	2,600	2,000	6,000	27,000	12,600	16,700	5,620	6,490	27,000
21	18.800	7,180	2,400	2,600	2,000	7,900	28,500	11,700	13,300	4,790	5,620	21,700
25	15,700	6,950	2,200	2,600	2,000	12,300	27,400	10,700	9,810	4,790	5,410	17,400
26	15,300	6,490	2,400	2,600	1,900	31,200	24,700	8,960	8,420	4,400	4,790	14,700
27	11,700	6,490	2,600	2,600	2,000	61,300	22,500	8,420	9,520	4,210	4,790	12,600
28	12,800	6,490	2,600	2,600	2,000	73,800	21,000	7,660	9,520	4,210	4,400	11,700
29	12,000	6,720	2,600	2,300		73,400	20,300	8,690	11,300	4,030	4,210	10,700
30	10,700	7,180	2,600	2,300		84,200	21,000	10,700	13,000	4,030	4,790	10,100
31	10,100		2,600	2,200		92,400		9,810		10,700	6,950	

Daily discharge, in second-feet, of Connecticut River at Sunderland, Mass., for the years ending Sept. 30, 1904–1918—Continued.

Day.	Oct.	Nov.	Dec.	Jan.	Feb.	Mar.	Apr.	May.	June.	July.	Aug.	Sept.
1905–6.												
1	8,960	6,050	15,300	18,100	15,300	16,700	26,200	22,500	36,900	10,700	8,690	5,830
2	8,160	6,050	12,000	17,000	14,300	15,000	22,500	22,500	28,100	17,400	8,420	4,790
3	8,420	6,050	13,300	14,000	14,000	14,700	19,200	26,600	22,500	16,700	7,660	4,210
4	7,910	6,490	37,300	13,300	9,500	24,700	19,200	31,500	21,000	15,000	7,910	3,670
5	7,660	7,180	30,800	14,700	9,800	30,000	24,000	30,000	19,900	14,000	7,910	3,500
6	7,180	7,660	24,300	15,700	10,000	24,000	30,000	27,000	18,100	13,000	6,950	4,030
7	6,720	9,520	20,300	14,700	10,000	19,900	31,500	26,200	17,800	11,000	6,270	4,400
8	6,270	12,300	17,000	14,000	10,000	17,400	27,400	25,500	19,500	9,240	6,490	4,210
9	5,620	11,700	15,300	13,300	10,000	16,700	24,700	24,000	24,300	8,160	5,620	4,030
10	6,050	10,400	13,300	13,000	10,000	14,300	23,200	28,900	25,500	7,420	4,790	4,030
11	6,050	9,520	12,600	12,300	8,600	10,100	22,500	32,700	24,700	7,420	4,400	3,670
12	6,270	8,960	10,700	12,600	8,600	8,960	25,500	31,200	22,500	7,420	4,210	3,330
13	8,420	8,160	10,700	14,700	8,600	7,660	30,400	28,100	19,500	7,420	4,030	3,500
14	8,960	8,420	9,520	15,700	8,000	7,420	34,200	28,500	16,400	7,420	3,850	3,170
15	8,690	8,420	11,300	14,700	7,800	7,910	48,600	32,700	14,000	6,490	3,850	3,020
16	7,910	8,160	14,000	14,700	7,400	10,700	76,300	31,500	11,300	5,830	3,670	2,450
17	7,910	7,910	13,300	16,000	7,400	12,000	75,100	29,300	10,400	5,200	3,670	3,020
18	7,180	7,660	13,000	16,000	6,700	8,690	67,700	27,000	12,300	6,270	3,330	2,580
19	6,950	7,420	12,600	15,700	6,700	6,490	62,100	24,000	16,000	7,660	2,580	2,720
20	7,420	7,180	13,000	14,300	6,800	9,520	58,900	22,100	15,700	6,720	2,870	2,720
21	8,960	6,720	13,000	13,600	7,000	10,700	55,300	21,400	13,300	5,620	2,450	2,720
22	9,520	6,050	15,000	14,300	7,400	7,660	51,300	20,300	11,000	5,410	3,020	3,020
23	8,690	5,830	16,000	27,700	11,000	6,050	49,000	18,100	9,240	6,490	4,400	2,580
24	8,960	5,830	15,700	46,200	15,000	5,620	46,600	16,000	10,700	6,720	3,330	3,330
25	8,420	6,050	14,000	54,100	17,000	5,620	41,000	14,000	15,700	6,950	3,330	2,720
26	7,660	5,830	12,600	41,200	18,100	5,200	35,300	14,700	16,000	6,270	2,870	2,870
27	7,420	5,830	12,300	34,600	18,100	5,830	28,900	17,400	14,300	5,620	3,500	3,020
28	6,950	6,490	11,700	30,000	17,400	16,400	26,600	45,500	12,000	4,790	3,330	2,720
29	6,270	6,950	12,000	25,500	....	26,200	22,500	68,100	10,700	4,790	3,500	2,720
30	5,830	15,300	15,700	19,500	....	28,100	21,700	62,900	8,960	5,830	6,050	1,960
31	5,830	....	17,400	16,400	....	27,400	....	49,000	....	7,180	6,490	....
1906–7.												
1	2,720	6,050	7,910	5,000	5,400	4,400	55,300	49,400	12,600	12,600	6,490	2,080
2	2,200	5,830	7,420	5,000	5,400	4,800	49,400	52,500	11,300	13,300	7,180	2,080
3	2,580	5,620	6,050	5,000	4,800	4,400	40,000	54,100	14,700	14,300	8,420	2,200
4	2,870	5,200	6,500	7,000	5,400	4,800	33,800	53,700	19,900	15,300	9,520	2,320
5	2,720	4,790	6,500	20,300	5,600	4,800	30,800	53,300	17,800	16,000	9,240	4,030
6	2,450	4,790	6,300	19,200	5,600	5,000	32,300	48,200	20,600	14,700	8,960	6,490
7	1,730	4,400	5,800	17,400	5,400	4,800	30,800	40,400	20,300	12,300	8,420	6,950
8	3,020	4,400	5,600	18,800	5,400	4,800	27,700	35,300	17,800	10,100	7,910	8,960
9	2,720	4,210	5,000	17,000	5,400	4,800	25,500	32,300	15,300	8,160	7,180	7,910
10	3,170	4,030	5,600	14,700	4,800	4,400	24,000	28,900	13,600	7,420	6,720	6,950
11	3,330	3,850	5,600	14,000	5,200	4,000	22,500	27,700	13,000	7,180	6,050	6,270
12	3,670	4,030	5,600	13,300	5,200	4,800	21,000	27,700	11,700	7,420	5,200	6,050
13	3,500	4,590	5,400	10,700	5,000	5,200	21,000	27,000	10,700	11,000	4,790	6,490
14	4,030	4,790	5,400	11,700	4,800	5,400	21,700	24,700	9,520	10,400	4,030	6,270
15	5,200	4,790	5,200	11,000	4,800	5,600	24,000	21,700	7,910	8,960	3,850	6,950
16	5,410	4,790	4,800	9,500	4,400	5,600	26,200	20,300	7,420	8,420	4,030	6,950
17	4,790	4,400	5,200	8,700	3,700	4,800	25,500	28,500	6,950	8,420	3,670	6,720
18	4,210	4,400	5,400	7,900	4,400	7,400	24,000	36,100	6,490	8,160	3,330	5,830
19	4,030	6,720	5,400	7,000	4,400	10,100	23,200	33,400	6,270	7,910	3,500	4,990
20	4,400	12,300	5,400	6,000	4,400	15,300	21,700	30,000	6,050	6,490	2,580	4,790
21	7,420	13,300	5,200	6,500	4,200	20,300	19,500	26,200	6,270	6,050	2,870	4,400
22	6,720	13,600	5,200	6,500	4,200	21,700	18,400	22,800	8,960	6,490	3,020	4,030
23	6,720	14,000	4,400	6,300	4,000	27,400	18,100	19,900	13,600	5,830	3,170	4,030
24	6,490	13,300	5,000	5,800	3,700	49,800	22,500	17,400	12,600	5,620	3,020	8,420
25	6,270	11,300	5,000	5,600	4,000	40,000	42,000	15,300	10,700	7,910	2,200	9,520
26	6,720	9,520	5,000	5,400	4,000	28,500	48,200	14,000	8,420	8,960	2,720	9,240
27	7,180	8,960	5,000	4,800	4,000	22,100	55,300	13,300	7,660	7,420	2,080	8,420
28	7,910	8,960	5,000	5,800	4,000	26,200	62,500	16,400	7,910	6,720	2,080	7,910
29	7,910	8,960	5,000	5,800	....	43,100	55,300	16,700	7,910	6,050	2,720	8,960
30	7,660	9,240	4,600	5,800	....	57,300	50,500	15,300	9,520	5,620	2,580	27,000
31	6,720	....	4,800	5,600	....	60,900	....	13,600	....	5,830	2,200	....

Daily discharge, in second-feet, of Connecticut River at Sunderland, Mass., for the years ending Sept. 30, 1904–1918—Continued.

Day.	Oct.	Nov.	Dec.	Jan.	Feb.	Mar.	Apr.	May.	June.	July.	Aug.	Sept.
1907–8.												
1	25,100	36,900	10,400	27,000	8,700	9,500	47,800	56,100	24,700	3,670	2,720	2,720
2	22,500	30,000	8,960	23,200	7,300	11,000	41,600	60,900	20,600	3,330	2,720	2,580
3	20,300	47,000	8,420	19,500	9,000	10,000	35,700	53,300	19,900	3,330	2,720	2,320
4	18,800	49,800	8,420	17,000	8,200	11,000	30,000	48,000	19,500	3,330	2,450	2,200
5	22,500	45,500	8,160	16,000	6,900	9,800	24,700	44,300	16,700	3,670	2,720	2,200
6	18,400	38,800	8,690	14,600	7,000	9,500	22,100	38,000	13,300	3,500	3,020	2,080
7	16,700	70,200	8,960	15,000	6,500	8,800	24,700	32,700	11,300	4,030	4,790	2,080
8	16,700	71,000	8,690	15,300	5,600	8,800	27,700	40,800	9,240	3,670	5,410	1,840
9	25,500	54,500	8,960	16,000	5,200	10,000	36,100	46,200	8,960	3,330	7,910	1,840
10	27,400	42,300	9,520	16,400	9,000	11,000	38,400	45,500	7,660	3,330	8,160	2,080
11	25,800	36,100	41,600	12,300	7,000	9,700	35,000	42,300	6,950	3,020	7,420	1,960
12	25,800	30,800	51,300	10,100	6,500	10,000	36,100	39,200	6,490	2,720	6,490	1,960
13	27,700	26,200	47,400	11,000	6,400	11,000	35,300	35,300	6,270	2,720	5,620	1,730
14	25,500	23,600	38,400	13,300	6,400	13,300	32,300	32,300	6,050	2,450	6,050	1,730
15	23,200	21,000	30,400	12,600	7,000	25,000	28,500	36,500	5,410	2,450	5,410	1,620
16	20,600	18,400	24,700	11,300	35,000	34,600	28,500	36,100	5,830	3,170	4,790	1,510
17	18,100	17,000	21,700	10,100	59,700	40,000	28,900	30,800	8,160	2,450	4,400	1,730
18	15,300	15,300	19,500	9,500	51,300	32,300	27,700	26,200	15,300	2,200	4,400	1,730
19	13,600	15,000	16,700	8,400	36,900	25,500	28,500	23,200	16,400	1,960	5,620	1,730
20	12,300	14,300	16,000	8,400	27,400	23,200	31,500	20,600	14,000	2,450	4,790	1,730
21	10,700	14,000	14,700	7,900	19,200	19,500	29,600	19,200	11,000	2,450	4,400	1,730
22	11,300	13,600	14,000	7,400	17,000	18,800	25,500	19,500	8,690	2,720	4,400	1,730
23	10,400	14,000	14,000	7,900	15,000	21,000	22,500	24,700	7,420	5,200	4,400	1,620
24	9,810	13,300	21,400	8,400	12,600	28,500	23,600	21,400	6,490	5,200	4,400	1,510
25	9,240	12,600	35,000	10,700	11,700	36,100	25,100	17,400	5,830	4,790	4,408	1,510
26	8,420	13,300	33,800	10,100	9,500	41,200	28,500	15,300	5,620	8,420	3,850	1,510
27	8,160	13,300	28,100	10,100	10,400	44,700	34,600	14,700	4,990	6,270	3,670	1,080
28	12,000	12,600	24,700	10,400	11,000	53,700	40,800	14,000	4,790	4,400	3,330	1,730
29	45,500	12,000	25,500	10,100	11,300	58,900	48,600	12,300	4,030	4,790	3,170	1,730
30	60,500	11,700	26,200	8,700	....	63,700	53,300	12,600	4,030	3,330	3,020	1,510
31	48,600	....	28,500	8,300	....	56,500	....	20,600	....	2,720	2,720	....
1908–9.												
1	1,620	1,960	3,670	2,300	4,700	20,300	20,600	33,100	16,700	3,500	2,320	2,450
2	1,730	2,080	4,400	2,300	4,600	18,300	21,700	35,300	16,700	3,020	2,450	2,320
3	1,730	1,730	3,670	1,900	4,400	15,400	24,000	34,600	14,300	3,020	2,320	2,450
4	1,560	1,960	3,330	2,100	4,100	16,400	24,700	32,700	12,600	3,020	2,200	2,320
5	1,840	1,960	2,720	2,300	3,900	16,700	28,500	32,300	11,000	3,670	2,450	1,510
6	1,620	1,960	2,320	6,200	3,900	14,200	38,000	30,800	13,600	4,030	2,450	1,960
7	1,560	1,960	2,720	11,700	4,000	13,100	54,100	31,200	21,000	4,400	2,580	1,960
8	2,080	1,240	3,020	12,300	7,200	13,100	75,100	33,400	20,600	4,120	2,450	2,200
9	2,200	1,960	3,850	12,000	8,700	12,100	75,100	33,800	14,000	4,030	2,720	2,200
10	2,200	1,730	3,850	10,900	10,100	14,200	63,300	33,100	12,000	3,670	2,320	2,200
11	1,960	1,840	4,400	10,400	11,500	16,300	50,100	32,300	12,300	3,670	2,450	2,580
12	2,080	1,960	5,830	9,500	11,300	16,300	42,000	33,800	9,520	3,850	2,200	2,450
13	1,840	2,200	5,200	8,200	11,000	17,200	38,400	34,600	9,520	3,330	3,020	3,020
14	1,960	2,200	6,050	7,300	9,600	17,900	57,300	33,800	9,520	3,170	3,330	3,020
15	1,960	1,730	4,990	6,500	9,700	17,000	89,100	31,500	7,420	2,870	2,720	2,720
16	1,840	1,960	4,790	6,000	9,200	15,700	95,400	28,500	6,490	3,020	2,580	2,450
17	1,730	1,960	4,790	4,600	8,600	14,700	90,000	27,000	12,600	3,020	3,670	2,450
18	1,400	2,320	3,200	4,700	10,400	12,600	85,000	29,600	12,000	3,020	5,410	2,200
19	1,840	2,580	3,200	4,500	12,300	11,300	79,600	33,100	12,600	2,720	5,620	1,560
20	1,840	2,450	2,600	4,500	14,200	10,100	77,100	33,800	9,810	2,450	4,400	2,320
21	1,620	2,450	2,800	4,400	20,300	8,690	71,800	31,900	12,300	2,320	3,500	2,320
22	1,840	1,730	2,400	4,400	21,000	7,910	64,500	29,300	10,100	2,450	3,020	2,450
23	1,730	2,450	2,600	4,500	20,300	7,910	62,100	25,500	6,950	2,580	3,170	2,200
24	1,730	1,840	2,500	5,200	22,500	8,160	55,700	22,800	6,720	4,030	2,720	2,080
25	1,180	1,960	2,300	5,600	25,300	10,100	47,800	20,600	6,720	5,200	2,870	1,840
26	1,730	2,450	2,500	5,900	23,200	20,300	42,000	19,200	6,490	3,330	2,720	1,200
27	1,730	2,720	1,900	5,600	23,200	18,800	37,700	18,100	4,210	2,450	2,450	1,620
28	1,960	2,450	2,400	5,200	21,400	18,800	35,000	15,300	3,670	2,320	2,720	3,500
29	2,200	1,840	2,400	4,900	....	19,900	34,600	15,700	5,200	2,450	1,960	5,830
30	2,200	1,960	2,300	4,800	....	19,500	32,700	17,400	5,200	2,720	2,450	6,950
31	2,200	....	2,300	4,000	....	19,900	....	17,400	....	2,720	2,870	....

Daily discharge, in second-feet, of Connecticut River at Sunderland, Mass., for the years ending Sept. 30, 1904–1918—Continued.

Day.	Oct.	Nov.	Dec.	Jan.	Feb.	Mar.	Apr.	May.	June.	July.	Aug.	Sept.
1909–10.												
1	8,960	3,670	5,620	3,600	11,700	68,500	52,500	24,700	18,400	5,620	2,080	2,450
2	8,960	4,030	5,200	3,500	11,000	85,800	51,700	16,000	18,400	5,410	3,020	2,450
3	7,180	3,670	4,790	3,300	10,700	78,400	49,400	20,600	19,200	4,400	3,170	2,450
4	6,720	3,670	4,790	3,200	11,700	68,500	42,700	21,700	18,100	4,210	3,020	1,510
5	5,830	3,500	4,590	3,000	10,400	56,500	41,200	24,000	16,700	4,210	3,500	1,180
6	5,410	3,330	4,500	3,200	10,700	46,200	38,400	26,600	18,100	4,590	10,700	1,960
7	5,620	2,580	4,400	3,800	9,000	42,300	36,100	26,600	21,700	4,030	9,240	3,670
8	5,410	3,500	4,200	5,000	7,000	45,500	36,900	23,200	22,500	4,030	6,950	4,080
9	4,790	4,210	4,100	6,000	6,000	42,700	38,400	20,600	21,400	3,500	6,050	5,830
10	3,020	4,210	4,000	5,000	5,600	35,000	39,200	18,100	19,500	3,170	4,400	5,200
11	3,330	4,030	3,800	4,500	5,300	30,000	31,800	17,400	18,100	3,020	4,400	3,670
12	4,030	4,030	3,800	4,000	5,000	25,500	30,000	17,400	22,500	4,030	4,400	3,830
13	3,670	4,080	4,200	3,700	5,200	24,000	24,700	16,700	24,300	4,030	4,590	4,400
14	3,670	3,330	6,950	3,500	5,400	24,700	21,700	15,300	21,000	3,500	4,030	4,400
15	3,500	3,850	6,270	3,200	5,600	22,100	18,800	14,000	16,000	3,020	3,330	4,080
16	3,170	4,030	5,200	3,000	5,600	21,000	18,400	14,000	15,300	3,020	4,030	4,400
17	2,200	3,670	4,990	2,800	5,800	19,500	15,000	12,600	14,000	2,080	4,030	4,030
18	2,870	3,670	4,790	5,000	6,000	17,000	14,700	12,600	15,000	2,080	4,030	2,580
19	3,330	4,030	4,400	8,000	6,000	16,000	18,100	12,900	13,300	2,320	4,210	2,200
20	3,330	3,850	4,400	12,000	5,600	15,300	19,800	12,000	13,600	2,720	4,080	3,170
21	3,170	2,320	4,400	20,900	5,600	21,000	18,800	12,000	12,300	2,870	3,850	3,500
22	3,330	3,330	4,300	45,100	5,800	25,500	18,400	14,300	10,100	3,020	3,670	3,020
23	3,330	3,670	4,100	57,300	6,000	27,000	18,800	10,100	8,960	2,580	4,400	3,020
24	2,720	3,670	3,900	40,400	6,000	31,500	20,600	12,300	7,910	1,840	4,030	2,870
25	3,670	4,030	3,800	33,100	10,000	36,100	23,600	11,300	6,950	1,400	4,210	1,840
26	4,590	4,210	4,000	27,400	12,000	56,100	27,000	12,000	7,910	2,200	4,210	1,510
27	5,410	5,200	4,200	23,200	15,000	58,100	34,200	18,100	4,210	2,580	3,330	2,200
28	4,400	5,620	4,400	19,500	20,000	51,300	33,800	25,500	5,830	3,170	1,960	2,450
29	4,400	6,050	4,000	17,800	......	49,400	31,500	25,100	5,410	3,020	1,840	2,720
30	4,030	6,490	3,900	15,000	......	52,500	28,100	23,200	6,050	3,020	2,450	2,450
31	3,330	......	3,800	11,300	......	53,700	......	20,300	......	2,200	2,720	......
1910–11.												
1	5,200	4,590	3,670	5,800	6,100	5,000	22,800	47,000	5,830	4,400	4,590	8,160
2	5,410	4,590	4,030	6,400	5,800	5,200	16,000	51,300	6,050	2,720	4,590	7,180
3	4,790	4,790	4,030	10,000	5,200	5,400	12,600	53,700	6,050	1,840	4,400	6,950
4	4,400	5,620	2,450	30,000	4,600	5,900	12,600	50,100	4,990	1,510	4,400	4,030
5	4,400	6,490	2,320	26,000	4,200	4,000	11,000	44,700	5,830	1,960	3,850	3,020
6	4,400	9,520	2,500	19,600	4,600	4,600	12,600	39,200	6,270	2,200	2,200	4,790
7	4,400	8,960	3,800	17,000	4,800	4,800	34,200	31,200	6,490	2,450	2,200	5,200
8	4,400	8,420	3,000	15,000	4,800	5,000	49,400	22,800	7,910	2,320	2,870	4,400
9	2,870	7,910	2,600	13,700	4,600	5,000	49,400	21,400	6,720	1,510	3,170	5,830
10	2,450	6,950	2,300	12,000	4,500	6,000	41,200	20,600	8,960	1,290	2,870	11,700
11	4,030	6,050	1,800	10,700	4,000	6,900	34,200	20,300	6,050	2,200	2,450	7,910
12	4,590	5,620	1,600	9,200	2,600	4,000	32,300	19,200	3,670	3,020	2,200	6,490
13	4,400	4,590	2,500	9,000	3,300	3,300	32,700	18,400	9,520	2,720	1,240	6,050
14	4,030	3,330	2,500	8,600	3,700	5,200	37,700	18,800	9,240	2,720	1,060	5,410
15	3,330	5,200	2,500	8,300	3,900	6,000	48,600	13,300	8,160	2,580	1,960	5,620
16	1,730	4,990	2,300	8,000	4,000	6,300	58,900	18,000	7,420	1,290	1,960	5,200
17	1,620	4,790	2,000	7,700	4,000	6,400	53,300	12,000	6,270	1,240	2,200	3,850
18	2,320	4,790	1,700	7,400	4,000	6,500	47,400	11,000	6,720	2,200	2,080	6,490
19	2,320	4,790	1,700	7,200	2,800	5,200	40,800	12,600	6,950	2,320	1,960	5,830
20	2,720	3,330	2,000	6,800	3,000	4,200	38,800	9,810	6,270	2,450	1,080	5,620
21	2,720	3,500	2,300	6,400	3,200	6,500	37,700	8,960	5,620	2,200	1,080	5,620
22	2,450	4,590	2,300	6,000	4,000	7,000	37,300	8,960	5,830	2,200	1,840	5,410
23	1,730	5,200	2,300	6,000	4,300	7,100	35,000	7,660	5,830	1,400	1,840	4,400
24	1,840	4,400	2,300	6,200	4,400	7,100	30,800	7,420	5,200	1,180	2,200	2,870
25	3,020	2,720	2,200	6,600	4,000	6,300	33,400	6,490	3,330	2,080	2,200	2,720
26	3,020	3,020	2,200	7,000	3,400	6,400	35,300	6,270	2,450	2,450	3,330	3,330
27	3,020	1,620	2,700	7,300	4,000	6,720	38,400	6,950	4,590	2,720	2,450	4,990
28	2,870	2,200	3,600	7,200	4,800	36,100	39,200	4,790	4,590	3,020	2,720	4,990
29	3,670	2,870	4,700	6,600	......	30,800	47,000	3,020	4,400	2,580	3,850	4,590
30	2,870	3,330	5,600	6,600	......	31,200	49,400	4,400	4,790	1,960	5,620	5,410
31	2,720	......	5,600	7,000	......	27,700	......	6,270	......	3,020	7,910	......

Daily discharge, in second-feet, of Connecticut River at Sunderland, Mass., for the years ending Sept. 30, 1904–1918—Continued.

Day.	Oct.	Nov.	Dec.	Jan.	Feb.	Mar.	Apr.	May.	June.	July.	Aug.	Sept.
1911–12.												
1	4,790	11,300	19,200	18,800	5,200	4,800	45,500	32,300	47,000	2,450	2,720	4,210
2	7,420	12,000	19,500	20,300	5,200	4,600	54,100	28,100	50,100	5,200	3,500	3,330
3	9,520	11,300	19,500	19,500	5,200	3,300	47,000	24,300	49,800	5,200	3,850	5,200
4	8,960	10,100	13,300	19,500	4,690	4,000	36,100	21,400	49,800	4,400	2,720	6,050
5	13,300	8,690	12,600	18,800	3,800	4,000	29,300	21,400	44,300	1,960	2,080	6,050
6	13,300	9,240	8,960	18,400	4,600	4,200	38,000	11,700	37,700	2,320	2,870	5,620
7	14,700	9,520	7,910	18,100	4,400	4,400	60,500	19,500	31,900	1,730	3,670	5,620
8	13,600	13,300	8,960	15,000	4,800	4,400	78,000	21,000	28,500	1,730	3,670	4,790
9	12,000	12,600	9,520	12,100	4,800	4,600	82,500	20,600	25,500	4,030	3,670	4,210
10	12,000	12,600	10,100	10,500	4,800	4,200	69,800	21,000	15,300	4,030	3,500	4,790
11	9,240	11,700	10,100	9,000	2,700	4,400	56,500	19,500	17,400	3,670	3,330	5,200
12	9,520	10,700	11,700	5,600	3,100	5,200	48,200	18,100	14,700	3,850	3,330	4,790
13	7,910	12,000	14,000	4,000	4,300	5,800	43,500	19,200	13,300	4,030	4,990	4,790
14	6,950	12,600	16,700	3,800	4,200	9,900	42,300	19,900	13,300	2,720	5,200	4,400
15	7,180	12,600	24,600	4,000	4,000	9,500	41,200	20,300	12,600	2,450	4,990	3,020
16	5,410	13,000	27,700	4,460	3,900	22,500	43,900	21,000	12,000	2,870	5,200	6,050
17	5,830	12,600	26,600	4,600	3,800	25,500	58,900	33,100	7,660	2,450	5,410	11,000
18	6,950	12,600	24,700	4,600	3,200	19,200	64,100	39,600	10,100	2,450	4,030	9,520
19	41,600	19,500	20,300	4,700	3,900	30,400	65,300	32,700	8,960	3,170	5,200	9,240
20	47,800	15,300	17,000	4,800	4,460	26,200	66,900	27,000	8,960	4,400	6,050	8,690
21	34,600	16,000	12,600	6,500	4,900	24,700	61,300	25,100	8,160	3,020	5,830	12,600
22	34,600	17,000	9,240	7,000	5,400	19,200	54,100	30,000	7,420	4,790	4,790	13,600
23	30,000	14,300	16,000	7,000	5,400	16,000	49,000	36,500	4,790	3,020	4,030	17,400
24	28,500	12,000	33,800	6,600	5,500	15,300	49,400	32,700	6,490	3,330	3,670	17,400
25	24,000	12,300	33,800	5,600	4,800	15,000	47,800	27,700	6,950	3,500	3,170	14,000
26	20,300	12,700	32,700	5,000	4,900	14,700	46,200	24,300	6,720	4,400	2,720	9,810
27	18,100	11,700	30,000	4,500	5,200	14,000	43,100	19,500	6,490	4,400	3,170	7,910
28	20,300	10,100	24,700	4,000	5,200	14,000	39,200	19,500	6,490	3,020	3,670	6,950
29	19,500	12,600	19,500	5,000	5,100	19,900	38,100	17,000	6,050	2,580	4,990	7,180
30	6,950	19,900	15,300	5,200	....	40,800	34,200	22,100	2,720	2,450	5,620	5,410
31	10,100	....	14,000	5,200	....	42,300	....	41,600	....	1,960	6,490	....
1912–13.												
1	6,270	12,600	12,300	23,500	16,000	26,100	60,100	22,500	31,200	4,590	6,050	1,730
2	7,420	12,000	6,270	18,800	23,600	16,700	54,100	19,200	25,500	5,200	6,050	1,960
3	8,960	18,100	10,100	19,200	14,700	11,300	45,500	17,800	22,500	5,620	3,850	2,200
4	8,420	13,000	22,500	24,700	13,300	10,700	39,200	15,000	16,700	5,200	4,080	1,450
5	7,910	11,700	20,300	26,600	13,300	10,700	37,700	14,700	13,600	3,020	6,050	1,130
6	6,950	10,100	21,000	22,800	12,000	10,700	40,000	13,300	13,300	2,080	4,990	1,960
7	6,490	10,700	24,300	20,300	12,000	11,700	38,800	12,600	11,300	2,320	3,670	1,840
8	6,050	17,400	22,500	26,200	10,400	12,600	36,100	12,000	12,000	2,450	4,080	1,620
9	5,830	33,100	17,400	25,500	20,600	9,810	31,500	10,400	7,420	3,020	3,850	2,200
10	6,050	34,200	15,300	19,500	13,300	8,420	27,000	10,100	8,420	3,020	3,020	1,730
11	6,050	28,600	12,600	16,700	13,600	19,200	24,700	8,960	8,960	3,020	2,330	1,840
12	5,620	23,200	12,000	20,300	13,000	24,000	34,200	5,620	8,420	3,020	3,020	2,200
13	3,020	18,800	10,700	17,400	12,600	18,800	23,600	7,910	8,420	2,720	3,020	1,960
14	3,670	18,800	10,700	21,700	12,300	22,800	30,800	7,180	7,910	2,720	2,450	1,730
15	5,830	23,200	10,100	18,800	10,400	47,800	30,800	8,420	5,200	3,330	3,670	1,730
16	6,270	21,700	8,960	16,700	6,500	60,900	28,500	7,420	5,830	4,210	2,720	2,580
17	6,050	27,000	9,810	17,400	5,600	51,700	27,000	6,490	8,160	4,400	2,200	2,200
18	6,050	10,700	9,520	26,600	9,200	39,200	26,200	8,960	7,420	5,200	1,730	2,450
19	5,830	14,700	8,960	35,700	9,500	31,500	25,100	7,180	5,620	4,080	2,720	1,960
20	4,080	13,300	12,000	36,100	9,500	29,300	24,000	9,240	4,790	2,580	2,450	1,960
21	4,990	13,300	15,300	32,300	9,000	38,400	24,700	7,910	4,790	2,450	1,960	1,840
22	6,490	11,300	17,400	36,900	10,100	46,600	24,300	7,910	3,330	3,170	2,450	1,620
23	11,700	12,000	7,420	31,500	10,100	52,500	22,800	8,690	3,670	2,720	2,720	4,400
24	44,700	17,800	8,160	27,700	15,000	47,800	21,400	24,900	5,620	3,020	1,960	3,020
25	54,100	7,420	13,000	28,500	14,000	46,200	19,500	30,000	5,200	3,670	1,620	6,270
26	45,500	10,700	16,400	24,000	12,300	63,700	19,500	26,600	4,590	3,500	2,450	4,030
27	37,700	11,700	14,300	20,300	11,700	88,300	25,800	25,500	4,400	2,580	2,200	2,200
28	17,400	10,700	15,000	17,400	14,000	107,000	15,000	19,500	4,030	2,720	2,080	1,620
29	20,300	10,100	14,000	15,300	....	104,000	25,800	28,100	2,720	4,210	2,450	1,290
30	15,300	9,810	12,000	13,300	....	86,200	26,200	36,100	1,960	4,400	2,870	3,500
31	14,700	....	21,400	12,300	....	69,300	....	33,400	....	4,400	2,450	....

Daily discharge, in second-feet, of Connecticut River at Sunderland, Mass., for the years ending Sept. 30, 1904–1918—Continued.

Day.	Oct.	Nov.	Dec.	Jan.	Feb.	Mar.	Apr.	May.	June.	July.	Aug.	Sept.
1913–14.												
1	3,330	9,240	4,990	3,800	5,200	2,300	41,600	61,300	6,720	3,670	3,850	7,910
2	3,170	8,160	7,910	4,000	4,600	3,700	54,900	53,300	7,420	4,210	2,320	8,420
3	3,020	7,910	7,910	4,200	7,400	11,400	62,100	45,500	7,180	3,850	2,580	7,180
4	3,020	8,420	7,910	2,700	6,300	22,800	50,100	40,800	6,950	2,450	3,500	4,990
5	2,320	7,910	6,490	2,300	5,600	24,700	40,800	38,800	6,490	2,450	3,170	3,670
6	1,510	6,490	6,050	3,600	3,300	22,000	34,600	45,500	8,960	3,330	3,020	2,450
7	4,400	6,050	5,200	3,600	4,000	18,000	30,800	44,700	10,400	8,160	3,850	2,320
8	4,030	5,620	10,100	4,000	2,600	13,600	32,700	42,300	6,490	7,660	3,330	2,870
9	4,400	3,850	15,000	3,800	2,200	11,500	50,500	39,600	7,660	6,270	1,960	3,670
10	3,670	30,000	11,300	3,800	3,700	10,000	59,700	39,600	8,690	6,490	2,200	4,030
11	3,330	18,800	8,960	3,500	3,500	8,600	52,500	28,500	7,910	6,950	3,330	3,500
12	2,200	16,700	9,810	2,700	3,000	8,000	56,100	31,500	6,720	5,830	3,670	3,330
13	2,080	14,000	8,690	3,800	2,900	8,600	55,300	35,300	5,620	4,590	3,500	3,170
14	3,330	8,960	6,270	5,200	2,700	9,200	48,600	36,500	3,020	6,050	3,170	3,170
15	4,030	10,100	6,270	4,800	1,400	7,000	43,500	31,900	3,170	6,050	3,330	3,330
16	4,400	11,000	8,420	4,200	2,100	5,200	41,600	26,600	4,790	4,990	1,730	3,670
17	3,670	5,830	8,690	2,900	2,600	9,000	39,200	24,300	4,400	5,200	2,080	3,670
18	3,670	6,490	8,160	2,700	2,800	9,400	40,000	16,000	3,850	3,850	2,720	3,670
19	2,200	7,420	7,910	1,500	3,000	11,300	47,800	18,100	4,030	2,320	3,020	4,400
20	2,080	6,050	11,700	3,000	2,800	11,900	70,600	16,700	3,670	2,720	2,720	2,870
21	4,210	6,950	7,910	3,000	2,200	10,900	88,300	15,000	2,580	4,400	3,170	1,510
22	4,400	7,420	6,050	3,300	2,100	7,900	87,500	14,700	2,870	3,850	4,030	2,870
23	8,160	9,520	6,050	3,600	1,500	8,600	80,000	14,700	4,210	3,850	2,720	2,450
24	7,660	9,520	6,270	3,500	2,200	9,400	70,600	16,000	3,500	3,500	2,580	2,450
25	8,960	9,810	4,030	3,800	2,100	9,400	60,500	8,160	3,670	3,170	4,080	2,580
26	12,300	8,160	4,030	4,000	2,100	10,000	50,900	10,400	3,670	1,840	3,670	2,450
27	16,700	6,950	6,490	4,600	2,100	11,000	49,400	7,910	3,330	2,320	3,500	1,730
28	17,000	6,950	6,950	3,450	2,800	39,200	52,900	6,950	2,320	3,850	3,330	1,280
29	15,300	8,420	6,800	3,200	....	50,500	52,500	8,960	2,450	3,670	2,720	2,450
30	13,300	4,990	6,000	4,600	....	47,400	58,100	8,960	3,850	3,500	3,170	2,320
31	11,300	....	5,200	6,000	....	40,800	....	6,050	....	3,500	6,050	....
1914–15.												
1	2,200	1,730	5,620	3,330	6,050	41,600	7,660	22,100	6,490	3,670	13,300	8,960
2	2,450	1,450	5,620	3,170	6,050	32,700	7,660	24,700	5,620	8,690	19,200	9,240
3	2,580	2,080	4,790	2,870	5,830	28,500	6,950	24,000	4,590	16,000	14,300	9,520
4	3,020	2,720	6,050	2,580	5,830	23,600	7,910	23,200	4,990	18,400	22,500	6,950
5	2,200	2,870	7,180	2,320	5,830	18,100	8,160	18,400	5,200	17,800	54,900	3,020
6	3,500	2,720	6,270	1,960	6,950	13,300	9,810	16,700	3,330	12,000	36,100	2,450
7	3,020	2,720	6,050	3,020	7,660	14,700	9,810	13,600	3,330	14,300	27,400	4,210
8	2,720	1,290	7,660	7,910	7,910	15,000	10,100	12,600	4,210	11,700	25,100	5,830
9	2,450	1,510	6,050	7,420	7,420	12,600	15,000	12,000	3,330	65,300	18,100	5,410
10	2,870	2,720	4,990	6,950	6,720	12,600	18,800	13,000	3,670	54,100	19,200	5,620
11	1,620	3,330	4,400	6,050	5,830	12,300	33,800	12,600	3,500	44,700	18,800	4,790
12	1,400	3,330	4,400	5,830	5,200	12,000	53,300	12,000	3,330	32,700	17,400	3,670
13	2,320	3,670	3,020	6,050	4,590	11,300	57,300	11,000	1,740	34,600	18,800	4,030
14	2,200	3,670	3,670	4,030	4,400	8,690	51,700	9,240	2,200	16,400	17,800	5,200
15	2,200	1,730	5,410	3,850	4,210	8,160	42,700	8,420	3,170	13,300	16,000	5,200
16	2,080	2,320	7,420	3,850	12,000	8,960	36,900	7,420	3,670	15,300	12,000	4,400
17	2,200	6,270	6,270	3,170	18,100	8,160	31,900	8,420	4,400	14,700	13,300	4,400
18	1,730	6,720	5,200	2,580	18,800	8,690	26,200	8,420	5,200	16,000	12,300	4,400
19	1,400	6,950	4,400	7,180	15,700	8,960	24,000	6,490	6,270	9,240	10,400	3,170
20	2,450	4,990	3,330	13,000	12,600	7,660	22,500	8,420	4,400	13,300	7,910	2,870
21	2,720	4,790	2,080	12,300	10,700	5,620	21,000	10,100	5,410	20,300	7,180	4,400
22	3,020	4,030	2,870	11,700	8,960	6,720	19,500	8,960	6,950	16,400	7,910	7,180
23	3,170	4,030	3,020	8,160	9,240	6,950	17,000	3,330	8,420	14,000	24,000	7,180
24	3,020	4,990	2,580	7,910	9,520	6,950	19,500	4,400	5,830	14,700	14,700	6,490
25	1,960	4,590	1,960	7,420	43,100	8,420	11,300	7,180	4,590	14,000	16,700	7,180
26	2,200	3,170	1,330	7,180	63,700	11,300	8,960	7,910	5,200	9,520	18,100	7,180
27	3,500	2,720	1,740	6,950	70,200	13,600	14,700	6,720	2,720	6,270	16,700	6,490
28	3,330	3,330	2,080	6,720	55,700	10,700	20,300	6,050	2,450	18,800	15,300	6,270
29	3,330	2,320	2,450	6,490	....	10,700	21,000	6,720	4,030	10,400	8,160	6,490
30	2,720	3,020	2,870	6,490	....	10,700	22,500	4,400	3,670	13,300	10,100	8,420
31	2,200	....	3,330	6,270	....	9,520	....	3,170	....	13,300	11,300	....

Daily discharge, in second-feet, of Connecticut River at Sunderland, Mass., for the years ending Sept. 30, 1904–1918—Continued.

Day.	Oct.	Nov.	Dec.	Jan.	Feb.	Mar.	Apr.	May.	June.	July.	Aug.	Sept.
1915–16.												
1	6,720	4,790	9,520	15,300	47,200	22,500	58,100	36,500	13,300	12,600	9,520	4,590
2	5,620	6,050	9,240	14,000	41,200	20,300	63,700	33,800	14,000	10,000	8,160	4,790
3	4,210	6,720	10,700	10,700	37,700	18,400	72,600	32,300	14,700	10,100	5,200	3,170
4	3,670	6,720	10,700	11,700	33,400	16,700	62,100	30,000	17,000	16,700	4,990	2,450
5	6,490	7,420	6,490	12,600	31,200	16,400	52,900	29,300	15,700	16,000	5,200	3,850
6	6,270	7,420	4,400	11,300	24,300	7,910	49,800	27,000	17,400	20,600	2,720	5,200
7	6,950	6,950	7,660	13,000	25,500	12,300	44,700	27,000	21,700	21,400	3,850	4,210
8	7,420	3,500	6,490	11,700	25,100	12,600	40,400	18,800	21,700	16,400	4,790	4,080
9	6,270	5,620	5,830	8,960	17,000	12,300	33,800	23,200	22,800	11,300	4,990	2,870
10	6,050	4,790	5,410	8,420	18,400	11,000	32,300	21,400	21,400	11,000	9,240	2,080
11	7,660	4,590	8,960	10,100	17,400	11,000	30,800	19,900	22,800	13,300	9,810	3,020
12	6,950	5,830	7,660	9,520	17,000	8,690	33,100	17,100	24,700	11,000	16,000	4,590
13	5,410	5,620	5,200	9,810	11,700	6,950	38,000	19,500	26,200	10,100	17,400	4,590
14	5,410	4,210	5,830	7,910	11,300	11,000	37,300	10,100	25,800	12,600	10,700	4,080
15	6,050	4,400	4,990	9,520	13,600	10,300	36,900	8,420	24,000	11,300	8,160	4,790
16	6,490	7,420	4,590	7,180	11,700	9,520	36,900	14,700	18,100	6,050	7,910	14,000
17	5,620	7,910	4,400	6,490	14,300	9,810	40,000	20,600	21,000	7,420	5,830	11,700
18	4,590	10,100	4,400	10,700	17,000	11,000	40,800	42,700	29,300	10,100	5,200	8,420
19	6,050	9,240	18,100	11,000	13,300	5,830	42,000	44,700	26,600	7,910	6,050	6,270
20	5,620	11,300	11,000	7,180	8,760	5,200	43,900	39,600	28,100	5,830	3,170	7,180
21	5,620	11,700	12,000	7,180	4,210	8,420	43,500	34,200	29,900	9,810	3,330	6,270
22	5,830	8,690	11,700	8,960	10,400	8,420	41,600	30,800	24,700	7,910	4,590	6,720
23	7,180	11,300	11,700	16,700	12,600	8,960	44,700	25,100	24,000	6,950	4,400	6,270
24	4,030	9,240	9,810	18,100	14,700	8,960	56,100	22,800	17,400	6,490	5,200	3,330
25	3,020	8,420	8,420	17,400	17,000	10,400	58,900	21,700	16,000	12,000	5,620	6,490
26	4,400	8,160	23,600	17,000	34,200	5,620	52,900	21,000	15,300	12,000	3,850	5,950
27	4,790	8,960	34,600	23,600	40,000	6,950	49,000	18,400	16,000	19,500	2,320	6,050
28	5,620	7,420	31,500	39,000	33,100	16,000	44,700	12,600	16,000	17,400	3,330	5,620
29	5,410	4,400	25,500	61,700	26,600	24,000	42,300	9,810	14,300	15,700	4,790	5,620
30	6,490	7,180	19,900	56,900	....	33,100	38,000	12,600	12,600	10,700	5,830	8,420
31	4,400	....	17,400	49,800	....	53,700	....	13,000	....	9,520	5,200	....
1916–17.												
1	14,300	6,950	26,200	4,590	6,270	15,700	40,400	23,200	19,900	10,100	3,330	11,300
2	8,960	5,620	30,800	6,720	5,830	13,600	46,200	30,800	16,700	17,400	3,330	10,700
3	10,100	5,200	29,300	6,490	6,270	9,810	46,200	34,600	23,600	17,000	4,210	10,100
4	10,100	5,830	27,000	6,270	3,020	8,960	47,800	33,800	11,300	16,400	7,910	9,810
5	8,690	5,200	22,500	6,720	3,330	5,200	50,100	31,900	18,100	11,000	2,720	9,240
6	7,180	6,720	18,800	7,660	5,200	7,180	46,600	33,100	20,300	10,400	2,080	9,520
7	6,050	6,950	18,800	7,660	5,620	6,270	44,700	32,700	19,200	10,400	4,030	8,690
8	3,020	6,050	17,400	11,700	5,200	6,270	45,500	30,000	22,800	6,050	4,590	6,050
9	5,410	5,620	17,400	10,400	5,200	6,720	45,100	28,900	22,100	7,910	4,400	3,330
10	6,050	5,620	15,300	10,400	4,590	7,910	40,400	27,000	21,700	7,420	4,400	3,330
11	5,620	6,490	14,000	8,690	2,450	6,490	33,100	25,500	23,200	6,270	6,050	4,400
12	4,030	5,620	14,000	7,910	3,020	5,830	30,400	24,000	36,900	6,490	3,330	5,620
13	4,990	4,590	13,300	7,180	5,200	8,160	20,300	30,800	48,600	6,270	5,200	4,790
14	5,200	6,490	12,000	4,210	4,210	8,420	24,000	19,900	47,800	7,660	6,950	4,790
15	3,020	6,950	11,700	11,700	4,210	9,810	27,700	24,000	43,100	4,790	5,200	4,210
16	3,500	7,420	11,300	11,700	4,400	9,520	18,100	24,000	35,700	6,050	6,490	2,720
17	6,050	7,420	8,960	12,300	4,400	8,690	21,000	22,500	30,400	7,910	6,490	3,500
18	5,200	6,720	7,180	12,300	2,200	5,830	21,700	20,300	28,500	8,160	7,420	4,990
19	6,050	3,330	9,520	11,000	1,960	8,690	24,700	17,800	38,000	7,660	10,700	4,790
20	6,720	3,330	9,520	10,700	4,590	8,160	33,100	15,700	39,600	6,950	10,700	4,400
21	10,400	5,620	8,420	7,420	4,400	7,660	48,200	17,800	35,300	6,490	12,600	4,210
22	8,960	5,620	9,810	6,270	4,030	9,520	56,500	18,800	31,500	7,910	13,300	3,020
23	10,100	5,200	7,420	7,910	4,030	10,100	59,700	20,300	27,700	5,200	12,000	1,620
24	10,700	9,810	10,100	6,720	4,210	11,700	58,100	22,900	22,500	6,490	15,000	2,200
25	8,960	22,500	11,300	6,490	2,200	34,500	53,300	22,100	24,000	6,950	14,000	3,330
26	7,910	19,500	9,520	6,270	3,020	32,300	48,200	20,300	22,500	6,720	11,300	3,170
27	6,950	14,700	9,240	6,050	6,950	38,400	43,500	19,900	18,800	7,180	16,000	3,330
28	6,050	14,000	8,420	4,590	14,000	56,500	37,700	17,000	16,700	5,830	12,000	3,330
29	4,210	13,300	10,100	5,200	....	61,300	36,100	19,900	14,700	3,020	10,400	3,020
30	5,830	17,000	7,910	6,720	....	59,700	29,300	25,500	11,000	3,500	10,700	1,960
31	6,960	....	5,410	5,830	....	45,500	....	23,600	....	4,210	11,700	....

Daily discharge, in second-feet, of Connecticut River at Sunderland, Mass., for the years ending Sept. 30, 1904–1918—Continued.

Day.	Oct.	Nov.	Dec.	Jan.	Feb.	Mar.	Apr.	May.	June.	July.	Aug.	Sept.
1917–18.												
1	2,580	41,600	6,490	4,090	5,200	16,700	49,400	29,300	8,420	4,210	4,160	1,600
2	4,080	40,000	6,050	4,090	5,200	17,000	57,300	36,900	7,910	5,860	4,090	945
3	4,030	34,200	4,790	4,030	3,330	14,000	68,500	36,900	7,910	6,170	3,440	1,700
4	4,030	29,600	6,050	4,210	2,200	13,000	68,500	35,000	10,700	4,850	2,330	2,880
5	4,400	22,800	6,050	4,400	2,200	13,000	59,700	30,800	10,100	5,360	2,870	3,090
6	6,960	17,800	6,050	3,020	4,030	13,600	49,400	27,400	8,960	6,270	3,500	3,240
7	4,790	15,700	6,950	3,020	4,210	14,700	46,200	22,100	6,950	3,530	3,340	3,280
8	9,240	13,600	6,490	4,790	4,590	13,300	39,600	19,500	5,620	4,380	3,270	1,800
9	10,100	9,520	6,270	4,400	4,400	14,300	40,000	18,800	4,030	5,670	3,290	2,680
10	8,960	8,690	6,050	4,590	2,720	11,700	44,300	19,200	5,620	4,630	4,820	3,420
11	8,420	6,490	6,950	4,210	1,620	8,690	43,100	18,800	10,400	4,450	9,200	3,510
12	7,420	6,950	6,490	3,670	2,450	9,810	39,200	25,500	10,600	4,320	11,400	2,980
13	6,720	8,690	5,620	3,020	3,670	9,810	33,100	17,000	9,360	3,590	10,100	3,300
14	5,200	8,420	5,200	2,450	4,030	9,810	33,400	27,000	12,800	3,310	6,580	2,690
15	6,050	7,420	4,400	2,720	4,400	9,810	28,900	32,700	10,600	5,700	5,740	1,570
16	8,160	7,660	2,450	4,400	4,030	11,700	35,300	31,500	8,630	7,900	6,140	2,240
17	7,910	7,180	2,580	4,590	3,670	6,270	36,900	29,300	5,740	7,550	5,690	3,440
18	8,420	3,670	4,080	4,210	4,210	8,690	40,800	23,600	8,690	9,320	3,340	3,550
19	8,420	4,990	4,030	3,330	5,200	13,000	43,900	20,300	8,170	10,500	3,910	5,530
20	8,420	6,720	4,400	2,720	10,700	17,800	40,000	17,800	7,620	7,590	4,430	5,400
21	4,790	6,960	4,590	3,330	17,400	29,600	33,800	16,000	7,030	3,270	4,240	6,850
22	4,990	6,720	6,490	3,170	16,400	48,200	41,600	14,700	6,400	4,060	4,270	8,710
23	7,420	6,950	4,210	4,210	15,700	58,100	42,700	12,300	8,360	5,170	3,770	13,900
24	7,180	7,910	2,870	4,990	9,810	58,900	40,800	11,700	13,100	4,300	2,750	14,000
25	12,000	4,990	3,500	5,620	12,300	52,900	38,400	12,000	12,300	4,510	1,300	11,900
26	12,600	5,620	3,500	5,620	14,000	42,000	34,600	7,180	11,200	4,590	1,980	17,000
27	11,700	7,660	4,400	3,020	20,300	35,300	34,600	6,950	10,700	3,530	2,760	35,300
28	10,700	6,720	4,590	2,080	18,400	30,800	27,700	8,960	12,500	1,770	2,750	35,700
29	11,000	5,620	4,030	4,080	....	30,000	20,300	10,100	8,470	2,820	2,810	31,200
30	15,700	5,200	2,450	5,830	....	31,200	24,700	7,910	3,930	4,650	2,760	25,800
31	40,000	....	2,200	5,830	....	38,400	....	7,420	....	4,460	2,270	....

NOTE.—Stage-discharge relation affected by ice as follows: Dec. 11, 1904, to Mar. 26, 1905; Feb. 3 to Mar. 2, 1906; Dec. 4, 1906, to Mar. 20, 1907; Jan. 8 to Mar. 25, 1908; Dec. 18, 1908, to Mar. 16, 1909; Dec. 6–13 and Dec. 20, 1909, to Jan. 21, 1910; Feb. 7–28, and Dec. 6, 1910, to Mar. 26, 1911; Jan. 9 to Mar. 27, 1912; Feb. 5–26, 1913; Dec. 29, 1913, to Mar. 29, 1914; Dec. 22, 1914, to Feb. 26, 1915; Dec. 14, 1915, to Apr. 2, 1916; Dec. 16, 1916, to Mar. 25, 1917; Dec. 1, 1917, to Mar. 21, 1918; daily discharge for these periods determined from gage heights corrected for effect of ice by means of discharge measurements, observer's notes, weather records, and hydrographic comparison with other Connecticut River records.

Monthly discharge of Connecticut River at Sunderland, Mass., for the years ending Sept. 30, 1904–1918.

[Drainage area, 8,000 square miles.]

Month.	Discharge in second-feet.				Run-off (depth in inches on drainage area).
	Maximum.	Minimum.	Mean.	Per square mile.	
1904.					
April	69,300	21,700	37,900	4.74	5.29
May	62,100	13,300	30,700	3.84	4.43
June	17,400	3,500	7,300	.912	1.02
July	6,950	2,080	3,890	.486	.56
August	8,420	1,960	4,450	.556	.64
September	19,900	2,720	7,750	.969	1.08
1904–5.					
October	25,500	7,180	12,500	1.56	1.80
November	9,520	4,790	6,560	.820	.91
December	9,240	2,200	4,490	.561	.65
January	3,300	2,200	2,730	.341	.39
February	2,300	1,900	2,090	.261	.27
March	92,400	2,000	16,200	2.02	2.33
April	93,300	19,900	37,800	4.72	5.27
May	22,500	7,660	15,700	1.96	2.26
June	16,700	6,270	8,960	1.12	1.25
July	22,100	4,030	7,950	.994	1.15
August	27,000	4,210	9,570	1.20	1.38
September	42,300	7,180	20,600	2.58	2.88
The year	93,300	1,900	12,100	1.51	20.54

Monthly discharge of Connecticut River at Sunderland, Mass., for the years ending Sept. 30, 1904–1918—Continued.

Month.	Discharge in second-feet.				Run-off (depth in inches on drainage area).
	Maximum.	Minimum.	Mean.	Per square mile.	
1905–6.					
October	9,520	5,620	7,520	0.940	1.08
November	15,300	5,830	7,870	.984	1.10
December	37,300	9,520	15,300	1.91	2.20
January	54,100	12,300	19,900	2.49	2.87
February	18,100	6,700	10,700	1.32	1.38
March	30,000	5,200	13,800	1.72	1.98
April	76,300	19,200	37,600	4.70	5.24
May	68,100	14,000	29,000	3.26	4.17
June	36,900	8,960	17,300	2.16	2.41
July	17,400	4,790	8,260	1.03	1.19
August	8,690	2,450	4,820	.602	.69
September	5,830	1,960	3,350	.419	.47
The year	76,300	1,960	14,600	1.82	24.78
1906–7.					
October	7,910	1,730	4,720	.590	.68
November	14,000	3,850	7,170	.896	1.00
December	7,910	4,400	5,490	.686	.79
January	20,300	4,800	9,450	1.18	1.36
February	5,600	3,700	4,700	.588	.61
March	60,900	4,400	16,600	2.08	2.40
April	62,500	18,100	32,400	4.05	4.52
May	54,100	13,300	29,600	3.70	4.27
June	20,600	6,050	11,400	1.42	1.58
July	16,000	5,620	9,070	1.13	1.30
August	9,520	2,080	4,830	.604	.70
September	27,000	2,080	6,770	.846	.94
The year	62,500	1,730	11,900	1.49	20.15
1907–8.					
October	60,500	8,160	21,200	2.65	3.06
November	71,000	11,700	27,800	3.48	3.88
December	51,300	8,160	21,400	2.68	3.09
January	27,000	7,900	12,500	1.56	1.80
February	59,700	5,200	15,000	1.88	2.03
March	63,700	8,800	24,700	3.09	3.56
April	53,300	22,100	32,400	4.05	4.52
May	60,900	12,300	31,700	3.96	4.56
June	24,700	4,030	10,200	1.28	1.43
July	8,420	1,960	3,580	.448	.52
August	8,160	2,450	4,480	.560	.65
September	2,720	1,080	1,830	.229	.26
The year	63,700	1,080	17,300	2.16	29.36
1908–9.					
October	2,200	1,560	1,830	.229	.26
November	2,720	1,240	2,050	.256	.29
December	6,050	1,900	3,390	.424	.49
January	12,300	1,900	5,960	.745	.86
February	25,300	3,900	12,200	1.52	1.58
March	20,300	7,910	14,900	1.86	2.14
April	95,400	20,600	53,800	6.72	7.50
May	35,300	15,300	28,400	3.55	4.09
June	21,000	3,670	10,700	1.34	1.50
July	5,200	2,320	3,230	.404	.47
August	5,620	1,960	2,910	.364	.42
September	6,950	1,290	2,550	.319	.36
The year	95,400	1,240	11,800	1.48	19.96
1909–10.					
October	8,960	2,200	4,430	.554	.64
November	6,490	2,320	3,980	.498	.56
December	6,950	3,800	4,510	.564	.65
January	57,300	2,800	12,900	1.61	1.80
February	20,000	5,000	8,200	1.02	1.06
March	85,800	15,300	40,200	5.02	5.79
April	52,500	14,700	29,800	3.72	4.15
May	26,600	10,100	17,800	2.22	2.56
June	24,300	4,210	14,800	1.85	2.06
July	5,620	1,400	3,250	.406	.47
August	10,700	1,840	4,190	.524	.60
September	5,830	1,180	3,100	.388	.43
The year	85,800	1,180	12,300	1.54	20.83

Monthly discharge of Connecticut River at Sunderland, Mass., for the years ending Sept. 30, 1904–1918—Continued.

Month.	Discharge in second-feet.				Run-off (depth in inches on drainage area).
	Maximum.	Minimum.	Mean.	Per square mile.	
1910–11.					
October	5,410	1,620	3,350	0.419	0.48
November	9,520	1,620	4,960	.620	.69
December	5,600	1,600	2,810	.351	.40
January	30,000	5,800	10,000	1.25	1.44
February	6,100	2,600	4,160	.520	.54
March	36,100	3,300	8,960	1.12	1.29
April	58,900	11,000	35,700	4.46	4.98
May	53,700	3,020	19,400	2.42	2.79
June	9,520	2,450	6,070	.759	.85
July	4,400	1,180	2,250	.281	.32
August	7,910	1,060	2,850	.356	.41
September	11,700	2,720	5,470	.684	.76
The year	58,900	1,060	8,820	1.10	14.95
1911–12.					
October	47,800	4,790	16,300	2.04	2.35
November	19,900	8,690	12,700	1.59	1.77
December	33,800	7,910	18,200	2.28	2.63
January	20,300	3,800	9,100	1.14	1.31
February	5,500	2,700	4,500	.562	.61
March	43,500	3,300	14,100	1.76	2.03
April	82,500	29,300	51,000	6.38	7.12
May	41,600	11,700	24,800	3.10	3.57
June	50,100	2,720	18,700	2.34	2.61
July	5,200	1,730	3,210	.401	.46
August	6,490	2,080	4,130	.516	.59
September	17,400	3,020	7,630	.954	1.06
The year	82,500	1,730	15,300	1.91	26.11
1912–13.					
October	54,100	3,020	12,800	1.60	1.84
November	34,200	7,420	16,100	2.01	2.24
December	24,300	6,270	13,900	1.74	2.01
January	36,900	12,300	23,000	2.88	3.32
February	23,600	5,600	12,400	1.55	1.61
March	107,000	8,420	39,400	4.92	5.67
April	60,100	15,000	30,300	3.79	4.23
May	36,100	5,620	15,200	1.90	2.19
June	31,200	1,960	9,100	1.14	1.27
July	5,620	2,080	3,500	.438	.50
August	6,050	1,620	3,130	.391	.45
September	4,400	1,130	2,270	.284	.32
The year	107,000	1,130	15,100	1.89	25.65
1913–14.					
October	17,000	1,510	5,910	.739	.85
November	30,000	3,850	9,260	1.16	1.29
December	15,000	4,080	7,530	.941	1.08
January	6,000	1,500	3,650	.456	.53
February	1,400	1,400	3,170	.396	.41
March	50,500	2,300	15,300	1.91	2.20
April	88,300	30,800	53,500	6.69	7.46
May	61,300	6,050	26,900	3.36	3.87
June	10,400	2,320	5,220	.652	.73
July	8,160	1,840	4,340	.542	.62
August	6,050	1,730	3,160	.395	.46
September	8,420	1,290	3,480	.435	.49
The year	88,300	1,290	11,800	1.48	19.99

Monthly discharge of Connecticut River at Sunderland, Mass., for the years ending Sept. 30, 1904–1918—Continued.

Month.	Discharge in second-feet.				Run-off (depth in inches on drainage area).
	Maximum.	Minimum.	Mean.	Per square mile.	
1914–15.					
October	3,500	1,400	2,510	0.314	0.36
November	6,950	1,290	3,390	.424	.47
December	7,660	1,390	4,380	.541	.62
January	13,000	1,960	5,960	.745	.86
February	70,200	4,210	15,700	1.96	2.04
March	41,600	5,620	13,200	1.65	1.90
April	57,300	6,950	21,900	2.74	3.06
May	24,700	3,170	11,000	1.38	1.59
June	8,420	1,740	4,400	.550	.61
July	65,300	3,670	18,800	2.35	2.71
August	54,900	7,180	17,600	2.20	2.54
September	9,520	2,450	5,690	.711	.79
The year	70,200	1,290	10,300	1.29	17.55
1915–16.					
October	7,660	2,020	5,690	.711	.82
November	11,700	3,500	7,200	.900	1.00
December	34,600	4,400	11,500	1.44	1.66
January	61,700	6,490	16,900	2,11	2.43
February	47,200	4,210	21,700	2.71	2.92
March	53,700	5,200	13,700	1.71	1.97
April	72,600	30,800	45,400	5.68	6.34
May	44,700	8,420	23,800	2.98	3.44
June	29,300	12,600	20,400	2.55	2.84
July	21,400	5,830	11,900	1.49	1.72
August	17,400	2,320	6,370	.796	.92
September	14,000	2,080	5,650	.706	.79
The year	72,600	2,080	15,800	1.98	26.85
1916–17.					
October	14,300	3,020	7,010	.876	1.01
November	22,500	3,330	8,180	1.02	1.14
December	30,800	5,410	14,000	1.75	2.02
January	12,300	4,210	7,930	.991	1.14
February	14,000	1,960	4,640	.580	.60
March	61,300	5,200	17,200	2.15	2.48
April	59,700	18,100	39,300	4.91	5.48
May	34,600	15,700	24,500	3.06	3.53
June	48,600	11,000	26,400	3.30	3.68
July	17,400	3,020	7,930	.991	1.14
August	16,000	2,080	8,020	1.00	1.15
September	11,300	1,620	5,180	.648	.72
The year	61,300	1,620	14,200	1.78	24.09
1917–18.					
October	40,000	2,580	8,780	1.10	1.27
November	41,600	3,670	12,200	1.52	1.70
December	6,950	2,200	4,850	.606	.70
January	5,830	2,080	3,990	.499	.58
February	20,300	1,620	7,370	.921	.96
March	58,900	6,270	22,600	2.82	3.25
April	68,500	20,300	41,200	5.15	5.75
May	36,900	6,950	20,500	2.56	2.95
June	13,100	3,930	8,760	1.10	1.23
July	10,500	1,770	5,110	.639	.74
August	11,400	1,300	4,300	.538	.62
September	35,700	945	8,640	1.08	1.20
The year	68,500	945	12,300	1.54	20.95

Days of deficiency in discharge of Connecticut River at Sunderland, Mass., during the years ending Sept. 30, 1905–1918.

Discharge in second-feet per square mile.	Discharge in second-feet.	Theoretical horsepower per foot of fall.	Days of deficiency in discharge.													
			1904–5.	1905–6.	1906–7.	1907–8.	1908–9.	1909–10.	1910–11.	1911–12.	1912–13.	1913–14.	1914–15.	1915–16.	1916–17.	1917–18.
0.15	1,200	136	....	....	....	1	1	1	4	....	1	....	....	....	....	1
.2	1,600	182	....	....	....	6	8	4	11	....	3	6	6	....	....	3
.3	2,400	273	44	1	10	30	89	19	57	6	34	33	31	2	7	16
.4	3,200	364	87	21	26	52	152	56	105	29	67	83	72	9	19	43
.5	4,000	455	95	37	36	65	173	106	127	51	78	144	98	17	38	73
.6	4,800	545	113	51	79	82	197	172	183	95	95	173	125	46	68	128
.7	5,600	636	126	54	131	89	207	189	209	129	102	184	139	61	81	144
.8	6,400	727	144	85	168	99	216	213	245	142	124	206	166	92	117	170
.9	7,200	818	175	110	198	111	223	220	274	158	131	225	197	117	150	194
1.0	8,000	909	198	138	219	119	228	222	286	164	142	243	214	132	170	209
1.1	8,800	1,000	213	156	229	137	233	223	292	167	151	260	229	146	181	226
1.2	9,600	1,090	231	171	247	152	238	228	301	184	163	274	243	162	194	234
1.3	10,400	1,180	240	178	249	163	245	231	303	191	173	280	248	176	207	243
1.4	11,200	1,270	246	190	255	176	250	236	306	194	185	285	255	192	222	252
1.5	12,000	1,360	252	194	259	182	254	240	307	200	198	292	261	206	233	259
1.6	12,800	1,450	264	206	264	192	266	250	313	220	211	293	275	219	239	266
1.75	14,000	1,590	270	218	274	200	269	252	316	229	224	295	286	225	243	275
1.9	15,200	1,730	278	242	281	213	275	261	317	240	236	300	297	233	252	281
2.05	16,400	1,860	284	259	287	221	281	267	318	247	241	303	303	243	256	285
2.25	18,000	2,050	296	274	294	231	290	273	319	254	255	307	312	262	267	294
2.5	20,000	2,270	305	286	301	244	298	291	323	278	268	310	327	273	279	299
2.75	22,000	2,500	318	291	312	254	309	304	326	289	279	310	331	284	287	302
3.0	24,000	2,730	325	300	317	264	313	311	328	291	292	312	337	291	300	305
3.5	28,000	3,180	334	323	332	289	318	326	330	306	318	315	344	307	313	311
4.0	32,000	3,640	337	340	338	304	326	332	335	315	320	320	346	316	325	322
4.5	36,000	4,090	343	345	343	317	343	336	342	326	335	328	350	327	335	334
5.0	40,000	4,550	348	347	344	330	346	342	350	332	345	329	352	336	341	341
7.0	56,000	6,360	357	358	362	357	352	357	364	356	357	354	361	358	359	359
10.0	80,000	9,090	362	365	365	366	361	364	365	365	361	362	365	366	365	365
15.0	120,000	13,600	365	....	....	....	365	365	....	366	365	365	....	....	....	....

NOTE.—The above table gives the theoretical horsepower per foot of fall that may be developed at different rates of discharge, and shows the number of days on which the discharge and corresponding horsepower were respectively less than the amounts given in the columns for discharge and horsepower. In using this table allowance should be made for the various losses, the principal ones being the wheel loss, which may be as large as 20 per cent, and the head loss, which may be as large as 5 per cent.

PASSUMPSIC RIVER AT PIERCE'S MILLS, NEAR ST. JOHNSBURY, VT.

LOCATION.—At suspension footbridge just below Pierce's mills, 2 miles below mouth of Sheldon Branch, 4 miles above mouth of Moose River, and 5 miles north of St. Johnsbury, Caledonia County.

DRAINAGE AREA.—237 square miles.

RECORDS AVAILABLE.—May 26, 1909, to September 30, 1918.

GAGE.—Staff, in two sections; low-water section a vertical staff bolted to ledge just above bridge; high-water section an inclined staff bolted to ledge below bridge; read by W. I. Cox and Clinton G. Taylor.

DISCHARGE MEASUREMENTS.—Made from footbridge or by wading below the bridge.

CHANNEL AND CONTROL.—Channel composed of ledge rock partly covered with gravel and alluvial deposits. At high stages the control is probably at the dam near Centervale.

EXTREMES OF DISCHARGE.—Maximum stage recorded during year water over top of gage on mornings of October 31 and April 3 (discharge about 2,900 second-feet); minimum stage recorded, 1.2 feet at 6 p. m. August 25 and 5.30 p. m. August 31 (discharge, 71 second-feet).

1909–1918: Maximum stage recorded, 14.8 feet during the night of March 27, 1913, determined by leveling from flood marks (discharge not computed); minimum stage recorded, zero flow at various times due to water being held back by mills.

ICE.—River freezes over at the control, causing the stage-discharge relation to be seriously affected, ice jams occasionally form below the gage.

REGULATION.—There is a small diurnal fluctuation caused by the operation of Pierce's mills,[a] just above the station, and by other mills farther upstream. The effect of the diurnal fluctuation was studied by means of a portable automatic gage from August 16 to September 11, 1914. Although the results obtained from twice-a-day gage heights were found to be occasionally in error for individual days, the mean discharge for the period determined from twice-a-day gage heights and was found to be identical with that obtained from the hourly record

ACCURACY.—The stage-discharge relation practically permanent except when affected by ice. Rating curve fairly well defined below 2,000 second-feet. Gage read to quarter-tenths twice daily, except from December 20 to March 24 when it was read once a day. Daily discharge ascertained by applying mean daily gage height to rating table and making correction for effect of ice during the winter. Record good.

Discharge measurements of Passumpsic River at Pierce's mills, near St. Johnsbury, Vt., during the year ending Sept. 30, 1918.

Date.	Made by—	Gage height.	Discharge.	Date.	Made by—	Gage height.	Discharge.
		Feet.	*Sec.-ft.*			*Feet.*	*Sec.-ft.*
Oct. 10	M. R. Stackpole	2.40	396	Mar. 28	M. R. Stackpole	[b]2.87	407
Dec. 14	.....do	[b]2.30	210	Apr. 10	.....do	4.09	1,050
Jan. 28	.....do	[b]2.60	134	10	.....do	4.10	1,050
Mar. 4	.....do	[b]3.00	223	July 23	C. H. Pierce	1.54	138

[a] Pierce's mills not in operation during the summer of 1918.
[b] Stage-discharge relation affected by ice.

Daily discharge in second-feet, of Passumpsic River at Pierce's mills, near St. Johnsbury, Vt., for the year ending Sept. 30, 1918.

Day.	Oct.	Nov.	Dec.	Jan.	Feb.	Mar.	Apr.	May.	June.	July.	Aug.	Sept.
1	390	1,080	260	110	130	460	2,120	1,310	640	245	176	202
2	340	790	230	110	130	360	2,600	1,260	640	500	130	420
3	260	670	245	110	130	260	2,480	1,000	390	275	130	202
4	460	600	260	90	130	215	1,460	790	290	230	122	105
5	640	530	230	90	130	215	1,080	750	260	202	202	117
6	830	530	230	100	130	200	1,040	600	245	189	360	120
7	530	530	260	110	130	200	1,260	640	870	152	202	202
8	375	420	260	120	110	200	1,410	560	600	275	202	126
9	530	420	200	150	130	175	1,760	530	420	460	1,000	120
10	375	460	260	120	130	175	1,120	500	340	375	460	126
11	320	420	260	110	130	175	1,040	1,000	290	260	260	109
12	290	460	275	130	130	190	950	640	500	245	216	93
13	600	360	290	130	140	200	1,080	870	830	340	189	152
14	405	290	215	130	150	230	830	2,000	530	390	176	216
15	390	290	200	130	175	230	1,220	1,120	390	360	230	164
16	600	360	215	130	175	230	1,510	790	360	260	164	130
17	405	340	230	130	175	260	1,360	560	320	320	130	164
18	340	305	230	130	150	320	1,260	460	360	360	120	360
19	305	460	200	130	150	320	870	420	290	230	122	460
20	670	360	236	130	175	390	790	390	245	176	111	275
21	500	320	165	130	230	420	830	390	216	164	105	910
22	375	320	200	110	175	500	1,310	375	530	152	101	500
23	320	405	175	130	175	530	1,410	600	640	141	109	305
24	320	390	215	130	150	560	1,260	460	560	130	91	530
25	790	230	175	150	150	600	870	340	375	141	82	530
26	500	275	175	130	260	600	790	305	290	126	78	910
27	390	305	150	130	670	530	830	530	230	122	82	1,880
28	670	260	150	130	600	670	870	600	176	117	91	910
29	530	230	150	150	....	750	1,120	420	245	120	89	560
30	1,510	260	140	150	....	950	1,360	560	460	260	91	420
31	2,300	....	130	150	....	1,560	....	500	....	260	75	....

NOTE.—Stage-discharge relation affected by ice Nov. 27 to Mar. 29; daily discharge during this period determined from gage heights corrected for effect of ice by means of four discharge measurements, observer's notes, and weather records.

Monthly discharge of Passumpsic River at Pierce's mills, near St. Johnsbury, Vt., for the year ending Sept. 30, 1918.

[Drainage area, 237 square miles.]

Month.	Discharge in second-feet.				Run-off (depth in inches on drainage area).
	Maximum.	Minimum.	Mean.	Per square mile.	
October	2,300	260	557	2.35	2.71
November	1,080	230	422	1.78	1.99
December	290	130	213	.899	1.04
January	150	90	125	.527	.61
February	670	110	187	.790	.82
March	1,560	175	409	1.73	1.99
April	2,600	790	1,260	5.32	5.94
May	2,000	305	686	2.89	3.33
June	870	176	418	1.76	1.96
July	500	117	244	1.03	1.19
August	1,000	75	184	.776	.89
September	1,880	93	377	1.59	1.77
The year	2,600	75	424	1.79	24.24

WHITE RIVER AT WEST HARTFORD, VT.

LOCATION.—About 500 feet above highway bridge in village of West Hartford, Windsor County, and 7 miles above mouth.

DRAINAGE AREA.—687 square miles.

RECORDS AVAILABLE.—June 9, 1915, to September 30, 1918.

GAGE.—Inclined staff on left bank; read by F. P. Morse.

DISCHARGE MEASUREMENTS.—Made from cable 1,500 feet below the gage or by wading.

CHANNEL AND CONTROL.—Channel wide and of fairly uniform cross section at measuring section; bed covered with gravel and small boulders. Control formed by rock ledge 100 feet below the gage; well defined.

EXTREMES OF DISCHARGE.—Maximum stage recorded during year, 10.0 feet at 5 p. m. October 30 (discharge, by extension of rating curve, about 10,000 second-feet); minimum stage recorded 2.22 feet at 7 p. m. August 4 (discharge, by extension of rating curve, about 35 second-feet).

1915–1918: Maximum stage recorded, 11.1 feet at 6 p. m. June 12, 1917 (discharge, by extension of rating curve, about 11,700 second-feet); minimum stage recorded, 2.33 feet at 6 a. m. August 29, 1916 (discharge, by extension of rating curve, about 26 second-feet). The high water of March 27, 1913, reached a stage of 18.9 feet, as determined from reference point on scale platform opposite gage (discharge not determined).

ICE.—River freezes over at the gage; control usually remains partly open, although ice on the rocks and along the shore affects the stage-discharge relation.

REGULATION.—There are several power plants on the main stream and tributaries above the station, the nearest being that of the Sharon Power Co. at Sharon; when this plant is in operation it causes some diurnal fluctuation in discharge at low stages; this plant was operated only a short time, if at all, during the year. The effect of power plants farther upstream is eliminated by the large amount of pondage at Sharon.

ACCURACY.—Stage-discharge relation practically permanent except when affected by ice. Rating curve fairly well defined between 150 and 5,000 second-feet. Staff gage read to quarter-tenths twice daily. Daily discharge ascertained by applying mean daily gage height to rating table, and making correction for effect of ice during the winter. Records good.

Discharge measurements of White River at West Hartford, Vt., during the year ending Sept. 30, 1918.

Date.	Made by—	Gage height.	Discharge.	Date.	Made by—	Gage height.	Discharge.
		Feet.	*Sec.-ft.*			*Feet.*	*Sec.-ft.*
Dec. 19	M. R. Stackpole	a 3.83	428	Apr. 13	M. R. Stackpole	6.31	2,780
Jan. 22	.....do	a 4.15	303	July 28	H. W. Fear	2.96	165
Feb. 27	.....do	a 7.98	2,820	Aug. 27	J. W. Moulton	3.00	171
Mar. 21	.....do	a 7.36	2,430				

a Stage-discharge relation affected by ice.

Daily discharge, in second-feet, of White River at West Hartford, Vt., for the year ending Sept. 30, 1918.

[illegible]

Note.—Stage-discharge relation affected by ice from Nov. 25 to Mar. 28; daily discharge for this period determined from gage heights corrected for effect of ice by means of four discharge measurements, observer's notes, and weather records. Stage-discharge relation affected by backwater from logs May 26–28; correction estimated.

Monthly discharge of White River at West Hartford, Vt., for the year ending Sept. 30, 1918.

[Drainage area, [illegible] square miles.]

Month.	Discharge in second-feet.				Run-off (depth in inches on drainage area).
	Maximum.	Minimum.	Mean.	Per square mile.	
October	[illegible]	[illegible]	[illegible]	1.22	1.41
November	[illegible]	[illegible]	[illegible]	[illegible]	1.19
December	[illegible]	[illegible]	[illegible]	.386	.58
January	[illegible]	[illegible]	[illegible]	.361	.42
February	[illegible]	[illegible]	[illegible]	.986	1.03
March	[illegible]	[illegible]	[illegible]	2.46	2.84
April	[illegible]	[illegible]	[illegible]	5.76	6.43
May	[illegible]	[illegible]	[illegible]	2.26	2.61
June	[illegible]	[illegible]	[illegible]	.543	.94
July	[illegible]	[illegible]	[illegible]	.482	.57
August	[illegible]	[illegible]	[illegible]	.348	.40
September	[illegible]	[illegible]	640	.945	1.05
The year	[illegible]	64	983	1.43	19.44

ASHUELOT RIVER AT HINSDALE, N. H.

LOCATION.—At lower steel highway bridge, a quarter of a mile below dam of Fisk Paper Co. and 1¼ miles above mouth.

DRAINAGE AREA.—440 square miles.

RECORDS AVAILABLE.—February 22, 1907, to December 31, 1909, and July 11, 1914, to September 30, 1918.

GAGE.—Chain gage on downstream side of bridge; read by Teresa Golden.

DISCHARGE MEASUREMENTS.—Made from highway bridge.

CHANNEL AND CONTROL.—Bed covered with coarse gravel and boulders. Control is a short distance below gage and is practically permanent.

EXTREMES OF DISCHARGE.—Maximum stage recorded during year, 6.80 feet at 4 p. m. April 3 (discharge, from extension of rating curve, about 4,150 second-feet); minimum stage recorded, 2.18 feet at 4 p. m. August 11 (discharge, from extension of rating curve, about 20 second-feet).

1914-1918: Maximum stage recorded, 7.5 feet at 5 p. m. February 26, 1915 (discharge, from extension of rating curve, about 5,190 second-feet); minimum stage recorded, 2.0 feet at 4 p. m. October 4, 1914 (discharge, from extension of rating curve, about 10 second-feet).

ICE.—Ice forms below bridge on control, affecting stage-discharge relation for short periods.

REGULATION.—The mills immediately above station are operated continuously except for Sundays and holidays, but cause little fluctuation in stage. Several reservoirs and ponds on the river and tributaries have some effect on the distribution of flow.

ACCURACY.—Stage-discharge relation practically permanent except when affected by ice. Rating curve fairly well defined below 4,000 second-feet. Gage read to hundredths twice daily. Discharge ascertained by applying mean daily gage height to rating table and making correction for effect of ice during the winter. Records good.

Discharge measurements of Ashuelot River at Hinsdale, N. H., during the year ending Sept. 30, 1918.

Date.	Made by—	Gage height.	Discharge.	Date.	Made by—	Gage height.	Discharge.
		Feet.	*Sec.-ft.*			*Feet.*	*Sec.-ft.*
Jan. 4	M. R. Stackpole........	a 4.45	130	Mar. 20	M. R. Stackpole.......	4.40	999
Feb. 13	.....do..................	a 3.14	106	June 8	O. W. Hartwell........	3.53	349

a Stage-discharge relation affected by ice.

Daily discharge, in second-feet, of Ashuelot River at Hinsdale, N. H., for the year ending Sept. 30, 1918.

Day.	Oct.	Nov.	Dec.	Jan.	Feb.	Mar.	Apr.	May.	June.	July.	Aug.	Sept.
1	115	161	185	76	140	720	2,300	520	340	206	90	106
2	115	161	170	86	105	1,000	3,280	520	350	223	106	120
3	122	161	300	86	130	1,200	4,010	460	350	173	115	94
4	161	235	260	105	140	600	3,720	400	231	185	24	79
5	134	375	300	140	130	350	3,720	350	345	215	82	98
6	120	1,160	140	155	120	350	2,860	810	315	239	94	104
7	111	2,170	130	155	120	430	2,170	660	375	260	82	45
8	111	1,910	280	155	120	400	1,550	555	310	167	86	73
9	122	1,550	120	155	140	320	2,440	350	215	167	98	132
10	161	350	240	155	140	350	2,300	400	247	209	90	86
11	134	310	300	140	140	320	2,170	460	223	185	25	58
12	134	223	280	105	140	239	2,580	520	375	243	161	106
13	115	264	220	120	155	268	2,860	590	330	139	170	215
14	122	335	185	130	220	350	2,300	770	375	124	191	65
15	115	215	170	130	155	282	1,550	900	350	231	223	106
16	167	176	130	130	240	400	2,040	1,210	235	282	255	82
17	161	173	130	105	260	260	2,300	1,380	282	215	155	84
18	161	106	170	130	300	247	1,610	1,100	264	315	115	134
19	150	215	185	120	400	330	1,160	950	231	255	134	137
20	161	197	200	140	460	1,000	2,170	695	209	215	137	273
21	161	176	185	130	350	1,670	2,720	490	235	120	139	460
22	161	206	130	155	260	2,040	2,860	430	282	167	134	660
23	139	335	86	120	300	2,580	2,040	231	430	161	111	460
24	134	400	130	120	460	2,720	1,100	194	855	145	120	291
25	134	278	155	105	700	2,440	810	264	520	115	52	223
26	147	185	140	140	520	2,580	1,210	300	490	139	134	855
27	147	200	140	130	460	2,440	900	264	375	98	139	1,790
28	243	155	140	120	520	1,910	625	209	350	68	111	2,170
29	206	185	130	105	......	1,790	325	320	375	106	137	2,040
30	161	105	105	120	......	1,910	375	350	282	102	139	1,910
31	206	......	96	120	......	2,440	......	350	......	102	134	......

NOTE.—Stage-discharge relation affected by ice Nov. 26 to Mar. 11; daily discharge for this period determined from gage heights corrected for effect of ice by means of two discharge measurements, observer's notes, and weather records.

Monthly discharge of Ashuelot River at Hinsdale, N. H., for the year ending Sept, 30, 1918.

[Drainage area, 440 square miles.]

Month.	Discharge in second-feet.				Run-off (depth in inches on drainage area).
	Maximum.	Minimum.	Mean.	Per square mile.	
October	243	111	146	0.332	0.38
November	2,170	105	422	.959	1.07
December	300	86	178	.405	.47
January	155	76	125	.284	.33
February	700	105	262	.595	.62
March	2,720	239	1,090	2.48	2.86
April	4,010	325	2,070	4.70	5.24
May	1,380	194	549	1.25	1.44
June	855	209	338	.768	.86
July	315	68	180	.409	.47
August	255	24	122	.277	.32
September	2,170	45	435	.989	1.10
The year	4,010	24	492	1.12	15.16

MILLERS RIVER NEAR WINCHENDON, MASS.

Location.—At steel highway bridge known as Nolan's bridge, half a mile below mouth of Sip Pond Brook and 2 miles west of Winchendon, Worcester County.

Drainage area.—80.0 square miles.

Records available.—June 5, 1916, to September 30, 1918.

Gages.—Stevens continuous water-stage recorder on right bank below highway bridge installed July 4, 1917. Chain gage on downstream side of bridge installed June 5, 1916. Foxboro water-stage recorder used from June 5 to July 3, 1917; inspected by Franklin Epps.

Discharge measurements.—Made from the highway bridge or by wading.

Channel and control.—Bed covered with gravel and alluvial deposits. Control for low and medium stages is about 80 feet below gage. Clearly defined.

Extremes of discharge.—Maximum open-water stage during year, from water-stage recorder, 6.56 feet at 9.30 p. m. April 3 (discharge, 715 second-feet); a stage of 8.13 feet was recorded at 6 p. m. March 23, but the stage-discharge relation was affected by ice at the time; minimum stage during year, from water-stage recorder, 2.02 feet at 5 a. m. September 20 (discharge, practically zero; water held back by dams).

1916–1918: Maximum open-water stage recorded, 6.56 feet at 9.30 p. m. April 3, 1918 (discharge, 715 second-feet); minimum stage recorded September 20, 1918.

Ice.—Stage-discharge relation seriously affected by ice. Complete ice cover usually remains intact throughout the winter. Owing to large diurnal fluctuation caused by operation of power plants in the vicinity of Winchendon, water frequently overflows the ice.

Regulation.—Distribution of flow affected by operation of power plants at and below Winchendon and by storage in Lake Monomonac and other reservoirs.

Accuracy.—Stage-discharge relation somewhat shifting on account of gravel bar 80 feet below the gage. Two rating curves have been used, both well defined for periods covered. Operation of water-stage recorder satisfactory throughout the year except from December 29 to February 8, when clock frequently stopped on account of low temperatures. Daily discharge for open-water period ascertained by use of discharge integrator. Records good for open-water periods and when the water-stage recorder was in operation, but only fair for winter period.

Discharge measurements of Millers River at Winchendon, Mass., during the year ending Sept. 30, 1918.

Date.	Made by—	Gage height.	Discharge.	Date.	Made by—	Gage height.	Discharge.
		Feet.	*Sec.-ft.*			*Feet.*	*Sec.-ft.*
Dec. 9	M. R. Stackpole	[a] 3.31	49.5	Apr. 9	H. W. Fear	4.35	249
Jan. 5	.....do	[a] 4.70	79	July 18	.....do	3.54	130
Feb. 8	.....do	[a] 5.25	39.7	18	A. N. Weeks	3.31	104
Mar. 8	H. W. Fear	[a] 6.82	228	Aug. 20	J. W. Moulton	3.51	115
Apr. 4	.....do	6.32	658	28	H. W. Fear	2.63	13.9

[a] Stage-discharge relation affected by ice.

Daily discharge, in second-feet, of Millers River near Winchendon, Mass., for the year ending Sept. 30, 1918.

Day.	Oct.	Nov.	Dec.	Jan.	Feb.	Mar.	Apr.	May.	June.	July.	Aug.	Sept.
1	56	305	145	18	62	330	540	255	73	79	49	22
2	56	220	45	45	50	300	590	270	18	80	54	15
3	56	126	85	50	15	260	620	225	71	66	62	44
4	54	57	95	50	50	300	590	152	77	50	28	50
5	45	80	85	78	62	240	495	122	76	65	35	62
6	39	66	78	18	55	195	395	112	68	65	50	55
7	28	68	78	45	45	220	290	142	95	50	60	40
8	42	64	70	30	40	230	380	128	93	70	70	14
9	40	62	50	35	40	220	325	112	22	73	86	42
10	48	59	85	30	18	230	330	114	57	75	67	54
11	57	32	70	35	30	220	300	79	99	74	22	40
12	59	70	78	30	40	205	290	37	99	75	62	46
13	46	79	85	13	45	195	235	102	92	65	71	54
14	13	74	78	62	50	220	190	122	96	40	79	30
15	56	55	62	55	50	205	345	144	95	50	69	11
16	50	55	35	70	45	205	345	134	50	50	58	42
17	58	48	62	62	25	160	340	122	78	88	45	39
18	60	25	62	55	50	220	360	104	79	108	17	53
19	36	83	62	45	105	260	295	41	73	70	53	46
20	50	59	50	15	170	315	240	85	72	71	72	49
21	14	67	50	55	330	375	215	97	61	16	66	77
22	40	61	45	62	300	475	490	102	134	59	71	41
23	44	125	15	55	270	555	460	94	210	67	66	75
24	52	105	50	50	220	535	390	92	290	58	58	85
25	61	36	18	45	280	555	350	90	170	54	14	66
26	102	90	78	45	345	515	245	40	136	55	61	116
27	84	160	50	18	330	495	200	104	136	53	55	365
28	24	116	45	45	345	475	154	104	134	27	54	355
29	142	38	40	50	......	425	190	104	90	53	58	220
30	250	92	15	55	......	455	164	46	60	71	55	180
31	400	......	78	55	......	495	......	84	......	62	52	......

NOTE.—Stage-discharge relation affected by ice, Dec. 1 to Mar. 31; daily discharge for this period determined from gage heights corrected for effect of ice by means of four discharge measurements, observer's notes, and weather records, and comparison with record of flow of Millers River at Erving. Discharge estimated Oct. 15-21, May 25-26; June 15-16, July 5-8, 12-15, and Aug. 6-8, 30, by hydrograph comparison with records at other stations.

Monthly discharge of Millers River near Winchendon, Mass., for the year ending Sept. 30, 1918.

[Drainage area, 80.0 square miles.]

Month.	Discharge in second-feet.				Run-off (depth in inches on drainage area).
	Maximum.	Minimum.	Mean.	Per square mile.	
October	400	13	69.7	0.871	1.00
November	305	25	85.9	1.07	1.19
December	145	15	62.7	.784	.90
January	78	13	44.4	.555	.64
February	345	15	124	1.55	1.61
March	555	160	325	4.06	4.68
April	620	154	345	4.31	4.81
May	270	37	115	1.44	1.66
June	290	18	96.8	1.21	1.35
July	108	16	62.5	.781	.90
August	86	14	55.5	.694	.80
September	365	11	79.6	.995	1.11
The year	620	11	122	1.52	20.65

MILLERS RIVER AT ERVING, MASS.

LOCATION—A quarter of a mile below dam at Erving, Franklin County, 8 miles above confluence of Millers River with Connecticut River, and below all important tributaries.

DRAINAGE AREA.—372 square miles.

RECORDS AVAILABLE.—August 1, 1914, to September 30, 1918.

GAGES.—Vertical staff attached to downstream end of factory; read by Arthur Lemire. Water-stage recorder installed in gage house on right bank July 1, 1915; gage heights referred to gage datum by a hook gage inside the well.

DISCHARGE MEASUREMENTS.—Made from cable near gage or by wading.

CHANNEL AND CONTROL.—Bed covered with coarse gravel and boulders. Control section is a short distance below the gage; practically permanent.

EXTREMES OF DISCHARGE.—Maximum open-water stage during year, from water-stage recorder, 4.63 feet at 7 a. m. April 3 (discharge, 3,090 second-feet); a stage of 5.97 feet was recorded at 8.30 a. m. February 27, but the stage-discharge relation was affected by ice; minimum stage, from water-stage recorder, 1.0 foot at 10 a. m. August 4 (discharge, 9 second-feet).

1914–1918: Maximum open-water stage recorded, 5.6 feet at 4 p. m. February 25, 1915 (discharge, 5,160 second-feet); see also preceding paragraph; minimum discharge, practically zero at various times during 1915, and at 3.30 p. m. October 29, 1916, when water was held back by dams above the gage.

ICE.—River freezes over below the gage at various times during the winter; ice considerably broken by rising and falling stages due to operation of power plants; stage-discharge relation seriously affected.

REGULATION.—Distribution of flow affected by operation of various power plants and storage reservoirs above the station.

ACCURACY.—Stage-discharge relation practically permanent except when affected by ice. Rating curve well defined below 4,000 second-feet. Staff gage read to hundredths twice daily. Daily discharge ascertained by use of discharge integrator, except for periods when continuous gage-height record was not obtained, and then the staff-gage records were used with corrections as determined by various comparisons with the water-stage recorder. Records good, except for times of ice effect, for which they are fair.

Discharge measurements of Millers River at Erving, Mass., during the year ending Sept. 30, 1918.

Date.	Made by—	Gage height.	Discharge.	Date.	Made by—	Gage height.	Discharge.
		Feet.	*Sec.-ft.*			*Feet.*	*Sec.-ft.*
Dec. 8	M. R. Stackpole	[a]3.37	766	Mar. 19	M. R. Stackpole	[a]3.70	1,230
Jan. 8	....do	[a]4.00	243	June 17	H. W. Fear	2.86	657
Feb. 10	....do	[a]3.84	200	July 17	A. N. Weeks	2.42	437

[a] Stage-discharge relation affected by ice.

Daily discharge, in second-feet, of Millers River at Erving, Mass., for the year ending Sept. 30, 1918.

Day.	Oct.	Nov.	Dec.	Jan.	Feb.	Mar.	Apr.	May.	June.	July.	Aug.	Sept.
1	260	1,610	280	165	260	1,550	2,240	900	370	265	180	88
2	70	1,220	120	150	220	1,550	2,510	1,200	250	200	150	31
3	225	790	350	180	55	820	2,810	1,100	350	255	110	35
4	215	580	400	220	220	1,150	2,610	960	260	148	14	148
5	210	620	450	260	260	780	2,150	720	210	200	270	132
6	215	495	420	40	220	630	1,830	820	170	215	186	132
7	140	470	400	180	180	780	1,500	700	280	100	132	100
8	290	460	400	95	200	740	1,450	650	420	270	140	31
9	140	440	120	135	220	660	1,500	610	360	225	124	138
10	190	340	350	120	200	1,150	1,450	590	300	200	180	135
11	235	290	400	150	180	1,050	1,400	570	410	220	126	124
12	280	480	350	120	200	1,000	1,300	630	530	235	188	128
13	255	240	170	70	180	950	1 250	550	540	250	160	130
14	150	390	300	260	200	950	1,050	580	590	160	175	125
15	275	405	260	180	220	900	1,500	800	500	330	185	40
16	255	335	75	300	350	860	1,650	770	340	225	230	146
17	290	345	260	220	220	570	1,600	640	400	305	240	146
18	315	120	240	260	280	950	1,500	560	375	350	42	130
19	215	340	260	95	350	950	1,500	330	290	370	190	124
20	280	280	220	120	570	1,260	1,300	360	270	385	170	230
21	145	310	200	220	950	1,490	1,100	400	295	185	138	450
22	225	370	180	260	1,500	1,910	1,600	520	640	265	134	330
23	235	445	20	220	1,560	2,420	1,800	450	950	210	172	320
24	260	510	220	220	950	2,610	1,700	390	1,100	205	152	265
25	275	440	55	180	1,000	2,510	1,450	430	960	200	50	295
26	430	370	220	150	1,150	2,510	1,250	350	590	182	114	385
27	270	300	200	55	1,620	2,240	1,050	410	540	132	116	1,180
28	355	285	200	180	1,370	1,910	900	350	465	31	130	1,340
29	475	270	180	220	....	1,830	840	370	385	152	143	1,080
30	800	285	55	240	....	1,830	860	420	180	230	145	850
31	1,730	....	220	240	....	1,910	....	270	....	176	205	....

NOTE.—Stage-discharge relation affected by ice Dec. 1 to Mar. 19; daily discharge for this period determined from gage heights corrected for effect of ice by means of four discharge measurements, observer's notes, and weather records. Discharge estimated May 8-13, 26-28, June 4-10, and July 7, by comparison with records at other stations in the Millers River basin.

Monthly discharge of Millers River at Erving, Mass., for the year ending Sept. 30, 1918.

[Drainage area, 372 square miles.]

Month.	Discharge in second-feet.				Run-off (depth in inches on drainage area).
	Maximum.	Minimum.	Mean.	Per square mile.	
October	1,730	70	313	0.841	0.97
November	1,610	120	461	1.24	1.38
December	450	20	244	.656	.76
January	300	40	178	.478	.55
February	1,620	55	532	1.43	1.49
March	2,610	570	1,370	3.68	4.24
April	2,810	840	1,550	4.17	4.65
May	1,200	270	594	1.60	1.84
June	1,140	170	444	1.19	1.33
July	385	31	222	.597	.69
August	270	14	150	.403	.47
September	1,340	31	293	.788	.88
The year	2,810	14	528	1.42	19.24

SIP POND BROOK NEAR WINCHENDON, MASS.

LOCATION.—About 500 feet above highway bridge a quarter of a mile below Massachusetts-New Hampshire State line, 1½ miles below outlet of Sip Pond, and 3 miles northwest of Winchendon, Worcester County.

DRAINAGE AREA.—18.8 square miles.

RECORDS AVAILABLE.—May 29, 1916, to September 30, 1918.

GAGES.—Gurley 7-day water-stage recorder installed June 26, 1917, and vertical staff gage installed June 9, 1917, on left bank, 500 feet above highway bridge. Inclined staff gage on right bank 50 feet above highway bridge, used May 29 to June 29, and December 13, 1916, to June 26, 1917. Stevens 8-day water-stage recorder at same site and datum used June 30 to December 12, 1916. Gages read by W. G. Greenall and Hazel Greenall. All gages at same datum, but owing to slope of stream and different control section, present gage reads higher than those previously used.

DISCHARGE MEASUREMENTS.—Made from footbridge 15 feet below vertical staff gage or by wading.

CHANNEL AND CONTROL.—Bed rough, covered with boulders. Control clearly defined. Considerable aquatic vegetation in channel below inclined staff gage during summer.

EXTREMES OF DISCHARGE.—Maximum discharge during year, 221 second-feet, occurred at noon April 3; minimum discharge, 4 second feet, occurred at 2 p. m. August 25.

1916–1918: Maximum discharge during period, about 294 second-feet, occurred at 6 p. m., March 28, 1917; minimum discharge, August 25, 1918.

REGULATION.—The distribution of flow is considerably affected by operation of mills at State Line, N. H., and by storage in Pearly Pond and Sip Pond.

ACCURACY.—Stage-discharge relation practically permanent for present site. Rating curve well defined below 200 second-feet. Operation of water-stage recorder satisfactory, except during winter, when it was affected by ice in gage well. Daily discharge determined by use of discharge integrator, except during winter. Open-water records excellent; winter records fair.

Discharge measurements of Sip Pond Brook near Winchendon, Mass., during the year ending Sept. 30, 1918.

Date.	Made by—	Gage height.	Discharge.	Date.	Made by—	Gage height.	Discharge.
		Feet.	*Sec.-ft.*			*Feet.*	*Sec.-ft.*
Dec. 10	M. R. Stackpole	[a] 5.68	14.5	Apr. 4	H. W. Fear	8.07	188
Jan. 5	....do	[a] 6.04	18.8	9	....do	7.14	96
Feb. 7	....do	[a] 5.44	8.4	July 18	....do	5.77	20.3
Mar. 8	H. W. Fear	[a] 6.67	44.1	Aug. 21	J. W. Moulton	5.06	6.0

[a] Stage-discharge relation affected by ice.

Daily discharge, in second-feet, of Sip Pond Brook near Winchendon, Mass., for the year ending Sept. 30, 1918.

Day.	Oct.	Nov.	Dec.	Jan.	Feb.	Mar.	Apr.	May.	June.	July.	Aug.	Sept.
1	13	85	27	19	10	53	156	42	19	14	12	12
2	13	60	14	19	10	50	180	55	14	21	11	9.4
3	13	51	30	19	7	48	205	55	16	20	11	9.4
4	13	43	26	19	10	56	188	45	16	12	7.1	8.6
5	13	38	27	19	9	50	162	32	16	16	11	9.4
6	13	31	29	11	8	48	122	36	14	16	12	8.9
7	10	26	29	18	8	45	102	28	19	9.2	12	9.5
8	13	25	21	24	8	42	99	26	18	14	11	5.8
9	13	23	11	18	8	42	92	24	10	14	8.2	9.5
10	13	23	19	15	6	32	90	21	17	13	9.2	8.0
11	14	16	19	15	8	30	80	21	17	13	7.5	7.6
12	13	20	18	15	9	35	68	14	19	13	10	7.2
13	13	22	15	11	10	38	66	21	20	16	10	8.1
14	10	19	16	12	10	40	62	25	21	10	7.7	6.5
15	16	15	19	13	13	42	78	33	24	13	11	6.2
16	18	17	12	13	14	47	82	35	13	13	10	5.7
17	14	17	16	13	10	53	80	29	19	13	9.0	8.6
18	12	11	18	12	13	64	85	24	17	14	6.7	11
19	11	19	18	12	19	65	81	16	10	14	8.5	12
20	12	20	18	10	22	65	70	23	16	12	9.1	19
21	11	20	18	11	24	67	63	21	18	9.0	9.0	22
22	14	21	16	11	22	116	99	24	32	14	9.6	18
23	14	22	11	10	20	140	100	25	75	12	9.5	22
24	14	22	19	10	16	134	88	26	74	12	9.1	22
25	19	16	10	10	40	138	75	24	53	12	4.2	18
26	18	28	18	10	69	134	65	20	35	13	10	34
27	17	30	18	8	80	120	56	24	28	11	9.7	110
28	15	23	16	10	62	104	48	24	24	7.1	8.8	120
29	21	17	16	10	....	93	47	21	21	10	10	77
30	33	19	10	10	....	108	42	21	14	11	8.2	63
31	72	....	18	10	....	130	....	21	....	11	9.0	....

NOTE.—Stage-discharge relation affected by ice Dec. 10 to Mar. 14, and extreme cold also affected operation of water-stage recorder for short periods; daily discharge during this period determined from gage heights corrected for effect of ice by means of four discharge measurements, observer's notes, and weather records.

Monthly discharge of Sip Pond Brook near Winchendon, Mass., for the year ending Sept. 30, 1918.

[Drainage area, 18.8 square miles.]

Month.	Discharge in second-feet.				Run-off (depth in inches on drainage area).
	Maximum.	Minimum.	Mean.	Per square mile.	
October	72	10	16.4	0.872	1.01
November	85	11	26.6	1.41	1.57
December	30	10	18.5	.984	1.13
January	24	8	13.5	.718	.83
February	80	6	19.5	1.04	1.08
March	140	30	71.9	3.82	4.40
April	205	42	94.0	5.00	5.58
May	55	14	27.6	1.47	1.70
June	75	10	23.6	1.26	1.41
July	20	7.1	13.0	.691	.80
August	12	4.2	9.39	.499	58
September	120	5.7	22.9	1.22	1.36
The year	205	4.2	29.7	1.58	21.45

PRIEST BROOK NEAR WINCHENDON, MASS.

Location.—At highway bridge 3 miles above confluence of Priest Brook with Millers River and 3½ miles west of Winchendon, Worcester County.

Drainage area.—18.8 square miles.

Records available.—May 25, 1916, to September 30, 1917, and July 18 to September 30, 1918.

Gage.—Sloping staff on left bank 200 feet below highway bridge; read by R. D. Hutchinson.

Discharge measurements.—Made from highway bridge or by wading.

Channel and control.—Channel above the station is straight, with fairly uniform section and gravel bottom. Control formed by the foundation of an old dam 30 feet below the gage; practically permanent.

Extremes of discharge.—Maximum stage recorded during the period covered by records, 4.88 feet at 7 a. m. March 28 and 29, 1917 (discharge, 306 second-feet); minimum stage recorded during periods, 2.11 feet at 7 a. m. August 26, 1918 (discharge, 1.3 second-feet).

Regulation.—Flow not appreciably affected by regulation.

Accuracy.—Stage-discharge relation practically permanent. Rating curve well defined below 200 second-feet. Gage read to hundredths twice daily. Daily discharge ascertained by applying mean daily gage height to rating table. Records good.

Discharge measurements of Priest Brook near Winchendon, Mass., during the year ending Sept. 30, 1918.

Date.	Made by—	Gage height.	Discharge.	Date.	Made by—	Gage height.	Discharge.
		Feet.	*Sec.-ft.*			*Feet.*	*Sec.-ft.*
Oct. 13	M. R. Stackpole........	2.91	15.4	Aug. 20	J. W. Moulton........	2.18	1.6
July 18	A. N. Weeks...........	2.84	15.3				

Daily discharge, in second-feet, of Priest Brook near Winchendon, Mass., for the year ending Sept. 30, 1918.

Day.	July.	Aug.	Sept.	Day.	July.	Aug.	Sept.	Day.	July	Aug.	Sept.
1		2.5	2.4	11		2.5	1.6	21	4.0	1.5	35
2		2.1	2.4	12		2.5	1.5	22	3.4	1.5	25
3		2.0	2.0	13		2.0	3.6	23	3.2	1.4	20
4		1.9	1.8	14		2.2	2.6	24	2.8	1.4	20
5		2.0	1.6	15		4.8	2.0	25	2.8	1.4	21
6		4.6	1.5	16		2.7	2.0	26	3.2	1.3	31
7		1.9	1.3	17		2.0	2.0	27	2.5	1.5	165
8		3.2	1.4	18	13	1.7	2.6	28	2.2	1.5	123
9		2.1	1.8	19	7.3	1.7	7.9	29	40	1.8	78
10		2.2	1.8	20	4.6	1.6	20	30	16	2.0	60
								31	2.8	1.9	

Monthly discharge of Priest Brook near Winchendon, Mass., for the year ending Sept. 30, 1918.

[Drainage area, 18.8 square miles.]

Month.	Discharge in second-feet.				Run-off (depth in inches on drainage area).
	Maximum.	Minimum.	Mean.	Per square mile.	
July 18-31	40	2.2	7.70	0.410	0.21
August	4.8	1.3	2.11	.112	.13
September	165	1.3	21.4	1.14	1.27

EAST BRANCH OF TULLY RIVER NEAR ATHOL, MASS.

LOCATION.—At highway bridge half a mile below mouth of Lawrence Brook and 3½ miles north of Athol, Worcester County.

DRAINAGE AREA.—50.2 square miles.

RECORDS AVAILABLE.—June 13, 1916, to September 30, 1918.

GAGE.—Vertical staff on downstream side of right abutment; read by W. A. Thompson.

DISCHARGE MEASUREMENTS.—Made from highway bridge or by wading.

CHANNEL AND CONTROL.—Two channels under bridge, one channel above; about 200 feet below the gage channel is divided by an island, and the control sections are formed by rocks and boulders in the two channels, probably permanent.

EXTREMES OF DISCHARGE.—Maximum stage recorded during the year, 3.35 feet at 7 a. m. April 3 (discharge, 588 second-feet); minimum stage recorded, 0.24 foot at 7 a. m. August 29 (discharge, 2.5 second-feet).

1916–1918: Maximum stage recorded, 3.76 feet at 1 p. m. March 28, 1917 (discharge, 780 second-feet); minimum stage recorded, August 29, 1918.

ICE.—River freezes slightly along banks, and stage-discharge relation is affected for short periods.

DIVERSIONS.—About half a mile below the station water is diverted through a canal into Packard Pond. A discharge measurement July 19, 1918, showed a flow of 10.5 second-feet diverted through the canal. On August 28, canal was dry.

REGULATION.—Flow not seriously affected by regulation.

ACCURACY.—Stage-discharge relation practically permanent, except for short periods when affected by ice. Rating curve well defined below 300 second-feet. Gage read to hundredths twice daily, except from December 9 to March 31, when it was read once daily. Daily discharge ascertained by applying mean daily gage height to rating table and making corrections for effect of ice during winter. Records good.

Discharge measurements of East Branch of Tully River near Athol, Mass., during the year ending Sept. 30, 1918.

Date.	Made by—	Gage height.	Discharge.	Date.	Made by—	Gage height.	Discharge.
		Feet.	*Sec.-ft.*			*Feet.*	*Sec.-ft.*
Jan. 7	M. R. Stackpole	[a] 1.12	24.1	July 19	C. H. Pierce	1.31	44.3
Feb. 9	.....do	[a] .96	18.3	Aug. 28	H. W. Fear	.26	2.9

[a] Stage-discharge relation affected by ice.

Daily discharge, in second-feet, of East Branch of Tully River near Athol, Mass., for the year ending Sept. 30, 1918.

Day.	Oct.	Nov.	Dec.	Jan.	Feb.	Mar.	Apr.	May.	June.	July.	Aug.	Sept.
1	14	365	39	23	24	197	421	149	36	35	9.5	10
2	14	251	45	23	25	183	485	172	33	31	8.2	10
3	14	197	48	24	20	170	565	172	27	25	6.7	9.5
4	13	157	46	25	24	145	505	149	22	20	5.8	9.2
5	13	128	45	26	20	149	389	127	19	17	4.9	8.2
6	19	112	43	26	20	127	316	110	18	16	4.9	5.2
7	26	96	42	24	23	145	282	96	21	15	4.4	3.8
8	24	83	35	24	24	149	276	93	46	16	3.8	4.4
9	24	72	34	22	18	130	248	89	37	15	4.9	6.7
10	22	65	34	21	19	120	263	73	37	14	6.1	4.9
11	20	61	33	21	19	113	246	70	35	14	8.2	4.1
12	20	60	33	28	18	104	218	72	41	14	9.8	3.8
13	34	55	33	28	18	99	197	66	77	22	9.2	7.3
14	45	49	35	31	21	99	193	101	76	23	9.5	9.5
15	40	45	33	34	24	90	260	149	62	25	18	12
16	42	45	32	34	31	93	269	125	46	20	18	11
17	39	43	31	31	34	88	254	97	36	19	16	10
18	34	41	32	29	37	104	248	79	29	43	12	12
19	28	41	34	27	40	123	243	66	23	50	8.5	27
20	31	39	35	27	76	161	207	58	19	38	7.0	49
21	42	38	37	29	96	207	190	50	16	29	6.4	108
22	38	42	39	24	134	309	289	56	86	28	4.4	107
23	33	71	40	26	149	429	298	56	232	18	3.6	80
24	32	76	39	23	165	437	269	48	200	14	3.1	63
25	64	76	36	24	149	429	226	42	145	12	3.4	53
26	70	57	34	22	174	437	193	45	103	9.8	2.9	72
27	59	45	32	24	202	421	165	45	79	8.8	3.1	320
28	76	40	28	22	202	437	147	40	60	7.6	2.9	309
29	94	35	26	25	......	429	132	35	49	6.4	3.1	215
30	117	32	24	25	......	337	125	36	42	6.7	5.2	163
31	425	......	23	25	......	302	......	37	......	11	4.9	......

NOTE.—Stage-discharge relation affected by ice Dec. 9–20, and Dec. 26 to Feb. 19; daily discharge during these periods determined from gage heights corrected for effect of ice by means of two discharge measurements, observer's notes, and weather records.

Monthly discharge of East Branch of Tully River near Athol, Mass., for the year ending Sept. 30, 1918.

[Drainage area, 50.2 square miles.]

Month.	Discharge in second-feet.				Run-off (depth in inches on drainage area).
	Maximum.	Minimum.	Mean.	Per square mile.	
October	425	13	50.5	1.00	1.15
November	365	32	83.9	1.67	1.86
December	48	23	35.5	.707	.82
January	34	21	25.7	.512	.59
February	202	18	64.5	1.28	1.33
March	437	88	218	4.34	5.00
April	565	125	271	5.40	6.02
May	172	35	84.0	1.67	1.92
June	232	16	58.4	1.16	1.29
July	50	6.4	19.9	.396	.46
August	18	2.9	7.05	.140	.16
September	320	3.8	56.8	1.13	1.26
The year	565	2.9	81.0	1.61	21.86

MOSS BROOK AT WENDELL DEPOT, MASS.

LOCATION.—A quarter of a mile above confluence with Millers River and a quarter of a mile from Wendell Depot, Franklin County.

DRAINAGE AREA.—12.2 square miles.

RECORDS AVAILABLE.—June 7, 1916, to September 30, 1918. From June 4 to October 16, 1909, records were obtained at a station near the mouth of the stream, and from April 25 to August 27, 1910, at a weir a short distance below the present location.

GAGE.—Sloping staff on left bank; read by C. M. Porter.

DISCHARGE MEASUREMENTS.—Made by wading.

CHANNEL AND CONTROL.—Bed composed principally of ledge rock and boulders. Control practically permanent.

EXTREMES OF DISCHARGE.—Maximum stage recorded during the year, 2.87 feet at 9 a. m. March 24 (discharge, 106 second-feet); minimum stage recorded, 0.85 foot at 9 a. m. August 26 (discharge, 0.9 second-foot).

1916–1918: Maximum stage recorded, 3.52 feet at 12.45 p. m. March 28, 1917 (discharge, by extension of rating curve, about 187 second-feet); minimum stage recorded, 0.85 foot at 9 a. m. August 26, 1918 (discharge, 0.9 second-foot).

ICE.—Stage-discharge relation slightly affected by ice.

REGULATION.—Flow not affected by regulation.

ACCURACY.—Stage-discharge relation changed by ice action, February 12–13; two rating curves used during the year, well defined below 60 second-feet. Gage read to hundredths twice daily, except from December 13 to April 8, when it was read once daily. Daily discharge ascertained by applying mean daily gage height to rating table, and making corrections for effect of ice during the winter. Records good.

Discharge measurements of Moss Brook at Wendell Depot, Mass., during the year ending Sept. 30, 1918.

Date.	Made by—	Gage height.	Discharge.	Date.	Made by—	Gage height.	Discharge.
		Feet.	*Sec.-ft.*			*Feet.*	*Sec.-ft.*
Dec. 8	M. R. Stackpole	a1.33	6.7	Feb. 9	M. R. Stackpole	a1.34	6.2
Jan. 8	do	a1.32	4.8	Aug. 28	H. W. Fear	.87	1.0

a Stage-discharge relation affected by ice.

Daily discharge, in second-feet, of Moss Brook at Wendell Depot, Mass., for the year ending Sept. 30, 1918.

Day.	Oct.	Nov.	Dec.	Jan.	Feb.	Mar.	Apr.	May.	June.	July.	Aug.	Sept.
1	2.1	53	5.5	4.5	4.5	46	92	47	8.2	4.8	2.2	3.4
2	2.0	25	9.5	4.5	4.5	44	101	47	6.8	5.5	1.6	2.0
3	2.0	19	10	4.5	5	42	98	35	5.1	4.4	1.4	1.7
4	1.9	16	9.5	4.5	4.5	40	78	30	4.4	3.5	1.4	1.3
5	2.7	13	9	4.5	4.5	38	66	27	4.1	3.1	1.5	1.2
6	7.8	12	8.5	4.5	4.5	37	55	22	6.8	2.8	1.4	1.3
7	4.5	12	7.5	5	4	34	51	20	12	2.6	1.3	1.3
8	3.6	11	6.5	4.5	4.5	32	48	17	16	2.4	1.2	1.3
9	3.3	10	7	4	6	30	47	16	9.7	2.3	1.6	2.0
10	2.9	10	6.5	4	4.5	28	52	19	7.6	2.2	2.3	1.4
11	2.7	9.4	6.5	4	5	26	45	21	6	2.2	1.7	1.2
12	2.6	8.9	6	8.5	6	25	41	17	15	2.2	1.4	1.1
13	11	9.4	5.5	8	6.8	21	40	16	17	3.1	1.4	6
14	7.8	9.4	5	7	7.9	20	44	58	11	5.5	1.6	3.2
15	6.3	8.4	5.5	6.5	9.7	23	65	42	8.2	4.3	2.1	1.8
16	7.5	8.4	5	6	9.7	23	63	30	6	3.4	1.7	1.6
17	5.7	7.8	5	6	9.3	22	54	21	4.1	4.8	1.4	1.6
18	4.6	7.3	7	6	9.7	28	52	17	2.7	11	1.3	2.1
19	4.3	7.8	8	6	14	37	47	14	2.3	5.7	1.2	2.4
20	6.1	7.3	8	5.5	34	55	39	12	2.1	3.4	1.1	5.7
21	6.1	7.5	9	5	32	62	52	14	1.9	2.3	1.0	15
22	5.0	8.9	10	5	30	73	68	14	46	2.3	1.0	9
23	4.3	15	8	5	30	89	62	13	28	2.2	1.0	5.3
24	5.0	13	8	5	28	106	49	10	20	2.0	1.0	4.6
25	12	12	7.5	5	34	80	42	9	13	1.8	1.0	3.8
26	9.4	10	7	5	66	84	34	13	9	1.8	1.1	27
27	6.8	8.5	6.5	5	68	63	30	10	7.1	1.7	1.7	46
28	21	7	6	5	57	52	28	9	5.3	1.6	1.0	29
29	20	5.5	5	4.5	......	59	25	8.2	5.1	1.4	2.7	14
30	39	5	4	4.5	......	69	27	9.3	3.9	1.8	1.3	10
31	91	......	4	4.5	......	84	......	8.8	......	3.2	1.3	......

NOTE.—Stage-discharge relation affected by ice Nov. 26 to Feb. 12, and Mar. 7-11; daily discharge during these periods determined from gage heights corrected for effect of ice by means of three discharge measurements, observer's notes, and weather records.

Monthly discharge of Moss Brook at Wendell Depot, Mass., for the year ending Sept. 30, 1918.

[Drainage area, 12.2 square miles.]

Month.	Discharge in second-feet.				Run-off (depth in inches on drainage area).
	Maximum.	Minimum.	Mean.	Per square mile.	
October	91	1.9	10.0	0.820	0.95
November	53	5	11.9	.976	1.09
December	10	4	6.97	.571	.66
January	8.5	4	5.21	.427	.49
February	68	4	18	1.48	1.54
March	106	20	47.5	3.89	4.48
April	101	25	53.2	4.36	4.86
May	53	8.2	20.7	1.70	1.96
June	46	1.9	9.81	.804	.90
July	11	1.4	3.27	.265	.31
August	2.7	1.0	1.45	.118	.14
September	46	1.1	6.91	.566	.63
The year	106	1.0	16.2	1.33	18.01

DEERFIELD RIVER AT CHARLEMONT, MASS.

Location.—About 1 mile below village of Charlemont, Franklin County.

Drainage area.—362 square miles.

Records available.—June 19, 1913, to September 30, 1918.

Gages.—Friez water-stage recorder on left bank, referred to gage datum by a hook gage inside the well; an inclined staff gage is used for auxiliary readings.

Discharge measurements.—Made from cable or by wading.

Channel and control.—Bed covered with coarse gravel and boulders. Section fairly uniform. Control practically permanent.

Extremes of discharge.—Maximum open-water stage during year, from water-stage recorder, 9.25 feet at 9 a. m. March 22 (discharge, 15,300 second-feet); a stage of 11.75 feet was recorded at noon March 21, but the water was held back by an ice jam; minimum stage during year, from water-stage recorder, 1.40 feet at 7 a. m. July 7 (discharge, 32 second-feet).

1913–1918: Maximum stage recorded, 15.7 feet on July 8, 1915 (discharge, by extension of rating curve, about 45,000 second-feet); minimum stage recorded, 1.35 feet September 21 and November 3, 1914 (discharge, 23 second-feet).

Ice.—River usually frozen over during the greater part of the winter; ice jams occasionally form below the gage, causing several feet of backwater.

Regulation.—Flow during low and medium stages largely regulated by a storage reservoir at Somerset, Vt. Several power plants above the station cause diurnal fluctuation.

Accuracy.—Stage-discharge relation practically permanent except when affected by ice. Rating curve well defined. Operation of water-stage recorder satisfactory, except for short periods as shown in the footnote to the daily-discharge table. Daily discharge ascertained by use of discharge integrator. Records good.

Discharge measurements of Deerfield River at Charlemont, Mass., during the year ending Sept. 30, 1918.

Date.	Made by—	Gage height.	Discharge.	Date.	Made by—	Gage height.	Discharge.
		Feet.	*Sec.-ft.*			*Feet.*	*Sec.-ft.*
Jan. 11	M. R. Stackpole	[a] 4.56	430	July 16	A. N. Weeks	2.38	436
Feb. 12	do	[a] 4.54	309	Sept. 6	H. W. Fear	1.90	169
Mar. 18	do	[a] 5.23	868				

[a] Stage-discharge relation affected by ice.

Daily discharge, in second-feet, of Deerfield River at Charlemont, Mass., for the year ending Sept. 30, 1918.

Day.	Oct.	Nov.	Dec.	Jan.	Feb.	Mar.	Apr.	May.	June.	July.	Aug.	Sept.
1	420	1,420	460	280	370	1,250	4,800	2,400	410	170	370	180
2	360	980	265	280	310	1,050	6,500	1,840	200	360	355	140
3	360	640	640	400	75	780	5,400	1,200	325	225	250	320
4	340	500	560	400	135	780	3,150	970	230	91	126	270
5	320	580	560	370	310	720	2,150	610	174	170	450	225
6	375	480	440	60	500	960	1,540	740	180	114	480	210
7	460	440	440	400	370	1,350	1,900	600	440	46	440	190
8	255	350	260	540	370	1,050	2,200	530	770	174	400	61
9	220	420	100	460	370	720	3,000	540	325	205	1,000	200
10	260	220	440	440	220	640	3,000	560	405	260	350	225
11	340	205	560	460	50	720	1,740	1,460	340	290	142	255
12	180	510	640	400	135	640	1,300	980	405	250	240	275
13	460	460	720	75	310	720	1,040	830	750	335	186	340
14	420	405	880	310	260	720	1,000	2,960	590	140	186	270
15	345	430	500	560	370	640	1,700	1,740	410	425	300	100
16	620	470	440	720	310	540	2,200	1,140	168	340	250	205
17	420	300	500	640	220	500	2,700	820	260	275	240	190
18	325	85	560	440	310	780	3,300	700	240	325	225	172
19	245	270	640	135	370	880	2,300	430	220	225	180	300
20	225	305	720	260	4,200	1,250	1,720	570	215	190	240	315
21	310	290	640	260	3,600	3,000	2,300	650	230	60	255	600
22	235	410	310	310	1,850	4,450	4,850	590	1,360	178	295	700
23	285	480	75	640	1,250	3,950	3,350	580	1,200	240	275	290
24	465	380	310	640	960	2,750	2,800	540	830	230	260	255
25	1,960	215	135	560	960	2,750	1,800	360	550	245	138	340
26	940	410	340	440	1,600	2,450	1,400	310	365	280	455	3,500
27	580	450	310	75	1,600	1,700	1,200	650	220	210	320	1,100
28	640	450	340	100	1,250	1,320	1,200	590	165	79	300	650
29	810	250	340	370	......	1,500	1,450	520	140	270	280	475
30	5,400	480	50	440	......	2,250	2,000	405	86	250	270	420
31	3,950	......	240	440	......	3,150	......	630	......	330	200	......

NOTE.—Stage-discharge relation affected by ice from Dec. 3 to Mar. 21; daily discharge for this period determined from gage heights corrected for effect of ice by three discharge measurements, observer's notes and weather records, and comparison with records at New England Power Co.'s plant No. 4 at Shelburne Falls. Water-stage recorder not in operation Apr. 28 to May 1; Aug. 8-10, 28; and Sept. 27-28; discharge for these periods estimated by comparison with records at other stations.

Monthly discharge of Deerfield River at Charlemont, Mass., for the year ending Sept. 30, 1918.

[Drainage area, 362 square miles.]

Month.	Observed discharge (second-feet).			Gain or loss in storage at Somerset, Vt. (millions of cubic feet).	Discharge corrected for storage (second-feet).		Run-off (depth in inches on drainage area).
	Maximum.	Minimum.	Mean.		Mean.	Per square mile.	
October	5,400	180	727	+103	765	2.11	2.43
November	1,420	85	443	−166	379	1.05	1.17
December	880	50	433	−508	243	.671	.77
January	720	60	384	−446	217	.599	.69
February	4,200	50	808	− 55	785	2.17	2.26
March	4,450	500	1,480	+269	1,580	4.36	5.03
April	6,500	1,000	2,500	+620	2,740	7.57	8.45
May	2,950	310	885	+387	1,030	2.85	3.29
June	1,360	86	407	+176	475	1.31	1.46
July	425	46	225	−299	113	.312	.36
August	1,000	126	305	−536	105	.290	.33
September	3,500	61	426	0	426	1.18	1.32
The year	6,500	46	749	−455	735	2.03	27.56

NOTE.—The increase (+) or decrease (−) of water held in storage at Somerset, Vt., during the month has been computed by engineers of the Geological Survey from data of storage increase or decrease furnished by the company operating the reservoir.

WARE RIVER AT GIBBS CROSSING, MASS.

LOCATION.—Between highway and electric railway bridges at Gibbs Crossing, three-quarters of a mile above mouth of Beaver Brook and 3 miles below Ware, Hampshire County.

DRAINAGE AREA.—201 square miles.

RECORDS AVAILABLE.—August 20, 1912, to September 30, 1918.

GAGES.—Barrett & Lawrence water-stage recorder on the right bank referred to gage datum by a hook gage inside of well; an inclined staff gage is used for auxiliary readings.

DISCHARGE MEASUREMENTS.—Made from the electric railway bridge or by wading.

CHANNEL AND CONTROL.—Bed rough and subject to a growth of aquatic vegetation during summer. Control free from weeds and at ordinary stages well defined at a section near the gage; shifts occasionally; at high stages the control is probably at the dam at Thorndike, 4 miles below the gage.

EXTREMES OF DISCHARGE.—Maximum open-water stage during year, from water-stage recorder, 3.84 feet at 12 noon March 23 (discharge, 1,260 second-feet); a stage of 8.85 feet was recorded at 10 a. m. February 27, but the water was held back by an ice jam; minimum stage during year, from water-stage recorder, 1.38 feet at 4 a. m. July 29 (discharge, 21 second-feet).

1912–1918: Maximum open-water stage recorded, 5.9 feet on March 2, 1914 (discharge, 2,770 second-feet); minimum stage recorded, 1.20 feet on October 26, 1914 (discharge, 5 second-feet).

ICE.—River freezes over, and the stage-discharge relation is seriously affected by the ice; the large diurnal fluctuation in flow breaks up the ice and causes a variable backwater effect.

REGULATION.—Flow affected by operation of mills at Ware, which at low stages causes a large variation in discharge on days when the mills are in operation and a low discharge on Sundays and holidays.

ACCURACY.—Slight changes in the stage-discharge relation occurred during the year. Rating curve fairly well defined. The operation of water-stage recorder was satisfactory, except for short periods as shown in footnote to daily-discharge table. Daily discharge ascertained by use of discharge integrator. Records good.

Discharge measurements of Ware River at Gibbs Crossing, Mass., during the year ending Sept. 30, 1918.

Date.	Made by—	Gage height.	Discharge.	Date.	Made by—	Gage height.	Discharge.
		Feet.	*Sec.-ft.*			*Feet.*	*Sec.-ft.*
Nov. 8	H. W. Fear	2.42	256	Mar. 15	M. R. Stackpole	3.10	528
27	....do	2.24	196	June 6	A. N. Weeks	2.22	168
Dec. 19	....do	a 3.55	198	July 6	....do	1.70	51
Jan. 29	....do	a 3.61	142	7	....do	1.47	29.4
Feb. 27	....do	a 8.80	1,320				

a Stage-discharge relation affected by ice.

Daily discharge, in second-feet, of Ware River at Gibbs Crossing, Mass., for the year ending Sept. 30, 1918.

Day.	Oct.	Nov.	Dec.	Jan.	Feb.	Mar.	Apr.	May.	June.	July.	Aug.	Sept.
1	56	455	184	29	76	890	720	560	122	156	82	22
2	61	310	83	66	52	700	710	670	104	128	74	32
3	60	240	245	86	40	490	780	580	178	128	87	60
4	55	225	280	96	64	380	780	480	110	43	21	46
5	53	215	205	110	80	540	720	400	148	136	112	70
6	51	200	198	39	100	980	610	415	124	84	70	120
7	32	190	210	70	90	1,000	500	380	164	37	83	72
8	56	174	164	105	90	800	510	375	148	156	65	23
9	90	156	100	80	48	650	490	345	168	82	70	64
10	100	120	170	86	37	600	470	325	205	61	52	64
11	82	96	200	155	88	540	440	275	190	110	60	75
12	74	178	210	155	140	510	440	245	210	132	162	80
13	55	150	190	105	210	670	440	280	205	80	96	73
14	35	156	180	260	110	620	425	295	190	26	88	67
15	52	172	125	260	175	550	680	295	160	138	64	30
16	172	148	82	195	280	480	630	290	150	91	132	63
17	130	112	190	165	230	530	580	265	188	140	83	100
18	90	54	145	37	215	800	545	180	130	120	21	124
19	70	100	135	45	190	790	550	170	140	142	102	118
20	55	126	94	56	380	720	480	235	136	92	94	110
21	34	134	115	76	1,000	850	470	230	79	20	60	480
22	80	132	88	165	790	990	790	205	300	124	60	370
23	94	158	52	220	540	1,100	760	205	490	57	67	275
24	124	164	135	115	380	1,120	700	170	400	66	50	260
25	140	130	41	120	300	1,080	600	128	280	90	18	180
26	170	198	56	50	790	1,000	490	108	250	61	35	200
27	162	152	86	37	1,110	880	445	205	200	41	39	790
28	156	122	120	80	540	760	385	178	170	16	39	830
29	230	67	94	120	......	700	430	184	112	60	68	445
30	310	116	35	135	......	670	405	87	89	64	50	360
31	420	......	115	76	......	660	......	210	......	160	40	......

NOTE.—Stage-discharge relation affected by ice from Dec. 10 to Mar. 5; discharge for this period determined from gage heights corrected for effect of ice by means of three discharge measurements, observer's notes, and weather records. Daily discharge Oct. 19-20, Nov. 5-7, and Dec. 1-2, estimated by means of hydrograph comparisons with records in adjacent drainage basins.

Monthly discharge of Ware River at Gibbs Crossing, Mass., for the year ending Sept. 30, 1918.

[Drainage area, 201 square miles.]

Month.	Discharge in second-feet.				Run-off (depth in inches on drainage area).
	Maximum.	Minimum.	Mean.	Per square mile.	
October	420	32	108	0.537	0.62
November	455	54	165	.821	.92
December	280	35	140	.697	.80
January	260	29	109	.542	.62
February	1,110	37	291	1.45	1.51
March	1,120	380	743	3.70	4.27
April	790	385	566	2.82	3.15
May	670	87	289	1.44	1.66
June	490	79	185	.920	1.03
July	160	16	91.9	.457	.53
August	162	18	69.2	.344	.40
September	830	22	187	.930	1.04
The year	1,120	16	245	1.22	16.55

SWIFT RIVER AT WEST WARE, MASS.

Location.—About 1,000 feet below old wooden dam opposite West Ware station of Boston & Albany Railroad, 6 miles downstream from Enfield, Franklin County, and 3 miles below confluence of East and West branches of Swift River.

Drainage area.—186 square miles.

Records available.—July 15, 1910, to September 30, 1918.

Gages.—Barrett & Lawrence water-stage recorder on left bank, referred to gage datum by means of a hook gage inside the well; an inclined staff gage is used for auxiliary readings. Prior to August 25, 1912, a chain gage on footbridge 600 feet upstream from the present station was used.

Discharge measurements.—Made from cable or by wading.

Channel and control.—Bed consists of gravel and alluvial deposits; some aquatic vegetation in channel during summer. Control subject to slight changes at high-water periods; at high stages the control is probably at the dam at Bondsville, 4 miles below the gage.

Extremes of discharge.—Maximum open-water stage during year, from water-stage recorder, 5.86 feet at noon March 25 (discharge, 1,100 second-feet); a stage of 7.2 feet was recorded at 8 a. m. March 2, but the water was held back by an ice jam; minimum stage during year, from water-stage recorder, 1.73 feet at 8 p. m. August 27 (discharge, 53 second-feet).

1910–1918: Maximum stage recorded, 9.1 feet on February 26, 1915 (discharge, by extension of rating curve, 2,240 second-feet); minimum stage recorded, 1.36 feet on September 22, 1914 (discharge, 22 second-feet).

Ice.—River usually freezes over, and the stage-discharge relation is somewhat affected by the ice.

Regulation.—Operation of mills at Enfield, 6 miles above the station, affects distribution of flow at low and medium stages, but has only a slight effect when the mean daily discharge is over 200 second-feet.

Accuracy.—Stage-discharge relation unchanged during the year except when affected by ice. Rating curve fairly well defined below 1,200 second-feet. Daily discharge ascertained by applying to rating table mean daily gage height determined by inspecting recorder graph. Records only fair during the period affected by ice, but are good for rest of year.

Discharge measurements of Swift River at West Ware, Mass., during the year ending Sept. 30, 1918.

Date.	Made by—	Gage height.	Discharge.	Date.	Made by—	Gage height.	Discharge.
		Feet.	*Sec.-ft.*			*Feet.*	*Sec.-ft.*
Dec. 21	H. W. Fear	a 2.26	98	May 9	H. W. Fear	3.22	325
Jan. 31	....do	a 3.42	101	June 5	A. N. Weeks	2.35	138
Mar. 6	....do	a 6.17	638	July 5	O. W. Hartwell	2.36	139

a Stage-discharge relation affected by ice.

Daily discharge, in second-feet, of Swift River at West Ware, Mass., for the year ending Sept. 30, 1918.

Day.	Oct.	Nov.	Dec.	Jan.	Feb.	Mar.	Apr.	May.	June.	July.	Aug.	Sept.
1	97	478	146	70	100	790	715	491	124	134	80	60
2	97	491	184	70	90	820	745	491	123	120	84	70
3	91	440	205	70	86	760	790	504	146	118	76	67
4	88	385	219	78	84	700	825	478	135	84	69	64
5	84	281	200	84	84	670	790	416	139	113	87	64
6	84	234	174	78	84	640	745	380	130	120	90	68
7	83	198	174	84	90	640	685	358	153	113	79	70
8	76	174	174	94	90	610	640	349	168	139	77	65
9	91	154	137	98	90	580	612	312	146	123	81	80
10	94	146	150	110	98	570	584	237	156	103	81	80
11	90	137	155	110	98	560	570	272	158	92	68	81
12	90	137	160	115	90	560	556	270	174	104	92	71
13	110	132	120	130	84	570	556	261	192	92	94	79
14	98	139	110	150	105	580	543	256	200	79	79	75
15	98	130	125	135	145	600	543	270	198	97	83	69
16	104	129	130	140	240	610	584	277	178	97	83	77
17	106	124	140	130	230	610	612	274	158	101	74	81
18	121	115	130	130	200	626	626	256	147	103	75	77
19	113	127	115	120	260	626	612	241	146	103	76	84
20	109	116	120	120	340	670	598	223	137	100	79	97
21	113	123	120	120	430	730	570	209	124	90	74	115
22	116	129	125	130	530	825	584	202	205	101	71	95
23	115	154	130	135	560	965	640	198	358	88	71	123
24	112	174	130	130	580	1,080	670	188	428	87	71	116
25	129	202	115	120	580	1,080	670	178	392	83	63	118
26	142	200	115	120	500	1,040	612	174	320	81	59	151
27	140	190	130	110	730	1,000	556	174	243	75	55	351
28	156	188	115	110	760	860	517	154	188	70	60	478
29	174	174	105	100	....	760	491	154	160	75	70	428
30	200	151	90	100	....	670	491	144	146	80	70	347
31	336	....	78	100	....	600	....	134	....	81	65	....

NOTE.—Stage-discharge relation affected by ice from Dec. 10 to Mar. 10; discharge for this period determined from gage heights corrected for effect of ice by means of three discharge measurements, observer's notes, and weather records. Pipe to gage well partly clogged Apr. 23 to June 2; gage heights determined by comparison with readings on inclined staff. Daily discharge June 26, July 6, 27–30, Aug. 28–31, and Sept. 1, 7–8, estimated by hydrograph comparisons with records in adjacent drainage basins.

Monthly discharge of Swift River at West Ware, Mass., for the year ending Sept. 30, 1918.

[Drainage area, 186 square miles.]

Month.	Discharge in second-feet.				Run-off (depth in inches on drainage area).
	Maximum.	Minimum.	Mean.	Per square mile.	
October	336	76	118	0.634	0.73
November	491	115	198	1.06	1.18
December	219	78	139	.747	.86
January	150	70	109	.586	.68
February	760	84	263	1.41	1.47
March	1,080	560	723	3.89	4.48
April	825	491	624	3.35	3.74
May	504	134	275	1.48	1.71
June	428	123	189	1.02	1.14
July	139	70	98.3	.528	.61
August	94	55	75.4	.405	.47
September	478	60	127	.683	.76
The year	1,080	55	244	1.31	17.83

QUABOAG RIVER AT WEST BRIMFIELD, MASS.

Location.—At two-span highway bridge in Hampden County near West Brimfield station of Boston & Albany Railroad, one-third of a mile above mouth of Blodgett Mill Brook.

Drainage area.—150 square miles.

Records available.—August 23, 1909, to September 30, 1918.

Gages.—Stevens continuous water-stage recorder at downstream end of center pier of bridge, referred to gage datum by means of a hook gage inside of well; a vertical staff is used for auxiliary readings. Prior to August 19, 1912, a vertical staff on upstream side of right abutment of bridge, at same datum as present gage, was used.

Discharge measurements.—Made from highway bridge or by wading near bridge.

Channel and control.—Stream bed covered with boulders, gravel, and alluvial deposits. Slight shifts in control have occurred at infrequent intervals.

Extremes of discharge.—Maximum open-water stage during year, from water-stage recorder, 3.59 feet at 11.30 a. m. March 14 and 10 a. m. March 22 (discharge, 756 second-feet); a stage of 6.07 feet was recorded at 9 a. m. March 1, but the water was held back by an ice jam; minimum stage during year from water-stage recorder, 1.51 feet at 11.15 a. m. September 15 (discharge, 5.5 second-feet).

1909–1918: Maximum stage recorded, 4.9 feet on March 1, 1910 (discharge, 1,660 second-feet); minimum stage recorded, 1.40 feet on September 17 and 18, 1910 (discharge, 2.5 second-feet).

Ice.—River freezes over and the stage-discharge relation is affected by the ice; the diurnal fluctuation in flow breaks up the ice and causes a variable backwater effect.

Regulation.—Flow affected by operation of power plants at West Warren, 3 miles above station, which at low stages causes a large variation in discharge on days when the mills are in operation and a low discharge on Sundays and holidays.

Accuracy.—A slight change in stage-discharge relation occurred during the year. Rating curves well defined. Operation of water-stage recorder satisfactory except for short periods as shown in the footnote to daily-discharge table. Daily discharge ascertained by discharge integrator. Records good, except for periods affected by ice, for which they are fair.

Discharge measurements of Quaboag River at West Brimfield, Mass., during the year ending Sept. 30, 1918.

Date.	Made by—	Gage height.	Discharge.	Date.	Made by—	Gage height.	Discharge.
		Feet.	*Sec.-ft.*			*Feet.*	*Sec.-ft.*
Nov. 9	H. W. Fear	2.28	129	Mar. 15	M. R. Stackpole	3.26	555
Dec. 20	.....do	[a]3.12	166	June 6	A. N. Weeks	[b]2.46	143
Jan. 8	.....do	[a]3.36	70	July 7	.....do	2.17	86
30	.....do	[a]3.70	91	Sept. 10	H. W. Fear	2.19	90
Feb. 26	.....do	[a]5.73	975				

[a] Stage-discharge relation affected by ice. [b] Stage-discharge relation affected by débris.

Daily discharge, in second-feet, of Quaboag River at West Brimfield, Mass., for the year ending Sept. 30, 1918.

Day.	Oct.	Nov.	Dec.	Jan.	Feb.	Mar.	Apr.	May.	June.	July.	Aug.	Sept.
1	73	210	75	50	55	830	540	370	100	102	87	45
2	58	200	110	55	55	560	520	380	98	87	79	50
3	56	186	135	55	50	345	500	375	104	90	63	48
4	60	166	165	65	55	275	510	355	77	84	74	46
5	58	160	135	55	65	590	470	340	76	100	90	46
6	50	160	110	50	65	830	460	305	91	105	75	48
7	50	150	110	55	55	930	450	285	102	90	75	52
8	77	132	85	55	50	790	400	270	100	97	73	50
9	67	144	65	50	75	690	390	255	102	76	70	55
10	62	114	85	50	75	630	385	250	114	73	53	55
11	62	120	95	50	55	590	380	225	110	72	66	45
12	59	128	110	110	55	560	405	225	136	70	91	47
13	64	118	95	110	50	720	380	210	182	54	72	55
14	69	116	65	150	65	650	370	150	172	64	74	44
15	92	120	55	120	95	550	355	150	160	91	90	20
16	80	114	65	135	235	430	345	170	154	71	72	61
17	72	96	75	165	165	540	335	180	148	64	62	46
18	70	85	85	135	150	560	355	130	128	90	61	53
19	80	104	85	135	150	580	330	150	114	96	72	60
20	66	100	75	120	420	580	320	140	100	86	57	70
21	56	110	75	150	660	610	345	120	100	66	50	85
22	90	114	55	120	530	630	415	134	225	94	48	70
23	72	146	50	95	365	620	395	150	220	73	47	83
24	90	122	85	75	275	620	385	130	182	88	45	90
25	144	100	50	75	235	630	365	116	154	91	42	84
26	126	100	75	65	530	620	355	130	144	85	52	114
27	91	100	95	50	760	580	340	140	130	65	45	198
28	120	85	75	65	500	590	325	130	118	65	48	152
29	116	80	85	65	....	570	300	114	110	81	52	140
30	190	75	75	55	....	550	290	132	106	66	52	142
31	250	....	65	55	....	550	....	136	....	75	47	....

NOTE.—Stage-discharge relation affected by ice Dec. 11 to Mar. 6; daily discharge for this period determined from gage heights corrected for effect of ice by means of four discharge measurements, observer's notes, and weather records. Stage-discharge relation slightly affected by débris from about June 1 to July 7; correction estimated from results of one discharge measurement. Daily discharge Nov. 26 to Dec. 10, Aug. 22-31, and Sept. 1-10, 19-22, estimated by hydrograph comparisons with records in adjacent drainage basins.

Monthly discharge of Quaboag River at West Brimfield, Mass., for the year ending Sept. 30, 1918.

[Drainage area, 150 square miles.]

Month.	Discharge in second-feet.				Run-off (depth in inches on drainage area).
	Maximum.	Minimum.	Mean.	Per square mile.	
October	250	50	86.1	0.574	0.66
November	210	75	125	.833	.93
December	165	50	86.0	.573	.66
January	165	50	85.2	.568	.65
February	760	50	211	1.41	1.47
March	930	275	606	4.04	4.66
April	540	290	390	2.60	2.90
May	380	114	205	1.37	1.58
June	225	76	129	.860	.96
July	105	54	81.0	.540	.62
August	91	42	64.0	.427	.49
September	198	20	71.8	.479	.53
The year	930	20	178	1.19	16.11

WESTFIELD RIVER AT KNIGHTVILLE, MASS.

Location.—At single-span steel highway bridge known locally as Pitcher Bridget in Knightville, Hampshire County, 1 mile north of outlet of Norwich Lake and 3 miles above confluence with Middle Branch of Westfield River.

Drainage area.—162 square miles.

Records available.—August 26, 1909, to September 30, 1918.

Gage.—Chain attached to downstream side of highway bridge; read by J. A. Burr.

Discharge measurements.—Made from highway bridge or by wading.

Channel and control.—Bed consists of boulders and ledge rock; control fairly permanent.

Extremes of discharge.—Maximum open-water stage recorded during yeare 4.61 feet at 6 p. m. April 2 (discharge, 1,880 second-feet); a stage of 6.5 feet was recorded at 4.30 p. m. February 20, but the water was held back by an ice jam; minimum stage recorded, 0.70 foot at 7 a. m. August 26 (discharge, 15 second-feet).

1909–1918: Maximum open-water stage recorded, 8.9 feet on March 27, 1913 (discharge, by extension of rating curve, about 5,100 second-feet); a gage height of 9.4 feet was recorded at 9.15 a. m. January 22, 1910, but channel was probably obstructed by ice at that time; minimum stage recorded, 0.60 foot on August 10, 1913 (discharge, 4 second-feet).

Ice.—Ice usually forms in the river early in the winter and seriously affects the stage-discharge relation.

Regulation.—Flow not seriously affected by regulation.

Accuracy.—The stage-discharge relation changed slightly during high water of April 1–3; individual discharge measurements have at times appeared erratic, the rough and irregular channel causing difficulty in securing accurate discharge measurements. Rating curve fairly well defined below 2,500 second-feet. Gage read to hundredths twice daily. Daily discharge ascertained by applying daily gage height to rating table and making corrections for effect of ice during winter. Records good.

Discharge measurements of Westfield River at Knightville, Mass., during the year ending Sept. 30, 1918.

Date.	Made by—	Gage height.	Discharge.	Date.	Made by—	Gage height.	Discharge.
		Feet.	*Sec.-ft.*			*Feet.*	*Sec.-ft.*
Dec. 22	H. W. Fear	[a] 2.35	83	Mar 16	M. R. Stackpole	[a] 3.60	309
Feb. 2	.....do	[a] 2.65	52	July 11[b]	A. N. Weeks	1.16	50
Mar. 1	.....do	[a] 5.90	984				

[a] Stage-discharge relation affected by ice. [b] Results uncertain.

Daily discharge, in second-feet, of Westfield River at Knightville, Mass., for the year ending Sept. 30, 1918.

Day.	Oct.	Nov.	Dec.	Jan.	Feb.	Mar.	Apr.	May.	June.	July.	Aug.	Sept.
1	21	435	113	56	56	600	1,440	512	153	52	31	49
2	21	285	150	56	45	540	1,780	655	130	115	28	52
3	21	215	157	64	45	540	1,440	485	113	84	25	36
4	20	167	143	64	40	440	1,200	350	92	68	25	28
5	21	152	134	64	35	490	910	310	82	55	29	24
6	57	143	125	50	27	660	715	292	84	49	28	21
7	63	123	105	70	31	600	715	275	130	60	28	20
8	38	119	86	70	27	540	655	240	156	50	29	20
9	35	117	86	80	27	540	780	225	108	44	29	28
10	33	113	96	70	27	600	655	202	97	100	28	24
11	29	109	84	70	35	540	595	370	87	45	51	23
12	30	105	64	145	27	490	512	310	163	50	49	23
13	92	96	60	170	86	490	485	225	210	61	37	27
14	98	91	105	170	145	390	540	1,050	141	85	34	35
15	77	87	105	170	145	350	780	485	93	139	42	35
16	71	94	96	170	170	300	655	350	77	92	31	34
17	68	94	86	145	145	520	625	275	68	67	27	25
18	58	92	86	145	170	1,050	780	225	64	106	23	29
19	55	85	80	125	145	1,200	568	205	63	79	21	64
20	50	81	80	125	900	1,350	460	173	56	67	20	82
21	47	91	86	145	1,350	1,690	568	153	48	59	19	175
22	45	172	80	125	980	1,690	1,360	183	540	49	19	146
23	45	345	80	125	660	1,600	845	199	460	40	19	92
24	105	200	86	145	540	1,280	780	163	188	38	19	67
25	845	115	80	125	350	1,280	540	148	136	34	17	59
26	265	94	70	105	660	1,120	435	210	109	32	16	512
27	139	94	56	105	1,100	845	390	275	84	27	18	910
28	125	94	56	105	660	780	350	188	68	25	17	258
29	192	87	70	86	....	845	330	130	48	28	22	158
30	910	87	80	64	....	980	485	136	39	42	42	113
31	1,200	....	64	64	....	1,200	....	163	....	49	34	....

NOTE.—Stage-discharge relation affected by ice Dec. 7 to Mar. 20; discharge for this period determined from gage heights corrected for effect of ice by means of four discharge measurements, observer's notes, and weather records.

Monthly discharge of Westfield River at Knightville, Mass., for the year ending Sept. 30, 1918.

[Drainage area, 162 square miles.]

Month.	Discharge in second-feet.				Run-off (depth in inches on drainage area).
	Maximum.	Minimum.	Mean.	Per square mile.	
October	1,200	20	157	0.969	1.12
November	435	81	139	.858	.96
December	157	56	91.9	.567	.65
January	170	50	106	.654	.75
February	1,350	27	308	1.90	1.98
March	1,690	300	824	5.09	5.87
April	1,780	330	746	4.60	5.13
May	1,050	130	296	1.83	2.11
June	540	39	130	.802	.90
July	139	27	61.0	.377	.43
August	51	16	27.6	.170	.20
September	910	20	106	.654	.73
The year	1,780	16	248	1.53	20.83

WESTFIELD RIVER NEAR WESTFIELD, MASS.

Location.—At Trap Rock crossing, 3 miles east of Westfield, Hampden County, 1 mile below mouth of Big Brook, and 2 miles below mouth of Westfield Little River.

Drainage area.—496 square miles.

Records available.—June 27, 1914, to September 30, 1918.

Gages.—Stevens continuous water-stage recorder on right bank, referred to gage datum by means of a hook gage inside the well; an inclined staff gage is used for auxiliary readings.

Discharge measurements.—Made from cable or by wading.

Channel and control.—Bed covered with gravel and alluvial deposits. Riffle of boulders about 200 feet below gage forms control at low and medium stages; at high stages control is probably formed by crest of storage dam at Mittineaugue 3 miles below the station.

Extremes of discharge.—Maximum stage during year, from water-stage recorder, 11.10 feet at 11 p. m. October 30 (discharge, 7,900 second-feet); minimum stage during year, from water-stage recorder, 3.18 feet at 9 p. m. August 24 (discharge, 88 second-feet).

1914–1918: Maximum stage recorded, 17.4 feet on August 4, 1915 (discharge, by extension of rating curve, about 17,400 second-feet); minimum stage recorded, 3.02 feet on September 24, 1914 (discharge, 46 second-feet).

Ice.—Stage-discharge relation affected by ice for short periods during the winter.

Diversions.—Water is diverted from Westfield Little River and carried to Springfield for municipal use.

Regulation.—Operating of several power plants above the station causes some diurnal fluctuation of flow; the nearest dam is at Westfield.

Accuracy.—Stage-discharge relation practically permanent except when affected by ice. Rating curve well defined below 7,500 second-feet. Operation of water-stage recorder satisfactory except for short periods as shown in the footnote to the daily-discharge table. Daily discharge ascertained by discharge integrator. Records good.

Discharge measurements of Westfield River near Westfield, Mass., during the year ending Sept. 30, 1918.

Date.	Made by—	Gage height.	Discharge.	Date.	Made by—	Gage height.	Discharge.
		Feet.	*Sec.-ft.*			*Feet.*	*Sec.-ft.*
Nov. 9	H. W. Fear	4.18	461	Feb. 28	H. W. Fear	6.22	1,900
Dec. 20	.....do	3.75	285	July 9	O. W. Hartwell	3.80	288
Jan. 7	.....do	a 3.51	153	10	.....do	3.53	190
Feb. 1	.....do	3.72	275				

a Stage-discharge relation affected by ice.

Daily discharge, in second-feet, of Westfield River near Westfield, Mass., for the year ending Sept. 30, 1918.

Day.	Oct.	Nov.	Dec.	Jan.	Feb.	Mar.	Apr.	May.	June.	July.	Aug.	Sept.
1	168	1,480	410	210	250	2,750	3,850	1,500	450	200	160	192
2	136	970	430	210	290	2,800	4,900	1,800	340	235	150	148
3	132	820	330	220	260	2,100	3,750	1,850	390	315	145	174
4	150	610	385	225	250	1,600	3,100	1,000	340	290	140	205
5	140	590	385	225	270	1,500	2,400	880	230	205	166	184
6	188	500	305	220	270	1,900	2,000	870	280	275	200	180
7	205	465	310	210	250	2,300	1,750	860	340	250	205	180
8	245	400	220	210	250	2,050	1,500	850	610	180	176	164
9	285	415	315	230	250	1,700	1,500	780	385	260	160	192
10	210	440	250	175	230	1,740	1,900	670	370	250	160	160
11	195	310	290	230	230	1,480	1,480	790	300	220	230	150
12	210	340	250	430	325	1,280	1,280	760	330	215	215	158
13	260	370	260	400	265	1,460	1,200	720	660	240	186	168
14	290	360	250	580	280	1,700	1,260	1,910	530	280	200	160
15	335	400	315	560	450	1,480	2,050	1,760	395	345	220	150
16	360	370	250	530	560	1,360	1,700	1,120	300	415	170	158
17	220	290	230	500	590	1,300	1,520	870	255	365	200	200
18	235	300	290	480	620	1,980	1,540	730	245	385	210	215
19	265	350	290	450	560	2,150	1,520	610	225	340	190	210
20	250	360	290	430	2,350	3,000	1,240	620	225	325	132	285
21	200	245	300	430	4,050	3,800	1,240	530	170	360	126	550
22	170	385	300	440	2,700	4,650	3,500	570	690	220	122	620
23	205	565	270	420	2,350	4,450	2,350	600	1,220	250	120	475
24	470	600	270	430	1,640	3,100	1,880	550	660	192	110	345
25	1,920	445	290	420	1,300	3,000	1,500	450	490	176	130	300
26	900	400	335	400		2,900	1,250	470	395	190	130	1,350
27	530	345	330	360	2,350	2,150	1,100	900	320	142	130	1,700
28	770	260	300	350	1,900	1,850	1,020	600	285	140	124	900
29	755	225	250	330	....	2,150	960	420	240	155	134	600
30	2,550	275	250	310	....	2,500	1,300	440	230	195	156	440
31	3,600	....	230	290	....	3,100	....	480	....	240	132	....

NOTE.—Stage-discharge relation affected by ice Jan. 7-14 and Feb. 5-7; corrections for these periods based on one discharge measurement and comparison with records at Knightville. Water-stage recorder not operating satisfactorily Dec. 28-31; Jan. 1-5, 16-31; Mar. 15-16, 28-30; Apr. 2-6, 29-30; May 1-6, 27-31; June 1; July 29-31; Aug. 1-3; Sept. 26-30; and discharge estimated by hydrograph comparison with records at Knightville.

Monthly discharge of Westfield River near Westfield, Mass., for the year ending Sept. 30, 1918.

[Drainage area, 496 square miles.]

Month.	Observed discharge (second-feet).			Diversion from Westfield Little River (millions of gallons).	Total discharge (second-feet).		Run-off (depth in inches on drainage area).
	Maximum.	Minimum.	Mean.		Mean.	Per square mile.	
October	3,600	132	534	397.9	554	1.12	1.29
November	1,480	225	463	393.3	483	.974	1.09
December	430	220	296	396.2	316	.637	.73
January	580	175	352	449.8	374	.754	.87
February	4,050	230	1,000	411.6	1,020	2 06	2.14
March	4,650	1,280	2,300	436.8	2,320	4.68	5.40
April	4,900	960	1,920	400.8	1,940	3.91	4.36
May	1,910	420	854	431.7	876	1.77	2.04
June	1,220	170	397	423.0	419	.845	.94
July	415	140	253	429.9	274	.552	.64
August	230	110	162	429.1	183	.369	.43
September	1,700	148	364	395.3	384	.774	.86
The year	4,900	110	738	4,997.4	759	1.53	20.79

NOTE.—Effect of storage in Borden Brook reservoir not taken into account in computing the total discharge.

MIDDLE BRANCH OF WESTFIELD RIVER AT GOSS HEIGHTS, MASS.

Location.—At highway bridge in Goss Heights, Hampshire County, 1½ miles above village of Huntington and half a mile above confluence of Middle and North branches of Westfield River.

Drainage area.—53 square miles.

Records available.—July 14, 1910, to September 30, 1918.

Gages.—Gurley 7-day water-stage recorder on upstream side of bridge abutment on right bank, referred to gage datum by means of a hook gage inside of well; an inclined staff is used for auxiliary readings. Prior to September 8, 1912, a chain gage on upstream side of bridge was used.

Discharge measurements.—Made from highway bridge or by wading.

Channel and control.—Bed covered with coarse gravel and boulders. A shift in control has occurred at various times.

Extremes of discharge.—Maximum open-water stage during year, from water-stage recorder, 3.65 feet at 9 p. m. March 22 (discharge, 1,220 second-feet); a stage of 5.54 feet was recorded at 7 p. m. March 6, but the water was held back by an ice jam; minimum stage during year, from water-stage recorder, 0.76 foot at 2 a. m. August 18 (discharge, 4.8 second-feet).

1910–1918: Maximum open-water stage recorded, 7.33 feet at 9 a. m., July 8, 1915 (discharge, by extension of rating curve, 4,500 second-feet); a gage height of 7.7 feet was recorded February 26, 1916, but channel was obstructed by ice at that time; minimum stage recorded 0.70 foot on October 26–27, 1914 (discharge practically zero flow).

Ice.—River usually frozen over during the greater part of the winter; ice jams occasionally form below the gage, causing several feet of backwater.

Regulation.—Flow somewhat affected at times by operation of small power plant about 2 miles above station.

Accuracy.—Stage-discharge relation unchanged during the year except when affected by ice (December to March). Rating curve fairly well defined below 1,000 second-feet. Daily discharge ascertained by applying to rating table mean daily gage height determined by inspecting recorder graph, except for periods as noted in footnote to daily-discharge table. Open-water records good; winter records fair.

Discharge measurements of Middle Branch of Westfield River at Goss Heights, Mass., during the year ending Sept. 30, 1918.

Date.	Made by—	Gage height.	Discharge.	Date.	Made by—	Gage height.	Discharge.
		Feet.	*Sec.-ft.*			*Feet.*	*Sec.-ft.*
Nov. 28	H. W. Fear	[a]1.09	19.6	Mar. 16	M. R. Stackpole	[a]3.19	169
Dec. 22	do	[a]1.80	27.4	Apr. 16	O. W. Hartwell	1.81	193
Feb. 2	do	[a]2.24	18.4	July 10	A. N. Weeks	.89	11.7

[a] Stage-discharge relation affected by ice.

Daily discharge, in second-feet, of Middle Branch of Westfield River at Goss Heights, Mass., for the year ending Sept. 30, 1918.

Day.	Oct.	Nov.	Dec.	Jan.	Feb.	Mar.	Apr.	May.	June.	July.	Aug.	Sept.
1	7.5	81	26	7	19	400	592	231	33	17	12	9.0
2	7.0	48	38	6	18	305	705	186	26	24	11	12
3	7.0	38	28	8	18	180	510	126	20	22	10	7.0
4	7.0	34	26	8	14	180	350	112	18	18	10	6.0
5	7.0	26	23	8	11	115	231	90	17	20	10	6.0
6	9.5	24	20	7	8	240	175	86	17	18	12	6.5
7	12	22	16	11	11	240	175	95	29	20	11	6.5
8	9.0	20	18	11	8	165	165	79	32	17	11	6.5
9	8.5	19	20	16	6	150	200	68	23	14	11	7.5
10	7.0	20	20	12	8	150	219	60	19	13	11	7.0
11	6.5	20	21	14	11	150	132	104	17	10	11	6.5
12	8.5	18	12	26	6	135	109	84	35	11	11	7.0
13	18	18	14	40	18	135	95	68	44	17	10	6.5
14	12	19	18	44	37	135	165	400	27	20	10	8.0
15	9.5	18	20	44	50	86	240	182	20	28	16	8.5
16	10	16	26	44	68	165	189	101	17	20	10	8.0
17	10	17	24	35	50	240	165	72	14	20	6.0	7.0
18	10	17	21	34	68	400	193	61	13	22	5.0	8.0
19	9	16	23	32	50	693	148	54	12	20	5.5	12
20	8	14	24	28	260	765	112	44	10	18	5.5	22
21	7	14	23	35	620	885	482	38	10	16	6.0	56
22	7	20	24	32	300	855	450	45	132	14	6.5	28
23	6	35	18	32	180	658	256	47	70	13	7.0	16
24	21	32	20	35	130	455	189	37	40	11	8.5	12
25	160	28	24	28	80	455	139	33	26	11	8.5	11
26	28	24	14	25	220	360	112	45	20	11	8.0	145
27	20	19	16	25	480	240	98	47	17	11	7.0	219
28	43	19	14	25	180	200	95	41	15	11	6.5	63
29	28	19	11	22	......	260	90	32	16	11	6.0	37
30	296	18	9	20	......	375	114	33	16	14	6.0	25
31	278	......	8	20	......	455	......	34	......	16	6.0	......

NOTE.—Stage-discharge relation affected by ice from Nov. 26 to Mar. 18; discharge for this period determined from gage heights corrected for effect of ice by means of four discharge measurements, observer's notes, and weather records. Operation of water-stage recorder not satisfactory Oct. 19–25, May 12–13, and July 19–23; daily discharge for these periods estimated by comparison with records at Knightville.

Monthly discharge of Middle Branch of Westfield River at Goss Heights, Mass., for the year ending Sept. 30, 1918.

[Drainage area, 53 square miles.]

Month.	Discharge in second-feet.				Run-off (depth in inches on drainage area).
	Maximum.	Minimum.	Mean.	Per square mile.	
October	296	6	34.7	0.655	0.76
November	81	14	24.4	.460	.51
December	30	8	19.7	.372	.43
January	44	6	23.7	.447	.52
February	620	6	105	1.98	2.06
March	885	86	330	6.23	7.18
April	705	90	230	4.34	4.84
May	400	32	88.2	1.66	1.91
June	132	10	26.8	.506	.56
July	28	10	16.4	.309	.36
August	16	5	8.87	.167	.19
September	219	6	26.0	.491	.55
The year	885	5	77.6	1.46	19.87

WESTFIELD LITTLE RIVER NEAR WESTFIELD, MASS.

Location.—At diversion dam of Springfield waterworks, in the town of Russell, Hampden County, 3 miles below the confluence of Pebble and Borden brooks) and about 3 miles west of Westfield. Originally (July, 1905, to December, 1909, a short distance below Borden Brook near Cobble Mountain.

Drainage area.—43 square miles at original site; 48 square miles at present site.

Records available.—July 13, 1905, to September 30, 1918.

Determination of discharge.—At the original site below Borden Brook (used 1905–1909) the discharge was determined by methods commonly employed at current-meter gaging stations. From August, 1906, to September, 1907, a 30-foot weir was maintained a short distance below the gage.[1] Since March 1, 1910, high-water flow determined from continuous record of head on concrete diversion dam (crest length, 155.4 feet), for which coefficients have been deduced from experiments at Cornell University; low-water flow—less than 163 second-feet—determined from continuous record of head on a 12-foot sharp-crested weir without end contractions, the crest being 2.55 feet below that of the dam. Water diverted to city of Springfield is measured by a 54-inch Venturi meter, using continuous record chart. Daily record corrected for storage in a reservoir on Borden Brook about 5 miles above station, but owing to the time required for water to reach the dam and the natural storage along the stream the record as corrected does not represent exactly the natural flow of the stream at all times.

Extremes of discharge.—Maximum discharge for 24 hours recorded during year. 641 second-feet, March 22; minimum discharge for 24 hours recorded, apparently zero from July 23 to 29, inclusive, when the water released from the reservoir was equal to or greater than the total flow at the diversion dam.

1909–1918: Maximum discharge for 24 hours, 1,490 second-feet, March 28, 1914; minimum discharge, apparently zero at various times when the water released from the reservoir was equal to or greater than the total flow at the diversion dam.

Diversions.—Record of water diverted at station for municipal supply of Springfield included in records as published.

Cooperation.—Data collected and compiled under the direction of E. E. Lochridge, chief engineer, board of water commissioners, Springfield, Mass.

Daily discharge, in second-feet, of Westfield Little River near Westfield, Mass., for the year ending Sept. 30, 1918.

Day.	Oct.	Nov.	Dec.	Jan.	Feb.	Mar.	Apr.	May.	June.	July.	Aug.	Sept.
1	9.2	150	22.6	23.2	22.1	54.9	248	162	21.2	14.3	8.6	15.5
2	9	84.9	27	17.6	21	85.3	279	138	18	23.3	5.9	8.6
3	11.3	58.6	26.6	16.4	20.5	27.4	307	109	15.8	19.1	8.9	9.8
4	9.5	51.1	26	12.2	31.3	45.4	278	91.3	12.9	16.3	9.8	14.6
5	15.1	42.7	24.3	32.8	19.9	15.9	173	80.3	15.3	13.9	16.2	13.5
6	15.3	38.6	23	38	29.8	33	148	68.7	29.9	12.5	15.6	10.9
7	13.5	34.7	18.7	20.6	52.7	32	109	63.7	53.8	12.2	14.5	8.6
8	10.3	31.1	28.5	17.8	69.9	21	111	61	44.9	12.2	9.1	9.7
9	10.8	30.3	19.6	17.7	23.6	18.5	122	51	21.4	11.9	8.6	17.2
10	11	26.4	20.9	17.1	19.6	20.3	153	45.7	20.2	10.6	13.3	14.1
11	11.2	24.9	18.9	20.3	30.2	13.1	127	41.6	17.2	10.1	28.2	10.9
12	15.8	24.9	17	38.9	35.1	15.7	113	37	66.6	10	14.8	11.8
13	36.2	22.2	29	48.8	67.1	18.7	99.2	44.5	76.9	9.7	9.6	16.2
14	18	22.2	19.2	67.9	102	20.8	142	140	44.5	10	14.7	11.4
15	14.3	21.8	21	65.4	92.4	15.5	185	101	34.1	10.5	15.6	4
16	12.3	20.2	49.6	62.5	82.9	12.1	147	75.8	23.6	11	15.1	8.9
17	11.5	21.6	31.1	46.6	76.5	15.6	124	58.1	17.2	6.1	13.2	9.6
18	17.8	20.4	28.6	39.9	61.5	20.4	137	45.1	15.3	7.4	8	29.6
19	14.4	18.7	29.8	35.3	65.6	26.1	125	38.4	10.2	6.5	9.4	18
20	17.8	20.3	31	37.5	456	35	105	34.2	12.4	6.5	8.6	33.1
21	13.4	20.9	32.2	35.1	295	52.7	218	30.2	18.5	1.3	9	61.1
22	6.1	37	21.7	33.4	185	64.1	310	29.1	111	1.3	9.4	34.9
23	11.3	69.6	21.1	30.8	134	48.9	216	28.7	70.9	......	8.7	21.1
24	153	51.4	21.2	30.5	121	34.2	141	25.5	44.9	......	9.2	16.2
25	190	26.8	20.9	39	140	25.6	121	24.8	29.3	......	9.1	14.9
26	65.2	19.4	20.3	35.6	314	25.7	109	38.1	18.9	......	8.2	165
27	50.4	19	30.3	25.9	319	19	90.9	33	16.1	......	8.7	201
28	138	20.6	19.2	33	215	15.7	80.6	30.4	12.4	......	8.5	85.8
29	74.3	19.1	17.5	35.9	......	15.8	75.6	24.4	12.7	......	9.9	46.9
30	428	19.9	16.3	28.4	......	18.2	87	22.7	15.9	10.6	9.4	34.4
31	317	......	17.7	23	......	22.1	......	22.1	......	26.5	14.2	......

Note.—Discharge determined by subtracting from the total flow at the diversion dam the quantity of water apparently released from Borden Brook reservoir, or by adding the quantity of water apparently stored in the reservoir, as indicated by elevation of water surface in reservoir. As no allowance has been made for evaporation and seepage from the reservoir, the results show the natural flow at the diversion dam only approximately. For days when no discharge records are given, the apparent storage release was equal to or greater than the total flow at the diversion dam.

[1] Results obtained by weir and current-meter methods are compared in U. S. Geol. Survey Water-Supply Papers 201, pp. 105-110, and 241, pp. 164-168.

Monthly discharge of Westfield Little River near Westfield, Mass., for the year ending Sept. 30, 1918.

[Drainage area, 48.5 square miles.]

Month.	Discharge in second-feet.				Run-off (depth in inches on drainage area).
	Maximum.	Minimum.	Mean.	Per square mile.	
October	428	6.1	55.8	1.15	1.33
November	150	18.7	35.6	.735	.820
December	49.6	17.0	24.2	.499	.575
January	67.9	12.2	33.1	.683	.787
February	456	19.6	111	2.29	2.38
March	641	121	271	5.58	6.44
April	310	75.6	156	3.22	3.59
May	162	22.1	57.9	1.19	1.38
June	111	10.2	30.8	.636	.710
July	26.5	(a)	8.82	.182	.210
August	28.2	5.9	11.4	.234	.270
September	201	4.0	31.9	.658	.734
The year	641	(a)	68.7	1.42	19.23

[a] On certain days the apparent storage release from Borden Brook reservoir was equal to or greater than the total flow at the diversion dam.

BORDEN BROOK NEAR WESTFIELD, MASS.

Location.—At the outlet of Borden Brook reservoir in town of Granville, Hampden County, 2 miles above confluence of Borden and Pebble brooks, and 8 miles west of Westfield.

Drainage area.—8 square miles.

Records available.—January 1, 1910, to September 30, 1918.

Determination of discharge.—Flow determined from a continuous record of the head on a 5-foot sharp-crested weir without end contractions. The results are then corrected for the apparent gain or loss in stored water in the reservoir, but no allowance is made for evaporation.

Extremes of discharge.—Maximum 24-hour flow recorded during year, 309 second-feet on March 4; minimum apparent flow, 0.0 second-foot at various times when the apparent storage release was equal to or greater than the measured flow at the weir.

1912–1918: Maximum 24-hour flow recorded, 309 second-foot on March 4, 1918; minimum apparent flow, 0. 0 second-foot.

Cooperation.—Records furnished by the Board of Water Commissioners of Springfield through E. E. Lochridge, chief engineer.

Daily discharge, in second-feet, of Borden Brook near Westfield, Mass., for the year ending Sept. 30, 1918.

Day.	Dec.	Jan.	Feb.	Mar.	Apr.	May.	June.	July.
1		23.1		116	43.9	16.2		
2		17.6		65.0	44.6	17.9		1.9
3				54.2	45.4	16.2		
4		3.0	10.8	309	31.1	15.0		
5		17.5		20.6	20.4	13.9		
6	1.2		29.7	46.3	31.0	12.2		
7			8.1	40.2	6.4	11.6		
8	9.3		.7	29.5	12.8	11.6		
9				46.5	12.8	11.6		
10				41.5	15.0	11.0		
11			10.8	28.0	16.2	11.5		
12		1.4	9.3	28.9	15.0	10.5	17.2	
13	10.8	9.3	10.8	13.6	13.9	8.6		
14			20.1	33.5	16.3	7.6		
15			20.1	24.9	20.4			
16	29.4	10.8		12.3	19.8	8.8		
17	10.8			32.8	17.9	7.6	.2	
18	9.3			28.9	17.9	5.8	.2	
19	9.3		9.3	46.6	17.9	5.0		
20	10.8		10.8	60.7	16.7	3.6		
21	10.8		9.3	82.9	51.4	1.7		
22				101	38.6	1.1	19.4	
23				72.2	38.9	1.1	8.0	
24			10.8	49.9	31.0	1.1	8.0	
25		10.8	29.4	20.7	14.5	.9		
26		9.3		40.9	17.3	.7		
27	9.3		41.8	21.7	16.2			
28			30.9	20.6	13.9			
29				18.6	12.2			
30				21.6	12.8		5.0	1.4
31				32.0				

NOTE.—Discharge determined by subtracting from the quantity of water passing over the weir the quantity apparently released from the reservoir, or by adding the quantity apparently stored in the reservoir, as indicated by elevation of water surface in reservoir. As no allowance has been made for evaporation and seepage from the reservoir, the results show the natural flow at the outlet of the reservoir only approximately. For days for which discharge is not given, the quantity apparently released from storage was equal to or greater than the quantity passing over the weir.

Monthly discharge of Borden Brook near Westfield, Mass., for the year ending Sept. 30, 1918.

Month.	Discharge in second-feet.				Run-off (depth in inches on drainage area).
	Maximum.	Minimum.	Mean.	Per square mile.	
October			0.00	0.000	0.00
November			.00	.000	.00
December	29.4		3.58	.448	.52
January	23.1		3.32	.415	.48
February	41.8		9.38	1.17	1.22
March	309	12.3	50.4	6.30	7.26
April	51.4	6.4	22.7	2.84	3.17
May	17.9		6.83	.854	.98
June	19.4		1.93	.241	.27
July	1.9		.11	.014	.02
August			.00	.000	.00
September			.00	.000	.00
The year	309		8.20	1.02	13.92

FARMINGTON RIVER NEAR NEW BOSTON, MASS.

Location.—At highway bridge a quarter of a mile below Clam River and 1 mile south of New Boston, Berkshire County.

Drainage area.—92.7 square miles.

Records available.—May 27, 1913, to September 30, 1918.

Gages.—Barrett & Lawrence water-stage recorder on left bank, downstream side of bridge, referred to gage datum by a hook gage inside the well; a vertical staff on bridge abutment is used for auxiliary readings.

Discharge measurements.—Made from a cable or by wading.

Channel and control.—Channel rocky and filled with boulders. Control practically permanent.

Extremes of discharge.—Maximum open-water stage during year, from water-stage recorder, 5.54 feet at 10 p. m. March 22 (discharge, 1,010 second-feet); a stage of 8.9 feet was recorded at 4 p. m. February 20, but the water was held back by an ice jam; minimum stage during year from water-stage recorder, 2.47 feet at 4 p. m. November 19 (discharge, 14 second-feet).

1913–1918: Maximum open-water stage from water-stage recorder, 7.64 feet on October 26, 1913 (discharge, by extension of rating curve, about 3,200 second-feet); minimum stage from water-stage recorder, 2.22 feet on August 27, 1913 (discharge, 4.4 second-feet).

Ice.—River frozen over during greater part of winter; stage-discharge relation seriously affected. Ice jams occasionally form below the gage causing several feet of backwater.

Regulation.—Flow affected by storage in Otis reservoir, about five miles above New Boston, and by operation of a woodworking shop just above the station.

Accuracy.—Stage-discharge relation practically permanent except when affected by ice. Rating curve well defined below 1,700 second-feet. Operation of water-stage recorder satisfactory except for short periods as shown in footnote to the daily-discharge table. Daily discharge ascertained by applying to rating table mean daily gage height determined by inspecting recorder graph and making corrections for effect of ice during winter. Open-water records good; winter records fair.

Discharge measurements of Farmington River near New Boston, Mass., during the year ending Sept. 30, 1918.

Date.	Made by—	Gage height.	Discharge.	Date.	Made by—	Gage height.	Discharge.
		Feet.	Sec.-ft.			Feet.	Sec.-ft.
Jan. 5	H. W. Fear	[a]3.96	18.3	Mar. 5	H. W. Fear	[a]6.11	218
Feb. 6	do	[a]3.40	24.3	July 12	O. W. Hartwell	3.24	84

[a] Stage-discharge relation affected by ice.

Daily discharge, in second-feet, of Farmington River near New Boston, Mass., for the year ending Sept. 30, 1918.

Day.	Oct.	Nov.	Dec.	Jan.	Feb.	Mar.	Apr.	May.	June.	July.	Aug.	Sept.
1	77	185	44	9	16	500	478	254	53	61	71	131
2	76	141	60	9	16	455	550	238	48	76	71	90
3	75	91	54	11	14	375	550	185	41	71	70	85
4	77	80	49	16	14	300	500	162	40	64	71	85
5	84	65	44	19	19	240	375	131	44	62	100	84
6	99	61	40	19	22	210	286	131	44	73	76	81
7	91	56	40	22	22	270	238	131	108	98	71	77
8	85	41	44	29	29	395	210	122	114	99	70	77
9	82	40	40	29	11	500	224	106	76	87	65	76
10	78	40	36	29	9	500	269	93	65	86	80	74
11	76	29	26	36	9	430	238	96	52	85	173	75
12	78	33	26	49	9	356	197	102	86	82	141	90
13	105	32	29	60	44	337	185	197	162	80	122	131
14	90	31	32	77	54	302	254	395	94	94	105	131
15	84	31	36	90	49	238	337	254	73	100	122	131
16	80	29	40	84	98	238	302	185	63	59	99	122
17	80	30	44	71	90	254	254	141	50	54	59	122
18	75	20	40	65	71	320	286	105	53	60	44	118
19	74	16	40	65	49	356	254	76	56	54	53	99
20	77	24	36	60	210	500	210	74	53	60	66	106
21	70	25	29	60	285	625	356	68	53	100	93	151
22	62	37	22	60	335	840	600	71	173	102	93	82
23	46	68	16	49	300	770	435	78	162	102	96	99
24	68	58	14	40	240	600	375	75	120	104	107	48
25	162	42	19	36	160	550	269	71	88	106	107	46
26	84	40	11	36	710	455	224	116	82	141	116	264
27	66	40	14	32	500	337	185	99	76	131	141	395
28	106	36	11	25	270	286	173	93	66	131	116	197
29	86	34	9	19	......	269	162	77	62	131	100	131
30	286	29	9	16	......	302	162	74	47	141	98	94
31	395	......	9	16	......	375	......	70	......	131	118	......

NOTE.—Stage-discharge relation affected by ice Dec. 5 to Mar. 8; discharge for this period determined from gage heights corrected for effect of ice by means of three discharge measurements, observer's notes, and weather records. Operation of water-stage recorder unsatisfactory Mar. 11, May 5-7, 11-13, 21-22, and July 8-11; discharge estimated.

Monthly discharge of Farmington River near New Boston, Mass., for the year ending Sept. 30, 1918.

[Drainage area 92.7 square miles.]

Month.	Discharge in second-feet.				Run-off (depth in inches on drainage area).
	Maximum.	Minimum.	Mean.	Per square mile.	
October	395	46	99.2	1.07	1.23
November	185	16	49.5	.534	.60
December	60	9	31.1	.335	.39
January	90	9	39.9	.430	.50
February	710	9	131	1.41	1.47
March	840	210	403	4.35	5.02
April	600	162	305	3.29	3.67
May	395	68	131	1.41	1.63
June	173	40	76.8	.829	.92
July	141	54	91.1	.983	1.13
August	173	44	94.0	1.01	1.16
September	395	46	116	1.25	1.40
The year	840	9	131	1.41	19.12

HOUSATONIC RIVER BASIN.

HOUSATONIC RIVER NEAR GREAT BARRINGTON, MASS.

LOCATION.—At highway bridge, a quarter of a mile northeast of Van Deusenville station of New York, New Haven & Hartford Railroad (Berkshire division) and 2 miles north of Great Barrington, Berkshire County.

DRAINAGE AREA.—280 square miles.

RECORDS AVAILABLE.—May 17, 1913, to September 30, 1918.

GAGE.—Inclined staff attached to concrete anchorages on downstream side of left abutment of highway bridge; vertical high-water section attached to bridge abutment; read by Martin Love.

DISCHARGE MEASUREMENTS.—Made from upstream side of highway bridge or by wading.

CHANNEL AND CONTROL.—Bed composed of sand and gravel. Control for high stages is not well defined. At low stages control is at well-defined riffle a few hundred feet below the gage.

EXTREMES OF DISCHARGE.—Maximum open-water stage recorded during year, 5.22 feet at 8 a. m. March 23 (discharge, 2,670 second-feet); minimum stage recorded, 0.2 foot at 8 a. m. July 28 (discharge, 2 second-feet).

1913–1918: Maximum stage recorded, 8.0 feet on March 31, 1916 (discharge, by extension of rating curve about 5,300 second-feet). Zero flow recorded at various times caused by storage of water at dams above.

ICE.—Stage-discharge relation affected by ice for short periods during the winter.

REGULATION.—Storage above dam of a paper mill about a mile above station causes low flow on Sundays and holidays.

ACCURACY.—Stage-discharge relation practically permanent during the year, except as affected by ice for a few days in December and January. Rating curve well defined below 2,000 second-feet. Gage read to quarter-tenths twice daily. Daily discharge ascertained by applying mean daily gage height to rating table. Records good.

Discharge measurements of Housatonic River near Great Barrington, Mass., during the year ending Sept. 30, 1918.

Date.	Made by—	Gage height.	Discharge.	Date.	Made by—	Gage height.	Discharge.
		Feet.	*Sec.-ft.*			*Feet.*	*Sec.-ft.*
Jan. 3	H. W. Fear	a 1.69	183	Mar. 2	H. W. Fear	3.48	1,220
Feb. 4	.....do	1.10	67	July 13	A. N. Weeks	1.34	107

a Stage-discharge relation affected by ice.

Daily discharge, in second-feet, of Housatonic River near Great Barrington, Mass., for the year ending Sept. 30, 1918.

Day.	Oct.	Nov.	Dec.	Jan.	Feb.	Mar.	Apr.	May.	June.	July.	Aug.	Sept.
1	103	780	215	155	135	1,570	1,340	720	358	215	200	87
2	51	512	46	118	105	1,810	1,570	720	155	215	140	46
3	83	335	185	185	11	1,200	1,810	720	170	232	155	170
4	120	270	200	155	51	1,270	1,810	600	200	77	18	145
5	118	270	185	74	105	1,060	1,410	335	252	95	215	132
6	16	290	185	64	97	1,060	1,200	540	358	232	130	125
7	58	215	232	81	132	1,490	885	512	312	132	120	200
8	110	155	170	85	135	1,410	920	660	405	250	89	15
9	135	215	31	103	85	1,130	990	512	97	250	97	59
10	185	185	200	118	58	920	1,060	430	430	215	101	120
11	135	101	215	142	97	990	1,060	512	270	185	48	108
12	200	120	155	380	152	780	920	97	335	130	130	132
13	95	200	170	77	132	920	780	312	335	108	108	142
14	43	215	215	155	97	920	720	1,340	430	28	93	140
15	85	145	120	81	200	885	920	1,410	250	97	232	19
16	97	155	85	135	145	815	885	1,060	105	155	335	155
17	128	118	66	155	77	660	780	780	270	130	215	155
18	105	14	185	83	250	1,060	750	1,410	290	120	13	145
19	125	155	185	87	335	1,060	630	458	250	128	43	120
20	118	118	155	87	780	1,200	1,060	380	170	105	155	125
21	70	170	142	105	1,200	1,970	600	485	290	145	185	132
22	103	215	145	28	1,490	2,130	1,490	430	250	120	155	132
23	95	185	29	97	1,340	2,650	1,570	458	105	101	142	185
24	155	120	130	142	812	2,050	1,340	485	335	170	125	155
25	215	34	130	145	990	1,970	1,130	485	312	185	24	155
26	185	170	270	110	1,340	1,810	990	250	405	120	56	105
27	170	130	170	21	1,490	1,490	815	312	358	120	58	405
28	76	185	170	145	1,270	1,270	458	458	335	2.6	77	690
29	132	68	115	101	....	990	430	430	200	110	118	405
30	250	95	49	97	....	1,060	405	170	8	118	145	512
31	720	....	458	145	....	1,060	....	250	....	150	89	....

NOTE.—Stage-discharge relation affected by ice from Dec. 26 to Jan. 10. Discharge for this period determined from gage heights corrected for effect of ice by means of one discharge measurement, observer's notes, and weather records.

Monthly discharge of Housatonic River near Great Barrington, Mass., for the year ending Sept. 30, 1918.

[Drainage area, 280 square miles.]

Month.	Discharge in second-feet.				Run-off (depth in inches on drainage area).
	Maximum.	Minimum.	Mean.	Per square mile.	
October	720	16	138	0.493	0.5
November	780	14	198	.707	.
December	458	29	162	.579	.
January	380	21	118	.421	.
February	1,490	11	449	1.60	1.
March	2,650	600	1,310	4.68	5.
April	1,810	405	1,020	3.64	4.
May	1,410	97	572	2.04	2.
June	430	8	267	.954	1.
July	250	2.6	143	.511	.
August	335	13	123	.439	.
September	690	15	173	.618	.
The year	2,650	2.6	389	1.39	

HOUSATONIC RIVER AT FALLS VILLAGE, CONN.

LOCATION.—Half a mile below power plant of Connecticut Power Co. at Falls Village, Litchfield County, and 23 miles north of Gaylordsville.

DRAINAGE AREA.—644 square miles.

RECORDS AVAILABLE.—July 11, 1912, to September 30, 1918.

GAGES.—Stevens continuous water-stage recorder on left bank, referred to gage datum by hook gage inside the well; a vertical staff on river bank 25 feet upstream and chain gage 300 feet upstream are used for auxiliary readings.

DISCHARGE MEASUREMENTS.—Made from cable 150 feet above gage or by wading.

CHANNEL AND CONTROL.—Channel deep and fairly uniform in cross-section; one channel at all times. Control not clearly defined except at low stages; probably permanent.

EXTREMES OF DISCHARGE.—Maximum open-water stage during year, from water-stage recorder, 8.22 feet at 9 p. m. March 23 (discharge, 4,220 second-feet); a stage of 9.60 feet was recorded at 11 p. m. February 26, but the water was held back by an ice jam; minimum stage, from water-stage recorder, 0.56 foot at 7 a. m. September 11 (discharge, 21 second-feet).

1912–1918: Maximum stage recorded, 13.3 feet on March 29, 1914 (discharge, 8,830 second-feet); minimum stage recorded, zero flow at various times owing to storage of water above power plant.

ICE.—Stage-discharge relation seriously affected by ice.

REGULATION.—Flow at low water completely regulated by power plant at Falls Village.

ACCURACY.—Stage-discharge relation practically permanent, except when affected by ice. Rating curve well defined between 200 and 7,000 second-feet. Operation of the water-stage recorder satisfactory. Daily discharge ascertained by using discharge integrator, and making corrections for ice during the winter. Records good.

Discharge measurements of Housatonic River at Falls Village, Conn., during the year ending Sept. 30, 1918.

Date.	Made by—	Gage height.	Discharge.	Date.	Made by—	Gage height.	Discharge.
		Feet.	*Sec.-ft.*			*Feet.*	*Sec.-ft.*
Jan. 4	H. W. Fear	[a]3.02	465	Mar. 4	H. W. Fear	[a]8.49	2,760
Feb. 5	.....do	[a]2.83	336	July 13	A. N. Weeks	2.43	599

[a] Stage-discharge relation affected by ice.

Daily discharge, in second-feet, of Housatonic River at Falls Village, Conn., for the year ending Sept. 30, 1918.

Day.	Oct.	Nov.	Dec.	Jan.	Feb.	Mar.	Apr.	May.	June.	July.	Aug.	Sept.
1	240	1,780	610	60	270	3,100	1,940	1,540	880	300	210	285
2	182	1,320	540	140	320	3,400	2,150	1,640	375	310	230	350
3	182	1,100	670	220	180	3,200	2,400	1,660	690	325	116	200
4	184	395	560	230	290	2,600	2,550	1,340	590	215	57	192
5	200	710	455	170	230	2,000	2,500	1,220	415	390	215	260
6	198	630	415	25	250	1,950	2,150	1,340	440	420	215	196
7	150	470	340	240	240	2,300	1,720	1,160	660	57	215	230
8	230	340	310	220	260	2,200	1,600	1,100	790	410	255	29
9	200	400	285	210	360	2,600	1,460	1,000	460	570	192	166
10	170	550	400	200	170	3,300	1,480	990	710	420	178	162
11	210	210	400	190	150	3,100	1,640	820	540	325	46	160
12	315	315	340	360	300	2,650	1,560	480	510	320	230	225
13	290	295	340	160	540	2,450	1,300	960	810	250	280	196
14	178	300	460	320	460	2,650	1,280	1,300	850	51	240	142
15	275	295	380	310	800	2,600	1,660	2,250	500	235	240	186
16	215	305	360	450	1,150	2,300	1,700	2,150	350	280	245	198
17	260	350	360	440	700	2,000	1,680	1,700	810	315	470	192
18	192	112	360	390	1,050	150	1,600	1,300	600	260	59	220
19	220	305	340	280	850	300	1,700	1,060	460	265	200	275
20	265	300	350	100	1,600	600	1,420	1,220	405	210	200	405
21	75	290	290	370	2,700	3,150	1,540	890	300	80	160	370
22	210	330	270	180	2,600	3,800	2,500	850	310	240	160	290
23	230	375	190	250	2,060	,100	2,850	860	480	225	162	425
24	290	485	320	250	,550	,900	2,800	890	810	270	110	495
25	600	230	140	210	,600	,500	2,500	890	510	270	59	370
26	370	275	260	340	2,900	3,100	2,140	510	580	320	176	680
27	350	340	250	200	3,500	2,750	1,780	1,140	540	265	190	1,580
28	250	285	260	330	3,200	2,350	1,460	1,000	580	96	172	1,780
29	570	56	260	330	......	2,000	1,350	940	480	184	176	1,280
30	700	365	50	260	......	1,900	1,280	510	210	240	200	1,080
31	570	......	190	300	......	1,850	......	1,020	......	240	240	......

NOTE.—Stage-discharge relation affected by ice Dec. 11 to Mar. 9; daily discharge for this period determined from gage heights corrected for effect of ice by means of three discharge measurements, observer's notes, weather records, and study of power plant records at Falls Village.

Monthly discharge of Housatonic River at Falls Village, Conn., for the year ending Sept. 30, 1918.

[Drainage area, 644 square miles.]

Month.	Discharge in second-feet.				Run-off (depth in inches on drainage area).
	Maximum.	Minimum.	Mean.	Per square mile.	
October	700	75	277	0.430	0.50
November	1,780	56	450	.700	.78
December	670	50	347	.539	.62
January	450	25	248	.385	.44
February	3,500	150	1,080	1.68	1.75
March	4,100	1,850	2,700	4.19	4.83
April	2,850	1,280	1,860	2.89	3.22
May	2,260	480	1,150	1.79	2.06
June	880	210	555	.862	.96
July	570	51	270	.419	.48
August	470	46	189	.293	.34
September	1,780	29	437	.679	.76
The year	4,100	25	795	1.23	16.74

HUDSON RIVER BASIN.

HUDSON RIVER NEAR INDIAN LAKE, N. Y.

LOCATION.—About 1 mile below mouth of Cedar River, 1½ miles above mouth of Indian River, and 6 miles northeast of Indian Lake village, Hamilton County.

DRAINAGE AREA.—418 square miles (measured on topographic maps).

RECORDS AVAILABLE.—August 30, 1916, to September 30, 1918.

GAGE.—Gurley printing water-stage recorder on right bank; inspected by John A. Bolton.

DISCHARGE MEASUREMENTS.—Made from cable about 100 yards below gage or by wading.

CHANNEL AND CONTROL.—Solid ledge overlain with coarse gravel; probably permanent.

EXTREMES OF DISCHARGE.—Maximum stage during year, from water-stage recorder, 8.08 feet at 6.30 a. m. May 19 (discharge, 8,960 second-feet); minimum stage, February 7 (discharge, 80 second-feet).

1916–1918: Maximum stage, from water-stage recorder, 9.87 feet at 11 a. m. June 12, 1917 (discharge, 13,500 second-feet); minimum stage from water-stage recorder 1.43 feet from 11 a. m. September 11 to 8 a. m. September 13, 1916 (discharge, 56 second-feet).

ICE.—Stage-discharge relation affected by ice.

REGULATION.—Large diurnal fluctuation due to logging operations during the spring months. Seasonal distribution of flow slightly affected by storage.

ACCURACY.—Stage-discharge relation practically permanent; affected by logs during October and November and by ice from December to March. Rating curve fairly well defined between 75 and 600 second-feet and well defined between 600 and 6,000 second-feet. Operation of water-stage recorder satisfactory. Daily discharge ascertained by applying mean daily gage height to rating table except when fluctuation required mean of hourly discharge. Records good.

Discharge measurements of Hudson River near Indian Lake, N. Y., during the year ending Sept. 30, 1918.

Date.	Made by—	Gage height.	Discharge.	Date.	Made by—	Gage height.	Discharge.
		Feet.	*Sec.-ft.*			*Feet.*	*Sec.-ft.*
Dec. 12[a]	E. D. Burchard	2.52	175	Apr. 29[c]	J. W. Moulton	4.34	1,830
Jan. 7[a]	J. W. Moulton	2.90	111	30	E. D. Burchard	3.14	987
31[a]	E. D. Burchard	3.07	183	30	J. W. Moulton	3.21	1,070
Feb. 27[a]	J. W. Moulton	4.84	851	June 21	.....do	2.22	352
Mar. 22[b]	.....do	4.97	1,070	21	.....do	2.22	338
Apr. 3[c]	.....do	6.37	4,910	July 14	.....do	2.78	696

[a] Measurement made through complete ice cover.
[b] Measurement made through partial ice cover.
[c] Log jam on the control.

Daily discharge, in second-feet, of Hudson River near Indian Lake, N. Y., for the year ending Sept. 30, 1918.

Day.	Oct.	Nov.	Dec.	Jan.	Feb.	Mar.	Apr.	May.	June.	July.	Aug.	Sept.
1	309	3,630	260	130	100	900	3,000	4,730	295	313	216	195
2	416	2,820	280	130	95	850	3,800	2,920	1,350	524	229	224
3	379	1,990	280	130	120	800	5,000	4,060	688	482	234	236
4	319	1,420	280	120	140	750	4,800	3,170	1,320	562	211	290
5	366	1,200	260	120	120	700	4,400	2,030	445	595	195	252
6	565	990	240	120	90	650	3,430	1,450	1,140	500	187	247
7	595	990	240	110	80	650	3,000	1,810	884	428	171	379
8	595	838	220	140	85	600	2,800	2,930	1,190	383	167	434
9	565	568	200	120	90	600	2,800	2,400	1,530	351	175	372
10	506	429	200	150	100	600	2,660	2,850	3,730	356	238	306
11	449	595	200	180	120	650	2,280	2,210	1,280	405	440	247
12	368	924	190	180	150	650	1,920	2,870	1,270	530	440	211
13	477	595	170	200	200	1,000	1,750	1,520	1,540	665	367	208
14	535	355	170	220	200	1,000	1,640	2,280	1,540	735	315	211
15	595	291	170	220	240	1,000	1,390	2,370	1,640	735	252	199
16	660	582	160	280	240	900	2,040	1,890	1,400	595	238	224
17	800	683	160	280	240	900	2,600	1,550	890	530	183	280
18	730	506	160	280	280	900	3,400	530	665	500	157	361
19	628	506	150	280	300	900	3,200	2,750	506	446	146	688
20	695	595	150	280	340	900	2,400	440	405	399	142	772
21	875	389	150	280	380	950	1,900	1,350	372	356	135	735
22	912	344	150	260	440	1,100	2,200	341	356	315	128	810
23	800	320	150	260	500	1,400	4,600	1,260	341	276	132	772
24	730	280	150	260	550	1,900	2,600	280	351	247	125	700
25	875	280	150	240	550	2,200	1,600	1,240	378	229	122	770
26	950	260	150	220	600	2,400	1,609	346	367	211	115	735
27	912	240	140	200	850	2,400	850	1,130	315	191	109	1,080
28	1,030	240	140	170	900	2,200	1,200	522	295	171	102	1,290
29	1,110	260	130	170	....	2,200	2,800	1,410	285	160	102	1,290
30	2,290	260	130	170	....	2,200	2,100	367	285	203	102	1,240
31	4,710	....	130	130	....	2,800	....	1,420	....	247	105	....

NOTE.—Discharge Nov. 23 to Apr. 4 estimated, because of ice, and discharge Apr. 18–30 estimated, because of logs on the control, from discharge measurements, weather records, study of recorder graph, and comparison with similar studies for Hudson River at North Creek.

Monthly discharge of Hudson River near Indian Lake, N. Y., for the year ending Sept. 30, 1918.

[Drainage area, 418 square miles.]

Month.	Discharge in second-feet.				Run-off (depth in inches on drainage area).
	Maximum.	Minimum.	Mean.	Per square mile.	
October	4,710	309	831	1.99	2.29
November	3,630	240	779	1.86	2.08
December	280	130	184	.440	.51
January	280	110	195	.467	.54
February	900	80	289	.691	.72
March	2,800	600	1,210	2.89	3.33
April	5,000	850	2,660	6.36	7.10
May	4,730	280	1,820	4.35	5.02
June	3,730	285	902	2.16	2.41
July	735	160	408	.976	1.13
August	440	102	193	.462	.53
September	1,290	195	528	1.26	1.41
The year	5,000	80	834	2.00	27.07

HUDSON RIVER AT NORTH CREEK, N. Y.

Location.—At two-span steel highway bridge in village of North Creek, Warren County, immediately above mouth of North Creek.

Drainage area.—804 square miles.

Records available.—September 21, 1907, to September 30, 1918.

Gage.—Chain at upstream side of left span of the bridge; read by William Alexander.

Discharge measurements.—Made from the upstream side of the highway bridge.

Channel and control.—Heavy gravel; fairly permanent.

Extremes of discharge.—Maximum stage recorded during year, 7.65 feet at 6 p. m. April 3 (discharge, 11,100 second-feet); minimum stage, 2.25 feet at 8 a. m. July 24 (discharge, 302 second-feet).

1907–1918: Maximum stage recorded 12.0 feet during the evening of March 27, 1913 (discharge about 30,000 second-feet); minimum stage, 2.05 feet at 7.05 a. m. September 30, 1913 (discharge, 168 second-feet).

Ice.—Stage-discharge relation affected by ice.

Regulation.—The numerous lakes and ponds in the basin of the upper Hudson have a decided effect on the low-water flow; especially the reservoir at Indian Lake. Many of the reservoirs are used to make flood waves in the spring in connection with log driving.

Accuracy.—Stage-discharge relation practically permanent; affected by ice from December to March, inclusive. Rating curve well defined between 250 and 6,000 second-feet. Gage read to half-tenths twice daily. Daily discharge ascertained by applying mean daily gage height to rating table. Open-water records good; winter records fair.

Discharge measurements of Hudson River at North Creek, N. Y., during the year ending Sept. 30, 1918.

Date.	Made by—	Gage height.	Discharge.	Date.	Made by—	Gage height.	Discharge.
		Feet.	*Sec.-ft.*			*Feet.*	*Sec.-ft.*
Dec. 12[a]	E. D. Burchard	4.22	899	Apr. 4	J. W. Moulton	6.22	6,880
Jan. 5[a]	J. W. Moulton	4.40	599	May 2	E. D. Burchard	4.15	2,460
Feb. 1[b]	E. D. Burchard	4.64	626	June 20	J. W. Moulton	2.66	588
28[b]	J. W. Moulton	5.54	1,520	July 13	do	3.76	1,770
Mar. 24[b]	do	7.10	2,710				

[a] Measurement made through incomplete ice cover. [b] Measurement made through complete ice cover.

Daily discharge, in second-feet, of Hudson River at North Creek, N. Y., for the year ending Sept. 30, 1918.

Day.	Oct.	Nov.	Dec.	Jan.	Feb.	Mar.	Apr.	May.	June.	July.	Aug.	Sept.
1	990	5,840	750	700	550	1,400	4,890	6,340	610	404	790	990
2	1,100	4,010	750	650	500	1,400	7,400	5,360	610	610	790	1,040
3	1,100	2,870	800	650	550	1,500	10,000	6,090	610	570	790	1,100
4	1,100	2,140	800	650	550	1,600	7,680	3,420	5,360	745	790	940
5	990	1,760	750	650	500	1,600	6,600	3,050	530	790	1,160	790
6	890	1,480	700	650	480	1,600	4,890	1,910	3,230	700	990	790
7	840	1,350	700	750	440	1,000	4,890	1,550	460	610	890	890
8	790	1,350	650	800	440	1,000	4,440	4,660	1,910	530	790	990
9	790	1,280	650	750	440	1,100	4,890	4,220	1,550	530	940	1,100
10	700	1,220	650	800	460	1,100	4,440	2,870	6,340	530	990	990
11	745	890	950	850	500	1,200	4,010	3,610	2,370	610	1,100	990
12	790	890	1,100	850	460	1,400	3,230	2,060	3,050	745	1,100	940
13	940	940	1,000	850	600	2,200	2,870	2,870	2,700	1,840	990	890
14	940	890	1,000	900	600	2,200	2.530	2,060	2,370	1.620	990	890
15	890	495	1,000	900	650	2,000	2,700	1,830	2,530	1,620	890	700
16	940	700	1,000	1,000	650	1,900	3,230	4,440	1,980	1,040	840	530
17	990	700	1,000	1,000	650	1,900	4,440	2,130	1,220	940	890	570
18	1,040	700	1,000	1,000	700	1,900	5,360	1,350	940	890	790	610
19	890	700	1,000	1,000	700	2,000	4,890	1,760	745	745	890	990
20	990	700	1,000	950	800	2,000	4,890	940	570	610	990	1,100
21	1,220	655	800	900	850	2,200	4,010	700	530	610	990	1,100
22	1,220	570	700	900	950	2,200	3,230	700	530	530	890	990
23	1,100	530	700	850	1,100	2,600	8,520	790	530	330	940	990
24	1,100	530	700	800	1,100	2,800	5,600	990	530	319	940	890
25	1,420	460	750	850	1,100	3,200	4,220	655	530	700	890	890
26	1,620	460	750	850	1,200	4,000	4,440	570	530	790	890	940
27	1,550	460	800	750	1,400	5,000	2,060	790	460	655	890	1,620
28	1,690	500	800	650	1,600	5,500	1,620	700	460	700	890	1,760
29	1,760	500	750	650	....	5,360	3,230	1,160	378	700	890	1,690
30	2,870	700	750	650	....	4,890	4,890	790	2,210	790	840	1,480
31	7,400	....	700	600	....	4,440	....	2,210	....	890	840	....

NOTE.—Discharge Nov. 26 to Mar. 28 estimated, because of ice, from discharge measurements, weather records, study of recorder graph and comparison with similar studies for Hudson River near Indian Lake.

Monthly discharge of Hudson River at North Creek, N. Y., for year ending Sept. 30, 1918.

[Drainage area, 804 square miles.]

Month.	Discharge in second-feet.				Run-off (depth in inches on drainage area).
	Maximum.	Minimum.	Mean.	Per square mile.	
October	7,400	700	1,340	1.67	1.91
November	5,840	460	1,210	1.50	1.67
December	1,100	650	821	1.02	1.18
January	1,000	600	800	.995	1.14
February	1,600	440	734	.913	.95
March	5,500	1,000	2,390	2.97	3.42
April	10,000	1,620	4,670	5.81	6.48
May	6,340	570	2,340	2.91	3.36
June	6,340	378	1,550	1.92	2.14
July	1,840	319	764	.950	1.10
August	1,160	790	912	1.13	1.30
September	1,760	530	1,010	1.26	1.41
The year	10,000	319	1,540	1.92	26.06

HUDSON RIVER AT THURMAN, N. Y.

Location.—At Delaware & Hudson Railroad bridge near Thurman railroad station, Warren County, half a mile below mouth of Schroon River, and 13 miles above mouth of Sacandaga River.

Drainage area.—1,550 square miles.

Records available.—September 1, 1907, to September 30, 1918.

Gage.—Chain at upstream side near center of left span; read by S. H. Spencer.

Discharge measurements.—Made from upstream side of bridge.

Channel and control.—Sand and gravel; fairly permanent. Logs occasionally lodge on a small island on the control.

Extremes of discharge.—Maximum stage recorded during year, 7.28 feet in the afternoon, April 23 (discharge, 14,800 second-feet); minimum stage recorded, 2.4 feet in the morning, July 28 (discharge, 680 second-feet).

1907–1918: Maximum stage (determined by leveling from flood marks), 12.5 feet during the late evening of March 27, 1913 (discharge about 46,000 second-feet); minimum stage recorded, 2.12 feet at 8.55 a. m. and 6.20 p. m. September 30, 1913 (discharge about 290 second-feet).

Ice.—Stage-discharge relation seriously affected by ice. Discharge determined from records at North Creek and Riverbank.

Regulation.—Discharge is regulated to some extent by the storage reservoirs at Indian Lake and Schroon Lake and the mills on Schroon River.

Accuracy.—Stage-discharge relation practically permanent; affected by ice during large part of the period from December to March, inclusive, and by logs during parts of June, July, and September. Rating curve well defined between 550 and 20,000 second-feet. Gage read to hundredths twice daily. Daily discharge ascertained by applying mean daily gage height to rating table. Records good; winter estimates fair.

Cooperation.—Gage heights furnished by the International Paper Co.

Discharge measurements of Hudson River at Thurman, N. Y., during the year ending Sept. 30, 1918.

Date.	Made by—	Gage height.	Discharge.	Date.	Made by—	Gage height.	Discharge.
		Feet.	*Sec.-ft.*			*Feet.*	*Sec.-ft.*
Dec. 16[a]	E. D. Burchard	5.16	1,570	June 20	J. W. Moulton	3.14	1,560
May 3	J. W. Moulton	5.41	8,050	July 12	....do	2.82	985

[a] Measurement made through complete ice cover.

Daily discharge, in second-feet, of Hudson River at Thurman, N. Y., for the year ending Sept. 30, 1918.

Day.	Oct.	Nov.	Dec.	Apr.	May.	June.	July.	Aug.	Sept.
1	1,460	7,760	1,550	11,400	6,780	2,040	850	1,220	1,380
2	1,460	6,170	1,460	12,100	7,430	3,590	1,300	1,080	1,460
3	1,550	4,420	1,460	14,100	6,780	2,150	1,500	1,150	1,380
4	1,380	4,140	1,940	14,100	6,470	4,420	950	1,080	1,380
5	1,550	3,590	1,380	12,500	5,000	950	1,100	1,460	1,080
6	1,550	3,200	1,550	11,000	4,710	1,550	1,100	1,300	1,080
7	1,220	2,960	1,500	10,600	3,860	1,380	950	1,220	1,080
8	1,380	2,840	1,460	9,500	5,580	1,220	950	1,220	1,080
9	1,080	2,480	1,400	9,860	4,710	1,380	850	1,220	1,550
10	1,020	1,940	1,500	9,860	5,290	4,710	850	1,220	1,380
11	850	1,740	1,600	8,790	5,290	3,590	850	1,150	1,300
12	1,080	1,940	1,800	7,430	5,870	3,080	800	1,460	1,300
13	1,300	2,150	1,800	7,430	4,710	4,710	1,700	1,380	1,220
14	1,460	1,740	1,700	7,100	6,470	4,140	2,200	1,300	1,460
15	1,460	1,740	1,600	6,470	8,100	4,140	2,200	1,300	2,150
16	1,300	1,940	1,600	7,430	5,580	3,860	1,700	1,150	850
17	1,460	2,040	1,500	7,100	4,140	2,600	1,300	1,080	905
18	1,640	1,940	1,500	7,760	3,860	2,370	1,300	905	1,220
19	1,380	1,640	1,500	11,400	3,590	2,150	1,200	1,020	1,300
20	1,300	1,640	1,400	7,760	3,460	1,840	1,200	1,220	1,550
21	1,460	1,740	1,400	8,790	3,330	1,740	1,000	1,380	2,040
22	1,940	1,460	1,200	7,430	2,840	2,150	850	1,080	1,740
23	1,840	1,460	1,100	11,000	2,480	1,220	800	1,220	1,550
24	1,460	1,460	1,100	9,500	2,260	1,220	750	1,300	1,550
25	1,940	1,300	1,100	6,170	3,860	1,500	850	1,380	1,380
26	2,260	1,080	1,100	7,100	1,550	1,400	1,220	1,150	1,550
27	2,150	1,020	1,100	5,290	1,940	1,500	1,020	1,150	2,800
28	2,260	905	1,100	5,000	2,260	1,300	680	1,080	2,600
29	2,480	1,640	1,100	9,140	4,140	1,300	1,220	1,150	2,400
30	3,590	2,150	1,100	8,790	2,150	700	1,220	1,150	2,200
31	8,790	....	1,000	....	5,000	....	1,380	1,080	....

NOTE.—Discharge Dec. 9-31 estimated, because of ice, from one discharge measurement, weather records, and study of recorder graph. Determinations of discharge, June 25 to July 24, and Sept. 27-30, somewhat uncertain because of logs on the control.

Monthly discharge of Hudson River at Thurman, N. Y., for the year ending Sept. 30, 1918.

[Drainage area, 1,550 square miles.]

Month.	Discharge in second-feet.				Run-off (depth in inches on drainage area).
	Maximum.	Minimum.	Mean.	Per square mile.	
October	8,790	850	1,840	1.19	1.37
November	7,760	905	2,410	1.55	1.73
December	1,940	1,000	1,410	.910	1.05
January	....	....	1,160	.748	.86
February	....	....	940	.606	.63
March	....	....	3,620	2.34	2.70
April	14,100	5,000	9,060	5.85	6.53
May	8,100	1,550	4,500	2.90	3.34
June	4,710	805	2,330	1.50	1.67
July	2,200	680	1,160	.748	.86
August	1,460	905	1,200	.774	.89
September	2,800	850	1,530	.987	1.10
The year	14,100	....	2,600	1.68	22.73

HUDSON RIVER AT SPIER FALLS, N. Y.

LOCATION.—Half a mile below Spier Falls dam, Saratoga County, and 11½ miles below mouth of Sacandaga River.

DRAINAGE AREA.—2,800 square miles (measured on topographic maps).

RECORDS AVAILABLE.—October 7, 1912, to September, 30, 1918.

GAGE.—Gurley 2-day water-stage recorder in a brick shelter 5 feet square on the right bank about half a mile below the Spier Falls dam. Recorder inspected by T. F. Malone, chief operator of power plant.

DISCHARGE MEASUREMENTS.—Made from a cable about 1,000 feet downstream from the gage.

CHANNEL AND CONTROL.—Coarse gravel and boulders; probably permanent.

EXTREMES OF DISCHARGE.—Maximum stage during year from water-stage recorder, 12.16 feet at 8 a. m. April 4 (discharge, 34,500 second-feet); minimum stage from water-stage recorder, 0.93 foot at 7 a. m. September 1 (discharge, 140 second-feet).

1912–1918: Maximum stage from water-stage recorder, 18.59 feet at 12.25 a. m. March 28, 1913 (discharge about 89,100 second-feet); minimum stage, −0.12 foot at 4 p. m. September 23, 1917, observed during current-meter measurement (discharge, about 5.5 second-feet).

ICE.—Stage-discharge relation not affected by ice, except for a short time during extremely cold periods.

REGULATION.—Large diurnal fluctuation in discharge due to the operation of the Spier Falls power plant. Seasonal flow affected by storage at Indian Lake and many small lakes and reservoirs in the upper part of the drainage.

ACCURACY.—Stage-discharge relation practically permanent; affected by ice February 2 to 16. Rating curve well defined for all stages except about 9 feet, where the rating curve may be 4 or 5 per cent large. Operation of the water-stage recorder satisfactory throughout the year. Daily discharge ascertained by averaging the results obtained by applying hourly gage heights to rating table. Records good.

COOPERATION.—Water-stage recorder inspected by an employee of the Adirondack Electric Power Corporation.

Discharge measurements of Hudson River at Spier Falls, N. Y., during the year ending Sept. 30, 1918.

Date.	Made by—	Gage height.	Discharge.
		Feet.	*Sec.-ft.*
Jan. 3[a]	J. W. Moulton	2.84	1,150
Feb. 2[b]	E. D. Burchard	2.85	1,400
June 18	J. W. Moulton	4.67	4,990

[a] Measurement made through complete ice cover. [b] Measurement made through incomplete ice cover.

Daily discharge, in second-feet, of Hudson River at Spier Falls, N. Y., for the year ending Sept. 30, 1918.

Day.	Oct.	Nov.	Dec.	Jan.	Feb.	Mar.	Apr.	May.	June.	July.	Aug.	Sept.
1	1,980	17,700	3,160	1,390	1,330	5,780	21,500	13,500	5,200	1,810	1,480	905
2	1,780	15,200	1,020	1,820	2,000	5,350	22,900	13,800	3,800	1,700	1,540	1,940
3	1,470	12,600	2,720	1,480	1,240	5,100	31,400	12,600	4,570	2,780	1,120	2,350
4	1,860	10,200	1,960	1,480	1,380	4,860	32,300	13,200	3,750	1,590	1,390	1,790
5	1,940	8,350	2,220	1,770	1,620	4,750	27,900	11,400	3,870	1,260	1,580	1,690
6	2,680	6,950	2,090	727	1,730	4,900	24,200	8,780	2,840	1,510	1,490	1,650
7	1,510	5,840	2,310	1,760	1,460	4,340	21,700	7,830	3,400	1,510	1,670	1,530
8	1,860	5,160	2,110	1,540	1,350	3,770	20,100	8,320	3,470	1,910	1,430	922
9	1,810	4,560	610	1,330	1,650	3,130	21,800	9,600	4,010	1,480	1,590	1.980
10	1,600	3,600	1,530	1,300	661	3,090	22,200	10,200	6,500	1,250	1,060	2,060
11	1,640	3,370	1,580	1,430	1,460	3,860	21,100	10,200	6,570	1,410	1,600	1,700
12	1,630	4,140	1,780	1,440	1,160	2,850	18,900	8,660	4,900	1,850	1,710	1,430
13	2,030	3,350	2,130	1,700	1,430	2,770	16,700	9,190	6,700	2,080	1,880	1,770
14	1,940	2,980	2,490	1,460	1,380	3,660	14,800	12,800	6,990	2,690	1,680	1,850
15	3,250	2,880	2,060	1,310	1,730	4,140	14,000	15,100	6,460	3,690	1,570	725
16	2,920	3,040	2,110	1,920	1,660	3,560	14,300	13,400	5,720	2,880	1,840	1,970
17	2,990	2,910	2,030	1,840	1,490	3,230	15,200	11,700	4,680	2,470	1,050	1,360
18	3,020	2,160	2,450	1,780	1,850	4,230	17,200	9,730	3,930	2,410	1,230	1,410
19	2,630	3,280	2,180	1,690	1,790	4,750	20,100	8,280	3,550	2,730	1,810	1,960
20	2,760	2,520	2,170	1,050	2,400	5,620	19,000	8,440	3,170	2,240	1,450	2,410
21	2,280	2,450	2,480	1,890	2,900	7,030	17,200	6,360	2,840	1,400	1,380	3,390
22	4,080	2,260	2,580	1,790	3,150	9,230	18,200	6,210	2,420	1,690	1,430	1,660
23	3,270	3,270	1,170	1,730	3,810	12,500	19,400	4,860	1,830	1,480	1,440	2,310
24	2,770	3,470	1,990	1,660	4,140	13,500	20,100	5,100	2,970	1,310	606	2,340
25	3,240	2,670	1,820	1,660	4,220	15,200	16,300	3,870	2,530	1,330	1,410	2,230
26	4,020	2,990	1,870	2,150	4,480	16,200	16,100	4,310	2,390	1,440	1,690	2,620
27	4,280	2,010	2,290	740	5,490	15,700	12,600	4,040	2,100	1,480	1,400	5,410
28	4,130	2,490	1,980	2,170	6,150	14,800	11,800	4,610	2,030	1,290	1,440	6,100
29	5,070	1,130	2,070	1,690	....	15,100	10,800	4,230	1,720	1,420	1,350	5,350
30	10,200	2,510	1,140	1,820	....	16,700	13,400	4,440	1,490	1,590	1,200	4,650
31	17,200	....	3,060	1,480	....	18,300	....	5,200	....	1,530	1,520	....

NOTE.—Discharge Jan. 1 to Feb. 15 estimated, because of ice, by comparison with discharge computed from power-house records.

Monthly discharge of Hudson River at Spier Falls, N. Y., for the year ending Sept. 30, 1918.

[Drainage area, 2,800 square miles.]

Month.	Discharge in second-feet.				Run-off (depth in inches on drainage area).
	Maximum.	Minimum.	Mean.	Per square mile.	
October	17,200	1,470	3,350	1.20	1.38
November	17,700	1,130	4,870	1.74	1.94
December	3,160	610	2,040	.729	.84
January	2,170	727	1,580	.564	.65
February	6,150	661	2,320	.829	.86
March	18,300	2,770	7,680	2.74	3.16
April	32,300	10,800	19,100	6.82	7.61
May	15,100	3,870	8,710	3.11	3.59
June	6,990	1,490	3,880	1.39	1.55
July	3,690	1,250	1,850	.661	.76
August	1,880	606	1,450	.518	.60
September	6,100	725	2,310	.825	.92
The year	32,300	606	4,920	1.76	23.86

HUDSON RIVER AT MECHANICVILLE, N. Y.

LOCATION.—At Duncan dam of West Virginia Pulp & Paper Co. in Mechanicville, Saratoga County, 3,700 feet above mouth of Anthony Kill, 1¼ miles below mouth of Hoosic River, and 19 miles above mouth of Mohawk River at Cohoes.

DRAINAGE AREA.—4,500 square miles.

RECORDS AVAILABLE.—1888 to 1918.

GAGE.—Water-stage recorder at the dam, installed in 1910; previous to that date staff gage.

COMPUTATIONS OF DISCHARGE.—Discharge over spillway determined from a rating curve based on United States Geological Survey coefficients for dams of ogee section; discharge through turbines computed from records of their operation; discharge at lock and through Barge canal turbines at lock computed from records of the number of lockages per day.

EXTREMES OF DISCHARGE.—Maximum daily discharge during year, 35,500 second-feet April 3; minimum daily discharge, 576 second-feet, Sunday, January 20.

1888–1918: Maximum discharge recorded, 120,000 second-feet at 6 a. m. March 28, 1913.[1] The plant is occasionally shut down and the flow of the river stored in the pond so that the discharge below the station becomes practically zero.

DIVERSIONS.—Water diverted above this station into the Champlain canal. No correction made for this diversion. During 1915 a Barge canal lock, through the Duncan dam, was completed and put into operation. Water used at the lock is included in the record.

COOPERATION.—Discharge over the spillway and through turbines of the West Virginia Pulp & Paper Co. furnished by Mr. W. J. Barnes, engineer of the company.

Daily discharge, in second-feet, of Hudson River at Mechanicville, N. Y., for the year ending Sept. 30, 1918.

Day.	Oct.	Nov.	Dec.	Jan.	Feb.	Mar.	Apr.	May.	June.	July.	Aug.	Sept.
1	1,190	19,800	3,720	1,870	1,620	8,370	25,800	15,000	7,070	2,330	1,640	631
2	1,430	16,800	4,170	1,870	1,060	7,430	30,600	17,800	5,510	2,650	1,350	1,050
3	1,750	14,300	3,000	1,830	638	8,050	35,500	14,800	5,960	2,460	1,740	2,600
4	2,120	11,300	3,840	1,810	685	6,840	35,200	15,000	4,540	1,460	587	2,520
5	2,010	10,200	3,250	1,810	638	6,040	30,800	13,400	4,750	3,020	1,220	2,140
6	2,020	8,600	3,180	1,160	584	6,980	26,500	11,200	4,280	2,290	1,670	2,170
7	1,640	7,580	3,040	1,610	1,340	6,510	23,200	9,700	4,060	1,430	1,650	1,790
8	2,440	6,460	2,840	1,600	1,850	5,390	22,500	9,500	5,050	1,990	1,520	1,190
9	1,940	6,120	2,500	1,610	1,780	5,680	24,300	10,600	5,720	2,990	1,410	1,120
10	1,980	5,270	2,250	1,620	587	6,150	25,200	11,400	5,660	2,580	1,420	1,740
11	1,600	2,800	1,900	1,540	614	5,940	23,900	11,100	7,320	2,090	1,160	2,140
12	1,530	4,910	1,880	1,500	1,050	5,250	21,200	10,700	6,060	1,820	1,200	1,890
13	1,960	4,750	1,950	795	749	5,820	9,200	11,300	6,670	2,250	2,040	2,040
14	1,570	4,520	2,040	1,400	2,520	6,640	17,800	15,200	7,140	1,710	2,010	1,580
15	2,740	4,020	2,480	1,220	4,060	6,190	16,900	16,900	6,590	3,930	1,940	1,080
16	3,490	3,760	2,670	1,420	4,210	5,740	16,000	15,800	5,490	4,140	1,710	1,360
17	3,280	3,720	2,830	1,130	1,200	8,150	17,200	13,700	5,320	3,640	1,680	2,180
18	3,390	3,160	2,480	606	3,570	9,920	18,800	11,700	5,130	3,790	988	2,090
19	3,440	3,710	2,520	606	5,840	11,700	20,500	9,450	4,770	3,400	1,190	1,780
20	3,340	3,870	2,710	576	22,400	14,200	20,900	11,200	4,020	3,120	1,670	1,780
21	2,310	3,580	2,810	741	9,610	16,400	19,500	8,430	3,300	1,830	1,670	3,360
22	3,320	3,760	2,850	1,760	7,230	18,600	23,200	8,070	4,070	2,240	1,650	3,290
23	4,300	3,920	2,080	1,940	6,960	20,900	22,400	6,710	3,380	2,160	1,630	2,700
24	3,590	5,440	3,120	2,050	6,350	21,100	24,200	6,710	3,940	1,770	1,600	3,120
25	3,770	4,450	2,220	2,010	7,830	22,700	19,600	5,480	4,790	1,350	788	2,690
26	4,440	4,680	2,700	1,720	16,000	22,500	17,500	5,560	4,130	1,260	1,040	5,230
27	4,650	4,090	2,360	1,140	11,300	20,600	15,600	6,200	3,980	1,280	1,300	12,100
28	4,790	2,940	2,170	1,250	9,950	19,100	14,100	6,450	3,100	810	1,470	9,740
29	5,310	2,700	2,000	1,900	....	19,000	12,700	5,990	2,820	1,720	1,520	8,970
30	12,000	2,600	1,350	1,480	....	20,600	14,400	5,750	1,600	2,170	1,630	7,800
31	20,100	....	1,790	1,110	....	22,400	....	6,090	....	1,920	1,460	....

[1] Highest known flood prior to this time occurred in April, 1869, calculated discharge, 70,000 second-feet. See U. S. Geological Survey Water-Supply Paper 65, p. 51, and report of U. S. Board of Engineers on Deep Waterways, Part I, pp. 377–388.

Monthly discharge of Hudson River at Mechanicville, N. Y., for the year ending Sept. 30, 1918.

[Drainage area, 4,500 square miles.]

Month.	Discharge in second-feet.				Run-off (depth in inches on drainage area).
	Maximum.	Minimum.	Mean.	Per square mile.	
October	20,100	1,190	3,660	0.813	0.94
November	19,800	2,600	6,130	1.36	1.52
December	4,170	1,350	2,600	.578	.67
January	2,060	576	1,440	.320	.37
February	22,400	584	4,720	1.05	1.09
March	22,700	5,250	12,000	2.67	3.08
April	35,500	12,700	21,800	4.84	5.40
May	17,800	5,480	10,500	2.33	2.69
June	7,320	1,600	4,870	1.08	1.20
July	4,140	810	2,310	.513	.59
August	2,040	587	1,470	.327	.38
September	12,100	631	3,130	.696	.78
The year	35,500	576	6,210	1.38	18.71

INDIAN LAKE RESERVOIR AT INDIAN LAKE, N. Y.

Location.—At the masonry storage dam at outlet of Indian Lake, 2 miles south of Indian Lake village, Hamilton County and 7½ miles above confluence of Indian River with the Hudson.

Drainage area.—131 square miles, including about 9.3 square miles of water surface of Indian Lake at the elevation of crest of spillway (measured on topographic maps).

Records available.—Records of stage and gate openings from July, 1900, to September 30, 1918.

Gages.—Elevation of water surface in reservoir is determined by chain gage on the crest of the dam near the gate house. Gage installed November 17, 1911, to replace staff gage previously maintained at the same point. Mean elevation of crest of spillway is at gage height 33.38 feet. Widths of sluice gate openings determined by gage scales at sides of gate stems inside gate house. Gages read by Lester Savarie.

Extremes of stage.—Maximum elevation of water surface in reservoir, 34.2 feet July 16, 17, and 18; minimum elevation, 5.15 feet February 25–26.

1900–1918: Maximum elevation recorded, 38.8 feet March 28, 1913; minimum elevation, 2.0 feet March 9 to 18, 1907, and January 3 to 17, 1910.

Regulation.—At ordinary stages the discharge is completely regulated by the operation of the sluice gates. Water is held in storage until needed to supplement the flow of the upper Hudson during the low-water period. This storage capacity of about 4.7 billion cubic feet provides for a discharge of about 600 second-feet for a period of 90 days. For record of discharge see Indian River near Indian Lake, N. Y., pages 146–147.

Daily gage height, in feet, of Indian Lake reservoir at Indian Lake, N. Y., for the year ending Sept. 30, 1918.

Day.	Oct.	Nov.	Dec.	Jan.	Feb.	Mar.	Apr.	May.	June.	July.	Aug.	Sept.
1	17.95	21.8	23.65	16.7	9.5	6.2	12.15	27.85	33.5	33.65	32.2	22.5
2	17.65	22.1	23.6	16.4	9.25	6.5	13.3	28.25	33.6	33.65	32.0	22.15
3	17.85	22.3	23.55	16.15	9.0	6.4	14.55	28.55	33.65	33.65	31.75	21.8
4	17.1	22.5	23.5	15.9	8.75	6.4	15.45	28.85	33.65	33.7	31.4	21.6
5	16.95	22.65	23.45	15.65	8.5	6.35	16.0	29.05	33.65	33.7	31.1	21.45
6	17.0	22.75	23.35	15.4	8.25	6.6	16.55	29.1	33.85	33.75	30.8	21.2
7	17.05	22.85	23.25	15.15	8.0	6.9	16.95	29.3	34.0	33.8	30.55	21.0
8	17.1	22.9	23.15	14.9	7.7	7.2	17.45	29.55	34.0	33.85	30.3	20.7
9	17.15	23.05	23.05	14.65	7.5	7.5	18.1	29.85	34.0	33.9	30.0	20.35
10	17.15	23.15	22.9	14.4	7.3	7.8	18.7	30.05	34.0	33.95	29.85	20.0
11	17.1	23.2	22.7	14.1	7.1	8.1	19.1	30.3	34.0	34.0	29.55	19.7
12	17.1	23.25	22.45	13.85	6.9	8.05	19.5	30.55	33.8	34.05	29.25	19.4
13	17.15	23.3	22.2	13.65	6.7	8.0	19.8	30.8	33.75	34.1	29.0	19.1
14	17.3	23.35	21.95	13.45	6.5	7.95	20.05	31.2	33.65	34.1	28.75	18.85
15	17.45	23.4	21.65	13.25	6.3	7.9	20.3	31.5	33.6	34.15	28.55	18.75
16	17.6	23.4	21.3	13.05	6.1	7.85	20.75	31.7	33.6	34.2	28.25	18.65
17	17.7	23.45	20.9	12.65	5.9	7.8	21.4	31.9	33.65	34.2	28.0	18.55
18	17.8	23.5	20.55	12.45	5.7	7.75	22.2	32.05	33.6	34.2	27.65	18.5
19	17.9	23.55	20.2	12.25	5.6	7.9	22.9	32.15	33.55	34.15	27.25	18.55
20	18.0	23.65	19.9	12.05	5.5	8.2	23.4	32.25	33.5	34.15	26.85	18.55
21	18.2	23.75	19.65	11.85	5.4	8.5	23.75	32.4	33.5	34.1	26.45	18.65
22	18.3	23.85	19.4	11.6	5.3	8.7	24.55	32.5	33.5	34.0	26.1	18.75
23	18.45	23.95	19.2	11.4	5.25	8.85	25.15	32.6	33.5	34.0	25.65	18.8
24	18.5	24.0	19.0	11.2	5.2	9.15	25.55	32.7	33.5	33.9	25.2	18.85
25	18.85	24.05	18.8	11.0	5.15	9.6	25.8	32.85	33.55	33.6	24.9	18.9
26	19.0	24.05	18.5	10.85	5.15	10.0	26.05	32.9	33.55	33.5	24.45	19.05
27	19.15	24.05	18.2	10.65	5.5	10.4	26.2	33.0	33.5	33.35	24.05	19.35
28	19.35	24.05	17.9	10.45	5.9	10.7	26.65	33.1	33.6	33.05	23.65	19.5
29	19.55	23.9	17.6	10.25	....	11.0	27.0	33.2	33.6	32.85	23.3	19.7
30	20.35	23.75	17.3	10.0	....	11.3	27.45	33.3	33.6	32.5	23.0	19.85
31	21.25	....	17.0	9.75	....	11.55	....	33.4	....	32.35	22.75	....

Gate openings, in inches, at Indian Lake reservoir at Indian Lake for the year ending Sept. 30, 1918.

From—		To—		Sluice gate A open.	Sluice gate B open.
Date.	Hour.	Date.	Hour.		
				Inches.	*Inches.*
Sept. 12	6 a. m.	Oct. 5	6 a. m.	....	48
Sept. 15	5 p. m.	Oct. 6	4 p. m.	60	....
Oct. 10	5 p. m.	Oct. 13	3 p. m.	60	....
Nov. 28	6 p. m.	Dec. 21	6 a. m.	60	....
Dec. 11	6 a. m.	Feb. 27	7 a. m.	....	48
Dec. 25	6 a. m.	Feb. 27	7 a. m.	30	....
Mar. 3	7 a. m.	Mar. 5	6 p. m	30	....
Mar. 3	7 a. m.	Mar. 5	6 p. m.	....	48
Mar. 11	5 p. m.	Mar. 19	1 p. m.	60	....
Mar. 11	5 p. m.	Mar. 19	1 p. m.	....	45
Apr. 20	1 p. m.	Apr. 20	9 p. m.	60	....
Apr. 20	9 p. m.	Apr. 21	7 a. m.	30	....
Apr. 21	7 a. m.	Apr. 21	1 p. m.	60	....
Apr. 23	3 p. m.	Apr. 23	11 p. m.	60	....
Apr. 24	10 p. m.	Apr. 26	5 a. m.	60	....
Apr. 26	1 p. m.	Apr. 27	11 a. m.	60	....
May 5	7 p. m.	May 6	7 p. m.	60	....
July 24	9 a. m.	July 25	6 p. m.	....	54
July 25	6 p. m.	July 27	5 p. m.	....	30
July 27	5 p. m.	Sept. 14	4 p. m.	....	54
Aug. 18	7 a. m.	Sept. 3	11 a. m.	60	....
Sept. 7	5 p. m.	Sept. 20	6 p. m.	60	....

NOTE.—The main logway was open 15 feet during the following periods: June 10, 7 a. m. to 10 a. m.; June 12, 7 a. m. to 6 p. m.; June 13, 10 a. m. to 2 p. m.; June 14, 9 a. m. to 6 p. m.; June 15, 2 p. m. to 6 p. m. It was also open 1 foot in width from 7 p. m. Aug. 3 to 7 a. m. Aug. 18.

INDIAN RIVER NEAR INDIAN LAKE, N. Y.

Location.—Three-fourths of a mile below State dam at the outlet of Indian Lake, : miles south of Indian Lake village, Hamilton County, 1 mile above mouth of Big Brook, and 6¼ miles above mouth.

Drainage area.—132 square miles (measured on topographic maps).

Records available.—July 1, 1912, to June 30, 1914; June 5, 1915, to September 30, 1918; also miscellaneous measurements in 1911.

Gage.—Gurley repeating-hydrograph water-stage recorder; installed August 30, 1916, in a standard wooden shelter on the right bank about three-fourths mile below the dam, at same datum as staff gage previously used. The staff gage is still in place and is used for checking the recorder. Recorder inspected by Lester Savarie.

Discharge measurements.—Made from cable or by wading at the head of the rapids about 150 feet below the gage.

Extremes of discharge.—Maximum stage, from water-stage recorder, 4.85 feet at 4 a. m. June 12 (discharge, 1,450 second-feet); minimum stage, from water-stage recorder, 0.07 foot at 12 p. m. September 30 (discharge, about 0.7 second-foot).

1900–1918: Maximum stage recorded; 7.8 feet March 28, 1913 (discharge, 3,460 second-feet); minimum stage that of September 30, 1918.

Channel and control.—The gage is at the side of a pool about 500 feet wide, called the "lower frog pond." The reef of coarse gravel at the outlet of this pool forms the control and is permanent.

Winter flow.—Stage-discharge relation not affected by ice.

Regulation.—Discharge at this station is regulated by the operation of gates at the dam.

Accuracy.—Stage-discharge relation permanent; not affected by ice. Rating curve well defined between 15 and 1,500 second-feet. Daily discharge for days on which no changes were made in the sluice gate openings at Indian Lake dam ascertained by applying to rating table; mean daily gage height determined by inspecting recorder graph; discharge for days on which gate openings are changed is mean of 24 hourly determinations.

Discharge measurements of Indian River at Indian Lake, N. Y., during the year ending Sept. 30, 1918.

Date.	Made by—	Gage height.	Discharge.
		Feet.	*Sec.-ft.*
June 22[a]	J. W. Moulton	1.51	85.8
July 15[a]	.....do	1.40	91.3

[a] Logs on the control.

Daily discharge, in second-feet, of Indian River near Indian Lake, N. Y., for the year ending Sept. 30, 1918.

Day.	Oct.	Nov.	Dec.	Jan.	Feb.	Mar.	Apr.	May.	June.	July.	Aug.	Sept.
1	623	4	278	600	402	4	8	9	24	18	564	725
2	603	3	278	600	388	3	9	10	25	26	564	725
3	603	2	275	600	368	203	6	10	26	30	575	599
4	603	2	272	600	363	338	4	10	26	36	623	453
5	603	2	272	560	358	265	1	65	27	49	623	453
6	150	2	270	550	353	4	2	292	31	40	623	453
7	4	1	270	545	347	3	2	258	54	42	623	473
8	2	1	270	545	338	3	2	18	67	40	623	684
9	2	1	270	545	325	3	5	13	74	46	603	684
10	39	1	270	526	319	3	4	13	499	50	603	664
11	200	1	592	526	313	75	3	13	152	60	603	664
12	205	1	725	526	307	313	2	12	874	75	584	664
13	140	1	725	500	304	313	2	14	428	86	603	643
14	4	1	725	500	301	310	4	16	795	90	603	433
15	2	1	725	500	298	307	4	15	566	90	584	220
16	2	1	725	480	298	316	3	15	110	90	584	217
17	2	1	725	480	295	310	3	15	190	100	584	217
18	1	2	725	480	292	307	2	16	95	95	668	214
19	1	2	725	480	289	180	2	16	95	90	832	212
20	2	2	704	480	286	11	115	18	90	100	810	187
21	2	2	544	460	284	11	155	19	90	90	810	6
22	2	2	436	460	284	9	7	19	90	85	788	2
23	1	2	436	460	280	9	93	18	90	80	788	1
24	2	2	436	440	280	6	24	17	90	448	767	1
25	4	2	570	440	280	5	278	19	90	570	767	1
26	2	2	623	440	280	5	178	22	173	353	767	1
27	2	2	623	420	88	3	160	22	18	405	746	2
28	3	64	623	420	4	2	7	22	14	584	746	1
29	3	281	600	420	......	3	7	22	13	584	725	1
30	15	281	600	460	......	47	7	23	12	564	725	1
31	7	......	600	400	......	130	......	24	......	564	746	......

NOTE.—Discharge Dec. 29 to Jan. 6, and Jan. 13 to 31 estimated, for lack of gage-height record, from study of recorder graph and examination of record of operation of gates at Indian Lake dam. Discharge June 16 to July 25 estimated, because of logs on the control, from discharge measurements and study of recorder graph.

Monthly discharge of Indian River near Indian Lake, N. Y., for the year ending Sept. 30, 1918.

[Drainage area, 132 square miles.]

Month.	Discharge in second-feet.				Run-off (depth in inches on drainage area).
	Maximum.	Minimum.	Mean.	Per square mile.	
October	623	1	124	0.939	1.08
November	281	1	22.4	.170	.19
December	725	270	513	3.89	4.48
January	600	400	496	3.76	4.34
February	402	4	297	2.25	2.34
March	338	2	113	.856	.99
April	278	2	36.7	.278	.31
May	292	9	34.7	.263	.30
June	874	12	161	1.22	1.36
July	584	18	180	1.36	1.57
August	832	564	673	5.10	5.88
September	725	1	320	2.42	2.70
The year	874	1	248	1.88	25.54

SCHROON RIVER AT RIVERBANK, N. Y.

LOCATION.—At steel highway bridge near Riverbank post office, Warren County, near Tumblehead Falls, 9 miles below Schroon Lake, and 9 miles above Warrensburg.

DRAINAGE AREA.—534 square miles.

RECORDS AVAILABLE.—September 2, 1907, to September 30, 1918.

GAGE.—Chain, on upstream side of bridge; read by J. H. Roberts.

DISCHARGE MEASUREMENTS.—Made from the upstream side of bridge.

CHANNEL AND CONTROL.—Gravel; occasionally shifting. Logs become lodged on the control for a portion of nearly every year.

EXTREMES OF DISCHARGE.—Maximum stage recorded during year, 7.25 feet at 9 a. m. and 4 p. m. April 4 (discharge, 5,820 second-feet); minimum stage recorded, 1.16 feet at 4 p. m. October 10 (discharge, 89 second-feet).

1907–1918: Maximum stage recorded, 10.7 feet at 5 p. m. March 28, 1913 (discharge about 13,500 second-feet); minimum stage recorded, 0.85 foot at 5 p. m. October 17, 1909 (discharge about 28 second-feet).

ICE.—Stage-discharge relation affected by ice.

REGULATION.—Flow affected by storage in Schroon and Brant lakes.

ACCURACY.—Stage-discharge relation probably permanent during year, except as affected by ice for a large part of the period from December to March and by logs on the control for a short period in May and June. Rating curve well defined between 150 and 4,000 second-feet. Gage read to hundredths twice daily. Daily discharge ascertained by applying mean daily gage height to rating table. Open-channel records good; other records fair.

Discharge measurements of Schroon River at Riverbank, N. Y., during the year ending Sept. 30, 1918.

Date.	Made by—	Gage height.	Discharge.	Date.	Made by—	Gage height.	Discharge.
		Feet.	*Sec.-ft.*			*Feet.*	*Sec.-ft.*
Dec. 15[a]	E. D. Burchard	3.08	394	Apr. 10[c]	J. W. Moulton	6.07	3,660
Jan. 9[a]	J. W. Moulton	2.41	357	May 3	.....do	4.52	2,050
28[b]	E. D. Burchard	2.34	207	June 19[c]	.....do	3.86	1,090
Mar. 2[b]	J. W. Moulton	2.85	324	July 12	.....do	1.54	179
25[c]	.....do	4.86	1,380	12	.....do	1.54	180
Apr. 1[c]	.....do	6.02	3,040				

[a] Measurement made through incomplete ice cover.
[b] Measurement made through complete ice cover.
[c] Gage height affected by logs on the control.

Daily discharge, in second-feet, of Schroon River at Riverbank, N. Y., for the year ending Sept. 30, 1918.

Day.	Oct.	Nov.	Dec.	Jan.	Feb.	Mar.	Apr.	May.	June.	July.	Aug.	Sept.
1	248	1,060	535	280	200	300	3,000	2,040	990	585	201	186
2	216	1,290	535	280	200	320	3,600	2,150	900	156	201	172
3	264	1,290	490	260	190	340	5,000	2,040	1,100	180	201	158
4	201	1,210	512	260	190	340	5,750	1,940	750	130	186	156
5	201	1,210	490	260	200	360	5,570	1,560	650	135	186	156
6	216	1,130	468	260	190	380	4,950	1,640	600	140	172	153
7	201	1,060	468	260	200	400	4,320	1,600	600	153	172	158
8	125	1,060	460	260	200	420	4,170	1,600	900	140	158	186
9	93	990	440	260	200	440	4,020	1,700	600	132	156	158
10	89	920	440	240	200	480	3,880	1,600	1,200	156	148	180
11	148	860	420	240	190	500	3,740	1,500	980	167	186	167
12	216	800	420	240	180	550	3,470	1,500	500	172	186	164
13	251	800	400	240	180	550	3,210	1,700	400	172	186	172
14	298	860	400	260	180	600	2,960	2,000	1,000	186	186	172
15	298	860	400	260	180	600	2,840	2,200	1,000	232	201	169
16	298	920	400	260	170	600	2,840	2,200	460	232	186	167
17	298	860	380	260	150	550	2,960	2,000	1,000	264	186	490
18	298	800	380	260	150	600	3,080	2,000	1,100	298	158	662
19	298	800	360	260	150	650	3,080	1,800	1,100	296	160	540
20	298	745	360	260	160	800	3,080	1,600	920	264	153	232
21	298	690	340	260	170	800	2,840	1,500	920	264	148	201
22	264	718	340	260	190	800	2,840	1,300	990	264	145	186
23	232	745	320	240	200	900	2,840	1,000	407	264	153	369
24	248	662	320	240	220	1,100	2,840	1,200	407	248	153	369
25	216	635	320	240	240	1,400	2,840	850	535	248	142	369
26	216	610	320	220	260	1,600	2,600	800	535	232	142	407
27	248	665	320	220	280	1,900	2,370	800	512	216	140	298
28	232	560	300	200	280	2,200	2,150	750	535	216	145	351
29	264	535	300	200	....	2,400	1,740	750	298	232	140	351
30	490	512	280	200	....	2,400	1,840	750	153	232	142	369
31	216	....	280	200	....	2,600	....	800	....	216	132	....

NOTE.—Discharge Dec. 8 to Apr. 3 estimated, because of ice, and discharge May 7 to June 19 estimated, because of logs, from discharge measurements, weather records, study of recorder graph, and comparison with similar studies for Hudson River at North Creek.

Monthly discharge of Schroon River at Riverbank, N. Y., for the year ending Sept. 30, 1918.

[Drainage area, 534 square miles.]

Month.	Discharge in second-feet.				Run-off (depth in inches on drainage area).
	Maximum.	Minimum.	Mean.	Per square mile.	
October	490	89	241	0.451	0.52
November	1,290	512	862	1.62	1.81
December	535	280	394	.738	.85
January	280	200	246	.461	.53
February	280	150	196	.367	.38
March	2,600	300	899	1.68	1.94
April	5,750	1,740	3,350	6.28	7.01
May	2,200	750	1,510	2.83	3.26
June	1,200	153	724	1.35	1.51
July	585	130	219	.410	.47
August	201	132	166	.311	.36
September	662	153	262	.492	.55
The year	5,750	89	755	1.41	19.19

SACANDAGA RIVER NEAR HOPE, N. Y.

LOCATION.—About 1½ miles below junction of East and West branches, 3½ miles above Hope post office, Hamilton County, and 12 miles above Northville.

DRAINAGE AREA.—494 square miles (measured on topographic maps).

RECORDS AVAILABLE.—September 15, 1911, to September 30, 1918.

Gage.—Staff in two sections, the lower inclined, the upper vertical; read by Melvin Willis.

Discharge measurements.—Made from a cable about 100 feet below the gage or by wading.

Channel and control.—Rocky; probably permanent.

Extremes of discharge.—Maximum stage recorded during year, 6.7 feet at 5.55 p. m. October 30 (discharge, 8,490 second-feet); minimum stage recorded, 1.28 feet at 6.30 p. m. August 28 and 7.20 a. m. August 29 (discharge, 37 second-feet).

1911–1918: Maximum stage recorded, 10.0 feet at 5.30 p. m. March 27, 1913 (discharge, 24,800 second-feet); minimum stage recorded, 1.17 feet at 7.55 a. m. September 30, 1913 (discharge about 20 second-feet).

Ice.—Stage-discharge relation affected by ice.

Accuracy.—Stage-discharge relation permanent; affected by ice for a large part of the period December to March, inclusive. Rating curve well defined between 60 and 10,000 second-feet. Gage read to half-tenths twice daily. Daily discharge ascertained by applying mean daily gage height to rating table. Open-water records good; winter records fair.

Discharge measurements of Sacandaga River near Hope, N. Y., during the year ending Sept. 30, 1918.

Date.	Made by—	Gage height.	Discharge.
		Feet.	*Sec.-ft.*
Jan. 8[a]	E. D. Burchard	2.62	203
29[a]	J. W. Moulton	2.70	203
30[a]	.....do	2.72	201

[a] Measurement made through complete ice cover.

Daily discharge, in second-feet, of Sacandaga River near Hope, N. Y., for the year ending Sept. 30, 1918.

Day.	Oct.	Nov.	Dec.	Jan.	Feb.	Mar.	Apr.	May.	June.	July.	Aug.	Sept.
1	202	3,540	240	200	200	1,330	4,480	2,740	660	310	114	590
2	164	2,740	240	200	200	1,220	5,790	2,740	660	320	106	335
3	147	2,230	220	190	200	1,160	7,530	2,930	590	273	101	175
4	230	1,810	240	190	200	1,010	6,350	2,930	590	255	89	111
5	400	1,440	220	180	200	910	5,250	2,560	558	230	81	111
6	525	1,220	220	180	200	820	3,760	2,080	525	217	83	202
7	410	1,110	220	180	200	910	3,540	1,810	910	221	79	154
8	370	1,010	220	200	200	910	4,480	1,810	1,110	213	73	141
9	335	910	220	220	200	820	6,070	1,680	1,010	213	161	132
10	310	820	220	220	200	740	5,790	1,560	1,010	320	141	138
11	264	700	240	260	220	700	3,990	1,680	820	273	128	164
12	380	625	240	280	240	660	4,230	1,560	910	273	122	182
13	910	558	260	320	260	740	2,390	4,230	1,110	255	111	186
14	780	525	260	280	320	780	2,230	5,520	1,160	400	96	213
15	820	495	260	240	400	910	2,740	3,990	960	365	89	205
16	1,010	495	260	240	500	820	3,330	3,130	780	350	83	175
17	1,010	495	260	240	700	820	4,230	2,560	660	330	75	175
18	960	465	240	240	850	865	5,250	2,080	590	305	71	242
19	1,010	443	240	260	1,000	910	4,990	1,560	465	273	68	280
20	1,330	421	260	260	1,100	1,110	3,760	1,330	400	255	61	280
21	1,160	410	260	260	1,300	1,440	3,330	1,290	355	230	59	310
22	910	380	260	240	1,300	2,560	3,130	1,160	454	213	56	340
23	820	360	260	240	1,300	2,740	3,330	1,110	465	182	52	360
24	910	340	240	240	1,220	2,390	3,330	1,110	443	161	48	330
25	1,560	320	240	220	1,220	2,930	3,130	1,010	375	141	45	310
26	1,560	320	220	220	1,440	2,740	2,740	1,010	340	132	44	310
27	1,330	300	230	260	1,440	2,390	2,560	910	315	116	40	292
28	1,940	280	220	220	1,440	2,230	2,560	820	292	116	38	315
29	1,810	260	200	200	......	2,230	2,560	780	255	108	39	335
30	1,810	260	200	200	......	3,130	2,740	820	238	122	48	350
31	5,790	......	200	200	......	3,540	......	740	......	136	45	......

Note.—Discharge Nov. 22 to Feb. 28 estimated, because of ice, from discharge measurements, weather records, study of recorder graph, and comparison with similar studies for Sacandaga River near Hadley.

Monthly discharge of Sacandaga River near Hope, N. Y., for the year ending Sept. 30, 1918.

[Drainage area, 494 square miles.]

Month.	Discharge in second-feet.				Run-off (depth in inches on drainage area).
	Maximum.	Minimum.	Mean.	Per square mile.	
October	5,790	147	1,010	2.06	2.38
November	3,540	260	843	1.71	1.91
December	260	200	235	.476	.55
January	320	180	226	.457	.53
February	1,440	200	652	1.32	1.38
March	3,540	660	1,500	3.04	3.50
April	7,530	2,230	3,990	8.08	9.02
May	5,820	740	1,970	3.99	4.60
June	1,160	238	634	1.28	1.43
July	400	108	235	.476	.55
August	161	38	78.9	.160	.18
September	590	111	244	.494	.55
The year	7,530	38	965	1.95	26.58

SACANDAGA RIVER AT HADLEY, N. Y.

LOCATION.—Half a mile west of railroad station at Hadley, Saratoga County, 1 mile above mouth of river, and 4½ miles below site of proposed storage dam at Conklingville.

DRAINAGE AREA.—1,060 square miles (measured on topographic maps).

RECORDS AVAILABLE.—January 1, 1911, to September 30, 1918. September 13, 1907, to December 31, 1910, at upper bridge station; September 24, 1909, to midsummer of 1911 at lower bridge station.

GAGE.—Gurley water-stage recorder in a concrete shelter on the left bank, about one-half mile west of railroad station at Hadley; installed January 6, 1916, replacing a Barrett & Lawrence water-stage recorder. Recorder inspected by J. F. Kelly.

DISCHARGE MEASUREMENTS.—Made from a cable about 30 feet above the gage, or by wading under the cable or about three-fourths of a mile above gage.

CHANNEL AND CONTROL.—Very rough, but permanent.

EXTREMES OF DISCHARGE.—Maximum stage during year, from water-stage recorder, 8.8 feet from 1 to 4 a. m. April 4 (discharge, 13,900 second-feet); minimum stage, from water-stage recorder, 2.36 feet at 10 p. m. August 28 (discharge, 92 second-feet).

1911–1918: Maximum stage, from water-stage recorder, 12.36 feet from 11 a. m. till noon March 28, 1913 (discharge, from 35,500 second-feet); minimum stage, from water-stage recorder, 2.25 feet all day September 15, 1913 (discharge about 61 second-feet).

ICE.—Stage-discharge relation seriously affected by ice.

ACCURACY.—Stage-discharge relation permanent; affected by ice during a large part of period from December to March, inclusive. Rating curve well defined between 150 and 20,000 second-feet. Operation of water-stage recorder satisfactory throughout the year. Daily discharge ascertained by applying to the rating table mean daily gage height determined by inspecting recorder graph. Open-water records excellent; winter records fair.

Discharge measurements of Sacandaga River at Hadley, N. Y., for the year ending Sept. 30, 1918.

Date.	Made by—	Gage height.	Discharge.	Date.	Made by—	Gage height.	Discharge.
		Feet.	*Sec.-ft.*			*Feet.*	*Sec.-ft.*
Dec. 11[a]	E. D. Burchard	5.63	486	Apr. 2	J. W. Moulton	7.82	10,200
Jan. 4[b]	J. W. Moulton	3.61	410	25	....do	6.91	7,400
29[b]	E. D. Burchard	3.44	437	26	E. D. Burchard	6.74	6,630
Mar. 1[a]	J. W. Moulton	8.52	3,750	July 11	J. W. Moulton	3.29	607
9[a]	....do	5.48	1,850	11	....do	3.31	589
21[a]	....do	5.72	3,190				

[a] Incomplete ice cover or ice jam on control. [b] Complete ice cover on control.

Daily discharge, in second-feet, of Sacandaga River at Hadley, N. Y., for the year ending Sept. 30, 1918.

Day.	Oct.	Nov.	Dec.	Jan.	Feb.	Mar.	Apr.	May.	June.	July.	Aug.	Sept.
1	218	7,430	420	420	420	3,800	8,700	4,580	1,800	539	250	214
2	250	7,130	440	420	440	3,800	10,400	4,580	1,620	601	250	631
3	268	6,140	460	420	440	2,800	12,600	4,460	1,330	714	232	545
4	278	5,080	480	400	420	2,200	13,500	4,460	1,100	637	222	383
5	334	4,100	460	380	440	1,800	11,500	4,340	986	582	210	302
6	608	3,250	440	340	420	1,600	9,710	4,100	947	506	201	283
7	730	2,600	460	300	420	1,700	8,700	3,660	1,040	481	197	307
8	668	2,110	460	260	420	1,800	8,050	3,350	1,780	486	184	389
9	582	1,660	440	240	420	1,860	8,700	2,960	1,740	455	222	344
10	506	1,520	460	240	380	1,900	9,370	2,780	1,560	461	263	286
11	474	1,880	440	260	360	2,000	9,710	2,960	1,760	553	328	252
12	443	1,240	550	300	400	1,600	8,700	2,870	1,950	615	317	234
13	594	1,130	550	320	600	1,500	7,430	3,060	2,780	660	307	227
14	1,150	1,020	550	320	800	1,700	6,410	4,700	2,870	746	292	263
15	1,060	956	600	340	800	1,900	5,730	5,880	2,600	996	273	344
16	1,290	901	600	400	1,000	2,000	5,730	6,000	2,110	1,090	245	366
17	1,480	882	600	380	1,300	1,600	6,270	5,470	1,650	966	222	334
18	1,230	847	600	420	1,600	1,700	6,980	4,700	1,340	956	218	336
19	1,110	795	550	440	2,000	2,000	7,740	3,880	1,100	976	189	443
20	1,160	778	550	480	2,400	2,400	7,740	3,950	919	847	176	566
21	1,650	780	550	800	2,600	3,200	7,430	2,780	787	706	161	663
22	1,530	750	550	440	2,600	4,400	7,280	2,430	821	601	149	795
23	1,270	750	350	400	2,600	5,730	7,430	2,170	1,090	493	146	795
24	1,170	750	550	400	2,400	6,980	7,430	1,880	1,200	436	138	683
25	1,480	750	500	400	2,400	7,740	6,980	1,600	1,140	412	135	630
26	2,110	650	480	440	2,800	7,740	6,550	1,520	976	401	124	1,040
27	2,110	600	480	460	3,400	7,430	5,860	1,600	821	355	118	2,840
28	1,960	550	480	440	3,800	7,740	5,210	1,770	714	297	101	3,150
29	2,600	500	480	440	....	7,430	4,700	1,720	622	283	101	2,690
30	4,440	440	460	420	....	7,740	4,580	1,650	559	263	107	2,190
31	6,550	....	440	420	....	7,740	....	2,030	....	250	121	....

NOTE.—Discharge Nov. 22 to Mar. 22 estimated, because of ice, from discharge measurements, weather records, study of graph, and comparison with similar studies for Sacandaga River near Hope.

Monthly discharge of Sacandaga River near Hadley, N. Y., for the year ending Sept. 30, 1918.

[Drainage area, 1,060 square miles.]

Month.	Discharge in second-feet.				Run-off (depth in inches on drainage area).
	Maximum.	Minimum.	Mean.	Per square mile.	
October	6,550	218	1,330	1.25	1.44
November	7,430	440	1,910	1.80	2.01
December	600	420	504	.475	.55
January	500	240	382	.360	.42
February	3,800	360	1,360	1.28	1.33
March	7,740	1,500	3,720	3.51	4.05
April	13,500	4,580	7,900	7.45	8.31
May	6,000	1,520	3,330	3.14	3.62
June	2,870	559	1,390	1.31	1.46
July	1,090	250	591	.558	.64
August	323	101	200	.189	.22
September	3,150	214	751	.708	.79
The year	13,500	101	1,940	1.83	24.84

HOOSIC RIVER NEAR EAGLE BRIDGE, N. Y.

LOCATION.—Half a mile below Walloomsac River and 1½ miles above Owl Kill and Eagle Bridge, Rensselaer County.

DRAINAGE AREA.—512 square miles (measured on topographic maps).

RECORDS AVAILABLE.—August 13, 1910, to September 30, 1918. September 25, 1903, to December 31, 1908, at Buskirk, 4 miles below present station.

GAGE.—Chain gage on the left bank near the farmhouse of James Russell, about 1½ miles above Eagle Bridge, installed September 4, 1918. From August 17, 1914, to September 3, 1918, an inclined staff gage on the left bank about 50 feet above the chain gage. From August 13, 1910, to August 16, 1914, chain gage on the left bank about 450 feet above the present chain gage. Gage read by Mrs. Viola Davis, Mrs. Volney Russell, and Mrs. J. E. Sherman.

DISCHARGE MEASUREMENTS.—Made from cable half a mile below gage or by wading.

CHANNEL AND CONTROL.—Gravel; somewhat shifting.

EXTREMES OF DISCHARGE.—Maximum stage recorded during year, 12.8 feet at 5 p. m. February 15 (discharge about 11,300 second-feet); minimum stage recorded, 2.1 feet at 7.30 a. m. September 8 (discharge about 40 second-feet).

1910–1918: Maximum stage not recorded, as gage used prior to August 17, 1914, could not be reached at high stages; minimum stage recorded, 6.1 feet at 5 p. m. September 14, 1913 (discharge practically zero).

ICE.—Stage-discharge relation affected by ice.

REGULATION.—Flow affected by storage on Walloomsac River and at Hoosick Falls about 2 miles above gage.

ACCURACY.—Stage-discharge relation probably permanent between dates of shifting; affected by ice during a large part of the period December to March, inclusive. Rating curve well defined between 75 and 7,000 second-feet. Gage read to quarter-tenths twice daily. Daily discharge ascertained by applying mean daily gage height to rating table. Records good except for periods of low water when semidaily gage heights may not indicate the true mean, and those for periods when the stage-discharge relation is affected by ice, which are fair.

Discharge measurements of Hoosic River near Eagle Bridge, N. Y., during the year ending Sept. 30, 1918.

Date.	Made by—	Gage height.	Discharge.	Date.	Made by—	Gage height.	Discharge.
		Feet.	*Sec.-ft.*			*Feet.*	*Sec.-ft.*
Dec. 28[a]	J. W. Moulton	3.80	201	May 20	J. W. Moulton	4.52	1,040
Jan. 7[a]	E. D. Burchard	4.10	133	June 19	M. H. Carson	3.14	288
28[a]	J. W. Moulton	4.68	199	19	do	3.21	288
Apr. 1	E. D. Burchard	6.19	2,830	19	E. D. Burchard	3.21	294
1	do	6.08	2,630	Sept. 4	do	[b] 2.86	181
May 20	J. W. Moulton	4.54	1,040	4	do	[b] 2.85	178

[a] Measurement made under complete ice cover. [b] Observed on chain gage installed this day.

Daily discharge, in second-feet, of Hoosic River near Eagle Bridge, N. Y., for the year ending Sept. 30, 1918.

Day.	Oct.	Nov.	Dec.	Jan.	Feb.	Mar.	Apr.	May.	June.	July.	Aug.	Sept.
1	133	1,390	445	130	200	1,770	2,810	1,670	450	340	156	320
2	162	940	370	110	200	1,570	4,300	2,100	428	320	132	268
3	130	860	498	100	65	1,470	4,150	1,570	450	360	108	188
4	159	555	370	130	130	1,020	2,690	1,280	340	208	88	185
5	182	645	280	95	220	1,020	1,990	1,100	302	302	82	136
6	152	498	348	110	120	2,570	1,570	870	320	250	142	150
7	200	370	445	220	110	1,990	1,670	835	500	136	110	112
8	193	420	370	280	100	1,100	1,880	765	582	285	132	65
9	152	395	272	120	170	870	2,210	765	428	250	168	97
10	179	302	440	280	70	835	2,450	1,470	500	250	199	108
11	133	348	480	240	160	555	1,880	980	340	340	150	116
12	268	420	440	220	200	800	1,570	940	450	302	145	82
13	216	325	360	190	460	2,100	1,280	905	640	340	142	124
14	248	325	380	260	600	1,770	1,280	3,590	582	217	140	85
15	182	325	280	300	7,000	1,370	1,670	2,450	475	428	130	68
16	260	280	360	220	4,400	980	1,770	1,770	340	220	128	110
17	208	348	420	280	2,200	1,570	1,770	1,280	405	285	120	128
18	200	248	420	240	1,700	1,570	2,330	1,100	268	320	72	130
19	248	348	420	200	2,200	2,330	1,880	940	268	285	80	190
20	280	280	340	140	9,000	2,690	1,470	1,020	250	235	132	208
21	280	260	190	200	8,870	3,450	1,570	765	250	185	140	555
22	280	325	340	240	2,100	4,450	3,730	835	450	199	128	640
23	204	498	170	260	1,990	4,150	2,690	300	1,020	170	91	405
24	220	420	120	320	1,990	2,690	2,450	610	905	170	70	285
25	370	280	200	220	1,470	3,590	1,880	640	582	132	92	360
26	470	445	190	280	7,070	2,330	1,570	730	450	145	104	1,770
27	260	470	180	95	2,570	1,770	1,280	730	475	140	126	3,190
28	302	395	240	240	1,990	1,280	1,190	640	302	86	100	1,370
29	470	280	130	190	......	1,370	1,190	555	268	110	120	800
30	325	302	65	260	......	1,670	1,190	582	170	130	130	730
31	3,330	......	130	180	......	2,100	......	528	......	130	110	......

NOTE.—Discharge Dec. 10 to Feb. 20 estimated, because of ice, from discharge measurements, weather records, and study of recorder graph. Discharge Sept. 4 to 30 determined from gage heights observed on new chain gage.

Monthly discharge of Hoosic River near Eagle Bridge, N. Y., for the year ending Sept. 30, 1918.

[Drainage area, 512 square miles.]

Month.	Discharge in second-feet.				Run-off (depth in inches on drainage area).
	Maximum.	Minimum.	Mean.	Per square mile.	
October	3,330	130	336	0.656	0.76
November	1,390	248	448	.866	.97
December	498	65	313	.611	.70
January	320	95	205	.400	.46
February	9,000	65	1,870	3.65	3.80
March	4,450	555	1,900	3.71	4.28
April	4,800	1,190	2,050	4.00	4.46
May	3,590	528	1,120	2.19	2.52
June	1,020	170	440	.859	.96
July	428	86	235	.459	.53
August	199	70	121	.236	.27
September	3,190	65	432	.844	.94
The year	9,000	65	779	1.52	20.65

MOHAWK RIVER AT VISCHER FERRY DAM, N. Y.

Location.—At Vischer Ferry dam of Barge canal (Lock No. 7), 1 mile above Stony Creek and Vischer Ferry, 7 miles below Schenectady, Schenectady County, and 11 miles above mouth.

Drainage area.—3,430 square miles (measured on topographic maps).

Records available.—June 24, 1913, to September 30, 1918.

Gage.—Stevens water-gage recorder (showing head on crest of spillway) in the southerly corner of the basin near upper end of Barge canal lock, installed August 18, 1916. Inclined staff gage at foot of an old bridge abutment about 100 feet above Vischer Ferry, read June 24 to December 16, 1913, and May 24 to June 2, 1914; staff gage in masonry of outer lock wall, just above upper gates, read March 30 to May 23, 1914, and March 30 to August 17, 1916. Datum of staff gage 12.1 feet lower than that of recorder. Gurley water-stage recorder in the northerly (out stream) corner of the basin, used December 17, 1913, to March 29, 1914, and May 24, 1914, to February 23, 1916. This gage was destroyed by ice April 2, 1916, and the record from February 24 to April 2 was lost with it. Water-stage recorder inspected by engineers from the Albany office of the United States Geological Survey; staff gage read by lock tenders.

Discharge measurements.—Made by wading below the dam at low water during 1913–14. During the spring of 1915 the Crescent dam (next downstream) was closed, making further measurement impossible. No provision for measurements at medium and high stages.

Channel and control.—The control is the crest of the spillway.

Extremes of discharge.—Maximum stage during year, from water-stage recorder, 4.00 feet at 7 a. m. October 31 (discharge, 50,200 second-feet); minimum stage, from water-stage recorder, 0.29 foot at 6.45 p. m. October 14 (discharge, 670 second-feet).

1913–1918: Maximum stage recorded, 7.6 feet just before noon March 28, 1914, determined by leveling from flood marks (discharge estimated by New York State engineer about 140,000 second-feet). This stage lasted but a few moments and was caused by the breaking of an ice jam near Schenectady. Minimum stage from water-stage recorder 0.18 foot from 4 a. m. to 5 a. m. and 4 p. m. to 6 p. m. October 31, 1914 (discharge about 290 second-feet).

Diversions.—Water was diverted into Erie canal at temporary lock in north end of dam prior to December, 1914. Measurements of this diversion were made at bridge 48, about a mile downstream, but no allowance for the diversion was made in computing the flow.

Barge canal lock No. 7 at the south end of dam was put in operation May 15, 1915. The following tables of discharge include the flow over the spillway and through the lock and water wheels.

Accuracy.—Stage-discharge relation practically permanent; probably not affected by ice. Rating curve fairly well defined by discharge measurements between 350 and 2,500 second-feet; above 2,500 second-feet, based on theoretic coefficients. Operation of water-stage recorder satisfactory during periods of record. Daily discharge determined by use of discharge integrator. Records good for periods of low water when the water-stage recorder was in operation; fair for other periods.

Cooperation.—Recorder inspected by an employee of the State superintendent of public works.

Daily discharge, in second-feet, of Mohawk River at Vischer Ferry dam, N. Y., for the year ending Sept. 30, 1918.

Day.	Oct.	Nov.	Dec.	Mar.	Apr.	May.	June.	July.	Aug.	Sept.
1	1,200	23,900	3,930		15,800		2,940	2,490	1,080	
2	1,740	14,900	4,810		18,200		2,790	4,800	1,250	
3	1,570	9,830	5,570		18,900		2,020	2,690	1,510	
4	1,740	7,280	4,860		15,700		2,080	2,490	1,490	
5	2,390	5,740			12,800		2,020	1,960	1,450	
6	2,670	5,120			10,500		2,180	2,700	1,510	
7	2,800	5,020			9,790		2,610	2,130	1,580	2,520
8	1,720	5,270			9,680		3,480	1,660	2,730	1,260
9	1,810	4,470			15,080		2,850	1,950	1,490	1,180
10	2,130	3,750			19,600		3,340	2,280	3,570	1,110
11	2,340	3,920			15,800		4,630	3,980	2,490	1,320
12	1,910	3,740			13,200		11,100	2,610		1,890
13	2,680	3,770			11,460		8,900	3,040		
14	2,440	3,620			12,500		6,670	3,600		
15	3,410	3,410			16,700		3,730	5,180		
16	6,380	3,310			14,600		3,590	5,240		
17	4,310	2,870			13,660		3,440	2,950		1,790
18	4,120	3,220			16,700	4,540		3,350		3,420
19	3,290	2,990		20,400	19,200	4,000	2,340	3,040		2,880
20	6,830	3,100		22,100	15,700	3,720	2,150	2,440	1,120	2,960
21	8,070	3,230		26,600	12,600	4,920	2,080	2,380	1,380	3,830
22	6,490	3,790		28,700	16,800	4,270	3,260	1,820	1,260	3,730
23	4,480	8,840		36,000		4,580	2,770	2,180	1,140	2,560
24	5,870	6,680		29,500		3,560	2,590	1,650	1,130	3,620
25	10,200	4,820		22,300		3,660	3,150	1,710	1,160	2,910
26	8,880	2,820		19,300		2,800	2,520	1,480		7,370
27	6,520	2,470		14,400		5,120	2,340	1,420		
28	7,120	3,690		11,100		5,480	2,020	1,250		
29	8,980	2,340		11,000		4,470	2,540	1,160		
30	23,200	2,650		13,400		3,780	1,560	1,180		
31	43,900			14,900		3,300		1,020		

Note.—No discharge record Dec. 5 to Mar. 18, Apr. 23 to May 17, June 18, Aug. 12 to 19, Aug. 26 to Sept. 6, Sept. 13–16, and 27–30.

Monthly discharge of Mohawk River at Vischer Ferry dam, N. Y., for the year ending Sept. 30, 1918.

[Drainage area, 3,430 square miles.]

Month.	Discharge in second-feet.				Run-off (depth in inches on drainage area).
	Maximum.	Minimum.	Mean.	Per square mile.	
October	43,900	1,200	6,170	1.80	2.08
November	28,900	2,340	5,350	1.56	1.74
December	5,800	1,550	2,900	.845	.97
January	....	1,570	1,890	.551	.64
February	34,600	1,390	6,930	2.02	2.10
March	36,000	6,380	15,400	4.49	5.18
April	22,900	6,980	14,100	4.11	4.59
May	17,300	2,800	5,810	1.70	1.96
June	11,100	1,560	3,340	.974	1.09
July	5,240	1,020	2,480	.723	.83
August	3,570	1,010	1,490	.435	.50
September	12,300	1,110	3,130	.912	1.02
The year	43,900	1,010	5,730	1.67	22.70

NOTE.—Above table completed by using discharge from Crescent dam station on days when no record is available.

MOHAWK RIVER AT CRESCENT DAM, N. Y.

LOCATION.—At Crescent dam of Barge canal, about 3 miles above mouth of river at Cohoes, Albany County.

DRAINAGE AREA.—3,490 square miles (measured on topographic maps by State engineer department).

RECORDS AVAILABLE.—December 1, 1917, to September 30, 1918.

GAGE.—Gurley 7-day water-stage recorder on left bank about 50 feet above guard gate at head of Waterford flight of locks, about 200 yards from left end of spillway; inspected by operator from Barge canal power house at the dam.

DISCHARGE MEASUREMENTS.—Made from steel highway bridge at Crescent, about 1½ miles upstream.

CHANNEL AND CONTROL.—The control is the crest of the spillway.

DIVERSIONS.—Water is diverted at this point for canal purposes through Lock 6 and through the power plant located at this lock. The following tables of discharge include the flow through Lock 6 and through the power plant.

REGULATION.—Seasonal distribution of flow regulated by the Delta reservoir on the upper Mohawk, and by Hinckley reservoir on West Canada Creek. Large diurnal fluctuations during low water caused by operation of movable dams upstream.

ACCURACY.—Stage-discharge relation permanent; probably not affected by ice. Rating curve well defined between 5,000 and 50,000 second-feet. Record from water-stage recorder satisfactory. Records good.

COOPERATION.—Station established and maintained by the United States Geological Survey in cooperation with the State engineer and surveyor. Recorder inspected by an employee of the State superintendent of public works.

No discharge measurements made at station during year.

Daily discharge, in second-feet, of Mohawk River at Crescent dam, N. Y., for the year ending Sept. 30, 1918.

Day.	Dec.	Jan.	Feb.	Mar.	Apr.	May.	June.	July.	Aug.	Sept.
1	5,400	....	1,670	16,700	18,700	8,680	4,110	2,150	1,270	2,760
2	5,560	....	1,620	16,000	21,500	10,300	3,710	4,770	1,280	1,790
3	5,640	....	1,620	13,500	22,900	8,360	3,090	2,670	1,340	1,630
4	6,040	....	1,480	13,000	20,100	5,960	3,020	2,840	1,370	1,640
5	4,890	2,180	1,530	10,500	15,300	6,080	2,940	2,120	1,390	1,690
6	5,800	2,130	1,970	9,210	12,700	4,930	2,950	2,420	1,300	1,980
7	4,040	2,070	1,870	11,500	11,700	5,000	3,300	2,160	1,430	2,890
8	3,710	1,920	1,670	11,600	11,500	5,220	4,110	1,430	2,360	2,270
9	2,530	1,770	1,570	9,220	16,800	3,690	3,160	1,850	1,430	2,110
10	1,550	1,670	1,530	7,590	25,100	3,420	3,580	2,140	2,600	1,880
11	1,550	1,620	1,480	6,450	20,100	5,000	4,160	3,020	2,020	2,040
12	2,250	1,820	1,390	6,860	16,700	5,000	7,410	2,720	1,690	2,190
13	2,420	1,570	1,430	8,310	13,900	6,500	11,200	2,950	1,680	2,370
14	2,370	1,570	2,020	18,000	13,300	17,300	8,510	3,580	1,420	2,580
15	1,940	1,770	4,570	15,300	18,000	12,700	5,170	4,680	1,630	2,580
16	1,840	1,670	5,540	9,940	16,700	8,660	4,090	5,050	1,270	2,310
17	1,990	2,070	5,460	7,490	15,300	5,280	3,710	3,220	1,390	2,030
18	2,250	1,970	5,000	18,000	17,300	4,960	3,550	3,240	1,090	3,910
19	2,470	1,670	4,360	24,300	22,900	4,290	2,750	3,360	1,010	3,770
20	2,470	1,720	12,600	26,500	18,700	3,900	2,330	2,840	1,120	3,710
21	2,470	2,070	34,600	31,800	14,500	5,100	2,620	2,420	1,080	3,460
22	2,250	1,720	17,100	35,000	18,000	4,520	3,410	1,870	1,080	3,660
23	....	1,720	12,000	44,800	22,900	4,800	3,160	2,370	1,070	2,720
24	....	1,620	9,210	39,800	18,700	3,970	2,790	2,030	1,210	3,230
25	....	1,620	8,690	29,500	14,700	3,910	3,460	2,020	1,470	2,770
26	....	1,770	10,500	23,600	11,700	3,550	2,650	1,860	1,250	6,330
27	....	1,820	22,900	17,300	8,390	4,900	2,600	1,740	1,050	12,300
28	2,530	2,020	18,700	12,700	7,480	6,280	2,110	1,560	1,020	7,830
29	2,530	2,070	....	11,700	6,980	5,160	2,380	1,440	1,220	4,480
30	....	1,720	....	14,700	7,480	4,580	1,830	1,440	1,340	3,580
31	....	1,720	....	17,300	....	3,980	....	1,310	1,870	....

NOTE.—Mean daily discharge estimated Dec. 23–27, 2,420 second-feet; 30–31, 2,330 second-feet; Jan. 1–4, 2,310 second-feet; Dec. 9–10, Feb. 1–2, Sept. 11–14, as shown in table, from hydrograph of staff gage readings; no automatic record.

Monthly discharge of Mohawk River at Crescent dam, N. Y., for the year ending Sept. 30, 1918.

[Drainage area, 3,490 sqare miles.]

Month.	Discharge in second-feet.				Run-off (depth in inches on drainage area).
	Maximum.	Minimum.	Mean.	Per square mile.	
December	6,040	1,550	3,010	0.862	0.99
January	....	1,570	1,890	.542	.62
February	34,600	1,390	6,930	1.99	2.07
March	44,800	6,360	17,300	4.96	5.72
April	25,100	6,980	16,000	4.58	5.11
May	17,300	3,420	6,000	1.72	1.98
June	11,200	1,830	3,800	1.09	1.22
July	5,050	1,310	2,560	.734	.85
August	2,600	1,010	1,410	.404	.47
September	12,300	1,630	3,280	.940	1.05

DELAWARE RIVER BASIN.

EAST BRANCH OF DELAWARE RIVER AT FISH EDDY, N. Y.

LOCATION.—At railway bridge in village of Fish Eddy, Delaware County, 4 miles below mouth of Beaver Kill and 5½ miles above confluence of East and West branches.

DRAINAGE AREA.—790 square miles (measured on Post Route map).

RECORDS AVAILABLE.—November 19, 1912, to September 30, 1918. Records were obtained at Hancock, about 4 miles below from October 14, 1902, to December 31, 1912.

Gage.—Staff, in two sections, on downstream end of left pier of railroad bridge; read by J. P. Lyons.

Discharge measurements.—Made from the highway bridge about 200 feet above the gage or by wading.

Channel and control.—Coarse gravel; occasionally shifting.

Extremes of discharge.—Maximum open-water stage recorded during year, 15.4 feet at 3 p. m., October 30 (discharge, about 27,400 second-feet); minimum stage recorded, 1.70 feet several times in August and September (discharge, 141 second feet 1912–1918: Maximum stage, 17.4 feet during the afternoon of March 27, 1913, determined by leveling from flood marks (discharge, about 33,500 second-feet); minimum stage recorded, 1.64 feet at 5 p. m., October 12, 14, 15, 1914 (discharge, 97 second-feet)..

Ice.—Stage-discharge relation seriously affected by ice.

Accuracy.—Stage-discharge relation apparently permanent, except for two or three months immediately after the spring flood; affected by ice during a large part of the period from December to March, inclusive. Rating curve well defined between 200 and 20,000 second-feet. Gage read twice daily. Open-water records good; winter records fair.

Discharge measurements of East Branch of Delaware River at Fish Eddy, N. Y., during the year ending Sept. 30, 1918.

Date.	Made by—	Gage height.	Discharge.	Date.	Made by—	Gage height.	Discharge.
		Feet.	*Sec.-ft.*			*Feet.*	*Sec.-ft.*
Oct. 15	E. D. Burchard	2.95	702	Mar. 9	E. D. Burchard	5.13	2,670
Dec. 20[a]	C. C. Covert	4.92	590	June 5	do	3.55	1,120
Jan. 14[b]	do	3.85	456	Aug. 15	do	2.08	243
Feb. 9[a]	E. D. Burchard	3.50	250				

[a] Measurement made through incomplete ice cover. [b] Measurement made through complete ice cover.

Daily discharge, in second-feet, of East Branch of Delaware River at Fish Eddy, N. Y., for the year ending Sept. 30, 1918.

Day.	Oct.	Nov.	Dec.	Jan.	Feb.	Mar.	Apr.	May.	June.	July.	Aug.	Sept.
1	308	7,360	1,080	360	340	4,560	2,210	2,210	2,210	530	228	340
2	300	5,620	1,080	340	340	4,390	2,210	2,100	1,890	480	228	385
3	300	3,910	1,000	320	340	4,390	2,100	1,410	1,590	480	228	300
4	300	3,760	1,000	300	300	4,390	2,100	1,160	1,320	430	228	245
5	408	3,760	1,080	300	280	3,610	2,100	1,160	1,160	385	228	228
6	320	3,460	1,160	300	260	3,760	1,990	1,160	920	385	228	183
7	390	2,430	1,160	300	220	3,610	1,990	1,160	850	385	213	168
8	300	1,690	1,160	300	240	3,320	1,990	1,160	850	385	213	163
9	360	1,500	1,160	300	280	2,920	2,920	1,000	780	385	198	141
10	281	1,160	1,200	300	220	3,050	3,460	1,000	745	408	198	141
11	920	1,080	1,100	300	200	2,550	2,920	920	1,590	480	198	141
12	1,690	1,080	1,000	340	220	2,320	2,790	920	2,430	480	198	141
13	1,320	1,000	1,000	400	300	2,320	3,050	1,000	1,790	320	228	141
14	960	920	900	550	500	3,050	3,320	1,160	1,240	281	228	141
15	710	850	900	500	1,000	2,790	3,320	1,500	1,160	281	245	141
16	650	780	800	440	3,400	2,320	3,910	1,320	1,040	300	228	141
17	590	780	750	460	2,400	2,320	4,730	1,240	850	408	198	168
18	590	780	650	440	1,500	3,460	4,900	1,080	780	480	183	168
19	650	650	600	480	1,000	3,610	5,620	1,080	710	430	174	198
20	2,100	590	550	420	5,500	5,620	5,620	2,210	710	385	168	262
21	1,790	590	550	400	4,900	6,000	5,810	1,500	710	340	168	455
22	1,590	710	500	440	3,760	7,970	6,000	1,320	1,320	340	154	620
23	1,240	2,320	500	420	2,790	7,760	5,440	1,320	1,160	320	154	430
24	1,080	1,890	500	440	2,550	7,160	4,900	1,240	1,040	300	141	385
25	3,910	1,790	500	340	2,430	7,160	4,230	1,160	960	300	141	455
26	2,920	1,690	480	320	2,550	6,380	3,460	2,100	780	300	141	430
27	2,790	1,500	440	320	2,670	4,070	2,920	2,670	710	281	141	3,610
28	4,560	1,160	380	360	3,610	2,790	2,550	2,580	710	262	141	1,890
29	3,050	1,160	380	360	......	2,430	2,320	2,320	590	228	141	1,500
30	17,500	885	380	340	......	2,320	2,100	2,100	530	228	141	1,000
31	14,500	......	380	360	......	2,320	......	2,320	......	228	168	......

Note.—Discharge Dec. 10 to Feb. 20 estimated, because of ice, from discharge measurements, weather records, study of recorder graph, and comparison with similar studies for the station at Hale Eddy.

Monthly discharge of East Branch of Delaware River at Fish Eddy, N. Y., for the year ending Sept. 30, 1918.

[Drainage area, 790 square miles.]

Month.	Discharge in second-feet.				Run-off (depth in inches on drainage area).
	Maximum.	Minimum.	Mean.	Per square mile.	
October	14,500	281	2,200	2.79	3.22
November	7,360	590	1,900	2.41	2.69
December	1,200	380	785	.994	1.15
January	550	300	373	.472	.54
February	5,500	200	1,580	2.00	2.08
March	7,970	2,320	4,020	5.09	5.87
April	6,000	1,990	3,430	4.34	4.84
May	2,670	990	1,500	1.90	2.19
June	2,430	530	1,100	1.39	1.55
July	530	228	360	.456	.53
August	245	141	189	.239	.28
September	3,610	141	490	.620	.69
The year	14,500	141	1,490	1.89	25.63

DELAWARE RIVER AT PORT JERVIS, N. Y.

Location.—At toll bridge at Port Jervis, Orange County, 1 mile above Neversink River and 6 miles below Mongaup River.

Drainage area.—3,250 square miles.

Records available.—October 12, 1904, to September 30, 1918.

Gage.—Staff, in two sections; the upper section vertical and attached to downstream end of left abutment; the lower section inclined, about 30 feet downstream. Prior to June 20, 1914, a chain gage on the bridge was used; read by Mrs. Bella Fuller.

Discharge measurements.—Made from the highway bridge or by wading.

Channel and control.—Gravel; occasionally shifting.

Extremes of discharge.—Maximum stage recorded during year, 12.3 feet at 8 a. m. October 31 (discharge, 61,600 second-feet); minimum stage recorded, 1.1 feet, 8 a. m. August 26 and 5 p. m. August 28 (discharge, 390 second-feet).

1904–1918: Maximum stage recorded, 16.0 feet at 8 a. m. March 28, 1914 (discharge, 92,700 second-feet); minimum stage recorded, 0.60 foot at 8 a. m. September 22 and 23, 1908 (discharge, 175 second-feet).

Ice.—Stage-discharge relation somewhat affected by ice.

Accuracy.—Stage-discharge relation practically permanent between dates of shifting; affected by ice during large part of January and February. Rating curve well defined between 1,000 and 30,000 second-feet. Gage read to hundredths twice daily from October 1 to December 31, and to tenths once daily, January 1 to September 30. Daily discharge ascertained by applying mean daily gage height to rating table. Open-water records good; winter records fair.

Cooperation.—Gage heights, October 1 to June 30, furnished by United States Weather Bureau.

Discharge measurements of Delaware River at Port Jervis, N. Y., during the year ending Sept. 30, 1918.

Date.	Made by—	Gage height.	Discharge.	Date.	Made by—	Gage height.	Discharge.
		Feet.	*Sec.-ft.*			*Feet.*	*Sec.-ft.*
Oct. 17	E. D. Burchard	2.37	1,800	June 8	J. W. Moulton	3.10	3,320
Feb. 8[a]	C. C. Covert	3.19	1,170	Aug. 13	E. D. Burchard	1.50	650
Mar. 12	E. D. Burchard	4.82	9,450	13	.....do	1.53	657
12	.....do	4.80	9,540				

[a] Measurement made through incomplete ice cover.

Daily discharge, in second-feet, of Delaware River at Port Jervis, N. Y., for the year ending Sept. 30, 1918.

Day.	Oct.	Nov.	Dec.	Jan.	Feb.	Mar.	Apr.	May.	June.	July.	Aug.	Sept.
1	685	33,500	2,920	1,200	1,200	14,100	6,700	7,810	7,060	2,070	780	830
2	685	19,200	3,160	1,200	1,100	28,200	7,430	7,060	6,700	1,720	880	880
3	685	13,100	3,160	1,200	1,000	18,600	8,200	6,700	5,360	2,070	780	1,110
4	990	10,360	3,160	1,200	1,000	14,100	9,010	5,680	3,910	1,890	732	985
5	780	8,200	2,920	1,200	1,000	11,600	8,600	5,360	3,910	1,640	780	780
6	780	7,060	2,470	1,000	1,000	11,600	8,200	5,050	3,650	1,240	685	780
7	880	6,010	2,070	1,000	1,000	20,500	6,010	4,750	3,650	1,240	642	685
8	990	5,680	1,720	950	1,200	14,100	5,360	4,460	3,400	1,390	685	642
9	1,110	4,750	1,390	1,200	1,200	11,600	5,050	3,650	3,650	1,550	685	600
10	990	3,910	1,720	1,300	1,200	10,300	9,840	3,400	2,920	1,390	732	525
11	880	3,910	2,070	1,300	1,200	12,100	10,300	3,160	2,470	1,390	732	490
12	780	3,650	2,920	1,400	1,000	9,010	9,010	2,920	2,690	1,470	685	490
13	1,110	3,160	2,690	1,600	1,200	7,810	8,600	4,460	3,160	1,470	780	490
14	3,650	2,920	2,470	1,600	1,600	14,100	8,200	7,430	5,360	1,550	780	562
15	2,470	2,690	2,260	1,700	2,400	15,100	13,100	6,350	4,180	1,720	990	880
16	2,070	2,690	2,000	1,700	3,600	12,100	16,200	5,050	3,650	1,980	990	780
17	1,890	2,690	2,000	1,900	8,500	11,200	14,100	4,460	2,920	1,640	780	685
18	1,890	2,470	1,900	1,500	8,000	15,100	15,100	3,910	2,470	1,550	685	685
19	1,720	2,260	1,700	1,500	7,000	16,200	16,800	3,650	2,260	1,550	562	880
20	1,720	2,260	1,600	1,300	11,600	18,600	13,600	3,650	2,070	1,550	490	1,050
21	4,460	2,070	1,600	1,200	35,000	20,500	11,600	6,010	1,890	1,550	455	1,640
22	3,910	2,070	1,600	1,000	29,000	21,800	19,200	5,360	2,070	1,240	422	2,690
23	3,400	4,460	1,600	1,000	15,100	23,900	21,200	5,360	5,360	1,180	390	2,260
24	2,920	4,180	1,600	1,600	10,700	19,800	16,200	4,750	4,460	990	390	1,890
25	4,460	3,650	1,700	1,500	8,200	15,100	13,600	3,910	3,400	880	390	1,550
26	9,010	3,400	1,600	1,200	8,600	13,100	11,200	3,910	2,920	780	390	1,550
27	7,060	3,160	1,700	1,200	35,000	11,200	9,010	4,460	2,470	780	455	6,700
28	6,010	2,920	1,600	1,100	24,600	9,010	7,810	6,010	2,070	685	390	7,430
29	7,060	2,690	1,500	1,100	....	7,430	6,700	6,010	1,890	685	455	5,050
30	9,420	2,470	1,400	1,100	....	7,060	6,350	6,010	1,720	685	455	3,650
31	61,600	....	1,300	1,100	....	6,700	....	8,200	....	880	455	....

NOTE.—Discharge Dec. 10 to Feb. 19 estimated, because of ice, from discharge measurements, weather records, study of recorder graph, and comparison with similar studies for stations on the East and West branches.

Monthly discharge of Delaware River at Port Jarvis, N. Y., for the year ending Sept. 30, 1918.

[Drainage area, 3,250 square miles.]

Month.	Discharge in second-feet.				Run-off (depth in inches on drainage area).
	Maximum.	Minimum.	Mean.	Per square mile.	
October	61,600	685	4,710	1.45	1.67
November	33,500	2,070	5,720	1.76	1.96
December	3,160	1,300	2,030	.624	.72
January	1,900	950	1,290	.397	.46
February	35,000	1,000	7,980	2.45	2.55
March	28,200	6,700	14,200	4.38	5.05
April	21.200	5,050	10,700	3.30	3.68
May	8,200	2,920	5,130	1.58	1.82
June	7,060	1,720	3,460	1.06	1.18
July	2,070	685	1,370	.422	.49
August	990	390	629	.194	.22
September	7,430	490	1,640	.505	.56
The year	61,600	390	4,880	1.50	20.36

DELAWARE RIVER AT RIEGELSVILLE, N. J.

LOCATION.—At toll suspension bridge between Riegelsville, N. J., and Riegelsville, Pa., 600 feet above Musconetcong River and 9 miles below Lehigh River.

DRAINAGE AREA.—6,430 square miles.

Records available.—July 3, 1906, to September 30, 1918.

Gage.—Staff in three sections installed November 14, 1914, on left bank (New Jersey side) at upstream side of bridge; lower section inclined, middle and upper sections vertical. Prior to November 14, 1914, chain gage attached to upstream side of bridge. Gage read by Herbert J. Bernholz.

Discharge measurements.—Made from bridge.

Channel and control.—Large boulders; practically permanent.

Extremes of discharge.—Maximum stage recorded during year, 18.4 feet at 4 p. m. October 31 (discharge, 90,700 second-feet); minimum stage recorded, 1.95 feet, August 28 (discharge, 1,420 second-feet).

1906–1918: Maximum stage[1] recorded, 25 feet March 28, 1913 (discharge, 144,000 second-feet); minimum stage recorded, 1.55 feet 8 a. m. Sept. 20, 1908 (discharge, 870 second-feet).

Ice.—Stage-discharge relation affected by ice during severe winters only.

Diversions.—The Delaware division of the Pennsylvania canal diverts about 250 second-feet from Lehigh River near its mouth from about the last of March to the middle of December each year.

Accuracy.—Stage-discharge relation practically permanent; affected by ice to some extent during December, January, and February. Rating curve well defined. Gage read to quarter-tenths twice a day. Daily discharge obtained by applying mean daily gage height to rating table. Records good.

No current-meter measurements were made during the year.

Daily discharge, in second-feet, of Delaware River at Riegelsville, N. J., for the year ending Sept. 30, 1918.

Day.	Oct.	Nov.	Dec.	Jan.	Feb.	Mar.	Apr.	May.	June.	July.	Aug.	Sept.
1	1,990	62,400	5,610	2,340	3,390	44,200	12,000	13,900	14,600	4,710	3,880	4,140
2	1,990	34,500	5,610	2,340	3,390	32,700	11,600	14,600	12,400	4,140	2,940	2,340
3	1,990	24,000	5,920	2,340	3,390	31,600	12,400	13,900	10,200	4,140	2,730	2,530
4	1,990	18,800	5,610	2,340	3,390	29,300	15,000	12,700	8,820	4,140	2,530	2,440
5	1,990	15,000	5,610	2,160	3,390	25,600	14,600	12,000	7,490	3,880	2,340	2,340
6	2,160	12,400	5,010	2,160	3,390	23,500	13,100	10,900	7,490	3,880	2,530	2,440
7	2,080	10,900	4,420	2,160	3,390	33,300	11,200	10,200	7,490	3,390	2,340	2,340
8	1,990	9,840	3,880	2,080	3,390	35,700	9,840	9,500	8,150	3,160	2,160	2,250
9	2,340	8,820	2,730	2,160	3,280	24,500	10,900	8,820	7,820	3,160	2,160	1,990
10	2,250	8,150	2,160	2,160	3,390	25,600	16,300	8,480	6,850	2,940	1,990	1,990
11	2,160	7,490	2,840	1,990	3,390	22,600	21,600	8,480	7,490	3,160	1,990	1,820
12	2,160	6,850	2,940	7,820	3,390	20,700	19,700	8,480	6,850	3,050	1,990	1,820
13	2,630	6,540	2,940	8,820	3,630	18,800	18,400	8,480	7,490	3,050	2,340	1,990
14	2,940	6,230	2,940	7,490	5,920	23,500	19,700	10,200	7,820	3,390	3,160	1,990
15	3,390	5,610	3,160	7,170	10,500	36,900	23,500	12,700	8,820	3,880	3,390	1,990
16	4,140	5,610	3,390	5,610	13,100	30,400	29,300	12,700	7,490	3,630	2,940	2,160
17	3,390	5,010	3,630	5,310	12,000	25,600	31,000	10,900	6,540	3,390	2,840	1,990
18	3,160	4,710	3,880	5,010	11,600	25,000	31,600	9,500	5,610	3,630	2,340	1,990
19	2,940	4,710	3,880	4,710	13,900	28,800	34,500	8,480	5,010	3,390	2,080	1,990
20	3,390	4,420	3,880	4,420	56,700	28,800	35,100	8,150	4,420	2,940	1,820	2,440
21	4,140	4,420	4,140	4,710	65,300	32,100	35,100	8,480	4,140	2,730	1,820	3,630
22	5,310	4,710	4,140	5,010	46,500	33,300	38,100	10,900	5,920	2,730	1,660	4,710
23	5,610	5,610	4,420	4,710	27,700	34,500	46,400	10,900	8,150	2,530	1,660	4,710
24	5,310	7,490	4,140	5,010	22,600	31,000	38,100	10,200	9,160	2,530	1,580	4,420
25	8,150	8,150	3,880	4,710	20,700	26,600	33,300	9,160	7,490	2,630	1,580	3,880
26	10,500	6,850	3,390	3,880	66,800	22,600	24,500	8,480	6,230	2,530	1,500	3,390
27	11,600	5,010	3,160	3,880	59,500	19,700	20,700	7,820	5,310	2,340	1,500	4,420
28	9,160	4,710	2,940	3,880	52,500	17,100	18,000	17,500	5,010	2,340	1,420	7,490
29	9,840	4,420	2,530	3,880	....	14,200	16,300	12,400	4,420	2,340	1,500	7,820
30	13,900	4,710	2,530	3,390	....	13,100	14,600	11,200	3,880	2,340	1,580	7,170
31	73,300	....	2,340	3,390	....	12,400	....	12,700	....	4,710	1,660	....

Note.—Discharge interpolated Feb. 5–7 as gage was read to top of ice. Stage-discharge relation probably affected by ice to some extent in December and January but no correction made therefor. Gage not read Feb. 22; discharge interpolated.

[1] It has been estimated that the flood of Oct. 10–11, 1903, reached a stage of 41.5 feet with a corresponding discharge of 275,000 second-feet.

Monthly discharge of Delaware River at Riegelsville, N. J., for the year ending Sept. 30, 1918.

[Drainage area, 6,430 square miles.]

Month.	Discharge in second-feet.				Run-off (depth in inches on drainage area).
	Maximum.	Minimum.	Mean.	Per square mile.	
October	73,300	1,990	6,710	1.06	1.24
November	62,400	4,420	10,600	1.68	1.87
December	5,920	2,160	3,800	.600	.69
January	8,820	1,990	4,100	.638	.74
February	66,800	3,280	18,900	2.94	3.06
March	44,200	12,400	26,600	4.15	4.78
April	46,400	9,840	22,500	3.55	3.96
May	17,500	8,150	10,700	1.71	1.97
June	14,600	3,880	7,290	1.17	1.30
July	4,710	2,340	3,250	.541	.62
August	3,880	1,420	2,190	.376	.43
September	7,820	1,820	3,220	.537	.60
The year	73,300	1,420	9,880	1.57	21.26

NOTE.—To allow for water diverted by the canal, 230 second-feet was added to the daily discharge, Oct. 1 to Dec. 9 and Mar. 16 to Sept. 30, before computing discharge per square mile: first three columns of table therefore indicate actual quantity of water flowing in the river; the two remaining columns represent the total run-off from drainage area above Riegelsville, including the discharge of the canal.

BEAVER KILL AT COOKS FALLS, N. Y.

LOCATION.—At covered highway bridge in Cooks Falls, Delaware County.

DRAINAGE AREA.—236 square miles (measured on Post Route and topographic maps).

RECORDS AVAILABLE.—July 25, 1913, to September 30, 1918.

GAGE.—Vertical staff, in two sections, bolted to rock on left bank under the bridge; read by Ralph Rosa and H. B. Couch.

DISCHARGE MEASUREMENTS.—Made from the bridge or by wading a short distance downstream.

CHANNEL AND CONTROL.—Coarse gravel, boulders, and solid ledge; practically permanent.

EXTREMES OF DISCHARGE.—Maximum stage recorded during year, 12.4 feet at 5 p. m. October 30 (discharge, about 9,700 second-feet); minimum stage recorded, 0.84 foot at 7 a. m. and 3 p. m. August 24 (discharge, 41 second-feet).

1913-1918: Maximum stage recorded, 12.4 feet at 5 p. m. October 30, 1917 (discharge, about 9,700 second-feet); minimum stage recorded, 0.70 foot from 7 a. m. October 12 to 7 a. m. October 13, 1916 (discharge, 30 second-feet).

ICE.—Stage-discharge relation somewhat affected by ice.

ACCURACY.—Stage-discharge relation practically permanent; affected by ice during parts of the period from December to March, inclusive. Rating curve well defined between 50 and 4,500 second-feet. Gage read to half-tenths twice daily. Daily discharge ascertained by applying mean daily gage height to rating table. Open-water records good; winter records fair.

Discharge measurements of Beaver Kill at Cooks Falls, N. Y., during the year ending Sept. 30, 1918.

Date.	Made by—	Gage height.	Discharge.	Date.	Made by—	Gage height.	Discharge.
		Feet.	*Sec.-ft.*			*Feet.*	*Sec.-ft.*
Oct. 16	E. D. Burchard	2.32	366	Mar. 11	E. D. Burchard	3.39	820
Nov. 22	C. C. Covert	2.05	270	June 7	J. W. Moulton	2.32	316
Dec. 20[a]	.....do	2.20	201	Aug. 15	E. D. Burchard	1.39	129
Jan. 14[a]	.....do	3.10	207	15	.....do	1.39	128
Feb. 9[b]	E. D. Burchard	2.28	107				

[a] Measurement made through complete ice cover. [b] Measurement made through incomplete ice cover.

Daily discharge, in second-feet, of Beaver Kill at Cooks Falls, N. Y., for the year ending Sept. 30, 1918.

Day.	Oct.	Nov.	Dec.	Jan.	Feb.	Mar.	Apr.	May.	June.	July.	Aug.	Sept.
1	186	1,730	371	200	130		1,330	805	455	197	80	244
2	186	1,400	355	190	130		1,800	705	371	197	72	132
3	175	1,080	355	190	120		1,800	615	325	164	65	80
4	244	805	325	190	120		1,940	570	296	175	62	67
5	269	705	310	190	120	1,370	1,260	530	296	164	89	59
6	208	615	282	190	110		1,020	490	269	146	76	56
7	186	282	256	200	110		910	455	355	146	64	59
8	175	404	244	190	110		805	371	325	146	59	56
9	220	325	232	190	110	805	1,400	355	256	142	59	56
10	164	310	220	200	110	830	1,460	355	256	164	59	51
11	154	296	200	200		805	1,260	355	232	175	128	54
12	310	282	200	200		755	1,080	340	355	164	120	54
13	244	296	200	200		855	910	355	340	186	101	75
14	340	296	200	200		1,020	1,020	1,020	282	256	76	58
15	355	296	200	200		755	1,330	660	256	310	130	52
16	310	269	190	200		705	1,400	490	220	186	91	51
17	256	282	200	200		855	1,200	455	208	164	78	48
18	232	325	200	200		1,260	1,940	420	197	164	62	55
19	232	310	200	200		1,730	1,400	387	197	142	59	132
20	530	296	200	190		2,240	1,080	387	175	130	55	110
21	490	282	200	180	584	2,720	1,800	420	164	118	48	310
22	325	404	200	180		3,310	2,720	387	855	112	46	175
23	282	1,140	190	170		2,960	1,730	387	404	105	43	140
24	530	615	186	170		2,160	1,400	355	325	100	41	124
25	910	371	197	170		1,940	1,140	340	256	98	122	112
26	570	355	197	160		1,660	910	455	232	94	64	530
27	1,590	340	208	160		1,400	805	420	197	89	51	910
28	1,260	325	197	160		1,020	705	387	186	82	46	490
29	1,940	340	200	150		910	705	325	186	85	72	325
30	7,110	387	200	140		1,260	805	455	175	83	64	269
31	2,400	......	200	130		1,260	......	530	......	92	43	......

NOTE.—Discharge Dec. 11-23 and Dec. 29 to Mar. 8 estimated, because of ice, from discharge measurements, weather records, study of recorder graph and comparison with similar studies for East Branch of Delaware River at Fish Eddy. Braced figures show mean discharge for periods included.

Monthly discharge of Beaver Kill at Cooks Falls, N. Y., for the year ending Sept. 30, 1918.

[Drainage area, 236 square miles]

Month.	Discharge in second-feet.				Run-off (depth in inches on drainage area).
	Maximum.	Minimum.	Mean.	Per square mile.	
October	7,110	154	722	3.06	3.53
November	1,730	269	505	2.14	2.39
December	371	186	230	.975	1.12
January	200	130	184	.780	.90
February	....	....	417	1.77	1.84
March	3,310	705	1,420	6.02	6.94
April	2,720	705	1,300	5.51	6.15
May	1,020	325	470	1.99	2.29
June	855	164	288	1.22	1.36
July	310	82	148	.627	.72
August	130	41	71.6	.303	.35
September	910	48	164	.695	.78
The year	7,110	41	493	2.09	28.37

WEST BRANCH OF DELAWARE RIVER AT HALE EDDY, N. Y.

LOCATION.—At highway bridge in village of Hale Eddy, Delaware County, 8 miles below power dam of Deposit Electric Co. and 8½ miles above junction with East Branch of Delaware River.

DRAINAGE AREA.—611 square miles (measured on Post Route map).

RECORDS AVAILABLE.—November 15, 1912, to September 30, 1918. Records obtained at Hancock, about 7 miles below, from October 15, 1902, to December 31, 1912.

GAGE.—Vertical staff in four sections, attached to rocks near right abutment of bridge and to abutment; read by William Seeley and W. J. Shanly.

Discharge measurements.—Made from cable, installed in July, 1916, about 400 feet below gage. Previous measurements made from highway bridge or by wading.

Channel and control.—Coarse gravel and boulders; practically permanent.

Extremes of discharge.—Maximum stage recorded during year, 13.4 feet at 4 p. m. February 20 (stage-discharge relation affected by ice, discharge not determined); minimum stage recorded, 1.5 feet several times in August (discharge, 65 second-feet).

1912–1918: Maximum stage recorded,[1] 15.3 at 5 p. m. March 27, 1913 (discharge, about 25,000 second-feet); minimum stage recorded, 1.0 foot at 6 p. m. September 21, 1913 (discharge, 34 second-feet).

Ice.—Stage-discharge relation seriously affected by ice.

Accuracy.—Stage-discharge relation practically permanent. Rating curve well defined between 300 and 18,000 second-feet. Gage read to half-tenths twice daily. Daily discharge ascertained by applying mean daily gage height to rating table. Open-water records good; winter records fair.

Discharge measurements of West Branch of Delaware River at Hale Eddy, N. Y., during the year ending Sept. 30, 1918.

Date.	Made by—	Gage height.	Discharge.	Date.	Made by—	Gage height.	Discharge.
		Feet.	*Sec.-ft.*			*Feet.*	*Sec.-ft.*
Oct. 15	E. D. Burchard	2.81	484	Mar. 9	E. D. Burchard	4.71	1,860
Dec. 21[a]	C. C. Covert	3.14	225	June 5	J. W. Moulton	3.56	883
Jan. 15[a]	.....do	3.53	270	5	.....do	3.58	875
Feb. 9[a]	.....do	3.20	212	Aug. 14	E. D. Burchard	1.62	94
Mar. 9	E. D. Burchard	4.72	1,850	14	.....do	1.61	92.5

[a] Measurement made through complete ice cover.

Daily discharge, in second-feet, of West Branch of Delaware River at Hale Eddy, N. Y., for the year ending Sept. 30, 1918.

Day.	Oct.	Nov.	Dec.	Jan.	Feb.	Mar.	Apr.	May.	June.	July.	Aug.	Sept.
1	135	5,900	580	100	260	7,650	1,260	1,180	1,580	388	150	101
2	155	3,800	605	100	240	4,960	1,580	1,180	1,180	555	142	130
3	130	2,670	455	120	240	3,800	1,850	1,110	1,110	660	118	232
4	170	2,140	455	170	240	3,540	1,940	900	1,040	480	118	232
5	250	1,760	410	85	240	2,560	1,580	900	900	432	110	170
6	325	1,420	388	40	240	4,080	1,260	780	840	410	89	250
7	325	1,260	325	90	220	4,660	1,260	780	900	432	85	268
8	305	1,110	305	110	220	3,030	1,110	780	1,040	410	69	215
9	200	970	300	130	220	1,940	1,850	660	840	345	85	200
10	232	840	300	130	220	1,940	2,240	555	720	305	105	155
11	215	780	300	160	240	1,760	1,940	530	605	345	170	170
12	250	720	300	360	300	1,580	1,940	505	2,560	388	118	150
13	1,110	660	280	260	420	2,790	1,940	1,110	2,340	388	105	161
14	720	555	280	260	800	2,670	1,940	1,940	1,420	455	85	142
15	505	555	260	280	1,300	2,140	3,150	1,760	970	530	95	150
16	555	480	260	280	2,000	1,760	3,030	1,340	970	505	130	118
17	480	455	240	280	2,400	1,760	2,560	1,110	840	432	142	130
18	365	455	240	280	2,400	2,790	2,340	900	840	455	130	215
19	365	410	240	260	2,600	3,280	2,560	840	605	455	118	250
20	1,500	410	240	260	2,600	3,540	2,340	1,340	505	410	110	285
21	1,180	410	220	280	2,600	4,360	2,340	2,040	505	388	105	720
22	720	480	240	280	2,560	4,660	3,030	1,670	1,850	345	110	780
23	720	900	240	280	2,670	3,030	2,910	1,940	1,420	325	89	720
24	900	840	300	280	2,670	2,560	2,340	1,580	1,040	285	69	840
25	2,340	480	200	280	2,910	2,340	2,140	1,260	840	250	69	1,260
26	2,140	388	300	260	10,900	2,040	1,850	1,580	605	232	75	2,340
27	1,340	345	300	260	3,800	1,760	1,580	1,850	605	170	81	2,560
28	2,140	432	200	260	3,540	1,760	1,420	2,140	505	101	81	2,240
29	1,940	505	170	260	....	1,420	1,180	2,040	455	95	95	2,040
30	11,600	455	150	260	....	1,260	1,180	2,140	455	118	105	1,340
31	12,800	....	90	260	....	1,180	....	1,850	....	250	95	....

Note.—Discharge Dec. 9 to Feb. 21 estimated, because of ice, from discharge measurements, weathe records, study of recorder graph, and comparison with similar studies for the station at Fish Eddy.

[1] The observer states that on Oct. 10, 1893, the water rose to an elevation indicated by a nail in a tree near the gage. This nail is at gage height 20.3 feet. No data available indicating whether the present rating is applicable to this gage height.

Monthly discharge of West Branch of Delaware River at Hale Eddy, N. Y., for the year ending Sept. 30, 1918.

[Draining area, 611 square miles.]

Month.	Discharge in second-feet.				Run-off (depth in inches on drainage area).
	Maximum.	Minimum.	Mean.	Per square mile.	
October	12,800	130	1,490	2.44	2.81
November	5,900	345	1,090	1.78	1.99
December	605	90	296	.484	.56
January	360	40	217	.355	.41
February	10,900	220	1,750	2.86	2.98
March	7,650	1,180	2,860	4.68	5.40
April	3,150	1,110	1,990	3.26	3.64
May	2,140	505	1,300	2.13	2.46
June	2,560	455	1,000	1.64	1.83
July	660	95	366	.599	.69
August	170	69	105	.172	.20
September	2,560	101	619	1.01	1.13
The year	12,800	40	1,080	1.77	24.10

SUSQUEHANNA RIVER BASIN.

SUSQUEHANNA RIVER AT CONKLIN, N. Y.

Location.—At steel highway bridge just below Conklin, Broome County, 5 miles below Big Snake Creek and 8 miles above Chenango River.

Drainage area.—2,350 square miles.

Records available.—November 13, 1912, to September 30, 1918. Records were obtained at Binghamton, 8 miles below, from July 31, 1901, to December 31, 1912.

Gage.—Stevens water-stage recorder on left bank, just below the bridge, installed October 4, 1914. Prior to that date, staff in two sections, the lower section inclined, the upper vertical, attached to left abutment. Water-stage recorder inspected by George W. Marvin.

Discharge measurements.—Made from the bridge or by wading.

Channel and control.—Coarse gravel and boulders; probably permanent.

Extremes of discharge.—Maximum stage during year, from water-stage recorder, 12.87 feet at 10.30 a. m. March 1 (discharge, about 25,900 feet), minimum stage from water-stage recorder, 2.40 feet October 1–5 (discharge, 470 second-feet).

1912–1918: Maximum stage recorded 19.74 feet at the former station in Binghamton, at 7.40 a. m., March 2, 1902 (discharge, about 62,500 second-feet); minimum stage recorded, 1.32 feet at 8.20 a. m. and 4 p. m. September 16, 1913 (discharge, 106 second-feet).

Ice.—Stage-discharge relation affected by ice.

Accuracy.—Stage-discharge relation practically permament, except when affected by ice (a large part of the period from January to March, inclusive).. Rating curve well defined between 250 and 55,000 second-feet. Operation of the water-stage recorder fairly satisfactory. Daily discharge ascertained by applying mean daily gage height to rating table, except for days when the mean gage height would not give the discharge within 1 per cent when the discharge is the mean of 24 hourly determinations. Gage heights determined by inspecting recorder graph or by taking mean of two observations per day. Open-water records good; winter records fair.

Discharge measurements of Susquehanna River at Conklin, N. Y., during the year ending Sept. 30, 1918.

Day.	Made by—	Gage height.	Discharge.	Date.	Made by—	Gage height.	Discharge.
		Feet.	*Sec.-ft.*			*Feet.*	*Sec.-ft.*
Jan. 17[a]	C. C. Covert	5.06	811	Mar. 19	C. C. Covert	8.45	11,000
Feb 11[a]	.....do	4.25	959	Apr. 26	.....do	6.12	5,740
Mar. 3[b]	.....do	11.1	10,600	June. 4	J. W. Moulton	4.50	2,620
8[b]	E. D. Burchard	9.83	11,200	Aug 16	E. D. Burchard	2.73	672

[a] Measurement made through complete ice cover. [b] Measurement made through incomplete ice cover.

Daily discharge, in second-feet, of Susquehanna River at Conklin, N. Y., for the year ending Sept. 30, 1918.

Day.	Oct.	Nov.	Dec.	Jan.	Feb.	Mar.	Apr.	May.	June.	July.	Aug.	Sept.
1	506	8,280	1,700	900	800	12,000	4,840	5,170	5,380	1,570	607	800
2	500	6,860	1,800	900	800	11,000	5,170	5,720	4,140	2,000	572	1,840
3	470	5,380	1,800	850	800	9,000	5,720	5,170	3,330	2,510	558	1,350
4	506	4,640	1,700	850	950	7,000	6,860	4,530	2,750	1,880	524	979
5	537	3,860	1,600	800	950	5,500	6,170	4,230	2,360	1,690	512	775
6	726	3,500	1,500	800	1,000	7,000	4,640	3,770	2,210	1,520	530	882
7	1,010	3,160	1,400	750	850	10,000	3,950	3,500	2,510	1,330	512	826
8	1,080	2,830	1,300	750	900	11,000	3,590	3,240	3,950	1,200	506	698
9	1,020	3,590	1,100	700	900	8,500	4,980	2,990	3,680	1,100	488	712
10	938	2,360	1,200	700	960	8,000	7,100	2,590	2,590	1,150	500	642
11	890	2,360	1,200	700	950	8,000	6,630	2,440	2,280	1,300	530	600
12	1,060	2,510	1,200	700	1,000	7,500	5,720	2,280	4,680	1,890	530	544
13	1,520	2,510	1,200	650	1,600	7,000	5,380	5,460	5,720	2,360	635	680
14	2,140	2,360	1,200	700	2,400	12,000	6,570	13,700	4,430	1,940	726	733
15	2,000	2,070	1,200	700	6,500	13,000	11,500	10,500	3,420	1,750	768	670
16	1,750	1,350	1,100	750	8,500	10,000	12,800	6,860	2,750	1,880	691	677
17	2,070	1,810	1,100	750	10,000	8,500	10,500	4,840	2,280	1,630	663	712
18	1,810	1,810	1,100	750	9,500	9,500	10,800	3,950	2,000	1,460	558	818
19	1,690	1,810	1,100	800	8,000	12,000	10,500	3,330	1,690	1,750	530	914
20	3,330	1,750	1,100	800	6,500	14,000	8,280	3,080	1,520	1,570	530	1,300
21	4,530	1,880	1,100	850	6,500	15,500	6,860	6,130	1,350	1,270	530	2,590
22	3,680	1,810	1,100	850	6,500	16,800	9,500	5,280	2,830	1,200	530	2,440
23	2,990	2,990	1,100	850	6,500	16,100	11,300	5,720	3,950	1,060	530	1,940
24	3,640	3,500	1,100	800	6,500	13,100	9,740	4,740	3,420	1,010	530	1,570
25	6,860	2,910	1,100	800	7,000	10,200	8,280	3,590	2,590	997	530	1,400
26	6,170	2,210	1,100	800	7,500	8,760	6,860	4,640	2,140	890	530	3,930
27	4,840	1,940	1,100	800	8,000	7,500	5,720	7,330	1,750	803	530	7,100
28	5,500	1,750	1,000	900	9,500	6,400	5,060	7,330	1,460	726	530	6,400
29	6,860	1,690	1,000	850	....	5,380	4,330	5,280	1,330	656	530	4,640
30	20,400	1,700	1,000	750	....	4,950	4,530	7,500	1,250	663	530	3,240
31	28,000	....	950	700	....	4,840	....	6,170	....	663	530	....

NOTE.—Discharge Oct. 31 to Nov. 10 estimated, for lack of gage-height record, from study of recorder graph and comparison with record of flow of Chenango River near Chenango Forks. Discharge Nov. 30 to Mar. 20 estimated, because of ice, from discharge measurements, weather records, study of recorder graph, and comparison with similar studies for Chenango River near Chenango Forks.

Monthly discharge of Susquehanna River at Conklin, N. Y., for the year ending Sept. 30, 1918.

[Drainage area, 2,350 square miles.]

Month.	Discharge in second-feet.				Run-off (depth in inches on drainage area).
	Maximum.	Minimum.	Mean.	Per square mile.	
October	28,000	470	3,840	1.63	1.88
November	8,280	1,350	2,870	1.22	1.36
December	1,800	950	1,230	.523	.60
January	900	650	782	.333	.38
February	10,000	800	4,350	1.85	1.93
March	16,800	4,840	9,680	4.12	4.75
April	12,800	3,590	7,130	3.03	3.38
May	13,700	2,280	5,200	2.21	2.55
June	5,720	1,250	2,860	1.22	1.36
July	2,510	656	1,400	.596	.69
August	768	488	558	.238	.27
September	7,100	544	1,750	.744	83
The year	28,000	470	3,460	1.47	19.98

CHENANGO RIVER NEAR CHENANGO FORKS, N. Y.

Location.—About 1½ miles below Tioughnioga River, 2 miles by road below Chenango Forks post office, Broome County, and 11½ miles above Binghamton and mouth.

Drainage area.—1,380 square miles; area from which water is diverted not included. See "Diversions."

Records available.—November 11, 1912, to September 30, 1918. Records were obtained at Binghamton July 31, 1901, to December 31, 1911.

Gage.—Stevens water-stage recorder on the left bank on the farm of Erastus Ingraham.

Discharge measurements.—Made from cable about 100 feet above the gage or by wading.

Channel and control.—Sand, gravel, and small cobble stones; practically permanent.

Extremes of discharge.—Maximum stage during year, from water-stage recorder, 10.75 feet at noon May 14 (discharge, about 22,000 second-feet); minimum stage recorded, 2.40 feet at 4 p. m. August 4 and 7 a. m. August 5 (discharge, 170 second-feet).

1901–1918: Maximum stage recorded, 12.18 feet from noon until 1 p. m. April 2, 1916 (discharge, 27,900 second-feet); minimum stage recorded, 4.6 feet at the former station in Binghamton at 8 a. m. August 29, 1909 (discharge, about 10 second-feet).

Ice.—Stage-discharge relation affected by ice.

Diversions.—The run-off from 87.3 square miles at head of Chenango River and from 15.7 square miles at head of Tioughnioga River is stored in reservoirs and, except for discharge over the spillways, is diverted out of the drainage area into the Erie canal. The drainage area for Chenango River does not include these two areas.

Accuracy.—Stage-discharge relation practically permanent except when affected by ice (a large part of the period from January to March, inclusive). Rating curve well defined between 120 and 35,000 second-feet. Operation of the water-stage recorder fairly satisfactory throughout the year. Daily discharges ascertained by applying to rating table mean daily gage height, determined by inspecting recorder graph, or for days of considerable fluctuation by averaging the hourly discharge. Open-water records good; winter records fair.

Discharge measurements of Chenango River near Chenango Forks, N. Y., during the year ending Sept. 30, 1918.

Date.	Made by—	Gage height.	Discharge.	Date.	Made by—	Gage height.	Discharge.
		Feet.	*Sec.-ft.*			*Feet.*	*Sec.-ft.*
Oct. 14	E. D. Burchard	4.02	1,820	Mar. 7[a]	E. D. Burchard	8.98	10,600
Dec. 16[a]	C. C. Covert	3.94	838	22	C. C. Covert	9.08	14,800
Jan. 16[b]	.....do	5.06	640	Apr. 26	.....do	4.72	3,100
Feb. 11[b]	.....do	4.29	595	June 3	J. W. Moulton	3.87	1,680
Mar. 2[a]	.....do	9.35	8,880	Aug. 16	E. D. Burchard	3.01	559

[a] Measurement made through incomplete ice cover. [b] Measurement made through complete ice cover.

Daily discharge, in second-feet, of Chenango River near Chenango Forks, N. Y., for the year ending Sept. 30, 1918.

Day.	Oct.	Nov.	Dec.	Jan.	Feb.	Mar.	Apr.	May.	June.	July.	Aug.	Sept.
1	740	8,800	1,560	650	380	8,000	3,550	3,260	3,160	3,160	496	338
2	750	5,920	2,180	600	380	9,000	3,860	2,970	2,100	1,860	414	360
3	740	4,720	1,630	550	380	7,000	4,280	2,270	1,620	1,600	322	360
4	750	3,800	1,520	550	400	4,200	4,500	2,360	1,700	1,400	232	360
5	1,170	3,160	1,420	460	420	3,400	3,160	2,100	1,170	1,200	246	360
6	1,430	2,790	1,280	460	440	6,500	2,520	1,860	1,300	950	398	360
7	1,250	2,610	1,080	440	480	10,000	2,180	1,660	3,260	850	398	360
8	1,030	2,270	994	420	500	7,500	2,100	1,550	2,880	750	446	360
9	1,380	2,020	809	409	559	7,000	4,970	1,410	1,769	700	338	360
10	1,310	1,940	900	400	550	7,000	4,960	1,380	1,520	750	487	360
11	1,090	1,780	1,000	400	600	7,500	4,060	1,570	1,560	2,000	555	360
12	1,040	1,660	1,100	420	700	8,000	3,750	1,530	3,810	2,930	487	360
13	2,790	1,520	1,200	440	1,100	10,000	3,550	3,960	3,580	1,860	487	414
14	2,020	1,380	1,100	550	1,600	17,800	5,030	5,640	2,440	1,520	860	574
15	1,670	1,280	1,000	650	2,600	16,600	8,800	3,350	1,860	1,530	660	740
16	3,160	1,270	950	650	3,800	11,800	7,100	2,440	1,490	1,300	438	772
17	2,270	1,270	900	500	4,200	10,900	5,430	1,940	1,270	1,200	438	740
18	1,720	1,180	850	480	3,600	13,400	7,650	1,660	1,140	1,400	438	882
19	1,670	1,170	850	440	5,500	11,200	6,440	1,590	1,010	1,100	438	970
20	5,210	1,140	850	360	9,500	13,000	4,500	1,300	904	900	438	1,780
21	4,180	1,120	800	360	9,000	14,200	4,060	5,680	827	750	414	2,610
22	2,880	1,300	850	380	8,000	14,280	5,550	3,160	2,190	650	360	1,600
23	2,360	1,260	850	380	8,000	12,700	5,430	3,160	1,940	574	322	1,700
24	3,140	1,410	850	380	7,000	9,400	4,720	2,520	1,570	772	322	2,180
25	7,080	1,720	850	380	7,000	7,100	3,860	2,700	1,250	882	322	3,160
26	5,070	1,340	850	380	10,000	5,800	3,160	2,790	1,020	700	322	3,960
27	3,650	1,250	850	380	9,500	4,840	2,790	4,170	871	555	322	4,060
28	4,900	1,250	800	380	8,500	3,960	2,360	3,750	761	360	322	2,610
29	4,840	1,230	800	380	....	3,550	2,100	3,160	710	622	322	1,660
30	11,600	1,250	750	380	....	3,550	2,700	2,970	982	504	322	1,350
31	14,280	....	700	380	....	3,550	....	3,350	....	504	322	....

NOTE.—Discharge Dec. 9 to Mar. 13 estimated, because of ice, from discharge measurements, weather records, study of recorder graph, and comparison with similar studies for Susquehanna River at Conklin. Discharge May 18 to June 10 and July 23 to Sept. 30 determined from semidaily observations on the staff gage, discharge July 3–7 and 16–22 estimated by comparison of recorder graph with that for the Susquehanna River at Conklin.

Monthly discharge of Chenango River near Chenango Forks, N. Y., for the year ending Sept. 30, 1918.

[Drainage area, 1,380 square miles.]

Month.	Discharge in second-feet.				Run-off (depth in inches on drainage area).
	Maximum.	Minimum.	Mean.	Per square mile.	
October	14,200	740	3,130	2.27	2.62
November	8,800	1,120	2,160	1.57	1.75
December	2,180	700	1,040	.754	.87
January	650	360	452	.328	.38
February	10,000	380	3,740	2.71	2.82
March	17,800	3,400	8,790	6.37	7.34
April	8,800	2,100	4,300	3.12	3.48
May	5,680	1,300	2,680	1.94	2.24
June	3,810	710	1,720	1.25	1.40
July	3,160	360	1,160	.841	.97
August	860	232	409	.296	.34
September	4,060	338	1,200	.870	.97
The year	17,800	232	2,560	1.86	25.18

CHEMUNG RIVER AT CHEMUNG, N. Y.

LOCATION.—At highway bridge about midway between Chemung, Chemung County, N. Y., and Willawana, Pa., half a mile upstream from State line and 10 miles above mouth.

DRAINAGE AREA.—2,440 square miles.

RECORDS AVAILABLE.—September 11, 1903, to September 30, 1918.

GAGE.—Tape gage at the upstream side of the right span of the bridge; read by D. L. Orcutt.

DISCHARGE MEASUREMENTS.—Made from the bridge or by wading.

CHANNEL AND CONTROL.—Sand and gravel; occasionally shifting.

EXTREMES OF DISCHARGE.—Maximum stage recorded during year, 17.96 feet at 7 a. m. March 15 (discharge, about 67,000 second-feet); minimum stage recorded 1.64 feet at 6.30 a. m. August 30 (discharge, 146 second-feet).

1903–1918: Maximum stage recorded, that of March 15, 1918; minimum stage recorded, 1.47 feet at 7 a. m. August 14, 1911 (discharge, about 49 second-feet).

ICE.—Stage-discharge relation affected by ice.

REGULATION.—Power is developed above the station, the largest plant being at Elmira, N. Y.

ACCURACY.—Stage-discharge relation probably permanent between dates of shift; affected by ice for a large part of the period from December to March, inclusive. Rating curve well defined between 200 and 45,000 second-feet. Gage read to hundredths twice daily. Daily discharge ascertained by applying mean daily gage height to rating table. Open-water record good; winter record fair.

Discharge measurements of Chemung River at Chemung, N. Y., during the year ending Sept. 30, 1918.

Date.	Made by—	Gage height.	Discharge.	Date.	Made by—	Gage height.	Discharge.
		Feet.	*Sec.-ft.*			*Feet.*	*Sec.-ft.*
Oct. 18	E. D. Burchard	3.17	1,230	Mar. 20	C. C. Covert	5.91	5,200
Dec. 24[a]	C. C. Covert	3.46	1,010	Apr. 28	.....do	4.16	2,500
Feb. 10[b]	E. D. Burchard	3.28	344	June 1	E. D. Burchard	4.85	3,710
Mar. 6	.....do	5.19	4,420	July 19	.....do	2.08	336

[a] Measurement made through incomplete ice cover. [b] Measurement made through complete ice cover.

Daily discharge, in second-feet, of Chemung River at Chemung, N. Y., for the year ending Sept. 30, 1918.

Day.	Oct.	Nov.	Dec.	Jan.	Feb.	Mar.	Apr.	May.	June.	July.	Aug.	Sept.
1	602	7,850	870	700	400	25,700	1,860	2,290	3,650	630	299	168
2	588	5,760	960	650	380	18,000	1,860	2,600	2,600	581	282	192
3	588	4,650	870	650	360	10,400	2,000	2,140	2,000	339	255	208
4	567	3,840	1,000	600	340	5,700	3,100	1,860	1,540	518	250	227
5	710	3,280	870	600	260	4,440	2,760	1,540	1,420	490	343	200
6	1,050	2,760	790	600	280	5,080	2,140	1,480	1,250	470	288	200
7	960	2,440	750	600	320	10,400	1,860	1,480	2,760	451	451	208
8	870	2,290	670	600	320	4,860	1,730	1,300	3,100	401	354	338
9	750	2,000	490	550	320	4,440	3,280	1,250	1,730	377	321	288
10	870	1,860	500	550	340	7,280	4,240	1,150	1,300	377	299	266
11	830	1,730	700	550	380	8,440	3,460	1,200	1,200	377	389	208
12	670	1,540	850	650	480	5,080	3,460	1,360	2,000	407	630	200
13	1,730	1,420	800	500	16,800	11,400	3,280	1,420	3,100	438	532	232
14	1,860	1,300	850	550	12,400	38,200	5,530	2,600	1,860	389	401	255
15	1,300	1,200	1,000	600	,400	54,900	20,400	2,440	1,360	343	360	525
16	1,420	1,150	1,000	600	11,000	12,400	33,100	1,730	1,150	343	302	383
17	1,600	1,150	1,000	600	[illegible],840	8,440	23,000	1,480	960	332	288	432
18	1,200	1,150	900	600	,600	6,490	22,500	1,250	830	310	266	870
19	1,050	1,050	900	600	,140	,300	12,400	1,150	750	299	236	1,200
20	16,800	1,000	850	550	1 ,200	,000	7,560	1,050	670	299	204	1,480
21	7,010	960	800	500	17,600	6,490	6,000	2,000	602	288	196	5,760
22	4,240	1,000	800	500	[illegible],440	6,490	7,560	2,760	2,000	282	184	2,600
23	3,100	1,200	850	500	,460	5,760	6,240	3,460	3,280	266	184	1,730
24	3,650	1,300	1,000	480	,100	4,440	5,530	4,440	2,000	266	184	1,300
25	24,300	1,100	1,100	500	,100	3,650	4,440	2,760	1,420	266	172	1,050
26	17,200	710	1,300	460	9,700	3,180	3,460	3,460	1,100	432	168	1,360
27	16,800	670	1,500	460	11,400	2,760	2,930	4,800	870	419	154	2,140
28	18,000	790	1,300	440	6,750	,440	2,600	5,300	750	343	157	1,600
29	13,100	790	1,000	460	....	,290	2,140	3,460	790	299	164	1,300
30	17,200	830	800	420	....	,000	2,140	8,440	670	288	154	1,000
31	13,800	....	750	400	....	,000	....	6,240	....	277	161	....

NOTE.—Discharge Dec. 10 to Feb. 12 estimated, because of ice, from discharge measurements, weather records, study of recorder graph, and comparison with similar studies for near-by streams.

Monthly discharge of Chemung River at Chemung, N. Y., for the year ending Sept. 30, 1918.

[Drainage area, 2,440 square miles.]

Month.	Discharge in second-feet.				Run-off (depth in inches on drainage area).
	Maximum.	Minimum.	Mean.	Per square mile.	
October	24,300	567	5,630	2.31	2.66
November	7,850	670	1,960	.804	.90
December	1,500	490	898	.368	.42
January	700	400	550	.225	.26
February	19,200	260	5,150	2.11	2.20
March	54,900	2,000	9,500	3.89	4.49
April	33,100	1,730	6,750	2.77	3.09
May	8,440	1,050	2,580	1.06	1.22
June	3,650	602	1,620	.663	.74
July	630	266	380	.156	.18
August	630	154	278	.114	.13
September	5,760	168	931	.382	.43
The year	54,900	154	3,000	1.23	16.72

COHOCTON RIVER NEAR CAMPBELL, N. Y.

Location.—At highway bridge known locally as Red Bridge, nearly 2 miles upstream from Campbell, Steuben County, and midway between Campbell and Savona.

Drainage area.—Not determined.

Records available.—July 11, 1918, to Sept. 30, 1918.

Gage.—Standard chain gage fastened to the downstream handrail of the bridge near the left abutment; read by Miss Dora Wood.

Discharge measurements.—Made from bridge or by wading.

Channel and control.—Firmly bedded gravel, not likely to shift.

Ice.—Stage-discharge relation probably affected by ice.

Regulation.—Seasonal distribution of flow is probably not affected by operation of small reservoirs above.

Cooperation.—Station established by the Lamoka Electric Power Co. under the direction of the United States Geological Survey; maintained by the Survey in cooperation with the power company and the State of New York.

Discharge measurements of Cohocton River near Campbell, N. Y., during the year ending Sept. 30, 1918.

Date.	Made by—	Gage height.	Discharge.	Date.	Made by—	Gage height.	Discharge.
		Feet.	*Sec.-ft.*			*Feet.*	*Sec.-ft.*
July 17	E. D. Burchard	0.82	94.2	July 19	E. D. Burchard	0.85	106
17	do	.82	91.3	Aug. 18	C. C. Covert	.72	68.8

Daily gage height, in feet, of Cohocton River near Campbell, N. Y., for the year ending Sept. 30, 1918.

Day.	July.	Aug.	Sept.	Day.	July.	Aug.	Sept.	Day.	July.	Aug.	Sept.
1		0.91	0.86	11	0.95	0.84	0.74	21	0.75	0.73	1.86
2		.81	.71	12	1.03	.83	.70	22	.75	.71	1.57
3		.81	.71	13	.97	.76	.82	23	.78	.77	1.37
4		.92	.70	14	.89	.77	.88	24	.83	.72	1.31
5		.81	.70	15	.83	.80	.76	25	1.23	.70	1.26
6		.83	.82	16	.87	.76	.73	26	1.04	.72	1.46
7		.89	.91	17	.86	.76	.98	27	.88	.71	1.42
8		.84	.78	18	.84	.73	1.10	28	.91	.70	1.31
9		.81	.70	19	.85	.74	1.41	29	.84	.70	1.22
10		.85	.68	20	.83	.72	2.07	30	.99	.73	1.13
								31	.98	.73	

MUD CREEK AT SAVONA, N. Y.

Location.—On farm of L. R. Travis in Savona, Steuben County, half a mile above mouth.

Drainage area.—Not determined.

Records available.—July 8 to September 30, 1918.

Gage.—Vertical staff fastened to timber planted in concrete at the water's edge on the left bank 150 feet upstream from farm bridge; read by L. R. Travis.

Discharge measurements.—Made by wading at the gage or from farm bridge.

Channel and control.—Fairly well compacted gravel; not likely to shift. Considerable grass grows in stream bed. Control probably submerged by backwater from the Cohocton River during extreme floods.

Ice.—Stage-discharge relation affected by ice.

REGULATION.—Operation of grist mills at Bradford, 7 miles upstream, causes some diurnal fluctuation in flow.

COOPERATION.—Station established by the Lamoka Electric Power Co. under the direction of the United States Geological Survey; maintained by the Survey in cooperation with the power company and the State of New York.

Discharge measurements of Mud Creek at Savona, N. Y., during the year ending Sept. 30, 1918.

Date.	Made by—	Gage height.	Discharge.
		Feet.	*Sec.-ft.*
July 19	E. D. Burchard	3.53	18.4
Aug. 18	C. C. Covert	3.49	14.3

Daily gage height, in feet, of Mud Creek at Savona, N. Y., for the year ending Sept. 30, 1918.

Day.	July.	Aug.	Sept.	Day.	July.	Aug.	Sept.	Day.	July.	Aug.	Sept.
1	......	3.54	3.60	11	3.59	3.52	3.47	21	3.56	3.52	4.05
2	......	3.52	3.46	12	3.66	3.50	3.47	22	3.50	3.52	3.70
3	......	3.50	3.48	13	3.60	3.50	3.58	23	3.51	3.66	3.55
4	......	3.58	3.53	14	3.62	3.62	3.48	24	3.72	3.48	3.56
5	......	3.54	3.50	15	3.54	3.51	3.42	25	4.04	3.46	3.56
6	......	3.58	3.48	16	3.54	3.60	3.40	26	3.76	3.47	3.76
7	......	3.56	3.50	17	3.54	3.63	3.50	27	3.60	3.60	3.68
8	3.54	3.52	3.52	18	3.54	3.50	3.59	28	3.54	3.49	3.59
9	3.56	3.54	3.47	19	3.52	3.48	3.47	29	3.52	3.50	3.57
10	3.63	3.62	3.48	20	3.58	3.50	4.26	30	3.62	3.50	3.48
								31	2.62	3.48	......

TIOGA RIVER NEAR ERWINS, N. Y.

LOCATION.—At highway bridge, a quarter of a mile below mouth of Canisteo River, near village of Erwins, Steuben County, and 3 miles above junction of Tioga and Cohocton rivers to form Chemung River at town of Painted Post.

DRAINAGE AREA.—1,320 square miles (furnished by Robert O. Hayt).

RECORDS AVAILABLE.—July 12, 1918, to September 30, 1918.

GAGE.—Chain near left abutment, downstream side of bridge; graduated and read to quarter-tenths twice daily by Miss Jane Sexton.

DISCHARGE MEASUREMENTS.—Made from bridge or by wading near the control, 100 yards downstream.

CHANNEL AND CONTROL.—Well-compacted gravel, probably permanent.

EXTREMES OF DISCHARGE.—Maximum stage recorded during period, 6.00 feet at 5.30 p. m. September 20 (discharge, 6,160 second-feet); minimum stage recorded, 0.92 foot August 30 (discharge, 54 second-feet).

ICE.—Stage-discharge relation affected by ice.

REGULATION.—There is no considerable storage to interfere with the seasonal flow.

ACCURACY.—Stage-discharge relation believed to be fairly permanent. Rating curve well defined for stages recorded.

COOPERATION.—Station established by the Lamoka Power Co., under the direction of the United States Geological Survey. Maintained by the Survey in cooperation with the power company and the State of New York.

Discharge measurements of Tioga River near Erwins, N. Y., during the year ending Sept. 30, 1918.

Date.	Made by—	Gage height.	Discharge.
		Feet.	*Sec.-ft.*
July 17	E. D. Burchard	1.15	125
17	....do	1.15	124
Aug. 17	C. C. Covert	1.28	143

Daily discharge, in second-feet, of Tioga River near Erwins, N. Y., for the year ending Sept. 30, 1918.

Day.	July.	Aug.	Sept.	Day.	July.	Aug.	Sept.	Day.	July.	Aug.	Sept.
1		138	50	11		548	106	21	118	106	2,340
2		112	90	12	124	513	112	22	127	109	1,380
3		82	121	13	142	306	146	23	118	103	980
4		97	100	14	130	265	432	24	112	88	830
5		146	121	15	118	220	294	25	106	80	660
6		562	112	16	100	180	220	26	154	70	980
7		200	240	17	109	180	390	27	138	65	1,240
8		190	205	18	118	138	1,100	28	112	60	940
9	79	154	170	19	97	121	900	29	82	60	700
10		230	121	20	106	109	3,920	30	94	54	590
								31	94	50	

NOTE.—Daily discharge estimated because of no gage-height record Aug. 25 to 29 and 31 to Sept. 3, inclusive.

Monthly discharge of Tioga River near Erwins, N. Y., for the year ending Sept. 30, 1918.

[Drainage area, 1,320 square miles.]

Month.	Discharge in second-feet.				Run-off (depth in inches on drainage area).
	Maximum.	Minimum.	Mean.	Per square mile.	
August	562	50	172	0.130	0.15
September	3,920	50	653	.495	.55

PATUXENT RIVER BASIN.

PATUXENT RIVER NEAR BURTONSVILLE, MD.

LOCATION.—At Columbia turnpike bridge, 1½ miles northeast of Burtonsville, Montgomery County, and about 4 miles northwest of Laurel.

DRAINAGE AREA.—127 square miles.

RECORDS AVAILABLE.—July 21, 1911, to June 15, 1912 (records furnished by United States Engineer Office); July 21, 1913, to September 30, 1918.

GAGE.—Stevens water-stage recorder referred to a staff gage in three sections on left bank about 80 feet below highway bridge; prior to July 23, 1914, a vertical staff fastened to left side of bridge pier; datum of recorder is 1.29 feet below that of gage on pier. Recorder inspected by Columbus Brashears and Arthur Beall.

DISCHARGE MEASUREMENTS.—Made from bridge or by wading.

CHANNEL AND CONTROL.—Banks are lined with trees and brush and are overflowed at stage of about 10 feet. Control is a flat gravel bar about 300 feet below bridge. Current is swift under bridge, but sluggish below bridge to control.

EXTREMES OF DISCHARGE.—Maximum stage during year, 8.68 feet at 12.30 a. m. January 14 (discharge, 2,190 second-feet); minimum stage, 1.69 feet August 25, 26, 27, and 28 (discharge, 47 second-feet).

1911–1918: Maximum stage recorded, 14.6 feet about 9 a. m. January 12, 1915 (discharge, from poorly defined rating curve, 5,100 second-feet); minimum stage, 0.18 foot August 25, 1911 (discharge, 6 second-feet).

ICE.—Stage-discharge relation affected by ice during severe winters only.

ACCURACY.—Stage-discharge relation affected by ice December 10 to January 11, January 12–14, and January 20 to February 12. Rating curve well defined between 50 and 200 second-feet and fairly well defined above 200 second-feet. Operation of water-stage recorder satisfactory throughout the year, except for period November 7–10. Daily discharge ascertained by use of discharge integrator, by hourly method, and by use of mean dailyg age height obtained by inspecting recorder graph. Records excellent.

Discharge measurements of Patuxent River near Burtonsville, Md., during the year ending Sept. 30, 1918.

Date.	Made by—	Gage height.	Discharge.	Date.	Made by—	Gage height.	Discharge.
		Feet.	*Sec.-ft.*			*Feet.*	*Sec.-ft.*
Nov. 6	G. C. Stevens	2.13	72.0	Dec. 17	G. C. Stevens	a 2.66	62.3
12	Parker and Horton	2.06	63.4	Apr. 6	Stevens and Hoyt	2.20	87.3

a Stage-discharge relation affected by ice.

Daily discharge, in second-feet, of Patuxent River near Burtonsville, Md., for the year ending Sept. 30, 1918.

Day.	Oct.	Nov.	Dec.	Jan.	Feb.	Mar.	Apr.	May.	June.	July.	Aug.	Sept.
1	55	76	75	19	84	117	100	200	103	63	53	162
2	54	72	55	23	76	103	97	151	144	58	45	49
3	53	76	50	23	76	98	94	131	162	51	43	41
4	51	76	48	23	69	92	94	126	92	51	41	39
5	50	72	45	23	62	177	87	122	83	50	42	38
6	49	72	43	23	69	130	84	112	78	49	42	43
7	49	70	42	23	108	153	81	105	89	47	40	43
8	49	67	55	28	369	117	82	102	80	45	270	38
9	51	63	62	23	270	107	229	94	72	44	55	42
10	55	59	92	28	190	126	1,050	107	70	41	47	40
11	55	55	100	49	357	87	607	260	76	41	47	36
12	53	62	84	1,810	844	92	468	108	75	41	95	37
13	78	65	69	291	1,620	121	520	103	68	121	171	37
14	56	63	69	219	405	312	393	117	63	78	72	37
15	53	65	69	190	1,150	346	270	92	66	58	62	32
16	50	65	62	190	357	162	219	87	65	47	40	34
17	49	65	69	200	171	126	200	84	62	43	32	32
18	47	65	62	200	148	110	180	80	61	44	34	186
19	49	63	55	190	323	97	162	76	59	47	36	84
20	121	62	62	171	944	89	157	72	56	47	34	62
21	89	61	62	171	229	700	638	126	56	41	30	162
22	69	63	62	162	144	430	393	577	68	38	28	69
23	82	61	49	171	162	200	239	131	63	35	26	50
24	468	55	55	162	162	151	200	108	59	35	24	47
25	135	49	43	144	153	157	201	103	59	38	22	41
26	76	47	43	126	323	130	177	102	62	36	22	41
27	92	49	32	108	153	117	162	190	63	35	22	41
28	223	49	38	108	124	114	153	97	61	35	22	41
29	108	55	15	92	....	108	146	260	56	36	40	36
30	507	56	23	100	....	105	149	124	56	74	47	37
31	153	....	19	92	....	102	....	146	....	76	171	....

NOTE.—Discharge estimated Nov. 7–10, account no record. Dec. 10 to Jan. 11, Jan. 15–17, and Jan. 20 to Feb. 12, discharge estimated as in table, because of ice, from discharge measurement, study of gage-height graph, and weather records.

Monthly discharge of Patuxent River near Burtonsville, Md., for the year ending Sept. 30, 1918.

Month.	Discharge in second-feet.				Run-off (depth in inches on drainage area).
	Maximum.	Minimum.	Mean.	Per square mile.	
October	507	47	104	0.819	0.94
November	76	47	62.6	.494	.55
December	100	15	55.1	.434	.50
January	1,810	19	167	1.31	1.51
February	1,620	62	326	2.57	2.68
March	700	87	164	1.29	1.49
April	1,050	81	254	2.00	2.23
May	577	72	136	1.07	1.23
June	162	56	74.2	.584	.65
July	121	85	49.7	.391	.45
August	270	22	56.6	.446	.51
September	186	32	55.9	.440	.49
The year	1,810	15	124	.976	13.23

POTOMAC RIVER BASIN.

POTOMAC RIVER AT POINT OF ROCKS, MD.

Location.—At steel highway bridge at Point of Rocks, Frederick County, about one-third mile below Catoctin Creek and 6 miles above Monocacy River.

Drainage area.—9,650 square miles.

Records available.—February 17, 1895, to September 30, 1918.

Gage.—Chain, attached to downstream side of left span of bridge; read by G. H. Hickman. Datum constant since September 2, 1902; prior to this date datum was 0.45 foot higher than at present. Sea-level elevation of gage datum, 200.54 feet.

Discharge measurements.—Made from the bridge.

Channel and control.—Practically permanent. The control is a ledge a few hundred feet below the station, the ledge extending completely across the river except for one relatively unimportant channel.

Extremes of discharge.—Maximum stage recorded during year, 17.1 feet at 6 p. m. April 22 (discharge, 115,000 second-feet); minimum stage recorded, 0.49 foot at 3 p. m. October 1 (discharge, 770 second-feet).

1895–1918: Maximum stage recorded, 29 feet on March 2, 1902 (discharge, 219,000 second-feet); minimum stage, 0.38 foot on September 10, 1914 (discharge, 540 second-feet).

The crest of the flood of June 2, 1889, as determined by the U. S. Army Engineers from high-water marks, reached a stage of 40.2 feet (discharge, 325,000 second-feet).

Ice.—Stage-discharge relation seldom affected by ice.

Diversions.—The Chesapeake & Ohio Canal parallels the Potomac on the Maryland side. The average discharge of the canal is 75 to 100 second-feet. The discharge in not included in the following tables:

Regulation.—Fluctuation at extremely low stages has been noted and is probably caused by the operation of power plants on the upper Potomac and tributaries.

Accuracy.—Stage-discharge relation practically permanent; affected by ice from December 12 to February 11. Rating curve well defined except at extremely low water. Gage read to hundredths once daily; during high water read oftener. Daily discharge ascertained by applying daily gage height to rating table. Records excellent except those for extremely low stages, which are fair.

The following discharge measurement was made by G. C. Stevens and M. I. Walters:
October 3, 1918: Gage height, 0.70 foot; discharge, 1,120 second-feet.

Daily discharge, in second-feet, of Potomac River at Point of Rocks, Md., for the year ending Sept. 30, 1918.

Day.	Oct.	Nov.	Dec.	Feb.	Mar.	Apr.	May.	June.	July.	Aug.	Sept.
1	770	18,600	1,990	....	400	5,020	16,300	5,020	2,940	3,240	2,940
2	1,080	12,500	2,250	....	800	6,520	14,600	5,750	2,660	2,940	2,800
3	1,190	9,070	2,250	....	100	5,750	11,000	5,020	3,240	3,540	2,380
4	1,060	6,920	2,520	....	20,300	5,750	9,070	4,840	4,040	3,390	3,700
5	945	5,380	2,660	....	18,200	5,380	8,620	4,840	4,840	3,240	3,240
6	835	4,840	2,380	....	1 700	4,500	8,180	5,380	4,500	2,520	2,940
7	1,510	4,500	2,250	....	1 000	4,170	7,330	4,500	4,670	2,800	3,700
8	1,290	4,010	2,120	....	070	4,010	6,520	4,170	4,500	3,090	3,640
9	1,190	4,010	2,120	....	5,070	7,320	6,130	2,520	4,170	3,390	3,590
10	1,100	3,090	2,250	....	9,530	29,400	5,750	2,380	4,010	2,660	3,090
11	835	3,090	2,250	....	12,000	60,600	6,520	2,520	4,010	2,250	2,940
12	945	3,540	....	63,900	11,000	50,800	6,130	2,940	3,860	2,120	2,800
13	1,220	2,940	....	105,000	13,500	27,500	5,380	2,800	3,090	2,940	2,380
14	1,260	2,800	....	80,500	15,700	35,600	6,920	2,660	3,240	3,090	2,120
15	1,060	2,660	....	68,000	43,000	93,000	5,750	2,520	3,700	3,540	1,990
16	900	2,800	....	48,400	28,800	111,000	5,750	2,380	3,390	3,090	2,120
17	1,260	2,940	....	40,000	26,100	97,100	5,380	2,250	3,240	3,240	2,250
18	1,220	1,910	....	33,500	1 800	93,800	4,840	2,520	2,940	2,940	2,520
19	1,100	1,540	....	23,600	19,500	80,500	4,190	2,520	2,800	2,800	4,500
20	1,340	1,390	....	30,100	16,500	54,000	3,540	2,940	2,520	2,660	6,520
21	1,030	1,260	....	64,700	9,530	37,100	3,240	2,800	2,940	2,940	6,920
22	945	1,510	....	55,600	11,000	110,000	2,940	3,240	2,520	2,520	6,720
23	965	1,680	....	38,500	8,620	95,400	3,540	3,860	2,520	2,380	6,520
24	11,500	1,790	....	21,100	9,070	35,600	3,090	3,090	2,380	2,520	5,380
25	22,900	2,120	....	9,530	12,000	33,500	2,660	2,800	2,120	2,380	5,020
26	18,600	1,940	....	22,300	520	28,800	3,540	2,660	2,940	2,250	5,750
27	9,530	1,540	....	33,500	520	20,400	3,090	2,520	3,240	2,660	5,380
28	7,750	1,290	....	19,800	070	22,900	2,940	2,660	2,940	2,940	5,020
29	7,320	1,480	....	....	6,520	15,700	3,090	2,520	3,090	3,240	4,760
30	15,200	1,760	....	....	6,500	14,100	4,330	2,730	4,330	3,240	4,500
31	30,100	....	....	....	5,750	....	5,380	....	2,660	3,700	....

NOTE.—Discharge estimated, on account of ice, from a study of weather records and daily gage-height graph as follows: Dec. 12-31, 2,700 second-feet; Jan. 1-31, 2,500 second-feet; Feb. 1-11, 3,200 second-feet. Discharge interpolated May 5 and 19, June 30, July 4, and Sept. 8, 22, and 29; discharge estimated Apr. 9.

Monthly discharge of Potomac River at Point of Rocks, Md., for the year ending Sept. 30, 1918.

Month.	Discharge in second-feet.				Run-off (depth in inches on drainage area).
	Maximum.	Minimum.	Mean.	Per square mile.	
October	30,100	770	4,770	0.494	0.57
November	18,600	1,260	3,830	.397	.44
December	....	....	2,550	.264	.30
January	....	....	2,500	.259	.30
February	....	....	28,300	2.93	3.05
March	43,000	4,500	13,600	1.41	1.63
April	111,000	4,010	39,800	4.12	4.60
May	16,300	2,660	5,990	.621	.72
June	5,750	2,250	3,310	.343	.38
July	4,840	2,120	3,360	.348	.40
August	3,700	2,120	2,910	.302	.35
September	6,920	1,990	3,940	.408	.46
The year	111,000	770	9,390	.973	13.20

MONOCACY RIVER NEAR FREDERICK, MD.

LOCATION.—At Ceresville bridge on toll road leading from Frederick, Frederick County, to Mount Pleasant, about 3,000 feet below Tuscarora Creek (entering from right), 2,000 feet above Israel Creek (entering from left), and 3 miles northeast of Frederick.

DRAINAGE AREA.—660 square miles.

RECORDS AVAILABLE.—August 4, 1896, to September 30, 1918.

GAGE.—Chain attached to downstream side of right span of bridge; read by Eugene L. Derr.

DISCHARGE MEASUREMENTS.—Made from the bridge or by wading.

CHANNEL AND CONTROL.—Bed composed of gravel and boulders; shifting during very high floods. Control not well defined. Banks lined with trees and brush; subject to overflow at high stages.

EXTREMES OF DISCHARGE.—Maximum stage recorded during the year, 22.1 feet at 5.20 p. m. February 20 (discharge, 14,300 second-feet); minimum stage recorded, 3.85 feet September 16 and 17 (discharge, 54 second-feet).

1896–1918: Maximum stage recorded, 27.2 feet at 11 a. m. January 13, 1915 (discharge, determined from rating curve used for 1916, 19,000 second-feet); minimum stage, 3.54 feet several days in October, 1910 (discharge, 15 second-feet).

ICE.—Stage-discharge relation affected by ice during severe winters only.

ACCURACY.—Stage-discharge relation affected by ice from December 9 to February 11. Rating curve well defined between 200 and 15,000 second-feet. Discharge measurements made during high water of March, 1917, indicate that rating curves used prior to 1916 gave results about 20 per cent too large at high stages. Gage read to half-tenths once daily; oftener during high water. Daily discharge ascertained by applying gage height to rating table. Records good.

The following discharge measurement was made by G. C. Stevens:

January 3, 1918: Gage height, 5.45 feet; discharge, 166 second-feet. Stage-discharge relation affected by ice.

Daily discharge, in second-feet, of Monocacy River near Frederick, Md., for the year ending Sept. 30, 1918.

Day.	Oct.	Nov.	Dec.	Feb.	Mar.	Apr.	May.	June.	July.	Aug.	Sept.
1	204	2,060	326	....	2,560	454	932	494	218	310	247
2	178	1,640	326	....	1,910	415	882	415	218	191	165
3	178	1,260	310	....	1,710	378	784	882	191	128	140
4	178	1,030	294	....	1,320	343	667	474	178	116	128
5	178	882	262	....	1,710	415	600	343	165	191	116
6	178	784	262	....	1,640	378	535	343	152	1,200	116
7	178	736	262	....	1,450	415	535	310	140	578	140
8	165	644	262	....	1,320	982	494	310	152	191	140
9	152	600	....	....	1,450	3,060	474	278	128	165	93
10	152	514	....	....	1,570	7,010	434	278	140	152	93
11	152	494	....	....	1,450	8,830	434	247	128	204	93
12	152	474	....	1,320	1,260	3,590	415	247	128	191	93
13	232	454	....	4,820	982	5,580	434	218	191	165	93
14	232	434	....	9,390	3,440	4,390	434	232	326	152	72
15	218	396	....	8,010	3,590	3,830	415	218	218	140	63
16	204	378	....	9,480	1,570	2,700	494	218	191	128	54
17	178	360	....	3,140	1,320	2,270	415	191	152	116	54
18	178	343	....	2,920	1,140	1,570	378	204	152	128	116
19	165	326	....	2,120	982	1,570	360	191	140	93	116
20	360	294	....	13,700	882	1,450	310	204	140	72	165
21	474	294	....	8,830	832	5,410	326	191	140	72	278
22	474	343	....	5,500	982	5,240	5,580	360	128	93	378
23	474	326	....	1,840	832	2,410	1,030	310	116	72	310
24	4,730	326	....	1,570	736	1,710	622	262	116	72	218
25	8,280	326	....	2,990	713	1,640	556	218	140	72	165
26	3,060	310	....	11,800	600	1,200	494	204	535	72	140
27	1,030	294	....	4,070	535	1,090	434	178	378	72	93
28	4,150	278	....	2,990	556	982	378	191	165	93	93
29	1,450	262	....	....	494	882	600	165	140	128	93
30	12,400	262	....	....	494	784	415	218	116	191	72
31	4,310	....	....	....	454	....	982	....	713	191	....

NOTE.—Discharge estimated, on account of ice, from discharge measurement, weather records, and a study of gage-height graph, as follows: Dec. 9–31, 270 second-feet; Jan. 1–12, 185 second-feet; Jan. 13–25, 590 second-feet; Jan. 26–Feb. 11, 460 second-feet.

Monthly discharge of Monocacy River near Frederick, Md., for the year ending Sept. 30, 1918.

Month.	Discharge in second-feet.				Run-off in inches.
	Maximum.	Minimum.	Mean.	Per square mile.	
October	12,400	152	1,440	2.18	2.51
November	3,060	262	604	.915	1.02
December			275	.417	.48
January			408	.618	.71
February	13,700		3,560	5.39	5.61
March	3,590	454	1,310	1.98	2.28
April	8,830	343	2,370	3.59	4.00
May	5,580	310	705	1.07	1.23
June	882	165	286	.433	.48
July	713	116	198	.300	.35
August	1,200	72	185	.280	.32
September	378	54	138	.209	.23
The year	13,700	54	935	1.42	19.22

RAPPAHANNOCK RIVER BASIN.

RAPPAHANNOCK RIVER NEAR FREDERICKSBURG, VA.

LOCATION.—At rear of McWhirt farm, 1½ miles above dam of Spottsylvania Power Co. and 3½ miles above Fredericksburg, Spottsylvania County.

DRAINAGE AREA.—1,590 square miles.

RECORDS AVAILABLE.—September 19, 1907, to September 30, 1918.

GAGE.—Vertical staff on right bank; installed November 4, 1913, to replace chain gage destroyed October 31, 1913. Original gage was a vertical staff which was destroyed February 14, 1908, and replaced February 20, 1908, by a chain gage under the cable. All three gages at practically the same location and referred to same datum. Gage read by Charles Perry.

DISCHARGE MEASUREMENTS.—Made from cable at gage. At extremely low water measurements can be made by wading or from a bridge over the power canal below the dam.

CHANNEL AND CONTROL.—Bed composed of boulders; somewhat rough. One channel. Banks wooded; water overflows right bank at stage about 15 feet and left bank at about 12 feet. Current sluggish at extremely low water. Control is a rocky section a few hundred feet below the gage; practically permanent.

EXTREMES OF DISCHARGE.—Maximum stage during the year, 11.45 feet at noon April 11 (discharge, 38,500 second-feet); minimum stage recorded, 0.73 foot at 3 p. m. September 17 (discharge, 191 second-feet).

1907–1918: Maximum stage recorded, 11.45 feet at noon April 11, 1918 (discharge, 38,500 second-feet); minimum stage recorded, 0.30 foot at 3 p. m. August 21, 1914 (discharge, 72 second-feet).

ICE.—Ice forms near gage but seldom in sufficient quantity at control to affect stage-discharge relation.

ACCURACY.—Stage-discharge relation practically permanent. Rating curve well defined except for extremely high and low stages. Gage read to hundredths twice daily. Daily discharge ascertained by applying mean daily gage height to rating table. Records good except for winter months. Comparison with records for other stations indicates that the winter records of the Rappahannock are not subject to large errors.

Daily discharge, in second-feet, of Rappahannock River near Fredericksburg, Va., for the year ending Sept. 30, 1918.

Day.	Oct.	Nov.	Dec.	Feb.	Mar.	Apr.	May.	June.	July.	Aug.	Sept.
1	342	2,920	1,040		2,080	1,220	3,100	1,480	1,420	729	2,400
2	282	2,080	860		1,920	1,220	2,740	1,560	1,220	494	
3	270	1,770	616		1,920	1,220	2,570	2,570	750	410	
4	260	1,420	569		1,700	1,220	2,080	2,920	687	440	
5	245	1,280	518		1,770	1,220	1,920	2,570	645	470	
6	276	1,160	502		2,740	1,160	1,770	1,840	578	395	
7	282	975	470		3,100	1,100	1,700	1,560	598	355	
8	250	918	534		2,570	1,040	1,480	1,420	560	2,920	
9	294	918	1,420		2,920	1,220	1,480	1,420	502	2,080	
10	329	750	918		2,400	32,500	1,480	729	502	750	
11	455	740	510		2,000	38,500	1,420	636	440	1,280	355
12	410	708	425		1,770	15,900	1,480	740	425	2,740	342
13	329	698			1,840	6,770	1,480	645	455	2,570	329
14	369	656			5,910	5,630	1,770	607	626	2,920	311
15	342	645			4,610	5,630	1,700	550	805	1,620	355
16	329	626			3,290	4,610	1,560	588	542	975	311
17	305	626			2,920	3,920	1,420	542	455	750	195
18	305	588			2,400	3,920	1,420	502	425	750	369
19	276	569			2,240	3,920	1,420	486	494	349	1,770
20	478	550			1,920	4,140	1,350	470	502	676	2,240
21	666	534			1,920	21,600	1,100	470	470	607	2,920
22	534	550			2,740	23,100	975	502	470	486	2,570
23	418	569			2,240	8,010	918	542	425	382	2,080
24	2,240	550			2,000	4,140	860	636	355	362	1,560
25	2,570	569			2,080	4,370	918	687	395	480	1,350
26	1,350	494			2,080	4,370	860	805	410	598	1,220
27	918	462		2,740	1,620	3,700	860	2,080	329	455	1,160
28	3,920	486		918	1,480	3,290	805	2,000	362	349	831
29	2,080	588			1,350	3,100	918	2,000	230	3,290	502
30	4,850	598			1,280	2,920	1,350	1,840	204	2,240	369
31	8,340				1,280		1,420		1,480	1,920	

NOTE.—Daily discharge estimated, on account of ice, from a study of gage heights, weather records, and comparison with near-by streams, as follows: Dec. 13–31, 400 second-feet; Jan. 1–31, 1,200 second-feet; Feb. 1–11, 3,300 second-feet; and on account of no gage readings, Feb. 12–26, 6,800 second-feet, and Sept. 2–10, 800 second-feet. Discharge interpolated Aug. 25 and Sept. 28.

Monthly discharge of Rappahannock River near Fredericksburg, Va., for the year ending Sept. 30, 1918.

Month.	Discharge in second-feet.				Run-off (depth in inches on drainage area).
	Maximum.	Minimum.	Mean.	Per square mile.	
October	8,340	245	1,110	0.698	0.80
November	2,920	462	866	.545	.61
December			516	.325	.37
January			1,200	.755	.87
February			5,200	3.27	3.40
March	5,910	1,280	2,320	1.46	1.68
April	38,500	1,040	7,160	4.50	5.02
May	3,100	805	1,490	.937	1.08
June	2,920	470	1,180	.742	.83
July	1,480	204	573	.360	.42
August	3,290	349	1,120	.704	.81
September	2,920	195	1,020	.642	.72
The year	38,500	195	1,950	1.23	16.61

MISCELLANEOUS MEASUREMENTS.

Miscellaneous discharge measurements in north Atlantic coast drainage basin during the year ending Sept. 30, 1918.

Date.	Stream.	Tributary to—	Locality.	Discharge.
				Sec.-ft.
May 19	Pond Brook	Pemigewasset River (via Bakers River).	Outlet of lower Baker Pond.	37.9
July 19	Canal	Diversion from East branch of Tully River.	Above Packard Pond	10.5

INDEX.

STREAM-GAGING STATIONS

AND

PUBLICATIONS RELATING TO WATER RESOURCES

PART I. NORTH ATLANTIC SLOPE BASINS

STREAM-GAGING STATIONS AND PUBLICATIONS RELATING TO WATER RESOURCES.

Part I. NORTH ATLANTIC SLOPE BASINS.

INTRODUCTION.

Investigation of water resources by the United States Geological Survey has consisted in large part of measurements of the volume of flow of streams and studies of the conditions affecting that flow, but it has comprised also investigation of such closely allied subjects as irrigation, water storage, water powers, underground waters, and quality of waters. Most of the results of these investigations have been published in the series of water-supply papers, but some have appeared in the bulletins, professional papers, monographs, and annual reports.

The results of stream-flow measurements are now published annually in 12 parts, each part covering an area whose boundaries coincide with natural drainage features as indicated below.

Part I. North Atlantic slope basins.
II. South Atlantic slope and eastern Gulf of Mexico basins.
III. Ohio River basin.
IV. St. Lawrence River basin.
V. Upper Mississippi River and Hudson Bay basins.
VI. Missouri River basin.
VII. Lower Mississippi River basin.
VIII. Western Gulf of Mexico basins.
IX. Colorado River basin.
X. Great Basin.
XI. Pacific slope basins in California.
XII. North Pacific slope basins, in three volumes:
A, Pacific slope basins in Washington and upper Columbia River basin.
B, Snake River basin.
C, Lower Columbia River basin and Pacific slope básins in Oregon.

This appendix contains, in addition to the list of gaging stations and the annotated list of publications relating specifically to the section, a similar list of reports that are of general interest in many sections and cover a wide range of hydrologic subjects; also brief references to reports published by State and other organizations (p. xxiv).

HOW GOVERNMENT REPORTS MAY BE OBTAINED OR CONSULTED.

Water-supply papers and other publications of the United States Geological Survey containing data in regard to the water resources of the United States may be obtained or consulted as indicated below.

1. Copies may be obtained free of charge by applying to the Director of the Geological Survey, Washington, D. C. The edition printed for free distribution is, however, small and is soon exhausted.
2. Copies may be purchased at nominal cost from the Superintendent of Documents, Government Printing Office, Washington, D. C., who will on application furnish lists giving prices.
3. Sets of the reports may be consulted in the libraries of the principal cities in the United States.
4. Complete sets are available for consultation in the local offices of the water-resources branch of the Geological Survey as follows:

Boston, Mass., 2500 Customhouse.
Albany, N. Y., 704 Journal Building.
Harrisburg, Pa., care of Water Supply Commission.
Asheville, N. C., 32-35 Broadway.
Chattanooga, Tenn., Temple Court Building.
Madison, Wis., c/o Railroad Commission of Wisconsin.
Chicago, Ill., 1404 Kimball Building.
Ames, Iowa, care of State Highway Commission.
Topeka, Kans., 25 Federal Building.
Austin, Tex., Capitol Building.
Helena, Mont., Montana National Bank Building.
Denver, Colo., 403 New Post Office Building.
Tucson, Ariz., University of Arizona.
Salt Lake City, Utah, 421 Federal Building.
Boise, Idaho, 615 Idaho Building.
Idaho Falls, Idaho, 228 Federal Building.
Tacoma, Wash., 406 Federal Building.
Portland, Oreg., 606 Post Office Building.
San Francisco, Calif., 328 Customhouse.
Los Angeles, Calif., 619 Federal Building.
Honolulu, Hawaii, 14 Capitol Building.

A list of the Geological Survey's publications may be obtained by applying to the Director, United States Geological Survey, Washington, D. C.

STREAM-FLOW REPORTS.

Stream-flow records have been obtained at more than 4,510 points in the United States, and the data obtained have been published in the reports indicated in the following table:

Stream-flow data in reports of the United States Geological Survey.

[A—Annual Report; B—Bulletin; W—Water-Supply Paper.]

Report.	Character of data.	Year.
10th A, pt. 2	Descriptive information only	
11th A, pt. 2	Monthly discharge and descriptive information	1884 to Sept., 1890.
12th A, pt. 2	....do	1884 to June 30, 1891.
13th A, pt. 3	Mean discharge in second-feet	1884 to Dec. 31, 1892.
14th A, pt. 2	Monthly discharge (long-time records, 1871 to 1893)	1888 to Dec. 31, 1893.
B 131	Descriptions, measurements, gage heights, and ratings	1893 and 1894.
16th A, pt. 2	Descriptive information only	
B 140	Descriptions, measurements, gage heights, ratings, and monthly discharge (also many data covering earlier years).	1895.
W 11	Gage heights (also gage heights for earlier years)	1896.
18th A, pt. 4	Descriptions, measurements, ratings, and monthly discharge (also similar data for some earlier years).	1895 and 1896.
W 15	Descriptions, measurements, and gage heights, eastern United States, eastern Mississippi River, and Missouri River above junction with Kansas.	1897.
W 16	Descriptions, measurements, and gage heights, western Mississippi River below junction of Missouri and Platte, and western United States.	1897.
19th A, pt. 4	Descriptions, measurements, ratings, and monthly discharge (also some long-time records).	1897.
W 27	Measurements, ratings, and gage heights, eastern United States, eastern Mississippi River, and Missouri River.	1898.
W 28	Measurements, ratings, and gage heights, Arkansas River and western United States.	1898.
20th A, pt. 4	Monthly discharge (also for many earlier years)	1898.
W 35 to 39	Descriptions, measurements, gage heights, and ratings	1899.
21st A, pt. 4	Monthly discharge	1899.
W 47 to 52	Descriptions, measurements, gage heights, and ratings	1900.
22d A, pt. 4	Monthly discharge	1900.
W 65, 66	Descriptions, measurements, gage heights, and ratings	1901.
W 75	Monthly discharge	1901.
W 82 to 85	Complete data	1902.
W 97 to 100	....do	1903.
W 124 to 135	....do	1904.
W 165 to 178	....do	1905.
W 201 to 214	....do	1906.
W 241 to 252	....do	1907–8.
W 261 to 272	....do	1909.
W 281 to 292	....do	1910.
W 301 to 312	....do	1911.
W 321 to 332	....do	1912.
W 351 to 362	....do	1913.
W 381 to 394	....do	1914.
W 401 to 414	....do	1915.
W 431 to 444	....do	1916.
W 451 to 464	....do	1917.
W 471 to 484	....do	1918.

NOTE.—No data regarding stream flow are given in the 15th and 17th annual reports.

The records at most of the stations discussed in these reports extend over a series of years, and miscellaneous measurements at many points other than regular gaging stations have been made each year. An index of the reports containing records obtained prior to 1904 has been published in Water-Supply Paper 119.

The following table gives, by years and drainage basin, the numbers of papers on surface-water supply published from 1899 to 1918. The data for any particular station will be found in the reports covering the years during which the station was maintained. For example, data for 1902 to 1918 for any station in the area covered by Part III are published in Water-Supply Papers 83, 98, 128, 169. 205, 243, 263, 283, 303, 323, 353, 383, 403, 433, 453, and 473, which contain records for the Ohio River basin for those years.

Numbers of water-supply papers containing results of stream measurements, 1899–1918.

Year.	I	II	III	IV	V	VI	VII	VIII	IX	X	XI	XII		
												North Pacific slope basins.		
	North Atlantic slope basins (St. John River to York River).	South Atlantic and eastern Gulf of Mexico basins (James River to the Mississippi).	Ohio River basin.	St. Lawrence River and Great Lakes basins.	Hudson Bay and upper Mississippi River basin.	Missouri River basin.	Lower Mississippi River basin.	Western Gulf of Mexico basins.	Colorado River basin.	Great Basin.	Pacific slope basins in California.	Pacific slope basins in Washington and upper Columbia River.	Snake River basin.	Lower Columbia River and Pacific slope basins in Oregon.
1899 [a]	35	[b] 35, 36	36	36	36	[c] 36, 37	37	37	[d] 37, 38	38, [e] 39	38, [f] 39	38	38	38
1900 [g]	47, [h] 48	48	48, [i] 49	49	49	49, [j] 50	50	50	50	51	51	51	51	51
1901	65, 75	65, 75	65, 75	65, 75	[k] 65, 66, 75	66, 75	[k] 65, 66, 75	66, 75	66, 75	66, 75	66, 75	66, 75	66, 75	66, 75
1902	82	[b] 82, 83	83	[l] 82, 83	[k] 83, 85	84	[k] 83, 84	84	85	85	85	85	85	85
1903	97	[b] 97, 98	98	97	[k] 98, 99, [m] 100	99	[k] 98, 99	99	100	100	100	100	100	100
1904	[n] 124, [o] 125, [p] 126	[p] 126, 127	128	129	[k] 128, 130	130, [q] 131	[k] 128, 131	132	133	133, [r] 134	134	135	135	135
1905	[n] 165, [o] 166, [p] 167	[p] 167, 168	169	170	171	172	[k] 169, 173	174	175, [s] 177	176, [r] 177	177	178	178	[t] 177, 178
1906	[n] 201, [o] 202, [p] 203	[p] 203, 204	205	206	207	208	[k] 205, 209	210	211	212, [r] 213	213	214	214	214
1907–8	241	242	243	244	245	246	247	248	249	250, [r] 251	251	252	252	252
1909	261	262	263	264	265	266	267	268	269	270, [r] 271	271	272	272	272
1910	281	282	283	284	285	286	287	288	289	290	291	292	292	292
1911	301	302	303	304	305	306	307	308	309	310	311	312	312	312
1912	321	322	323	324	325	326	327	328	329	330	331	332A	332B	332C
1913	351	352	353	354	355	356	357	358	359	360	361	362A	362B	362C
1914	381	382	383	384	385	386	387	388	389	390	391	392	393	394
1915	401	402	403	404	405	406	407	408	409	410	411	412	413	414
1916	431	432	433	434	435	436	437	438	439	440	441	442	443	444
1917	451	452	453	454	455	456	457	458	459	460	461	462	463	464
1918	471	472	473	474	475	476	477	478	479	480	481	482	483	484

[a] Rating tables and index to Water-Supply Papers 35–39 contained in Water-Supply Paper 39. Tables of monthly discharge for 1899 in Twenty-first Annual Report, Part IV.
[b] James River only.
[c] Gallatin River.
[d] Green and Gunnison rivers and Grand River above junction with Gunnison.
[e] Mohave River only.
[f] Kings and Kern rivers and south Pacific slope basins.
[g] Rating tables and index to Water-Supply Papers 47–52 and data on precipitation, wells, and irrigation in California and Utah contained in Water-Supply Paper 52. Tables of monthly discharge for 1900 in Twenty-second Annual Report, Part IV.
[h] Wissahickon and Schuylkill rivers to James River.
[i] Scioto River.
[j] Loup and Platte rivers near Columbus, Nebr., and all tributaries below junction with Platte.
[k] Tributaries of Mississippi from east.
[l] Lake Ontario and tributaries of St. Lawrence River.
[m] Hudson Bay only.
[n] New England rivers only.
[o] Hudson River to Delaware River, inclusive.
[p] Susquehanna River to Yadkin River, inclusive.
[q] Platte and Kansas rivers.
[r] Great Basin in California, except Truckee and Carson river basins.
[s] Below junction with Gila.
[t] Rogue, Umpqua, and Siletz rivers only.

In these papers and in the following lists the stations are arranged in downstream order. The main stem of any river is determined by measuring or estimating its drainage area—that is, the headwater stream having the largest drainage area is considered the continuation of the main stream, and lake surfaces and local changes in name are disregarded. All stations from the source to the mouth of the main stem of the river are presented first, and the tributaries in regular order from source to mouth follow, the streams in each tributary basin being listed before those of the next basin below.

In exception to this rule the records for Mississippi River are given in four parts, as indicated on page III, and the records for large lakes are taken up in order of streams around the rim of the lake.

PRINCIPAL STREAMS.

The principal streams flowing into the Atlantic Ocean between St. John River, Maine-New Brunswick, and York River, Virginia, are the St. Croix, Machias, Union, Penobscot, Kennebec, Androscoggin, Saco, Merrimack, Mystic, Blackstone, Connecticut, Hudson, Delaware, Susquehanna, Potomac, and Rappahannock. The streams drain wholly or in part the States of Connecticut, Delaware, Maine, Maryland, Massachusetts, New Jersey, New Hampshire, New York, Pennsylvania, Rhode Island, Vermont, Virginia, and West Virginia.

GAGING STATIONS.[1]

NOTE.—Dash after date indicates that station was being maintained September 30, 1918. Period after a date indicates discontinuance.

ST. JOHN RIVER BASIN.

St. John River near Dickey, Maine, 1910–11.
St. John River at Fort Kent, Maine, 1905–1915.
St. John River at Van Buren, Maine, 1908–
 Allagash River near Allagash, Maine, 1910–11.
 St. Francis River at St. Francis, Maine, 1910–11.
 Fish River at Wallagrass, Maine, 1903–1908; 1911.
 Madawaska River at St. Rose du Degele, Quebec, 1910–11.
 Aroostook River at Fort Fairfield, Maine, 1903–1910.

ST. CROIX RIVER BASIN.

St. Croix River near Woodland (Spragues Falls), Maine, 1902–1911.
St. Croix River at Baring, Maine, 1914.
 West Branch of St. Criox River at Baileyville, Maine, 1910–1912.

MACHIAS RIVER BASIN.

Machias River at Whitneyville, Maine, 1903–

[1] St. John River to York River, inclusive.

UNION RIVER BASIN.

Union River, West Branch (head of Union River), at Amherst, Maine, 1909–
Union River, West Branch, near Mariaville, Maine, 1909.
Union River at Ellsworth, Maine, 1909.
East Branch of Union River near Waltham, Maine, 1909.
Webb Brook at Waltham, Maine, 1909.
Green Lake (head of Reeds Brook) at Green Lake, Maine, 1909–1912.
Reeds Brook (Green Lake Stream) at Lakewood, Maine, 1909–1913.
Branch Lake (head of Branch Lake Stream) near Ellsworth, Maine, 1909–1915.
Branch Lake Stream near Ellsworth, Maine, 1909–1914.

PENOBSCOT RIVER BASIN.

Penobscot River, West Branch (head of Penobscot River), at Millinocket, Maine, 1901-
Penobscot River, West Branch, near Medway, Maine, 1916–
Penobscot River at West Enfield, Maine, 1901–
Penobscot River at Sunkhaze rips, near Costigan, Maine, 1899–1900.
East Branch of Penobscot River at Grand Lake dam, Maine, 1912.
East Branch of Penobscot River at Grindstone, Maine, 1902–
Mattawamkeag River at Mattawamkeag, Maine, 1902–
Piscataquis River near Foxcroft, Maine, 1902–
Passadumkeag River at Lowell, Maine, 1915–
Cold Stream Pond (head of Cold Stream), Maine, 1900–1911 (record of opening and closing of pond).
Cold Stream at Enfield, Maine, 1904–1906.
Kenduskeag Stream near Bangor, Maine, 1908–
Orland River:
Phillips Lake outlet near East Holden, Maine, 1904–1908.

ST. GEORGE RIVER BASIN.

St. George River at Union, Maine, 1913–14.

KENNEBEC RIVER BASIN.

Moose River (head of Kennebec River) near Rockwood, Maine, 1902–1908; 1910–1912.
Moosehead Lake (on Kennebec River) at Greenville, Maine, 1903–1906 (stage only).
Moosehead Lake at east outlet, Maine (stage only), 1895–
Kennebec River at The Forks, Maine, 1901–
Kennebec River at Bingham, Maine, 1907–1910.
Kennebec River at North Anson, Maine, 1901–1907.
Kennebec River at Waterville, Maine, 1892–
Kennebec River at Gardiner, Maine, 1785–1910 (record of opening and closing of navigation).
Roach River at Roach River, Maine, 1901–1908.
Dead River near The Forks, Maine, 1901–1907; 1910–
Carrabassett River at North Anson, Maine, 1901–1907.
Sandy River near Farmington, Maine, 1910–1915.
Sandy River near Madison, Maine, 1904–1908.
Sebasticook River at Pittsfield, Maine, 1908–
Messalonskee Stream at Waterville, Maine, 1903–1905.
Cobbosseecontee Lake (on Cobbosseecontee Stream), Maine, 1839–1911 (dates of opening and closing).
Cobbosseecontee Stream at Gardiner, Maine, 1890–1915.

ANDROSCOGGIN RIVER BASIN.

Rangeley Lake (head of Androscoggin River), Maine, 1879-1911 (dates of opening and closing).
Androscoggin River at Errol dam, N. H., 1905-
Androscoggin River at Berlin, N. H., 1913-
Androscoggin River at Gorham, N. H., 1903 (fragmentary).
Androscoggin River at Shelburne, N. H., 1903-1907; 1910.
Androscoggin River at Rumford Falls, Maine, 1892-1903; 1905-
Androscoggin River at Dixfield, Maine, 1902-1908.
    Magalloway River at Aziscohos dam, Maine, 1912-
    Auburn Lake, Maine, 1890-1911 (date of opening).
    Little Androscoggin River at Bisco Falls, near South Paris, Maine, 1913-

PRESUMPSCOT RIVER BASIN.

Presumpscot River at outlet of Sebago Lake, Maine, 1887-

SACO RIVER BASIN.

Saco River near Center Conway, N. H., 1903-1912.
Saco River at Cornish, Maine, 1916-
Saco River at West Buxton, Maine, 1907-
    Ossipee River at Cornish, Maine, 1916-

MERRIMACK RIVER BASIN.

Pemigewasset River (head of Merrimack River) at Plymouth, N. H., 1886-1913.
Merrimack River at Franklin Junction, N. H., 1903-
Merrimack River at Garvins Falls, N. H., 1904-1915.
Merrimack River at Lowell, Mass., 1848-1861; 1866-1916.
Merrimack River at Lawrence, Mass., 1880-
    Middle Branch of Pemigewasset River at North Woodstock, N. H., 1911-12.
    Smith River near Bristol, N. H., 1918-
    Lake Winnepesaukee at Lakeport, N. H., 1860-1911. (Stage only.)
    Contoocook River at Elmwood, N. H., 1918-
    Contoocook River at West Hopkinton, N. H., 1903-1907.
        Blackwater River near Contoocook, N. H., 1918-
    Suncook River at North Chichester, N. H., 1918-
    Suncook River at East Pembroke, N. H., 1904-5.
    Souhegan River at Merrimack, N. H., 1909-
    Nashua River:
        South Branch of Nashua River, Clinton, Mass., 1896-
    Concord River at Lowell, Mass., 1901-1916.
        Sudbury River at Framingham, Mass., 1875-
            Lake Cochituate at Cochituate, Mass., 1863-

MYSTIC RIVER BASIN.

Mystic Lake (on Mystic River) near Boston, Mass., 1878-1897.

CHARLES RIVER BASIN.

Charles River at Waltham, Mass., 1903-1909.

TAUNTON RIVER BASIN.

Matfield River (head of Taunton River) at Elmwood, Mass., 1909-10.
    Satucket River near Elmwood, Mass., 1909-10.

PROVIDENCE RIVER BASIN.

Providence River:
Seekonk River:
Tenmile River near Rumford, R. I., 1909.
Blackstone River at Woonsocket, R. I., 1904-5.
Blackstone River at Albion, R. I., 1914-1916.
Blackstone River at Berkeley, R. I., 1901-2.
Branch River at Branch Village, R. I., 1909-10; 1912-13.
Woonasquatucket River at Olneyville, R. I., 1910.

PAWTUXET RIVER BASIN.

Pawtuxet River at Harris, R. I., 1909.

PAWCATUCK RIVER BASIN.

Pawcatuck River:
Wood River at Hope Valley, R. I., 1909-10.

THAMES RIVER BASIN.

Thames River:
Quinebaug River:
Shetucket River at Willimantic, Conn., 1904-5.

CONNECTICUT RIVER BASIN.

Connecticut River at First Lake, near Pittsburg, N. H., 1917-
Connecticut River at Orford, N. H., 1900-
Connecticut River at Sunderland, Mass., 1904-
Connecticut River at Holyoke, Mass., 1880-1899.
Connecticut River at Hartford, Conn., 1896-1908.
Israel River above South Branch, near Jefferson Highlands, N. H., 1903-1906.
Israel River below South Branch, at Jefferson Highlands, N. H., 1903-1907.
Passumpsic River at Pierce's mills, near St. Johnsbury, Vt., 1909-
Passumpsic River at St. Johnsbury Center, Vt., 1903.
Ammonoosuc River at Bretton Woods, N. H., 1903-1907.
Zealand River near Twin Mountain, N. H., 1903-1907.
Little River at Twin Mountain, N. H., 1904-5.
White River at Sharon, Vt., 1903-1904; 1909-1913.
White River at West Hartford, Vt., 1915-
Ashuelot River at Winchester, N. H., 1903-1904.
Ashuelot River at Hinsdale, N. H., 1907-1909; 1914-
Millers River at Wendell Depot, Mass., 1909-1913.
Millers River near Winchenden, Mass., 1916-
Millers River at Erving, Mass., 1914-
Sip Pond Brook near Winchenden, Mass., 1916-
Priest Brook near Winchenden, Mass., 1916-
Otter River near Gardner, Mass., 1916-1917.
East Branch of Tully River near Athol, Mass., 1916-
Moss Brook at Wendell Depot, Mass., 1909-10; 1916-
Deerfield River at Hoosac Tunnel, Mass., 1909-1913.
Deerfield River at Charlemont, Mass., 1913-
Deerfield River at Shelburne Falls, Mass., 1907-1913.
Deerfield River at Deerfield, Mass., 1904-5.
Ware River (head of Chicopee River) at Ware, Mass., 1904-1911.

Connecticut River tributaries—Continued.
  Ware River at Gibbs Crossing, Mass., 1912–
    Burnshirt River near Templeton, Mass., 1909.
    Swift River at West Ware, Mass., 1910–
    Quaboag River at West Warren, Mass., 1903–1907.
    Quaboag River at West Brimfield, Mass., 1909–
  Westfield River at Knightville, Mass., 1909–
  Westfield River at Russell, Mass., 1904–5.
  Westfield River near Westfield, Mass., 1914–
    Middle Branch of Westfield River at Goss Heights, Mass., 1910–
    West Branch of Westfield River at Chester, Mass., 1915.
    Westfield Little River near Westfield, Mass., 1905–
      Borden Brook near Westfield, Mass., 1910–
  Farmington River near New Boston, Mass., 1913–
  Salmon River at Leesville, Conn., 1905–6.

HOUSATONIC RIVER BASIN.

Housatonic River near Great Barrington, Mass., 1913–
Housatonic River at Falls Village, Conn., 1912–
Housatonic River at Gaylordsville, Conn., 1900–1914.
  Tenmile River at Dover Plains, N. Y., 1901–1903.
  Pomperaug River at Bennetts Bridge, Conn., 1913–1916.

MIANUS RIVER BASIN.

Mianus River at Bedford, N. Y., 1903.
Mianus River near Stamford, Conn., 1903.

BYRAM RIVER BASIN.

Byram River, West Branch (head of Byram River), near Port Chester, N. Y., 1903.
Byram River at Pemberwick, Conn., 1903.
  East Branch of Byram River near Greenwich, Conn., 1903.
    Middle Branch of Byram River near Riverville, Conn., 1903.

HUDSON RIVER BASIN.

Hudson River near Indian Lake, N. Y., 1916–
Hudson River at North Creek, N. Y., 1907–
Hudson River at Thurman, N. Y., 1907–
Hudson River at Corinth, N. Y., 1904–1912.
Hudson River at Spier Falls, N. Y., 1912–
Hudson River at Fort Edward, N. Y., 1899–1908.
Hudson River at Mechanicville, N. Y., 1890–
  Cedar River near Indian Lake, N. Y., 1911–1917.
  Indian Lake reservoir near Indian Lake, N. Y., 1900–
  Indian River near Indian Lake, N. Y., 1912–1914; 1915–
  Schroon Lake (on Schroon River) at Pottersville, N. Y., 1908–1911.
  Schroon River at Riverbank, N. Y., 1907–
  Schroon River at Warrensburg, N. Y., 1895–1902.
  Sacandaga River at Wells, N. Y., 1907–1911.
  Sacandaga River near Hope, N. Y., 1911–
  Sacandaga River at Northville, N. Y., 1907–1910.
  Sacandaga River near Hadley, N. Y., 1907–1910.
  Sacandaga River (at cable) at Hadley, N. Y., 1911–

Hudson River tributaries—Continued.
Sacandaga River at Union Bag & Paper Co.'s mill at Hadley, N. Y., 1909–1911.
West Branch of Sacandaga River at Whitehouse, N. Y., 1910.
West Branch of Sacandaga River at Blackbridge, near Wells, N. Y., 1911–1916.
Batten Kill at Battenville, N. Y., 1908.
Fish Creek at Burgoyne, N. Y., 1905; 1908.
Hoosic River near Eagle Bridge, N. Y., 1910–
Hoosic River at Buskirk, N. Y., 1903–1908.
Mohawk River at Ridge Mills, near Rome, N. Y., 1898–1900.
Mohawk River at Utica, N. Y., 1901–1903.
Mohawk River at Little Falls, N. Y., 1898–1909; 1912.
Mohawk River at Rocky Rift dam, near Indian Castle, N. Y., 1901.
Mohawk River at Tribes Hill, N. Y., 1912.
Mohawk River at Schenectady, N. Y., 1899–1901.
Mohawk River at Rexford Flats, N. Y., 1898–1901.
Mohawk River at Vischer Ferry dam, N. Y., 1913–
Mohawk River at Dunsbach Ferry, N. Y., 1898–1909.
Mohawk River at Crescent Dam, N. Y., 1918–
Ninemile Creek at Stittville, N. Y., 1898–99.
Oriskany Creek at Coleman, N. Y., 1904–1906.
Oriskany Creek at Wood-road bridge, near Oriskany, N. Y., 1901–1904.
Oriskany Creek at State dam, near Oriskany, N. Y., 1898–1900.
Saquoit Creek at New York Mills, N. Y., 1898–1900.
Nail Creek at Utica, N. Y., 1904.
Reels Creek near Deerfield, N. Y., 1901–1904.
Reels Creek at Utica, N. Y., 1901–2.
Johnson Brook at Deerfield, N. Y., 1903–1905.
Starch Factory Creek at New Hartford, N. Y., 1903–1906.
Graefenberg Creek at New Hartford, N. Y., 1903–1906.
Sylvan Glen Creek at New Hartford, N. Y., 1903–1906.
West Canada Creek at Wilmurt, N. Y., 1912–13.
West Canada Creek at Twin Rock bridge, near Trenton Falls, N. Y., 1900–1909.
West Canada Creek at Poland, N. Y., 1913.
West Canada Creek at Middleville, N. Y., 1898–1901.
West Canada Creek at Kast Bridge, N. Y., 1905–1909; 1912–13.
East Canada Creek at Dolgeville, N. Y., 1898–1909; 1912.
Caroga Creek 3 miles above junction with Mohawk River, N. Y., 1898–99.
Cayadutta Creek at Johnstown, N. Y., 1899–1900.
Schoharie Creek at Prattsville, N. Y., 1902–1913.
Schoharie Creek at Schoharie Falls, above Mill Point, N. Y., 1900–1901.
Schoharie Creek at Mill Point, N. Y., 1900–1903.
Schoharie Creek at Fort Hunter, N. Y., 1898–1901.
Schoharie Creek at Erie Canal aqueduct, below Fort Hunter, N. Y., 1900.
Alplaus Kill near Charlton, N. Y., 1913–1916.
Quacken Kill at Quacken Kill, N. Y., 1894.
Normans Kill at Frenchs Mill, N. Y., 1891.
Kinderhook Creek at Wilsons dam, near Garfield, N. Y., 1892–1894.
Kinderhook Creek at East Nassau, N. Y., 1892–1894.
Kinderhook Creek at Rossman, N. Y., 1906–1909; 1911–1914.
Catskill Creek at South Cairo, N. Y., 1901–1907.
Esopus Creek at Olivebridge, N. Y., 1903–4.
Esopus Creek near Olivebridge, N. Y., 1906–1913.
Esopus Creek at Kingston, N. Y., 1901–1909.
Esopus Creek at Mount Marion, N. Y., 1907–1913.

Hudson River tributaries—Continued.
Rondout Creek at Rosendale, N. Y., 1901–1903; 1906–1913.
Diversion to Delaware and Hudson canal at Rosendale, N. Y., 1901–1903; 1906.
Wallkill River at Newpaltz, N. Y., 1901–1903.
Wappinger Creek at Wappinger Falls, N. Y., 1903–1905.
Fishkill Creek at Glenham, N. Y., 1901–1903.
Foundry Brook at Cold Spring, N. Y., 1902–3.
Croton River at Croton dam, near Croton Lake, N. Y., 1870–1899.

PASSAIC RIVER BASIN.

Passaic River at Millington, N. J., 1903–1906.
Passaic River near Chatham, N. J., 1902–1911.
Passaic River at Two Bridges (Mountain View), N. J., 1901–1903.
Rockaway River at Boonton, N. J., 1903–4.
Pompton River at Pompton Plains, N. J., 1903–4.
Pompton River at Two Bridges (Mountain View), N. J., 1901–1903.
Ramapo River near Mahwah, N. J., 1903–1906; 1908.
Wanaque River at Wanaque, N. J., 1903–1905.

RARITAN RIVER BASIN.

Raritan River, South Branch (head of Raritan River), at Stanton, N. J., 1903–1906.
Raritan River at Finderne, N. J., 1903–1907.
Raritan River at Boundbrook, N. J., 1903–1909.
North Branch of Raritan River at Pluckemin, N. J., 1903–1906.
Millstone River at Millstone, N. J., 1903–4.

DELAWARE RIVER BASIN.

Delaware River, East Branch (head of Delaware River), at Fish Eddy, N. Y., 1912–
Delaware River, East Branch, at Hancock, N. Y., 1902–1912.
Delaware River at Port Jervis, N. Y., 1904–
Delaware River at Riegelsville, N. J., 1906–
Delaware River at Lambertville, N. J., 1897–1908.
Beaver Kill at Cooks Falls, N. Y., 1913–
West Branch of Delaware River at Hale Eddy, N. Y., 1912–
West Branch of Delaware River at Hancock, N. Y., 1902–1912.
Mongaup River near Rio, N. Y., 1909–1913.
Neversink River at Godeffroy, N. Y., 1903; 1909–10; 1911–1914.
Neversink River at Port Jervis, N. Y., 1902–3.
Paulins Kill at Columbia, N. J., 1908–9.
Lehigh River at South Bethlehem, Pa., 1902–1905; 1909–1913.
Lehigh River at Easton, Pa., 1909.
Musconetcong River at Asbury, N. J., 1903.
Musconetcong River near Bloomsbury, N. J., 1903–1907.
Tohickon Creek at Point Pleasant, Pa., 1883–1889; 1901–1913.
Neshaminy Creek below Forks, Pa., 1884–1913.
Schuylkill River near Philadelphia, Pa., 1898–1912.
Perkiomen Creek near Frederick, Pa., 1884–1913.
Wissahickon Creek near Philadelphia, Pa., 1897–1902; 1905–6.

SUSQUEHANNA RIVER BASIN.

Susquehanna River at Colliersville, N. Y., 1907–8.
Susquehanna River at Conklin, N. Y., 1912–
Susquehanna River at Binghamton, N. Y., 1901–1912.

Susquehanna River at Wysox, Pa., 1908–9.
Susquehanna River at Wilkes-Barre, Pa., 1899–1913.
Susquehanna River at Danville, Pa., 1899–1913.
Susquehanna River at Harrisburg, Pa., 1891–1913.
Susquehanna River at McCall Ferry, Pa., 1902–1909.
Chenango River at South Oxford, N. Y., 1903.
Chenango River near Greene, N. Y., 1908.
Chenango River near Chenango Forks, N. Y., 1912–
Chenango River at Binghamton, N. Y., 1901–1912.
Eaton Brook, Madison County, N. Y., 1835.
Madison Brook, Madison County, N. Y., 1835.
Tioughnioga River at Chenango Forks, N. Y., 1903.
Cayuta Creek at Waverly, N. Y., 1898–1902. (Data in Water-Supply Paper 109, only.)
Chemung River at Chemung, N. Y., 1903– (Data for period prior to 1905 published in Water-Supply Paper 109.)
Cohocton River near Campbell, N. Y., 1918–
Mud Creek at Savona, N. Y., 1918–
Tioga River near Erwins, N. Y., 1918–
West Branch of Susquehanna River at Williamsport, Pa., 1895–1913.
West Branch of Susquehanna River at Allenwood, Pa., 1899–1902.
Juniata River at Newport, Pa., 1899–1913.
Broad Creek at Mill Green, Md., 1905–1909.
Octoraro Creek at Rowlandsville, Md., 1896–1899.
Deer Creek near Churchville, Md., 1905–1909.

GUNPOWDER RIVER BASIN.

Gunpowder Falls at Glencoe, Md., 1905–1909.
Little Gunpowder Falls near Belair, Md., 1905–1909.

PATAPSCO RIVER BASIN.

Patapsco River at Woodstock, Md., 1896–1909.

PATUXENT RIVER BASIN.

Patuxent River near Burtonsville, Md., 1911–12; 1913–
Patuxent River at Laurel, Md., 1896–1898.

POTOMAC RIVER BASIN.

Potomac River, North Branch (head of Potomac River), at Piedmont, W. Va., 1899–1906.
Potomac River, North Branch, at Cumberland, Md., 1894–1897.
Potomac River at Great Cacapon, W. Va., 1895.
Potomac River at Point of Rocks, Md., 1895–
Potomac River at Great Falls, Md., 1886–1891.
Potomac River at Chain Bridge, near Washington, D. C., 1892–1895.
Savage River at Bloomington, Md., 1905–6.
Georges Creek at Westernport, Md., 1905–6.
Wills Creek near Cumberland, Md., 1905–6.
South Branch of Potomac River near Springfield, W. Va., 1894–1896; 1899–1906.
Opequan Creek near Martinsburg, W. Va., 1905–6.
Tuscarora Creek at Martinsburg, W. Va., 1905.
Antietam Creek near Sharpsburg, Md., 1897–1905.

Potomac River tributaries—Continued.

- North River (head of South Fork of Shenandoah River, which is continuation of main stream) at Port Republic, Va., 1895–1899.
- South Fork of Shenandoah River near Front Royal, Va., 1899–1906.
- Shenandoah River at Millville, W. Va., 1895–1909.
  - Cooks Creek at Mount Crawford, Va., 1905–6.
  - Middle River:
    - Lewis Creek near Staunton, Va., 1905–6.
  - South River at Basic City, Va., 1905–6.
  - South River at Port Republic, Va., 1895–1899.
  - Elk Run at Elkton, Va., 1905–6.
  - Hawksbill Creek near Luray, Va., 1905–6.
  - North Fork of Shenandoah River near Riverton, Va., 1899–1906.
- Passage Creek at Buckton, Va., 1905–6.
- Monocacy River near Frederick, Md., 1896–
- Goose Creek near Leesburg, Va., 1909–1912.
- Rock Creek at Zoological Park, D. C., 1897–1900.
- Rock Creek at Lyons Mill, D. C., 1892–1894.
- Occoquan Creek near Occoquan, Va., 1913–1916.

RAPPAHANNOCK RIVER BASIN.

Rappahannock River near Fredericksburg, Va., 1907–

REPORTS ON WATER RESOURCES OF NORTH ATLANTIC COAST.[1]

PUBLICATIONS OF UNITED STATES GEOLOGICAL SURVEY.

WATER-SUPPLY PAPERS.

Water-supply papers are distributed free by the Geological Survey as long as its stock lasts. An asterisk (*) indicates that this stock has been exhausted. Many of the papers marked in this way may, however, be purchased (at price noted) from the SUPERINTENDENT OF DOCUMENTS, WASHINGTON, D. C. Omission of the price indicates that the report is not obtainable from Government sources. Water-supply papers are of octavo size.

*24. Water resources of the State of New York, Part I, by G. W. Rafter. 1899. 99 pp., 13 pls. 15c.

Describes the principal rivers of New York and their more important tributaries, and gives data on temperature, precipitation, evaporation, and stream flow.

*25. Water resources of the State of New York, Part II, by G. W. Rafter. 1899. 100 pp., 12 pls. 15c.

Contains discussion of water storage projects on Genesee and Hudson rivers, power development at Niagara Falls, descriptions and early history of State canals, and a chapter on the use and value of the water power of the streams and canals; also brief discussion of the water yields of sand areas of Long Island.

*44. Profiles of rivers in the United States, by Henry Gannett. 1901. 100 pp., 11, pls. 15c.

Gives elevations and distances along rivers of the United States, also brief descriptions of many of the streams, including St. Croix, Penobscot, Kennebec, Androscoggin, Saco, Merrimack, Connecticut, Housatonic, Hudson, Mohawk, Delaware, Lehigh, Schuylkill, Susquehanna, Juniata, Potomac, and James rivers.

*57. Preliminary list of deep borings in the United States, Part I (Alabama-Montana), by N. H. Darton. 1902. 60 pp. (See No. 149.) 5c.

*61. Preliminary list of deep borings in the United States, Part II (Nebraska-Wyoming), by N. H. Darton. 1902. 67 pp. 5c.

Nos. 57 and 61 contain information as to depth, diameter, yield, and head of water in borings more than 400 feet deep; under head "Remarks" give information concerning temperature, quality of water, purposes of boring, etc. The lists are arranged by States, and the States are arranged alphabetically. Revised edition published in 1905 as Water-Supply Paper 149 (q. v.).

*69. Water powers of the State of Maine, by H. A. Pressey. 1902. 124 pp., 14 pls. 20c.

Discusses briefly the geology and forests of Maine and in somewhat greater detail the drainage areas, lake storage, and water powers of the St. Croix, Penobscot, Kennebec, Androscoggin, Presumpscot, Saco, and St. John rivers, and the minor coastal streams; mentions also developed tidal powers.

72. Sewage pollution in the metropolitan area near New York City and its effect on inland water resources, by M. O. Leighton. 1902. 75 pp., 8 pls. 10c.

Defines "normal" and "polluted" waters and discusses the water of Raritan, Passaic, and Hudson rivers and their tributaries and the damage resulting from pollution.

76. Observations on the flow of rivers in the vicinity of New York City, by H. A. Pressey. 1903. 108 pp., 13 pls. 15c.

Describes methods of measuring stream flow in open channels and under ice, and the quality of the river water as determined by tests of turbidity, color, alkalinity, and permanent hardness. The streams considered are Catskill, Esopus, Rondout, and Fishkill creeks, and Wallkill, Tenmile, and Housatonic rivers.

---

[1] For stream-measurement reports see tables on pages IV, V, VI.

79. Normal and polluted waters in northeastern United States, by M. O. Leighton. 1903. 192 pp. 10c.

Defines essential qualities of water for various uses, the impurities in rain, surface, and underground waters, the meaning and importance of sanitary analyses, and the principal sources of pollution; chiefly "a review of the more readily available records" of examination of water supplies derived from streams in the Merrimack, Connecticut, Housatonic, Delaware, and Ohio River basins; contains many analyses.

88. The Passaic flood of 1902, by G. B. Hollister and M. O. Leighton. 1903. 56 pp. 15 pls. 15c.

Describes the topography of the area drained by the Passaic and its principal tributaries; discusses flood flow and losses caused by the floods, and makes comparison with previous floods; suggests construction of dam at Mountain View to control flood flow. See also No. 92.

92. The Passaic flood of 1903, by M. O. Leighton. 1904. 48 pp., 7 pls. 5c.

Discusses flood damages and preventive measures. See No. 88.

102. Contributions to the hydrology of eastern United States, 1903; M. L. Fuller, geologist in charge. 1904. 522 pp. 30c.

Contains brief reports on the wells and springs of the New England States and New York. The reports comprise tabulated well records giving information as to location, owner, depth, yield, head, etc., supplemented by notes as to elevation above sea, material penetrated, temperature, use, and quality; many miscellaneous analyses.

*103. A review of the laws forbidding pollution of inland waters in the United States, by E. B. Goodell. 1904. 120 pp. Superseded by 152.

Cites statutory restrictions of water pollution.

106. Water resources of the Philadelphia district, by Florence Bascom. 1904. 75 pp., 4 pls. 5c.

Describes the physiography, stratigraphic geology, rainfall, streams, ponds, springs, deep and artesian wells, and public water supplies of the area mapped on the Germantown, Norristown, Philadelphia, and Chester atlas sheets of the United States Geological Survey; compares quality of Delaware and Schuylkill River waters.

108. Quality of water in the Susquehanna River drainage basin, by M. O. Leighton, with an introductory chapter on physiographic features, by G. B. Hollister. 1904. 76 pp., 4 pls. 15c.

109. Hydrography of the Susquehanna River drainage basin, by J. C. Hoyt and R. H. Anderson. 1905. 215 pp., 29 pls. 25c.

The scope of No. 108 is sufficiently indicated by its title. No. 109 describes the physical features of the area drained by the Susquehanna and its tributaries, contains the results of measurements of flow at the gaging stations, and discusses precipitation, floods, low water, and water power.

*110. Contributions to the hydrology of eastern United States, 1904; M. L. Fuller, geologist in charge. 1905. 211 pp., 5 pls. 10c.

Contains brief reports on water resources, surface and underground, of districts in the North Atlantic slope drainage basins, as shown by the following list:

Drilled wells of the Triassic area of the Connecticut Valley, by W. H. C. Pynchon.

Triassic rocks of the Connecticut Valley as a source of water supply, by M. L. Fuller. Scope indicated by title.

Water resources of the Taconic quadrangle, New York, Massachusetts, and Vermont, by F. B. Taylor. Discusses rainfall, drainage, water powers, lakes and ponds, underground waters, and mineral springs; also quality of spring water as indicated by chemical and sanitary analyses of Sand Spring, near Williamstown.

Water resources of the Watkins Glen quadrangle, New York, by Ralph S. Tarr. Discusses the use of the surface and underground waters for municipal supplies and their quality as indicated by examination of Sixmile and Fall creeks, and sanitary analyses of well water at Ithaca.

Water resources of the central and southwestern highlands of New Jersey, by Laurence La Forge. Treats of population, industries, climate, and soils, lakes, ponds, swamps and rivers, mineral springs (with analyses), water power, and the Morris canal; present and prospective sources and quality of municipal supplies.

Water resources of the Chambersburg and Mercersburg quadrangles, Pennsylvania, by George W. Stose. Describes streams and springs.

Water resources of the Curwensville, Patton, Ebensburg, and Barnesboro quadrangles, Pennsylvania, by F. G. Clapp. Treats briefly of surface and underground waters and their use for municipal supplies; gives analyses of waters at Cresson Springs.

Water resources of the Accident and Grantsville quadrangles, Maryland, by G. C. Martin.

Water resources of the Frostburg and Flintstone quadrangles, Maryland and West Virginia, by G. C. Martin.

*114. Underground waters of eastern United States; M. L. Fuller, geologist in charge. 1905. 285 pp., 18 pls. 25c.

Contains brief reports on water supplies of the North Atlantic States as follows:
Maine, by W. S. Bayley.
New Hampshire, by M. L. Fuller.
Vermont, by G. H. Perkins.
Massachusetts and Rhode Island, by W. O. Crosby.
Connecticut, by H. E. Gregory.
New York, by F. B. Weeks.
New Jersey, by G. N. Knapp.
Pennsylvania, by M. L. Fuller.
Delaware, by N. H. Darton.
Maryland, by N. H. Darton and M. L. Fuller.
District of Columbia, by N. H. Darton and M. L. Fuller.
Virginia, by N. H. Darton and M. L. Fuller.

Each of these reports discusses the resources of the public and private water supplies and related subjects, and gives list of pertinent publications; mineral springs are listed and sales of mineral water are reported.

*122. Relation of the law to underground waters, by D. W. Johnson. 1905. 55 pp. 5c.

Cites legislative acts relating to ground waters in New Jersey.

140. Field measurements of the rate of movement of underground waters, by C. S. Slichter. 1905. 122 pp., 15 pls. 15c.

Contains chapter on measurement of rate of underflow on Long Island, N. Y.

144. The normal distribution of chlorine in the natural waters of New York and New England, by D. D. Jackson. 1905. 31 pp., 5 pls. 10c.

Discusses common salt in coast and inland waters, salt as an index to pollution of streams and wells, the solutions and methods used in chlorine determinations, and the use of the normal chlorine map; gives charts and tables for chlorine in the New England States and New York.

145. Contributions to the hydrology of eastern United States, 1905; M. L. Fuller, geologist in charge. 1905. 220 pp., 6 pls. 10c.

Contains several brief reports relating chiefly to areas in the North Atlantic slope drainage basins, as follows:

Water resources of the Portsmouth-York region, New Hampshire and Maine, by George Otis Smith. Gives results of investigations made for the War Department to determine water supplies available for forts at mouth of harbor.

Water supply from glacial gravels near Augusta, Maine, by George Otis Smith. Describes the Silver Lake system of ponds near Augusta and the series of springs at the head of Spring Brook.

Water resources of the Pawpaw and Hancock quadrangles, West Virginia, Maryland, and Pennsylvania, by George W. Stose and George C. Martin. Describes rocks, springs, and streams in the areas at the northernmost bend of the Potomac; discusses history of development, character of water (with analysis), flow, and origin of Berkeley Springs.

Water of a gravel-filled valley near Tully, N. Y., by George B. Hollister. Describes character of the sands and gravels, the volume of the springs issuing from them, deposits of tufa, the waters of the lakes, and the composition of the spring and lake waters; analyses.

147. Destructive floods in United States in 1904, by E. C. Murphy and others. 206 pp., 18 pls. 15c.

Describes floods on Susquehanna and Mohawk rivers and near Johnstown, Pa.

*149. Preliminary list of deep borings in the United States, second edition, with additions, by N. H. Darton. 1905. 175 pp. 10c.

Gives by States (and within the States by counties), location, depth, diameter, yield, height of water, and other available information, concerning wells 400 feet or more in depth; includes all wells listed in Water-Supply Papers 57 and 61; mentions also principal publications relating to deep borings.

*152. A review of the laws forbidding pollution of inland waters in the United States (second edition), by E. B. Goodell. 1905. 149 pp. 10c.

Cites statutory restrictions of water pollution.

*155. Fluctuations of the water level in wells, with special reference to Long Island, New York, by A. C. Veatch. 1906. 83 pp., 9 pls. 25c.

Includes general discussion of fluctuation due to rainfall and evaporation, barometric changes, temperature changes, changes in rivers, changes in lake level, tidal changes, effects of settlement, irrigation, dams, underground-water developments, and to indeterminate causes.

*162. Destructive floods in the United States in 1905, with a discussion of flood discharge and frequency and an index to flood literature, by E. C. Murphy and others. 1906. 105 pp., 4 pls. 15c.

Contains accounts of floods in North Atlantic slope drainage basins as follows: Flood on Poquonnock River, Connecticut, by T. W. Norcross; flood on the Unadilla and Chenango rivers, New York, by R. E. Horton and C. C. Covert; also estimates of flood discharge and frequency on Kennebec, Androscoggin, Merrimack, Connecticut, Hudson, Passaic, Raritan, Delaware, Susquehanna, and Potomac rivers; gives index to literature on floods on American streams.

*185. Investigations on the purification of Boston sewage, with a history of the sewage-disposal problem, by C.-E. A. Winslow and E. B. Phelps. 1906. 163 pp. 25c.

Discusses composition, disposal, purification, and treatment of sewage and sewage-disposal practice in England, Germany, and the United States; describes character of crude sewage at Boston, removal of suspended matter, treatment in septic tanks, and purification in intermittent sand filtration and coarse material; gives bibliography.

*192. The Potomac River basin (Geographic history; rainfall and stream flow; pollution, typhoid fever, and character of water; relation of soils and forest cover to quality and quantity of surface water; effect of industrial wastes on fishes), by H. N. Parker, Bailey Willis, R. H. Bolster, W. W. Ashe, and M. C. Marsh. 1907. 364 pp., 10 pls. 60c.

Scope indicated by title.

*198. Water resources of the Kennebec River basin, Maine, by H. K. Barrows, with a section on the quality of Kennebec River water, by G. C. Whipple. 1907. 235 pp., 7 pls. 30c.

Describes physical characteristics and geology of the basin, the flow of the streams, evaporation, floods, developed and undeveloped water powers, water storage, log driving, and lumbering; under quality of water discusses effect of tides, pollution, and the epidemic of typhoid fever in 1902–3; contains gazetteer of rivers, lakes, and ponds.

*223. Underground waters of southern Maine, by F. G. Clapp, with records of deep wells, by W. S. Bayley. 1909. 268 pp., 24 pls. 55c.

Describes physiography, rivers, water-bearing rocks, amount, source, and temperature of the ground waters, recovery of waters by springs, collecting galleries and tunnels, and wells; discusses well-drilling methods, municipal water supplies, and the chemical composition of the ground waters; gives details for each county.

232. Underground-water resources of Connecticut, by H. E. Gregory, with a study of the occurrence of water in crystalline rocks, by E. E. Ellis. 1909. 200 pp., 5 pls. 20c.

Describes physiographic features, drainage, forests, climate, population and industries, and rocks; circulation, amount, temperature, and contamination of ground water; discusses the ground waters of the crystalline rocks, the Triassic sandstones and traps, and the glacial drift; the quality of the ground waters (with analyses); well construction; temperature, volume, character, uses, and production of spring waters.

*236. The quality of surface waters in the United States, Part I, Analyses of waters east of the one hundredth meridan, by R. B. Dole. 1909. 123 pp. 10c.

Describes collection of samples, method of examination, preparation of solutions, accuracy of estimates, and expression of analytical results; gives results of analyses of waters of Androscoggin, Hudson, Raritan, Delaware, Susquehanna, Lehigh, Potomac, and Shenandoah rivers.

*258. Underground-water papers, 1910, by M. L. Fuller, F. G. Clapp, G. C. Matson, Samuel Sanford, and H. C. Wolff. 1911. 123 pp., 2 pls. 15c.

Contains four brief reports pertaining especially to districts in the North Atlantic slope drainage area:

Occurrence and composition of well waters in the slates of Maine, by F. G. Clapp. Analyses.
Occurrence and composition of well waters in the granites of New England, by F. G. Clapp. Discusses proportion of successful wells and water supply and depth. Analyses.
Composition of mineral springs in Maine, by F. G. Clapp.
Saline artesian waters of the Atlantic Costal Plain, by Samuel Sanford
Underground waters near Manassas, Va., by F. G. Clapp.

279. Water resources of the Penobscot River basin, Maine, by H. K. Barrows, and C. C. Babb. 1912. 285 pp., 19 pls. 65c.

Describes the topography, drainage, geology, forests, population, industries, transportation lines, and precipitation in the basin; gives results of investigations of stream flow at gaging stations; discusses relation of run-off to precipitation, evaporation, floods, low water, developed, and undeveloped water powers, storage, log driving, and lumbering; contains gazetteer of rivers, lakes, and ponds.

364. Water analyses from the laboratory of the United States Geological Survey, tabulated by F. W. Clarke, chief chemist. 1914. 40 pp.

Contains analyses of spring and well waters in Maine, District of Columbia, and Virginia.

374. Ground water in the Hartford, Stamford, Salisbury, Willimantic, and Saybrook areas, Connecticut, by H. E. Gregory and A. J. Ellis. 1916. 150 pp., 13 pls. 30c.

Describes occurrence of ground water, methods of developing, and requirements for municipal use. Gives, by towns, a description of the surface and ground water and of the public water supply, and records of wells and springs.

397. Ground water in the Waterbury area, Connecticut, by A. J. Ellis, under direction of H. E. Gregory. 1916. 73 pp., 4 pls. 15c.

Describes the geology of the area, the occurrence of ground water, its use for private and municipal supply, and methods of developing. Discusses under towns the population and industries, topography, water-bearing formations, surface and ground water, and public supplies, and gives records of wells and springs.

415. Surface waters of Massachusetts, by C. H. Pierce and H. J. Dean. 1916. 433 pp., 12 pls. 45c.

A compilation of available stream-flow data, including the classic records collected on the Merrimack at Lowell and Lawrence, on the Connecticut at Holyoke, and on the Cochituate at Sudbury by the Metropolitan Water and Sewerage Board, as well as records covering shorter periods; prepared in cooperation with the Commonwealth of Massachusetts. Contains a gazetteer of streams, lakes, and ponds.

424. Surface waters of Vermont, by C. H. Pierce. 1917. 218 pp., 14 pls.

A compilation of available stream-flow data; prepared in cooperation with the Commonwealth of Vermont. Contains a gazetteer of streams, lakes, and ponds.

ANNUAL REPORTS.

Each of the papers contained in the annual reports was also issued in separate form.

Annual reports are distributed free by the Geological Survey as long as its stock lasts. An asterisk (*) indicates that this stock has been exhausted. Many of the papers so marked, however, may be purchased from the SUPERINTENDENT OF DOCUMENTS, WASHINGTON, D. C.

*Sixth Annual Report of the United States Geological Survey, 1884–85, J. W. Powell, Director. 1885. xxix, 570 pp., 65 pls. Cloth $2.00. Contains:

* Seacoast swamps of the eastern United States, by N. S. Shaler. pp. 353–398. Describes the coast swamps of New England; discusses economic problems connected with marine swamps; gives a detailed account of selected areas of salt marsh lands, and a list of the principal areas of salt marshes between Hudson River and Portland, Maine.

*Tenth Annual Report of the United States Geological Survey, 1888–89, J. W. Powell, Director. 1890. 2 parts. *Pt. I—Geology, xv, 774 pp., 98 pls. Cloth $2.35. Contains:

* General account of the fresh-water morasses of the United States, with a description of the Dismal Swamp district of Virginia and North Carolina, by N. S. Shaler, pp. 255–339, Pls. 6 to 19. Scope indicated by title.

Fourteenth Annual Report of the United States Geological Survey, 1892–93, J. W. Powell, Director. 1893. (Pt. II, 1894.) 2 parts. *Pt. II.—Accompanying papers, xx, 597 pp., 73 pls. Cloth $2.10. Contains:

* The potable waters of the eastern United States, by W. J. McGee, pp. 1 to 47. Discusses cistern water, stream waters, and ground waters, including mineral springs and artesian wells.

PROFESSIONAL PAPERS.

Professional papers are distributed free by the Geological Survey as long as its stock lasts. An asterisk (*) indicates that this stock has been exhausted. Many of the papers marked with an asterisk may, however, be purchased from the SUPERINTENDENT OF DOCUMENTS, WASHINGTON, D. C. Professional papers are of quarto size.

*44. Underground-water resources of Long Island, N. Y., by A. C. Veatch, C. S. Slichter, Isaiah Bowman, W. O. Crosby, and R. E. Horton. 1906. 394 pp., 34 pls. $1.25.

Describes the geologic formations, the source of the ground waters, and requisite conditions for flowing wells; the springs, streams, ponds, and lakes; artesian and deep wells; fluctuation of ground-water table; blowing wells; waterworks; discusses measurements of velocity of underflow, the results of sizing and filtration tests, and the utilization of stream waters; gives well records and notes (with chemical analyses) concerning representative wells.

BULLETINS.

An asterisk (*) indicates that the Geological Survey's stock of the paper is exhausted. Many of the papers so marked may be purchased from the SUPERINTENDENT OF DOCUMENTS, WASHINGTON, D. C.

*138. Artesian well prospects in the Atlantic Coastal Plain region, by N. H. Darton. 1896. 232 pp., 19 pls.

Describes the general geologic structure of the Atlantic Coastal Plain region and summarizes the conditions affecting subterranean water in the Coastal Plain; discusses the general geologic relations in New York, southern New Jersey, Delaware, Maryland, District of Columbia, Virginia, North Carolina, South Carolina, and eastern Georgia; gives for each of the States a list of the deep wells and discusses well prospects. The notes on the wells that follow the tabulated lists contain many well sections and analyses of the waters.

*264. Record of deep well drilling for 1904, by M. L. Fuller, E. F. Lines, and A. C. Veatch. 1905. 106 pp. 10c.

Discusses the importance of accurate well records to the driller, to owners of oil, gas, and water wells, and to the geologist; describes the general methods of work; gives tabulated records of wells in Connecticut, Maine, Massachusetts, New Hampshire, New Jersey, New York, Pennsylvania, Rhode Island, and Virginia, and detailed records of wells at Pleasantville and Atlantic Highlands, N. J., and Tully, N. Y. These wells were selected because they give definite stratigraphic information.

*298. Record of deep well drilling for 1905, by M. L. Fuller and Samuel Sanford. 1906. 299 pp. 25c.

Gives an account of progress in the collection of well records and samples; contains tabulated records of wells in Connecticut, Delaware, Maine, Maryland, Massachusetts, New Hampshire, New Jersey, New York, Pennsylvania, Rhode Island, Vermont, and Virginia, and detailed records of wells in Newcastle County, Del.; Cumberland County, Maine; Anne Arundel, St. Mary, and Talbot counties, Md.; Hampshire County, Mass.; Monmouth County, N. J., Saratoga County, N. Y.; and Lycoming and Somerset counties, Pa. The wells of which detailed sections are given were selected because they afford valuable stratigraphic information.

*531. Contributions to economic geology, 1911, Part II, Mineral fuels; M. R. Campbell, geologist in charge. 1913. 361 pp. 24 pls. 45c.

Issued also in separate chapters. The following papers contain information on ground water.

*(d) Geologic structure of the Punxsutawney, Curwensville, Houtzdale, Barnesboro, and Patton quadrangles, central Pennsylvania, by G. H. Ashley and M. R. Campbell (pp. 69–89, Pls. VII–VIII). Discusses the geologic structure of the five quadrangles named and includes a map showing structure contours. It contains a brief statement in regard to shallow and deep wells and artesian prospects (pp. 88–89). The ground water in the Barnesboro and Patton quadrangles is also briefly described in Geologic Folio 189, and the ground water in these two quadrangles and in the Curwensville quadrangle is briefly described in Water-Supply Paper 110.

GEOLOGIC FOLIOS.

Under the plan adopted for the preparation of a geologic map of the United States the entire area is divided into small quadrangles, bounded by certain meridians and parallels, and these quadrangles, which number several thousand, are separately surveyed and mapped.[2] The unit of survey is also the unit of publication, and the maps and description of each quadrangle are issued in the form of a folio. When all the folios are completed they will constitute the Geologic Atlas of the United States.

A folio is designated by the name of the principal town or of a prominent natural feature within the quadrangle. Each folio includes maps showing the topography, geology, underground structure, and mineral deposits of the area mapped and several pages of descriptive text. The text explains the maps and describes the topographic and geologic features of the country and its mineral products. The topographic map shows roads, railroads, waterways, and, by contour lines, the shapes of the hills and valleys and the height above sea level of all points in the quadrangle. The areal-geology map shows the distribution of the various rocks at the surface. The structural-geology map shows the relations of the rocks to one another underground. The economic-geology map indicates the location of mineral deposits that are commercially valuable. The artesian-water maps show the depth to underground-water horizons. Economic-geology and artesian-water maps are included in folios if the conditions in the areas mapped warrant their publication. The folios are of special interest to students of geography and geology and are valuable as guides in the development and utilization of mineral resources.

Folios 1 to 163, inclusive, are published in only one form (18 by 22 inches), called the library edition. Some of the folios that bear numbers higher than 163 are published also in an octavo edition (6 by 9 inches). Owing to a fire in the Geological Survey building May 18, 1913, the stock of geologic folios was more or less damaged by fire and water, but the folios that are usable are sold at the uniform price of 5 cents each, with no reduction for wholesale orders. This rate applies to folios in stock from 1 to 184, inclusive (except reprints), also to the library edition of Folio 186. The library edition of Folios 185, 187, and higher numbers sells for 25 cents a copy, except that some folios which contain an unusually large amount of matter sell at higher prices. The octavo edition of Folio 185 and higher numbers sell for 50 cents a copy, except Folio 193, which sells for 75 cents a copy. A discount of 40 per cent is allowed on an order for folios or for folios together with topographic maps amounting to $5 or more at the retail rate.

All the folios contain descriptions of the drainage of the quadrangles. The folios in the following list contain also brief discussions of the underground waters in connection with the economic resources of the areas and more or less information concerning the utilization of the water resources.

An asterisk (*) indicates that the stock of the folio is exhausted.

*13. Fredericksburg, Virginia-Maryland. 1894. 5c.

*23. Nomini, Maryland-Virginia. 1896. 5c.

*70. Washington, District of Columbia-Maryland-Virginia. 1901.

*83. New York City (Paterson, Harlem, Staten Island, and Brooklyn quadrangles), New York-New Jersey. 1902.

Discusses the present and future water supply of New York City.

*136. St. Marys, Maryland-Virginia. 1906. 5c.

Discusses artesian wells.

*137. Dover, Delaware-Maryland-New Jersey. 1906. 5c.

Describes the shallow and deep wells used as sources of water supply; gives section of well at Middletown, Del.

---

[2] Index maps showing areas in the North Atlantic slope basins covered by topographic maps and by geologic folios will be mailed on receipt of request addressed to the Director, U. S. Geological Survey, Washington, D. C.

*149. **Penobscot Bay, Maine. 1907. 5c.**
Describes the wells and springs; gives analysis of spring water from North Bluehill.

152. **Patuxent, Maryland-District of Columbia. 1907. 5c.**
Discusses the springs, shallow wells, and artesian wells.

*157. **Passaic, New Jersey-New York. 1908.**
Discusses the underground water of the quadrangle, including the cities of Newark, Hoboken, Jersey City, Paterson, Elizabeth, Passaic, Plainfield, Rahway, and Perth Amboy, and a portion of the city of New York; gives a list of the deep borings in the New Jersey portion of the quadrangle, and notes concerning wells on Staten Island, Long Island, Hoffman Island, and Governors Island.

158. **Rockland, Maine. 1908. 5c.**
Describes the water supply in Knox County, Maine, of which Rockland is the principal city; discusses the water obtained from wells drilled in limestone and granite, and the city water supply of Camden, Rockport, Rockland, and Thomaston.

*160. **Accident-Grantsville, Maryland-Pennsylvania-West Virginia. 1908. 5c.**
Under "Mineral Resources" the folio describes Youghiogheny and Castleman rivers, Savage River, and Georges Creek, and the spring waters; notes possibility of obtaining artesian water.

*161. **Franklin Furnace, New Jersey. 1908.**
Describes the streams, water powers, and ground waters of a district in northwestern New Jersey, mainly in Sussex County but including also a small part of Morris County; gives tabulated list of water powers and of bored wells.

*162. **Philadelphia (Norristown, Germantown, Chester, and Philadelphia quadrangles), Pennsylvania-New Jersey-Delaware. 1909.**
Describes the underground waters of the Piedmont Plateau and the Coastal Plain and gives a tabulated list of wells; discusses the water supply of Philadelphia and Camden, also suburban towns; gives analysis of filtered water of Pickering Creek.

*167. **Trenton, New Jersey-Pennsylvania.[3] 1909. 5c.**
Describes streams tributary to Raritan and Delaware rivers (including estimates of capacity with and without storage) and the springs and wells; discusses also the public water supply of Trenton and suburban towns.

169. **Watkins Glen-Catatonk, New York. 1909. 5c.**
Describes the rivers, which include tributaries of the Susquehanna and the St. Lawrence, the lakes and swamps, and, under "Economic geology," springs and shallow and deep wells; discusses also water supply at Ithaca.

*170. **Mercersburg-Chambersburg, Pennsylvania.[4] 1909. [4]5c.**
Describes the underground waters, including limestone springs, sandstone springs, and wells, and mentions briefly the sources of the water supplies of the principal towns.

182. **Choptank, Maryland. 1912.[4] 5c.**
The Choptank quadrangle includes the entire width of Chesapeake Bay and portions of many large estuaries.

189. **Barnesboro-Patton, Pennsylvania. 1913. 25c.**
Discusses the water supply of various towns in the quadrangle.

191. **Raritan, New Jersey.[5] 1914.**
Discusses briefly the surface and ground waters of the quadrangle, the quality, and the utilization of streams for power; gives analysis of water from Raritan River and from Schooley Mountain Spring near Hackettstown.

192. **Eastport, Maine. 1914. 25c.**
Includes brief account of the water supply of the quadrangle and of the utilization of streams for power.

204. **Tolchester, Maryland. 1917. 25c.**
Discusses shallow and artesian wells.

---

[3] Octavo edition only.
[4] Issued in two editions—library (18 by 22 inches) and octavo (6 by 9 inches). Specify edition desired.
[5] Issued in two editions—library (18 by 22 inches), 25c., and octavo (6 by 9 inches), 50c. Specify edition desired.

MISCELLANEOUS REPORTS.

Other Federal bureaus and State and other organizations have from time to time published reports relating to the water resources of various sections of the country. Notable among those pertaining to the North Atlantic States are the reports of the Maine State Water Storage Commission (Augusta), the New Hampshire Forestry Commission (Concord), the Metropolitan Water and Sewerage Board (Boston, Mass.), the New York State Water-Supply Commission (Albany), the New York State Conservation Commission (Albany), the New York State engineer and surveyor (Albany), the various commissions on water supply of New York City, the Geological Survey of New Jersey (Trenton), State boards of health, and the Tenth Census (vol. 16).

The following reports deserve special mention:

Water power of Maine, by Walter Wells, Augusta, 1869.

Hydrology of the State of New York, by G. W. Rafter: New York State Museum Bull. 85, 1905.

Hydrography of Virginia, by N. C. Grover and R. H. Bolster: Virginia Geol. Survey Bull. 3, 1906.

Underground-water resources of the Coastal Plain province of Virginia, by Samuel Sanford: Virginia Geol. Survey Bull. 5, 1913.

Surface water supply of Virginia, by G. C. Stevens: Virginia Geol. Survey Bull. 10, 1916.

Many of these reports can be obtained by applying to the several commissions, and most of them can be consulted in the public libraries of the larger cities.

GEOLOGICAL SURVEY HYDROLOGIC REPORTS OF GENERAL INTEREST.

The following list comprises reports that are not readily classifiable by drainage basins and that cover a wide range of hydrologic investigations:

WATER-SUPPLY PAPERS.

*1. Pumping water for irrigation, by H. M. Wilson. 1896. 57 pp., 9 pls.

Describes pumps and motive powers, windmills, water wheels, and various kinds of engines; also storage reservoirs to retain pumped water until needed for irrigation.

*3. Sewage irrigation, by G. W. Rafter. 1897. 100 pp., 4 pls. 10c. (See Water-Supply Paper 22.)

Discusses methods of sewage disposal by intermittent filtration and by irrigation; describes utilization of sewage in Germany, England, and France, and sewage purification in the United States.

*8. Windmills for irrigation, by E. C. Murphy. 1897. 49 pp., 8 pls. 10c.

Gives results of experimental tests of windmills during the summer of 1896 in the vicinity of Garden, Kans.; describes instruments and methods and draws conclusions.

*14. New tests of certain pumps and water lifts used in irrigation, by O. P. Hood. 1898. 91 pp., 1 pl. 10c.

Discusses efficiency of pumps and water lifts of various types.

*20. Experiments with windmills, by T. O. Perry. 1899. 97 pp., 12 pls. 15c.

Includes tables and descriptions of wind wheels, compares wheels of several types, and discusses results.

*22. Sewage irrigation, Part II, by G. W. Rafter. 1899. 100 pp., 7 pls. 15c.

Gives résumé of Water-Supply Paper No. 3; discusses pollution of certain streams, experiments on purification of factory wastes in Massachusetts, value of commercial fertilizers, and describes American sewage-disposal plants by States; contains bibliography of publications relating to sewage utilization and disposal.

*41. The windmill: Its efficiency and economic use, Part I, by E. C. Murphy. 1901. 72 pp., 14 pls.

*42. The windmill: Its efficiency and economic use, Part II, by E. C. Murphy. 1901. 75 pp., 2 pls. 10c.

Nos. 41 and 42 give details of results of experimental tests with windmills of various types.

*43. Conveyance of water in irrigation canals, flumes, and pipes, by Samuel Fortier. 1901. 86 pp., 15 pls. 15c.

*56. Methods of stream measurement. 1901. 51 pp., 12 pls. 15c.

Describes the methods used by the Survey in 1901–2. See also Nos. 64, 94, and 95.

*64. Accuracy of stream measurements, by E. C. Murphy. 1902. 99 pp., 4 pls. (See No. 95.) 10c.

Describes methods of measuring velocity of water and of measuring and computing stream flow and compares results obtained with the different instruments and methods; describes also experiments and results at the Cornell University hydraulic laboratory. A second, enlarged edition published as Water-Supply Paper 95.

*67. The motions of underground waters, by C. S. Slichter. 1902. 106 pp., 8 pls. 15c.

Discusses origin, depth, and amount of underground waters; permeability of rocks and porosity of soils; causes, rates, and laws of motions of underground water; surface and deep zones of flow, and recovery of waters by open wells and artesian and deep wells; treats of the shape and position of the water table; gives simple methods of measuring yield of flowing wells; describes artesian wells at Savannah, Ga.

*80. The relation of rainfall to run-off, by G. W. Rafter. 1903. 104 pp. 10c.
Treats of measurements of rainfall and laws and measurements of stream flow; gives rainfall, run-off, and evaporation formulas; discusses effect of forests on rainfall and run-off.

87. Irrigation in India (second edition), by H. M. Wilson. 1903. 238 pp., 27 pls. 25c.
First edition was published in Part II of the Twelfth Annual Report.

93. Proceedings of first conference of engineers of the Reclamation Service, with accompanying papers, compiled by F. H. Newell, chief engineer. 1904. 361 pp. 25c.
Contains the following papers of more or less general interest:
Limits of an irrigation project, by D. W. Ross.
Relation of Federal and State laws to irrigation, by Morris Bien.
Electrical transmission of power for pumping, by H. A. Storrs.
Correct design and stability of high masonry dams, by Geo. Y. Wisner.
Irrigation surveys and the use of the plane table, by J. B. Lippincott.
The use of alkaline waters for irrigation, by Thomas H. Means.

*94. Hydrographic manual of the United States Geological Survey, prepared by E. C. Murphy, J. C. Hoyt, and G. B. Hollister. 1904. 76 pp., 3 pls. 10c.
Gives instruction for field and office work relating to measurements of stream flow by current meters. See also No. 95.

*95. Accuracy of stream measurements (second enlarged edition), by E. C. Murphy. 1904. 169 pp., 6 pls.
Describes methods of measuring and computing stream flow and compares results derived from different instruments and methods. See also No. 94.

*103. A review of the laws forbidding pollution of inland waters in the United States, by E. B. Goodell. 1904. 120 pp. (See No. 152.)
Explains the legal principles under which antipollution statutes become operative, quotes court decisions to show authority for various deductions, and classifies according to scope the statutes enacted in the different States.

*110. Contributions to the hydrology of eastern United States, 1904; M. L. Fuller, geologist in charge. 1905. 211 pp., 5 pls. 10c.
Contains the following reports of general interest. The scope of each paper is indicated by its title.
Description of underflow meter used in measuring the velocity and direction of underground water, by Charles S. Slichter.
The California or "stovepipe" method of well construction, by Charles S. Slichter.
Approximate methods of measuring the yield of flowing wells, by Charles S. Slichter.
Corrections necessary in accurate determinations of flow from vertical well casings, from notes furnished by A. N. Talbot.
Experiments relating to problems of well contamination at Quitman, Ga., by S. W. McCallie.

113. The disposal of strawboard and oil-well wastes, by R. L. Sackett and Isaiah Bowman. 1905. 52 pp., 4 pls. 5c.
The first paper discusses the pollution of streams by sewage and by trade wastes, describes the manufacture of strawboard, and gives results of various experiments in disposing of the waste. The second paper describes briefly the topography, drainage, and geology of the region about Marion, Ind., and the contamination of rock wells and of streams by waste oil and brine.

*114. Underground waters of eastern United States; M. L. Fuller, geologist in charge. 1905. 285 pp., 18 pls. 25c.
Contains report on "Occurrence of underground waters," by M. L. Fuller, discussing sources, amount, and temperature of waters, permeability and storage capacity of rocks, water-bearing formations, recovery of water by springs, wells, and pumps, essential condition of artesian flows and general conditions affecting underground waters in eastern United States.

115. River surveys and profiles made during 1903, arranged by W. C. Hall and J. C. Hoyt. 1905. 115 pp., 4 pls. 10c.
Contains results of surveys made to determine location of undeveloped power sites.

119. Index to the hydrographic progress reports of the United States Geological Survey, 1888 to 1903, by J. C. Hoyt and B. D. Wood. 1905. 253 pp. 15c.
Scope indicated by title.

120. Bibliographic review and index of papers relating to underground waters published by the United States Geological Survey, 1879–1904, by M. L. Fuller. 1905. 128 pp. 10c.

Scope indicated by title.

*122. Relation of the law to underground waters, by D. W. Johnson. 1905. 55 pp. 5c.

Defines and classifies underground waters, gives common-law rules relating to their use, and cites State legislative acts affecting them.

140. Field measurements of the rate of movement of underground waters, by C. S. Slichter. 1905. 122 pp., 15 pls. 15c.

Discusses the capacity of sand to transmit water, describes measurements of underflow in Rio Hondo, San Gabriel, and Mohave River valleys, Calif., and on Long Island, N. Y.; gives results of tests of wells and pumping plants, and describes stovepipe method of well construction.

143. Experiments on steel-concrete pipes on a working scale, by J. H. Quinton. 1905. 61 pp., 4 pls.

Scope indicated by title.

145. Contributions to the hydrology of eastern United States, 1905; M. L. Fuller, geologist in charge. 1905. 220 pp., 6 pls. 10c.

Contains brief reports of general interest as follows:

Drainage of ponds into drilled wells, by Robert E. Horton. Discusses efficiency, cost, and capacity of drainage wells, and gives statistics of such wells in southern Michigan.

Construction of so-called fountain and geyser springs, by Myron L. Fuller.

A convenient gage for determining low artesian heads, by Myron L. Fuller.

146. Proceedings of second conference of engineers of the Reclamation Service, with accompanying papers, compiled by F. H. Newell, chief engineer. 1905. 267 pp. 15c.

Contains brief account of the organization of the hydrographic [water-resources] branch and the Reclamation Service, reports of conferences and committees, circulars of instruction, and many brief reports on subjects closely related to reclamation, and a bibliography of techincal papers by members of the service. Of the papers read at the conference those listed below (scope indicated by title) are of more or less general interest:

Proposed State code of water laws, by Morris Bien.
Power engineering applied to irrigation problems, by O. H. Ensign.
Estimates on tunneling in irrigation projects, by A. L. Fellows.
Collection of stream-gaging data, by N. C. Grover.
Diamond-drill methods, by G. A. Hammond.
Mean-velocity and area curves, by F. W. Hanna.
Importance of general hydrographic data concerning basins of streams gaged, by R. E. Horton.
Effect of aquatic vegetation on stream flow, by R. E. Horton.
Sanitary regulations governing construction camps, by M. O. Leighton.
Necessity of draining irrigated land, by Thos. H. Means.
Alkali soils, by Thos. H. Means.
Cost of stream-gaging work, by E. C. Murphy.
Equipment of a cable gaging station, by E. C. Murphy.
Silting of reservoirs, by W. M. Reed.
Farm-unit classification, by D. W. Ross.
Cost of power for pumping irrigating water, by H. A. Storrs.
Records of flow at current-meter gaging stations during the frozen season, by F. H. Tillinghast.

147. Destructive floods in United States in 1904, by E. C. Murphy and others. 206 pp., 18 pls. 15c.

Contains a brief account of "A method of computing cross-section area of waterways," including formulas for maximum discharge and area of cross section.

*150. Weir experiments, coefficients, and formulas, by R. E. Horton. 1906. 189 pp., 38 pls. (See Water-Supply Paper 200.) 15c.

Scope indicated by title.

151. Field assay of water, by M. O. Leighton. 1905. 77 pp., 4 pls. 10c.

Dicusses methods, instruments, and reagents used in determining turbidity, color, iron, chlorides, and hardness, in connection with studies of the quality of water in various parts of the United States.

*152. A review of the laws forbidding pollution of inland waters in the United States (second edition), by E. B. Goodell. 1905. 149 pp. 10c.
Scope indicated by title.

*160. Underground-water papers, 1906; M. L. Fuller, geologist in charge. 1906. 104 pp., 1 pl.
Gives account of work in 1905, lists of publications relating to underground waters, and contains the following brief reports of general interest:
Significance of the term "artesian," by Myron L. Fuller.
Representation of wells and springs on maps, by Myron L. Fuller.
Total amount of free water in the earth's crust, by Myron L. Fuller.
Use of fluorescein in the study of underground waters, by R. B. Dole.
Problems of water contamination, by Isaiah Bowman.
Instances of improvement of water in wells, by Myron L. Fuller.

*162. Destructive floods in the United States in 1905, with a discussion of flood discharge and frequency and an index to flood literature, by E. C. Murphy and others. 1906. 105 pp., 4 pls. 15c.

*163. Bibliographic review and index of underground-water literature published in the United States in 1905, by M. L. Fuller, F. G. Clapp, and B. L. Johnson. 1906. 130 pp. 15c.
Scope indicated by title.

*179. Prevention of stream pollution by distillery refuse, based on investigations at Lynchburg, Ohio, by Herman Stabler. 1906. 34 pp., 1 pl. 10c.
Describes grain distillation, treatment of slop, sources, character, and effects of effluents on streams; discusses filtration, precipitation, fermentation, and evaporation methods of disposal of wastes without pollution.

*180. Turbine water-wheel tests and power tables, by R. E. Horton. 1906. 134 pp., 2 pls. 20c.
Scope indicated by title.

*186. Stream pollution by acid-iron wastes, a report based on investigations made at Shelby, Ohio, by Herman Stabler. 1906. 36 pp., 1 pl.
Gives history of pollution by acid-iron wastes at Shelby, Ohio, and resulting litigation; discusses effect of acid-iron liquors on sewage purification processes, recovery of copperas from acid-iron wastes, and other processes for disposal of pickling liquor.

*187. Determination of stream flow during the frozen season, by H. K. Barrows and R. E. Horton. 1907. 93 pp., 1 pl. 15c.
Scope indicated by title.

*189. The prevention of stream pollution by strawboard waste, by E. B. Phelps. 1906. 29 pp., 2 pls. 5c.
Describes manufacture of strawboard, present and proposed methods of disposal of waste liquors, laboratory investigations of precipitation and sedimentation, and field studies of amounts and character of water used, raw material and finished product, and mechanical filtration.

*194. Pollution of Illinois and Mississippi rivers by Chicago sewage (a digest of the testimony taken in the case of the State of Missouri *v.* the State of Illinois and the Sanitary district of Chicago), by M. O. Leighton. 1907. 369 pp., 2 pls. 40c.
Scope indicated by amplification of title.

*200. Weir experiments, coefficients, and formulas (revision of paper No. 150), by R. E. Horton. 1907. 195 pp., 38 pls. 35c.
Scope indicated by title.

*226. The pollution of streams by sulphite pulp waste, a study of possible remedies, by E. B. Phelps. 1909. 37 pp., 1 pl. 10c.
Describes manufacture of sulphite pulp, the waste liquors, and the experimental work leading to suggestions as to methods of preventing stream pollution.

*229. The disinfection of sewage and sewage filter effluents, with a chapter on the putrescibility and stability of sewage effluents, by E. B. Phelps. 1909. 91 pp., 1 pl. 15c.
Scope indicated by title.

*234. Papers on the conservation of water resources. 1909. 96 pp., 2 pls. 15c.

Contains the following papers, whose scope is indicated by their titles: Distribution of rainfall, by Henry Gannett; Floods, by M. O. Leighton; Developed water powers, compiled under the direction of W. M. Steuart, with discussion by M. O. Leighton; Undeveloped water powers, by M. O. Leighton; Irrigation, by F. H. Newell; Underground waters, by W. C. Mendenhall; Denudation, by R. B. Dole, and Herman Stabler; Control of catchment areas, by H. N. Parker.

*235. The purification of some textile and other factory wastes, by Herman Stabler and G. H. Pratt. 1909. 76 pp. 10c.

Discusses waste waters from wool scouring, bleaching and dyeing cotton yarn, bleaching cotton piece goods, and manufacture of oleomargarine, fertilizer, and glue.

*236. The quality of surface waters in the United States, Part I.—Analyses of waters east of the one hundredth meridian, by R. B. Dole. 1909. 123 pp. 10c.

Describes collection of samples, method of examination, preparation of solutions, accuracy of estimates, and expression of analytical results.

238. The public utility of water powers and their governmental regulation, by René Tavernier and M. O. Leighton. 1910. 161 pp. 15c.

Discusses hydraulic power and irrigation, French, Italian, and Swiss legislation relative to the development of water powers, and laws proposed in the French parliament; reviews work of bureau of hydraulics and agricultural improvement of the French department of agriculture, and gives résumé of Federal and State water-power legislation in the United States.

*255. Underground waters for farm use, by M. L. Fuller. 1910. 58 pp., 17 pls. 15c.

Discusses rocks as sources of water supply and the relative safety of supplies from different materials; springs, and their protection; open or dug and deep wells, their location, yield, relative cost, protection, and safety; advantages and disadvantages of cisterns and combination wells and cisterns.

*257. Well-drilling methods, by Isaiah Bowman. 1911. 139 pp., 4 pls. 15c.

Discusses amount, distribution, and disposal of rainfall, water-bearing rocks, amount of underground water and artesian conditions, and oil and gas bearing formations; gives history of well drilling in Asia, Europe, and the United States; describes in detail the various methods and the machinery used; discusses loss of tools and geologic difficulties; contamination of well waters and methods of prevention; tests of capacity and measurement of depth; and costs of sinking wells.

*258. Underground-water papers, 1910, by M. L. Fuller, F. G. Clapp, G. C. Matson, Samuel Sanford, and H. C. Wolff. 1911. 123 pp., 2 pls. 15c.

Contains the following papers (scope indicated by titles) of general interest:
Drainage by wells, by M. L. Fuller.
Freezing of wells and related phenomena, by M. L. Fuller.
Pollution of underground waters in limestone, by G. C. Matson.
Protection of shallow wells in sandy deposits, by M. L. Fuller.
Magnetic wells, by M. L. Fuller.

259. The underground waters of southwestern Ohio, by M. L. Fuller and F. G. Clapp, with a discussion of the chemical character of the waters, by R. B. Dole. 1912. 228 pp., 9 pls. 35c.

Describes the topography, climate, and geology of the region, the water-bearing formations, the source, mode of occurrence, and head of the waters, and municipal supplies; gives details by counties; discusses in supplement, under chemical character, method of analysis and expression of results, mineral constituents, effect of the constituents on waters for domestic, industrial, or medicinal uses, methods of purification, and chemical composition; many analyses and field assays. The matter in the supplement was also published in Water-Supply Paper 254 (The underground waters of north-central Indiana).

274. Some stream waters of the western United States, with chapters on sediment carried by the Rio Grande and the industrial application of water analyses, by Herman Stabler. 1911. 188 pp. 15c.

Describes collection of samples, plan of analytical work, and methods of analyses; discusses soap-consuming power of waters, water softening, boiler waters, and water for irrigation.

280. Gaging stations maintained by the United States Geological Survey, 1888–1910, and Survey publications relating to water resources, compiled by B. D. Wood. 1912. 102 pp. 10c.

315. The purification of public water supplies, by G. A. Johnson. 1913. 84 pp., 8 pls. 10c.

Discusses ground, lake, and river waters as public supplies, development of waterworks systems in the United States, water consumption, and typhoid fever; describes methods of filtration and sterilization of water and municipal water softening.

334. The Ohio Valley flood of March-April, 1913 (including comparisons with some earlier floods), by A. H. Horton and H. J. Jackson. 1913. 96 pp., 22 pls. 20c.

Although relating specifically to floods in the Ohio Valley, this report discusses also the causes of floods and the prevention of damage by floods.

337. The effects of ice on stream flow, by William Glenn Hoyt. 1913. 77 pp., 7 pls. 15c.

Discusses methods of measuring the winter flow of streams.

345. Contributions to the hydrology of the United States, 1914. N. C. Grover, chief hydraulic engineer. 1915. 225 pp., 17 pls. 30c. Contains:

*(e) A method of determining the daily discharge of rivers of variable slope, by M. R. Hall, W. E. Hall, and C. H. Pierce, pp. 53-65. 5c. Scope indicated by title.

364. Water analyses from the laboratory of the United States Geological Survey, tabulated by F. W. Clarke, chief chemist. 1914. 40 pp. 5c.

Contains analyses of waters from rivers, lakes, wells, and springs in various parts of the United States, including analyses of the geyser water of Yellowstone National Park, hot springs in Montana, brines from Death Valley, water from the Gulf of Mexico, and mine waters from Tennessee, Michigan, Missouri and Oklahoma, Montana, Colorado and Utah, Nevada and Arizona, and California.

371. Equipment for current-meter gaging stations, by G. J. Lyon. 1915. 64 pp., 37 pls. 20c.

Describes methods of installing recording and other gages and of constructing gage wells, shelters, and structures for making discharge measurements and artificial controls.

375. Contributions to the hydrology of the United States, 1915. N. C. Grover, chief hydraulic engineer. 1916. 181 pp., 9 pls. Contains:

(c) Relation of stream gaging to the science of hydraulics, by C. H. Pierce and R. W. Davenport, pp. 77-84.

(e) A method for correcting river discharge for changing stage, by B. E. Jones, pp. 117-130.

(f) Conditions requiring the use of automatic gages in obtaining stream-flow records, by C. H. Pierce, pp. 131-139.

*400. Contributions to the hydrology of the United States, 1916. N. C. Grover, chief hydraulic engineer. Contains:

(a) The people's interest in water-power resources, by G. O. Smith, pp. 1-8.

*(c) The measurement of silt-laden streams, by Raymond C. Pierce, pp. 39-51.

(d) Accuracy of stream-flow data, by N. C. Grover and J. C. Hoyt, pp. 53-59.

416. The divining rod, a history of water witching, with a bibliography, by Arthur J. Ellis. 1917. 39 pp. 10c.

A brief paper published "merely to furnish a reply to the numerous inquiries that are continually being received from all parts of the country" as to the efficacy of the divining rod for locating underground water.

*425. Contributions to the hydrology of the United States, 1917. N. C. Grover, chief hydraulic engineer. 1918. Contains:

(c) Hydraulic conversion tables and convenient equivalents, pp. 71-94. 1917.

427. Bibliography and index of the publications of the United States Geological Survey relating to ground water, by O. E. Meinzer. 1918. 169 pp., 1 pl.

Includes publications prepared, in whole or part, by the Geological Survey that treat any phase of the subject of ground water or any subject directly applicable to ground water. Illustrated by maps showing reports that cover specific areas more or less thoroughly.

PROFESSIONAL PAPERS.

*72. Denudation and erosion in the southern Appalachian region and the Monongahela basin, by L. C. Glenn. 1911. 137 pp., 21 pls. 35c.

Describes the topography, geology, drainage, forests, climate, population, and transportation facilities of the region, the relation of agriculture, lumbering, mining, and power development to erosion and denudation, and the nature, effects, and remedies of erosion; gives details of conditions in Holston, Nolichucky, French Broad, Little Tennessee, and Hiwassee river basins, along Tennessee River proper, and in the basins of the Coosa-Alabama system, Chattahoochee, Savannah, Saluda, Broad, Catawba, Yadkin, New, and Monongahela rivers.

*86. The transportation of débris by running water, by G. K. Gilbert, based on experiments made with the assistance of E. C. Murphy. 1914. 263 pp., 3 pls. 70c.

The results of an investigation which was carried on in a specially equipped laboratory at Berkeley, Calif., and was undertaken for the purpose of learning "the laws which control the movement of bed load and especially to determine how the quantity of load is related to the stream's slope and discharge and to the degree of comminution of the débris."

A highly technical report.

105. Hydraulic mining débris in the Sierra Nevada, by G. K. Gilbert. 1917. 154 pp., 34 pls.

Presents the results of an investigation undertaken by the United States Geological Survey in response to a memorial from the California Miners' Association asking that a particular study be made of portions of the Sacramento and San Joaquin valleys affected by detritus from torrential streams. The report deals largely with geologic and physiographic aspects of the subject, traces the physical effects, past and future, of the hydraulic mining of earlier decades, the similar effects which certain other industries induce through stimulation of the erosion of the soil, and the influence of the restriction of the area of inundation by the construction of levees. Suggests cooperation by several interests for the control of the streams now carrying heavy loads of débris.

BULLETINS.

*32. Lists and analyses of the mineral springs of the United States (a preliminary study), by A. C. Peale. 1886. 235 pp.

Defines mineral waters, lists the springs by States, and gives tables of analyses so far as available.

*264. Record of deep-well drilling for 1904, by M. L. Fuller, E. F. Lines, and A. C. Veatch. 1905. 106 pp. 10c.

*298. Record of deep-well drilling for 1905, by M. L. Fuller and Samuel Sanford. 1906. 299 pp. 25c.

Bulletins 264 and 298 discuss the importance of accurate well records to the driller, to owners of oil, gas, and water wells, and to the geologist; describes the general methods of work; gives tabulated records of wells by States, and detailed records selected as affording valuable stratigraphic information.

*319. Summary of the controlling factors of artesian flows, by Myron L. Fuller. 1908. 44 pp., 7 pls. 10c.

Describes underground reservoirs, the sources of underground waters, the confining agents, the primary and modifying factors of artesian circulation, the essential and modifying factors of artesian flow, and typical artesian systems.

*479. The geochemical interpretation of water analyses, by Chase Palmer. 1911. 31 pp. 5c.

Discusses the expression of chemical analyses, the chemical character of water, and the properties of natural waters; gives a classification of waters based on property values and reacting values, and discusses the character of the waters of certain rivers as interpreted directly from the results of analyses; discusses also the relation of water properties to geologic formations, silica in river water, and the character of the water of the Mississippi and the Great Lakes and St. Lawrence River as indicated by chemical analyses.

*616. The data of geochemistry (third edition), by F. W. Clarke. 1916. 821 pp. 45c.

Earlier editions were published as Bulletins 330 and 491. Contains a discussion of the statement and interpretation of water analyses and a chapter on "Mineral wells and springs" (pp. 179-216). Discusses the definition and classification of mineral waters, changes in the composition of water, deposits of calcareous, ocherous and siliceous materials made by water, vadose and juvenile waters, and thermal springs in relation to volcanism. Describes the different kinds of ground water and gives typical analyses. Includes a brief bibliography of papers containing water analyses.

ANNUAL REPORTS.

*Fifth Annual Report of the United States Geological Survey, 1883-84, J. W. Powell, Director. 1885. xxxvi, 469 pp., 58 pls. $2.25. Contains:

*The requisite and qualifying conditions of artesian wells, by T. C. Chamberlain, pp. 125 to 173, Pl. 21. Scope indicated by title.

*Twelfth Annual Report of the United States Geological Survey, 1890-91, J. W. Powell, Director. 1891. 2 parts. *Pt. II—Irrigation, xviii, 576 pp., 93 pls. $2. Contains:

*Irrigation in India, by H. M. Wilson, pp. 363-561, Pls. 107 to 146. See Water-Supply Paper 87.

Thirteenth Annual Report of the United States Geological Survey, 1891-92, J. W. Powell, Director. 1892. (Pts. II and III, 1893.) 3 parts. *Pt. III—Irrigation, xi, 486 pp., 77 pls. $1.85. Contains:

*American irrigation engineering, by H. M. Wilson, C. E., pp. 101-349, Pls. 111 to 146. Discusses the economic aspects of irrigation, alkaline drainage, silt, and sedimentation; gives brief history and legislation; describes canals; discusses water storage at reservoirs of the California and other projects, subsurface sources of supply, pumping, and subirrigation.

Fourteenth Annual Report of the United States Geological Survey, 1892-93, J. W. Powell, Director. 1893. (Pt. II, 1894). 2 parts. *Pt. II—Accompanying papers, xx, 597 pp., 73 pls. $2.10. Contains:

*The potable waters of the eastern United States, by W. J. McGee, pp. 1 to 47. Discusses cistern water, stream waters, and ground waters, including mineral springs and artesian wells.

*Natural mineral waters of the United States, by A. C. Peale, pp. 49-88, Pls. 3 and 4. Discusses the origin and flow of mineral springs, the source of mineralization, thermal springs, the chemical composition and analysis of spring waters, geographic distribution, and the utilization of mineral waters; gives a list of American mineral spring resorts; contains also some analyses.

Nineteenth Annual Report of the United States Geological Survey, 1897-98, Charles D. Walcott, Director. 1898. (Parts II, III, and V, 1899.) 6 parts in 7 vols. and separate case for maps with Pt. V. *Pt. II—Papers chiefly of a theoretic nature, v. 958 pp., 172 pls. $2.65. Contains:

*Principles and conditions of the movements of ground water, by F. H. King, pp. 59-294, Pls. 6 to 16. Discusses the amount of water stored in sandstone, in soil, and in other rocks; the depth to which ground water penetrates; gravitational, thermal, and capillary movements of ground waters, and the configuration of the ground-water surface; gives the results of experimental investigations on the flow of air and water through rigid porous media and through sands, sandstones, and silts; discusses results obtained by other investigators, and summarizes results of observations; discusses also rate of flow of water through sand and rock, the growth of rivers, rate of filtration through soil, interference of wells, etc.

*Theoretical investigation of the motion of ground waters, by C. S. Slichter, pp. 295-384, Pl. 17. Scope indicated by title.

INDEX BY AREAS AND SUBJECTS.

[A—Annual Reports; M—Monograph; B—Bulletin; P—Professional Paper; W—Water-Supply Paper; GF—Geologic folio. For titles see preceding pages.]

[1] Many of the reports contain brief subject bibliographies. See abstracts.

[2] Many analyses of river, spring, and well waters are scattered through publications, as noted in abstracts.

INDEX OF STREAMS.

O

DEPARTMENT OF THE INTERIOR
JOHN BARTON PAYNE, Secretary

UNITED STATES GEOLOGICAL SURVE
GEORGE OTIS SMITH, Director

ACE WATER SUPPLY OF UNITED STATES

1918

I. SOUTH ATLANTIC SLOPE AND EA GULF OF MEXICO BASINS

NATHAN C. GROVER, Chief Hydraulic Engineer

DEPARTMENT OF THE INTERIOR
JOHN BARTON PAYNE, Secretary

UNITED STATES GEOLOGICAL SURVEY
GEORGE OTIS SMITH, Director

Water-Supply Paper 472

SURFACE WATER SUPPLY OF THE UNITED STATES 1918

II. SOUTH ATLANTIC SLOPE AND EASTERN GULF OF MEXICO BASINS

NATHAN C. GROVER, Chief Hydraulic Engineer
GUY C. STEVENS and C. G. PAULSEN
District Engineers

WASHINGTON
GOVERNMENT PRINTING OFFICE
1920

CONTENTS.

ILLUSTRATIONS.

SURFACE WATER SUPPLY OF SOUTH ATLANTIC SLOPE AND EASTERN GULF OF MEXICO DRAINAGE BASINS, 1918.

AUTHORIZATION AND SCOPE OF WORK.

This volume is one of a series of 14 reports presenting results of measurements of flow made on streams in the United States during the year ending September 30, 1918.

The data presented in these reports were collected by the United States Geological Survey under the following authority contained in the organic law (20 Stat. L., p. 394):

Provided, That this officer [the Director] shall have the direction of the Geological Survey and the classification of public lands and examination of the geological structure, mineral resources, and products of the national domain.

The work was begun in 1888 in connection with special studies relating to irrigation in the arid west. Since the fiscal year ending June 30, 1895, successive sundry civil bills passed by Congress have carried the following item and appropriations:

For gaging the streams and determining the water supply of the United States, and for the investigation of underground currents and artesian wells, and for the preparation of reports upon the best methods of utilizing the water resources.

Annual appropriations for the fiscal years ended June 30, 1895–1919.

1895	$12,500
1896	20,000
1897 to 1900, inclusive	50,000
1901 to 1902, inclusive	100,000
1903 to 1906, inclusive	200,000
1907	150,000
1908 to 1910, inclusive	100,000
1911 to 1917, inclusive	150,000
1918	175,000
1919	148,244.10

In the execution of the work many private and State organizations have cooperated, either by furnishing data or by assisting in collecting data. Acknowledgments for cooperation of the first kind are made in connection with the description of each station affected; cooperation of the second kind is acknowledged on page 9.

Measurements of stream flow have been made at about 4,510 points in the United States and also at many points in Alaska and the Hawaiian Islands. In July, 1918, 1,180 gaging stations were being maintained by the Survey and the cooperating organizations. Many miscellaneous discharge measurements are made at other points. In connection with this work data were also collected in regard to precipitation, evaporation, storage reservoirs, river profiles, and water power in many sections of the country and will be made available in water-supply papers from time to time. Information in regard to publications relating to water resources is presented in the appendix to this report.

DEFINITION OF TERMS.

The volume of water flowing in a stream—the "run-off" or "discharge"—is expressed in various terms, each of which has become associated with a certain class of work. These terms may be divided into two groups—(1) those that represent a rate of flow, as second-feet, gallons per minute, miners' inches, and discharge in second-feet per square mile, and (2) those that represent the actual quantity of water, as run-off in depth in inches, acre-feet, and millions of cubic feet. The principal terms used in this series of reports are second-feet, second-feet per square mile, run-off in inches, and acre-feet. They may be defined as follows:

"Second-feet" is an abbreviation for "cubic feet per second." A second-foot is the rate of discharge of water flowing in a channel of rectangular cross section 1 foot wide and 1 foot deep at an average velocity of 1 foot per second. It is generally used as a fundamental unit from which others are computed.

"Second-feet per square mile" is the average number of cubic feet of water flowing per second from each square mile of area drained, on the assumption that the run-off is distributed uniformly both as regards time and area.

"Run-off (depth in inches)" is the depth to which an area would be covered if all the water flowing from it in a given period were uniformly distributed on the surface. It is used for comparing run-off with rainfall, which is usually expressed in depth of inches.

An "acre-foot," equivalent to 43,560 cubic feet, is the quantity required to cover an acre to the depth of 1 foot. The term is commonly used in connection with storage for irrigation.

The following terms not in common use are here defined:

"Stage-discharge relation;" an abbreviation for the term "relation of gage height to discharge."

"Control;" a term used to designate the section or sections of the stream channel below the gage which determine the stage-discharge relation at the gage. It should be noted that the control may not be the same section or sections at all stages.

The "point of zero flow" for a gaging station is that point on the gage—the gage height—to which the surface of the river falls when the discharge is reduced to zero.

EXPLANATION OF DATA.

The data presented in this report cover the year beginning October 1, 1917, and ending September 30, 1918. At the beginning of January in most parts of the United States much of the precipitation in the preceding three months is stored as ground water, in the form of snow or ice, or in ponds, lakes, and swamps, and this stored water passes off in the streams during the spring break-up. At the end of September, on the other hand, the only stored water available for run-off is possibly a small quantity in the ground; therefore the run-off for the year beginning October 1 is practically all derived from precipitation within that year.

The base data collected at gaging stations consist of records of stage, measurements of discharge, and general information used to supplement the gage heights and discharge measurements in determining the daily flow. The records of stage are obtained either from direct readings on a staff gage or from a water-stage recorder that gives a continuous record of the fluctuations. Measurements of discharge are made with a current meter. (See Pls. I, II.) The general methods are outlined in standard textbooks on the measurement of river discharge.

From the discharge measurements rating tables are prepared that give the discharge for any stage, and these rating tables, when applied to gage heights, give the discharge from which the daily, monthly, and yearly means of discharge are determined.

The data presented for each gaging station in the area covered by this report comprise a description of the station, a table giving results of discharge measurements, a table showing the daily discharge of the stream, and a table of monthly and yearly discharge and run-off.

If the base data are insufficient to determine the daily discharge, tables giving daily gage heights and results of discharge measurements are published.

The description of the station gives, in addition to statements regarding location and equipment, information in regard to any conditions that may affect the constancy of the stage-discharge relation, covering such subjects as the occurrence of ice, the use of the stream for log driving, shifting of control, and the cause and effect of backwater; it gives also information as to diversions that decrease the flow at the gage, artificial regulation, maximum and minimum recorded stages, and the accuracy of the records.

The table of daily discharge gives, in general, the discharge in second-feet corresponding to the mean of the gage heights read each day. At stations on streams subject to sudden or rapid diurnal fluctuation the discharge obtained from the rating table and the mean daily gage height may not be the true mean discharge for the day. If such stations are equipped with water-stage recorders the mean daily discharge may be obtained by averaging discharge at regular intervals during the day, or by using the discharge integrator, an instrument operating on the principle of the planimeter and containing as an essential element the rating curve of the station.

In the table of monthly discharge the column headed "Maximum" gives the mean flow for the day when the mean gage height was highest. As the gage height is the mean for the day it does not indicate correctly the stage when the water surface was at crest height, and the corresponding discharge was consequently larger than given in the maximum column. Likewise, in the column headed "Minimum" the quantity given is the mean flow for the day when the mean gage height was lowest. The column headed "Mean" is the average flow in cubic feet for each second during the month. On this average flow computations recorded in the remaining columns, which are defined on page 6, are based.

ACCURACY OF FIELD DATA AND COMPUTED RECORDS.

The accuracy of stream-flow data depends primarily (1) on the permanence of the stage-discharge relation and (2) on the accuracy of observation of stage, measurements of flow, and interpretation of records.

A paragraph in the description of the station gives information regarding the (1) permanence of the stage-discharge relation, (2) precision with which the discharge rating curve is defined, (3) refinement of gage readings, (4) frequency of gage readings, and (5) methods of applying daily gage height to the rating table to obtain the daily discharge.[1]

For the rating tables "well defined" indicates, in general, that the rating is probably accurate within 5 per cent; "fairly well defined," within 10 per cent; "poorly defined," within 15 to 25 per cent. These notes are very general and are based on the plotting of the individual measurements with reference to the mean rating curve.

The monthly means for any station may represent with high accuracy the quantity of water flowing past the gage, but the figures showing discharge per square mile and depth of run-off in inches may be subject to gross errors caused by the inclusion of large noncontributing districts in the measured drainage area, by lack of information concerning water diverted for irrigation or other use, or by inability to interpret the effect of artificial regulation of the flow of the river above the station. "Second-feet per square mile" and "Run-off (depth in inches)" are therefore not computed if such errors appear probable. The computations are also omitted for stations on

[1] For a more detailed discussion of the accuracy of records see Grover, N. C., and Hoyt, J. C. Accuracy of stream-flow data: U. S. Geol. Survey Water-Supply Paper 400, pp. 53-59, 1916.

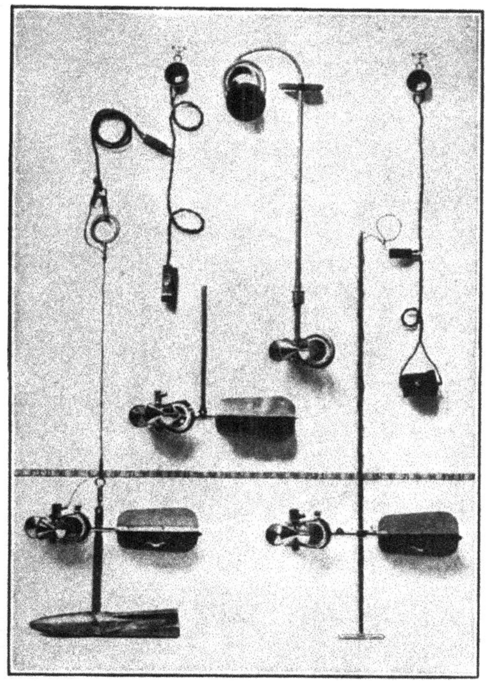

A. PRICE CURRENT METERS.

B. TYPICAL GAGING STATION.

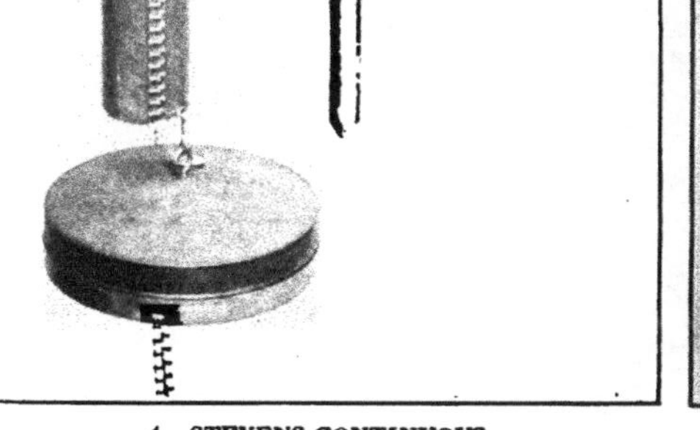

A. STEVENS CONTINUOUS.

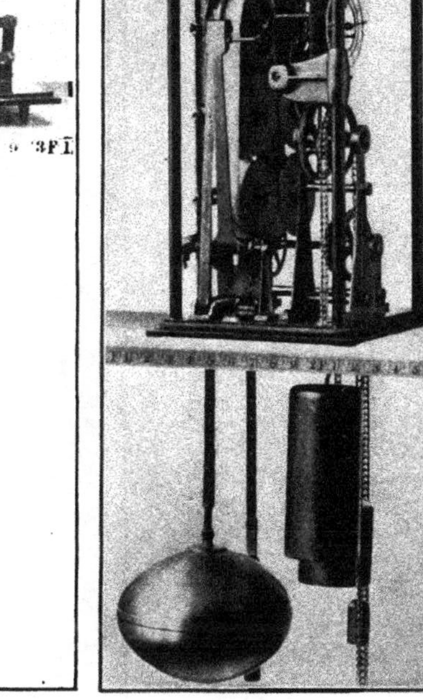

B. GURLEY PRINTING.

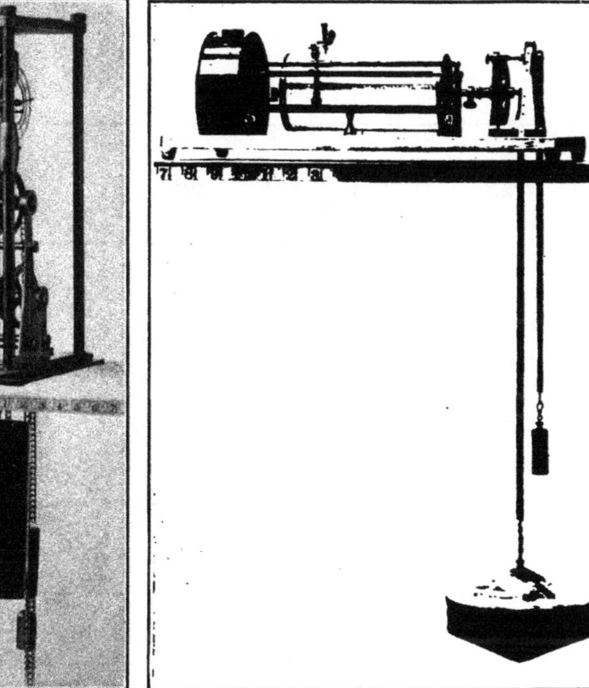

C. FRIEZ.

WATER-STAGE RECORDERS.

streams draining areas in which the annual rainfall is less than 20 inches. All figures representing "second-feet per square mile" and "run-off (depth in inches)" previously published by the Survey should be used with caution because of possible inherent but unknown sources of error.

The table of monthly discharge gives only a general idea of the flow at the station and should not be used for other than preliminary estimates; the tables of daily discharge allow more detailed studies of the variation in flow. It should be borne in mind, however, that the observations in each succeeding year may be expected to throw new light on data previously published.

COOPERATION.

Special acknowledgements are due for financial assistance rendered by the following corporations and individuals: Virginia Railway & Power Co., Alabama Geological Survey, United States Weather Bureau, Tallassee Power Co., Central Georgia Power Co., Columbus Power Co., Georgia Railway & Power Co., Alabama Power Co., Juliette Milling Co., and Rhodhiss Manufacturing Co.

DIVISION OF WORK.

Data for the stations in the James and Roanoke drainage basins were collected and prepared for publication under the direction of G. C. Stevens, district engineer, assisted by B. L. Hopkins, A. G. Fiedler, B. J. Peterson, and J. W. Moulton.

The data for all drainage basins south of Roanoke River were collected and prepared for publication under the direction of C. G. Paulsen, district engineer, assisted by B. J. Peterson, A. H. Condron, L. J. Hall, and Miss E. M. Tiller.

GAGING-STATION RECORDS.

JAMES RIVER BASIN.

JAMES RIVER AT BUCHANAN, VA.

LOCATION.—At highway bridge near Chesapeake & Ohio Railway station at Buchanan, Botetourt County.

DRAINAGE AREA.—2,060 square miles.

RECORDS AVAILABLE.—August 18, 1895, to September 30, 1918.

GAGE.—Chain gage attached to highway bridge, installed November 21, 1903, to replace original wire gage read from August 18, 1895, to that date; read by D. D. Booze for United States Weather Bureau. Datum of gage lowered 2 feet April 3, 1897, to avoid negative readings. A span of the bridge and the gage were destroyed by flood on the night of March 27, 1913. A temporary gage was used from April 22 to September 15, 1913, when a new chain gage was installed.

DISCHARGE MEASUREMENTS.—Made from downstream side of two-span highway bridge, or by wading.

Channel and control.—Bed under bridge in composed of rock overlain with a thick deposit of mud. Banks high; not overflowed except in extreme floods. Control of boulders and gravel several hundred feet below station. Stage-discharge relation not permanent.

Extremes of discharge.—Maximum stage recorded during year, 17.0 feet March 14 (discharge, 54,700 second-feet); minimum stage, 1.9 feet several days in October (discharge, 340 second-feet).

1895–1918: Maximum stage recorded, 31 feet during the night of March 27, 1913 (determined by levels from flood marks October 2, 1914; discharge not determined); minimum stage, 1.2 feet (present gage datum) April 17 and May 2, 1898 (discharge, 260 second-feet).

Ice.—Stage-discharge relation affected by ice during the severe winter of 1917–18.

Accuracy.—Stage-discharge assumed permanent during the year; affected by ice December 11 to February 10. Rating curve fairly well defined below 4,000 second-feet, and poorly defined above. Gage read to tenths once daily. Daily discharge ascertained by applying daily gage height to rating table, except for period of ice effect. Records fair for open water and poor for winter.

Cooperation.—Since July 15, 1906, gage-height records have been furnished by United States Weather Bureau.

The following discharge measurement was made by B. L. Hopkins and A. G. Fiedler:

May 29, 1918: Gage height, 4.48 feet; discharge, 3,240 second-feet.

Daily discharge, in second-feet, of James River at Buchanan, Va., for the year ending Sept. 30, 1918.

Day.	Oct.	Nov.	Dec.	Feb.	Mar.	Apr.	May.	June.	July.	Aug.	Sept.
1	340	975	390	2,000	3,880	2,560	4,560	2,560	5,040	1,080	975
2	340	975	390	2,000	3,470	2,220	3,670	2,330	5,540	975	880
3	340	880	390	1,400	3,270	1,920	3,080	2,220	2,900	880	795
4	340	795	390	1,200	3,080	1,780	2,900	2,220	2,560	795	715
5	340	715	390	1,200	2,900	1,650	2,720	2,070	1,920	715	975
6	340	640	390	1,000	6,320	1,650	2,390	2,070	1,520	640	2,390
7	340	570	390	1,000	7,100	1,520	2,220	1,920	1,400	640	2,220
8	340	505	390	1,000	9,900	1,520	2,070	1,780	1,290	975	2,330
9	340	505	390	3,000	7,100	3,080	1,920	1,400	1,180	795	1,920
10	340	505	390	6,000	5,040	14,700	1,920	1,400	1,180	795	2,070
11	340	505	......	16,100	4,100	15,000	1,780	1,290	1,180	715	1,650
12	340	505	......	16,500	3,470	12,400	2,070	1,290	975	795	1,400
13	340	505	......	16,100	2,900	10,200	1,920	1,080	880	795	1,180
14	340	445	......	18,700	54,700	7,640	2,070	1,080	880	715	975
15	340	445	......	18,300	19,500	6,060	1,780	975	795	715	880
16	340	445	......	18,000	9,040	5,290	1,780	975	715	640	880
17	340	445	......	13,000	6,060	4,800	1,650	1,650	715	640	795
18	340	445	......	6,320	4,560	4,560	1,650	2,330	975	640	1,180
19	340	445	......	3,670	4,100	4,100	1,650	3,880	1,520	1,080	2,070
20	390	445	......	4,560	3,670	6,580	1,520	2,560	1,780	975	1,920
21	390	445	......	6,060	6,580	14,700	1,520	2,070	1,400	880	1,780
22	390	390	......	7,100	13,300	26,600	1,400	1,650	1,180	795	1,650
23	340	390	......	6,580	9,900	13,000	1,400	1,650	1,080	795	1,520
24	340	390	......	5,540	7,100	9,320	3,270	1,520	975	715	1,400
25	340	390	......	5,040	5,800	6,840	4,560	1,520	880	715	1,180
26	340	390	......	7,100	4,560	6,060	4,100	5,540	795	640	1,080
27	390	390	......	10,500	4,100	6,580	3,880	14,000	715	640	975
28	390	390	......	5,040	3,670	8,760	3,670	10,200	640	880	880
29	390	390	......	......	3,080	6,580	3,470	7,640	640	1,080	795
30	1,080	390	......	......	2,900	5,040	4,560	5,540	570	880	715
31	975	......	......	......	2,720	......	2,560	......	1,180	795	......

Note.—Discharge estimated, because of ice, from weather records and comparison with records at other stations as follows: Dec. 11–20, 350 second-feet; Dec. 21–31, 500 second-feet; Jan. 1–15, 350 second-feet; Jan. 16–31, 1,200 second-feet; Feb. 1–10, as in table.

Monthly discharge of James River at Buchanan, Va., for the year ending Sept. 30, 1918.

[Drainage area, 2,060 square miles.]

Month.	Discharge in second-feet.				Run-off (depth in inches on drainage area).
	Maximum.	Minimum.	Mean.	Per square mile.	
October	1,080	340	394	0.191	0.22
November	975	390	522	.253	.28
December	....	....	416	.202	.23
January	....	....	789	.383	.44
February	18,700	1,000	7,280	3.53	3.68
March	54,700	2,720	7,350	3.57	4.12
April	26,600	1,520	7,090	3.44	3.84
May	4,560	1,400	2,570	1.25	1.44
June	14,000	975	2,950	1.43	1.60
July	5,540	570	1,450	.704	.81
August	1,080	640	800	.388	.45
September	2,390	715	1,340	.650	.73
The year	54,700	340	2,710	1.32	17.84

JAMES RIVER AT CARTERSVILLE, VA.

Location.—At highway bridge between Pemberton and Cartersville, Cumberland County, about 50 miles above Richmond. Willis River enters from the south about a mile above station, and Rivanna River from the north about 7 miles above.

Drainage area.—6,230 square miles.

Records available.—January 1, 1899, to September 30, 1918.

Gage.—Chain on downstream side and near Cartersville end of bridge; read by B. W. Palmore. Wire gage used previous to July 24, 1903.

Discharge measurements.—Made from bridge.

Channel and control.—Bed composed of rocks and sand; shifts somewhat during floods. Banks high; left bank is overflowed at a stage of about 20 feet.

Extremes of discharge.—Maximum stage recorded during year 17.0 feet at 9.30 a. m. April 22 (discharge, 52,800 second-feet); minimum stage, 0.75 foot at 9.30 a. m. October 3 (discharge, 910 second-feet).

1899–1918: Maximum stage recorded, 26.7 feet at 6 p. m. December 30, 1901 (discharge about 106,000 second-feet); minimum stage, 0.5 foot October 3, 1914 (discharge, 800 second-feet). A discharge of 603 second-feet (gage height 0.42 foot) was measured September 8, 1897, but gage-height record corresponding to this measurement is probably subject to error.

Ice.—Stage-discharge relation affected by ice during the winter of 1917–18.

Accuracy.—Stage-discharge relation practically permanent during year; affected by ice December 12 to February 10. Rating curve well defined between 1,300 and 40,000 second-feet, and is extended for high stages. Gage read to hundredths twice daily. Daily discharge ascertained by applying mean daily gage height to rating table, except during period of ice effect. Records good for open water periods and fair for winter period.

The following discharge measurement was made by B. J. Peterson and A. G. Fiedler:

June 24, 1918: Gage height, 2.10 feet; discharge, 3,350 second-feet.

Daily discharge, in second-feet, of James River at Cartersville, Va., for the year ending Sept. 30, 1918.

Day.	Oct.	Nov.	Dec.	Jan.	Feb.	Mar.	Apr.	May.	June.	July.	Aug.	Sept.
1	1,150	5,460	1,630	1,310	4,600	13,100	7,870	16,400	6,390	13,100	3,440	2,790
2	1,150	5,240	1,550	1,310	5,020	10,000	8,650	14,600	5,920	10,500	2,630	3,400
3	1,020	3,790	1,470	1,310	5,240	9,460	8,130	13,100	5,240	9,510	2,580	2,540
4	1,230	3,400	1,230	1,310	5,580	7,870	7,610	9,190	4,600	8,380	2,510	3,440
5	1,150	3,020	1,310	1,310	5,920	8,390	5,240	8,390	3,790	4,770	2,300	2,790
6	1,470	2,470	1,390	1,310	6,150	8,390	5,240	7,360	4,190	4,010	2,060	2,560
7	1,390	2,300	1,390	1,310	6,630	8,920	4,810	6,870	3,790	3,750	2,270	4,230
8	1,150	2,130	1,470	1,310	7,630	12,800	4,190	6,390	3,400	3,420	1,860	4,270
9	1,050	2,040	1,630	1,310	8,130	14,300	4,810	5,920	3,400	2,630	1,930	3,500
10	1,150	1,880	1,710	1,310	10,000	12,500	30,700	5,460	3,590	2,510	1,660	3,610
11	1,470	1,960	1,470	1,310	13,400	9,730	44,100	5,460	3,400	2,330	2,440	3,590
12	1,470	1,790	1,470	1,310	33,700	8,920	35,200	5,690	3,210	2,100	3,440	3,380
13	1,310	1,630	1,470	1,310	37,400	10,800	30,700	5,690	2,650	2,270	2,980	3,040
14	1,310	1,470	1,550	1,470	39,000	19,600	27,500	6,630	2,300	2,490	3,440	2,630
15	1,470	1,630	1,310	1,790	39,700	48,600	20,000	7,110	2,040	2,160	2,610	2,330
16	1,310	1,630	1,310	2,650	34,000	32,200	20,000	6,870	1,880	1,760	1,980	2,100
17	1,050	1,630	1,310	6,390	30,700	25,400	21,600	6,390	1,550	1,710	1,660	1,860
18	1,230	1,630	1,310	5,920	26,400	13,700	19,600	5,920	1,630	2,580	1,880	1,940
19	1,150	1,630	1,470	5,240	23,700	10,800	16,100	5,690	1,960	3,210	2,300	2,590
20	2,300	1,390	1,790	4,600	17,100	8,650	15,200	5,020	2,130	2,380	2,560	3,400
21	2,130	1,230	1,960	4,190	14,000	9,460	38,600	4,600	2,300	3,690	2,270	7,140
22	1,790	1,630	1,960	3,790	13,700	20,600	50,700	4,810	5,460	3,120	2,670	7,900
23	1,630	1,630	1,960	3,400	12,800	23,700	44,900	5,020	3,790	2,790	2,790	5,510
24	2,130	1,470	1,960	3,210	12,500	21,300	29,300	5,460	3,210	2,610	2,870	4,270
25	2,650	1,470	1,960	3,020	10,800	17,700	20,600	8,390	2,470	2,160	2,270	3,830
26	1,960	1,470	1,960	3,020	9,460	14,900	18,000	7,110	6,390	1,990	1,960	3,170
27	1,790	1,310	1,880	3,210	9,730	12,500	14,900	5,240	9,190	2,060	1,490	2,680
28	2,130	1,230	1,790	3,590	14,000	10,800	15,800	4,810	22,000	2,160	1,600	2,470
29	2,300	1,390	1,710	3,790	....	9,460	15,500	5,920	11,700	2,440	1,600	2,270
30	3,990	1,470	1,630	3,990	....	8,390	14,300	8,920	10,000	2,400	1,930	2,030
31	11,100	....	1,470	4,390	....	8,130	....	7,360	....	2,870	2,380	....

NOTE.—Daily discharge Dec. 12 to Feb. 10 estimated because of ice from observer's notes, weather records, and comparison with records at other stations.

Monthly discharge of James River at Cartersville, Va., for the year ending Sept. 30, 1918.

[Drainage area, 6,230 square miles.]

Month.	Discharge in second-feet.				Run-off (depth in inches on drainage area).
	Maximum.	Minimum.	Mean.	Per square mile.	
October	11,100	1,020	1,920	0.308	0.36
November	5,460	1,230	2,080	.334	.37
December	1,930	1,230	1,600	.257	.30
January	6,390	1,310	2,730	.438	.50
February	39,700	4,600	16,300	2.62	2.73
March	48,600	7,870	14,600	2.34	2.70
April	50,700	4,190	20,000	3.21	3.58
May	16,400	4,600	7,150	1.15	1.33
June	22,000	1,550	4,790	.769	.86
July	13,100	1,710	3,670	.589	.68
August	3,440	1,490	2,320	.372	.43
September	7,900	1,860	3,370	.541	.60
The year	50,700	1,020	6,620	1.06	14.44

ROANOKE RIVER BASIN.

ROANOKE RIVER AT ROANOKE, VA.

LOCATION.—At Walnut Street highway bridge in Roanoke, Roanoke County.

DRAINAGE AREA.—388 square miles.

RECORDS AVAILABLE.—July 10, 1896, to July 15, 1906; May 7, 1907, to September 30, 1918.

GAGE.—Chain on downstream side of Walnut Street bridge; read by employees of Roanoke Railway & Electric Co. Wire gage used previous to November 28, 1903.

DISCHARGE MEASUREMENTS.—Made from downstream side of Walnut Street bridge or by wading.

CHANNEL AND CONTROL.—Bed composed of coarse gravel and small boulders. Banks may be overflowed at extreme flood stages. Control, loose boulders.

EXTREMES OF DISCHARGE.—Maximum stage recorded during year, 7.5 feet 8 a. m. June 26 (discharge not determined); minimum stage, 0.49 foot December 29–31 (affected by ice); minimum open-water stage, 0.55 foot November 18.

1896–1918: Maximum stage recorded, 14.34 feet August 6, 1901 (discharge, 16,900 second-feet); minimum stage recorded, zero, on morning of December 23, 1909, when flow was retarded by freezing.

ICE.—Stage-discharge relation seriously affected by ice during the winter of 1917–18.

ACCURACY.—Current-meter measurements indicate that stage-discharge relation changed during the year; affected by ice from about December 9 to February 1. Gage read to tenths or half-tenths once daily. Daily discharge not ascertained owing to lack of current-meter measurements to define change in stage-discharge relation.

COOPERATION.—Gage-height record furnished by Roanoke Railway & Electric Co. J. W. Hancock, general manager.

Discharge measurements of Roanoke River at Roanoke, Va., during the year ending Sept. 30, 1918.

Date.	Made by—	Gage height.	Discharge.
		Feet.	*Sec.-ft.*
May 28	Hopkins and Fiedler	1.96	607
28	....do	1.96	603

Daily gage height, in feet, of Roanoke River at Roanoke, Va., for the year ending Sept. 30, 1918.

Day.	Oct.	Nov.	Dec.	Feb.	Mar.	Apr.	May.	June.	July.	Aug.	Sept.
1	0.62	0.95	0.72	....	1.25	....	2.70	1.40	3.40	3.90	2.00
2	.62	.82	.72	....	1.25	1.35	2.40	1.30	2.30	2.40	1.70
3	.60	.69	.67	1.35	1.20	1.25	2.15	1.25	1.80	1.95	1.50
4	.59	.65	.67	1.35	1.15	1.25	2.00	1.30	1.60	1.70	1.35
5	.57	.65	.67	1.35	1.25	1.15	1.85	1.20	1.48	1.45	1.30
6	.57	.57	.67	1.35	1.40	1.05	1.75	1.12	1.40	1.35	1.35
7	.58	.67	.67	.95	1.85	1.15	1.65	1.15	1.30	1.30	1.50
8	.60	.67	.67	1.05	2.15	1.15	1.65	1.13	1.28	1.35	1.40
9	.61	.62	.97	1.15	1.90	1.85	1.60	1.10	1.28	1.20	2.45
10	.62	.60	.97	3.30	1.80	2.80	1.55	1.08	1.25	1.15	1.80
11	.70	.67	.97	3.45	1.70	3.40	1.55	1.02	1.20	1.10	1.60
12	.62	.62	.77	3.40	1.60	3.20	1.45	1.00	1.10	1.08	1.50
13	.57	.62	.57	4.05	1.75	2.95	1.45	1.00	1.30	1.05	1.40
14	.62	.62	.57	3.15	2.05	2.65	1.55	.97	1.22	1.20	1.30
15	.61	.62	.57	2.85	1.85	2.85	1.50	.95	1.10	1.05	1.24
16	.60	.62	.77	2.95	1.65	2.55	1.45	.92	1.00	1.30	1.18
17	.58	.62	.72	2.30	1.50	2.25	1.45	1.25	1.13	1.15	1.11
18	.59	.55	.72	1.95	1.45	2.05	1.37	1.50	1.50	1.15	2.50
19	.58	.62	.67	1.75	1.35	1.85	1.55	4.20	2.35	2.80	2.05
20	.65	.62	.57	1.75	1.30	1.85	1.65	2.05	1.85	2.00	1.60
21	.60	.62	.52	1.90	2.10	4.20	2.05	1.50	1.55	1.50	1.90
22	.59	.62	.52	1.70	2.75	4.00	4.40	1.42	1.40	1.35	1.75
23	.58	.62	.52	1.70	2.25	2.85	2.25	1.30	1.30	1.25	1.65
24	.68	.62	.51	1.55	2.35	2.45	1.95	1.18	1.25	1.15	1.45
25	.74	.62	.51	1.45	2.45	2.15	1.70	1.10	1.15	1.10	1.38
26	.67	.62	.51	1.40	2.15	1.95	1.60	7.50	1.85	1.05	1.32
27	.67	.62	.50	1.40	1.95	2.55	1.55	2.95	1.35	1.30	1.28
28	.67	.62	.50	1.35	1.80	2.45	2.30	2.30	1.30	2.35	1.22
29	.69	.62	.49	....	1.60	2.25	1.75	1.90	1.70	1.65	1.16
30	.97	.72	.49	....	....	2.05	1.80	2.05	1.40	1.40	1.12
31	1.07	....	.49	....	....	....	1.52	....	1.50	1.30	....

ROANOKE RIVER AT OLD GASTON, N. C.

Location.—At bridge of Roanoke Railway Co. at Old Gaston, Northampton County, about three-fourths mile below mouth of Indian Creek, 1¼ miles north of Thelma, and 2¼ miles above mouth of Deep Creek.

Drainage area.—8,350 square miles.

Records available.—December 7, 1911, to September 30, 1918.

Gage.—Chain gage attached to outside of guard timber on downstream side of second span from right end of deck railroad bridge; read by R. A. Howell.

Discharge measurements.—Made from downstream side of bridge to which gage is attached. Measuring section broken by 11 bridge piers.

Channel and control.—Channel fairly permanent; point of control, about a mile below gage, is of rock and probably permanent. Left bank subject to overflow in extreme floods, but a fair determination can be made of the overflow discharge around the bridge.

Extremes of discharge.—Maximum stage recorded during year, 10.7 feet in the morning of April 23 (discharge, 72,300 second-feet); minimum stage. 1.0 foot October 6 (discharge, 900 second-feet).

1911–1918: Maximum stage recorded, 16.6 feet at 7 a, m. March 18, 1912 (discharge, 210,000 second-feet); minimum stage, 0.95 foot at 6 a. m. October 1. 1914 (discharge, 790 second-feet).

Ice.—Ice formed to considerable thickness at this station during the winter of 1917–18 and the stage-discharge relation was seriously affected.

Regulation.—During periods of low water there are variations in flow, probably due to weekly (Sunday) shutdown of large power plants farther up stream. These variations are observable at power plants at Roanoke Rapids and Weldon on Tuesdays or Wednesdays.

ACCURACY.—Stage-discharge relation practically permanent; affected by ice from December 12 to January 29. Rating curve well defined below 33,300 second-feet, and fairly well defined to 180,000 second-feet. Gage read to tenths once daily. Daily discharge ascertained by applying daily gage height to rating table. Records good for open water periods and fair for periods of ice effect.

The following discharge measurement was made by B. J. Peterson and A. G. Fiedler: June 21, 1918: Gage height, 2.36 feet; discharge, 4,980 second-feet.

Daily discharge, in second-feet, of Roanoke River at Old Gaston, N. C., for the year ending Sept. 30, 1918.

Day.	Oct.	Nov.	Dec.	Jan.	Feb.	Mar.	Apr.	May.	June.	July.	Aug.	Sept.
1	2,160	11,900	4,430	900	28,200	6,240	6,240	11,400	3,740	5,140	8,210	3,740
2	2,160	9,060	4,080	900	21,400	5,500	4,780	17,200	5,500	4,430	7,010	5,780
3	1,620	5,870	3,740	900	15,300	4,430	4,430	13,000	5,140	3,410	7,800	2,770
4	2,160	3,740	6,620	900	13,000	4,430	5,140	11,400	3,740	2,160	7,400	3,410
5	1,130	3,410	5,870	900	11,400	5,500	6,240	10,400	3,410	3,090	5,140	2,770
6	900	3,090	3,410	900	10,900	7,010	5,870	9,060	4,080	2,770	3,410	2,160
7	1,880	2,770	3,090	1,130	9,500	7,010	5,500	7,010	3,740	4,430	2,770	2,460
8	1,370	2,160	3,410	1,620	11,400	7,400	6,240	6,620	4,430	3,740	2,160	9,060
9	2,160	1,880	1,620	1,620	32,400	10,400	7,010	5,500	3,410	3,410	2,460	5,870
10	2,460	1,620	1,620	1,880	19,900	9,960	34,200	4,780	2,770	2,770	2,160	4,780
11	3,090	3,410	1,370	2,460	14,700	9,500	49,300	7,010	3,090	2,160	1,620	5,500
12	2,770	2,770	1,250	2,770	14,700	8,210	47,300	6,240	2,460	2,460	1,370	4,780
13	3,090	2,160	1,370	18,500	14,200	7,010	36,000	8,630	3,090	3,090	3,740	4,430
14	3,410	1,880	900	19,900	11,400	6,240	22,800	16,600	4,430	6,240	5,500	4,080
15	2,460	1,620	900	26,600	11,900	5,870	17,200	24,300	3,410	3,740	3,740	3,410
16	2,160	1,370	900	18,500	10,900	5,500	8,630	14,700	3,090	3,410	3,410	2,160
17	1,620	1,370	900	13,600	9,500	5,140	7,400	10,400	2,770	2,460	3,090	2,770
18	2,160	2,460	900	10,900	8,630	6,240	7,010	9,060	2,160	2,770	3,410	2,160
19	1,620	2,160	2,160	8,630	8,210	5,500	6,240	8,210	1,880	3,090	4,780	1,620
20	2,770	2,160	3,090	8,210	7,400	5,140	5,870	3,740	2,160	6,240	9,500	1,370
21	3,090	3,090	3,090	7,400	7,010	4,780	22,800	5,500	4,780	11,400	8,630	5,140
22	2,770	2,460	3,090	4,430	7,400	4,430	60,800	5,140	3,410	9,060	7,400	9,960
23	2,460	2,160	1,620	4,430	7,800	7,010	72,300	4,780	4,080	7,400	5,870	7,010
24	3,410	2,460	1,370	4,080	7,010	9,060	70,800	5,140	3,740	4,430	3,740	5,500
25	3,090	2,160	1,370	4,780	5,140	7,400	29,000	9,500	3,090	3,090	3,410	4,780
26	2,460	2,770	1,130	4,430	6,620	9,960	13,000	11,900	4,430	3,410	2,460	3,740
27	2,160	3,090	1,130	4,780	6,240	9,060	15,900	9,960	11,400	3,740	2,160	3,410
28	1,370	3,090	900	9,960	7,010	9,060	22,800	10,900	13,000	3,740	3,410	3,740
29	1,620	3,410	900	18,500	....	8,210	17,200	8,210	11,900	3,090	4,780	3,410
30	4,780	3,740	900	37,800	....	7,010	13,600	7,010	11,400	3,410	4,430	2,770
31	8,210	....	900	38,800	....	6,620	....	4,430	....	6,620	6,240	....

NOTE.—Discharge estimated, because of ice, as in table for Dec. 12–13, 18–28, 30–31, and Jan. 1–29, from observer's notes, weather records, and comparison with records at other stations.

Monthly discharge of Roanoke River at Old Gaston, N. C., for the year ending Sept. 30, 1918.

[Drainage area, 8,350 square miles.]

Month.	Discharge in second-feet.				Run-off (depth in inches on drainage area).
	Maximum.	Minimum.	Mean.	Per square mile.	
October	8,210	900	2,530	0.303	0.35
November	11,900	1,370	3,180	.381	.43
December	6,620	900	2,190	.262	.30
January	38,800	900	9,070	1.09	1.26
February	32,400	5,140	12,100	1.45	1.51
March	10,400	4,430	6,930	.830	.96
April	72,300	4,430	21,100	2.53	2.82
May	24,300	3,740	9,280	1.11	1.28
June	13,000	1,880	4,660	.558	.62
July	11,400	2,160	4,210	.504	.58
August	9,500	1,370	4,500	.516	.63
September	9,960	1,370	4,150	.497	.55
The year	72,300	900	6,940	.831	11.29

PEEDEE RIVER BASIN.

YADKIN RIVER AT DONNAHA, N. C.

LOCATION.—At toll bridge in Donnaha, Forsyth County, on road between Donnaha and East Bend, a quarter of a mile west of Donnaha railroad station, 6 miles downstream from Ararat River, which enters from the left, and 60 miles upstream from gaging station at Salisbury, N. C.

DRAINAGE AREA.—1,600 square miles.

RECORDS AVAILABLE.—April 11, 1913, to September 30, 1918.

GAGE.—Vertical gage in four sections on left bank, 150 feet downstream from left end of toll bridge; read twice daily to tenths by J. F. Goolsby. Section of gage below 10 feet was carried away by ice in February, 1918. Gage heights below 10 feet, after gage went out, obtained by measuring down from 12.5-foot mark on gage.

DISCHARGE MEASUREMENTS.—Prior to flood in July, 1916, measurements were made from the toll bridge; bridge washed out in July, 1916; no measurements after that date.

CHANNEL AND CONTROL.—Bed composed of sand and bedrock; probably permanent. Current slightly obstructed by two old steel trusses lying about 150 and 400 feet, respectively, below bridge; obstruction probably permanent. Control is a rock ledge extending across river and forming a shoal about 450 feet below gage.

EXTREMES OF STAGE.—Maximum stage recorded during year, 10.4 feet at 8 a. m. April 29 (discharge not determined); minimum stage recorded, 5.0 feet at 4 p. m. January 6 and 8 a. m. and 5 p. m. January 7 and 8 (discharge not determined).

1913–1918: Maximum stage recorded, 40.0 feet at 8 a. m. July 16, 1916, determined by observer, who measured from flood marks down to water surface at a lower stage (discharge not determined); minimum stage, 4.65 feet at 4 p. m. September 30, 1914 (discharge, 678 second-feet).

ICE.—Never enough to affect stage-discharge relation.

DIVERSIONS.—None.

REGULATION.—None, except for a few small mill dams on tributaries.

Data inadequate for determination of discharge.

No discharge measurements were made at this station during the year.

Daily gage height, in feet, of Yadkin River at Donnaha, N. C., for the year ending Sept. 30, 1918.

Day.	Oct.	Nov.	Dec.	Jan.	Feb.	Mar.	Apr.	May.	June.	July.	Aug.	Sept.
1	5.2	5.1	5.2	5.3	5.6	5.4	5.3	6.2	5.4	5.4	6.5	5.9
2	5.2	5.1	5.2	5.3	5.6	5.4	5.3	5.7	5.4	5.3	6.4	5.8
3	5.2	5.1	5.1	5.2	5.8	5.4	5.3	5.4	5.4	5.3	5.8	5.6
4	5.1	5.1	5.2	5.2	5.8	5.4	5.3	5.4	5.4	5.3	5.6	5.5
5	5.2	5.1	5.2	5.1	5.8	5.8	5.4	5.4	5.3	5.3	5.5	5.4
6	5.2	5.1	5.1	5.0	5.6	5.8	5.4	5.4	5.3	5.4	5.4	5.5
7	5.2	5.1	5.1	5.0	5.8	5.6	5.4	5.4	5.3	5.4	5.4	5.5
8	5.2	5.1	5.1	5.0	5.8	5.5	5.4	5.4	5.3	5.4	5.4	5.6
9	5.1	5.1	5.2	5.2	5.6	5.8	5.3	5.4	5.3	5.4	5.4	5.5
10	5.2	5.1	5.2	5.2	5.8	6.2	5.3	5.4	5.3	5.4	6.1	5.6
11	5.2	5.2	5.4	5.6	6.0	6.0	5.3	5.4	5.3	5.3	7.0	6.0
12	5.2	5.8	5.4	9.0	5.7	5.8	5.3	5.4	5.2	5.3	7.2	5.8
13	5.2	8.5	5.5	9.6	5.6	5.8	5.4	5.4	5.2	5.3	8.0	5.6
14	6.2	7.8	5.4	8.7	5.6	5.6	5.4	5.6	5.2	5.3	6.8	6.6
15	6.5	6.2	5.4	7.9	5.8	5.4	5.4	6.2	5.2	5.3	6.1	6.4
16	5.8	5.6	5.6	7.4	5.6	5.4	5.4	5.6	6.0	5.3	5.6	6.0
17	5.4	5.5	5.6	6.8	5.6	5.4	5.4	5.5	6.9	5.3	5.6	5.8
18	5.4	5.4	5.5	6.6	5.6	5.4	5.4	5.4	9.5	5.3	5.4	5.6
19	5.3	5.4	5.4	6.4	5.5	5.4	5.3	5.4	8.0	5.3	5.4	5.6
20	5.3	5.3	5.4	6.4	5.6	5.4	5.3	5.4	6.6	5.3	5.4	5.6
21	5.3	5.2	5.4	6.2	5.5	5.4	5.3	5.4	5.8	5.3	5.4	5.6
22	5.2	5.2	5.4	6.0	5.5	5.3	5.4	5.4	5.8	5.3	5.3	5.6
23	5.2	5.2	5.5	5.8	5.4	5.3	5.4	5.4	5.6	5.4	5.4	5.5
24	5.2	5.2	5.6	5.6	5.4	5.3	5.5	5.4	5.5	6.2	5.6	5.5
25	5.2	5.2	5.6	5.8	5.4	5.4	6.0	5.4	5.4	8.8	5.5	5.4
26	5.2	5.2	5.6	5.6	5.4	5.4	6.6	5.4	5.4	7.5	5.5	5.4
27	5.2	5.2	5.6	5.6	5.4	5.4	5.8	5.4	5.4	6.2	5.5	5.4
28	5.2	5.2	5.6	5.6	5.4	5.4	6.9	5.4	5.4	6.9	5.6	5.4
29	5.2	5.2	5.6	5.4	....	5.4	10.0	5.4	5.4	6.4	5.6	5.4
30	5.2	5.2	5.6	5.6	....	5.4	8.1	5.4	5.3	7.6	6.0	5.4
31	5.1	....	5.4	5.7	....	5.4	....	5.4	....	6.9	6.2	....

NOTE.—Gage heights after February, 1918, when gage below 10 feet was carried out by ice, obtained by measuring down from 12.5-foot mark; may be somewhat in error.

YADKIN RIVER NEAR SALISBURY, N. C.

LOCATION.—At highway bridge known as Piedmont toll bridge, 1,000 feet upstream from Southern Railway bridge, 4 miles east of Spencer, 5 miles downstream from mouth of South Yadkin River, 6 miles east of Salisbury, Rowan County, and 26 miles upstream from American Aluminum Co.'s hydroelectric plant near Whitney, N. C.

DRAINAGE AREA.—3,400 square miles.

RECORDS AVAILABLE.—September 24, 1895, to December 31, 1909; September 1, 1911, to September 30, 1918.

GAGE.—Chain gage attached to highway bridge; read by J. T. Yarbrough. From the date of establishment to May 31, 1899, the gage was at the Southern Railway bridge, and from the latter date it was at the highway bridge until moved back to the railroad bridge early in 1903, where it remained until the end of 1905. Since January 1, 1906, the gage has been at the highway bridge at the datum originally established there in 1899. The last gage at the railroad bridge read the same as the gage at the highway bridge at gage height 3.2 feet, but not for higher and lower stages. Datum of the original gage at the railroad bridge somewhat uncertain.

DISCHARGE MEASUREMENTS.—Made from highway bridge. During the time that gage was at railroad bridge most of the measurements were made from that bridge. During flood of July, 1916, water rose over floor of highway bridge, making it necessary to use railroad bridge.

CHANNEL AND CONTROL.—Channel wide; bed rather rough. Control is a rock ledge about 500 feet below bridge extending entirely across river.

EXTREMES OF DISCHARGE.—Maximum stage recorded during year, 7.55 feet at 7 a. m. April 22 (discharge, 24,300 second-feet); minimum stage recorded, 1.75 feet at 7 a. m. Dec. 13 (discharge, 1,250 second-feet).

1895–1909; 1911–1918: Maximum stage recorded, 23.8 feet at 1 a. m July 18, 1916 (discharge, 121,000 second-feet); minimum stage, 1.2 feet September 20, October 5, November 22 and 26, 1897 (discharge, 900 second-feet).

ICE.—Never enough to affect stage-discharge relation.

DIVERSIONS.—None.

REGULATION.—Flow during low stages may be slightly affected by developed powers on the river and tributaries above.

ACCURACY.—Stage-discharge relation practically permanent. Rating curve well defined below 20,000 second-feet and fairly well defined between 20,000 and 121,000 second-feet. Gage read to half-tenths twice daily; during high water read oftener. Daily discharge ascertained by applying mean daily gage height to rating table. Records good.

Discharge measurements of Yadkin River near Salisbury, N. C., during the year ending Sept. 30, 1918.

Date.	Made by—	Gage height.	Discharge.
		Feet.	*Sec.-ft.*
Feb. 28	L. J. Hall	2.46	3,030
May 26	C. G. Paulsen	3.27	5,240

Daily discharge, in second-feet, of Yadkin River near Salisbury, N. C., for the year ending Sept. 30, 1918.

Day.	Oct.	Nov.	Dec.	Jan.	Feb.	Mar.	Apr.	May.	June.	July.	Aug.	Sept.
1	2,290	3,790	2,290	1,630	10,000	2,660	2,410	5,050	2,170	2,060	6,100	3,210
2	1,840	2,800	2,060	1,630	7,220	2,540	2,290	4,720	2,170	2,170	8,000	2,930
3	1,730	2,170	2,170	1,840	6,100	2,660	2,290	4,090	2,170	2,410	3,790	1,940
4	1,630	2,170	1,940	1,940	6,100	2,660	2,290	3,500	1,940	1,730	2,660	1,730
5	1,840	2,290	1,940	2,060	5,390	2,800	2,290	3,360	2,170	1,730	2,410	1,630
6	1,730	2,060	2,060	2,060	4,400	2,800	2,290	3,070	2,170	1,530	2,060	1,730
7	1,840	2,060	1,940	2,660	4,090	2,800	2,060	2,930	2,170	1,530	1,940	2,800
8	2,060	2,060	2,060	2,660	4,400	2,540	2,540	2,800	2,540	1,730	1,730	3,500
9	1,630	2,060	2,290	2,800	4,090	2,540	6,460	3,790	2,170	1,940	1,940	3,790
10	2,170	2,060	2,290	2,540	4,720	2,800	10,000	3,500	2,060	2,290	2,660	3,210
11	2,660	1,840	1,630	2,290	4,400	3.070	6,100	3,070	2,170	2,060	2,170	2,660
12	2,170	2,170	1,840	12,000	4,400	2,660	4,720	2,800	1,940	1,730	3,070	2,170
13	1,840	1,840	1,630	14,700	3,790	2,540	3,500	3,070	1,730	1,630	2,410	1,940
14	1,840	2,170	1,730	8,800	4,090	2,540	3,370	5,740	1,730	1,530	2,060	1,730
15	2,060	2,170	1,940	7,220	3,500	2,540	3,210	5,390	1,730	1,630	1,730	1,630
16	1,730	2,060	1,840	9,200	3,210	2,290	2,930	3,790	1,530	1,530	4,400	1,730
17	1,840	1,840	2,170	7,220	3,790	2,290	3,360	3,210	2,170	1,530	2,660	1,630
18	1,840	1,840	2,170	4,720	3,790	2,800	3,070	3,210	4,400	1,940	6,840	1,730
19	1,840	2,060	2,410	4,090	3,210	2,540	3,360	2,540	3,500	4,400	14,700	2,170
20	2,170	2,660	2,410	3,210	3,500	2,540	10,400	2,800	2,800	3,500	8,800	3,790
21	4,400	1,940	2,290	2,930	4,090	3,360	20,500	3,360	2,540	1,940	4,090	5,390
22	2,800	1,940	2,410	2,540	3,500	5,050	22,000	3,210	2,410	1,940	2,540	4,400
23	2,060	1,840	2,800	2,800	3,070	4,400	10,000	3,360	2,410	1,730	2,170	2,660
24	1,940	1,840	2,540	2,930	3,070	3,790	6,460	3,070	2,660	4,400	1,940	2,060
25	1,840	1,840	2,290	2,800	3,210	4,090	4,720	3,360	2,170	3,360	1,730	1,940
26	2,060	1,940	2,290	3,360	2,800	3,790	5,050	6,100	2,170	3,070	2,170	1,730
27	1,840	1,730	2,540	5,740	2,930	3,070	4,090	5,050	2,930	3,500	2,170	2,800
28	1,940	1,840	2,290	10,800	2,800	2,660	4,400	3,070	2,410	3,070	2,410	3,070
29	2,060	2,290	2,540	14,700	....	2,660	4,090	2,660	1,940	4,090	3,500	2,170
30	2,660	1,940	1,630	18,000	....	2,540	4,090	2,660	1,730	4,720	2,660	2,060
31	5,740	....	1,340	16,000	....	2,290	....	2,410	....	5,050	2,060	....

Monthly discharge of Yadkin River near Salisbury, N. C., for the year ending Sept. 30, 1918.

[Drainage area, 3,400 square miles.]

Month.	Discharge in second-feet.				Run-off (depth in inches on drainage area).
	Maximum.	Minimum.	Mean.	Per square mile.	
October	5,740	1,630	2,200	0.647	0.75
November	3,790	1,730	2,110	.621	.69
December	2,800	1,340	2,120	.624	.72
January	18,000	1,630	5,740	1.69	1.95
February	10,000	2,800	4,270	1.26	1.31
March	5,050	2,290	2,910	.856	.99
April	22,000	2,060	5,480	1.61	1.80
May	6,100	2,410	3,570	1.05	1.21
June	4,400	1,530	2,290	.674	.75
July	5,050	1,530	2,500	.735	.85
August	14,700	1,730	3,530	1.04	1.20
September	5,390	1,630	2,530	.744	.83
The year	22,000	1,340	3,270	.962	13.05

SANTEE RIVER BASIN.

CATAWBA RIVER AT RHODHISS, N. C.

LOCATION.—At new highway bridge 1,000 feet below dam of Rhodhiss Manufacturing Co., 1 mile from Carolina & North Western Railroad station in Rhodhiss, Caldwell County. The tailrace of the company's cotton mills empties into river 300 feet upstream from gage.

DRAINAGE AREA.—1,180 square miles (determined by Rhodhiss Manufacturing Co.).

RECORDS AVAILABLE.—April 13, 1917, to September 30, 1918.

GAGE.—Chain gage attached to upstream side of highway bridge; read by H. C. Cobb.

DISCHARGE MEASUREMENTS.—Made from the bridge.

CHANNEL AND CONTROL.—Bed composed of rock; probably permanent.

EXTREMES OF DISCHARGE.—Maximum stage recorded during year, 5.8 feet at 8.30 a. m. January 29 (discharge, 10,100 second-feet); minimum stage recorded, 1.3 feet at 9.30 a. m. December 30 (discharge, 600 second-feet).

1917–1918: Maximum stage recorded, 8.58 feet at 7 a. m. September 1, 1917 (discharge, 18,800 second-feet); minimum stage, that of January 29, 1918.

ICE.—Stage-discharge relation not affected by ice.

REGULATION.—Slight fluctuations at low stages caused by operation of power plant of the Rhodhiss Manufacturing Co.

ACCURACY.—Stage-discharge relation probably permanent. Rating curve fairly well defined between 700 and 1,300 second-feet and well defined between 1,300 and 10,000 second-feet; extended above 10,000 second-feet. Gage read to half-tenths twice daily. Daily discharge ascertained by applying mean daily gage height to rating table. Records good except those below 1,000 second-feet which are subject to error owing to regulation caused by operation of power plant, and those above 10,000 second-feet, which are fair.

Discharge measurements of Catawba River at Rhodhiss, N. C., during the year ending Sept. 30, 1918.

Date.	Made by—	Gage height.	Discharge.
		Feet.	*Sec.-ft.*
Apr. 9	Babb and Hollar	5.32	8,540
May 27	C. G. Paulsen	2.59	2,130
Sept. 14	C. C. Babb	1.64	891

Daily discharge, in second-feet, of Catawba River at Rhodhiss, N. C., for the year ending Sept. 30, 1918.

Day.	Oct.	Nov.	Dec.	Jan.	Feb.	Mar.	Apr.	May.	June.	July.	Aug.	Sept.
1	1,160	1,160	900	858	4,830	1,330	1,210	2,110	1,330	4,590	2,110	950
2	1,050	1,050	815	1,050	3,880	1,610	1,270	1,690	1,270	2,290	1,770	858
3	950	1,050	900	1,050	3,660	1,330	1,210	1,610	1,210	1,540	1,210	815
4	950	1,020	900	1,000	3,440	1,210	1,160	1,330	1,160	1,330	1,160	815
5	950	1,000	900	1,900	2,470	1,270	1,160	1,270	1,330	1,330	1,000	950
6	1,000	1,050	900	1,050	2,470	1,210	1,100	1,400	1,210	1,100	950	1,000
7	858	1,000	858	1,540	2,110	1,210	1,000	1,330	1,460	1,100	1,000	1,000
8	900	950	705	1,330	2,020	1,270	4,350	1,460	1,400	1,100	1,210	1,160
9	900	950	705	1,160	2,110	1,210	7,980	1,770	1,280	1,270	1,620	1,610
10	900	950	705	1,050	2,110	1,400	3,880	1,610	1,160	1,000	2,020	1,210
11	900	858	705	1,100	2,290	1,270	2,840	1,270	1,210	950	2,020	1,050
12	1,050	950	858	6,360	2,020	1,210	2,290	1,210	1,100	950	2,020	815
13	858	1,100	930	2,470	1,860	1,160	1,860	1,270	1,210	950	1,050	900
14	778	1,210	1,000	2,110	1,770	1,210	1,860	2,840	1,100	858	900	815
15	900	1,000	1,000	2,290	1,610	1,160	1,770	2,110	1,050	950	900	815
16	900	1,000	778	2,110	1,940	1,100	1,540	1,610	975	815	900	815
17	900	1,000	1,000	1,860	2,110	1,100	1,610	1,540	900	900	1,380	815
18	950	950	1,000	1,690	1,860	1,210	1,540	1,460	1,690	900	1,860	1,770
19	1,460	1,000	1,050	1,400	1,770	1,160	1,610	1,540	1,270	1,000	3,230	1,400
20	4,350	950	1,000	1,610	2,110	1,210	2,290	1,690	1,460	950	1,770	1,770
21	1,770	900	1,100	1,330	2,290	1,610	4,590	2,290	1,400	975	1,270	1,270
22	1,460	950	1,000	1,330	2,110	1,270	3,660	2,290	2,840	1,000	1,050	1,140
23	1,210	950	950	1,270	1,860	1,210	2,650	2,290	2,120	1,000	1,000	1,000
24	1,000	950	900	1,400	1,770	2,110	2,290	2,110	1,400	1,460	900	1,000
25	1,000	858	950	1,330	1,610	1,860	2,020	2,020	1,460	1,460	925	858
26	950	900	1,160	1,540	1,540	1,540	1,940	1,940	1,400	1,460	950	858
27	950	858	1,270	1,460	1,470	1,400	1,860	2,110	1,160	1,330	1,100	1,460
28	900	815	1,100	6,620	1,400	1,270	1,690	1,940	1,160	1,720	1,160	1,100
29	1,000	900	1,000	9,160	....	1,270	1,610	1,610	1,160	2,110	1,100	858
30	2,290	950	600	9,160	....	1,210	1,610	1,460	2,880	2,470	1,050	858
31	1,270	....	900	7,700	....	1,210	....	1,460	....	2,470	1,050	....

NOTE.—Discharge interpolated for the following days: Nov. 4; Dec. 8, 13; Feb. 26, 27; June 2, 9, 16, 23, 30; July 7, 21, 28; Aug. 9, 11, 17, 25; Sept. 1, 15, and 22. Accuracy of records for the following days affected to some extent by regulation above gage: Oct. 1-20; Nov. 19-30; Dec. 1-31; Jan. 1-11.

Monthly discharge of Catawba River at Rhodhiss, N. C., for the year ending Sept. 30, 1918.

[Drainage area, 1,180 square miles.]

Month.	Discharge in second-feet.				Run-off (depth in inches on drainage area).
	Maximum.	Minimum.	Mean.	Per square mile.	
October	4,350	778	1,180	1.00	1.15
November	1,210	815	974	.825	.92
December	1,270	600	921	.781	.90
January	9,160	858	2,460	2.08	2.40
February	4,830	1,400	2,230	1.89	1.97
March	2,110	1,100	1,320	1.12	1.29
April	7,980	1,000	2,250	1.91	2.13
May	2,840	1,210	1,730	1.47	1.70
June	2,880	900	1,390	1.18	1.32
July	4,590	815	1,400	1.19	1.37
August	3,230	900	1,340	1.14	1.31
September	1,770	815	1,060	.898	1.00
The year	9,160	600	1,520	1.29	17.46

SAVANNAH RIVER BASIN.

CHATTOOGA RIVER NEAR TALLULAH FALLS, GA.

Location.—About 300 feet above mouth of Camp Creek, 5½ miles above junction with Tallulah River and 8 miles east of Tallulah Falls, Rabun County.

Drainage area.—256 square miles (measured on topographic maps).

Records available.—January 1, 1917, to January 28, 1918; September 25–30, 1918.

Gage.—Gurley 7-day recording gage installed on right bank August 17, 1917. On the same date a new vertical staff gage was installed about 30 feet upstream to which all recording gage records are referred. Prior to August 17, 1917, readings were taken from an old vertical staff gage at same location as new staff gage and set at same datum. Gage read by employees of Georgia Railway & Power Co.

Discharge measurements.—Made from cable at gage.

Channel and control.—Section under cable may shift somewhat but stage-discharge relation is kept permanent by a solid rock shoal about 100 feet below gage.

Extremes of discharge.—Maximum mean daily stage recorded during year, 4.44 feet January 28 (discharge, 2,690 second-feet); minimum mean daily stage recorded, 0.78 foot September 26 and 30 (discharge, 313 second-feet).

1917–1918: Maximum mean daily stage recorded January 1, 1917, to January 28, 1918, and September 25–30, 1918, 12.2 feet March 24, 1917 (discharge, about 12,000 second-feet); minimum mean daily stage recorded, 0.78 foot September 26 and 30, 1918 (discharge, 313 second-feet).

Ice.—Stage-discharge relation not affected by ice.

Accuracy.—Stage-discharge relation permanent. Rating curve well defined between 280 and 2,000 second-feet. Operation of recording gage satisfactory except for the period January 29 to September 24 for which there is no record owing to the gage well having been partly filled with sand. Daily discharge ascertained by applying to rating table mean daily gage height obtained by inspecting gage-height graph. Records excellent.

Cooperation.—Gage-height record furnished by Georgia Railway & Power Co.

Discharge measurements of Chattooga River near Tallulah Falls, Ga., during the year ending Sept. 30, 1918.

Date.	Made by—	Gage height.	Discharge.
		Feet.	*Sec.-ft.*
Oct. 11	Nelson and Wills [a]	0.98	383
May 9	Paulsen and Condron	1.57	642
Aug. 23	C. G. Paulsen	.80	321

[a] Employees of Georgia Railway & Power Co.

Daily discharge, in second-feet, of Chattooga River near Tallulah Falls, Ga., for the year ending Sept. 30, 1918.

Day.	Oct.	Nov.	Dec.	Jan.	Sept.	Day.	Oct.	Nov.	Dec.	Jan.	Sept.
1	438	510	386	358		16	376	422	406	840	
2	410	479	386	358		17	376	414	414	705	
3	394	470	386	351		18	376	414	422	605	
4	390	466	390	347		19	1,660	414	418	551	
5	390	462	390	347		20	1,120	418	406	551	
	383	450	390	361		21	755	410	406	502	
	383	446	390	458		22	630	398	406	515	
8	379	442	383	383		23	569	394	402	488	
9	410	438	394	354		24	520	383	394	488	
10	430	430	383	347		25	510	383	386	502	317
11	390	438	383	556		26	497	383	383	533	313
12	383	438	410	1,400		27	497	383	376	705	379
13	376	458	406	705		28	497	383	368	2,690	344
14	376	506	462	454		29	528	386	361		317
15	376	442	430	1,020		30	680	386	361		313
						31	578		361		

Note.—No gage-height record Jan. 29 to Sept. 24.

Monthly discharge of Chattooga River near Tallulah Falls, Ga., for the year ending Sept. 30, 1918.

[Drainage area, 256 square miles.]

Month.	Discharge in second-feet.				Run-off (depth in inches on drainage area).
	Maximum.	Minimum.	Mean.	Per square mile.	
October	1,660	376	519	2.03	2.34
November	510	383	428	1.67	1.86
December	462	361	395	1.54	1.78
January 1-28	2,690	347	624	2.44	2.54
September 25-30	379	313	330	1.29	0.29

TALLULAH RIVER NEAR SEED, GA.

Location.—One-fourth mile upstream from head of Rabun Lake, 1 mile downstream from Bridge Creek, 5 miles north of Seed, Rabun County, 6 miles due west of Lakemont railroad station, and 10 miles upstream from Rabun (Mathis) dam.

Drainage area.—127 square miles (measured on topographic maps).

Records available.—January 6, 1916, to September 30, 1918.

Gage.—A staff gage in three sections on right bank; read by employees of Georgia Railway & Power Co.

Discharge measurements.—Made from cable and car about 200 feet upstream for low and medium stages. Flood measurements made from suspension footbridge 1 mile downstream from gage.

Channel and control.—Bed composed of rock, sand, and gravel; rather rough, but permanent. Control is a ledge, which extends across river and over which water drops sharply, about 250 feet downstream from gage; probably permanent. Point of zero flow, gage height −0.5 foot.

Extremes of discharge.—Maximum daily stage recorded during year, 4.21 feet January 28 (discharge, 3,020 second-feet); minimum daily stage recorded, 0.68 foot August 18 (discharge, 70 second-feet).

1916–1918: Maximum stage recorded, 8.2 feet at 6 p. m. July 9, 1916 (discharge, 8,010 second-feet); minimum mean daily stage recorded, that of August 18, 1918.

Ice.—Never enough to affect stage-discharge relation.

Accuracy.—Stage-discharge relation permanent. Rating curve well defined between 100 and 5,500 second-feet. Gage read to hundredths three times daily. Daily discharge ascertained by applying mean daily gage height to rating table. Records good.

Discharge measurements of Tallulah River near Seed, Ga., during the year ending Sept. 30, 1918.

Date.	Made by—	Gage height.	Discharge.
		Feet.	*Sec.-ft.*
Oct. 12	Wills [a] and Nelson [a]	1.04	174
May 9	Paulsen and Condron	1.44	365
Aug. 24	C. G. Paulsen	.88	119

[a] Employees of Georgia Railway & Power Co.

Daily discharge, in second-feet, of Tallulah River near Seed, Ga., for the year ending Sept. 30, 1918.

Day.	Oct.	Nov.	Dec.	Jan.	Feb.	Mar.	Apr.	May.	June.	July.	Aug.	Sept.
1	250	286	192	192	825	400	316	495	250	346	183	112
2	225	274	183	183	745	388	322	430	240	235	230	109
3	215	250	175	187	705	376	298	406	240	210	187	127
4	200	240	192	179	595	394	316	382	304	196	162	121
5	200	230	179	183	528	382	292	364	262	192	155	205
6	192	225	175	322	495	418	274	352	316	183	144	144
7	179	220	171	328	495	430	462	340	280	171	138	388
8	179	210	230	256	462	382	1,080	462	245	171	134	200
9	262	205	179	210	462	364	668	388	250	162	134	235
10	200	200	175	205	462	495	528	630	240	158	138	166
11	187	200	175	240	418	382	462	430	268	151	196	144
12	175	210	200	668	418	364	462	406	240	148	138	134
13	171	210	196	382	406	370	406	400	280	148	127	131
14	166	286	220	382	394	358	388	400	230	141	121	118
15	166	225	200	668	400	340	370	364	205	141	115	118
16	162	210	200	406	560	322	400	346	200	141	124	115
17	158	205	210	340	630	334	394	340	200	138	171	112
18	166	200	205	304	495	316	495	334	280	192	70	245
19	1,220	196	215	280	668	316	412	364	240	220	158	148
20	495	200	205	280	430	495	406	358	210	187	138	322
21	352	205	200	250	668	412	462	322	352	220	151	196
22	298	192	200	286	560	370	400	352	316	210	118	155
23	268	183	192	250	495	352	370	376	235	205	112	141
24	240	183	183	256	495	376	364	370	215	179	115	131
25	230	179	183	262	462	352	358	340	205	245	121	124
26	220	179	215	280	462	328	462	322	256	215	121	124
27	215	179	200	668	418	322	382	304	225	230	112	148
28	210	179	187	3,020	412	316	376	286	215	192	121	131
29	220	210	187	1,600	....	304	394	280	215	205	134	144
30	495	200	138	1,600	....	298	560	262	286	210	118	151
31	322	....	162	1,310	....	292	....	256	....	200	115	....

Monthly discharge of Tallulah River near Seed, Ga., for the year ending Sept. 30, 1918.

[Drainage area, 127 square miles.]

Month.	Discharge in second-feet.				Run-off (depth in inches on drainage area).
	Maximum.	Minimum.	Mean.	Per square mile.	
October	1,220	158	266	2.09	2.41
November	286	179	212	1.67	1.86
December	230	138	191	1.50	1.73
January	3,020	179	515	4.06	4.68
February	825	394	520	4.09	4.26
March	495	292	366	2.88	3.32
April	1,080	274	429	3.38	3.77
May	630	256	370	2.91	3.36
June	352	200	250	1.97	2.20
July	346	138	192	1.51	1.74
August	230	70	139	1.09	1.26
September	388	109	161	1.27	1.42
The year	3,020	70	300	2.36	32.01

TALLULAH RIVER NEAR LAKEMONT, GA.

Location.—One-fourth mile downstream from Rabun dam (originally called Mathis dam), 1 mile upstream from mouth of Tiger Creek, and 1½ miles from Lakemont, Rabun County.

Drainage area.—Not measured.

Records available.—January 13, 1916, to September 30, 1918.

Gage.—A Barrett & Lawrence water-stage recorder, with 10-foot range of stage, at rock-filled log crib, originally a bridge abutment, on left bank of river; referred to vertical staff gage 20 feet upstream.

Discharge measurements.—Made from cable 5 feet downstream from gage.

Channel and control.—Bed rough and rocky, necessitating careful work in making discharge measurements. Control is a rock shoal 50 feet downstream from gage. Part of shoal is loose rock, and high water in last part of 1915 changed stage-discharge relation by changing the position of these rocks.

Extremes of discharge.—Maximum stage during year from water-stage recorder, 4.00 feet at 1.50 p. m. March 7 (discharge, 1,500 second-feet); minimum discharge, somewhat less than 5 second-feet, during periods when sluice gates in dam were closed.

1916–1918: Maximum stage recorded, 10.4 feet at 8.30 p. m. July 9, 1916 (discharge, 10,900 second-feet); minimum flow somewhat less than 5 second-feet at certain times when sluice gates at storage dam one-fourth mile upstream were shut and no water passed over crest of dam.

Ice.—Never enough to affect stage-discharge relation.

Diversions.—None.

Regulation.—The Rabun dam, one-fourth mile upstream, makes a very large reservoir which is used solely for storage in operating the great hydroelectric plant 7 miles downstream. Water is impounded or let loose at will of operators; consequently fluctuations are great, sudden, and frequent.

Accuracy.—Stage-discharge relation practically permanent. Rating curve well defined between 50 and 4,000 second-feet. Operation of water-stage recorder not entirely satisfactory on account of poor attention by observer. Daily discharge ascertained by use of discharge integrator. Records fair.

Discharge measurements of Tallulah River near Lakemont, Ga., during the year ending Sept. 30, 1918.

Date.	Made by—	Gage height.	Discharge.
		Feet.	*Sec.-ft.*
May 10	Paulsen and Condron	2.93	883
Aug. 25	C. G. Paulsen	−.21	a 4.6

a Sluice gates in Rabun dam closed when this measurement was made.

Daily discharge, in second-feet, of Tallulah River near Lakemont, Ga., for the year ending Sept. 30, 1918.

Day.	Oct.	Nov.	Dec.	Jan.	Feb.	Mar.	Apr.	May.	June.	July.	Aug.	Sept.
1	227	405	220	240	330	961	355	460	340	270	40	44
2	419	373	78	285	30	306	580	540	151	250	5	114
3	232	212	404	455	26	76	576	543	565	176	80	133
4	430	102	245	310	91	622	572	172	530	182	37	145
5	420	624	196	100	404	627	590	96	578	375	99	......
6	132	664	196	37	376	620	238	513	527	207	360	......
7	84	678	180	204	424	692	112	490	465	175	333	......
8	415	641	......	233	415	653	226	560	140	425	363	44
9	385	598	......	248	55	256	314	565	116	560	409	47
10	490	280	......	262	33	58	248	487	626	510	136	23
11	500	73	......	120	568	617	220	208	475	573	43	54
12	485	435	......	38	623	641	312	110	415	569	140	136
13	165	490	282	79	670	650	106	504	450	213	155	118
14	87	405	310	300	580	680	24	506	416	116	136	117
15	450	495	188	338	660	670	264	468	173	500	82	64
16	400	522	82	457	130	322	486	270	134	525	91	165
17	505	243	295	374	28	100	518	350	420	490	67	176
18	487	79	288	145	310	615	490	170	445	485	30	238
19	293	480	172	62	405	586	459	85	490	412	37	172
20	180	405	220	41	450	665	251	510	430	188	39	86
21	60	470	280	32	459	716	24	385	317	65	34	5
22	197	460	150	31	455	672	318	390	118	305	8	5
23	180	390	......	290	141	228	278	447	110	370	18	5
24	264	180	......	450	24	57	225	311	290	285	19	58
25	359	84	......	330	445	475	420	182	425	260	7	134
26	322	416	......	35	645	630	104	122	445	240	128	116
27	210	420	......	28	792	600	151	395	424	6	135	1[illegible]0
28	90	490	......	25	866	750	19	405	400	6	192	98
29	455	355	......	148	......	710	390	440	131	6	60	82
30	315	465	90	159	......	275	425	490	103	50	139	142
31	350	......	200	193	......	139	......	500	......	290	74	......

NOTE.—Gage-height record incomplete Dec. 31, Jan. 1, 24, 25 and June 1; discharge estimated for part of the day. No gage-height record Dec. 8-12, 23-29, and Sept. 5-7.

Monthly discharge, in second-feet, of Tallulah River, near Lakemont, Ga., for the year ending Sept. 30, 1918.

Month.	Maximum.	Minimum.	Mean.	Month.	Maximum.	Minimum.	Mean.
October	505	60	309	April	590	19	310
November	678	73	398	May	565	85	377
January	457	25	195	June	626	103	355
February	866	24	373	July	573	6	298
March	961	57	505	August	409	5	113

TIGER CREEK AT LAKEMONT, GA.

LOCATION.—100 feet from old Mathis post office, 100 feet upstream from Tallulah Falls Railway bridge, 600 feet downstream from Phillips's grist mill dam, 800 feet upstream from junction of creek with Tallulah River, and one-fourth mile downstream from Lakemont post office, Rabun County.

DRAINAGE AREA.—29 square miles (measured on topographic maps); revised since publication in Water-Supply Paper 432.

RECORDS AVAILABLE.—January 11, 1916, to September 30, 1918.

GAGE.—Staff gage in two sections on right bank; read by employee of Georgia Railway & Power Co.

DISCHARGE MEASUREMENTS.—Made from cable one-fourth mile upstream from gage in front of Lakemont railroad station.

CHANNEL AND CONTROL.—Bed rocky and rough at gage. Under gaging cable bed is sandy and shifting. Control is solid rock shoal just below gage; permanent. Backwater from very high floods on Tallulah River probably affects stage-discharge relation. This condition arises very infrequently, however.

Extremes of discharge.—Maximum mean daily stage during year, 3.01 feet January 28 (discharge, 518 second-feet); minimum mean daily stage, 1.19 feet September 26 (discharge, 31 second-feet).

1916–1918: Maximum stage about 7.0 feet (over top of gage) at 9 p. m. July 9, 1916 (discharge not determined); minimum mean daily stage that of September 26, 1918.

Ice.—Never enough to affect stage-discharge relation.

Diversions.—None.

Regulation.—Phillips' mill, which is infrequently operated, can cause considerable variation in stage. The gage is read only when mill is not running. As the pond above dam has practically no storage the gage heights are an accurate indication of natural flow.

Accuracy.—Stage-discharge relation practically permanent; not affected by ice. Rating curve well defined below 600 second-feet; above this point it is an extension. Gage read to half-tenths four times daily—at 6 a. m., noon, 6 p. m., and midnight. Daily discharge ascertained by applying mean daily gage height to rating table. Records good.

Cooperation.—Gage-height record furnished by Georgia Railway & Power Co.

Discharge measurements of Tiger Creek at Lakemont, Ga., during the year ending Sept. 30, 1918.

[Made by C. G. Paulsen.]

Date.	Gage height.	Discharge.
	Feet.	*Sec.-ft.*
May 10	1.67	108
Aug. 25	1.16	28.0

Daily discharge, in second-feet, of Tiger Creek at Lakemont, Ga., for the year ending Sept. 30, 1918.

Day.	Oct.	Nov.	Dec.	Jan.	Feb.	Mar.	Apr.	May.	June.	July.	Aug.	Sept.
1	50	47	41	48	105	66	55	73	55	46	41	33
2	46	43	39	39	107	66	57	68	54	43	51	32
3	43	42	39	39	100	65	57	63	52	42	43	37
4	42	42	38	43	83	71	55	58	61	42	40	35
5	41	42	37	39	71	66	55	58	55	40	38	47
6	40	41	37	57	71	63	54	58	57	38	36	41
7	40	41	38	45	73	68	98	55	55	37	34	42
8	40	40	42	43	71	63	95	66	52	37	33	41
9	51	39	41	43	69	65	105	91	52	37	33	41
10	43	38	46	43	71	66	87	100	51	36	33	38
11	42	37	43	109	65	61	81	73	48	34	37	34
12	40	40	48	89	66	58	71	68	57	34	34	33
13	40	87	46	65	65	58	68	69	52	34	33	33
14	40	61	43	60	68	57	63	65	50	34	33	32
15	39	51	42	89	69	55	63	61	48	34	33	32
16	39	48	41	69	105	55	68	58	47	34	33	32
17	39	43	42	60	100	58	65	58	47	34	73	34
18	43	42	43	52	85	55	79	57	47	35	39	46
19	162	42	43	54	122	54	68	75	47	39	46	45
20	63	45	41	52	127	60	73	68	47	36	38	48
21	55	43	41	51	105	65	73	60	63	48	34	42
22	54	42	41	50	89	63	68	98	50	42	33	34
23	52	42	40	50	81	60	65	73	47	43	33	34
24	50	42	42	52	75	63	61	135	46	45	37	32
25	48	42	40	54	71	60	60	79	43	41	35	32
26	48	41	43	54	71	57	69	69	50	47	34	31
27	48	39	43	98	68	55	63	65	45	63	34	33
28	48	37	41	518	68	55	63	61	42	46	39	32
29	48	41	41	186	......	54	61	58	42	42	37	37
30	52	41	39	259	......	54	83	57	47	43	35	36
31	47	......	41	137	......	52	......	57	......	42	34	......

Monthly discharge of Tiger Creek at Lakemont, Ga., for the year ending Sept. 30, 1918.

[Drainage area 29 square miles.][a]

Month.	Discharge in second-feet.				Run-off (depth in inches on drainage area).
	Maximum.	Minimum.	Mean.	Per square mile.	
October	162	39	49.5	1.71	1.97
November	87	37	44.0	1.52	1.70
December	48	37	41.4	1.43	1.65
January	518	39	85.4	2.94	3.39
February	127	65	82.9	2.86	2.98
March	71	52	60.3	2.08	2.40
April	105	54	69.4	2.39	2.67
May	135	55	69.5	2.40	2.77
June	63	42	50.3	1.73	1.93
July	63	34	40.3	1.39	1.60
August	73	33	37.6	1.30	1.50
September	48	31	36.6	1.26	1.41
The year	518	31	55.4	1.91	25.97

[a] Revised since publication in Water-Supply Paper 432.

ALTAMAHA RIVER BASIN.

OCMULGEE RIVER AT JULIETTE, GA.

LOCATION.—1 mile below Juliette railroad station, 1 mile below Juliette cotton mills, which are on left side of river opposite Juliette, 2½ miles below mouth of Towaliga River, and 20 miles upstream from Macon, Ga. Ocmulgee River forms line between Jones and Monroe counties.

DRAINAGE AREA.—2,100 square miles (measured from Post Route map of Georgia).

RECORDS AVAILABLE.—June 3, 1916, to September 30, 1918.

GAGE.—Stevens continuous water-stage recorder on left bank of river, referenced to a staff gage inside concrete well.

DISCHARGE MEASUREMENTS.—Made from a cable about 150 feet upstream from gage.

CHANNEL AND CONTROL.—Bed composed of sand and solid rock at gage section. Banks high; subject to overflow at about gage height 15 feet. A rock shoal about one-half mile downstream forms a control which keeps stage-discharge relation permanent.

EXTREMES OF DISCHARGE.—Maximum stage recorded during year, from water-stage recorder, 14.16 feet at 9 p. m. December 30 (discharge, 15,300 second-feet); minimum stage, from water-stage recorder, 3.06 feet at 2 a. m. June 17 (discharge, 430 second-feet).

1916–1918: Maximum stage from water-stage recorder, 26.4 feet at 3 p. m. July 10, 1916 (discharge, 42,400 second-feet); minimum stage from water-stage recorder, that of June 17, 1918.

Maximum stage of which there is any record, 32.0 feet during flood of 1886 (discharge determined from extension of rating curve, about 55,800 second-feet). This stage was determined with wye level from marks pointed out by local residents and is not reliable.

ICE.—Stage-discharge relation not affected by ice.

DIVERSIONS.—None.

REGULATION.—There is considerable regulation from three separate sources. Greatest fluctuations are caused by operation of the hydroelectric plant about 30 miles upstream near Jackson, Ga. Minor diurnal fluctuations are caused by operation of Juliette mills, 1 mile upstream and the hydroelectric plant on Towaliga River at High Falls, about 15 miles away.

ACCURACY.—Stage-discharge relation probably permanent, but some trouble was caused during the year by obstructions in intake pipe to gage well. Rating curve fairly well defined between 600 and 45,000 second-feet. Operation of water-stage recorder satisfactory. Slight errors in gage-height graph, due to lag in stage, caused by obstruction in intake pipe, are compensating, because there is considerable diurnal fluctuation. Discharge determined by use of discharge integrator. Records good.

Discharge measurements of Ocmulgee River at Juliette, Ga., during the year ending September 30, 1918.

Date.	Made by—	Gage height.	Discharge.	Date.	Made by—	Gage height.	Discharge.
		Feet.	*Sec.-ft.*			*Feet.*	*Sec.-ft.*
Feb. 23	C. G. Paulsen	4.80	1,540	June 5	A. H. Condron	4.00	914
Apr. 12	....do	4.68	1,680	Aug. 7	....do	4.48	1,200
May 27	A. H. Condron	4.08	1,130				

Daily discharge, in second-feet, of Ocmulgee River at Juliette, Ga., for the year ending September 30, 1918.

Day.	Oct.	Nov.	Dec.	Jan.	Feb.	Mar.	Apr.	May.	June.	July.	Aug.	Sept.
1	7,130	1,450	1,320	1,390	9,140	2,170	1,540	2,550	1,520	1,290	1,380	980
2	3,610	1,510	940	1,420	7,100	2,000	1,970	2,270	1,000	1,230	3,150	870
3	2,440	1,370	1,100	1,460	5,900	1,290	1,920	1,990	1,250	1,200	2,480	1,440
4	2,430	970	1,410	1,490	5,530	1,690	1,870	1,740	1,550	1,030	1,620	1,950
5	2,000	1,150	1,400	1,380	3,930	2,120	1,860	1,110	1,440	920	1,470	1,860
6	1,620	1,520	1,500	940	2,880	2,180	1,700	1,500	1,440	1,080	1,510	1,860
7	1,100	1,510	1,440	1,040	2,850	2,350	1,150	1,890	1,540	700	1,440	1,600
8	1,300	1,550	1,460	1,370	2,750	2,280	2,180	1,890	1,520	780	1,430	1,010
9	1,550	1,580	1,130	1,330	2,580	2,050	1,990	1,810	840	1,340	1,460	1,110
10	1,540	1,440	1,100	1,250	1,670	1,170	2,000	1,880	1,000	1,220	1,440	1,670
11	1,470	970	1,440	1,280	2,090	1,640	2,000	1,650	1,460	1,120	950	1,640
12	1,490	1,170	1,490	....	2,490	2,140	1,930	1,100	1,360	1,120	1,060	1,560
13	1,410	1,340	1,390	....	2,610	2,200	1,730	1,390	1,350	1,070	1,380	1,480
14	1,020	1,330	1,360	....	2,480	2,170	1,200	1,740	1,310	680	1,390	1,500
15	1,120	1,320	1,400	....	2,500	2,180	1,500	1,740	1,210	800	1,360	860
16	1,510	1,340	1,070	....	2,290	1,970	1,860	1,810	730	1,170	1,380	1,010
17	1,500	1,290	1,170	....	1,680	1,240	1,890	1,790	860	1,140	1,270	1,450
18	1,540	1,000	1,610	....	2,370	1,670	1,860	1,530	1,240	1,250	855	1,320
19	1,560	1,150	1,480	....	2,700	2,200	1,920	1,040	1,220	1,320	985	1,240
20	1,400	1,800	1,420	1,260	2,640	2,380	1,750	1,370	1,230	1,300	1,330	1,210
21	931	1,500	1,520	1,650	2,630	2,300	1,400	1,740	1,180	840	1,330	1,000
22	1,220	1,390	1,400	2,170	2,550	2,200	1,530	1,780	1,090	880	1,350	710
23	1,540	1,410	920	2,270	2,320	1,990	1,850	1,830	700	1,280	1,350	700
24	1,520	1,360	680	2,160	1,600	1,270	1,870	1,950	840	1,290	1,300	920
25	1,510	970	828	2,520	2,160	1,690	1,860	1,610	1,180	1,330	830	880
26	1,680	1,140	920	2,370	2,620	2,230	2,040	1,060	1,180	1,700	750	910
27	1,480	1,430	1,230	1,280	2,450	2,250	1,900	1,310	1,200	1,000	1,230	900
28	975	1,420	1,300	2,070	2,260	2,250	1,510	1,690	1,190	870	1,310	870
29	1,200	1,360	1,260	7,620	....	2,320	1,930	1,650	1,220	930	2,700	730
30	1,650	1,320	890	12,400	....	2,080	2,400	1,660	1,050	1,380	1,440	740
31	1,480	....	940	12,800	....	1,250	....	1,640	....	1,620	1,290	....

NOTE.—Discharge, Jan. 12–19, estimated, by comparison with records for Ocmulgee River at Jackson, as 1,570 second-feet.

Monthly discharge, in second-feet, of Ocmulgee River at Juliette, Ga., for the year ending September 30, 1918.

Month.	Maximum.	Minimum.	Mean.	Month.	Maximum.	Minimum.	Mean.
October	7,130	931	1,740	May	2,550	1,040	1,670
November	1,800	970	1,340	June	1,550	700	1,200
December	1,610	680	1,240	July	1,700	680	1,130
January	12,800	940	2,500	August	3,150	750	1,430
February	9,140	1,600	3,100	September	1,950	700	1,200
March	2,380	1,170	1,970				
April	2,400	1,150	1,800	The year.	12,800	680	1,680

OCONEE RIVER NEAR GREENSBORO, GA.

Location.—At highway bridge 1½ miles downstream from Town Creek, 4 miles upstream from mouth of Apalachee River, and 5 miles west of Greensboro, Greene County, on road to Madison, Ga.

Drainage area.—1,100 square miles.

Records available.—July 25, 1903, to September 30, 1918.

Gage.—Standard chain gage attached to bridge; read by F. M. Chambers to December, 1917, and by N. T. Oakes from January to September, 1918.

Discharge measurements.—Made from downstream side of bridge.

Channel and control.—Bed composed chiefly of sand; slightly shifting. Control section not known.

Extremes of discharge.—Maximum stage recorded during year, 14.1 feet at 4 p. m. January 31 (discharge, 8,260 second-feet); minimum stage, 0.2 foot in forenoon of July 15 (discharge, 141 second-feet).

1903–1918: Maximum stage recorded, 35.4 feet August 26, 1908 (discharge not determined). Discharge for this stage published in Water-Supply Papers 382 and 402, and determinations of discharges for stages above 13 feet prior to 1913, as published in previous water-supply papers, are too small, the error increasing with the stage.

Minimum stage recorded, 0.2 foot in forenoon of July 15, 1918 (discharge, 141 second-feet).

Ice.—None.

Diversions.—None.

Regulation.—Considerable diurnal fluctuation caused by operation of power plants.

Accuracy.—Stage-discharge relation practically permanent during the year. Rating curve well defined between 225 and 6,000 second-feet. Gage read to tenths twice daily. Daily discharge ascertained by applying mean daily gage height to rating table. Records good.

Discharge measurements of Oconee River near Greensboro, Ga., during the year ending Sept. 30, 1918.

Date.	Made by—	Gage height.	Discharge.
		Feet.	*Sec.-ft.*
Apr. 11	C. G. Paulsen	3.79	1,280
June 15	A. H. Condron	1.36	425

Daily discharge, in second-feet, of Oconee River near Greensboro, Ga., for the year ending Sept. 30, 1918.

Day.	Oct.	Nov.	Dec.	Jan.	Feb.	Mar.	Apr.	May.	June.	July.	Aug.	Sept.
1	2,330	890	520	432	7,500	925	582	1,150	352	1,110	1,920	1,110
2	1,280	520	490	314	3,470	890	750	1,110	404	890	3,590	550
3	715	614	490	490	2,550	855	750	960	352	680	4,480	460
4	680	490	490	550	2,020	925	614	820	550	520	5,130	432
5	550	490	490	582	1,720	1,280	582	750	490	314	6,180	750
6	520	490	550	490	1,640	995	715	715	404	378	2,550	1,920
7	460	490	550	614	1,500	960	550	785	1,540	378	995	1,150
8	490	550	550	820	1,460	960	1,150	785	960	460	890	890
9	614	490	490	715	1,280	890	2,120	1,110	520	1,920	820	647
10	680	490	520	614	1,190	890	1,820	2,440	647	460	820	550
11	520	490	490	614	1,230	890	1,230	1,070	820	432	582	550
12	614	432	520	4,220	1,150	890	1,110	855	614	432	785	520
13	550	614	520	4,740	1,190	820	960	820	582	326	995	432
14	404	750	460	4,870	1,190	820	890	1,070	490	228	680	432
15	404	680	490	4,220	1,190	820	890	1,030	490	252	680	352
16	378	680	490	4,220	1,030	820	855	855	404	276	582	288
17	432	550	550	4,220	1,150	750	820	750	550	314	432	404
18	432	550	490	1,920	1,280	890	960	750	750	314	378	404
19	490	490	490	1,460	1,230	890	995	550	614	614	352	404
20	750	750	550	1,110	1,360	820	890	855	490	2,550	520	432
21	750	647	550	1,110	1,320	995	890	785	582	2,080	404	680
22	614	614	582	1,190	1,230	1,110	960	715	550	1,280	432	614
23	520	550	550	2,120	1,070	960	890	715	378	960	432	520
24	550	550	432	2,220	1,030	1,720	785	680	680	1,720	432	550
25	460	550	614	1,920	1,030	1,190	750	647	460	2,660	378	550
26	460	404	614	1,820	1,030	890	890	404	1,820	1,540	352	520
27	582	432	750	1,680	960	820	1,030	582	1,360	1,320	326	432
28	490	490	582	1,590	925	820	1,030	680	890	925	520	490
29	550	490	614	4,610	......	785	1,030	614	550	1,680	680	264
30	820	550	432	5,840	......	750	1,110	432	614	2,280	890	288
31	1,320	......	404	7,700	......	550	......	378	......	3,830	995	......

Monthly discharge of Oconee River near Greensboro, Ga., for the year ending Sept. 30, 1918.

[Drainage area, 1,100 square miles.]

Month.	Discharge in second-feet.				Run-off (depth in inches on drainage area).
	Maximum.	Minimum.	Mean.	Per square mile.	
October	2,330	378	658	0.598	0.69
November	890	404	559	.508	.57
December	750	404	526	.478	.55
January	7,700	314	2,230	2.03	2.34
February	7,500	925	1,600	1.45	1.53
March	1,720	550	922	.838	.97
April	2,120	550	953	.866	.97
May	2,440	378	834	.758	.87
June	1,820	352	664	.604	.67
July	3,830	228	1,070	.973	1.12
August	6,180	326	1,260	1.15	1.33
September	1,920	264	586	.533	.59
The year	7,700	228	987	.897	12.18

OCONEE RIVER AT FRALEYS FERRY, NEAR MILLEDGEVILLE, GA.

LOCATION.—At Fraleys Ferry, in Baldwin County, 4 miles downstream from mouth of Little River, and 6 miles upstream from Milledgeville.

DRAINAGE AREA.—2,840 square miles.

RECORDS AVAILABLE.—May 23, 1906, to December 31, 1908; October 6, 1909, to September 30, 1918.

GAGE.—A combination sloping and vertical rod gage on left bank. Low-water section, inclined, is 75 feet upstream from ferry cable and extends to 8.5 feet; vertical section, 8.5 to 10 feet, at same site. High-water section, 10 to 20 feet, attached to tree 75 feet upstream from inclined section. Read by H. A. Taylor and B. L. Butts.

DISCHARGE MEASUREMENTS.—Made from ferryboat.

CHANNEL AND CONTROL.—Sand and shifting at measuring section. Control formed by a rock ledge extending across river 200 feet downstream; fairly permanent.

EXTREMES OF DISCHARGE.—No record of maximum stage (water over top of gage); minimum stage recorded, 4.3 feet at 7 a. m. July 15 and 5 p. m. July 16 (discharge, 400 second-feet).

1906–1918: Maximum stage recorded May 23, 1906, to December 31, 1908, and October 6, 1909, to September 30, 1918, about 24.6 feet March 17, 1913 (discharge determined from extension of rating curve, about 49,700 second-feet); minimum stage recorded, July 15 and 16, 1918.

ICE.—None.

DIVERSIONS.—None.

REGULATION.—Operation of power plants a great distance upstream can cause only slight fluctuations.

ACCURACY.—Stage-discharge relation permanent during the year. Rating curve very well defined below 2,000 second-feet, fairly well defined between 2,000 and 5,500 second-feet, and extended above 5,500 second-feet. Gage read to half-tenths twice daily. Daily discharge ascertained by applying mean daily gage-height to rating table. Records good up to 5,500 second-feet; above that point subject to error.

Discharge measurements of Oconee River at Fraleys Ferry, near Milledgeville, Ga., during the year ending Sept. 30, 1918.

Date.	Made by—	Gage height.	Discharge.
		Feet.	*Sec.-ft.*
Mar. 15	C. G. Paulsen	5.73	1,700
June 6	A. H. Condron	5.10	1,030
Aug. 8	.....do	5.57	1,480

Daily discharge, in second-feet, of Oconee River at Fraleys Ferry, near Milledgeville, Ga., for the year ending Sept. 30, 1918.

Day.	Nov.	Dec.	Jan.	Feb.	Mar.	Apr.	May.	June.	July.	Aug.	Sept.
1	......	1,540	1,270	......	2,210	1,540	3,160	710	1,320	4,030	2,360
2	......	1,430	1,320	......	2,210	2,060	2,680	792	1,790	3,670	1,790
3	......	1,430	1,320	......	2,210	1,790	2,210	710	1,430	9,240	1,270
4	1,540	1,430	1,380	7,970	2,060	1,790	1,790	632	972	5,940	880
5	1,120	1,430	1,430	5,940	2,060	1,660	1,540	880	835	5,940	835
6	1,120	1,320	1,540	4,410	2,060	1,540	1,540	1,120	632	5,740	2,060
7	1,120	1,430	1,790	3,500	2,060	1,540	1,540	3,330	710	1,540	2,360
8	1,020	1,920	1,790	3,000	2,060	1,540	1,540	3,160	710	1,540	2,360
9	1,020	1,790	1,920	2,680	2,060	3,000	1,790	1,790	4,030	1,270	1,120
10	1,540	1,540	1,790	2,680	2,060	2,680	3,330	1,170	4,030	1,270	1,270
11	2,840	1,540	2,060	2,520	2,060	2,060	3,000	1,540	1,270	1,540	1,070
12	2,360	1,430	7,140	2,680	2,060	2,680	1,790	1,540	925	1,380	835
13	3,670	1,380	7,140	3,160	1,790	2,210	1,540	1,660	710	1,380	792
14	3,330	1,430	7,140	3,670	1,790	2,060	1,790	1,430	670	1,380	710
15	3,160	1,320	7,760	3,330	1,790	1,790	2,360	1,070	460	972	670
16	3,000	1,320	7,140	3,000	1,660	1,790	2,060	880	430	1,120	525
17	2,840	1,430	6,140	3,000	1,790	1,540	1,540	1,380	632	1,790	632
18	2,680	1,430	4,030	3,000	2,060	1,540	1,540	2,360	835	880	710
19	2,520	1,430	3,330	3,000	2,060	2,060	1,320	1,540	1,790	670	670
20	3,160	1,430	3,000	3,000	2,060	2,360	1,380	1,170	3,000	1,170	835
21	2,060	1,430	2,680	3,160	2,060	3,160	1,430	1,020	4,600	972	880
22	1,540	1,430	4,030	3,000	2,210	2,360	1,380	925	3,000	670	972
23	1,540	1,430	4,410	3,000	2,210	2,060	1,220	880	1,920	670	880
24	1,320	1,430	4,030	2,840	1,790	1,660	1,270	710	3,000	595	792
25	1,220	1,430	3,670	2,680	2,680	1,660	1,220	632	4,030	670	792
26	1,220	1,790	3,330	2,520	2,360	2,060	1,170	1,120	4,980	670	750
27	1,220	2,210	3,160	2,520	1,540	2,360	972	2,840	3,000	595	792
28	1,320	2,060	3,160	2,360	1,540	2,680	925	2,060	2,360	632	835
29	1,320	1,540	......	......	1,540	2,210	925	1,380	1,790	1,070	792
30	1,540	1,380	......	......	1,660	2,680	880	1,540	1,430	1,790	632
31	......	1,270	......	......	1,430	......	880	......	4,220	1,660	......

NOTE.—Water overtopped the gage Dec. 29 to Feb. 3; discharge above 9,700 second-feet. No record Oct. 1 to Nov. 3.

Monthly discharge of Oconee River at Fraleys Ferry, near Milledgeville, Ga., for the year year ending Sept. 30, 1918.

[Drainage area, 2,840 square miles.]

Month.	Discharge in second-feet.				Run-off (depth in inches on drainage area).
	Maximum.	Minimum.	Mean.	Per square mile.	
November 4-30	3,670	1,020	1,940	0.683	0.69
December	2,210	1,270	1,510	.532	.61
January 1-28	7,760	1,270	3,530	1.24	1.29
February 4-28	7,970	2,360	3,300	1.16	1.08
March	2,680	1,430	1,970	.694	.80
April	3,160	1,540	2,070	.729	.81
May	3,330	880	1,670	.588	.68
June	3,330	632	1,400	.493	.55
July	4,980	430	1,980	.697	.80
August	9,240	595	2,010	.708	.82
September	2,360	525	1,060	.373	.42

APALACHICOLA RIVER BASIN.

CHATTAHOOCHEE RIVER NEAR GAINESVILLE, GA.

LOCATION.—At Clarke's covered wooden highway bridge, 500 feet downstream from Gainesville & Northwestern Railway bridge, 4 miles northeast of Gainesville, Hall County, 6 miles upstream from Dunlap dam of Georgia Railway & Power Co. and about 12 miles above mouth of Chestatee River.

DRAINAGE AREA.—Not measured.

RECORDS AVAILABLE.—January 1, 1917, to January 31, 1918, when station was discontinued.

GAGE.—Vertical staff, enamel-faced, attached to the upstream side of the wooden bridge; read by A. E. Maynard.

DISCHARGE MEASUREMENTS.—Made from boat a short distance below gage.

CHANNEL AND CONTROL.—Bed fairly permanent. Banks subject to overflow at a stage of about 12 feet. Backwater from Dunlap dam, 6 miles downstream, probably affects stage-discharge relation.

EXTREMES OF STAGE.—Maximum mean daily stage recorded, 7.85 feet January 12; minimum mean daily stage, 0.34 foot December 12.

1917–1918: Maximum mean daily stage recorded, 12.93 feet March 24, 1917; minimum mean daily stage recorded, 0.34 foot December 12, 1917.

ICE.—Stage-discharge relation not affected by ice.

REGULATION.—Owing to probable backwater effect from Dunlap dam, gage-height record should be used with caution.

COOPERATION.—Gage-height record furnished by the Georgia Railway & Power Co.

Data inadequate for determination of discharge.

Discharge measurements of Chattahooche River near Gainesville, Ga., during 1918.

[Made by C. G. Paulsen.]

Date.	Gage height.	Discharge.
	Feet.	*Sec.-ft.*
Aug. 26	0.76	396
Oct. 1	.74	364

Daily gage height, in feet, of Chattahoochee River near Gainesville, Ga., for the period Oct. 1, 1917, to Jan. 31, 1918.

Day.	Oct.	Nov.	Dec.	Jan.	Day.	Oct.	Nov.	Dec.	Jan.
1	2.25	1.55	0.95	0.80	16	0.75	1.12	2.55	6.45
2	1.62	1.50	.98	2.50	17	.73	1.03	2.35	3.05
3	1.05	1.60	.93	1.70	18	.68	1.07	2.05	2.83
4	1.05	1.10	.95	2.65	19	1.97	.87	1.82	1.45
5	1.05	1.25	.99	3.10	20	2.65	1.00	1.58	1.15
6	1.00	.95	.80	1.25	21	2.15	1.08	1.62	1.35
7	1.28	1.00	.98	1.68	22	1.25	.95	1.14	3.00
8	1.00	.95	.81	1.13	23	.85	.91	1.27	2.45
9	1.00	.92	1.25	.95	24	.88	.85	1.16	1.95
10	1.10	.94	1.10	1.10	25	.98	1.00	.93	2.45
11	1.00	.97	2.00	2.40	26	.90	.98	1.21	1.95
12	.75	.97	.34	7.85	27	1.10	1.25	1.25	1.52
13	.80	1.10	1.83	5.25	28	1.17	.97	1.05	4.15
14	1.05	1.45	3.05	5.30	29	1.05	.92	.96	7.40
15	.85	1.10	2.86	5.75	30	2.10	1.05	1.00	5.50
					31	1.35	....	1.05	5.65

CHATTAHOOCHEE RIVER NEAR NORCROSS, GA.

Location.—At Medlock's bridge, 1½ miles upstream from mouth of John Creek, 4½ miles north of Norcross, Gwinnett County, and about 5 miles above Suwanee Creek. The river forms the boundary between Gwinnett and Milton counties.

Drainage area.—1,170 square miles.

Records available.—January 9, 1903, to September 30, 1918.

Gage.—Chain gage on toll bridge, read by W. O. Medlock. January 1 to September 30, 1916, a Dexter water-stage recorder on right bank, just above bridge, and referred to chain gage without change in datum, was also used for recording stages below 7 feet.

Discharge measurements.—Made from downstream side of bridge.

Channel and control.—Bed sandy; shifts. Low-water control is a rock shoal about 2½ miles downstream; at higher stages shifting clay banks and other conditions may cause changes in the stage-discharge relation.

Extremes of discharge.—Maximum stage recorded during year, 10.4 feet at 6 p. m. January 29 (discharge, 10,800 second-feet); minimum stage recorded, 1.15 feet at 7 a. m. August 24 (discharge, 522 second-feet).

1903–1918: Maximum stage recorded, 21.4 feet at 2.30 p. m. December 30, 1915 (discharge, 36,200 second-feet); minimum stage recorded, 1.02 feet October 21, 1911 (discharge, 294 second-feet).

Ice.—Never enough to affect stage-discharge relation.

Regulation.—Diurnal fluctuation is caused by operation of hydroelectric plants on Chattahoochee and Chestatee rivers near Gainesville, Ga. Discharge January 1 to September 30, 1916, determined from records of water-stage recorder, agree very closely with that obtained by using mean daily gage-heights from two readings of chain gage per day. Errors in mean monthly discharge obtained by using records from chain gage varied from —1.6 per cent for February and May to +1.4 per cent for June. This study indicates that for medium and high stages, estimates of discharge for former years, as computed from records of the chain gage, are probably not seriously in error owing to diurnal fluctuation in stage. The effect on the accuracy of records for low stages has not been determined.

Accuracy.—Stage-discharge relation practically permanent during the year. Rating curve well defined between 700 and 10,000 second-feet. Gage read to hundredths twice daily. Daily discharge ascertained by applying mean daily gage height to rating table. Records good.

Discharge measurements of Chattahooche River near Norcross, Ga., during the year ending Sept. 30, 1918.

Date.	Made by—	Gage height.	Discharge.	Date.	Made by—	Gage height.	Discharge.
		Feet.	*Sec.-ft.*			*Feet.*	*Sec.-ft.*
Jan. 28	C. G. Paulsen	3.68	2,450	June 12	A. H. Condron	2.26	1,260
Mar. 9	.....do	2.81	1,800	July 11	.....do	1.77	930
Apr. 9	.....do	6.34	5,440	Sept. 5	.....do	3.22	2,200
9	.....do	5.87	4,890	5	.....do	3.12	2,150
May 18	A. H. Condron	2.48	1,540				

Daily discharge, in second-feet, of Chattahoochee River near Norcross, Ga., for the year ending Sept. 30, 1918.

Day.	Oct.	Nov.	Dec.	Jan.	Feb.	Mar.	Apr.	May.	June.	July.	Aug.	Sept.
1	1,920	1,560	1,230	1,150	4,670	1,830	1,390	2,460	1,080	1,740	1,740	800
2	1,650	1,390	1,230	1,080	3,440	1,740	1,390	2,370	1,080	1,310	3,870	730
3	1,560	1,310	1,150	1,230	3,240	1,740	1,390	2,010	1,010	1,080	4,310	730
4	1,390	1,310	1,230	1,150	2,940	1,740	1,470	1,830	1,080	940	2,100	2,840
5	1,310	1,230	1,390	1,150	2,550	1,740	1,470	1,740	1,150	870	1,650	2,100
6	1,310	1,310	1,310	1,230	2,280	1,740	1,390	1,560	1,310	870	1,150	1.650
7	1,310	1,230	1,230	1,470	2,190	1,740	1,560	1,470	1,920	870	1,230	1,390
8	1,150	1,230	1,230	1,560	2,100	1,560	4,090	1,560	1,740	765	1,080	1,150
9	1,310	1,230	1,230	1,470	2,100	1,650	5,030	1,920	1.310	870	870	1,230
10	1,310	1,230	1,230	1,310	1,920	1,740	2,740	1,650	1,150	800	905	1,080
11	1,310	1,230	1,230	2,460	1,920	1,650	2,100	1,390	1,150	765	1,920	905
12	1,310	1,230	1,830	9,060	1,830	1,560	2,010	1,470	1,150	730	1,010	870
13	1,230	1,310	2,010	5,150	1,830	1,560	1,830	1,470	1,150	730	975	870
14	1,230	1,310	1,830	2,460	1,740	1,560	1,740	1,830	1,310	730	975	800
15	1,080	1,390	1,920	5,270	2,100	1,390	1,470	1,560	1,080	670	975	765
16	1,150	1,310	2,010	4,790	2,640	1,390	1,650	1,390	1,010	730	800	730
17	1,150	1,230	1,650	2,640	3,650	1,560	1,830	1,310	940	730	800	670
18	1,150	1,310	1,740	2,190	3,440	1,560	1,830	1,310	1,010	800	800	765
19	1,390	1,150	1,560	1,920	2,640	1,390	1,830	1,310	1,310	1,230	765	800
20	1,920	1,230	1,560	1,830	2,840	1,470	1,830	1,470	1,230	2,190	765	1,310
21	1,830	1,230	1,310	1,740	2,940	1,740	1,740	1,390	1,390	1,740	730	1,390
22	1,560	1,230	1,230	2,280	2,640	1,740	1,740	1,560	1,390	1,150	730	1,230
23	1,470	1,230	1,310	2,100	2,370	1,560	1,650	1,390	1,390	1,390	730	1,080
24	1,390	1,230	1,230	2,100	2,190	1,650	1,650	1,470	1,150	1,310	730	1,010
25	1,230	1,150	1,310	2,010	2,100	1,560	1,470	1,470	1,080	1,390	730	905
26	1,230	1,080	1,390	2,280	2,010	1,560	1,830	1,230	1,920	2,370	905	800
27	1,230	1,010	1,230	2,100	1,920	1,470	2,370	1,310	1,470	2,460	870	800
28	1,230	1,230	1,310	3,440	1,920	1,390	1,920	1,310	1,150	1,830	975	835
29	1,390	1,230	1,310	10,000	......	1,390	2,010	1,230	1,080	1,560	940	800
30	1,740	1,230	1,230	8,020	......	1,390	2,100	1,150	1,740	1,560	800	800
31	1,470	......	1,150	7,760	......	1,470	......	1,080	......	2,100	800	......

Monthly discharge of Chattahoochee River near Norcross, Ga., for the year ending Sept. 30, 1918.

[Drainage area, 1,170 square miles.]

Month.	Discharge in second-feet.				Run-off (depth in inches on drainage area).
	Maximum.	Minimum.	Mean.	Per square mile.	
October	1,920	1,080	1,380	1.18	1.36
November	1,560	1,010	1,250	1.07	1.19
December	2,010	1,150	1,410	1.21	1.40
January	10,000	1,080	3,050	2.61	3.01
February	4,670	1,740	2,510	2.15	2.24
March	1,830	1,390	1,590	1.36	1.57
April	5,030	1,390	1,950	1.67	1.86
May	2,460	1,080	1,540	1.32	1.52
June	1,920	940	1,260	1.08	1.20
July	2,460	670	1,230	1.05	1.21
August	4,310	730	1,210	1.03	1.19
September	2,840	670	1,060	0.906	1.01
The year	10,000	670	1,620	1.38	18.76

CHATTAHOOCHEE RIVER AT WEST POINT, GA.

Location.—At West Point waterworks pumping plant just below Oseligee Creek, one-fourth mile east of Alabama-Georgia State line, in Troup County, and 1 mile upstream from West Point railroad station. Prior to October 20, 1912, station was at Montgomery Street Bridge in West Point.

Drainage area.—3,300 square miles.

Records available.—July 30, 1896, to September 30, 1918.

Gage.—Staff gage on left bank. By using a telescope the observer reads gage from pump house on right bank. October 20, 1912, to 1915, the gage was a vertical staff in two sections, a low-water section (0 to 6 feet) on right side of river and a high-water section on left side at same site as present gage and directly across river from low-water section. Datum of gage 0.2 foot above that of present gage. Prior to October 20, 1912, a chain gage at the Montgomery Street Bridge in West Point was used. Gage read by J. H. Miller.

Discharge measurements.—Made from Montgomery Street Bridge 1 mile downstream. No tributaries enter between gage and bridge.

Channel and control.—Bed rough and rocky; fairly permanent. Banks subject to overflow at high stages. Control is a rock ledge extending across river just below gage, and is probably not affected by Langdale dam 5 miles downstream. The old chain gage was abandoned in 1912 because of backwater from this dam.

Extremes of discharge.—Maximum mean daily stage, 16.3 feet January 12 (discharge, 34,800 second-feet); minimum mean daily stage, 2.2 feet July 16 (discharge, 1,300 second-feet).

1896–1918: Maximum stage recorded (old gage), 25.0 feet December 30, 1901 (discharge, 88,600 second-feet); minimum stage (old gage), 0.8 foot September 18–21, 1896 (discharge, 780 second-feet).

Ice.—None.

Diversions.—None.

Regulation.—Operation of power plants a great distance upstream causes some diurnal fluctuation, but a mean of three daily readings is probably very accurate.

Accuracy.—Stage-discharge relation permanent during the year. Rating curve well defined between 1,700 and 30,000 second-feet. Gage read to tenths three times daily; during high water read oftener. Daily discharge ascertained by applying mean daily gage height to rating table. Records good.

Discharge measurements of Chattahoochee River at West Point, Ga., during the year ending Sept. 30, 1918.

Date.	Made by—	Gage height.	Discharge.	Date.	Made by—	Gage height.	Discharge.
		Feet.	*Sec.-ft.*			*Feet.*	*Sec.-ft.*
Mar. 19	C. G. Paulsen	3.54	3,110	Aug. 1	A. H. Condron	3.95	3,970
Apr. 26	Paulsen and Cox	5.65	8,140	Sept. 26	do	2.68	1,770
June 19	A. H. Condron	3.42	2,870				

Daily discharge, in second-feet, of Chattahoochee River at West Point, Ga., for the year ending Sept. 30, 1918.

Day.	Oct.	Nov.	Dec.	Jan.	Feb.	Mar.	Apr.	May.	June.	July.	Aug.	Sept.
1	13,200	3,630	2,510	2,350	20,500	4,060	3,030	13,500	2,350	7,000	3,840	2,680
2	7,750	2,680	2,680	2,200	14,800	4,060	3,030	9,750	2,200	4,770	9,500	2,850
3	5,010	2,850	2,510	2,350	11,800	3,840	3,030	7,750	2,060	3,420	18,800	2,060
4	4,060	2,850	2,350	2,350	9,750	3,840	3,030	6,000	2,060	2,850	16,000	2,060
5	3,840	2,510	2,350	2,200	8,500	3,630	2,850	5,010	2,060	2,350	9,000	4,530
6	3,420	2,350	2,510	2,680	7,250	3,630	3,030	4,530	2,060	2,060	4,770	6,250
7	3,220	2,510	2,680	3,220	6,250	3,840	3,840	4,060	2,510	1,930	3,220	3,840
8	3,030	2,510	2,510	3,220	6,000	3,630	12,200	3,840	3,220	1,800	2,680	2,850
9	2,850	2,510	2,850	3,220	5,250	3,630	20,200	3,840	4,060	2,060	2,350	2,680
10	2,680	2,510	2,350	3,220	5,250	3,630	17,000	5,010	4,060	2,060	2,200	1,930
11	2,850	2,510	2,350	3,420	5,010	3,420	12,200	5,010	5,010	1,800	4,060	1,930
12	2,850	2,350	2,510	34,800	4,530	3,420	7,750	3,840	3,630	1,680	3,030	1,930
13	2,850	2,680	2,680	26,800	4,770	3,420	5,500	3,840	2,850	1,470	3,420	1,800
14	2,850	2,850	2,350	18,500	5,010	3,630	5,010	7,750	2,680	1,470	1,930	1,680
15	2,680	2,680	2,350	14,000	5,750	3,420	4,530	6,500	2,510	1,380	1,930	1,680
16	2,510	2,680	2,510	14,500	5,750	3,220	4,060	4,530	2,350	1,300	1,800	1,570
17	2,510	2,510	2,510	12,000	6,000	3,030	3,840	3,840	2,200	1,470	1,800	1,470
18	2,680	2,680	2,350	8,750	6,500	3,030	4,290	3,420	2,350	1,470	1,930	1,470
19	2,680	2,350	2,350	6,500	7,000	3,030	4,530	3,420	2,850	1,680	1,800	1,470
20	3,630	2,680	2,510	5,250	7,000	3,420	4,060	3,030	2,060	2,510	1,800	1,800
21	3,420	3,220	2,510	5,500	5,500	3,630	4,290	3,030	2,350	2,850	1,680	2,680
22	3,030	3,420	2,510	9,500	6,000	3,420	4,290	3,220	2,350	3,840	1,570	2,850
23	3,220	2,850	2,510	12,200	5,750	3,420	3,840	3,220	2,510	3,220	1,570	2,350
24	2,680	2,510	2,510	7,750	5,010	3,420	3,630	3,420	2,350	2,850	1,380	2,060
25	2,680	2,510	2,510	6,750	5,010	3,420	3,420	3,840	2,200	4,060	1,470	1,800
26	2,680	2,350	2,680	5,750	4,530	3,220	7,250	3,030	2,350	4,530	1,380	1,680
27	2,680	2,350	2,680	5,250	4,290	3,220	12,200	2,850	2,850	3,630	1,570	1,680
28	2,680	2,350	2,680	6,250	4,060	3,030	8,500	2,510	3,030	10,200	1,380	1,570
29	2,510	2,510	2,680	17,200	....	3,030	9,750	2,510	2,850	6,500	2,200	1,800
30	2,680	2,510	2,510	22,000	....	2,850	13,200	2,510	5,250	4,770	2,060	1,800
31	3,630	....	2,510	26,800	....	2,850	....	2,350	....	4,060	2,510	....

Monthly discharge of Chattahoochee River at West Point, Ga., for the year ending Sept. 30, 1918.

[Drainage area, 3,300 square miles.]

Month.	Discharge in second-feet.				Run-off (depth in inches on drainage area).
	Maximum.	Minimum.	Mean.	Per square mile.	
October	13,200	2,510	3,530	1.07	1.23
November	3,630	2,350	2,650	.803	.90
December	2,850	2,350	2,520	.764	.88
January	34,800	2,200	9,560	2.90	3.34
February	20,500	4,060	6,890	2.09	2.18
March	4,060	2,850	3,430	1.04	1.20
April	20,200	2,850	6,580	1.99	2.22
May	13,500	2,350	4,550	1.38	1.59
June	5,250	2,060	2,770	.839	.94
July	10,200	1,300	3,130	.948	1.09
August	18,800	1,380	3,700	1.12	1.29
September	6,250	1,470	2,290	.694	.77
The year	34,800	1,300	4,290	1.30	17.63

CHESTATEE RIVER AT NEW BRIDGE, GA.

Location.—Just below dam of Georgia Railway & Power Co., at New Bridge, Lumpkin County, 2 miles above mouth of Yellow Creek, 10 miles by direct route above confluence with Chattahoochee River, and 14 miles northwest of Gainesville.

Drainage area.—Not measured.

Records available.—January 1, 1917, to August 31, 1918, when station was discontinued.

Gage.—Vertical staff in tail race of Georgia Railway & Power Co.'s power plant; read to hundredths twice daily by J. M. Hulsey.

Discharge measurements.—Made from boat at a section 800 feet below gage.

Channel and control.—Bed of river rough and rocky.

Extremes of stage.—Maximum mean daily stage recorded during year, 3.25 feet January 29 and 30; minimum mean daily stage recorded, zero, May 7.

1917 and 1918: Maximum mean daily stage recorded, 5.2 feet March 4, 1917; minimum mean daily stage recorded, zero, May 7, 1918.

Ice.—Stage-discharge relation not affected by ice.

Regulation.—Owing to large diurnal fluctuations caused by operation of the power plant of the Georgia Railway & Power Co., gage heights should be used with caution. Also owing to the fact that the gage is located in the tail race, the stage-discharge relationship is not permanent when water is flowing over dam.

Cooperation.—Gage-height record furnished by Georgia Railway & Power Co.

Data inadequate for determination of discharge.

The following discharge measurement was made by C. G. Paulsen:

October 1, 1918: Gage height, 1.02 feet; discharge, 197 second-feet.

Daily gage-height, in second-feet, of Chestatee River at New Bridge, Ga., for the year ending Sept. 30, 1918.

Day.	Oct.	Nov.	Dec.	Jan.	Feb.	Mar.	Apr.	May.	June.	July.	Aug.
1	1.70	1.55	1.40	1.40	2.65	1.90	2.15	2.15	2.10	1.95	2.15
2	1.60	1.55	1.40	1.45	2.55	1.85	2.15	2.10	2.05	1.85	2.05
3	1.55	1.55	1.40	1.50	2.40	1.80	2.15	2.10	1.75	2.05	2.00
4	1.50	1.50	1.40	1.40	2.30	1.80	2.15	2.10	2.05	1.65	2.00
5	1.50	1.50	1.40	1.40	2.20	1.80	2.10	2.10	2.00	1.80	2.05
6	1.50	1.45	1.40	1.40	2.10	1.80	2.15	2.15	2.10	1.70	2.00
7	1.50	1.45	1.40	1.85	2.10	2.10	2.25	.00	2.00	1.80	1.85
8	1.50	1.40	1.40	1.40	2.00	2.10	2.85	2.10	1.65	1.85	1.90
9	1.45	1.40	1.40	1.40	2.00	2.10	2.50	2.20	1.75	2.00	1.80
10	1.50	1.45	.70	1.40	2.00	2.10	2.20	2.10	1.55	1.90	1.85
11	1.50	1.45	1.40	2.25	1.70	2.05	2.10	2.10	1.80	1.80	1.95
12	1.45	1.50	1.15	2.80	1.70	2.05	2.10	2.10	2.10	1.85	1.85
13	.70	1.50	1.30	1.80	2.00	2.05	2.10	2.10	2.10	1.85	1.90
14	1.40	1.45	.80	2.05	2.05	2.05	1.05	2.10	2.00	1.75	1.85
15	.75	1.10	1.55	2.90	2.40	2.00	1.15	2.10	1.50	1.75	1.90
16	1.45	1.10	1.50	2.45	2.55	1.85	2.20	2.05	2.10	1.85	1.75
17	1.40	1.40	1.40	2.10	2.85	2.00	2.20	2.10	2.40	1.90	1.70
18	.70	.70	1.40	2.10	2.45	2.10	2.20	2.00	1.80	1.90	1.75
19	2.00	1.45	1.45	2.10	2.40	2.10	2.15	2.10	1.60	2.05	1.80
20	1.90	1.15	1.45	1.80	2.55	1.80	2.15	2.15	1.80	2.10	1.65
21	1.65	1.10	1.40	2.00	2.45	2.10	2.20	2.10	1.90	2.00	1.60
22	1.55	1.40	1.50	1.80	2.35	2.05	2.20	2.10	2.10	1.90	1.70
23	1.50	1.15	1.45	1.90	2.20	1.95	2.15	2.10	2.10	2.05	1.80
24	1.50	1.15	1.40	1.80	2.15	2.05	2.10	2.10	1.80	2.25	1.70
25	1.50	1.40	1.60	1.80	2.10	2.05	2.45	2.10	1.80	2.10	1.85
26	1.50	1.40	1.70	1.90	2.10	2.10	2.65	2.10	2.05	2.10	1.65
27	1.40	1.15	1.50	2.00	2.05	2.10	2.30	1.95	2.00	2.10	1.90
28	.70	1.40	1.45	2.80	1.90	2.05	2.20	1.85	2.00	2.20	1.85
29	1.50	1.40	1.45	3.25	......	2.05	2.10	1.75	1.85	2.25	1.80
30	1.65	1.40	1.45	3.25	......	2.05	2.10	2.00	2.05	2.20	2.00
31	1.55	......	1.40	3.15	......	2.05	......	2.05	......	......	2.00

FLINT RIVER NEAR WOODBURY, GA.

LOCATION.—At Macon & Birmingham Railroad bridge one-fourth mile downstream from mouth of Elkins Creek, one-third mile upstream from mouth of Cane Creek, and 3 miles east of Woodbury, Pike County.

DRAINAGE AREA.—1,090 square miles.

RECORDS AVAILABLE.—March 29, 1900, to September 30, 1918.

GAGE.—Chain gage attached to guard rail on downstream side of Macon & Birmingham Railroad bridge; installed May 24, 1918. Prior to that date gage was a vertical staff in four sections on left bank about 300 feet above present gage. Gages set to same datum. Slope between gages negligible at low and medium stages. Zero of gage, 660 feet above sea level. Gage read by E. T. Riggins.

DISCHARGE MEASUREMENTS.—Made from downstream side of railroad bridge which does not make a right angle with the current.

CHANNEL AND CONTROL.—Bottom consists chiefly of rock; rough; current irregular. Control formed by a shoal 1 mile downstream; shifts occasionally.

EXTREMES OF DISCHARGE.—Maximum stage recorded during year, 6.0 feet at 7 a. m. January 31 (discharge, 8,320 second-feet); minimum stage recorded, −0.4 foot at 7 a. m. July 23 (discharge, 127 second-feet).

1900–1918: Maximum stage recorded, 16.2 feet March 15, 1913 (discharge, 35,300 second-feet); minimum stage, −0.4 foot October 8–10, 1911 (discharge, 86 second-feet).

ICE.—None.

DIVERSIONS.—None.

REGULATION.—Some slight diurnal fluctuations caused by operation of small mills on tributary streams.

ACCURACY.—Stage-discharge relation practically permanent. Rating curve well defined between 200 and 4,000 second-feet and fairly well defined between 4,000 and 24,000 second-feet. Gage read twice daily to tenths up to May 24 and to hundredths after that date. Daily discharge ascertained by applying mean daily gage height to rating table. Records good.

Discharge measurements of Flint River near Woodbury, Ga., during the year ending Sept. 30, 1918.

Date.	Made by—	Gage height.	Discharge.	Date.	Made by—	Gage height.	Discharge.
		Feet.	*Sec.-ft.*			*Feet.*	*Sec.-ft.*
Feb. 27	C. G. Paulsen	1.18	1,030	May 24	A. H. Condron	0.65	566
Mar. 26	do	.78	680	July 16	Paulsen and Condron	−.05	232
May 2	Paulsen and Condron	2.95	3,100				

Daily discharge, in second-feet, of Flint River near Woodbury, Ga., for the year ending Sept. 30, 1918.

Day.	Oct.	Nov.	Dec.	Jan.	Feb.	Mar.	Apr.	May.	June.	July.	Aug.	Sept.
1	7,530	690	610	610	6,580	1,040	610	3,900	370	950	1,400	770
2	5,000	610	610	610	6,040	950	610	2,850	370	770	2,420	1,310
3	2,170	610	610	610	5,170	860	690	2,050	325	540	3,900	860
4	1,220	610	610	610	4,050	860	690	1,600	325	420	3,150	860
5	860	540	610	610	2,570	860	610	1,220	325	325	2,710	770
6	690	540	690	690	2,300	860	610	1,040	420	325	1,710	950
7	690	540	610	1,040	1,820	860	860	950	540	285	1,040	1,220
8	610	540	690	950	1,600	950	3,000	860	540	285	690	950
9	610	540	770	860	1,400	950	4,050	860	610	420	540	540
10	610	540	690	860	1,310	950	5,340	770	610	325	420	420
11	610	540	690	1,040	1,220	950	3,450	770	1,040	285	540	420
12	540	540	770	5,000	1,220	860	2,300	690	1,400	285	610	370
13	540	540	770	4,200	1,310	860	1,600	690	1,040	285	540	325
14	540	540	770	4,050	1,220	770	1,220	770	770	285	480	325
15	540	540	690	4,680	1,400	770	1,040	1,040	540	250	370	325
16	540	540	690	4,050	1,500	770	950	1,130	420	250	370	285
17	540	540	690	3,000	2,300	770	860	860	420	215	370	285
18	480	540	690	2,050	1,820	690	860	770	370	250	370	285
19	540	540	690	1,600	1,600	690	860	690	370	325	480	250
20	610	690	690	1,400	1,400	690	950	540	420	540	540	370
21	690	860	610	1,220	1,600	690	1,040	540	420	480	480	420
22	610	770	610	1,600	1,400	770	860	540	420	540	370	420
23	610	770	610	2,050	1,220	770	860	540	420	152	325	370
24	540	770	610	2,170	1,220	690	690	540	370	180	285	325
25	540	690	610	2,170	1,220	690	690	610	325	480	420	325
26	540	610	950	2,050	1,130	690	1,820	540	325	770	325	325
27	540	610	1,040	1,710	1,040	690	2,570	480	325	1,600	285	325
28	540	610	860	1,400	1,040	610	3,150	420	370	1,710	285	325
29	540	610	770	4,680	......	610	3,300	420	370	1,820	420	325
30	860	610	690	6,040	......	610	3,600	420	480	2,300	690	325
31	770	......	610	8,320	......	610	......	370	......	1,600	860	......

Monthly discharge of Flint River near Woodbury, Ga., for the year ending Sept. 30, 1918.

[Drainage area, 1,690 square miles.]

Month.	Discharge in second-feet.				Run-off (depth in inches on drainage area).
	Maximum.	Minimum.	Mean.	Per square mile.	
October	7,530	480	1,040	0.954	1.10
November	860	540	607	.557	.62
December	1,040	610	697	.639	.74
January	8,320	610	2,320	2.13	2.46
February	6,580	1,040	2,060	1.89	1.97
March	1,040	610	787	.722	.83
April	5,340	610	1,660	1.52	1.70
May	3,900	370	951	.872	1.01
June	1,400	325	502	.461	.51
July	2,300	152	621	.570	.66
August	3,900	285	884	.811	.94
September	1,310	250	512	.470	.52
The year	8,320	152	1,050	.963	13.06

FLINT RIVER NEAR CULLODEN, GA.

Location.—At Grays Ferry, in Upson County, 1½ miles upstream from mouth of Auchumpkee Creek and 14 miles southwest of Culloden.

Drainage area.—2,000 square miles.

Records available.—July 1, 1911, to September 30, 1918.

Gage.—A vertical staff in four sections on left bank at old ferry landing; read by Lonie Williams until March 1, 1918; thereafter by Arthur Preston.

DISCHARGE MEASUREMENTS.—Made from row boat held in place by a small galvanized cable stretched taut across river.

CHANNEL AND CONTROL.—Channel sandy and shifting at gage. Control is a rock ledge one-half mile downstream; fairly permanent.

EXTREMES OF DISCHARGE.—Maximum stage recorded during year, 10.1 feet at 7 a. m. January 31 (discharge, 13,500 second-feet); minimum stage, 1.23 feet at 7 a. m. July 19 (discharge, 205 second-feet).

1911–1918: Maximum stage recorded, 33.3 feet during night of July 9, 1916 (discharge not determined); minimum stage, 1.0 foot October 8, 1911 (discharge, 165 second-feet).

ICE.—None.

DIVERSIONS.—None.

ACCURACY.—Stage-discharge relation practically permanent. Rating curve well defined below 4,000 second-feet. Above 4,000 second-feet rating curve is an extension. Gage read twice daily to tenths. Daily discharge ascertained by applying mean daily gage height to rating table. Low-water records good; determinations above 4,000 second-feet subject to error.

Discharge measurements of Flint River near Culloden, Ga., during the year ending Sept. 30, 1918.

Date.	Made by—	Gage height.	Discharge.	Date.	Made by—	Gage height.	Discharge.
		Feet.	*Sec.-ft.*			*Feet.*	*Sec.-ft.*
Mar. 27	C. G. Paulsen	2.45	976	July 17	Paulsen and Condron	1.37	284
May 3	Paulsen and Condron	4.72	3,440	Aug. 16	A. H. Condron	1.72	487
23	A. H. Condron	2.20	806	25	....do	1.30	526

Daily discharge, in second-feet, of Flint River near Culloden, Ga., for the year ending Sept. 30, 1918.

Day.	Oct.	Nov.	Dec.	Jan.	Feb.	Mar.	Apr.	May.	June.	July.	Aug.	Sept.
1	9,600	1,040	922	960	9,220	1,470	960	6,750	530	848	1,880	1,240
2	7,130	998	960	960	9,600	1,420	960	4,940	530	1,080	1,990	1,340
3	3,690	960	960	960	11,200	1,380	1,040	3,290	530	848	5,460	1,120
4	2,100	885	960	960	7,510	1,380	1,040	2,540	500	595	3,840	1,420
5	1,420	885	960	960	5,110	1,290	998	2,100	500	530	3,550	1,240
6	1,160	885	960	998	3,840	1,240	885	1,570	595	440	1,990	848
7	1,040	885	960	1,570	3,030	1,420	960	1,380	772	410	1,470	1,380
8	960	885	1,080	1,420	2,540	1,420	3,840	1,290	960	350	1,080	1,380
9	960	885	1,200	1,380	2,320	1,290	4,460	1,200	848	350	810	1,040
10	1,040	885	1,200	1,340	2,210	1,420	5,460	1,160	922	500	665	735
11	960	885	1,120	1,290	2,210	1,420	4,780	1,040	998	500	630	595
12	922	810	1,040	7,700	2,100	1,240	3.030	998	1,380	320	700	530
13	885	885	1,120	6,560	2,320	1,200	2,320	998	1,570	350	848	500
14	885	810	1,120	5.280	2,100	1,200	1,770	1,040	1,200	290	665	470
15	810	885	1,120	6,560	2,100	1,200	1,380	1,160	922	290	595	440
16	810	885	1,200	5,820	2,320	1,080	1,240	1,340	960	215	470	380
17	810	848	1,120	4,460	3,030	1,120	1,200	1,240	562	265	595	410
18	810	810	1,040	2,100	3,030	1,160	1,200	1,080	562	215	440	350
19	810	810	1,040	1,880	2,430	1,080	1,240	998	562	240	440	350
20	848	1,290	1,040	2,210	2,540	1,200	1,340	960	530	1,080	665	320
21	960	1,290	1,040	2,100	3,030	1,240	1,670	848	562	922	735	470
22	960	1,200	1,040	2,100	2,770	1,200	1,770	810	530	810	562	530
23	885	1,080	960	3,030	2,210	1,240	1,200	810	562	810	470	470
24	885	998	960	3,690	1,990	1,160	1,080	810	500	530	380	440
25	810	960	960	2,900	1,990	1,120	1,040	848	440	500	440	470
26	810	885	960	2,650	1,670	1,040	2,900	848	410	922	440	410
27	810	885	1,040	2,430	1,570	1,040	3,420	810	440	998	350	410
28	810	885	1,160	2,320	1,470	960	3,550	810	470	2,650	735	380
29	810	885	1,200	7,510	......	960	4,460	700	470	1,880	772	470
30	998	885	1,200	7,700	......	960	5,460	630	530	2,540	665	440
31	1,160	......	1,120	12,500	......	960	......	595	......	2,430	1,200	......

Monthly discharge of Flint River near Culloden, Ga., for the year ending Sept. 30, 1918.

[Drainage area, 2,000 square miles.]

Month.	Discharge in second-feet.				Run-off (depth in inches on drainage area).
	Maximum.	Minimum.	Mean.	Per square mile.	
October	9,600	810	1,530	0.765	0.88
November	1,290	810	935	.468	.52
December	1,200	922	1,060	.530	.61
January	12,500	960	3,360	1.68	1.94
February	11,200	1,470	3,480	1.74	1.81
March	1,470	960	1,210	.605	.70
April	5,460	885	2,220	1.11	1.24
May	6,750	595	1,470	.735	.85
June	1,570	410	695	.348	.39
July	2,650	215	797	.398	.46
August	5,460	350	1,150	.575	.66
September	1,420	320	686	.343	.38
The year	12,500	215	1,540	.770	10.44

FLINT RIVER AT ALBANY, GA.

LOCATION.—At Dougherty County highway bridge in Albany, 700 feet below Atlantic Coast Line Railroad bridge and 2 miles downstream from mouth of Muckafoonee Creek.

DRAINAGE AREA.—5,000 square miles.

RECORDS AVAILABLE.—April 10, 1893, to September 30, 1918 (United States Weather Bureau gage heights). Discharge measurements were begun by the Geological Survey in 1901, and determinations of daily discharge have been made from January 1, 1902, to September 30, 1915.

GAGE.—Chain gage, installed at the bridge April 20, 1904; read once daily by D. W. Brosnan. Original staff gage was washed out in 1898. It was again damaged in 1902, and on June 18 of that year a new gage was installed by the United States Weather Bureau at a datum 0.75 foot lower than that of the former gage. All gage heights published for 1902 by the United States Weather Bureau and the United States Geological Survey refer to the new datum. Present gage conforms with the United States Weather Bureau gage.

DISCHARGE MEASUREMENTS.—Fairly accurate measurements can be made at the section at the Atlantic Coast Line bridge, although it is very rough and train-switching in the yard interferes with the work. The section at the Georgia Northern Railway bridge, 1 mile above, at which measurements are sometimes made, is considered better, especially for medium and low stages.

CHANNEL AND CONTROL.—Channel at and below gage may shift slightly, but control is such that conditions of flow are practically permanent except for changes caused by dredging below gage. The river overflows both banks, but only under the approaches to the bridge.

EXTREMES OF DISCHARGE.—Maximum stage recorded during year, 12.3 feet at 7 a. m. February 8 and 9 (discharge not determined); minimum stage recorded, −0.8 foot at 7 a. m. September 21–23 (discharge not determined).

1902–1918: Maximum stage recorded, 30.3 feet at 7 a. m. March 21, 1913 (discharge, 53,700 second-feet); minimum stage, −1.1 feet October 9 to 12, 1911 (discharge, 1,110 second-feet).

ICE.—Stage-discharge relation not affected by ice.

DIVERSIONS.—None.

REGULATION.—Power developments on Muckalee Creek, which joins Flint River about 2 miles above the station, cause considerable diurnal fluctuation, especially at low stages. It is probable that the flow is also affected by other power plants farther up the river.

ACCURACY.—Discharge measurements made in 1918 indicate a decided change in the stage-discharge relation as expressed by the curve used from 1912 to 1915. This change was caused by dredging operations carried on by the United States Army Engineers during the summer of 1915. Discharge records for 1915 as published in Water Supply Paper 402 were determined from the old rating and should, therefore, be used with caution. Determination of discharge for 1918 is not possible until additional current-meter measurements can be obtained.

Discharge measurements of Flint River at Albany, Ga., during the year ending Sept. 30, 1918.

[Made by A. H. Condron.]

Date.	Gage height.	Discharge.
	Feet.	*Sec.-ft.*
June 7	0.40	2,420
June 24	.45	2,410

Daily gage height, in feet, of Flint River at Albany, Ga., for the year ending Sept. 30, 1918.

Day.	Oct.	Nov.	Dec.	Jan.	Feb.	Mar.	Apr.	May.	June.	July.	Aug.	Sept.
1	3.4	0.4	0.5	1.6	5.2	3.0	1.1	6.4	-0.3	1.4	3.6	0.8
2	4.5	.3	1.2	1.6	5.5	2.5	1.1	7.3	- .4	1.2	4.6	.8
3	5.6	.3	1.6	1.9	6.4	2.4	1.3	8.2	- .2	1.0	4.6	.8
4	6.1	.3	.8	1.1	8.5	2.4	1.4	8.6	.2	-.6	4.0	1.2
5	6.5	.2	.6	.8	9.5	2.1	1.3	8.8	.4	-.4	4.5	1.6
6	6.9	.1	.6	.9	11.1	2.0	1.4	8.8	.2	-.3	4.9	1.8
7	6.5	.2	.4	.8	12.1	2.0	1.4	8.0	.3	-.1	4.6	1.8
8	4.3	.1	.3	1.5	12.3	2.1	1.5	6.2	.4	-.1	4.6	2.1
9	3.1	.1	.5	2.0	12.3	1.8	1.1	3.8	.8	.4	3.7	1.9
10	.9	.4	.6	1.9	10.7	1.8	1.7	3.0	1.0	.2	3.4	1.8
11	.8	.3	.5	2.0	8.4	1.9	2.9	2.3	.7	.3	2.8	1.7
12	.9	.2	1.4	2.5	5.8	1.6	4.0	2.8	.6	.1	2.2	1.7
13	.7	.0	1.5	2.8	4.4	1.5	5.1	1.5	.6	-.2	1.8	1.2
14	.7	.3	1.1	3.3	4.3	2.0	5.5	1.2	.7	-.3	1.4	.7
15	.4	.0	1.1	4.4	4.3	1.4	5.6	1.4	1.0	-.4	.8	.4
16	.3	.0	1.4	5.0	4.6	1.4	4.3	1.5	.9	-.4	.7	.0
17	.3	.1	1.3	6.0	4.5	1.2	2.3	1.6	.6	-.5	1.0	-.2
18	.4	.0	1.5	6.5	4.5	1.4	1.8	1.6	.3	-.6	2.2	.0
19	.5	-.1	1.4	7.1	4.7	1.6	1.6	2.2	.5	-.5	3.6	-.5
20	.6	.0	.9	7.0	4.9	1.7	1.4	2.6	.1	.0	3.1	-.7
21	.4	.3	.6	7.0	5.6	1.4	2.0	2.3	.1	.2	2.0	-.8
22	.1	.7	.9	6.3	4.9	1.3	2.3	1.2	.1	.2	1.6	-.8
23	.1	.7	.8	5.0	4.5	1.5	2.3	.6	.0	.4	1.7	-.8
24	.2	.9	.5	4.4	4.3	1.4	2.7	.3	-.2	1.6	1.2	-.4
25	.0	1.2	.3	4.3	4.3	2.0	1.8	.2	.0	2.4	.7	-.1
26	.0	1.6	1.0	4.6	3.8	1.8	1.6	.2	.0	2.6	.3	-.2
27	.4	1.2	1.2	4.6	3.3	1.2	2.0	.2	.1	2.0	.1	-.3
28	.1	.7	1.5	4.6	2.9	1.4	3.0	.2	.2	1.8	.2	-.3
29	.0	.2	1.9	4.4	......	1.5	4.8	.2	.6	1.8	.4	-.2
30	.1	.3	1.8	4.1	......	1.3	5.3	-.1	1.0	1.9	.4	.0
31	.2	......	1.9	4.5	......	1.2	......	-.1	......	2.6	.6	......

LITTLE POTATO (TOBLER) CREEK NEAR YATESVILLE, GA.

LOCATION.—At Tobler mills, 1 mile downstream from Macon & Birmingham Railroad bridge, 2 miles north of Yatesville, Upson County, and 15 miles upstream from junction of creek with Flint River.

DRAINAGE AREA.—Not measured.

RECORDS AVAILABLE.—November 4, 1914, to September 30, 1918, when station was discontinued.

GAGE.—Vertical staff on right bank just below penstock of Tobler mills; read by J. K. Sanders.

DISCHARGE MEASUREMENTS.—Made from steel highway bridge across mill pond, about 600 feet above gage, during medium and high stages; by wading during low stages.

Channel and control.—Bed composed of boulders and solid rock. Control formed by solid rock shoal; permanent.

Extremes of stage.—Maximum stage recorded during year, 1.8 feet at 7.30 a. m. and 4.30 p. m. January 31 (discharge not determined); minimum stage recorded, 0.4 foot at 5.30 a. m. July 26 (discharge not determined).

1914–1918: Maximum stage recorded, 3.3 feet at 5.30 a. m. July 8 and 5 p. m. July 18, 1916 (discharge not determined); minimum stage, 0.3 foot at 6 a. m. September 29, 1915 (discharge not determined).

Ice.—None.

Diversions.—None.

Regulation.—Operation of Tobler mill causes large fluctuations in stage. Gage is read in morning before operation of mill in order to obtain readings that more nearly represent the normal stage.

Accuracy.—Stage-discharge relation permanent; not affected by ice. Owing to storage in mill pond, gage heights do not indicate the mean stage for the day accurately, particularly at low water. Therefore the gage-height record should be used with caution.

The following discharge measurement was made by C. G. Paulsen:

February 27, 1918: Gage height, 0.80 foot; discharge, 28.5 second-feet.

Daily gage-height, in feet, of Little Potato (Tobler) Creek near Yatesville, Ga., for the year ending Sept. 30, 1918.

Day.	Oct.	Nov.	Dec.	Jan.	Feb.	Mar.	Apr.	May.	June.	July.	Aug.	Sept.
1	0.6	0.7	0.7	0.7	1.0	0.8	0.8	0.9	0.68	0.7	0.8	0.7
2	.6	.7	.7	.7	1.1	.8	.8	.85	.7	.7	1.15	.7
3	.6	.7	.7	.7	1.2	.8	.85	.8	.7	.7	1.2	1.1
4	.6	.7	.7	.7	1.2	.8	.75	.8	.7	.7	.9	1.0
5	.6	.7	.7	.7	1.0	.8	.72	.7	.7	.7	.8	.85
6	.6	.7	.75	.7	.9	.8	.7	.7	.7	.7	.8	.88
7	.6	.7	.8	.7	.95	.8	.8	.7	.7	.7	.8	.65
8	.6	.7	.8	.8	.8	.8	.8	.7	.7	.7	.7	.7
9	.6	.7	.7	.8	.8	.8	.8	.68	.7	.7	.7	.7
10	.6	.7	.7	.8	.8	.8	.8	.72	.72	.7	.75	.7
11	.6	.7	.7	.82	.8	.8	.8	.7	.72	.7	.8	.7
12	.7	.7	.7	1.4	.8	.8	.8	.7	.7	.7	.8	.7
13	.7	.7	.7	.8	.9	.8	.75	.68	.7	.7	.8	.7
14	.7	.7	.7	.8	.8	.8	.8	.6	.7	.7	.8	.7
15	.7	.7	.7	1.0	.8	.8	.8	.75	.7	.7	.75	.7
16	.7	.7	.7	.9	.8	.8	.8	.7	.7	.7	.7	.7
17	.7	.7	.7	.85	.85	.8	.75	.7	.7	.7	.7	.7
18	.7	.7	.75	.8	.85	.8	.78	.7	.7	.72	.7	.7
19	.7	.7	.62	.8	.85	.8	.8	.7	.7	.75	.8	.72
20	.7	.8	.68	.8	.95	.85	.8	.7	.7	.82	.85	.75
21	.7	.8	.65	.8	.95	.8	.8	.7	.7	.9	.8	.72
22	.7	.8	.6	.8	.9	.8	.8	.7	.7	.9	.8	.7
23	.7	.8	.6	.8	.9	.8	.8	.68	.7	.9	.8	.7
24	.7	.8	.6	.8	.8	.8	.8	.65	.7	.9	.75	.7
25	.7	.8	.7	.8	.8	.8	.8	.65	.7	.65	.7	.7
26	.7	.8	.7	.8	.8	.8	1.1	.65	.7	.62	.7	.7
27	.7	.8	.6	.8	.8	.78	.8	.65	.7	.85	.75	.7
28	.75	.7	.6	.8	.8	.8	.9	.65	.7	.82	1.2	.7
29	.75	.7	.6	1.6	......	.8	.9	.65	.7	1.0	.8	.7
30	......	.7	.6	1.05	......	.8	.9	.65	.7	.9	.85	.7
31	......	......	.6	1.8	......	.8	......	.65	......	.9	.9	......

ESCAMBIA RIVER BASIN

CONECUH RIVER AT BECK, ALA.

Location.—At Simmons Bridge at Beck, Covington County, 8 miles west of Andalusia, a station on Central of Georgia Railway and Louisville & Nashville Railroad, and 12 miles downstream from mouth of Patsaliga Creek.

Drainage area.—1,290 square miles.

Records available.—1891 to 1898 (gage heights by United States Weather Bureau and discharge measurements by United States Geological Survey); 1899 to 1904 incomplete records of gage heights; continuous records January 1, 1905, to September 30, 1918.

Gage.—Chain gage attached to upstream side of wagon bridge; read once daily to tenths, except Sundays, from October 1, 1917, to January 31, 1918, by A. W. Lambert, and from February 1 to September 30, 1918, by C. E. Raley.

Discharge measurements.—Made from bridge.

Channel and control.—Channel cut in soft bedrock; practically permanent. Both banks subject to overflow at very high stages. Location of control not known.

Extremes of discharge.—Maximum stage recorded, 29.1 feet at 8 a. m. October 3 (discharge, 15,100 second-feet); minimum stage recorded, 0.9 foot at 8 a. m. July 15 and 19 (discharge, 208 second-feet).

1904–1918: Maximum stage (no gage height) March 18, 1913 (discharge, 26,000 second-feet, estimated by comparison with Pea River at Pera, Ala.); minimum stage, 0.7 foot October 4, 1904 (discharge, 187 second-feet).

Ice.—Stage-discharge relation not affected by ice.

Diversions.—None.

Regulation.—Flow may at times be affected by logging operations.

Accuracy.—Stage-discharge relation practically permanent. Rating curve, substantiated by one additional discharge measurement made subsequent to 1918, is fairly well defined between 225 and 7,000 second-feet above which it is extended. Daily discharge ascertained by applying mean daily gage height to rating table. Records fair.

The following discharge measurement was made by A. H. Condron:

June 22, 1918: Gage height, 1.96 feet; discharge, 366 second-feet.

Daily discharge, in second-feet, of Conecuh River at Beck, Ala., for the year ending Sept. 30, 1918.

Day.	Oct.	Nov.	Dec.	Jan.	Feb.	Mar.	Apr.	May.	June.	July.	Aug.	Sept.
1	14,300	996	955	795	3,630	1,260	585	6,630	395	278	834	1,150
2	14,000	955	955	834	3,240	1,170	720	6,120	395	278	874	1,040
3	15,100	874	955	795	3,970	1,120	1,680	5,380	395	247	720	955
4	15,000	874	955	757	4,700	1,080	1,120	4,030	395	262	758	914
5	12,800	874	996	720	4,870	1,040	1,300	3,940	443	278	795	1,040
6	10,200	834	996	1,010	4,870	955	1,400	3,860	524	262	617	874
7	7,260	795	874	1,300	4,700	996	1,840	4,190	443	270	496	650
8	4,310	757	996	1,260	4,250	955	2,270	3,630	955	278	443	618
9	2,970	684	1,100	1,260	3,860	955	1,780	2,540	914	278	395	585
10	2,540	684	1,210	1,300	3,520	955	1,590	1,890	874	262	373	554
11	2,050	702	1,210	1,300	3,190	955	1,540	1,440	757	220	343	496
12	1,730	720	1,350	1,730	2,750	874	1,440	1,180	617	220	313	524
13	1,540	684	1,300	1,610	2,540	874	1,400	914	524	220	525	496
14	1,420	684	1,170	1,490	2,160	834	1,280	3,130	496	214	373	496
15	1,300	650	1,170	3,350	2,210	795	1,170	2,910	469	208	352	414
16	1,210	684	1,100	3,190	1,890	757	1,040	2,270	496	220	352	332
17	1,120	650	1,040	2,700	1,920	720	955	2,000	524	220	352	352
18	1,120	650	955	2,320	1,940	684	1,080	1,640	496	220	385	352
19	1,040	650	914	2,050	1,890	720	955	1,450	418	208	418	313
20	996	795	834	2,160	1,830	720	955	1,260	395	247	834	332
21	955	720	834	2,270	1,940	720	1,020	1,080	373	332	395	395
22	914	757	795	2,700	1,730	720	1,080	955	352	418	395	354
23	834	720	776	955	1,730	684	1,040	874	342	332	332	313
24	834	795	757	2,160	1,660	684	914	720	332	332	352	313
25	834	776	757	2,000	1,590	684	914	650	295	395	332	295
26	795	757	1,040	1,890	1,540	684	914	602	278	332	313	278
27	796	684	1,040	1,810	1,400	650	834	554	352	469	278	295
28	758	720	996	1,730	1,300	650	1,280	554	332	512	496	332
29	720	1,040	955	1,780	....	585	1,730	524	295	554	1,040	314
30	2,000	914	914	1,890	....	617	3,240	469	286	524	955	295
31	1,040	....	874	3,970	....	601	....	443	....	469	1,260	....

Note.—Daily discharge interpolated for Sundays when gage was not read.

Monthly discharge of Conecuh River at Beck, Ala., for the year ending Sept. 30, 1918.

[Drainage area, 1,290 square miles.]

Month.	Discharge in second-feet.				Run-off (depth in inches on drainage area).
	Maximum.	Minimum.	Mean.	Per square mile.	
October	15,100	720	3,950	3.06	3.53
November	1,040	650	769	.596	.66
December	1,350	757	993	.770	.89
January	3,970	720	1,780	1.38	1.59
February	4,870	1,300	2,740	2.12	2.21
March	1,260	585	829	.643	.74
April	3,240	585	1,300	1.00	1.12
May	6,630	443	2,190	1.70	1.96
June	955	278	472	.366	.41
July	554	208	308	.239	.28
August	1,260	278	539	.418	.48
September	1,150	278	522	.405	.45
The year	15,100	208	1,360	1.05	14.33

MOBILE RIVER BASIN.

OOSTANAULA RIVER AT RESACA, GA.

LOCATION.—At Western & Atlantic (now Nashville, Chattanooga & St. Louis) Railroad bridge in Resaca, Gordon County, 400 feet upstream from Dixie highway bridge, 1 mile above Camp Creek, and 3 miles below junction of Conasauga and Coosawattee rivers, which form the Oostanaula.

DRAINAGE AREA.—1,610 square miles.

RECORDS AVAILABLE.—1891 to 1898 (gage heights by the United States Weather Bureau and discharge measurements and gage heights by the United States Geological Survey); 1899 to 1904, partial records of gage heights; continuous records, January 1, 1905, to September 30, 1918.

GAGE.—Heavy vertical timber attached to the downstream side of midstream pier of railroad bridge.

DISCHARGE MEASUREMENTS.—Made from the Dixie highway bridge or by wading.

CHANNEL AND CONTROL.—Bed composed of sand; somewhat shifting. Right bank a high bluff; not subject to overflow; left bank high but is overflowed at very high stages. Though the location of control is not exactly known, the fact that station rating has shown very little change in the past indicates that the control is practically permanent.

EXTREMES OF DISCHARGE.—Maximum stage recorded during year, 23.3 feet February 1 (discharge, 19,900 second-feet); minimum stage recorded, 1.3 feet November 26, December 2 and 4 (discharge, 390 second-feet).

1896–1918: Maximum stage recorded,[1] 31.7 feet March 15, 1909 (discharge 39,200 second-feet); minimum stage, 0.95 foot during discharge measurement, September 26, 1904 (discharge, 273 second-feet).

ICE.—Stage-discharge relation not affected by ice.

DIVERSIONS.—None.

REGULATION.—Practically none from the few small mills upstream.

[1] Gage-height records not obtained during the following periods: May 1 to July 31, 1896; May 1 to October 31, 1899; July 1 to October 31, 1900; May 1 to November 12, 1901, and January 1, 1902, to December 31, 1904.

ACCURACY.—Stage-discharge relation practically permanent. Rating curve well defined between 450 and 8,000 second-feet, above which curve is extended tangent. Gage read to tenths once daily. Gage heights at low water subject to error because of poor conditions of lower part of gage; therefore records at low stage should be used with caution. Daily discharge ascertained by applying mean daily gage height to rating table. Records fair.

The following discharge measurement was made by C. G. Paulsen:

April 17, 1918: Gage height, 8.13 feet; discharge, 4,650 second-feet.

Daily discharge, in second-feet, of Oostanaula River at Resaca, Ga., for the year ending Sept. 30, 1918.

Day.	Oct.	Nov.	Dec.	Jan.	Feb.	Mar.	Apr.	May.	June.	July.	Aug.	Sept.
1	1,760	2,200	452	640	19,900	2,420	820	4,890	820	1,380	2,650	522
2	1,080	2,040	390	600	18,900	1,830	820	4,020	820	1,080	1,440	452
3	1,140	1,760	420	640	16,600	1,760	870	3,270	870	870	1,380	640
4	1,080	1,380	390	600	8,590	1,380	820	2,420	820	600	1,080	2,420
5	870	1,140	2,120	640	4,540	1,080	820	1,690	820	600	772	2,040
6	820	772	1,080	1,690	2,880	1,140	2,120	1,140	870	640	640	1,760
7	870	640	640	1,760	1,690	2,040	1,440	1,080	1,690	600	870	1,380
8	820	560	600	1,080	1,380	1,830	9,070	1,080	3,190	600	772	1,080
9	870	600	640	640	1,760	1,760	12,600	1,760	2,880	640	600	870
10	820	560	600	452	2,120	2,120	11,600	1,140	2,500	560	486	772
11	870	600	640	3,270	2,420	1,900	9,560	1,080	2,040	420	420	726
12	820	522	600	11,500	2,420	1,690	3,190	1,080	1,600	420	420	682
13	920	560	640	10,100	2,500	1,140	2,500	1,760	3,270	452	452	640
14	600	486	600	6,690	2,420	820	1,690	6,690	2,420	420	420	600
15	640	486	640	9,660	2,420	820	1,080	5,780	1,690	420	420	522
16	600	420	600	8,890	2,880	870	4,110	4,980	1,440	452	2,880	522
17	640	452	640	5,160	8,590	820	4,890	3,190	1,080	420	3,190	420
18	560	420	600	3,600	4,890	820	4,890	2,420	1,080	420	2,420	522
19	1,440	452	640	2,500	8,690	870	3,270	1,140	2,500	1,440	1,760	640
20	3,190	560	560	2,040	6,780	870	2,500	2,120	1,140	3,270	1,140	820
21	2,120	452	600	2,120	5,780	1,690	5,330	3,190	1,080	2,800	820	972
22	820	420	560	2,420	4,890	1,690	5,780	2,800	820	2,420	682	870
23	820	452	600	3,270	4,720	1,440	2,500	2,500	640	1,760	640	870
24	600	420	560	2,800	4,450	1,080	1,690	1,690	600	1,080	600	772
25	640	420	600	2,500	4,280	920	1,080	1,380	1,380	1,030	560	600
26	600	390	600	1,690	4,110	870	2,500	1,140	2,500	972	560	560
27	640	420	640	3,680	3,190	820	3,190	1,080	1,690	870	452	420
28	600	452	600	8,590	2,880	820	2,420	1,080	1,080	820	682	522
29	640	452	640	13,600	......	870	4,980	1,030	870	2,500	870	640
30	1,690	420	600	15,500	......	870	5,870	870	1,030	2,800	726	600
31	1,140	......	640	17,600	......	820	......	820	......	2,800	640	......

Monthly discharge of Oostanaula River at Resaca, Ga., for the year ending Sept. 30, 1918.

[Drainage area, 1,610 square miles.]

Month.	Discharge in second-feet.				Run-off (depth in inches on drainage area).
	Maximum.	Minimum.	Mean.	Per square mile.	
October	3,190	600	991	0.616	0.71
November	2,200	390	697	.433	.48
December	2,120	390	649	.403	.46
January	17,600	452	4,700	2.92	3.37
February	19,900	1,380	5,600	3.48	3.62
March	2,420	820	1,290	.801	.92
April	12,600	820	3,800	2.36	2.63
May	6,690	820	2,270	1.41	1.63
June	3,270	600	1,510	.938	1.05
July	3,270	420	1,150	.714	.82
August	3,190	420	1,010	.627	.72
September	2,420	420	829	.515	.57
The year	19,900	390	2,010	1.25	16.98

COOSA RIVER AT CHILDERSBURG, ALA.

Location.—At Central of Georgia Railway bridge half a mile west of Childersburg, Talladega County, 35 miles above site of lock 12, and 75.3 miles above Wetumpka.

Drainage area.—8,390 square miles (determined by Alabama Power Co.).

Records available.—February 22, 1914, to September 30, 1918.

Gage.—Gurley printing water-stage recorder attached to downstream end of second pier from right bank of river, installed on May 5, 1914. Prior to that date readings were taken from a vertical staff gage fastened to upstream side of same pier to which the Gurley gage is now attached. Datum of Gurley gage is about 0.1 foot higher than that of the staff gage. This difference in datum is believed constant since 1914. All records from 1915 to 1918 are referred to datum of Gurley gage. Sea-level elevation of zero of staff gage is 421.00 feet (United States Army Engineers' datum).

Discharge measurements.—Made from the bridge.

Channel and control.—Channel straight for half a mile below gage. Left bank high; right bank subject to overflow at extreme high stages. Control not well defined; bed of stream probably permanent.

Extremes of discharge.—Maximum stage during year from water-stage recorder, 16.1 feet from 4 p. m. January 31 to 1 a. m. February 1 (discharge, 68,700 second-feet); minimum stage, 1.3 feet September 15 (discharge, 2,840 second-feet).

1914–1918: Maximum stage from water-stage recorder, 24.7 feet from 3 to 9 and 11 to 12 p. m. July 11, 1916 (discharge not determined owing to lack of data for extending rating curve); minimum discharge, 2,370 second-feet, September 20, 1914.

Regulation.—None.

Accuracy.—Stage-discharge relation practically permanent. Rating curve based on four discharge measurements made in 1918 and is well defined between 3,000 and 20,000 second-feet; extended above 20,000 second-feet. Operation of water-stage recorder satisfactory except for periods indicated in footnote to daily-discharge table. Daily discharge ascertained by applying to rating table mean daily gage height obtained by averaging hourly gage height or, for days of large variations in stage, by averaging the discharge for intervals of the day. Record good except those above 25,000 second-feet, which should be used with caution.

Cooperation.—Gage-height record furnished by the Alabama Power Co.

Discharge measurements of Coosa River at Childersburg, Ala., during the year ending Sept. 30, 1918.

Date.	Made by—	Gage height.	Discharge.
		Feet.	*Sec.-feet.*
Apr. 23	Paulsen and Hoyt	5.38	15,000
July 24	C. G. Paulsen	2.90	6,820

Daily discharge, in second-feet, of Coosa River at Childersburg, Ala., for the year ending Sept. 30, 1918.

Day.	Oct.	Nov.	Dec.	Jan.	Feb.	Mar.	Apr.	May.	June.	July.	Aug.	Sept.
1	21,200	5,370	4,150	3,920	67,500	11,000	5,240	31,700	5,240	8,550	9,190	3,800
2	19,300	5,900	4,150	3,920	65,800	10,000	5,110	31,200	5,110	6,600	7,330	4,260
3	13,000	5,630	4,150	3,690	64,000	9,520	5,110	30,800	4,980	7,040	6,900	4,380
4	8,550	5,240	4,260	3,580	58,800	9,190	5,110	25,400	4,860	7,180	6,460	3,920
5	6,460	4,620	4,150	3,580	53,000	8,550	5,110	18,900	4,740	6,040	6,040	15,500
6	5,370	4,260	4,150	4,040	45,000	7,930	5,110	14,800	4,620	5,110	7,930	17,400
7	4,860	4,150	4,740	4,620	32,200	7,630	6,460	12,600	4,620	4,500	6,750	19,300
8	4,620	4,150	5,630	4,620	20,400	7,500	17,900	11,200	4,980	4,150	5,110	6,750
9	4,380	4,150	5,370	4,860	16,200	7,400	29,900	10,200	5,630	3,920	4,500	5,370
10	4,380	4,150	4,500	5,370	14,800	7,330	40,200	9,520	6,600	3,690	4,380	4,500
11	4,150	3,920	4,380	10,100	14,000	7,330	44,400	8,870	7,630	3,690	5,500	3,920
12	4,040	3,800	4,380	27,200	13,000	7,180	44,400	9,190	6,600	3,650	5,110	3,690
13	3,920	3,800	4,380	34,600	13,000	7,040	38,100	9,850	6,460	3,600	4,260	3,470
14	3,800	3,800	4,380	39,700	12,600	7,040	27,600	15,900	6,040	3,550	4,740	3,360
15	3,920	3,690	4,150	47,200	11,900	6,900	16,600	20,000	5,900	3,500	4,860	2,840
16	3,920	3,690	3,920	47,200	11,200	6,460	12,200	20,000	5,630	3,500	4,150	3,470
17	3,800	3,800	3,920	42,300	13,300	6,460	11,200	16,600	5,240	3,470	3,800	3,250
18	3,690	3,800	3,920	36,600	17,400	6,320	11,600	14,400	4,860	3,470	3,690	3,040
19	4,150	3,800	3,920	27,600	21,200	6,040	12,600	12,200	4,500	3,470	3,690	3,250
20	5,240	4,500	3,920	19,300	23,700	5,900	13,000	9,850	4,380	4,150	3,690	3,250
21	5,760	5,370	3,920	15,100	25,800	6,180	13,000	8,550	4,860	4,980	3,920	3,040
22	6,900	5,110	3,920	21,600	25,400	6,180	12,200	8,240	5,900	6,180	3,800	3,040
23	7,040	4,860	3,920	20,000	22,800	6,180	15,100	7,930	5,630	7,180	3,360	3,140
24	6,180	4,620	4,040	17,400	20,000	6,600	15,900	7,930	5,240	6,750	3,250	3,580
25	5,240	4,380	4,150	15,900	17,400	6,600	14,000	8,240	4,860	6,460	3,250	3,690
26	4,500	4,150	4,150	14,400	15,100	6,320	11,900	7,630	5,240	6,180	3,360	3,690
27	4,260	3,920	4,150	13,000	13,000	6,040	10,200	7,040	4,860	10,200	3,250	3,470
28	4,040	3,800	4,150	14,400	12,000	6,040	13,300	6,750	4,620	12,200	3,140	5,630
29	4,040	4,040	4,040	35,900	....	5,760	21,200	6,180	6,040	9,520	3,040	6,750
30	4,620	4,260	3,920	53,000	....	5,500	29,400	5,760	8,550	9,520	3,040	4,980
31	4,860	....	3,920	66,900	....	5,240	....	5,500	....	10,500	3,140	....

NOTE.—Water-stage recorder did not operate satisfactorily Feb. 27 to Mar. 2, Mar. 8, 9, July 12-16, and Sept. 15-21; discharge estimated by comparison with records of stage at Riverside except that for July 12-16 which was estimated, and Sept. 15-21 which was determined from daily readings of staff gage reduced to datum of Gurley gage.

Monthly discharge of Coosa River at Childersburg, Ala., for the year ending Sept. 30, 1918.

[Drainage area, 8,390 square miles.]

Month.	Discharge in second-feet.				Run-off (depth in inches on drainage area).
	Maximum.	Minimum.	Mean.	Per square mile.	
October	21,200	3,690	6,140	0.732	0.84
November	5,900	3,690	4,360	.520	.58
December	5,630	3,920	4,220	.503	.58
January	66,900	3,580	21,300	2.54	2.93
February	67,500	11,200	26,400	3.15	3.28
March	11,000	5,240	7,080	.844	.97
April	44,400	5,110	17,100	2.04	2.28
May	31,700	5,500	13,300	1.59	1.83
June	8,550	4,380	5,480	.653	.73
July	12,200	3,470	5,890	.702	.81
August	9,190	3,040	4,670	.557	.64
September	19,300	2,840	5,320	.634	.71
The year	67,500	2,840	10,000	1.19	16.18

ETOWAH RIVER NEAR ROME, GA.

LOCATION.—At Freemans Ferry, a railroad stop on Nashville, Chattanooga & St. Louis Railway branch line from Kingston to Rome, Ga., 1 mile downstream from mouth of Dikes Creek and 5 miles upstream from Rome, Floyd County, where Etowah and Oostanaula rivers unite to form Coosa River.

DRAINAGE AREA.—1,800 square miles.

Records available.—August 17, 1904, to September 30, 1918.

Gage.—Vertical staff in three sections on left bank, 250 feet downstream from ferry; read by R. M. Pattillo.

Discharge measurements.—Made from boat held in place by ferry cable. Measurements can not be made at high water.

Channel and control.—Bed composed of rock, boulders, and gravel; practically permanent. Banks subject to overflow at extremely high stages. A shoal immediately below gage forms control.

Extremes of discharge.—Maximum stage recorded during year, 14.8 feet at 7 a. m. April 9 (discharge, obtained from extension or rating curve, 23,400 second-feet); minimum stage recorded, 1.55 feet at 7 a. m. and 6 p. m. September 26–27 (discharge, 668 second-feet).

1904–1918: Maximum stage recorded, 27.0 feet at 12 p. m. July 11, 1916 (discharge, 45,400 second-feet); prior to 1909 high-water rating was not defined and estimates of discharge based on an extension of the rating curve are considerably too large as shown by later measurements; minimum stage recorded, 1.2 feet October 10 and 24, 1904 (discharge, 360 second-feet).

Ice.—Stage-discharge relation not affected by ice.

Regulation.—The operation of a few saw mills upstream apparently has no effect on flow.

Accuracy.—Stage-discharge relation practically permanent. Rating curve well defined below 4,000 second-feet and extended tangent above that point. Gage read to half-tenths twice daily. Daily discharge ascertained by applying mean daily gage height to rating table. Records good below 4,000 second-feet; determinations above that point subject to error because of impossibility of obtaining flood discharge measurements.

The following discharge measurement was made by C. G. Paulsen:

March 13, 1918: Gage height, 2.50; discharge, 1,680 second-feet.

Daily discharge, in second-feet, of Etowah River near Rome, Ga., for the year ending Sept. 30, 1918.

Day.	Oct.	Nov.	Dec.	Jan.	Feb.	Mar.	Apr.	May.	June.	July.	Aug.	Sept.
1	4,360	1,360	1,200	1,090	11,200	2,060	1,360	11,200	1,200	2,200	3,300	895
2	3,300	1,250	1,140	1,090	6,880	2,060	1,360	9,040	1,140	1,920	2,960	1,540
3	2,640	1,200	1,140	1,090	4,540	1,920	1,300	7,240	1,090	1,790	5,800	4,900
4	2,340	1,140	1,090	1,090	3,820	1,790	1,300	4,000	1,300	1,790	3,640	5,800
5	2,060	1,140	1,090	1,040	3,640	1,660	1,250	3,640	1,790	1,660	2,490	3,640
6	1,790	1,090	1,090	1,040	3,470	1,660	1,250	3,300	1,300	1,600	2,340	2,060
7	1,480	1,090	1,090	990	3,470	1,600	1,790	3,300	1,200	1,540	2,340	1,300
8	1,250	1,090	1,200	990	3,300	1,600	16,600	3,300	1,200	1,420	2,200	1,090
9	1,040	1,090	1,200	1,540	2,960	1,540	21,300	2,960	1,140	1,300	2,200	990
10	942	1,090	1,090	1,250	2,800	1,540	9,760	2,960	1,090	1,200	2,060	990
11	895	1,090	1,090	2,340	2,640	1,540	4,360	2,960	1,090	1,200	1,920	942
12	848	1,090	1,090	15,200	2,640	1,540	3,300	2,800	990	1,090	1,790	895
13	800	1,040	1,040	9,040	2,490	1,600	2,960	2,340	2,340	990	1,660	895
14	800	1,040	1,040	3,470	2,490	1,600	2,960	3,640	2,060	895	1,480	848
15	755	1,040	1,040	7,240	2,340	1,600	2,800	2,960	1,790	848	1,250	848
16	755	990	1,090	7,240	2,200	1,540	2,800	2,640	1,540	800	1,090	800
17	710	990	1,090	5,440	3,640	1,540	2,640	2,490	1,420	755	2,340	800
18	710	990	1,090	4,180	3,820	1,540	2,640	2,340	2,200	710	3,130	755
19	2,490	990	1,090	3,640	2,960	1,540	2,640	2,200	1,790	2,340	1,600	755
20	2,200	990	1,040	3,640	2,800	1,540	2,960	2,060	1,660	2,200	1,420	2,200
21	1,600	942	990	3,300	2,640	1,540	4,720	1,920	1,600	2,060	1,300	1,540
22	1,420	942	990	3,130	2,490	1,540	2,960	1,790	1,540	2,060	1,200	990
23	1,420	942	1,090	2,640	2,340	1,540	2,640	1,790	1,420	1,920	1,090	800
24	1,360	895	1,140	2,490	2,340	1,480	2,490	1,790	1,360	3,640	990	710
25	1,300	895	1,200	2,340	2,340	1,480	2,340	1,660	1,300	3,640	895	710
26	1,250	895	1,200	2,340	2,340	1,420	6,700	1,660	2,200	2,340	848	668
27	1,200	1,090	1,140	2,340	2,200	1,420	7,420	1,540	1,660	5,440	800	668
28	1,200	1,090	1,090	5,800	2,060	1,420	5,800	1,420	1,420	7,240	942	1,540
29	1,200	1,090	1,090	17,300	....	1,420	5,440	1,420	1,300	5,440	2,490	1,300
30	1,140	1,200	1,090	16,600	....	1,360	7,960	1,300	3,130	4,000	1,540	1,200
31	1,090	....	1,090	18,400	....	1,360	....	1,200	....	3,640	1,040	....

Monthly discharge of Etowah River near Rome, Ga., for the year ending Sept. 30, 1918.

[Drainage area, 1,800 square miles.]

Month.	Discharge in second-feet.				Run-off (depth in inches on drainage area).
	Maximum.	Minimum.	Mean.	Per square mile.	
October	4,380	710	1,500	0.833	0.96
November	1,380	895	1,060	.589	.56
December	1,200	990	1,100	.611	.70
January	18,400	990	4,820	2.68	3.09
February	11,200	2,080	3,320	1.84	1.92
March	2,060	1,360	1,580	.878	1.01
April	21,300	1,250	4,530	2.52	2.81
May	11,200	1,200	3,060	1.70	1.96
June	3,130	990	1,540	.856	.96
July	7,240	710	2,250	1.25	1.44
August	5,800	800	1,940	1.08	1.24
September	5,800	666	1,440	.800	.89
The year	21,300	666	2,340	1.30	17.64

TALLAPOOSA RIVER AT STURDEVANT, ALA.

LOCATION.—At bridge of Central of Georgia Railway one-fourth mile west of Sturdevant, Tallapoosa County, and 5 miles below mouth of Hillabee Creek.

DRAINAGE AREA.—2,460 square miles (2,500 square miles used in computing table of monthly means, published in Water-Supply Papers 322 and 352 for years 1912 and 1913).

RECORDS AVAILABLE.—July 19, 1900, to September 30, 1918.

GAGE.—Vertical staff on right bank about 2,000 feet upstream from bridge; installed August 20, 1906; read by A. L. Stowe. Original gage, a staff attached to pier of railroad bridge, was read until July 10, 1905, when the present gage was substituted for the chain gage because it was impossible to obtain an observer for chain gage. From August 21, 1906, to September 30, 1915, readings on the present staff gage were reduced to datum of original gage by means of comparative readings; since October 1, 1915, gage heights have been obtained from readings on the present staff gage without reference to datum of old gage, which has been removed.

DISCHARGE MEASUREMENTS.—Made from a plank walk resting on lower members of deck of railroad bridge.

CHANNEL AND CONTROL.—Bed rough and rocky; permanent. At extreme high stage water overflows banks. Control is a series of rock ledges and shoals below gage; permanent.

EXTREMES OF DISCHARGE.—Maximum stage recorded during year, 16.4 feet January 12 (discharge, 39,900 second-feet); minimum stage recorded, 0.2 foot July 17 and August 25 (discharge, 585 second-feet).

1900–1918: Maximum stage recorded, 22.5 feet at 5. p. m. December 29, 1915 (discharge, 58,200 second-feet); minimum stage, −0.2 foot (old datum) October 25–29, 1904 (discharge, 250 second-feet).

ICE.—Stage-discharge relation not affected by ice.

REGULATION.—Practically none.

ACCURACY.—Stage-discharge relation permanent. Rating curve well defined between 500 and 20,000 second-feet; extended above that point. Gage read to hundredths twice daily. Daily discharge ascertained by applying mean daily gage height to rating table. Records good.

Discharge measurements of Tallapoosa River at Sturdevant, Ala., during the year ending Sept. 30, 1918.

Date.	Made by—	Gage height.	Discharge.	Date.	Made by—	Gage height.	Discharge.
		Feet.	*Sec.-ft.*			*Feet.*	*Sec.-ft.*
Mar. 21	C. G. Paulsen..........	2.51	2,380	June 20	A. H. Condron.........	1.77	1,670
Apr. 24	Paulsen and Hoyt......	2.10	1,960	Sept. 27	.....do..................	.72	870

Daily discharge, in second-feet, of Tallapoosa River at Sturdevant, Ala., for the year ending Sept. 30, 1918.

Day.	Oct.	Nov.	Dec.	Jan.	Feb.	Mar.	Apr.	May.	June.	July.	Aug.	Sept.
1	8,800	2,170	1,860	1,410	12,500	2,920	2,060	12,800	1,290	5,150	4,390	2,280
2	4,770	1,860	1,670	1,370	8,800	2,920	2,060	9,320	1,250	3,210	4,960	1,670
3	3,210	1,670	1,580	1,330	8,080	2,780	2,060	5,550	1,180	2,060	13,600	1,580
4	2,520	1,490	1,580	1,410	6,880	2,780	1,960	4,210	1,140	1,410	7,360	6,180
5	2,280	1,490	1,580	1,370	5,550	2,650	1,960	3,530	1,220	1,220	4,580	4,030
6	2,170	1,490	1,580	1,760	4,960	2,650	2,170	2,780	1,860	1,110	2,780	2,060
7	1,960	1,410	1,580	2,520	4,390	2,650	3,210	2,520	2,400	920	1,960	1,410
8	1,760	1,410	1,760	1,330	4,210	2,650	6,880	2,520	2,520	980	1,580	1,410
9	1,670	1,370	1,760	2,170	4,030	2,520	7,600	2,400	2,920	1,110	1,370	1,330
10	2,170	1,370	1,580	1,960	3,860	2,520	7,840	2,280	3,370	1,290	1,580	1,080
11	2,060	1,370	1,580	14,700	3,530	2,520	6,180	2,170	6,180	860	5,150	980
12	1,580	1,370	1,760	39,300	3,530	2,400	3,860	2,060	4,580	800	2,060	950
13	1,490	1,410	1,860	15,900	3,860	2,400	2,920	2,520	4,390	710	1,330	890
14	1,490	1,490	1,860	10,600	4,390	2,400	2,650	5,750	2,170	655	1,370	830
15	1,410	1,410	1,760	9,320	8,560	2,400	2,400	4,210	1,580	630	1,110	800
16	1,410	1,410	1,760	9,580	5,550	2,280	2,400	3,370	1,410	608	1,010	860
17	1,860	1,410	1,670	7,120	7,120	2,170	2,280	2,520	1,290	608	1,080	800
18	1,410	1,330	1,670	5,150	5,750	2,170	2,520	2,280	2,170	630	950	740
19	1,490	1,370	1,580	3,860	4,210	2,170	2,520	2,060	1,670	710	1,290	655
20	1,580	3,060	1,580	3,530	4,770	2,280	2,280	1,960	1,410	1,490	1,330	710
21	1,490	3,860	1,580	4,580	4,770	2,400	2,170	1,860	1,670	1,410	1,110	2,080
22	1,490	2,650	1,490	11,400	4,210	2,400	2,060	1,960	1,760	1,220	950	2,280
23	1,410	2,060	1,490	11,700	4,030	2,400	2,060	2,060	1,410	1,180	800	2,170
24	1,410	1,860	1,490	7,600	3,690	2,280	1,960	2,170	1,490	1,370	710	1,760
25	1,370	1,580	1,580	5,550	3,530	2,280	1,860	2,400	1,110	4,580	740	1,040
26	1,330	1,490	1,860	4,580	3,370	2,060	2,780	2,280	1,180	4,770	800	880
27	1,330	1,490	1,760	3,860	3,210	2,060	3,060	2,060	1,490	4,770	655	860
28	1,370	1,410	1,670	6,400	3,060	2,060	3,060	1,580	1,330	14,700	950	1,040
29	1,760	2,170	1,580	15,300	....	1,960	6,400	1,410	1,860	10,900	1,580	2,780
30	4,210	1,960	1,580	15,600	....	1,960	10,900	1,370	4,770	4,390	1,140	3,060
31	2,780	....	1,410	17,700	....	1,960	....	1,330	....	3,060	1,370	....

Monthly discharge of Tallapoosa River at Sturdevant, Ala., for the year ending Sept. 30, 1918.

[Drainage area, 2,460 square miles.]

Month.	Discharge in second-feet.				Run-off (depth in inches on drainage area).
	Maximum.	Minimum.	Mean.	Per square mile.	
October	8,800	1,330	2,160	0.878	1.01
November	3,860	1,330	1,730	.703	.78
December	1,860	1,410	1,650	.671	.77
January	39,300	1,330	7,740	3.15	3.63
February	12,500	3,060	5,160	2.10	2.19
March	2,920	1,960	2,390	.972	1.12
April	10,900	1,860	3,470	1.41	1.57
May	12,800	1,330	3,140	1.28	1.48
June	6,180	1,110	2,140	.870	.97
July	14,700	608	2,530	1.03	1.19
August	13,600	655	2,310	.939	1.08
September	6,180	655	1,640	.667	.74
The year	39,300	608	3,000	1.22	16.53

MISCELLANEOUS MEASUREMENTS.

Miscellaneous discharge measurements in south Atlantic and eastern Gulf of Mexico drainage basins during the year ending September 30, 1918.

Streams draining into south Atlantic Ocean.

Date.	Stream.	Tributary to—	Locality.	Gage height.	Discharge.
				Feet.	*Sec.-ft.*
June 19	Roanoke River	Atlantic Ocean	Former gaging station at Southern Railway bridge at Randolph, Va.	a 4.40	1,430
29	Cape Fear River	....do	Highway bridge at Fayetteville, N. C.	a 6.20	1,660
29	Lower Little River	Cape Fear River	Highway bridge at Manchester, N. C.		213
July 1	....do	....do	Lamont's bridge, 4 miles upstream from Manchester, N. C.		188
1	Rockfish Creek	....do	Rockfish bridge, half a mile upstream from mouth of Little Rockfish Creek, N.C.		254
1	Little Rockfish Creek	Rockfish Creek	Rockfish bridge, half a mile above mouth.		73
June 29	Beaver Creek	Little Rockfish Creek	Just below Beaver Lake, at bridge on Fayetteville-Carthage road, N. C.		10.2
29	Catawba River	Wateree River	Highway bridge at Bridgewater, N. C.	8.26	333
27	Linville River	Catawba River	One mile above mouth at Bridgewater, N. C.	2.56	125
Sept. 18	Intake canal to John P. King's cotton mill.	Diverts from Savannah River.	At Augusta, Ga.		822
19	Tailrace of the Sutherland cotton mill.	Savannah River	....do		158

Streams draining into eastern Gulf of Mexico.

Date.	Stream.	Tributary to—	Locality.	Gage height.	Discharge.
Aug. 27	Big Potato Creek	Flint River	At Nelson's highway bridge, 6 miles west of Thomaston, Ga.		36.2
July 31	Tallapoosa River	Alabama River	Former gaging station at Milstead, Ala.	3.98	4,640
Sept. 8	Etowah River	Coosa River	Former gaging station at Ball Ground, Ga.	2.60	453
7	Chamblee Creek	Etowah River	Half a mile above mouth, near Canton, Ga.		5.8

a United States Weather Bureau gage.

INDEX.

STREAM-GAGING STATIONS

AND

PUBLICATIONS RELATING TO WATER RESOURCES

PART II. SOUTH ATLANTIC SLOPE AND EASTERN GULF OF MEXICO BASINS

STREAM-GAGING STATIONS AND PUBLICATIONS RELATING TO WATER RESOURCES.

INTRODUCTION.

Investigation of water resources by the United States Geological Survey has consisted in large part of measurements of the volume of flow of streams and studies of the conditions affecting that flow, but it has comprised also investigation of such closely allied subjects as irrigation, water storage, water powers, ground waters, and quality of waters. Most of the results of these investigations have been published in the series of water-supply papers, but some have appeared in the bulletins, professional papers, monographs, and annual reports.

The results of stream-flow measurements are now published annually in 12 parts, each part covering an area whose boundaries coincide with natural drainage features as indicated below:

Part I. North Atlantic slope basins.
II. South Atlantic slope and eastern Gulf of Mexico basins.
III. Ohio River basin.
IV. St. Lawrence River basin.
V. Upper Mississippi River and Hudson Bay basins.
VI. Missouri River basin.
VII. Lower Mississippi River basin.
VIII. Western Gulf of Mexico basins.
IX. Colorado River basin.
X. Great Basin.
XI. Pacific slope basins in California.
XII. North Pacific slope basins; in three volumes:
A. Pacific slope basins in Washington and upper Columbia River basin.
B. Snake River basin.
C. Lower Columbia River basin and Pacific slope basins in Oregon.

HOW GOVERNMENT REPORTS MAY BE OBTAINED OR CONSULTED.

Water-supply papers and other publications of the United States Geological Survey containing data in regard to the water resources of the United States may be obtained or consulted as indicated below:

1. Copies may be obtained free of charge by applying to the Director of the Geological Survey, Washington, D. C. The edition printed for free distribution is, however, small, and is soon exhausted.

2. Copies may be purchased at nominal cost from the Superintendent of Documents, Government Printing Office, Washington, D. C., who will on application furnish lists giving prices.

3. Sets of the reports may be consulted in the libraries of the principal cities in the United States.

4. Complete sets are available for consultation in the local offices of the water-resources branch of the Geological Survey, as follows:

Boston, Mass., 2500 Customhouse.
Albany, N. Y., 704 Journal Building.
Atlanta, Ga., Post Office Building.
Madison, Wis., Capitol Building, care of Railroad Commission of Wisconsin.
Helena, Mont., Montana National Bank Building.
Topeka, Kans., 23 Federal Building.
Denver, Colo., 403 New Post Office Building.
Salt Lake City, Utah, 313 Federal Building.
Boise, Idaho, 615 Idaho Building.
Tucson, Ariz., University of Arizona.
Austin, Tex., Capitol Building.
Portland, Oreg., 606 Post Office Building.
Tacoma, Wash., 406 Federal Building.
San Francisco, Cal., 328 Customhouse.
Los Angeles, Cal., 619 Federal Building.
Honolulu, Hawaii, 25 Capitol Building.

A list of the Geological Survey's publications may be obtained by applying to the Director of the United States Geological Survey, Washington, D. C.

STREAM-FLOW REPORTS.

Stream-flow records have been obtained at more than 4,500 points in the United States, and the data obtained have been published in the reports tabulated below:

Stream-flow data in reports of the United States Geological Survey.

[A=Annual Report; B=Bulletin; W=Water-Supply Paper.]

Report.	Character of data.	Year.
10th A, pt. 2	Descriptive information only	
11th A, pt. 2	Monthly discharge and descriptive information	1884 to Sept., 1890.
12th A, pt. 2	.....do	1884 to June 30, 1891.
13th A, pt. 3	Mean discharge in second-feet	1884 to Dec. 31, 1892.
14th A, pt. 2	Monthly discharge (long-time records, 1871 to 1893)	1888 to Dec. 31, 1893.
B 131	Descriptions, measurements, gage heights, and ratings	1893 and 1894.
16th A, pt. 2	Descriptive information only	
B 140	Descriptions, measurements, gage heights, ratings, and monthly discharge (also many data covering earlier years).	1895.
W 11	Gage heights (also gage heights for earlier years)	1896.
18th A, pt. 4	Descriptions, measurements, ratings, and monthly discharge (also similar data for some earlier years).	1895 and 1896.
W 15	Descriptions, measurements, and gage heights, eastern United States, eastern Mississippi River, and Missouri River above junction with Kansas.	1897.
W 16	Descriptions, measurements, and gage heights, western Mississippi River below junction of Missouri and Platte, and western United States.	1897.
19th A, pt. 4	Descriptions, measurements, ratings, and monthly discharge (also some long-time records).	1897.
W 27	Measurements, ratings, and gage heights, eastern United States, eastern Mississippi River, and Missouri River.	1898.

Stream-flow data in reports of the United States Geological Survey—Continued.

Report.	Character of data.	Year.
W 28	Measurements, ratings, and gage heights, Arkansas River and western United States.	1898.
20th A, pt. 4	Monthly discharge (also for many earlier years)	1898.
W 35 to 39	Descriptions, measurements, gage heights, and ratings	1899.
21st A, pt. 4	Monthly discharge	1899.
W 47 to 52	Descriptions, measurements, gage heights, and ratings	1900.
22d A, pt. 4	Monthly discharge	1900.
W 65, 66	Descriptions, measurements, gage heights, and ratings	1901.
W 75	Monthly discharge	1901.
W 82 to 85	Complete data	1902.
W 97 to 100	do	1903.
W 124 to 135	do	1904.
W 165 to 178	do	1905.
W 201 to 214	do	1906.
W 241 to 252	do	1907-8
W 261 to 272	do	1909.
W 281 to 292	do	1910.
W 301 to 312	do	1911.
W 321 to 332	do	1912.
W 351 to 362	do	1913.
W 381 to 394	do	1914.
W 401 to 414	do	1915.
W 431 to 444	do	1916.
W 451 to 464	do	1917.
W 471 to 484	do	1918.

NOTE.—No data regarding stream flow are given in the 15th and 17th annual reports.

The records at most of the stations discussed in these reports extend over a series of years, and miscellaneous measurements at many points other than regular gaging stations have been made each year. An index of the reports containing records obtained prior to 1904 has been published in Water-Supply Paper 119.

The following table gives, by years and drainage basins, the numbers of the papers on surface-water supply published from 1899 to 1918. The data for any particular station will in general be found in the reports covering the years during which the station was maintained. For example, data for Machias River at Whitneyville, Me., 1903 to 1918, are published in Water-Supply Papers 97, 124, 165, 201, 241, 261, 281, 301, 321, 351, 381, 401, 431, 451, and 471, which contain records for the New England streams from 1903 to 1918. Results of miscellaneous measurements are published by drainage basins.

In these papers and in the following lists the stations are arranged in downstream order. The main stem of any river is determined by measuring or estimating its drainage area—that is, the headwater stream having the largest drainage area is considered the continuation of the main stream, and local changes in name and lake surface are disregarded. All stations from the source to the mouth of the main stem of the river are presented first, and the tributaries in regular order from source to mouth follow, the streams in each tributary basin being listed before those of the next basin below.

In exception to this rule the records for Mississippi River are given in four parts, as indicated on page III, and the records for large lakes are presented in order of streams around the rim of the lake.

Number of water-supply papers containing results of stream measurements, 1899–1918.

Year.	I	II	III	IV	V	VI	VII	VIII	IX	X	XI	XII		
												North Pacific slope basins.		
	North Atlantic slope basins (St. John River to York River).	South Atlantic slope and eastern Gulf of Mexico basins (James River to the Mississippi).	Ohio River basin.	St. Lawrence River basin.	Hudson Bay and upper Mississippi River basins.	Missouri River basin.	Lower Mississippi River basin.	Western Gulf of Mexico basins.	Colorado River basin.	Great Basin.	Pacific slope basins in California.	Pacific slope basins in Washington and upper Columbia River basin.	Snake River basin.	Lower Columbia River basin and Pacific slope basins in Oregon.
1899 [a]	35	[b] 35, 36	36	36	36	[c] 36, 37	37	37	[d] 37, 38	38, [e] 39	38, [f] 39	38	38	38
1900 [g]	47, [h] 48	48	48, [i] 49	49	49	49, [s]/[j] 50	50	50	50	51	51	51	51	51
1901	65, 75	65, 75	65, 75	65, 75	[k] 65, 66, 75	66, 75	[k] 65, 66, 75	66, 75	66, 75	66, 75	66, 75	66, 75	66, 75	66, 75
1902	82	[b] 82, 83	83	[l] 82, 83	[k] 83, 85	84	[k] 83, 84	84	85	85	85	85	85	85
1903	97	[b] 97, 98	98	97	[k] 98, 99, [m] 100	99	[k] 98, 99	99	100	100	100	100	100	100
1904	[n] 124, [o] 125, [p] 126	[p] 126, 127	128	129	[k] 128, 130	130, [q] 131	[k] 128, 131	132	133	133, [r] 134	134	135	135	135
1905	[n] 165, [o] 166, [p] 167	[p] 167, 168	169	170	171	172	[k] 169, 173	174	175, [s]/[e] 177	176, [r] 177	177	178	178	[s]/[t] 177, 178
1906	[n] 201, [o] 202, [p] 203	[p] 203, 204	205	206	207	208	[k] 205, 209	210	211	212, [r] 213	213	214	214	214
1907–8	241	242	243	244	245	246	247	248	249	250, [r] 251	251	252	252	252
1909	261	262	263	264	265	266	267	268	269	270, [r] 271	271	272	272	272
1910	281	282	283	284	285	286	287	288	289	290	291	292	292	292
1911	301	302	303	304	305	306	307	308	309	310	311	312	312	312
1912	321	322	323	324	325	326	327	328	329	330	331	332–A	332–B	332–C
1913	351	352	353	354	355	356	357	358	359	360	361	362–A	362–B	362–C
1914	381	382	383	384	385	386	387	388	389	390	391	392	393	394
1915	401	402	403	404	405	406	407	408	409	410	411	412	413	414
1916	431	432	433	434	435	436	437	438	439	440	441	442	443	444
1917	451	452	453	454	455	456	457	458	459	460	461	462	463	464
1918	471	472	473	474	475	476	477	478	479	480	481	482	483	484

[a] Rating tables and index to Water-Supply Papers 35–39 contained in Water-Supply Paper 39. Tables of monthly discharge for 1899 in Twenty-first Annual Report, Part IV.
[b] James River only.
[c] Gallatin River.
[d] Green and Gunnison rivers and Grand River above junction with Gunnison.
[e] Mohave River only.
[f] Kings and Kern rivers and south Pacific slope basins.
[g] Rating tables and index to Water-Supply Papers 47–52 and data on precipitation, wells, and irrigation in California and Utah contained in Water-Supply Paper 52. Tables of monthly discharge for 1900 in Twenty-second Annual Report, Part IV.
[h] Wissahickon and Schuylkill rivers to James River.
[i] Scioto River.
[j] Loup and Platte rivers near Columbus, Nebr., and all tributaries below junction with Platte.
[k] Tributaries of Mississippi from east.
[l] Lake Ontario and tributaries to St. Lawrence River proper.
[m] Hudson Bay only.
[n] New England Rivers only.
[o] Hudson River to Delaware River, inclusive.
[p] Susquehanna River to Yadkin River, inclusive.
[q] Platte and Kansas rivers.
[r] Great Basin in California except Truckee and Carson river basins.
[s] Below junction with Gila.
[t] Rogue, Umpqua, and Siletz rivers only.

PRINCIPAL STREAMS.

The south Atlantic slope and eastern Gulf of Mexico drainage basins include streams flowing into the Atlantic Ocean and Gulf of Mexico from York River, Va., to Pearl River, Miss., inclusive. The principal streams in this division are James, Roanoke, Cape Fear, Yadkin, Santee, Savannah, Altamaha, Apalachicola, Chotawhatchee, Mobile, and Pearl. The streams drain wholly or in part the States of Alabama, Florida, Georgia, Mississippi, North Carolina, South Carolina, and Virginia.

In addition to the annotated list of publications relating specifically to the section, these pages contain a similar list of reports that are of general interest in many sections and cover a wide range of hydrologic subjects, and also brief references to reports published by State and other organizations. (See p. XVII.)

GAGING STATIONS.

NOTE.—Dash after a date indicates that station was being maintained September 30, 1918; period after a date indicates discontinuance. Tributaries are indicated by indention.

JAMES RIVER BASIN.

Jackson River (head of James) at Covington, Va., 1907-8.
James River at Buchanan, Va., 1895-
James River at Holcomb Rock, Va., 1900-1915.
James River at Cartersville, Va., 1899-
    Cowpasture River near Clifton Forge, Va., 1907-8.
    North River near Glasgow, Va., 1895-1905.
    Appomattox River at Mattoax, Va., 1900-1905.

ROANOKE RIVER BASIN.

Roanoke River at Roanoke, Va., 1896-
Roanoke River at Randolph, Va., 1900-1906.
Roanoke River above Dan River, at Clarksville, Va., 1895-1898.
Roanoke River at Old Gaston, N. C., 1911-
Roanoke River near Weldon, N. C., 1912.
Roanoke River at Neal, N. C., 1896-1903.
    Tinker Creek at Roanoke, Va., 1907-8.
    Back Creek near Roanoke, Va., 1907-8.
    Dan River at Madison, N. C., 1903-1908.
    Dan River at South Boston, Va., 1900-1907.
    Dan River at Clarksville, Va., 1895-1898.
        Banister River at Houston, Va., 1904-5.

TAR RIVER BASIN.

Tar River near Tarboro, N. C., 1896-1900.

NEUSE RIVER BASIN.

Neuse River near Selma, N. C., 1896-1900.

CAPE FEAR RIVER BASIN.

Haw River (head of Cape Fear River) near Moncure, N. C., 1898–99.
Cape Fear River near Fayetteville, N. C., 1889–1903.
Deep River near Cumnock, N. C., 1900–1902.
Deep River near Moncure, N. C., 1898–99.
Rockfish Creek near Brunt, N. C., 1902–3.

YADKIN (OR PEEDEE) RIVER BASIN.

Yadkin River (head of Peedee River) at North Wilkesboro, N. C., 1903–1909.
Yadkin River at Siloam, N. C., 1900–1901.
Yadkin River at Donnaha, N. C., 1913–
Yadkin River near Salisbury, N. C., 1895–1909; 1911–
Yadkin River near Norwood, N. C., 1896–1899.
Yadkin River near Peedee, N. C., 1906–1912.
Peedee River at Cheraw, S. C., 1909–1912.

SANTEE RIVER BASIN.

Catawba River (head of Santee River) at Old Fort, N. C., 1907.
Catawba River near Morganton, N. C., 1900; 1903–1909.
Catawba River at Rhodhiss, N. C., 1917–
Catawba River at Catawba, N. C., 1896–1902.
Catawba River near Catawba, S. C., 1903–1905.
Catawba River near Rock Hill, S. C., 1895–1903.
Wateree River (lower part of Catawba) near Camden, S. C., 1903–1910.
Mill Creek at Old Fort, N. C., 1907.
Linville River at Fonta Flora, N. C., 1907–8.
Linville River near Bridgewater, N. C., 1900.
John River at Collettsville, N. C., 1907.
John River near Morganton, N. C., 1900–1901.
Broad River (of the Carolinas), head of Congaree River, at Uree, N. C., 1907–1909.
Broad River (of the Carolinas) at Dellinger, S. C., 1900–1901.
Broad River (of the Carolinas) near Gaffney, S. C., 1896–1899.
Broad River (of the Carolinas) at Alston, S. C., 1896–1907.
Green River near Saluda, N. C., 1907–1909.
Second Broad River near Logans Store, N. C., 1907–8.
Saluda River near Waterloo, S. C., 1896–1905.
Saluda River near Ninety Six, S. C., 1905.

EDISTO RIVER BASIN.

Four Hole Creek near Ridgeville, S. C., 1914–1917.

SAVANNAH RIVER BASIN.

Chattooga River (head of Savannah River) near Clayton, Ga., 1907–8.
Chattooga River near Tallulah Falls, Ga., 1917–
Tugaloo River (continuation of Chattooga River) near Toccoa, Ga., 1907–8.
Tugaloo River near Madison, S. C., 1898–1901; 1903–1910.
Savannah River near Calhoun Falls, S. C., 1896–1903.
Savannah River at Woodlawn, S. C., 1905–1910.
Savannah River at Augusta, Ga., 1884–1906.
Stekoa Creek near Clayton, Ga., 1907–8.
Tallulah River near Seed, Ga., 1916–
Tallulah River near Lakemont, Ga., 1916–

Savannah River at Augusta, Ga., 1899-1906—Continued.
Tallulah River at Mathis, Ga., 1912-1916.
Tallulah River at Tallulah Falls, Ga., 1900-1901; 1904-1912.
Tiger Creek at Lakemont, Ga., 1916-
Chauga River near Madison, S. C., 1907.
Seneca River near Clemson College, S. C., 1903-1905.
Broad River (of Georgia) near Carlton, Ga., 1897-1913.

OGEECHEE RIVER BASIN.

Ogeechee River near Millen, Ga., 1903.
Williamsons Swamp Creek near Davisboro, Ga., 1903-4.
Canoochee River near Groveland, Ga., 1903-1907.

ALTAMAHA RIVER BASIN.

South River (head of Ocmulgee River, which is head of Altamaha River) near Lithonia, Ga., 1903-4.
Ocmulgee River near Jackson, Ga., 1906-1915.
Ocmulgee River near Flovilla, Ga., 1901-1905.
Ocmulgee River at Juliette, Ga., 1916-
Ocmulgee River at Macon, Ga., 1893-1913.
Yellow River at Almon, Ga., 1897; 1899-1901.
Alcovy River near Covington, Ga., 1901-1904.
Alcovy River near Stewart, Ga., 1905-6.
Towaliga River near Juliette, Ga., 1899-1901.
Oconee River at Barnett Shoals, near Watkinsville, Ga., 1902.
Oconee River near Greensboro, Ga., 1903-
Oconee River at Carey, Ga., 1896-1898.
Oconee River at Fraleys Ferry, near Milledgeville, Ga., 1906-1908; 1909-
Oconee River at Milledgeville, Ga., 1903-1905.
Oconee River at Dublin, Ga., 1894-1913.
Middle Oconee River near Athens, Ga., 1901-2.
Apalachee River near Buckhead, Ga., 1901-1908.
Ohoopee River near Reidsville, Ga., 1903-1907.

ST. JOHNS RIVER BASIN.

Silver Spring near Silver Springs, Fla., 1906-7.

FLORIDA EVERGLADES DRAINAGE CANALS.

North New River canal near Fort Lauderdale, Fla., 1913.
North New River canal near Rita, Fla., 1913.
South New River canal near Zona, Fla., 1913.
South New River canal near Rita, Fla., 1913.
Miami canal near Miami, Fla., 1913.

SUWANNEE RIVER BASIN.

Suwannee River near White Springs, Fla., 1906-1908.

APALACHICOLA RIVER BASIN.

Chattahoochee River (head of Apalachicola River) near Ariel, Ga., 1907-1909.
Chattahoochee River near Leaf, Ga., 1907.
Chattahoochee River near Gainsville, Ga., 1901-1903; 1917-18.
Chattahoochee River near Buford, Ga., 1901.
Chattahoochee River near Norcross, Ga., 1903-

112130°—20—W S P 472——5

Chattahoochee River at Oakdale, Ga., 1895–1904.
Chattahoochee River at West Point, Ga., 1896–1910; 1912–
Chattahoochee River at Columbus, Ga., 1912.
Chattahoochee River at Alaga, Ala., 1908–1912.
Soque River near Demorest, Ga., 1904–1909.
Chestatee River at New Bridge, Ga., 1917–18.
Sweetwater Creek near Austell, Ga., 1904–5; 1913.
Flint River near Molina, Ga., 1897–98.
Flint River near Woodbury, Ga., 1900–
Flint River near Musella, Ga., 1907.
Flint River near Culloden, Ga., 1911–
Flint River near Montezuma, Ga., 1905–1909; 1911–12.
Flint River at Albany, Ga., 1897–
Flint River at Bainbridge, Ga., 1908–1913.
Little Potato (Tobler) Creek near Yatesville, Ga., 1914–1918.
Kinchafoonee Creek near Leesburg, Ga., 1905–1909.
Kinchafoonee Creek near Albany, Ga., 1903.
Muckalee Creek near Albany, Ga., 1903.
Ichawaynochaway Creek at Milford, Ga., 1905–1907.
Chipola River at Altha, Fla., 1912–13.

CHOCTAWHATCHEE RIVER BASIN.

Choctawhatchee River near Newton, Ala., 1906–1908; 1911–12.
Choctawhatchee River near Geneva, Ala., 1904.
Double Bridges Creek at Geneva, Ala., 1904.
Pea River at Pera, Ala, 1904–1913.
Pea River at Elba, Ala. 1906.

ESCAMBIA RIVER BASIN.

Conecuh River at Beck, Ala., 1904–

MOBILE RIVER BASIN.

Cartecay River (head of Mobile River) near Cartecay, Ga., 1904–5; 1907.
Coosawattee River (continuation of Cartecay River) at Carters, Ga., 1896–1908.
Oostanaula River (continuation of Coosawattee River) at Resaca, Ga., 1892–1901; 1905–
Coosa River (continuation of Oostanaula River) at Rome, Ga., 1897–1903.
Coosa River at Lock No. 4, above Riverside, Ala., 1890–1901.
Coosa River at Riverside, Ala., 1896–1916.
Coosa River at Lock No. 5, near Riverside, Ala., 1892–1899.
Coosa River at Childersburg, Ala., 1914–
Coosa River at Lock No. 12, near Clanton, Ala., 1912–1914.
Coosa River at Lock No. 18, near Wetumpka, Ala., 1912–1914.
Coosa River near Wetumpka, Ala., 1896–1898.
Alabama River (continuation of Coosa River) at Montgomery, Ala., 1899–1903.
Alabama River at Selma, Ala., 1899–1913.
Ellijay River at Ellijay, Ga., 1907.
Conasauga River at Beaverdale, Ga., 1907–8.
Etowah River near Ball Ground, Ga., 1907–1915.
Etowah River at Canton, Ga., 1892–1905.
Etowah River near Rome, Ga., 1904–
Etowah River at Rome, Ga., 1903.
Amicalola River near Potts Mountain, Ga., 1907–8; 1910–1913.

REPORTS ON WATER RESOURCES OF THE SOUTH ATLANTIC AND EASTERN GULF STATES.

WATER-SUPPLY PAPERS.

Water-supply papers are distributed free by the Geological Survey as long as its stock lasts. An asterisk (*) indicates that this stock has been exhausted. Many of the papers marked in this way may, however, be purchased (at price noted) from the SUPERINTENDENT OF DOCUMENTS, WASHINGTON, D. C. Omission of the price indicates that the report is not obtainable from Government sources. Water-supply papers are of octavo size.

*44. Profiles of rivers in the United States, by Henry Gannett. 1901. 100 pp., 11 pls. 15c.

Gives elevations and distances along rivers of the United States, and brief descriptions of many of the streams, including Roanoke, Cape Fear, Peedee, Santee, Savannah, Oconee, Apalachicola, Chattahoochee, Coosa, Tallapoosa, and Black Warrior rivers.

*57. Preliminary list of deep borings in the United States, Part I (Alabama-Montana), by N. H. Darton. 1902. 60 pp. 5c.

*61. Preliminary list of deep borings in the United States, Part II (Nebraska-Wyoming), by N. H. Darton. 1902. 67 pp. 5c.

A second, revised edition of Nos. 57 and 61 was published in 1905 as Water-Supply Paper 149 (q. v.).

62. Hydrography of the southern Appalachian Mountain region, Part I, by H. A. Pressey. 1902. 95 pp., 25 pls. 15c.

63. Hydrography of the southern Appalachian Mountain region, Part II, by H. A. Pressey. 1902. pp. 96–190, pls. 26–44. 15c.

Nos. 62 and 63 describe in a general way the mountains, rivers, climate, forests, soil, vegetation, and mineral resources of the southern Appalachian Mountains, and then discuss in detail the drainage basins, giving for each an account of the physical features, rainfall, forests, minerals, transportation, discharge measurements, and water powers. Most of the streams described are tributary through Tennessee River to the Ohio, but Part II (No. 63) includes also descriptions of several streams in the South Atlantic slope and eastern Gulf of Mexico drainage basins.

*67. The motions of underground waters, by C. S. Slichter. 1902. 106 pp., 8 pls. 15c.

Describes artesian wells at Savannah, Ga.

96. Destructive floods in the United States in 1903, by E. C. Murphy. 1904. 81 pp., 13 pls. 15c.

Contains an account of flood on tributaries of Broad River (of the Carolinas) in Spartanburg County, S. C.

101. Underground waters of southern Louisiana, by G. D. Harris, with discussions of their uses for water supplies and for rice irrigation, by M. L. Fuller. 1904. 98 pp., 11 pls. 20c.

Describes the geology and ground-water conditions of the area, gives data in regard to artesian wells, and outlines methods of well drilling, pumping, and rice irrigation. Includes 23 analyses of ground water.

102. Contributions to the hydrology of eastern United States, 1903; M. L. Fuller, geologist in charge. 1904. 522 pp. 30c.

Contains brief reports on municipal water supplies, wells, and springs of Georgia, Florida, Alabama, and Mississippi. The reports comprise tabulated well records, giving information as to location, owner, depth, yield, head, etc., supplemented by notes as to elevation above sea, materials penetrated, temperature, use, and quality; many miscellaneous analyses.

*103. A review of the laws forbidding pollution of inland waters in the United States, by E. B. Goodell. 1904. 120 pp. Superseded by 152.

Cites statutory restrictions of water pollution in Alabama, Florida, Georgia, Mississippi, North Carolina, and Virginia.

*107. Water powers of Alabama, with an appendix on stream measurements in Mississippi, by B. M. Hall. 1904. 253 pp., 9 pls. 20c.

Contains gage heights, rating tables, and estimates of monthly discharge at stations on Tallapoosa, Coosa, Alabama, Cahaba, Black Warrior, and Tombigbee rivers and their tributaries; gives estimates and short descriptions of water powers.

110. Contributions to the hydrology of eastern United States, 1904; M. L. Fuller, geologist in charge. 1905. 211 pp., 5 pls. 10c.

Contains reports as follows:

Experiment relating to problems of well contamination at Quitman, Ga., by S. W. McCallie. Scope indicated by title.

Water resources of the Cowee and Pisgah quadrangles, North Carolina, by Hoyt S. Gale. Discusses drainage, springs, and mineral waters of one of the units of the geologic atlas of the United States.

*114. Underground waters of eastern United States; M. L. Fuller, geologist in charge. 1905. 285 pp., 18 pls. 25c.

Contains brief reports relating to south Atlantic slope and eastern Gulf of Mexico drainage areas, as follows:

Virginia, by N. H. Darton and M. L. Fuller.
North Carolina, by M. L. Fuller.
South Carolina, by L. C. Glenn.
Georgia, by S. W. McCallie.
Florida, by M. L. Fuller.
Alabama, by A. E. Smith.
Mississippi, by L. C. Johnson.

Each of these reports describes the geology of the area in its relation to water supplies, notes the principal mineral springs, and gives list of pertinent publications.

115. River surveys and profiles made during 1903, arranged by W. C. Hall and J. C. Hoyt. 1905. 115 pp., 4 pls. 10c.

Contains results of surveys made to determine location of undeveloped power sites. Gives elevations and distances along Catawba, Tallulah, Chattooga, Tugaloo, Savannah, Broad, Ocmulgee, Yellow, South, Alcovy, Towaliga, and Chattahoochee rivers.

145. Contributions to the hydrology of eastern United States, 1905; M. L. Fuller, geologist in charge. 1905. 220 pp., 6 pls. 10c.

Contains "Notes on certain hot springs of the southern United States," by Walter Harvey Weed, including the "Warm springs of Georgia." Describes the location of the springs, the geologic conditions, and the composition of the waters (with analyses); estimates discharge.

*149. Preliminary list of deep borings in the United States, second edition with additions, by N. H. Darton. 1905. 175 pp. 10c.

Gives by States (and within the States by counties) location, depth, diameter, yield, height of water, and other valuable information concerning wells 400 feet or more in depth. Includes all wells listed in Water-Supply Papers 57 and 61; mentions also principal publications relating to deep borings.

*152. A review of the laws forbidding pollution of inland waters in the United States (second edition), by E. B. Goodell. 1905. 149 pp. 10c.

Cites statutory restrictions of water pollution in Alabama, Georgia, Florida, Mississippi, North Carolina, and Virginia.

159. Sumary of the underground-water resources of Mississippi, by A. F. Crider and L. C. Johnson. 1906. 86 pp., 6 pls. 20c.

Describes geography, topography, and general geology of the State; discusses the source, depth of penetration, rate of percolation, and recovery of ground waters; artesian requisites, and special conditions in the Coastal Plain formation; gives notes on wells by counties, deep well records, and selected records in detail; treats of sanitary aspects of wells and gives analyses.

*160. Underground-water papers, 1906; M. L. Fuller, geologist in charge. 1906. 104 pp., 1 pl.

Contains brief report entitled "Peculiar mineral waters from crystalline rocks of Georgia," by Myron L. Fuller, discussing origin of certain mineral springs and wells near Austell; gives analyses.

*162. Destructive floods in the United States in 1905, with a discussion of flood discharge and frequency and an index to flood literature, by E. C. Murphy and others. 1906. 105 pp., 4 pls. 15c.

Gives estimates of flood discharge and frequency on Cape Fear, Savannah, Alabama, and Black Warrior rivers.

*197. Water resources of Georgia, by B. M. and M. R. Hall. 1907. 342 pp., 1 pl. 50c.

Describes topographic and geologic features of the State; discusses by drainage basins, stream flow, river surveys, and water powers.

236. The quality of surface waters in the United States: Part I, Analyses of water. east of the one hundredth meridian, by R. B. Dole. 1909. 123 pp. 10c.

Describes collection of samples, methods of examination, preparation of solutions, accuracy of estimates, and expression of analytical results: gives results of analyses of waters of James, Roanoke, Dan, Neuse, Cape Fear, Peedee, Wateree, Saluda, Savannah, Ocmulgee, Oconee, Chattahoochee, Flint, Oostanaula, Alabama, Cahaba, Tombigbee, and Pearl rivers.

*258. Underground water papers, 1910; by M. L. Fuller, F. G. Clapp, G. C. Matson, Samuel Sanford, and H. C. Wolff. 1911. 123 pp., 2 pls. 15c. Contains:

Saline artesian waters of the Atlantic coastal plain, by Samuel Sanford. Discusses briefly the geology of the coastal plain, the artesian waters, the occurrence and character of the salt waters, the causes of salinity, and lateral changes in salinity.

*319. Geology and ground waters of Florida, by G. C. Matson and Samuel Sanford. 1913. 445 pp., 17 pls. 60c.

Describes the characteristic upland, lowland, and coastal features of the State—the springs, lakes, caverns, sand dunes, coral reefs, bars, inlets, tidal runways, pine lands, swamps, keys, and ocean currents; discusses in detail the stratigraphic position, lithologic character, thickness, physiographic expression, structure, and areal distribution of the geologic formations; treats of the source, amount, depth, circulation, and recovery of ground waters, the artesian waters, and public water supplies; and gives details concerning source, quality, and development of the water supplies by counties. Discusses briefly the quality of the well waters.

341. Underground waters of the coastal plain of Georgia, by L. W. Stephenson and J. O. Veatch, and a discussion of the quality of the waters, by R. B. Dole. 1915. 539 pp., 21 pls. 50c.

Describes the physiographic features of the State, the geologic provinces, the areal distribution, stratigraphic position, and lithologic character of the rocks belonging to the geologic systems; discusses the source and amount of the ground waters, the uses of the springs and shallow and artesian wells, and the distribution of the ground waters in the rocks of the various formations; gives details concerning each county. The chapter on the chemical character of the waters describes standards for classification and the general requisites of waters for miscellaneous industrial uses and for domestic use; treats also of methods of purifying water and of the relation of quality to geographic position, to water-bearing stratum, and to depth.

364. Water analyses from the laboratory of the United States Geological Survey, tabulated by F. W. Clarke, chief chemist. 1914. 40 pp. 5c.

Contains analyses of spring and well waters in Virginia, North Carolina, South Carolina, and Florida, and of water from the Gulf of Mexico.

ANNUAL REPORTS.

Each of the papers contained in the annual reports was also issued in separate form.

Annual reports are distributed free by the Geological Survey as long as its stock lasts. An asterisk (*) indicates that this stock has been exhausted. Many of the papers so marked, however, may be purchased from the SUPERINTENDENT OF DOCUMENTS, WASHINGTON, D. C.

*Tenth Annual Report of the United States Geological Survey, 1888-89, J. W. Powell, Director. 1890. 2 parts. *Pt. I. Geology, xv, 774 pp., 98 pls. $2.35. Contains:

*General account of the fresh-water morasses of the United States, with a description of the Dismal Swamp district of Virginia and North Carolina, by N. S. Shaler, pp. 235-339, pls. 6-19. Scope indicated by title.

Fourteenth Annual Report of the United States Geological Survey, 1892-93, J. W. Powell, Director. 1893. (Pt. II, 1894.) 2 parts. *Pt. II. Accompanying papers, xx, 597 pp., 73 pls. $2.10. Contains:

*Potable waters of eastern United States, by W. J. McGee, pp. 1-47. Discusses cistern water, stream waters, and ground waters, including mineral springs and artesian wells.

PROFESSIONAL PAPERS.

Professional papers are distributed free by the Geological Survey as long as its stock lasts. An asterisk (*) indicates that this stock has been exhausted. Many of the papers marked with an asterisk may, however, be purchased from the SUPERINTENDENT OF DOCUMENTS, WASHINGTON, D. C. Professional papers are of quarto size.

*37. The Southern Appalachian forests, by H. B. Ayers and W. W. Ashe. 1905. 291 pp., 37 pls. 80c.

Describes the relief, drainage, climate, natural resources, scenery, and water supply of the southern Appalachian forests, the trees, shrubs, and rate of growth; gives details concerning forests by drainage basins, including New, Holston (southern tributaries of South Fork only), Watauga, Nolichucky, French Broad, Pigeon, Little Tennessee, Hiwassee, Tallulah-Chattooga, Toxaway, Saluda and First and Second Broad rivers, Catawba and Yadkin rivers, describing many of the tributaries of each of the master streams.

*72. Denudation and erosion in the southern Appalachian region and the Monongahela basin, by L. C. Glenn. 1911. 137 pp., 21 pls. 35c.

Describes the topography, geology, drainage, forests, climate, and population, and transportation facilities of the region, the relation of agriculture, lumbering, mining, and power development to erosion and denudation, and the nature, effects, and remedies of erosion; gives details of conditions in Holston, Nolichucky, French Broad, Little Tennessee, and Hiwassee River basins, along Tennessee River proper, and in the basins of the Coosa-Alabama system, Chattahoochee, Savannah, Saluda, Broad, Catawba, Yadkin, New, and Monongahela rivers.

*90. Shorter contributions to general geology, 1914; David White, chief geologist. 1915. 199 pp., 21 pls. 40c.

Issued also in separate chapters. The following paper relates in part to ground water:
(h) A deep well at Charleston, S. C., by L. W. Stephenson, with a report on the mineralogy of the water, by Chase Palmer (pp. 69-94).

BULLETINS.

An asterisk (*) indicates that the Geological Survey's stock of the paper is exhausted. Many of the papers so marked may be purchased from the SUPERINTENDENT OF DOCUMENTS, WASHINGTON, D. C. Bulletins are of octavo size.

*138. Artesian-well prospects in the Atlantic Coastal Plain region, by N. H. Darton. 1896. 232 pp., 19 pls.

Describes the general geologic structure of the Atlantic Coastal Plain region and summarizes the conditions affecting subterranean water in the Coastal Plain; discusses the general geologic relations in New York, southern New Jersey, Delaware, Maryland, District of Columbia, Virginia, North Carolina, South Carolina, and eastern Georgia; gives for each of the States a list of the deep wells and discusses well prospects. The notes on the wells that follow the tabulated lists contain many sections and analyses of the waters.

*264. Record of deep-well drilling for 1904, by M. L. Fuller, E. F. Lines, and A. C. Veatch. 1905. 106 pp. 10c.

Discusses the importance of accurate well records to the driller, to owners of oil, gas, and water wells, and to the geologist; describes the general methods of work; gives tabulated records of wells in Alabama, Florida, Georgia, Mississippi, and North Carolina, and detailed records of wells in Hancock and Jackson counties, Mississippi. These wells were selected because they give definite stratigraphic information.

*298. Record of deep-well drilling for 1905, by M. L. Fuller and Samuel Sanford. 1906. 299 pp. 25c.

Gives an account of progress in the collection of well records and samples; contains tabulated records of wells in Alabama, Florida, Georgia, Mississippi, North Carolina, South Carolina, and Virginia; and detailed records of wells in Madison, Marengo, and Mobile counties, Alabama; Duval, Escambia, Sumter, and Volusia counties, Florida; Chatham, Decatur, Fulton, Pierce, and Tattnall counties, Georgia; Lenoir, New Hanover, and Moore counties, North Carolina; Hancock, Harrison, Jackson, Jones, Marshall, Newton, and Panola counties, Mississippi; and Aiken, Barnwell, Charleston, Hampton, Lee, and Orangeburg counties, South Carolina. The wells of which detailed sections are given were selected because they afford valuable stratigraphic information.

GEOLOGIC FOLIOS.

Under the plan adopted for the preparation of a geologic map of the United States the entire area is divided into small quadrangles, bounded by certain meridians and parallels, and these quadrangles, which number several thousand, are separately surveyed and mapped.[1] The unit of survey is also the unit of publication, and the maps and description of each quadrangle are issued in the form of a folio. When all the folios are completed they will constitute the Geologic Atlas of the United States.

A folio is designated by the name of the principal town or of a prominent natural feature within the quadrangle. Each folio includes maps showing the topography, geology, underground structure, and mineral deposits of the area mapped and several pages of descriptive text. The text explains the maps and describes the topographic and geologic features of the country and its mineral products. The topographic map shows roads, railroads, waterways, and, by contour lines, the shapes of the hills and valleys and the height above sea level of all points in the quadrangle. The areal-geology map shows the distribution of the various rocks at the surface. The structural-geology map shows the relations of the rocks to one another underground. The economic-geology map indicates the location of mineral deposits that are commercially valuable. The artesian-water map shows the depth to underground-water horizons. Economic-geology and artesian-water maps are included in folios if the conditions in the areas mapped warrant their publication. The folios are of special interest to students of geography and geology and are valuable as guides in the development and utilization of mineral resources.

The folios numbered from 1 to 163, inclusive, are published in only one form (18 by 22 inches), called the library edition. Some of the folios that bear numbers higher than 163 are published also in an octavo edition (6 by 9 inches). Owing to a fire in the Geological Survey building May 18, 1913, the stock of geologic folios was more or less damaged by fire and water, but many of the folios are usable. The damaged folios are sold at the uniform price of 5 cents each, with no reduction for wholesale orders. This rate applies to folios in stock from 1 to 184, inclusive (except reprints), also to the library edition of folio 186. The library edition of folios 185, 187, and higher numbers sells for 25 cents a copy, except that some folios which contain an unusually large amount of matter sell at higher prices. The octavo edition of folio 185 and higher numbers sells for 50 cents a copy, except folio 193, which sells for 75 cents a copy. A discount of 40 per cent is allowed on an order for folios or for folios together with topographic maps amounting to $5 or more at the retail rate.

All the folios contain descriptions of the drainage of the quadrangles. The folios in the following list contain also brief discussions of the ground waters in connection with the economic resources of the areas and more or less information concerning the utilization of the water resources.

*80. Norfolk, Virginia-North Carolina.

Describes the plains, Dismal Swamp, and the tidal marshes; discusses the reclamation of swamp lands and gives an account of the ground waters; gives sections of wells near Norfolk and at Fort Monroe, and analyses of waters from the test boring at Norfolk and the boring at Lambert Point.

90. Cranberry, North Carolina-Tennessee. 5c.

*124. Mount Mitchell, North Carolina-Tennessee.

[1] Index maps showing areas in the South Atlantic States covered by topographic maps and by geologic folios will be mailed on receipt of request addressed to the Director, U. S. Geological Survey, Washington, D. C.

*147. Pisgah, North Carolina-South Carolina.
*175. Birmingham, Alabama.[1] 5c.
187. Ellijay, Georgia-North Carolina-Tennessee.[2] 25c.

MISCELLANEOUS REPORTS.

Other Federal bureaus and State and other organizations have from time to time published reports relating to the water resources of the various sections of the country. Notable among those pertaining to the South Atlantic States are the reports of the State surveys of North Carolina, Georgia, Florida, and Alabama, and the Tenth Census (vol. 16).

The following reports deserve special mention:

Hydrography of Virginia, by N. C. Grover and R. H. Bolster: Virginia Geol. Survey Bull. 3, 1906.

Underground waters of the Coastal Plain province of Virginia, by Samuel Sanford: Virginia Geol. Survey Bull. 5, 1913.

Surface water supply of Virginia, by G. C. Stevens: Virginia Geol. Survey Bull. 10, 1916.

A preliminary report on the water powers of Georgia, by B. M. Hall: Georgia Geol. Survey Bull. 3-A, 1896.

A preliminary report on the artesian-well system of Georgia, by S. W. McCallie: Georgia Geol. Survey Bull. 7, 1898.

A preliminary report on the underground waters of Georgia, by S. W. McCallie: Georgia Geol. Survey Bull. 15, 1908.

Second report on the water powers of Georgia, by B. M. Hall and M. R. Hall: Georgia Geol. Survey Bull. 16, 1908.

A preliminary report on the mineral springs of Georgia, by S. W. McCallie: Georgia Geol. Survey Bull. 20, 1913.

Reports on condition of water supply at Savannah, Ga. Mayor of Savannah Ann. Rept., 1915.

Contains the following papers submitted by the United States Geological Survey:

Preliminary report on Savannah water supply, by L. W. Stephenson and R. B. Dole. Pp. 1-14.

The water supply of Savannah, Ga., by R. B. Dole. Pp. 15-89.

These papers discuss the yield and head of the artesian wells of Savannah, the consumption of water, the sanitary and chemical quality of the water, and the cost of operation. They give the results of fluorescein tests and several analyses of surface and ground waters. They conclude with recommendations for future development.

A preliminary report on the underground water supply of central Florida, by E. H. Sellards: Florida Geol. Survey Bull. 1, 1908.

Underground waters of Mississippi; a preliminary report by W. N. Logan and W. R. Perkins: Mississippi Agr. Exper. Sta. Bull. 89, 1905.

Report of the Secretary of Agriculture in relation to the forests, rivers, and mountains of the Southern Appalachian region: 57th Cong., 1st sess., S. Doc. 84, 1902.

Underground water resources of Alabama, by E. A. Smith. Montgomery, Ala., 1907.

Preliminary report on part of the water powers of Alabama, by B. M. Hall: Alabama Geol. Survey Bull. 7, 1903.

Papers on the water power in North Carolina, a preliminary report by George F. Swain, J. A. Holmes, and E. W. Myers: North Carolina Geol. Survey Bull. 8, 1899.

The Coastal Plain of North Carolina, by W. B. Clark, B. L. Miller, L. W. Stephenson, B. L. Johnson, and H. N. Parker: North Carolina Geol. and Econ. Survey Rept., vol. 3, 1912.

Many of these reports can be obtained by applying to the several organizations, and most of them can be consulted in the public libraries of the larger cities.

---

[1] Octavo edition only. [2] Octavo edition, 50c.

GEOLOGICAL SURVEY HYDROLOGIC REPORTS OF GENERAL INTEREST.

The following list comprises reports not readily classifiable by drainage basins and covering a wide range of hydrologic investigations:

WATER-SUPPLY PAPERS.

*1. Pumping water for irrigation, by H. M. Wilson. 1896. 57 pp., 9 pls.

Describes pumps and motive powers, windmills, water wheels, and various kinds of engines; also storage reservoirs to retain pumped water until needed for irrigation.

*3. Sewage irrigation, by G. W. Rafter. 1897. 100 pp., 4 pls. 10c. (See Water Supply Paper 22.)

Discusses methods of sewage disposal by intermittent filtration and by irrigation; describes utilization of sewage in Germany, England, and France and sewage purification in the United States.

*8. Windmills for irrigation, by E. C. Murphy. 1897. 49 pp., 8 pls. 10c.

Gives results of experimental tests of windmills during the summer of 1896 in the vicinity of Garden, Kans.; describes instruments and methods and draws conclusions.

*14. New tests of certain pumps and water lifts used in irrigation, by O. P. Hood 1898. 91 pp., 1 pl.

Discusses efficiency of pumps and water lifts of various types.

*20. Experiments with windmills, by T. O. Perry. 1899. 97 pp., 12 pls. 15c.

Includes tables and descriptions of wind wheels, compares wheels of several types, and discusses results.

*22. Sewage irrigation, Part II, by G. W. Rafter. 1899. 100 pp., 7 pls. 15c.

Gives résumé of Water-Supply Paper No. 3; discusses pollution of certain streams, experiments on purification of factory wastes in Massachusetts, value of commercial fertilizers, and describes American sewage-disposal plants by States; contains bibliography of publications relating to sewage, utilization, and disposal.

*41. The windmill; its efficiency and economic use, Part I, by E. C. Murphy. 1901. 72 pp., 14 pls. 5c.

*42. The windmill; its efficiency and economic use, Part II, by E. C. Murphy. 1901. 75 pp. (73–147), 2 pls. (15–16). 10c.

Nos. 41 and 42 give details of results of experimental tests with windmills of various types.

*43. Conveyance of water in irrigation canals, flumes, and pipes, by Samuel Fortier. 1901. 86 pp., 15 pls. 15c.

*56. Methods of stream measurement. 1901. 51 pp., 12 pls. 15c.

Describes the methods used by the Survey in 1901–2. (See also Nos. 64, 94, and 95.)

*64. Accuracy of stream measurements, by E. C. Murphy. 1902. 99 pp., 4 pls. (See No. 95.) 10c.

Describes methods of measuring velocity of water and of measuring and computing stream flow and compares results obtained with the different instruments and methods; describes also experiments and results at the Cornell University hydraulic laboratory. A second, enlarged edition published as Water-Supply Paper 95.

*67. The motions of underground waters, by C. S. Slichter, 1902. 106 pp., 8 pls 15c.

Discusses origin, depth, and amount of ground waters; permeability of rocks and porosity of soils; causes, rates, and laws of motions of ground water; surface and deep zones of flow, and recovery of waters by open wells and artesian and deep wells; treats of the shape and position of the water table; gives simple methods of measuring yield of flowing wells.

72. Sewage pollution in the metropolitan area near New York City and its effect on inland water resources, by M. O. Leighton. 1902. 75 pp., 8 pls. 10c.

Defines "normal" and "polluted" waters and discusses the damage resulting from pollution.

*80. The relation of rainfall to run-off, by G. W. Rafter. 1903. 104 pp. 10c.

Treats of measurements of rainfall and laws and measurements of stream flow; gives rainfall run-off, and evaporation formulas; discusses effects of forests on rainfall and run-off.

87. Irrigation in India (second edition), by H. M. Wilson. 1903. 238 pp., 27 pls. 25c.

First edition was published in Part II of the Twelfth Annual Report.

93. Proceedings of first conference of engineers of the Reclamation Service, with accompanying papers, compiled by F. H. Newell, chief engineer. 1904. 361 pp. 25c. [Requests for this report should be addressed to the U. S. Reclamation Service.]

Contains the following papers of more or less general interest:
Limits of an irrigation project, by D. W. Ross.
Relation of Federal and State laws to irrigation, by Morris Bien.
Electrical transmission of power for pumping, by H. A. Storrs.
Correct design and stability of high masonry dams, by Geo. Y. Wisner.
Irrigation surveys and the use of the plane table, by J. B. Lippincott.
The use of alkaline waters for irrigation, by Thomas H. Means.

*94. Hydrographic manual of the United States Geological Survey, prepared by E. C. Murphy, J. C. Hoyt, and G. B. Hollister. 1904. 76 pp., 3 pls. 10c.

Gives instruction for field and office work relating to measurements of stream flow by current meters. See also No. 95.

*95. Accuracy of stream measurement (second, enlarged edition), by E. C. Murphy. 1904. 169 pp., 6 pls.

Describes methods of measuring and computing stream flow and compares results derived from different instruments and methods. See also No. 94.

*103. A review of the laws forbidding pollution of inland waters in the United States, by E. B. Goodell. 1904. 120 pp. Superseded by No. 152, q. v.

Explains the legal principles under which antipollution statutes become operative, quotes court decisions to show authority for various deductions, and classifies according to scope the statutes enacted in the different States.

110. Contributions to the hydrology of eastern United States, 1904; M. L. Fuller, geologist in charge. 1905. 211 pp., 5 pls. 10c.

Contains the following reports of general interest. The scope of each paper is indicated by its title.

Description of underflow meter used in measuring the velocity and direction of underground water, by Charles S. Slichter.

The California or "stovepipe" method of well construction, by Charles S. Slichter.

Approximate methods of measuring the yield of flowing wells, by Charles S. Slichter.

Corrections necessary in accurate determinations of flow from vertical well casings, from notes furnished by A. N. Talbot.

Experiment relating to problems of well contamination at Quitman, Ga., by S. W. McCallie.

113. The disposal of strawboard and oil-well wastes, by R. L. Sackett and Isaiah Bowman. 1905. 52 pp., 4 pls. 5c.

The first paper discusses the pollution of streams by sewage and by trade wastes, describes the manufacture of strawboard and gives results of various experiments in disposing of the waste. The second paper describes briefly the topography, drainage, and geology of the region about Marion, Ind., and the contamination of rock wells and of streams by waste oil and brine.

*114. Underground waters of eastern United States; M. L. Fuller, geologist in charge. 1905. 285 pp., 18 pls. 25c.

Contains report on "Occurrence of underground waters," by M. L. Fuller, discussing sources, amount, and temperature of waters; permeability and storage capacity of rocks, water-bearing formations; recovery of water by springs, wells, and pumps; essential conditions of artesian flows; and general conditions affecting underground waters in eastern United States.

119. Index to the hydrographic progress reports of the United States Geological Survey, 1888 to 1903, by J. C. Hoyt and B. D. Wood. 1905. 253 pp. 15c.

120. Bibliographic review and index of papers relating to underground waters published by the United States Geological Survey, 1879-1904, by M. L. Fuller. 1905. 128 pp. 10c.

*122. Relation of the law to underground waters, by D. W. Johnson. 1905. 55 pp. 5c.

Defines and classifies underground waters, gives common-law rules relating to their use, and cites States legislative acts affecting them.

140. Field measurements of the rate of movement of underground waters, by C. S. Slichter. 1905. 122 pp., 15 pls. 15c.

Discusses the capacity of sand to transmit water; describes measurements of underflow in Rio Hondo, San Gabriel, and Mohave River valleys, Calif., and on Long Island, N. Y.; gives results of tests of wells and pumping plants, and describes stovepipe method of well construction.

143. Experiments on steel-concrete pipes on a working scale, by J. H. Quinton. 1905. 61 pp., 4 pls. 5c.

Scope indicated by title.

145. Contributions to the hydrology of eastern United States, 1905; M. L. Fuller, geologist in charge. 1905. 220 pp., 6 pls. 10c.

Contains brief reports of general interest as follows:

Drainage of ponds into drilled wells, by Robert E. Horton. Discusses efficiency, cost, and capacity of drainage wells and gives statistics of such wells in southern Michigan.

Construction of so-called fountain and geyser springs, by Myron L. Fuller.

A convenient gage for determining low artesian heads, by Myron L. Fuller.

146. Proceedings of second conference of engineers of the Reclamation Service, with accompanying papers, compiled by F. H. Newell, Chief Engineer. 1905. 267 pp. 15c. [Inquiries concerning this report should be addressed to the Reclamation Service.]

Contains brief account of the organization of the hydrographic [water resources] branch and the Reclamation Service, reports of conferences and committees, circulars of instruction, and many brief reports on subjects closely related to reclamation, and a bibliography of technical papers by members of the service. Of the papers read at the conference those listed below (scope indicated by title) are of more or less general interest:

Proposed State code of water laws, by Morris Bien.
Power engineering applied to irrigation problems, by O. H. Ensign.
Estimates on tunneling in irrigation projects, by A. L. Fellows.
Collection of stream-gaging data, by N. C. Grover.
Diamond-drill methods, by G. A. Hammond.
Mean-velocity and area curves, by F. W. Hanna.
Importance of general hydrographic data concerning basins of streams gaged, by R. E. Horton.
Effect of aquatic vegetation on stream flow, by R. E. Horton.
Sanitary regulations governing construction camps, by M. O. Leighton.
Necessity of draining irrigated land, by Thos. H. Means.
Alkali soils, by Thos. H. Means.
Cost of stream-gaging work, by E. C. Murphy.
Equipment of a cable gaging station, by E. C. Murphy.
Silting of reservoirs, by W. M. Reed.
Farm-unit classification, by D. W. Ross.
Cost of power for pumping irrigating water, by H. A. Storrs.
Records of flow at current-meter gaging stations during the frozen season, by F. H. Tillinghast.

147. Destructive floods in United States in 1904, by E. C. Murphy and others. 1905. 206 pp., 18 pls. 15c.

Contains a brief account of "A method of computing cross-section area of waterways," including formulas for maximum discharge and areas of cross section.

*150. Weir experiments, coefficients, and formulas, by R. E. Horton. 1906. 189 pp., 38 pls. (See Water-Supply Paper 200.) 15c.

Scope indicated by title.

151. Field assay of water, by M. O. Leighton. 1905. 77 pp., 4 pls.

Discusses methods, instruments, and reagents used in determining turbidity, color, iron, chlorides, and hardness in connection with the studies of the quality of water in various parts of the United States.

*152. A review of the laws forbidding pollution of inland waters in the United States, second edition, by E. B. Goodell. 1905. 149 pp. 10c.

Scope indicated by title.

*155. Fluctuations of the water level in wells, with special reference to Long Island, N. Y., by A. C. Veatch. 1906. 83 pp., 9 pls. 25c.

Includes general discussion of fluctuations due to rainfall and evaporation, barometric changes, temperature changes, changes in rivers, changes in lake level, tidal changes, effects of settlement, irrigation, dams, underground-water development, and to indeterminate causes.

*160. Underground-water papers, 1906; M. L. Fuller, geologist in charge. 1906. 104 pp., 1 pl.

Gives account of work in 1905, lists publications relating to underground waters, and contains the following brief reports of general interest:

Significance of the term "artesian," by Myron L. Fuller.
Representation of wells and springs on maps, by Myron L. Fuller.
Total amount of free water in the earth's crust, by Myron L. Fuller.
Use of fluorescein in the study of underground waters, by R. B. Dole.
Problems of water contamination, by Isaiah Bowman.
Instances of improvement of water in wells, by Myron L. Fuller.

*162. Destructive floods in the United States in 1905, with a discussion of flood discharge and frequency and an index to flood literature, by E. C. Murphy and others. 1906. 105 pp., 4 pls. 15c.

*163. Bibliographic review and index of underground-water literature published in the United States in 1905, by M. L. Fuller, F. G. Clapp, and B. L. Johnson. 1906. 130 pp. 15c.

Scope indicated by title.

*179. Prevention of stream pollution by distillery refuse, based on investigations at Lynchburg, Ohio, by Herman Stabler. 1906. 34 pp., 1 pl. 10c.

Describes grain distillation, treatment of slop, sources, character, and effects of effluents on streams; discusses filtration, precipitation, fermentation, and evaporation methods of disposal of wastes without pollution.

*180. Turbine water-wheel tests and power tables, by R. E. Horton. 1906. 134 pp., 2 pls. 20c.

Scope indicated by title.

*185. Investigations on the purification of Boston sewage, * * * with a history of the sewage-disposal problem, by C.-E. A. Winslow and E. B. Phelps. 1906. 163 pp. 25c.

Discusses composition, disposal, purification, and treatment of sewages and tendencies in sewage-disposal practice in England, Germany, and the United States; describes character of crude sewage at Boston, removal of suspended matter, treatment in septic tanks, and purification in intermittent sand filtration and coarse material; gives bibliography.

*186. Stream polution by acid-iron wastes, a report based on investigations made at Shelby, Ohio, by Herman Stabler. 1906. 36 pp., 1 pl.

Gives history of pollution by acid-iron wastes at Shelby, Ohio, and resulting litigation; discusses effect of acid-iron liquors on sewage purification processes, recovery of copperas from acid-iron wastes, and other processes for removal of pickling liquor.

*187. Determination of stream flow during the frozen season, by H. K. Barrows and R. E. Horton. 1907. 93 pp., 1 pl. 15c.

Scope indicated by title.

*189. The prevention of stream polution by strawboard waste, by E. B. Phelps. 1906. 29 pp., 2 pls.

Describes manufacture of strawboard, present and proposed methods of disposal of waste liquors, laboratory investigations of precipitation and sedimentation, and field studies of amounts and character of water used, raw material and finished product, and mechanical filtration.

*194. Pollution of Illinois and Mississippi rivers by Chicago sewage (a digest of the testimony taken in the case of the State of Missouri v. the State of Illinois and the Sanitary District of Chicago), by M. O. Leighton. 1907. 369 pp., 2 pls.

Scope indicated by amplification of title.

*200. Weir experiments, coefficients, and formulas (revision of paper No. 150), by R E. Horton. 1907. 195 pp., 38 pls. 35c.

Scope indicated by title.

*226. The pollution of streams by sulphite-pulp waste, a study of possible remedies, by E. B. Phelps. 1909. 37 pp., 1 pl. 10c.

Describes manufacture of sulphite pulp, the waste liquors, and the experimental work leading to suggestions as to methods of preventing stream pollution.

*229. The disinfection of sewage and sewage filter effluents, with a chapter on the putrescibility and stability of sewage effluents, by E. B. Phelps. 1909. 91 pp., 1 pl. 15c.

Scope indicated by title.

*234. Papers on the conservation of water resources. 1909. 96 pp., 2 pls. 15c

Contains the following papers, whose scope is indicated by their titles: Distribution of rainfall, by Henry Gannett; Floods, by M. O. Leighton; Developed water powers, compiled under the direction of W. M. Steuart, with discussion by M. O. Leighton; Undeveloped water powers, by M. O. Leighton; Irrigation, by F. H. Newell; Underground waters, by W. C. Mendenhall; Denudation, by R. B. Dole and Herman Stabler; Control of catchment areas, by H. N. Parker.

*235. The purification of some textile and other factory wastes, by Herman Stabler and G. H. Pratt.. 1909. 76 pp. 10c.

Discusses waste waters from wool scouring, bleaching and dyeing cotton yarn, bleaching cotton piece goods, and manufacture of oleomargarine, fertilizer, and glue.

236. The quality of surface waters in the United States: Part I, Analyses of waters east of the one hundredth meridian, by R. B. Dole. 1909. 123 pp. 10c.

Describes collection of samples, methods of examination, preparation of solutions, accuracy of estimate, and expression of analytical results.

238. The public utility of water powers and their governmental regulation, by René Tavernier and M. O. Leighton. 1910. 161 pp. 15c.

Discusses hydraulic power and irrigation, French, Italian, and Swiss legislation relative to the development of water powers, and laws proposed in the French Parliament; reviews work of bureau of hydraulics and agricultural improvement of the French department of agriculture and gives résumé of Federal and State water-power legislation in the United States.

*255. Underground waters for farm use, by M. L. Fuller. 1910. 58 pp., 17 pls. 15c.

Discusses rocks as sources of water supply and the relative safety of supplies from different materials, springs and their protection; open or dug and deep wells, their location, yield, relative cost, protection, and safety; advantages and disadvantages of cisterns and combination wells and cisterns.

*257. Well-drilling methods, by Isaiah Bowman. 1911. 139 pp., 4 pls. 15c.

Discusses amount, distribution, and disposal of rainfall, water-bearing rocks, amount of ground water, artesian conditions, and oil and gas bearing formation; gives history of well drilling in Asia, Europe, and the United States; describes in detail the various methods and the machinery used; discusses loss of tools and geologic difficulties; contamination of well waters and methods of prevention; tests of capacity and measurement of depth; and costs of sinking wells.

*258. Underground water-papers, 1910, by M. L. Fuller, F. G. Clapp, G. C. Matson, Samuel Sanford, and H. C. Wolff. 1911. 123 pp., 2 pls. 15c.

Contains the following papers (scope indicated by titles) of general interest:
Drainage by wells, by M. L. Fuller.
Freezing of wells and related phenomena, by M. L. Fuller.
Pollution of underground waters in limestone, by G. C. Matson.
Protection of shallow wells in sandy deposits, by M. L. Fuller.
Magnetic wells, by M. L. Fuller.

274. Some stream waters of the western United States, with chapters on sediment carried by the Rio Grande and the industrial application of water analyses, by Herman Stabler. 1911. 188 pp. 15c.

Describes collection of samples, plan of analytical work, and methods of analyses; discusses soap-consuming power of waters, water softening, boiler waters, and water for irrigation.

*315. The purification of public water supplies, by G. A. Johnson. 1913. 84 pp., 8 pls. 10c.

Discusses ground, lake, and river waters as public supplies; development of waterworks systems in the United States, water consumption, and typhoid fever; describes methods of filtration and sterilization of water, and municipal water softening.

334. The Ohio Valley flood of March–April, 1913 (including comparisons with some earlier floods), by A. H. Horton and H. J. Jackson. 1913. 96 pp., 22 pls. 20c.

Although relating specifically to floods in the Ohio Valley, this report discusses also the causes of floods and the prevention of damage by floods.

337. The effects of ice on stream flow, by William Glenn Hoyt. 1913. 77 pp., 7 pls. 15c.

Discusses methods of measuring the winter flow of streams.

*345. Contributions to the hydrology of the United States, 1914. N. C. Grover, chief hydraulic engineer. 1915. 225 pp., 7 pls. 30c. Contains:

*(e) A method of determining the daily discharge of rivers of variable slope, by M. R. Hall, W. E. Hall, and C. H. Pierce, pp. 53-65. Scope indicated by title.

364. Water analyses from the laboratory of the United States Geological Survey, tabulated by F. W. Clarke, chief chemist. 1914. 40 pp. 5c.

Contains analyses of waters from rivers, lakes, wells, and springs in various parts of the United States, including analyses of the geyser water of Yellowstone National Park, hot springs in Montana, brines from Death Valley, water from the Gulf of Mexico, and mine waters from Tennessee, Michigan, Missouri and Oklahoma, Montana, Colorado and Utah, Nevada, and Arizona and California.

371. Equipment for current-meter gaging stations, by G. J. Lyon. 1915. 64 pp., 37 pls. 20c.

Describes methods of installing automatic and other gages and of constructing gage wells, shelters, and structures for making discharge measurements and artificial controls.

*375. Contributions to the hydrology of the United States, 1915. N. C. Grover, chief hydraulic engineer. 1916. 181 pp., 9 pls. 15c.

Contains three papers presented at the conference of engineers of the water-resources branch in December, 1914, as follows:

*(c) Relation of stream gaging to the science of hydraulics, by C. H. Pierce and R. W. Davenport, pp. 77-84.

(e) A method for correcting river discharge for a changing stage, by B. E. Jones, pp. 117-130.

(f) Conditions requiring the use of automatic gages in obtaining records of stream flow, by C. H. Pierce, pp. 131-139.

*400. Contributions to the hydrology of the United States, 1916. N. C. Grover, chief hydraulic engineer. 1917. 108 pp., 7 pls. Contains:

(a) The people's interest in water-power resources, by G. O. Smith, pp. 1-8.
*(c) The measurement of silt-laden streams, by R. C. Pierce, pp. 39-51.
(d) Accuracy of stream-flow data, by N. C. Grover, and J. C. Hoyt, pp. 53-59.

416. The divining rod, a history of water witching, with a bibliography, by A. J. Ellis. 1917. 59 pp. 10c.

A brief paper published "merely to furnish a reply to the numerous inquiries that are continually being received from all parts of the country" as to the efficacy of the divining rod for locating underground water.

425. Contributions to the hydrology of the United States, 1917. N. C. Grover, chief hydraulic engineer. 1918. Contains:

*(c) Hydraulic conversion tables and convenient equivalents, pp. 71-94. 1917.

427. Bibliography and index of the publications of the United States Geological Survey relating to ground water, by O. E. Meinzer. 1918. 169 pp., 1 pl.

Includes publications prepared, in whole or part, by the Geological Survey that treat any phase of the subject of ground water or any subject directly applicable to ground water. Illustrated by map showing reports that cover specific areas more or less thoroughly.

ANNUAL REPORTS.

*Fifth Annual Report of the United States Geological Survey, 1883-84, J. W. Powell, Director. 1885. xxxvi, 469 pp., 58 pls. $2.25. Contains:

*The requisite and qualifying conditions of artesian wells, by T. C. Chamberlin, pp. 125-173, pl. 21. Scope indicated by title.

*Twelfth Annual Report of the United States Geological Survey, 1890-91, J. W. Powell, Director. 1891. 2 parts. *Pt. II, Irrigation, xviii, 576 pp., 93 pls. $2. Contains:

*Irrigation in India, by H. M. Wilson, pp. 363-561, pls. 107 to 146. (See Water-Supply Paper 87.)

Thirteenth Annual Report of the United States Geological Survey, 1891-92, J. W. Powell, Director. 1892. (Pts. II and III, 1893.) 3 parts. *Pt. III, Irrigation, pp. xi, 486, 77 plates. $1.85. Contains:

*American irrigation engineering, by H. M. Wilson, pp. 101-349, pls. 111 to 146. Discusses economic aspects of irrigation, alkaline drainage, silt, and sedimentation; gives brief history of legislation; describes perennial canals in Idaho-California, Wyoming, and Arizona; discusses water storage at reservoirs of the California and other projects, subsurface sources of supply, pumping, and subirrigation.

Fourteenth Annual Report of the United States Geological Survey, 1892-93, J. W. Powell, Director. 1893. (Pt. II, 1894.) 2 parts. *Pt. II, Accompanying papers, pp. xx, 597, 73 pls. $2.10. Contains:

*Natural mineral waters of the United States, by A. C. Peale, pp. 49-88, pls. 3 and 4. Discusses the origin and flow of mineral springs, the source of mineralization, thermal springs, the chemical composition and analysis of spring waters, geographic distribution, and the utilization of mineral waters; gives a list of American mineral spring resorts; contains also some analyses.

Nineteenth Annual Report of the United States Geological Survey, 1897-98, Charles D. Walcott, Director. 1898. (Parts II, III, and V, 1899.) 6 parts in 7 vols. and separate case for maps with Pt. V. *Pt. II, papers chiefly of a theoretic nature, pp. v, 958, 172 plates. $2.65. Contains:

*Principles and conditions of the movements of ground water, by F. H. King, pp. 59-294, pls. 6 to 16. Discusses the amount of water stored in sandstone, in soil and in other rocks, the depth to which ground water penetrates; gravitational, thermal, and capillary movements of ground waters, and the configuration of the ground-water surface; gives the results of experimental investigations on the flow of air and water through a rigid, porous medium and through sands, sandstones, and silts; discusses results obtained by other investigators, and summarizes results of observations; discusses also rate of flow of water through sand and rock, the growth of rivers, rate of filtration through soil, interference of wells, etc..

*Theoretical investigation of the motion of ground waters, by C. S. Slichter, pp. 295-384, pl. 17. Scope indicated by title.

PROFESSIONAL PAPERS.

86. The transportation of débris by running water, by G. K. Gilbert, based on experiments made with the assistance of E. C. Murphy. 1914. 263 pp., 3 pls. 70c.

The results of an investigation which was carried on in a specially equipped laboratory at Berkeley, Calif., and was undertaken for the purpose of learning "the laws which control the movement of bed load and especially to determine how the quantity of load is related to the stream slope and discharge and to the degree of comminution of the débris."

105. Hydraulic-mining débris in the Sierra Nevada, by G. K. Gilbert. 154 pp., 34 pls. 1917. 50c.

Presents the results of an investigation undertaken by the United States Geological Survey in response to a memorial from the California Miners' Association asking that a particular study be made of portions of the Sacramento and San Joaquin valleys affected by detritus from torrential streams. The report deals largely with geologic and physiographic aspects of the subject, traces the physical effects, past and future, of the hydraulic mining of earlier decades, the similar effects which certain other industries induce through stimulation of the erosion of the soil, and the influence of the restriction of the area of inundation by the construction of levees. Suggests cooperation by several interests for the control of the streams now carrying heavy loads of débris.

BULLETINS.

*32. Lists and analyses of the mineral springs of the United States (a preliminary study), by A. C. Peale. 1886. 235 pp.

Defines mineral waters, lists the springs by States, and gives tables of analyses.

*319. Summary of the controlling factors of artesian flows, by Myron L. Fuller. 1908. 44 pp., 7 pls. 10c.

Describes underground reservoirs, the sources of ground waters, the confining agents, the primary and modifying factors of artesian circulation, the essential and modifying factors of artesian flow, and typical artesian systems.

*479. The geochemical interpretation of water analyses, by Chase Palmer. 1911. 31 pp. 5c.

Discusses the expression of chemical analyses, the chemical character of water and the properties of natural water; gives a classification of waters based on property values and reacting values, and discusses the character of the waters of certain rivers as interpreted directly from the results of analyses; discusses also the relation of water properties to geologic formations, silica in river water, and the character of the water of the Mississippi and the Great Lakes and St. Lawrence River as indicated by chemical analyses.

616. The data of geochemistry (third edition), by F. W. Clarke. 1916. 821 pp. 45c.

Earlier editions were published as Bulletins 330 and 491. Contains a discussion of the statement and interpretation of water analyses and a chapter on "Mineral wells and springs" (pp. 179-216). Discusses the definition and classification of mineral waters, changes in the composition of water, deposits of calcareous, ocherous, and siliceous materials made by water, vadose and juvenile waters, and thermal springs in relation to volcanism. Describes the different kinds of ground water and gives typical analyses. Includes a brief bibliography of papers containing water analyses.

112130°—20—wsp 472——6

INDEX BY AREAS AND SUBJECTS.

[A—Annual Reports; M—Monograph; B—Bulletin; P—Professional Paper; W—Water-Supply Paper; G F—Geologic folio.]

[1] Many of the reports contain brief subject bibliographies. See abstracts.
[2] Many analyses of river, spring, and well waters are scattered through publications, as noted in abstracts.

INDEX OF STREAMS.

ADDITIONAL COPIES
OF THIS PUBLICATION MAY BE PROCURED FROM
THE SUPERINTENDENT OF DOCUMENTS
GOVERNMENT PRINTING OFFICE
WASHINGTON, D. C.
AT
10 CENTS PER COPY
▽

DEPARTMENT OF THE INTERIOR

ALBERT B. FALL, Secretary

NITED STATES GEOLOGICAL SURVE

GEORGE OTIS SMITH, Director

ACE WATER SUPPLY OF UNITED STATES

1918

PART III. OHIO RIVER BASIN

NATHAN C. GROVER, Chief Hydraulic Engineer

LBERT H. HORTON and C. G. PAULSEN, District Engineer

DEPARTMENT OF THE INTERIOR
ALBERT B. FALL, Secretary

UNITED STATES GEOLOGICAL SURVEY
GEORGE OTIS SMITH, Director

Water-Supply Paper 473

SURFACE WATER SUPPLY OF THE UNITED STATES

1918

PART III. OHIO RIVER BASIN

NATHAN C. GROVER, Chief Hydraulic Engineer
ALBERT H. HORTON and C. G. PAULSEN, District Engineers

Prepared in cooperation with
THE STATES OF ILLINOIS AND KENTUCKY

WASHINGTON
GOVERNMENT PRINTING OFFICE
1922

ADDITIONAL COPIES
OF THIS PUBLICATION MAY BE PROCURED FROM
THE SUPERINTENDENT OF DOCUMENTS
GOVERNMENT PRINTING OFFICE
WASHINGTON, D. C.
AT
10 CENTS PER COPY

CONTENTS.

ILLUSTRATIONS.

SURFACE WATER SUPPLY OF OHIO RIVER BASIN, 1918.

AUTHORIZATION AND SCOPE OF WORK.

This volume is one of a series of 14 reports presenting results of measurements of flow made on streams in the United States during the year ending September 30, 1918.

The data presented in these reports were collected by the United States Geological Survey under the following authority contained in the organic law (20 Stat. L., p. 394):

Provided, That this officer [the Director] shall have the direction of the Geological Survey and the classification of public lands and examination of the geological structure, mineral resources, and products of the national domain.

The work was begun in 1888 in connection with special studies relating to irrigation in the arid West. Since the fiscal year ending June 30, 1895, successive sundry civil bills passed by Congress have carried the following item and appropriations:

For gaging the streams and determining the water supply of the United States, and for the investigation of underground currents and artesian wells, and for the preparation of reports upon the best methods of utilizing the water resources.

Annual appropriations for the fiscal years ended June 30, 1895–1919.

1895	$12,500.00
1896	20,000.00
1897 to 1900, inclusive	50,000.00
1901 to 1902, inclusive	100,000.00
1903 to 1906, inclusive	200,000.00
1907	150,000.00
1908 to 1910, inclusive	100,000.00
1911 to 1917, inclusive	150,000.00
1918	175,000.00
1919	148,244.10

In the execution of the work many private and State organizations have cooperated either by furnishing data or by assisting in collecting data. Acknowledgments for cooperation of the first kind are made in connection with the description of each station affected; cooperation of the second kind is acknowledged on page 5.

Measurements of stream flow have been made at about 4,500 points in the United States and also at many points in Alaska and the Hawaiian Islands. In July, 1918, 1,180 gaging stations were being maintained by the Survey and the cooperating organizations. Many miscellaneous discharge measurements are made at other points. In connection with this work data were also collected in regard to precipitation, evaporation, storage reservoirs, river profiles, and water power in many sections of the country and will be made available in water-supply papers from time to time.

DEFINITION OF TERMS.

The volume of water flowing in a stream—the "run-off" or "discharge"—is expressed in various terms, each of which has become associated with a certain class of work. These terms may be divided into two groups—(1) those that represent a rate of flow, as second-feet, gallons per minute, miners' inches, and discharge in second-feet per square mile, and (2) those that represent the actual quantity of water, as run-off in depth in inches, acre-feet, and millions of cubic feet. The principal terms used in this series of reports are second-feet, second-feet per square mile, run-off in inches, and acre-feet. They may be defined as follows:

"Second-feet" is an abbreviation for "cubic feet per second." A second-foot is the rate of discharge of water flowing in a channel of rectangular cross section 1 foot wide and 1 foot deep at an average velocity of 1 foot per second. It is generally used as a fundamental unit from which others are computed.

"Second-feet per square mile" is the average number of cubic feet of water flowing per second from each square mile of area drained, on the assumption that the run-off is distributed uniformly both as regards time and area.

"Run-off in inches" is the depth to which an area would be covered if all the water flowing from it in a given period were uniformly distributed on the surface. It is used for comparing run-off with rainfall, which is usually expressed in depth of inches.

An "acre-foot," equivalent to 43,560 cubic feet, is the quantity required to cover an acre to the depth of 1 foot. The term is commonly used in connection with storage for irrigation.

The following terms not in common use are here defined:

"Stage-discharge relation," an abbreviation for the term "relation of gage height to discharge."

"Control," a term used to designate the section or sections of the stream below the gage which determines the stage-discharge relation at the gage. It should be noted that the control may not be the same section or sections at all stages.

The "point of zero flow" for a gaging station is that point on the gage—the gage height—at which water ceases to flow over the gage.

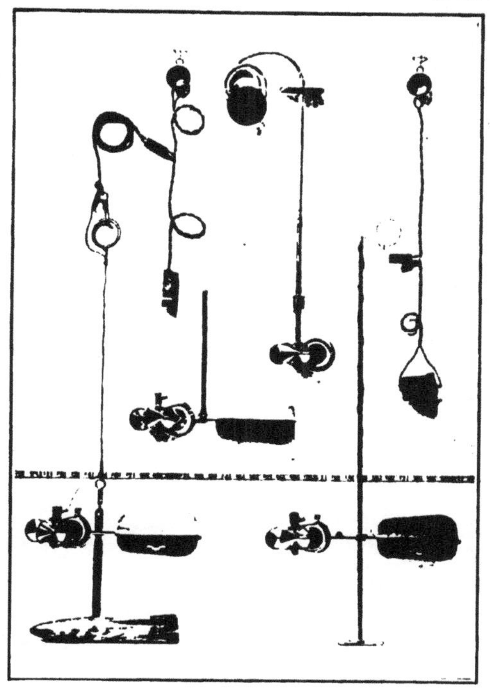

A. PRICE CURRENT METERS.

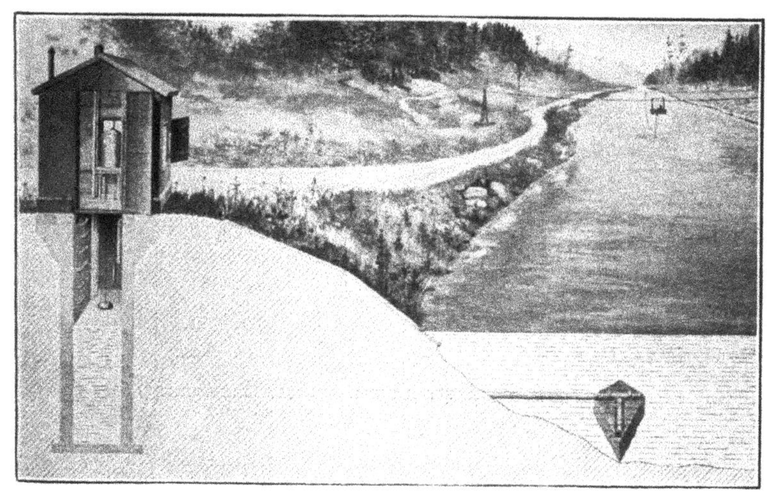

B. TYPICAL GAGING STATION.

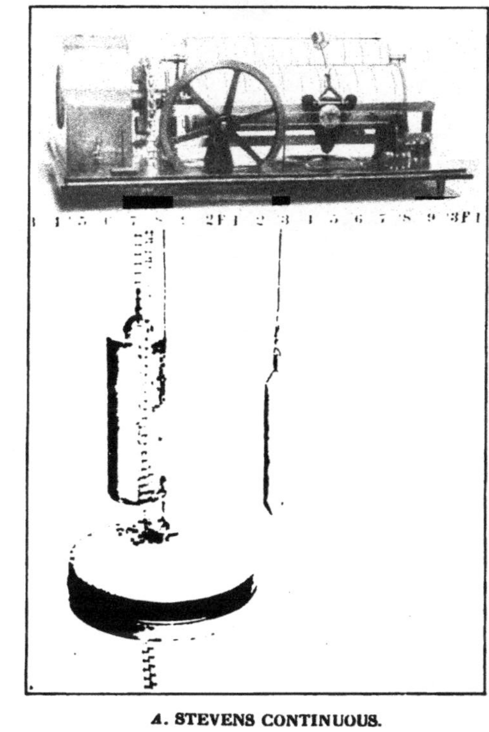

A. STEVENS CONTINUOUS.

B. GURLEY PRINTING.

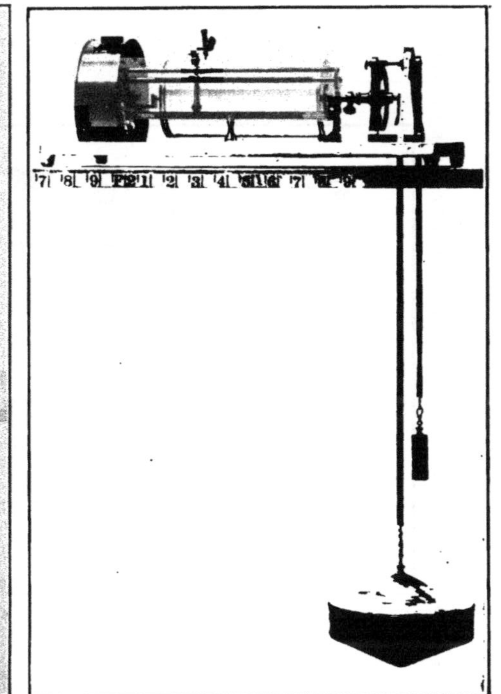

C. FRIEZ.

WATER-STAGE RECORDERS.

EXPLANATION OF DATA.

The data presented in this report cover the year beginning October 1, 1917, and ending September 30, 1918. At the beginning of January in most parts of the United States much of the precipitation in the preceding three months is stored as ground water in the form of snow or ice, or in ponds, lakes, and swamps, and this stored water passes off in the streams during the spring break-up. At the end of September, on the other hand, the only stored water available for run-off is possibly a small quantity in the ground; therefore the run-off for the year beginning October 1 is practically all derived from precipitation within that year.

The base data collected at gaging stations consist of records of stage, measurements of discharge, and general information used to supplement the gage heights and discharge measurements in determining the daily flow. The records of stage are obtained either from direct readings on a staff gage or from a water-stage recorder that gives a continuous record of the fluctuations. Measurements of discharge are made with a current meter. (See Pls. I, II.) The general methods are outlined in standard textbooks on the measurement of river discharge.

From the discharge measurements rating tables are prepared that give the discharge for any stage, and these rating tables, when applied to gage heights, give the discharge from which the daily, monthly, and yearly mean discharge is determined.

The data presented for each gaging station in the area covered by this report comprise a description of the station, a table giving results of discharge measurements, a table showing the daily discharge of the stream, and a table of monthly and yearly discharge and run-off.

If the base data are insufficient to determine the daily discharge, tables giving daily gage heights and results of discharge measurements are published.

The description of the station gives, in addition to statements regarding location and equipment, information in regard to any conditions that may affect the permanence of the stage-discharge relation, covering such subjects as the occurrence of ice, the use of the stream for log driving, shifting of control, and the cause and effect of backwater; it gives also information as to diversions that decrease the flow at the gage, artificial regulation, maximum and minimum recorded stages, and the accuracy of the records.

The table of daily discharge gives, in general, the discharge in second-feet corresponding to the mean of the gage heights read each day. At stations on streams subject to sudden or rapid diurnal fluctuation the discharge obtained from the rating table and the mean daily gage height may not be the true mean discharge for the day. If such stations are equipped with water-stage recorders the mean daily discharge may be obtained by averaging discharge at regular intervals during the day, or by using the discharge integrator, an instrument operating on the principle of the planimeter and containing as an essential element the rating curve of the station.

In the table of monthly discharge the column headed "Maximum" gives the mean flow for the day when the mean gage height was highest. As the gage height is the mean for the day it does not indicate correctly the stage when the water surface was at crest height, and the corresponding discharge was consequently larger than given in the maximum column. Likewise, in the column headed "Minimum" the quantity given is the mean flow for the day when the mean gage height was lowest. The column headed "Mean" is the average flow in cubic feet for each second during the month. On this average flow computations recorded in the remaining columns, which are defined on page 2, are based.

ACCURACY OF FIELD DATA AND COMPUTED RESULTS.

The accuracy of stream-flow data depends primarily (1) on the permanence of the stage-discharge relation and (2) on the accuracy of observation of stage, measurements of flow, and interpretation of records.

A paragraph in the description of the station gives information regarding the (1) permanence of the stage-discharge relation, (2) precision with which the discharge rating curve is defined, (3) refinement of gage readings, (4) frequency of gage readings, and (5) methods of applying daily gage heights to the rating table to obtain the daily discharge.[1]

For the rating tables "well defined" indicates, in general, that the rating is probably accurate within 5 per cent; "fairly well defined," within 10 per cent; "poorly defined," within 15 to 25 per cent. These notes are very general and are based on the plotting of the individual measurements with reference to the mean rating curve.

The monthly means for any station may represent with high accuracy the quantity of water flowing past the gage, but the figures showing discharge per square mile and depth of run-off in inches may be subject to gross errors caused by the inclusion of large noncontributing districts in the measured drainage area, by lack of information concerning water diverted for irrigation or other use, or by inability to interpret the effect of artificial regulation of the flow of the river above the station. "Second-feet per square mile" and "run-off in inches" are therefore not computed if such errors appear probable. The computations are also omitted for stations on

[1] For a more detailed discussion of the accuracy of stream-flow data see Grover, N. C., and Hoyt, J. C. Accuracy of stream-flow data: U. S. Geol. Survey Water-Supply Paper 400, pp. 53-59, 1916.

streams draining areas in which the annual rainfall is less than 20 inches. All figures representing "second-feet per square mile" and "run-off in inches" previously published by the Survey should be used with caution because of possible inherent sources of error not known to the Survey.

The table of monthly discharge gives only a general idea of the flow at the station and should not be used for other than preliminary estimates; the tables of daily discharge allow more detailed studies of the variation in flow. It should be borne in mind, however, that the observations in each succeeding year may be expected to throw new light on data previously published.

COOPERATION.

Data for Allegheny River at Red House, N. Y., were collected in cooperation with the State of New York.

Work in Illinois during the year ending September 30, 1918, was carried on in cooperation with the State through the division of waterways of the Department of Public Works.

Work in Kentucky was done in cooperation with the State Geological Survey, J. B. Hoeing, State geologist.

The United States Engineer Corps cooperated in the maintenance of 9 gaging stations in the Ohio River basin and furnished base data for 30 additional stations.

Financial assistance was also rendered by the Alabama Geological Survey and the Tennessee Power Co.

DIVISION OF WORK.

Data for Allegheny River at Red House, N. Y., were collected and prepared for publication under the direction of C. C. Covert, district engineer, assisted by O. W. Hartwell, E. D. Burchard, and J. W. Moulton.

Data for the Ohio River basin, except those for the Allegheny at Red House, N. Y., for stations in Illinois, and for the basin of Tennessee River, were collected and prepared for publication under the direction of A. H. Horton, district engineer, assisted by B. J. Peterson, B. L. Hopkins, and B. L. Bigwood.

Data for stations in Illinois in the Ohio basin were collected and prepared for publication under direction of W. G. Hoyt, district engineer, assisted by H. C. Beckman.

Field data for stations in the Tennessee River basin were collected and prepared for publication under the direction of C. G. Paulsen, district engineer, assisted by B. J. Peterson, A. H. Condron, L. J. Hall, and Miss E. M. Tiller.

The records were assembled and reviewed by B. J. Peterson.

GAGING-STATION RECORDS.

ALLEGHENY RIVER BASIN.

ALLEGHENY RIVER AT RED HOUSE, N. Y.

LOCATION.—At highway bridge in Red House, Cattaraugus County, 5 miles below Salamanca and 13 miles above boundary line between New York and Pennsylvania. Conewango Creek, outlet of Chautauqua Lake, enters the Allegheny in Pennsylvania 30 miles below station.

DRAINAGE AREA.—1,640 square miles.

RECORDS AVAILABLE.—September 4, 1903, to September 30, 1918.

GAGE.—Gurley seven-day water-stage recorder on left bank just below highway bridge, installed September 3, 1917. Prior to that date, chain gage attached to upstream side of bridge near left end. Recorder inspected by W. E. Coe.

DISCHARGE MEASUREMENTS.—Made from downstream side of bridge.

CHANNEL AND CONTROL.—Coarse gravel; shifts occasionally.

EXTREMES OF DISCHARGE.—Maximum stage during year from water stage-recorder, 11.70 feet at 5 a. m. March 15 (discharge, 30,000 second-feet); minimum stage, 3.1 feet from 10 a. m. to 5 p. m. July 24 (discharge, 260 second-feet).

1903–1918: Maximum stage recorded, 12.7 feet March 26, 1913 (discharge, about 40,000 second-feet); minimum stage recorded, 2.7 feet on several days in December, 1908 (discharge, about 100 second-feet).

ICE.—Stage-discharge relation somewhat affected by ice.

REGULATION.—Low-water flow may be slightly affected by the operation of several small power plants above Salamanca. A storage reservoir on the divide between Oil Creek, tributary to Allegheny River, and Black Creek, tributary to Genesee River, was formerly used for supplying water to the Erie canal system through the abandoned Genesee River canal and Genesee River. This reservoir is no longer used for canal purposes, and all the water is turned into Allegheny River through Olean Creek.

ACCURACY.—Stage-discharge relation practically permanent between dates of shifting; affected by ice during most of the period from December to February. Rating curve well defined between 300 and 900 second-feet, and between 6,000 and 15,000 second-feet. Operation of water-stage recorder satisfactory. Daily discharge ascertained by applying to rating table the mean daily gage height obtained by inspecting gage-height graph. Open-water records good; others fair.

Discharge measurements of Allegheny River at Red House, N. Y., during the year ending Sept. 30, 1918.

Date.	Made by—	Gage height.	Discharge.	Date.	Made by—	Gage height.	Discharge.
		Feet.	*Sec.-ft.*			*Feet.*	*Sec.-ft.*
Dec. 21	E. D. Burchard	a 4.37	958	Mar. 20	E. D. Burchard	6.26	6,170
Jan. 21	.....do	a 4.47	374	May 28	J. W. Moulton	5.96	5,300
Feb. 28	.....do	7.37	9,900	June 21	E. D. Burchard	3.58	657
28	.....do	7.30	9,560	Aug. 22	.....do	3.32	408

a Stage-discharge relation affected by ice.

Daily discharge, in second-feet, of Allegheny River at Red House, N. Y., for the year ending Sept. 30, 1918.

Day.	Oct.	Nov.	Dec.	Jan.	Feb.	Mar.	Apr.	May.	June.	July.	Aug.	Sept.
1	340	15,000	1,460	800	360	10,600	1,740	1,660	2,240	992	635	538
2	402	10,200	1,870	750	340	13,300	1,710	1,580	2,110	918	481	675
3	410	7,260	1,480	650	380	13,300	1,640	1,460	1,860	836	392	625
4	655	4,930	1,420	600	360	11,000	2,160	1,310	1,490	727	463	556
5	1,050	3,940	1,350	480	340	8,230	2,250	1,240	1,370	655	1,540	585
6	1,010	3,180	1,170	550	340	9,250	2,060	1,240	1,220	625	1,360	1,550
7	998	2,780	998	550	360	10,200	1,860	1,240	1,240	547	894	1,370
8	878	2,460	775	700	380	7,440	1,770	1,220	1,460	509	780	980
9	844	2,160	700	750	420	6,540	1,920	1,210	1,160	490	1,040	802
10	867	1,990	700	600	460	11,600	2,110	1,460	980	490	942	696
11	786	1,830	750	400	800	9,600	1,970	2,480	905	528	906	605
12	786	1,680	800	360	1,700	8,070	2,060	2,980	1,180	518	1,100	576
13	1,170	1,540	850	320	6,500	9,250	2,110	2,820	1,920	490	1,070	1,250
14	1,440	1,410	850	360	9,000	23,400	3,660	2,290	1,580	445	870	2,330
15	1,290	1,290	850	380	9,600	28,400	5,290	2,020	1,160	400	980	1,640
16	2,070	1,250	800	380	8,900	21,800	5,290	1,940	1,000	378	1,020	1,480
17	1,920	1,210	800	380	5,830	15,100	5,420	1,800	942	362	859	3,090
18	1,480	1,170	750	380	3,310	9,600	5,560	1,770	942	340	696	3,660
19	1,790	1,100	750	380	3,540	7,140	5,560	1,740	848	325	595	2,620
20	7,510	1,040	750	380	18,800	5,970	5,290	1,360	696	299	518	3,420
21	7,560	1,010	1,100	380	16,800	5,420	4,770	1,460	685	292	461	4,640
22	6,170	1,080	1,700	380	12,900	5,160	4,270	2,290	2,580	280	427	4,020
23	4,550	1,280	1,700	360	8,900	4,640	3,780	3,810	4,820	266	409	3,200
24	5,080	1,280	1,900	360	8,230	4,140	3,540	4,520	3,730	292	392	2,580
25	10,200	1,180	2,800	360	6,990	3,660	3,200	4,400	2,600	1,140	370	2,220
26	10,200	1,170	2,600	380	10,900	3,200	2,680	5,560	2,040	1,020	332	2,110
27	12,800	938	2,000	400	11,000	2,840	2,350	7,140	1,690	675	325	2,310
28	17,800	1,080	1,600	400	9,600	2,540	2,110	5,160	1,420	500	299	2,200
29	18,800	1,100	1,400	400	....	2,270	1,890	3,660	1,210	409	378	1,860
30	23,800	1,130	1,200	380	....	2,040	1,770	2,860	1,080	716	566	1,580
31	21,800	....	950	360	....	2,040	....	2,660	....	665	538	....

NOTE.—Discharge Dec. 9 to Feb. 14 estimated, because of ice, from discharge measurements, weather records, study of gage-height graph and comparison with similar studies for near-by streams.

Monthly discharge of Allegheny River at Red House, N. Y., for the year ending Sept. 30, 1918.

[Drainage area, 1,640 square miles.]

Month.	Discharge in second-feet.				Run-off in inches.
	Maximum.	Minimum.	Mean.	Per square mile.	
October	23,800	340	5,370	3.27	3.77
November	15,000	938	2,620	1.60	1.78
December	2,800	700	1,250	.762	.88
January	800	320	462	.282	.33
February	18,800	340	5,610	3.42	3.56
March	28,400	2,040	8,960	5.46	6.30
April	5,560	1,640	3,060	1.87	2.09
May	7,140	1,210	2,510	1.53	1.76
June	4,820	685	1,610	.982	1.10
July	1,140	266	553	.337	.39
August	1,540	299	699	.426	.49
September	4,640	538	1,860	1.13	1.26
The year	28,400	266	2,860	1.74	23.71

MONONGAHELA RIVER BASIN.

TYGART RIVER NEAR DAILEY, W. VA.

LOCATION.—At Burnt Bridge, on Staunton-Parkersburg pike 1 mile northeast of Dailey, Randolph County, and 2 miles south of Beverly, on Western Maryland Railway, Stalnaker Run enters river on right 1,000 feet below station.

DRAINAGE AREA.—194 square miles (measured on topographic maps).

RECORDS AVAILABLE.—April 20, 1915, to September 30, 1918.

GAGE.—Vertical staff on face of right abutment of bridge near downstream end; read by Charles W. Chenoweth.

DISCHARGE MEASUREMENTS.—Made from bridge or by wading.

CHANNEL AND CONTROL.—Channel straight for 100 feet above bridge; curves slightly to right below bridge. Bed composed of small boulders. Banks sandy. Right bank high; left bank low; large overflow through meadows at high stages. Control probably permanent. Point of zero flow, September 26, 1917, at gage height 0.2 foot ±0.1 foot.

EXTREMES OF STAGE.—Maximum stage recorded during year, 15.9 feet at 5 p. m. March 13; minimum stage recorded, 0.68 foot October 7, 8, and 9.

1915–1918: Maximum stage recorded same as for 1918. Highest known flood reached a stage represented by gage height about 16 feet. Minimum stage recorded, 0.6 foot September 6, 1916.

ICE.—Stage-discharge relation affected by ice during severe winters.

ACCURACY.—Stage-discharge relation probably permanent except as affected by ice. Rating curve not fully developed. Gage read to hundredths twice daily. Records good.

The following discharge measurement was made by B. L. Hopkins:

May 3, 1918: Gage height, 2.10 feet; discharge, 216 second-feet.

Daily gage height, in feet, of Tygart River near Dailey, W. Va., for the year ending Sept. 30, 1918.

Day.	Oct.	Nov.	Dec.	Jan.	Feb.	Mar.	Apr.	May.	June.	July.	Aug.	Sept.
1	0.80	2.35	2.66	2.50	2.70	3.00	1.70	2.40	1.85	3.57	1.10	1.94
2	.77	1.80	2.29	....	2.40	2.88	1.68	2.25	1.72	2.60	.90	1.45
3	.74	1.58	2.05	....	2.30	2.58	1.99	2.10	1.62	2.38	.89	1.32
4	.72	1.44	1.98	....	2.20	2.32	2.22	2.02	1.52	1.90	.87	1.16
5	.71	1.37	1.82	2.50	2.60	2.85	....	1.92	1.42	1.69	.85	1.00
6	.70	1.30	1.68	1.65	2.70	3.28	....	1.82	1.30	1.60	.82	1.24
7	.66	1.23	1.58	1.95	3.35	5.62	....	2.05	1.90	1.48	.85	1.08
8	.68	1.86	1.49	3.09	3.95	4.78	2.24	8.10	1.95	1.32	1.08	1.25
9	.68	1.24	1.45	....	5.92	3.55	5.68	5.35	1.50	1.28	1.00	1.18
10	.69	1.10	1.45	....	7.70	3.25	4.35	3.45	1.40	1.60	1.18	1.08
11	.70	1.06	....	....	4.60	3.10	3.62	2.88	1.28	1.68	1.57	1.02
12	.72	1.05	....	3.40	4.38	2.88	3.38	2.55	1.15	1.32	1.78	.96
13	.74	1.04	....	3.40	4.70	12.55	2.58	3.71	1.12	1.11	2.00	.97
14	.80	1.04	....	....	4.00	10.95	3.12	7.64	1.06	1.00	2.38	1.58
15	.84	1.01	1.45	....	4.80	8.05	5.80	4.25	.98	1.00	1.95	1.36
16	.81	.98	1.45	3.20	4.00	5.14	4.85	3.25	.95	1.10	1.60	1.17
17	.78	.96	....	....	3.15	3.55	3.95	3.00	1.78	1.10	1.42	1.38
18	.76	.92	....	....	2.45	2.92	3.22	2.46	2.45	1.10	1.78	2.56
19	.76	.91	....	3.20	2.50	2.68	2.72	2.05	2.02	1.91	2.81	2.86
20	1.52	.91	....	....	7.70	2.55	2.60	2.08	1.58	1.55	1.94	2.22
21	1.54	.90	....	....	6.05	2.45	2.88	2.15	1.42	1.35	1.57	2.70
22	1.30	.92	1.50	1.80	3.50	2.18	3.22	1.98	1.50	1.18	1.38	2.49
23	1.14	.97	1.42	....	3.10	2.79	2.92	3.68	1.38	.99	1.24	1.93
24	1.14	.98	3.16	....	2.75	2.56	3.00	3.58	1.28	.94	1.15	1.70
25	1.36	1.10	4.68	1.80	4.30	2.55	3.75	5.90	1.41	.90	1.06	1.58
26	1.48	1.10	3.58	....	9.60	2.45	3.61	8.91	3.34	1.04	1.01	1.40
27	1.80	1.10	2.97	6.45	5.15	2.28	3.38	4.18	2.15	1.08	.95	1.34
28	2.40	1.10	2.70	10.18	3.52	2.20	3.08	3.55	1.78	.90	.90	1.24
29	1.86	2.16	2.50	8.18	....	2.05	2.75	2.98	8.75	.88	.88	1.18
30	2.20	2.30	2.50	4.45	....	2.00	2.52	2.44	4.88	.89	.88	1.12
31	2.40	....	2.50	3.32	....	2.00	....	2.15	....	1.28	2.12	....

NOTE.—Stage-discharge relation affected by ice Dec. 10 to Jan. 29. Gage not read Dec. 11–14, 17–21, Jan. 2–4, 9–11, 14, 15, 17, 18, 20, 21, 23, 24, 26. Gage heights Apr. 5-7 withheld because of observer's error in making readings.

TYGART RIVER AT BELINGTON, W. VA.

Location.—At highway bridge at Belington, Barbour County, a quarter of a mile above mouth of Mill Creek.

Drainage area.—390 square miles.

Records available.—June 5, 1907, to September 30, 1918.

Gage.—Chain gage attached to the upstream side of highway bridge to left of center of the river; read by S. A. Campbell. Sea-level elevation of zero of gage, 1,679.89 feet.

Discharge measurements.—Made from upstream side of bridge or by wading.

Channel and control.—Channel straight above and below bridge. Bed composed of firm, coarse gravel. Banks high. Control slightly shifting.

Extremes of discharge.—Maximum stage recorded during year, 19.42 feet at 7.30 a. m. March 14 (discharge, 17,400 second-feet); minimum stage recorded, 1.90 feet at 7 a. m. October 9 (discharge, 8 second-feet).

1907–1918: Maximum stage recorded, 21.48 feet March 13, 1917 (discharge, 20,100 second-feet); minimum stage recorded, 1.70 feet October 2, 1914 (discharge, 3 second-feet).

Ice.—Stage-discharge relation affected by ice during severe winters.

Accuracy.—The change in rating curve indicated by discharge measurements made in 1920 probably was caused by the high water in March, 1918. Stage-discharge relation also affected by ice. Rating curve used for open-water periods October 1, 1917, to March 15, 1918, fairly well defined between 13 and 300 second-feet and well defined between 300 and 4,000 second-feet; curve extended beyond these limits. Curve used March 16 to September 30, 1918, fairly well defined between 50 and 150 second-feet and well defined between 150 and 3,000 second-feet; extended beyond these limits. Gage read to hundredths once daily. Owing to indistinct figures at footmarks on gage scale, some of the gage readings were in error by multiples of half a foot. Records for these days were interpreted by comparison with records for stations at Dailey, Fetterman, Midvale, and Hall. Daily discharge for open-water periods ascertained by applying daily gage height to rating table; for period of ice effect estimated by means of observer's notes, weather records and comparison with records for other stations. Open-water records fair; those for period of ice effect, roughly approximate.

Records of discharge for years ending September 30, 1916 and 1917, as given in following tables supersede those published in previous reports owing to revision based on comparison of discharge at Belington with that at Dailey and the combined discharge at stations at Belington, Midvale, and Hall with that at Fetterman.

The following discharge measurement was made by B. L. Hopkins:

August 29,1918: Gage height, 4.38 feet; discharge, 823 second-feet.

[illegible]	[illegible]	[illegible]	[illegible]	155	[illegible]	[illegible]
[illegible]	[illegible]	[illegible]	[illegible]	155	[illegible]	[illegible]
[illegible]	[illegible]	[illegible]	[illegible]	119	[illegible]	[illegible]
[illegible]	[illegible]	[illegible]	[illegible]	222	[illegible]	[illegible]
[illegible]	[illegible]	[illegible]	[illegible]	350	[illegible]	[illegible]
[illegible]	[illegible]	[illegible]	104	182	[illegible]	[illegible]
[illegible]	[illegible]	[illegible]	99	210	[illegible]	[illegible]
[illegible]	[illegible]	[illegible]	70		[illegible]	[illegible]
[illegible]	[illegible]	[illegible]	67	310	[illegible]	[illegible]
[illegible]	[illegible]	212	79	365	[illegible]	[illegible]
[illegible]	[illegible]	138	72	[illegible]	[illegible]	[illegible]
[illegible]	[illegible]	[illegible],000	61		[illegible]	[illegible]
[illegible]	[illegible]	[illegible],380	60		[illegible]	[illegible]
[illegible]	140	6,380	240	[illegible]	[illegible]	[illegible]
[illegible]	[illegible]	3,430	396		[illegible]	[illegible]
[illegible]	......	2,350	......			[illegible]

Daily discharge, in second-feet, of Tygart River at Belington, W. Va., for the years ending Sept. 30, 1916–1918—Continued.

Day.	Oct.	Nov.	Dec.	Jan.	Feb.	Mar.	Apr.	May.	June.	July.	Aug.	Sept.
1917–18.												
1	14	440	645	250	1,270	1,270	321	644	407	1,450	85	213
2	14	310	590	250	985	1,200	282	498	321	904	63	407
3	13	205	490	240	700	940	321	452	246	570	62	213
4	14	128	372	210	700	880	694	385	207	430	43	156
5	13	119	330	210	490	1,000	644	302	156	407	40	106
6	14	110	270	210	820	1,550	522	321	73	264	87	87
7	19	95	219	220	2,750	1,830	430	210	183	264	108	87
8	13	82	179	270	3,010	4,230	430	990	797	230	186	98
9	8	70	150	340	7,380	1,760	2,790	3,460	475	130	178	85
10	13	46	120	490	4,930	1,760	2,870	1,660	246	110	146	82
11	13	48	100	700	3,550	1,830	2,000	960	154	80	264	65
12	17	46	80	760	2,190	1,060	1,780	670	110	82	264	48
13	13	49	70	1,450	2,110	5,370	1,440	520	92	92	282	59
14	20	48	60	1,400	2,190	17,400	1,100	3,050	90	59	407	106
15	15	48	60	1,240	1,620	12,400	3,810	1,440	63	73	694	119
16	15	48	50	1,130	2,350	4,010	2,870	1,130	59	73	407	90
17	13	36	50	1,300	1,900	1,920	2,230	730	63	82	213	106
18	13	40	40	910	800	1,190	1,370	660	1,440	59	119	282
19	18	30	40	600	610	797	959	342	694	108	321	1,250
20	30	30	30	420	5,590	644	644	321	363	282	498	694
21	77	30	40	330	1,900	546	595	302	342	142	246	797
22	158	34	60	290	1,130	644	1,130	342	522	106	170	745
23	75	35	80	260	1,130	850	1,310	1,560	363	73	112	452
24	55	40	90	230	618	694	1,020	1,500	321	58	90	321
25	99	36	100	230	2,830	694	959	2,800	213	47	78	204
26	252	46	1,020	210	5,590	644	1,370	6,420	904	42	60	183
27	270	49	1,130	1,170	7,260	595	1,130	2,940	1,130	32	53	132
28	672	58	660	3,840	2,190	595	850	1,620	183	32	46	94
29	418	65	430	7,680	......	452	797	1,450	1,500	54	34	94
30	290	645	270	5,370	......	385	694	745	2,900	59	53	98
31	672	......	250	2,190	......	321	......	570	......	66	98	......

NOTE.—Discharge estimated because of ice, Dec. 6–16, 1915, Jan. 16–21, and Dec. 10–20, 1916, Jan. 13–20, Feb. 2–19, and Dec. 9, 1917, to Jan. 29, 1918. Discharge for following days estimated by comparison with records of flow for stations at Dailey, Fetterman, Midvale, and Hall: Feb. 18, 19, Apr. 14, May 7–13, 16–18, 24–29, June 30 and July 1, 1918. Discharge Nov. 5 and 15, 1917, and Feb. 2, 1918, interpolated. Braced figures show mean discharge for periods indicated.

Monthly discharge of Tygart River at Belington, W. Va., for the years ending Sept. 30, 1916–1918.

[Drainage area, 390 square miles.]

Month.	Discharge in second-feet.				Run-off in inches.
	Maximum.	Minimum.	Mean.	Per square mile.	
1915–16.					
October	5,260	77	577	1.48	1.71
November	4,530	55	603	1.55	1.73
December	8,100	....	1,330	3.41	3.93
January	9,920	540	1,990	5.10	5.88
February	6,250	590	1,610	4.13	4.45
March	6,250	672	1,900	4.87	5.62
April	2,350	540	908	2.33	2.60
May	1,480	270	576	1.48	1.71
June	1,900	172	439	1.13	1.26
July	2,670	....	394	1.01	1.16
August	645	50	211	.541	.62
September	2,040	22	289	.741	.83
The year	9,920	22	902	2.31	31.50

Monthly discharge of Tygart River at Belington, W. Va., for the years ending Sept. 30, 1916–1918—Continued.

Month.	Discharge in second-feet.				Run-off in inches.
	Maximum.	Minimum.	Mean.	Per square mile.	
1916–17.					
October	700	70	217	0.556	0.64
November	760	49	142	.364	.41
December	6,030		809	2.07	2.39
January	9,060	418	1,580	4.05	4.67
February			1,450	3.72	3.87
March	20,100	540	3,750	9.62	11.09
April	2,510	195	669	1.72	1.92
May	7,260	158	1,110	2.85	3.29
June	1,340	60	330	.846	.94
July	565	22	175	.449	.52
August	182	12	42.7	.109	.13
September	540	14	96.0	.246	.27
The year	20,100	12	866	2.22	30.14
1917–18.					
October	672	8	108	0.277	0.32
November	645	30	102	.262	.29
December	1,130	30	260	.667	.77
January	7,680	210	1,110	2.85	3.29
February	7,380	490	2,450	6.28	6.54
March	17,400	321	2,240	5.74	6.62
April	3,810	282	1,260	3.23	3.60
May	6,420	210	1,260	3.23	3.72
June	2,900	59	487	1.25	1.40
July	1,450	32	208	.533	.61
August	694	34	178	.456	.53
September	1,250	48	249	.638	.71
The year	17,400	8	815	2.09	28.40

TYGART RIVER AT FETTERMAN, W. VA.

Location.—At highway bridge at Fetterman, Taylor County, three-fourths mile above mouth of Otter Creek.

Drainage area.—1,340 square miles.

Records available.—June 3, 1907, to September 30, 1918.

Gage.—Chain gage attached to downstream side of highway bridge; read by Joseph Weaver. Sea-level elevation of zero of gage, 957.86 feet.

Discharge measurements.—Made from downstream side of the bridge or by wading.

Channel and control.—Channel straight above and below bridge. Both banks high. Control practically permanent.

Extremes of discharge.—Maximum stage recorded during year, 24.1 feet at midnight March 13 (discharge, about 45,400 second-feet); minimum stage recorded, 3.15 feet October 8–11 (discharge, 58 second-feet).

1907–1918: Maximum stage recorded, 29.1 feet July 25, 1912 (discharge, 57,600 second-feet); minimum stage recorded, 2.30 feet October 27–28, and November 4–10, 1912 (discharge, 12 second-feet).

Ice.—Stage-discharge relation affected by ice during severe winters.

Accuracy.—Stage-discharge relation practically permanent, except as affected by ice. Rating curve well defined between 80 and 23,000 second-feet, poorly defined below 80 second-feet; extended above 23,000 second-feet. Gage read to half-tenths twice daily. Daily discharge ascertained by applying mean daily gage height to rating table except for periods of ice effect. Records good except those for periods of ice effect which are poor.

The following discharge measurement was made by B. L. Hopkins:

April 27, 1918: Gage height, 5.75 feet; discharge, 3,040 second-feet.

Daily discharge, in second-feet, of Tygart River at Fetterman, W. Va., for the year ending Sept. 30, 1918.

Day.	Oct.	Nov.	Dec.	Jan.	Feb.	Mar.	Apr.	May.	June.	July.	Aug.	Sept.
1	65	1,450	1,850	750	3,910	4,290	920	1,930	1,610	2,790	348	2,100
2	65	1,100	1,610	650	2,610	3,340	865	1,690	1,100	1,850	255	2,020
3	65	920	1,450	600	2,020	2,970	865	1,450	865	1,450	348	1,030
4	65	760	1,230	650	2,270	2,610	1,300	1,300	710	975	255	665
5	65	578	975	680	1,690	3,150	1,770	1,160	620	865	244	535
6	65	455	810	710	1,380	4,670	1,380	1,030	535	1,030	200	455
7	65	380	710	760	3,150	6,000	1,300	975	3,340	665	155	380
8	58	315	620	920	7,940	7,560	1,380	6,380	3,340	535	138	315
9	58	315	500	1,200	11,300	6,380	7,940	9,700	1,690	455	120	285
10	58	285	350	1,700	17,600	6,000	9,120	5,240	1,030	380	155	200
11	58	285	300	1,990	11,100	5,620	7,560	3,530	710	348	620	255
12	65	285	250	2,610	6,780	5,240	7,750	2,610	535	315	455	200
13	65	200	200	4,860	6,000	16,200	5,810	2,270	455	380	810	228
14	65	200	200	4,670	5,430	41,600	5,620	7,160	380	328	810	255
15	65	200	200	4,290	5,240	28,800	6,000	8,340	315	315	1,380	200
16	65	155	200	3,910	5,240	13,400	7,360	4,480	255	303	1,100	191
17	90	155	180	4,480	3,910	8,190	5,620	2,970	1,610	255	710	348
18	90	155	160	3,150	2,610	,910	3,720	2,100	3,150	255	495	865
19	138	155	160	2,100	2,270	,790	2,790	1,530	2,440	315	380	1,530
20	535	155	170	1,450	16,400	,100	2,270	1,770	1,380	255	455	2,610
21	380	155	200	1,160	15,700	1,930	2,610	1,450	1,030	267	665	2,610
22	455	138	267	1,030	6,780	2,790	5,240	1,230	865	418	455	2,610
23	535	138	348	920	3,910	2,610	4,100	1,300	1,690	315	380	1,770
24	535	138	535	760	3,340	2,270	3,150	2,610	1,030	255	285	1,160
25	620	148	920	760	6,780	2,100	2,790	7,560	710	418	228	760
26	1,100	228	3,530	710	22,900	1,930	2,610	20,200	1,030	255	200	620
27	1,230	348	3,910	4,100	18,800	1,850	2,970	13,400	3,150	200	178	535
28	1,380	440	2,270	21,900	7,560	975	3,340	5,430	1,690	178	155	455
29	1,690	665	1,450	25,600	....	1,380	2,970	4,100	2,020	155	200	380
30	1,610	1,930	920	15,900	....	1,160	2,270	2,610	4,480	620	178	315
31	1,300	....	860	7,160	....	1,030	....	2,270	....	810	578	....

NOTE.—Discharge estimated because of ice Dec. 9-21, 30, 31, Jan. 1-5, 8-24, by means of observer's notes weather records, and comparison with records of flow at other stations on this river.

Monthly discharge of Tygart River at Fetterman, W. Va., for the year ending Sept. 30, 1918.

[Drainage area, 1,340 square miles.]

Month.	Discharge in second-feet.				Run-off in inches.
	Maximum.	Minimum.	Mean.	Per square mile.	
October	1,690	58	410	0.306	0.35
November	1,930	138	428	.319	.36
December	3,910	160	882	.658	.76
January	25,600	600	3,940	2.94	3.39
February	22,900	1,380	7,310	5.46	5.69
March	41,600	975	6,220	4.64	5.35
April	9,120	865	3,780	2.82	3.15
May	20,200	975	4,190	3.13	3.61
June	4,480	255	1,460	1.09	1.22
July	2,790	155	579	.432	.50
August	1,380	120	417	.311	.36
September	2,610	191	863	.644	.72
The year	41,600	58	2,510	1.87	25.46

MONONGAHELA RIVER AT LOCK 15, HOULT, W. VA.

LOCATION.—At Lock 15, at Hoult, 2½ miles below county highway bridge at Fairmont, Marion County, and 4 miles below mouth of West Fork. Buffalo Creek enters on left three-fourths mile above station.

DRAINAGE AREA.—2,430 square miles (measured on topographic maps).

RECORDS AVAILABLE.—October 1, 1914, to September 30, 1918. Upper and lower gages at Lock 15 have been read under direction of United States Engineer Corps since May 1, 1904.

GAGE.—Upper vertical staff gage at lock. Lower section is set in recess in left lock wall just above upper gate; upper section, 61.5 feet from face of right lock wall, directly opposite lower section, was used until January 29, 1918, when it was carried away by ice. Read by Charles R. Hall, lockmaster.

DISCHARGE MEASUREMENTS.—Made from bridge at Fairmont or by wading on crest of dam at the lock. Flow of Buffalo Creek is added to discharge measured at bridge.

CHANNEL AND CONTROL.—One channel at all stages; straight half a mile above and below bridge. Control for station is crest of dam; permanent. Point of zero flow, gage height 6.9 feet, elevation of crest of dam. Leakage through lock and occasional opening of valves of lock may affect stage at which flow would be zero.

EXTREMES OF DISCHARGE.—Maximum stage recorded during year, 21.1 feet at 5 a. m. March 14 (discharge, 91,200 second-feet); minimum stage recorded, 6.96 feet at 6 p. m. July 29, due to opening valves. Minimum stage recorded under normal conditions, 7.02 feet at 5 p. m. October 2 (discharge, 55 second-feet).

1915–1918: Maximum stage recorded same as for 1918; minimum stage recorded, 6.10 feet July 31, 1916, due to opening valves. Minimum stage recorded under normal conditions, 7.00 feet September 26, 1917 (discharge, 47 second-feet). Flood of 1888, before dam No. 15 was built, reached a stage represented by gage height of about 26 feet.

ICE.—Stage-discharge relation affected by ice when ice in pool above dam forms close to crest of dam.

DIVERSIONS.—Leakage through lock and water used for lockages. See "Accuracy."

REGULATION.—None under normal conditions. Pool No. 15 may be lowered at times in the interest of navigation.

ACCURACY.—Stage-discharge relation permanent except for effect of operations at lock and change in leakage through lock, the change depending on which gates are open; slightly affected by ice. Rating curve well defined to 62,000 second-feet. Gage read to hundredths twice daily. Daily discharge ascertained by applying mean daily gage height to rating table, and adding amount of water used for lockage. Rating table makes allowance based on measurement for leakage through upper gates, for under normal conditions upper gates are closed; gage reader records number of lockages and length of time upper gates are open. Daily discharge corrected for effect of lockage and change in leakage when upper gates at lock are open. Records good.

The following discharge measurement was made by B. L. Hopkins:

May 10, 1918: Gage height, 10.07 feet; discharge, 7,860 second-feet.

Daily discharge, in second-feet, of Monongahela River at Lock 15, Hoult. W. Va., for the year ending Sept. 30, 1918.

Day.	Oct.	Nov.	Dec.	Jan.	Feb.	Mar.	Apr.	May.	June.	July.	Aug.	Sept.
1	97	2,740	2,360	980	5,	6,620	1,290	3,140	2,750	3,750	900	3,020
2	93	2,120	2,230	810		5,170	1,180	2,740	1,780	2,240	488	3,000
3	95	1,570	2,230	730		4,340	1,370	2 120	1,280	1,900	287	1,680
4	89	1,030	1,780	810	880	3,720	2,000	1 790	949	1,300	255	995
5	104	836	1,470	810	2,[illegible]	5,520	2,730	1,570	821	1,060	297	654
6	93	598	1,280	963	2,000	8,270	2,120	1,380	624	1,110	253	510
7	98	566	984	827	6,250	6,250	1,780	1,370	4,040	869	192	454
8	112	510	844	1,370	14,000	8,710	2,120	13,500	6,650	622	178	385
9	107	468	650	2,000	22,000	7,020	14,500	14,500	3,280	466	161	337
10	96	411	455	2,000	25,900	9,680	15,000	7,820	1,680	136	168	300
11	94	411	400	2,120	17,200	8,710	12,000	5,190	1,120	362	303	282
12	134	373	380	3,000	40,100	6,250	17,200	4,180	810	324	401	287
13	144	320	300	6,620	9,630	16,600	13,500	3,140	561	322	523	273
14	161	291	300	5,890	8,280	77,000	9,630	7,000	455	331	744	322
15	147	282	300	5,170	9,170	46,500	8,710	11,000	384	305	1,580	320
16	142	292	300	7,820	9,630	19,700	9,190	6,260	313	288	1,470	302
17	138	284	273	6,620	7,000	9,170	7,000	4,180	493	282	1,010	340
18	140	273	246	4,500	4,360	5,880	5,180	3,150	3,870	264	631	591
19	201	273	246	2,860	3,870	4,340	4,180	2,230	3,150	284	444	1,370
20	1,780	264	264	2,120	28,800	3,160	2,860	1,700	1,900	269	375	2,860
21	1,470	236	255	1,780	26,000	2,600	3,280	8,140	1,180	255	635	2,880
22	892	210	340	1,280	10,700	3,590	5,880	1,690	878	282	506	3,000
23	810	220	552	1,130	6,260	3,870	6,250	2,470	1,570	292	422	2,480
24	682	213	666	980	5,050	3,150	4,830	3,000	1,780	246	335	1,890
25	1,570	210	1,280	980	7,120	2,770	3,870	11,500	1,060	340	249	1,090
26	2,870	219	3,870	980	32,200	3,280	3,570	30,800	759	340	227	844
27	2,370	228	5,880	1,570	28,700	3,000	3,720	21,400	3,610	262	210	604
28	2,120	246	3,720	12,500	11,500	2,470	5,000	8,740	2,480	274	206	446
29	2,250	312	2,230	44,000	....	1,780	4,340	9,180	1,590	168	198	393
30	2,010	1,580	1,130	24,600	....	1,470	3,720	4,660	3,720	236	218	360
31	2,470	....	1,020	10,600	....	1,370	....	3,880	....	1,790	487	....

NOTE.—Daily discharge, Jan. 8-10, estimated because of ice, by comparison with flow of stations on Tygart River. Wickets open Mar. 21, May 6, and July 29; discharge estimated.

Monthly discharge of Monongahela River at Lock 15, Hoult, W. Va., for the year ending Sept. 30, 1918.

[Drainage area, 2,430 square miles.]

Month.	Discharge in second-feet.				Run-off in inches.
	Maximum.	Minimum.	Mean.	Per square mile.	
October	2,870	89	761	0.313	0.36
November	2,740	210	586	.241	.27
December	5,880	246	1,230	.506	.58
January	44,000	730	5,110	2.10	2.42
February	32,200	2,000	11,600	4.77	4.97
March	77,000	1,370	9,420	3.88	4.47
April	17,200	1,180	5,930	2.44	2.72
May	30,800	1,370	6,400	2.63	3.03
June	6,650	313	1,850	.761	.85
July	3,750	168	686	.282	.33
August	1,580	161	463	.191	.22
September	3,020	273	1,080	.444	.50
The year	77,000	89	3,710	1.53	20.72

MIDDLE FORK AT MIDVALE, W. VA.

LOCATION.—A third of a mile above Midvale railroad station on Coal & Coke Railway, two-thirds mile below post office at Ellamore, Randolph County. Laurel Creek enters river on right 1¾ miles above station.

DRAINAGE AREA.—122 square miles (measured on topographic maps).

RECORDS AVAILABLE.—May 3, 1915, to September 30, 1918.

GAGE.—Vertical and inclined staff on right bank; read by Anna Riley.

DISCHARGE MEASUREMENTS.—Made from cable or by wading.

CHANNEL AND CONTROL.—One channel at all stages; straight 300 feet above and 100 feet below cable section. Banks are high and in most places wooded. Control slightly shifting.

EXTREMES OF STAGE.—Maximum stage recorded during year, 16.1 feet at 7.30 a. m. January 28 (stage-discharge relation affected by ice); minimum stage recorded, 1.25 feet at 7 a. m. August 8 and 30.

1915–1918: Maximum stage recorded same as for 1918; minimum stage recorded, 1.12 feet August 29, 1917 (discharge, 2.6 second-feet).

Floods of 1888 and 1912 reached gage height of about 18 feet.

ICE.—Stage-discharge relation affected by ice during severe winters.

ACCURACY.—The change in rating curve indicated by discharge measurements made during 1918 and 1920 probably was caused by the high water in January, 1918. Stage-discharge relation seriously affected by ice. New rating curve not fully developed. Gage read to hundredths twice daily. Records good.

The following discharge measurement was made by B. L. Hopkins:

May 4, 1918: Gage height, 2.55 feet; discharge, 184 second-feet.

Daily gage height, in feet, of Middle Fork at Midvale, W. Va., for the year ending Sept. 30, 1918.

Day.	Oct.	Nov.	Dec.	Jan.	Feb.	Mar.	Apr.	May.	June.	July.	Aug.	Sept.
1	1.36	2.40	3.09	....	3.	3.	2.	2.	2.28	2.68	1.62	2.32
2	1.29	2.16	3.00	....	2.	3.	2.	2.	2.09	2.43	1.49	1.88
3	1.26	2.04	2.74	....	2.	3.	2.	2.	2.02	2.16	1.42	1.66
4	1.26	1.92	2.56	....	2.22	3.66	2.26	2.88	1.90	2.02	1.36	1.66
5	1.29	1.83	2.37	....	2.[illegible]	3.[illegible]	2.[illegible]	2.[illegible]	1.80	1.88	1.32	1.60
6	1.31	1.76	2.23	....	3.	3.	2.	2.	1.74	1.84	1.28	1.56
7	1.34	1.72	2.16	....	3.	5.	2.	2.	2.64	1.72	1.27	1.60
8	1.34	1.62	2.07	....	4.	4.	2.	4.	2.44	1.68	1.58	1.54
9	1.35	1.63	2.17	....	6.	3.34	4.	4.36	2.08	1.62	1.60	1.48
10	1.36	1.64	2.24	....	6.24	3.[illegible]	4.44	3.[illegible]	1.93	1.56	1.82	1.42
11	1.38	1.58	2.01	....	4.	3.58	3.	3.	1.82	1.52	1.76	1.37
12	1.42	1.60	2.04	6.85	4.	3.48	3.	3.	1.74	1.46	1.63	1.36
13	1.52	1.64	2.02	....	4.	10.34	3.	2.	1.70	1.48	2.55	1.38
14	1.60	1.58	2.00	....	3.89	7.99	2.76	5.26	1.65	1.58	2.10	1.36
15	1.52	1.52	1.98	6.34	3.[illegible]	7.15	4.[illegible]	4.[illegible]	1.56	1.48	1.92	1.34
16	1.48	1.56	1.92	5.86	3.	5.00	4.	3.	1.51	1.42	1.75	1.30
17	1.40	1.54	1.85	5.58	3.	4.14	3.	3.	1.64	1.40	1.60	1.47
18	1.40	1.52	1.90	5.11	2.	3.47	3.	2.	2.54	1.64	1.53	3.00
19	1.48	1.52	1.89	4.46	2.58	3.13	3.38	2.58	2.06	1.60	2.08	2.70
20	1.94	1.52	1.93	4.14	7.[illegible]	2.88	2.[illegible]	2.[illegible]	1.88	1.70	1.79	2.60
21	1.82	1.52	1.99	3.76	5.	2.	3.	2.44	1.72	1.52	1.63	3.16
22	1.66	1.55	2.12	4.15	4.	3.	3.	2.48	2.18	1.42	1.52	2.50
23	1.60	1.62	2.18	4.04	2.	3.	3.	3.42	2.08	1.36	1.43	2.24
24	1.62	1.73	2.38	3.78	3.03	2.96	3.36	3.24	1.88	1.32	1.44	2.03
25	1.80	1.58	6.06	3.73	3.[illegible]	3.[illegible]	3.[illegible]	6.63	1.80	1.28	1.42	1.88
26	2.12	1.58	7.44	4.68	8.72	2.	3.	6.	4.79	1.44	1.36	1.78
27	2.91	1.62	....	11.60	5.28	2.	3.	4.	3.22	1.40	1.32	1.71
28	1.82	1.84	....	16.00	4.17	2.	3.	3.	2.60	1.36	1.30	1.63
29	1.40	3.43	....	7.52	....	2.	3.	3.	2.84	1.30	1.26	1.60
30	3.02	3.24	....	4.86	....	2.94	2.26	2.50	2.72	1.32	1.27	1.56
31	2.72	....	....	3.98	....	2.[illegible]	....	2.[illegible]	....	1.58	1.86	....

NOTE.—Gage heights Oct. 28, 29, and Apr. 14, are apparently 1 foot too low. Stage-discharge relation affected by ice Dec. 9 to Jan. 28. Gage not read Dec. 27-31, Jan 1-11, 13 and 14.

BUCKHANNON RIVER AT HALL, W. VA.

LOCATION.—About 500 feet below ruins of an old milldam, a quarter of a mile above post office and county highway bridge at Hall, Barbour County, 1 mile from Baltimore & Ohio Railroad station. Pecks Run enters river on left 1 mile below station.

DRAINAGE AREA.—277 square miles (measured on topographic maps).

RECORDS AVAILABLE.—June 7, 1907, to May 25, 1909; April 15, 1915, to September 30, 1918.

GAGE.—Vertical and inclined staff on right bank used since April 15, 1915; read by James Newcomb. From June 7, 1907, to May 25, 1909, a chain gage at county highway bridge one-quarter of a mile below was used.

DISCHARGE MEASUREMENTS.—Made from county highway bridge.

CHANNEL AND CONTROL.—Gage is about midway between beginning and end of rapids having approximately 10-foot fall. Bed of stream in rapids composed of large boulders, rocks, and gravel; practically permanent. Banks are high and wooded and are not overflowed except into an old mill race on left bank.

EXTREMES OF DISCHARGE.—Maximum stage recorded during year, 14.7 feet at 7 a. m. March 14 (discharge not determined); minimum stage recorded, 1.70 feet at 6 a. m. October 7 (discharge, 8 second-feet).

1907–1909: Maximum stage recorded, 13.8 feet (gage at highway bridge) February 6, 1908 (discharge not determined); minimum stage recorded, 1.40 feet during several days in October and November, 1908 (discharge not determined).

1915–1918: Maximum and minimum stages occurred during year ending September 30, 1918.

Highest flood known reported to have reached a gage height of about 14 feet in 1888, referred to datum of present gage.

ICE.—Stage-discharge relation affected by ice during severe winters.

DIVERSIONS.—No water diverted above station except small quantity which may flow around gage through abandoned mill race above ordinary low stages and which is included in flow measured at county highway bridge.

ACCURACY.—Stage-discharge relation permanent except as affected by ice, December 28 to January 28 and February 3–8. Rating curve well defined between 40 and 4,500 second-feet; extended beyond these limits. Gage read to hundredths twice daily. Daily discharge ascertained by applying mean daily gage height to rating table, except for periods of ice effect for which it was ascertained by means of observer's notes, weather records, and comparison with records for other stations in this basin. Records good except those for periods of ice effect.

The following discharge measurement was made by B. L. Hopkins:

May 6, 1918: Gage height, 2.63 feet; discharge, 252 second-feet.

Daily discharge, in second-feet, of Buckhannon River at Hall, W. Va., for the year ending Sept. 30, 1918.

Day.	Oct.	Nov.	Dec.	Jan.	Feb.	Mar.	Apr.	May.	June.	July.	Aug.	Sept.
1	17	396	628	180	1,070	920	218	535	482	410	180	722
2	12	270	628	180	675	820	208	442	311	450	114	474
3	15	208	490	170	410	675	260	367	228	297	78	228
4	11	166	396	150	244	490	426	311	194	228	57	154
5	12	138	324	150	194	1,070	352	270	162	189	40	124
6	10	111	265	150	244	1,070	297	233	130	162	37	104
7	9	98	218	170	410	1,280	284	218	410	130	45	93
8	12	88	175	210	675	1,720	410	1,960	535	111	50	83
9	10	78	162	260	1,720	1,280	2,400	1,610	304	98	73	73
10	10	71	98	340	3,600	1,500	1 840	970	199	93	114	65
11	10	71	86	490	1,840	1,120	1,440	770	154	88	98	56
12	13	63	86	580	1,280	820	1,960	580	127	69	130	47
13	13	63	78	1,070	1,020	6,660	1,440	535	93	57	218	45
14	23	53	70	870	920	12,200	1,120	1,720	83	61	490	63
15	21	57	65	1,280	970	6,770	1,220	1,500	71	50	304	45
16	18	51	63	1,960	1,120	2,920	1,170	970	65	37	194	36
17	24	43	61	1,070	770	1,390	970	628	170	50	127	47
18	27	45	73	675	580	870	820	458	284	53	117	78
19	30	43	69	490	490	628	628	360	324	51	83	770
20	98	47	63	410	3,220	474	490	442	180	43	76	628
21	194	39	57	330	3,600	490	770	284	138	45	83	770
22	150	45	86	300	1,390	820	1,120	213	194	43	65	675
23	104	36	104	240	870	628	920	580	249	50	51	450
24	98	47	127	220	820	535	770	628	170	47	40	304
25	170	50	338	220	1,170	580	675	1,960	130	40	37	228
26	450	51	2,180	170	5,790	628	628	6,660	580	31	33	162
27	338	47	1,340	1,070	4,170	490	970	2,620	675	23	43	124
28	466	57	675	2,620	1,580	426	1,280	2,400	374	26	26	101
29	396	450	304	6,880	....	338	920	1,280	442	43	30	88
30	338	722	218	4,070	....	270	675	970	490	63	24	88
31	580	....	180	1,500	....	249	....	675	....	65	162	....

NOTE.—Discharge, Dec. 14 and 15, estimated; gage not read. Discharge Dec. 28 to Jan. 28 and Feb. 3-8 estimated because of ice, from observer's notes, study of weather records, and comparison with records for other stations in basin.

Monthly discharge of Buckhannon River at Hall, W. Va., for the year ending Sept. 30, 1918.

[Drainage area, 277 square miles.]

Month.	Discharge in second-feet.				Run-off in inches.
	Maximum.	Minimum.	Mean.	Per square mile.	
October	580	9	119	0.430	0.50
November	722	36	123	.444	.50
December	2,180	57	313	1.13	1.30
January	6,880	150	919	3.32	3.83
February	5,790	194	1,460	5.27	5.49
March	12,200	249	1,620	5.85	6.74
April	2,400	208	889	3.21	3.58
May	6,660	213	1,070	3.86	4.45
June	675	65	265	.957	1.07
July	450	23	103	.372	.43
August	490	24	104	.375	.43
September	770	36	231	.834	.93
The year	12,200	9	596	2.15	29.25

WEST FORK AT BUTCHERVILLE, W. VA.

Location.—At Weston & Clarksburg Electric Railway Co.'s trolley bridge, a quarter of a mile upstream from Butcherville, Lewis County, 3 miles north of Weston. Freemans Creek enters river on left 1 mile below station.

Drainage area.—181 square miles (measured on topographic maps).

Records available.—April 8, 1915, to September 30, 1918.

Gage.—Chain gage fastened to upstream side of trolley bridge near center of span; read by Bess Ervin.

Discharge measurements.—Made from bridge or by wading.

Channel and control.—One channel except at extreme high stages, when river overflows right bank and a little water passes through two small culverts in trolley embankment; straight for 500 feet above and curved for 1,000 feet below station. Stream bed composed of sand and gravel. Control is rock ledge; probably permanent. Growth of aquatic plants may cause backwater at gage during summer.

Extremes of stage.—Maximum stage recorded during year, 24.0 feet at 4.30 p. m. March 13; minimum stage recorded, 3.28 feet at 8.30 a. m. August 14.

1915–1918: Maximum and minimum stages same as for 1918. Highest flood known is reported to have reached a stage represented by gage height of about 27 feet in 1888. Dam since washed out may have increased height of this flood.

Ice.—Stage-discharge relation affected by ice during severe winters.

Accuracy.—Stage-discharge relation probably permanent; seriously affected by ice in December and January. Measurements of flow do not indicate noteworthy backwater from growth of aquatic plants. Rating curve not fully developed. Gage read to hundredths twice daily. Records excellent.

The following discharge measurement was made by B. L. Hopkins:

May 8, 1918: Gage height, 7.20 feet; discharge, [illegible]71 second-feet.

Daily gage height, in feet, of West Fork at Butcherville, W. Va., for the year ending Sept. 30, 1918.

Day.	Oct.	Nov.	Dec.	Jan.	Feb.	Mar.	Apr.	May.	June.	July.	Aug.	Sept.
1	3.74	5.64	5.50	4.57	5.74	6.02	4.36	5.48	5.56	5.36	4.82	7.44
2	3.66	5.28	5.34	......	5.19	5.72	4.90	5.22	5.28	5.22	4.63	5.36
3	3.62	4.90	5.25	......	5.11	5.48	5.07	5.06	5.04	5.07	4.38	5.06
4	8.64	4.72	5.08	4.51	5.04	5.50	5.90	5.90	4.81	5.20	4.22	4.70
5	3.62	4.64	4.94	......	4.92	8.26	5.54	4.78	4.64	5.04	4.06	4.26
6	3.60	4.42	4.84	......	4.84	6.81	5.26	4.69	4.46	4.89	3.94	4.14
7	3.56	4.30	4.68	8.16	6.59	6.20	5.04	4.64	7.90	4.56	3.86	4.15
8	3.56	4.28	4.50	7.50	7.84	5.80	7.32	6.92	5.66	4.30	3.76	4.14
9	3.58	4.24	4.48	6.24	11.22	6.54	10.81	5.43	4.93	4.16	3.64	4.06
10	3.60	4.25	4.40	5.62	8.86	7.83	7.84	5.32	4.74	4.05	3.53	4.04
11	3.65	4.22	4.34	5.18	7.02	7.05	9.84	5.32	4.64	3.98	3.48	4.00
12	3.80	4.20	......	7.78	......	6.30	9.76	5.20	4.50	4.02	3.40	4.34
13	4.10	4.14	......	7.13	......	21.28	8.48	5.43	4.32	4.28	3.34	4.91
14	4.10	4.10	......	6.70	......	......	7.14	6.69	4.15	4.14	3.62	4.46
15	4.12	4.07	4.21	8.25	......	12.40	6.56	6.00	4.00	4.00	4.65	4.27
16	4.01	4.10	......	8.00	......	7.80	5.96	5.44	3.99	3.89	4.44	4.20
17	3.94	4.08	......	7.27	......	6.42	5.74	5.16	4.32	3.86	4.22	4.14
18	3.94	4.05	4.07	6.56	5.37	5.99	5.65	4.91	5.28	3.84	4.10	4.49
19	4.42	4.02	4.11	5.64	5.69	5.46	5.54	4.76	4.76	3.71	3.96	4.34
20	6.22	3.49	......	5.16	15.00	5.24	6.02	4.66	4.50	4.04	3.78	4.56
21	5.28	4.04	......	4.98	9.42	6.40	8.00	4.42	4.36	3.97	3.76	5.86
22	4.87	4.04	4.32	4.86	6.54	6.63	8.42	4.24	4.50	3.78	3.72	4.70
23	4.88	4.06	4.48	4.79	5.64	6.22	5.96	5.60	4.48	3.60	3.66	4.74
24	6.08	4.02	4.78	4.71	6.38	5.68	5.72	6.19	4.32	3.42	3.62	4.68
25	7.72	3.98	6.92	4.62	7.58	6.76	5.47	11.62	4.19	3.36	3.56	4.48
26	6.55	3.97	6.66	4.96	16.30	6.16	5.26	16.98	7.48	3.44	3.46	4.32
27	5.70	4.04	5.78	8.46	9.66	5.74	9.03	8.98	6.62	3.54	3.44	4.22
28	5.61	4.62	5.48	14.62	6.58	5.39	7.14	13.86	5.50	3.61	3.88	4.16
29	5.43	5.48	5.25	15.78	......	5.13	6.24	8.29	4.85	3.64	3.64	4.10
30	5.44	5.62	4.87	7.44	......	4.87	5.78	6.26	4.44	3.72	3.48	4.04
31	5.62	......	4.60	6.48	......	4.58	......	5.90	......	5.02	5.22	......

NOTE.—Gage not read Dec. 12-14, 16, 17, 20, 21, Jan. 2, 3, 5, 6, Feb. 12-17. Gage height at 5 p. m. Mar. 14, 17.32 feet; gage not read in morning.

WEST FORK AT ENTERPRISE, W. VA.

LOCATION.—At highway bridge at Enterprise, Harrison County, three-fourths mile above mouth of Bingamon Creek.

DRAINAGE AREA.—750 square miles.

RECORDS AVAILABLE.—June 2, 1907, to September 30, 1918, when station was discontinued.

GAGE.—Chain gage attached to bridge; read by R. M. Wharton. Sea-level elevation of zero of gage, 869.91 feet.

DISCHARGE MEASUREMENTS.—Made from downstream side of bridge.

CHANNEL AND CONTROL.—Channel at measuring section broken by one pier; smooth rock bottom; straight above and below. Control practically permanent.

EXTREMES OF STAGE.—Maximum stage recorded during year, 21.75 feet at 7.50 a. m. March 14; minimum stage recorded, 0.98 foot at 7.15 a. m. July 22.

1907-1918: Maximum stage recorded, 25.35 feet January 22, 1917 (discharge not determined); minimum stage recorded, 0.6 foot September 10, 14, and 25, 1908 (discharge, 12 second-feet). Flood of 1888 reached stage represented by about 33 feet referred to datum of present gage.

ICE.—Stage-discharge relation affected by ice during severe winters.

ACCURACY.—Stage-discharge relation practically permanent; probably affected by ice during greater part of December and January. A measurement made October 2, 1917, indicates a marked change in rating curve at low stages, or that operation of mill at the dam at Worthington about 3 miles below gage affects the gage readings. The gates of the milldam were open December 5-12, 1908, in order to drain the pond, but no effect was apparent on the gage readings. This may have been due to unreliable gage readings. The low-water discharge for this station as published in previous water-supply papers may at times be in error; this condition should be considered in using the data. Gage read to half-tenths once daily.. Data inadequate for determination of daily discharge. Records uncertain.

The following discharge measurement was made by B. L. Hopkins:

May 9, 1918: Gage height, 5.90 feet; discharge, 3,010 second-feet.

Daily gage height, in feet, of West Fork at Enterprise, W. Va., for the year ending Sept. 30, 1918.

Day.	Oct.	Nov.	Dec.	Jan.	Feb.	Mar.	Apr.	May.	June.	July.	Aug.	Sept.
1	1.10	3.25	2.52	3.70	7.27	4.10	2.20	3.20	3.30	2.48	2.38	
2	1.10	2.55				3.73	2.20	1.87		2.13	2.11	2.00
3	1.03	2.28	2.23				2.30	1.80	2.30	2.12	1.77	2.55
4	1.10		2.20	2.23	5.37	3.87	2.35	1.82	2.05	1.93		2.40
5	1.12	2.03	2.15		4.85	3.90	2.30		1.90	1.90	1.42	2.13
6	1.07	1.80	1.95		5.23	4.00	2.32	1.97	1.82	1.85	1.32	1.73
7		1.75	1.85	4.10	6.80	4.25		2.30	5.16		1.27	1.60
8	1.02	1.72	1.82	5.22	7.82	4.03	4.23	10.17	6.60	1.62	1.22	
9	1.05	1.62			9.88	3.95	6.95	6.12		1.50	1.10	1.43
10	1.08	1.60	1.67				6.47	4.12	2.68	1.42	1.08	1.40
11	1.12			4.58	5.20	4.10	7.01	3.77	2.30	1.40		1.38
12	1.25	1.55		5.25	4.92	4.20	9.27		2.08	1.28	1.41	1.35
13	1.23	1.56			4.90	15.77	8.25	2.87	1.87	1.25	1.85	1.37
14		1.55	1.41		4.85	21.75		4.00	1.72		1.60	1.55
15	1.30	1.50	1.40	5.70	4.70	10.15	4.65	4.18	1.62	1.43	3.06	
16	1.22	1.53		8.60	5.10	6.46	4.98	3.45		1.42	2.35	1.50
17	1.13	1.54	1.47				4.60	2.92	1.50	1.40	1.85	2.40
18	1.07			4.85	5.85	5.20	4.45	2.56	3.08	1.40		1.92
19	2.15	1.52		4.60	6.40	4.87	4.27		2.20	1.30	1.60	2.10
20	3.05	1.50			13.65	4.65		2.15	2.18	1.32	1.42	2.87
21		1.50	1.56		5.85	4.85		2.02	1.87		1.35	2.75
22	2.46	1.48		3.32	4.87	4.15	4.45	1.98	1.82	.98	1.27	
23	3.23	1.60		3.20	3.85	3.90	4.30	2.12		1.18	1.30	2.37
24	3.55	1.75	2.06				4.10	2.82	2.37	1.13	1.18	2.07
25	3.50			3.27	4.50	3.75	3.70	6.40	2.10	1.22		2.00
26	3.10	1.95			13.60	3.68	3.45		1.87	1.52	1.22	1.75
27	3.20	2.00	4.35		8.57	3.25	3.00	3.85	1.80	1.35	1.20	1.80
28		2.45	3.95	10,37	7.23	2.90		4.86	1.92		1.27	1.70
29	2.80	2.35		24.65		2.60	1.95	7.30	1.95	1.12	1.27	
30	3.10	2.23		11.20		2.40	3.10	4.20		1.21	1.30	1.70
31	3.13							3.70		3.13	1.92	

NOTE.—Gage not read on days for which no gage-height is given.

ELK CREEK NEAR CLARKSBURG, W. VA.

LOCATION.—At a footbridge near Clarksburg, Harrison County, 300 feet above Turkey Run and 6 miles above mouth of creek.

DRAINAGE AREA.—107 square miles (determined by Pittsburgh Flood Commission).

RECORDS AVAILABLE.—October 11, 1910, to September 30, 1918, when station was discontinued.

GAGE.—Vertical gage in two sections consisting of enameled gage scale attached to cypress backing. Section below 6.73 feet attached to downstream end of right abutment of footbridge. Upper section, 6.73 to 16.9 feet attached to an oak tree 5 feet downstream from low-water section. Prior to October 1, 1917, the gage was a wooden staff at downstream end of right abutment, braced to the oak tree. Sea-level elevation of zero of gage, 955.01 feet. Gage read by E. H. Smith.

DISCHARGE MEASUREMENTS.—Made from footbridge or by wading at section about 200 feet below bridge.

CHANNEL AND CONTROL.—Rocky and practically permanent. Banks high and not subject to overflow. Point of zero flow, about gage height 0.9 foot.

EXTREMES OF STAGE.—Maximum stage recorded during year, 14.4 feet at 6 p. m. January 28 (stage-discharge relation affected by ice jam); minimum stage recorded, 1.15 feet at 10 a. m. October 2 and 3.

1911–1918: Maximum stage recorded, 15.0 feet July 25, 1912; minimum stage recorded, 0.8 foot August 21–24, 1911.

ICE.—Stage-discharge relation affected by ice during severe winters.

ACCURACY.—Stage-discharge relation practically permanent, except as affected by ice during greater part of December and January. Gage read to half-tenths daily. Data inadequate for determining daily discharge. Records good.

Discharge measurements of Elk Creek near Clarksburg, W. Va., during the year ending Sept. 30, 1918.

Date.	Made by—	Gage height.	Discharge.
		Feet.	*Sec.-ft.*
Oct. 1	Peterson and Hopkins	1.20	0.6
May 7	B. L. Hopkins	1.76	32.5
8	....do	4.27	1,040

Daily gage height, in feet, of Elk Creek near Clarksburg, W. Va., for the year ending Sept. 30, 1918.

Day.	Oct.	Nov.	Dec.	Jan.	Feb.	Mar.	Apr.	May.	June.	July.	Aug.	Sept.
1	1.25	2.00	1.90	....	2.50	2.50	1.95	1.95	2.10	2.00	1.80	2.95
2	1.15	1.90	1.90	....	2.20	2.50	1.95	1.90	2.00	2.10	1.60	2.20
3	1.15	1.80	1.85	....	2.30	2.80	2.00	1.80	1.85	1.85	.50	1.95
4	1.30	1.75	1.80	1.70	2.05	2.20	2.40	1.75	1.75	1.75	.45	1.85
5	1.30	1.70	1.75	....	2.00	3.40	2.20	1.70	1.70	1.70	.40	1.70
6	1.30	1.65	1.70	....	2.00	2.90	2.10	1.70	1.65	1.65	1.40	1.65
7	1.30	1.60	1.65	3.40	3.90	2.70	2.00	1.70	3.20	1.65	1.35	1.60
8	1.25	1.60	1.60	2.40	3.00	2.60	2.20	6.20	2.90	1.60	1.35	1.60
9	1.30	1.55	1.70	2.00	4.40	2.50	4.70	3.00	2.40	1.55	1.30	1.55
10	1.30	1.55	1.70	1.90	4.40	3.20	3.40	2.54	2.10	1.53	1.30	1.50
11	1.30	1.50	1.70	1.90	2.90	2.70	3.60	2.53	1.90	1.50	1.30	1.45
12	1.30	1.50	1.65	4.10	2.70	2.50	4.20	2.25	1.80	1.55	1.30	1.45
13	1.40	1.50	1.65	3.00	2.60	7.40	8.70	2.15	1.70	1.53	1.90	1.90
14	1.40	1.50	1.60	2.80	2.50	5.40	3.10	2.70	1.65	1.50	1.95	1.80
15	1.40	1.50	1.60	5.00	3.40	4.20	2.70	2.40	1.60	1.50	2.70	1.70
16	1.40	1.50	1.55	4.00	2.70	2.90	2.45	2.20	1.56	1.58	2.10	1.60
17	1.35	1.45	1.55	2.80	2.60	2.60	2.35	2.00	2.80	1.50	1.80	1.70
18	1.35	1.45	1.55	2.20	2.40	2.40	2.25	1.90	2.60	1.50	1.65	2.40
19	1.50	1.40	1.55	2.05	2.30	2.20	2.10	1.85	2.40	1.65	1.55	2.20
20	2.40	1.40	1.55	2.00	7.60	2.10	2.00	1.76	2.00	1.55	1.50	2.40
21	2.00	1.40	1.60	1.95	3.60	2.10	2.40	1.74	1.80	1.50	1.45	2.60
22	1.80	1.40	1.60	....	3.00	2.60	2.70	1.70	2.00	1.45	1.40	2.25
23	1.60	1.40	1.60	....	2.60	2.40	2.50	1.90	2.40	1.40	1.40	2.00
24	2.00	1.40	1.70	....	2.40	2.20	2.40	1.90	1.90	1.50	1.35	1.90
25	2.20	1.40	2.00	1.90	3.40	2.45	2.20	5.95	1.75	1.45	1.35	1.80
26	2.30	1.40	3.00	....	7.30	2.30	2.10	3.90	1.80	1.40	1.30	1.70
27	2.00	1.45	2.00	....	3.40	2.20	2.05	2.73	1.90	1.40	1.30	1.65
28	1.90	1.50	1.90	10.85	2.80	2.10	2.00	2.55	1.90	1.35	1.40	1.60
29	1.80	1.70	1.80	5.80	....	2.00	2.00	2.20	1.80	1.35	1.45	1.55
30	2.20	1.80	1.75	3.90	....	1.95	2.00	3.20	2.30	1.35	1.40	1.50
31	2.35	....	1.70	2.80	....	1.90	....	2.30	....	2.20	2.15	....

NOTE.—Gage not read Jan. 1–3, 5, 6, 22–24, 26, and 27.

BUFFALO CREEK AT BARRACKVILLE, W. VA.

LOCATION.—At steel highway bridge 1,000 feet above covered highway bridge at Barrackville, Marion County, 2¾ miles northwest of Fairmont. Finch's Run enters on left 1,600 feet below station.

DRAINAGE AREA.—115 square miles (measured on topographic maps).

RECORDS AVAILABLE.—June 3, 1907, to December 31, 1908; May 8, 1915, to September 30, 1918.

GAGE.—Chain gage fastened to downstream handrail of bridge; read by E. M. Beall.

DISCHARGE MEASUREMENTS.—Made from highway bridge or by wading.

CHANNEL AND CONTROL.—One channel at all stages; straight about 100 feet above and below station. Banks high. Stream bed rocky; some gravel. Control not permanent.

EXTREMES OF DISCHARGE.—Maximum stage recorded during year, 11.02 feet at 6.45 a. m. February 26 (discharge, about 4,850 second-feet); minimum stage recorded 0.52 foot at 6.35 a. m. and 3.50 p. m. August 21 and at 6.50 a. m. August 22 (discharge, 0.2 second-foot).

1907–1908; 1915–1918: Maximum stage recorded, 14.22 feet January 22, 1917 (discharge, about 6,800 second-feet); no flow during greater part of September, October, and November, 1908. Flood of July, 1912, reached a stage represented by about 16 feet on present gage.

ICE.—Stage-discharge relation affected by ice during severe winters.

ACCURACY.—The change in rating curve indicated by two discharge measurements made in 1918 probably was caused by the high water of February, 1918. Stage-discharge relation seriously affected by ice. Rating curve used October 1 to February 25 well defined below 1,600 second-feet; above 1,600 second-feet the curve is an extension. New rating curve used February 26 to September 30 fairly well defined between 100 and 400 second-feet; poorly defined below 100 second-feet and extended above 400 second-feet on basis of form of previous curve. Gage read to hundredths twice daily. Daily discharge ascertained by applying mean daily gage height to rating table except for periods of ice effect and periods when gage was not read. Prior to February 26, open-water records good; for periods of ice effect poor: after February 26, records fair except for days when gage was not read, for which they are poor.

Discharge measurements of Buffalo Creek at Barrackville, W. Va., during the year ending Sept. 30, 1918.

Date.	Made by—	Gage height.	Dis-charge.
		Feet.	*Sec.-ft.*
Oct. 3	Peterson and Hopkins	0.67	0.6
May 9	B. L. Hopkins	2.58	314
11	do	1.85	146

Daily discharge, in second-feet, of Buffalo Creek at Barrackville, W. Va., for the year ending Sept. 30, 1918.

Day.	Oct.	Nov.	Dec.	Jan.	Feb.	Mar.	Apr.	May.	June.	July.	Aug.	Sept.
1	0.4	84	38		90	207	26	129	102	8.4	14	97
2	.4	56	30		91	166	30	97	67	5.4	8.4	20
3	.4	42	28		105	120	62	88	50	6.2	5.4	9.8
4	.9	32	26		94	102	147	76	43	5.0	3.0	5.4
5	1.1	27	22		71	815	96	68	32	3.8	2.5	4.6
6	.7	23	19		73	318	73	60	25	4.2	2.0	3.4
7	.7	20	18		1,640	252	60	54	21	2.6	1.5	2.8
8	.9	19	14		1,040	166		2,000	33	2.3	1.3	2.5
9	.5	15	13		2,800	156		360	24	2.2	1.4	2.3
10	.5	11			1,310	405	640	207	18	2.1	1.1	2.0
11	.7	19			540	207		120	13	2.0	2.8	1.7
12	4.1	17			392	176		111	10	2.1	2.2	1.7
13	21	17			409	375		129	7.0	1.9	1.8	1.3
14	68	12			330	1,400	166	435	5.4	1.6	4.2	1.1
15	76	5.9		44	873	625	166	218	4.6	1.3	2.4	1.0
16	4.5	4.8	9		345	264	147	147	3.8	1.2	1.6	1.0
17	3.8	4.8			135	229	129	120	3.0	1.3	1.1	1.4
18	3.3	4.8			94	186	111	86	4.2	1.4	1.0	1.8
19	[illegible]	4.5			258	111	99	67	3.4	1.4	.7	1.6
20	245	4.5			2,500	99	80	60	4.2	5.0	.4	3.0
21	30	4.3			330	83	96	72	8.4	2.8	.2	5.8
22	42	4.1			159	73	96	59	5.8	2.8	.3	5.0
23	26	4.5			108	64	96	1,020	3.8	5.4	.7	3.4
24	55	4.5	106		97	57	86	229	3.8	2.4	.7	3.0
25	520	4.5	245		202	53	89	1,180	2.9	2.0	.8	3.0
26	193	4.5	130		3,490	50	78	815	2.8	1.7	.6	2.7
27	86	4.5	91		470	46	89	240	2.8	1.4	1.0	2.3
28	159	5.9	73		290	40	99	715	2.6	1.3	1.2	2.1
29	105	12	54		......	36	97	240	2.3	1.1	.9	1.9
30	108	35	37	873	......	32	129	218	13	2.6	1.0	1.7
31	143	......	26	392	......	29	......	156	......	13	3.8	......

NOTE.—Discharge estimated, because of ice effect, Dec. 10-23, 29-31; Jan. 1-29, by means of observer's notes, weather records, and comparison with records at other stations in the Monongahela basin. Discharge, Apr. 8-13 and Sept. 24-25, estimated because of lack of gage readings, by means of weather records and comparison with records at other stations in the Monongahela basin. Braced figures show mean discharge for periods indicated.

Monthly discharge of Buffalo Creek at Barrackville, W. Va., for the year ending Sept. 30, 1918.

[Drainage area, 115 square miles.]

Month.	Discharge in second-feet.				Run-off in inches.
	Maximum.	Minimum.	Mean.	Per square mile.	
October	520	0.4	63.7	0.554	0.64
November	84	4.1	16.9	.147	.16
December	245	....	35.4	.308	.36
January	873	....	82.0	.713	.82
February	3,490	71	655	5.70	5.94
March	1,400	29	224	1.95	2.25
April	....	26	206	1.79	2.00
May	2,000	54	309	2.69	3.10
June	102	2.3	17.4	.151	.17
July	13	1.1	3.16	.027	.03
August	14	.2	2.26	.020	.02
September	97	1.0	6.54	.057	.06
The year	3,490	.2	132	1.15	15.55

CHEAT RIVER NEAR PARSONS, W. VA.

LOCATION.—At Moss highway bridge, 2 miles north of Parsons, Tucker County, 2 miles below junction with Shavers Fork, and 5 miles below junction of Dry Fork and Blackwater River.

DRAINAGE AREA.—716 square miles (determined by Hydro-Electric Co. of West Virginia).

RECORDS AVAILABLE.—January 1, 1913, to September 30, 1918.

GAGE.—Chain gage near center of bridge on downstream guard rail; read by Mrs. E. C. Linger.

DISCHARGE MEASUREMENTS.—Made from downstream side of highway bridge.

CHANNEL AND CONTROL.—Rocky and probably permanent. Water is swift and turbulent at high stages.

EXTREMES OF DISCHARGE.—Maximum stage recorded during year, 15.6 feet at 4 p. m. March 13 (discharge, about 33,000 second-feet); minimum stage recorded, 1.72 feet at 6 p. m. October 8 (discharge, 51 second-feet).

1913–1918: Maximum stage recorded, 17.98 feet March 12, 1917 (discharge, about 40,000 second-feet); minimum stage recorded, 1.52 feet September 6, 1917 (discharge, 29 second-feet).

ICE.—Stage-discharge relation affected by ice during severe winters.

REGULATION.—Some regulation above at various pulp mills and sawmills. Effect probably compensating, so that two gage readings a day give correct basis for determining daily discharge.

ACCURACY.—Stage-discharge relation probably permanent, except as affected by ice. Rating curve fairly well defined between 65 and 1,000 second-feet and well defined between 1,000 and 5,500 second-feet; beyond these limits curve is an extension and may be considerably in error. Gage read to quarter-tenths twice daily. Daily discharge ascertained by applying mean daily gage height to rating table, except for period of ice effect. Open-water records fair, except for extremely high stages; winter records roughly approximate.

The following discharge measurement was made by B. L. Hopkins:

May 2, 1918: Gage height, 3.55 feet; discharge, 1,280 second-feet.

Daily discharge, in second-feet, of Cheat River near Parsons, W. Va., for the year ending Sept. 30, 1918.

Day.	Oct.	Nov.	Dec.	Jan.	Feb.	Mar.	Apr.	May.	June.	July.	Aug.	Sept.
1	670	1,370	950		1,500	3,850	590	1,250	1,150	2,830	378	1,370
2	141	1,100	1,100		1,200	4,020	760	1,200	1,000	2,050	290	1,200
3	124	900	900		1,200	3,000	1,150	1,050	760	1,370	244	1,050
4	124	715	850		900	2,050	1,560	850	520	1,050	195	950
5	96	590	715		670	3,000	1,250	805	420	760	138	420
6	75	520	630		715	3,340	1,000	715	1,500	715	127	800
7	61	450	520		2,510	7,200	850	555	2,350	590	134	690
8	54	390	450		3,680	5,290	1,100	3,850	1,770	485	450	410
9	85	420			8,620	3,340	5,880	3,340	950	420	485	360
10	68	378			9,100	4,370	4,370	2,200	590	354	290	340
11	61	320			5,680	3,680	3,000	1,770	485	310	590	330
12	120	300			5,880	2,830	2,350	1,440	390	280	760	330
13	117	271		270	9,100	23,700	2,050	1,200	336	290	670	310
14	102	240			6,300	17,800	2,670	3,680	310	336	850	700
15	148	175			7,660	13,200	5,480	2,670	286	285	1,370	490
16	117	179	245		6,300	6,300	6,970	1,770	235	244	900	276
17	93	183			4,020	3,850	7,660	1,500	540	244	520	850
18	88	187			2,510	2,670	6,740	1,150	2,830	253	420	3,850
19	105	191			2,670	2,050	3,850	1,000	2,050	590	1,630	3,000
20	2,050	171			19,000	1,700	2,510	1,500	1,200	520	1,200	1,560
21	1,370	211			7,430	1,770	5,100	2,050	805	450	760	2,510
22	420	191			3,680	2,670	4,730	3,000	1,440	310	520	1,770
23	310	171			2,670	2,200	3,340	2,200	1,770	253	420	1,200
24	310	152			2,350	1,630	2,670	2,200	1,150	215	366	850
25	342	134	2,510		3,340	1,440	3,000	5,100	760	300	342	590
26	235	117	1,910	3,000	19,300	1,150	2,830	7,660	2,670	336	290	450
27	2,200	102	1,630	8,620	6,740	950	2,350	4,190	2,670	360	240	450
28	2,670	88	1,100	5,880	4,020	850	1,910	3,850	1,630	235	219	450
29	1,700	590	555	4,730	......	760	1,500	2,050	11,200	187	148	366
30	805	670	360	3,000	......	670	1,310	2,200	4,020	191	138	325
31	2,830	......	300	2,050	......	590	......	1,500	......	187	330	......

NOTE.—Discharge estimated because of ice effect, Dec. 9–25 and Dec. 30 to Jan. 26 by means of observer's notes, weather records, and comparison with records for other stations in this river basin. Discharge, Nov. 16–18 and July 20, interpolated, and Sept 6–15, estimated by comparison with records of flow for Shavers Fork at Parsons; observer's gage readings in error.

Monthly discharge of Cheat River near Parsons, W. Va., for the year ending Sept. 30, 1918.

[Drainage area, 716 square miles.]

Month.	Discharge in second-feet.				Run-off in inches.
	Maximum.	Minimum.	Mean.	Per square mile.	
October	2,830	54	571	0.797	0.92
November	1,370	88	383	.535	.60
December	2,510		594	.830	.96
January	8,620		1,100	1.54	1.78
February	19,300	670	5,310	7.42	7.73
March	23,700	590	4,260	5.95	6.86
April	7,660	590	3,020	4.22	4.71
May	7,660	555	2,240	3.13	3.61
June	11,200	235	1,590	2.22	2.48
July	2,830	187	548	.765	.88
August	1,630	127	497	.694	.80
September	3,850		942	1.32	1.47
The year	23,700	54	1,730	2.42	32.80

CHEAT RIVER AT ROWLESBURG, W. VA.

Location.—At Baltimore & Ohio Railroad bridge at Rowlesburg, Preston County, 300 feet above mouth of Salt Lick Creek.

Drainage area.—960 square miles (includes drainage area of Salt Lick Creek).

Records available.—July 19, 1912, to September 30, 1918. The United States Weather Bureau has collected gage-height records since 1884.

Gage.—Mott tape gage attached to upstream side of bridge; read by J. F. Pierce.

Discharge measurements.—Made from upstream side of bridge. Salt Lick Creek is measured separately and the discharge added to that measured at bridge.

Channel and control.—Channel is curved above and below bridge. Control consists of small boulders; probably permanent. Salt Lick Creek enters between the control and the gage. Stage at which flow would be zero was about 0.45 foot in September, 1917.

Extremes of stage.—Maximum stage recorded during year, 13.6 feet during night of March 13; minimum stage recorded, 2.0 feet October 4–10.

1912–1918: Maximum stage recorded, 14.7 feet at 5 p. m. March 12, 1917; minimum stage recorded, 1.4 feet October 6–8, 1914.

The highest stage of which there is any record occurred, according to the records of the United States Weather Bureau, on July 10, 1888, when the water reached a stage of 22 feet.

Ice.—Stage-discharge relation affected by ice during severe winters.

Accuracy.—Stage-discharge relation probably permanent, except as affected by ice. Rating curve not developed. Gage read to tenths daily. Records fair.

Cooperation.—Gage-height record furnished by the United States Weather Bureau.

The following measurement was made by B. L. Hopkins:

April 26, 1918: Gage height, 4.45 feet; discharge, 3,410 second-feet.

Daily gage height, in feet, of Cheat River at Rowlesburg, W. Va., for the year ending Sept. 30, 1918.

Day.	Oct.	Nov.	Dec.	Jan.	Feb.	Mar.	Apr.	May.	June.	July.	Aug.	Sept.
1	2.2	3.9	3.3	2.9	3.8	4.7	2.9	3.4	3.4	4.4	2.3	3.2
2	2.1	3.6	3.7	2.9	3.4	5.0	2.9	3.4	3.2	4.2	2.5	3.4
3	2.2	3.3	3.5	2.9	3.3	4.6	2.9	3.4	3.0	3.6	2.4	2.9
4	2.0	3.1	3.2	2.9	3.3	4.1	3.4	3.2	2.9	3.3	2.3	2.7
5	2.0	3.0	3.2	2.9	3.3	4.1	3.5	3.1	2.8	3.0	2.3	2.6
6	2.0	2.9	3.2	2.9	3.3	4.7	3.3	3.0	2.7	3.1	2.3	2.5
7	2.0	2.8	3.2	2.9	3.8	4.6	3.0	3.0	3.6	3.0	2.1	2.4
8	2.0	2.8	3.0	2.9	5.3	5.5	3.1	3.9	4.4	2.8	2.1	2.7
9	2.0	2.8	3.0	2.9	6.2	4.6	5.6	5.3	3.7	2.7	2.9	2.5
10	2.0	2.8	3.0	2.9	8.3	5.1	5.5	4.3	3.2	2.6	2.6	2.4
11	2.1	2.6	3.0	2.9	6.4	5.0	4.9	4.0	2.9	2.6	2.6	2.4
12	2.1	2.5	3.0	2.9	6.5	4.4	4.5	3.7	2.8	2.5	3.2	2.3
13	2.1	2.6	3.0	2.9	7.3	9.0	4.2	3.5	2.7	2.5	3.0	2.4
14	2.1	2.6	3.0	2.9	6.5	12.0	4.0	5.2	2.6	2.5	2.9	2.3
15	2.2	2.5	3.0	2.9	6.6	8.6	6.6	4.8	2.6	2.5	3.0	2.6
16	2.1	2.5	3.0	4.3	6.6	6.5	6.9	4.1	2.5	2.5	3.4	2.5
17	2.1	2.4	3.0	4.2	5.2	5.2	6.4	3.8	3.4	2.4	2.9	2.5
18	2.1	2.4	3.0	4.2	4.4	4.5	6.2	3.5	5.6	2.4	2.7	3.3
19	2.3	2.4	3.0	4.2	4.0	4.1	5.2	3.3	3.9	2.5	2.5	5.1
20	2.5	2.4	2.8	4.2	9.7	3.8	4.4	3.3	3.8	2.8	3.6	4.0
21	3.7	2.3	2.8	4.2	7.5	3.6	4.6	3.8	3.2	2.9	3.5	3.8
22	3.0	2.4	2.8	4.2	5.3	4.2	5.9	3.7	3.1	2.7	2.8	4.2
23	2.7	2.4	2.8	4.2	4.4	4.1	5.1	4.2	3.9	2.5	2.4	3.6
24	3.3	2.4	2.8	4.2	4.1	3.8	4.5	4.0	3.5	2.4	2.5	3.3
25	2.8	2.4	3.9	4.2	4.2	3.4	4.2	3.8	3.2	2.3	2.4	3.1
26	3.0	2.4	5.0	3.8	9.6	3.6	4.4	7.8	3.0	2.3	2.4	2.6
27	3.5	2.4	3.8	5.1	7.1	3.5	4.2	5.3	4.4	2.4	2.4	2.8
28	4.5	2.4	3.6	6.6	5.4	3.3	3.9	4.4	3.7	2.5	2.3	2.7
29	4.3	2.4	3.2	6.0	....	3.1	3.8	4.2	3.4	2.3	2.3	2.7
30	3.7	3.1	2.8	4.9	....	3.1	3.6	3.7	5.4	2.3	2.2	2.6
31	4.7	....	2.9	4.3	....	3.0	....	3.7	....	2.4	2.3	....

NOTE.—Stage-discharge relation affected by ice Dec. 10-24 and Dec. 30 to Jan. 26.

CHEAT RIVER NEAR MORGANTOWN, W. VA.

LOCATION.—At highway bridge at Uneva, Monongalia County, 10 miles above mouth of river. Parallel of 39° 40′ crosses the river at this bridge.

DRAINAGE AREA.—1,380 square miles.

RECORDS AVAILABLE.—July 8 to December 30, 1899; July 1 to December 29, 1900; August 21, 1902, to December 31, 1905; November 18, 1908, to December 31, 1917. Bridge and gage were torn out by an ice jam February 9, 1918.

GAGE.—Chain gage attached to bridge; read by C. F. Baker.

DISCHARGE MEASUREMENTS.—Made from upstream side of bridge or by wading.

CHANNEL AND CONTROL.—Probably permanent.

EXTREMES OF DISCHARGE.—Maximum stage recorded during period, October 1 to December 31, 1917, 4.70 feet at 8 a. m. and 5 p. m. October 28 (discharge, 5,550 second-feet): minimum stage recorded, 1.90 feet at 8 a. m. October 5 (discharge, 135 second-feet).

ICE.—Stage-discharge relation seriously affected by ice during severe winters. Ice forms sometimes to a thickness of several inches, and large ice jams may occur when this ice breaks up.

ACCURACY.—Stage-discharge relation practically permanent except as affected by ice. Rating curve fairly well defined above 175 second-feet. Gage read to hundredths twice daily. Daily discharge ascertained by applying mean daily gage height to rating table except for period of ice effect. Records good except those for periods of ice effect, which are poor.

Discharge measurements of Cheat River near Morgantown, W. Va., during the year ending Sept. 30, 1918.

[Made by Peterson and Hopkins.]

Date.	Gage height.	Discharge.
	Feet.	*Sec.-ft.*
Oct. 4	2.06	192
5	2.06	196

Daily discharge, in second-feet, of Cheat River near Morgantown, W. Va., for the period Oct. 1 to Dec. 31, 1917.

Day	Oct.	Nov.	Dec.	Day.	Oct.	Nov.	Dec.	Day.	Oct.	Nov.	Dec.
1	195	2,860	1,630	11	163	600		21	2,450	397	
2	215	2,080	2,320	12	195	600		22	1,080	384	400
3	261	1,530	2,080	13	249	560		23	835	397	
4	210	1,290	1,450	14	237	520		24	640	397	
5	135	1,080	1.360	15	243	520	400	25	1,220	378	
6	180	950	1,360	16	273	480		26	1,730	371	
7	151	835	1,220	17	237	452		27	2,860	297	1,300
8	163	730	1,010	18	215	431		28	5,550	315	
9	167	685	640	19	200	410		29	4,510	640	
10	167	685	400	20	1,220	397		30	2,860	1,290	
								31	4,900	......	

NOTE.—Discharge, Dec. 10-31, estimated because of ice, on basis of observer's notes and weather records. Gage read Jan. 1-28, but data inadequate for determining discharge.

Monthly discharge of Cheat River near Morgantown, W. Va., for the period Oct. 1 to Dec. 31, 1917.

[Drainage area, 1,380 square miles.]

Month.	Discharge in second-feet.				Run-off in inches.
	Maximum.	Minimum.	Mean.	Per square mile.	
October	5,550	135	1,090	0.790	0.91
November	2,860	297	752	.545	.61
December	....	....	938	.680	.78

BLACKWATER RIVER AT HENDRICKS, W. VA.

LOCATION.—At highway bridge at Hendricks, Tucker County, an eighth of a mile above mouth of river.

DRAINAGE AREA.—148 square miles (determined by West Virginia Development Co.)

RECORDS AVAILABLE.—October 13, 1911, to September 30, 1918, when station was discontinued.

GAGE.—Chain gage attached to upstream side of bridge; read by William Cochran and J. W. Garrett.

DISCHARGE MEASUREMENTS.—Made from bridge or by wading.

CHANNEL AND CONTROL.—Bed composed of coarse gravel and stones; very rough. Control shifting. Right bank high. Left bank subject to overflow at high stages.

EXTREMES OF STAGE.—Maximum stage recorded during year, 6.42 feet at 8 a. m. February 26; minimum stage recorded, 1.56 feet at 7 a. m. and 5 p. m. September 30.

1911–1918: Maximum stage recorded, 8.37 feet March 12, 1917; minimum stage recorded, 1.49 feet October 15, 1916. Maximum flood occurred July 10, 1888; stage unknown.

ICE.—Stage-discharge relation affected by ice during severe winters.

ACCURACY.—Stage-discharge relation not permanent; affected by ice during greater part of December and January. Gage read to hundredths twice daily. Data inadequate for determination of discharge.

Discharge measurements of Blackwater River at Hendricks, W. Va., during the year ending Sept. 30, 1918.

[Made by B. L. Hopkins.]

Date.	Gage height.	Discharge.
	Feet.	*Sec.-ft.*
Apr. 30	2.35	202
May 12	2.28	172
13	2.26	163

Daily gage height, in feet, of Blackwater River at Hendricks, W. Va., for the year ending Sept. 30, 1918.

Day.	Oct.	Nov.	Dec.	Jan.	Feb.	Mar.	Apr.	May.	June.	July.	Aug.	Sept.
1	1.71	2.74	2.65	2.46	2.26	3.42	2.12	2.34	2.18	2.22	1.94	2.18
2	1.68	2.38	2.50	2.46	2.21	3.34	2.12	2.34	2.10	2.08	1.87	1.98
3	1.66	2.28	2.50	2.46	2.18	3.24	2.32	2.30	2.04	2.00	1.80	1.86
4	1.66	2.20	2.46	2.46	3.18	3.26	2.62	2.24	2.00	1.90	1.78	1.84
5	1.77	2.19	2.26	2.46	2.16	3.54	2.40	2.14	2.00	1.89	1.76	1.84
6	1.80	2.18	2.21	2.62	2.25	3.55	2.20	2.14	2.04	2.21	1.72	1.84
7	1.78	2.18	2.17	2.70	4.14	3.42	2.18	2.16	3.41	2.05	1.83	1.82
8	1.79	2.13	2.16	2.67	3.76	3.41	2.36	3.11	2.72	1.95	2.04	1.80
9	1.77	2.12	2.26	2.64	3.17	2.88	3.59	2.72	2.17	1.94	2.02	1.77
10	1.75	2.12	2.26	2.64	3.66	3.60	2.83	2.40	2.12	1.90	1.86	1.75
11	1.74	2.10	2.26	2.76	3.04	2.96	2.76	2.36	2.10	1.85	2.78	1.73
12	1.80	2.08	2.26	3.32	3.22	3.16	2.62	2.28	2.02	1.84	2.42	1.76
13	1.78	2.10	2.26	3.34	3.24	5.56	2.70	2.36	2.02	1.88	2.19	1.78
14	1.74	2.12	2.14	3.14	3.18	5.37	3.00	3.12	1.99	1.94	2.31	1.76
15	1.74	2.09	2.14	3.76	3.26	4.80	3.76	2.58	1.95	1.92	2.20	1.75
16	1.72	2.07	2.14	3.81	3.10	3.96	4.16	2.38	1.94	1.85	2.02	1.72
17	1.70	2.07	2.11	3.81	3.20	3.01	4.25	2.31	1.99	1.88	1.96	2.50
18	1.68	2.04	2.11	3.47	3.44	2.82	4.02	2.22	2.58	1.88	1.92	3.85
19	2.05	2.03	2.11	3.47	3.97	2.62	3.22	2.20	2.14	1.92	1.88	2.95
20	2.28	2.00	2.11	3.47	6.04	2.50	2.92	2.34	2.05	1.90	1.90	2.52
21	2.00	2.00	2.11	3.02	4.04	2.78	3.75	2.55	2.00	1.85	1.87	2.66
22	1.98	2.02	2.11	3.02	3.84	2.90	3.55	2.32	2.27	1.80	1.81	2.30
23	2.02	2.06	2.28	3.02	3.70	2.62	2.98	2.34	2.36	1.80	1.74	2.10
24	2.20	2.14	2.44	3.02	3.75	2.49	2.66	2.26	2.12	1.88	1.95	2.01
25	2.12	2.10	2.60	3.01	4.08	2.37	2.79	2.97	2.04	2.06	1.79	1.98
26	2.02	2.08	2.70	3.35	5.66	2.32	2.64	3.02	2.16	1.90	1.77	1.97
27	2.26	2.08	2.62	4.45	3.62	2.22	2.52	2.56	2.12	1.84	1.80	1.89
28	2.19	2.16	2.65	4.12	3.31	2.20	2.32	2.48	2.04	1.78	1.77	1.80
29	2.64	2.60	2.60	3.65	....	2.18	2.34	2.33	2.04	1.76	1.62	1.91
30	3.73	2.70	2.47	2.89	....	2.18	2.31	2.24	2.17	1.78	1.77	1.56
31	2.99	....	2.46	2.44	....	2.12	....	2.18	....	1.78	1.88	....

SHAVERS FORK AT PARSONS, W. VA.

LOCATION.—At steel highway bridge 600 feet northwest of railroad station at Parsons, Tucker County, and half a mile above confluence with Dry Fork.

DRAINAGE AREA.—210 square miles (determined by Pittsburgh Flood Commission).

RECORDS AVAILABLE.—October 14, 1910, to September 30, 1918.

Gage.—Standard chain gage attached to bridge, read by R. W. Evans. Sea-level elevation of zero of gage, 1,631.70 feet.

Discharge measurements.—Made from downstream side of bridge or by wading.

Channel and control.—Channel rocky. Control, coarse gravel and rocks; probably permanent.

Extremes of discharge.—Maximum stage recorded during year, 8.40 feet at 4.30 p. m. February 26 (discharge, 8,700 second-feet); minimum stage recorded, 2.60 feet at 9 a. m. October 14 (discharge, 25 second-feet).

1910–1918: Maximum stage recorded, 9.90 feet January 30, 1912, and March 12, 1917 (discharge, 12,300 second-feet); minimum discharge recorded, 1 second-foot October 1, 1914 (gage height, 2.0 feet). High waters of 1888 and 1907 reached a stage represented by approximately 12.5 feet, referred to present gage datum.

Ice.—Stage-discharge relation affected by ice during severe winters.

Regulation.—Flow at low stages may be affected by storage of water at pulp mill dam about three-fourths mile above station.

Accuracy.—Stage-discharge relation practically permanent except as affected by ice December 28 to January 27. Rating curve well defined between 40 and 10,000 second-feet; extended beyond these limits. Gage read to hundredths twice daily. Daily discharge ascertained by applying mean daily gage height to rating table except for period of ice effect for which it was ascertained by means of observer's notes, weather records, and comparison with records at other stations. Open-water records fair; records for period of ice effect poor.

The following discharge measurements were made by B. L. Hopkins:

April 30, 1918: Gage height, 3.84 feet; discharge, 461 second-feet.

May 1, 1918: Gage height, 3.79 feet; discharge, 445 second-feet.

Daily discharge, in second-feet, of Shavers Fork at Parsons, W. Va., for the year ending Sept. 30, 1918.

Day.	Oct.	Nov.	Dec.	Jan.	Feb.	Mar.	Apr.	May.	June.	July.	Aug.	Sept.
1	72	905	400	140	800	295	274	480	504	2,070	146	1,260
2	74	643	500	130	852	260	225	365	365	1,140	135	380
3	34	652	400	120	776	173	480	365	309	634	135	204
4	34	365	365	110	700	1,260	700	365	215	425	135	225
5	38	135	425	100	605	1,140	440	295	225	351	61	281
6	56	146	425	110	662	1,140	380	281	165	260	44	464
7	56	146	253	120	800	2,830	295	295	173	365	61	225
8	34	120	225	130	1,320	1,790	365	1,790	562	183	76	135
9	39	135	210	120	3,170	700	3,350	1,590	480	150	80	158
10	38	135	170	120	1,390	700	1,520	905	199	154	80	173
11	60	135	160	130	1,520	800	1,210	681	154	123	46	143
12	56	87	140	130	1,600	800	905	520	135	110	46	135
13	46	135	110	140	2,510	4,870	652	624	111	80	295	99
14	25	92	100	140	2,830	5,270	520	1,790	87	80	260	365
15	55	104	100	140	3,530	3,530	1,930	1,260	61	135	588	295
16	46	80	82	140	4,670	1,660	2,220	800	80	72	605	129
17	58	82	110	140	3,530	1,260	1,930	700	225	80	135	225
18	38	104	110	130	1,020	852	1,590	837	905	102	173	1,020
19	65	104	104	130	905	1,020	1,020	605	960	80	960	700
20	1,790	70	110	130	6,330	1,660	800	750	520	85	456	960
21	974	72	104	120	3,900	1,140	905	700	154	80	267	1,390
22	158	63	82	120	1,930	1,520	1,930	496	852	154	173	1,390
23	135	61	135	120	1,260	750	1,140	700	700	120	135	440
24	116	70	139	110	1,020	700	960	900	562	116	99	169
25	135	70	400	110	1,080	800	700	2,510	480	128	80	410
26	120	61	1,390	200	8,220	700	1,020	4,090	2,070	104	80	267
27	135	55	1,020	300	2,220	546	905	1,590	1,390	70	80	246
28	135	46	700	2,830	520	425	800	1,520	624	61	76	267
29	571	200	300	2,670	......	295	681	800	2,830	80	80	253
30	750	300	170	1,660	......	199	546	605	1,020	80	80	154
31	960	......	150	852	......	199	......	554	......	104	80	......

NOTE.—Discharge, Oct. 21, Feb. 3, Apr 11, June 13 and 14, interpolated, Dec. 9–15 estimated, because gage was not read; Nov. 29 to Dec 3 and Dec. 25, estimated by comparison with records for Cheat River near Parsons (observer's readings were in error). Discharge, Dec. 28 to Jan. 27, estimated because of ice, from observer's notes, study of weather records and comparison with records for Cheat River near Parsons.

Monthly discharge of Shavers Fork at Parsons, W. Va., for the year ending Sept. 30, 1918.

[Drainage area, 210 square miles.]

Month.	Discharge in second-feet.				Run-off in inches.
	Maximum.	Minimum.	Mean.	Per square mile.	
October	1,790	25	223	1.06	1.22
November	905	46	179	.852	.95
December	1,390	82	293	1.40	1.61
January	2,830	100	376	1.79	2.06
February	8,220	520	2,130	10.1	10.52
March	5,270	173	1,270	6.05	6.98
April	3,350	225	1,010	4.81	5.37
May	4,090	281	947	4.51	5.20
June	2,830	61	571	2.72	3.04
July	2,070	61	251	1.20	1.38
August	960	44	185	.881	1.02
September	1,390	99	419	2.00	2.23
The year	8,220	25	644	3.07	41.58

BIG SANDY CREEK AT ROCKVILLE, W. VA.

Location.—At highway bridge at Rockville, Preston County, 5 miles above mouth of creek and 6 miles below Bruceton Mills.

Drainage area.—202 square miles (determined by West Virginia Development Co.).

Records available.—May 7, 1909, to March 31, 1918, when station was discontinued.

Gage.—Chain gage attached to downstream side of bridge; read by Levi Zweyer.

Discharge measurements.—Made from bridge or by wading.

Channel and control.—Channel bed composed of boulders and bedrock. Control practically permanent.

Extremes of discharge.—Maximum stage recorded during period, 12.8 feet at 7 a. m. February 26 (discharge, about 11,900 second-feet); minimum stage recorded, 3.30 feet at 6 p. m. October 3 and 4 (discharge, 12 second-feet).

Ice.—Stage-discharge relation affected by ice during severe winters.

Regulation.—Gristmills at Rockville, Clifton Mills, and Bruceton Mills operated by water power, may produce fluctuations in stage during low water.

Accuracy.—Stage-discharge relation practically permanent except as affected by ice. Rating curve well defined between 10 and 8,000 second-feet. Gage read to hundredths twice daily. Daily discharge ascertained by applying mean daily gage height to rating table except for period of ice effect. Records fair except those for periods of ice effect, which are poor.

The following discharge measurement was made by Peterson and Hopkins:

October 6, 1917: Gage height, 3.45 feet; discharge, 13.7 second-feet.

Daily discharge, in second-feet, of Big Sandy Creek at Rockville, W. Va., for the period Oct. 1, 1917, to Mar. 31, 1918.

Day.	Oct.	Nov.	Dec.	Jan.	Feb.	Mar.	Day.	Oct.	Nov.	Dec.	Jan.	Feb.	Mar.
1	34	346	299			798	16	39	81	90		2,500	796
2	19	284	330			633	17	29	88	88		1,200	503
3	13	284	269			523	18	27	88	85		739	429
4	12	191	284			465	19	61	81	88		1,690	346
5	16	169	191		140	862	20	503	64	103		7,970	299
6	18	150	150			739	21	241	64	142		2,360	269
7	16	150	150			684	22	169	81	191		931	241
8	27	133	133	130		543	23	126	85	299	130	798	229
9	17	110	120		4,400	633	24	133	103	346		739	215
10	16	118	120		2,640	1,200	25	465	88			862	191
11	19	110	110		2,360	862	26	586	48			8,810	169
12	32	103	110		3,080	633	27	1,010	81	226		1,950	150
13	68	103	100		7,340	1,100	28	931	103			1,010	142
14	52	96	90		2,220	4,060	29	633	118			....	133
15	61	81	90		6,920	1,690	30	739	169			....	133
							31	465	....			....	126

NOTE.—Discharge, Dec. 9-16 and Dec. 25 to Feb. 8, estimated because of ice effect, by means of observer's notes and weather records.

Monthly discharge of Big Sandy Creek at Rockville, W. Va., for the period Oct. 1, 1917, to Mar. 31, 1918.

[Drainage area, 202 square miles.]

Month.	Discharge in second-feet.				Run-off in inches.
	Maximum.	Minimum.	Mean.	Per square mile.	
October	1,010	12	212	1.05	1.21
November	346	48	126	.624	.70
December	....	85	179	.886	1.02
January	....	....	130	.644	.74
February	8,810	....	2,200	10.9	11.35
March	4,060	126	641	3.17	3.66

LITTLE BEAVER CREEK BASIN.

LITTLE BEAVER CREEK NEAR EAST LIVERPOOL, OHIO.

LOCATION.—At steel highway bridge known as Grimms Bridge, 4 miles above mouth of creek and 4 miles northeast of East Liverpool, Columbiana County. North Fork enters creek on left about 3 miles above station.

DRAINAGE AREA.—505 square miles (measured on topographic maps).

RECORDS AVAILABLE.—May 17, 1915, to September 30, 1918.

GAGE.—Chain gage fastened to downstream side of highway bridge; read by G. W. Garn and Bessie Garn.

DISCHARGE MEASUREMENTS.—Made from bridge or by wading.

CHANNEL AND CONTROL.—One channel at all stages; at extreme high stages water flows around both bridge abutments. Channel straight for 100 feet above and 300 feet below station. Rapids about 600 feet below bridge act as primary control; probably permanent. Point of zero flow, gage height, 0.10±0.2 foot.

EXTREMES OF STAGE.—Maximum stage recorded, 11.2 feet at 8 a. m. February 20: minimum stage recorded, 1.78 feet at 6 p. m. August 22 and 7 a. m. August 26.

1915-1918: Maximum and minimum stages recorded same as for 1918 above. Highest known flood reached a stage represented by gage height about 20 feet.

ICE.—Stage-discharge relation affected by ice and ice jams during severe winters.

ACCURACY.—Stage-discharge relation permanent except as affected by ice. Rating curve not fully developed. Gage read to hundredths twice daily. Records excellent.

The following discharge measurement was made by Peterson and Hopkins:

October 9, 1917: Gage height, 2.50 feet; discharge, 70.8 second-feet.

Daily gage height, in feet, of Little Beaver Creek near East Liverpool, Ohio, for the year ending Sept. 30, 1918.

Day.	Oct.	Nov.	Dec.	Jan.	Feb.	Mar.	Apr.	May.	June.	July.	Aug.	Sept.
1	2.20	4.86	3.13		2.96	6.20	3.13	4.15	4.98	2.72	2.44	2.72
2	2.20	4.07	3.10			6.44	3.19	3.87	4.25	2.42	2.26	2.46
3	2.32	4.04	3.04			5.22	3.27	3.68	3.77	2.46	2.17	2.40
4	2.87	3.82	2.92	3.92	3.00	4.64	3.59	3.58	3.50	2.44	2.23	2.49
5	3.07	3.67	2.88			4.92	3.34	3.45	3.32	2.41	2.23	2.46
6	2.78	3.58	2.85			5.00	3.18	3.35	3.48	2.32	2.10	2.89
7	2.62	3.48	2.84	3.34		4.94	3.10	3.40	3.41	2.28	1.92	2.56
8	2.53	3.42	2.51		3.25	4.60	4.40	3.32	3.25	2.20	2.32	2.37
9	2.50	3.30	2.57		7.02	4.92	4.36	3.23	3.10	2.20	2.85	2.25
10	2.48	3.25	2.80		7.15	6.35	4.02	3.38	3.08	2.18	2.70	2.16
11	2.40	3.21	2.86	3.31	6.78	5.12	4.32	3.52	3.08	2.15	2.58	2.10
12	2.38	3.14	2.85		7.80	4.95	5.72	4.08	3.12	3.22	2.38	2.14
13	2.62	3.15	2.86		8.74	4.72	5.68	6.26	3.11	3.10	2.30	2.46
14	2.60	3.15	2.86	3.15	6.69	5.68	5.65	5.99	2.94	2.78	2.12	2.32
15	2.52	3.15	2.86		7.92	5.76	4.86	4.88	2.83	2.58	2.05	2.80
16	2.47	3.15	2.96		6.03	4.72	4.42	4.44	2.79	2.40	2.09	2.40
17	2.44	3.15	2.95		5.00	4.48	4.62	3.95	2.70	2.31	2.16	2.59
18	2.42	3.14	2.90	3.05	5.18	4.28	5.28	3.78	2.65	2.24	2.02	2.68
19	4.74	2.98	2.98		5.62	4.08	4.34	3.57	2.58	2.22	1.90	2.50
20	5.68	2.92	3.23		10.40	3.92	4.00	3.44	2.52	2.26	1.88	2.52
21	4.08	2.92	3.70	3.00	6.65	3.84	4.12	3.62	2.56	2.20	1.95	2.60
22	3.58	2.91	4.42		5.29	3.77	4.32	3.56	2.76	2.16	1.79	2.43
23	3.50	2.98	4.16		4.70	3.62	4.12	4.00	2.92	2.15	1.83	2.36
24	6.12	3.00			4.62	3.52	4.38	3.86	2.72	2.22	1.86	2.28
25	7.95	2.90		2.95	4.70	3.42	4.16	4.93	2.62	2.40	1.86	2.24
26	5.92	2.86			5.72	3.36	3.98	5.72	2.54	2.46	1.80	2.16
27	5.00	2.90			4.75	3.26	3.80	4.31	2.48	2.36	1.82	2.12
28	4.86	2.90	4.68	2.95	4.40	3.26	4.75	3.85	2.44	2.32	1.83	2.08
29	4.54	2.88				3.24	4.91	3.57	2.45	2.15	2.00	2.04
30	6.18	2.95				3.19	4.60	4.04	2.38	2.65	2.54	2.03
31	5.76		3.94			3.17		5.90		2.50	2.48	

NOTE.—Stage-discharge relation affected by ice Dec. 10 to Feb. 9; gage read once or twice a week Dec. 24 to Feb. 7.

YELLOW CREEK BASIN.

YELLOW CREEK AT HAMMONDSVILLE, OHIO.

LOCATION.—At covered highway bridge on Steubenville Pike, a fifth of a mile southwest of Hammondsville, Jefferson County. North Fork enters on left 1,000 feet below station.

DRAINAGE AREA.—169 square miles (measured on topographic maps).

RECORDS AVAILABLE.—May 13, 1915, to September 30, 1918.

GAGE.—Chain gage on downstream side of bridge about 25 feet from left end; read by W. J. Sprague.

DISCHARGE MEASUREMENTS.—Made from bridge or by wading.

CHANNEL AND CONTROL.—One channel, but at extreme high stages stream flows around both abutments; straight 1,000 feet above and curved 100 feet below station. Control not permanent. Point of zero flow, gage height about 1.4 feet September, 1915 and 1916, and October, 1917.

EXTREMES OF STAGE.—Maximum stage recorded during year, 9.61 feet at 8.45 a. m. February 20; minimum stage recorded, 1.28 feet at 7.10 p. m. August 28.

1915–1918: Maximum stage recorded, 10.75 feet December 29, 1915; minimum stage same as for 1918 above. Highest known flood reached a stage represented by gage height about 16 feet.

ICE.—Stage-discharge relation affected by ice during severe winters.

ACCURACY.—Stage-discharge relation practically permanent except as affected by ice during December, January, and February. Rating curve not fully developed. Gage read to hundredths twice daily. Records good.

The following discharge measurement was made by Peterson and Hopkins:

October 8, 1917: Gage height, 2.17 feet; discharge, 16.1 second-feet.

Daily gage height, in feet, of Yellow Creek at Hammondsville, Ohio, for the year ending Sept. 30, 1918.

Day.	Oct.	Nov.	Dec.	Jan.	Feb.	Mar.	Apr.	May.	June.	July.	Aug.	Sept.
1	2.10	3.79	2.50	4.47	2.65	3.72	2.44	3.16	3.48	1.92	1.52	2.[illegible]
2	2.09	3.54	2.52	4.70	......	3.52	2.46	3.02	3.00	1.86	1.50	2.35
3	2.09	3.31	2.46	4.74	......	3.36	2.47	2.94	2.72	1.84	1.48	2.04
4	2.27	3.10	2.45	4.77	2.42	3.28	2.52	2.90	2.65	1.81	1.46	1.74
5	2.50	2.98	2.42	4.82	......	3.65	2.38	2.88	2.55	1.75	1.42	2.28
6	2.36	2.94	2.40	4.86	......	3.56	2.26	2.88	2.48	1.71	1.36	2.40
7	2.24	2.89	2.35	4.88	......	3.51	2.18	2.90	2.50	1.66	1.30	2.06
8	2.23	2.81	2.42	4.88	2.46	3.44	3.40	2.91	2.44	1.62	1.54	1.89
9	2.20	2.75	2.84	4.82	5.84	3.32	3.79	2.89	2.37	1.59	1.81	1.81
10	2.18	2.73	2.72	4.73	5.44	3.86	3.06	2.89	2.35	1.56	1.90	1.74
11	2.16	2.71	......	4.60	5.16	3.65	3.42	2.99	2.42	1.60	2.27	1.68
12	2.21	2.68	......	4.56	6.14	3.63	4.62	3.04	2.33	1.87	2.22	1.90
13	2.22	2.66	......	4.28	6.39	3.50	4.20	3.83	2.24	2.02	2.18	1.99
14	2.20	2.50	2.58	3.69	4.53	3.96	3.96	4.63	2.18	1.92	2.14	1.96
15	2.26	2.58	......	3.46	5.64	3.80	3.62	3.90	2.18	1.82	2.06	1.90
16	2.26	2.60	......	3.30	3.78	3.51	3.40	3.47	2.10	1.71	2.00	1.90
17	2.24	2.58	3.30	3.20	3.13	3.40	3.38	3.17	2.03	1.64	1.94	2.02
18	2.20	2.58	......	3.16	3.18	3.22	3.30	2.99	1.98	1.60	1.86	2.07
19	4.75	2.54	......	......	4.34	3.06	3.00	2.88	1.92	1.60	1.79	2.01
20	4.02	2.51	......	......	7.92	2.94	2.88	2.82	1.90	1.60	1.70	1.98
21	3.40	2.82	3.38	3.05	5.65	2.89	3.34	2.76	1.96	1.56	1.62	2.16
22	3.04	2.57	4.54	......	6.79	2.80	3.43	2.81	1.99	1.54	1.52	2.10
23	2.85	2.58	4.46	......	6.92	2.72	3.20	5.21	1.98	1.52	1.46	1.97
24	5.10	2.52	4.16	......	4.66	2.66	3.36	4.05	1.94	1.50	1.41	1.81
25	5.36	2.34	4.02	3.00	3.76	2.70	3.22	4.28	1.93	1.70	1.36	1.75
26	4.18	2.42	3.96	......	4.91	2.62	3.11	4.18	1.92	1.66	1.32	1.76
27	4.04	2.46	3.84	......	4.12	2.54	3.04	3.82	1.92	1.60	1.31	1.72
28	3.97	2.54	3.82	2.98	3.80	2.54	2.90	3.23	1.88	1.56	1.29	1.70
29	3.79	2.52	3.85	......	......	2.50	3.82	3.07	1.84	1.52	1.93	1.68
30	4.79	2.54	3.93	......	......	2.46	3.42	4.72	1.84	1.55	2.54	1.64
31	4.10	......	4.20	......	......	2.44	......	4.60	......	1.53	3.03	......

NOTE.—Gage not read on days for which no gage heights are given.

MIDDLE ISLAND CREEK BASIN.

MIDDLE ISLAND CREEK AT LITTLE, W. VA.

LOCATION.—At highway bridge at Little, 6 miles southeast of Friendly, Tyler County. Stewart Run enters on left 500 feet below station.

DRAINAGE AREA.—458 square miles (measured on topographic maps).

RECORDS AVAILABLE.—May 7, 1915, to September 30, 1918.

GAGE.—Vertical and inclined staff on left bank immediately below the bridge; read to tenths twice daily by E. F. Weigand.

DISCHARGE MEASUREMENTS.—Made from bridge or by wading. Stay wire is used for measurements at high stages.

CHANNEL AND CONTROL.—One channel at all stages; straight for about 400 feet above and 250 feet below station. Primary control is at foundation of old milldam 250 feet below station; composed of bedrock, foundation timbers, small deposit of rock and sand; probably permanent. Point of zero flow, gage height 1.4 feet ±0.2 foot.

EXTREMES OF STAGE.—Maximum stage recorded during year, 18.7 feet at 7 a. m. March 14; minimum stage, 1.74 feet January 13, August 29 and 30.

1915–1918: Maximum stage recorded, 22.22 feet at 5 p. m. January 22, 1917; minimum stage, 1.74 feet January 13, August 29 and 30, 1918. Highest known flood occurred in August, 1875; gage height about 33.5 feet.

ICE.—Stage-discharge relation affected by ice during winters.

COOPERATION.—Base data furnished by United States Engineer Corps.

Determination of discharge withheld until additional data are obtained.

No discharge measurements were made at this station during the year.

Daily gage height, in feet, of Middle Island Creek at Little, W. Va., for the year ending Sept. 30, 1918.

Day.	Oct.	Nov.	Dec.	Jan.	Feb.	Mar.	Apr.	May.	June.	July.	Aug.	Sept.
1	2.04	3.99	3.19	2.64	4.84	3.94	2.64	3.84	3.49	2.54	2.29	2.89
2	2.04	3.44	2.94	2.39	3.84	3.69	2.74	3.44	3.19	2.49	2.59	3.79
3	2.04	3.24	2.89	2.14	4.14	3.49	3.19	3.39	2.84	2.44	2.69	3.29
4	2.04	3.19	2.74	2.04	3.99	3.34	4.24	3.29	2.74	2.34	2.44	2.89
5	2.04	3.04	2.69	2.04	3.69	4.79	3.79	3.09	2.69	2.29	2.29	2.84
6	2.04	2.99	2.64	1.94	3.54	4.54	3.54	2.99	2.64	2.19	2.14	2.69
7	2.04	2.84	2.54	2.19	6.69	3.89	3.34	2.84	6.74	2.14	2.09	2.64
8	2.04	2.79	2.59	2.69	11.44	3.59	4.99	15.26	5.69	2.04	2.04	2.54
9	2.04	2.69	2.59	2.39	12.55	3.54	8.59	6.29	3.34	2.04	2.64	2.44
10	2.04	2.54	2.49	2.24	9.64	4.14	5.44	4.14	3.09	2.04	3.69	2.39
11	2.04	2.54	2.44	2.04	5.49	3.99	6.84	3.79	2.89	2.04	3.24	2.34
12	2.19	2.84	2.34	1.84	4.84	3.74	8.44	3.84	2.79	2.19	2.99	2.29
13	2.34	2.44	2.29	1.74	6.29	8.99	5.39	5.79	2.64	2.49	2.49	2.24
14	2.79	2.44	2.24	2.14	5.04	16.70	4.54	6.14	2.64	2.59	2.34	2.19
15	2.74	2.39	2.09	2.64	6.64	7.24	4.09	4.19	2.49	2.49	2.39	2.14
16	2.74	2.34	2.04	3.74	6.34	4.24	3.69	3.74	2.44	2.39	2.49	2.09
17	2.54	2.34	2.04	4.34	4.44	3.99	3.64	3.49	2.44	2.34	2.39	2.09
18	2.59	2.34	2.04	4.19	3.69	3.64	3.54	3.19	2.59	2.29	2.29	2.44
19	3.89	2.34	2.09	3.94	5.84	3.39	3.39	2.99	3.14	2.19	2.14	2.49
20	7.12	2.34	2.29	3.59	11.84	3.24	3.24	2.89	2.74	2.14	2.04	2.44
21	3.59	2.24	3.04	3.39	6.99	3.19	3.59	2.69	2.54	2.44	2.04	2.69
22	3.24	2.24	3.79	3.24	4.39	3.04	4.09	2.59	2.39	2.64	2.04	2.49
23	2.94	2.34	3.29	3.14	3.84	2.89	3.84	3.74	3.34	2.54	2.04	2.69
24	4.09	2.34	3.14	3.04	3.99	2.84	3.69	3.59	2.89	2.44	2.04	2.79
25	7.47	2.34	3.94	3.29	4.64	2.79	3.64	8.34	2.69	2.34	1.99	2.59
26	6.62	2.24	4.34	3.44	11.54	2.89	3.84	10.64	2.59	2.29	1.94	2.49
27	3.62	2.34	4.04	3.69	5.64	2.89	3.79	4.54	2.49	2.24	1.84	2.39
28	5.29	2.54	3.64	4.49	4.19	2.84	3.84	4.94	2.74	2.09	1.84	2.29
29	3.84	3.09	3.44	11.54	......	2.74	3.74	3.69	2.84	2.09	1.79	2.24
30	5.34	3.34	3.04	9.14	......	2.74	3.94	3.34	2.84	2.39	1.74	2.14
31	4.64	......	2.89	6.64	......	2.64	......	3.09	2.69	2.24	1.89	......

NOTE.—Stage-discharge relation may have been affected by ice during part of December, January, and February.

LITTLE MUSKINGUM RIVER BASIN.

LITTLE MUSKINGUM RIVER AT FAY, OHIO.

LOCATION.—A mile northwest of Fay, Washington County, Ohio, 7 miles from St. Marys, W. Va., and 12 miles from Marietta, Ohio. Bear Run enters on left half a mile above station. Covered highway bridge crosses river just above Bear Run.

DRAINAGE AREA.—259 square miles (measured on topographic maps).

RECORDS AVAILABLE.—May 14, 1915, to September 30, 1918.

GAGE.—Inclined and vertical staff on right bank about 400 feet below suspension footbridge; read by G. I. Smith.

DISCHARGE MEASUREMENTS.—Made from suspension bridge or by wading.

CHANNEL AND CONTROL.—One channel at all stages; straight several hundred feet above and below bridge. Overflow at gage height about 13 feet; wide overflow at maximum stages. Bed of stream composed of mud, sand, rock and gravel; primary control at ford 50 feet below gage compact sand and gravel; fairly permanent. Point of zero flow, gage height 0.7 ± 0.2 foot May, 1915.

EXTREMES OF STAGE.—Maximum stage recorded during year, 15.50 feet at 8 a. m. February 10 and 5 p. m. February 20; minimum stage, 1.17 feet at 5 p. m. October 2 and 8 a. m. October 3.

1915–1918: Maximum stage recorded, 21.5 feet at 5 p. m. January 22, 1917; minimum stage, 1.17 feet at 5 p. m. October 2 and 8 a. m. October 3, 1917.

Highest flood known reached a stage represented by gage height about 23 feet.

ICE.—Stage-discharge relation affected by ice in severe winters.

COOPERATION.—Base data furnished by United States Engineer Corps.

Data inadequate for determination of discharge.

The following discharge measurement was made by Shick and Quattlebaum:

February 20, 1918: Gage height, 15.50 feet; discharge, 6,430 second-feet.

Daily gage height, in feet, of Little Muskingum River at Fay, Ohio, for the year ending Sept. 30, 1918.

Day.	Oct.	Nov.	Dec.	Jan.	Feb.	Mar.	Apr.	May.	June.	July.	Aug.	Sept.
1	1.20	3.32	2.46	2.49	3.22	3.40	2.12	2.70	2.70	1.56	1.33	3.44
2	1.18	2.90	2.29	2.34	2.92	3.24	2.46	2.58	2.45	1.48	1.32	2.25
3	1.18	2.61	2.17	2.25	2.80	3.02	4.12	2.44	2.28	1.41	1.32	2.04
4	1.20	2.40	2.09	2.18	2.82	2.94	3.37	2.33	2.14	1.40	1.33	1.75
5	1.24	2.26	2.02	2.14	2.66	3.45	2.86	2.26	2.06	1.40	1.32	1.66
6	1.26	2.17	1.98	2.20	2.59	3.34	2.60	2.22	2.04	1.39	1.30	1.62
7	1.28	2.10	1.90	3.02	5.40	3.14	2.69	2.32	2.18	1.34	1.30	1.56
8	1.35	2.06	1.83	3.35	9.40	2.96	3.69	13.58	2.45	1.33	1.35	1.48
9	1.34	1.99	1.68	2.94	14.05	3.05	4.55	4.40	2.09	1.40	1.40	1.42
10	1.30	1.92	1.65	2.69	13.15	4.30	3.96	3.58	1.98	1.38	1.58	1.42
11	1.30	1.88	1.69	2.55	7.75	3.69	7.30	4.18	1.88	1.37	3.38	1.40
12	1.45	1.83	1.58	3.01	8.45	3.34	9.18	4.10	1.84	1.36	2.01	1.42
13	1.47	1.84	1.53	3.29	9.30	10.75	4.90	7.08	1.74	1.38	1.74	1.71
14	1.56	1.78	1.55	2.97	5.24	12.45	3.82	8.30	1.69	1.48	1.56	1.80
15	1.56	1.73	1.62	2.85	8.02	4.68	3.35	4.10	1.65	1.58	1.42	1.56
16	1.46	1.68	1.62	2.87	4.75	3.64	3.11	3.34	1.60	1.52	1.42	1.52
17	1.44	1.64	1.62	2.70	3.63	3.26	3.10	3.02	1.61	1.56	1.42	2.76
18	1.44	1.62	1.64	2.52	3.12	3.06	3.09	2.82	1.58	1.46	1.36	2.65
19	3.27	1.62	1.63	2.42	5.62	2.84	2.83	2.64	1.60	1.42	1.32	2.15
20	5.30	1.62	1.95	2.26	15.05	2.72	2.85	2.52	1.56	1.37	1.30	2.14
21	2.94	1.60	3.74	2.18	4.96	2.63	3.31	2.50	1.53	1.34	1.28	2.30
22	2.42	1.56	5.11	2.23	3.55	2.56	3.19	2.46	1.56	1.32	1.27	2.00
23	2.33	1.54	4.20	2.23	3.25	2.48	2.90	2.84	1.52	1.30	1.41	1.65
24	3.41	1.54	3.78	2.27	3.61	2.40	2.78	2.73	1.52	1.30	1.29	1.58
25	8.03	1.56	4.72	2.24	4.15	2.34	2.68	12.55	1.50	1.30	1.24	1.40
26	6.10	1.52	4.10	2.26	11.00	2.29	2.61	4.54	1.48	1.32	1.23	1.23
27	4.16	1.54	3.37	2.40	4.75	2.22	2.57	5.25	1.50	1.72	1.31	1.23
28	5.45	2.10	3.09	2.77	3.78	2.14	2.55	4.30	1.48	1.50	1.30	1.30
29	3.72	2.64	2.66	4.20	....	2.10	2.68	3.56	1.44	1.44	1.29	1.22
30	6.87	2.62	2.71	4.26	....	2.06	2.89	3.11	1.39	1.43	1.28	1.20
31	4.32	....	2.81	3.72	....	2.04	....	2.93	....	1.36	4.10	....

NOTE.—Stage-discharge relation may have been affected by ice during part of December, January, and February.

MUSKINGUM RIVER BASIN.

MUSKINGUM RIVER AT FRAZIER, OHIO.

LOCATION.—At highway bridge at Frazier, Muskingum County, 4½ miles below Zanesville. Brush Creek enters on right one-third mile below gage.

DRAINAGE AREA.—7,160 square miles (revised measurement).

RECORDS AVAILABLE.—June 1, 1915, to September 30, 1918.

GAGE.—Staff near upper corner of right abutment of bridge; read by D. A. Burns Sea-level elevation of zero of gage, 663.29 feet.

DISCHARGE MEASUREMENTS.—Made from upstream side of bridge or by wading on crest of dam No. 9, about 4 miles below gage. Leakage past dam, through lock and power plants, should be included with flow over crest. The measurement of August 12, 1916, made by wading on crest of dam, includes the flow over crest (620 second-feet); discharge through upper gate of lock (5 second-feet); and discharge through headgate of Carter's mill (47 second-feet).

CHANNEL AND CONTROL.—River straight above and below. Control is crest of dam No. 9 at Philo, about 4 miles below gage. Except for leakage through lock and dam and leakage and flow to flour mill at left end of dam, and leakage and flow through gate at right end of dam leading to old canal for supply to railroad pumping station, the gage height of the crest of the dam, 8.83 feet, is the point at which flow would be zero.

EXTREMES OF STAGE.—Maximum stage recorded during year, 22.9 feet at 6 a. m. February 15; minimum stage, 9.1 feet August 21 and 22. Flood of March, 1913, reached a stage of 49.1 feet; highest stage ever recorded.

ICE.—Stage-discharge relation affected by ice jams at times.

Regulation.—Leakage through the lock and the power plants at dam No. 9 and the operation of power plants at dams Nos. 9 and 10 may affect the low-water flow to some extent.

Accuracy.—Stage-discharge relation permanent, except as the relation may be affected by leakage through dam No. 9, through the gates of the power plants and through the lock, and by the operation of the power plants at dam No. 9; probably not affected by ice. The flow from the area between the measuring section and the crest of dam No. 9 may be sufficient at times to affect the stage-discharge relation. This area, however, is small, and such conditions would be of rare occurrence and of small effect. Gage read twice daily to tenths. Records good.

Cooperation.—Base data furnished by the United States Engineer Corps.

No discharge measurements were made at this station during the year.

Daily gage height, in feet, of Muskingum River at Frazier, Ohio, for the year ending Sept. 30, 1918.

Day.	Oct.	Nov.	Dec.	Jan.	Feb.	Mar.	Apr.	May.	June.	July.	Aug.	Sept.
1	9.3	11.95	9.4	9.5	9.5	13.45	10.1	10.95	11.7	9.5	9.8	10.15
2	9.3	11.45	9.4	9.2	9.5	14.95	10.3	10.75	11.8	9.5	9.7	9.85
3	9.3	11.05	9.5	9.2	9.5	15.05	10.5	10.6	11.25	9.5	9.5	9.6
4	9.3	10.9	9.5	9.2	9.5	13.35	10.35	10.45	11.0	9.5	9.4	9.5
5	9.3	10.5	9.55	9.2	9.5	12.85	10.2	10.35	10.65	9.5	9.3	9.4
6	9.3	10.4	9.6	9.2	9.5	12.6	10.25	10.0	10.45	9.5	9.3	9.55
7	9.3	10.25	9.6	9.2	9.5	12.45	10.3	10.05	10.35	9.5	9.2	9.9
8	9.3	9.9	9.5	9.2	9.5	12.15	10.35	10.45	10.2	9.5	9.2	9.75
9	9.3	10.0	9.4	9.5	10.15	11.9	10.25	10.1	10.35	9.4	9.15	9.6
10	9.3	9.95	9.4	9.5	12.4	12.3	10.5	10.2	10.25	9.4	9.3	9.5
11	9.3	10.0	9.2	9.5	14.9	12.9	10.75	10.35	10.0	9.4	9.4	9.4
12	9.3	10.0	9.2	9.5	16.9	12.6	11.2	10.45	10.0	9.4	9.5	9.4
13	9.3	9.8	9.2	9.5	19.65	12.45	11.85	13.55	10.0	9.5	9.5	9.4
14	9.3	9.8	9.0	9.5	20.6	14.2	12.15	17.0	9.9	9.5	9.4	9.3
15	9.3	9.8	9.0	9.5	21.95	14.0	11.9	15.95	9.9	9.4	9.4	9.55
16	9.3	9.8	9.0	9.5	20.65	13.7	11.5	14.55	8	9.4	9.4	9.7
17	9.3	9.7	9.0	9.5	18.35	12.85	11.5	13.65	9	9.3	9.3	9.65
18	9.3	9.7	9.0	9.5	15.45	12.7	10.85	12.6	8	9.2	9.3	9.65
19	9.35	9.6	9.0	9.5	13.8	11.7	10.7	11.6	9.7	9.2	9.2	9.85
20	9.4	9.7	9.0	9.5	17.8	11.4	10.8	11.5	9.7	9.2	9.2	10.0
21	9.95	9.4	9.0	9.5	18.4	11.05	10.8	11.0	9.6	9.2	9.1	9.8
22	10.5	9.4	9.1	9.5	17.4	10.8	10.9	10.9	9.6	9.2	9.1	9.8
23	10.45	9.5	9.3	9.5	15.95	10.8	11.15	11.45	9.6	9.2	9.2	9.65
24	10.25	9.4	10.2	9.5	13.95	10.6	11.05	12.95	9.6	9.2	9.2	9.6
25	10.45	9.5	11.1	9.5	13.5	10.6	11.0	12.45	9.6	9.2	9.25	9.5
26	11.45	9.3	11.0	9.5	15.8	10.45	10.9	13.25	9.6	9.4	9.35	9.5
27	11.5	9.3	11.0	9.5	14.4	10.5	10.8	12.7	9.6	9.5	9.3	9.4
28	11.45	9.3	10.5	9.5	13.8	10.4	10.55	12.45	9.6	9.4	9.3	9.4
29	11.15	9.25	10.5	9.5	....	10.3	10.75	12.55	9.5	9.5	9.2	9.5
30	11.4	9.4	9.6	9.5	....	10.2	10.95	11.8	9.5	9.5	9.2	9.4
31	11.8	....	9.6	9.5	....	10.2	....	11.5	....	9.55	9.55	....

MUSKINGUM RIVER AT BEVERLY, OHIO.

Location.—At Lock 4 at Beverly, Washington County. Wolf Creek enters on right immediately above station.

Drainage area.—7,700 square miles (United States Engineer Corps).

Records available.—June 1, 1915, to September 30, 1918.

Gage.—Ceramic tile gage, graduated to tenths of a foot, on lower buttress of river wall of Lock 4, about 1,000 feet above the measuring section. Sea-level elevation of zero of gage, 602.60 feet (United States Engineer Corps).

Discharge measurements.—Made from upstream side of highway bridge 1,000 feet below gage.

Channel and control.—Bed of stream gravel and masonry débris of old bridge piers; probably permanent. Stream curves slightly to the left from 1,000 feet above to 1,000 feet below the section. Control is crest of dam No. 3, 10.8 miles below. At gage height 5.2 feet or crest of dam No. 3 flow would be zero provided there was no leakage through dam, lock, or power plant at dam.

Extremes of stage.—Maximum stage recorded during year, 24.4 feet at 6 p. m. February 15; minimum stage, 3.3 feet October 1–3.

Flood of March, 1913, reached a stage of 46.55 feet, the highest stage ever recorded.

Ice.—Stage-discharge relation not affected by ice.

Regulation.—Leakage through dam No. 3, lock, and the power plant at the dam may affect the low-water flow to some extent.

Accuracy.—Stage-discharge relation practically permanent; not affected by ice. Dam No. 3, about 11 miles below, the control for the gage, leaks so that the water falls below the crest during low water. Change in this leakage, leakage and flow through the power plant, leakage through lock, and inflow into pool 3 below the measuring section may all affect the stage-discharge relation at low and medium stages. When the stage of the Ohio at Marietta is about 39 feet or more, the stage-discharge relation is affected by backwater. Records of daily discharge withheld for additional information. Gage read twice daily to tenths. Records good, except as may be affected by described conditions at low and medium stages.

Cooperation.—Base data furnished by United States Engineer Corps.

No discharge measurements were made at this station during the year.

Daily gage height, in feet, of Muskingum River at Beverly, Ohio, for the year ending Sept. 30, 1918.

Day.	Oct.	Nov.	Dec.	Jan.	Feb.	Mar.	Apr.	May.	June.	July.	Aug.	Sept
1	3.3	9.55	5.6		0	12.15	6.	7.85	9.15	5.6	5.85	6.35
2	3.3	9.1	5.6		0	13.3	6.	7.7	9.3	5.6	5.9	6.15
3	3.3	8.35	5.6		0	14.35	7.6	7.55	8.85	5.6	5.75	6.1
4	3.45	7.85	5.6	6.	6.0	12.75	7.6	7.35	8.2	5.6	5.6	5.8
5	3.5	7.2	5.7	6.6 45	6.0	11.3	6.25	6.95	7.7	5.6	5.0	5.6
6	3.5	6.9	5.7	6.4	6.0	.0	6.7	6.7	7.3	5.5	4.65	5.5
7	3.5	6.6	5.7	6.8	6.1	11.6	6.7	6.6	7.5	5.5	4.35	5.8
8	3.6	6.5	5.7	6.65	7.7	10.2	7.1	7.25	7.0	5.4	4.55	6.1
9	3.55	6.35	5.55	6.45	13.0	65	7.05	7.0	6.8	5.3	4.6	5.9
10	3.9	6.2	5.4	6.4	15.45	5	7.0	7.1	6.7	5.2	4.75	5.65
11	4.1	6.1	5.4	6.	16.85	10.75	8.9	7.	6.55	5.2	5.35	5.5
12	4.2	6.0	5.4	6.	18.95	11.15	10.4	7.	6.35	5.2	5.7	5.45
13	4.3	5.9	5.4	6.3	20.6	14.7	9.7	12.	6.2	5.2	5.65	5.4
14	4.3	5.9	5.4	6.3	21.35	15.15	10.2	17.	6.2	5.2	5.5	5.3
15	4.3	5.9	5.4	6.3	24.2	13.35	9.95	16.8	6.1	5.2	5.4	5.3
16	4.55	5.8	5.4	6.3	22.9	12.7	9.3	14.15	6.0	5.2	5.2	5.5
17	4.95	5.7	5.4	6.2	19.9	11.45	8.6	12.55	6.0	5.2	4.9	6.95
18	5.0	5.7	5.3	6.2	16.35	10.45	8.1	11.25	5.9	5.1	4.65	7.0
19	6.25	5.7	5.3	6.2	14.5	9.6	7.75	9.6	5.8	5.0	4.45	6.55
20	5.85	5.7	5.45	6.2	18.3	8.85	7.6	8.65	5.7	5.0	4.25	6.35
21	5.5	5.7	5.75	6.1	19.6	8.3	7.7	8.7	5.7	4.9	3.85	6.15
22	6.8	5.6	6.05	6.1	17.35	8.05	7.85	8.15	5.7	4.45	3.6	6.1
23	6.9	5.6	6.2	6.1	16.3	7.8	8.1	8.9	5.6	4.45	3.45	6.1
24	6.75	5.6	6.8	6.0	13.8	7.55	8.2	11.1	5.6	4.6	3.45	5.95
25	7.7	5.6	8.4	6.0	12.1	7.5	8.05	11.0	5.6	4.6	3.4	5.75
26	8.25		8.35	0	18.65	7.3	7.8	11.55	5.6	4.9	3.4	5.7
27	9.05		8.0	6.0	13.95	7.1	7.6	11.8	5.6	5.4	3.4	5.6
28	8.95		8.0	6.0	12.85	7.0	7.45	10.65	5.6	5.5	3.6	5.5
29	8.25		7.8	6.0	....	6.9	7.3	10.55	5.6	5.3	3.95	5.4
30	9.45	[illegible]	7.05	6.0	....	6.8	7.65	10.0	5.6	5.65	4.25	5.3
31	8.75	....	6.7	6.0	....	6.7	....	8.9	....	5.7	4.85	....

LITTLE KANAWHA RIVER BASIN.

LITTLE KANAWHA RIVER AT GLENVILLE, W. VA.

Location.—At three-span steel highway bridge at Glenville, Gilmer County. Stewart Creek enters on right 1½ miles above station.

Drainage area.—385 square miles (measured on topographic maps).

Records available.—June 1, 1915, to September 30, 1918.

Gage.—Vertical and inclined staff attached to upstream side of right pier of bridge; read by Hollie Gainor. Gage was established by the United States Weather Bureau September 10, 1900 (read daily to tenths at 8 a. m.), repaired and its datum lowered 2.5 feet on June 1, 1915.

Discharge measurements.—Made from bridge or by wading.

Channel and control.—One channel at all stages; straight for 100 feet above and 150 feet below station. Bed of river composed of mud, rock, sand, and gravel; control is probably fairly permanent. Point of zero flow, gage height about 1.0 foot June and September, 1915.

Extremes of stage.—Maximum stage recorded during year, 31.7 feet at 5.40 p. m., March 13; minimum stage, 1.35 feet at 6 p. m. July 23.

1915–1918: Maximum and minimum recorded stages same as those for year ending September 30, 1918.

Ice.—Stage-discharge relation affected by ice during severe winters.

Accuracy.—Stage-discharge relation practically permanent; probably affected by ice during periods in December and January. Gage read to half-tenths twice daily. Data inadequate for determination of discharge.

Cooperation.—Base data furnished by United States Engineer Corps.

No discharge measurements were made at this station during the year.

Daily gage height, in feet, of Little Kanawha River at Glenville, W. Va., for the year ending Sept. 30, 1918.

Day.	Oct.	Nov.	Dec.	Jan.	Feb.	Mar.	Apr.	May.	June.	July.	Aug.	Sept.
1	1.52	4.05	4.60	2.88	4.92	4.92	3.42	4.45	3.92	4.20	3.68	6.40
2	1.48	3.98	4.30	2.72	4.28	4.65	3.32	4.32	3.15	4.10	2.98	3.95
3	1.48	3.88	4.20	2.62	4.28	4.55	4.18	4.28	2.95	3.20	2.90	3.88
4	1.62	3.78	3.98	2.55	4.28	4.72	4.38	4.15	2.75	2.85	2.82	3.68
5	1.58	3.68	3.55	2.50	4.12	7.15	4.12	4.00	2.38	2.68	2.78	3.38
6	1.52	3.52	3.30	2.98	4.08	5.30	3.78	3.75	2.50	2.52	2.65	3.08
7	1.48	3.42	3.12	6.55	5.30	7.20	3.72	3.62	2.68	2.45	2.45	2.80
8	1.48	3.32	2.98	6.02	5.35	6.40	7.08	4.52	2.58	2.40	2.30	2.55
9	1.58	3.22	2.82	4.30	8.45	5.38	11.05	4.92	2.58	2.25	2.18	2.48
10	1.62	3.18	2.70	3.68	8.85	6.45	6.85	4.52	2.48	2.08	2.10	2.65
11	1.60	2.95	2.58	3.30	6.15	5.02	7.78	4.32	2.40	1.98	2.05	2.18
12	1.58	2.80	2.48	6.00	5.15	4.78	9.80	3.92	2.38	1.92	5.20	2.10
13	1.52	2.58	2.38	5.40	6.02	24.40	7.50	3.95	2.38	1.85	4.70	2.08
14	1.50	2.42	2.30	5.05	4.60	27.75	5.95	5.70	2.38	1.82	4.50	2.05
15	1.42	2.32	2.25	7.30	5.78	14.05	5.18	5.18	2.40	1.78	3.55	2.00
16	1.42	2.22	2.20	7.80	4.90	6.60	5.00	4.60	2.35	1.75	3.12	2.05
17	1.40	2.12	2.15	5.58	4.30	5.40	4.88	4.18	4.75	1.80	2.98	2.85
18	1.48	2.02	2.08	4.85	4.20	4.72	4.78	3.88	5.00	1.75	2.88	3.22
19	2.95	1.92	2.02	4.22	4.12	4.35	4.45	3.72	3.85	1.70	2.80	3.12
20	4.18	1.85	2.08	4.08	14.55	4.15	5.35	3.62	3.35	1.65	2.72	4.05
21	3.72	1.80	2.12	3.88	7.35	4.78	6.95	3.48	3.25	1.58	2.62	3.75
22	3.62	1.82	2.05	3.72	5.35	5.08	6.08	3.45	3.10	1.52	2.55	3.82
23	3.45	1.80	2.12	3.62	4.68	4.60	5.20	5.08	3.00	1.48	2.48	3.20
24	3.78	1.72	2.22	3.48	5.30	4.30	4.88	4.40	2.35	1.52	2.38	3.05
25	6.55	1.65	4.40	3.32	7.90	4.82	4.48	14.92	2.20	1.52	2.28	2.92
26	4.85	1.38	5.88	3.38	19.65	4.80	4.32	23.65	7.10	2.42	2.22	2.82
27	4.35	1.58	4.55	7.50	9.35	4.45	7.50	6.28	4.80	2.38	2.18	2.68
28	4.25	1.70	3.80	15.90	5.48	4.10	6.20	11.50	3.85	2.30	2.22	2.52
29	4.15	3.10	3.40	19.15	....	3.88	5.25	5.50	3.05	2.25	2.20	2.42
30	4.20	4.58	3.20	7.22	....	3.72	4.70	4.80	2.95	3.85	2.22	2.32
31	4.12	....	3.00	5.45	....	3.55	....	4.32	....	6.25	5.10	....

LITTLE KANAWHA RIVER AT LOCK 4, PALESTINE, W. VA.

Location.—At Lock 4, Palestine, Wirt County, 30 miles from Parkersburg by Little Kanawha Railroad. Reedy Creek enters from left 1 mile above gage.

Drainage area.—1,500 square miles (measured on map prepared by United States Geological Survey; scale, 1 :500,000).

Records available.—April 25, 1915, to September 30, 1918. The upper and lower gages at the lock have been read under direction of the United States Engineer Corps, since November 5, 1905.

Gage.—Upper gage at lock; vertical staff on right bank bolted to right side of river wall of lock just above upper gates; an inclined section of gage extends above top of lock wall; read by James Burton, lockmaster.

Discharge measurements.—Made at cable about 1,200 feet below gage or by wading on crest of dam.

Channel and control.—One channel at all stages. Crest of dam No. 4 is the control for the gage; lowest point in crest of dam is at 9.4 feet gage height, which is the point of zero flow except for leakage through dam, lock gates, and valves. Backwater submerges dam No. 4 during extreme floods on Ohio River.

Extremes of stage.—Maximum stage recorded during year, 25.8 feet at 8 a. m. March 14; minimum stage, 9.45 feet August 27 and 28.

1915–1918: Maximum stage recorded, that of March 14, 1918; minimum stage 9.40 feet at 6 p. m. September 21, 1915.

Highest headwater as reported by lockmaster occurred in 1897, and was equivalent to a gage height of about 30 feet on the lower gage, which corresponds to a reading of about 24.4 feet on upper gage, assuming fall of 1 foot at dam.

Ice.—Stage-discharge relation probably not affected by ice.

Regulation.—Flow may be affected at times by the manipulation of the pool above dam No. 5, about 9.5 miles above dam No. 4, and the occasional use of flashboards on dam No. 4.

Accuracy.—Stage-discharge relation practically permanent; not affected by ice during year. Variable leakage through lock and dam may affect the stage-discharge relation at low stages. Data inadequate for determining daily discharge. Gage read to hundredths twice daily.

Cooperation.—Base data furnished by United States Engineer Corps.

The following discharge measurement was made by H. S. Shick:

March 15, 1918·Gage height, 19.87 feet; discharge, 31,200 second-feet.

Daily gage height, in feet, of Little Kanawha River at Lock 4, Palestine, W. Va., for the year ending Sept. 30, 1918.

Day.	Oct.	Nov.	Dec.	Jan.	Feb.	Mar.	Apr.	May.	June.	July.	Aug.	Sept.
1	9.60	11.04	10.63	10.06	11.72	11.55	10.20	11.05	10.41	10.29	10.72	9.76
2	9.61	10.60	10.59	10.04	11.04	11.16	10.15	10.85	10.25	10.12	10.47	10.96
3	9.58	10.36	10.44	9.98	10.98	10.94	10.50	10.58	10.08	10.09	11.06	10.51
4	9.57	10.21	10.30	9.96	10.90	10.76	10.82	10.46	9.99	10.08	9.91	10.15
5	9.57	10.08	10.19	9.96	10.30	11.84	10.79	10.36	9.92	9.98	9.80	9.93
6	9.57	10.00	10.12	9.97	10.40	12.56	10.41	10.27	9.88	9.80	9.66	9.82
7	9.56	9.94	10.02	11.84	11.28	12.28	10.32	10.20	9.88	9.76	9.48	9.76
8	9.54	9.88	9.97	12.04	12.26	12.54	12.24	10.20	10.00	9.68	9.50	9.68
9	9.54	9.84	9.99	11.82	13.65	11.85	15.12	10.19	10.30	9.62	9.50	9.65
10	9.55	9.82	9.91	11.06	14.84	11.64	13.62	10.59	10.20	9.60	9.54	9.61
11	9.53	9.77	9.82	10.72	13.15	11.78	13.25	10.56	10.01	9.58	9.58	9.55
12	9.52	9.76	9.77	10.70	12.28	11.20	15.50	10.58	9.90	9.60	9.88	9.56
13	9.52	9.74	9.75	10.68	11.85	18.94	14.08	10.76	9.76	9.60	9.66	9.59
14	9.56	9.70	9.64	11.18	11.70	25.10	12.82	11.12	9.74	9.54	9.72	9.55
15	9.66	9.68	9.60	11.32	12.20	20.10	11.89	11.24	9.64	9.52	10.38	9.55
16	9.55	9.69	9.60	11.90	12.33	14.64	11.36	11.00	9.62	9.52	10.10	9.52
17	9.62	9.69	9.65	12.15	11.40	12.08	11.18	10.62	9.60	9.48	9.93	9.59
18	9.58	9.68	9.66	11.36	10.92	11.40	11.07	10.46	10.50	9.46	9.79	9.51
19	11.55	9.64	9.66	10.88	11.18	11.04	10.95	10.74	10.16	9.47	9.73	9.56
20	12.10	9.62	9.66	10.53	15.24	10.78	10.80	10.16	9.99	9.46	9.65	9.60
21	11.15	9.60	9.74	10.31	13.77	10.57	12.98	10.02	9.85	9.47	9.60	9.61
22	10.56	9.59	9.81	10.18	12.44	11.14	12.70	10.00	9.81	9.46	9.56	9.78
23	10.28	9.60	....	10.03	11.41	10.87	11.90	10.14	8.90	9.46	9.54	9.86
24	10.54	9.62	....	10.00	11.62	10.59	11.33	10.95	9.91	9.52	9.50	9.96
25	12.04	9.62	....	10.00	12.50	10.66	11.04	10.62	11.06	10.05	9.50	9.86
26	12.32	9.60	....	10.09	16.86	11.54	10.76	15.27	11.35	10.29	9.52	9.79
27	11.36	9.59	....	10.61	16.41	10.90	10.95	15.86	10.58	10.00	9.46	9.74
28	11.26	9.60	....	12.56	12.75	10.52	12.32	12.00	10.26	9.82	9.46	9.66
29	11.00	9.94	....	18.20	....	10.44	11.41	12.72	10.08	9.76	9.47	9.61
30	11.54	10.25	10.22	15.86	....	10.34	11.36	11.16	....	10.26	9.54	9.59
31	11.40	....	10.10	12.35	....	10.72	....	10.62	....	....	9.83	....

NOTE.—Gage not read Dec. 23-29.

SOUTH FORK OF HUGHES RIVER AT MACFARLAN, W. VA.

LOCATION.—About 80 feet above highway bridge half a mile east of Macfarlan, Ritchie County. Dutchman Run enters river on left 3,000 feet below station.

DRAINAGE AREA.—210 square miles (measured on topographic maps).

RECORDS AVAILABLE.—May 17, 1915, to September 30, 1918.

GAGE.—Vertical staff on right bank; read by A. H. Reynolds.

DISCHARGE MEASUREMENTS.—Made from bridge or by wading.

CHANNEL AND CONTROL.—One channel at all stages; straight 300 feet above and 1,500 feet below bridge. Bed of stream rock and mud. Control probably fairly permanent.

EXTREMES OF STAGE.—Maximum stage recorded during year 24.0 feet at 6 p. m. March 13; minimum stage, 2.00 feet September 16 and 22.

1915–1918: Maximum stage recorded, 25.7 feet at 8 a. m. January 22, 1917; minimum stage 1.50 feet June 28, 29, July 2, and July 24, 1915.

Highest flood known reached a stage represented by gage height about 29 feet.

ICE.—Stage-discharge relation affected by ice during severe winters.

ACCURACY.—Stage-discharge relation practically permanent; probably affected by ice part of December and January. Gage read twice daily to hundredths.

COOPERATION.—Base data furnished by United States Engineer Corps.

The following discharge measurement was made by H. S. Shick:

March 14, 1918: Gage height, 8.51 feet; discharge, 2,200 second-feet.

Daily gage height, in feet, of South Fork of Hughes River at Macfarlan, W. Va., for the year ending Sept. 30, 1918.

Day.	Oct.	Nov.	Dec.	Jan.	Feb.	Mar.	Apr.	May.	June.	July	Aug.	Sept.
1	2.60	4.00	3.30	2.45	6.40	3.90	2.86	4.15	3.00	2.40	2.43	2.80
2	2.60	3.78	3.30	2.94	6.40	3.86	2.85	3.70	3.00	2.40	2.53	2.69
3	2.60	3.70	3.10	2.94	4.30	3.41	4.48	3.50	2.68	2.38	2.56	2.59
4	2.60	3.69	3.10	2.94	4.30	3.70	4.46	3.47	2.60	2.38	2.33	2.44
5	2.62	3.55	2.90	2.92	4.20	4.71	4.39	3.26	2.47	2.38	2.30	2.35
6	2.64	3.06	2.90	2.90	4.20	4.38	4.13	3.10	2.50	2.30	2.29	2.30
7	2.64	2.99	2.90	6.10	7.60	3.97	3.15	3.08	2.60	3.30	2.20	2.39
8	2.56	2.93	2.80	6.45	7.88	3.71	7.25	3.06	2.72	3.29	2.27	2.29
9	2.55	2.90	2.90	4.95	11.17	3.53	6.98	3.19	3.02	2.71	3.40	2.22
10	2.56	2.86	2.90	3.60	8.45	3.71	4.80	3.06	2.90	2.40	3.40	2.20
11	2.58	2.80	2.90	3.70	4.63	4.33	7.70	3.03	2.60	2.40	3.28	2.18
12	2.57	2.70	2.90	3.70	5.10	5.09	8.25	3.00	2.56	2.42	3.26	2.14
13	2.70	2.70	2.90	3.70	5.15	20.17	5.81	3.55	2.50	2.50	3.28	2.15
14	2.70	2.70	2.90	3.70	5.65	12.75	4.67	4.78	2.55	2.30	3.35	2.08
15	2.70	2.70	2.90	5.10	6.00	7.10	4.10	4.27	2.60	2.32	3.27	2.05
16	2.82	2.70	2.90	4.50	5.05	5.92	3.78	3.93	2.60	2.35	3.33	2.10
17	2.76	2.60	2.90	4.40	4.26	4.10	3.66	3.05	2.60	2.37	3.15	2.25
18	2.67	2.60	2.90	4.[illegible]0	3.87	3.76	3.69	2.86	2.45	2.36	2.20	2.20
19	6.65	2.60	2.90	4.20	3.65	3.48	3.46	2.78	2.30	2.30	2.20	2.20
20	6.63	2.60	2.90	3.20	12.80	3.42	3.35	2.70	2.30	2.34	2.20	2.45
21	5.88	2.00	2.80	3.20	4.36	3.28	5.15	2.69	2.28	2.79	2.20	2.40
22	4.83	2.60	2.80	3.30	4.30	3.24	4.61	2.60	2.30	2.50	2.20	2.02
23	4.32	2.60	3.50	3.30	4.34	3.20	4.10	3.30	2.36	2.42	2.13	2.15
24	4.47	2.60	3.50	3.30	4.47	3.24	3.77	3.28	2.39	2.37	2.10	2.23
25	7.10	2.60	3.73	3.00	4.20	3.78	3.67	8.98	2.46	2.38	2.12	2.35
26	5.45	2.60	3.60	3.09	11.41	3.48	3.75	7.50	2.50	4.80	2.06	2.41
27	5.10	2.60	3.60	3.90	6.55	3.12	4.31	4.30	2.49	3.98	2.10	2.50
28	5.02	2.60	3.43	9.85	4.41	3.20	4.34	4.15	2.59	3.88	2.10	2.50
29	5.35	3.45	3.20	13.40	......	3.04	4.37	3.49	2.53	3.72	2.10	2.29
30	5.55	3.60	3.10	7.60	......	2.97	4.56	3.46	2.41	3.38	2.10	2.20
31	4.70	......	3.10	6.50	......	2.85	......	2.97	......	2.46	2.35	......

HUGHES RIVER AT CISKO, W. VA.

Location.—At Cisko, 1 mile below junction of North and South forks and 6 miles south of Petroleum, Ritchie County.

Drainage area.—453 square miles (measured on topographic maps).

Records available.—May 29, 1915, to September 30, 1918.

Gage.—Vertical and inclined staff on right bank; read by S. J. Enoch.

Discharge measurements.—Made from cable 40 feet below gage or by wading at the same section.

Channel and control.—One channel at all stages; straight for about 150 feet above and 500 feet below cable section. Bed of river is sand, gravel, mud, and boulders; control is probably permanent.

Extremes of stage.—Maximum stage recorded during year, 27.1 feet at 2 a. m. March 14; minimum stage, 2.21 feet August 26 and 27.

1915–1918: Maximum stage recorded, 30.25 feet at 3 p. m. January 22, 1917; minimum, 2.14 feet October 14 and 15, 1916.

Highest known flood previous to installation of gage reached a stage represented by gage height about 30 feet.

Ice.—Stage-discharge relation affected by ice during winters.

Accuracy.—Stage-discharge relation probably permanent; probably affected by ice December, January, and February. Stages of Ohio River at Parkersburg of about 40 feet or more will probably cause backwater at the gage.

Data inadequate for determination of discharge.

Cooperation.—Base data furnished by United States Engineer Corps.

No discharge measurements were made at this station during the year.

Daily gage height, in feet, of Hughes River at Cisko, W. Va., for the year ending Sept. 30, 1918.

Day.	Oct.	Nov.	Dec.	Jan.	Feb.	Mar.	Apr.	May.	June.	July.	Aug.	Sept.
1	2.38	4.97	3.98	3.60	6.62	4.82	3.42	4.83	3.71	2.94	2.80	4.42
2	2.40	4.35	3.78	3.48	6.04	4.59[a]	3.54	4.41	3.56	2.80	3.00	3.85
3	2.34	4.02	3.60	3.41	5.76	4.38	4.98	4.10	3.31	2.76	2.78	3.42
4	2.40	3.79	3.48	3.40	5.25	4.20	5.72	3.93	3.16	2.72	2.71	3.26
5	2.43	3.62	3.38	3.37	5.30	5.80	4.61	3.78	3.03	2.71	2.58	3.03
6	2.36	3.50	3.28	3.33	5.00	5.58	4.13	3.70	2.96	2.65	2.51	2.86
7	2.34	3.40	3.21	5.78	6.84	4.98	4.06	3.54	8.00	2.56	2.44	2.74
8	2.22	3.33	3.18	7.44	13.62	4.56	8.30	6.14	5.20	2.46	2.47	2.70
9	2.42	3.25	3.36	5.50	14.30	4.48	10.10	4.68	4.08	2.44	3.94	2.55
10	2.32	3.18	3.22	4.84	15.55	6.94	6.08	4.06	3.62	2.35	3.72	2.48
11	2.34	3.12	3.08	4.52	8.70	5.40	8.06	3.93	3.36	2.34	4.23	2.56
12	2.48	3.14	2.97	4.50	6.32	4.80	11.62	3.79	3.15	2.35	4.16	2.46
13	2.72	3.00	2.85	4.74	6.62	20.30	6.32	4.94	3.02	3.08	4.11	2.64
14	3.00	3.02	2.78	4.72	6.03	20.95	5.78	6.64	2.98	3.28	3.33	2.46
15	3.15	2.94	2.86	5.01	7.93	9.34	5.01	4.92	2.74	3.00	3.00	2.42
16	3.07	2.94	2.76	6.78	6.56	6.22	4.59	4.26	2.70	2.84	2.88	2.60
17	2.86	2.90	2.81	6.20	5.22	5.04	4.44	3.92	2.68	2.78	2.66	2.66
18	2.78	2.84	2.68	5.28	4.52	4.64	4.48	3.67	2.60	2.78	2.58	2.76
19	6.15	2.95	2.87	4.70	5.50	4.33	4.23	3.50	2.46	2.58	2.56	2.95
20	8.82	2.70	2.71	4.36	15.45	4.10	4.16	3.40	2.98	2.68	2.44	3.00
21	4.96	2.76	3.00	4.11	7.21	3.99	6.00	3.30	2.83	2.64	2.38	3.04
22	4.23	2.76	3.96	3.88	5.29	3.94	5.73	3.21	2.88	3.03	2.34	3.04
23	3.84	2.72	4.34	3.77	4.78	3.84	4.94	3.70	3.39	2.81	2.48	3.06
24	4.35	2.70	4.16	3.71	5.67	3.70	4.50	4.09	3.30	2.64	2.38	2.92
25	9.73	2.66	4.68	3.75	6.48	4.13	4.58	5.50	3.10	2.56	2.24	2.82
26	7.20	2.74	5.71	3.69	12.99	4.30	4.70	6.80	3.12	3.60	2.21	2.74
27	5.30	2.74	4.72	4.08	6.89	4.06	4.93	5.15	3.07	3.72	2.21	2.63
28	6.20	2.86	4.52	7.42	5.36	3.76	5.28	4.64	3.21	3.31	2.28	2.54
29	5.38	3.07	4.12	15.88	....	3.62	4.82	4.28	3.13	3.00	2.33	2.50
30	7.65	3.92	3.84	10.05	....	3.52	4.97	3.88	2.95	2.80	2.42	2.52
31	6.70	....	3.78	7.80	....	3.46	....	3.92	....	2.69	4.22	....

HOCKING RIVER BASIN.

HOCKING RIVER AT ATHENS, OHIO.

LOCATION.—At single-span highway bridge at Mill Street, three-fourths of a mile from business section of Athens, Athens County. Margaret Creek enters on right 3½ miles above station.

DRAINAGE AREA.—944 square miles (measured on topographic maps).

RECORDS AVAILABLE.—May 3, 1915, to September 30, 1918.

GAGE.—Vertical and inclined staff at downstream end of right abutment; read by W. A. Casley.

DISCHARGE MEASUREMENTS.—Made from bridge or by wading.

CHANNEL AND CONTROL.—Channel straight about 700 feet above and below station. Left bank overflows at gage height 17 feet and water passes around bridge. Bed of stream rocky with sand deposits near both banks. Ruins of old milldam 300 feet below gage act as control. Stage-discharge relation will change as dam decays.

EXTREMES OF STAGE.—Maximum stage recorded during year, 17.9 feet at 5 p. m., March 14; minimum stage, 2.65 feet August 22 and 23.

1915–1918: Maximum stage recorded, 17.9 feet at 5 p. m. December 18, 1915, and 5 p. m. March 14, 1918 (discharge, 12,600 second-feet); minimum stage 2.65 feet August 22 and 23, 1918.

Highest flood known reached a stage represented by gage height about 26 feet.

ICE.—Stage-discharge relation probably not materially affected by ice except during extremely cold weather.

ACCURACY.—Stage-discharge relation practically permanent; affected by ice part of December and January. Gage read to half-tenths twice daily.

COOPERATION.—Base data furnished by United States Engineer Corps.

No discharge measurements were made at this station during the year.

Daily gage height, in feet, of Hocking River at Athens, Ohio, for the year ending Sept. 30, 1918.

Day.	Oct.	Nov.	Dec.	Jan.	Feb.	Mar.	Apr.	May.	June.	July.	Aug.	Sept.
1	2.90	3.55	3.10	3.57	3.70	5.53	3.60	3.93	3.43	3.00	2.95	3.23
2	2.93	3.30	3.10	3.47	3.50	4.93	3.75	3.83	3.33	3.10	2.83	3.05
3	2.95	3.17	3.05	3.67	3.45	4.63	5.30	3.67	3.25	3.00	2.80	2.95
4	2.97	3.10	3.05	....	3.45	4.47	5.15	3.57	3.23	2.95	2.77	2.90
5	3.07	3.07	3.00	....	3.45	4.67	4.47	3.53	3.17	2.90	2.75	2.83
6	3.00	3.05	2.95	3.53	3.55	4.67	4.15	3.45	3.15	2.85	2.75	2.80
7	3.00	3.00	2.95	4.87	4.13	4.53	3.93	3.43	4.15	2.83	2.75	2.85
8	3.00	3.00	3.00	5.33	5.65	4.37	4.15	3.80	3.50	2.80	2.73	2.80
9	3.15	2.97	2.90	4.45	11.75	4.30	4.43	3.80	3.23	2.80	2.70	2.80
10	3.17	2.95	2.85	4.27	14.45	5.03	4.17	3.60	3.13	2.75	2.77	2.75
11	3.05	2.95	2.85	4.07	14.55	4.60	5.33	3.45	3.37	2.75	2.87	2.75
12	2.95	2.95	2.85	4.00	14.55	4.47	9.50	4.65	3.17	2.83	2.90	2.77
13	2.90	2.95	2.80	3.95	15.40	12.35	7.55	8.35	3.07	2.85	3.25	2.80
14	2.95	2.95	2.80	3.95	14.10	17.62	5.80	11.25	3.00	2.85	3.10	2.83
15	2.95	3.03	2.80	3.97	11.75	14.95	4.80	6.45	3.00	2.85	2.87	2.85
16	2.90	3.05	2.80	3.93	8.25	8.20	4.40	4.87	3.05	2.80	2.77	2.80
17	2.90	2.95	2.80	3.80	5.97	5.70	4.30	4.27	3.15	2.80	2.75	2.97
18	2.85	2.90	2.80	3.70	4.93	4.97	4.45	4.00	2.97	2.80	2.70	3.10
19	3.85	2.90	2.87	3.67	6.63	4.65	4.10	3.87	2.95	2.85	2.70	2.97
20	3.50	2.90	2.93	3.55	14.00	4.37	3.93	3.90	2.90	2.80	2.70	3.00
21	3.25	2.90	3.05	3.55	9.35	4.33	4.45	5.03	2.90	2.80	2.70	3.00
22	3.07	2.90	3.25	3.45	6.35	4.23	5.97	4.65	2.90	2.75	2.65	2.95
23	3.00	2.95	3.65	3.45	5.00	4.08	5.03	3.75	2.85	2.75	2.67	2.87
24	3.03	2.95	3.55	3.40	4.77	3.93	4.55	3.67	2.85	2.75	2.75	2.83
25	3.43	2.90	3.13	3.40	4.53	4.03	4.43	4.05	2.90	2.90	2.75	2.75
26	3.40	2.90	4.60	3.35	12.65	4.03	4.20	4.47	2.95	3.30	2.80	2.75
27	3.27	2.90	3.90	3.35	10.98	3.90	4.23	3.77	2.95	3.35	2.85	2.75
28	3.33	2.93	3.60	3.[illegible]0	7.75	3.80	4.10	4.00	2.90	3.20	2.87	2.75
29	3.47	3.00	3.83	4.20	....	3.73	4.07	3.75	2.85	3.10	2.95	2.70
30	4.10	3.03	3.73	4.05	....	3.65	4.10	3.70	2.85	3.13	2.93	2.70
31	4.23	....	3.67	3.80	....	3.60	....	3.47	....	3.07	2.90	....

KANAWHA RIVER BASIN.

NEW RIVER AT EGGLESTON, VA.

Location.—At highway bridge at Eggleston, Giles County.

Drainage area.—2,920 square miles.

Records available.—October 1, 1914, to September 30, 1918.

Gage.—Chain gage attached to downstream side of bridge; read by J. A. Bishop.

Discharge measurements.—Made from upstream side of bridge.

Channel and control.—Stream bed composed of rock covered with silt. Primary control is rock ledge about 1¼ miles below gage; permanent.

Extremes of discharge.—Maximum stage recorded during year, 8.93 feet at 8 a. m. June 26 (discharge, 16,300 second-feet); minimum stage recorded, 2.58 feet at 8 a. m. December 9 and 8 a. m. and 5 p. m. December 10 (discharge, 815 second-feet); minimum discharge may have occurred during periods of ice effect in December and January.

1914–1918: Maximum stage recorded, 39.5 feet July 16, 1916 (discharge, about 152,000 second-feet); minimum stage recorded, 2.37 feet August 29, 1917 (discharge, 652 second-feet). The flood of 1878 reached a stage of about 40 feet on present gage.

Ice.—Stage-discharge relation affected by ice during severe winters.

Accuracy.—Stage-discharge relation practically permanent except as affected by ice. Rating curve well defined between 1,200 and 45,000 second-feet; extended beyond these limits. Gage read to hundredths twice daily. Daily discharge ascertained by applying mean daily gage height to rating table except for periods of ice effect. Records good except those for periods of ice effect, which are poor.

Discharge measurements of New River at Eggleston, Va., during the year ending Sept. 30, 1918.

Date.	Made by—	Gage height.	Discharge.
		Feet.	*Sec.-ft.*
Jan. 22	B. L. Hopkins	[a] 3.20	1,160
May 27	Hopkins and Fiedler	4.62	3,810

[a] Stage-discharge relation affected by ice.

Daily discharge, in second-feet, of New River at Eggleston, Va., for the year ending Sept. 30, 1918.

Day.	Oct.	Nov.	Dec.	Jan.	Feb.	Mar.	Apr.	May.	June.	July.	Aug.	Sept.
1	980	2,120	1,400		8,580	4,120	2,940	5,690	3,710	4,120	7,480	1,970
2	980	1,740	1,340		8,020	3,310	3,510	5,930	3,310	4,550	7,210	1,970
3	1,340	1,740	1,340		5,450	2,440	3,310	4,990	1,970	3,710	4,330	1,400
4	1,220	1,600	1,600		3,120	2,120	3,310	4,770	1,970	3,510	3,710	1,900
5	1,170	1,220	1,530			2,600	2,940	4,550	2,600	2,600	2,280	1,900
6	1,220	1,170	1,460		2,800	3,510	2,770	2,770	1,970	3,310	1,970	2,600
7	1,120	1,400	1,400			3,910	2,280	3,310	2,120	2,600	2,280	3,710
8	980	1,400	1,400			3,510	2,440	3,510	2,440	1,820	1,820	3,910
9	1,070	1,400	815		8,300	3,510	3,710	3,310	1,820	1,600	1,970	5,930
10	1,340	1,460	815		10,000	3,510	10,000	4,120	1,740	1,670	2,120	5,450
11	3,120	1,460			11,600	2,770	8,300	4,550	1,970	1,970	2,280	3,310
12	1,340	1,170			10,400	2,770	7,750	3,910	1,970	2,440	1,820	2,600
13	1,280	937			9,160	2,940	7,750	3,310	1,970	2,280	1,900	2,600
14	1,340	1,400			8,300	3,910	5,450	3,310	1,970	2,280	1,970	1,970
15	1,280	1,400		1,100	7,750	3,120	4,330	5,220	1,820	1,740	1,970	1,900
16	1,400	1,460			6,430	2,600	5,690	4,990	1,400	1,970	2,280	1,820
17	1,400	1,460			4,990	2,770	4,770	5,220	1,400	1,900	2,600	2,120
18	1,400	1,280			4,550	1,970	3,510	4,550	3,120	2,280	2,940	2,280
19	1,280	1,280			5,450	1,740	3,710	4,120	7,750	2,600	3,910	3,310
20	1,340	1,340			4,550	2,440	4,770	4,120	5,690	2,770	4,550	2,600
21	1,340	1,400	1,400		3,510	3,710	6,180	4,770	4,550	2,280	3,510	2,280
22	3,510	1,400			2,770	7,750	10,400	5,690	3,510	1,740	2,770	2,600
23	2,940	1,670			4,770	6,180	7,750	5,930	4,550	2,120	1,900	2,440
24	1,900	1,530			3,710	5,450	5,690	5,450	3,510	2,120	2,770	2,280
25	1,400	1,530			2,600	5,450	5,450	5,220	4,330	2,120	2,280	1,900
26	1,400	1,070			2,440	7,210	4,770	4,550	15,900	4,120	2,120	2,120
27	1,280	1,400			3,120	6,430	4,770	4,120	8,300	3,910	2,440	1,970
28	1,280	1,280			3,510	5,930	4,330	4,330	6,430	3,120	2,120	1,600
29	1,400	1,400			......	5,690	3,510	4,120	4,770	3,120	2,120	1,900
30	1,970	1,400		13,700	......	4,550	4,770	3,510	4,550	3,510	1,900	1,600
31	4,550	......		10,400	......	4,120	......	3,710	......	4,120	1,670	......

NOTE.—Discharge, Dec. 11 to Jan. 29, and Feb. 5-8, estimated because of ice, by means of observer's notes, weather records, one current-meter measurement, and comparison with gage-height record for New River at Radford where ice effect was not so pronounced.

Monthly discharge of New River at Eggleston, Va., for the year ending Sept. 30, 1918.

[Drainage area, 2,920 square miles.]

Month.	Discharge in second-feet.				Run-off in inches.
	Maximum.	Minimum.	Mean.	Per square mile.	
October	4,550	980	1,600	0.548	0.63
November	2,120	937	1,420	.486	.54
December	......	......	1,370	.469	.54
January	13,700	......	1,810	.620	.72
February	11,600	2,440	5,510	1.89	1.97
March	7,750	1,740	3,940	1.35	1.56
April	10,400	2,280	5,030	1.72	1.92
May	5,930	2,770	4,440	1.52	1.75
June	15,900	1,400	3,770	1.29	1.44
July	4,550	1,660	2,710	.928	1.07
August	7,480	1,670	2,810	.962	1.11
September	5,930	1,400	2,530	.866	.97
The year	15,900	......	3,060	1.05	14.22

KANAWHA RIVER AT LOCK 2, MONTGOMERY, W. VA.

Location.—At Lock 2, three-fourths of a mile below Chesapeake & Ohio Railway station at Montgomery, Fayette County. Morris Creek enters on left 300 feet below the gage.

Drainage area.—8,470 square miles.

Records available.—June 22, 1915, to September 30, 1918. Upper and lower gages at the lock have been read since December, 1887, under the direction of the Corps of Engineers, United States Army.

Gage.—Upper gage at lock, vertical and inclined staff on right bank, short distance above upper lock gates; vertical section fastened to land wall of lock, inclined section at upstream end of paved slope; read by George Meyers, lockmaster. A chain gage fastened to downstream handrail near center of toll bridge at Montgomery is used in referring water surface at bridge when making discharge measurements.

Discharge measurements.—Made from bridge at Montgomery or by wading on the crest of the dam.

Channel and control.—One channel at all stages; straight for 300 feet above and 800 feet below bridge. Bed of river composed of rock, sand, and mud. The dam at Lock No. 2 is control for all stages, as there is a fall of about 2 feet at the dam at the maximum stage. Except for the leakage through the dam and lock, point of zero flow is at lowest point in crest of dam, which is 17.9 feet above zero of upper gage.

Extremes of discharge.—Maximum stage recorded during year, 37.0 feet at 5 p. m. March 13 (discharge, 140,000 second-feet); minimum stage, 18.20 feet at 5 p. m. October 12 (discharge, 1,030 second-feet).

Highest stage recorded occurred May 23, 1901, at 6 a. m.: upper gage 49.65 feet. lower gage 47.70 feet (discharge, about 250,000 second-feet).

Ice.—Stage-discharge relation not affected by ice.

Leakage.—At about gage height 19 feet on upper gage, leakage through the dam amounts to about 500 second-feet. Leakage through the lock gates amounts to about 110 and 260 second-feet, depending upon which of the two gates is closed

Accuracy.—Stage-discharge relation practically permanent except as may be affected by change in leakage through lock and dam; not affected by ice. Rating curve well defined throughout. Gage read twice daily to hundredths. Daily discharge ascertained by applying mean daily gage height to rating table, which is adjusted for leakage through dam and lock gates. Records good.

Cooperation.—Base data furnished by United States Engineer Corps.

No discharge measurements were made at this station during the year.

Daily discharge, in second-feet, of Kanawha River at Lock 2, Montgomery, W. Va., for the year ending Sept. 30, 1918.

Day.	Oct.	Nov.	Dec.	Jan.	Feb.	Mar.	Apr.	May.	June.	July.	Aug.	Sept.
1	1,970	12,800	5,800	3,270	30,800	20,500	10,300	16,800	8,800	20,500	7,400	3,430
2	1,970	9,800	5,900	3,950	23,000	18,800	9,300	15,300	7,300	16,800	9,300	5,300
3	1,900	5,800	4,800	3,270	18,800	18,300	7,800	14,800	6,300	14,300	11,300	5,800
4	1,640	5,300	4,580	2,860	16,800	14,900	8,800	13,300	5,800	11,300	7,300	3,950
5	1,640	4,350	4,130	2,880	14,300	18,300	8,800	12,300	4,150	9,300	5,300	2,860
6	1,770	3,950	3,590	2,740	11,800	28,600	8,800	10,800	4,800	7,300	4,350	2,990
7	1,770	3,270	3,430	5,800	12,300	45,900	7,800	9,800	4,900	4,800	3,590	2,990
8	1,580	2,740	3,430	5,800	17,800	48,400	11,800	9,300	4,800	5,800	2,740	4,800
9	1,580	2,740	3,590	4,800	23,600	30,100	39,000	10,800	5,300	4,800	2,740	5,300
10	1,510	2,990	2,990	4,580	55,400	22,400	51,000	9,800	4,800	3,580	2,640	7,800
11	1,450	2,740	2,530	3,590	59,800	19,900	41,500	10,800	4,150	3,270	3,270	8,800
12	1,270	2,640	2,270	5,300	51,900	16,800	37,300	10,800	3,270	3,950	2,990	5,800
13	1,900	2,530	2,120	11,300	45,800	75,800	33,100	10,800	3,590	3,770	3,130	4,580
14	2,270	2,530	2,190	10,800	47,600	120,000	27,800	19,900	2,990	4,150	4,800	3,770
15	1,970	2,350	2,270	8,800	44,100	82,900	32,300	21,700	2,530	3,770	4,350	3,590
16	2,190	2,530	2,530	8,300	45,000	40,700	30,800	19,400	2,640	3,770	4,350	3,270
17	2,190	2,530	2,530	7,900	37,300	25,700	29,300	16,300	2,740	3,270	4,150	3,270
18	1,640	2,440	2,350	7,300	23,600	18,800	25,000	13,900	3,130	3,270	5,300	2,860
19	1,510	2,530	2,120	4,800	16,800	14,800	21,100	11,300	3,950	3,950	4,580	8,300
20	2,040	2,350	2,270	4,580	17,800	12,300	17,800	10,800	10,400	6,800	9,300	12,800
21	5,300	2,350	2,350	3,770	39,800	11,300	27,100	10,300	11,800	7,800	7,800	10,400
22	4,150	2,120	2,860	3,590	31,600	15,800	37,300	12,800	7,800	6,300	6,300	9,800
23	4,350	2,040	2,740	3,590	20,500	27,800	35,600	14,800	6,300	4,800	4,150	8,300
24	3,590	2,440	2,440	3,130	17,800	20,500	26,400	14,300	8,800	3,590	3,950	6,800
25	5,300	2,440	3,270	3,270	16,300	20,500	21,300	12,800	7,800	3,590	3,270	4,800
26	4,350	2,350	7,800	3,590	21,100	21,700	19,400	15,800	23,600	5,300	3,270	4,580
27	4,580	2,190	8,300	4,800	43,300	19,900	18,800	16,800	64,200	4,350	3,430	3,770
28	6,800	2,190	6,300	50,200	30,100	17,800	24,300	16,300	27,100	6,300	3,270	3,590
29	6,300	3,590	6,300	85,600	....	15,800	22,400	14,300	18,300	6,800	3,590	3,430
30	5,800	5,800	3,950	59,800	....	12,800	17,300	12,300	19,900	5,800	3,770	3,130
31	8,300	....	3,590	42,400	....	11,300	....	9,800	....	6,300	3,590	....

Monthly discharge of Kanawha River at Lock 2, Montgomery, W. Va., for the year ending Sept. 30, 1918.

[Drainage area, 8,470 square miles.]

Month.	Discharge in second-feet.				Run-off in inches.
	Maximum.	Minimum.	Mean.	Per square mile.	
October	8,300	1,270	3,050	0.360	0.42
November	12,800	2,040	3,550	.419	.47
December	8,300	2,120	3,720	.439	.51
January	85,600	2,740	12,100	1.43	1.65
February	59,800	11,800	29,900	3.52	3.66
March	120,000	11,300	28,700	3.39	3.91
April	51,000	7,800	23,700	2.80	3.12
May	21,700	9,300	13,500	1.59	1.83
June	64,200	2,530	9,720	1.15	1.28
July	20,500	3,270	6,430	.759	.88
August	11,300	2,640	4,810	.568	.65
September	12,800	2,860	5,260	.621	.69
The year	120,000	1,270	11,900	1.40	19.07

GREENBRIER RIVER AT ALDERSON, W. VA.

LOCATION.—At reinforced-concrete arch highway bridge at Alderson, Monroe County, half a mile above mouth of Muddy Creek.

DRAINAGE AREA.—1,340 square miles.

RECORDS AVAILABLE.—July 30, 1895, to June 30, 1906; May 10, 1907, to September 30, 1918.

GAGE.—Chain gage attached to downstream side of bridge near center of second span from left side of river; read by W. J. Hancock.

DISCHARGE MEASUREMENTS.—Made from bridge or by wading.

CHANNEL AND CONTROL.—The channel and control are composed of coarse gravel and are practically permanent.

EXTREMES OF STAGE.—Maximum stage recorded during year, 22 feet during night March 13–14; minimum stage recorded, 1.70 feet at 6 p. m. October 10 and 8 a. m. and 6 p. m. October 11.

1895–1918: Maximum stage recorded same as for 1918 above; minimum discharge recorded, 46 second-feet September 30 to October 6, October 17, 24, 27–31, and November 7, 10, 11, 1904 (gage height, 1.40 feet).

ICE.—Stage-discharge relation occasionally affected by ice for short periods during severe winters.

ACCURACY.—Stage-discharge relation changed during year; may have been slightly affected by ice during December and January. New rating curve not fully developed. Gage read to hundredths twice daily. Records excellent.

Discharge measurements of Greenbrier River at Alderson, W. Va., during the year ending Sept. 30, 1918.

[Made by B. L. Hopkins.]

Date.	Gage height.	Discharge.	Date.	Gage height.	Discharge.
	Feet.	*Sec.-ft.*		*Feet.*	*Sec.-ft.*
Feb. 15	6.95	11,800	Apr. 15	6.16	8,840
15	7.15	12,400	May 15	3.97	2,750
16	8.49	16,400	June 23	3.64	2,060
18	4.64	5,330			

Daily gage height, in feet, of Greenbrier River at Alderson, W. Va., for the year ending Sept. 30, 1918.

Day.	Oct.	Nov.	Dec.	Jan.	Feb.	Mar.	Apr.	May.	June.	July.	Aug.	Sept.
1	1.96	3.48	2.16	2.77	3.78	4.70	3.35	4.15	3.21	5.16	3.58	4.36
2	1.90	2.96	2.15	2.66	3.38	5.05	3.26	3.90	3.06	4.72	3.34	3.52
3	1.85	2.70	2.18	2.53	3.25	4.48	3.28	3.72	2.94	4.14	3.03	3.21
4	1.84	2.55	2.15	2.65	3.18	3.90	3.46	3.58	2.85	3.70	2.83	2.96
5	1.82	2.42	2.14	2.58	3.08	4.75	3.41	3.48	2.75	3.39	2.69	2.79
6	1.77	2.34	2.15	2.55	2.98	5.49	3.31	3.38	2.76	3.18	2.57	2.88
7	1.75	2.27	2.13	2.52	2.99	6.40	3.24	3.30	2.84	3.03	2.52	3.16
8	1.72	2.22	2.09	2.49	3.93	6.05	3.36	3.29	2.98	2.90	2.43	3.22
9	1.72	2.18	2.06	2.41	4.10	4.70	7.40	3.34	2.89	2.80	2.37	3.11
10	1.71	2.14	1.92	2.39	6.53	4.28	6.75	3.85	2.79	2.74	2.33	2.96
11	1.70	2.10	1.87	2.44	7.08	4.12	5.85	3.64	2.68	2.66	2.32	2.83
12	1.73	2.06	1.83	2.53	6.70	3.82	5.68	3.50	2.54	2.64	2.37	2.72
13	1.74	2.04	2.06	2.68	7.00	8.02	5.24	3.42	2.45	2.70	2.42	2.64
14	1.74	2.02	2.04	2.70	8.00	18.62	5.00	3.52	2.44	2.72	2.64	2.58
15	1.74	1.99	2.06	2.88	7.12	8.52	6.18	3.80	2.41	2.66	3.13	2.56
16	1.74	1.98	2.10	2.82	8.14	6.36	6.28	3.90	2.38	2.58	2.91	2.65
17	1.74	1.96	2.09	2.81	5.78	5.35	5.58	3.68	2.62	2.63	2.65	2.58
18	1.74	1.94	2.08	2.74	4.58	4.58	5.02	3.50	2.86	2.69	2.64	2.71
19	1.74	1.92	2.03	2.68	4.02	4.18	4.64	3.64	4.00	2.93	2.87	5.73
20	2.04	1.91	2.08	2.61	6.08	3.91	4.35	3.52	4.00	3.60	3.15	4.25
21	2.67	1.90	2.06	2.65	8.30	3.85	6.45	3.62	3.43	3.39	3.16	3.77
22	2.56	1.90	2.05	2.57	5.45	3.88	7.05	3.65	3.25	3.03	2.89	3.73
23	2.34	1.88	2.17	2.54	4.42	5.30	5.55	4.00	3.62	2.82	2.73	3.56
24	2.25	1.87	2.25	2.49	3.98	4.58	4.82	3.95	3.36	2.68	2.64	3.33
25	2.22	1.84	2.40	2.48	3.88	4.46	4.55	3.76	3.15	2.59	2.89	3.14
26	2.16	1.84	2.56	2.47	7.28	4.38	4.55	5.20	9.00	2.52	3.03	3.00
27	2.18	1.84	2.83	2.62	7.92	4.15	5.44	5.22	6.50	2.99	2.81	2.88
28	2.27	1.83	3.01	5.36	5.26	3.92	5.66	4.34	4.68	2.76	2.68	2.79
29	2.40	1.82	2.94	6.63	......	3.72	4.86	3.90	4.76	2.61	2.64	2.71
30	3.68	1.95	2.86	5.18	......	3.56	4.38	3.70	6.33	2.58	2.62	2.64
31	4.40	......	2.31	4.26	......	3.42	......	3.45	......	3.10	2.59	......

LITTLE COAL RIVER AT McCORKLE, W. VA.

LOCATION.—At McCorkle, Lincoln County, on Coal River branch of Chesapeake & Ohio Railway. Cobb Creek enters river on left 400 feet below station.

DRAINAGE AREA.—375 square miles (measured on topographic maps).

RECORDS AVAILABLE.—July 23, 1915, to September 30, 1918.

GAGE.—Vertical and inclined staff on left bank just below McCorkle Hotel; read by F. M. Priestly.

DISCHARGE MEASUREMENTS.—Made from cable 40 feet above inclined section of gage or by wading.

CHANNEL AND CONTROL.—One channel at all stages; slightly curved above and below cable section. Bed of stream composed of loose sand; but control is probably fairly permanent. Flow of Cobb Creek affects stage at gage and should be included in station.

EXTREMES OF STAGE.—Maximum stage recorded during year, 24.0 feet at 6 p. m., January 28; minimum stage, 1.69 feet July 29. Highest known flood August 9, 1916, reached a stage of 28.57 feet (discharge, roughly, 24,000 second-feet).

ICE.—Stage-discharge relation affected by ice during severe winters.

ACCURACY.—Changes in stage-discharge relation may be caused by floods; ice effect during part of December and January. Gage read to half-tenths twice daily.

COOPERATION.—Base data furnished by United States Engineer Corps.

No discharge measurements were made at this station during the year.

Daily gage height, in feet, of Little Coal River at McCorkle, W. Va., for the year ending Sept. 30, 1918.

Day.	Oct.	Nov.	Dec.	Jan.	Feb.	Mar.	Apr.	May.	June.	July.	Aug.	Sept.
1	2.70	3.02	3.28	3.15	5.54	3.72	3.28	3.39	2.62	2.72	2.74	2.74
2	2.52	2.92	3.38	3.10	4.54	3.52	3.22	3.29	2.54	2.62	2.54	2.69
3	2.35	2.88	3.30	3.00	3.96	3.36	3.24	3.14	3.06	2.54	2.36	2.56
4	2.40	2.80	3.15	3.00	3.66	3.32	3.49	3.09	2.76	2.44	2.29	2.52
5	2.42	2.72	3.02	3.00	3.56	3.94	3.44	3.06	2.62	2.42	2.24	2.34
6	2.38	2.65	2.95	3.12	5.74	4.84	3.36	3.00	2.59	2.28	2.19	2.34
7	2.30	2.60	2.86	5.80	5.19	5.94	3.42	2.99	3.66	2.22	2.14	2.36
8	2.31	2.56	2.85	5.30	3.39	5.36	7.53	3.06	3.64	2.19	2.16	2.49
9	2.22	2.52	2.85	4.25	3.92	4.49	9.20	3.49	3.06	2.16	2.06	2.42
10	2.15	2.49	2.68	3.65	4.86	4.06	6.02	3.29	2.86	2.14	2.24	2.32
11	2.20	2.45	2.68	3.42	4.49	3.69	4.86	3.22	2.69	2.16	2.64	2.22
12	2.42	2.45	2.70	3.55	4.14	3.56	5.69	3.22	2.62	2.24	2.39	2.17
13	2.35	2.45	2.78	4.90	3.82	9.57	5.86	3.24	2.44	2.29	2.44	2.16
14	2.40	2.41	2.78	5.25	3.59	8.45	4.96	3.82	2.36	2.14	2.34	2.10
15	2.34	2.39	2.72	6.02	3.64	6.42	4.42	3.86	2.34	2.06	2.39	2.09
16	2.28	2.32	2.68	5.62	3.64	4.72	4.06	3.62	2.29	1.99	2.42	2.04
17	2.22	2.30	2.65	6.05	3.59	4.12	3.86	3.38	2.24	2.09	2.36	2.09
18	2.20	2.30	2.65	5.00	3.44	3.84	3.74	3.19	2.46	2.09	2.34	2.14
19	3.62	2.28	2.60	4.45	3.36	3.59	3.56	3.12	2.32	2.16	2.34	2.26
20	3.95	2.25	2.60	4.00	5.52	3.46	3.52	3.34	2.24	2.29	2.29	2.46
21	3.35	2.24	2.62	3.70	5.44	3.39	3.64	5.09	2.16	2.14	2.26	2.46
22	3.05	2.20	2.70	3.60	4.42	3.39	4.09	4.02	2.14	2.12	2.22	2.42
23	2.92	2.20	2.68	3.60	3.99	3.34	3.96	3.59	2.12	2.02	2.19	2.39
24	3.05	2.20	2.78	3.55	3.74	3.32	4.16	3.32	2.06	1.89	2.14	2.34
25	3.25	2.25	3.02	3.58	3.62	4.86	3.99	3.12	2.09	1.84	2.14	2.26
26	3.38	2.25	2.88	3.52	3.99	4.92	3.86	3.09	3.89	1.82	2.06	2.22
27	3.35	2.28	3.15	4.90	4.19	4.26	3.79	3.04	3.29	1.76	1.94	2.16
28	3.18	2.45	3.50	20.27	3.96	3.86	3.62	2.94	2.96	1.74	2.02	2.14
29	3.02	2.72	3.32	9.00	......	3.64	3.54	2.94	2.76	1.69	2.00	2.14
30	3.15	3.20	3.30	6.95	......	3.46	3.46	2.82	2.69	1.92	1.96	2.14
31	3.18	......	3.28	6.40	......	3.36	......	2.74	......	2.79	1.99	......

RACCOON CREEK BASIN.

RACCOON CREEK AT ADAMSVILLE, OHIO.

Location.—About 200 feet above covered highway bridge at Adamsville, Gallia County, 5 miles southwest of Hocking Valley Railroad station at Bidwell. Indian Creek enters on right 1½ miles above station.

Drainage area.—537 square miles (measured on topographic maps).

Records available.—June 25, 1915, to September 30, 1918.

Gage.—Vertical and inclined staff on left bank 200 feet above bridge; read by Irene Call.

Discharge measurements.—Made from covered highway bridge or by wading.

Channel and control.—Straight for about 500 feet above and 600 feet below bridge. Bed of stream composed of mud, sand, and gravel. Principal control at ruins of old milldam, 1,200 feet below bridge; probably permanent.

Extremes of stage.—Maximum stage recorded during year, 18.49 feet at 5 p. m. March 15; minimum stage, 1.81 feet at 7 a. m., September 1.

1915–1918: Maximum stage recorded that of March 15, 1918; minimum stage, 1.75 feet at 7 a. m. September 26, 1917 (discharge, 18 second-feet).

High-water marks indicate maximum stage of about 24.5 feet.

Ice.—Stage-discharge relation affected by ice during severe winters.

Accuracy.—Stage-discharge relation practically permanent; affected by ice part of December and January. Gage read to hundredths twice daily.

Cooperation.—Base data furnished by United States Engineer Corps.

No discharge measurements were made at this station during the year.

Daily gage height, in feet, of Raccoon Creek at Adamsville, Ohio, for the year ending Sept. 30, 1918.

Day.	Oct.	Nov.	Dec.	Jan.	Feb.	Mar.	Apr.	May.	June.	July.	Aug.	Sept.
1	1.98	3.55	2.26	2.58	4.66	10.71	3.12	4.88	3.06	3.58	2.24	1.82
2	2.05	3.06	2.26	2.46	4.15	7.20	3.14	4.28	2.72	3.60	2.18	2.02
3	2.08	2.70	2.35	2.38	4.06	5.88	3.38	3.96	2.66	3.75	2.12	2.13
4	2.08	2.55	2.30	2.35	4.10	5.72	6.69	3.58	2.55	3.42	2.09	2.11
5	2.10	2.49	2.32	2.33	3.55	5.75	6.28	3.67	2.56	3.02	2.16	2.13
6	2.26	2.50	2.30	2.75	3.55	5.36	5.12	3.26	2.68	2.37	2.18	2.06
7	2.00	2.40	2.25	5.37	5.98	5.28	4.02	3.12	2.88	2.18	2.14	2.32
8	1.98	2.28	2.30	5.02	7.28	4.76	5.00	3.06	3.03	2.26	2.11	2.30
9	2.08	2.35	2.13	4.78	12.02	4.45	5.13	3.04	2.90	2.23	2.12	2.08
10	2.03	2.30	2.10	4.48	13.86	5.58	5.43	3.04	2.61	2.10	2.16	2.04
11	2.00	2.40	2.10	4.15	14.96	5.18	5.98	3.04	2.36	2.30	2.09	2.38
12	2.10	2.48	2.12	3.00	16.00	5.41	9.96	3.66	2.23	2.18	2.10	2.41
13	2.11	2.42	2.08	2.80	15.66	13.55	9.97	7.53	2.55	2.12	2.08	2.10
14	2.12	2.28	2.06	2.68	15.61	17.51	4.45	11.26	2.40	2.08	2.04	2.02
15	2.00	2.21	....	2.58	15.24	18.47	4.88	11.00	2.24	2.14	2.05	2.00
16	2.08	2.28	....	2.61	14.28	17.88	5.78	9.78	2.22	2.10	2.13	1.98
17	2.09	2.25	2.12	2.45	11.71	16.05	6.48	8.21	2.33	2.10	2.06	2.55
18	2.18	2.15	2.10	2.38	7.06	8.25	5.11	6.95	2.60	2.22	2.04	2.36
19	4.66	2.10	2.13	2.30	6.96	5.78	4.80	5.55	2.26	2.18	2.04	2.11
20	3.86	2.12	2.10	2.50	12.70	5.11	4.68	4.68	2.18	2.20	2.06	2.06
21	3.05	2.23	2.04	2.76	13.58	4.70	5.28	4.92	2.23	2.10	2.03	2.04
22	2.78	2.08	2.03	2.35	14.25	4.40	6.10	4.40	2.09	2.06	2.02	2.05
23	2.48	2.08	2.46	2.24	12.26	4.29	6.16	4.16	2.05	2.13	2.00	2.07
24	2.48	2.28	2.12	2.20	7.76	3.66	5.35	3.80	2.22	2.15	2.02	2.05
25	2.13	2.08	2.10	2.20	6.12	4.00	5.15	3.28	4.62	2.11	2.04	1.99
26	2.40	2.10	2.18	2.32	12.35	3.98	5.46	3.08	7.16	2.26	2.05	1.94
27	2.60	2.32	3.44	3.60	11.81	3.75	5.62	3.18	5.28	2.20	1.97	1.90
28	2.66	2.13	3.50	4.23	11.56	3.58	5.83	3.06	3.83	2.56	....	1.84
29	2.60	2.18	3.38	5.32	....	3.38	5.58	3.08	3.53	2.18	....	2.02
30	4.06	2.25	3.15	5.56	....	3.30	5.23	3.72	3.69	2.18	2.00	2.01
31	3.90	....	2.62	5.00	....	3.18	....	3.55	....	2.18	2.03	....

Note.—Gage not read Dec. 15–16; gage readings in error Aug. 28–29.

GUYANDOT RIVER BASIN.

GUYANDOT RIVER AT WILBER, W. VA.

Location.—At site of Hutchinson Lumber Co.'s suspension bridge at Wilber, three fourths mile below Manbar, Logan County. Rich Creek enters river on left 600 feet above station.

Drainage area.—791 square miles (measured on map of West Virginia: scale, 1:500,000).

Records available.—July 13, 1915, to September 30, 1918.

Gage.—Vertical and inclined staff on right bank; read by Allie Smith. Vertical section fastened to downstream corner of right timber crib pier; inclined section is about 10 feet downstream. Gage washed out by flood on January 28, 1918; replaced March 6.

Discharge measurements.—Made from cable installed between towers of former bridge in February, 1916, or by wading.

Channel and control.—Channel straight for about 1,000 feet above and 500 feet below station. Bed of river composed of solid rock, boulders, and mud; control probably permanent. Point of zero flow, gage height 0.00 ± 0.5 foot.

Extremes of stage.—Maximum stage recorded during year, 24.8 feet at 4 p. m. January 28; minimum stage, 1.60 feet October 9 and 11.

1915–1918: Maximum stage recorded that of January 28, 1918; minimum stage 1.10 feet September 26, 1917.

Ice.—Stage-discharge relation not affected by ice except during severe winters.

Accuracy.—Stage-discharge relation probably permanent; affected by ice during part of December and January. Gage read to tenths twice daily; records fair.

Cooperation.—Base data furnished by United States Engineer Corps.

No discharge measurements were made at this station during the year.

Daily gage height, in feet, of Guyandot River at Wilber, W. Va., for the year ending Sept. 30, 1918.

Day.	Oct.	Nov.	Dec.	Jan.	Feb.	Mar.	Apr.	May.	June.	July.	Aug.	Sept.
1	3.02	3.40	3.40	3.80	......	......	4.22	4.55	3.20	7.35	3.80	4.90
2	2.80	3.52	3.90	3.80	......	......	4.80	4.25	3.00	6.70	3.60	5.70
3	2.60	3.30	3.82	3.80	......	......	4.70	4.05	3.05	6.40	3.50	5.40
4	2.20	3.02	3.80	3.80	......	......	4.38	3.80	3.20	5.85	4.70	4.80
5	2.02	2.85	3.60	3.80	......	......	4.10	3.75	3.40	5.25	4.40	4.05
6	2.00	2.52	3.40	3.80	......	7.61	3.70	3.70	2.70	4.75	4.10	3.80
7	1.80	2.45	3.00	4.42	......	13.61	4.00	3.65	2.60	4.35	3.80	3.40
8	1.65	2.42	3.00	4.45	......	8.60	8.10	3.55	2.70	3.95	3.80	3.40
9	1.60	2.40	3.00	4.60	......	7.20	10.95	3.45	2.60	3.90	3.90	3.40
10	1.80	2.22	2.85	4.65	......	7.15	8.00	3.40	2.60	3.60	4.30	3.40
11	1.60	2.10	2.80	4.80	......	5.55	6.25	3.55	2.60	3.52	4.50	3.60
12	2.02	2.00	2.80	5.00	......	5.48	6.10	3.70	2.60	3.20	4.30	3.70
13	1.90	2.00	2.65	5.00	......	6.45	6.15	3.55	2.60	3.00	4.05	3.60
14	1.80	1.82	2.60	5.00	......	8.51	5.75	3.30	2.60	2.90	4.00	3.40
15	2.05	1.80	2.45	5.00	......	7.75	5.50	3.20	2.60	2.90	3.90	3.40
16	2.02	2.02	2.25	5.00	......	6.60	5.10	3.15	2.60	2.90	4.05	3.20
17	1.92	2.00	2.25	5.00	......	5.90	4.95	3.20	2.61	2.90	3.90	3.20
18	1.82	1.85	2.10	4.60	......	5.56	4.80	3.20	2.70	2.90	3.70	3.10
19	2.20	1.80	2.05	4.50	......	5.45	4.55	3.25	2.60	2.90	3.70	3.00
20	2.42	1.65	2.00	4.80	......	5.25	4.60	3.85	2.40	2.90	4.30	3.00
21	2.80	1.65	1.85	4.20	......	4.40	7.90	4.40	2.40	2.90	4.90	3.25
22	3.00	1.80	1.82	4.00	......	5.10	8.10	4.82	2.20	3.05	4.80	3.30
23	3.02	1.80	2.00	3.80	......	5.65	7.32	5.30	2.20	3.30	4.60	3.15
24	2.92	1.85	2.18	3.40	......	6.10	7.62	4.90	2.15	3.70	4.40	3.00
25	2.82	2.00	2.85	3.40	......	7.85	6.22	4.45	2.10	4.10	4.15	2.95
26	2.90	2.05	2.72	3.40	......	7.65	6.55	3.90	2.10	4.35	4.00	2.90
27	3.60	2.05	2.60	5.40	......	7.38	5.65	3.80	3.40	4.70	4.50	2.90
28	3.70	2.08	2.40	21.40	......	6.70	4.88	3.60	4.85	5.00	5.15	2.90
29	3.20	2.35	3.00	......	......	6.11	4.75	3.45	8.60	4.75	5.25	2.90
30	3.10	3.00	3.80	......	......	5.10	4.65	3.40	7.90	4.65	4.70	2.90
31	3.20	......	3.80	......	......	4.90	......	3.30	......	4.15	4.40	......

GUYANDOT RIVER AT BRANCHLAND, W. VA.

Location.—At highway bridge at Branchland, Lincoln County. Fourmile Creek enters river on left 20 feet above bridge.

Drainage area.—1,230 square miles (measured on map of West Virginia; scale, 1:500,000).

Records available.—July 8, 1915, to September 30, 1918.

Gage.—Chain gage fastened to handrail on upstream side of bridge near center of main span; read by John A. Broaddus.

Discharge measurements.—Made from bridge or by wading.

Channel and control.—Bed of stream is composed of rock, gravel, sand, and mud and is fairly permanent; character of control not determined.

Extremes of stage.—Maximum stage recorded during year, 39.24 feet at 7.20 a. m. January 29; minimum stage, 2.72 feet at 7 a. m. June 22.

1915–1918: Maximum stage recorded, that of January 29, 1918; minimum stage, that of June 22, 1918.

Highest known flood reached a stage of about 44 feet by present gage.

Ice.—Stage-discharge relation affected by ice during cold winters.

Accuracy.—Stage-discharge relation may change during floods; affected by ice part of December and January. Gage read to hundredths twice daily.

Cooperation.—Base data furnished by United States Engineer Corps.

No discharge measurements were made at this station during the year.

Daily gage height, in feet, of Guyandot River at Branchland, W. Va., for the year ending Sept. 30, 1918.

Day.	Oct.	Nov.	Dec.	Jan.	Feb.	Mar.	Apr.	May.	June.	July.	Aug.	Sept.
1	4.23	4.78	4.78	4.18	14.76	6.54	5.67	5.42	3.60	4.00	4.35	4.22
2	3.89	4.60	4.96	4.62	10.32	5.87	4.96	5.13	3.49	3.96	3.64	4.09
3	3.67	4.37	4.94	4.64	8.02	5.38	4.94	4.87	3.71	3.82	3.38	3.78
4	3.54	4.10	4.82	4.49	7.06	5.20	5.14	4.62	3.42	3.70	3.24	3.68
5	3.46	3.94	4.56	4.32	5.91	6.42	4.97	4.47	3.22	3.51	3.12	3.50
6	3.40	3.84	4.38	5.39	5.46	12.43	4.81	4.42	3.19	3.38	2.99	3.36
7	3.36	3.70	4.21	7.29	5.76	15.57	4.78	4.31	3.68	3.22	2.93	3.42
8	3.28	3.66	4.14	8.53	5.86	19.60	9.82	4.30	3.99	3.12	2.84	4.15
9	3.24	3.60	3.88	7.39	6.96	12.21	20.18	4.25	3.58	3.08	3.22	3.92
10	3.22	3.56	3.74	6.54	8.66	8.98	15.47	4.26	3.40	3.10	3.72	3.76
11	3.21	3.52	3.86	5.70	11.34	7.56	12.77	4.22	3.30	3.05	3.43	3.75
12	3.23	3.48	3.85	5.35	9.04	6.78	....	4.26	3.18	3.07	3.34	3.62
13	3.28	3.40	3.82	5.89	7.44	12.02	....	5.46	3.11	3.03	3.52	3.49
14	3.28	3.42	3.85	6.76	6.56	16.20	....	5.46	3.02	2.98	3.40	3.40
15	3.28	3.40	3.81	7.62	6.06	12.68	7.60	6.05	2.96	2.91	3.29	3.29
16	3.30	3.38	3.78	7.28	5.89	9.77	7.04	6.22	2.89	2.86	3.40	3.24
17	3.37	3.36	3.74	6.64	5.63	7.44	6.48	5.36	2.85	2.84	3.50	3.19
18	3.30	3.34	3.60	6.28	5.30	6.42	6.15	4.86	2.81	2.87	3.38	3.20
19	5.18	3.30	3.55	5.56	5.10	5.82	5.81	4.53	2.88	2.94	3.56	3.21
20	5.74	3.28	3.48	4.92	8.32	5.34	5.56	4.58	2.79	2.89	3.83	3.22
21	5.03	3.26	3.50	4.69	11.86	5.32	7.18	5.52	2.75	2.86	4.18	3.34
22	4.20	3.23	3.55	4.25	9.86	5.86	10.79	6.04	2.83	2.88	3.70	3.64
23	4.20	3.24	3.70	4.49	7.66	6.22	9.37	5.11	2.98	2.95	3.46	3.69
24	4.26	3.24	3.86	4.37	6.49	6.22	7.98	4.68	2.97	2.94	3.27	3.70
25	4.38	3.22	4.28	4.34	6.02	9.54	6.74	4.28	3.64	2.87	3.08	3.58
26	4.42	3.19	4.61	4.44	6.44	12.96	6.04	4.08	4.65	2.83	3.04	3.43
27	4.44	3.19	4.72	4.86	6.68	10.33	5.65	3.94	6.88	2.78	2.99	3.32
28	4.64	3.31	5.04	24.62	6.90	8.86	5.60	3.98	5.72	3.08	3.06	3.22
29	4.52	3.58	5.02	37.82	....	7.67	5.84	4.08	4.70	2.94	3.39	3.14
30	4.83	4.10	3.43	21.36	....	6.87	5.69	4.00	4.22	3.05	3.35	3.09
31	4.90	....	4.22	17.46	....	6.40	....	3.79	....	4.20	3.54	....

Note.—Gage not read Apr. 12-14.

MUD RIVER AT YATES, W. VA.

Location.—About 200 feet above highway bridge at Yates, Cabell County, 2 miles above Howell milldam, and 15 miles from Huntington.

Drainage area.—318 square miles (measured on topographic maps).

Records available.—July 19, 1915, to September 30, 1918.

Gage.—Vertical and inclined staff on left bank; read by C. J. McDonie.

Discharge measurements.—Made from single-span steel highway bridge below gage.

Channel and control.—One channel up to high stages, when right bank is overflowed around right abutment; straight for about 50 feet above and 75 feet below bridge. Primary control at ford, about 100 feet below gage; fairly permanent.

Extremes of stage.—Maximum stage recorded during year, 20.0 feet at 5.30 p. m. March 14; minimum stage 1.20 feet at 6 p. m. September 30.

Highest flood known reached a gage height of about 23 feet by present gage.

Ice.—Stage-discharge relation affected by ice during severe winters.

Accuracy.—Stage-discharge relation probably permanent; affected by ice part of December and January. Gage read to hundredths twice daily.

Cooperation.—Base data furnished by United States Engineer Corps.

No discharge measurements were made at this station during the year.

Daily gage height, in feet, of Mud River at Yates, W. Va., for the year ending Sept. 30, 1918.

Day	Oct.	Nov.	Dec.	Jan.	Feb.	Mar.	Apr.	May	June.	July.	Aug.	Sept.
1	1.79	3.60	3.51	3.69	8.40	3.98	3.14	3.20	2.39	3.16	1.46	1.49
2	1.83	3.14	3.08	3.59	5.80	3.75	3.04	3.04	2.31	2.98	1.44	1.48
3	1.86	2.85	2.85	3.52	5.49	3.55	4.92	2.92	2.22	2.56	1.42	1.54
4	1.88	2.72	2.84	3.49	5.49	3.46	4.34	2.82	2.20	2.35	1.43	1.54
5	1.81	2.58	2.62	3.25	5.28	6.68	3.65	2.74	2.35	2.20	1.44	1.48
6	1.66	2.55	2.52	4.00	4.82	6.50	3.22	2.65	2.29	2.14	1.42	1.55
7	1.66	2.42	2.49	8.80	5.98	6.09	4.22	2.59	2.96	2.08	1.40	1.58
8	1.64	2.42	2.50	9.77	7.62	5.10	8.80	2.52	4.66	2.04	1.41	1.52
9	1.58	2.32	2.95	6.28	10.95	4.50	12.60	2.48	3.38	1.96	1.42	1.49
10	1.47	2.28	2.64	4.90	9.25	4.20	7.18	2.74	2.83	1.98	1.54	1.46
11	1.50	2.25	2.60	4.45	6.90	3.81	7.30	2.82	2.59	1.92	1.48	1.44
12	1.64	2.18	2.49	4.80	5.78	4.52	9.45	3.06	2.42	1.98	1.44	1.44
13	1.81	2.19	2.38	5.02	5.34	15.05	8.85	6.00	2.30	1.91	1.44	1.44
14	1.86	2.21	2.32	4.40	4.70	19.40	6.25	8.85	2.20	2.04	1.46	1.37
15	1.96	2.18	2.31	5.20	4.94	16.55	5.09	5.32	2.13	2.04	1.49	1.34
16	2.04	2.14	2.29	7.28	4.40	7.85	4.55	3.84	2.07	1.92	1.44	1.33
17	2.09	2.12	2.24	6.20	3.98	5.45	4.14	3.40	2.04	1.89	1.46	1.38
18	2.04	2.14	2.21	4.95	3.60	4.74	4.02	3.11	2.04	1.96	1.54	1.34
19	7.20	2.16	2.29	4.74	3.52	4.18	3.70	2.92	2.00	1.81	1.47	1.30
20	7.53	2.02	2.32	4.25	8.52	3.85	3.64	2.82	1.94	1.86	1.43	1.40
21	4.90	2.04	2.52	3.94	9.70	3.65	6.68	5.83	2.04	1.80	1.39	1.40
22	3.37	2.06	2.66	3.89	5.75	3.64	5.41	3.81	2.06	1.82	1.38	1.36
23	2.98	2.05	2.66	3.82	4.58	3.52	4.22	3.25	2.00	1.80	1.38	1.52
24	3.14	2.05	2.66	3.58	4.12	3.46	4.03	2.88	1.94	1.73	1.54	1.52
25	3.79	1.99	4.80	3.40	3.98	4.41	3.86	2.74	3.18	1.60	1.50	1.46
26	3.77	1.90	6.80	3.60	5.80	6.30	3.78	2.68	6.84	1.56	1.49	1.42
27	3.32	2.06	4.99	4.60	5.32	5.08	3.62	2.63	4.33	1.54	1.48	1.36
28	3.25	2.51	4.69	11.12	4.35	3.90	3.50	3.46	3.24	1.54	1.46	1.32
29	2.99	3.74	4.20	16.10	....	3.58	3.38	3.12	2.78	1.46	1.42	1.30
30	5.18	3.77	3.79	17.10	....	3.38	3.38	2.77	2.52	1.62	1.39	1.24
31	4.29	....	3.62	10.50	....	3.22	....	2.58	....	1.50	1.44	....

TWELVEPOLE CREEK BASIN.

TWELVEPOLE CREEK AT WAYNE, W. VA.

Location.—At highway bridge 500 feet above railroad bridge of East Lynne branch of Norfolk & Western Railway at Wayne, Wayne County, three-fourths mile below junction of East and West forks.

Drainage area.—291 square miles (measured on topographic maps).

Records available.—July 1, 1915, to September 30, 1918.

Gage.—Chain gage attached to upstream handrail about 90 feet from left abutment; read by Byron Smith.

Discharge measurements.—Made from highway bridge or by wading.

Channel and control.—Straight for about 80 feet above and 1,200 feet below bridge. Bed of stream composed of rock and sand. Principal control is Sampson's milldam; probably permanent, but at low stages the operation of the mill may affect the discharge relation.

Extremes of stage.—Maximum stage, recorded during year, 20.48 feet at midnight January 28; minimum stage, 1.24 feet August 7 and 9.

Highest flood known reached a stage represented by gage height about 25 feet.

Ice.—Stage-discharge relation probably not materially affected by ice.

Regulation.—None, except for backwater caused during low-water periods by operation of small power plant at Sampson's mill about a mile below gage.

Accuracy.—Stage-discharge relation probably permanent; slightly affected by ice part of December and January. Operation of power plant at dam about a mile below gage may have slight effect upon stage-discharge relation at low stages, but this effect, if any, is small as the plant is only operated occasionally for a few hours at a time. Gage read to hundredths twice daily.

Cooperation.—Base data furnished by United States Engineer Corps.

No discharge measurements were made at this station during the year.

Daily gage height, in feet, of Twelvepole Creek at Wayne, W. Va., for the year ending Sept. 30, 1918.

Day.	Oct.	Nov.	Dec.	Jan.	Feb.	Mar.	Apr.	May.	June.	July.	Aug.	Sept.
1	2.27	3.21	2.72	....	5.64	4.32	3.57	3.19	2.04	3.56	1.44	1.87
2	2.20	2.84	2.19	....	4.99	8.94	3.40	3.00	2.22	3.11	1.44	1.87
3	2.17	2.71	2.67	....	4.54	3.68	4.77	2.92	3.25	2.94	1.40	2.04
4	2.13	2.63	2.58	....	4.35	4.10	4.70	2.86	2.75	2.68	1.38	2.78
5	1.98	2.58	2.54	3.79	4.28	6.12	3.92	2.77	2.64	2.34	1.31	1.70
6	1.70	2.50	2.44	7.14	4.36	5.26	3.74	2.62	3.16	2.02	1.28	1.68
7	1.65	2.43	2.31	....	5.09	4.89	4.50	2.54	4.20	1.95	1.25	1.62
8	1.56	2.39	2.18	....	6.14	4.52	11.47	2.51	4.70	1.89	1.26	1.59
9	1.50	2.34	2.50	....	7.64	4.12	9.18	2.49	3.45	1.91	1.24	1.54
10	1.45	2.28	....	....	7.24	3.86	6.82	2.47	3.18	1.94	1.42	1.46
11	1.51	2.18	....	....	5.69	3.79	5.87	2.60	2.86	1.87	1.44	1.40
12	1.81	2.11	....	....	5.54	3.72	5.70	2.99	2.41	1.85	1.38	1.42
13	1.75	2.05	....	....	5.84	12.44	5.44	6.42	2.20	1.82	2.49	1.56
14	2.08	2.05	....	....	5.60	9.79	5.20	7.87	2.06	1.80	1.94	1.58
15	2.00	2.11	....	....	5.38	6.46	4.90	5.40	1.93	1.77	1.73	1.49
16	1.90	2.07	2.50	....	5.12	5.59	4.48	4.82	1.81	1.78	1.64	1.47
17	1.83	2.06	....	....	4.83	4.86	4.00	4.47	1.75	1.76	1.59	1.43
18	1.79	2.14	....	....	4.16	4.36	3.64	3.87	1.71	1.79	2.02	1.41
19	8.10	2.08	....	....	3.36	4.12	3.14	3.14	1.72	1.77	1.99	1.38
20	5.58	1.98	....	....	11.04	3.84	3.00	2.92	1.75	1.75	1.84	1.83
21	3.46	1.93	....	....	7.59	3.76	5.00	2.98	1.86	1.72	1.74	1.61
22	3.36	1.89	....	....	6.36	3.94	4.60	2.95	1.83	1.63	1.68	1.53
23	3.34	1.91	2.53	....	5.04	3.72	4.57	2.92	1.82	1.50	1.72	1.66
24	3.56	1.95	2.81	....	4.26	4.19	4.40	2.81	1.75	1.43	1.64	1.62
25	3.37	1.97	2.96	....	4.62	7.94	4.24	2.78	5.36	1.51	1.55	1.57
26	3.29	1.94	3.00	....	4.68	7.29	3.91	3.00	7.38	1.47	1.44	1.49
27	3.17	1.96	3.18	4.74	4.56	6.21	3.70	3.10	5.11	1.45	1.43	1.46
28	4.40	2.59	4.98	14.59	4.50	4.86	3.12	3.20	3.70	1.55	1.39	1.41
29	4.50	2.78	4.36	16.32	....	4.22	3.47	2.84	3.05	1.52	1.41	1.35
30	5.40	2.87	3.96	9.02	....	4.00	3.35	2.34	2.88	1.55	1.70	1.29
31	4.40	....	....	6.12	....	3.78	....	2.04	....	1.49	2.04	....

Note.—Gage not read Dec. 10-15, 17-22, Jan. 1-4, and 7-26.

BIG SANDY RIVER BASIN.

LEVISA FORK AT THELMA, KY.

Location.—At Chesapeake & Ohio Railway bridge at Thelma, Johnston County, 2 miles below Paintsville. Buffalo Creek enters on right half a mile above station.

Drainage area.—2,090 square miles (measured by United States Engineer Corps).

Records available.—June 1, 1915, to September 30, 1918.

Gage.—Vertical staff gage attached to right shore pier of bridge, portion of gage above 24 feet is cut in masonry steps on upper end of right abutment; read by John Stambaugh. Sea-level elevation of gage, 561.82 feet (United States Engineer Corps).

Discharge measurements.—Made from boardwalk constructed on the lower downstream chord of bridge.

Channel and control.—Channel straight half a mile above and 300 feet below gage. Bed of stream sandy. Remains of cofferdams around piers, and piles at measuring section. Primary control about 2,400 feet downstream composed of rock which extends three-fourths of the way across stream; remainder is firm sand, fairly permanent.

Extremes of stage.—Maximum stage recorded during year, 40.7 feet at 6 p. m. January 29; minimum stage, 1.3 feet August 25 and 26.

Ice.—Stage-discharge relation probably not affected by ice except during extremely cold periods.

Regulation.—Splash dams on tributaries and in main stream about 50 miles above used by timber companies may affect low-water flow to some extent.

Accuracy.—Stage-discharge relation may change during high water; affected by ice during part of December and January. Gage read to half-tenths twice daily until May 31, 1918, and once daily thereafter

Cooperation.—Base data furnished by United States Engineer Corps.

No discharge measurements were made at this station during the year.

Daily gage height, in feet, of Levisa Fork at Thelma, Ky., for the year ending Sept. 30, 1918.

Day.	Oct.	Nov.	Dec.	Jan.	Feb.	Mar	Apr	May.	June.	July.	Aug.	Sept.
1	4.05	4.05	2.25	4.50	16.50	5.40	5.00	5.45	4.1	5.3	5.9	3.9
2	3.00	4.15	2.30	4.28	11.70	5.08	4.62	5.05	3.9	5.0	4.8	4.0
3	2.88	3.75	2.45	4.55	8.50	4.90	4.45	4.68	3.5	4.2	3.5	3.5
4	2.65	3.40	2.80	4.65	7.65	4.60	5.75	4.22	3.2	4.0	3.0	3.1
5	2.48	3.12	2.80	4.52	6.00	8.55	6.95	4.00	2.8	3.5	2.6	3.0
6	2.30	2.95	2.70	4.12	5.55	10.15	6.48	3.92	2.8	3.0	2.4	3.2
7	2.18	2.78	2.62	6.05	5.65	12.98	6.10	3.68	2.8	2.35	2.1	3.2
8	2.08	2.65	2.62	9.55	5.85	17.25	12.70	3.52	3.0	2.5	2.0	4.1
9	2.02	2.55	2.72	9.45	6.70	12.88	21.75	3.50	2.9	2.5	2.0	3.5
10	2.00	2.48	2.75	7.70	7.75	9.75	17.80	3.42	2.7	2.4	1.9	3.2
11	1.92	2.40	2.80	5.75	7.75	8.85	11.55	3.55	2.6	2.3	1.9	3.3
12	1.98	2.40	2.80	5.60	7.60	8.30	10.15	3.88	2.4	2.3	2.0	3.0
13	2.20	2.35	2.80	5.68	6.55	7.88	8.55	4.95	2.3	2.2	2.0	2.8
14	2.20	2.30	2.80	7.80	6.55	8.05	7.30	8.40	2.3	2.1	2.0	2.4
15	2.12	2.28	2.85	7.30	5.50	8.62	6.48	8.90	2.0	2.0	2.1	2.4
16	2.05	2.20	2.90	7.90	5.05	7.90	5.85	8.25	1.9	1.9	2.0	2.4
17	2.05	2.12	2.90	8.10	4.95	6.55	5.58	5.70	1.75	1.8	1.9	2.3
18	2.00	2.10	2.90	7.95	4.65	6.00	5.35	4.80	2.7	1.7	1.8	2.2
19	5.38	2.10	2.90	6.78	4.48	5.30	5.30	5.10	6.3	1.7	1.7	2.2
20	8.50	2.10	2.55	5.60	10.85	4.95	5.30	5.58	3.0	2.8	1.65	2.1
21	5.75	2.05	2.45	5.08	12.00	5.10	5.75	8.50	2.5	3.0	1.6	2.0
22	4.70	2.00	2.65	4.62	10.40	5.50	8.45	8.45	4.5	2.8	1.4	2.0
23	3.96	2.00	3.02	4.50	8.25	8.70	9.75	7.05	6.0	2.5	1.4	2.2
24	3.70	2.00	3.25	4.28	6.98	9.25	7.95	8.78	5.5	2.6	1.35	2.2
25	3.50	2.00	3.70	4.28	6.20	12.15	6.80	7.45	5.0	2.3	1.3	2.1
26	3.30	1.95	4.52	4.38	6.12	13.45	6.05	6.25	12.6	2.5	1.3	2.1
27	3.15	1.95	4.98	7.65	5.90	11.50	6.15	5.85	13.0	2.2	1.35	2.1
28	3.08	2.02	5.25	29.00	5.68	8.60	6.45	5.05	8.3	2.5	1.35	2.0
29	2.98	2.18	4.10	40.60	....	7.15	6.42	4.65	6.0	2.6	2.1	1.9
30	3.80	2.20	3.58	28.35	....	6.10	6.10	3.95	5.9	3.6	2.1	1.9
31	4.55	....	4.08	22.00	....	5.40	....	4.22	....	4.0	2.3	....

TUG FORK AT KERMIT, W. VA.

Location.—About 150 feet above United Fuel Gas Co.'s ferry at Kermit, Mingo County. Marrowbone Creek enters on right 2 miles below gage.

Drainage area.—1,240 square miles (measured by United States Engineer Corps).

Records available.—June 1, 1915, to September 30, 1918.

Gage.—Vertical staff gage in three sections attached to trees on right bank of river; 0–20 feet, 160 feet above cable; 20–38 feet, 130 feet below cable; and 38 to 48 feet at cable; read by C. C. Preece. Sea-level elevation of zero of gage, 574.77 feet (United States Engineer Corps).

Discharge measurements.—Made from car on ferry cable or by wading under cable.

Channel and control.—Channel straight above and below, bed of stream sandy; control about 150 feet below cable composed of solid rock which extends half way across from left bank and loose rock placed in river for fording, probably permanent.

Extremes of stage.—Maximum stage recorded during year, 38.8 feet, January 29; minimum stage, 2.00 feet October 11, November 26 and 27.

Ice.—Stage-discharge relation seldom affected by ice.

Accuracy.—Stage-discharge relation practically permanent; probably affected by ice during part of December and January. Gage read to hundredths twice daily until May 31, 1918, and once daily thereafter.

No discharge measurements were made at this station during the year.

Daily gage height, in feet, of Tug Fork at Kermit, W. Va., for the year ending Sept. 30, 1918.

Day.	Oct.	Nov.	Dec.	Jan.	Feb.	Mar.	Apr.	May.	June.	July.	Aug.	Sept.
1	3.42	3.85	3.35	4.40	13.5	5.98	5.90	5.65	3.40	4.50	4.70	3.15
2	3.35	3.76	3.82	4.05	10.05	5.62	5.56	5.30	3.20	4.10	4.00	3.12
3	2.60	3.45	3.85	3.98	8.40	5.15	5.48	4.94	3.00	3.80	3.32	3.12
4	2.40	3.39	3.65	3.80	7.41	4.90	5.56	4.71	2.95	3.40	2.90	3.10
5	2.38	3.10	3.45	3.74	5.94	9.26	5.60	4.58	2.62	3.10	2.70	2.80
6	2.29	2.88	3.25	3.82	5.74	11.20	5.35	4.41	2.50	2.90	2.50	2.30
7	2.16	2.79	3.15	6.20	6.00	15.55	5.25	4.29	2.75	2.70	2.40	4.15
8	2.30	2.68	3.30	8.20	5.82	15.85	11.25	4.08	3.10	2.65	2.20	4.28
9	2.10	2.60	3.18	7.68	5.85	10.80	19.55	4.00	3.10	2.80	2.10	3.80
10	2.35	2.50	3.00	6.55	8.22	8.90	14.05	3.97	3.15	2.62	3.60	3.60
11	2.00	2.45	2.90	6.3	9.40	8.32	10.60	3.89	2.85	2.60	2.58	3.40
12	2.16	2.40	3.10	7.12	8.15	7.88	9.65	3.88	2.60	2.55	2.50	3.30
13	2.32	2.39	4.02	6.90	7.45	8.82	9.30	4.10	2.40	2.40	3.00	2.95
14	2.26	2.36	4.18	9.15	6.45	9.42	8.75	5.35	2.30	2.35	2.48	2.75
15	2.31	2.30	4.00	8.50	6.02	9.12	8.05	5.90	2.20	2.32	2.25	2.60
16	2.36	2.30	3.65	8.38	5.65	7.98	7.25	5.25	2.12	2.20	2.75	2.55
17	2.26	2.25	3.20	7.80	5.45	6.95	6.70	4.70	2.10	2.20	2.75	2.42
18	2.19	2.20	2.88	7.45	5.16	6.32	6.42	4.25	2.35	2.28	2.70	2.50
19	4.60	2.18	2.82	6.80	4.96	5.82	6.22	3.92	2.12	2.30	2.60	2.45
20	5.35	2.11	2.80	6.18	8.55	5.34	6.01	4.00	2.00	2.48	2.55	2.50
21	4.05	2.10	2.80	5.65	10.22	5.65	6.41	5.20	2.10	3.00	3.50	3.20
22	3.61	2.10	2.80	5.55	8.68	9.80	10.30	5.52	2.35	2.57	2.92	3.10
23	3.40	2.10	3.40	5.45	7.45	9.75	8.86	4.75	2.60	2.40	2.55	3.50
24	3.50	2.10	3.25	5.2	6.72	9.80	7.70	4.35	3.32	2.30	2.32	3.35
25	3.58	2.08	3.36	5.22	6.04	14.40	6.84	4.82	3.00	2.28	2.15	3.00
26	3.62	2.00	4.22	5.35	6.00	13.65	6.25	4.38	5.90	2.20	2.15	2.78
27	3.55	2.00	4.20	6.65	6.10	10.95	5.95	3.92	11.35	2.25	2.02	2.60
28	3.42	2.20	4.34	31.85	6.48	9.14	6.20	3.71	6.60	2.20	3.00	2.50
29	3.36	3.05	3.62	34.00	....	8.15	6.26	4.65	4.95	2.55	3.60	2.40
30	3.76	3.10	4.72	15.25	....	7.00	6.00	3.82	4.20	2.45	3.20	2.30
31	3.66	....	4.40	19.75	....	6.36	....	3.38	....	6.00	3.22	....

BLAINE CREEK AT YATESVILLE, KY.

LOCATION.—At covered highway bridge one-fourth mile above Yatesville, Lawrence County. Morgan Branch enters on left 2 miles above station.

DRAINAGE AREA.—216 square miles (United States Engineer Corps).

RECORDS AVAILABLE.—June 1, 1915, to September 30, 1918.

GAGE.—Vertical staff gage in two sections attached to elm tree on right bank about 50 feet above bridge; read by Hattie M. Carter.

DISCHARGE MEASUREMENTS.—Made from board walk constructed on inside of bridge near top of siding. Wading measurements are made under bridge.

CHANNEL AND CONTROL.—Stream curved above and straight below bridge, right bank is overflowed at high stages, stream bed compact sand and gravel; control composed of bedrock extending halfway across stream, sand and gravel rest of way, probably permanent.

EXTREMES OF DISCHARGE.—Maximum stage recorded during year, 15.0 feet at 6 p. m. January 28 (discharge, 5,960 second-feet); minimum stage recorded, 0.90 foot October 8 and 9 (discharge, 10 second-feet.)

ICE.—Stage-discharge relation seldom affected by ice.

ACCURACY.—Stage-discharge relation probably permanent; not affected by ice. Rating curve well defined between 20 and 4,000 second-feet; extended beyond these limits. Gage read twice daily to hundredths. Daily discharge ascertained by applying mean daily gage height to rating table. Records fair.

COOPERATION.—Base data furnished by United States Engineer Corps.

No discharge measurements were made at this station during the year.

Daily discharge, in second-feet, of Blaine Creek at Yatesville, Ky., for the year ending Sept. 30, 1918.

Day.	Oct.	Nov.	Dec.	Jan.	Feb.	Mar.	Apr.	May.	June.	July.	Aug.	Sept.
1	252	252	24	800	525	277	192	150	34	290	54	24
2	150	204	43	730	375	227	150	132	96	227	43	21
3	96	114	30	465	331	204	555	110	54	114	34	21
4	66	88	39	375	317	304	555	100	96	66	30	24
5	27	88	42	405	331	465	345	96	60	60	27	21
6	17	66	44	465					34	54	24	19
7	15	60	39	525					80	43	21	19
8	12	78	43	465					264	66	24	17
9	12	48	38	525	345	465	264	86	192	60	27	17
10	15	34	59	405	1,[illegible]	[illegible]	1,[illegible]	[illegible]	54	54	43	21
11	24	48	80	465		252	495	74	34	43	34	19
12	80	38	30	465		252	405	82	30	48	27	19
13	38	30	28	405		880	360	360	27	43	30	17
14	96	48	30	435	800	880	290	1,040	27	38	27	21
15	66	38	33	695	[illegible]	1,200	264	405	24	34	34	27
16	66	38	35	960	227	525	239	239	21	30	27	17
17	80	48	28	880	181	405	239	264	19	34	38	21
18	48	30	35	800	170	277	264	495	21	30	54	43
19	2,770	38	33	625	264	304	204	252	21	34	43	21
20	1,200	30	28	360	1,710	277	192	465	21	27	34	54
21	465	27	33	360	880	317	435	465	17	28	30	27
22	331	24	123	405	660	405	375	239	96	27	27	24
23	252	24	192	405	590	304	277	160	114	30	24	27
24	123	24	204	405	405	317	239	141	38	29	24	27
25	123	24	405	405	375	800	215	105	34	30	21	21
26	123	27	960	405	465	525	192	123	1,400	38	21	19
27	27	24	800	1,200	465	405	204	141	345	660	24	21
28	43	36	660	4,420	331	304	181	88	192	114	27	17
29	88	41	730	4,070	......	252	192	88	132	66	21	17
30	317	30	800	1,120	......	239	170	88	114	80	24	21
31	304	......	695	695	......	215	......	60	......	54	27	......

Monthly discharge of Blaine Creek at Yatesville, Ky., for the year ending Sept. 30, 1918.

[Drainage area, 216 square miles.]

Month.	Discharge in second-feet.				Run-off in inches.
	Maximum.	Minimum.	Mean.	Per square mile.	
October	2,770	12	236	1.09	1.26
November	252	24	56.5	.262	.29
December	960	24	205	.949	1.09
January	4,420	360	811	3.75	4.32
February	1,710	170	538	2.49	2.59
March	1,200	204	411	1.90	2.19
April	1,530	150	367	1.70	1.90
May	1,040	60	205	.949	1.09
June	1,400	17	123	.569	.63
July	660	27	82.3	.381	.44
August	54	21	30.5	.141	.16
September	54	17	22.8	.106	.12
The year	4,420	12	256	1.19	16.08

SCIOTO RIVER BASIN.

SCIOTO RIVER AT WAVERLY, OHIO.

Location.—At Norfolk & Western Railway bridge 1 mile southeast of Waverly, Pike County.

Drainage area.—5,730 square miles (United States Engineer Corps).

Records available.—March 23, 1916, to September 30, 1918.

Gage.—Chain gage fastened to downstream side of bridge; read by W. G. Johnston. Sea-level elevation of zero of gage, 542.00 feet (United States Engineer Corps).

Discharge measurements.—Made from downstream side of bridge to which gage is attached, or from highway bridge 2,000 feet below gage.

Channel and control.—For stages over 12 feet the river spreads over the bottom lands, but all water passes under the bridge.

Extremes of stage.—Maximum stage during year, 18.16 feet at 4.10 p. m. February 15; minimum stage, 0.77 foot at 7 a. m. August 26.

1916–1918: Maximum stage recorded, 21.9 feet March 29, 1916 (discharge, 97,800 second-feet); minimum stage, 0.77 foot at 7 a. m. August 26, 1918.

Ice.—Stage-discharge relation not affected by ice except during severe winters.

Accuracy.—Stage-discharge relation probably permanent, but no current-meter measurements have been made since October 18, 1916, to check the rating curve; ice effect during part of December, January, and February. Gage read to hundredths twice daily.

Cooperation.—Gage-height record furnished by United States Engineer Corps.

Daily gage height, in feet, of Scioto River at Waverly, Ohio, for the year ending Sept. 30, 1918.

Day.	Oct.	Nov.	Dec.	Jan.	Feb.	Mar.	Apr.	May.	June.	July.	Aug.	Sept.
1	......	2.15	1.17	1.79	2.82	6.18	2.91	4.23	2.65	1.66	......	2.71
2	......	2.24	1.18	1.79	2.84	6.11	3.03	3.80	2.45	1.81	......	2.55
3	1.06	2.39	1.19	1.79	2.86	6.36	3.88	3.39	2.34	1.90	......	2.11
4	.89	2.29	......	1.79	2.86	6.71	4.01	3.18	2.25	1.66	1.35	2.25
5	.84	2.02	......	1.79	2.76	6.66	4.11	2.96	2.09	1.57	1.26	1.93
6	.83	1.71	......	1.97	3.31	6.76	......	2.78	2.08	1.55	1.25	1.57
7	.93	1.54	1.18	2.25	4.61	6.16	......	2.71	2.44	1.39	1.22	1.77
8	1.02	1.56	1.21	......	6.61	5.84	......	2.59	2.36	1.33	1.14	1.90
9	1.05	1.48	1.26	3.17	8.96	5.18	......	2.59	2.36	1.51	1.10	2.01
10	.95	1.46	1.39	3.17	11.81	5.66	......	2.59	2.27	1.52	1.07	1.86
11	.96	1.41	1.39	3.39	10.51	6.14	......	2.57	3.87	1.49	1.82	1.65
12	1.01	1.33	1.39	......	13.06	5.74	4.40	2.79	2.80	1.44	1.79	1.54
13	1.00	1.26	1.39	3.42	14.76	11.20	4.13	7.77	2.27	1.41	1.57	1.42
14	.99	1.28	1.39	3.39	17.14	17.91	3.87	10.70	2.15	1.39	1.25	1.33
15	.96	......	1.39	......	18.11	15.11	3.68	14.30	2.10	1.33	1.19	1.55
16	.96	1.31	1.39	2.97	16.88	10.30	3.35	11.60	1.93	1.27	1.25	1.51
17	.89	1.23	1.39	2.97	13.56	8.14	3.19	6.90	1.72	1.23	1.46	1.86
18	.97	1.11	1.39	2.97	7.86	6.51	3.30	5.85	1.59	1.19	2.24	2.56
19	1.02	.97	1.39	2.97	7.66	5.31	3.09	4.85	1.55	1.17	1.94	3.11
20	.87	1.13	1.39	2.97	8.26	4.54	2.93	4.05	1.56	1.19	1.15	3.23
21	1.11	1.06	......	2.97	9.01	4.21	3.05	4.03	1.58	1.18	.95	3.15
22	1.42	1.07	......	2.97	8.41	4.08	4.45	3.87	1.56	1.21	.98	3.02
23	1.44	1.08	......	2.93	9.56	3.94	4.50	3.83	1.55	1.22	.93	2.91
24	1.49	1.09	......	2.79	10.01	3.76	4.10	3.77	1.49	1.23	.90	2.58
25	1.48	1.09	1.79	2.77	10.27	3.68	3.90	3.70	1.49	1.35	.79	2.26
26	1.43	1.07	1.79	2.76	10.38	3.56	3.85	3.55	3.34	1.30	.78	2.08
27	1.48	1.03	1.79	2.76	10.61	3.28	3.95	3.45	2.93	1.27	.94	1.92
28	1.57	1.17	1.79	2.76	8.86	3.13	......	4.10	2.70	1.50	1.29	1.93
29	1.66	1.17	1.79	2.76	......	2.98	3.90	3.29	1.84	1.76	1.45	1.78
30	1.81	1.15	1.79	2.84	......	2.92	......	2.99	......	......	1.55	1.67
31	2.08	......	1.79	2.85	......	2.89	......	2.86	......	......	1.60	......

LITTLE MIAMI RIVER BASIN.

LITTLE MIAMI RIVER AT MIAMIVILLE, OHIO.

Location.—At two-span steel highway bridge one-third mile southeast of Miamiville, Clermont County.

Drainage area.—1,200 square miles.

Records available.—June 21, 1915, to September 30, 1918.

Gage.—Chain gage attached to downstream side of bridge; read by J. M. Barrere.

Discharge measurements.—Made from downstream side of bridge, except at low stages, when they are made by wading.

Channel and control.—Channel clean of vegetation, except at high stages. Control probably permanent.

Extremes of stage.—Maximum stage recorded during year, 10.88 feet at 4 p. m. February 12; minimum stage, 1.30 feet at 6.38 p. m. November 19.

Regulation.—Low-water flow regulated to some extent by operation of flour mill at Fosters crossing about 11 miles upstream.

Accuracy.—Stage-discharge relation probably permanent; affected by ice during December, January, and February. Gage read to hundredths twice daily.

Cooperation.—Base data furnished by United States Army Engineers.

No discharge measurements were made at this station during the year.

82287—22—wsp 473——5

Daily gage height, in feet, of Little Miami River at Miamiville, Ohio, for the year ending Sept. 30, 1918.

Day.	Oct.	Nov.	Dec.	Jan.	Feb.	Mar.	Apr.	May.	June.	July.	Aug.	Sept.
1	1.46	1.97	1.63	1.91	2.22	3.92	2.37	2.75	2.22	2.77	1.98	3.32
2	1.37	1.90	1.69	1.91	2.39	3.62	2.49	2.61	2.08	2.17	1.83	2.76
3	1.47	1.75	1.81	1.95	2.41	3.57	3.43	2.51	2.11	1.77	1.74	2.56
4	1.51	1.71	1.72	1.89	2.39	3.47	3.42	2.43	2.02	1.62	1.63	2.67
5	1.43	1.82	1.66	1.91	2.56	4.37	2.88	2.35	2.04	1.90	1.73	2.70
6	1.49	1.71	1.59	2.96	2.57	4.07	2.70	2.37	2.71	1.67	1.65	2.46
7	1.53	1.69	1.58	4.23	2.36	3.62	2.58	2.29	2.83	1.56	1.55	2.30
8	1.60	1.68	1.71	3.38	2.63	3.38	2.60	2.30	2.42	1.93	1.52	2.05
9	1.32	1.66	1.81	2.79	5.38	3.31	2.49	2.30	2.35	1.70	1.54	2.06
10	1.34	1.51	1.32	2.62	6.83	3.24	2.43	2.29	2.21	1.62	1.62	1.91
11	1.44	1.75	1.35	2.65	6.78	3.08	2.48	2.36	2.73	1.67	1.57	1.84
12	1.53	1.69	1.50	2.51	9.98	2.94	2.61	5.42	2.40	1.67	1.57	1.94
13	1.47	1.56	1.59	2.39	9.88	8.39	2.71	8.22	2.05	1.60	1.79	2.38
14	1.74	1.52	1.81	2.31	6.83	7.07	2.60	5.52	1.95	1.62	1.65	2.22
15	1.49	1.65	1.44	2.35	7.18	4.97	2.53	4.42	1.86	1.71	1.55	2.00
16	1.42	1.65	1.58	2.42	5.08	4.17	2.46	3.82	1.83	1.64	1.51	4.31
17	1.53	1.52	1.84	2.27	4.27	3.72	2.65	3.42	1.88	1.71	1.55	5.06
18	1.51	1.40	1.60	2.09	3.77	3.52	2.84	3.17	1.93	1.67	......	3.76
19	1.62	1.31	1.99	2.24	4.87	3.27	2.65	......	1.70	1.64	1.65	3.51
20	2.28	1.44	1.92	2.23	7.52	3.06	2.70	3.47	1.71	1.58	1.57	3.41
21	1.84	1.58	1.89	2.31	5.77	2.96	2.86	3.06	1.72	1.60	1.56	3.26
22	1.84	1.62	1.78	2.36	4.12	2.87	3.14	2.72	1.71	1.78	1.49	2.92
23	1.76	1.53	1.95	2.31	3.57	2.84	2.91	2.63	1.70	2.97	1.32	2.70
24	1.71	1.59	2.05	2.30	3.52	2.83	2.78	2.51	1.83	2.53	1.46	2.50
25	1.73	1.54	2.15	2.26	3.67	2.89	2.74	2.40	3.31	2.20	1.45	2.38
26	1.65	1.65	2.22	2.11	6.52	2.76	3.72	2.36	3.00	2.19	1.49	2.33
27	1.74	1.56	2.09	2.24	5.22	2.63	3.72	2.32	2.26	2.05	1.40	2.21
28	1.69	1.53	2.05	2.42	4.17	2.52	3.30	2.61	2.11	2.81	2.62	2.05
29	1.63	1.59	2.01	2.29	......	2.50	3.15	2.49	1.99	4.11	2.91	1.86
30	2.01	1.73	1.91	2.23	......	2.41	2.99	2.36	2.11	2.68	3.06	2.01
31	2.08	......	1.87	2.14	......	2.32	......	2.43	......	2.13	3.65	......

NOTE.—Gage not read May 19 and Aug. 18.

LITTLE MIAMI RIVER AT PLAINVILLE, OHIO.

LOCATION.—At steel highway bridge half a mile above Pennsylvania Railroad station at Plainville, Hamilton County.

DRAINAGE AREA.—1,680 square miles.

RECORDS AVAILABLE.—July 10, 1914, to September 30, 1915; August 18, to September 30, 1918.

GAGE.—Chain gage attached to downstream side of bridge.

DISCHARGE MEASUREMENTS.—Made from downstream side of bridge or by wading.

CHANNEL AND CONTROL.—Bed composed of heavy gravel and rock covered with layer of mud. Control is at a riffle about 600 feet below gage.

COOPERATION.—Base data furnished by United States Engineer Corps.

Data inadequate for determination of discharge.

Discharge measurements of Little Miami River at Plainville, Ohio, during the year ending Sept. 30, 1918.

[Made by U. S. Army Engineers.]

Date.	Gage height.	Discharge.
	Feet.	*Sec.-ft.*
Aug. 30	6.9	797
Sept. 11	5.85	161
13	6.1	202

Daily gage height, in feet, of Little Miami River at Plainville, Ohio, for the year ending Sept. 30, 1918.

Day.	Aug.	Sept.	Day.	Aug.	Sept.	Day.	Aug.	Sept.
1		8.75	11		6.05	21	5.30	7.30
2		8.50	12		5.80	22	5.30	7.05
3		7.95	13		5.95	23	5.30	6.75
4		8.70	14		6.10	24	5.30	6.40
5		8.80	15		6.06	25	5.30	6.25
6		8.60	16		8.40	26	5.30	6.00
7		8.00	17		9.30	27	5.30	5.85
8		7.10	18	5.40	8.20	28	7.65	5.80
9		6.60	19	5.40	7.80	29	7.40	5.75
10		6.20	20	5.30	7.20	30	7.20	5.70
						31	10.10	....

EAST FORK OF LITTLE MIAMI RIVER AT PERINTOWN, OHIO.

Location.—At single-span steel highway bridge at Perintown, Clermont County, 5 miles above junction of East Fork and Little Miami River.

Drainage area.—459 square miles.

Records available.—May 7, 1915, to September 30, 1918.

Gage.—Chain gage attached to downstream side of bridge; read by G. W. Taylor.

Discharge measurements.—Made from downstream side of bridge except at low stages when they are made by wading.

Channel and control.—Bed of river mostly rock; banks covered with trees and brush above a stage of about 5 feet; control rock and gravel; probably permanent.

Extremes of stage.—Maximum stage recorded during year, 18.2 feet at 8 p. m. February 11; minimum stage, −0.18 foot October 3–6.

1915–1918: Maximum stage recorded, 18.6 feet at noon December 27, 1916 (discharge, about 21,300 second-feet); minimum stage, −0.18 foot October 3–6, 1917.

Ice.—Stage-discharge relation affected by ice during severe winters.

Accuracy.—Stage-discharge relation probably permanent; affected by ice during part of December, January, and February. Gage read to hundredths twice daily.

Cooperation.—Base data furnished by United States Engineer Corps.

No discharge measurements were made at this station during the year.

Daily gage height, in feet, of East Fork of Little Miami River at Perintown, Ohio, for the year ending Sept. 30, 1918.

Day.	Oct.	Nov.	Dec.	Jan.	Feb.	Mar.	Apr.	May.	June.	July.	Aug.	Sept.
1	−0.16	0.53	0.00	0.48	0.66	1.79	0.86	1.56	0.42	2.73	0.92	2.74
2	− .17	.42	.00	.46	.80	1.54	.81	1.23	.43	2.40	.68	1.84
3	− .18	.37	.00	.31	.80	1.38	2.65	1.06	.42	1.50	.46	1.21
4	− .18	.31	.00	.16	.80	1.61	2.70	.95	.40	1.10	.38	1.16
5	− .18	.27	.02	.17	.80	2.60	1.88	.84	.40	.85	.33	1.31
6	− .18	.23	.01	2.01	.80	2.41	1.31	.80	3.40	.68	.29	1.34
7	− .16	.20	.00	6.60	.94	1.92	1.15	.78	1.76	.56	.25	1.14
8	− .16	.18	.00	6.40	1.30	1.63	1.04	.74	1.92	.49	.21	.93
9	− .16	.15	.00	6.20	12.80	1.72	.95	.68	1.47	.44	.15	.74
10	− .16	.11	.00	4.10	16.70	1.97	.94	.70	1.16	.38	.13	.61
11	− .16	.10	.00	2.30	17.35	1.72	.94	.77	.84	.33	.12	.50
12	− .15	.08	.00	1.38	15.65	1.41	.96	4.90	.68	.31	.31	.46
13	− .14	.05	.00	1.30	10.15	8.00	.96	12.85	.56	.29	.58	1.01
14	− .14	.04	.00	1.30	5.00	9.40	.96	4.70	.48	.24	.47	.79
15	− .14	.02	.00	1.24	4.60	3.95	.92	2.68	.40	.19	.49	.59
16	− .14	.02	− .04	1.24	3.25	2.33	.87	2.10	.33	.18	.61	.69
17	− .14	.00	− .04	1.22	2.25	1.76	1.07	1.66	.26	.16	.37	.81
18	− .14	.00	− .03	.80	1.80	1.54	1.55	1.45	.20	.14	.19	1.18
19	.03	.00	.00	.70	1.97	1.41	1.05	1.28	.18	.14	.17	1.27
20	.19	− .01	.02	.70	5.60	1.28	1.10	1.15	.18	.14	.16	.98
21	.14	− .04	.10	.66	3.80	1.18	3.65	1.05	.18	.14	.14	.71
22	.07	− .06	.22	.60	1.90	1.16	3.20	.97	.16	.41	.11	.60
23	.01	− .06	.31	.60	1.70	1.14	1.96	1.06	.13	.31	.07	.55
24	.00	− .06	.44	.60	1.53	1.30	1.72	1.19	.12	.21	.03	.48
25	.00	− .06	.22	.56	2.13	1.55	2.24	1.09	3.63	.39	− .01	.42
26	− .03	− .06	.24	.56	5.30	1.58	3.85	.97	5.38	1.00	− .02	.32
27	− .06	− .07	.39	.56	2.85	1.45	4.02	.83	2.48	1.03	.05	.24
28	.00	− .08	.54	.56	2.05	1.25	2.82	.73	1.91	1.31	2.84	.18
29	.07	− .03	.54	.56	....	1.13	2.10	.64	1.64	1.52	1.87	.19
30	.12	.00	.52	.56	....	1.02	1.88	.58	2.95	.79	2.37	.16
31	.65	....	.52	.56	....	.93	....	.50	....	.87	5.33	....

LICKING RIVER BASIN.

LICKING RIVER AT FARMERS, KY.

Location.—About 100 feet below Chesapeake & Ohio Railway bridge and 300 feet below two-span steel highway bridge three-fourths of a mile west of Farmers, Rowan County.

Drainage area.—768 square miles (measured by United States Engineer Corps).

Records available.—July 20, 1915, to September 30, 1918.

Gage.—Combination vertical staff and slope gage on east bank of river; read by Mrs. S. P. Cassity.

Discharge measurements.—Made from downstream side of two-span highway bridge 300 feet above gage.

Channel and control.—Bed of stream solid rock, straight above and below gage. Control is a rock reef about 1 mile below gage.

Extremes of stage.—Maximum stage recorded during year, 21.3 feet at 7 a. m. January 30; minimum stage, 1.25 feet August 15 and 16.

1915–1918: Maximum stage recorded, 25.6 feet at 7 a. m. January 22, 1917; minimum stage, 1.1 feet August 17 and 18, 1917.

Ice.—Stage-discharge relation not affected by ice except during severe winters.

Regulation.—The flow at low stages may be affected by storage of water for use of a sawmill at a movable dam a short distance above the gage. Dam is submerged at gage height 5 feet.

Accuracy.—Stage-discharge relation probably permanent; affected by ice during part of December and January. Gage read to half-tenths twice daily; not checked since August 4, 1917.

Cooperation.—Base data furnished by United States Engineer Corps.

No discharge measurements were made at this station during the year.

Daily gage height, in feet, of Licking River at Farmers, Ky., for the year ending Sept. 30, 1918.

Day.	Oct.	Nov.	Dec.	Jan.	Feb.	Mar.	Apr.	May.	June.	July.	Aug.	Sept.
1	2.80	4.32	2.72	3.55	18.60	4.32	3.65	4.68	2.50	2.42	1.78	1.85
2	2.70	3.55	2.60	3.38	13.05	4.05	3.92	4.20	2.42	2.38	1.72	1.80
3	2.12	2.90	2.38	3.82	9.75	3.88	5.10	3.82	2.35	2.28	1.68	1.92
4	1.88	2.78	2.20	3.82	8.65	3.75	5.32	3.68	2.45	2.22	1.62	2.18
5	1.78	2.60	2.05	3.88	8.28	5.02	4.60	3.18	2.32	2.12	1.50	2.30
6	1.72	2.45	2.05	3.30	7.25	6.48	3.62	3.05	2.50	2.08	1.45	2.05
7	1.62	2.38	1.98	8.35	6.80	7.18	3.70	2.98	2.82	1.98	1.52	2.05
8	1.50	2.20	2.05	8.62	10.02	6.50	5.30	2.85	2.72	2.20	1.50	2.10
9	1.42	2.18	2.12	7.92	16.15	6.48	8.45	2.70	2.68	2.45	1.40	2.00
10	1.55	2.50	2.22	7.15	17.20	4.55	9.40	2.98	2.58	2.72	1.35	1.92
11	1.52	2.50	2.28	6.22	14.85	4.65	7.15	2.90	2.48	2.78	1.30	1.82
12	1.62	2.35	2.12	6.05	12.72	4.38	5.75	3.08	2.32	2.52	1.35	1.72
13	1.72	1.95	2.22	6.00	8.70	7.15	4.80	6.60	2.20	2.10	1.32	1.82
14	1.88	1.88	2.12	5.50	5.45	14.00	3.90	13.20	2.05	1.68	1.30	1.72
15	1.95	1.78	2.10	7.22	5.05	8.82	3.95	8.78	1.92	1.55	1.25	1.60
16	1.82	1.70	2.00	9.95	4.88	6.45	3.88	6.58	1.82	1.78	1.25	1.80
17	1.72	1.65	1.95	8.68	4.78	5.32	3.70	5.10	1.72	1.68	1.30	2.48
18	1.72	1.60	1.98	7.45	4.72	4.70	3.42	4.20	1.68	1.82	1.35	2.00
19	3.70	1.50	2.00	6.68	5.05	4.88	3.28	4.02	1.78	1.88	1.70	2.18
20	7.38	1.48	2.25	6.28	8.58	4.88	3.60	3.90	1.92	1.68	1.62	2.15
21	1.68	1.70	2.58	6.38	11.38	4.78	11.25	5.40	2.05	1.58	1.55	2.00
22	4.85	1.68	2.82	5.70	10.58	4.68	7.75	5.80	2.82	1.48	1.48	2.08
23	3.28	1.68	2.72	5.18	6.25	3.68	5.48	6.20	2.38	1.38	1.42	1.98
24	2.80	1.58	2.85	5.25	5.42	4.32	4.85	6.05	2.25	1.38	1.38	1.88
25	2.75	1.52	4.10	5.32	5.12	5.75	4.55	6.45	2.80	1.48	1.32	1.78
26	2.45	1.60	5.25	5.65	5.52	7.65	10.18	6.00	5.65	1.52	1.28	1.68
27	2.18	1.45	5.42	8.25	5.38	6.72	12.28	4.30	3.25	1.62	1.30	1.58
28	2.48	1.38	4.65	13.80	4.95	5.48	6.58	4.00	2.20	2.15	1.42	1.52
29	2.68	2.20	4.00	20.70	......	4.98	5.75	3.65	......	1.45	1.72	1.48
30	3.52	2.68	3.42	21.28	......	4.20	4.98	3.15	1.98	3.00	1.72	1.40
31	4.52	......	3.52	21.12	......	3.88	......	2.72	......	2.65	1.82	......

Note.—Gage not read June 29.

LICKING RIVER AT CATAWBA, KY.

Location.—About 200 feet below Catawba ford, one-fourth mile north of Catawba, Pendleton County. Kinkaid Creek enters from right, 1,000 feet below gage.

Drainage area.—3,300 square miles.

Records available.—July 14, 1916, to September 30, 1918.

Gage.—Combination slope and vertical staff on south bank of river about 200 feet below the ford; read by G. A. Frank. Elevation of zero of gage is 498.37 feet above sea level, which corresponds approximately to 69 feet on the United States Weather Bureau gage on Ohio River at Cincinnati, Ohio.

Discharge measurements.—Made from cable about 500 feet upstream from gage.

Channel and control.—Bed of river at cable is mostly ledge rock. The banks are heavily wooded above an elevation of about 7 feet on the gage. The control is a rock bar just below the mouth of Kinkaid Creek; probably permanent.

Extremes of stage.—Maximum stage recorded during year, 35.00 feet at 6 a. m. February 10; minimum stage, 0.80 foot at 6 a. m September 29.

1916–1918: Maximum stage recorded, that of February 10, 1918; minimum stage 0.80 foot September 28, 1917, and September 29, 1918.

Ice.—Stage-discharge relation affected by ice during severe winters.

Accuracy.—Stage-discharge relation probably permanent; probably affected by ice during part of December and January. Gage read to hundredths twice daily. Gage has not been checked since August 2, 1917.

Cooperation.—Base data furnished by United States Engineer Corps.

No discharge measurements were made at this station during the year.

Daily gage height, in feet, of Licking River at Catawba, Ky., for the year ending Sept. 30, 1918.

Day.	Oct.	Nov.	Dec.	Jan.	Feb.	Mar.	Apr.	May.	June.	July.	Aug.	Sept.
1	2.38	5.12	3.60	5.40	17.50	6.05	4.18	5.95	3.18	3.60	3.78	3.40
2	3.12	4.55	3.82	5.38	16.65	5.45	4.02	5.22	3.00	3.38	3.40	2.65
3	2.60	4.15	3.75	5.30	16.00	5.10	8.60	4.75	2.75	3.48	2.75	2.68
4	2.20	3.62	3.48	5.15	13.55	4.78	9.65	4.42	2.52	2.98	2.32	2.50
5	2.00	3.25	3.20	5.10	10.15	4.80	7.82	4.10	2.30	2.75	2.00	2.92
6	1.78	3.02	2.90	9.32	8.85	4.95	6.10	3.85	3.10	2.52	1.85	2.62
7	1.70	2.82	2.72	12.50	10.05	7.15	5.05	3.62	3.30	2.28	1.65	2.32
8	1.65	2.60	2.68	11.05	18.45	6.80	4.52	3.48	3.00	2.15	1.48	2.25
9	1.58	2.45	3.48	8.95	31.80	6.75	4.92	3.70	2.80	1.95	1.35	1.92
10	1.50	2.35	3.50	7.18	34.00	6.48	6.20	4.48	2.95	1.85	1.28	1.80
11	1.38	2.25	3.58	6.40	25.45	5.85	7.10	3.80	2.95	1.78	1.28	1.72
12	1.45	2.18	3.48	7.95	17.78	5.40	6.85	4.70	2.60	1.70	1.25	2.15
13	1.40	2.10	3.35	7.10	13.90	12.10	5.78	12.15	2.32	1.62	1.08	2.68
14	1.40	2.00	3.12	8.05	9.95	20.15	5.02	12.80	2.15	1.60	1.10	2.10
15	1.38	1.95	3.10	8.15	9.18	16.10	4.62	11.72	1.98	1.58	1.10	1.75
16	1.32	1.92	3.02	12.02	9.15	11.75	4.25	9.90	1.85	1.52	1.05	1.65
17	1.30	1.90	3.00	15.60	7.95	8.25	4.05	7.80	1.75	1.42	.95	1.68
18	1.58	1.90	2.80	15.25	6.62	6.78	4.20	5.82	1.62	1.38	1.00	1.58
19	1.72	1.88	2.82	12.75	7.72	5.82	4.22	5.10	1.48	1.35	1.30	1.50
20	1.88	1.82	2.95	10.20	17.50	5.22	4.00	5.30	1.42	1.95	1.38	1.38
21	5.80	1.80	4.08	9.40	14.02	4.88	7.40	8.65	1.40	1.62	1.25	1.22
22	6.18	1.72	6.38	8.70	12.40	4.55	11.10	10.00	1.38	1.45	1.12	1.18
23	5.42	1.68	6.08	7.58	10.78	4.42	10.75	6.72	1.68	1.35	1.10	1.15
24	3.95	1.62	6.00	7.20	8.60	4.30	7.25	6.00	1.70	1.30	1.05	1.32
25	3.38	1.58	5.62	6.72	6.73	4.75	7.50	5.65	2.50	1.22	1.00	1.18
26	2.95	1.68	4.90	6.75	8.68	5.40	7.50	4.88	6.18	1.85	1.02	1.12
27	2.72	1.78	5.25	7.05	7.02	6.35	10.25	5.50	5.80	3.22	1.10	1.05
28	2.60	1.55	5.72	9.65	6.70	6.60	11.45	5.08	5.42	3.20	2.15	.90
29	2.58	1.70	5.15	16.10	....	5.75	9.55	4.55	5.65	2.60	1.65	.85
30	3.48	3.20	4.98	19.48	....	4.95	7.02	3.82	4.05	9.98	2.05	.92
31	4.65	....	5.75	18.70	....	4.50	....	3.50	....	5.25	3.20	....

SOUTH FORK OF LICKING RIVER AT HAYES, KY.

LOCATION.—At two-span steel highway bridge at Hayes, Pendleton County, 2½ miles south of Falmouth.

DRAINAGE AREA.—922 square miles (measured by United States Engineer Corps).

RECORDS AVAILABLE.—July 7, 1916, to September 30, 1918.

GAGE.—Chain gage attached to downstream handrail of bridge: read by J. K. Frazer. Sea-level elevation of zero of gage, 540.10 feet.

DISCHARGE MEASUREMENTS.—Made from upstream side of bridge.

CHANNEL AND CONTROL.—Bed of river composed of ledge rock; banks lined with vegetation. Control about 800 feet below gage; probably permanent. Backwater begins to affect the stage-discharge relation at this station when the main Licking River reaches a stage of about 28 feet on the gage at Falmouth.

EXTREMES OF STAGE.—Maximum stage recorded during year, 15.9 feet February 9; minimum stage, 0.24 foot October 5.

1916–1918: Maximum stage recorded, that of February 9, 1918; minimum stage, 0.20 foot at 6 a. m. September 6, 1917.

ICE.—Stage-discharge relation not affected by ice except during severe winters.

ACCURACY.—Stage-discharge relation probably permanent, except as may be affected by ice part of December and January. Rating curve not fully developed. Gage read to hundredths twice daily. As gage has not been checked since August 2, 1917, readings may be too large owing to elongation of gage chain.

COOPERATION.—Base data furnished by United States Engineer Corps.

No discharge measurements were made at this station during the year.

Daily gage height, in feet, of South Fork of Licking River at Hayes, Ky., for the year ending Sept. 30, 1918.

Day.	Oct.	Nov.	Dec.	Jan.	Feb.	Mar.	Apr.	May.	June.	July.	Aug.	Sept.
1	0.41	0.76	0.93	2.63	6.08	2.49	1.54	2.12	1.47	1.13	2.55	2.35
2	.35	1.09	1.15	2.24	5.35	2.34	1.47	1.90	1.36	1.02	1.92	1.66
3	.29	1.26	1.34	2.16	5.10	2.18	3.33	1.74	1.22	1.01	1.58	1.45
4	.26	1.08	1.19	2.10	4.60	2.10	3.27	1.60	1.15	1.02	1.37	1.71
5	.24	.98	1.05	2.03	4.12	2.09	2.47	1.48	1.08	.97	1.19	2.06
6	.26	1.00	.99	4.92	4.22	2.18	2.03	1.39	2.12	.93	1.04	1.57
7	.47	.96	.93	5.18	5.85	2.87	1.80	1.33	1.80	.83	.88	1.43
8	.47	.84	1.15	5.68	11.72	2.57	1.67	1.24	1.82	.86	82	1.29
9	.39	.74	.97	4.05	14.10	2.38	1.54	1.31	1.39	.80	.72	1.09
10	.32	.70	1.16	3.04	10.72	2.11	1.51	1.24	1.21	.80	.74	1.01
11	.26	.75	1.11	2.58	7.00	2.00	1.51	1.32	1.06	.75	.71	.91
12	.33	.71	1.02	3.05	6.95	1.96	1.46	1.94	.95	.70	.67	1.06
13	.35	.62	1.13	3.85	5.15	5.88	1.39	5.15	.95	.64	.57	1.72
14	.36	.64	1.02	3.55	4.28	7.22	1.32	5.75	.89	.67	.77	1.07
15	.33	.60	.86	3.62	4.05	4.80	1.25	4.25	.82	.66	.67	.81
16	.30	.55	.83	3.62	4.22	3.68	1.23	3.11	.77	.54	.55	.86
17	.28	.53	.85	7.38	3.65	3.03	1.22	2.60	.73	.55	.44	.96
18	.27	.59	.82	6.08	3.11	2.73	1.14	2.31	.54	.57	.38	.81
19	.28	.53	.85	4.92	3.77	2.46	1.19	2.03	.49	1.03	1.10	.73
20	.33	.47	1.07	4.12	7.25	2.25	1.19	3.01	.46	1.13	.96	.64
21	.31	.49	1.72	3.80	6.02	2.12	2.38	5.88	.41	.78	.82	.66
22	.27	.47	2.67	3.58	4.38	1.98	2.86	5.98	.39	.67	.77	.66
23	.29	.50	2.61	3.29	3.58	1.89	3.14	3.95	.41	.63	.73	.75
24	.31	.41	3.11	3.30	3.14	1.81	2.40	3.08	.34	.58	.70	.76
25	.29	.42	3.00	3.13	2.85	1.93	3.24	2.62	.86	.56	.66	.60
26	.27	.49	3.00	2.99	3.72	2.22	3.16	2.28	1.58	.76	.58	.59
27	.26	.42	3.02	3.03	2.97	2.19	2.94	2.03	1.37	.72	.53	.49
28	.43	.45	2.48	3.78	2.71	1.99	3.08	1.96	2.15	.67	1.41	.39
29	.41	.52	1.97	7.88	....	1.85	2.73	1.92	1.94	.59	1.71	.39
30	.70	.59	1.99	9.18	....	1.71	2.40	1.70	1.33	6.62	1.33	.39
31	.81	....	2.14	7.28	....	1.62	....	1.69	....	3.82	2.21	....

MIAMI RIVER BASIN.

MIAMI RIVER AT VENICE, OHIO.

Location.—About 400 feet downstream from boundary line between Hamilton and Butler counties, at single-span highway bridge three-fourths of a mile southeast of Venice, Butler County. Indian Creek enters from right 1.4 miles above station.

Drainage area.—3,790 square miles (measured by United States Army Engineers).

Records available.—June 14, 1915, to September 30, 1918.

Gage.—Chain gage fastened to downstream side of bridge; read by H. B. Watson.

Discharge measurements.—Made from downstream side of bridge.

Channel and control.—The control for medium stages is the remains of an old milldam about 1¼ miles below the gage. For stages below about 3 feet a riffle is formed by an unstable gravel bar under the bridge. This bar scours out during high water and reforms at low stages. All water flows under the bridge for stages less than 25 feet.

Extremes of stage.—Maximum stage recorded during year, 18.72 feet at 7 a. m. February 13; minimum stage, 1.50 feet at 6 p. m. August 25.

1915–1918: Maximum stage recorded, 23.1 feet February 1, 1916 (discharge, 52,300 second-feet); minimum stage, 1.31 feet September 5, 1916.

The highest known stage corresponds to about 38 feet on the gage during the flood of 1913.

Diversions.—The Miami & Erie canal is fed by water taken from Miami River at Middletown and Miamisburg, Ohio. The canal at Lindenwald, near the point where it leaves the drainage basin, has a flow of about 100 second-feet, which is a considerable part of the low-water flow of Miami River.

Regulation.—The flow during low stages is probably regulated to a large extent by power plants in Hamilton.

Accuracy.—Stage-discharge relation practically permanent except for possible slight changes at low stage because of shifts in the gravel bar at the bridge; probably affected by ice during part of December, January, and February. Gage read to hundredths twice daily.

Cooperation.—Gage-height record furnished by United States Engineer Corps.

No discharge measurements were made at this station during the year.

Daily gage height, in feet, of Miami River at Venice, Ohio, for the year ending Sept. 30, 1918.

Day.	Oct.	Nov.	Dec.	Jan.	Feb.	Mar.	Apr.	May.	June.	July.	Aug.	Sept.
1	1.59	3.60	1.89	3.11		6.00	2.84	5.00	2.90	2.40	2.86	5.07
2	1.62	3.17	1.80	3.36		8.58	3.00	4.30	2.76	2.46	2.52	3.53
3	1.60	2.72	1.76	3.61		6.64	4.15	3.85	2.60	2.28	2.30	2.84
4	1.60	2.46	1.82			6.18	4.27	3.56	2.54	2.10	2.14	2.70
5	1.62	2.33	1.83			7.22	3.72	3.28	2.46	2.00	2.08	2.77
6	1.63		1.84	5.36		7.08	3.46	3.14	2.90	1.94	1.98	3.08
7	1.56	2.10	1.82	7.43		6.23	3.14	3.05	3.32	1.92	1.89	3.02
8	1.57	2.05	1.78	5.16		5.40	3.04	3.17	3.20	1.90	1.86	2.78
9	1.59	1.97	1.82	4.17	7.42	5.12	2.90	3.17	2.82	1.87	1.81	2.50
10	1.56	1.96		3.82	8.13	5.07	2.80	3.12	2.64	1.82	1.78	2.30
11	1.58	1.90		3.67	9.78	5.82	2.82	3.00	3.42	1.78	1.75	2.16
12	1.62	1.84			14.88	5.17	2.82	9.90	2.79	1.74	1.74	3.92
13	1.60	1.86			18.41	7.51	2.84	16.22	2.42	1.70		3.94
14	1.56	1.83			15.80	9.98	2.80	12.28	3.24	1.69	1.84	3.48
15	1.55	1.82			13.82	9.60	2.63	8.38	2.14	1.65	1.90	3.14
16	1.58	1.82			10.46	7.23	2.67	6.72	2.09	1.64	1.80	5.85
17	1.60	1.80			7.58	6.04	3.95	5.63	1.98	1.62	1.76	5.93
18	1.62	1.76	2.11		6.46	5.24	4.26	4.86	1.98	1.62	1.66	5.05
19	2.36	1.74	2.11		7.53	4.76	3.56	4.38	1.92	1.62	1.63	5.78
20	2.44	1.77	1.68		11.21	4.46	3.33	4.50	1.84	1.61	1.64	6.14
21	2.04	1.78	1.78		8.52	4.13	3.62	4.12	1.78	1.58	1.62	5.09
22	1.96	1.77	1.80		6.38	3.94	3.40	3.80	1.74	2.34	1.58	4.36
23	1.83	1.80	1.82		5.70	3.80	3.26	3.72	1.70	3.02	1.56	3.78
24	1.76	1.78	2.08		5.09	3.70	3.20	3.66	1.76	4.46	1.52	3.49
25	1.75	1.70	2.38		5.06	3.84	3.10	3.66	2.19	3.31	1.50	3.26
26	1.76	1.78	2.45		9.78	3.52	4.46	3.24	2.22	2.90	1.72	3.11
27	1.70	1.76	2.36		6.46	3.29	6.63	3.18	1.86	2.88	3.28	2.97
28	1.63	1.84	2.25		5.59	3.19	6.39	3.18	2.20	2.40	3.65	2.81
29	1.67	1.85	1.66			3.06	7.92	2.97	2.12	3.66	4.00	2.73
30	5.36	1.86	2.07			2.98	6.06	2.92	2.13	4.72	4.16	2.58
31	4.46		2.23			2.95		2.90		3.92	6.39	

Note.—Gage not read Dec. 10–17, Jan. 4–5, 12–31, and Feb. 1–8.

WHITEWATER RIVER AT BROOKVILLE, IND.

Location.—At two-span steel highway bridge three-fourths mile south of Brookville, Franklin County, and 2,000 feet below junction of East and West forks of Whitewater River.

Drainage area.—1,180 square miles.

Records available.—June 8, 1915, to September 30, 1918.

Gage.—Chain gage fastened to downstream side of bridge; read by H. Koemer and Raymond Logan.

Discharge measurements.—Made from downstream side of bridge.

Channel and control.—Control about 500 feet below gage is probably permanent.

Extremes of stage.—Maximum stage recorded during year, 9.53 feet at 7 a. m. May 13; minimum stage, 0.94 foot at 6 p. m. August 24.

1915–1918: Maximum stage recorded, 17.18 feet January 31, 1916 (discharge, about 54,000 second-feet); minimum stage, 0.94 foot at 6 p. m. August 24, 1918.

Regulation.—Flow regulated to some extent by the Thompson-Norris strawboard mill at Brookville. Water is diverted from the West Fork about 10 miles above station and flows down the old Whitewater canal to the mill and is returned to the river a few hundred feet above junction of the East and West forks.

Accuracy.—Stage-discharge relation practically permanent; probably affected by ice during part of December and January. Gage read to hundredths twice daily.

Cooperation.—Gage-height record furnished by United States Engineer Corps.

No discharge measurements were made at this station during the year.

Daily gage height, in feet, of Whitewater River at Brookville, Ind., for the year ending Sept. 30, 1918.

Day.	Oct.	Nov.	Dec.	Jan.	Feb.	Mar.	Apr.	May.	June.	July.	Aug.	Sept.
1	1.23	2.13	1.32	1.31	1.41	3.54	1.84	2.88	1.67	2.21	1.28	1.82
2	1.23	1.82	1.29	1.29	1.34	3.86	2.09	2.62	1.62	1.73	1.25	1.54
3	1.19	1.69	1.29	1.27	1.37	2.97	3.15	2.43	1.57	1.54	1.17	1.43
4	1.21	1.61	1.24	1.29	1.35	3.44	2.54	2.33	1.57	1.44	1.16	1.84
5	1.23	1.56	1.27	1.31	1.49	4.07	2.61	2.22	1.82	1.38	1.12	2.17
6	1.24	1.53	1.27	3.39	1.47	4.22	2.39	2.08	2.11	1.36	1.10	1.97
7	1.21	1.49	1.28	3.33	1.48	2.87	2.27	2.49	1.87	1.32	1.08	1.62
8	1.17	1.51	1.25	2.28	1.78	2.48	2.12	2.34	1.74	1.28	1.04	1.50
9	1.19	1.47	1.29	2.00	5.52	2.54	2.00	2.14	1.61	1.24	1.12	1.43
10	1.15	1.46	1.25	1.92	4.27	2.67	1.96	2.06	1.53	1.23	1.03	1.35
11	1.19	1.45	1.23	1.80	5.52	2.55	2.00	2.08	1.62	1.22	1.06	1.31
12	1.21	1.41	1.23	1.52	7.12	2.67	2.02	6.93	1.63	1.28	1.12	4.05
13	1.21	1.28	1.17	1.48	6.32	4.48	1.93	8.18	1.49	1.20	1.08	2.10
14	1.22	1.25	1.17	1.59	4.27	4.22	1.79	4.58	1.44	1.12	1.14	1.72
15	1.19	1.25	1.21	1.63	6.32	3.18	1.76	3.65	1.42	1.23	1.35	1.56
16	1.21	1.33	1.25	1.61	4.77	2.86	1.76	2.98	1.40	1.19	1.38	3.86
17	1.15	1.35	1.23	1.54	2.52	2.60	3.49	2.86	1.40	1.31	1.15	3.19
18	1.23	1.34	1.23	1.42	2.66	2.40	3.89	2.65	1.34	1.34	1.17	2.46
19	1.76	1.30	1.24	1.49	2.58	2.40	2.70	2.52	1.35	1.28	1.18	3.80
20	1.68	1.32	1.29	1.56	6.12	2.26	2.48	2.45	1.33	1.22	1.08	3.08
21	1.51	1.29	1.60	1.45	3.56	2.18	3.75	2.30	1.29	1.16	1.02	2.38
22	1.41	1.27	1.49	1.46	2.46	2.12	3.19	2.19	1.29	1.37	1.03	2.13
23	1.32	1.29	1.72	1.47	2.46	2.15	2.85	2.18	1.24	1.69	1.08	1.91
24	1.33	1.32	1.73	1.56	2.49	2.17	2.77	2.08	1.30	1.50	.98	1.75
25	1.33	1.29	1.74	1.45	3.60	2.35	2.64	1.99	1.66	1.47	1.01	1.68
26	1.30	1.28	1.68	1.51	3.97	2.16	2.68	1.92	1.71	1.57	1.86	1.62
27	1.27	1.23	1.65	1.43	3.68	2.12	4.28	1.89	1.49	1.97	1.40	1.49
28	1.29	1.26	1.78	1.52	2.84	1.94	4.51	1.80	2.42	2.00	2.58	1.52
29	3.18	1.37	1.62	1.54	....	1.87	4.08	1.74	2.05	1.75	1.87	1.45
30	3.40	1.34	1.29	1.53	....	1.86	3.31	1.81	1.81	1.78	3.38	1.41
31	2.36	....	1.33	1.47	....	1.84	....	1.77	....	1.42	2.59	....

KENTUCKY RIVER BASIN.

DIX RIVER NEAR BURGIN, KY.

LOCATION.—At covered wooden highway bridge on Burgin and Buena Vista pike, 3¾ miles due east of Burgin, Mercer County. Kennedy's mill is a quarter of a mile above station.

DRAINAGE AREA.—395 square miles (86 per cent measured on topographic maps and 14 per cent on map of Kentucky, compiled by United States Geological Survey; scale, 1:500,000).

RECORDS AVAILABLE.—July 2, 1910, to July 16, 1911; October 1, 1911, to September 30, 1918.

GAGE.—Staff gage attached to right upstream wing wall of bridge near face of abutment; read by Frank Martin. Soundings taken at the measuring section indicate that the zero of the gage as replaced by the observer on February 15, 1913, is approximately 0.2 foot below zero of gage installed when station was established. Gage readings subsequent to February 15, 1913, refer to a datum which is about 0.2 foot below datum of original gage.

DISCHARGE MEASUREMENTS.—Made from upstream side of bridge, from a boat, or by wading.

CHANNEL AND CONTROL.—Probably permanent except during extreme floods. At stages above low water the growth of foliage on trees and brush at the control may affect the stage-discharge relation to a small extent.

EXTREMES OF DISCHARGE.—Maximum stage recorded during year, 19.0 feet at 5 p. m. January 28 (discharge, 14,500 second-feet); minimum stage recorded, 2.60 feet at 6 a. m. June 19 (discharge, 0.8 second-foot).

1910–1918: Maximum stage about 30 feet; date unknown. Minimum stage same as for 1918.

ICE.—Ice forms only during severe winters.

ACCURACY.—Stage-discharge relation practically permanent; not affected by ice during the year. Rating table well defined up to 455 second-feet and fairly well defined between 455 and 12,000 second-feet, above 12,000 second-feet, curve is an extension. Gage read twice daily to quarter-tenths. Daily discharge ascertained by applying mean daily gage height to rating table. Records good.

The following discharge measurement was made by Hopkins and Kidwell:

June 11, 1918: Gage height, 3.14 feet; discharge, 12.9 second-feet.

Daily discharge, in second-feet, of Dix River near Burgin, Ky., for the year ending Sept. 30, 1918.

Day.	Oct.	Nov.	Dec.	Jan.	Feb.	Mar.	Apr.	May.	June.	July.	Aug.	Sept.
1	85	318	49	374	1,150	374	100	190	36	92	44	4.5
2	54	190	66	345	735	331	116	133	25	75	38	4.5
3	47	142	75	305	608	244	133	116	28	40	29	2.6
4	27	108	66	331	875	280	1,530	100	17	22	14	2.6
5	17	88	54	331	438	529	875	40	16	14	12	3.6
6	11	75	49	359	438	875	389	49	11	7.4	7.4	3.6
7	9.0	65	47	3,600	875	825	306	62	22	4.5	8.2	3.6
8	14	58	49	1,400	2,170	389	233	54	22	7.4	14	3.6
9	9.4	53	44	691	3,310	374	455	44	14	23	14	3.6
10	9.4	47	40	421	2,590	318	359	33	11	40	9.4	3.6
11	11	44	36	389	1,600	268	211	116	11	22	6.0	3.6
12	9.4	40	31	455	1,090	292	211	151	11	18	4.5	3.6
13	7.4	40	38	1,090	1,030	331	190	305	9.4	11	4.1	3.6
14	7.8	36	42	1,150	735	331	151	2,590	8.2	6.0	3.0	3.6
15	12	36	38	1,940	691	305	142	1,150	6.8	3.6	3.6	3.6
16	14	36	31	4,500	735	244	116	331	5.1	2.6	3.6	3.6
17	16	36	36	3,220	491	211	124	280	3.0	3.6	7.4	3.6
18	23	33	42	1,660	389	180	100	160	1.5	1.5	4.5	3.6
19	318	31	47	875	359	170	97	160	1.4	1.5	5.1	3.6
20	405	28	78	649	3,220	151	78	222	1.5	2.0	4.1	3.6
21	256	27	133	825	3,800	151	649	244	2.6	2.0	3.0	3.6
22	180	25	151	730	1,270	151	1,030	389	2.6	4.5	2.4	3.6
23	116	23	389	635	780	151	405	280	2.6	7.4	1.5	3.6
24	82	20	389	545	649	151	280	190	2.6	75	2.0	3.6
25	62	17	491	455	491	190	268	133	4.5	491	2.0	3.6
26	51	16	875	608	529	389	233	108	7.4	359	1.7	3.6
27	47	16	608	5,490	608	318	405	78	7.4	305	2.4	3.6
28	44	20	491	12,700	455	211	374	66	14	66	3.6	3.6
29	42	27	345	10,300	......	190	268	54	25	70	2.6	3.6
30	491	44	318	2,420	......	151	233	47	14	62	5.1	3.6
31	649	......	374	1,800	......	116	......	40	......	49	8.2	......

NOTE—Gage not read Jan. 22-24; discharge interpolated.

Monthly discharge of Dix River near Burgin, Ky., for the year ending Sept. 30, 1918.

[Drainage area, 395 square miles.]

Month.	Discharge in second-feet.				Run-off in inches.
	Maximum.	Minimum.	Mean.	Per square mile.	
October	649	7.4	101	0.256	0.30
November	318	16	58	.147	.16
December	875	31	178	.451	.52
January	12,700	305	1,950	4.94	5.70
February	3,800	359	1,150	2.91	3.03
March	875	116	296	.749	.86
April	1,530	78	335	.848	.95
May	2,590	33	255	.646	.74
June	36	1.4	11.3	.029	.03
July	491	1.5	60.9	.154	.18
August	44	1.5	8.72	.022	.03
September	4.5	2.6	3.59	.0091	.01
The year	12,700	1.4	364	.922	12.51

ELKHORN CREEK AT FORKS OF ELKHORN, KY.

LOCATION.—At footbridge at Forks of Elkhorn, Franklin County, three-fourths mile below forks of stream and 5 miles northeast of Frankfort.

DRAINAGE AREA.—415 square miles (measured by United States Engineer Corps).

RECORDS AVAILABLE.—April 26, 1915, to September 30, 1918.

GAGE.—Vertical staff in two sections on left bank; section reading 0 to 5 feet attached to elm tree 40 feet below bridge, other section attached to sycamore tree about 20 feet below bridge; read by L. I. McDaniel.

DISCHARGE MEASUREMENTS.—Made from footbridge.

CHANNEL AND CONTROL.—Bed of stream loose stone and bedrock; probably permanent. Control short distance below gage, composed of solid rock and boulders; permanent.

EXTREMES OF DISCHARGE.—Maximum stage recorded during year, 9.7 feet at 6 p. m. February 9 (discharge, 8,730 second-feet); minimum stage, 0.2 foot for long periods (discharge, 49 second-feet).

ICE.—Stage-discharge relation probably not affected by ice except during severe winters.

ACCURACY.—Stage-discharge relation probably permanent; not affected by ice during year. Rating curve well defined, 65 to 18,000 second-feet and fairly well defined at other stages. Gage read to tenths twice daily; record only fair. Daily discharge ascertained by applying mean daily gage height to rating table. Records fair.

COOPERATION.—Base data furnished by United States Engineer Corps.

No discharge measurements were made at this station during the year.

Daily discharge, in second-feet, of Elkhorn Creek at Forks of Elkhorn, Ky., for the year ending Sept. 30, 1918.

Day.	Oct.	Nov.	Dec.	Jan.	Feb.	Mar.	Apr.	May.	June.	July.	Aug.	Sept.
1	49	49	49	135	455	266	135	86	109	128	109	81
2	49	49	49	135	335	222	135	86	109	103	86	66
3	49	49	49	135	335	213	164	86	109	81	81	97
4	49	49	49	135	335	213	149	86	109	76	66	109
5	49	49	49	149	335	213	135	86	109	76	66	109
6	49	49	49	370	335	213	135	86	109	76	66	109
7	49	49	49	455	1,540	232	135	86	135	76	66	86
8	49	49	49	570	6,000	213	135	86	135	109	57	71
9	49	49	49	278	8,190	204	135	86	135	97	49	66
10	49	49	49	180	6,000	172	135	86	97	76	49	66
11	49	49	49	164	2,980	156	135	142	86	76	49	66
12	49	49	49	164	2,420	149	135	1,800	86	76	49	66
13	49	49	49	164	2,060	910	135	1,710	86	76	49	66
14	49	49	49	164	1,220	910	135	1,540	81	66	49	66
15	49	49	49	196	1,540	320	128	765	66	66	49	66
16	49	49	49	455	1,060	232	109	455	66	57	49	49
17	49	49	49	570	605	180	97	266	62	49	49	49
18	49	49	49	370	455	135	86	222	53	49	49	49
19	49	49	49	335	1,220	135	86	149	49	97	49	49
20	49	49	49	335	3,920	135	109	135	49	62	49	49
21	49	49	86	335	2,420	135	180	1,970	49	49	49	49
22	49	49	97	335	1,460	135	122	1,620	49	49	49	49
23	49	49	109	213	835	135	86	730	49	49	49	49
24	49	49	109	196	600	135	86	305	49	86	49	49
25	49	49	135	180	540	135	86	232	180	122	49	49
26	49	49	135	180	370	135	109	172	149	122	49	49
27	49	49	135	335	335	135	86	135	128	116	49	49
28	49	49	135	1,140	305	135	86	135	103	109	49	49
29	49	49	135	1,970	......	135	86	135	86	122	49	49
30	49	49	135	1,380	......	135	86	135	122	232	57	49
31	49	......	135	835	......	135	......	135	......	149	97	......

Monthly discharge of Elkhorn Creek at Forks of Elkhorn, Ky., for the year ending Sept. 30, 1918.

[Drainage area, 415 square miles.]

Month.	Discharge in second-feet.				Run-off in inches.
	Maximum.	Minimum.	Mean.	Per square mile.	
October	49	49	49	0.118	0.14
November	49	49	49	.118	.13
December	135	49	75	.181	.21
January	1,970	135	405	.976	1.13
February	8,190	305	1,720	4.14	4.31
March	910	135	223	.537	.62
April	180	86	119	.287	.32
May	1,970	86	443	1.07	1.23
June	180	49	93.5	.225	.25
July	232	49	89.6	.216	.25
August	109	49	57.4	.138	.16
September	109	49	64.2	.155	.17
The year	8,190	49	273	.658	8.92

EAGLE CREEK AT GLENCOE, KY.

Location.—At county highway bridge half a mile south of Glencoe, Gallatin County.

Drainage area.—445 square miles (United States Engineer Corps).

Records available.—April 29, 1915, to September 30, 1918.

Gage.—Vertical staff attached to upstream side of first pier from left abutment of bridge; read by Athaleen Connelly and Elphia Connelly.

Discharge measurements.—Made from bridge.

Channel and control.—Bed of stream sand and loose stone; probably permanent. Small island covered with trees about 250 feet below bridge. Point of control not determined.

Extremes of discharge.—Maximum stage recorded during year, 17.0 feet at 5 p. m. February 9 (discharge, 17,600 second-feet); minimum stage, 0.1 foot October 3–11 and 14–18 (discharge, 1 second-foot).

Ice.—Stage-discharge relation probably not affected by ice except in very cold winters.

Accuracy—Stage-discharge relation probably permanent; probably not seriously affected by ice during year. Rating curve fairly well defined below 15,000 second-feet, extended above this limit. Gage read twice daily to tenths. Daily discharge ascertained by applying mean daily gage height to rating table. Records fair except those for December and January which may be too large because stage-discharge relation may have been affected by ice.

Cooperation.—Gage-height record furnished by United States Engineer Corps.

The following discharge measurement was made by Hopkins and Kidwell:

June 21, 1918: Gage height, 0.61 foot; discharge, 5.2 second-feet.

Daily discharge, in second-feet, of Eagle Creek at Glencoe, Ky., for the year ending Sept. 30, 1918.

Day.	Oct.	Nov.	Dec.	Jan.	Feb.	Mar.	Apr.	May.	June.	July.	Aug.	Sept.
1	2	87	2	30	155	260	113	183	69	225	260	314
2	2	46	12	30	155	193	100	155	81	120	193	155
3	1	40	46	30	155	173	2,010	140	87	69	138	148
4	1	14	40	30	155	155	1,200	126	51	51	81	488
5	1	9	30	30	155	204	345	126	35	35	57	314
6	1	6	21	1,920	155	193	225	113	30	30	40	362
7	1	5	14	4,390	1,200	173	173	100	51	30	35	204
8	1	5	14	695	4,740	155	155	100	133	30	21	113
9	1	5	14	286	14,400	155	148	94	248	20	14	81
10	1	4	14	155	2,580	133	133	87	164	14	9	57
11	1	4	14	155	330	113	126	94	120	14	6	51
12	2	4	14	155	1,500	113	126	1,280	81	12	6	51
13	2	4	14	155	1,500	2,010	113	5,200	57	14	21	113
14	1	3	14	155	204	6,020	113	2,190	35	8	12	214
15	1	3	14	155	155	870	100	488	30	6	9	140
16	1	3	14	155	155	362	87	314	26	6	9	106
17	1	2	14	155	155	248	193	225	21	6	6	63
18	1	2	14	155	155	204	148	214	14	5	6	51
19	4	2	14	155	236	183	106	173	14	5	6	51
20	3	2	14	155	7,150	164	87	173	9	5	5	40
21	6	2	14	155	1,130	155	810	173	6	5	5	40
22	4	2	645	155	400	140	565	300	6	5	5	30
23	6	2	420	155	273	140	260	248	6	5	4	21
24	6	2	420	155	248	140	193	183	6	21	4	14
25	4	2	420	155	214	330	1,660	155	8	8	4	14
26	4	2	420	155	930	248	780	120	100	6	4	9
27	3	2	420	155	600	183	1,200	106	286	345	8	6
28	3	2	420	155	314	155	465	81	183	204	138	5
29	2	2	420	155	......	133	300	75	300	120	126	5
30	286	2	420	155	......	126	248	57	204	2,980	46	5
31	155	......	420	155	......	126	......	46	......	810	695	......

NOTE.—Gage washed out Dec. 23 to Jan. 1; discharge estimated from weather records and comparison with records for other streams.

Monthly discharge of Eagle Creek at Glencoe, Ky., for the year ending Sept. 30, 1918.

[Drainage area, 445 square miles.]

Month.	Discharge in second-feet.				Run-off in inches.
	Maximum.	Minimum.	Mean.	Per square mile.	
October	286	1	16.4	0.037	0.04
November	87	2	9.0	.020	.02
December	....	2	154	.346	.40
January	4,390	30	350	.787	.91
February	14,400	155	1,410	3.17	3.30
March	6,020	113	450	1.01	1.16
April	2,010	87	408	.917	1.02
May	5,200	46	423	.951	1.10
June	300	6	82	.184	.21
July	2,980	5	168	.378	.44
August	695	4	63.1	.142	.16
September	488	5	109	.245	.27
The year	14,400	1	296	.665	9.03

GREEN RIVER BASIN.

GREEN RIVER AT MUNFORDVILLE, KY.

LOCATION.—At toll highway bridge at Munfordville, Hart County, 1 mile above Louisville & Nashville Railroad bridge.

DRAINAGE AREA.—1,790 square miles (measured on map of Kentucky compiled by United States Geological Survey; scale, 1:500,000).

RECORDS AVAILABLE.—February 27, 1915, to September 30, 1918.

GAGE.—Chain gage attached to upstream handrail of bridge; read by Chester Williams.

DISCHARGE MEASUREMENTS.—Made from upstream side of bridge or by wading 100 feet below the bridge.

CHANNEL AND CONTROL.—The control for low stages is at a riffle used as a ford immediately below the bridge and is believed to be permanent; control at high stages is also believed to be permanent. Discharge relation may be affected to some extent at high stages by differences in the foliage on the brush and trees in the flood plain.

EXTREMES OF DISCHARGE.—Maximum stage recorded during year, 33.04 feet at 5.25 a. m. January 31 (discharge, 28,500 second-feet); minimum stage, 2.65 feet at 5.30 a. m. July 18 (discharge, 72 second-feet).

1915–1918: Maximum stage recorded, 44.48 feet at 5.20 a. m. December 18, 1915 (discharge, 42,400 second-feet); minimum stage, that of July 18, 1918.

Highest known stage about 54 feet; date unknown.

ICE.—Ice seldom forms at this station.

ACCURACY.—Stage-discharge relation practically permanent; affected by ice during parts of December and January. Rating curve well defined below and fairly well defined above 1,700 second-feet. Gage read to hundredths twice daily. Daily discharge ascertained by applying mean daily gage height to rating table. Records good.

COOPERATION.—Station maintained in cooperation with the Kentucky Geological Survey, J. B. Hoeing, State geologist.

Discharge measurements of Green River at Munfordville, Ky., during the year ending Sept. 30 1918.

Date.	Made by—	Gage height.	Discharge.
		Feet.	*Sec.-ft.*
Apr. 13	B. L. Hopkins	4.31	1,250
June 19	Hopkins and Kidwell	2.99	204

Daily discharge, in second-feet, of Green River at Munfordville, Ky., for the year ending Sept. 30, 1918.

Day.	Oct.	Nov.	Dec.	Jan.	Feb.	Mar.	Apr.	May.	June.	July.	Aug.	Sept.
1	490	1,800	358	800	15,700	2,180	788	1,200	475	397	712	230
2	429	1,200	365	700	5,400	1,880	1,580	975	445	305	638	312
3	365	975	397	700	4,050	1,880	6,960	712	712	230	413	275
4	208	788	365	700	3,150	1,880	7,800	638	1,120	245	328	320
5	260	712	365	700	2,250	2,250	5,320	825	638	230	320	445
6	238	675	350	800	2,320	4,120	3,060	788	592	208	305	675
7	215	638	350	3,380	3,220	4,950	2,480	712	675	136	189	568
8	202	512	350	5,880	5,880	3,520	1,950	600	825	95	176	429
9	189	468	335	5,400	8,140	3,150	1,580	675	560	189	176	365
10	189	437	298	3,300	8,740	2,180	1,200	1,120	482	290	176	290
11	208	421	275	1,950	7,720	2,020	1,280	2,250	405	222	176	268
12	290	421	270	1,900	5,250	1,950	1,200	2,020	268	156	170	215
13	358	389	270	1,900	4,280	1,580	1,050	2,780	252	128	136	156
14	320	373	270	2,250	3,750	1,420	975	4,420	290	141	186	136
15	275	358	270	2,920	3,220	1,280	938	6,620	268	170	136	136
16	282	350	260	5,960	2,700	1,120	1,200	4,280	238	150	132	123
17	413	342	260	6,440	2,480	1,050	975	2,180	202	141	150	176
18	305	320	260	6,530	2,180	1,280	1,050	1,280	202	132	170	222
19	245	312	270	4,500	1,950	1,050	1,280	1,280	189	132	429	128
20	260	312	280	2,780	7,040	938	1,880	1,580	170	132	675	123
21	268	298	300	2,180	13,000	938	2,400	2,020	170	132	275	102
22	938	298	500	1,720	14,900	938	3,450	3,220	170	132	230	90
23	373	298	700	1,580	7,550	900	2,480	4,500	170	132	196	90
24	335	282	1,050	1,580	4,580	1,050	1,880	2,020	170	132	202	90
25	320	268	1,580	1,500	3,520	1,420	1,650	2,020	170	320	141	90
26	328	268	2,400	1,500	2,850	1,500	1,720	1,880	429	552	141	82
27	328	252	2,700	7,640	2,700	1,350	2,020	1,050	712	1,500	150	90
28	358	260	2,320	16,700	2,480	1,200	1,880	825	397	2,850	150	90
29	900	320	1,420	25,400	....	1,120	1,500	788	475	1,500	105	90
30	1,050	335	1,050	27,900	....	938	1,350	592	712	3,000	141	90
31	1,580	....	900	27,900	....	862	....	512	....	1,500	100	....

NOTE.—Discharge Dec. 12-13, 31, Jan. 1-6, and 13, estimated because of ice effect.

Monthly discharge of Green River at Munfordville, Ky., for the year ending Sept. 30, 1918.

[Drainage area, 1,790 square miles.]

Month.	Discharge in second-feet.				Run-off in inches.
	Maximum.	Minimum.	Mean.	Per square mile.	
October	1,580	189	407	0.227	0.26
November	1,800	252	489	.273	.30
December	2,700	260	682	.381	.44
January	27,900	700	5,650	3.16	3.64
February	15,700	1,950	5,400	3.02	3.14
March	4,950	862	1,740	.972	1.12
April	7,800	788	2,160	1.21	1.35
May	6,620	512	1,820	1.02	1.18
June	1,120	170	419	.234	.26
July	3,000	95	503	.281	.32
August	712	100	244	.136	.16
September	675	82	217	.121	.14
The year	27,900	82	1,620	.905	12.31

WABASH RIVER BASIN.

VERMILION RIVER NEAR DANVILLE, ILL.

Location.—In sec. 22, T. 19 N., R. 11 W., at Chicago & Eastern Illinois Railroad bridge 3 miles south of Danville, Vermilion County, 1¾ miles above Stony Creek, and 3 miles below mouth of North Fork.

Drainage area.—1,280 square miles.

Records available.—November 12, 1914, to September 30, 1918.

Gage.—Chain gage attached to downstream side of bridge; read by William Taylor.

Discharge measurements.—Made from downstream side of bridge or by wading.

Channel and control.—Soft mud and sand; may shift.

Extremes of discharge.—Maximum stage recorded during year, 11.0 feet at 6 a. m. April 27 (discharge, 5,920 second-feet); minimum stage recorded, 2.21 feet October 17 (discharge, 26 second-feet).

1915–1918: Maximum stage recorded, 18.9 feet January 31, 1916 (discharge, 12,800 second-feet); minimum stage recorded, 2.00 feet November 20 and 23 to 25, 1915 (discharge, 15 second-feet).

Accuracy.—Stage-discharge relation assumed to have changed February 13; affected by ice December 7 to February 13. Rating curves used prior and subsequent to February 13 fairly well defined above 50 second-feet. Gage read to hundredths once daily; readings somewhat unreliable. Daily discharge ascertained by applying daily gage height to rating table except for period of ice effect. Records fair except for very low stages and period of ice effect, for which they are poor.

Discharge measurements of Vermilion River near Danville, Ill., during the year ending Sept. 30, 1918.

[Made by H. C. Beckman.]

Date.	Gage height.	Discharge.	Date.	Gage height.	Discharge.
	Feet.	*Sec.-ft.*		*Feet.*	*Sec.-ft.*
Nov. 20	2.60	86	June 26	3.50	421
20	2.60	92	Aug. 30	2.47	69

Daily discharge, in second-feet, of Vermilion River near Danville, Ill., for the year ending Sept. 30, 1918.

Day.	Oct.	Nov.	Dec.	Jan.	Feb.	Mar.	Apr.	May.	June.	July.	Aug.	Sept.
1	52	373	71			1,530	98	4,480	720	1,160	94	......
2	43	329	71			1,380	126	4,000	669	1,460	81	......
3	37	250	69			1,300	152	3,520	619	1,530	73	......
4	36	216	66			1,230	179	2,880	594	1,600	63	430
5	35	179	64	35		1,160	201	2,580	570	1,380	53	669
6	32	154	62			1,100	216	2,430	546	1,300	47	876
7	31	144			230	1,040	302	2,360	669	1,230	44	1,160
8	30	134				985	343	2,280	771	1,230	42	985
9	30	125				930	408	2,200	876	1,380	39	771
10	35	118				876	430	2,130	930	1,460	36	644
11	37	116				823	475	2,130	985	1,530	......	475
12	33	107				771	498	2,060	930	1,380	......	302
13	31	98	65			720	522	2,130	876	1,300	......	235
14	30	92			2,660	644	570	2,280	771	1,230	......	186
15	29	88			2,640	594	594	2,430	594	1,160	......	179
16	27	84			2,620	546	619	2,500	570	1,040	......	172
17	26	80			2,600	522	669	2,660	522	985	......	152
18	32	78			2,580	498	930	2,730	475	876	......	152
19	54	77			2,500	475	1,100	2,730	430	823	......	149
20	62	75			2,430	475	1,230	2,500	386	720	......	146
21	66	73		30	2,360	452	1,530	2,360	322	619	......	146
22	71	71			2,360	430	1,980	2,280	282	498	......	152
23	62	73			2,280	430	2,730	2,130	322	386	......	159
24	52	78			2,200	386	3,360	1,900	386	322	......	166
25	48	82			2,130	343	4,720	1,680	452	262	......	172
26	59	86	55		2,060	322	5,280	1,530	546	216	......	179
27	71	84			1,980	282	5,920	1,380	594	172	......	186
28	114	80			1,830	246	5,600	1,230	771	152	......	193
29	269	77			......	223	5,120	1,040	876	140	......	160
30	396	73			......	186	4,880	985	1,040	123	......	126
31	418	......			......	81	......	876	......	108	......	......

Note.—Discharge interpolated Feb. 15–17, Mar. 10, and Sept. 29, because of no gage-height record; discharge, Dec. 7 to Feb. 13, estimated because of ice, from gage heights, observer's notes, and weather records. Braced figures show mean discharge for periods indicated.

Monthly discharge of Vermilion River near Danville, Ill., for the year ending Sept. 30, 1918.

[Drainage area, 1,280 square miles.]

Month.	Discharge in second-feet.				Run-off in inches.
	Maximum.	Minimum.	Mean.	Per square mile.	
October	418	26	75.7	0.059	0.07
November	373	71	123	.096	.11
December			61.9	.048	.06
January			31.6	.025	.03
February	2,660		1,360	1.06	1.10
March	1,530	81	677	.529	.61
April	5,920	98	1,690	1.32	1.47
May	4,480	876	2,270	1.77	2.04
June	1,040	282	636	.497	.55
July	1,600	108	896	.700	.81
August 1-10	94	36	57.2	.045	.02
September 4-30	1,160	126	345	.270	.27

EMBARRASS RIVER AT STE. MARIE, ILL.

LOCATION.—In sec. 30, T. 6 N., R. 14 W., at highway bridge at north end of Main Street, Ste. Marie, Jasper County, 450 feet downstream from Cincinnati, Indianapolis & Western Railroad bridge and 2½ miles upstream from mouth of Hickory (or North Fork) Creek.

DRAINAGE AREA.—1,540 square miles.

RECORDS AVAILABLE.—October 20, 1909, to December 31, 1912; August 24, 1914, to September 30, 1918.

GAGE.—Chain gage attached to bridge; read by V. C. Wuerth.

DISCHARGE MEASUREMENTS.—Made from downstream side of highway bridge at ordinary stages; during high water made also from downstream side of five wooden trestles on Cincinnati, Indianapolis & Western Railroad bridge, northwest of highway bridge.

CHANNEL AND CONTROL.—Measuring section is at a pool; control is about 1,800 feet below gage; may shift.

EXTREMES OF DISCHARGE.—Maximum stage recorded during year, 18.6 feet February 14 and April 30 at 4 p. m. (discharge, 7,240 second-feet); minimum stage recorded, 1.97 feet October 15 to 17 (discharge, 55 second-feet).

1909–1912; 1914–1918: Maximum stage recorded 21.2 feet June 6, 1917 (discharge, 14,000 second-feet); minimum stage recorded, 1.1 feet September 5 to 9 and October 19, 1914 (discharge, 1 second-foot).

Flood of spring of 1908 reached a height of 22.5 feet on the present gage.

ACCURACY.—Stage-discharge relation changed during high water in February; seriously affected by ice December 6 to February 12. Rating curve used to February 13 fairly well defined; curve used after that date fairly well defined between 102 and 5,030 second-feet; above 5,030 second-feet it is based on an extension of curve for main river channel and estimated overflow. Gage read to hundredths once daily. Daily discharge ascertained by applying daily gage height to rating table except for period of ice effect. Open-water records good, except for very high stages and for extremely low stages in August, for which they are fair; records for period of ice effect, poor.

Discharge measurements of Embarrass River at Ste. Marie, Ill., during the year ending Sept. 30, 1918.

[Made by H. C. Beckman.]

Date.	Gage height.	Discharge.
	Feet.	*Sec.-ft.*
Oct. 10	2.05	63
June 24	2.62	143
Aug. 29	2.82	188

Daily discharge, in second-feet, of Embarrass River at Ste. Marie, Ill., for the year ending Sept. 30, 1918.

Day.	Oct.	Nov.	Dec.	Jan.	Feb.	Mar.	Apr.	May.	June.	July.	Aug.	Sept.
1	91	1,060	136			1,530	386	6,780	365	2,860	238	1,500
2	91	1,060	136			2,740	365	6,040	642	2,740	220	344
3	91	1,080	136			2,300	365	5,440	692	3,100	211	742
4	85	780	130			2,100	344	4,560	545	2,060	194	817
5	79	661	124			1,900	894	3,180	498	1,820	365	4,740
6	73	554		270		1,700	1,100	2,380	452	1,640	169	5,510
7	67	458			1,550	950	868	1,980	430	1,010	161	3,810
8	65	422				842	817	1,820	692	3,580	145	2,180
9	62	369				792	742	1,940	452	5,650	145	1,670
10	60	335				717	667	2,100	408	6,220	138	1,280
11	57	335				617	598	2,020	365	5,300	130	980
12	57	335				545	521	4,260	365	3,860	123	2,260
13	55	287	70		6,580	545	521	6,220	324	3,060	109	1,670
14	55	272			7,240	545	475	5,230	285	1,820	102	692
15	55	257			6,680	521	452	4,410	285	1,320	98	498
16	55	242		150	6,130	521	408	3,180	238	1,100	116	521
17	56	227			5,880	521	1,280	3,020	220	894	98	2,820
18	57	212			4,110	521	3,340	2,580	202	792	89	2,820
19	73	212			2,780	475	2,740	2,780	194	767	81	2,020
20	85	212			2,140	452	3,220	2,540	186	717	498	3,020
21	110	198			1,980	430	5,750	1,940	177	642	408	1,900
22	257	184			1,530	408	6,220	1,530	169	521	211	1,320
23	335	184			1,280	386	5,720	1,280	153	452	123	798
24	212	184			1,280	365	5,040	1,160	145	452	98	667
25	184	177			1,160	365	4,620	1,070	920	408	177	545
26	184	170	225	55	1,070	408	4,560	980	3,810	365	169	521
27	170	170			980	452	4,980	868	2,820	344	194	408
28	184	156			920	430	5,440	667	1,600	304	169	386
29	257	150			....	365	6,680	593	3,060	304	116	365
30	955	143			....	344	7,240	452	4,460	304	123	324
31	805	....			....	324	....	452	....	247	2,180	....

NOTE.—Discharge, Dec. 6 to Feb. 12, estimated because of ice, from gage heights, observer's notes, and weather records. Braced figures show mean discharge for periods indicated.

Monthly discharge of Embarrass River at Ste. Marie, Ill., for the year ending Sept. 30, 1918.

[Drainage area, 1,540 square miles.]

Month.	Discharge in second-feet.				Run-off in inches.
	Maximum.	Minimum.	Mean.	Per square mile.	
October	955	55	162	0.105	0.12
November	1,060	143	368	.239	.27
December			135	.088	.10
January	....	....	155	.101	.12
February	7,240	....	2,510	1.63	1.70
March	2,740	324	810	.526	.61
April	7,240	344	2,540	1.65	1.84
May	6,780	452	2,690	1.75	2.02
June	4,460	145	838	.544	.61
July	6,220	247	1,760	1.14	1.31
August	2,180	81	238	.155	.18
September	5,510	324	1,570	1.02	1.14
The year	7,240	55	1,140	.740	10.02

WEST BRANCH OF WHITE RIVER NEAR NOBLESVILLE, IND.

LOCATION.—In sec. 16, T. 19 N., R. 5 E. second principal meridian, at steel highway bridge known as Conners Bridge, 4½ miles northeast of Noblesville, Hamilton County.

DRAINAGE AREA.—900 square miles (measured on map compiled by United States Geological Survey; scale, 1 : 500,000).

RECORDS AVAILABLE.—May 13, 1915, to September 30, 1918.

GAGE.—Chain gage attached to upstream side of bridge; read by Marvin Scearce.

DISCHARGE MEASUREMENTS.—Made from downstream side of bridge or by wading.

CHANNEL AND CONTROL.—Coarse sand and gravel, strewn with boulders; probably permanent.

EXTREMES OF DISCHARGE.—Maximum stage recorded during period of records, 15.0 feet at 7 a. m. February 1, 1916 (discharge, 18,700 second-feet); minimum stage, 1.08 feet at 5 p. m. August 15, 1918 (discharge, 85 second-feet).

During the flood of March, 1913, the water reached a stage of about 21.5 feet on the present gage (discharge not known).

ICE.—Stage-discharge relation affected by ice during severe winters.

ACCURACY.—Stage-discharge relation practically permanent, except for periods of ice effect and from July 8 to August 31, 1915, when there probably was backwater from obstructions. Rating curve used July 8 to August 31, 1915, poorly defined; curve used for remainder of time well defined between 290 and 11,000 second-feet and fairly well defined beyond these limits. Gage read to hundredths twice daily. Daily discharge ascertained by applying mean daily gage height to rating tables, except for periods of ice effect and periods when gage was not read. Records good except for periods of ice effect and for July 8 to August 31, 1915, for which they are poor.

COOPERATION.—Gage-height record furnished by Noblesville Heat, Light & Power Co., Noblesville, Ind.

Discharge measurements of West Branch of White River near Noblesville, Ind., during the year ending Sept. 30, 1918.

[Made by H. C. Beckman.]

Date.	Gage height.	Discharge.	Date.	Gage height.	Discharge.
	Feet.	*Sec.-ft.*		*Feet.*	*Sec.-ft.*
May 15	3.37	1,020	June 21	1.47	148
16	2.92	743	July 2	1.73	215
17	2.62	589			

Daily discharge, in second-feet, of West Branch of White River near Noblesville, Ind., for the years ending Sept. 30, 1915–1918.

Day.	May.	June.	July.	Aug.	Sept.	Day.	May.	June.	July.	Aug.	Sept.
1915.						1915.					
1		1,560	273	345	414	16	330	1,900	1,650	971	239
2		1,480	291	312	392	17	506	1,080	2,660	693	223
3		1,640	255	345	350	18	350	734	2,260	650	239
4		1,320	273	2,460	330	19	273	612	1,420	527	436
5		940	291	3,080	330	20	255	584	1,730	527	371
6		734	273	1,080	330	21	291	532	1,280	1,730	310
7		642	255	875	330	22	310	459	1,020	2,860	273
8		800	2,860	737	310	23	330	371	828	2,460	239
9		1,010	4,810	450	291	24	273	310	875	1,570	209
10		436	4,260	450	291	25	239	291	527	923	209
11		532	1,730	608	273	26	255	273	450	828	209
12		459	1,140	1,730	255	27	255	330	414	693	734
13	150	459	923	3,410	239	28	414	414	379	608	1,810
14	160	330	737	1,980	239	29	2,170	310	379	527	1,160
15	150	2,080	1,140	1,020	239	30	1,990	273	414	450	702
						31	1,900		379	414	

Day.	Oct.	Nov.	Dec.	Jan.	Feb.	Mar.	Apr.	May.	June.	July.	Aug.	Sept.
1915-16.												
1	612	291	506	1,720	17,700	642	1,720	584	557	459	150	160
2	557	273	482	10,300	10,500	584	1,240	532	506	1,010	150	150
3	557	255	459	15,700	3,300	584	1,160	557	482	1,010	141	141
4	482	239	436	10,500	1,810	459	1,160	767	557	940	160	141
5	414	239	414	4,330	1,640	459	940	870	671	905	195	141
6	392	255	371	4,210	1,640	612	940	734	767	835	209	150
7	350	239	350	2,560	1,160	940	905	905	2,360	273	195	239
8	310	239	350	1,640	1,080	1,320	767	1,010	3,190	255	183	171
9	330	223	350	1,560	940	940	671	2,260	1,990	255	255	171
10	273	223	330	1,400	506	870	642	1,900	1,720	239	291	150
11	255	223	330	1,560	557	800	642	1,240	1,480	239	330	150
12	239	310	350	3,080	584	734	612	905	1,160	239	392	132
13	239	330	371	6,090	1,320	642	584	702	905	223	273	124
14	414	330	310	5,790	940	702	532	870	835	273	209	132
15	940	310	414	1,990	940	702	506	734	940	350	195	124
16	940	291	350	1,240	1,160	642	482	671	870	273	183	124
17	734	255	506	940	1,010	642	459	557	835	223	171	132
18	800	310	1,560	1,810	1,240	584	459	482	835	223	506	124
19	1,720	2,080	1,560	3,740	1,160	584	459	436	870	239	330	124
20	1,240	2,970	1,080	4,450	940	557	459	392	734	223	209	124
21	1,010	2,260	532	3,740	870	557	532	392	2,170	371	171	115
22	940	1,560	800	2,760	800	584	642	459	2,760	532	160	115
23	734	1,240	584	2,560	940	940	671	767	2,560	273	532	115
24	506	1,010	506	1,900	1,320	1,240	612	642	1,560	223	506	115
25	459	835	584	1,320	1,400	1,320	557	459	671	209	414	124
26	414	702	1,560	1,240	1,400	1,160	532	414	734	183	255	124
27	371	642	1,080	1,320	940	4,090	584	392	767	183	414	132
28	350	612	1,080	2,170	870	5,220	734	371	642	150	255	150
29	310	612	1,080	2,860	734	4,830	734	459	557	160	239	141
30	310	506	1,080	8,120		3,190	734	734	506	150	183	150
31	291		612	16,300		1,990		905		150	171	

Daily discharge, in second-feet, of West Branch of White River near Noblesville, Ind., for the years ending Sept. 30, 1915-1918—Continued.

Day.	Oct.	Nov.	Dec	Jan.	Feb.	Mar.	Apr.	May.	June.	July.	Aug.	Sept.
1916-17.												
1	141	124	141	506		642	2,260	1,990	2,860	1,010	310	160
2	141	124	150	459		532	2,970	2,860	1,900	1,320	278	160
3	141	124	150	414		506	1,900	2,860	1,320	940	247	160
4	132	124	141	436		532	2,460	3,080	1,160	557	215	160
5	132	124	150	642	650	532	2,660	3,300	1,640	371	183	160
6	124	124	150	4,830		482	4,090	3,970	2,460	392	183	150
7	124	124	150	4,570		414	4,960	2,170	3,080	414	183	150
8	124	115	150	2,760		371	3,740	1,810	1,900	414	171	273
9	115	132	141	1,560		584	2,170	1,640	1,640	436	171	160
10	115	124	141	1,240		940	1,240	1,320	3,080	482	150	160
11	115	115	150	1,080		1,010	1,160	1,240	2,970	482	171	160
12	115	132	310	905		2,660	1,010	1,160	1,990	436	160	160
13	115	141	1,080	702		4,450	835	1,010	1,320	459	160	160
14	124	132	584	532		8,760	702	905	1,240	532	160	160
15	124	132	459		670	7,480	671	671	940	2,080	171	160
16	124	132				3,740	642	612	800	1,900	171	160
17	124	124				1,990	557	584	734	1,560	160	141
18	124	132				1,900	642	557	612	1,560	150	124
19	132	612				1,640	1,080	506	557	905	150	124
20	150	612				1,010	1,720	482	506	642	141	124
21	150	612			2,970	1,320	1,480	506	532	557	141	124
22	150	835		630	2,660	2,320	1,320	642	482	436	150	132
23	141	3,080	390		1,010	,400	1,080	1,160	459	371	141	132
24	141	1,900			671	,460	1,480	1,010	459	330	150	141
25	141	1,240			612	660	1,990	800	436	310	141	132
26	132	124			557	1,990	2,560	702	436	612	141	132
27	124	132			835	2,260	2,970	1,640	905	506	141	132
28	132	132			800	2,360	2,260	2,080	2,080	436	132	132
29	124	124			......	1,160	1,900	2,660	1,560	404	141	124
30	124	150			......	1,240	1,480	3,080	1,240	373	150	124
31	124	......			......	1,010	......	3,300	......	341	160	......
1917-18.												
1	124	940	150			557	532	1,010	371	223	141	160
2	124	734	557			557	1,900	800	255	195	132	132
3	124	532	490			702	2,660	671	239	171	124	124
4	124	506	424			1,240	2,660	557	239	171	115	150
5	124	482	357	240	170	584	1,640	506	223	160	108	940
6	124	436	291			642	1,240	436	209	150	108	584
7	124	350				734	940	414	223	141	100	330
8	124	255				940	734	392	291	132	94	209
9	124	223				1,240	584	350	273	132	94	160
10	124	209				1,010	532	350	223	132	100	150
11	124	195			8,600	1,010	506	330	436	124	100	160
12	124	209			5,360	1,320	459	392	436	124	94	183
13	124	209	250		6,690	2,360	436	436	291	124	94	506
14	124	195			4,700	3,740	392	1,640	239	115	94	255
15	124	183		160	3,410	3,300	350	1,010	209	115	87	209
16	124	171			2,560	2,070	350	734	183	124	94	209
17	124	171			1,990	835	734	584	171	132	108	506
18	183	171			1,320	835	870	506	160	150	115	702
19	835	160			905	835	1,010	459	160	141	108	506
20	800	160			800	734	767	436	150	132	100	436
21	584	160			1,080	612	940	392	150	124	94	350
22	310	160			1,320	557	1,080	392	141	115	183	273
23	209	160			734	482	1,080	414	141	115	141	239
24	195	150			734	482	800	414	132	132	115	209
25	209	150			702	482	642	506	141	115	115	195
26	209	150	280	130	702	532	532	532	141	115	100	183
27	209	150			671	506	767	532	141	124	100	160
28	223	150			612	482	2,260	482	255	115	100	150
29	642	150			......	436	1,560	371	255	239	209	132
30	1,160	150			......	436	1,160	459	239	291	371	141
31	1,160	......			......	371	......	532	......	171	209	......

NOTE.—Discharge estimated Jan. 15-19, 1916, Dec. 16, 1916, to Jan. 3, 1917, Jan. 15 to Feb. 20, 1917, and Dec. 7, 1917, to Feb. 10, 1918, because of ice, from gage heights, observer's notes, and weather records; interpolated for July 29 to Aug. 4, Oct. 1-6, and Dec. 3-5, 1917, and Aug. 5, 1918, because of no gage-height record. Braced figures show mean discharge for periods indicated.

Monthly discharge of West Branch of White River near Noblesville, Ind., for the years ending Sept. 30, 1915–1918.

[Drainage area, 900 square miles.]

Month.	Discharge in second-feet.				Run-off in inches.
	Maximum.	Minimum.	Mean.	Per square mile.	
1915.					
May 13–31	2,170	150	558	0.620	0.44
June	2,080	273	763	.848	.95
July	4,810	255	1,170	1.30	1.50
August	3,410	312	1,140	1.27	1.46
September	1,810	209	399	.443	.49
1915–16.					
October	1,720	239	564	.627	.72
November	2,970	223	662	.736	.82
December	1,560	310	657	.730	.84
January	16,800	940	4,160	4.62	5.33
February	17,700	506	2,050	2.28	2.46
March	5,220	459	1,260	1.40	1.61
April	1,720	459	722	.802	.89
May	2,260	371	745	.828	.95
June	3,190	482	1,170	1.30	1.45
July	1,010	150	364	.404	.47
August	532	141	259	.288	.33
September	239	115	140	.156	.17
The year	17,700	115	1,060	1.18	16.04
1916–17.					
October	150	115	130	.144	.17
November	3,080	115	394	.438	.49
December	1,080	141	332	.369	.43
January	4,830	414	1,010	1.12	1.29
February	2,970	....	833	.926	.96
March	8,760	371	1,910	2.12	2.44
April	4,960	557	1,870	2.08	2.32
May	3,970	482	1,620	1.80	2.08
June	3,080	436	1,380	1.53	1.71
July	2,080	310	696	.773	.89
August	310	132	170	.189	.22
September	273	124	149	.166	.19
The year	8,760	115	874	.971	13.19
1917–18.					
October	1,160	124	291	.323	.37
November	940	150	267	.297	.33
December	557	150	285	.317	.37
January	....	....	175	.194	.22
February	8,600	....	1,590	1.77	1.84
March	3,740	371	988	1.10	1.27
April	2,660	350	1,000	1.11	1.24
May	1,640	330	550	.611	.70
June	436	132	224	.249	.28
July	291	115	147	.163	.19
August	371	87	124	.138	.16
September	940	124	288	.320	.36
The year	8,600	87	486	.540	7.33

LITTLE WABASH RIVER AT WILCOX, ILL.

LOCATION.—In SW. ¼ sec. 3, T. 2 N., R. 8 E., at highway bridge at Wilcox, Clay County, 6 miles southeast of Clay City and a quarter of a mile below mouth of Big Muddy Creek.

DRAINAGE AREA.—1,130 square miles.

RECORDS AVAILABLE.—August 22, 1914, to September 30, 1918.

GAGE.—Chain gage attached to bridge; read by Mrs. Kate Holman.

DISCHARGE MEASUREMENTS.—At ordinary stages made from downstream side of highway bridge, which is at a pool; during high water made also from bridge across drainage ditch and overflow section about half a mile east of highway bridge.

Channel and control.—Heavy clay, probably permanent; control section is about 100 feet below bridge. Point of zero flow was determined August 22, 1914, to be at a stage represented by a gage height about 1.2 feet.

Extremes of discharge.—Maximum stage recorded during year, 22.8 feet at 6 a. m. April 23 (discharge, 6,770 second-feet); minimum stage recorded, 1.81 feet at 6 a. m. October 6 (discharge, 5.2 second-feet).

1914–1918: Maximum stage prevailed August 22, 1915 (gage inaccessible, discharge estimated at 10,000 second-feet); minimum stage recorded, 1.70 feet August 23, 1914 (discharge, 4 second-feet).

Accuracy.—Stage-discharge relation practically permanent; seriously affected by ice December 9 to February 8. Rating curve well defined between 63 and 420 second-feet, fairly well defined below 63 second-feet and between 420 and 3,360 second-feet, and poorly defined above 3,360 second-feet. Gage read to hundredths once daily. Daily discharge ascertained by applying daily gage height to rating table, except for period of ice effect. Records good except for very high stages and for period of ice effect, for which they are poor.

Discharge measurements of Little Wabash River at Wilcox, Ill., during the year ending Sept. 30, 1918.

[Made by H. C. Beckman.]

Date.	Gage height.	Discharge.
	Feet.	*Sec.-ft.*
June 22	2.47	30.2
22	2.47	29.9

Daily discharge, in second-feet, of Little Wabash River at Wilcox, Ill., for the year ending Sept. 30, 1918.

Day.	Oct.	Nov.	Dec.	Jan.	Feb.	Mar.	Apr.	May	June.	July.	Aug.	Sept.
1	7.0	390	21			809	56	4,070	98	1,140	13	1,360
2	7.0	347	21			1,460	46	3,880	108	971	11	1,740
3	5.6	184	20			1,620	46	3,640	108	406	11	1,060
4	6.0	108	18			1,080	56	2,900	98	161	11	436
5	6.0	71	16		10	773	56	1,420	80	161	13	1,600
6	5.2	52	16	135		1,080	80	468	80	139	12	3,160
7	6.6	43	16			845	63	308	71	184	12	3,760
8	6.6	37	16			484	63	280	63	468	10	3,820
9	7.0	32			1,480	319	63	452	76	791	10	3,360
10	7.0	27			3,360	232	56	2,900	134	2,410	12	2,900
11	7.0	26			3,940	196	49	4,140	134	2,590	13	1,560
12	7.0	24			4,140	161	49	4,350	84	2,590	13	484
13	7.0	24			4,210	144	49	5,650	76	2,560	11	308
14	6.0	21			4,560	134	49	5,260	76	1,770	10	172
15	6.0	21	12		4,700	484	46	4,350	84	1,500	10	208
16	6.0	25		100	4,000	256	49	3,580	46	1,220	10	184
17	6.0	32			3,310	172	516	4,070	39	755		580
18	8.0	28			2,620	134	971	2,620	39	614		2,560
19	8.0	24			1,920	113	5,050	1,360	39	144		2,500
20	18.0	20			719	103	4,700	971	26	139		2,260
21	28.0	20			719	95	6,450	2,470	26	84	200	3,310
22	39.0	18			564	87	6,690	3,520	22	84		2,940
23	49.0	16			406	79	6,770	2,590	30	80		1,920
24	12.0	15			268	71	6,530	1,100	30	98		631
25	56.0	15			232	62	6,370	452	30	220		308
26	12.0	13	35	9	232	54	5,970	347	10	134	701	232
27	12.0	14			232	46	5,490	232	618	103	1,240	791
28	12.0	16			220	71	4,280	184	935	71	773	150
29	63.0	24			......	63	4,840	161	935	12	172	150
30	76.0	22			......	63	4,490	150	791	14	71	139
31	648.0	......			......	46	......	118	......	12	172	......

Note.—Discharge interpolated Oct. 20–22, Nov. 9, Dec. 3, Feb. 16–18, and Mar. 21–25, and estimated Aug. 17–25, because of no gage-height record. Discharge, Dec. 9 to Feb. 8, estimated because of ice, from gage heights, observer's notes, and weather records. Braced figures show mean discharges for periods indicated.

Monthly discharge of Little Wabash River at Wilcox, Ill., for the year ending Sept. 30, 1918.

[Drainage area, 1,130 square miles.]

Month.	Discharge in second-feet.				Run-off in inches.
	Maximum.	Minimum.	Mean.	Per square mile.	
October	648	5.2	37.1	0.033	0.04
November	390	13	57.0	.050	.06
December	....	....	21.7	.019	.02
January	....	....	79.0	.070	.08
February	4,700	....	1,500	1.33	1.38
March	1,620	46	366	.324	.37
April	6,770	46	2,330	2.06	2.30
May	5,650	118	2,190	1.94	2.24
June	935	10	167	.148	.17
July	2,590	12	698	.618	.71
August	1,240	10	165	.146	.17
September	3,820	139	1,490	1.32	1.47
The year	6,770	5.2	749	.663	9.01

SKILLET FORK AT WAYNE CITY, ILL.

Location.—In sec. 18, T. 2 S., R. 6 E., at Southern Railway bridge 1 mile east of Wayne City, Wayne County, and 4 miles below mouth of Horse Creek.

Drainage area.—481 square miles.

Records available.—August 16, 1908, to December 31, 1912; June 22, 1914, to September 30, 1918.

Gage.—Chain gage attached to bridge; read by J. C. Taylor.

Discharge measurements.—Made from downstream side of bridge; in high water also from downstream side of wooden trestle about 1 mile east of main channel. Low-water measurements made by wading below gage.

Channel and control.—Channel practically permanent; rough. Control is remains of rock dam at bridge section. Point of zero flow was determined August 20, 1914, to be at a stage represented by a gage height of 1.6 feet.

Extremes of discharge.—Maximum stage recorded during year, 21.2 feet at 1 p. m. May 13 (discharge, 5,400 second-feet); minimum stage recorded, 2.00 feet October 1 to 17 and August 9 to 12. Minimum discharge, 0.7 second-foot, August 9 to 12,

1908–1912; 1914–1918: Maximum stage recorded, 23.1 feet August 22, 1915 (discharge, 9,350 second-feet, supersedes figure previously published); zero flow existed for 54 days in September to December, inclusive, of 1908.

Diversions.—About 30,000 gallons of water a day are pumped from river above gage into service tank of Southern Railway.

Accuracy.—Stage-discharge relation practically permanent; affected by ice December 9 to February 10. Rating curves fairly well defined between 15 and 5,000 second-feet, and poorly defined beyond these limits. Gage read to hundredths once daily. Daily discharge ascertained by applying daily gage height to rating table except for period of ice effect. Records good, except for high stages and for period of ice effect, for which they are poor.

Discharge measurements of Skillet Fork at Wayne City, Ill., during the year ending Sept 30, 1918.

[Made by H. C. Beckman.]

Date.	Gage height.	Discharge.
	Feet.	*Sec.-ft.*
June 20	2.32	4.8
20	2.31	4.6

Daily discharge, in second-feet, of Skillet Fork at Wayne City, Ill., for the year ending Sept. 30, 1918.

Day.	Oct.	Nov.	Dec.	Jan.	Feb.	Mar.	Apr.	May.	June.	July.	Aug.	Sept.
1	0.8	59	22			355	9.0	4,050	16	95	3.8	13
2	.8	37	9.5			1,340	9.0	3,020	15	46	3.8	8.2
3	.8	34	9.5			770	8.2	788	15	26	1.3	8.2
4	.8	18	8.7			551	6.6	192	15	14	1.3	34
5	.8	15	7.9			406	6.6	88	15	9.0	1.2	395
6	.8	12	7.5	8	350	564	5.0	59	635	6.6	1.2	551
7	.8	9.5	6.3			318	16	41	806	5.0	1.2	752
8	.8	9.5	4.2			185	14	37	80	4.0	1.2	220
9	.8	7.5				125	9.0	538	34	3.2	.7	44
10	.8	7.5				92	7.4	260	28	2.3	.7	24
11	.8	7.5			3,670	56	5.8	125	13	2.3	.7	14
12	.8	7.5			3,600	52	5.8	2,700	13	2.3	.7	11
13	.8	7.5			3,950	32	5.0	5,400	5.0	5.0	1.9	9.0
14	.8	7.5			4,050	30	5.0	4,980	5.0	16	1.3	32
15	.8	7.5	2		3,600	28	5.0	4,250	5.0	11	1.3	18
16	.8	5.5			2,780	26	5.0	3,810	4.8	8.2	1.3	25
17	.8	5.5			605	22	291	1,800	4.8	4.8	1.3	245
18	3.0	5.5			207	21	3,020	283	4.8	4.5	9.0	463
19	5.5	5.5			125	21	3,100	365	4.5	3.8	40	395
20	5.5	5.5			132	18	3,190	475	4.5	2.3	22	318
21	2.2	5.5		1	102	14	3,670	1,420	6.6	2.3	15	551
22	2.2	5.5			98	14	3,530	1,470	5.0	2.3	15	512
23	2.2	4.2			73	11	3,350	1,420	5.0	2.3	9.0	125
24	2.2	4.2			59	11	1,620	428	3.2	2.3	9.0	40
25	2.2	4.2			56	15	2,180	155	3.2	1.9	8.2	28
26	2.2	4.2	20		52	17	4,150	110	2.8	1.8	26	18
27	2.2	9.5			52	11	4,350	80	2.4	1.8	44	14
28	22	9.5			245	10	4,250	31	2.4	3.5	140	8.6
29	200	37			......	9	4,850	25	207	3.2	88	7.0
30	252	62			......	9	4,350	22	102	25	33	5.4
31	185	......			......	9	......	21	......	5.4	21	......

NOTE.—Discharge, Dec. 9 to Feb. 10, estimated, because of ice, from gage heights, observer's notes, and weather records. Braced figures show mean discharge for periods indicated.

Monthly discharge of Skillet Fork at Wayne City, Ill., for the year ending Sept. 30, 1918.

[Drainage area, 481 square miles.]

Month.	Discharge in second-feet.				Run-off in inches.
	Maximum.	Minimum.	Mean.	Per square mile.	
October	252	0.8	22.6	0.047	0.05
November	62	4.2	14.0	.029	.03
December	......	......	10.3	.021	.02
January	......	......	3.26	.0068	.008
February	4,050	......	963	2.00	2.08
March	1,340	9.0	166	.345	.40
April	4,850	5.0	1,530	3.18	3.55
May	5,400	21	1,240	2.58	2.97
June	806	2.4	68.8	.143	.16
July	95	1.8	10.4	.022	.03
August	140	.7	16.2	.034	.04
September	752	5.4	163	.339	.38
The year	5,400	.7	345	.717	9.72

CUMBERLAND RIVER BASIN.

CUMBERLAND RIVER AT CUMBERLAND FALLS, KY.

LOCATION.—At Cumberland Falls post office, Whitley County, 400 feet above falls, 13 miles from Parkers Lake post office and Cumberland Falls railroad station. McCreary County, on Queen & Crescent Route.

DRAINAGE AREA.—2,040 square miles (measured on maps of Kentucky and Tennessee prepared by the United States Geological Survey; scale, 1:500,000).

RECORDS AVAILABLE.—August 15, 1907, to December 10, 1911; April 1, 1915, to September 30, 1918.

GAGE.—Staff, inclined and vertical, on right bank, 400 feet above brink of falls, established April 3, 1915; read by Alice Brunson. An inclined and vertical staff gage was established in August, 1907, by Viele, Blackwell & Buck, on right bank about 300 feet above site of Survey gage; this gage was read twice daily until March 18, 1911, and once daily from March 19 to December 10, 1911, by H. C. Brunson; nothing is left of it except the bench mark to which it was referred. A staff gage reading to about 6 feet was installed in 1914 on a large boulder in the river near the left bank, practically opposite the site of the gage established in August, 1907; no readings of this gage are available.

DISCHARGE MEASUREMENTS.—Made from cable about 600 feet above gage. A reference on left bank near cable is used to determine depths when soundings can not be made.

CHANNEL AND CONTROL.—Solid rock; permanent. At high stages the edge of the falls serves as control, there being a vertical drop of about 68 feet at the falls at low water.

EXTREMES OF DISCHARGE.—Maximum stage recorded during year, 12.50 feet at 7.30 a. m. January 28 (discharge, 59,600 second-feet); minimum stage, 1.22 feet at 5.30 p. m. October 18 (discharge, 78 second-feet).

1907–1911; 1915–1918: Maximum stage recorded, that of January 28, 1918; minimum stage, 55 second-feet October 3-7 and 23–27, 1908.

Highest known stage prior to 1918 corresponds to about 12 feet on Survey gage; lowest stage, according to William Taylor, a local resident in September, 1916, occurred in 1902, when entire flow of river was confined in a channel 7 feet wide, 1 foot deep, flowing fast; under these conditions, the discharge would probably be about 30 second-feet.

ICE.—Stage-discharge relation not affected by ice.

REGULATION.—Low-water flow may be affected to a small extent by operation of power plant at Williamsburg, about 25 miles above the station.

ACCURACY.—Stage-discharge relation permanent; affected by ice December 13–20. Rating curve well defined. Gage read to hundredths twice daily. Daily discharge ascertained by applying mean daily gage height to rating table. Records good.

The following discharge measurement was made by Hopkins and Kidwell:

June 13, 1918: Gage height, 1.50 feet; discharge, 320 second-feet.

Daily discharge, in second-feet, of Cumberland River at Cumberland Falls, Ky., for the year ending Sept. 30, 1918.

Day.	Oct.	Nov.	Dec.	Jan.	Feb.	Mar.	Apr.	May.	June.	July.	Aug.	Sept.
1	500	1,890	180	1,110	42,100	2,920	2,000	2,920	920	4,450	830	890
2	404	1,460	180	1,080	24,200	2,560	1,890	2,560	785	7,260	604	954
3	338	1,060	199	1,050	9,000	2,330	2,680	2,220	785	3,570	476	1,270
4	256	860	218	1,020	4,450	2,000	5,750	2,000	658	1,890	428	860
5	199	714	228	1,020	3,440	2,110	6,470	1,780	565	1,360	317	971
6	159	617	218	1,090	2,680	2,440	5,410	1,560	591	988	266	1,460
7	138	526	208	2,440	2,560	3,570	4,760	1,360	658	714	237	1,670
8	124	452	218	4,150	2,560	6,100	23,000	1,270	617	604	218	1,360
9	110	404	218	4,150	2,440	6,860	24,900	1,670	440	658	199	1,090
10	102	359	218	2,560	2,220	4,760	19,200	2,000	416	845	180	728
11	94	338	208	1,890	2,220	4,760	10,400	2,000	370	686	275	526
12	94	317	218	2,000	2,330	4,150	5,410	1,460	317	500	428	428
13	94	296	210	3,440	2,330	3,570	4,150	3,300	286	428	380	359
14	86	275	210	4,760	2,110	3,180	3,050	18,000	275	338	275	306
15	86	256	200	6,100	1,890	3,050	2,560	16,800	237	296	237	266
16	82	246	200	7,260	1,780	3,050	2,330	9,460	199	256	228	237
17	82	237	200	6,860	1,780	2,680	2,330	4,150	180	218	218	237
18	82	228	200	4,760	1,670	2,330	2,440	2,920	266	180	208	218
19	275	218	200	3,050	1,460	2,110	2,800	2,560	208	190	237	199
20	6,470	208	250	2,220	2,800	1,890	2,920	5,080	180	199	266	208
21	3,570	199	306	1,890	6,860	2,110	3,850	6,470	1,040	199	338	180
22	2,110	190	359	1,670	6,860	2,800	8,110	4,760	3,300	180	286	218
23	1,360	180	476	1,460	5,080	2,920	7,680	8,110	2,330	180	218	218
24	920	180	604	1,460	3,850	2,800	5,080	7,260	2,000	166	166	199
25	686	180	742	1,460	3,050	4,450	3,570	5,080	1,460	275	138	190
26	578	190	1,560	1,780	3,050	5,080	3,440	3,570	1,890	237	124	173
27	476	180	2,560	19,800	3,300	5,080	3,570	2,560	3,050	180	180	173
28	428	180	2,560	57,500	3,050	3,850	3,180	2,110	2,330	190	218	173
29	380	173	2,330	54,700	....	3,180	2,920	1,780	1,560	246	199	173
30	552	180	1,740	56,100	....	2,560	3,050	1,560	2,330	275	338	166
31	630	....	1,140	56,800	....	2,220	....	1,160	....	742	565	....

NOTE.—Discharge, Dec. 13-20, estimated because of ice; Dec. 30 and Jan. 1-3, interpolated.

Monthly discharge of Cumberland River at Cumberland Falls, Ky., for the year ending Sept. 30, 1918.

[Drainage area, 2,040 square miles.]

Month.	Discharge in second-feet.				Run-off in inches.
	Maximum.	Minimum.	Mean.	Per square mile.	
October	6,470	82	692	0.339	0.39
November	1,890	173	426	.209	.23
December	2,560	180	599	.294	.34
January	57,500	1,020	10,200	5.00	5.76
February	42,100	1,460	5,400	2.65	2.76
March	6,860	1,890	3,340	1.64	1.89
April	24,900	1,890	5,960	2.92	3.26
May	18,000	1,160	4,180	2.05	2.36
June	3,300	180	1,010	.495	.55
July	7,260	166	919	.450	.52
August	830	124	299	.147	.17
September	1,670	166	537	.263	.29
The year	57,500	82	2,790	1.37	18.52

CUMBERLAND RIVER AT BURNSIDE, KY.

Location.—Below mouth of South Fork of Cumberland River at Burnside, Pulaski County.

Drainage area.—4,890 square miles (measured on maps of Kentucky and Tennessee, prepared by United States Geological Survey; scale, 1:500,000).

Records available.— October 1, 1914, to September 30, 1918.

Gage.—Vertical staff in two sections on piers of toll bridge across South Fork of Cumberland River about 700 feet above mouth; installed in July, 1914, by United States Weather Bureau; readings on this gage by the Weather Bureau began January 1, 1915; sea-level elevation of zero, 589.53 feet (Smith Shoals Survey datum, United States Engineer Corps); datum same as that of gage which was marked on the rails of inclines 1 and 2 leading from the South Fork to the warehouse, about 500 feet below the present gage, and which was established in 1884 and read daily until January 1, 1915; upper part of old gage, reading from 54 to 71 feet, was spiked to office of Col. Cole. The United States Weather Bureau [1] reports that "the old river gage was changed on several unknown dates and by amounts that are uncertain, so that readings prior to January 1, 1915, are not comparable by from 0.1 to 0.7 foot." New gage is read for the United States Geological Survey by L. M. Cheeley.

Discharge measurements.—Flow of South Fork is measured from the highway bridge; the Cumberland above the South Fork is measured from a boat, from the Queen & Crescent Railroad bridge, or by means of floats, the method used depending on the stage; flow below the South Fork is the combined flow of both streams.

Channel and control.—Channel considered permanent except for deposits of mud, which are washed away at high stages. Low-water control is crest of dam No. 21, 28 miles below Burnside; gage height of crest of dam, 1.47 feet. The dam is a recently built concrete structure, and probably little or no water leaks through dam or lock.

Extremes of discharge.—Maximum stage recorded during year, 69.5 feet at 1 a. m. January 29 (discharge, roughly, 157,000 second-feet); minimum stage, 2.07 feet October 11 (discharge, 289 second-feet).

1915–1918: Maximum stage recorded, that of January 29, 1918; minimum stage 1.97 feet July 13 and 14, 1917, due to lowering of pool to flood steamer off bar below lock.

The stage of January 29, 1918, is the maximum stage since December 15, 1884, the date of establishment of the United States Weather Bureau gage.

Ice.—Stage-discharge relation seldom affected by ice.

Regulation.—Stage at low water will be affected by any manipulation of the level of pool No. 21 at the lock.

Accuracy.—Stage-discharge relation practically permanent; affected by ice during parts of December and January. Rating curve fairly well defined to 30,000 second feet (gage height approximately, 20 feet); above 30,000 second-feet curve is an extension and may be considerably in error. Gage read to hundredths twice daily. Daily discharge ascertained by applying mean daily gage height to rating table. At low stages discharge relation may be affected by water entering between the gage and the dam owing to heavy local showers in the basins of the small intervening tributaries. Records good for discharge of less than 30,000 second-feet.

---

[1] Daily river stages, pt. 12, p. 29.

Discharge measurements of Cumberland River at Burnside, Ky., during the year ending Sept. 30, 1918.

Date.	Made by—	Gage height.	Discharge.
		Feet.	*Sec.-ft.*
Apr. 11	Peterson and Hopkins	17.57	24,200
June 17	Hopkins and Kidwell	2.44	552

Daily discharge, in second-feet, of Cumberland River at Burnside, Ky., for the year ending Sept. 30, 1918.

Day.	Oct.	Nov.	Dec.	Jan.	Feb.	Mar.	Apr.	May.	June.	July.	Aug.	Sept.
1	1,780	4,080	524	3,000	69,100	6,260	4,540	7,540	2,580	5,460	1,890	970
2	1,430	3,730	695	3,000	44,000	5,680	4,190	6,260	2,120	6,260	2,000	1,430
3	1,030	3,040	800	2,500	25,000	5,340	4,880	5,460	2,120	5,920	1,430	1,780
4	912	2,580	912	2,500	13,400	5,340	14,100	4,760	1,890	4,080	1,200	2,920
5	695	2,120	970	2,500	9,600	7,540	16,600	4,300	1,660	2,580	912	4,300
6	636	1,780	912	2,500	7,260	11,200	12,900	3,730	1,660	1,890	745	7,400
7	569	1,540	912	4,880	6,740	13,000	10,200	3,380	1,660	1,370	695	6,030
8	508	1,430	912	9,900	7,820	13,700	53,300	3,270	1,780	1,140	533	3,840
9	441	1,320	912	9,150	9,450	12,400	61,100	3,960	1,660	1,080	550	2,460
10	378	1,200	912	7,260	10,400	11,500	39,000	7,000	1,320	1,140	533	1,780
11	303	1,030	900	4,800	8,550	9,450	23,700	5,220	1,140	1,370	524	1,260
12	325	970	900	6,000	7,400	8,550	15,200	4,760	1,030	1,200	490	1,140
13	378	912	850	11,000	6,620	7,540	9,900	7,540	912	970	490	1,030
14	425	912	800	10,000	6,140	6,870	7,820	36,800	800	800	490	695
15	362	855	750	18,900	5,680	6,260	6,030	37,000	695	607	465	626
16	378	855	650	25,400	5,460	5,800	5,460	20,500	645	533	449	533
17	362	800	650	19,100	5,340	5,570	5,800	10,500	569	533	449	533
18	645	800	650	12,200	4,880	5,000	6,740	7,400	533	490	490	490
19	5,110	695	650	9,000	4,420	4,420	6,870	5,800	578	490	745	516
20	6,870	696	700	7,000	14,800	4,190	6,260	6,140	695	433	550	607
21	11,500	645	800	5,500	29,000	4,420	8,850	17,000	745	370	533	607
22	5,570	645	900	5,000	22,400	4,880	18,900	14,600	2,240	401	533	578
23	3,960	645	1,780	4,500	14,300	5,340	16,100	16,100	4,420	441	524	533
24	2,700	616	2,120	5,340	10,500	5,460	12,200	30,300	3,380	441	449	533
25	2,120	569	2,350	4,540	8,250	11,000	9,000	15,500	3,040	457	409	533
26	1,540	542	4,540	4,420	7,540	12,000	7,820	9,900	2,810	490	370	465
27	1,430	524	5,460	27,900	7,260	10,200	11,700	7,130	3,840	490	385	449
28	1,260	524	6,140	115,000	6,870	8,700	12,500	5,900	3,960	626	516	385
29	1,540	508	5,460	149,000	....	7,000	9,900	4,650	3,500	1,030	645	332
30	2,460	482	4,650	101,000	....	5,800	8,100	3,730	3,620	1,430	578	310
31	3,270	....	3,500	91,200	....	5,000	....	3,040	....	1,540	695	....

NOTE.—Discharge, Dec. 11-22, 31, Jan. 1-6, 11-14, and 21-23, estimated because of ice effect.

Monthly discharge of Cumberland River at Burnside, Ky., for the year ending Sept. 30, 1918.

[Drainage area, 4,890 square miles.]

Month.	Discharge in second-feet.				Run-off in inches.
	Maximum	Minimum.	Mean.	Per square mile.	
October	11,500	303	1,960	0.401	0.46
November	4,080	482	1,230	.252	.28
December	6,140	524	1,730	.354	.41
January	149,000	2,500	22,100	4.52	5.21
February	69,100	4,420	13,500	2.76	2.87
March	13,700	4,190	7,590	1.55	1.79
April	61,100	4,190	14,300	2.92	3.26
May	37,000	3,040	10,300	2.11	2.43
June	4,420	533	1,920	.393	.44
July	6,260	370	1,490	.305	.35
August	2,000	370	686	.140	.16
September	7,400	310	1,500	.307	.34
The year	149,000	303	6,490	1.33	18.00

SOUTH FORK OF CUMBERLAND RIVER AT NEVELSVILLE, KY.

LOCATION.—One-fourth mile below Turkey Creek ferry on Greenwood-Monticello pike, 1 mile from Nevelsville, McCreary County. Little South Fork enters on left 1¾ miles above station.

DRAINAGE AREA.—1,260 square miles (measured on maps of Kentucky and Tennessee prepared by United States Geological Survey; scale, 1:500,000).

RECORDS AVAILABLE.—March 10, 1915, to September 30, 1918.

GAGE.—Vertical staff gage in 5 sections bolted to rock ledges on left bank; read by Ben Whitehead. A reference gage for use in referencing soundings at the measuring section, is attached to a tree on the left bank 110 feet below cable.

DISCHARGE MEASUREMENTS.—Made from cable about 2,000 feet below gage or by wading at low stages.

CHANNEL AND CONTROL.—Channel straight above and below; bed, compact gravel. Low-water control is partly the bed of the river below gage and partly a gravel bar about 2 miles below gage. Both are probably permanent. High-water control is bed of stream for several miles below gage, and may be slightly affected by foliage along the banks.

EXTREMES OF DISCHARGE.—Maximum stage recorded during year, 51.4 feet at 5.30 p. m. January 28 (discharge, roughly, 84,300 second-feet); minimum stage, 1.84 feet at 5.30 p. m. August 16 and 26 (discharge, 88 second-feet).

1915–1918: Maximum stage recorded, that of January 28, 1918; minimum stage 1. 82 feet at 5.30 a. m. July 13, 1917 (discharge, 64 second-feet).

ICE.—Stage-discharge relation seldom affected by ice.

REGULATION.—Operation of a small power plant short distance above gage may affect flow at extreme low water.

ACCURACY.—Stage-discharge relation probably permanent; affected by ice during parts of December and January. Rating curve well defined between 500 and 25,000 second-feet, and fairly well defined below 500 second-feet; extended above 25,000 second-feet. Gage read to hundredths twice daily. Daily discharge ascertained by applying mean daily gage height to rating table. Records good.

The following discharge measurement was made by Hopkins and Kidwell:

June 15, 1918: Gage height, 2.43 feet; discharge, 235 second-feet.

Daily discharge, in second-feet, of South Fork of Cumberland River at Nevelsville, Ky., for the year ending Sept. 30, 1918.

Day.	Oct.	Nov.	Dec.	Jan.	Feb.	Mar.	Apr.	May.	June.	July.	Aug.	Sept.
1	500	1,450	261	750	7,930	1,660	1,210	2,810	740	261	850	410
2	371	1,090	296	750	5,860	1,520	1,150	2,150	685	333	525	550
3	296	850	602	680	4,190	1,390	2,080	1,800	630	333	410	500
4	244	685	575	670	3,210	1,270	8,840	1,520	602	314	278	850
5	195	550	475	660	2,500	1,590	5,980	1,330	550	244	218	1,210
6	190	500	430	900	2,010	1,590	3,740	1,150	575	195	179	1,730
7	177	452	410	1,800	2,150	2,430	4,190	1,030	712	165	170	1,270
8	151	430	390	2,650	2,150	1,940	35,300	970	685	156	156	685
9	140	390	390	2,000	1,940	2,150	22,900	970	575	212	140	500
10	134	371	270	1,600	1,730	1,870	8,580	1,030	500	314	130	352
11	134	352	330	1,200	1,940	1,940	4,970	850	410	296	118	278
12	147	333	350	2,100	1,520	1,800	3,650	850	352	231	110	225
13	151	314	350	4,400	1,520	1,660	2,810	2,730	296	179	98	195
14	147	314	330	2,600	1,450	1,450	2,080	12,900	261	151	95	170
15	140	296	310	5,100	1,330	1,450	1,800	6,220	222	134	92	156
16	136	296	290	7,900	1,330	1,390	1,730	3,210	201	126	88	142
17	134	296	270	4,300	1,270	1,330	3,740	2,500	179	120	108	138
18	130	278	260	2,800	1,210	1,210	4,670	2,290	177	114	187	132
19	2,570	261	260	2,000	1,150	1,150	3,470	1,520	278	114	174	130
20	5,630	261	270	1,500	4,280	1,150	2,970	2,150	244	118	352	134
21	2,430	244	330	1,350	7,800	1,520	5,190	2,730	575	118	333	130
22	1,330	244	370	1,200	4,870	1,520	7,300	2,890	795	110	278	126
23	910	238	480	1,150	3,470	1,390	4,870	6,100	1,210	118	218	122
24	685	231	520	1,100	2,730	1,800	3,470	11,600	575	170	145	114
25	525	225	680	1,050	2,220	2,810	2,650	4,770	430	170	122	114
26	452	218	1,500	1,300	2,290	2,730	3,050	2,810	390	147	100	114
27	296	209	1,400	16,300	2,150	2,500	5,410	2,010	371	138	170	114
28	158	201	1,300	53,100	1,800	2,010	4,190	1,730	314	333	278	122
29	390	212	1,100	31,100	....	1,730	3,470	1,330	278	575	238	151
30	740	234	900	15,400	....	1,390	3,210	1,030	261	550	261	179
31	1,520	....	750	18,800	....	1,270	....	850	....	740	278	....

NOTE.—Discharge, Dec. 10 to Jan. 26, estimated because of ice effect.

Monthly discharge of South Fork of Cumberland River at Nevelsville, Ky., for the year ending Sept. 30, 1918.

[Drainage area, 1,260 square miles].

Month.	Discharge in second-feet.				Run-off in inches.
	Maximum.	Minimum.	Mean.	Per square mile.	
October	5,630	130	682	0.541	0.62
November	1,450	201	401	.318	.35
December	1,500	260	532	.422	.49
January	53,100	660	6,070	4.82	5.56
February	7,930	1,150	2,790	2.21	2.30
March	2,810	1,150	1,700	1.35	1.56
April	35,300	1,150	5,620	4.46	4.98
May	12,900	850	2,830	2.25	2.59
June	1,210	177	469	.372	.42
July	740	114	235	.187	.22
August	850	88	223	.177	.20
September	1,730	114	368	.292	.33
The year	53,100	88	1,820	1.44	19.62

CANEY FORK NEAR ROCK ISLAND, TENN.

Location.—About 100 feet downstream from power house of Tennessee Power Co., half a mile downstream from mouth of Collins River, and 1 mile northwest of Rock Island, Warren County.

Drainage area.—1,640 square miles (measured from Post Route map).

Records available.—November 14, 1911, to September 30, 1918.

Gage.—Bristol water-stage recorder known as gage No. 3, 100 feet downstream from power house and about half a mile downstream from Rock Island dam. This gage has been used since January 1, 1917. From March 26 to December 31, 1916, a Bristol water-stage recorder installed March 26, 1916, at site of staff gage known as gage B (No. 2), half a mile upstream from gage No. 3 and 300 feet downstream from Rock Island dam, was used. The closing of sluice gates in dam on December 8, 1916, and diversion of flow through tunnel on December 12 made gage B useless after December 7, 1916. Prior to March 26, 1916, daily mean stage was determined from a water-stage recorder known by the Billesby Co., as gage A, 400 feet upstream from gage B, just above point at which dam is now built; date of installation of recorder not known. Backwater from dam began to affect stage-discharge relation at gage A on March 26, 1916.

Discharge measurements.—Formerly made from cable at gage B or from sluice ways in dam. No discharge measurements have been made since closing of the sluiceways on December 8, 1916.

Channel and control.—Bed of stream above and below gage consists chiefly of solid rock; probably permanent.

Extremes of discharge.—Maximum mean daily stage from water-stage recorder, 16.28 feet January 28 (discharge, about 44,900 second-feet); minimum mean daily stage, zero on gage July 27, 28, and September 1, 26 (discharge, 140 second-feet).

1911–1918: Maximum stage recorded, 13.2 feet April 2, 1912 (discharge, 107,000 second-feet); minimum stage recorded, same as for 1918.

Diversions.—None.

Regulation.—Considerable fluctuation caused by storage in reservoir and operation of plant.

Accuracy.—Stage-discharge relation practically permanent. Rating curve is fairly well defined between 300 and 9,000 second-feet, above which it is an extension. Daily discharge ascertained by applying to rating table mean daily gage height obtained by inspecting gage-height graph. Records good except for extreme high and low stages.

Cooperation.—Gage-height record furnished by Tennessee Power Co.

No discharge measurements were made at this station during the year.

Daily discharge, in second-feet, of Caney Fork near Rock Island, Tenn., for the year ending Sept. 30, 1918.

Day.	Oct.	Nov.	Dec.	Jan.	Feb.	Mar.	Apr.	May.	June.	July.	Aug.	Sept.
1	1,370	1,950	700	1,610	22,100	1,730	1,570	2,950	1,060	1,490	760	140
2	1,410	1,690	730	1,610	16,100	3,880	2,080	5,690	1,370	880	578	155
3	1,410	1,530	650	1,610	16,900	1,370	1,410	2,520	1,230	1,090	625	212
4	1,340	1,370	915	1,120	13,400	1,770	1,060	1,950	1,120	1,260	386	222
5	1,260	1,260	1,450	1,450	8,760	3,320	2,220	2,420	1,370	950	850	564
6	1,300	1,260	1,490	1,530	9,430	1,950	3,450	3,450	1,340	1,020	850	1,020
7	1,200	1,260	1,300	3,320	9,430	1,410	5,260	2,420	1,200	950	850	1,300
8	1,060	1,200	1,200	3,880	6,940	1,410	32,100	1,610	1,120	610	850	850
9	915	1,300	1,300	3,880	6,410	4,190	28,100	1,860	790	760	850	510
10	820	950	1,650	1,860	7,510	1,650	18,500	1,860	985	630	880	470
11	760	950	1,450	2,730	3,450	1,530	10,800	1,260	850	850	546	486
12	730	880	1,450	8,440	4,190	2,000	6,160	1,260	1,200	820	790	438
13	675	880	1,490	9,090	4,030	2,040	2,730	9,770	1,200	650	730	519
14	578	850	1,490	6,670	3,730	2,620	4,520	25,300	1,230	430	366	615
15	450	820	1,450	13,000	5,060	2,420	2,730	9,090	1,060	390	555	287
6	450	850	1,530	14,200	4,690	1,410	4,350	7,510	880	332	478	730
17	450	880	1,530	8,120	7,810	1,490	16,900	2,840	820	280	332	700
18	450	880	1,450	6,940	9,090	1,576	16,500	3,590	880	578	225	790
19	1,450	850	1,300	5,690	7,810	1,730	11,500	1,200	730	418	280	675
20	8,440	790	1,370	2,840	12,200	1,900	11,200	1,690	610	410	262	301
21	2,620	675	1,410	2,080	14,200	5,920	21,700	3,070	354	564	173	298
22	2,130	600	1,490	1,530	13,400	3,590	16,900	3,070	434	573	248	175
23	1,370	625	2,950	3,320	10,800	3,730	8,760	3,450	700	519	188	188
24	1,370	600	1,610	2,220	6,940	3,450	6,940	11,900	985	490	192	306
25	1,230	650	2,730	1,610	4,520	7,510	6,160	5,060	985	386	256	250
26	1,200	675	1,370	4,690	3,320	6,160	4,520	2,620	1,120	537	250	140
27	1,200	650	2,520	27,300	2,180	4,030	11,200	1,610	1,020	140	591	615
28	1,120	650	2,420	44,900	3,190	4,190	4,350	2,420	1,090	140	532	346
29	950	650	1,570	39,700	....	2,220	6,940	1,090	2,950	336	410	175
30	2,320	820	1,860	35,300	....	1,820	6,940	1,300	1,770	336	242	700
31	2,130	....	1,610	34,100	....	1,900	....	1,230	....	430	210	....

Monthly discharge of Caney Fork near Rock Island, Tenn., for the year ending Sept. 30. 1918.

[Drainage area, 1,640 square miles.]

Month.	Discharge in second-feet.				Run-off in inches.
	Maximum.	Minimum.	Mean.	Per square mile.	
October	8,440	450	1,420	0.866	1.00
November	1,950	600	966	.589	.66
December	2,950	650	1,530	.933	1.08
January	44,900	1,120	9,560	5.82	6.71
February	22,100	2,180	8,490	5.18	5.39
March	7,510	1,370	2,770	1.69	1.95
April	32,100	1,060	9,250	5.64	6.29
May	25,300	1,090	4,100	2.50	2.88
June	2,950	354	1,080	.659	.74
July	1,490	140	621	.379	.44
August	880	173	495	.302	.35
September	1,300	140	473	.288	.32
The year	44,900	140	3,360	2.05	27.81

COLLINS RIVER NEAR ROWLAND, TENN.

Location.—At Hennessee's iron highway bridge, 1 mile below Mountain Creek, 2½ miles northwest of Rowland, Warren County, 5 miles southwest of Rock Island, and 8 miles upstream, by river, from junction with Caney Fork.

Drainage area.—800 square miles (measured by Tennessee Power Co.).

Records available.—April 1, 1916, to September 30, 1918.

Gage.—Chain gage on downstream side of bridge at middle of second span from right bank; read by Joe Keathley. Sea-level elevation of zero of gage, 795.86 feet.

Discharge measurements.—Made from upstream side of bridge or by wading.

Channel and control.—Bed composed of rock, boulders, and sand. Channel fairly straight for a considerable distance above and below gage. Right bank is a steep rock bluff; left bank is low and subject to overflow above a stage of 8 feet. A series of rock and boulder riffles beginning just below bridge forms the control, probably permanent.

Extremes of discharge.—Maximum stage recorded during year, 12.17 feet at 5 p. m. January 28 (discharge, 23,200 second-feet); minimum stage recorded, 1.02 feet at 7 p. m. September 15 (discharge, 92 second-feet).

1916–1918: Maximum stage recorded, 14.1 feet at noon March 4, 1917 (discharge, 28,900 second-feet); minimum stage recorded same as for 1918.

By means of levels the elevation of marks of the flood of 1854 (exact date unknown), obtained from old residents nearby, indicates that stage rose to 32.6 feet (discharge estimated at 82,200 second-feet). Elevation of marks of the flood of 1902 (exact date unknown), obtained in the same manner, indicates that stage rose to 27.2 feet (estimated discharge, 66,000 second-feet).

Ice.—Stage-discharge relation not affected by ice.

Diversions.—None.

Regulation.—Small mills upstream probably cause some diurnal fluctuation.

Accuracy.—Stage-discharge relation practically permanent; not affected by ice. Rating curve well defined below 8,000 second-feet; above that point curve is an extension. Gage read to hundredths twice daily; during high water read oftener. Daily discharge ascertained by applying mean daily gage height to rating table. Records good except above stage of overflow (about 8 feet, discharge, 11,300 second-feet) when they are subject to error.

Cooperation.—Gage-height record furnished by the Tennessee Power Co.

Discharge measurements of Collins River near Rowland, Tenn., during the year ending Sept. 30, 1918.

[Made by L. J. Hall.]

Date.	Gage height.	Discharge.
	Feet.	*Sec.-ft.*
Nov. 7	1.60	440
Feb. 19	3.16	2,100
Aug. 21	1.10	123

Daily discharge, in second-feet, of Collins River near Rowland, Tenn., for the year ending Sept. 30, 1918.

Day.	Oct	Nov.	Dec.	Jan.	Feb.	Mar.	Apr.	May.	June.	July.	Aug.	Sept.
1	840	890	510	582	6,760	1,320	735	1,460	636	358	249	398
2	690	753	609	618	4,220	1,180	717	1,250	564	318	235	318
3	564	663	600	600	3,300	1,020	726	1,100	502	286	235	242
4	494	591	555	564	2,720	953	762	1,010	510	263	228	207
5	454	555	537	546	2,240	920	910	900	470	263	207	242
6	414	502	510	636	2,000	910	860	880	446	263	207	221
7	358	462	478	1,450	2,050	880	2,020	800	486	256	180	228
8	382	510	510	1,340	1,950	840	19,300	860	546	263	193	228
9	318	446	780	1,180	1,730	900	14,000	800	494	214	180	186
10	334	430	800	1,030	1,660	986	5,440	820	438	228	167	160
11	318	422	708	1,050	1,600	1,030	3,300	840	414	228	167	144
12	302	430	690	2,610	1,570	964	2,450	1,180	382	228	173	144
13	302	462	627	2,650	1,640	986	1,960	4,300	342	207	167	160
14	278	494	600	2,170	1,480	910	1,710	8,120	326	221	167	167
15	294	462	573	5,070	1,380	900	1,470	4,000	326	214	148	104
16	270	430	546	4,460	1,590	840	2,430	2,430	286	221	148	144
17	242	430	528	3,040	3,060	780	3,860	1,830	278	214	156	128
18	256	398	486	2,330	2,600	771	4,980	1,450	286	228	148	136
19	1,100	374	537	1,850	2,140	744	3,860	1,220	278	256	140	128
20	1,610	398	582	1,630	4,800	762	3,510	1,070	663	286	167	124
21	663	382	654	1,380	4,080	762	4,460	1,350	860	270	173	112
22	790	374	744	1,270	2,960	780	4,000	1,290	564	256	186	97
23	627	398	810	1,180	2,420	762	2,910	1,840	422	256	173	140
24	555	374	840	1,100	2,070	870	2,250	3,650	366	242	160	136
25	486	302	830	1,120	1,820	1,210	1,850	2,270	358	214	152	124
26	454	318	820	1,530	1,720	1,140	1,610	1,490	326	207	152	136
27	438	302	880	[illegible]320	1,610	1,050	1,780	1,250	318	263	160	173
28	438	318	800	2,700	,470	942	1,590	,080	318	256	173	186
29	430	350	726	1 300	......	830	1,820	880	510	242	152	173
30	1,460	398	672	500	......	810	1,690	762	422	318	148	173
31	1,110	......	636	300	......	735	......	708	......	256	342	......

Monthly discharge of Collins River near Rowland, Tenn., for the year ending Sept. 30, 1918.

[Drainage area, 800 square miles.]

Month.	Discharge in second-feet.				Run-off in inches.
	Maximum.	Minimum.	Mean.	Per square mile.	
October	1,610	242	557	0.696	0.80
November	890	302	454	.568	.63
December	880	478	651	.814	.94
January	21,700	546	3,940	4.92	5.67
February	6,760	1,380	2,450	3.06	3.19
March	1,320	735	919	1.15	1.33
April	19,300	717	3,300	4.12	4.60
May	8,120	708	1,700	2.12	2.44
June	860	278	438	.548	.61
July	358	207	251	.314	.36
August	342	140	182	.228	.26
September	398	97	175	.219	.24
The year	21,700	97	1,240	1.55	21.07

TENNESSEE RIVER BASIN.

FRENCH BROAD RIVER AT ASHEVILLE, N. C.

LOCATION.—At new concrete highway bridge which replaced old Smith's Bridge, washed out July 16, 1916, 1 mile below Southern Railway station at Asheville, Buncombe County, and 2 miles below mouth of Swannanoa River.

DRAINAGE AREA.—987 square miles.

RECORDS AVAILABLE.—January 1, 1905, to July 16, 1916; January 1, 1917, to September 30, 1918. Records were obtained at Bingham School Bridge, about 3 miles below Asheville, from September 17, 1895, to December 31, 1901.

GAGE.—Vertical staff, graduations from −2.0 to 14.7 feet stamped on right downstream face of third pier from right bank. The original gages, a vertical staff attached to one of the bridge piers of the old Smith's Bridge and an auxiliary chain gage (for obtaining readings below zero) attached to that bridge, were used until the flood in July, 1916. All gages set to same datum. From January 1 to November 21, 1917, readings were obtained from a temporary staff gage set at different datum; readings reduced to datum of present gage. Gage read by O. S. Snook.

DISCHARGE MEASUREMENTS.—Made from highway bridge.

CHANNEL AND CONTROL.—Bed composed chiefly of rock; practically permanent. Control formed by rock shoal and concrete piers of Southern Railway bridge; permanent, though piers of bridge may become choked with débris during extreme floods, causing backwater at gage for short periods.

EXTREMES OF DISCHARGE.—Maximum stage recorded during year, 5.0 feet at 5 p. m. January 28 (discharge, 12,200 second-feet); minimum stage recorded, −0.6 foot December 21, September 3–4 and 16–18 (discharge, 760 second-feet).

1905–1918: Maximum stage, 24.13 feet July 16, 1916, determined by levels from flood marks November 21, 1917 (discharge not determined; stage-discharge relation probably affected by backwater from drift lodged against the Southern Railway bridge). Maximum stage recorded before or after the flood in July, 1916, 7.8 feet January 23, 1906 (discharge, 25,800 second-feet). Minimum stage recorded, −0.7 foot September 16 and 20, 1907 (discharge, 380 second-feet).

ICE.—Stage-discharge relation seldom affected by ice.

DIVERSIONS.—None.

REGULATION.—Slight diurnal fluctuation may be caused by operation of small mills upstream.

ACCURACY.—Stage-discharge relation practically permanent, except as affected by ice during December and January. Rating curve well defined below 10,800 second-feet. Gage read to tenths once daily. Daily discharge ascertained by applying daily gage height to rating table except for periods of ice effect. Records good.

COOPERATION.—Gage-height record furnished by United States Weather Bureau.

Discharge measurements of French Broad River at Asheville, N. C., during the year ending Sept. 30, 1918.

Date.	Made by—	Gage height.	Discharge.
		Feet.	*Sec.-ft.*
Mar. 1	L. J. Hall	0.48	1,8[illegible]
May 28	C. G. Paulsen	.40	1,7[illegible]

Daily discharge, in second-feet, of French Broad River at Asheville, N. C., for the year ending Sept. 30, 1918.

Day.	Oct.	Nov.	Dec.	Jan.	Feb.	Mar.	Apr.	May.	June.	July.	Aug.	Sept.
1	1,640	1,750	1,020	....	5,750	2,100	1,420	2,360	1,310	2,100	1,860	1,020
2	1,420	1,420	1,020	1,420	4,060	1,860	1,420	1,980	1,210	1,530	1,530	840
3	1,420	,530	930	1,310	4,050	1,750	1,420	1,750	1,210	1,310	1,530	760
4	1,310	,310	930	,110	3,480	1,640	1,420	1,750	2,970	1,210	1,110	760
5	1,310	,310	930	,110	2,500	1,750	1,420	1,530	1,640	1,110	1,020	840
6	1,310	1,420	930	1,210	1,980	1,750	1,420	1,530	1,210	1,110	1,020	1,020
7	1,210	1,310	840	2,360	1,980	1,750	1,750	1,530	2,500	1,110	930	1,110
8	1,210	1,210	840	100	1,980	1,750	1,980	1,530	1,860	1,020	930	1,210
9	1,310	1, 0	1,020	530	1,980	1,640	5,990	2,100	1,530	1,210	930	1,310
10	1,420	1, 10	1,020	....	1,980	1,640	3,670	1,980	1,530	1,110	1,530	1,110
11	1,310	1,210	930	....	1,980	1,750	2,650	1,750	1,310	1,110	1,420	1,020
12	1,310	1,210	930	4,630	2,100	1,640	2,500	1,640	1,310	1,110	1,530	1,020
13	1,310	1, 1	840	....	2,100	1,640	2,100	1,530	1,310	1,110	1,110	930
14	1,110	1, 1	840	....	1,980	1,640	1,980	2,360	1,210	1,110	930	840
15	1,020	1, 1	840	2,500	1,980	1,640	1,860	2,100	1,210	1,020	840	840
16	1,110	1,210	840	3,670	1,980	1,530	1,860	1,640	1,110	930	930	760
17	1,020	1,110	760	2,500	2,100	1,530	1,860	1,530	1,110	930	1,210	760
18	1,020	1,110	760	2,360	500	1,530	1,750	1,640	1,750	1,210	1,420	760
19	1,640	1,210	760	1,860	360	1,530	1,750	1,530	1,210	1,020	2,500	1,110
20	5,750	1,110	760	....	,670	1,420	1,860	1,860	1,210	930	1,750	1,210
21	2,500	1,110	760	....	3,670	1,980	1,980	2,360	1,640	930	1,210	1,310
22	2,100	1,110	930	....	2,130	1,750	1,980	1,980	3,670	930	1,020	1,210
23	1,860	1,020	1,020	....	500	1,530	1,860	2,100	1,860	1,750	930	930
24	1,530	1,020	1,020	....	360	2,230	1,750	1,980	1,640	1,210	930	840
25	1,310	1,020	1,020	....	,100	2,100	1,640	2,360	1,310	1,110	930	840
26	1,420	1,020	1,020	....	2,230	1,860	1,640	2,360	1,640	1,020	840	840
27	1,310	1,020	1,110	3,300	2,230	1,640	1,860	2,360	1,530	1,310	930	1,310
28	1,310	1,020	1,020	6,500	2,100	'640	,750	,860	1,310	1,420	840	1,020
29	1,420	930	1,020	10,400	....	530	,640	,750	1,210	1,530	1,210	930
30	2,230	930	840	8,900	....	530	,640	,530	1,530	1,420	1,210	840
31	2,230	....	840	8,620	....	420	....	,530	....	2,360	1,020	....

NOTE.—River frozen Dec. 14-20, 30-31, Jan. 1, 10, 11, 13, 14, and 20-26; gage not read. Discharge Dec. 14-20, 30, and 31 estimated.

Monthly discharge of French Broad River at Asheville, N. C., for the year ending Sept. 30, 1918.

[Drainage area, 987 square miles.]

Month.	Discharge in second-feet.				Run-off in inches.
	Maximum.	Minimum.	Mean.	Per square mile.	
October	5,750	1,020	1,590	1.61	1.86
November	1,750	930	1,200	1.22	1.36
December	1,110	760	914	.926	1.07
February	5,750	1,980	2,600	2.63	2.74
March	2,230	1,420	1,700	1.72	1.98
April	5,990	1,420	1,990	2.02	2.25
May	2,360	1,530	1,860	1.88	2.17
June	3,670	1,110	1,570	1.59	1.77
July	2,360	930	1,240	1.26	1.45
August	2,500	840	1,200	1.22	1.41
September	1,310	760	977	.990	1.10

TENNESSEE RIVER AT CHATTANOOGA, TENN.

Location.—At Walnut Street Bridge in Chattanooga, Hamilton County, just below Chattanooga Island, 3 miles above mouth of Chattanooga Creek, 4 miles below mouth of Chickamauga Creek, 33 miles above Hales Bar dam, 188 miles below junction of French Broad and Holston rivers, and 464 miles above mouth of Tennessee River.

Drainage area.—21,400 square miles (measured on topographic maps).

Records available.—April 1, 1874, to October 21, 1913; March 1, 1915, to September 30, 1918, when station was discontinued.

Gages.—Two gages, 7 miles apart and set to same datum, are used at this station to determine variation in slope of water surface caused by operation of power plant and locks at Hales Bar dam, as the station is within influence of backwater from the dam. Gage No. 1 consists of a sloping section of iron (railroad T rail) bolted to rock and a vertical timber attached to the rock cliff on left bank about 200 feet upstream from Walnut Street Bridge; read by L. M. Andress from October 1, 1917, to February 9, 1918, and by J. B. Miller after that date. Gage No. 2 is a vertical staff gage in three sections, fastened to trees on left bank about 100 feet above Cincinnati Southern Railroad bridge 7 miles upstream from Chattanooga; read by Floyd Gooden from October 1 to November 10, 1917, and by M. M. Swafford from March 1 to September 30, 1918. Prior to October 21, 1913, gage No. 1 was used alone, but on that date backwater from Hales Bar dam began to affect stage-discharge relation, and the station was abandoned until March 1, 1915, when gage No. 2 was installed.

Discharge measurements.—Made from downstream footway of Walnut Street Bridge.

Channel and control.—Channel practically permanent. Control now formed by the Hales Bar lock and dam and power plant.

Extremes of discharge.—Maximum stage recorded during year, 42.45 feet (gage No. 1) at 7 a. m. February 2 (discharge, 270,000 second-feet); minimum discharge recorded, 8,000 second-feet December 17.

1874–1918: Maximum stage recorded, 54.0 feet at 7 a. m. March 1, 1875 (discharge, 361,000 second-feet); minimum stage recorded, zero on gage September 11–14, 1881, and September 19, 1883 (discharge, 4,800 second-feet).

Ice.—Stage-discharge relation not affected by ice.

Diversions.—None.

Regulation.—See "Accuracy."

Accuracy.—Stage-discharge relation affected by changes in slope of water surface caused by operation of power plant at Hales Bar dam and by rising and falling stages. Discharge determined by slope method (see Water-Supply Paper 345) except for periods indicated in footnote to daily discharge table. Rating curve well defined between 11,500 and 363,000 second-feet. Gages are read to hundredths twice daily, but means are subject to error due to diurnal fluctuations. Records fair.

Discharge measurements of Tennessee River at Chattanooga, Tenn., during the year ending Sept. 30, 1918.

Date.	Made by—	Gage height.		Discharge.	Date.	Made by—	Gage height.		Discharge.
		Gage No. 1.	Gage No. 2.				Gage No. 1.	Gage No. 2.	
		Feet.	*Feet.*	*Sec.-ft.*			*Feet.*	*Feet.*	*Sec.-ft.*
Nov. 14	L. J. Hall.......	8.00	9.87	11,500	Feb. 5	C. G. Paulsen....	22.14	25.07	99,500
Feb. 2	Paulsen and Hall	42.10	....	266,000	Mar. 29	L. J. Hall.......	13.23	16.73	51,700

Daily discharge, in second-feet, of Tennessee River at Chattanooga, Tenn., for the year ending Sept. 30, 1918.

Day.	Oct.	Nov.	Dec.	Jan.	Feb.	Mar.	Apr.	May.	June.	July.	Aug.	Sept.
1...	23,800	16,200	8,400	11,600	263,000	37,500	30,400	40,700	23,400	30,500	21,900	11,900
2...	20,900	17,400	8,800	12,200	266,000	36,000	27,200	38,600	22,500	30,700	23,100	11,200
3...	17,900	16,400	8,400	11,900	200,000	32,200	26,100	36,700	19,000	27,200	21,000	14,300
4...	16,200	15,900	8,800	11,900	167,000	30,900	26,100	35,200	17,600	27,600	18,200	17,800
5...	14,200	15,300	9,200	11,900	104,000	27,800	32,100	32,600	16,400	25,600	15,800	18,900
6...	14,200	15,100	8,800	11,600	71,500	27,100	30,600	28,600	18,900	22,800	14,200	14,600
7...	13,700	15,300	8,800	11,900	59,200	26,600	30,300	27,800	21,200	20,500	14,000	13,000
8...	13,000	14,700	9,800	15,400	53,100	26,600	56,200	27,000	25,200	19,600	13,600	12,300
9...	12,600	14,500	10,400	18,200	48,900	26,600	83,200	25,400	28,500	17,200	13,500	12,500
10...	12,200	14,500	9,800	19,700	45,900	28,600	73,500	26,400	22,700	16,400	13,000	13,400
11...	12,200	....	10,400	21,400	42,900	33,700	67,600	30,300	20,700	17,900	12,700	13,600
12...	12,700	....	10,800	30,400	42,900	39,500	72,800	36,200	19,300	17,700	12,500	14,500
13...	12,900	....	8,800	29,200	41,700	35,200	63,900	40,300	17,500	16,000	14,000	14,900
14...	12,200	....	8,800	34,800	42,900	34,200	51,800	54,300	17,800	15,000	17,900	14,000
15...	11,800	....	8,800	34,200	42,900	31,800	39,800	70,800	19,500	14,400	16,000	13,100
16...	12,400	....	8,400	47,700	41,700	29,500	37,800	62,000	16,200	13,500	14,500	12,800
17...	12,600	....	8,000	53,900	47,700	28,600	45,100	50,800	15,300	13,400	14,000	12,400
18...	11,600	....	8,800	52,700	53,400	29,500	51,000	43,100	15,500	13,900	13,000	12,200
19...	12,200	....	11,600	45,900	54,400	27,900	55,700	36,200	15,800	14,500	12,900	12,700
20...	14,800	....	11,300	37,100	53,400	26,400	57,000	35,600	18,400	14,500	14,800	12,600
21...	20,700	....	11,000	30,300	56,600	26,600	64,800	32,200	18,800	15,100	17,000	12,700
22...	25,000	....	10,500	26,200	65,500	26,900	69,500	32,800	21,300	14,200	18,300	14,000
23...	21,700	....	10,000	23,700	65,500	28,900	66,300	34,500	44,000	13,800	15,300	15,800
24...	20,100	....	9,700	20,700	58,000	31,800	59,300	42,200	56,600	14,200	13,500	15,700
25...	17,800	....	9,290	21,400	51,400	37,800	53,000	45,100	57,000	14,000	12,700	15,100
26...	15,700	....	12,100	23,400	46,500	45,200	52,200	37,500	39,900	14,300	11,900	14,800
27...	15,300	....	12,900	25,100	42,300	57,900	50,600	30,300	31,900	16,000	11,600	14,200
28...	14,900	....	14,000	47,700	39,400	59,100	49,700	29,100	29,200	17,700	11,500	13,500
29...	14,000	....	14,200	143,000	....	52,700	44,300	28,500	26,700	16,500	11,200	13,300
30...	15,100	....	14,300	182,000	....	44,700	41,900	27,900	32,400	18,500	11,600	13,000
31...	14,600	....	14,000	232,000	....	36,800	....	24,300	....	20,600	11,900	....

NOTE.—Discharge record Dec. 1-18, Jan. 6-29, and Feb. 4-28 furnished by Tennessee Power Co.; discharge determined from the gage-height record for the company's gage below Hales Bar dam, the discharge thus obtained being corrected for increase or decrease in storage in order to obtain the natural flow. Discharge for other periods obtained by slope method.

Monthly discharge of Tennessee River at Chattanooga, Tenn., for the year ending Sept. 30, 1918.

[Drainage area, 21,400 square miles.]

Month.	Discharge in second-feet.				Run-off in inches.
	Maximum.	Minimum.	Mean.	Per square mile.	
October	25,000	11,600	15,500	0.724	0.83
November 1-10	17,400	14,500	15,500	.724	.27
December	14,300	8,000	10,300	.481	.55
January	232,000	11,600	41,900	1.96	2.26
February	266,000	39,400	77,400	3.62	3.77
March	59,100	26,400	34,300	1.60	1.84
April	83,200	26,100	50,400	2.36	2.63
May	70,800	24,300	36,900	1.72	1.98
June	57,000	15,300	25,000	1.17	1.30
July	30,700	13,400	18,200	.850	.98
August	23,100	11,200	14,700	.687	.79
September	18,900	11,200	13,800	.645	.72

TENNESSEE RIVER AT FLORENCE, ALA.

LOCATION.—At Southern Railway bridge at lower end of Pattons Island, just below Little Muscle Shoals, 1 mile south of Florence, Lauderdale County, 3 miles above upper end of Sevenmile Island, 208 miles below Chattanooga, Tenn., and 256 miles above mouth of river.

DRAINAGE AREA.—30,800 square miles.

RECORDS AVAILABLE.—November 7, 1871, to September 30, 1918.

GAGE.—Rod gage consisting of four sections of steel, three-eighths inch by 7½ inches, attached to right face of stone draw pier, which has batter of 1 inch to the foot. These sections form one continuous gage, graduated from −1.92 to 33.5 feet; read by R. E. Coburn. Zero of gage is 400.85 feet above sea level. For description of gages used prior to September 30, 1913, see Water-Supply Paper 353, page 151.

DISCHARGE MEASUREMENTS.—Prior to May, 1918, made from downstream side of highway section (the low-level or through section) of 17-span combined railway and highway bridge. Special care was necessary to counteract effect of obstruction of current by piers. During summer of 1918 measurements were made from boat at a section three-quarters of a mile below gage.

CHANNEL AND CONTROL.—Bed rocky, rough, and uneven; probably permanent. Control is practically permanent.

EXTREMES OF DISCHARGE.—Maximum stage recorded during year, 22.0 feet, afternoon of February 6 (discharge, 276,000 second-feet); minimum stage recorded, −0.3 foot, afternoon of September 1 (discharge, 10,400 second-feet).

1871–1918: Maximum stage recorded, 32.5 feet at 10 and 12 p. m. March 19, 1897 (discharge, 444,000 second-feet; supersedes figure previously published); minimum stage recorded, −0.8 foot September 18, 1878 (discharge, 7,350 second-feet).

ICE.—Stage-discharge relation not affected by ice.

DIVERSIONS.—None.

REGULATION.—Operation of power plant at Hales Bar lock and dam, 175 miles upstream, may cause some diurnal fluctuation in low-stage flow.

ACCURACY.—Stage-discharge relation practically permanent. Rating curve is well defined above 12,000 second-feet. Gage read to tenths twice daily; oftener during high water. Daily discharge ascertained by applying mean daily gage height to rating table. Records good.

COOPERATION.—Gage-height record furnished by Mississippi River Commission.

Discharge measurements of Tennessee River at Florence, Ala., during the year ending Sept. 30, 1918.

Date.	Made by—	Gage height.	Discharge.	Date.	Made by—	Gage height.	Discharge.
		Feet.	*Sec.-ft.*			*Feet.*	*Sec.-ft.*
Oct. 30	L. J. Hall	1.07	17,100	May 18	Hall and Wright	8.35	81,600
Nov. 25	....do	.25	12,800	20	....do	6.15	57,400
Feb. 3	Paulsen and Hall	19.55	230,000	21	....do	5.37	50,000
Apr. 1	L. J. Hall	5.47	52,500	June 24	Hall and Adams	2.10	23,400
5	....do	3.48	36,400	July 31	L. J. Hall	1.74	20,900
May 14	Paulsen and Adams	8.56	83,900	Aug. 1	....do	1.72	20,200
17	Hall and Wright	9.08	91,300	2	....do	1.96	21,900

NOTE.—Measurements made in May, June, July, and August were made from boat at a section three-fourths mile below gage.

Daily discharge, in second-feet, of Tennessee River at Florence, Ala., for the year ending Sept. 30, 1918.

Day.	Oct.	Nov.	Dec.	Jan.	Feb.	Mar.	Apr.	May.	June.	July.	Aug.	Sept.
1	25,400	19,600	12,700	17,200	228,000	57,900	53,400	75,500	36,000	42,800	21,600	10,900
2	25,400	20,200	12,700	16,700	240,000	53,400	48,000	69,300	35,200	41,200	21,600	10,900
3	24,600	20,200	12,700	16,700	240,000	51,600	42,800	60,600	32,600	38,600	23,100	13,200
4	26,100	19,000	12,700	17,200	253,000	48,000	37,800	57,000	31,000	36,000	23,800	14,700
5	25,400	20,200	13,200	16,200	269,000	44,600	36,000	53,400	29,300	34,400	23,800	13,700
6	23,100	19,000	13,200	14,700	274,000	42,800	34,400	50,700	27,700	31,000	23,100	15,200
7	20,200	18,400	13,700	14,700	263,000	39,400	39,400	46,200	24,600	29,300	20,200	17,800
8	17,800	17,800	13,700	14,700	244,000	38,600	83,200	45,400	23,100	27,700	18,400	19,000
9	16,700	16,700	13,700	16,200	184,000	37,800	127,000	42,800	23,100	26,100	16,200	16,700
10	15,700	16,700	13,200	17,200	133,000	37,800	154,000	41,200	23,100	21,600	14,700	14,700
11	14,700	16,200	13,700	21,600	90,200	36,000	154,000	42,000	24,600	19,600	14,700	13,700
12	13,700	16,200	14,700	27,700	70,300	36,000	138,000	39,400	24,600	19,000	14,200	13,700
13	13,200	15,700	15,200	31,000	64,300	37,800	114,000	58,800	24,600	18,400	3,700	13,700
14	12,700	15,200	15,200	42,000	60,600	41,200	96,200	83,200	23,800	18,400	3,700	14,700
15	12,200	14,700	16,700	55,200	58,800	44,600	87,800	83,200	23,100	19,000	3,200	15,200
16	12,200	14,200	14,700	68,300	58,800	42,800	74,400	85,400	20,200	17,800	13,700	15,200
17	11,800	14,200	13,700	76,600	63,300	41,200	72,300	90,200	18,400	16,700	16,700	15,200
18	11,800	13,700	12,700	77,700	74,400	37,800	92,600	83,200	18,400	16,700	16,700	13,700
19	11,800	13,200	11,800	81,000	83,200	36,000	103,000	68,300	18,400	17,800	16,700	12,700
20	12,700	13,200	11,800	74,400	86,600	35,200	99,800	58,800	17,800	17,200	14,700	11,800
21	12,700	13,200	11,800	64,300	95,000	36,000	99,800	51,600	17,200	16,700	13,700	11,800
22	15,700	13,200	12,200	56,100	93,800	37,800	101,000	44,600	19,000	17,800	13,200	11,800
23	19,000	12,700	13,700	52,500	93,800	36,000	102,000	42,800	22,400	19,000	13,700	11,400
24	20,200	12,200	13,700	49,800	90,200	35,200	98,600	44,600	24,600	17,800	17,200	11,400
25	23,100	12,200	13,200	42,000	86,600	36,000	90,200	45,400	26,900	18,400	18,400	10,900
26	23,100	12,700	14,200	36,000	78,800	39,400	81,000	46,200	49,800	17,800	16,700	12,700
27	21,600	12,700	14,700	37,800	70,300	41,200	72,300	51,600	56,200	17,200	14,700	15,700
28	19,000	12,700	15,200	58,800	62,400	46,200	68,300	48,900	51,600	20,200	13,700	15,700
29	17,800	12,700	15,200	90,200	....	57,000	70,300	42,800	43,700	20,200	12,700	15,200
30	16,700	12,700	15,200	152,000	....	62,400	76,600	39,400	48,000	19,000	12,200	14,700
31	17,800	....	15,700	208,000	....	58,800	....	36,000	....	20,200	11,800	.....

Monthly discharge of Tennessee River at Florence, Ala., for the year ending Sept. 30, 1918.

[Drainage area, 30,800 square miles.]

Month.	Discharge in second-feet.				Run-off in inches.
	Maximum.	Minimum.	Mean.	Per square mile.	
October	26,100	11,800	17,900	0.581	0.67
November	20,200	12,200	15,400	.500	.56
December	16,700	1,800	13,800	.448	.52
January	208,000	4,700	50,500	1.64	1.89
February	274,000	58,800	132,000	4.29	4.47
March	62,400	35,200	42,800	1.39	1.60
April	154,000	34,400	84,900	2.76	3.08
May	90,200	36,000	55,800	1.81	2.09
June	55,200	17,200	28,600	.929	1.04
July	42,800	16,700	23,000	.747	.86
August	23,800	11,800	16,500	.536	.62
September	19,000	10,900	13,900	.451	.50
The year	274,000	10,900	40,600	1.32	17.90

SOUTH FORK OF HOLSTON RIVER AT BLUFF CITY, TENN.

LOCATION.—At highway bridge at Bluff City, Sullivan County, 300 feet below Virginia & Southwestern Railway bridge, 1 mile below mouth of Indian Creek, and 10 miles above mouth of Watauga River.

DRAINAGE AREA.—828 square miles.

RECORDS AVAILABLE.—July 17, 1900, to September 30, 1918.

GAGE.—Vertical staff attached to downstream side of bridge pier nearest the right bank; read by W. C. Massengill.

DISCHARGE MEASUREMENTS.—Made from downstream side of highway bridge; or from railroad bridge 300 feet above, where section is much better. At low stages the current becomes sluggish.

CHANNEL AND CONTROL.—Bed of river very rough. Control consists of a shallow ledge; probably permanent. Depth and velocity of current very irregular.

EXTREMES OF DISCHARGE.—Maximum stage recorded during year, 9.1 feet January 29 (discharge, 15,100 second-feet); minimum stage recorded, zero on gage December 12 and 13 (discharge, 185 second-feet).

1900–1918: Maximum stage recorded, 11.45 feet February 28, 1902 (discharge, 33,000 second-feet); minimum stage recorded, —0.1 foot October 16 to 19, 21 to 26, 28 to 31, and November 1, 1904 (discharge, 150 second-feet).

ICE.—Stage-discharge relation not affected by ice.

REGULATION.—Operation of small mills upstream causes some diurnal fluctuation.

ACCURACY.—Stage-discharge relation practically permanent. Rating curve fairly well defined below 6,000 second-feet. Gage read to tenths once daily. Daily discharge ascertained by applying daily gage height to rating table. Records good, except for stages below 800 second-feet, for which they are only fair owing to poor definition of rating curve at low stages.

COOPERATION.—Gage-height record furnished by United States Weather Bureau.

The following discharge measurement was made by L. J. Hall:

February 25, 1918: Gage height, 1.61 feet; discharge, 1,150 second-feet.

Daily discharge, in second-feet, of South Fork of Holston River at Bluff City, Tenn., for the year ending Sept. 30, 1918.

Day.	Oct.	Nov.	Dec.	Jan.	Feb.	Mar.	Apr.	May.	June.	July.	Aug.	Sept.
1	245	590	475	325	5,990	1,480	1,280	1,480	650	1,100	785	370
2	212	475	370	370	4,390	1,280	1,100	1,380	590	940	530	325
3	212	370	420	370	3,320	1,280	1,020	1,190	590	785	370	325
4	245	370	325	325	2,760	1,020	1,020	1,100	590	715	285	285
5	245	325	370	325	2,250	1,190	860	1,020	590	650	285	245
6	245	285	325	325	2,010	1,480	860	940	530	530	285	245
7	245	285	285	1,190	1,900	1,680	940	785	530	420	285	285
8	245	285	285	1,100	1,790	2,500	1,190	1,100	530	370	650	325
9	285	285	212	100	1,900	1,900	3,610	1,280	590	590	475	785
10	325	285	212	785	3,180	1,680	3,610	1,680	530	650	650	650
11	285	212	212	590	3,760	1,580	2,500	1,380	420	530	475	420
12	285	245	185	2,250	2,760	1,380	2,010	2,500	370	475	370	325
13	370	285	185	2,250	,370	,280	1,680	2,370	370	420	325	370
14	325	285	245	,480	,130	,790	1,480	2,630	370	370	325	420
15	245	370	370	,680	,790	,680	1,380	2,250	325	325	325	325
16	245	325	325	1,380	1,680	1,480	1,380	1,680	325	325	285	325
17	245	285	285	1,100	1,580	1,280	1,190	1,480	325	370	370	285
18	245	245	212	940	,380	1,100	1,380	1,190	245	325	370	370
19	285	245	212	860	,190	1,020	1,480	1,100	1,900	420	420	530
20	860	212	245	785	,190	1,020	1,480	1,020	1,020	650	370	530
21	715	245	245	715	1,580	2,010	2,900	940	785	420	370	475
22	530	245	285	715	1,190	3,760	2,500	1,190	1,680	420	325	475
23	420	245	245	715	1,280	,630	,010	1,190	2,130	370	285	420
24	370	245	285	860	1,190	2,500	,680	940	1,280	475	245	370
25	370	245	285	590	1,190	3,910	,380	1,020	785	370	245	325
26	325	245	370	590	1,190	4,060	1,280	940	1,280	285	245	285
27	325	245	325	6,180	1,900	3,040	1,190	860	2,250	245	285	325
28	285	245	420	12,500	,680	2,250	1,100	860	1,280	420	370	370
29	285	285	475	15,100	......	1,900	1,020	860	1,020	420	325	325
30	370	325	370	7,830	......	1,580	1,020	860	860	370	325	285
31	715	......	370	9,410	......	1,380	......	785	......	475	325	......

Monthly discharge of South Fork of Holston River at Bluff City, Tenn., for the year ending Sept. 30, 1918

[Drainage area, 828 square miles.]

Month.	Discharge in second-feet.				Run-off in inches.
	Maximum.	Minimum.	Mean.	Per square mile.	
October	860	212	342	0.413	0.48
November	590	212	294	.355	.40
December	475	185	304	.367	.42
January	15,100	325	2,410	2.91	3.36
February	5,990	1,190	2,160	2.61	2.72
March	4,060	1,020	1,870	2.26	2.61
April	3,610	860	1,580	1.91	2.13
May	2,630	785	1,290	1.56	1.80
June	2,250	245	825	.996	1.11
July	1,100	245	491	.593	.68
August	785	245	374	.452	.52
September	785	245	380	.459	.51
The year	15,100	185	1,020	1.23	16.74

HOLSTON RIVER NEAR ROGERSVILLE, TENN.

Location.—At Virginia & Southwestern Railway bridge near Austins Mill, Hawkins County, half a mile below the county highway bridge, 2 miles downstream from mouth of Dodson Creek, 3 miles south of Rogersville, and 11 miles northeast of Bulls Gap, Tenn.

Drainage area.—3,060 square miles.

Records available.—March 10, 1902 (daily-discharge record beginning January 1, 1904), to September 30, 1918.

Gage.—Vertical staff attached to right side of bridge pier nearest right bank.

Discharge measurements.—Made from steel highway bridge about half a mile upstream from gage.

Channel and control.—Bed of stream composed of solid rock, boulders, and gravel. Right bank high and not subject to overflow; left bank high but subject to overflow at extremely high stages. Control formed by rock shoals below bridge; practically permanent.

Extremes of discharge.—Maximum stage during year, 20.0 feet at crest on January 29 (discharge, about 70,900 second-feet); minimum stage recorded, 1.3 feet December 24 to 26 (discharge, 680 second-feet).

1904–1918: Maximum stage recorded, 19.1 feet March 28, 1913 (discharge, about 67,000 second-feet); minimum stage recorded, 1.0 foot October 23 to November 3, 1904 (discharge, 490 second-feet).

Ice.—Stage-discharge relation seldom affected by ice.

Regulation.—Some diurnal fluctuation caused by Austin's mill power plant and by several other small plants situated on tributaries. The effect is negligible except in extreme low water.

Accuracy.—Stage-discharge relation practically permanent; probably not affected by ice although river was frozen over January 13 to 27. Rating curve well defined below 33,000 second-feet; extended above that point. Below 10,000 second-feet it coincides with curve used from 1911 to 1915; above 10,000 second-feet revised and slightly changed as a result of flood data obtained in March, 1917. Gage read to tenths once daily (morning) except during period of ice cover when no readings were made. Daily discharge ascertained by applying daily gage height to rating table. Records fair.

Discharge measurements of Holston River near Rogersville, Tenn., during the year ending Sept. 30, 1918.

[Made by L. J. Hall.]

Date.	Gage height.	Discharge.
	Feet.	*Sec.-ft.*
Nov. 19	1.51	1,020
Feb. 23	3.13	4,410

Daily discharge, in second-feet, of Holston River near Rogersville, Tenn., for the year ending Sept. 30, 1918.

Day.	Oct.	Nov.	Dec.	Jan.	Feb.	Mar.	Apr.	May.	June.	July.	Aug.	Sept.
1	1,570	1,570	850	1,390	22,400	4,500	3,970	3,720	3,000	6,230	1,950	4,230
2	1,210	1,760	1,210	1,390	15,000	3,970	3,720	4,770	2,350	6,530	2,350	3,720
3	850	1,570	1,390	1,390	11,300	3,720	3,470	3,970	2,150	4,500	1,950	3,470
4	850	1,570	1,210	1,390	9,850	3,470	3,720	3,720	2,150	3,470	1,570	1,390
5	850	1,390	1,390	1,390	7,790	3,000	3,720	3,470	2,350	2,780	1,570	1,210
6	850	1,390	1,210	1,390	6,230	2,780	3,000	3,000	2,150	2,350	1,390	1,030
7	850	1,390	1,030	1,760	6,230	4,230	2,780	3,000	2,150	2,150	1,390	1,030
8	850	1,210	1,030	2,350	6,230	4,770	3,970	3,000	2,150	1,950	1,390	1,570
9	850	1,030	1,210	3,720	5,930	6,530	7,150	7,470	2,150	1,950	1,390	1,760
10	850	1,030	1,390	3,230	6,230	5,340	12,400	7,470	2,150	2,150	2,150	1,950
11	850	1,030	1,210	3,230	9,850	6,230	9,490	6,230	1,760	2,150	1,950	1,950
12	850	1,030	1,030	3,720	9,490	5,340	7,150	5,630	1,760	1,760	1,760	1,570
13	1,030	1,030	1,030	....	7,150	4,500	5,630	6,530	1,570	1,760	1,570	1,570
14	1,210	1,030	1,030	....	6,530	4,500	5,340	7,150	1,570	1,570	1,390	1,760
15	1,570	1,030	850	....	5,930	5,340	4,500	7,470	1,390	1,570	1,390	1,760
16	1,030	1,030	850	....	5,630	5,050	3,970	6,530	1,390	1,390	1,760	1,570
17	1,030	1,030	850	....	5,630	4,500	3,970	5,050	1,390	1,390	1,390	1,570
18	850	1,210	850	....	5,050	4,230	7,150	4,230	1,390	1,570	1,390	1,570
19	850	1,030	850	....	4,230	3,720	6,230	3,970	1,950	1,570	1,570	1,760
20	1,030	1,030	850	....	4,770	3,230	5,050	4,500	3,470	1,950	1,570	1,760
21	3,740	1,030	850	....	5,050	3,970	5,930	3,970	3,720	1,950	1,570	2,350
22	2,350	850	850	....	4,770	6,530	9,490	3,720	5,930	1,760	1,570	2,150
23	1,950	850	850	....	4,770	8,120	7,470	5,050	8,800	1,570	1,390	1,950
24	1,570	850	680	....	4,230	6,530	6,530	3,970	6,530	1,570	1,210	1,760
25	1,570	850	680	....	3,970	8,460	5,340	3,720	4,230	1,760	1,030	1,760
26	1,390	850	680	....	3,970	11,000	4,770	3,720	3,720	1,760	1,210	1,570
27	1,210	850	850	....	3,970	9,850	4,230	4,230	7,790	1,760	1,570	1,390
28	1,210	850	1,760	39,900	4,770	7,470	3,970	3,720	7,150	1,950	1,950	1,570
29	1,210	850	1,760	58,600	....	5,630	3,720	3,970	4,770	1,760	1,570	1,570
30	1,390	850	1,390	34,600	....	5,050	3,470	3,470	3,720	1,570	1,570	1,390
31	1,390	....	1,390	34,600	....	4,500	....	3,470	....	1,760	1,950	....

NOTE.—No record Jan. 13–27.

Monthly discharge of Holston River near Rogersville, Tenn., for the year ending Sept. 30, 1918.

[Drainage area, 3,060 square miles.]

Month.	Discharge in second-feet.				Run-off in inches.
	Maximum.	Minimum.	Mean.	Per square mile.	
October	3,470	850	1,240	0.405	0.47
November	1,760	850	1,100	.359	.40
December	1,760	680	1,070	.350	.40
February	22,400	3,970	7,030	2.30	2.40
March	11,000	2,780	5,360	1.75	2.02
April	12,400	2,780	5,380	1.76	1.96
May	7,470	3,000	4,640	1.52	1.75
June	8,800	1,390	3,220	1.05	1.17
July	6,530	1,390	2,260	.739	.85
August	2,350	1,030	1,590	.520	.60
September	4,230	1,030	1,860	.608	.68

TOCCOA RIVER NEAR DIAL, GA.

LOCATION.—About 2,600 feet above Shallow Ford, 1 mile above Rock Creek, 2½ miles below Big Creek, 3½ miles below Noontootley Creek, 4 miles northwest of Dial, Fannin County, and 12 miles by river above gaging station at Morganton.

DRAINAGE AREA.—175 square miles (measured on topographic maps).

RECORDS AVAILABLE.—January 1, 1913, to September 30, 1918.

GAGE.—Bristol water-stage recorder. Sea-level elevation of zero of auxiliary staff gage, 1,781.13 feet.

DISCHARGE MEASUREMENTS.—Made from cable about 1,000 feet upstream from gage.

CHANNEL AND CONTROL.—Bed of stream consists of gravel and boulders; fairly smooth. Left bank is overflowed at a stage of about 12 feet. Control is formed by the head of rapids just below gage; practically permanent.

EXTREMES OF DISCHARGE.—Maximum mean daily stage during year from water stage recorder, 3.85 feet January 28 (discharge, 1,880 second-feet); minimum mean daily stage, 0.80 foot September 2 and 16 (discharge, 140 second-feet).

1913–1918: Maximum stage recorded, 10.0 feet at 6 p. m. July 9, 1916 (discharge, 9,200 second-feet); minimum stage recorded, 0.55 foot October 13, 29, and 30, 1914 (discharge, 109 second-feet).

DIVERSIONS.—None.

REGULATION.—Slight diurnal fluctuations are caused by operation of small mills upstream.

ACCURACY.—Stage-discharge relation practically permanent. Rating curve well defined below 4,000 second-feet. Operation of water-stage recorder was satisfactory throughout the year. Daily discharge ascertained by applying to rating table mean daily gage height obtained by inspecting gage-height graph. Records good.

COOPERATION.—Gage-height record furnished by Tennessee Power Co.

No discharge measurements were made at this station during the year.

Daily discharge, in second-feet, of Toccoa River near Dial, Ga., for the year ending Sept. 30, 1918.

Day.	Oct.	Nov.	Dec.	Jan.	Feb.	Mar.	Apr.	May.	June.	July.	Aug.	Sept.
1	312	312	225	210	785	428	334	656	374	382	330	148
2	278	295	225	210	755	424	323	610	370	330	455	140
3	278	295	225	210	700	419	330	585	402	312	330	192
4	278	278	278	195	595	410	354	560	442	302	312	171
5	278	278	242	195	545	406	323	545	406	295	298	210
6	260	260	225	312	545	406	316	536	455	284	267	180
7	242	260	210	260	570	432	845	522	424	284	260	195
8	242	260	242	225	522	390	1,030	700	394	295	242	210
9	260	260	195	225	478	419	785	536	390	281	232	201
10	260	242	278	225	455	446	620	522	362	267	338	162
11	242	260	195	1,200	432	390	565	496	342	260	312	162
12	242	260	225	595	455	378	514	482	370	253	288	165
13	225	260	278	410	432	394	464	755	370	242	213	165
14	225	260	260	432	432	386	455	620	330	239	213	152
15	225	260	295	815	432	370	437	550	320	232	242	142
16	225	242	260	478	700	354	645	527	306	225	320	140
17	210	242	210	410	672	362	595	504	428	242	302	148
18	242	242	210	370	595	358	672	504	509	610	260	250
19	785	225	210	330	700	350	565	532	398	575	195	267
20	500	260	210	330	672	398	532	509	362	342	195	354
21	390	242	210	312	620	386	545	565	460	330	180	216
22	350	242	225	350	570	358	500	610	374	330	165	180
23	330	242	210	312	522	362	478	555	330	309	165	165
24	312	225	210	295	522	378	460	500	302	330	180	155
25	312	225	210	312	500	354	585	491	700	414	162	152
26	312	225	225	350	500	338	1,030	468	545	398	160	160
27	295	225	242	545	455	334	700	432	402	386	171	168
28	295	225	210	1,880	455	330	662	410	350	386	165	165
29	330	278	210	1,270	......	330	635	410	437	350	171	183
30	595	242	165	1,500	......	330	728	402	442	390	162	168
31	390	......	210	1,100	......	330	......	382	......	370	165	......

Monthly discharge of Toccoa River near Dial, Ga., for the year ending Sept. 30, 1918.

[Drainage area, 175 square miles.]

Month.	Discharge in second-feet.				Run-off in inches.
	Maximum.	Minimum.	Mean.	Per square mile.	
October	785	210	314	1.79	2.06
November	312	225	254	1.45	1.62
December	295	165	227	1.30	1.50
January	1,880	195	512	2.93	3.38
February	785	432	558	3.19	3.32
March	446	330	379	2.17	2.50
April	1,030	316	568	3.25	3.63
May	755	382	531	3.03	3.49
June	700	302	403	2.30	2.57
July	[illegible]10	225	330	1.89	2.18
August	455	160	240	1.37	1.58
September	354	140	182	1.04	1.16
The year	1,880	140	374	2.14	28.99

TOCCOA RIVER NEAR MORGANTON, GA.

Location.—At Morganton highway bridge on road from Blueridge, Ga., to Morganton, half a mile downstream from mouth of Star Creek, 2 miles west of Morganton post office, Fannin County, 4 miles east of Blueridge, 12 miles downstream from Dial gaging station, 14 miles upstream from Georgia-Tennessee State line at Copperhill, Tenn., and 28 miles upstream from gaging station on Ocoee River at Emf, Tenn. At State line name of river is changed from Toccoa to Ocoee.

Drainage area.—231 square miles (measured on topographic maps).

Records available.—November 25, 1898, to March 31, 1903, and April 1, 1913, to September 30, 1918. Records 1898 to 1903 published in Water-Supply Paper 197 under "Toccoa River near Blueridge, Ga."

Gage.—Bristol water-stage recorder on right bank 200 feet downstream from bridge and 150 feet downstream from the old vertical staff which was used from 1898 to 1903; zeros of both gages, 1,544.50 feet above sea level, but on account of slope in water surface readings of the two gages do not agree for all stages. The water-stage recorder was installed in 1914 (exact date not recorded). A rod gage has been placed at site of automatic gage. Observer visits gage every day and checks record sheet with rod reading.

Discharge measurements.—Made from cable 1,800 feet downstream from gage.

Channel and control.—Bed composed of gravel and boulders. Left bank subject to overflow at about gage height 15 feet; right bank not subject to overflow. Low-water control is a low shoal or riffle just below gage; subject to small shifts occasionally; high-water control formed by combination of shoals and banks; practically permanent.

Extremes of discharge.—Maximum mean daily stage during year from water-stage recorder, 5.3 feet January 28 (discharge, 2,220 second-feet); minimum mean daily stage, 2.3 feet October 12 and December 11 (discharge, 196 second-feet).

1913–1918: Maximum stage recorded, 13.0 feet at 9. p. m. July 9, 1916 (discharge, 13,900 second-feet); minimum stage recorded, 1.8 feet September 10, 14 to 17, 29, 30, and October 1, 1914 (discharge, 129 second-feet).

DIVERSIONS.—None.

REGULATION.—Slight diurnal fluctuations, probably caused by operation of small mills upstream.

ACCURACY.—Stage-discharge relation permanent during year. Rating curve well defined. Daily discharge ascertained by applying to rating table the mean daily gage height obtained by inspecting gage-height graph. Records good.

COOPERATION.—Gage-height record furnished by Tennessee Power Co.

Discharge measurements of Toccoa River near Morganton, Ga., during the year ending Sept. 30, 1918.

Date.	Made by—	Gage height.	Discharge.
		Feet.	*Sec.-ft.*
Nov. 21	L. J. Hall	2.56	281
Feb. 8	C. G. Paulsen	3.21	593

Daily discharge, in second-feet, of Toccoa River near Morganton, Ga., for the year ending Sept. 30, 1918.

Day.	Oct.	Nov.	Dec.	Jan.	Feb.	Mar.	Apr.	May.	June.	July.	Aug.	Sept.
1	331	310	272	254	982	570	434	827	529	461	369	272
2	290	290	272	272	908	540	424	743	512	383	494	254
3	290	290	254	254	870	529	404	716	529	354	378	306
4	310	290	310	254	764	512	434	683	594	354	354	298
5	310	290	290	254	696	494	393	670	540	340	336	327
6	254	272	272	378	668	483	393	657	594	331	323	306
7	238	272	272	354	696	546	1,050	696	576	345	323	290
8	238	254	272	272	600	489	1,340	1,120	512	340	319	290
9	238	254	254	272	570	494	975	764	500	323	302	323
10	223	254	238	254	540	558	736	696	494	310	364	268
11	210	254	196	1,300	540	483	663	683	483	298	440	254
12	196	254	290	908	540	483	600	663	517	290	298	261
13	254	254	310	512	540	483	558	1,040	517	290	290	261
14	272	254	272	834	540	483	540	885	456	286	290	248
15	272	254	272	945	540	456	529	764	434	286	290	245
16	254	254	290	600	1,060	440	729	716	429	272	350	238
17	238	254	272	483	945	440	736	696	650	283	419	235
18	238	254	290	429	764	440	777	683	750	709	388	354
19	238	238	272	404	764	429	689	709	512	856	354	350
20	512	254	272	378	908	540	683	736	500	451	290	478
21	354	254	254	378	798	483	689	805	594	383	276	323
22	310	254	254	429	729	440	600	856	517	388	268	261
23	290	254	272	378	663	434	588	805	445	331	265	254
24	272	254	254	378	663	478	552	696	429	378	310	248
25	272	254	254	378	663	451	696	670	975	424	283	245
26	254	254	254	429	632	429	1,260	650	827	472	268	272
27	254	254	254	663	600	419	870	600	523	429	290	283
28	272	254	272	2,220	570	419	812	576	456	512	290	279
29	290	290	254	1,500	....	404	798	564	523	540	279	283
30	540	290	254	1,940	....	398	922	552	570	540	272	261
31	331	....	254	1,420	....	419	....	540	....	456	286	....

NOTE.—Gage heights Dec. 29-31, doubtful; discharge estimated.

Monthly discharge of Toccoa River near Morganton, Ga., for the year ending Sept. 30, 1918.

[Drainage area, 231 square miles.]

Month.	Discharge in second-feet.				Run-off in inches.
	Maximum.	Minimum.	Mean.	Per square mile.	
October	540	196	285	1.23	1.42
November	310	238	264	1.14	1.27
December	310	196	267	1.16	1.34
January	2,220	254	636	2.75	3.17
February	1,060	540	705	3.05	3.18
March	570	398	473	2.05	2.36
April	1,340	393	696	3.01	3.36
May	1,120	540	725	3.14	3.62
June	975	429	550	2.38	2.66
July	856	272	400	1.73	1.99
August	494	265	324	1.40	1.61
September	478	235	286	1.24	1.38
The year	2,220	196	466	2.02	27.36

OCOEE RIVER AT McHARGE, TENN.

Location.—At county highway bridge at Rogers Ferry, Polk County, half a mile below McHarge railroad siding, half a mile below mouth of Potato Creek, and 2½ miles below Copperhill.

Drainage area.—451 square miles (measured on topographic maps).

Records available.—April 24, 1917, to June 6, 1918.

Gage.—Vertical staff bolted to left downstream side of concrete bridge pier on left bank; read by B. V. Karaivanoff.

Discharge measurements.—Made from downstream side of bridge.

Channel and control.—Left bank subject to overflow at extreme stages, but all water will always pass under bridge. Channel straight for about 300 feet above and 700 feet below gage. Control consists of rock riffle about 300 feet below gage; practically permanent.

Extremes of discharge.—Maximum stage recorded during year, and period of records, 7.1 feet at 7 p. m. January 28, 1918 (discharge not determined); minimum stage recorded, 0.5 foot December 19–23 and 25, 1917 (discharge, 340 second-feet).

Ice.—Stage-discharge relation not affected by ice.

Accuracy.—Stage-discharge relation permanent. Rating curve well defined between 400 and 2,000 second-feet; extended above 2,000 second-feet. Gage read to half-tenths twice daily; oftener during high water. Daily discharge ascertained by applying mean daily gage height to rating table. Records good.

Discharge measurements of Ocoee River at McHarge, Tenn., during the year ending Sept. 30, 1918.

Date.	Made by—	Gage height.	Discharge.
		Feet.	*Sec.-ft.*
Nov. 22	L. J. Hall	0.67	441
Feb. 8	Paulsen and Hall	1.68	1,150

Daily discharge, in second-feet, of Ocoee River at McHarge, Tenn., for the period Oct. 1, 1917, to June 6, 1918.

Day.	Oct.	Nov.	Dec.	Jan.	Feb.	Mar.	Apr.	May	June.
1	770	630	475	445	1,610	990	630	1,610	805
2	665	565	445	445	1,710	915	700	1,160	770
3	630	565	418	418	1,810	915	840	1,250	878
4	915	535	475	445	1,430	878	840	1,160	952
5	565	535	445	445	1,250	878	700	1,070	915
6	535	505	445	505	1,250	878	630	1,030	878
7	535	505	418	630	1,250	1,160	1,610	990	....
8	505	305	445	505	1,120	952	2,370	2,130	....
9	535	505	445	475	1,070	878	2,020	2,020	....
10	565	505	445	445	1,070	990	1,610	1,810	....
11	505	505	445	1,910	952	915	1,250	1,430	....
12	505	505	445	2,250	990	840	1,160	1,070	....
13	445	475	445	1,070	952	878	990	1,710	....
14	445	505	418	1,430	915	840	990	1,810	....
15	445	475	390	2,760	1,160	805	990	1,250	....
16	445	505	390	1,250	1,610	770	1,430	1,160	....
17	445	505	365	1,030	2,370	770	2,250	1,160	....
18	445	445	365	840	1,520	770	1,430	1,070	....
19	1,430	445	340	770	1,520	840	1,250	1,340	....
20	1,250	445	340	735	1,810	915	1,160	1,250	....
21	770	445	340	630	1,610	915	1,250	1,340	....
22	700	445	340	952	1,430	770	1,160	1,160	....
23	630	445	340	700	1,430	770	1,070	1,610	....
24	630	418	365	735	1,160	770	952	1,160	....
25	565	418	340	665	1,250	770	915	1,070	....
26	505	418	365	735	1,250	735	2,130	990	....
27	365	445	365	990	1,160	735	1,520	990	....
28	505	445	390	4,540	990	700	1,250	915	....
29	505	475	390	3,830	....	665	1,340	840	....
30	840	565	475	5,290	....	630	1,120	840	....
31	665	....	505	3,410	....	630	....	840	....

Monthly discharge of Ocoee River at McHarge, Tenn., for the period Oct. 1, 1917, to May 31, 1918.

[Drainage area, 451 square miles.]

Month.	Discharge in second-feet.				Run-off in inches.
	Maximum.	Minimum.	Mean.	Per square mile.	
October	1,430	445	628	1.39	1.60
November	630	418	489	1.08	1.20
December	505	340	407	.902	1.04
January	5,290	418	1,330	2.95	3.40
February	2,370	915	1,340	2.97	3.09
March	1,160	630	834	1.85	2.13
April	2,370	630	1,250	2.77	3.09
May	2,130	840	1,270	2.82	3.25

OCOEE RIVER AT EMF, TENN.

LOCATION.—About 600 feet below Tennessee Power Co's. plant No. 2, known as "Caney Creek plant," half a mile upstream from Emf post office, Polk County, 1¼ miles below mouth of Goforth Creek, and 8 miles upstream from Parksville, Tenn.

DRAINAGE AREA.—530 square miles (determined by Tennessee Power Co.).

RECORDS AVAILABLE.—January 1, 1913, to September 30, 1918.

GAGE.—Bristol water-stage recorder on left bank; checked daily with a staff gage which is bolted to rock near the recorder. Sea-level elevation of zero of staff gage, 830.00 feet.

DISCHARGE MEASUREMENTS.—Made from suspension footbridge 1,000 feet downstream from gage. Prior to August 29, 1917, made from a cable 2,000 feet below gage, and a few of the early measurements were made from boat.

CHANNEL AND CONTROL.—Bed of stream for several hundred feet below gage is composed of boulders, gravel, and solid rock. Banks high; subject to small overflow. Control is formed by a shoal and island 700 feet downstream from gage; practically permanent.

EXTREMES OF DISCHARGE.—Maximum mean daily stage during year from water-stage recorder, 7.5 feet January 30 (discharge, 7,730 second-feet); minimum mean daily stage 2.89 feet December 11 (discharge, 288 second-feet).

1913–1918: Maximum stage recorded, 13.7 feet at 12.30 a. m. July 10, 1916 (discharge, 21,400 second-feet); minimum stage recorded, 2.77 feet September 15 to 17, 1914 (discharge, 285 second-feet).

DIVERSIONS.—None.

REGULATION.—The operation of plant No. 2 causes considerable fluctuation at times, but as a rule, this plant runs on a steady load, the quantity of water used depending largely on stage of river. Storage at diversion dam very small. When plant is shut down water overflows dam in a short time, so that periods of fluctuation will be short.

ACCURACY.—Stage-discharge relation practically permanent. Rating curve well defined between 400 and 8,000 second-feet; above 8,000 second-feet curve is extended as a tangent. Daily discharge ascertained by applying to rating table mean daily gage height determined by inspecting gage-height graph. Records excellent.

COOPERATION.—Gage-height record furnished by Tennessee Power Co.

Discharge measurements of Ocoee River at Emf, Tenn., during the year ending Sept. 30, 1918.

Date.	Made by—	Gage height.	Discharge.	Date.	Made by—	Gage height.	Discharge.
		Feet.	*Sec.-ft.*			*Feet.*	*Sec.-ft.*
Oct. 12	L. J. Hall	3.28	561	June 20	L. J. Hall	3.88	1,070
Nov. 16	.....do	3.25	509	20	.....do	3.83	[a] 1,060
Feb. 6	Paulsen and Hall	4.20	1,400	Aug. 17	.....do	3.51	730
Apr. 18	.....do	4.34	1,540				

[a] Measurement made at old cable section which was abandoned in August, 1917. This measurement indicates that results obtained at cable section are somewhat too large.

Daily discharge, in second-feet, of Ocoee River at Emf, Tenn., for the year ending Sept. 30, 1918.

Day.	Oct.	Nov.	Dec.	Jan.	Feb.	Mar.	Apr.	May.	June.	July.	Aug.	Sept.
1	681	706	487	523	2,640	1,140	793	1,890	922	1,480	942	538
2	600	584	530	487	2,120	1,080	829	1,530	848	1,010	811	501
3	615	631	466	472	2,040	1,060	811	1,420	903	820	875	479
4	607	664	459	415	1,660	1,050	829	1,300	961	767	811	600
5	689	664	472	472	1,430	1,050	838	1,130	1,060	758	732	681
6	600	592	472	561	1,380	1,040	750	1,180	1,040	715	615	592
7	545	545	453	681	1,360	1,250	1,410	1,170	1,160	732	664	523
8	545	545	459	592	1,290	1,140	3,110	1,820	1,090	838	581	530
9	545	545	538	494	1,230	990	2,730	1,550	913	732	600	538
10	561	545	538	434	1,280	1,160	1,760	1,340	866	681	568	538
11	545	545	288	2,920	1,210	1,070	1,060	1,250	857	664	732	487
12	545	545	391	4,220	1,180	980	1,240	1,160	866	623	732	472
13	530	545	600	1,270	1,180	951	1,140	1,720	1,070	615	561	568
14	538	545	440	1,690	1,140	1,000	1,090	2,370	838	600	545	516
15	545	530	409	3,110	1,260	970	1,030	1,510	884	600	545	459
16	545	530	538	1,620	2,460	903	1,400	1,320	741	584	607	440
17	538	516	538	1,140	3,300	857	1,820	1,320	732	584	932	440
18	530	472	623	922	1,960	942	1,590	1,300	1,310	776	961	523
19	1,720	508	631	802	1,630	884	1,490	1,200	1,130	1,690	767	561
20	1,760	479	623	793	2,280	961	1,400	1,300	1,030	1,100	656	802
21	866	487	623	723	1,890	1,040	1,700	1,440	1,100	913	561	767
22	838	479	631	961	1,630	922	1,420	1,530	1,170	857	323	576
23	767	479	538	811	1,500	903	1,280	2,120	913	884	508	523
24	723	472	545	776	1,300	1,020	1,210	1,280	793	932	353	472
25	706	545	447	750	1,340	980	1,140	1,610	913	922	576	447
26	681	530	459	784	1,340	922	2,120	1,330	2,040	1,030	538	472
27	584	472	487	1,240	1,250	857	1,890	1,200	1,130	1,060	494	648
28	600	434	466	4,420	1,290	820	1,500	1,070	903	1,090	568	568
29	508	440	472	4,950	......	802	1,500	990	2,370	1,520	553	530
30	951	466	434	7,730	......	793	1,760	970	2,920	1,300	516	487
31	884	......	434	5,060	......	776	......	1,400	......	1,250	553	......

Monthly discharge of Ocoee River at Emf, Tenn., for the year ending Sept. 30, 1918.

[Drainage area, 530 square miles.]

Month.	Discharge in second-feet.				Run-off in inches.
	Maximum.	Minimum.	Mean.	Per square mile.	
October	1,760	508	706	1.33	1.53
November	706	434	535	1.01	1.13
December	631	288	500	.943	1.09
January	7,730	415	1,670	3.15	3.63
February	3,300	1,140	1,630	3.08	3.21
March	1,250	776	978	1.85	2.13
April	3,110	750	1,420	2.68	2.99
May	2,370	970	1,410	2.66	3.07
June	2,920	732	1,120	2.11	2.35
July	1,690	584	908	1.71	1.97
August	961	494	650	1.23	1.42
September	802	440	543	1.02	1.14
The year	7,730	288	1,000	1.89	25.66

BIG BEAR RIVER NEAR RED BAY, ALA.

LOCATION.—At Norman Bridge, 2½ miles east of Red Bay, Franklin County, 3 miles east of Mississippi State line, 4 miles below mouth of Blue Creek, and 35 miles above junction with Tennessee River.

DRAINAGE AREA.—254 square miles (measured on map; scale, 1: 500,000).

RECORDS AVAILABLE.—August 24, 1913, to September 30, 1918.

GAGE.—Vertical staff attached to a sweet gum tree on left bank 25 feet upstream from bridge; installed April 10, 1918. Zero of this gage is 0.66 foot below zero of old gage as originally installed, but owing to settlement of old gage, the 8-foot marks on both gages are at the same elevation. Both gages attached to same tree. See paragraph under "Gage" in Water-Supply Paper 453 for additional information as to settlement of old gage. Gage read by Ed. Bullen.

CHANNEL AND CONTROL.—Bed composed of gravel; probably shifting. During extreme low water current is sluggish and irregular. Left bank subject to overflow at stages above 12 feet. Control is a gravel bar 100 feet downstream; practically permanent.

EXTREMES OF DISCHARGE.—Maximum stage recorded during year, 13.3 feet at 1 p. m. April 9 (discharge, 3,810 second-feet); minimum stage recorded, 1.2 feet August 15–17 and September 17 (discharge, 10 second-feet).

1913–1918: Maximum stage recorded, 14.2 feet at 7 p. m. July 9, 1916, referred to original datum of gage installed August 24, 1913, or 14.86 feet referred to datum of gage installed April 10, 1918 (discharge, 4,700 second-feet; figure previously published is in error owing to erroneous extension of rating curve); minimum discharge, 10 second-feet August 15–17 and September 17, 1918.

ICE.—Stage-discharge relation not affected by ice.

ACCURACY.—Stage-discharge relation practically permanent. Rating curve well defined between 80 and 4,000 second-feet; poorly defined below 80 second-feet. Gage read to tenths once daily. Daily discharge ascertained by applying daily gage height to rating table. Records October to January should be used with caution owing to uncertainty in regard to corrections applied to gage heights (see paragraph under "Gage"). Records February to September good, except those below 80 second-feet, which are only fair.

Discharge measurements of Big Bear River near Red Bay, Ala., during the year ending Sept. 30, 1918.

[Made by L. J. Hall.]

Date.	Gage height.	Discharge.	Date.	Gage height.	Discharge.
	Feet.	*Sec.-ft.*		*Feet.*	*Sec.-ft.*
Apr. 8	13.25	3,900	May 16	5.18	848
9	11.89	2,940	16	4.92	784
10	7.34	1,470	July 27	1.91	111
10	6.46	1,200	28	2.66	258
11	5.26	870			

NOTE.—Gage heights of above measurements referred to datum of staff gage installed Apr. 10, 1918.

Daily discharge, in second-feet, of Big Bear River near Red Bay, Ala., for the year ending Sept. 30, 1918.

Day.	Oct.	Nov.	Dec.	Jan.	Feb.	Mar.	Apr.	May.	June.	July.	Aug.	Sept.
1	130	130	148	46	1,610	263	130	646	130	671	78	32
2	95	78	112	46	906	243	130	481	130	305	62	46
3	62	62	95	46	671	223	130	435	263	166	46	32
4	62	62	95	46	550	204	130	347	148	130	46	46
5	62	62	78	46	413	204	112	305	112	95	46	284
6	46	62	78	62	347	185	112	263	598	78	32	130
7	32	46	112	130	326	263	148	243	527	62	32	32
8	46	46	148	148	326	284	3,440	223	263	46	32	46
9	62	46	130	112	305	243	3,810	1,190	185	46	32	46
10	46	46	112	95	284	223	1,610	722	148	46	20	46
11	32	46	112	130	284	223	906	435	130	32	20	32
12	32	62	112	305	326	204	646	369	112	32	20	32
13	32	78	95	130	458	185	527	1,620	95	32	20	32
14	32	62	112	263	369	185	435	3,620	95	32	20	20
15	32	62	112	458	305	166	413	1,770	78	32	10	20
16	32	46	112	800	326	166	369	906	78	32	10	20
17	32	46	130	504	646	166	435	598	78	32	10	10
18	32	46	130	326	696	148	1,950	481	62	46	130	20
19	46	46	112	284	574	130	2,260	391	62	130	263	32
20	62	62	112	263	598	148	960	326	46	112	130	32
21	62	62	112	223	574	166	1,310	284	46	78	78	20
22	62	62	95	204	550	204	1,070	223	527	130	46	20
23	62	46	95	166	550	185	933	204	263	95	46	20
24	46	46	95	130	527	166	550	185	148	78	32	20
25	46	46	95	148	435	166	458	166	78	78	32	20
26	46	46	78	185	391	166	391	166	95	46	20	20
27	62	46	78	347	347	148	347	148	481	46	20	46
28	62	32	78	800	305	148	413	148	284	263	62	284
29	46	78	78	1,130	....	130	722	130	185	223	46	204
30	62	166	78	1,980	....	130	748	166	263	263	32	130
31	112	....	62	2,450	....	130	....	148	....	112	32	....

Monthly discharge of Big Bear River near Red Bay, Ala., for the year ending Sept. 30, 1918.

[Drainage area, 254 square miles.]

Month.	Discharge in second-feet.				Run-off in inches.
	Maximum.	Minimum.	Mean.	Per square mile.	
October	130	32	54	0.213	0.25
November	166	32	60.9	.240	.27
December	148	62	103	.406	.47
January	2,450	46	387	1.52	1.75
February	1,610	284	500	1.97	2.05
March	284	130	187	.736	.85
April	3,810	112	853	3.36	3.71
May	3,620	130	559	2.20	2.54
June	598	46	190	.748	.83
July	671	32	115.	.453	.52
August	263	10	48.5	.191	.22
September	284	10	59.1	.233	.26
The year	3,810	10	257	1.01	13.76

MISCELLANEOUS MEASUREMENTS.

The results of measurements of flow of streams in the Ohio River basin at points other than regular gaging stations are presented in the following table:

Miscellaneous measurements in the Ohio River drainage basin in the year ending Sept. 30, 1918.

Date.	Stream.	Tributary to—	Locality.	Gage height.	Discharge.
1918.				*Feet.*	*Sec.-ft.*
Feb. 12	Hiwassee River....	Tennessee River..	Old gaging station at Reliance, Tenn.	2.20	2,8[illegible]0
Apr. 6	Tuscumbia Spring.	.....do............	Weir 1 mile above pumping station of Government nitrate plant No. 1 at Sheffield, Tenn.	....	61.7
6	.....do............	.....do............	.....do........................	....	60.6
May 20	.....do............	.....do............	.....do........................	1.52	175
Apr. 19	Ocoee River.......	Hiwassee River...	Old gaging station at Parksville, Tenn.	5.85	2,530

INDEX.

82287—22—WSP 473——9

DEPARTMENT OF THE INTERIOR
JOHN BARTON PAYNE, Secretary

UNITED STATES GEOLOGICAL SURVE
GEORGE OTIS SMITH, Director

ACE WATER SUPPLY OF UNITED STATES

1918

RT IV. ST. LAWRENCE RIVER BAS

NATHAN C. GROVER, Chief Hydraulic Engineer

W. G. HOYT, A. H. HORTON, C. C. COVERT, and
C. H. PIERCE, District Engineers

DEPARTMENT OF THE INTERIOR
JOHN BARTON PAYNE, Secretary

UNITED STATES GEOLOGICAL SURVEY
GEORGE OTIS SMITH, Director

Water-Supply Paper 474

SURFACE WATER SUPPLY OF THE UNITED STATES

1918

PART IV. ST. LAWRENCE RIVER BASIN

NATHAN C. GROVER, Chief Hydraulic Engineer
W. G. HOYT, A. H. HORTON, C. C. COVERT, and
C. H. PIERCE, District Engineers

Prepared in cooperation with the
STATES OF WISCONSIN, NEW YORK, AND VERMONT

WASHINGTON
GOVERNMENT PRINTING OFFICE
1920

ADDITIONAL COPIES
OF THIS PUBLICATION MAY BE PROCURED FROM
THE SUPERINTENDENT OF DOCUMENTS
GOVERNMENT PRINTING OFFICE
WASHINGTON, D. C.
AT
15 CENTS PER COPY

CONTENTS.

ILLUSTRATIONS.

SURFACE WATER SUPPLY OF ST. LAWRENCE RIVER BASIN, 1918.

AUTHORIZATION AND SCOPE OF WORK.

This volume is one of a series of 14 reports presenting results of measurements of flow made on streams in the United States during the year ending September 30, 1918.

The data presented in these reports were collected by the United States Geological Survey under the following authority contained in the organic law (20 Stat. L., p. 394):

Provided, That this officer [the Director] shall have the direction of the Geological Survey and the classification of public lands and examination of the geological structure, mineral resources, and products of the national domain.

The work was begun in 1886 in connection with special studies relating to irrigation in the arid west. Since the fiscal year ending June 30, 1895, successive sundry civil bills passed by Congress have carried the following item and appropriations:

For gaging the streams and determining the water supply of the United States, and for the investigation of underground currents and artesian wells, and for the preparation of reports upon the best methods of utilizing the water resources.

Annual appropriations for the fiscal years ended June 30, 1895-1919.

1895	$12,500
1896	20,000
1897 to 1900, inclusive	50,000
1901 to 1902, inclusive	100,000
1903 to 1906, inclusive	200,000
1907	150,000
1908 to 1910, inclusive	100,000
1911 to 1917, inclusive	150,000
1918	175,000
1919	148,244.10

In the execution of the work many private and State organizations have cooperated, either by furnishing data or by assisting in collecting data. Acknowledgments for cooperation of the first kind are made in connection with the description of each station affected· cooperation of the second kind is acknowledged on page 9.

Measurements of stream flow have been made at about 4,500 points in the United States and also at many points in Alaska and the Hawaiian Islands. In July, 1918, 1,180 gaging stations were being maintained by the Survey and the cooperating organizations. Many miscellaneous discharge measurements are made at other points. In connection with this work data were also collected in regard to precipitation, evaporation, storage reservoirs, river profiles, and water power in many sections of the country and will be made available in water-supply papers from time to time. Information in regard to publications relating to water resources is presented in the appendix to this report.

DEFINITION OF TERMS.

The volume of water flowing in a stream—the "run-off" or "discharge"—is expressed in various terms, each of which has become associated with a certain class of work. These terms may be divided into two groups—(1) those that represent a rate of flow, as second-feet, gallons per minute, miners' inches, and discharge in second-feet per square mile, and (2) those that represent the actual quantity of water, as run-off in depth in inches, acre-feet, and millions of cubic feet. The principal terms used in this series of reports are second-feet, second-feet per square mile, run-off in inches, and acre-feet. They may be defined as follows:

"Second-feet" is an abbreviation for "cubic feet per second." A second-foot is the rate of discharge of water flowing in a channel of rectangular cross section 1 foot wide and 1 foot deep at an average velocity of 1 foot per second. It is generally used as a fundamental unit from which others are computed.

"Second-feet per square mile" is the average number of cubic feet of water flowing per second from each square mile of area drained, on the assumption that the run-off is distributed uniformly both as regards time and area.

"Run-off (depth in inches)" is the depth to which an area would be covered if all the water flowing from it in a given period were uniformly distributed on the surface. It is used for comparing run-off with rainfall, which is usually expressed in depth in inches.

An "acre-foot," equivalent to 43,560 cubic feet, is the quantity required to cover an acre to the depth of 1 foot. The term is commonly used in connection with storage for irrigation.

The following terms not in common use are here defined:

"Stage-discharge relation," an abbreviation for the term "relation of gage height to discharge."

"Control," a term used to designate the section or sections of the stream below the gage which determine the stage-discharge relation at the gage. It should be noted that the control may not be the same section or sections at all stages.

The "point of zero flow" for a given gaging station is that point on the gage—the gage height—to which the surface of the river would fall if there were no flow.

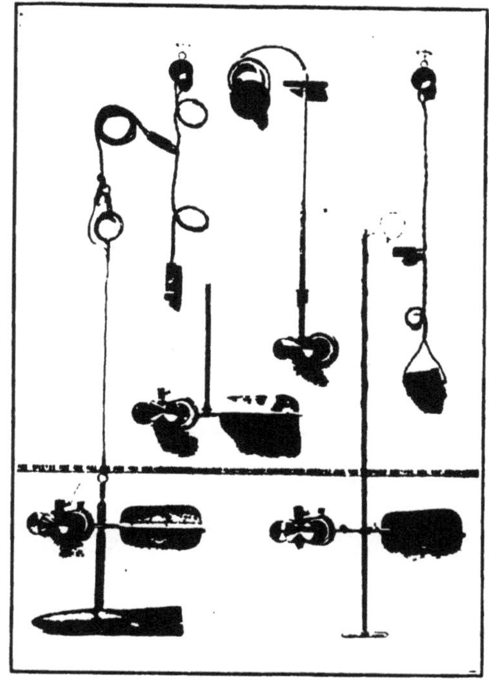

A. PRICE CURRENT METERS.

B. TYPICAL GAGING STATION.

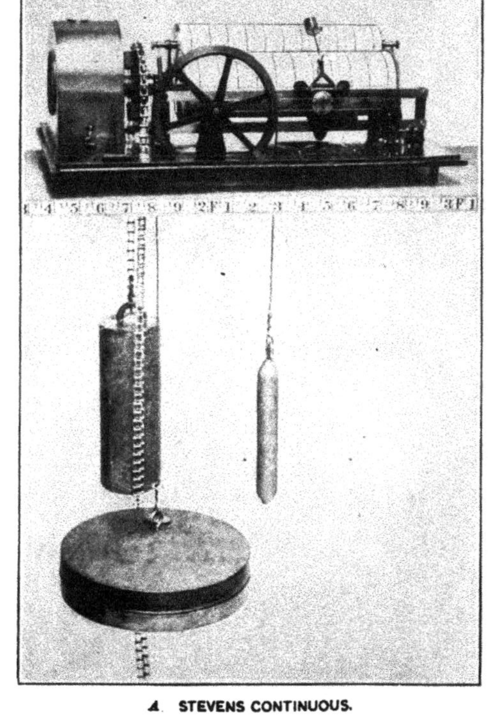

A. STEVENS CONTINUOUS.

B. GURLEY PRINTING.

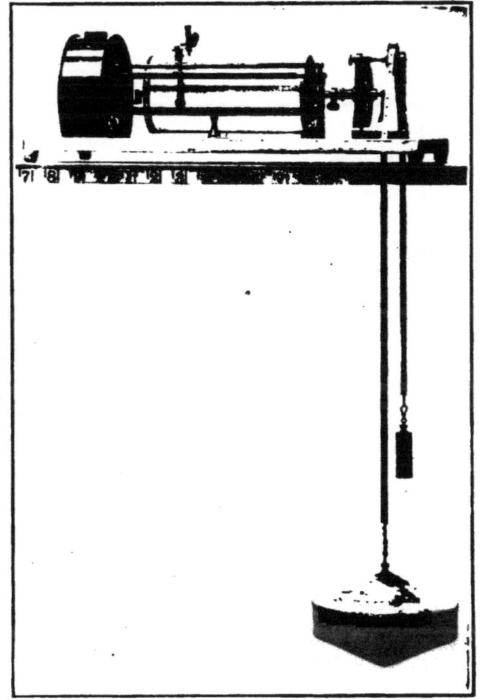

C. FRIEZ.

WATER-STAGE RECORDERS.

EXPLANATION OF DATA.

The data presented in this report cover the year beginning October 1, 1917, and ending September 30, 1918. At the beginning of January in most parts of the United States much of the precipitation in the preceding three months is stored as ground water in the form of snow or ice, or in ponds, lakes, and swamps, and this stored water passes off in the streams during the spring break-up. At the end of September, on the other hand, the only stored water available for run-off is possibly a small quantity in the ground; therefore the run-off for the year beginning October 1 is practically all derived from precipitation within that year.

The base data collected at gaging stations consist of records of stage, measurements of discharge, and general information used to supplement the gage heights and discharge measurements in determining the daily flow. The records of stage are obtained either from direct readings on a staff gage or from a water-stage recorder that gives a continuous record of the fluctuations. Measurements of discharge are made with a current meter. (See Pls. I, II.) The general methods are outlined in standard textbooks on the measurement of river discharge.

From the discharge measurements rating tables are prepared that give the discharge for any stage, and these rating tables, when applied to gage heights, give the discharge from which the daily, monthly, and yearly mean discharge is determined.

The data presented for each gaging station in the area covered by this report comprise a description of the station, a table giving results of discharge measurements, a table showing the daily discharge of the stream, and a table of monthly and yearly discharge and run-off.

If the base data are insufficient to determine the daily discharge, tables giving daily gage heights and results of discharge measurements are published.

The description of the station gives, in addition to statements regarding location and equipment, information in regard to any conditions that may affect the constancy of the discharge relation, covering such subjects as the occurrence of ice, the use of the stream for log driving, shifting of control, and the cause and effect of backwater; it gives also information as to diversions that decrease the flow at the gage, artificial regulation, maximum and minimum recorded stages, and the accuracy of the records.

The table of daily discharge gives, in general, the discharge in second-feet corresponding to the mean of the gage heights read each day. At stations on streams subject to sudden or rapid diurnal fluctuation the discharge obtained from the rating table and the mean daily gage height may not be the true mean discharge for the day. If such stations are equipped with water-stage recorders the mean daily discharge may be obtained by averaging discharge at regular intervals during the day, or by using the discharge integrator, an instrument operating on the principle of the planimeter and containing as an essential element the rating curve of the station.

In the table of monthly discharge the column headed "Maximum" gives the mean flow for the day when the mean gage height was highest. As the gage height is the mean for the day it does not indicate correctly the stage when the water surface was at crest height, and the corresponding discharge was consequently larger than given in the maximum column. Likewise, in the column headed "Minimum" the quantity given is the mean flow for the day when the mean gage height was lowest. The column headed "Mean" is the average flow in cubic feet for each second during the month. On this average flow computations recorded in the remaining columns, which are defined on page 6, are based.

ACCURACY OF FIELD DATA AND COMPUTED RESULTS.

The accuracy of stream-flow data depends primarily (1) on the permanence of the discharge relation and (2) on the accuracy of observation of stage, measurements of flow and interpretation of records.

A paragraph in the description of the station or footnotes added to the tables gives information regarding the (1) permanence of the stage-discharge relation, (2) precision with which the discharge rating curve is defined, (3) refinement of gage readings, (4) frequency of gage readings, and (5) methods of applying daily gage heights to the rating table to obtain the daily discharge.[1]

For the rating tables "well defined" indicates, in general, that the rating is probably accurate within 5 per cent; "fairly well defined," within 10 per cent; "poorly defined," within 15 to 25 per cent. These notes are very general and are based on the plotting of the individual measurements with reference to the mean rating curve.

The monthly means for any station may represent with high accuracy the quantity of water flowing past the gage, but the figures showing discharge per square mile and depth of run-off in inches may be subject to gross errors caused by the inclusion of large noncontributing districts in the measured drainage area, by lack of information concerning water diverted for irrigation or other use, or by inability to interpret the effect of artificial regulation of the flow of the river above the station. "Second-feet per square mile" and "Run-off (depth in inches)" are therefore not computed if such errors appear probable. The computations are also omitted for stations on

[1] For a more detailed discussion of the accuracy of stream-flow data see Grover, N. C., and Hoyt, J. C. Accuracy of stream-flow data: U. S. Geol. Survey Water-Supply Paper 400, pp. 53-59, 1916.

streams drainage areas in which the annual rainfall is less than 20 inches. All figures representing "second-feet per square mile" and "run-off (depth in inches)" previously published by the Survey should be used with caution because of possible inherent sources of error not known to the Survey.

The table of monthly discharge gives only a general idea of the flow at the station and should not be used for other than preliminary estimates; the tables of daily discharge allow more detailed studies of the variation in flow. It should be borne in mind, however, that the observations in each succeeding year may be expected to throw new light on data previously published.

COOPERATION.

The work in Wisconsin during the year ending September 30, 1918, was done in cooperation with the Railroad Commission of Wisconsin, C. M. Larson, chief engineer, and at certain stations with the following organizations: Menominee & Marinette Light & Traction Co., Edward Daniel, general manager (Menominee River below Koss, Mich.); Corps of Engineers, United States Army (Wolf River at New London, Fox River at Berlin, and Fox River at Rapide Croche dam); United States Indian Office (Wolf River at Keshena).

The station on Little Calumet River at Harvey, Ill., was maintained in cooperation with division of waterways of the Illinois Department of Public Works and Buildings, W. L. Sackett, director.

The gage reader for Huron River at Flat Rock, Mich., was paid by Gardner S. Williams.

Work in the State of New York has been conducted under cooperative agreements with the State engineer and surveyor and since July 1, 1911, with the division of inland waters of the State Conservation Commission as provided by an act of the State legislature.

The water-stage recorder on Genessee River at Rochester, N. Y., was inspected by an employee of the Rochester Railway & Light Co.

The water-stage recorder on Raquette River at Piercefield, N. Y., was inspected by an employee of the International Paper Co.

The work in Vermont has been carried on in cooperation with the State of Vermont, Horace F. Graham, governor, and Herbert M. McIntosh, State engineer, and at certain stations in cooperation with the following organizations and individuals: Vermont Marble Co. (Otter Creek at Middlebury); the department of civil engineering of Norwich University (Dog River at Northfield); Newport Electric Light Co. (Clyde River at West Derby).

DIVISION OF WORK.

The data for stations in the Lake Superior and Lake Michigan drainage basins in Wisconsin and Illinois were collected and prepared for publication under the direction of W. G. Hoyt, district engineer, assisted by S. B. Soulé, H. C. Beckman, L. L. Smith, T. G. Bedford, A. M. Wahl, and H. S. Wahl.

Data for stations in the St. Lawrence drainage basin in New York were collected and prepared for publication under the direction of C. C. Covert, district engineer, assisted by O. W. Hartwell, E. D. Burchard, J. W. Moulton, Max H. Carson, and W. A. James.

Data for stations in Vermont were collected and prepared for publication under the direction of C. H. Pierce, district engineer, assisted by O. W. Hartwell, H. W. Fear, M. R. Stackpole, J. W. Moulton, and Hope Hearn.

The manuscript was assembled by B. J. Peterson.

GAGING-STATION RECORDS.

STREAMS TRIBUTARY TO LAKE SUPERIOR.

BAD RIVER NEAR ODANAH, WIS.

LOCATION.—In sec. 25, T. 47 N., R. 3 W., 8 miles upstream from Odanah, Ashland County, 12 miles above mouth. Potato River enters from right about 8 miles above station.

DRAINAGE AREA.—607 square miles (measured on map issued by Wisconsin Geological and Natural History Survey, edition of 1911; scale, 1 inch=6 miles).

RECORDS AVAILABLE.—July 31, 1914, to September 30, 1918.

GAGE.—Stevens continuous water-stage recorder, installed March 31, 1915, over a wooden well, just above the first falls in the river above the mouth; a Gurley water-stage recorder at the same site was used July 31, 1914, to March 31, 1915.

DISCHARGE MEASUREMENTS.—Made from a cable about 700 feet upstream from the gage.

CHANNEL AND CONTROL.—Bed sand and gravel. Rock outcrops at the beginning of rapids about 200 feet below the gage form a permanent control. During log-driving periods logs may collect on the outcrop and cause backwater at the gage. Right bank high, not subject to overflow; left bank of medium height and may be overflowed during extremely high water.

EXTREMES OF DISCHARGE.—Maximum stage recorded during year, 5.61 feet at 9 p. m. June 1 (discharge 8,590 second-feet); minimum open-water stage 0.82 foot, afternoon of August 27, (discharge about 88 second-feet). Discharge during January and February may have been slightly less than 88 second-feet.

1914-1918: Maximum stage recorded 6.66 feet at 1 a. m., April 22, 1916 (discharge 12,200 second-feet); minimum stage recorded that of August 27, 1918.

ICE.—Stage-discharge relation seriously affected by ice.

REGULATION.—A number of small reservoirs are operated during the early spring and summer as an aid to log driving. During such periods the stage may fluctuate rapidly.

ACCURACY.—Stage-discharge relation fairly permanent, except when affected by ice; rating curve well defined between 80 and 7,270 second-feet; above 7,270 second-feet extended and may be subject to considerable error. Operation of water-stage recorder satisfactory except during winter period. Daily discharge ascertained as follows: October 1–15, by use of discharge integrator; October 16 to December 2, and March 22 to September 30 by applying to rating table mean daily gage height obtained by planimeter from recorder graph, except April 18–20, which was interpolated; December 2 to March 21, determined, because of ice, from discharge measurements, and comparisons with records of flow in adjacent drainage basins. Open-water records good; winter records roughly approximate.

Discharge measurements of Bad River near Odanah, Wis., during the year ending Sept. 30, 1918.

Date.	Made by—	Gage height.	Discharge.	Date.	Made by—	Gage height.	Discharge.
		Feet.	*Sec.-ft.*			*Feet.*	*Sec.-ft.*
Dec. 20[a]	T. G. Bedford..........	1.60	123	Apr. 27[c]	T. G. Bedford..........	1.40	376
Jan. 21[b]	.....do......	1.82	106	Aug. 3[d]	S. B. Soule.............	.88	116

[a] Made through complete ice cover at the gage section. Measured discharge probably too low because of low velocity in measuring section.
[b] Complete ice cover at control and measuring section.
[c] Made at cable section; a few logs lodged on control.
[d] Made by wading.

Daily discharge, in second-feet, of Bad River near Odanah, Wis., for the year ending Sept. 30, 1918.

Day.	Oct.	Nov.	Dec.	Jan.	Feb.	Mar.	Apr.	May.	June.	July.	Aug.	Sept.
1	280	649	305				1590	1350	6960	183	139	209
2	250	601	294				1460	1200	6340	177	129	188
3	250	568	275				1180	950	3730	172	116	167
4	240	542					1050	930	2520	188	120	153
5	230	518				100	990	750	1780	236	120	139
6	230	510					1010	800	1530	188	158	112
7	245	494					1120	750	1120	167	258	112
8	255	486					1150	840	930	144	264	100
9	270	470					970	820	1250	139	253	108
10	270	463					910	1470	1340	139	253	112
11	270	442	190				770	1850	990	134	253	129
12	270	421					712	1530	910	129	219	153
13	280	407					780	1430	658	125	264	193
14	320	407					730	1130	577	129	299	183
15	350	400			100		780	1050	394	134	253	167
16	435	380		110		440	810	980	368	134	214	153
17	526	380					900	990	342	144	183	158
18	940	368					900	850	269	139	158	158
19	1590	368					900	1160	247	134	139	177
20	1660	329					900	1780	247	134	129	247
21	1590	361					900	1780	203	129	125	374
22	1370	348				1850	1050	1780	193	129	125	361
23	1160	348				1850	910	1920	183	129	116	305
24	1030	348				1850	890	1640	177	129	116	264
25	930	342				1660	790	1590	158	129	100	219
26	860	329	140			1400	685	2860	153	129	96	193
27	880	323				1240	435	3420	158	116	96	183
28	830	317				1250	496	2860	158	139	108	172
29	800	311			....	1260	830	2360	148	172	153	153
30	760	311			....	1140	1300	1780	153	158	158	139
31	694	....			....	1370	....	2200	....	153	198	....

NOTE.—Stage-discharge relation affected by ice Dec. 3 to Mar. 21; discharge Apr. 18–20 interpolated. Braced figures show mean discharge for period included.

Monthly discharge of Bad River near Odanah, Wis., for the year ending Sept. 30, 1918.

[Drainage area, 607 square miles.]

Month.	Discharge in second-feet.				Run-off (depth in inches on drainage area).
	Maximum.	Minimum.	Mean.	Per square mile.	
October	1,660	230	646	1.06	1.22
November	649	311	418	.689	.77
December			182	.300	.35
January			110	.181	.21
February			100	.165	.17
March			668	1.10	1.27
April	1,590	435	930	1.53	1.71
May	3,420	750	1,510	2.49	2.97
June	6,960	148	1,140	1.88	2.10
July	236	116	148	.244	.28
August	299	96	171	.282	.33
September	374	100	183	.301	.34
The year	6,960		519	.855	11.62

MONTREAL RIVER AT IRONWOOD, MICH.

Location.—At main highway bridge on State line between Hurley, Wis., and Ironwood, Mich., about 8 miles upstream from junction of West Branch, and 22 miles above mouth of river.

Drainage area.—About 73 square miles (measured on Hixon's County Atlas; scale, 1 inch = 6 miles).

Records available.—April 24 to September 30, 1918.

Gage.—Chain gage fastened to downstream side of highway bridge, read by W. A. Markert.

Discharge measurements.—Made from wooden bridge at lumber mill, one-fourth mile above gage, or by wading.

Channel and control.—Bed at and downstream from gage fairly heavy gravel; fairly permanent. Concrete retaining walls on both sides of the river below the gage prevent overflow at flood stages.

Extremes of discharge.—Maximum stage recorded, 3.1 feet, June 2 (discharge, about 455 second-feet); minimum stage recorded, 0.71 foot July 23 (discharge, about 2.9 second-feet).

Regulation.—Water stored in Pine Lake, in secs. 28, 29, 32, and 33, T. 44 N., R. 3 E., is used to increase the water supply for Ironwood and Hurley during periods of low flow; effect of this regulation on flow at station probably slight.

Accuracy.—Stage-discharge relation assumed fairly permanent except as affected by ice during winter months. Rating curve poorly defined below 275 second-feet, and extended above. Gage read to hundredths once daily. Daily discharge ascertained by applying gage height to rating table. Records probably fair.

Discharge measurements of Montreal River at Ironwood, Mich., during the year ending Sept. 30, 1918.

Date.	Made by—	Gage height.	Discharge.
		Feet.	*Sec.-ft.*
Apr. 24	W. G. Hoyt	1.68	74
June 8	T. G. Bedford	2.04	150
Aug. 23	S. B. Soulé	.94	6.4

Daily discharge, in second-feet, of Montreal River at Ironwood, Mich., for the year ending Sept. 30, 1918.

Day.	Apr.	May.	June.	July.	Aug.	Sept.	Day.	Apr.	May.	June.	July.	Aug.	Sept.
1.	......	106	425	152	6.1	14	16.	......	82	40	8.3	13	20
2.	......	115	455	14	7.8	14	17.	......	96	16	12	13	23
3.	......	100	335	13	6.6	13	18.	......	65	48	16	12	22
4.	......	89	191	15	6.1	13	19.	......	111	6.4	14	6.1	29
5.	......	89	204	19	6.6	3.9	20.	......	232	12	14	5.4	64
6.	......	78	133	13	8.3	7.5	21.	......	152	10	9.9	8.0	59
7.	......	204	65	14	7.8	8.3	22.	......	165	10	11	8.6	24
8.	......	204	152	11	14	7.8	23.	......	152	5.8	3.2	8.0	42
9.	......	204	96	9.9	16	7.8	24.	76	122	7.5	4.4	8.3	14
0.	......	218	91	9.5	18	14	25.	70	191	4.0	4.0	8.0	19
11.	......	275	113	8.3	14	2.9	26.	62	365	6.1	4.4	8.3	16
12.	......	165	41	7.2	13	19	27.	56	365	7.2	4.5	7.5	13
3	......	85	59	6.6	18	35	28.	58	350	7.5	4.4	9.9	13
14.	......	65	64	6.6	30	26	29.	191	410	16	4.5	7.5	9.5
15.	......	94	43	8.6	13	26	30.	178	335	16	4.4	24	9.0
							31.	......	260	......	4.7	17	......

NOTE.—Gage not read May 30 and Sept. 12; discharge interpolated.

Monthly discharge of Montreal River at Ironwood, Mich., for the year ending Sept. 30, 1918.

[Drainage area, 73 square miles.]

Month.	Discharge in second-feet.				Run-off (depth in inches on drainage area).
	Maximum.	Minimum.	Mean.	Per square mile.	
April 24-30	191	56	98.7	1.35	0.35
May	410	65	179.	2.45	2.82
June	455	4.0	89.3	1.22	1.36
July	152	3.2	13.9	.190	.22
August	30	5.4	11.3	.155	.18
September	64	2.9	19.6	.268	.30

WEST BRANCH OF MONTREAL RIVER AT GILE, WIS.

LOCATION.—In sec. 27, T. 46 N., R. 2 E., 800 feet upstream from highway bridge at Gile, Iron County, 2½ miles southwest of Hurley, Wis., and 4 miles upstream from junction of East and West branches.

DRAINAGE AREA.—About 70 square miles (measured on Hixon's County Atlas; scale, 1 inch=2 miles).

RECORDS AVAILABLE.—April 26 to September 30, 1918.

GAGE.—Standard sloping gage bolted to rock ledge on left bank of river, a few hundred feet upstream from pump house of Ottawa mine; read by Lyle Slender.

DISCHARGE MEASUREMENTS.—Made from downstream side of highway bridge 800 feet below gage or by wading.

CHANNEL AND CONTROL.—Control formed by permanent rock ledge across narrow section of stream about 15 feet below gage; fall at control about 4 feet.

EXTREMES OF DISCHARGE.—Maximum stage recorded during period, 5.65 feet, June 28 (discharge, about 377 second-feet); minimum stage recorded, 1.32 feet July 23 (discharge, 2.4 second-feet).

REGULATION.—None.

ACCURACY.—Stage-discharge relation permanent. Rating curve fairly well defined below 200 second-feet; extended above 200 second-feet. Gage read to hundredths once daily. Daily discharge ascertained by applying gage height to rating table. Records good for days when gage was read; records of discharge obtained by interpolation subject to error.

Discharge measurements of West Branch of Montreal River at Gile, Wis., during the year ending Sept. 30, 1918.

Date.	Made by—	Gage height.	Discharge.	Date.	Made by—	Gage height.	Discharge.
		Feet.	*Sec.-ft.*			*Feet.*	*Sec.-ft.*
Apr. 25..	W. G. Hoyt............	3.46	87	Aug. 23..	S. B. Soulé............	1.57	5.3
June 8...	T. G. Bedford..........	4.25	161	23..	.....do..................	1.57	5.4

Daily discharge, in second-feet, of West Branch of Montreal River at Gile, Wis., for the year ending Sept. 30, 1918.

Day.	Apr.	May.	June.	July.	Aug.	Sept.	Day.	Apr.	May.	June.	July.	Aug.	Sept.
1	......	184	368	24	2.4	11	16	......	131	46	3.7	11	19
2	......	158	359	21	2.5	11	17	......	112	38	3.7	9.4	18
3	......	136	350	21	2.5	11	18	......	104	54	3.3	8.8	36
4	......	122	334	22	2.5	14	19	......	144	48	3.1	8.3	19
5	......	115	270	22	2.5	14	20	......	184	41	3.0	5.0	36
6	......	108	240	20	4.0	14	21	......	198	32	2.8	4.0	54
7	......	117	212	16	7.0	15	22	......	198	30	2.6	4.8	48
8	......	117	158	11	12	13	23	......	226	31	2.4	4.8	41
9	......	150	147	9.9	14	11	24	......	198	31	2.8	5.1	34
10	......	184	136	8.3	14	11	25	......	198	32	2.6	4.2	28
11	......	212	122	5.9	14	12	26	72	274	32	2.6	3.3	30
12	......	191	104	5.6	14	18	27	65	350	29	2.5	3.6	26
13	......	170	82	4.8	13	23	28	100	368	25	2.9	3.7	22
14	......	136	65	4.4	14	22	29	146	334	21	3.3	5.9	20
15	......	122	54	4.0	14	20	30	184	302	22	3.3	11	19
							31	......	270	......	3.4	11	......

NOTE.—Gage not read Apr. 28, May 5, 9, 12, 19, 26, June 2, 9, 16, 23, 24, 30, July 3, 7, 14, 21, 28, Aug. 4, 11, 18, 25, 31, Sept. 1, 2, 8, 12, 15, 22, and 29; discharge interpolated.

Monthly discharge of West Branch of Montreal River at Gile, Wis., for the year ending Sept. 30, 1918.

[Drainage area, 70 square miles].

Month.	Discharge in second-feet.				Run-off (depth in inches on drainage area.)
	Maximum.	Minimum.	Mean.	Per square mile.	
April 26-30	184	65	113	1.61	0.30
May	368	104	188	2.69	3.10
June	368	21	117	1.67	1.86
July	24	2.4	8.0	.114	.13
August	14	2.4	7.6	.109	.13
September	54	11	22.8	.319	.36

STREAMS TRIBUTARY TO LAKE MICHIGAN.

MENOMINEE RIVER BELOW KOSS, MICH.

LOCATION.—In sec. 5, T. 33 N., R. 23 E., at "Grand Rapids," about 4 miles below Koss, Menominee County, Mich., and 3 miles west of Ingalls, Mich. Little Cedar River, draining an area entirely in Michigan, enters from the left about half a mile below the station.

DRAINAGE AREA.—3,790 square miles.

RECORDS AVAILABLE.—July 1, 1913, to September 30, 1918.

DISCHARGE.—The flow is computed by the Menominee & Marinette Light & Traction Co., of Menominee, Mich., as follows: Each hour the load on the generators is noted and gage heights are read of the head and tail-water to determine the head on the spillway of the dam and the acting head on the turbines. The flow through the turbines for each hour is taken from a table giving the discharge corresponding to load and head. The flow over the spillway is taken from a table computed from a weir formula. When water is wasted through the gates the magnitude and duration of the gate openings are noted and the quantity wasted determined from computed tables. The sum of the hourly discharge through the turbines and over the spillway, plus the quantity wasted through the gates, divided by the number of seconds in 24 hours, gives the average discharge in second-feet for the day. No account is taken of the water passing through the exciter turbine, nor waste over the "trash gate" at the power house. This amount is, however, relatively small.

EXTREMES OF DISCHARGE.—Maximum daily discharge during year, 15,000 second-feet May 30; minimum daily discharge, 1,160 second-feet February 3.

1913–1918: Maximum daily discharge recorded, 23,200 second-feet, April 23 and 25, 1916; minimum daily discharge recorded, 1,000 second-feet, June 14, 1914.

REGULATION.—Above the station are the following power plants: Sturgeon Falls, owned by Pennsylvania Iron Mining Co., 50 miles; Little Quinnesec, owned by Kimberly Clark Co., 57 miles; Upper Quinnesec, owned by Oliver Iron Mining Co., 62 miles; Twin Falls, owned by Peninsular Power Co. With the exception of the Kimberly Clark dam at Little Quinnesec, the dams furnish power for utility and mining uses so that the flow past the dams is comparatively uniform. The Kimberly Clark dam is used for paper mills and regulates the flow on Sundays and holidays. The effect of this regulation is noticeable at the station generally on Tuesdays. The monthly flow probably represents the natural flow.

ACCURACY.—No measurements have been made by the Survey engineers at this plant, but measurements made at Koss, Mich., in 1914, show a close comparison with the discharge as determined at the power house.

COOPERATION.—Daily-discharge records furnished monthly by Edward Daniell, general manager of the Menominee & Marinette Light & Traction Co.

Daily discharge, in second-feet, of Menominee River below Koss, Mich., for the year ending Sept. 30, 1918.

Day.	Oct.	Nov.	Dec.	Jan.	Feb.	Mar.	Apr.	May.	June.	July.	Aug.	Sept.
1	2,430	4,100	2,140	1,420	1,520	1,680	6,300	5,260	11,600	1,970	1,900	2,340
2	2,280	3,540	2,200	1,590	1,420	1,700	6,470	5,340	11,600	2,270	2,010	3,140
3	2,270	3,330	2,220	1,480	1,160	1,540	5,560	5,370	10,500	2,580	1,960	2,220
4	2,230	3,140	2,120	1,550	1,480	1,630	5,200	4,730	10,500	3,040	1,850	2,790
5	2,270	3,300	1,830	1,600	1,460	1,720	4,830	4,250	10,000	3,040	1,840	3,260
6	2,370	3,090	1,840	1,420	1,420	1,840	4,680	3,900	8,940	2,850	2,000	3,340
7	2,550	3,220	2,030	1,620	1,470	1,910	4,280	4,740	7,860	2,660	2,270	3,360
8	2,620	3,210	2,110	1,470	1,500	1,910	3,840	4,720	7,490	2,760	2,840	2,960
9	2,160	3,120	1,920	1,400	1,420	1,630	4,100	4,660	6,480	2,660	3,560	3,100
10	2,360	3,320	1,720	1,600	1,310	1,750	4,080	5,540	5,940	2,210	4,650	2,180
11	2,510	2,900	1,280	1,680	1,520	1,710	4,060	6,190	5,130	2,110	5,460	2,290
12	2,440	2,780	1,780	1,420	1,400	1,500	3,880	6,810	4,970	2,110	5,430	2,410
13	2,570	2,520	1,630	1,720	1,550	1,610	3,210	6,360	4,640	1,980	4,000	2,780
14	2,900	2,840	1,170	1,560	1,540	1,700	3,580	5,970	3,970	1,970	3,840	2,770
15	2,560	2,880	1,380	1,680	1,310	1,670	2,940	5,520	3,820	1,700	3,310	3,030
16	2,500	2,990	1,160	1,640	1,440	1,670	3,210	5,090	3,500	1,850	3,220	2,870
17	2,680	2,810	1,370	1,640	1,310	1,750	3,610	4,970	3,210	2,070	3,260	2,440
18	3,110	2,380	1,320	1,420	1,440	1,840	3,840	4,970	3,430	2,020	2,600	2,590
19	3,210	2,680	1,380	1,540	1,380	2,060	4,050	4,920	3,400	1,850	2,720	2,870
20	4,070	2,710	1,460	1,600	1,460	2,820	4,050	5,570	2,210	1,770	2,350	3,550
21	5,270	2,960	1,710	1,450	1,320	3,380	4,140	6,760	2,550	1,750	1,880	4,050
22	5,220	3,020	1,690	1,590	1,350	4,490	3,870	6,740	2,340	1,810	1,970	4,220
23	4,280	2,900	1,590	1,440	1,370	5,940	4,350	6,830	2,360	1,710	2,400	4,560
24	4,170	2,890	1,740	1,720	1,330	6,230	4,120	6,020	2,130	2,070	2,570	3,890
25	4,270	2,950	1,810	1,550	1,640	7,180	3,990	6,010	1,820	2,000	2,640	3,970
26	4,100	2,040	1,760	1,550	1,590	7,850	3,580	7,130	1,960	2,460	1,790	3,830
27	3,990	1,660	1,270	1,330	1,600	7,740	3,430	7,850	1,980	2,260	1,830	3,390
28	4,000	2,500	1,460	1,520	1,660	7,500	3,320	10,800	2,110	1,960	1,700	3,300
29	3,890	2,340	1,500	1,390	......	8,030	3,240	11,600	2,050	2,420	2,190	2,960
30	4,220	2,050	1,460	1,520	......	8,620	4,480	15,000	2,080	1,970	2,300	2,680
31	4,050	......	1,440	1,600	......	7,300	......	11,700	......	1,900	2,420	......

Monthly discharge of Menominee River below Koss, Mich., for the year ending Sept. 30, 1918.

[Drainage area, 3,790 square miles.]

Month.	Discharge in second-feet.				Run-off (depth in inches on drainage area).
	Maximum.	Minimum.	Mean.	Per square mile.	
October	5,270	2,160	3,210	0.847	0.98
November	4,100	1,660	2,870	.757	.84
December	2,220	1,160	1,660	.438	.50
January	1,720	1,330	1,540	.406	.47
February	1,660	1,160	1,440	.380	.40
March	8,620	1,500	3,550	.937	1.08
April	6,470	2,940	4,140	1.09	1.22
May	15,000	3,900	6,490	1.71	1.97
June	11,600	1,820	5,020	1.32	1.47
July	3,040	1,710	2,190	.578	.67
August	5,460	1,700	2,730	.720	.83
September	4,560	2,180	3,100	.818	.91
The year	15,000	1,160	3,170	.836	11.34

NOTE.—Monthly and yearly discharge computed by U. S. Geological Survey from daily discharge records furnished by the Menominee & Marinette Light & Traction Co.

PINE RIVER NEAR FLORENCE, WIS.

Location.—In secs. 23 and 26, T. 39 N., R. 17 E., at highway bridge 8 miles south west of Florence, Florence County, and 12 miles above mouth of river. Popple River enters from right about 200 feet above station.

Drainage area.—488 square miles (measured on map issued by Wisconsin Geological and Natural History Survey, edition of 1911; scale, 1 inch=6 miles).

Records available.—January 22, 1914, to September 30, 1918.

Gage.—Chain gage fastened to guardrail on upstream side of bridge; read by William Taft.

Discharge measurements.—Made from upstream side of bridge or by wading.

Channel and control.—Coarse gravel and stones; left bank high and not subjcet to overflow; extremely high water may overflow right bank around approach to bridge.

Extremes of discharge.—Maximum stage recorded during year, 5.80 feet May 30, 31, and June 1 (discharge, 1,720 second-feet; minimum recorded stage 1.50 feet July 18–20 (discharge, about 160 second-feet).

1914–1918: Maximum recorded stage, 9.25 feet at noon, April 23, 1916, (discharge approximately 4,520 second-feet); minimum recorded stage 1.6 feet, September 6 and 7, 1915 (discharge about 118 second-feet).

Ice.—Stage-discharge relation seriously affected by ice.

Regulation.—River not used for log driving during year. Gates of a dam below station remained open throughout the year.

Accuracy.—Stage-discharge relation practically permanent; rating curve fairly well defined between 250 and 1,840 second-feet; extension of curve below 250 and above 1,840 second-feet may be subject to considerable error. Gage read to half-tenths once daily. Daily discharge ascertained by applying daily gage height to rating table, except for period when stage-discharge relation was affected by ice, for which it was obtained from results of discharge measurements, observer's notes, and weather records. Records fair.

Discharge measurements of Pine River near Florence, Wis., during the year ending Sept. 30, 1918.

Date.	Made by—	Gage height.	Discharge.
		Feet.	*Sec.-ft.*
Dec. 17[a]	L. L. Smith	2.59	171
Jan. 16[a]	.....do	2.91	174
Apr. 22	T. G. Bedford	2.48	400

[a] Complete ice cover at control and measuring section.

Monthly discharge, in second-feet, of Pine River near Florence, Wis., for the year ending Sept. 30, 1918.

Day.	Oct.	Nov.	Dec.	Jan.	Feb.	Mar.	Apr.	May.	June.	July.	Aug.	Sept.
1	352	552					930	541	1720	292	198	575
2	319	518					890	507	1620	266	198	575
3	319	451			180		575	490	1570	266	198	541
4	319	385					507	507	1520	266	198	507
5	287	352					473	541	1340	266	198	439
6	287	352					439	575	1250	242	242	405
7	319	354					439	575	1090	220	318	374
8	319	336	195			300	422	610	930	220	610	346
9	319	319					405	680	890	220	890	346
10	352	319					405	750	820	209	1,090	318
11	352	287					405	820	758	198	970	305
12	368	287					405	785	750	198	930	292
13	368	287					405	785	715	188	855	292
14	385	287					374	785	680	178	715	292
15	385	287					374	750	680	178	575	292
16	418	287		175			374	715	575	178	507	292
17	484	272			195		405	715	541	169	374	318
18	552	256					405	785	507	160	374	346
19	905	256					422	785	473	160	374	405
20	905						439	820	439	160	346	473
21	905						439	855	374	178	318	541
22	869						439	890	292	198	292	507
23	833		170			760	439	930	292	220	266	473
24	833						473	1010	266	220	266	439
25	833	230					473	1210	266	242	266	439
26	797						473	1250	242	242	266	405
27	797						490	1250	242	242	292	374
28	725						507	1340	242	220	374	374
29	690				......		507	1470	266	220	645	374
30	655				......		541	1720	292	220	645	346
31	620	......			......		......	1720	......	209	610	......

NOTE.—Stage-discharge relation affected by ice Nov. 20 to Mar. 31. Braced figures show mean discharge for period included.

Monthly discharge of Pine River near Florence, Wis., for the year ending Sept. 30, 1918.

[Drainage area, 488 square miles.[a]]

Month.	Discharge in second-feet.				Run-off (depth in inches on drainage area).
	Maximum.	Minimum.	Mean.	Per square mile.	
October	905	287	644	1.11	1.28
November	552	....	299	.613	.68
December	....	....	182	.373	.43
January	....	....	175	.359	.41
February	....	....	192	.393	.41
March	....	....	537	1.10	1.27
April	930	374	476	.975	1.09
May	1,720	490	876	1.80	2.08
June	1,720	242	722	1.48	1.65
July	292	160	214	.439	.51
August	1090	198	465	.953	1.10
September	575	292	400	.820	.91
The year	1720	....	425	.871	11.82

[a] Revised since publication of 1916 report, on the assumption that Kentuck Lake discharges into Brule River instead of into Pine River.

PIKE RIVER AT AMBERG, WIS.

LOCATION.—In sec. 15, T. 35 N., R. 21 E., at Chicago, Milwaukee & St. Paul Railway bridge half a mile south of Amberg, Marinette County, immediately below the junction of two branches of Pike River and about 11 miles above mouth.

DRAINAGE AREA.—240 square miles (measured on map issued by Wisconsin Geological and Natural History Survey, edition of 1911; scale, 1 inch=6 miles.

RECORDS AVAILABLE.—February 26, 1914, to September 30, 1918.

GAGE.—Chain gage fastened to guardrail on upstream side of bridge; read by Frank Bunce.

DISCHARGE MEASUREMENTS.—Made from a highway bridge a quarter of a mile downstream from the bridge to which the gage is attached, or by wading.

CHANNEL AND CONTROL.—Solid rock and some loose granite boulders; channel permanent but very rough at gage. Banks medium high; not subject to overflow.

EXTREMES OF DISCHARGE.—Maximum stage recorded during year, 3.85 feet at 7.10 a. m., May 28 (discharge 841 second-feet); minimum discharge estimated 70 second-feet December 9–11, 30 and 31.

1914–1918: Maximum stage recorded, 4.65 feet at 8.10 p. m., July 14, 1914 (discharge, 1,200 second-feet); minimum open-water stage recorded, 1.55 feet September 7, 1915 (discharge 109 second-feet). Minimum discharge for winter periods estimated 70 second-feet December 9–11, 30, and 31, 1917.

REGULATION.—None.

ACCURACY.—Stage-discharge relation permanent except when affected by ice. Rating curve well defined between 180 and 1,120 second-feet. Gage read to quarter-tenths once daily. Daily discharge ascertained by applying daily gage height to rating table or for periods in which stage-discharge relation was affected by ice, from discharge measurements, observer's notes, and weather records. Open-water records good, except for extremely low stages, for which they are fair. Winter records fair.

Discharge measurements of Pike River at Amberg, Wis., during the year ending Sept. 30, 1918.

Date.	Made by—	Gage height.	Discharge.	Date.	Made by—	Gage height.	Discharge.
		Feet.	*Sec.-ft.*			*Feet.*	*Sec.-ft.*
Dec. 18[a]	L. L. Smith	1.73	112	Feb. 20[a]	L. L. Smith	2.14	101
Jan. 15[a]	.....do	1.97	117	Apr. 20..	T. G. Bedford	2.36	294

[a] Complete ice cover at control and measuring section.

Daily discharge, in second-feet, of Pike River at Amberg, Wis., for the year ending Sept. 30, 1918.

Day.	Oct.	Nov.	Dec.	Jan.	Feb.	Mar.	Apr.	May.	June.	July.	Aug.	Sept.
1	158	258	140	80	80	150	364	510	738	204	154	199
2	158	244	120	100	80	160	364	476	738	217	148	185
3	158	348	110	140	90	170	348	412	658	204	142	230
4	158	204	100	160	90	170	333	348	620	204	138	244
5	158	204	90	160	100	160	310	348	546	204	138	230
6	162	204	80	150	110	150	288	318	428	185	204	217
7	169	204	80	150	110	140	303	348	396	169	288	204
8	162	204	80	140	120	140	318	348	348	158	364	192
9	158	185	70	140	120	140	310	396	318	148	510	180
10	169	180	70	130	120	150	303	582	318	142	698	158
11	169	185	70	130	120	160	296	658	288	138	582	192
12	180	192	80	120	110	160	288	658	273	134	476	258
13	185	185	80	120	110	170	266	582	244	128	364	273
14	185	185	80	120	110	170	244	476	230	122	303	258
15	192	185	90	120	110	170	244	396	230	128	258	230
16	180	185	100	120	100	180	244	364	217	154	230	217
17	192	180	100	110	100	205	303	333	204	154	199	192
18	244	180	110	110	100	230	333	333	204	142	192	244
19	230	180	110	110	100	290	318	364	192	138	169	288
20	244	180	120	110	100	350	318	380	180	128	158	333
21	230	185	120	100	110	410	318	348	180	118	142	318
22	230	192	110	100	120	550	318	364	169	118	230	318
23	230	185	110	100	130	700	318	364	169	154	230	303
24	230	180	100	100	140	780	318	348	169	169	288	258
25	204	169	100	100	160	698	303	396	162	192	258	230
26	204	158	90	90	160	604	288	658	158	192	230	217
27	258	155	90	90	160	510	258	738	169	180	204	204
28	288	150	80	90	160	453	288	820	176	176	192	180
29	288	145	80	80	....	396	412	820	162	204	217	169
30	273	140	70	80	....	380	546	820	158	192	230	169
31	258	....	70	80	....	364	....	738	....	176	204	....

NOTE.—Stage-discharge relation affected by ice Nov. 27 to Mar. 24. Gage not read on every alternate day, Mar. 26 to Apr. 15; discharge interpolated.

Monthly discharge of Pike River at Amberg, Wis., for the year ending Sept. 30, 1918.

[Drainage area, 240 square miles.]

Month.	Discharge in second-feet.				Run-off (depth in inches on drainage area).
	Maximum.	Minimum.	Mean.	Per square mile.	
October	288	158	203	0.846	0.98
November	348	140	191	.796	.89
December	140	70	93.5	.390	.45
January	160	80	114	.475	.55
February	160	80	115	.479	.50
March	780	140	305	1.27	1.46
April	546	244	315	1.31	1.46
May	820	318	485	2.02	2.33
June	738	158	301	1.25	1.40
July	217	118	164	.683	.79
August	698	138	263	1.10	1.27
September	333	158	230	.958	1.07
The year	820	70	232	.967	13.15

PESHTIGO RIVER AT HIGH FALLS, NEAR CRIVITZ, WIS.

LOCATION.—In sec. 1, T. 32 N., R. 18 E., at High Falls, near Crivitz, Marinette County, about a quarter of a mile downstream from power house of Wisconsin Public Service Co., 1 mile upstream from Thunder River (coming in from right), and 15 miles by road northwest of Crivitz.

DRAINAGE AREA.—520[1] square miles (measured on map issued by Wisconsin Geological and Natural History Survey, edition of 1911; scale, 1 inch=6 miles).

RECORDS AVAILABLE.—October 1, 1912, to September 30, 1918.

GAGE.—Barrett and Lawrence water-stage recorder, set over a wooden well about 15 feet from the left bank and quarter of a mile downstream from power house; well is protected from floating logs by a large boulder.

DISCHARGE MEASUREMENTS.—Made from cable half a mile below gage. About 2 second-feet of seepage water enters the river below the gage but above the cable and is included in the determined discharge as published.

CHANNEL AND CONTROL.—Banks at control and measuring section are high and not subject to overflow. Control at low stages is a small gravel riffle about 50 feet downstream from the gage; at medium and high stages this control is apparently drowned out and is probably formed by some point farther downstream.

EXTREMES OF DISCHARGE.—Maximum mean daily discharge during the year, May 31, 2,140 second-feet. Minimum mean discharge 110 second-feet February 10.

1912–1918: Maximum stage, from water-stage recorder, 7.2 feet May 13, 1916 (discharge 3,480 second-feet); minimum stage, 1.1 feet at 5 p. m. March 21, 1915 (discharge, 54 second-feet). Owing to artificial regulation, extremes given do not represent the natural flow.

ICE.—Because of the relatively warm water in the large service reservior, ice does not form on the river in the vicinity of the gage. Open-water rating curve used throughout year.

REGULATION.—Flow controlled by operation of the power plant. Considerable diurnal fluctuation caused by the operation of the power plant and during log-driving season by the manipulation of the gates. The mean monthly flow does not represent the natural flow because of storage in the service reservoir.

ACCURACY.—Stage-discharge relation permanent; not affected by ice. Rating curve well defined between 145 and 3,980 second-feet. Daily discharge for periods when recording gage was in operation ascertained by averaging the results obtained by applying gage height for hourly or other regular interval to the rating table; discharge for periods when gage was not in operation (see footnote to table of daily discharge) obtained by adding 10 per cent to discharge indicated by records of power plant. Correction determined by study of records available from water-stage recorder. Records fair.

No discharge measurements were made at this station during the year.

[1] Revised since publication of Water-Supply Paper 434.

Daily discharge, in second-feet, of Peshtigo River at High Falls, near Crivitz, Wis., for the year ending Sept. 30, 1918.

Day.	Oct.	Nov.	Dec.	Jan.	Feb.	Mar.	Apr.	May.	June.	July.	Aug.	Sept.
1	456	399	464	116	292	236	622	708	1,800	660	456	440
2	462	418	179	288	274	288	657	583	1,360	615	485	335
3	455	496	410	316	170	216	670	569	2,060	475	435	730
4	380	236	424	342	252	262	573	590	1,630	169	310	735
5	355	484	460	338	346	287	656	381	1,310	347	565	700
6	330	428	399	127	402	445	667	580	988	373	650	590
7	124	451	388	309	282	318	410	697	1,210	230	680	600
8	380	418	527	344	236	375	650	685	956	477	620	290
9	354	399	292	339	245	388	680	711	613	455	525	575
10	370	407	415	348	110	174	678	661	940	422	445	775
11	327	202	435	322	214	348	667	727	782	370	208	700
12	337	448	467	258	292	373	670	283	770	389	500	722
13	347	425	428	124	266	438	662	1,490	765	395	564	670
14	172	436	461	276	374	460	393	1,210	790	162	550	530
15	364	444	410	265	271	520	595	1,140	720	376	535	187
16	406	450	174	241	330	537	695	922	380	482	500	506
17	435	428	382	228	177	344	700	860	722	479	329	600
18	407	240	424	295	243	444	697	766	824	544	215	570
19	430	462	423	228	253	457	720	380	770	535	400	531
20	364	455	384	116	253	522	705	1,180	800	433	365	540
21	186	480	368	211	270	607	355	785	865	256	413	533
22	448	464	321	224	288	660	685	784	800	529	395	202
23	415	460	139	330	202	752	674	905	415	562	400	535
24	343	462	171	295	137	423	731	1,160	690	535	410	690
25	322	173	119	314	212	650	683	1,040	760	413	318	710
26	375	407	413	299	248	685	666	320	760	470	537	630
27	346	467	378	184	208	694	694	1,270	760	436	650	660
28	181	419	417	240	231	677	394	1,300	780	196	716	512
29	430	185	410	338	....	680	634	2,060	680	512	776	207
30	406	428	120	298	....	669	692	1,790	390	476	749	513
31	415	....	234	268	....	375	....	2,140	....	449	773	....

NOTE.—Records for following periods obtained from water-stage recorder: Oct. 5–7, 12, 13, 19, 20, 26, Nov. 2–7, Apr. 15–22, May 2–10, 26, June 2, 9–15, 19–22, 24–30, July 1–3, 7–12, Aug. 1–9, 12–23, Sept. 1–13 and 24–27. Daily discharge for other periods determined from records of power plant, as noted in paragraph under "Accuracy."

Monthly discharge of Peshtigo River at High Falls, near Crivitz, Wis., for the year ending Sept. 30, 1918.

[Drainage are, 520 square miles.]

Month.	Discharge in second-feet.				Run-off (depth in inches on drainage area).
	Maximum.	Minimum.	Mean.	Per square mile.	
October	462	124	359	0.690	0.80
November	496	173	402	.773	.86
December	527	119	356	.685	.79
January	348	116	265	.510	.59
February	402	110	253	.487	.51
March	752	174	461	.887	1.02
April	731	355	632	1.22	1.36
May	2,140	283	925	1.78	2.05
June	2,060	380	903	1.74	1.94
July	660	162	427	.821	.95
August	776	208	499	.960	1.11
September	775	187	550	1.06	1.18
The year	2,140	110	503	.967	13.16

OCONTO RIVER NEAR GILLETT, WIS.

Location.—In sec. 34, T. 28 N., R. 18 E., at highway bridge 2½ miles southeast of Gillett, Oconto County, and about 27 miles above mouth of river.

Drainage area.—678 square miles (measured on map issued by Wisconsin Geological and Natural History Survey, edition of 1911; scale, 1 inch=6 miles).

Records available.—June 7, 1906, to March 30, 1909; January 6, 1914, to September 30, 1918.

Gage.—Chain gage attached to iron railing on upstream side of bridge; read by Miss Nettie Gilbertson. Zero of gage used from January 6, 1914, to September 30, 1918, is 4 feet above that of gage used June 7, 1906, to March 31, 1909.

Discharge measurements.—Made from upstream side of bridge to which gage is fastened.

Channel and control.—Gravel; fairly permanent. Left bank of medium height and not subject to overflow; during extreme flood stages water may overflow right bank around the end of the bridge.

Extremes of discharge.—Maximum stage recorded during year, 4.45 feet at 3.30 p. m., May 30 (discharge, 2,510 second-feet); minimum discharge 230 second-feet, February 6-9.

1906-1918: Maximum stage recorded, 5.3 feet at 3.30 p. m., April 25, 1916 (discharge, 3,220 second-feet); minimum open-water discharge, 95 second-feet January 3 and 6, 1907.

Ice.—Stage-discharge relation seriously affected by ice.

Regulation.—A dam above the station stores water to float logs during the spring; except when dam is in operation flow at the gage is natural.

Accuracy.—Stage-discharge relation practically permanent, except as affected by ice. Rating curve well defined between 239 and 1,790 second-feet. Gage read to quarter-tenths once daily. Daily discharge obtained by applying daily gage height to rating table, except for period when stage-discharge relation was affected by ice, for which it was obtained by applying to rating table daily gage height corrected for effect of ice by means of discharge measurements, observer's notes, and weather records. Open-water records good except at highest flood stages, for which they are only fair; winter records fair.

Discharge measurements of Oconto River near Gillett, Wis., during the year ending Sept. 30, 1918.

Date.	Made by—	Gage height.	Discharge.	Date.	Made by—	Gage height.	Discharge.
		Feet.	*Sec.-ft.*			*Feet.*	*Sec.-ft.*
Dec. 19[a]	L. L. Smith	2.33	339	Feb. 21[a]	L. L. Smith	3.10	295
Jan. 17[a]	do	2.64	342	Apr. 19	T. G. Bedford	2.16	845

[a] Complete ice cover at control and measuring section.

Daily discharge, in second-feet, of Oconto River near Gillett, Wis., for the year ending Sept. 30, 1918.

Day.	Oct.	Nov.	Dec.	Jan.	Feb.	Mar.	Apr.	May.	June.	July.	Aug.	Sept.
1	446	670	340	295	270	305	992	1,020	2,320	468	515	468
2	446	670	330	300	260	305	960	960	2,090	468	515	492
3	446	670	320	300	250	305	1,020	1,020	1,940	540	468	468
4	468	670	310	300	240	305	992	1,160	1,720	565	382	424
5	446	642	305	305	240	305	1,020	930	1,570	590	424	424
6	424	615	300	305	230	310	780	930	1,640	615	424	446
7	424	565	290	310	230	325	1,290	1,090	1,430	590	424	446
8	424	565	290	310	230	320	780	1,290	1,290	590	424	424
9	446	565	280	315	230	320	870	1,430	1,020	468	446	446
10	468	565	270	320	235	320	810	1,290	992	515	468	492
11	468	565	270	320	240	330	752	1,360	1,290	492	565	492
12	468	565	270	325	240	340	698	1,860	960	515	565	515
13	468	540	270	325	260	350	698	2,020	960	492	565	540
14	468	540	270	335	270	360	698	1,860	780	468	540	565
15	468	515	270	340	280	370	725	1,720	725	424	515	515
16	468	515	280	340	280	390	698	1,640	615	424	468	468
17	468	565	290	340	260	410	780	1,430	615	424	468	492
18	468	492	320	330	240	440	840	1,720	615	492	590	515
19	468	515	340	320	260	460	900	1,790	590	515	515	540
20	515	492	330	310	270	470	810	1,640	565	515	468	540
21	515	492	325	305	290	615	780	1,500	382	492	446	515
22	515	515	325	305	290	1,020	810	1,430	382	492	424	515
23	515	515	320	305	300	2,020	870	1,430	342	468	446	540
24	515	492	310	305	300	2,390	870	1,290	424	446	403	565
25	515	424	305	305	305	2,090	900	1,500	468	468	403	540
26	540	403	305	305	310	2,020	960	1,860	492	492	446	565
27	565	390	305	305	320	1,870	1,360	2,090	492	515	446	515
29	615	380	305	300	825	1,720	840	2,160	468	540	492	492
29	565	360	300	290	....	1,720	870	2,470	468	515	565	492
30	590	340	290	290	....	1,290	810	2,470	615	515	540	492
31	590	....	290	290	....	1,020	....	2,320	....	515	515	....

NOTE.—Stage-discharge relation affected by ice Nov. 27 to Mar. 25. Gage not read Mar. 27; discharge interpolated.

Monthly discharge of Oconto River near Gillett, Wis., for the year ending Sept. 30, 1918.

[Drainage area, 678 square miles.]

Month.	Discharge in second-feet.				Run-off (depth in inches on drainage area).
	Maximum.	Minimum.	Mean.	Per square mile.	
October	615	424	490	0.723	0.83
November	670	340	527	.777	.87
December	340	270	301	.444	.51
January	340	290	311	.459	.53
February	325	230	266	.392	.41
March	2,390	305	800	1.18	1.36
April	1,360	698	873	1.29	1.44
May	2,470	930	1,570	2.32	2.68
June	2,320	342	942	1.39	1 55
July	615	424	504	.743	.86
August	590	382	480	.708	.82
September	565	424	498	.735	.82
The year	2,470	230	632	.932	12.68

FOX RIVER AT BERLIN, WIS.

Location.—In sec. 16, T. 17 N., R. 13 E., at government lock and dam about 2½ mile upstream from Berlin, Green Lake County.

Drainage area.—1,430 square miles (measured on map issued by the Wisconsin Geological and Natural History Survey, edition of 1911; scale, 1 inch=6 miles).

Records available.—1898 to September 30, 1918 (publication of records prior to Sept. 30, 1917, is held up pending collection of data relative to effect of ice on stage-discharge relation).

Gage.—Staff gage located in pool immediately below the dam. Read by United States Army Engineer.

Channel and control.—Sand and gravel, one channel at all stages. Both banks low and subject to overflow.

Discharge measurements.—Made from downstream side of Huron Street highway bridge in city of Berlin about 2½ miles downstream from gage. Rating curves for gage corrected for small inflow between the gage and measuring section.

Extremes of discharge.—Maximum mean daily discharge recorded during year, 6,050 second-feet, March 21–23; minimum mean daily discharge 480 second-feet January 1–3.

Ice.—Stage-discharge relation affected by ice.

Accuracy.—Stage-discharge relation practically permanent except for effect of ice. Rating curve well defined between 800 and 6,000 second-feet. Gage read three times daily, but generally noon reading alone is used in determination of daily discharge. Daily discharge ascertained by applying daily gage height to rating table, except for period when stage-discharge relation was affected by ice, for which it was obtained from results of one discharge measurement and observer's notes. Open-water records good; winter records roughly approximate.

Cooperation.—Records have been collected and computations of daily discharge made by United States Army Engineers. Open-water records obtained from rating curves based on discharge measurements made by United States Geological Survey.

Discharge measurements of Fox River at Berlin, Wis., during the period June 1, 1917, to Sept. 30, 1918.

Date.	Made by—	Gage height.	Discharge.	Date.	Made by—	Gage height.	Discharge.
1917.		Feet.	Sec.-ft.	1917.		Feet.	Sec.-ft.
June 7	R. B. Kilgore	10.37	1,960	Nov. 7	R. B. Kilgore	10.17	1,780
14	Kilgore and Kane	11.27	2,460	1918.			
July 25	....do	9.83	1,600	Jan. 18	Hoyt and Grover	a 8.75	609
Aug. 1	Hoyt and Kane	8.97	1,210	Mar. 28	W. G. Hoyt	13.92	5,080
28	Kilgore and Welsch	8.10	824	Apr. 5	T. G. Bedford	11.86	2,940

a Stage-discharge relation affected by ice; ice cover, 13 inches thick.

Note.—Discharge measured at Huron Street highway bridge. Discharge at gage obtained by applying a correction factor of 0.993 to the figures shown in the above table.

Daily discharge, in second-feet, of Fox River at Berlin, Wis., for the year ending Sept. 30, 1918.

Day.	Oct.	Nov.	Dec.	Jan.	Feb.	Mar.	Apr.	May.	June.	July.	Aug.	Sept.
1	940	1,460	940	480	700	940	3,920	1,800	3,080	940	735	675
2	905	1,460	905	480	700	980	3,620	1,740	3,000	975	765	675
3	905	1,520	865	480	700	1,060	3,350	1,740	2,910	905	735	675
4	905	1,570	865	510	740	1,200	3,170	1,680	2,830	905	735	615
5	905	1,680	765	510	700	1,350	3,000	1,570	2,670	975	765	615
6	905	1,850	800	540	660	1,600	2,830	1,460	2,600	940	735	615
7	905	1,800	700	540	660	1,800	2,750	1,420	2,520	940	675	645
8	830	1,740	700	540	700	2,000	2,670	1,320	2,380	905	735	590
9	865	1,680	800	540	700	2,200	2,520	1,270	2,310	865	735	590
10	865	1,620	800	540	740	2,200	2,450	1,740	2,240	800	765	590
11	865	1,570	800	570	740	2,200	2,310	1,910	2,170	800	735	645
12	865	1,460	800	570	740	2,300	2,170	2,040	2,040	800	765	645
13	865	1,420	750	600	740	2,500	2,100	2,100	1,910	800	735	645
14	865	1,360	800	600	780	2,700	1,980	2,100	1,850	765	705	645
15	865	1,320	800	600	780	2,900	1,850	2,040	1,680	800	765	645
16	865	1,270	800	600	780	3,100	1,740	1,910	1,520	800	705	645
17	865	1,220	800	600	780	3,340	1,620	1,850	1,420	765	735	645
18	865	1,180	800	600	780	3,700	1,570	2,040	1,320	765	735	675
19	905	1,140	840	630	780	4,420	1,520	2,240	1,220	765	735	645
20	940	1,140	840	630	820	5,790	1,420	2,830	1,180	765	675	645
21	905	1,140	840	630	820	6,060	1,460	2,450	1,140	735	645	645
22	905	1,100	840	630	820	6,060	1,680	3,530	1,100	735	645	645
23	975	1,060	880	630	820	6,050	1,740	4,120	1,020	735	675	675
24	975	1,020	880	630	820	5,920	1,740	4,020	975	675	675	675
25	1,020	1,020	880	630	860	5,920	1,800	3,820	940	765	645	645
26	1,100	1,020	750	660	900	5,520	1,800	3,530	905	800	675	645
27	1,220	975	750	660	900	5,270	1,740	3,440	865	765	645	615
28	1,270	975	700	660	940	5,030	1,680	3,350	905	735	645	645
29	1,360	975	750	660	......	4,790	1,740	3,260	865	800	645	615
30	1,360	940	750	660	......	4,560	1,800	3,170	905	800	675	615
31	1,420	......	750	700	......	4,230	......	3,080	......	735	675	......

Monthly discharge of Fox River at Berlin, Wis., for the year ending Sept. 30, 1918.

[Drainage area 1,430 square miles.]

Month.	Discharge in second-feet.				Run-off (depth in inches on drainage area).
	Maximum.	Minimum.	Mean.	Per square mile.	
October	1,420	830	974	0.681	0.79
November	1,850	940	1,320	.923	1.03
December	940	700	805	.563	.65
January	700	480	591	.413	.48
February	940	660	771	.539	.56
March	6,050	940	3,470	2.43	2.80
April	3,920	1,420	2,190	1.53	1.71
May	4,120	1,270	2,410	1.69	1.95
June	3,080	865	1,750	1.22	1.36
July	975	675	815	.570	.66
August	765	645	707	.494	.57
September	675	590	640	.448	.50
The year	6,050	480	1,370	.958	13.06

FOX RIVER AT RAPIDE CROCHE DAM, NEAR WRIGHTSTOWN, WIS.

Location.—At Rapide Croche dam, in sec. 4, T. 21 N., R. 19 E., about 2 miles upstream from Wrightstown, Brown County, 19 miles downstream from Lake Winnebago and 20 miles upstream from mouth of river at Green Bay.

Records available.—March 3, 1896 to September 30, 1918. Daily-discharge records for this station, 1896–1914, were published by the Wisconsin Railroad Commission in "Water Power Report to the Legislature, 1915." The records published in this report have since been found to be considerably in error and should not be used. See "Determination of flow."

Drainage area.—6,150 square miles (measured on map issued by Wisconsin Geological and Natural History Survey, edition of 1911; scale, 1 inch=6 miles).

Determination of discharge.—This dam is owned and operated by the United States Army Engineers to aid navigation and the flow is computed by the United States Army Engineers as follows: The dam is made of timber and is equipped with four needle sluice gates which are used only in times of high water. A vertical staff gage at the lower end of the canal leading to the lock and about a quarter of a mile below the dam is read five times daily—at 7 a. m., 9 a. m., noon, 3 p. m., and 6 p. m. The mean flow for the day is computed from a formula using the five gage heights for the day, assuming gradual changes in gage height between the readings, and weighting the different gage heights by elapsed time. Prior to 1917 determinations of daily discharge were based on tables derived from theoretical formulas for flow over a sharp-crested weir and through the sluice gates. During 1917 discharge measurements were made by engineers of the United States Geological Survey from a cable a short distance downstream from the dam. Seven measurements were made with the four sluices closed and eight with all sluices open. The measured discharge varied from 1,000 to 13,000 second-feet. Curves based on the discharge measurements show that the theoretical formulas previously used gave results ranging from about 850 second-feet too small at low stages, with the sluices closed, to 250 second-feet too large at high stages, with all sluices open. The deficiency of amounts in the old records as published is due to the fact that no allowance was made for leakage through the dam, which is now determined to be about 1,000 second-feet when water is at the crest of the dam and all gates are closed. Discharge measurements made by the United States Geological Survey in 1902 and 1903 at Wrightstown, about 2 miles below the dam, indicate that the leakage at the dam was apparently the same during 1902 and 1903 as in 1917. As Rapide Croche dam was built in 1878 and existed in 1902 as in 1917, it is considered necessary and proper to correct the old records for 1896–1917 to agree with the results of the current-meter measurements made in 1917. The recomputed records published in Water Supply Paper 454, are the old records corrected by means of the curves for 1917, each recomputation taking into consideration the relation between the old and new curves according to the number of sluices open. Corrections were applied to the semimonthly and monthly mean discharge.

Extremes of discharge.—Information relative to daily maximum and minimum, 1896–1917 may be obtained from the United States Army Engineer office, Milwaukee, Wis. During 1918, the maximum mean daily discharge was 16,300 second-feet May 25; minimum mean daily discharge, 1,330 second-feet October 22.

Regulation.—Flow regulated by Lake Winnebago, which has an area of 215 square miles, and also by dams between the outlet of Lake Winnebago and the station, the dams being operated for power development and to some extent in the interests of navigation. Under existing conditions, which, as regards storage, have been the same throughout the period covered by the records, the flow past the station is natural.

Accuracy.—Records good.

Cooperation.—Records collected and daily discharge computed by United States Army Engineers from curves developed by current-meter measurements made by engineers of the United States Geological Survey.

Daily discharge, in second-feet, of Fox River at Rapide Croche dam, near Wrightstown, Wis., for the year ending Sept. 30, 1918.

Day.	Oct.	Nov.	Dec.	Jan.	Feb.	Mar.	Apr.	May.	June.	July.	Aug.	Sept.
1	1,870	3,380	2,880	4,740	5,590	4,340	7,820	6,480	16,100	3,830	3,670	1,640
2	3,090	3,440	2,070	4,580	5,570	4,440	9,220	6,630	14,800	4,600	3,360	2,040
3	2,830	3,440	2,740	4,770	4,690	4,300	9,740	6,500	14,700	4,460	3,190	2,100
4	2,940	2,330	3,260	4,830	4,470	4,850	11,600	6,300	15,300	3,060	1,930	2,150
5	2,780	1,970	3,360	4,730	4,980	4,740	11,600	4,970	15,100	3,350	2,480	2,260
6	2,920	3,960	4,080	3,860	5,380	4,440	11,500	4,680	15,000	3,680	3,140	2,380
7	1,750	4,270	4,050	4,750	5,470	4,420	10,800	6,100	14,300	2,960	2,170	2,200
8	1,510	4,280	4,140	5,000	5,330	4,420	10,700	6,360	14,500	4,170	2,410	1,670
9	3,260	4,230	5,400	4,700	5,340	4,200	11,200	6,700	13,700	4,670	2,430	2,070
10	3,310	4,070	3,820	4,680	4,530	3,740	11,300	8,130	13,900	4,690	2,460	1,980
11	3,370	2,610	4,480	4,810	5,080	4,530	11,100	7,510	14,100	4,550	1,650	2,180
12	3,290	2,390	4,760	4,570	5,090	4,730	11,100	5,700	13,100	4,600	2,180	2,080
13	3,150	4,270	4,720	3,600	4,920	4,680	10,800	5,380	13,400	4,550	2,800	2,010
14	2,070	4,380	4,730	4,580	4,760	4,800	9,780	9,070	12,800	3,410	2,640	2,100
15	1,700	4,420	4,730	5,080	4,450	4,860	9,000	9,480	12,400	3,630	2,720	1,570
16	2,950	4,060	4,020	5,130	4,620	4,760	9,040	10,700	11,700	4,440	2,750	1,810
17	2,920	3,740	4,190	5,060	3,860	4,230	8,460	10,900	11,500	4,470	2,660	1,940
18	2,930	2,280	5,050	4,080	4,690	6,230	6,420	11,500	11,800	4,460	1,780	1,830
19	2,570	2,450	5,070	3,880	4,540	7,300	6,390	11,800	11,400	4,390	2,350	1,920
20	2,600	3,860	4,680	4,020	4,440	7,120	6,200	12,200	10,200	4,440	2,800	1,980
21	1,920	4,050	4,590	4,090	4,420	6,080	4,750	13,300	7,960	3,160	2,760	1,940
22	1,330	3,910	4,610	4,070	4,570	5,510	5,420	15,700	5,630	3,670	2,360	1,620
23	2,780	4,000	3,400	4,700	4,500	5,370	6,510	13,800	3,690	4,340	2,430	1,740
24	3,320	4,060	3,510	5,670	3,900	4,120	6,500	14,200	3,920	4,440	2,630	1,960
25	3,290	2,330	3,700	5,700	4,460	4,830	6,590	16,300	4,700	4,460	1,900	1,960
26	3,430	2,650	4,510	5,470	4,280	5,280	6,350	14,800	4,890	4,580	2,040	1,940
27	3,540	4,180	4,370	4,390	4,350	5,450	6,170	14,800	4,940	4,510	2,420	2,100
28	2,100	3,770	4,170	4,160	4,360	5,700	4,790	15,200	4,780	3,110	2,480	1,990
29	2,070	3,730	4,510	4,990	......	6,000	5,170	15,000	4,630	2,570	2,430	1,530
30	3,220	3,270	3,930	5,440	......	6,510	6,210	15,400	3,700	3,430	2,270	1,760
31	3,420	......	4,400	5,520	......	6,820	......	15,600	......	3,540	2,090	......

Monthly discharge of Fox River at Rapide Croche dam, near Wrightstown, Wis., for the year ending Sept. 30, 1918.

[Drainage area, 6,150 square miles.]

Month.	Discharge in second-feet.				Run-off (depth in inches on drainage area).
	Maximum.	Minimum.	Mean.	Per square mile.	
October	3,540	1,330	2,720	0.442	0.51
November	4,420	1,970	3,530	.574	.64
December	5,400	2,070	4,130	.672	.77
January	5,700	3,600	4,700	.764	.88
February	5,570	3,860	4,740	.771	.80
March	7,300	3,740	5,120	.833	.96
April	11,600	4,750	8,410	1.37	1.53
May	16,300	4,680	10,400	1.69	1.95
June	16,100	3,690	10,600	1.72	1.92
July	4,690	2,570	4,010	.652	.75
August	3,670	1,650	2,500	.407	.47
September	2,380	1,530	1,950	.317	.35
The year	16,300	1,330	5,220	.849	11.53

WOLF RIVER AT KESHENA, WIS.

Location.—In sec. 26, T. 28 N., R. 15 E., at highway bridge at Keshena, Shawano County, 3 miles below junction with West Branch of Wolf River, coming in from right.

Drainage area.—840 [a] square miles.

Records available.—May 9, 1907, to March 31, 1909; February 10, 1911, to September 30, 1918.

Gage.—Chain gage fastened to downstream side of new bridge December 9, 1914; May 9, 1907, to November 29, 1914, vertical staff gage fastened to downstream end of left abutment; both gages at same datum. Gage read by Jerome M. Beauprey.

Discharge measurements.—Made from the bridge.

Channel and control.—Gravel; smooth and practically permanent. Banks of medium height; overflow improbable.

Extremes of discharge.—Maximum stage recorded during year 4.88 feet at 4 p. m. May 28 (discharge, 2,530 second-feet); minimum discharge, about 315 second-feet, February 20.

1907–1909 and 1911–1918: Maximum discharge recorded, 3,910 second-feet, September 2, 1912; minimum discharge during open-water periods, 275 second-feet, September 26, 1908.

Ice.—Stage-discharge relation seriously affected by ice.

Regulation.—The river and its main tributaries above Keshena are controlled to some extent by logging dams.

Accuracy.—Stage-discharge relation permanent except for effect of ice. Rating curve well defined between 380 and 2,000 second-feet; above and below these limits curve is extended and subject to error. Gage read to quarter-tenths twice daily. Daily discharge ascertained by applying mean daily gage height to rating table, except for period when stage-discharge relation was affected by ice, for which it was ascertained by applying to rating table mean daily gage height corrected for effect of ice by means of discharge measurements, observer's notes, and weather records. Open-water records good, except those for extremely high and low stages, which are fair; winter records fair.

Discharge measurements of Wolf River at Keshena, Wis., during the year ending Sept. 30, 1918.

Date.	Made by—	Gage height.	Discharge.	Date.	Made by—	Gage height.	Discharge.
		Feet.	*Sec.-ft.*			*Feet.*	*Sec.-ft.*
Dec. 20[a]	L. L. Smith	2.26	461	Feb. 22[b]	L. L. Smith	2.89	389
Jan. 18[a]	.....do	2.70	390	Apr. 29	T. G. Bedford	2.98	1,290

[a] Revised since publication of Water-Supply Paper 454.
[b] Complete ice cover at control and measuring section.

Daily discharge, in second-feet, of Wolf River at Keshena, Wis., for the year ending Sept. 30, 1918.

Day.	Oct.	Nov.	Dec.	Jan.	Feb.	Mar.	Apr.	May.	June.	July.	Aug.	Sept.
1	630	715	490	430	350	475	1,110	1,160	2,190	806	672	853
2	590	715	480	415	360	480	950	1,050	2,190	760	760	1,000
3	552	715	475	435	350	490	760	1,000	1,850	901	630	901
4	590	672	460	430	360	495	950	950	1,530	760	552	672
5	630	672	430	395	325	505	950	853	1,530	760	515	760
6	552	715	430	400	350	510	950	853	1,460	715	515	853
7	515	715	435	435	350	510	1,050	853	1,460	853	552	950
8	497	715	435	390	330	510	1,050	901	1,400	760	672	853
9	590	672	435	385	330	510	1,000	950	1,400	715	950	760
10	672	672	440	375	335	510	806	1,790	1,280	760	1,220	760
11	672	760	440	410	335	510	806	1,920	1,160	672	1,160	672
12	715	672	445	385	325	510	760	1,850	1,110	630	760	590
13	630	590	445	340	325	565	853	1,400	1,050	672	1,050	672
14	552	590	430	365	330	605	760	1,220	950	515	1,050	760
15	552	672	445	350	330	625	760	1,280	950	672	1,000	590
16	590	715	475	360	320	670	806	1,000	950	672	950	672
17	590	590	475	365	320	810	806	950	1,050	672	1,050	715
18	672	590	470	390	330	860	853	1,160	1,050	552	760	901
19	672	552	465	375	325	910	901	1,280	901	590	1,000	672
20	760	552	460	375	315	960	853	1,340	760	590	806	715
21	715	715	460	335	355	1,020	901	1,000	715	630	715	672
22	672	760	430	350	390	1,380	901	1,050	806	590	715	760
23	672	590	430	325	400	1,310	853	1,110	760	672	760	806
24	715	590	420	345	415	1,250	806	1,220	672	715	1,000	853
25	760	540	450	375	445	1,190	853	1,160	672	715	1,050	853
26	760	535	445	365	455	1,130	715	1,460	806	672	1,000	853
27	901	530	445	365	460	1,100	806	2,120	672	672	901	590
28	1,000	515	390	365	470	1,070	853	2,330	853	715	1,000	590
29	1,220	505	390	350	....	1,190	1,280	1,590	901	760	1,050	590
30	1,000	495	395	365	....	1,400	1,220	2,060	806	715	950	590
31	760	....	400	365	....	1,110	....	2,060	....	715	1,000	....

NOTE.—Stage-discharge relation affected by ice Nov. 25 to Mar. 29.

Monthly discharge of Wolf River at Keshena, Wis., for the year ending Sept. 30, 1918.

[Drainage area, 840 square miles.[a]]

Month.	Discharge in second-feet.				Run-off (depth in inches on drainage area).
	Maximum.	Minimum.	Mean	Per square mile.	
October	1,220	497	690	0.821	0.95
November	760	495	635	.756	.84
December	490	390	442	.526	.61
January	435	325	378	.450	.52
February	470	315	360	.429	.45
March	1,400	475	812	.967	1.11
April	1,280	715	897	1.07	1.19
May	2,330	853	1,320	1.57	1.81
June	2,190	672	1,130	1.35	1.51
July	901	515	697	.830	.96
August	1,220	515	863	1.03	1.19
September	1,000	590	749	.892	1.00
The year	2,330	315	750	.893	12.14

[a] Revised since publication of Water-Supply Paper 454.

WOLF RIVER AT NEW LONDON, WIS.

Location.—In sec. 12, T. 22 N., R. 14 E., at Pearl Street highway bridge, New London, Waupaca County. Embarrass River enters from the right three-fourths of a mile above, and Little Wolf River, also from the right, 5 miles below the station.

Drainage area.—2,240 square miles (measured on map issued by Wisconsin Geological and Natural History Survey, edition of 1911; scale, 1 inch=6 miles).

Records available.—Gage heights March 1, 1899, to September 30, 1918; daily discharge determinations October 1, 1913, to September 30, 1918.

Gage.—Enameled steel gage, graduated from 1.0 to 13.0 feet, fastened to right hand downstream pier of Pearl Street Bridge. Datum of the gage raised 0.641 foot on March 1, 1911, according to United States Army Engineers; zero of gage is at an elevation of 748.874 feet above mean sea level, New York City datum.

Discharge measurements.—Made from the Shawano Street Bridge, two blocks below the gage.

Channel and control.—Sand, hardpan, and mud; not permanent; control not well defined. Both banks at the gage fairly high and not subject to overflow. During extreme flood stages it is reported that the water from the Embarrass River will flow across the city of New London and empty into channel of the Wolf River below gage.

Extremes of discharge.—Maximum stage recorded during year, 9.5 May 30 and 31 (discharge, 7,270 second-feet); minimum discharge, about 700 second-feet February 6-9.

1914-1918: Maximum discharge recorded, 9.7 feet April 4, 1916 (discharge, 8,960 second-feet); minimum discharge, that of February 6-9, 1918. The United States Army Engineers report a stage of 11.6 feet on April 16, 1888.

Ice.—Stage-discharge relation affected by ice.

Regulation.—Little if any diurnal fluctuation due to operation of power plants on the river above station, has been observed at the gage; monthly flow natural.

Accuracy.—Stage-discharge relation not permanent. Two rating curves used during 1918, one, applicable October 1 to November 25 and March 12 to September 30, fairly well defined between 20 and 2,750 second-feet; the other, applicable November 26 to March 11, fairly well defined between 810 and 9,280 second-feet; both curves poorly defined outside these limits. Gage read to tenths once daily. Daily discharge ascertained by applying daily gage height to rating table, except for period when stage-discharge relation was affected by ice, for which it was obtained by applying to rating table mean daily gage height corrected for effect of ice by means of discharge measurements, observer's notes, and weather records. Records fair.

Discharge measurements of Wolf River at New London, Wis., during the year ending Sept. 30, 1918.

Date.	Made by—	Gage height.	Discharge.	Date.	Made by—	Gage height.	Discharge.
		Feet.	*Sec.-ft.*			*Feet.*	*Sec.-ft.*
Dec. 21[a]	Hoyt and Smith	2.02	814	Apr. 30	T. G. Bedford	5.41	2,440
Jan. 19[a]	L. L. Smith	2.40	725	July 19	W. G. Hoyt	1.90	1,090
Feb. 23[a]	do	2.97	704				

[a] Complete ice cover at control and measuring section.

Daily discharge, in second-feet, of Wolf River at New London, Wis., for the year ending Sept. 30, 1918.

Day.	Oct.	Nov.	Dec.	Jan.	Feb.	Mar.	Apr.	May.	June.	July.	Aug.	Sept.
1	953	1,810	910	795	740	795	4,060	2,450	7,000	1,420	1,310	1,310
2	888	1,770	875	780	725	810	3,760	2,550	6,490	1,500	1,230	1,350
3	888	1,650	875	780	725	890	3,500	2,600	6,020	1,500	1,160	1,310
4	920	1,540	840	780	710	945	3,420	2,600	5,610	1,420	1,160	1,230
5	920	1,540	810	780	710	1,020	3,190	2,650	5,250	1,540	1,160	1,160
6	920	1,500	780	795	700	1,140	2,500	2,600	4,940	1,610	1,120	1,060
7	920	1,460	750	795	700	1,280	2,920	2,550	4,650	1,500	1,020	1,060
8	920	1,540	750	780	700	1,420	2,860	2,500	4,390	1,350	1,020	1,060
9	888	1,460	750	780	700	1,610	2,700	2,400	4,160	1,350	1,160	1,060
10	920	1,460	750	765	710	1,810	2,650	2,500	3,850	1,350	1,230	1,090
11	986	1,380	750	750	710	2,060	2,500	2,800	3,670	1,270	1,380	1,060
12	1,060	1,350	765	750	725	2,090	2,450	2,920	3,340	1,200	1,540	1,090
13	1,120	1,310	780	750	740	2,130	2,350	3,060	3,120	1,200	1,690	1,120
14	1,120	1,350	765	750	750	2,220	2,130	3,120	2,980	1,120	1,650	1,120
15	1,120	1,270	765	740	750	2,220	2,060	3,190	2,750	1,090	1,500	1,120
16	1,090	1,160	780	740	750	2,260	1,970	3,340	2,600	1,090	1,460	1,090
17	986	1,120	780	740	740	2,300	1,890	3,420	2,400	1,120	1,420	1,090
18	1,060	1,160	795	725	740	2,450	2,010	3,850	2,220	,120	,380	1,090
19	1,090	1,200	795	725	740	3,120	1,970	4,160	2,050	,060	,270	1,090
20	1,060	1,200	810	725	725	3,960	1,930	5,420	1,890	,020	,200	1,090
21	1,160	1,160	815	740	725	5,420	2,010	6,250	1,730	1,020	1,160	1,120
22	1,230	1,160	810	750	710	6,740	2,090	6,250	1,570	986	1,120	1,120
23	1,270	1,160	810	765	705	6,490	2,170	,020	1,460	953	,120	,200
24	1,270	1,200	810	780	725	6,740	2,220	,810	1,380	953	,120	,200
25	1,270	1,090	825	795	740	6,490	2,220	,020	1,310	1,090	,120	,230
26	1,380	980	825	795	750	6,020	2,130	6,250	1,310	1,060	1,160	1,230
27	1,540	980	810	795	780	6,020	2,090	6,490	1,350	1,020	1,230	1,200
28	1,570	960	810	780	780	5,610	2,090	6,740	1,350	1,090	1,270	1,120
29	1,690	945	795	765	......	5,090	2,130	7,000	1,310	1,120	1,270	1,060
30	1,730	945	795	750	......	4,650	2,300	7,270	1,350	1,160	1,310	1,020
31	1,770	......	795	740	......	4,390	......	7,270	......	1,200	1,310	......

NOTE.—Stage-discharge relation affected by ice Nov. 26 to Mar. 11.

Monthly discharge of Wolf River at New London, Wis., for the year ending Sept. 30, 1918.

[Drainage area, 2,240 square miles.]

Month.	Discharge in second-feet.				Run-off (depth in inches on drainage area).
	Maximum.	Minimum.	Mean.	Per square mile.	
October	1,770	888	1,150	0.513	0.59
November	1,810	945	1,290	.576	.64
December	910	750	799	.357	.41
January	795	725	764	.341	.39
February	780	700	729	.325	.34
March	6,740	795	3,230	1.44	1.66
April	4,060	1,890	2,480	1.11	1.24
May	7,270	2,400	4,260	1.90	2.19
June	7,000	1,310	3,120	1.39	1.55
July	1,610	953	1,210	.540	.62
August	1,690	1,020	1,270	.567	.65
September	1,350	1,020	1,140	.509	.57
The year	7,270	700	1,790	.799	10.85

LITTLE WOLF RIVER AT ROYALTON, WIS.

Location.—In sec. 1, T. 22 N., R. 13 E., at highway bridge in Royalton, Waupaca County, about 4 miles above mouth of river.

Drainage area.—485 square miles (measured on map issued by Wisconsin Geological and Natural History Survey, edition of 1911; scale, 1 inch=6 miles).

Records available.—January 13, 1914, to September 30, 1918.

Gage.—Sloping gage located on left bank of river, about 150 feet upstream from highway bridge, used since August 21, 1915. Chain gage fastened to upstream side of highway bridge was used until August 20, 1915. Datum of the sloping gage is 0.75 foot higher than that of the chain gage. Owing to change in slope, however, difference between the readings from the two gages is not constant.

Discharge measurements.—Made from a cable about 500 feet upstream from bridge.

Channel and control.—Bed at the gage section consists of heavy gravel and rock and is fairly permanent; at the measuring section, fine, smooth gravel. Neither bank is overflowed to any extent at flood stages.

Extremes of discharge.—Maximum stage recorded during year, 4.69 feet at 5.30 p. m. May 19 (discharge about 2,850 second-feet); minimum discharge about 132 second-feet February 2.

1914–1918: Maximum stage recorded, 7.5 feet at 7.15 p. m. June 7, 1914 (discharge, 5,350 second-feet); minimum discharge about 130 second-feet March 5 and 6, 1916, and January 23, 1917.

Ice.—Stage-discharge relation affected by ice.

Regulation.—The few power plants above the station have little storage, and no diurnal fluctuation has been observed at the gage.

Accuracy.—Stage-discharge relation fairly permanent throughout the year. Rating curve well defined between 209 and 1,570 second-feet. Gage read to quarter-tenths twice daily. Daily discharge ascertained by applying mean daily gage height to rating table, except for period when stage-discharge relation was affected by ice, for which it was obtained by applying to rating table mean daily gage height corrected for effect of ice by means of discharge measurements, observer's notes, and weather records. During winter period chain gage was read. Open-water records good, except those for high stages, which are fair; winter records fair.

Discharge measurements of Little Wolf River at Royalton, Wis., during the year ending Sept. 30, 1918.

Date.	Made by—	Gage height.	Discharge.	Date.	Made by—	Gage height.	Discharge.
		Feet.	*Sec.-ft.*			*Feet.*	*Sec.-ft.*
Dec. 21[a]	Hoyt and Smith	[b] 1.18	178	Apr. 30	T. G. Bedford	[c] 2.96	998
Jan. 19[a]	L. L. Smith	[b] 1.91	17	July 19	W. G. Hoyt	1.45	230
Feb. 25[a]	do	[b] 2.40	194				

[a] Complete ice cover at control and measuring section.
[b] Referred to chain gage.
[c] Referred to sloping gage; some uncertainty as to correct gage height as it was determined from reading of chain gage, correction being deduced from previous simultaneous reading of the two gages.

Daily discharge, in second-feet, of Little Wolf River at Royalton, Wis., for the year ending Sept. 30, 1918.

Day.	Oct.	Nov.	Dec.	Jan.	Feb.	Mar.	Apr.	May.	June.	July.	Aug.	Sept.
1	245	402	245	162	148	200	800	970	1,210	314	560	257
2	227	417	223	162	132	203	770	865	1,060	301	276	245
3	230	450	238	170	148	205	800	590	970	314	284	251
4	223	472	232	178	155	207	740	680	830	347	243	254
5	238	439	227	203	148	209	710	590	770	417	223	236
6	207	356	215	186	155	211	650	501	770	361	243	236
7	223	356	207	203	155	213	800	560	770	310	219	211
8	211	402	200	194	162	215	680	620	650	273	301	203
9	219	347	194	203	162	219	650	620	650	267	501	196
10	225	356	189	178	162	223	590	1,130	650	264	620	201
11	245	366	186	194	155	234	590	1,390	650	264	650	257
12	254	371	186	178	162	245	530	1,480	501	236	710	337
13	248	352	183	178	170	266	461	1,390	461	257	530	264
14	227	328	180	178	178	530	407	1,130	450	264	386	251
15	264	301	178	178	178	710	456	830	407	270	301	241
16	251	318	173	194	186	830	417	770	501	238	264	257
17	264	289	170	170	178	1,050	450	970	347	230	273	236
18	332	270	170	155	186	1,210	620	2,400	407	238	257	254
19	366	293	170	177	217	1,390	590	2,740	386	238	254	264
20	356	305	170	149	178	1,570	472	2,070	347	254	270	245
21	328	328	178	155	170	1,870	501	1,870	347	238	270	236
22	323	305	164	140	178	2,070	650	1,670	332	232	251	211
23	276	310	162	148	203	2,290	590	1,300	305	241	257	219
24	318	284	161	162	186	2,400	710	1,300	310	310	251	236
25	318	267	160	170	194	1,210	650	1,870	276	530	243	276
26	386	245	160	162	194	1,130	434	2,070	251	397	236	276
27	456	245	162	170	194	1,050	472	2,740	386	264	254	264
28	472	254	167	162	196	800	590	2,620	347	243	257	241
29	472	248	164	155	......	770	830	2,400	289	461	301	203
30	501	227	168	162	......	770	1,090	2,070	314	318	276	217
31	530	......	173	148	......	770	......	1,570	......	590	264	......

NOTE.—Stage-discharge relation affected by ice Dec. 6 to Mar. 24.

Monthly discharge of Little Wolf River at Royalton, Wis., for the year ending Sept. 30, 1918.

[Drainage area, 485 square miles.]

Month.	Discharge in second-feet.				Run-off (depth in inches on drainage area).
	Maximum.	Minimum.	Mean.	Per square mile.	
October	530	207	304	0.627	0.72
November	472	227	330	.680	.76
December	245	160	186	.384	.44
January	203	140	172	.355	.41
February	217	132	172	.355	.37
March	2,400	200	815	1.68	1.94
April	1,090	407	623	1.28	1.43
May	2,740	501	1,410	2.91	3.36
June	1,210	251	531	1.09	1.22
July	590	230	306	.631	.73
August	710	219	330	.680	.78
September	337	196	243	.501	.56
The year	2,740	132	455	.938	12.72

WAUPACA RIVER NEAR WAUPACA, WIS.

Location.—In sec. 34, T. 22 N., R. 12 E., at Waupaca County highway bridge, about 4 miles downstream from Waupaca, Wis.

Drainage area.—305 square miles (measured on map issued by Wisconsin Geological and Natural History Survey, edition of 1911; scale, 1 inch=6 miles).

Records available.—October 18, 1917, to September 30, 1918; June 28, 1916, to October 18, 1917, records were obtained at a station near Weyauwega, about a mile downstream from present site.

Gage.—Chain gage bolted to upstream handrail of bridge; read by Harry Radtke.

Discharge measurements.—Made from upstream side of bridge or by wading.

Channel and control.—Bed consists of fine gravel and clay, clean and free from vegetation. Control not well defined; may shift slightly. Right bank is high and will rarely be overflowed; left bank of medium height and will be overflowed in time of flood stage.

Ice.—Stage-discharge relation affected by ice.

Extremes of stage.—Maximum stage recorded during year 6.0 feet, March 19 (stage discharge relation affected by ice); minimum open-water stage recorded 1.57 feet September 30 (minimum discharge occurred probably during winter period).

Regulation.—The operation of power plants at and above Waupaca on the main stream and also several on the Crystal River may cause slight fluctuation during low stages. The pondage at the various plants is small and mean monthly discharge is believed to represent nearly the natural flow.

Data inadequate for determination of discharge.

Discharge measurements of Waupaca River near Waupaca, Wis., during the year ending Sept. 30, 1918.

Date.	Made by—	Gage height.	Discharge.	Date.	Made by—	Gage height.	Discharge.
		Feet.	*Sec.-ft.*			*Feet.*	*Sec.-ft.*
Oct. 19[a]	R. B. Kilgore	1.92	238	Feb. 26[b]	L. L. Smith	3.60	168
26[a]	do	2.06	289	Mar. 28	T. G. Bedford	2.19	327
Dec. 22[b]	L. L. Smith	2.66	179	June 6	do	2.06	299
Jan. 21[b]	do	3.07	138	July 19[a]	W. G. Hoyt	1.70	182

[a] Measurement made by wading. [b] Complete ice cover at control and measuring section.

Daily gage height, in feet, of Waupaca River near Waupaca, Wis., for the year ending Sept. 30, 1918.

Day.	Oct.	Nov.	Dec.	Jan.	Feb.	Mar.	Apr.	May.	June.	July.	Aug.	Sept.
1	......	1.86	1.69	2.75	3.4	3.6	2.1	2.1	2.35	1.92	1.68	1.75
2	......	1.78	1.76	2.75	3.3	3.7	2.06	1.96	2.25	1.93	1.76	1.69
3	......	1.85	1.68	3.0	3.4	4.0	2.0	1.99	2.15	1.89	1.80	1.72
4	......	1.86	1.78	2.85	3.4	4.5	1.99	1.84	2.1	1.81	1.78	1.68
5	......	1.84	1.90	2.8	3.3	4.7	1.98	1.82	2.05	1.93	1.71	1.58
6	......	1.96	3.6	2.95	3.4	4.4	1.92	1.81	2.1	1.91	1.71	1.62
7	......	1.83	2.85	2.85	3.4	4.2	2.0	1.88	2.1	1.88	1.71	1.65
8	......	1.80	2.7	3.0	3.4	4.0	2.0	1.81	2.0	1.89	1.92	1.68
9	......	1.83	2.65	3.0	3.5	3.0	1.95	1.90	2.1	1.75	1.98	1.62
10	......	1.84	2.1	2.85	3.5	2.45	1.88	2.45	2.1	1.83	1.98	1.72
11	......	1.78	1.97	3.0	3.5	3.5	1.93	2.6	2.05	1.83	1.88	1.99
12	......	1.75	1.98	2.9	3.5	3.9	1.86	2.45	1.99	1.82	2.2	1.85
13	......	1.79	2.1	3.1	3.5	4.3	1.86	2.25	1.92	1.80	2.2	1.91
14	......	1.74	2.06	3.0	3.5	4.4	1.87	2.1	1.91	1.85	1.99	1.78
15	......	1.76	2.0	3.1	3.4	4.2	1.98	2.0	1.91	1.76	1.90	1.77
16	......	1.69	2.06	3.1	3.5	4.2	1.84	1.99	1.84	1.89	1.80	1.73
17	......	1.72	2.05	3.1	3.5	4.0	1.77	1.94	1.86	1.84	1.86	1.79
18	......	1.75	2.0	3.1	3.5	4.5	1.90	3.2	1.87	1.75	1.79	1.82
19	1.78	1.69	2.0	3.1	3.5	6.0	1.87	2.8	1.85	1.75	1.76	1.73
20	1.80	1.77	1.99	3.0	3.6	5.6	1.91	2.8	1.85	1.67	1.76	1.70
21	1.79	1.72	2.05	3.1	3.5	4.7	1.93	2.4	1.86	1.75	1.73	1.80
22	1.80	1.76	2.66	3.2	3.5	3.6	1.96	2.55	1.81	1.71	1.70	1.75
23	1.83	1.74	2.65	2.95	3.4	2.9	2.0	2.5	1.84	1.75	1.71	1.69
24	1.84	2.0	2.7	3.2	3.5	2.6	1.89	2.3	1.81	1.78	1.72	1.75
25	1.81	1.76	2.5	3.3	3.6	2.4	1.87	3.6	1.86	1.83	1.74	1.69
26	1.82	1.68	2.55	3.3	3.5	2.3	1.79	3.4	1.83	1.87	1.75	1.62
27	1.96	1.78	2.6	3.2	3.6	2.25	1.86	3.5	1.87	1.80	1.69	1.62
28	2.1	1.75	2.7	3.3	3.6	2.1	1.90	3.0	1.97	1.75	1.57	1.64
29	2.0	1.74	2.65	3.2	......	2.15	2.3	2.65	1.85	1.85	2.05	1.66
30	1.96	1.66	2.7	3.4	......	2.15	2.2	2.55	1.81	1.88	1.87	1.57
31	2.1	......	2.75	3.3	......	2.1	......	2.35	......	1.78	1.94	......

NOTE.—Stage-discharge relation affected by ice Nov. 24, 25 and Dec. 4 to Mar. 22.

SHEBOYGAN RIVER NEAR SHEBOYGAN, WIS.

LOCATION.—In sec. 28, T. 15 N., R. 23 E., about 2 miles west of Sheboygan, Sheboygan County, and 2½ miles above mouth.

DRAINAGE AREA.—403 square miles (measured on map issued by Wisconsin Geological and Natural History Survey, edition of 1911; scale, 1 inch=6 miles).

RECORDS AVAILABLE.—June 30, 1916, to September 30, 1918.

GAGE.—Chain gage fastened to upstream side of bridge; read by Hattie Opgenorth.

DISCHARGE MEASUREMENTS.—Made from highway bridge or by wading; at extreme flood stages, from Chicago & North Western Railway bridge, one-third mile downstream.

CHANNEL AND CONTROL.—Control is a well-defined riffle about 200 feet below bridge. Bed of stream is heavy gravel; clear and free from aquatic grass. Banks are of medium height and are rarely overflowed.

EXTREMES OF STAGE.—1916–1918: Maximum stage recorded, 8.85 feet at 8.15 a.m., March 20, 1918. The stage on March 18 and 19, 1918 was somewhat higher, as the observer reports inability to read the gage due to overflow around approach. Minimum stage 1.68 feet at 7.15 p. m., September 13, 1918.

ICE.—Stage-discharge relation affected by ice.

REGULATION.—At low stages there is a small amount of diurnal fluctuation due to operation of small power plants above.

[Sta]ge-discharge relation apparently not permanent. Determination of daily dis[char]ge during year held up pending the making of additional discharge measurements.

Discharge measurements of Sheboygan River near Sheboygan, Wis., during the year ending Sept. 30, 1918.

Date.	Made by—	Gage height.	Discharge.	Date.	Made by—	Gage height.	Discharge.
		Feet.	*Sec.-ft.*			*Feet.*	*Sec.-ft.*
Dec. 20[a]	W. G. Hoyt	2.66	63	Mar. 27	T. G. Bedford	5.16	1,630
Jan. 17[a]	do	2.79	22	July 18	W. G. Hoyt	2.33	51

a Complete ice cover at control and measuring section.

Daily gage height, in feet, of Sheboygan River near Sheboygan, Wis., for the year ending Sept. 30, 1918.

Day.	Oct.	Nov.	Dec.	Jan.	Feb.	Mar.	Apr.	May.	June.	July.	Aug.	Sept.
1	2.18	2.94	2.44	2.34	3.22	5.95	4.02	2.78	3.31	2.35	2.29	1.88
2	2.21	2.79	2.32	2.60	3.45	5.95	3.95	2.77	3.26	2.51	2.45	2.13
3	2.22	2.84	2.46	2.26	3.35	6.25	3.68	2.79	3.18	2.44	2.37	1.99
4	2.21	2.85	2.42	2.34	......	7.35	3.30	2.81	3.06	2.54	2.46	2.08
5	2.44	2.80	2.42	2.98	3.02	7.75	3.35	2.73	3.00	2.38	2.20	2.09
6	2.28	2.77	......	3.02	3.80	7.35	2.86	2.59	2.94	2.43	2.48	2.01
7	2.15	2.77	2.42	2.38	3.40	7.30	2.75	2.99	2.81	2.49	2.33	1.91
8	2.06	2.74	40	36	.50	7.30	05	3.02	2.74	2.33	2.32	2.30
9	2.02	2.68	46	46	.60	......	89	2.77	2.87	2.29	2.45	2.08
10	2.17	2.74	42	66	.45	5.45	90	3.16	2.78	2.26	2.26	2.10
11	2.26	2.57	2.32	2.56	3.45	......	3.04	3.11	2.84	2.39	2.27	2.16
12	2.33	2.48	2.36	2.66	3.55	8.30	2.80	3.01	2.64	2.42	2.30	1.99
13	2.20	2.53	2.34	......	.65	60	69	2.95	2.59	2.52	2.31	1.92
14	2.24	2.53	2.50	2.64	.80	70	59	2.94	2.49	2.32	2.33	2.12
15	2.22	2.51	2.50	2.76	.70	00	61	2.86	2.55	2.26	2.26	2.08
16	2.16	2.45	2.56	2.70	3.60	8.80	2.58	2.85	2.74	2.29	2.36	2.08
17	2.28	2.42	2.56	3.78	3.50	8.84	2.61	2.77	2.49	2.33	2.85	1.94
18	2.37	2.46	2.38	14	.75	......	73	3.06	2.59	2.32	2.41	1.99
19	2.29	2.40	2.86	28	......	......	77	3.16	2.44	2.32	2.31	2.02
20	2.26	2.43	2.68	10	4.10	8.78	71	3.32	2.39	2.33	2.10	2.09
21	2.22	3.20	2.84	3.40	3.20	7.65	3.00	2.97	2.45	2.26	2.19	1.94
22	2.29	2.64	2.80	2.90	3.75	7.05	3.46	3.28	2.14	2.35	2.12	1.95
23	2.29	2.61	2.56	2.96	3.80	6.32	3.02	3.26	2.26	2.37	2.62	1.94
24	2.57	2.78	2.68	3.10	4.15	5.50	2.91	3.00	2.33	2.32	2.29	1.95
25	2.73	2.62	2.46	3.14	4.50	5.60	3.02	3.02	2.32	2.49	2.22	1.96
26	2.64	2.28	2.02	3.10	......	5.40	3.11	3.23	2.34	2.46	1.95	2.06
27	3.48	2.60	2.34	3.02	5.05	5.15	2.57	3.30	2.31	2.32	2.09	2.05
28	3.45	2.28	2.46	3.06	5.70	4.78	84	3.80	2.51	2.30	2.16	2.00
29	3.10	2.40	2.40	2.96	......	4.65	38	3.68	2.34	2.49	2.20	1.91
30	2.96	2.36	2.34	3.60	......	4.43	.28	3.50	2.55	2.28	2.26	2.06
31	2.72	......	2.36	3.32	......	4.28	......	3.34	......	2.47	2.56	......

NOTE.—Stage-discharge relation affected by ice Nov. 24 to Mar. 20.

MILWAUKEE RIVER NEAR MILWAUKEE, WIS.

Location.—In NW. ¼ sec. 5, T. 7 N., R. 22 E., immediately above an old quarry near north limits of Milwaukee, Milwaukee County, half a mile below concrete highway bridge and 1 mile above Mineral Spring road; 5½ miles above confluence of Milwaukee and Menominee rivers.

Drainage area.—661 square miles (measured on map issued by Wisconsin Geological and Natural History Survey, edition of 1911; scale, 1 inch=6 miles).

Records available.—April 30, 1914, to September 30, 1918.

Gage.—Inclined gage on concrete foundations on left bank of river; prior to April 18, 1918, chain gage fastened to cantilever arm supported by posts set in concrete foundations. Both gages at same datum. Gage read by Miss Bertha Kuehl.

Channel and control. Bed of channel at gage heavy gravel; about 200 feet below the gage is a rock outcrop with a 4-foot fall which forms the control and is fairly permanent, changing only during exceptionally heavy floods. Below the control the river flows in an artificial channel which at one time was a quarry. Left bank above and below the control high and not subject to overflow; right bank above control of medium height; below the control the right bank is artificial and of such height that overflow will rarely occur.

Discharge measurements.—Made by wading immediately above the gage section; at medium and high stages from a concrete highway bridge about a mile upstream from the gage.

Extremes of discharge.—Maximum stage during year, determined by levels to high-water mark, 9.00 feet, early in morning of March 20 (discharge, about 12,100 second-feet); minimum discharge about 45 second-feet, January 20 to February 2.

1914–1918: Maximum stage recorded, that of March 20, 1918; minimum stage recorded, 0.50 foot at 8.31 p. m., August 2, 1916 (discharge, about 26 second-feet).

Ice.—Stage-discharge relation affected by ice.

Regulation.—No diurnal fluctuation at the gage resulting from operation of small plants above.

Accuracy.—Stage discharge relation changed somewhat during the flood of March. Two rating curves used during year, both well defined between 88 and 3,710 second-feet. Gage read to quarter-tenths twice daily. Daily discharge ascertained by applying mean daily gage height to rating table, except for period when stage-discharge relation was affected by ice, for which it was obtained by applying to rating table mean daily gage height corrected for ice effect by means of discharge measurements, observer's notes, and weather records. Open-water records excellent, except those for extremely high and low stages, which are only good; winter records fair.

Discharge measurements of Milwaukee River near Milwaukee, Wis., during the year ending Sept. 30, 1918.

Date.	Made by—	Gage height.	Discharge.	Date.	Made by—	Gage height.	Discharge.
		Feet.	*Sec.-ft.*			*Feet.*	*Sec.-ft.*
Dec. 20[a]	W. G. Hoyt	1.50	141	Apr. 17[c]	T. G. Bedford	1.31	349
Jan. 17[a]	do	2.05	58	July 18	W. G. Hoyt	.65	91
Mar. 25[b]	Hoyt and Potts	8.25	10,400				

[a] Complete ice cover at control and measuring section.

[b] Velocity determined by timing movement of ice cakes and débris over a measured course 200 feet long at old bridge section 1,000 feet downstream from gage.

[c] Made at second highway bridge 1 mile upstream from gage.

Daily discharge, in second-feet, of Milwaukee River near Milwaukee, Wis., for the year ending Sept. 30, 1918.

Day.	Oct.	Nov.	Dec.	Jan.	Feb.	Mar.	Apr.	May.	June	July	Aug.	Sept.
1	150	777	230	130	45	1,270	860	860	770	117	95	66
2	117	734	294	120	45	1,310	860	728	495	120	91	51
3	127	650	307	115	50	1,360	770	568	389	127	86	78
4	127	610	247	110	55	1,360	685	460	347	117	93	82
5	127	532	195	95	60	1,180	645	389	330	127	66	70
6	146	494	115	90	65	1,270	605	365	305	136	80	58
7	130	532	110	85	70	1,680	605	371	285	125	91	66
8	127	494	105	80	75	1,790	770	447	244	107	78	70
9	117	460	100	70	80	1,360	815	495	258	102	62	64
10	154	394	90	65	90	1,270	645	568	240	100	60	91
11	210	373	90	60	100	1,180	568	728	240	78	60	104
12	247	367	95	60	110	1,790	495	605	215	104	91	93
13	288	360	100	60	115	2,260	434	495	206	100	125	117
14	247	360	100	60	130	2,380	402	447	180	95	117	117
15	215	353	105	60	145	2,630	383	383	159	107	95	100
16	205	327	110	60	150	2,760	389	335	146	120	102	91
17	225	327	115	55	160	3,150	347	276	136	93	117	109
18	353	301	120	50	170	4,410	421	335	102	93	109	91
19	294	270	125	50	185	8,260	568	860	82	95	84	91
20	264	282	130	45	210	12,100	645	1,040	93	130	86	78
21	205	288	145	45	240	10,300	860	950	109	117	84	80
22	210	288	165	45	270	7,450	1,130	1,220	117	58	51	91
23	394	294	185	45	290	4,860	1,130	1,130	136	72	48	86
24	820	320	190	45	360	3,430	950	995	133	58	84	95
25	952	294	205	45	425	2,400	685	770	117	55	72	78
26	1,270	360	190	45	735	1,920	530	530	93	55	48	62
27	1,360	347	185	45	1,090	1,500	460	728	95	51	53	78
28	1,360	294	170	45	1,180	1,310	530	995	80	82	60	70
29	1,360	294	160	45	....	1,080	728	1,040	86	95	55	80
30	1,180	301	150	45	....	995	905	995	86	93	66	72
31	908	....	145	45	....	905	....	905	....	91	80	....

NOTE.—Stage-discharge relation affected by ice Dec. 6 to Mar. 10. Gage washed out Mar. 19; discharge interpolated.

Monthly discharge of Milwaukee River near Milwaukee, Wis., for the year ending Sept. 30, 1918.

[Drainage area, 661 square miles.]

Month.	Discharge in second-feet.				Run-off (depth in inches on drainage area).
	Maximum.	Minimum.	Mean.	Per square mile.	
October	1,360	117	448	0.678	0.78
November	777	270	403	.610	.68
December	307	90	154	.233	.27
January	130	45	65.0	.098	.11
February	1,180	45	239	.362	.38
March	12,100	905	2,930	4.43	5.11
April	1,130	347	661	1.00	1.12
May	1,220	276	678	1.03	1.19
June	770	80	209	.316	.35
July	136	51	97.4	.147	.17
August	125	48	80.3	.121	.14
September	117	51	82.6	.125	.14
The year	12,100	45	508	.769	10.43

LITTLE CALUMET RIVER AT HARVEY, ILL.

Location.—In NW. ¼ sec. 9, T. 36 N., R. 14 E., at Illinois Central Railroad bridge 800 feet north of railroad station at One Hundred and Forty-seventh Street, Harvey, Cook County, 11 miles above mouth of river.

Drainage area.—570 square miles (measured on map issued by United States Geological Survey; scale, 1:500,000).

Records available.—Daily discharge, October 1, 1916, to September 30, 1918; daily gage heights, collected by Sanitary District of Chicago, June 10, 1907, to September 30, 1916.

Gage.—Vertical staff gage attached to bridge pier; read by Mrs. H. Wurtman.

Discharge measurements.—Made from downstream side of bridge during medium and high stages, or by wading during low stages.

Channel and control.—Bed of river composed of clay and gravel. Low-water control is at "The Rocks," about a mile below gage; bed of river, heavy gravel, somewhat shifting. Banks not subject to overflow.

Extremes of discharge.—Maximum stage recorded during year, 8.8 feet at 8 a. m and 4 p. m. February 15 (discharge not determined because of backwater from ice). Maximum open-water stage recorded, 7.1 feet at 8 a. m. and 4 p. m. March 1 (discharge, 1,680 second-feet); minimum discharge, probably somewhat less than 25 second-feet, occurred in January.

1910–1918: Maximum stage recorded, 13.4 feet March 6, 1908 (discharge not determined); minimum discharge, that in January, 1918.

Accuracy.—Stage-discharge relation probably permanent throughout the year; seriously affected by ice during the winter. Rating curve well defined above and fairly well defined below 125 second-feet. Gage read to hundredths once daily. Daily discharge ascertained by applying daily gage height to rating table. Records good for open-water periods; poor for winter.

Discharge measurements of Little Calumet River at Harvey, Ill., during the year ending Sept. 30, 1918.

[Made by H. C. Beckman.]

Date.	Gage height.	Discharge.	Date.	Gage height.	Discharge.
	Feet.	*Sec.-ft.*		*Feet.*	*Sec.-ft.*
Nov. 1	3.67	188	Sept. 18	3.10	68
Mar. 2	6.98	1,600	18	3.10	76
May 27	4.30	395			

Daily discharge, in second-feet, of Little Calumet River at Harvey, Ill., for the year ending Sept. 30, 1918.

Day.	Oct.	Nov.	Dec.	Jan.	Feb.	Mar.	Apr.	May.	June.	July.	Aug.	Sept.
1	70	182	109			1,680	472	378	530	361	147	70
2	68	195	109			1,620	433	344	452	311	119	65
3	68	182	111			1,510	414	311	414	280	113	71
4	74	182	109			1,400	378	280	396	280	96	73
5	72	182	109			1,290	344	280	396	280	85	94
6	77	170		85	30	1,290	328	265	378	296	77	91
7	74	170				1,190	311	265	378	311	68	87
8	71	158				1,090	280	280	361	328	65	85
9	70	147				1,090	265	265	344	311	65	84
10	71	138				1,090	250	280	328	311	65	77
11	71	134				995	236	344	296	296	62	84
12	74	127				905	208	311	265	280	62	91
13	74	119	80			905	195	361	236	250	59	91
14	77	115				1,340	170	361	222	236	56	94
15	77	113				1,340	156	328	195	208	56	91
16	77	113		40	1,130	1,090	145	296	170	208	53	84
17	113	125				995	136	280	136	182	65	77
18	147	129				905	236	250	123	158	125	73
19	158	127				905	222	236	105	136	81	74
20	170	117				860	195	650	98	125	71	74
21	170	113				816	236	414	91	113	65	74
22	170	117				773	311	361	87	98	62	73
23	170	119				731	296	344	84	87	58	70
24	182	109				731	265	328	82	81	98	68
25	170	111				690	265	414	79	76	84	66
26	182	107	130	25	1,520	650	265	396	77	101	74	65
27	182	109				610	265	378	77	115	71	65
28	182	109				570	280	361	98	147	64	65
29	182	111				530	452	452	91	182	58	65
30	170	109				510	378	690	101	182	62	68
31	182	......				472	......	650	......	170	65	......

NOTE.—Discharge Dec. 6 to Feb. 28 estimated, because of ice, from gage heights, observer's notes, and weather records. Braced figures show mean discharge for periods included.

Monthly discharge of Little Calumet River at Harvey, Ill., for the year ending Sept. 30, 1918.

[Drainage area, 570 square miles.]

Month.	Discharge in second-feet.				Run-off (depth in inches on drainage area).
	Maximum.	Minimum.	Mean.	Per square mile.	
October	182	68	119	0.209	0.24
November	195	107	135	.237	.26
December	....	....	102	.179	.21
January	....	....	49.2	.086	.10
February	....	....	849	1.49	1.55
March	1,680	472	986	1.73	1.99
April	472	136	280	.491	.55
May	690	236	360	.632	.73
June	530	77	223	.391	.44
July	361	76	210	.368	.42
August	147	53	75.8	.133	.15
September	94	65	77.0	.135	.15
The year	....	....	285	.500	6.79

GRAND RIVER AT GRAND RAPIDS, MICH.

Location.—At Fulton Street Bridge, Grand Rapids.

Drainage area.—4,900 square miles.

Records available.—March 12, 1901, to September 30, 1918.

Gage.—Staff, attached to bridge; read to tenths; occasionally, October 1, 1917, to February 10, and July 1 to August 5, 1918; twice daily, February 11 to June 30, except on Sundays. Gage read by Charles Darling and J. M. Knoll.

Discharge measurements.—Made from downstream side of bridge.

Extremes of stage.—Maximum stage recorded during year 16.2 feet at 8 a. m. and 5 p. m. March 18; minimum stage recorded, −1.8 feet several days in June, July and August.

Ice.—Stage-discharge relation somewhat affected by ice.

Regulation.—Operation of power plants above station may modify low-water flow.

Cooperation.—Records furnished by city engineer of Grand Rapids.

No discharge measurements made during the year.

Daily gage height, in feet, of Grand River at Grand Rapids, Mich., for the year ending Sept. 30, 1918.

Day.	Oct.	Nov.	Dec.	Jan.	Feb.	Mar.	Apr.	May.	June.	July.	Aug.
1	−1.0	1.6	−0.8		0.3	12.05	3.75	0.85	0.8	−1.8	
2		1.4		0.4	.4	12.0	3.45	1.0			
3			− .8	.3			3.45	.85	.3	−1.5	−1.8
4	−1.0		− .7	.4	.3	11.35	3.6	.8	.35		
5	−1.0	.6	− .7	.4	.4	11.0	3.5		.3		−1.8
6		.4	− .6		.3	10.95	3.4	.4		−1.6	
7			− .7	.3	.3	11.15		.45	− .4		
8		− .2	− .6	.3	.4	10.6	2.45	.4	− .4	−1.8	
9	−1.0	− .4		.3	.4	10.55	1.95	.3			
10			− .6	.4			1.75	.1	− .4	−1.6	
11			− .6	.3	.4	9.4	1.55	.2	− .3		
12		− .4	− .6	.3	.45	9.35	1.4		− .4		
13	−1.0		− .6		2.35	10.5	1.4	.6	− .38	−1.7	
14		− .4	− .6	.2	4.85	11.75		.95	− .35		
15			− .6	.3	7.65	13.0	.55	1.2	− .4	−1.8	
16		− .6		.3	8.95	14.35	.5	1.2			
17		− .8	− .7	.3	9.65	15.9	.4	.9	− .5	−1.6	
18	−1.0			.3	11.75	16.2	.55	.55	− .9	−1.8	
18		− .8	− .5	.4	12.75	15.6	1.4		− .85		
20		− .9	.2		14.3	14.6	.8	−.1	−1.0	−1.8	
21		−1.0		.4	14.5	13.7		.1	−1.0		
22	− .6	−1.0	.4	.4	14.5	12.8	1.55	−.1	−1.2	−1.7	
23		−1.0		.3	14.3	11.92	1.35	.3			
24		−1.0	.4	.3	13.45		1.0	.3	−1.25	−1.8	
25	.2			.3	12.55	10.15	.9	.2	−1.65		
26		−1.0	.4	.3	12.35	9.2	.9		−1.8	−1.8	
27	.8		.4		12.05	8.3	.8	.85	−1.8		
28		− .9		.3	12.05	7.1		1.85	−1.6		
29	1.2		.4	.4		5.8	.55	1.85	−1.7	−1.8	
30				.3		4.8	.95			−1.8	
31	1.4		.4	.3				.8		−1.8	

STREAMS TRIBUTARY TO LAKE HURON.

TITTABAWASSEE RIVER AT FREELAND, MICH.

Location.—At highway bridge at Freeland.

Drainage area.—2,530 square miles.

Records available.—August 22, 1903, to August 3, 1906; October 28, 1906, to December 31, 1909; January 1, 1912, to September 30, 1918

Cooperation.—Estimates of daily discharge were made and furnished by G. S. Williams, consulting engineer, Ann Arbor, Mich.

Daily discharge, in second-feet, of Tittabawassee River at Freeland, Mich., for the year ending Sept. 30, 1918.

Day.	Oct.	Nov.	Dec.	Jan.	Feb.	Mar.	Apr.	May.	June.	July.	Aug.	Sept.
1	620	1,110	1,244	828	967	4,500	5,275	2,270	4,800	930	675	700
2	636	1,110	2,025	821	967	4,700	5,060	2,246	3,285	1,140	646	730
3	646	1,098	1,985	838	967	4,700	4,800	2,230	2,875	1,080	620	760
4	646	1,080	1,905	838	967	4,205	4,250	2,230	2,400	1,050	566	786
5	675	1,088	1,921	821	967	3,905	8,520	2,230	1,785	1,020	566	786
6	700	990	1,921	787	967	3,800	3,285	2,105	1,705	1,002	566	815
7	700	960	1,905	770	967	3,620	3,285	2,065	1,600	990	582	930
8	700	930	1,985	770	948	3,330	3,031	2,025	2,270	930	566	990
9	690	882	1,093	762	928	3,255	2,700	2,025	1,235	930	566	930
10	675	870	1,020	758	928	3,225	2,400	1,985	1,221	845	582	900
11	690	870	928	758	948	3,480	2,270	1,945	1,200	815	592	845
12	700	882	1,000	750	983	3,620	2,025	1,905	1,182	786	608	815
13	712	918	1,032	750	1,112	3,905	1,865	1,865	1,170	760	592	815
14	700	930	1,130	770	1,244	4,825	1,825	1,825	1,170	730	592	821
15	706	930	1,300	794	1,308	5,790	1,825	1,825	1,166	700	608	845
16	712	900	1,390	814	1,855	5,520	1,865	1,865	1,140	700	620	815
17	730	918	1,410	821	2,330	5,490	1,865	1,865	1,020	690	646	786
18	730	930	1,300	828	2,300	5,790	1,825	1,825	930	675	646	760
19	748	900	1,244	838	2,275	6,180	1,825	1,801	900	658	658	748
20	748	900	1,112	838	2,290	7,650	2,400	1,785	845	658	675	700
21	760	942	967	866	2,100	10,000	4,100	1,745	815	646	700	663
22	786	930	948	928	2,125	9,600	4,250	1,785	786	646	700	646
23	815	930	928	928	2,250	8,200	4,400	1,825	760	690	700	636
24	900	1,300	928	928	2,430	7,400	4,250	1,985	760	815	700	620
25	930	1,441	910	928	2,670	5,870	3,475	2,875	748	845	700	620
26	930	1,432	891	948	3,055	5,790	2,610	4,050	730	960	690	592
27	942	1,428	871	967	3,855	5,600	2,315	7,109	730	990	680	582
28	990	1,390	861	983	4,390	5,500	2,306	9,075	700	990	685	582
29	1,020	1,365	858	967	....	5,450	2,270	8,700	730	900	690	566
30	1,050	1,300	838	948	....	5,400	2,270	7,735	730	845	690	566
31	1,098	....	838	967	....	5,300	....	6,930	....	760	700	....

Monthly discharge of Tittabawassee River at Freeland, Mich., for the year ending Sept. 30, 1918.

[Drainage area, 2,530 square miles.]

Month.	Discharge in second-feet.				Run-off (depth in inches on drainage area).
	Maximum.	Minimum.	Mean.	Per square mile.	
October	1,098	620	777	0.307	0.35
November	1,441	870	1,050	.415	.46
December	2,025	838	1,250	.494	.57
January	983	750	849	.336	.39
February	4,390	928	1,750	.692	.72
March	10,000	3,225	5,340	2.11	2.43
April	5,275	1,825	2,980	1.18	1.32
May	9,075	1,745	3,020	1.19	1.37
June	4,800	700	1,380	.545	.61
July	1,140	646	843	.333	.38
August	700	566	639	.253	.29
September	990	566	745	.294	.33
The year	10,000	566	1,720	.680	9.22

NOTE.—Monthly and yearly discharge computed by United States Geological Survey.

STREAMS TRIBUTARY TO LAKE ERIE.

HURON RIVER AT BARTON, MICH.

Location.—At dam and power plant of Eastern Michigan Edison Co. at Barton, near Ann Arbor, 4 miles above station at Geddes.

Drainage area.—723 square miles.

Records available.—January 1 to September 30, 1918.

Determination of discharge.—Flow computed from records of operation of power plant, the flow through under-sluice during floods, and the depth of flow over dam. The flow through the power house is determined from a calibration of the turbines by means of a specially constructed weir, the crest of which was formed by a ¼-inch by 5-inch milled plate, the discharge over the weir being computed by Bazin's formula for free overflow. The greater part of the flood water passes through under-sluices in the power-house foundations, and this flow is determined from a weir calibration of the sluices. Water flows over crest of dam only a few days during the year.

Cooperation.—Daily-discharge record furnished by G. S. Williams, consulting engineer, Ann Arbor, Mich.

Daily discharge, in second-feet, of Huron River at Barton, Mich., for the year ending Sept. 30, 1918.

Day.	Oct.	Nov.	Dec.	Jan.	Feb.	Mar.	Apr.	May.	June.	July.	Aug.	Sept.
1	164	478	222	196	150	2,499	914	518	255	177	91	70
2	155	420	205	189	155	2,602	922	523	245	106	88	85
3	162	406	280	192	136	2,686	941	442	256	98	83	69
4	166	385	203	217	145	2,568	899	516	211	68	18	133
5	158	419	228	156	153	2,370	857	442	214	111	92	160
6	182	376	231	168	150	2,185	786	433	207	117	79	112
7	163	331	221	175	164	1,939	778	459	186	70	89	134
8	215	317	211	186	160	1,811	812	403	180	108	84	119
9	143	346	152	188	165	1,729	733	393	198	99	85	168
10	134	326	220	179	202	1,720	660	412	194	101	49	113
11	170	313	179	177	242	1,487	532	411	178	102	40	177
12	171	314	206	167	575	1,765	608	418	174	97	87	143
13	146	314	191	103	862	2,459	564	508	166	112	92	111
14	161	278	219	183	1,338	5,841	521	581	163	59	97	131
15	185	340	160	163	2,424	4,138	538	502	160	98	97	139
16	169	264	210	156	1,642	3,603	426	441	143	153	90	175
17	194	313	190	158	1,378	3,497	505	452	149	77	74	147
18	217	305	217	159	1,326	3,382	545	458	162	108	48	151
19	235	290	194	149	1,928	3,286	594	415	145	104	87	152
20	266	298	189	145	2,197	2,822	551	426	158	109	96	141
21	262	272	261	146	2,249	2,555	567	346	136	48	126	160
22	285	289	277	140	1,914	2,197	576	346	135	100	92	102
23	297	273	315	144	1,668	2,142	891	309	44	105	81	146
24	364	278	326	143	1,661	1,759	464	294	135	95	68	175
25	368	250	246	146	2,467	1,577	501	331	129	97	18	130
26	364	273	312	151	3,806	1,346	482	226	107	109	72	137
27	373	255	243	117	3,194	1,335	503	264	119	94	35	142
28	413	221	213	167	2,776	1,205	426	284	100	22	61	151
29	458	254	218	187	......	1,145	504	222	98	96	54	108
30	515	266	220	151	......	981	489	272	65	120	68	129
31	476	......	232	146	......	917	......	281	......	94	68	......

Monthly discharge of Huron River at Barton, Mich., for the year ending Sept. 30, 1918.

[Drainage area, 723 square miles.]

Month.	Discharge in second-feet.				Run-off (depth in inches on drainage area).
	Maximum.	Minimum.	Mean.	Per square mile.	
October	515	134	249	0.344	0.40
November	478	221	315	.436	.49
December	326	152	226	.313	.36
January	217	103	163	.225	.26
February	3,806	136	1,260	1.74	2.01
March	5,841	917	2,310	3.20	3.69
April	941	426	636	.880	.98
May	581	222	398	.550	.63
June	256	44	160	.221	.25
July	177	22	98.5	.136	.16
August	126	18	74.5	.103	.12
September	177	69	134	.185	.21
The year	5,841	18	498	.689	9.56

NOTE.—Monthly and yearly discharge computed by United States Geological Survey.

HURON RIVER AT FLAT ROCK, MICH.

LOCATION.—At highway bridge at Flat Rock, 2,000 feet below crossing of Detroit, Toledo & Ironton Railway.

DRAINAGE AREA.—1,000 square miles.

RECORDS AVAILABLE.—August 6, 1904, to September 30, 1918.

GAGE.—Staff; read daily to tenths, occasionally to half tenths twice daily, by John Vincent.

DISCHARGE MEASUREMENTS.—Made from downstream side of bridge.

CHANNEL AND CONTROL.—Probably permanent.

EXTREMES OF STAGE.—Maximum stage during year above 11 feet (water over gage) March 15; minimum stage recorded, 0.9 foot, several days in July and August.

ICE.—Ice jams form below the station and cause backwater at the gage; in general the section above the station is kept open by the power plant.

REGULATION.—At ordinary stages flow of the river is controlled by a dam and power plant immediately above station, but operation of this plant is assumed to have little effect on diurnal fluctuations of stage.

No discharge measurements were made at this station during the year.

Daily gage height, in feet, Huron River at Flat Rock, Mich., for the year ending Sept. 30, 1918.

Day.	Oct.	Nov.	Dec.	Jan.	Feb.	Mar.	Apr.	May.	June.	July.	Aug.	Sept.
1	1.6	2.8	1.75	2.6	2.8	9.62	4.62	3.1	1.65	1.45	1.6	......
2	1.4	2.65	1.6	2.45	3.0	9.1	4.6	2.7	......	1.45	1.35	1.55
3	1.65	2.55	1.5	2.35	2.8	9.8	4.65	2.7	1.7	1.4	1.5	1.3
4	1.5	2.2	1.8	2.55	2.8	9.78	4.5	2.5	1.8	......	......	1.4
5	1.55	1.95	1.3	2.5	2.8	9.4	4.4	......	1.85	1.4	1.15	1.55
6	1.65	2.45	1.75	2.4	2.8	8.88	4.25	2.35	1.7	1.05	.95	1.6
7	1.5	2.1	1.6	2.2	2.8	8.7	4.0	2.35	2.0	......	1.2	1.65
8	1.35	2.1	1.6	2.35	2.8	8.38	3.7	2.6	1.55	1.2	1.35	......
9	1.6	1.9	1.8	2.6	2.9	7.78	3.6	2.1	......	1.2	1.35	1.5
10	1.4	1.9	1.75	2.45	2.8	7.7	3.5	2.2	1.55	1.5	1.25	1.6
11	1.4	1.9	1.65	2.55	2.75	7.4	3.4	2.5	1.6	1.4	......	1.6
12	1.55	1.75	2.15	2.8	3.7	7.05	2.9	2.5	1.45	1.35	1.4	1.6
13	1.4	2.0	1.9	2.8	5.15	7.1	3.0	2.6	1.45	1.3	1.4	1.5
14	1.6	1.9	2.0	2.3	7.0	9.12	2.8	2.8	1.4	......	1.2	1.45
15	1.35	1.85	1.95	2.2	8.4	......	2.75	2.9	1.45	1.0	1.3	......
16	1.6	1.9	1.8	3.0	8.75	9.3	2.85	2.35	......	1.05	1.4	1.4
17	1.6	1.8	1.7	2.8	9.3	8.8	2.19	2.5	......	1.25	1.45	1.45
18	1.7	1.8	2.2	2.6	8.8	8.52	2.9	2.3	......	1.3	......	1.7
19	1.65	1.75	2.3	2.7	8.25	8.4	3.1	......	......	1.35	1.45	1.6
20	2.0	2.2	2.25	2.6	8.55	8.28	2.95	2.1	......	1.35	1.0	1.6
21	1.6	1.9	2.5	2.4	9.25	8.06	2.9	2.1	......	......	1.05	1.6
22	1.65	2.0	2.7	2.5	9.75	7.8	3.05	1.9	......	1.5	1.45	......
23	2.0	1.9	2.6	2.5	8.95	7.5	3.1	1.7	......	1.35	1.35	1.3
24	1.95	1.65	2.1	2.4	8.62	7.1	2.25	1.75	1.4	1.53	1.6	1.45
25	2.05	1.5	2.8	2.7	8.3	6.72	2.5	2.25	1.4	1.5	......	1.6
26	2.0	1.6	2.65	2.6	8.52	6.25	2.1	......	1.35	1.5	1.45	1.45
27	2.2	1.7	2.55	2.6	9.58	6.02	2.0	1.9	1.3	1.45	1.05	1.55
28	2.2	1.8	2.5	2.4	10.4	5.6	2.0	1.6	1.5	......	1.2	1.55
29	2.7	1.8	2.5	2.5	......	4.25	2.6	2.05	1.35	1.2	1.45	......
30	3.05	1.85	2.2	2.7	......	5.1	3.25	......	1.4	1.05	1.5	1.45
31	3.0	......	2.05	2.85	......	4.7	......	1.8	......	1.5	1.4	......

CATTARAUGUS CREEK AT VERSAILLES, N. Y.

Location.—At three-span highway bridge in Versailles, Cattaraugus County, 2½ miles above mouth of Clear Creek, 6 miles below Gowanda, and 8 miles above mouth of stream.

Drainage area.—467 square miles (measured on post-route map).

Records available.—September 23, 1910, to September 30, 1918.

Gage.—Chain, on upstream side of right span of bridge; read by Charles Wilson.

Discharge measurements.—Made from the downstream side of bridge or by wading.

Channel and control.—Rock and gravel; shifting.

Extremes of discharge.—Maximum stage recorded during year, 12.0 feet at 8 a. m. February 23 (stage-discharge relation affected by ice, discharge not computed); minimum stage recorded during year, 4.35 feet several times in August (discharge about 49 second-feet).

1910–1918: Maximum open-water stage recorded, 11.6 feet at 5.40 p. m., March 25, 1913 (discharge, about 30,000 second-feet); minimum stage recorded 4.35 feet several times in August, 1918 (discharge, about 49 second-feet).

Ice.—Stage-discharge relation seriously affected by ice.

Accuracy.—Stage-discharge relation not permanent; affected by ice during much of the period from December to March. Gage read to half-tenths twice daily. Daily discharge ascertained by applying mean daily effective gage height to rating table. Records fair.

Discharge measurements of Cattaraugus Creek at Versailles, N. Y., during the year ending Sept. 30, 1918.

[Made by E. D. Burchard.]

Date.	Gage height.	Discharge.	Date.	Gage height.	Discharge.
1911.	*Feet.*	*Sec.-ft.*	1912.	*Feet.*	*Sec.-ft.*
Jan. 22[a]	6.43	232	Aug. 22	4.45	78.1
Mar. 1	6.18	1,950	22	4.50	78.4
May 29	4.99	333	22	4.60	117
29	4.99	347			

[a] Measurement made through complete ice cover.

Daily discharge, in second-feet, of Cattaraugus Creek at Versailles, N. Y., for the year ending Sept. 30, 1918.

Day.	Oct.	Nov.	Dec.	Jan.	Feb.	Mar.	Apr.	May.	June.	July.	Aug.	Sept.
1	190	1,900	1,400	380	220	1,800	480	300	280	240	180	120
2	170	1,500	1,000	340	220	1,100	650	280	240	220	150	85
3	190	1,300	650	340	220	1,500	550	240	200	180	140	85
4	1,600	1,100	600	360	220	950	500	240	200	150	140	100
5	1,200	1,000	500	380	220	2,400	460	240	200	150	160	160
6	1,000	900	480	340	220	4,000	400	240	220	120	130	300
7	800	900	420	320	220	1,800	380	280	200	130	85	170
8	380	750	280	320	220	1,000	380	380	200	110	100	110
9	380	700	280	320	380	1,200	420	280	170	140	130	110
10	340	700	320	320	1,500	2,000	340	320	200	220	150	65
11	280	650	380	320	3,200	1,400	380	500	180	280	140	100
12	300	600	400	320	1,700	1,400	420	380	500	200	180	85
13	500	550	480	320	2,000	3,400	420	460	440	160	140	140
14	850	500	550	300	2,200	16,000	900	700	320	150	120	180
15	900	500	550	300	2,600	4,000	800	400	260	100	80	150
16	1,100	550	600	260	1,500	1,400	600	300	200	120	55	170
17	650	550	650	260	1,000	1,200	480	300	200	140	65	440
18	420	500	750	240	800	1,200	800	280	200	120	80	360
19	1,100	550	850	240	900	1,100	600	240	170	110	65	320
20	2,100	500	1,200	240	2,600	1,400	460	1,000	160	100	80	550
21	900	480	2,400	240	1,500	1,400	400	550	160	100	75	420
22	700	550	2,200	240	2,200	1,200	550	340	340	85	80	380
23	650	750	1,500	240	4,400	950	600	900	420	95	110	360
24	1,600	650	1,500	240	4,400	750	550	440	300	95	80	340
25	3,600	500	3,400	240	3,900	700	500	340	240	340	65	440
26	2,800	500	1,760	240	7,000	600	400	950	180	160	80	300
27	5,500	380	1,000	240	2,000	600	380	650	180	130	65	440
28	6,000	550	800	220	1,800	600	320	420	160	220	65	500
29	8,500	500	550	220	......	550	300	300	170	160	95	320
30	10,000	550	500	220	......	550	300	300	180	800	160	260
31	3,400	......	440	220	......	500	......	280	......	280	85	......

NOTE.—Stage-discharge relation affected by ice Dec. 10, to Feb 25.

Monthly discharge of Cattaraugus Creek at Versailles, N. Y., for year ending Sept. 30, 1918.

[Drainage area, 467 square miles.]

Month.	Discharge in second-feet.				Run-off (depth in inches on drainage area).
	Maximum.	Minimum.	Mean.	Per square mile.	
October	10,000	170	1,870	4.00	4.61
November	1,900	380	720	1.54	1.72
December	3,400	280	914	1.96	2.26
January	380	220	282	.631	.73
February	4,400	220	1,760	3.78	3.94
March	16,000	500	1,890	4.05	4.67
April	900	300	491	1.05	1.17
May	1,000	240	414	.877	1.01
June	500	160	236	.505	.56
July	800	85	181	.388	.45
August	180	55	107	.229	.26
September	550	65	252	.540	.60
The year	16,000	55	756	1.62	21.98

STREAMS TRIBUTARY TO LAKE ONTARIO.

LITTLE TONAWANDA CREEK AT LINDEN, N. Y.

LOCATION.—At stone-arch highway bridge in Linden, Genesee County, about 3 miles above junction with Tonawanda Creek.

DRAINAGE AREA.—22.0 square miles (measured on topographic maps).

RECORDS AVAILABLE.—July 8, 1912, to September 30, 1918.

GAGE.—Vertical staff, on upstream side of right abutment. Lower 2 feet of enameled iron, graduated to hundredths of foot; upper 4 feet of bronze, graduated to half-tenths; read by C. L. Schenck.

DISCHARGE MEASUREMENTS.—Made from cable 1,000 feet above gage, or by wading near gage.

CHANNEL AND CONTROL.—A standard Francis weir, 2.01 feet long and 8 inches high, constructed under the upstream side of the bridge, formed the control until February 20, 1918, when it was entirely destroyed by ice and has not since been replaced. When the water overtopped this weir it flowed over a 2-inch plank about 13 feet long, including the 2 feet of weir. The section of the channel that forms the control since the destruction of the weir is of coarse gravel and boulders and is probably permanent between dates of shift.

EXTREMES OF DISCHARGE.—Maximum stage recorded during year, 7.45 feet at 8 p. m. February 19 (stage-discharge relation affected by ice; discharge not determined); minimum stage recorded, −0.46 foot at 8 p. m. August 20 (discharge, 0.5 second-foot).

1912–1918: Maximum stage determined by leveling from flood marks, 14.6 feet during the flood of April 22, 1916 (discharge about 2,400 second-feet); minimum stage recorded, 0.18 foot August 20 and 21, September 14–16, and October 8, 1913 (discharge, 0.43 second-foot).

ACCURACY.—Stage-discharge relation changed when weir was destroyed on February 20. Rating curve for weir in good condition, well defined up to 250 second-feet and fairly well defined between 250 and 750 feet; rating curve for period after the weir was destroyed fairly well defined. Gage read to hundredths twice daily. Daily discharge ascertained by applying mean daily gage height to rating table. Records good for period when weir was in good condition and fairly good for remainder of year.

Discharge measurements of Little Tonawanda Creek near Linden, N. Y., during the year ending Sept. 30, 1918.

Date.	Made by—	Gage height.	Discharge.	Date.	Made by—	Gage height.	Discharge.
		Feet.	*Sec.-ft.*			*Feet.*	*Sec.-ft.*
Mar. 4	E. D. Burchard........	0.26	41	Mar. 19	E. D. Burchard........	1.18	147
19	.....do................	.86	106	May 31	.....do................	— .24	6.8
19	.....do................	.94	116	31	.....do................	— .24	6.8
19	.....do................	1.02	128	July 23	C. C. Covert............	— .39	.70
19	.....do................	1.12	140	Aug. 21	E. D. Burchard........	— .47	.60

Daily discharge, in second-feet, of Little Tonawanda Creek at Linden, N. Y., for the year ending Sept. 30, 1918.

Day.	Oct.	Nov.	Dec.	Jan.	Feb.	Mar.	Apr.	May.	June	July	Aug.	Sept.
1	1.51	51	43	9.0	4.2	33	25	9.2	5.9	5.9	1.2	0.9
2	1.45	41	21	8.4	4.6	105	43	8.7	4.7	4.3	1.0	.8
3	1.51	34	12	7.8	3.6	52	25	8.2	4.0	3.2	1.0	.6
4	2.25	27	11	7.2	3.6	38	18	7.8	3.8	3.0	1.2	.6
5	2.86	24	10	6.6	3.48	79	15	7.4	3.8	2.7	1.3	1.3
6	3.28	21	9.7	6.6	3.6	203	14	6.6	5.9	2.1	1.0	1.2
7	2.38	19	8.4	7.2	4.6	50	13	7.4	7.4	2.1	.9	.8
8	2.12	16	8.7	6.6	5.1	56	16	8.2	5.1	2.1	.9	.8
9	2.25	16	6.1	6.6	6.1	32	14	7.0	4.3	2.7	1.9	.6
10	2.18	15	7.2	6.6	9.7	158	13	22	5.1	3.2	1.3	.6
11	2.12	13	9.0	7.2	25	77	14	17	4.3	2.7	3.2	.6
12	2.32	13	9.0	7.8	......	585	15	14	75	2.7	.9	.8
13	3.36	12	9.0	6.6	......	203	38	15	21	2.1	.8	1.5
14	7.8	11	9.0	6.4	......	740	44	12	16	1.9	.8	.9
15	8.4	12	9.0	6.1	......	97	25	9.2	11	1.9	.8	.9
16	9.7	13	8.1	6.1	......	63	15	8.2	7.4	1.9	.6	1.6
17	6.1	13	8.1	5.9	......	73	22	7.4	5.9	1.9	.6	2.1
18	5.6	13	8.1	5.6	......	65	80	6.6	5.1	1.6	.6	1.3
19	12	13	8.4	5.3	......	110	32	6.2	4.0	1.6	.6	1.6
20	19	12	13	5.6	......	108	21	5.9	3.8	1.6	.5	3.8
21	11	12	35	5.6	......	71	21	5.1	4.3	1.3	.6	3.0
22	7.2	17	39	5.1	......	60	25	5.1	8.2	1.3	1.3	2.1
23	7.8	18	23	5.1	......	42	19	6.6	7.4	1.3	.8	1.6
24	154	13	37	5.1	......	32	19	5.1	5.9	1.3	.8	2.7
25	164	12	59	4.6	......	30	17	5.9	4.3	1.3	.8	3.2
26	154	11	24	4.9	115	25	14	22	3.8	1.3	.6	3.2
27	135	10	18	4.9	88	21	13	15	3.5	1.2	.6	3.2
28	135	10	13	4.9	46	26	11	10	3.2	1.0	.6	3.2
29	144	10	12	4.6	......	25	10	7.8	3.0	2.1	.9	2.7
30	288	14	11	4.2	......	26	9.2	8.2	3.0	2.1	.8	2.4
31	83	......	9.7	4.4	......	25	......	6.6	......	1.5	1.0	......

NOTE.—Discharge Feb. 12-25 estimated at 141 second-feet because of ice.

Monthly discharge of Little Tonawanda Creek at Linden, N. Y., for the year ending Sept. 30, 1918.

[Drainage area, 22.0 square miles.]

Month.	Discharge in second-feet.				Run-off (depth in inches on drainage area).
	Maximum.	Minimum.	Mean.	Per square mile.	
October	288	1.45	44.6	2.03	2.34
November	51	10	17.2	.782	.87
December	59	6.1	16.9	.768	.86
January	9.0	4.2	6.1	.277	.32
February	....	3.48	82	3.73	3.88
March	740	21	107	4.86	5.60
April	80	9.2	22	1.00	1.12
May	22	5.1	9.39	.427	.49
June	75	3.0	8.34	.379	.42
July	5.9	1.0	2.16	.098	.11
August	3.2	.5	.964	.044	.05
September	3.8	.6	1.69	.077	.09
The year	740	.5	26.2	1.19	16.15

GENESEE RIVER AT SCIO, N. Y.

Location.—At steel highway bridge half a mile above Vandermark Creek, half a mile above Scio, Allegheny County, and a mile above Knight Creek.

Drainage area.—297 square miles (measured on maps issued by United States Geological Survey; scale, 1:500,000.)

Records available.—June 12, 1916, to September 30, 1918.

Gage.—Vertical staff attached to downstream face of left bridge abutment; read by Raymond Sisson until November 3, and by Miss Retta B. Potter, after that date.

Discharge measurements.—Made from downstream side of bridge or by wading.

Channel and control.—Coarse gravel; probably permanent.

Extremes of discharge.—Maximum stage recorded during the year, 9.0 feet at 8 a. m. March 14 (discharge, 10,400 second-feet); minimum discharge 34 second-feet, January 20.

1916–1918: Maximum stage recorded, that of March 14, 1918; minimum discharge recorded, 25 second-feet, August 25 and 26, 1916.

Ice.—Stage-discharge relation affected by ice.

Accuracy.—Stage-discharge relation practically permanent, except as affected by ice December 7 to February 13. Rating curve well defined between 25 and 5,500 second-feet. Gage read to hundredths twice daily; gage-height record unreliable, April 27 to May 22, and June 14–20. Daily discharge ascertained by applying mean daily gage height to rating table. Records good, except those for period of ice effect and for periods in which gage-height record was unreliable, which are fair.

Discharge measurements of Genesee River at Scio, N. Y., during the year ending Sept. 30, 1918.

Date.	Made by—	Gage height.	Discharge.	Date.	Made by—	Gage height.	Discharge.
		Feet.	*Sec.-ft.*			*Feet.*	*Sec.-ft.*
Dec. 22[a]	E. D. Burchard	1.83	186	June 21	E. D. Burchard	0.74	74
Jan. 19[a]	....do	2.05	55	21	do	.74	73
Mar. 5	....do	2.02	609	Aug. 23	....do	.69	56.7
May 27	J. W. Moulton	1.61	346	23	....do	.69	58.2

a Measurement made through complete ice cover.

Daily discharge, in second-feet, of Genesee River at Scio, N. Y., for the year ending Sept. 30, 1918.

Day.	Oct.	Nov.	Dec.	Jan.	Feb.	Mar.	Apr.	May.	June.	July.	Aug.	Sept.
1	61	1,150	361	120	46	6,300	572	460	465	345	61	178
2	74	2,680	312	100	46	1,360	545	440	279	322	41	74
3	64	1,310	265	120	46	1,150	920	360	218	300	41	71
4	91	690	198	95	46	780	850	340	200	279	41	66
5	265	545	178	85	46	660	660	340	238	258	440	74
6	121	386	158	75	46	1,680	600	360	322	238	147	264
7	98	438	180	75	46	780	572	400	518	218	87	147
8	88	438	120	75	46	690	600	360	415	200	87	116
9	118	386	140	70	60	750	720	320	518	200	147	113
10	101	336	140	70	160	815	750	320	518	132	102	113
11	88	336	160	65	380	750	850	320	415	74	218	116
12	202	288	120	70	1,300	1,150	815	500	415	61	147	116
13	312	242	120	65	1,800	1,490	780	650	279	41	147	147
14	190	265	140	65	1,310	10,000	690	550	200	61	147	164
15	361	242	160	60	2,800	2,300	780	440	150	41	147	147
16	490	220	140	65	1,310	1,070	1,490	340	120	41	147	141
17	312	242	160	60	1,150	885	1,490	300	85	41	116	300
18	251	198	160	150	990	750	1,880	260	60	61	116	258
19	1,580	265	140	55	750	720	1,880	340	60	41	116	218
20	2,680	158	140	34	8,070	720	1,990	600	60	41	87	2,540
21	1,150	178	100	38	990	720	1,780	550	77	41	87	750
22	850	220	180	42	850	750	1,580	500	322	41	87	518
23	990	242	240	46	780	750	1,230	1,310	279	41	61	440
24	2,100	265	500	48	815	750	750	780	258	41	61	390
25	2,100	312	440	46	720	720	720	440	200	61	61	345
26	1,880	312	220	46	4,560	630	600	518	200	61	41	300
27	3,570	336	150	46	1,150	600	550	440	200	41	41	300
28	4,130	312	140	46	815	630	500	390	181	41	39	238
29	3,440	336	110	46	......	600	460	465	181	41	43	238
30	2,920	312	130	16	......	572	440	465	238	61	74	218
31	1,680	......	120	46	......	572	......	390	......	61	119	......

NOTE.—Discharge, Dec. 7 to Feb. 13 estimated, because of ice, from discharge measurements, weather records, study of gage-height graph, and comparison with records for stations downstream. Discharge Apr. 27 to May 22, and June 14–20, estimated by comparison with records of flow at St. Helena.

Monthly discharge of Genesee Rivver at Scio, N. Y., for year ending Sept. 30, 1918.

[Drainage area, 297 square miles.]

Month.	Discharge in second-feet.				Run-off (depth in inches on drainage area).
	Maximum.	Minimum.	Mean.	Per square mile.	
October	4,130	61	1,040	3.50	4.04
November	2,680	158	455	1.53	1.71
December	500	100	188	.633	.73
January	150	34	67	.226	.26
February	8,070	46	1,110	3.74	3.90
March	10,000	572	1,360	4.58	5.28
April	1,990	440	935	3.15	3.51
May	1,310	260	460	1.55	1.79
June	518	60	256	.862	.96
July	345	41	114	.384	.44
August	440	39	106	.357	.41
September	2,540	66	303	1.02	1.14
The year	10,000	34	529	1.78	24.17

GENESEE RIVER AT ST. HELENA, N. Y.

Location.—At steel highway bridge in St. Helena, Wyoming County, about 5½ miles below Portageville and site of proposed storage dam of State of New York Conservation Commission, and 9½ miles above mouth of Canaseraga Creek

Drainage area.—1,030 square miles.

Records available.—August 14, 1908, to September 30, 1918.

Gage.—Stevens continuous water-stage recorder on left bank just below bridge and a chain gage fastened to the upstream side of the bridge; middle-span chain gage installed August 14, 1908; water-stage recorder installed August 24, 1911. Water-stage recorder inspected by C. S. De Golyer. Chain gage read by Herman Piper.

Discharge measurements.—Made from the bridge, or by wading.

Channel and control.—Gravel and rocks; frequently shifting.

Extremes of discharge.—Maximum stage recorded during year, 11.4 feet at 5 p. m. March 14 (discharge about 29,500 second-feet); minimum stage recorded, 2.00 feet at 7 a. m. July 26 and 6 p. m. August 30 (discharge, 40 second-feet).

1908–1918: Maximum stage, from water-stage recorder, 12.81 feet at 8 a. m. May 17, 1916 (discharge, 43,500 second-feet); minimum stage recorded, 1.70 feet at 5 p. m. October 5 and 8 a. m. October 17, 1913 (discharge, approximately 18 second-feet).

Ice.—Stage discharge relation somewhat affected by ice.

Accuracy.—Stage-discharge relation not permanent. Rating curve well defined between 75 and 2,000 second-feet and fairly well defined between 2,000 and 30,000 second-feet. Chain gage read to quarter-tenths twice daily. Daily discharge ascertained by applying mean daily gage heights to rating table, except for days of great range in stage, when it was determined by averaging the results obtained by applying gage heights for two-hour periods to rating table. Records fair.

Discharge measurements of Genesee River at St. Helena, N. Y., during the year ending Sept. 30, 1918.

Date.	Made by—	Gage height.	Discharge.	Date.	Made by—	Gage height.	Discharge.
		Feet.	*Sec.-ft.*			*Feet.*	*Sec.-ft.*
Oct. 29	C. S. De Golyer	7.68	10,800	Apr. 27	D. S. De Golyer	3.55	930
Nov. 2	E. D. Burchard	4.97	2,950	May 25	J. W. Moulton	3.44	774
14	D. S. De Golyer	3.24	690	30	C. S. De Golyer	3.16	588
Dec. 12[a]	do	3.52	379	June 27	do	2.76	319
Jan. 5[b]	do	3.87	238	July 13	E. D. Burchard	2.51	194
25[b]	do	3.84	146	13	do	2.50	191
Feb. 8[b]	do	3.68	153	25	C. S. De Golyer	2.15	71
13	do	7.53	9,860	Aug. 21	E. D. Burchard	2.40	144
Mar. 9	do	4.56	2,200	28	C. S. De Golyer	2.10	57.5
13	do	6.15	5,750	Sept. 20	do	3.23	579
15	do	9.78	19,300				

[a] Measurement made through partial ice cover. [b] Measurement made through complete ice cover.

Daily discharge, in second-feet, of Genesee River at St. Helena, N. Y., for the year ending Sept. 30, 1918.

Day.	Oct.	Nov.	Dec.	Jan.	Feb.	Mar.	Apr.	May.	June.	July.	Aug.	Sept.
1	245	3,980	760	220	150	5,320	790	555	520	257	132	109
2	268	2,950	1,040	200	190	5,320	882	628	425	262	140	103
3	261	2,420	670	190	170	4,100	882	555	401	199	126	182
4	511	1,980	590	200	160	2,100	980	488	346	182	115	136
5	895	1,640	590	200	140	2,260	930	455	329	186	225	182
6	805	1,440	475	260	240	5,850	745	443	351	154	451	214
7	630	1,290	392	240	190	3,660	665	488	384	154	293	335
8	510	1,120	309	240	170	1,960	665	555	431	91	209	282
9	573	1,000	221	280	170	2,180	790	443	384	170	190	214
10	622	940	240	190	180	10,400	980	520	346	150	186	209
11	489	868	320	240	240	3,050	790	590	329	190	178	166
12	447	760	360	260	850	3,050	745	590	745	182	281	228
13	820	670	240	220	15,400	6,420	882	665	835	174	257	218
14	931	670	240	220	4,810	26,000	2,480	930	530	147	204	346
15	796	630	300	260	10,400	14,800	5,530	705	407	149	182	329
16	1,540	806	220	220	3,450	3,840	4,060	555	335	134	278	329
17	1,150	590	320	240	1,680	2,830	2,830	443	329	126	228	373
18	823	510	280	220	1,140	2,830	5,010	395	292	123	190	1,130
19	1,090	590	280	280	1,300	3,020	2,650	384	247	122	166	1,060
20	9,170	550	240	150	9,000	2,830	1,930	455	232	111	154	835
21	3,830	630	380	240	4,810	2,830	1,590	882	228	112	140	2,100
22	2,180	630	650	300	1,810	2,830	1,860	745	419	106	143	1,060
23	1,690	760	750	220	1,360	2,150	1,590	1,240	1,030	109	129	808
24	4,470	670	750	190	1,540	1,590	1,470	1,180	628	103	122	650
25	8,820	380	1,200	130	1,420	1,350	1,300	745	455	100	115	628
26	7,040	447	1,100	190	11,500	1,180	1,030	745	362	97	110	605
27	10,700	332	650	170	4,100	1,030	882	1,130	308	143	104	808
28	12,000	440	440	170	3,050	930	745	835	247	122	98	781
29	10,800	428	360	160	......	882	665	590	257	109	103	628
30	12,300	496	320	170	......	835	628	555	242	122	115	507
31	6,500	......	260	180	......	835	......	665	......	136	103	......

NOTE.—Discharge Nov. 11 to July 13 and Aug. 31 to Sept. 20 determined from chain-gage heights. Discharge Dec. 10 to Feb. 12 estimated, because of ice, from discharge measurements, weather records, study of gage-height graph and comparison with records for stations at Scio and Jones Bridge. Discharge Feb. 20 estimated by comparison with station at Jones Bridge.

Monthly discharge of Genesee River at St. Helena, N. Y., for the year ending Sept. 30, 1918.

[Drainage area, 1,030 square miles.]

Month.	Discharge in second-feet.				Run-off (depth in inches on drainage area).
	Maximum.	Minimum.	Mean.	Per square mile.	
October	12,300	245	3,320	3.22	3.71
November	3,980	332	1,020	.990	1.10
December	1,200	220	482	.468	.54
January	300	130	215	.209	.24
February	15,400	140	2,840	2.76	2.87
March	26,000	835	4,140	4.02	4.64
April	5,530	628	1,570	1.52	1.70
May	1,240	384	650	.631	.73
June	1,030	228	412	.400	.45
July	262	91	146	.142	.16
August	451	98	176	.171	.20
September	2,100	103	520	.505	.56
The year	26,000	91	1,280	1.24	16.90

GENESEE RIVER AT JONES BRIDGE, NEAR MOUNT MORRIS, N. Y.

LOCATION.—At highway bridge known as Jones Bridge, 1½ miles below Canaseraga Creek, 1¾ miles above mouth of Beads Creek, 5 miles below Mount Morris, Livingston County, and 6 miles by river above Geneseo.

DRAINAGE AREA.—1,410 square miles.

RECORDS AVAILABLE.—May 22, 1903, to April 30, 1906; August 12, 1908, to December 31, 1913; July 12, 1915, to September 30, 1918.

GAGE.—Gurley seven-day water-stage recorder installed September 11, 1915, on the right bank about 60 feet downstream from the bridge. Prior to 1915, a chain gage fastened to upstream side of highway bridge was used. Datum of water-stage recorder is 2.73 feet higher than that for the former chain gage (540.00 feet Conservation Commission datum). Water-stage recorder inspected by Theron S. Trewer.

DISCHARGE MEASUREMENTS.—Made from footbridge erected on the lower chord of the truss at the upstream side of the bridge.

CHANNEL AND CONTROL.—Sandy clay; likely to shift, but as shown by current-meter measurements, fairly permanent in recent years.

EXTREMES OF DISCHARGE.—Maximum stage during year estimated from record 25.5 feet at 3.30 a. m. February 21 (stage-discharge relation affected by ice; discharge not determined); minimum stage, 0.45 foot at 1 a. m. July 25 (discharge 63 second-feet).

1903–1918 (not including periods of no record; see "Records available"): Maximum stage recorded 25.44 feet at noon May 17, 1916 (discharge, 54,500 second-feet); minimum stage recorded, 2.7 feet at 6 p. m. August 29, 1909 (discharge about 18 second-feet).

ICE.—Stage-discharge relation affected by ice.

REGULATION.—During extreme low water there is some diurnal fluctuation in flow caused by mills at Mount Morris.

ACCURACY.—Stage-discharge relation practically permanent during the year except as affected by ice December 8 to March 22. Rating curve well-defined between 150 and 7,000 second-feet and fairly well defined between 7,000 and 60,000 second-feet. Operation of the water-stage recorder satisfactory throughout the year. Daily discharge ascertained by applying to the rating table mean daily gage height determined by inspecting the recorder graph, or for days of considerable fluctuation by use of discharge integrator. Records good.

Discharge measurements of Genesee River at Jones Bridge, near Mount Morris, N. Y., during the year ending Sept. 30, 1918.

Date.	Made by—	Gage height.	Discharge.	Date.	Made by—	Gage height.	Discharge.
		Feet.	*Sec.-ft.*			*Feet.*	*Sec.-ft.*
Nov. 1	E. D. Burchard	11.12	6,040	Feb. 26[b]	E. D. Burchard	22.0	11,700
1	J. W. Moulton	10.12	5,320	27[b]	do	21.5	8,400
2	do	7.40	3,900	Mar. 2[b]	do	19.21	6,970
Dec. 19[a]	E. D. Burchard	2.98	530	4[b]	do	15.0	4,120
Jan. 16[a]	do	2.59	318	15	do	24.2	[c] 28,300
Feb. 11[b]	do	2.96	313	18	do	8.90	4,880
13[b]	do	12.4	3,700	19	do	7.08	3,770
14[b]	do	21.42	6,860	May 23	do	3.21	1,190
15[b]	do	21.60	8,450	July 12	do	1.36	292
16[b]	do	21.9	7,920	Aug. 21	do	.91	159

[a] Measurement made through complete ice cover.
[b] Ice jam on control.
[c] Includes overflow of 6,300 second-feet on left bank

Daily discharge, in second-feet, of Genesee River at Jones Bridge, near Mount Morris, N. Y., for the year ending Sept. 30, 1918.

Day.	Oct.	Nov.	Dec.	Jan.	Feb.	Mar.	Apr.	May.	June.	July.	Aug.	Sept.
1	305	7,050	892	440	280	5,000	1,200	892	840	365	165	140
2	353	4,200	1,330	440	280	7,000	1,200	892	690	357	162	204
3	365	3,240	1,080	420	280	6,000	1,300	865	615	327	155	213
4	502	2,750	865	420	280	4,200	1,360	815	565	278	126	225
5	1,080	2,280	815	380	260	3,600	1,420	790	515	305	162	210
6	1,030	2,020	690	360	240	4,800	1,140	765	506	238	413	238
7	740	1,840	590	320	260	5,500	1,000	740	535	155	425	258
8	640	1,660	650	320	260	3,000	1,080	1,000	540	273	295	319
9	690	1,480	600	300	300	4,570	1,140	840	590	258	285	302
10	740	1,300	600	300	320	8,310	1,420	865	515	275	298	278
11	615	1,220	600	300	440	10,700	1,170	1,140	492	269	229	271
12	560	1,140	600	320	1,600	9,700	1,170	1,030	740	255	281	258
13	857	1,080	600	320	5,800	12,500	1,220	1,000	1,250	235	353	229
14	1,060	975	600	320	7,500	21,600	2,790	1,280	840	190	281	264
15	840	948	550	320	8,000	22,200	6,790	1,200	665	236	248	369
16	1,540	920	550	320	7,500	12,100	4,970	948	535	223	241	369
17	1,320	920	550	320	5,500	7,980	3,760	815	466	216	316	425
18	920	865	550	320	3,800	4,500	5,740	740	466	188	245	1,470
19	1,120	840	500	320	3,200	3,500	4,270	665	399	167	248	892
20	10,500	815	500	320	8,500	3,800	2,820	690	341	164	238	867
21	5,080	790	650	400	9,000	3,700	2,280	1,220	349	135	213	2,570
22	2,680	815	900	420	6,500	3,600	2,410	1,280	461	136	181	1,420
23	1,960	920	1,000	380	4,800	3,170	2,410	1,250	1,120	133	168	1,030
24	4,740	920	1,100	340	3,600	2,380	2,020	1,720	920	130	133	865
25	13,700	740	1,500	340	3,400	2,000	1,840	1,140	690	216	140	765
26	10,800	665	2,000	340	6,500	1,700	1,600	1,440	535	145	......	715
27	13,200	615	1,500	300	7,500	1,600	1,300	1,840	448	131	......	892
28	16,500	615	1,000	300	6,000	1,400	1,110	1,480	399	153	......	1,000
29	15,100	690	750	320	......	1,400	1,080	1,030	365	181	......	790
30	17,300	715	550	320	......	1,300	920	948	323	163	......	640
31	14,100	......	500	320	......	1,300	......	948	......	164	140	......

NOTE.—Discharge Dec. 8 to Mar. 22 estimated, because of ice, from discharge measurements, weather records, study of gage height graph and comparison with records for St. Helena and Rochester. Discharge Aug. 26–30 estimated 140 second-feet.

Monthly discharge of Genesee River at Jones Bridge, near Mount Morris, N. Y., for year ending Sept. 30, 1918.

[Drainage area, 1,410 square miles.]

Month.	Discharge in second-feet.				Run-off (depth in inches on drainage area).
	Maximum.	Minimum.	Mean.	Per square mile.	
October	17,300	305	4,550	3.23	3.72
November	7,050	615	1,500	1.06	1.18
December	2,000	500	810	.575	.66
January	440	300	344	.244	.28
February	9,000	240	3,640	2.58	2.69
March	22,200	1,300	5,940	4.21	4.85
April	6,790	920	2,130	1.51	1.69
May	1,840	665	1,040	.738	.85
June	1,250	323	590	.418	.47
July	365	130	215	.152	.18
August	425	126	221	.157	.18
September	2,570	140	616	.437	.49
The year	22,200	126	1,790	1.27	17.74

GENESEE RIVER AT ROCHESTER, N. Y.

Location.—At Elmwood Avenue Bridge, at north end of South Park, 3¼ miles below mouth of Black Creek, 3½ miles above center of city of Rochester, Monroe County, and 7½ miles above mouth of river.

Drainage area.—2,360 square miles.

Records available.—February 9, 1904, to September 30, 1918. Fragmentary records prior to this period published in Water-Supply Papers 24, 65, and 97.

Gage.—Gurley water-stage recorder installed in December, 1910, in the pump house immediately below the bridge on the right bank. Recorder inspected by Geo. A. Bailey. Prior to December, 1910, a staff gage bolted to the downstream end of the first pier from the right abutment. Elevation of zero of gage 506.848 feet, barge canal datum, and 245.591 feet, Rochester city datum.

Discharge measurements.—Made from downstream side of the bridge. Prior to 1904, measurements and elevation of water surface taken in conjunction with the city of Rochester.

Channel and control.—Smooth gravel; practically permanent until May, 1918, when dredging operations for the barge canal were begun near the control. These operations were continued through the summer, causing a gradual change in the stage-discharge relation.

Extremes of discharge.—Maximum stage during year, from water-stage recorder, 10.97 feet at 9.15 p. m., March 16 (discharge, 27,900 second-feet); minimum discharge about 110 second-feet during afternoons of July 21 and 22.

1904–1918: Maximum stage, from water-stage recorder, 12.3 feet at midnight March 30, 1916 (discharge, 48,300 second-feet); minimum discharge, July 21 and 22, 1918.

Ice.—Stage-discharge relation affected by ice during a large part of the period from December to March, inclusive.

Accuracy.—Stage-discharge relation practically permanent until May 1 except as affected by ice December 10 to February 13; May 1 to September 30, a gradual change in stage-discharge relation was caused by dredging operations. Rating curve well defined between 2,000 and 44,000 second-feet. Operation of water-stage recorder satisfactory throughout the year. Mean daily gage height ascertained by averaging hourly gage heights. Daily discharge prior to May ascertained by applying mean daily gage height to rating table; May to September, by the shifting-control method. Records good except those for periods when the stage discharge relation was affected by ice or dredging on the control, which are fair.

Cooperation.—Water-stage recorder inspected by an employee of the Rochester Railway & Light Co.

Discharge measurements of Genesee River at Rochester, N. Y., during the year ending Sept. 30, 1918.

Date.	Made by—	Gage height.	Discharge.	Date	Made by—	Gage height.	Discharge.
		Feet.	*Sec.-ft.*			*Feet.*	*Sec.-ft.*
Nov. 3	J. W. Moulton	4.05	4,970	July 12	E. D. Burchard	1.14	764
Dec. 19[a]	E. D. Burchard	1.99	865	20	.....do	1.20	673
Jan. 16[b]	.....do	2.18	517	27	.....do	.76	664
Feb. 11[b]	.....do	1.63	400	31	.....do	.60	666
13[a]	.....do	8.36	7,720	Aug. 19	.....do	.49	597
Mar. 22	.....do	4.59	6,440	28	.....do	.40	512
May 22	.....do	2.36	1,680	Sept. 24	.....do	1.21	1,580
June 20	.....do	1.12	742				

[a] Measurement made through partial ice cover.
[b] Measurement made through complete ice cover.

Daily discharge, in second-feet, of Genesee River at Rochester, N. Y., for the year ending Sept. 30, 1918.

Day.	Oct.	Nov.	Dec.	Jan.	Feb.	Mar.	Apr.	May.	June.	July.	Aug.	Sept.
1	531	14,900	1,480	850	360	8,060	2,340	1,500	900	750	460	550
2	510	8,330	1,980	750	360	7,540	2,500	1,500	1,100	850	480	550
3	553	5,120	2,010	750	400	9,430	2,500	1,400	950	850	480	550
4	619	3,880	1,620	700	340	8,330	2,340	1,400	850	850	480	550
5	776	3,300	1,480	700	420	5,930	2,340	1,300	850	850	480	550
6	1,380	2,850	1,280	600	420	6,050	2,040	1,200	900	900	500	550
7	1,340	2,590	1,150	550	420	8,330	1,860	1,100	950	800	900	550
8	1,070	2,340	1,320	600	400	7,800	1,840	1,100	950	700	1,100	550
9	966	2,180	1,340	650	440	5,700	1,900	1,300	850	750	1,000	550
10	920	2,000	1,300	650	440	5,130	2,040	1,100	850	800	650	550
11	980	1,890	1,100	650	420	8,870	2,000	1,200	900	800	500	550
12	896	1,760	1,000	600	1,200	8,060	1,980	1,500	850	750	500	550
13	812	1,680	1,000	600	7,500	15,600	[illegible],060	1,500	1,100	800	500	550
14	1,110	1,580	1,000	600	11,200	19,200	,490	1,200	1,460	700	500	550
15	1,370	1,510	1,000	550	13,600	23,000	,900	850	1,400	750	750	550
16	1,250	1,450	1,000	550	14,000	27,200	5,960	1,200	1,100	800	700	550
17	1,920	1,440	950	550	12,100	25,100	5,240	1,700	950	700	650	550
18	1,620	1,400	900	550	8,060	14,900	,930	1,200	950	700	650	700
19	1,240	1,340	850	500	5,020	7,800	,030	1,000	900	650	600	1,600
20	3,680	1,330	800	500	8,870	6,530	,700	800	800	550	600	1,300
21	8,870	1,270	900	500	12,400	6,290	3,680	1,000	800	550	550	1,600
22	4,600	1,300	1,200	500	13,600	[illegible],050	3,400	1,400	950	300	550	2,800
23	2,760	1,410	1,700	500	11,200	,130	,490	1,800	800	480	500	2,200
24	2,500	1,550	1,800	480	6,290	,080	3,120	1,700	1,400	750	500	1,800
25	10,600	1,450	2,200	480	4,600	,490	2,850	1,700	1,400	800	500	1,400
26	14,900	1,200	3,000	460	7,800	3,120	2,500	1,100	1,200	800	500	1,400
27	13,300	1,040	3,200	440	11,200	2,850	2,180	1,500	900	650	550	1,300
28	14,300	1,060	2,200	420	11,800	2,680	1,960	2,200	800	650	550	1,500
29	15,300	1,070	1,800	420	......	2,500	1,800	1,600	900	950	550	1,800
30	16,000	1,190	1,200	420	......	2,420	1,650	1,500	800	600	550	1,500
31	17,000	......	950	420	......	2,340	......	1,400	......	650	550	......

NOTE.—Discharge Dec. 10 to February 13 estimated, because of ice, from discharge measurements, weather records, study of gage-height graph, and comparison with records for station upstream.

Monthly discharge of Genesee River at Rochester, N. Y., for year ending Sept. 30, 1918.

[Drainage area, 2,360 square miles.]

Month.	Discharge in second-feet.				Run-off (depth in inches on drainage area).
	Maximum.	Minimum.	Mean.	Per square mile.	
October	17,000	510	4,630	1.96	2.26
November	14,900	1,040	2,510	1.06	1.18
December	3,200	800	1,440	.610	.70
January	850	420	560	.239	.28
February	14,000	340	5,890	2.50	2.60
March	27,200	2,340	8,700	3.69	4.25
April	7,930	1,650	3,190	1.35	1.51
May	2,200	850	1,350	.572	.66
June	1,400	800	982	.416	.46
July	950	300	725	.307	.35
August	1,100	460	591	.250	.29
September	2,800	550	1,010	.428	.48
The year	27,200	300	2,610	1.11	15.02

CANASERAGA CREEK AT CUMMINSVILLE, N. Y.

Location.—At bridge on State road in Cumminsville, Livingston County, about a mile downstream from station formerly maintained near Dansville, 1½ miles below mouth of Mill Brook and 21 miles above mouth of creek.

Drainage area.—171 square miles (measured by State conservation commission).

Records available.—October 23, 1917, to September 30, 1918; July 21, 1910, to December 31, 1912, and July 10, 1915, to December 29, 1917, at station near Dansville.

Gage.—Vertical staff gage in three sections on downstream face of bridge pier; read to tenths daily by George Freed.

Discharge measurements.—Made from downstream side of bridge or by wading.

Channel and control.—Fairly well compacted gravel and small boulders may shift during severe floods, otherwise practically permanent.

Extremes of stage.—Maximum stage recorded during year, 5.2 feet at 3.30 p. m. February 12 (stage discharge relation affected by ice); minimum stage recorded during year, 0.7 foot several times in August and September.

Ice.—Stage-discharge relation affected by ice.

Data inadequate for determination of daily discharge.

Discharge measurements of Canaseraga Creek at Cumminsville, N. Y., during year ending Sept. 30, 1918.

Date.	Made by—	Gage height.	Discharge.	Date.	Made by—	Gage height.	Discharge.
		Feet.	*Sec.-ft.*			*Feet.*	*Sec.-ft.*
Oct. 20	E. D. Burchard	2.06	478	Feb. 15	E. D. Burchard	3.00	1,130
20	.....do	1.98	425	Mar. 18	.....do	1.63	289
23	.....do	1.38	135	21	.....do	1.70	326
25	.....do	3.05	1,140	May 26	.....do	1.45	183
25	.....do	2.92	1,020	31	J. W. Moulton	1.21	88
Dec. 20[a]	.....do	1.44	120	July 15	E. D. Burchard	.89	38.2
Jan. 17[a]	.....do	1.66	49	Aug. 23	.....do	.77	24.7
Feb. 12[b]	.....do	4.60	782				

a Measurement made through complete ice cover.
b Measurement made through partial ice cover.

Daily gage height, in feet, of Canaseraga Creek at Cumminsville, N. Y., for the year ending Sept. 30, 1918.

Day.	Oct.	Nov.	Dec.	Jan.	Feb.	Mar.	Apr.	May.	June.	July.	Aug.	Sept.
1	......	1.78	1.18	1.95	1.50	2.10	1.38	1.12	1.28	0.90	0.90	0.70
2	......	1.71	1.18	1.95	1.45	2.05	1.38	1.11	1.20	.90	.90	.70
3	......	1.57	1.18	2.10	1.45	2.00	1.38	1.10	1.10	.85	.90	.70
4	......	1.51	1.20	1.95	1.50	1.85	1.36	1.10	1.10	.80	.80	.70
5	......	1.52	1.19	1.90	1.50	1.90	1.30	1.10	1.10	.80	.85	.75
6	......	1.51	1.18	1.90	1.45	2.30	1.30	1.10	1.10	.80	.80	.80
7	......	1.48	1.18	1.80	1.54	1.95	1.34	1.08	1.10	.80	.89	.80
8	......	1.37	1.17	1.60	1.50	2.05	1.39	1.30	1.10	.80	.80	.80
9	......	1.37	1.31	1.60	1.50	1.90	1.39	1.20	1.10	.85	.80	.80
10	......	1.34	1.34	1.67	1.50	2.90	1.32	1.28	1.10	1.00	.80	.80
11	......	1.33	1.40	1.60	2.56	1.95	1.28	1.35	1.05	1.00	1.00	.80
12	......	1.32	1.46	1.85	4.43	2.70	1.29	1.30	1.40	.90	.95	.80
13	......	1.30	1.50	1.80	2.85	2.55	1.35	1.30	1.25	.90	.80	.90
14	......	1.28	1.70	1.60	2.20	4.00	1.80	1.30	1.20	.90	.80	.80
15	......	1.29	1.71	1.60	3.05	2.55	2.05	1.23	1.10	.90	.80	.80
16	......	1.26	1.70	1.60	2.00	1.90	1.90	1.20	1.00	.90	.80	.90
17	......	1.26	1.70	1.55	1.70	1.85	1.70	1.14	1.00	.85	.80	1.00
18	......	1.26	1.70	1.40	1.70	1.60	1.95	1.14	1.00	.90	.80	.90
19	......	1.27	1.71	1.45	2.90	1.60	1.72	1.13	.90	.90	.80	.80
20	......	1.23	1.50	1.50	3.90	1.64	1.55	1.24	.90	.90	.80	1.40
21	......	1.21	1.30	1.48	1.80	1.67	1.69	1.30	.90	.80	.70	1.10
22	......	1.27	1.30	1.50	1.70	1.83	1.59	1.30	1.00	.80	.70	.95
23	......	1.28	1.28	1.50	1.70	1.67	1.43	1.30	1.00	.80	.70	.90
24	......	1.25	1.33	1.50	1.70	1.55	1.38	1.40	1.00	.80	.70	.90
25	......	1.22	1.45	1.60	1.75	1.41	1.38	1.34	1.00	1.00	.70	.90
26	......	1.20	1.50	1.55	2.95	1.32	1.30	1.30	.95	.95	.70	.90
27	......	1.20	1.69	1.50	2.35	1.31	1.26	1.30	.90	.90	.70	.90
28	2.66	1.20	1.68	1.43	1.80	1.30	1.20	1.27	.90	.90	.70	.90
29	2.95	1.18	1.82	1.40	......	1.36	1.20	1.30	.90	.90	.70	.90
30	2.94	1.18	2.00	1.45	......	1.36	1.16	1.34	.90	1.00	.70	.90
31	2.35	......	2.00	1.50	......	1.38	......	1.30	......	.90	.80	......

NOTE.—Stage-discharge relation affected by ice during large part of period from December to February.

CANASERAGA CREEK AT GROVELAND STATION, N. Y.

LOCATION.—At highway bridge at Groveland Station, Livingston County. The creek is flowing through the improved channel at this point.

DRAINAGE AREA.—195 square miles measured by engineers of the New York State Conservation Commission.

RECORDS AVAILABLE.—August 5, 1915, to September 30, 1916, and March 1, 1917, to September 30, 1918.

GAGE.—Chain, near center of downstream side of bridge. Prior to March 30, 1916, inclined staff gage on right bank about 400 feet above the bridge, at practically the same datum (560.00 feet conservation commission datum); read by Thomas Maimone.

DISCHARGE MEASUREMENTS.—Made from highway bridge or by wading.

CHANNEL AND CONTROL.—Gravel; likely to shift.

EXTREMES OF DISCHARGE.—Maximum stage recorded during year, 19.01 feet at 7 a. m. February 13 (stage-discharge relation affected by ice, discharge not determined); minimum stage recorded, 6.3 feet at 6 p. m. August 20 and 30 (discharge about 22 second-feet).

1915–1918: Maximum open-water stage recorded 16.5 feet from 2 to 3 p. m. July 29, 1917 (discharge, 4,170 second-feet); minimum stage recorded, 6.3 feet at 6 p. m. August 20 and 30, 1918.

ICE.—Stage-discharge relation affected by ice; gage observations suspended during winter.

Accuracy.—Stage-discharge relation not permanent; affected by ice December to March and by shifting control during the rest of the year. Rating curve well defined between 35 and 3,000 second-feet. Gage read to half-tenths twice daily. Daily discharge ascertained by applying mean daily gage height to rating table, for the period previous to winter, and for the remainder of the year by the shifting-control method. Records fair.

Discharge measurements of Canaseraga Creek at Groveland Station, N. Y., during the year ending Sept. 30, 1918.

Date.	Made by—	Gage height.	Discharge.	Date.	Made by—	Gage height.	Discharge.
		Feet.	*Sec.-ft.*			*Feet.*	*Sec.-ft.*
Oct. 25	E. D. Burchard	12.59	1,200	Mar. 18	E. D. Burchard	8.91	314
25	.....do	12.50	1,190	21	.....do	9.30	394
31	.....do	10.42	678	May 24	.....do	7.61	118
31	.....do	10.30	637	June 23	.....do	7.30	88
Nov. 1	.....do	9.08	418	July 15	.....do	6.64	36
Mar. 16[a]	.....do	11.11	400	Aug. 24	.....do	6.52	29

[a] Slush ice in the current and flats below flooded, causing backwater.

Daily discharge, in second-feet, of Canaseraga Creek at Groveland Station, N. Y., for the year ending Sept. 30, 1918.

Day.	Oct.	Nov.	Dec.	Mar.	Apr.	May.	June.	July.	Aug.	Sept.
1	54	474	179		200	95	85	42	32	34
2	54	365	187		200	100	70	40	32	36
3	47	328	155		190	85	60	68	32	36
4	116	292	155		180	75	55	38	34	28
5	109	274	124		130	70	60	36	32	36
6	102	256	124		120	65	60	36	30	36
7	139	238	109		110	70	60	36	36	
8	124	204	179		120	170	55	36	28	
9	124	196			200	95	50	48	60	
10	95	196			140	120	55	70	32	
11	139	184			200	110	60	50	32	
12	109	171			140	120	200	42	32	
13	139	171			180	180	120	38	32	
14	95	155			650	140	90	36	28	
15	95	155			650	100	65	36	28	
16	139	163		400	420	95	60	34	28	
17	102	155		320	300	75	55	40	26	
18	83	147		300	550	65	44	42	26	36
19	460	155		300	320	65	44	36	28	80
20	536	139		360	260	100	44	32	22	130
21	256	155		380	240	190	48	32	28	65
22	204	155		400	220	170	95	32	28	46
23	171	163		300	200	170	75	32	28	40
24	975	139		240	190	110	65	40	28	40
25	1,610	139		220	190	85	60	65	28	36
26	1,090	155		200	150	260	48	42	26	35
27	1,320	139		170	110	300	50	86	28	40
28	1,000	139		190	100	160	44	34	24	38
29	1,130	124		170	90	150	44	34	24	36
30	1,490	139		200	90	120	42	40	22	36
31	675			200		110		35	24	

Note.—Discharge Dec. 9 to Mar. 15 not determined because of ice. Discharge Sept. 7-17 estimated at 36 second-feet.

Monthly discharge of Canaseraga Creek at Groveland Station, N. Y., for the year ending Sept. 30, 1918.

[Drainage area, 195 square miles.]

Month.	Discharge in second-feet.				Run-off (depth in inches on drainage area).
	Maximum.	Minimum.	Mean.	Per square mile.	
October	1,610	47	412	2.11	2.43
November	479	124	196	1.00	1.12
April	650	90	228	1.17	1.30
May	300	65	122	.626	.72
June	200	42	65.4	.335	.37
July	70	32	39.6	.203	.23
August	60	22	29.4	.151	.17
September	150	24	41.8	.214	.24

CANASERAGA CREEK AT SHAKERS CROSSING, N. Y.

LOCATION.—At highway bridge at Shakers Crossing, about a mile above mouth and 1½ miles northeast of Mount Morris, Livingston County.

DRAINAGE AREA.—347 square miles (measured by engineers of the New York State Conservation Commission).

RECORDS AVAILABLE.—Current-meter measurements 1904–1915; continuous record of gage height and occasional current-meter measurements July 13, 1915, to September 30, 1918.

GAGE.—Gurley seven-day water-stage recorder on the left bank, just below the bridge. Datum of gage same as that established on Genesee River at Jones Bridge near Mount Morris July 12, 1916 (540 feet conservation commission datum). Recorder inspected by Mrs. Wm. Russell.

DISCHARGE MEASUREMENTS.—Made from highway bridge or by wading.

CHANNEL AND CONTROL.—Firm gravel; not likely to shift; subject to backwater from Genesee River.

ICE.—Stage-discharge relation affected by ice.

EXTREMES OF STAGE.—Maximum stage during year, from water-stage recorder, 27.9 feet at 4 a. m. February 21; minimum stage from water-stage recorder, 7.86 at 6 p. m. August 31.

1915–1918: Maximum stage from water-stage recorder, 28.92 feet at 1 p. m. May 17, 1916; minimum stage from water-stage recorder 7.86 feet at 6 p. m. August 31, 1918.

Stage-discharge relation is affected by backwater from the Genesee River to such an extent that daily discharge has not been determined.

Discharge measurements of Canaseraga Creek at Shakers Crossing, N. Y., during the year ending Sept. 30, 1918.

Date.	Made by—	Gage height.	Discharge.	Date.	Made by—	Gage height.	Discharge.
		Feet.	*Sec.-ft.*			*Feet.*	*Sec.-ft.*
Nov. 1	E. D. Burchard	15.44	1,910	Feb. 14	E. D. Burchard	24.75	1,650
1	J. W. Moulton	14.74	1,623	Mar. 16	....do	22.82	5,620
2	....do	12.62	980	May 23	....do	9.79	421
Feb. 13[a]	E. D. Burchard	24.97	−1,640	July 15	....do	8.70	157

[a] Measurement shows flow upstream due to backwater flow from Genesee River caused by ice jam near Jones Bridge.

Daily gage height, in feet, of Canaseraga Creek at Shakers Crossing, N. Y., for the year ending Sept. 30, 1918.

Day.	Oct.	Nov.	Dec.	Jan.	Feb.	Mar.	Apr.	May.	June.	July.	Aug.	Sept.
1	8.63	14.90	9.98	9.48	....	20.15	9.71	9.36	9.10	8.63	8.22	8.29
2	8.65	12.32	9.88	9.46	....	22.61	9.77	9.33	8.94	8.61	8.18	8.29
3	8.68	11.26	9.52	9.52	....	21.15	9.75	9.26	8.85	8.54	8.16	8.23
4	8.96	10.71	9.48	9.42	....	17.98	9.66	9.24	8.75	8.40	8.00	8.26
5	9.31	10.29	9.41	9.42	....	16.48	9.53	9.20	8.75	8.52	8.20	8.18
6	9.17	10.11	9.42	9.31	9.29	18.95	9.40	9.18	8.75	8.41	8.32	8.35
7	8.98	10.01	9.32	9.41	9.35	18.16	9.31	9.13	8.79	8.42	8.26	8.36
8	8.95	9.88	9.45	9.40	9.39	14.30	9.32	9.77	8.74	8.46	8.20	8.40
9	9.14	9.74	9.48	9.35	9.50	11.64	9.67	9.42	8.65	8.37	8.36	8.48
10	9.02	9.56	8.88	9.48	9.83	16.28	9.51	9.63	8.69	8.59	8.36	8.45
11	8.90	9.42	9.72	9.74	10.81	18.39	....	10.10	8.70	8.61	8.39	8.45
12	8.93	9.52	9.78	9.52	15.94	18.36	....	9.68	9.49	8.54	8.41	8.37
13	9.39	9.46	9.85	9.51	20.97	19.90	9.70	9.68	9.38	8.50	8.50	8.35
14	9.25	9.42	9.89	9.52	24.19	24.02	11.94	9.82	9.01	8.49	8.46	8.40
15	9.16	9.42	9.88	9.55	24.53	26.68	15.01	9.53	8.90	8.49	8.41	8.61
16	9.66	9.58	9.72	9.56	24.03	22.14	12.99	9.33	8.76	8.37	8.40	8.51
17	9.32	9.70	9.68	9.70	20.97	16.76	11.78	9.22	8.79	8.32	8.44	8.96
18	9.03	9.54	9.68	9.82	17.65	13.47	14.04	9.08	8.74	8.27	8.42	9.02
19	9.96	9.58	9.62	9.60	15.81	11.96	12.35	9.02	8.70	8.21	8.41	8.56
20	17.43	9.55	9.62	9.53	25.23	12.20	10.89	9.17	8.63	8.16	8.30	8.95
21	13.01	9.62	10.02	9.58	26.88	12.19	10.38	9.90	8.62	8.16	8.20	9.95
22	10.45	9.56	10.45	9.55	23.56	12.02	10.70	9.47	8.98	8.05	8.15	8.64
23	9.79	9.52	9.90	9.50	21.94	11.45	10.50	9.70	9.09	8.07	8.14	8.75
24	13.20	9.50	10.17	9.47	17.59	....	10.14	9.57	8.97	8.05	8.16	8.91
25	20.48	9.42	10.94	9.49	16.75	....	9.92	9.20	8.80	8.53	8.20	8.71
26	18.38	9.46	10.87	9.51	23.40	10.08	9.68	10.53	8.72	8.22	8.22	8.70
27	19.96	9.49	10.10	9.55	23.90	9.86	9.49	10.79	8.64	8.16	8.15	8.74
28	22.26	9.52	9.80	9.50	21.54	9.82	9.32	9.78	8.64	8.14	8.09	8.75
29	21.12	9.65	9.50	9.58	....	9.84	9.20	9.38	8.63	8.20	8.07	8.70
30	22.69	9.72	9.32	9.58	....	9.74	9.28	9.50	8.64	8.17	8.15	8.69
31	20.01	....	9.48	....	....	9.72	....	9.31	....	8.20	8.09	....

NOTE.—Gage heights Oct. 20 and 21 estimated by comparison with records on Genesee River at Jones Bridge. Gage heights Nov. 16 to Dec. 18, and Dec. 29 to Jan. 16 from observations on staff gage.

KESHEQUA CREEK AT CRAIG COLONY, SONYEA, N. Y.

LOCATION.—About 200 feet downstream from private highway bridge on grounds of Craig Colony at Sonyea, Livingston County.

DRAINAGE AREA.—69 square miles (measured by engineers of the State conservation commission).

RECORDS AVAILABLE.—October 31, 1917, to September 30, 1918, at present site; July 22, 1910, to December 31, 1912, at a site about 200 feet upstream, and from August 29, 1915, to October 31, 1917, at a station about 1 mile downstream near the Delaware, Lackawanna & Western Railroad bridge.

GAGE.—Vertical staff gage in three sections on retaining wall on left bank just above the concrete weir; read by A. J. Porter.

DISCHARGE MEASUREMENTS.—Made from downstream side of the private highway bridge or by wading.

CHANNEL AND CONTROL.—Double-crested concrete weir built by Craig Colony for maintaining water level for their pumping plant; permanent.

EXTREMES OF DISCHARGE.—Maximum stage recorded during period of record at present station, 5.9 feet at 6.30 a. m. March 14 (discharge, about 3,700 second-feet); minimum stage recorded, 0.13 foot at 8 a. m. August 20 (discharge 0.7 second-foot).

ICE.—Stage-discharge relation slightly affected by ice.

ACCURACY.—Stage-discharge relation permanent, except when slightly affected by ice from December 10 to February 12 and by use of flashboards on downstream crest of dam, August 17–22. Rating curve well defined below 450 second-feet. Gage read to hundredths twice daily. Daily discharge, except for periods of backwater, determined by applying mean daily gage height to rating table. Records good.

Discharge measurements of Keshequa Creek at Craig Colony, Sonyea, N. Y., during the year ending Sept. 30, 1918.

Date.	Made by—	Gage height.	Discharge.	Date.	Made by—	Gage height.	Discharge.
		Feet.	*Sec.-ft.*			*Feet.*	*Sec.-ft.*
Oct. 26	E. D. Burchard	1.60	245	Feb. 15	E. D. Burchard	1.70	210
31	....do	1.33	151	Mar. 16	....do	1.30	156
Nov. 3	....do	1.00	68	May 24	J. W. Moulton	.64	21
Dec. 20[a]	....do	.87	22	June 23	E. D. Burchard	.52	14
Jan. 17[b]	....do	.66	11	July 15	....do	.30	3.4
Feb. 12[a]	....do	3.15	1,450	Aug. 21	....do	.19	1.3

[a] Measurement made through partial ice cover.
[b] Measurement made through complete ice cover.

Daily discharge, in second-feet, of Keshequa Creek at Craig Colony, Sonyea, N. Y., for the year ending Sept. 30, 1918.

Day.	Nov.	Dec.	Jan.	Feb.	Mar.	Apr.	May.	June.	July.	Aug.	Sept.
1	105	75	8	6	56	37	26	20	7.0	5.6	1.8
2	83	54	8	6	197	47	25	17	7.5	3.8	2.6
3	66	38	7	5	75	40	23	14	9.3	3.2	3.0
4	56	36	6	5	368	38	20	12	5.6	2.6	2.2
5	50	34	6	5	115	31	20	12	4.8	4.1	3.0
6	47	33	6	5	368	26	18	14	5.6	6.3	6.3
7	44	33	6	6	77	26	17	17	3.8	5.2	3.0
8	36	15	8	6	61	26	79	12	4.1	2.3	2.2
9	34	26	9	10	75	40	33	11	3.8	15	3.0
10	33	40	12	28	620	28	90	13	6.3	7.5	2.5
11	31	60	11	190	95	26	81	29	7.5	6.3	2.0
12	31	55	15	900	395	32	48	45	4.8	2.4	1.4
13	28	48	36	455	395	34	65	20	3.4	7.0	3.0
14	26	26	36	154	1,590	226	70	14	4.1	3.0	4.8
15	26	17	17	595	226	190	38	11	3.4	2.0	3.8
16	29	14	12	61	245	110	29	8.8	4.5	2.2	3.4
17	28	18	11	50	75	72	23	9.8	4.1	1.6	14
18	25	22	11	35	72	245	19	7.5	4.8	3.0	9.3
19	17	20	10	245	79	162	18	7.0	3.0	2.0	13
20	22	24	9	545	105	68	45	7.0	2.4	.8	21
21	23	110	9	36	112	60	44	7,9	2.2	1.3	15
22	28	90	9	29	112	128	23	28	2.6	1.4	5.2
23	28	32	8	83	75	72	40	12	2.6	1.4	6.3
24	22	46	8	43	51	61	23	12	3.0	1.1	4.8
25	21	110	8	68	51	50	21	7.9	15	1.0	7.5
26	24	26	8	425	47	41	207	7.5	5.9	.9	7.0
27	22	26	8	66	35	36	118	7.0	3.4	1.0	8.8
28	32	24	8	68	38	31	44	8.8	1.8	1.4	7.0
29	30	14	8	......	40	26	29	5.6	9.3	1.2	6.3
30	51	11	8	......	37	26	48	4.1	9.8	1.0	5.6
31	......	10	6	......	40	......	31	......	9.8	2.4	......

NOTE.—Discharge Dec. 10 to Feb. 12 estimated, because of ice, from discharge measurements, weather records, study of gage-height graph, and comparison with records for near-by streams.

Monthly discharge of Keshequa Creek at Craig Colony, Sonyea, N. Y., for the year ending Sept. 30, 1918.

[Drainage area, 69 square miles.]

Month.	Discharge in second-feet.				Run-off (depth in inches on drainage area).
	Maximum.	Minimum.	Mean.	Per square mile.	
November	105	17	36.6	0.534	0.6
December	110	10	38.3	.555	.64
January	36	6	10.7	.155	.16
February	900	5	146	2.12	2.21
March	1,590	35	191	2.77	3.1
April	245	26	65.8	.954	1.06
May	207	17	45.6	.661	.76
June	45	4.1	13.4	.194	.22
July	15	1.8	5.32	.077	.09
August	15	.8	3.23	.047	.05
September	21	1.4	5.96	.086	.10

OWASCO LAKE OUTLET NEAR AUBURN, N. Y.

LOCATION.—On farm of Charles H. Pearce, 2 miles below center of Auburn, Cayuga County, and 3¾ miles below State dam at outlet of Owasco Lake.

DRAINAGE AREA.—206 square miles (measured on topographic maps)

RECORDS AVAILABLE.—November 17, 1912, to September 30, 1918.

GAGE.—Gurley water-stage recorder in a concrete shelter on the left bank on the farm of Charles H. Pearce. Recorder inspected by Charles H. Pearce.

DISCHARGE MEASUREMENTS.—Made by wading directly opposite the gage in low water and from a cable at the same section in high water.

CHANNEL AND CONTROL.—A low concrete control has been constructed about 15 feet below the gage. Crest of control is 1 foot wide and the slopes of both upstream and downstream faces are ½:1. A small horizontal apron built on a level with the bed of the stream extends downstream 2½ feet from toe of dam. Mean elevation of the left end of the dam for a distance of 50 feet is at a gage height 1.28 feet; the remaining 50 feet of the crest of the dam is at a gage height 2.13 feet

EXTREMES OF DISCHARGE.—Maximum stage during year, from water-stage recorder, 3.5 feet at 3 a. m. March 17 (discharge 1,100 second-feet); minimum stage during year, from water-stage recorder, 1.48 feet at 10 and 11 p. m. October 7 (discharge 12 second-feet).

1912-1918: Maximum stage, 6.4 feet during period March 25–30, 1913, determined by leveling from flood marks (discharge, 2,750 second-feet); minimum stage from water-stage recorder, 1.41 feet at 1 a. m. October 15, 1915 (discharge, 5.6 second-feet).

ICE.—Stage-discharge relation seldom affected by ice.

DIVERSIONS.—An average flow of about 10 second-feet is pumped from Owasco Lake for the municipal water supply of Auburn. Proportion returning to stream above the gaging station is not known.

REGULATION.—Large diurnal fluctuation in flow during low-water periods due to operation of mills in Auburn; seasonal flow regulated at the State dam

ACCURACY.—Stage-discharge relation permanent except when affected by ice December 30 to January 10. Rating curve well defined between 1 and 1,700 second-feet. Operation of the water-stage recorder satisfactory throughout year, except as indicated in footnote to daily discharge table. Daily discharge ascertained by averaging the hourly discharge. Records good.

The following discharge measurement was made by E. D. Burchard July 11, 1918: Gage height, 2.43 feet; discharge, 254 second-feet.

Daily discharge, in second-feet, of Owasco Lake outlet near Auburn, N. Y., for the year ending Sept. 30, 1918.

Day.	Oct.	Nov.	Dec.	Jan.	Feb.	Mar.	Apr.	May.	June.	July.	Aug.	Sept.
1	181	534	271	......	a 175	434	484	439	414	......	194	134
2	184	509	267	......	a 170	420	471	407	393	......	205	149
3	176	522	261	......	125	353	468	394	401	......	195	157
4	181	525	274	......	154	317	458	389	402	......	166	156
5	188	507	263	......	167	340	422	332	369	......	191	159
6	158	499	267	......	174	352	434	284	337	......	205	147
7	48	474	261	......	139	427	361	298	332	......	204	141
8	155	458	256	......	139	506	237	284	286	......	209	132
9	209	a 455	254	......	130	526	239	269	206	......	196	149
10	212	a 445	245	......	114	558	202	214	190	......	a 190	150
11	212	a 435	251	145	135	796	301	265	181	188	a 185	145
12	273	a 425	263	172	160	937	335	265	192	166	a 180	155
13	213	a 420	250	124	157	923	336	276	171	168	a 180	177
14	205	a 412	196	203	156	921	429	203	179	160	a 180	188
15	219	a 404	249	150	169	931	524	203	161	168	176	96
16	211	a 396	244	168	160	946	539	267	160	171	194	137
17	211	388	a 225	156	162	921	560	203	184	162	168	181
18	211	330	a 220	161	181	876	596	205	147	161	185	167
19	244	322	214	158	184	784	643	196	171	a 165	175	159
20	206	324	205	195	199	748	640	194	175	171	177	188
21	64	315	206	198	175	718	612	174	242	181	175	169
22	232	303	194	176	189	689	622	194	315	191	183	77
23	359	309	202	171	170	591	628	348	258	178	179	124
24	376	298	202	178	173	a 578	626	396	235	184	163	80
25	428	296	193	153	214	a 580	580	390	226	a 185	174	137
26	457	285	202	181	262	a 570	522	373	......	a 190	a 175	134
27	468	289	202	175	304	a 560	495	398	......	a 190	173	126
28	452	287	149	152	351	a 550	526	402	......	a 190	167	116
29	492	278	202	140	......	a 530	516	407	......	a 190	167	92
30	537	264	a 202	146	......	515	480	402	......	a 195	162	138
31	545	......	a 200	131	......	504	......	414	......	a 195	172	......

a Estimated; no gage height record.

NOTE.—Discharge, Jan. 1-10, estimated 198 second-feet; June 26-30, 216 second-feet; July 1-10, 206 second feet.

Monthly discharge of Owasco Lake outlet near Auburn, N. Y., for the year ending Sept. 30, 1918.

[Drainage area, 206 square miles.]

Month.	Discharge in second-feet.				Run-off (depth in inches on drainage area).
	Maximum.	Minimum.	Mean.	Per square mile.	
October	545	48	268	1.30	1.50
November	534	264	390	1.89	2.11
December	274	149	229	1.11	1.28
January	203	124	174	.845	.97
February	351	114	178	.864	.90
March	946	317	626	3.04	3.51
April	643	202	476	2.31	2.58
May	439	174	306	1.49	1.72
June	414	147	247	1.20	1.34
July	....	160	187	.908	1.06
August	209	162	182	.883	1.02
September	188	77	142	.689	.77
The year	946	48	284	1.38	18.75

WEST BRANCH OF ONONDAGA CREEK AT SOUTH ONONDAGA, N. Y.

Location.—At highway bridge in South Onondaga, Onondaga County, about 1¾ miles above mouth of creek and 10 miles above Syracuse.

Drainage area.—20.8 square miles (measured on topographic maps).

Records available.—August 22, 1916, to June 30, 1918, when station was discontinued.

Gage.—Staff on downstream side of right abutment of bridge.

Discharge measurements.—Made from bridge or by wading.

Channel and control.—Fine and coarse gravel; probably shifting.

Extremes of stage.—Maximum stage recorded, 3.34 feet at 7.20 a. m., February 20; minimum stage recorded, 1 foot at 7.15 a. m. October 30.

1916–1918: Maximum stage recorded, 3.34 feet at 7.20 a. m. February 20, 1918; minimum stage recorded, 0.90 foot at 6.45 p. m. September 24 and 6.35 a. m. September 25, 1917.

Ice.—Stage-discharge relation probably affected by ice.

Data inadequate for determination of discharge.

The following discharge measurement was made by E. D. Burchard.

April 5, 1918: Gage height, 1.76 feet; discharge, 19 second-feet.

Daily gage height, in feet, of West Branch of Onondaga Creek at South Onondaga, N. Y., for the year ending Sept. 30, 1918.

Day.	Oct.	Nov.	Dec.	Feb.	Mar.	Apr.	May.	June.
1	1.18	1.62	1.60		1.84	1.88	1.73	1.37
2	1.70	1.52	1.47		1.86	1.89	1.67	1.33
3	1.40	1.45	1.38		1.88	1.86	1.61	1.30
4	1.48	1.39	1.35		1.79	1.80	1.65	1.31
5	1.41	1.37	1.34		2.11	1.79	1.59	1.31
6	1.38	1.33	1.27		2.63	1.75	1.58	1.38
7	1.14	1.36	1.36		2.06	1.74	1.55	1.69
8	1.59	1.28	1.16		1.95	1.81	1.53	1.41
9	1.29	1.28	1.22		1.83	2.45	1.52	1.34
10	1.15	1.27			2.53	1.93	1.59	1.44
11	1.11	1.27			2.11	1.98	1.60	1.34
12	1.16	1.27		2.34	2.24	1.95	1.57	1.86
13	1.20	1.24		2.49	2.43	2.39	1.79	1.65
14	1.13	1.18		2.26	3.03	2.15	1.78	1.51
15	1.24	1.19		3.47	2.75	1.93	1.58	1.44
16	1.15	1.25		2.72	2.46	1.85	1.51	1.35
17	1.09	1.26			2.49	1.84	1.47	1.33
18	1.60	1.21			2.21	2.34	1.43	1.32
19	1.68	1.27		2.04	2.19	2.04	1.40	1.29
20	1.82	1.30		3.15	2.21	1.89	1.57	1.28
21	1.31	1.26		2.17	2.22	1.92	1.78	1.28
22	1.24	1.39		2.41	2.17	2.60	1.54	1.51
23	1.25	1.38		2.06	2.27	1.96	1.61	1.67
24	1.50	1.30		1.76	1.97	1.93	1.45	1.45
25	2.17	1.20		2.01	1.97	1.83	1.45	1.37
26	1.77	1.24		3.14	1.94	1.78	1.54	1.33
27	1.51	1.20		2.35	1.92	1.73	1.50	1.28
28	1.65	1.23		2.05	1.85	1.69	1.45	1.29
29	1.59	1.24			1.89	1.63	1.46	1.49
30	1.96	1.32			1.88	1.64	1.44	1.80
31	1.79				1.88		1.39	

BLACK RIVER NEAR BOONVILLE, N. Y.

Location.—At highway bridge 1 mile above mouth of Sugar River, 2 miles northeast of Boonville, Oneida County, and 2 miles by river downstream from Hawkinsville.

Drainage area.—303 square miles (measured on topographic maps).

Records available.—February 16, 1911, to June 30, 1918.

Gage.—Chain, near center of left span, downstream side of bridge. Staff gage, graduated from 6 to 13 feet, on downstream side of right abutment, used for high water readings. Gage read by W. D. Charbonneau.

Discharge measurements.—Made from cable about half a mile above gage or by wading near cable.

Channel and control.—Rough; full of boulders; permanent.

Extremes of discharge.—Maximum stage recorded during year, 9.6 feet at 5 p. m. October 31 (discharge, 4,960 second-feet); minimum stage recorded, 2.40 feet at 5 p. m. August 26 (discharge about 5 second-feet).

1911–1918: Maximum stage about 12.5 feet during night of March 28, 1913 (determined by leveling from flood mark), discharge about 10,000 second-feet. Minimum stage recorded, that of August 26, 1918.

Ice.—Stage-discharge relation affected by ice.

Regulation and diversion.—The State dam at Forestport, about 8 miles upstream, provides a reservoir with a capacity of about 2 billion cubic feet. During the navigation season water is diverted westward from this reservoir through the Forestport feeder to a storage basin in Boonville. The Black River canal flows north from this basin, entering Black River at the foot of Lyons Falls. A spillway from the basin overflows into Mill Creek, a tributary of Black River. Water flowing through this spillway and through Black River canal returns to the river below the gaging station, thus passing around it. The Black River canal also flows south from Boonville, passing out of the Black River drainage basin and entering the summit level of the Erie Canal (or Barge Canal) at Rome.

Occasional discharge measurements have been made at three points to indicate the distribution of the diverted water. The water entering Boonville through the Forestport feeder has been measured at the highway bridge about a mile northeast of Boonville. During October, 1915, two water-stage recorders were installed on this canal to obtain a continuous record of the flow. This is published as a separate station—"Forestport feeder near Boonville, N. Y." The water flowing north from the basin through the Black River canal has been measured at the highway bridge just below the lock into this canal near the railroad station. The water flowing south from the basin has been measured at a private farm bridge about 1 mile southeast of Boonville. During September, 1915, two water-stage recorders were installed on this canal to obtain a continuous record of the flow. This is published as a separate station under the heading "Black River Canal, flowing south, near Boonville, N. Y."

Accuracy.—Stage-discharge relation practically permanent except as affected by ice December 10 to March 24. Rating curve well defined between 35 and 2,800 second-feet and fairly well defined between 2,800 and 4,500 second-feet. Gage read to hundredths twice daily. Daily discharge ascertained by applying mean daily gage height to rating table. Records good, except those for period of ice effect which are fair.

Discharge measurements of Black River near Boonville, N. Y., during the year ending Sept. 30, 1918.

Date.	Made by—	Gage height.	Discharge.	Date.	Made by—	Gage height.	Discharge.
		Feet.	*Sec.-ft.*			*Feet.*	*Sec.-ft.*
Dec. 19[a]	J. W. Moulton	5.67	318	Mar. 19[a]	J. W. Moulton	6.85	574
Jan. 11[a]	E. D. Burchard	4.69	170	Apr. 13	E. D. Burchard	6.70	1,400
Feb. 9[b]	J. W. Moulton	4.85	173	June 6	M. H. Carson	3.65	92
Mar. 14[a]	.....do	7.08	586				

[a] Measurement made through partial ice cover. [b] Measurement made through complete ice cover.

Daily discharge, in second-feet, of Black River near Boonville, N. Y., for the year ending Sept. 30, 1918.

Day.	Oct.	Nov.	Dec.	Jan.	Feb.	Mar.	Apr.	May.	June.	July.	Aug.	Sept.
1	90	4,140	680	220	120	1,200	2,500	1,290	145	194	28	119
2	78	1,940	630	220	110	850	2,385	1,210	154	250	42	90
3	127	1,370	558	220	120	800	2,500	1,135	136	216	46	66
4	250	1,210	450	220	120	850	2,385	1,210	111	205	49	72
5	305	1,060	335	200	100	900	2,160	1,210	97	154	49	90
6	490	920	227	190	90	800	2.160	1,060	97	63	44	70
7	735	795	227	190	130	650	2,160	920	430	66	36	174
8	680	855	238	190	160	550	2,270	855	920	154	24	227
9	605	680	250	180	180	600	2,270	795	630	558	28	194
10	580	535	260	170	200	700	2,620	785	305	855	56	184
11	580	335	280	220	240	600	2,385	795	194	795	84	227
12	558	275	300	440	300	500	1,740	795	164	535	70	275
13	795	250	300	280	460	480	1,455	1,060	154	174	56	305
14	1,140	194	320	300	480	600	1,740	1,545	97	145	61	410
15	1,140	512	320	340	550	600	1,945	1,370	90	535	46	535
16	1,140	1,540	320	280	550	850	2,060	1,210	63	430	44	450
17	1,060	1,540	320	240	460	1,200	1,945	855	56	262	49	410
18	855	1,210	320	200	440	1,200	1,740	920	40	205	59	680
19	795	1,140	320	190	550	1,000	1,545	795	36	154	70	795
20	795	855	320	200	650	800	1,840	680	38	127	59	990
21	795	735	300	180	900	1,600	1,945	735	35	104	46	1,140
22	795	735	280	200	1,100	2,400	1,740	855	205	111	33	855
23	735	630	280	200	1,100	2,400	1,545	795	470	63	27	795
24	795	512	260	180	1,200	2,200	1,545	680	370	66	21	795
25	795	450	260	150	1,200	2,160	1,370	680	290	49	11	735
26	795	430	240	140	1,400	2,060	1,370	680	262	30	7	795
27	855	450	220	140	1,700	2,160	1,210	795	227	40	10	795
28	920	450	220	120	1,900	2,380	990	1,060	164	44	26	735
29	1,940	512	220	95	....	2,270	1,060	855	84	49	53	690
30	3,750	535	220	100	....	2,160	1,210	680	275	30	70	605
31	4,820	....	200	110	....	2,380	....	227	....	36	84	....

NOTE.—Discharge Dec. 10 to Mar. 24 estimated, because of ice, from discharge measurements, weather records and study of gage-height graph.

Monthly discharge of Black River near Boonville, N. Y., for the year ending Sept. 30, 1918.

[Drainage area, 303 square miles.]

Month.	Discharge in second-feet.				Run-off (depth in inches on drainage area).
	Maximum.	Minimum.	Mean.	Per square mile.	
October	4,820	78	961	3.17	3.66
November	4,140	194	893	2.94	3.28
December	680	200	312	1.03	1.19
January	440	95	203	.670	.77
February	1,900	90	590	1.95	2.03
March	2,400	480	1,290	4.26	4.91
April	2,620	990	1,860	6.14	6.85
May	1,540	227	919	3.03	3.47
June	920	35	211	.696	.78
July	855	30	216	.713	.82
August	84	7	44.8	.148	.17
September	1,140	66	476	1.57	1.75
The year	4,820	7	663	2.19	29.68

NOTE.—Water diverted past this station by the Forestport feeder not included in the above table.

BLACK RIVER AT BLACK RIVER, N. Y.

Location.—About one-fourth mile below concrete-arch highway bridge and the power plant of Northern New York Utilities Co., and three-fourths mile below village of Black River, Jefferson County.

Drainage area.—1,870 square miles (measured on topographic maps).

Records available.—March 24, 1917, to September 30, 1918.

Gage.—Vertical staff, in two sections, spiked to large cedar tree on the left bank one-fourth mile below highway bridge; a low-water section fastened to rocks 10 feet upstream; read by Erwin W. Hart.

Discharge measurements.—Made from a cable 100 yards above the gage.

Channel and control.—Solid rock.

Extremes of discharge.—Maximum stage recorded, 12.3 feet at 8.40 a. m. April 4 (discharge, 16,300 second-feet); minimum discharge, 440 second-feet, January 20.

1917–1918: Maximum stage recorded 13.4 feet from 6 p. m., April 4, to 7 a. m., April 5, 1917 (discharge, 19,300 second-feet); minimum stage recorded, 1.05 feet at 2.45 p. m. Sunday, July 29, 1917, during a current-meter measurement (discharge about 16 second-feet).

Ice.—Stage-discharge relation affected by ice.

Regulation.—Seasonal distribution of flow is regulated by Beaver River flow, Fulton Chain Lakes, Forestport reservoir, and other storage reservoirs in the upper part of the drainage basin. Some diurnal fluctuation at low stages due to mills and power plants above the station.

Diversions.—Water is diverted from Black River into the Forestport feeder at Forestport. A part of this water returns to the river through various spillways and through the Black River canal (flowing north); the rest passes out of the drainage basin through the Black River canal (flowing south), the record at the station on Black River canal (flowing south) at Boonville indicates the amount of this diversion. See also "Regulation and diversion" in description of station on Black River near Boonville.

Accuracy.—Stage-discharge relation permanent except as affected by ice December 7 to February 19. Rating curve well defined between 500 and 18,000 second-feet. Gage read to tenths twice daily. Daily discharge ascertained by applying mean daily gage height to rating table. Records good except for days of low discharge when they may be poor.

Discharge measurements of Black River at Black River, N. Y., during the year ending Sept. 30, 1918.

Date.	Made by—	Gage height.	Discharge.	Date.	Made by—	Gage height.	Discharge.
		Feet.	*Sec.-ft.*			*Feet.*	*Sec.-ft.*
Jan. 14[a]	E. D. Burchard	5.78	1,340	Mar. 18	J. W. Moulton	6.20	3,930
Feb. 18[a]	J. W. Moulton	5.28	1,370	Apr. 6	E. D. Burchard	11.32	14,300
Mar. 14	.....do	6.20	3,760				

[a] Measurement made through partial ice cover.

Daily discharge, in second-feet, of Black River at Black River, N. Y., for the year ending Sept. 30, 1918.

Day.	Oct.	Nov.	Dec.	Jan.	Feb.	Mar.	Apr.	May.	June.	July.	Aug.	Sept.
1	1,570	10,400	2,360	1,400	1,800	7,420	12,600	5,840	2,730	1,470	1,100	690
2	2,010	12,600	2,240	2,400	1,700	7,240	13,100	6,520	2,600	1,680	1,100	745
3	,010	11,100	,480	2,200	1,200	6,700	14,600	7,060	2,730	3,390	1,100	690
4	,010	9,570	,360	1,800	1,400	6,010	16,000	6,880	1,900	3,250	1,100	950
5	,480	8,370	,240	2,400	1,600	5,510	15,300	6,350	1,900	2,120	1,020	1270
6	2,990	6,520	1,900	1,700	1,200	4,870	13,800	5,840	1,570	1,900	950	950
7	3,120	5,030	1,790	2,200	1,200	4,550	13,100	4,870	2,360	2,120	1,100	810
8	,250	4,550	1,500	,400	1,500	4,100	10,800	5,840	4,710	1,680	845	712
9	,860	3,950	1,300	,000	950	3,670	10,400	6,180	5,670	2,240	1,100	1,020
10	,990	3,670	1,500	,200	750	3,580	10,800	6,180	4,710	3,390	950	1,100
11	2,600	3,580	1,700	1,700	1,100	3,120	11,100	7,240	8,950	2,120	1,900	950
12	2,360	3,390	2,600	850	1,400	2,990	11,100	7,240	4,870	1,680	1,370	1,100
13	3,120	3,120	,400	850	1,700	3,250	9,990	7,610	5,510	1,900	950	1,100
14	4,550	2,860	,000	1,500	2,200	3,950	9,570	7,610	5,840	2,360	1,100	1,790
15	5,190	2,600	,000	1,300	3,400	4,100	9,170	7,610	5,350	3,120	950	810
16	5,510	2,360	1,600	1,400	3,600	3,950	8,570	8,370	4,870	3,670	880	650
17	4,870	2,730	2,200	1,800	4,200	3,670	8,770	7,610	4,250	2,990	1,180	560
18	4,400	2,600	,200	1,300	4,400	3,810	8,570	6,700	2,730	2,480	1,570	1,470
19	3,670	2,860	,000	650	4,600	4,550	9,370	5,840	2,360	2,600	1,370	3,120
20	6,180	3,120	,500	440	6,500	6,010	9,780	4,870	2,360	2,480	950	3,390
21	7,240	2,990	1,300	1,200	6,880	8,180	9,780	3,810	2,240	1,900	1,470	4,250
22	6,520	3,250	1,200	850	6,880	9,570	9,570	4,870	2,120	1,790	1,100	4,710
23	,840	3,250	1,000	1,700	5,840	10,800	9,780	5,030	1,790	1,680	1,370	4,400
24	,710	3,390	1,000	1,700	5,510	11,100	9,990	5,030	1,900	1,370	1,680	4,870
25	,840	2,600	1,300	,400	5,190	11,100	9,570	4,400	2,240	1,370	1,270	3,670
26	6,180	2,360	2,200	1,700	7,240	10,800	8,770	3,670	2,120	1,470	1,180	2,730
27	6,180	2,360	1,800	750	7,420	10,600	7,800	4,550	1,900	1,270	880	2,480
28	5,840	1,900	1,800	480	7,990	9,570	6,180	4,870	1,680	810	1,100	3,120
29	5,840	2,010	2,000	850	......	8,770	5,840	4,710	1,900	1,100	1,100	3,530
30	6,880	2,010	1,600	1,000	......	8,770	5,510	4,250	1,680	1,270	1,370	3,390
31	9,170	......	2,200	1,500	......	9,990	......	4,250	......	1,020	950	......

NOTE.—Discharge Dec. 7 to Feb. 19, estimated because of ice from discharge measurements, weather records study of gage-height graph, and comparison with records for Black River near Boonville.

Monthly discharge, of Black River at Black River, N. Y., for the year ending Sept. 30, 1918.

[Drainage area, 1,870 square miles.]

Month.	Discharge in second-feet.				Run-off (depth in inches on drainage area).
	Maximum.	Minimum.	Mean.	Per square mile.	
October	9,170	1,570	4,450	2.38	2.74
November	12,600	1,900	4,370	2.34	2.61
December	2,600	1,000	1,850	.989	1.14
January	2,400	440	1,410	.754	.87
February	7,990	750	3,550	1.90	1.98
March	11,100	2,990	6,520	3.49	4.02
April	16,000	5,510	10,300	5.51	6.15
May	8,370	3,670	5,860	3.13	3.61
June	5,840	1,570	3,080	1.65	1.84
July	3,670	810	2,050	1.10	1.27
August	1,900	845	1,160	.620	.71
September	4,870	560	2,030	1.09	1.22
The year	16,000	440	3,880	2.07	28.16

FORESTPORT FEEDER NEAR BOONVILLE, N. Y.

LOCATION.—At lower end of feeder, above point at which it enters basin at Boonville.

RECORDS AVAILABLE.—Occasional current-meter measurements 1900 and 1905–1915; continuous record October 30, 1915, to September 30, 1918.

GAGES.—Two Gurley seven-day water-stage recorders, with natural scale for gage heights. Gage No. 1 is at the downstream end of the left abutment of the steel highway bridge in the village of Hawkinsville; gage No. 2 is on the left bank just below a farm bridge, about a mile above the basin at Boonville; the gages are about 2.53 miles apart. These gages and the two gages on Black River canal (flowing south) near Boonville are all set at the same datum. Recorder at gage No. 1 is inspected by Mrs. Anna Zwahlen and Charles Nugent; that at gage No. 2 is inspected by Charles Nugent.

DISCHARGE MEASUREMENTS.—Made from steel highway bridge at gage No. 1 in Hawkinsville.

DETERMINATION OF DISCHARGE.—Daily discharge determined by Chezy formula. The coefficient, c, computed from each current-meter measurement, is plotted with reference to the date of measurement. A smooth curve drawn through the plotted points shows the variation of c through the season, and the coefficient for each day is taken off the curve. The other factors in the Chezy formula are obtained from gage-height records and the cross section of the canal.

DIVERSIONS.—A spillway takes water from the feeder just below gage No. 2, discharging it into Mill Creek, which enters Black River below the gaging station at Boonville. Other spillways above Hawkinsville discharge into Black River above the gaging station. There are no spillways between gage No. 1 and gage No. 2. The sum of the flow at this station and that of Black River near Boonville indicates the total run-off of Black River above the station near Boonville. The way in which water is diverted from Black River is briefly described under "Black River near Boonville" (pp. 66–67).

ICE.—There is usually no water in the canal during the winter, but water was observed in the canal several times during the winter of 1917–18, and occasional current-meter measurements of the discharge were made. See table of discharge measurements.

ACCURACY.—Records good except for days on which the discharge varies widely from the mean, for which they are fair.

Discharge measurements of Forestport feeder near Boonville, N. Y., during the year ending Sept. 30, 1918.

Date.	Made by—	Gage height (feet).		Discharge (second-feet).	Date.	Made by—	Gage height (feet).		Discharge (second-feet).
		Gage No. 1.	Gage No. 2.				Gage No. 1.	Gage No. 2.	
Oct. 25	O. W. Hartwell	3.254	1.934	239	June 27	J. W. Moulton	3.002	1.592	241
Nov. 13	E. D. Burchard	3.240	1.877	262	27	do	3.026	1.625	246
13	do	3.239	1.876	262	July 18	do	3.122	1.776	237
Feb. 9[a]	J. W. Moulton			60	18	do	3.124	1.779	243
Mar. 14[a]	do			21	Aug. 15	C. C. Covert	3.044	1.724	201
19[a]	do			23	Sept. 7	do	3.526	2.005	254
Apr. 13	M. H. Carson			40	20	O. W. Hartwell	3.627	2.057	291
June 6	do	3.222	1.858	281					

[a] Measurement made through complete ice cover.

Daily discharge, in second-feet, of Forestport feeder near Boonville, N. Y., for the year ending Sept. 30, 1918.

Day.	Oct.	Nov.	June.	July.	Aug.	Sept.	Day.	Oct.	Nov.	June.	July.	Aug.	Sept.
1	266	235	134	246	238	215	16	......	......	238	243	197	247
2	255	227	194	292	221	240	17	......	......	236	236	195	243
3	250	246	238	307	216	229	18	......	......	235	238	179	240
4	246	239	226	255	207	221	19	......	......	255	234	220	219
5	248	271	237	254	212	228	20	261	......	240	224	217	260
6	246	284	259	245	221	239	21	255	......	226	215	227	252
7	238	288	261	240	222	251	22	252	......	212	230	227	246
8	230	300	251	238	224	237	23	235	......	254	239	229	243
9	214	262	228	252	225	200	24	273	......	265	226	230	214
10	257	263	230	265	233	193	25	246	......	256	217	222	206
11	......	264	224	264	220	206	26	230	......	244	206	218	206
12	......	261	230	240	219	242	27	217	......	240	223	215	251
13	......	261	248	226	220	214	28	238	......	240	198	209	240
14	......	257	259	249	213	209	29	238	......	228	205	221	213
15	......	......	248	257	203	238	30	258	......	212	208	231	196
							31	250	......	......	235	238	......

NOTE.—Discharge, Oct. 11-19, estimated at 240 second-feet; Nov. 15-30, 250 second-feet.

Monthly discharge, in second-feet, of Forestport feeder near Boonville, N. Y., for the year ending Sept. 30, 1918.

Month.	Maximum.	Minimum.	Mean.	Month.	Maximum.	Minimum.	Mean.
October	273	214	244	July	307	198	239
November	300	227	255	August	238	179	218
June	265	134	235	September	260	193	228

BLACK RIVER CANAL (FLOWING SOUTH) NEAR BOONVILLE, N. Y.

LOCATION.—Slope station in summit level of Black River canal near Boonville, Oneida County.

RECORDS AVAILABLE.—Occasional discharge measurements 1900, 1905 to 1915. Continuous record September 16, 1915, to September 30, 1918.

GAGES.—Gurley seven-day water-stage recorders with natural scale for gage heights, 1.81 miles apart. These gages and two gages in the Forestport feeder near Boonville are all set at the same datum. Gage No. 1 is located on the right bank (opposite tow path) about 50 feet downstream from the collector's office in Boonville. Gage No. 2 is located on the right bank opposite tow path) about 300 yards above Lock 70 and 50 yards above the spillway from the canal in Lansing Kill. Recorders inspected by Philip Joynt and Charles Nugent.

DISCHARGE MEASUREMENTS.—Made from the steel and concrete highway bridge in the village of Boonville, a short distance below Gage No. 1.

DETERMINATION OF DISCHARGE.—Daily discharge determined by use of Chezy formula. The coefficient, *c*, computed from each current measurement is plotted with reference to date of measurement. A smooth curve, then drawn through the plotted points, shows the variation of *c* through the season and the coefficient for each day is taken off the curves. The other factors in Chezy formula are obtained from gage-height records and cross section of canal.

DIVERSIONS.—There are no diversions between gage No. 1 and gage No. 2. This station indicates the amount of water diverted from the Black River drainage into the Mohawk River drainage for canal purposes. For brief description of way in which water is diverted from Black River, see "Black River near Boonville."

GULATION.—Flow in the canal is regulated by the operation of the spillway and sluice gates at Lock 70 and also by discharge of Forestport feeder into the basin at Boonville.

t.—No flow in the canal during the frozen season.

CURACY.—Records good.

charge measurements of Black River canal (flowing south) near Boonville, N. Y., during the year ending Sept. 30, 1918.

ate.	Made by—	Gage height (feet).		Discharge. (sec.-ft.)	Date.	Made by—	Gage height (feet).		Discharge (sec.-ft.).
		Gage No. 1.	Gage No. 2.				Gage No. 1.	Gage No. 2.	
t. 26	O. W. Hartwell	1.465	1.200	151	June 27	J. W. Moulton	1.457	1.345	126
v. 13	E. D. Burchard	1.550	1.286	175	27	.....do	1.395	1.328	111
13	.....do	1.526	1.279	168	27	.....do	1.285	1.085	163
13	.....do	1.500	1.278	168	July 18	.....do	1.462	1.262	156
14	.....do	1.506	1.291	170	18	.....do	1.456	1.255	153
14	.....do	1.502	1.285	165	Aug. 16	C. C. Covert	1.486	1.196	164
ne 7	M. H. Carson	1.415	1.258	146	Sept. 20	O. W. Hartwell	1.62	1.29	168

ily discharge, in second-feet, of Black River canal (flowing south) near Boonville, N. Y., for the year ending Sept. 30, 1918.

Day.	Oct.	Nov.	June.	July.	Aug.	Sept.	Day.	Oct.	Nov.	June.	July.	Aug.	Sept.
	182	173	100	165	159	167	16			177	150	155	144
	184	182	217	177	163	173	17			173	149	158	160
	179	175	202	181	153	167	18			173	145	143	173
	179	177	205	154	166	159	19			173	143	162	158
	173	197	195	155	165	173	20			176	136	162	166
	179	199	227	140	150	155	21	148		178	133	162	159
	186	184	182	157	153	160	22	157		159	151	162	154
	179	199	194	153	160	161	23	166		170	167	169	144
	178	192	180	157	169	166	24	176		171	162	166	136
	183	184	179	160	168	156	25	171		165	162	169	142
	194	180	184	168	162	153	26	170		163	151	161	132
	173	180	188	140	158	183	27	171		151	165	157	166
	171	185	184	138	164	166	28	176		161	158	160	165
		181	195	153	157	158	29	186		157	156	164	143
			182	156	157	153	30	198		146	149	165	139
							31	182			152	172	

NOTE.—Discharge estimated as follows: Oct. 14–20, 175 second-feet; Nov. 15–30, 180 second-feet.

onthly discharge, in second-feet, of Black River canal (flowing south) near Boonville, N. Y., for the year ending Sept. 30, 1918.

Month.	Maximum.	Minimum.	Mean.	Month.	Maximum.	Minimum.	Mean.
ctober	198	148	176	July	181	133	154
ovember	199	173	182	August	172	143	161
ne	227	100	177	September	183	132	158

MOOSE RIVER AT MOOSE RIVER, N. Y.

Location.—In village of Moose River, Lewis County, about 3 miles downstream from McKeever, 5 miles below mouth of South Branch of Moose River and nearly 20 miles above junction of Black and Moose rivers at Lyons Falls.

Drainage area.—370 square miles (measured on topographic maps).

Records available.—June 5, 1900, to September 30, 1918.

Gage.—Staff in two sections on the left bank; read by H. W. Hoch. The gage datum was lowered 0.17 foot on February 28, 1903, and again 5.00 feet on January 1, 1913.

Discharge measurements.—Made from a cable a short distance below the gage.

Channel and control.—Cobblestones and boulders; fairly permanent. Current smooth, depth comparatively uniform.

Extremes of discharge.—Maximum stage recorded during year, 12.8 feet at 8 a. m. October 31 (discharge, 6,680 second-feet); minimum discharge 65 second-feet January 31.

1900–1918: Maximum stage recorded, 16.3 feet during the afternoon of March 27, 1913, determined by leveling from flood marks (discharge about 16,500 second-feet); minimum stage recorded 4.94 feet July 21, 23, 25, 26, and 27, 1913 (discharge about 42 second-feet).

Ice.—Stage-discharge relation affected by ice.

Regulation.—A timber dam at McKeever, 3 miles upstream, is used for power and for the regulation of flow during log driving. Seasonal flow affected by operation of the State dam at Old Forge. This regulation is indicated by a record from station "Middle Branch of Moose River at Old Forge."

Accuracy.—Stage-discharge relation practically permanent except as affected by ice December 8 to April 16. Rating curve fairly well defined between 100 and 5,500 second-feet. Gage read to half-tenths twice daily. Daily discharge ascertained by applying mean daily gage height to rating table. Records fairly good except for periods of ice effeet or low discharge, for which they are fair.

Discharge measurements of Moose River at Moose River, N. Y., during the year ending Sept. 30, 1918.

Date.	Made by—	Gage height.	Discharge.	Date.	Made by—	Gage height.	Discharge.
		Feet.	*Sec.-ft.*			*Feet.*	*Sec.-ft.*
Oct. 5	E. D. Burchard	6.61	488	Mar. 13[b]	J. W. Moulton	8.63	568
Dec. 18[a]	J. W. Moulton	6.50	277	Apr. 12	E. D. Burchard	9.08	1,910
Jan. 10[b]	E. D. Burchard	6.70	151	12	M. H. Carson	8.99	1,820
Feb. 8[b]	J. W. Moulten	8.0	284				

[a] Measurement made through partial ice cover.
[b] Measurement made through complete ice cover.

Daily discharge, in second-feet, of Moose River at Moose River, N. Y., for the year ending Sept. 30, 1918.

Day.	Oct.	Nov.	Dec.	Jan.	Feb.	Mar.	Apr.	May.	June.	July.	Aug.	Sept.
1	460	2,660	441	360	170	1,100	800	2,760	625	369	202	540
2	369	2,000	441	280	240	950	1,500	2,860	670	1,320	216	560
3	274	1,830	404	240	180	650	2,600	2,180	540	810	230	422
4	441	1,590	386	300	110	950	2,600	1,910	422	715	230	189
5	580	1,320	336	280	200	700	2,400	1,560	369	625	176	352
6	810	1,200	352	260	220	600	2,000	1,560	369	404	151	259
7	715	1,140	220	260	75	550	1,800	1,520	670	386	151	230
8	580	1,020	340	200	280	600	2,200	2,180	1,080	386	151	422
9	500	965	380	190	110	550	2,600	2,270	760	422	126	422
10	500	965	400	150	170	750	2,400	2,270	860	441	189	259
11	460	910	440	220	160	600	2,200	2,180	860	715	230	176
12	386	860	400	220	180	600	,900	1,830	860	760	176	336
13	965	860	550	220	200	550	1,500	2,180	1,200	810	202	386
14	1,080	760	340	179	360	600	1,200	3,170	1,020	860	164	460
15	860	715	360	240	360	550	1,500	2,460	860	810	151	441
16	1,260	625	440	260	400	600	1,900	1,910	760	670	259	352
17	1,140	670	420	240	380	500	3,060	1,670	670	580	336	860
18	860	670	280	360	400	600	3,170	1,260	625	670	320	1,260
19	715	670	440	260	340	600	3,060	1,260	500	625	202	1,260
20	1,830	670	340	180	550	550	2,560	1,020	404	540	151	1,140
21	1,380	540	280	280	700	750	2,360	1,260	404	386	164	1,200
22	1,200	500	280	440	1,110	850	2,860	1,140	500	336	189	1,380
23	1,080	500	280	240	950	1,200	2,860	1,080	670	289	289	1,140
24	965	500	240	100	750	1,200	2,460	810	810	274	274	1,140
25	860	500	240	220	850	1,100	2,180	860	580	259	230	1,080
26	1,140	460	180	150	850	1,000	1,910	860	580	230	259	1,020
27	1,080	500	420	180	1,100	900	1,830	860	369	216	259	1,680
28	1,200	441	400	170	1,100	750	1,910	1,080	320	244	259	1,260
29	1,590	404	320	360	....	700	2,090	910	336	230	230	1,140
30	2,660	500	300	70	....	700	2,360	810	176	244	259	910
31	5,170	....	280	65	....	700	....	715	....	274	230	....

NOTE.—Discharge Dec. 8 to Apr. 16 estimated, because of ice, from discharge measurements, weather records, study of gage-height graph, and comparison with records for Black River near Boonville.

Monthly discharge of Moose River at Moose River, N. Y., for the year ending Sept. 30, 1918.

[Drainage area, 370 square miles.]

Month.	Discharge in second-feet.				Run-off (depth in inches on drainage area).
	Maximum.	Minimum.	Mean.	Per square mile.	
October	5,170	274	1,070	2.89	3.33
November	2,660	404	900	2.43	2.71
December	550	180	353	.954	1.10
January	440	65	231	.624	.72
February	1,100	75	446	1.21	1.26
March	1,200	500	742	2.01	2.32
April	3,170	800	2,190	5.92	6.61
May	3,170	715	1,680	4.41	5.08
June	1,200	176	629	1.70	1.90
July	1,320	216	513	1.39	1.60
August	336	126	215	.581	.67
September	1,380	176	719	1.94	2.16
The year	5,170	65	802	2.17	29.46

MIDDLE BRANCH OF MOOSE RIVER AT OLD FORGE, N. Y.

Location.—About 300 feet below highway bridge and 400 feet below State dam at Old Forge, Herkimer County.

Drainage area.—51.5 square miles (measured on topographic maps).

Records available.—November 9, 1911, to September 30, 1918.

Gage.—Vertical staff on left bank, 300 feet below highway bridge; read by Jacob Edick.

Discharge measurements.—Made from highway bridge or by wading.

Channel and control.—Bed, near the gage, composed of stone and gravel. Control is rock ledge about 200 feet below gage; practically permanent.

Extremes of discharge.—Maximum stage recorded during year, 4.0 feet at 8 a. m. and 3.30 p. m. May 13 (discharge, 530 second-feet); minimum discharge, 16 second-feet June 23.

1911–1918: Maximum stage recorded, 6.3 feet on March 28, 1913 (stage-discharge relation affected by backwater from Moose River); discharge computed from records at dam, 760 second-feet.

Ice.—Stage-discharge relation not affected by ice.

Regulation.—Flow controlled by dam.

Accuracy.—Stage-discharge relation practically permanent between dates of shift; not affected by ice. Rating curve well defined from 20 to 400 second-feet. Gage read to hundredths twice daily. Daily discharge ascertained by applying to the rating table mean daily gage height weighted on days of changing gates, from records of gate opening at dam. Records good except those computed from gate openings at dam which are fair.

Discharge measurements of Middle Branch of Moose River at Old Forge, N. Y., during the year ending Sept. 30, 1918.

Date.	Made by—	Gage height.	Discharge.	Date.	Made by—	Gage height.	Discharge.
		Feet.	*Sec.-ft.*			*Feet.*	*Sec.-ft.*
Oct. 4	E. D. Burchard	1.81	97	May 11	J. W. Moulton	3.68	451
4	.....do	2.20	149	11	E. D. Burchard	3.79	493
5	.....do	2.42	182	11	J. W. Moulton	2.58	177
5	.....do	1.39	36	June 24	.....do	1.20	28
5	.....do	1.32	36	24	.....do	1.77	83
Apr. 11	M. H. Carson	2.40	137	24	.....do	2.33	163
11	E. D. Burchard	1.86	35	July 16	.....do	2.76	212
May 11	.....do	3.39	382				

Daily discharge, in second-feet, of Middle Branch of Moose River at Old Forge, N. Y., for the year ending Sept. 30, 1918.

Day.	Oct.	Nov.	Dec.	Jan.	Feb.	Mar.	Apr.	May.	June.	July.	Aug.	Sept.
1	98	280	49	58	143	130	130	232	63	34	34	104
2	98	311	51	56	136	130	136	232	53	25	34	104
3	98	299	52	58	136	130	106	232	35	31	28	104
4	98	290	52	57	136	130	115	232	27	32	27	104
5	98	290	52	57	136	130	115	232	38	33	27	104
6	98	280	54	57	136	130	125	223	126	34	27	104
7	98	280	57	57	136	130	115	214	86	40	28	104
8	98	280	57	57	136	130	115	214	58	42	28	98
9	98	280	56	58	136	130	125	290	63	36	29	98
10	98	280	58	58	136	130	135	378	220	36	31	98
11	98	280	58	56	136	130	135	378	311	43	33	98
12	98	280	58	56	136	130	135	378	241	74	32	98
13	98	270	58	56	130	130	135	451	86	200	31	98
14	98	260	59	56	130	130	150	530	63	223	31	98
15	104	250	63	60	130	130	150	503	74	298	32	110
16	104	270	63	60	130	130	150	402	74	324	30	98
17	104	250	63	59	130	123	165	280	53	272	29	98
18	104	250	63	59	130	123	165	184	41	200	29	104
19	98	165	63	59	130	123	135	141	35	36	27	104
20	98	58	63	59	130	123	167	141	50	42	75	104
21	98	54	61	57	130	123	178	111	126	44	173	104
22	98	51	61	57	130	123	324	86	74	58	173	104
23	98	54	60	57	130	123	324	86	16	63	173	104
24	98	55	60	57	130	130	324	74	53	63	173	104
25	104	56	60	57	130	130	324	63	311	58	165	98
26	104	55	60	57	130	130	298	74	241	58	165	98
27	104	50	60	57	130	130	298	63	24	58	165	104
28	143	48	60	56	130	130	248	46	21	58	116	98
29	165	48	58	56	......	130	232	175	21	58	98	98
30	181	48	58	104	......	130	232	53	28	58	98	98
31	214	......	58	143	......	130	......	63	......	86	98	......

NOTE.—Discharge Apr. 3-13, 19-28 and May 18 to July 12 determined from special rating curves based on discharge measurements made when logs were lodged on the control. Discharge Sept. 21-28 estimated because of logs on the control.

Monthly discharge of Middle Branch of Moose River at Old Forge, N. Y., for the year ending Sept. 30, 1918.

[Drainage area, 51.5 square miles.]

Month.	Discharge in second-feet.				Run-off (depth in inches on drainage area).
	Maximum.	Minimum.	Mean.	Per square mile.	
October	214	98	109	2.12	2.44
November	311	48	190	3.69	4.12
December	63	49	58	1.13	1.30
January	143	56	61.6	1.20	1.38
February	143	130	133	2.58	2.69
March	130	123	128	2.49	2.87
April	324	106	183	3.56	3.97
May	530	46	218	4.24	4.89
June	311	16	87.1	1.69	1.89
July	324	25	87.6	1.70	1.96
August	173	27	72.2	1.40	1.61
September	110	98	101	1.96	2.19
The year	530	16	119	2.31	31.31

BEAVER RIVER AT STATE DAM NEAR BEAVER RIVER, N. Y.

LOCATION.—At concrete storage dam at outlet of Beaver River flow, 7½ miles west of Beaver River post office, Herkimer County, and 7 miles above Beaver Lake at Number Four.

DRAINAGE AREA.—176 square miles (measured on topographic maps).

RECORDS AVAILABLE.—May 11, 1908, to September 30, 1918.

GAGES.—Elevation of water surface in the reservoir is determined by a staff gage in two sections, on the west corner of the gage house; read by James Dunbar, gate tender. The mean elevation of the crest of the spillway is at gage height 16.96 feet. Prior to September 28, 1913, elevation of water surface was determined by measuring the distance from the water surface to a reference point set at the elevation of the crest of the spillway. Widths of sluice gate openings determined by measuring on the gate stems the distances they have been raised.

DISCHARGE MEASUREMENTS.—Made from a temporary footbridge at the mouth of the outlet tunnel, below the gates.

DETERMINATION OF DISCHARGE.—Records include the discharge through one or more of four 4-foot circular sluice gates, when opened, the discharge over the spillway, and the discharge through the logway at the west end of the spillway. The sluice gates have been rated by current-meter measurements made at different elevations of the lake, but no measurements have been made of the discharge over the spillway or through the logway. Theoretic coefficients based on the experiments [1] in the hydraulic laboratory at Cornell University have been used to compute ratings for the spillway and logway.

REGULATION.—At ordinary stages the discharge of Beaver River is completely regulated by the operation of the sluice gates.

EXTREMES OF STAGE.—Maximum elevation of water surface in reservoir recorded during year, 18.5 feet on April 4 and 5; minimum stage recorded 7.85 feet at 8:35 a. m. February 13.

1908–1918: Maximum elevation of water surface in reservoir, 19.46 feet on March 29, 1913; minimum stage, 2.9 feet on September 29 and October 1, 1913.

EXTREMES OF DISCHARGE.—Maximum daily discharge during year, 1,900 second-feet on April 5; minimum discharge, zero, during periods when gates were closed and there was no flow over spillway.

1908–1918: Maximum discharge, 3,300 second feet on May 2, 1911.

ACCURACY.—Stage-discharge relation permanent. Probably not affected by ice. Rating curves for sluice gates well defined. Lake gage read to half-tenths once daily. The accuracy of these computations depends to a large extent on the care with which the gates were set to the recorded openings Records fairly good.

Monthly discharge of Beaver River at State dam near Beaver River, N. Y., for the year ending Sept. 30, 1918.

[Drainage area, 176 square miles.]

Month.	Discharge in second-feet.				Run-off (depth in inches on drainage area).
	Maximum.	Minimum.	Mean.	Per square mile.	
October	269	200	228	1.30	1.50
November	536	253	328	1.86	2.08
December	252	238	246	1.40	1.61
January	237	199	219	1.24	1.43
February	224	166	188	1.07	1.11
March	338	227	245	1.39	1.60
April	1,900	536	1,100	6.22	6.94
May	1,260	552	846	4.80	5.53
June	835	173	475	2.70	3.01
July	363	160	237	1.35	1.56
August	253	218	237	1.35	1.56
September	221	194	208	1.18	1.32
The year	1,900	160	380	2.16	29.35

[1] U. S. Geol. Survey Water-Supply Paper 200.

STREAMS TRIBUTARY TO ST. LAWRENCE RIVER.

EAST BRANCH OF OSWEGATCHIE RIVER AT NEWTON FALLS, N. Y.

LOCATION.—600 feet below lower dam of Newton Falls Paper Co., in Newton Falls, St. Lawrence County, 4 miles above mouth of Little River, and 10 miles below outlet of Cranberry Lake.

DRAINAGE AREA.—166 square miles (measured by engineers of the State of New York Conservation Commission).

RECORDS AVAILABLE.—October 6, 1912, to September 30, 1918.

GAGE.—Vertical staff on left bank about 600 feet above the lower dam; read by Henry Van Waldick.

DISCHARGE MEASUREMENTS.—Made by wading or from a cable 30 feet above gage.

CHANNEL AND CONTROL.—Small boulders and rock; covered with waste from pulp mill; permanent.

EXTREMES OF DISCHARGE.—Maximum stage recorded during year, 4.53 feet at 5.10 p. m. May 16 (discharge, 1,240 second-feet); minimum stage is reached nearly every Sunday during low-water period when paper mills shut down.

1912–1918: Maximum stage recorded, 6.1 feet at 5.15 p. m. March 28, 1913 (discharge, 2,200 second-feet).

ICE.—Stage-discharge relation affected by ice only for a short time during extremely cold weather.

REGULATION.—Some diurnal fluctuation in flow caused by the paper mills. Seasonal flow largely controlled by storage at Cranberry Lake.

ACCURACY.—Stage-discharge relation practically permanent; not affected by ice during year. Rating curve well defined between 20 and 1,200 second-feet. Gage read to hundredths twice daily. Daily discharge ascertained by applying to the rating table weighted mean gage heights based on observer's notes concerning operation of paper mills. Records good.

Discharge measurements of East Branch of Oswegatchie River at Newton Falls, N. Y., during the year ending Sept. 30, 1918.

Date.	Made by—	Gage height.	Discharge.	Date.	Made by—	Gage height.	Discharge.
		Feet.	*Sec.-ft.*			*Feet.*	*Sec.-ft.*
Feb. 12a	J. W. Moulton	2.63	399	June 25	J. W. Moulton	2.42	412
Apr. 7	E. D. Burchard	1.31	168	July 17	....do	2.09	318
7	....do	.85	94	17	....do	1.99	296
7	....do	1.05	117	17	....do	1.98	295
June 25	J. W. Moulton	2.78	508	17	....do	1.93	301
25	....do	2.66	473				

a Measurement made through incomplete ice cover.

Daily discharge, in second-feet, of East Branch of Oswegatchie River at Newton F N. Y., for the year ending Sept. 30, 1918.

Day.	Oct.	Nov.	Dec.	Jan.	Feb.	Mar.	Apr.	May.	June.	July.	Aug.	Sept.
1	148	350	363	22	363	430	460	622	416	326	304	[illegible]
2	363	315	180	363	363	430	588	810	242	293	304	[illegible]
3	338	223	214	376	22	22	460	1,030	416	20	350	[illegible]
4	338	22	272	326	416	460	506	1,030	402	20	326	[illegible]
5	315	338	272	338	416	430	522	538	445	326	326	[illegible]
6	350	338	232	22	402	460	416	852	326	20	326	[illegible]
7	22	252	252	338	45	445	152	506	389	20	293	[illegible]
8	430	188	232	350	402	326	522	506	416	376	350	[illegible]
9	402	293	232	338	416	293	490	506	304	262	326	[illegible]
10	389	163	658	338	22	22	460	571	588	475	350	[illegible]
11	376	180	852	315	430	430	416	694	894	293	137	[illegible]
12	350	223	214	315	416	460	430	554	852	376	326	[illegible]
13	376	205	223	22	430	445	402	894	1,120	350	460	[illegible]
14	171	223	196	326	193	430	144	938	1,120	20	389	[illegible]
15	376	252	445	338	460	416	460	1,220	1,070	304	363	[illegible]
16	350	252	22	315	445	430	506	1,220	588	304	350	[illegible]
17	338	242	554	326	22	22	554	1,070	810	315	376	[illegible]
18	282	232	363	315	460	338	554	938	554	282	137	[illegible]
19	315	350	283	326	445	445	389	770	430	272	304	[illegible]
20	338	304	389	350	460	350	363	770	402	262	326	[illegible]
21	326	293	338	338	430	445	20	588	293	20	326	[illegible]
22	293	272	363	338	460	490	522	522	242	232	338	[illegible]
23	363	293	87	350	460	430	445	522	202	282	304	[illegible]
24	272	304	350	363	22	152	402	490	350	282	363	[illegible]
25	262	205	363	350	475	430	506	490	326	304	130	[illegible]
26	262	223	376	350	430	338	389	282	304	272	416	[illegible]
27	272	223	522	22	445	445	376	588	326	262	376	[illegible]
28	22	376	363	350	460	460	20	460	315	242	326	[illegible]
29	75	283	338	304	......	445	152	490	304	252	363	[illegible]
30	272	363	99	252	......	460	350	522	293	326	376	[illegible]
31	262	......	163	252	......	202	......	475	......	293	338	......

Monthly discharge, of East Branch of Oswegatchie River at Newton Falls, N. Y., for the year ending Sept. 30, 1918.

[Drainage area, 166 square miles.]

Month.	Discharge in second-feet.				Run-off (depth in inches on drainage area).
	Maximum.	Minimum.	Mean.	Per square mile.	
October	430	22	292	1.76	[illegible]
November	376	22	259	1.56	[illegible]
December	852	22	316	1.90	[illegible]
January	376	22	291	1.75	[illegible]
February	475	22	350	2.11	[illegible]
March	490	22	367	2.21	[illegible]
April	554	20	399	2.40	[illegible]
May	1,220	282	692	4.17	[illegible]
June	1,120	202	491	2.96	[illegible]
July	475	20	248	1.49	[illegible]
August	460	130	325	1.96	[illegible]
September	522	20	338	2.04	[illegible]
The year	1,220	20	364	2.19	[illegible]

NOTE.—Table shows run-off as regulated at Cranberry Lake, and by paper mills at Newton Falls.

OSWEGATCHIE RIVER NEAR HEUVELTON, N. Y.

LOCATION.—2½ miles above Heuvelton, St. Lawrence County, 3 miles below Rensselaer Falls, and 7 miles above mouth of Indian River (outlet to Black Lake).

DRAINAGE AREA.—961 square miles (measured on topographic maps and map of State of New York, issued by United States Geological Survey).

RECORDS AVAILABLE.—June 23, 1916, to September 30, 1918.

GAGE.—Gurley seven-day water-stage recorder on the right bank, about 2½ miles above Heuvelton, installed September 16, 1916. Prior to this date gage height was determined by measuring the distance from a reference point to the water surface. Recorder inspected by George Todd.

CHANNEL AND CONTROL.—Solid rock.

EXTREMES OF DISCHARGE.—Maximum stage, from water-stage recorder, 6.6 feet from midnight to 8 p. m. April 4 (discharge, 9,220 second-feet); minimum stage from water-stage recorder 0.95 foot at 5 a. m. August 24 (discharge 340 second-feet).

1916–1918: Maximum stage from water-stage recorder, 7.6 feet from 9 to 12 a. m. March 30, 1917 (discharge, 11,700 second-feet); minimum stage from water-stage recorder, 0.91 foot at 11 p. m. October 16, 1916 (discharge 320 second-feet).

ICE.—Stage-discharge relation slightly affected by ice.

REGULATION.—Some diurnal fluctuation due to operation of mills at Rensselaer Falls and above. Seasonal flow regulated by storage in Cranberry Lake.

ACCURACY.—Stage-discharge relation permanent, except as affected by ice December 28 to March 7. Rating curve well defined between 400 and 15,000 second-feet. Operation of water-stage recorder satisfactory during the year. Daily discharge ascertained by applying mean daily gage height to rating table. Open-water records good; winter records fair.

Discharge measurements of Oswegatchie River at Heuvelton, N. Y., during the year ending Sept. 30, 1918.

Date.	Made by—	Gage height.	Discharge.	Date.	Made by—	Gage height.	Discharge.
		Feet.	Sec.-ft.			Feet.	Sec.-ft.
Dec. 20a	J. W. Moulton	1.47	675	Mar. 16a	J. W. Moulton	2.60	1,780
Jan. 12a	E. D. Burchard	1.50	656	Apr. 9	E. D. Burchard	4.46	4,830
Feb. 14b	J. W. Moulton	2.02	735	June 7	M. H. Carson	1.95	1,180

a Measurement made through incomplete ice cover.
b Measurement made through complete ice cover.

Daily discharge, in second-feet, of Oswegatchie River at Heuvelton, N. Y., for the year ending Sept. 30, 1918.

Day.	Oct.	Nov.	Dec.	Jan.	Feb.	Mar.	Apr.	May.	June.	July.	Aug.	Sept.
1	513	3,700	1,080	650	500	4,800	6,450	1,480	1,600	800	446	452
2	560	3,780	1,530	600	600	4,000	7,890	1,520	1,590	791	426	440
3	620	3,700	1,410	500	500	3,400	8,990	1,700	1,490	800	520	404
4	690	3,210	1,190	500	500	3,000	9,220	2,180	1,290	966	510	459
5	770	2,750	1,040	550	460	2,600	8,990	2,320	1,230	863	495	513
6	870	2,320	956	550	700	2,600	8,100	2,180	1,140	686	480	499
7	938	2,040	872	550	600	2,200	6,850	1,910	1,110	600	490	492
8	881	1,700	755	500	380	1,910	5,480	1,720	1,470	555	400	485
9	909	1,470	600	550	420	1,780	4,830	1,630	2,320	562	440	492
10	966	1,360	592	550	550	1,650	4,560	1,780	2,530	728	541	472
11	1,020	1,240	694	480	550	1,650	4,380	2,040	2,530	881	719	520
12	1,080	1,100	654	650	500	1,650	4,040	2,390	2,460	947	863	492
13	1,060	1,080	615	650	600	1,650	3,870	3,780	2,460	938	800	446
14	1,100	985	678	600	700	1,590	3,870	4,650	2,750	854	622	446
15	1,240	881	800	600	1,000	1,650	3,960	6,050	2,980	800	555	420
16	1,410	809	764	650	1,800	1,910	3,620	5,860	2,900	719	555	459
17	1,410	809	719	650	2,000	1,840	3,370	5,480	2,390	615	541	472
18	1,400	800	702	650	2,200	2,040	3,370	4,040	1,970	615	520	534
19	1,360	881	686	600	2,600	2,600	3,370	3,530	1,660	600	485	555
20	1,540	1,000	662	550	4,000	3,450	2,820	2,900	1,330	622	466	938
21	2,020	985	670	480	4,400	5,100	2,750	2,530	1,130	593	492	1,170
22	2,180	1,080	881	550	4,200	6,650	2,980	2,460	1,040	555	459	1,420
23	2,180	1,310	995	650	4,000	7,680	2,980	2,460	966	513	398	1,840
24	1,980	1,410	1,040	650	3,800	7,890	2,820	2,250	918	506	355	1,730
25	2,320	1,360	1,040	600	3,200	7,890	2,600	2,040	1,000	459	398	1,740
26	2,530	1,210	938	600	4,000	7,470	2,390	1,840	1,100	433	420	1,780
27	2,600	1,060	976	600	5,000	6,850	2,180	1,780	1,040	440	392	1,980
28	2,600	956	918	460	5,000	6,250	1,910	1,720	928	420	420	1,840
29	2,460	881	900	420	......	5,670	1,730	1,840	863	446	446	1,590
30	2,530	809	800	420	......	5,480	1,570	1,840	800	459	446	1,510
31	3,290	......	750	480	......	5,480	......	1,730	......	485	472	......

NOTE.—Discharge Dec. 28 to Mar. 7 estimated, because of ice, from discharge measurements, weather records and study of gage-height graph. Discharge Aug. 4-9 estimated by study of gage-height graph.

Monthly discharge of Oswegatchie River near Heuvelton, N. Y., for the year ending Sept. 30, 1918.

[Drainage area, 961 square miles.]

Month.	Discharge in second-feet.				Run-off (depth in inches on drainage area).
	Maximum.	Minimum.	Mean.	Per square mile.	
October	3,290	513	1,520	1.58	1.82
November	3,780	800	1,560	1.62	1.81
December	1,530	592	867	.902	1.04
January	650	420	564	.588	.68
February	5,000	380	1,960	2.04	2.12
March	7,890	1,590	3,890	4.04	4.66
April	9,220	1,570	4,400	4.58	5.11
May	6,050	1,480	2,630	2.74	3.16
June	2,980	800	1,630	1.70	1.90
July	966	420	653	.679	.78
August	863	355	502	.522	.60
September	1,980	404	886	.922	1.03
The year	9,220	355	1,750	1.82	24.71

WEST BRANCH OF OSWEGATCHIE RIVER NEAR HARRISVILLE, N. Y.

LOCATION.—At highway bridge near Geers Corners, 2½ miles downstream from Harrisville, Lewis County.

DRAINAGE AREA.—245 square miles (measured on topographic maps and map of New York, issued by United States Geological Survey; scale, 1:500,000).

RECORDS AVAILABLE.—July 1, 1916, to September 30, 1918.

GAGE.—Vertical staff in three sections on the right bank. One section graduated from 0.0 to 3.3 feet about 25 feet below bridge, and two sections graduated from 3.3 to 10.1 feet on downstream side of bridge abutment; read by Frank Osborne.

DISCHARGE MEASUREMENTS.—Made from cable 200 feet above the bridge, or by wading.

CHANNEL AND CONTROL.—Rocky and rough; probably permanent.

EXTREMES OF DISCHARGE.—Maximum stage recorded during year, 7.4 feet at 6 p. m. April 3 (discharge, 3,980 second-feet); minimum stage recorded, 1.1 feet at 7 a. m. August 28 and 29 (discharge 42 second-feet).

1916–1918: Maximum stage recorded 8.1 feet at 6.30 a. m. and 6 p. m. March 28, 1917 (discharge, 4,880 second-feet); minimum stage recorded 1.10 feet at 6 p. m. August 11, 1917, and 7 a. m. August 28 and 29, 1918 (discharge 42 second-feet).

ICE.—Stage-discharge relation probably not affected by ice.

REGULATION.—The pulp mill at Harrisville causes some diurnal fluctuation.

ACCURACY.—Stage-discharge relation practically permanent; not affected by ice. Rating curve well defined between 50 and 4,000 second-feet. Gage read to half-tenths twice daily. Daily discharge ascertained by applying mean daily gage height to rating table. Records good.

Discharge measurements of West Branch of Oswegatchie River near Harrisville, N. Y., during the year ending Sept. 30, 1918.

Date.	Made by—	Gage height.	Discharge.
		Feet.	*Sec.-ft.*
Feb. 12	J. W. Moulton	1.99	165
Apr. 8	E. D. Burchard	4.88	1,580
June 26	J. W. Moulton	2.63	339

Daily discharge, in second-feet, of West Branch of Oswegatchie River near Harrisville, N. Y., for the year ending Sept. 30, 1918.

Day.	Oct.	Nov.	Dec.	Jan.	Feb.	Mar.	Apr.	May.	June.	July.	Aug.	Sept.
1	158	1,960	305	106	106	1,560	1,800	560	480	220	195	54
2	245	1,640	335	115	91	1,210	2,890	650	422	440	170	70
3	220	1,350	305	106	106	1,090	3,980	650	405	370	124	58
4	275	1,090	245	106	106	970	3,300	600	352	245	106	58
5	335	850	232	91	98	800	2,690	600	320	275	106	79
6	405	650	220	68	91	750	2,130	560	275	245	124	91
7	388	560	245	77	106	650	1,640	520	520	245	77	74
8	460	422	170	85	91	560	1,560	600	1,090	220	106	77
9	480	405	158	77	85	480	1,640	600	1,210	245	195	54
10	480	370	170	79	77	480	1,800	560	1,030	370	320	63
11	405	352	170	91	115	440	1,640	650	910	335	245	66
12	305	370	170	124	124	370	1,420	800	850	305	158	56
13	370	335	158	98	146	405	1,280	1,150	970	320	124	70
14	520	275	170	79	220	370	1,210	1,720	1,090	275	135	70
15	560	220	170	158	440	370	1,210	1,800	970	275	91	68
16	560	220	195	115	480	370	1,210	1,490	750	220	66	91
17	480	260	182	106	480	370	1,350	1,210	650	195	63	106
18	520	275	207	106	560	405	1,350	1,030	520	209	68	275
19	520	305	158	115	650	440	1,350	850	440	195	79	320
20	750	405	170	106	1,210	600	1,210	750	352	170	70	460
21	970	370	195	124	1,490	850	1,090	700	320	146	68	750
22	1,030	405	207	98	1,350	1,350	1,090	650	388	124	68	850
23	850	460	195	124	1,350	1,800	1,150	560	422	106	51	650
24	750	440	195	146	1,210	1,960	1,090	480	480	124	60	700
25	850	370	195	124	1,210	1,960	970	440	422	115	58	800
26	1,090	335	207	124	1,800	1,960	910	460	335	146	63	800
27	1,090	335	260	124	1,800	1,640	800	560	305	146	56	600
28	910	305	195	98	1,720	1,350	700	650	245	106	56	560
29	700	275	170	79	......	1,280	650	700	275	98	54	560
30	850	290	170	98	......	1,280	600	650	245	195	58	520
31	1,420	......	115	106	......	1,420	......	560	......	245	56	......

Monthly discharge of West Branch of Oswegatchie River near Harrisville, N. Y., for the year ending Sept. 30, 1918.

[Drainage area, 245 square miles.]

Month.	Discharge in second-feet.				Run-off (depth in inches on drainage area).
	Maximum.	Minimum.	Mean.	Per square mile.	
October	1,420	158	611	2.50	2.88
November	1,960	220	530	2.16	2.41
December	335	115	201	.82	.95
January	158	68	105	.429	.38
February	1,800	77	618	2.52	2.63
March	1,960	370	953	3.89	4.48
April	3,980	600	1,520	6.22	6.94
May	1,800	440	766	3.13	3.61
June	1,210	245	568	2.32	2.59
July	440	98	223	.910	1.05
August	320	51	105	.429	.49
September	850	54	302	1.23	1.37
The year	3,980	51	540	2.20	29.77

RAQUETTE RIVER AT PIERCEFIELD, N. Y.

LOCATION.—Half a mile below dam of International Paper Co. at Piercefield, St. Lawrence County and three-fourths mile above head of Black Rapids.

DRAINAGE AREA.—723 square miles (all but 16 square miles measured on topographic maps).

RECORDS AVAILABLE.—August 20, 1908, to September 30, 1918.

GAGE.—Stevens water-stage recorder on right bank about one-half mile below dam. Prior to January 1, 1913, the following gages were used: August 20, 1908, to September 3, 1910, vertical staff fastened to an old pine stump; September 4 to December 31, 1910, chain fastened to same stump and having same datum; June 1, 1911, datum of the chain gage was lowered 2 feet. Water-stage recorder was set at this datum. Recorder inspected by M. O. Wood.

DISCHARGE MEASUREMENTS.—Made from a cable three-fourths mile below gage, just above Black Rapids.

CHANNEL AND CONTROL.—Channel opposite gage is a deep pond with no perceptible velocity. Control is at head of Black Rapids.

EXTREMES OF DISCHARGE.—Maximum stage during year, from water-stage recorder, 10.6 feet at 1 p. m. April 2 (discharge, 5,990 second-feet); minimum stage from water-stage recorder, 1.8 feet at 3 p. m. January 20 (discharge, 56 second-feet).

1908–1918: Maximum stage from water-stage recorder, 11.68 feet at 3 a. m. April 1, 1913 (discharge, 7,100 second-feet); minimum stage from water-stage recorder, 0.85 foot at 11 a. m. September 2, 1913 (discharge, about 10 second-feet).

ICE.—Rapids that form control rarely freeze and measurements made when the pond was covered with ice indicate that the stage-discharge relation was not affected.

REGULATION.—Large diurnal fluctuation in flow caused by dam during low and medium stages. Numerous lakes in the upper part of the drainage basin afford considerable storage, most of which is so controlled that the effect on the seasonal distribution of flow is large.

ACCURACY.—Stage-discharge relation practically permanent; not affected by ice. Rating curve well defined between 50 and 7,000 second-feet. Operation of the water-stage recorder satisfactory throughout the year. Daily discharge ascertained by use of discharge integrator. Records good.

COOPERATION.—Water-stage recorder inspected by an employee of the International Paper Co.

Discharge measurements of Raquette River at Piercefield, N. Y., during the year ending Sept. 30, 1918.

Date.	Made by—	Gage height.	Dis- charge.	Date.	Made by—	Gage height.	Dis- charge.
		Feet.	*Sec.-ft.*			*Feet.*	*Sec.-ft.*
Oct. 4	E. D. Burchard........	4.05	475	Mar. 12	J. W. Moulton.........	6.08	1,420
Feb. 7[a]	J. W. Moulton.........	4.21	387	May 10	.....do................	8.50	3,550

[a] Measurement made through incomplete ice cover.

Daily discharge, in second-feet of Raquette River at Piercefield, N. Y., for the year ending Sept. 30, 1918.

Day.	Oct.	Nov.	Dec.	Jan.	Feb.	Mar.	Apr.	May.	June.	July.	Aug.	Sept.
1	318	1,800	1,070	620	300	915	1,900	3,900	1,980	854	865	235
2	480	1,930	676	450	275	950	3,290	3,850	1,750	1,250	765	259
3	485	1,980	1,000	440	140	490	2,980	3,870	1,960	1,200	740	370
4	496	2,020	1,000	470	70	975	3,280	3,900	2,070	782	485	523
5	484	2,310	887	550	450	1,070	3,610	3,840	1,970	962	565	387
6	480	2,240	654	144	550	1,200	3,740	3,880	1,870	1,270	740	328
7	226	2,180	668	210	460	1,110	3,820	3,830	1,970	824	713	204
8	369	2,140	696	254	209	1,100	3,850	3,800	1,990	964	710	117
9	510	2,030	436	315	245	1,180	4,050	3,510	1,690	1,260	746	273
10	480	1,950	778	410	200	620	4,150	3,550	2,090	1,210	677	417
11	484	1,680	914	440	105	1,200	4,180	3,650	2,010	1,260	421	407
12	502	1,870	708	456	338	1,180	4,170	3,500	2,160	1,240	838	408
13	519	1,770	556	204	522	1,170	4,120	3,840	2,200	1,280	830	408
14	238	1,730	538	301	535	1,120	4,010	3,780	2,150	830	867	385
15	425	1,680	734	366	520	1,200	3,910	3,750	2,130	1,330	862	154
16	564	1,630	420	130	450	1,230	3,920	3,740	1,860	1,380	845	278
17	758	1,530	680	254	246	460	3,880	3,630	2,150	1,330	835	458
18	978	1,270	800	448	250	1,000	3,970	3,680	2,060	1,350	523	414
19	959	1,470	620	448	518	1,230	4,020	3,470	1,990	1,290	775	453
20	1,000	1,550	520	180	540	1,140	3,930	3,430	1,950	1,380	845	531
21	387	1,590	510	297	575	1,120	4,170	3,300	1,850	898	785	532
22	810	1,550	650	356	700	1,130	4,180	3,170	1,480	1,400	710	300
23	1,310	1,400	271	196	935	1,140	4,400	3,050	1,330	1,380	695	401
24	1,350	1,330	577	344	365	655	4,300	2,840	1,520	1,110	657	614
25	1,480	812	277	408	638	1,330	4,220	2,900	1,440	1,100	277	780
26	1,440	1,180	464	383	810	1,550	4,290	2,450	1,270	1,110	417	1,070
27	1,460	884	579	190	810	1,540	4,200	2,720	1,170	960	417	1,070
28	1,070	1,220	580	86	920	1,560	4,060	2,440	1,230	640	340	1,110
29	1,570	1,240	580	398	......	1,350	4,000	2,150	1,340	895	285	950
30	1,630	1,120	320	450	......	1,640	3,880	2,000	754	983	205	1,290
31	1,730	......	520	431	......	1,380	......	1,970	......	975	160	......

NOTE.—Discharge Dec. 16-22, Dec. 29 to Jan. 5, and Jan. 10-12 estimated for lack of gage-height record, from study of record for the periods Dec. 8-15 and Jan. 19-26.

Monthly discharge of Raquette River at Piercefield, N. Y., for the year ending Sept. 30, 1918

[Drainage area, 723 square miles.]

Month.	Discharge in second-feet.				Run-off (depth in inches on drainage area).
	Maximum.	Minimum.	Mean.	Per square mile.	
October	1,730	226	800	1.11	1.28
November	2,310	812	1,640	2.27	2.53
December	1,000	271	635	.878	1.01
January	620	86	343	.475	.55
February	935	70	453	.627	.65
March	1,640	460	1,130	1.56	1.80
April	4.400	1,900	3,880	5.37	5.99
May	3,900	1,970	3,340	4.62	5.33
June	2,200	754	1,780	2.46	2.74
July	1,400	640	1,120	1.55	1.79
August	867	160	632	.874	1.01
September	1,290	117	504	.697	.78
The year	4,400	70	1,360	1.88	25.46

ST. REGIS RIVER AT BRASHER CENTER, N. Y.

LOCATION.—Near steel highway bridge in Brasher Center, St. Lawrence County, 5 miles downstream from Brasher Falls, 6¼ miles below junction of East and West branches of St. Regis River, and about 12 miles above mouth.

DRAINAGE AREA.—621 square miles (measured on post-route map).

RECORDS AVAILABLE.—August 22, 1910, to November 10, 1917, when the station was discontinued.

GAGES.—Staff gage consisting of inclined and vertical sections, on right bank about 600 feet above bridge; installed June 24, 1916. Prior to this date, chain on right hand downstream side of bridge. Gages not at same datum; subject to different controls. Gage read by George Myers.

DISCHARGE MEASUREMENTS.—Made from a cable at the staff gage installed in June, 1916; previously made from the highway bridge or by wading.

CHANNEL AND CONTROL.—Small boulders and coarse gravel at cable; large boulders and gravel; very rough at bridge; both sections fairly permanent.

EXTREMES OF DISCHARGE.—1910–1917: Maximum stage recorded, 9.1 feet at 7 a. m. March 27, 1914 (discharge, 16,200 second-feet); minimum stage recorded 5.25 feet at 5 p. m. August 8, 1917 (discharge about 34 second-feet).

ICE.—Stage-discharge relation seriously affected by ice.

ACCURACY.—Stage-discharge relation practically permanent. Gage read to quarter-tenths twice dialy. Daily discharge ascertained by applying mean daily gage height to rating table. Records good.

Discharge measurements of St. Regis River at Brasher Center, N. Y., during the year ending Sept. 30, 1918.

Date.	Made by—	Gage height.	Dis-charge.	Date.	Made by—	Gage height.	Dis-charge.
		Feet.	*Sec.-ft.*			*Feet.*	*Sec.-ft.*
Oct. 2	J. W. Moulton	6.20	441	Mar. 17a	J. W. Moulton	6.67	545
3	E. D. Burchard	6.21	442	Apr. 10	E. D. Burchard	8.33	3,400

a Measurement made through incomplete ice cover.

Daily discharge, in second-feet, of St. Regis River at Brasher Center, N. Y., for the period Oct. 1 to Nov. 10, 1917.

Day.	Oct.	Nov.	Day.	Oct.	Nov.	Day.	Oct.	Nov.
1	404	1,880	12	625	....	22	930	....
2	510	1,520	13	685	....	23	930	....
3	529	1,240	14	705	....	24	810	....
4	586	1,050	15	930	....	25	1,120	....
5	655	930	16	810	....	26	1,380	....
6	810	810	17	705	....	27	1,310	....
7	930	705	18	605	....	28	1,180	....
8	810	625	19	625	....	29	1,240	....
9	705	529	20	990	....	30	1,590	....
10	685	438	21	990	....	31	1,960	....
11	625	....						

NOTE.—Mean discharge for October is 883 second-feet, or 1.42 second-feet per square mile, equivalent to a run-off of 1.64 inches from drainage area above station.

RICHELIEU RIVER AT FORT MONTGOMERY, ROUSES POINT, N. Y.

LOCATION.—Inside fort three-eighths mile south of international boundary, about one-half mile below outlet of Lake Champlain and 1 mile northeast of village of Rouses Point, Clinton County.

DRAINAGE AREA.—7,870 square miles, including 436 square miles of water surface (from Annual Report of New York State Engineer and Surveyor).

RECORDS AVAILABLE.—1875 to 1918.

GAGE.—Staff, inside the fort; read by Thomas Bourke. Elevation of gage zero 92.50 feet above mean sea level.

EXTREMES OF STAGE.—Maximum elevation recorded during year, 98.95 feet on April 11, 12, and 15; minimum elevation recorded, 93.65 feet at 10 a. m. September 10.

1869–1918: Maximum elevation recorded, 103.28 feet April, 1869; [1] minimum elevation recorded, 91.9 feet November 13, 1908.

COOPERATION.—Gage heights observed under direction of United States Engineer Corps and reported weekly to the United States Geological Survey.

[1] Hoyt, J. C., U. S. Geol. Survey Water-Supply Paper 97, p. 340. 1904.

Daily gage height, in feet, of Richelieu River at Fort Montgomery, N. Y., for the year ending Sept. 30, 1918.

Day.	Oct.	Nov.	Dec.	Jan.	Feb.	Mar.	Apr.	May.	June.	July.	Aug.	Sept.
1	1.2	2.45	2.15	1.45	1.2	2.25	4.9	5.75	4.25	2.95	1.9	1.4
2	1.2	2.6	2.0	1.45	1.2	2.35	5.25	5.9	4.1	2.9	1.9	1.45
3	1.4	2.7	2.1	1.4	1.2	2.4	5.7	5.75	4.05	2.9	1.75	1.55
4	1.3	2.7	2.05	1.4	1.2	2.45	6.0	5.55	4.0	2.85	1.8	1.3
5	1.2	2.8	2.1	1.4	1.2	2.4	6.15	5.6	3.9	2.8	1.85	1.4
6	1.25	2.9	2.0	1.4	1.2	2.45	6.2	5.6	3.85	2.8	1.65	1.3
7	1.35	2.7	1.95	1.35	1.2	2.45	6.3	5.45	3.9	2.75	1.7	1.3
8	1.4	2.75	1.95	1.35	1.2	2.5	6.3	5.45	3.75	2.7	1.65	1.3
9	1.25	2.75	1.9	1.3	1.2	2.5	6.25	5.25	3.65	2.75	1.7	1.3
10	1.25	2.7	1.95	1.3	1.2	2.45	6.25	6.0	3.65	2.65	1.8	1.15
11	1.35	2.7	1.85	1.3	1.25	2.6	6.45	5.1	3.6	2.65	2.1	1.2
12	1.35	2.6	1.85	1.3	1.25	2.6	6.45	5.05	3.75	2.6	1.85	1.5
13	1.55	2.6	1.85	1.3	1.2	2.55	6.35	5.15	3.55	2.55	1.9	1.25
14	1.45	2.6	1.9	1.3	1.25	2.6	6.4	5.15	3.6	2.5	1.85	1.2
15	1.6	2.6	1.9	1.3	1.6	2.6	6.45	5.1	3.55	2.5	1.8	1.2
16	1.4	2.45	1.85	1.3	1.3	2.65	6.4	5.5	3.6	2.5	1.7	1.2
17	1.35	2.5	1.8	1.3	1.3	2.6	6.4	5.1	3.5	2.45	1.7	1.2
18	1.55	2.6	1.8	1.3	1.3	2.6	6.25	5.1	3.5	2.4	1.7	1.2
19	1.8	2.4	1.8	1.3	1.7	2.6	6.35	5.05	3.4	2.4	1.65	1.3
20	1.45	2.5	1.75	1.25	1.6	2.75	6.35	5.05	3.35	2.35	1.7	1.3
21	1.5	2.3	1.75	1.25	1.6	2.8	6.25	4.75	3.45	2.3	1.65	1.35
22	1.55	2.3	1.65	1.25	1.65	2.95	6.25	4.75	3.3	2.3	1.6	1.4
23	1.6	2.25	1.7	1.25	1.65	3.15	6.25	4.65	3.2	2.3	1.65	1.5
24	1.55	2.3	1.7	1.25	1.7	3.4	6.25	4.6	3.15	2.2	1.55	1.5
25	1.7	2.2	1.6	1.25	1.7	3.6	6.05	4.55	3.15	2.25	1.55	1.65
26	1.65	2.1	1.55	1.25	1.95	3.8	6.15	4.4	3.1	2.25	1.6	1.75
27	1.65	2.2	1.55	1.25	2.05	3.95	6.1	4.4	3.1	2.15	1.45	1.95
28	1.75	2.2	1.6	1.2	2.15	4.05	6.0	4.2	3.2	1.95	1.55	2.35
29	1.8	2.15	1.45	1.2	....	4.2	6.05	4.25	3.1	2.0	1.8	2.2
30	1.9	2.2	1.5	1.2	....	4.4	5.85	4.25	3.0	2.0	1.35	2.3
31	2.2	.....	1.45	1.2	....	4.65	....	4.3	....	1.85	1.45	....

SARANAC RIVER NEAR PLATTSBURG, N. Y.

LOCATION.—At Indian Rapids power plant of Plattsburg Gas & Electric Co., 6 miles above mouth of river at Plattsburg, Clinton County.

DRAINAGE AREA.—607 square miles (measured on topographic maps).

RECORDS AVAILABLE.—March 27, 1903, to September 30, 1918.

GAGES.—Crest gage a vertical staff on the angle of the wing wall at the end of the racks; datum raised 0.76 foot August 20, 1906. Tailrace gage, a vertical staff spiked to timberwork dike between tailrace and river and about 50 feet below power house. Datum has changed slightly owing to settling of cribwork. Records of kilowatt output are obtained by a watt meter on switchboard at half-hour intervals. An inclined staff gage at the cable station, about one-fourth mile below the dam. Gages and watt meters read by power-house operators.

DISCHARGE MEASUREMENTS.—Made from a cable at head of Indian Rapids, one-fourth mile below dam, or, at low water, by wading under cable or in tailrace.

DISCHARGE RATING.—Records include flow over concrete spillway 171.25 feet in crest length, a rating for which has been prepared for use of coefficients[1] derived from experiments made in the hydraulic laboratory of Cornell University on a model section of the dam; the discharge through two power units equipped with 300-kilowatt generators which have been rated by current-meter measurements; and the discharge through two 5-foot waste gates when open. Occasional observations are made on the inclined staff gage at the cable as a check on the ratings of spillway and turbines.

[1] Horton, R. E., Weir experiments, coefficients, and formulas: U. S. Geol. Survey Water-Supply Paper 200, pp. 98–100, 1907.

EXTREMES OF DISCHARGE.—Maximum daily discharge during year, 5,600 second-feet April 3; minimum daily discharge, 200 second-feet August 4.

1908–1918: Maximum daily discharge recorded, 6,410 second-feet, April 20, 1914; minimum daily discharge recorded, 90 second-feet, September 28, 1914.

ICE.—The crest of the spillway is kept free from ice so that the stage-discharge relation is not affected.

REGULATION.—The lakes and ponds on the main stream and tributaries above the station have a water surface area of about 25.5 square miles. The actual storage afforded by these reservoirs has been largely increased by the State dam at Lower Saranac Lake, the operation of which affects the distribution of flow throughout the year.

ACCURACY.—Discharge measurements made during the year indicate that the ratings of spillway and turbines have not changed. Discharge over the spillway ascertained by applying to the rating table mean gage heights for 6-hour periods; discharge through the turbines ascertained by applying to their ratings the mean kilowatt output and head for 12-hour periods. Records fairly good.

COOPERATION.—Gage-height records and watt meter readings furnished by Plattsburg Gas & Electric Co., Herbert A. Stutchbury, superintendent.

The following discharge measurement was made by J. W. Moulton:

May 9, 1918: Gage height, 2.79 feet; discharge, 1,300 second-feet.

Daily discharge, in second-feet, of Saranac River near Plattsburg, N. Y., for the year ending Sept. 30, 1918.

Day.	Oct.	Nov.	Dec.	Jan.	Feb.	Mar.	Apr.	May.	June.	July.	Aug.	Sept.
1	740	1,040	440	330	450	1,550	3,500	1,750	1,040	700	350	420
2	880	940	440	440	640	2,600	4,900	2,000	1,300	700	290	470
3	1,080	660	520	410	440	1,500	5,600	1,700	920	620	300	390
4	940	820	410	520	860	1,200	4,000	1,650	800	580	200	370
5	880	760	370	450	410	1,100	3,200	1,800	820	700	250	360
6	920	700	360	480	420	920	2,700	1,600	700	620	310	390
7	720	740	300	540	700	900	2,450	1,600	1,240	900	290	620
8	760	760	260	340	840	800	2,600	1,550	1,300	740	220	600
9	520	780	230	520	440	760	2,500	1,500	1,060	540	520	600
10	460	740	420	560	620	620	2,000	1,300	920	840	780	580
11	500	660	310	470	880	780	1,800	1,250	860	780	900	580
12	430	800	280	580	470	820	1,650	1,400	1,000	720	1,180	560
13	560	800	470	540	640	780	1,600	1,300	1,060	740	1,220	620
14	490	800	450	810	580	960	1,800	1,450	1,080	440	1,180	700
15	620	780	470	750	580	820	1,850	1,250	960	700	940	600
16	520	720	410	680	660	840	2,100	1,300	920	580	720	600
17	480	760	560	460	920	620	2,050	1,350	880	400	620	560
18	600	660	430	390	840	900	2,100	1,240	880	480	520	640
19	520	620	300	560	640	860	1,950	1,250	840	580	600	900
20	560	520	370	280	760	900	1,850	960	800	580	390	900
21	640	500	390	520	2,200	1,450	1,800	1 000	800	460	480	1,080
22	660	480	370	310	1,500	2,050	2,200	740	780	540	500	1,220
23	540	410	290	300	1,240	2,900	2,200	820	840	580	490	1,040
24	540	270	370	240	960	2,300	2,050	920	820	520	490	1,020
25	620	225	260	330	1,020	2,300	1,850	820	800	1,140	430	1,200
26	880	290	480	380	1,550	2,300	1,750	1,040	720	840	420	1,300
27	700	260	370	700	2,000	2,000	1,700	1,200	680	600	370	1,600
28	680	320	320	1,050	1,900	1,900	1,500	1,400	700	370	360	1,600
29	880	500	470	410	......	2,050	1,350	960	720	400	420	1,250
30	900	500	440	320	......	2,500	1,700	940	680	310	400	1,180
31	1,220	......	460	460	......	2,800	......	900	......	310	380	......

Monthly discharge of Saranac River near Plattsburg, N. Y., for the year ending Sept. 30, 1918.

[Drainage area, 607 square miles.]

Month.	Discharge in second-feet.				Run-off (depth in inches on drainage area).
	Maximum.	Minimum.	Mean.	Per square mile.	
October	1,220	430	692	1.14	1.31
November	1,040	225	627	1.03	1.15
December	560	230	388	.639	.74
January	1,050	240	488	.804	.93
February	2,200	410	899	1.48	1.54
March	2,900	620	1,440	2.37	2.73
April	5,600	1,350	2,340	3.86	4.31
May	2,000	740	1,290	2.13	2.46
June	1,300	680	897	1.48	1.65
July	1,140	310	613	1.01	1.16
August	1,220	200	533	.878	1.01
September	1,600	360	798	1.31	1.46
The year	5,600	200	915	1.51	20.45

AUSABLE RIVER AT AUSABLE FORKS, N. Y.

LOCATION.—In village of Ausable Forks, Clinton County, immediately below junction of East and West branches and about 15 miles above mouth of river.

DRAINAGE AREA.—444 square miles (measured on topographic maps).

RECORDS AVAILABLE.—August 17, 1910, to September 30, 1918.

GAGE.—Chain on left bank 1,000 feet below junction of East and West branches; read by A. S. Baker.

DISCHARGE MEASUREMENTS.—Made from a cable about 1½ miles below gage, or by wading, either near the cable or a short distance above the gage.

CHANNEL AND CONTROL.—Stone and gravel, occasionally shifting. Channel divided by an island opposite the gage.

EXTREMES OF DISCHARGE.—Maximum stage recorded during year, 6.46 feet at 5.15 p. m., April 1, and 7 a. m., April 22 (discharge, 6,070 second-feet); minimum discharge, 80 second-feet, January 14 and 15 and February 1-3.

1910–1918: Maximum stage recorded, 10.2 feet in the evening of March 27, 1913 (discharge, roughly, 25,000 second-feet); minimum stage recorded, 3.0 feet at 7 a. m., July 21, 1912 (discharge, practically zero).

SPECIAL STUDY.—A portable water-stage recorder was installed at this station and a continuous gage-height record obtained July 11 to September 30, 1914, which showed a continual small fluctuation in stage. It was shown that determinations of monthly mean discharge based on semidaily gage heights are in error, as follows: July 11-31, 3.5 per cent; August, 1.1 per cent; September, 0.5 per cent. Some of the determinations of daily discharge showed greater errors, which were, however, largely compensating.

ICE.—Stage-discharge relation slightly affected by ice.

ACCURACY.—Stage-discharge relation probably permanent between dates of shifts. affected by ice December 10 to February 13. Rating curve fairly well defined between 175 and 3,000 second-feet. Gage read to hundredths twice daily. Daily discharge ascertained by applying mean daily gage height to rating table. Records good.

Discharge measurements of Ausable River at Ausable Forks, N. Y., during the year ending Sept. 30, 1918.

[Made by J. W. Moulton.]

Date.	Gage height.	Discharge.
	Feet.	*Sec.-ft.*
Jan. 10[a]	3.59	124
May 4	4.78	1,790
6	5.28	2,840

[a] Measurement made through incomplete ice cover.

Daily discharge, in second-feet, of Ausable River at Ausable Forks, N. Y., for the year ending Sept. 30, 1918.

Day.	Oct.	Nov.	Dec.	Jan.	Feb.	Mar.	Apr.	May.	June.	July.	Aug.	Sept.
1	371	1,440	234	220	80	890	4,210	3,690	656	345	250	1,230
2	417	1,010	221	220	80	668	5,600	3,320	588	998	196	567
3	371	751	189	260	80	1,060	3,950	3,690	679	436	170	436
4	379	599	183	260	85	557	2,490	1,830	557	336	164	221
5	1,090	546	208	220	100	515	1,730	1,440	345	294	142	227
6	739	455	202	160	110	398	1,350	2,160	319	302	121	611
7	597	436	183	130	100	362	1,530	2,720	465	407	121	679
8	455	388	170	120	95	371	1,620	2,950	1,230	484	142	426
9	398	407	157	120	100	407	2,380	1,440	1,940	526	5,310	362
10	345	362	180	120	110	354	1,530	1,260	998	526	2,600	294
11	319	311	180	110	110	319	1,260	2,720	515	505	2,050	234
12	398	328	190	100	140	336	1,120	1,350	1,130	536	1,940	177
13	1,010	302	200	90	200	426	1,010	1,620	1,530	515	1,620	170
14	578	264	220	80	407	417	1,200	3,070	1,180	634	567	929
15	567	280	200	80	864	407	2,270	1,730	813	536	465	436
16	955	264	200	90	800	336	1,620	1,200	567	407	302	354
17	567	227	220	100	505	319	2,600	929	484	336	257	679
18	484	272	220	110	436	526	2,600	851	388	319	227	1,180
19	465	280	200	110	668	788	1,830	764	328	302	189	1,260
20	903	280	160	140	3,190	788	1,440	1,040	257	250	164	702
21	727	280	160	130	942	1,260	1,440	1,030	264	214	196	1,100
22	588	272	160	120	903	2,050	5,030	764	214	214	177	1,210
23	484	311	180	110	890	3,070	2,490	800	328	189	177	1,070
24	515	311	200	110	788	2,160	2,600	702	567	164	177	1,040
25	864	202	220	120	714	1,730	1,440	588	546	153	164	1,180
26	788	208	240	130	3,070	1,350	1,350	825	407	153	183	1,350
27	1,070	221	240	130	2,160	1,040	1,350	825	354	102	177	2,490
28	1,260	208	220	110	1,620	903	1,830	764	311	132	164	1,530
29	1,620	208	220	110	......	1,040	1,730	1,260	272	110	189	984
30	2,400	208	240	100	......	1,350	3,690	903	242	234	183	813
31	3,070	......	220	85	......	1,830	......	714	......	436	183	......

NOTE.—Discharge Dec. 10 to Feb. 13, estimated because of ice from discharge measurements, weather records, and study of gage-height graph.

Monthly discharge of Ausable River at Ausable Forks, N. Y., for the year ending Sept. 30, 1918.

[Drainage area, 444 square miles.]

Month.	Discharge in second-feet.				Run-off (depth in inches on drainage area).
	Maximum.	Minimum.	Mean.	Per square mile.	
October	3,070	319	800	1.80	2.08
November	1,440	202	388	.874	.98
December	240	157	201	.453	.52
January	260	80	132	.298	.34
February	3,190	80	691	1.56	1.62
March	3,070	319	904	2.04	2.35
April	5,600	1,010	2,210	4.98	5.56
May	3,690	588	1,580	3.56	4.10
June	1,940	214	616	1.39	1.55
July	998	102	358	.806	.93
August	5,310	121	612	1.38	1.59
September	2,490	170	798	1.80	2.01
The year	5,600	80	772	1.74	23.63

LAKE GEORGE AT ROGERS ROCK, N. Y.

Location.—At boathouse in small bay on north side of steamboat landing at Rogers Rock, Essex County.

Drainage area.—Not measured.

Records available.—July 10, 1913, to September 30, 1918.

Gage.—Vertical staff fastened to a pile in the back end of the boathouse. Datum 3.15 feet [1] below crest of dam at outlet of lake; read once daily by George O. Cook.

Extremes of stage.—Maximum stage recorded during year, 4.2 feet May 20, 22, 27, 30, and June 3; minimum stage recorded, 1.55 feet February 16.

1913–1918: Maximum stage recorded, 4.98 feet on May 2, 1914; minimum stage recorded, 1.2 feet on November 21 and December 22, 1916.

Regulation.—The elevation of lake surface is regulated by the operation of gates and wheels at the dam at the outlet of the lake at Ticonderoga.

Cooperation.—Gage-height record furnished by International Paper Co.

[1] Determined by levels; supersedes the estimated figure previously published.

Daily gage height, in feet, of Lake George at Rogers Rock, N. Y., for the year ending Sept. 30, 1918.

Day.	Oct	Nov.	Dec.	Jan.	Feb.	Mar.	Apr.	May.	June.	July.	Aug.	Sept.
1			2.25	1.85	1.70	1.82	2.80	3.82	4.10	3.65	3.18	2.58
2				1.90	1.70	1.85	2.92	3.90	4.15	3.50	3.10	2.62
3				1.80	1.72	1.80	3.00	3.80	4.20	3.52	3.00	2.60
4	2.20	2.60	2.28	1.78	1.70	1.82	3.12	3.80	4.05	3.48	3.10	2.52
5	[illegible]	2.58	[illegible]	1.75	1.62	1.88	3.20	3.85	4.00	3.50	3.05	2.50
6				1.78	1.65	1.85	3.22	3.85	4.02	3.50	2.98	2.55
7				1.80	1.62	1.82	3.25	3.90	4.10	3.55	3.00	2.50
8				1.82	1.65	1.80	3.30	3.88	4.05	3.55	2.95	2.55
9	2.28	2.58	2.15	1.78	1.70	1.85	3.35	3.80	4.08	3.52	2.90	2.40
10	[illegible]	[illegible]	[illegible]	1.75	1.70	1.88	3.42	3.88	3.98	3.50	2.92	2.38
11				1.78	1.68	1.92	3.48	3.85	4.00	3.50	3.00	2.40
12				1.80	1.65	1.95	3.52	3.92	4.08	3.48	2.95	2.45
13				1.82	1.65	1.95	3.55	4.02	4.00	3.45	2.98	2.48
14	2.12	2.40	2.15	1.80	1.62	1.92	3.58	4.08	4.05	3.42	2.98	2.45
15	2.10	[illegible]	[illegible]	1.85	1.60	1.95	3.60	4.05	3.90	3.45	2.95	2.42
16				1.90	1.55	1.98	3.62	4.15	3.98	3.45	2.88	2.40
17				1.88	1.60	2.00	3.65	4.12	3.95	3.40	2.85	2.38
18				1.85	1.65	1.98	3.68	4.15	3.90	3.38	2.80	2.35
19	2.05	2.30	2.10	1.88	1.68	2.00	3.70	4.18	3.78	3.40	2.75	2.40
20	[illegible]	[illegible]	2.08	1.85	1.70	1.98	3.72	4.20	3.75	3.40	2.80	2.35
21				1.82	1.68	2.02	3.75	4.15	3.80	3.38	2.70	2.35
22				1.85	1.65	2.15	3.85	4.20	3.78	3.40	2.75	2.40
23				1.82	1.65	2.20	3.82	4.15	3.78	3.35	2.75	2.35
24	2.00	2.22	2.00	1.80	1.68	2.30	3.85	4.12	3.75	3.32	2.72	2.32
25	2.98	[illegible]	[illegible]	1.85	1.70	2.35	3.80	4.10	3.70	3.28	2.70	2.40
26				1.80	1.80	2.40	3.82	4.15	3.68	3.30	2.68	2.35
27				1.75	1.80	2.42	3.80	4.20	3.65	3.30	2.65	2.50
28				1.78	1.82	2.45	3.80	4.12	3.68	3.20	2.62	2.50
29		2.35		1.80	......	2.50	3.82	4.12	3.62	3.25	2.60	2.48
30	2.05	[illegible]	1.95	1.78	......	2.55	3.80	4.20	3.58	3.30	2.58	2.45
31	[illegible]	......	[illegible]	1.75	......	2.62	......	4.18	......	3.12	2.55	......

LAKE CHAMPLAIN AT BURLINGTON, VT.

Location.—On south side of roadway leading to dock of Champlain Transportation Co., at foot of King Street, Burlington.

Records available.—May 1, 1907, to September 30, 1918.

Gage.—Staff. Comparisons of gage readings indicate that zero of gage at Burlington is at practically the same elevation as that of gage at Fort Montgomery—92.5 feet above mean sea level. Gage read by employee of the Champlain Transportation Co.

Extremes of stage.—Maximum stage recorded during year, 6.78 feet on April 10 and 11; minimum stage recorded, 1.44 feet on September 14.

1907–1918: Maximum stage recorded, 8.20 feet on April 7, 1913; minimum stage recorded, −0.25 foot on December 4, 1908.

Ice.—Wider parts of Lake Champlain not usually frozen over until last part of January. Occasionally closure does not occur until February and in some years it lasts only for a few days. The northern end of the lake above the outlet is usually covered with ice from the middle of December to the middle of April.

Accuracy.—Gage read to hundredths once a day except Sundays from October 1 to December 21 and from March 25 to April 20; readings at irregular intervals during the rest of the year. Gage readings made when the lake is rough subject to inaccuracies due to wave action.

Cooperation.—Gage-height record furnished through the courtesy of Mr. D. A. Loomis, general manager of the Champlain Transportation Co.

Daily gage height, in feet, of Lake Champlain at Burlington, Vt., for the year ending Sept. 30, 1918.

Day.	Oct.	Nov.	Dec.	Jan.	Feb.	Mar.	Apr.	May.	June.	July.	Aug.	Sept.
1	1.48	3.10	2.38			2.68	5.30	6.14	4.48		2.18	
2	1.50	3.18				2.68	5.65	6.10		3.19		1.52
3	1.52	3.20	2.35				6.14	6.04				
4	1.56		2.33				6.49	6.02	4.29		2.08	
5	1.56	3.23	2.32			2.84	6.61		4.24	3.04		
6	1.58	3.21	2.30				6.63	5.88			1.92	
7		3.18	2.25					5.77	4.10		1.92	1.60
8	1.63	3.18	2.23				6.60	5.68				
9	1.64	3.15		1.58			6.65	5.58			2.09	
10	1.67	3.13	2.15				6.78		3.95	2.93		
11	1.67		2.13	1.58		2.98	6.78	5.45	3.90	2.89		1.50
12	1.68	3.05	2.08				6.75			2.84	2.14	
13	1.68	2.98	2.06			3.03	6.75	5.35	3.92			
14		2.95	2.06					5.40	3.95			1.44
15	1.74	2.90	2.03				6.65	5.48		2.78	2.10	
16	1.74	2.83					6.65	5.45			2.08	1.49
17	1.72	2.76	2.01				6.65	5.39	3.83	2.72		
18	1.70		2.00			2.99	6.70		3.78	2.71		
19	1.70	2.65	2.00			2.99	6.65		3.73		2.02	
20	1.73	2.62	1.98			2.99	6.61				1.98	1.67
21		2.58	1.98					5.15		2.60		1.70
22	1.79	2.58				3.35	6.48	5.04			1.92	
23	1.83	2.55					6.53	4.96		2.52	1.86	1.76
24	1.87	2.54						4.92	3.43	2.48		1.82
25	1.87				2.03	4.20		4.82	3.45			1.89
26	1.95	2.47				4.42	6.44		3.50	2.30		2.06
27	2.03	2.47			2.34	4.55	6.37	4.70		2.20	1.75	2.16
28		2.47				4.67		4.67				2.46
29	2.35	2.43				4.75		4.60		2.20		
30	2.70	2.40				4.87					1.54	2.76
31	3.00											

NOTE.—Thickness of ice 50 feet from dock: Jan. 9, 9½ inches; Jan. 18, 11½ inches; Jan. 21, 11½ inches; Jan. 28, 15½ inches; Feb. 4, 18½ inches; Feb. 11 and 18, 22 inches; Feb. 25, 23½ inches; Mar. 4, 22½ inches; Mar. 11, 21 inches; Mar. 18, 22½ inches; Mar. 25, 19 inches; Apr. 1, 13 inches; lake was frozen over Jan. 24 and was clear of ice again on Apr. 10.

OTTER CREEK AT MIDDLEBURY, VT.

LOCATION.—At railroad bridge half a mile south of railroad station at Middlebury, Addison County, 3½ miles below mouth of Middlebury River, and 3½ miles above mouth of New Haven River.

DRAINAGE AREA.—615 square miles.

RECORDS AVAILABLE.—April 1, 1903, to May 1, 1907, October 5, 1910, to September 30, 1918.

GAGE.—Chain; read by Almon Lovett.

DISCHARGE MEASUREMENTS.—Made from a boat just below railroad bridge, at the stone-arch highway bridge just above the dam, or by wading.

CHANNEL AND CONTROL.—Channel deep; current sluggish for several miles above the station. Control for low stages is gravel and boulder rips about 800 feet below gage, probably somewhat shifting; control at high stages is near the dam 800 feet farther downstream.

EXTREMES OF DISCHARGE.—Maximum stage recorded during year, 16.1 feet at 7.15 a. m. March 30 (discharge, 3,500 second-feet); minimum stage recorded, 11.75 feet at various times during the year (discharge, 202 second-feet).

1903–1907 and 1910–1918: Maximum stage recorded, 21.07 feet March 30, 1913 (discharge from extension of rating curve, about 8,000 second-feet); minimum open-water stage recorded, 11.45 feet September 15, 1913 (discharge, 138 second-feet). A somewhat lower discharge has possibly occurred at various times when the stage-discharge relation has been affected by ice.

ICE.—Ice forms to a considerable thickness at the gage and occasionally at the control, affecting the stage-discharge relation. Winter discharge ascertained by means of gage heights, current-meter measurements, observer's notes, and climatic records.

REGULATION.—Probably little if any effect from operation of power plants above the station. Considerable storage has been developed on tributaries near the headwaters.

ACCURACY.—Stage-discharge relation apparently permanent during the year, except when affected by ice. Rating curve well defined between 200 and 4,000 second-feet. Gage read to quarter-tenths once daily. Daily discharge ascertained by applying daily gage height to rating table, with corrections for ice from December 27 to March 23. Records good.

Discharge measurements of Otter Creek at Middlebury, Vt., during the year ending Sept. 30, 1918.

Date.	Made by—	Gage height.	Discharge.	Date.	Made by—	Gage height.	Discharge.
		Feet.	*Sec.-ft.*			*Feet.*	*Sec.-ft.*
Dec. 11	M. R. Stackpole	12.24	368	Apr. 2	M. R. Stackpole	15.82	3,270
Feb. 1	do	a12.42	278	July 27	H. W. Fear	12.10	320
Mar. 11	do	a13.25	592				

a Stage-discharge relation affected by ice.

Daily discharge, in second-feet, of Otter Creek at Middlebury, Vt., for the year ending Sept. 30, 1918.

Day.	Oct.	Nov.	Dec.	Jan.	Feb.	Mar.	Apr.	May.	June.	July.	Aug.	Sept.
1	232	2,510	360	220	280	2,400	3,230	1,140	810	320	283	320
2	283	2,510	501	210	280	2,500	3,320	1,700	670	360	248	360
3	265	2,330	360	210	280	2,400	3,320	1,790	555	403	232	403
4	265	2,960	450	220	280	2,200	3,230	1,610	450	340	232	360
5	360	1,610	426	250	220	1,800	3,140	1,360	426	301	202	320
6	403	1,190	403	220	220	1,350	3,140	1,030	426	320	248	301
7	403	917	320	210	250	1,100	3,050	955	450	340	265	320
8	403	775	403	220	250	880	2,960	917	610	360	248	301
9	403	670	301	210	230	740	2,870	1,030	555	450	381	248
10	381	610	202	280	250	660	2,690	880	528	450	501	265
11	340	610	360	280	250	580	2,600	1,190	610	450	360	301
12	320	450	301	300	230	520	2,510	1,150	670	501	283	301
13	403	501	265	280	250	520	2,420	1,150	955	670	283	301
14	450	475	320	280	320	660	2,060	2,600	1,110	381	320	320
15	426	475	360	220	400	1,200	1,970	2,510	880	810	403	301
16	475	403	381	300	500	1,100	1,970	2,330	670	640	403	248
17	426	450	265	320	1,250	740	1,970	2,150	475	501	360	301
18	426	450	320	320	1,100	740	1,970	1,700	450	501	301	426
19	403	320	360	320	960	1,100	1,970	1,190	403	475	248	528
20	450	403	403	320	960	1,700	1,970	992	403	450	232	555
21	501	426	403	220	2,300	2,100	1,880	1,070	360	381	283	640
22	381	426	381	220	2,200	2,300	1,970	1,110	360	320	248	775
23	403	501	360	230	2,200	2,500	2,060	955	535	320	265	880
24	403	705	283	230	1,950	2,690	2,150	810	1,360	340	248	640
25	501	640	283	260	2,100	2,780	2,060	670	1,440	320	248	1,030
26	740	450	283	300	2,500	2,780	1,970	670	1,030	320	217	775
27	740	340	300	340	2,400	2,960	1,790	640	670	320	202	2,330
28	775	381	280	340	2,400	3,050	1,520	740	528	283	248	2,240
29	670	340	360	280	....	3,140	1,360	810	450	265	265	1,970
30	955	283	280	250	....	3,500	1,360	810	450	202	265	1,880
31	2,690	....	220	280	....	3,320	....	810	....	283	283	....

NOTE.—Stage-discharge relation affected by ice Dec. 27 to Mar. 23. Determination of discharge for this period based on gage heights corrected for effect of ice by means of two discharge measurements, observer's notes, and weather records.

Monthly discharge of Otter Creek at Middlebury, Vt., for the year ending Sept. 30, 1918.

[Drainage area, 615 square miles.]

Month.	Discharge in second-feet.				Run-off (depth in inches on drainage area).
	Maximum.	Minimum.	Mean.	Per square mile.	
October	2,690	232	525	0.854	0.98
November	2,510	283	807	1.31	1.46
December	501	202	339	.551	.64
January	340	210	263	.428	.49
February	2,500	220	958	1.56	1.62
March	3,500	520	1,810	2.94	3.39
April	3,320	1,360	2,350	3.82	4.26
May	2,600	640	1,250	2.03	2.34
June	1,440	360	644	1.05	1.17
July	810	202	399	.649	.75
August	501	202	284	.462	.53
September	2,330	248	665	1.08	1.20
The year	3,500	202	854	1.39	18.83

WINOOSKI RIVER AT MONTPELIER, VT.

LOCATION.—1 mile downstream from Central Vermont Railway station in Montpelier, Washington County, about three-eighths mile above mouth of Dog River and 1¼ miles below mouth of Worcester branch.

DRAINAGE AREA.—420 square miles.

RECORDS AVAILABLE.—May 19, 1909, to September 30, 1918.

GAGE.—Gurley seven-day water-stage recorder on right bank, installed July 4, 1914, gage heights referred to datum by means of a hook gage inside the well; an outside staff gage is used for auxiliary readings; records June 16 to July 3, 1914, obtained from the staff gage. Chain gage at highway bridge just above the Central Vermont Railway station used from May 19, 1909, to June 30, 1914.

DISCHARGE MEASUREMENTS.—Made from a cable, or by wading.

CHANNEL AND CONTROL.—Channel deep and fairly uniform in section at the gage; control is formed by sharply defined rock outcrop about 500 feet below gage

EXTREMES OF DISCHARGE.—Maximum open-water stage during year, from water-stage recorder, 11.45 feet at 9 p. m. October 30 (discharge from extension of rating curve, 8,780 second-feet); minimum stage from water-stage recorder, 2.95 feet at 7 a. m. July 26 and 8 a. m. August 29 (discharge, 42 second-feet).

1909–1918: Maximum stage determined by leveling from flood marks preserved on building near present gage, 17.31 feet, April 7, 1912 (discharge not determined), minimum stage from water-stage recorder 1914–1918, 2.77 feet, August 13, 1914, and October 24, 1915 (discharge, 19 second-feet).

ICE.—Stage-discharge relation seriously affected by ice during the winter; discharge ascertained by means of gage heights, current-meter measurements, observer's notes, and climatic records.

REGULATION.—Operation of power plants on main stream and tributaries above station cause large diurnal fluctuations in stage.[1]

ACCURACY.—Stage-discharge relation practically permanent except when affected by ice. Rating curve well defined between 30 and 5,000 second-feet. Operation ter-stage recorder satisfactory during the year, except during part of October ovember, when it was temporarily removed for cleaning; Sanborn water-ecorder used November 16 to December 17. Daily discharge determined charge integrator, except for high stages and the period November 16 to 28, when mean daily gage heights were used. Open-water records good; ords fair.

[1] See fig. 1, p. 41, U. S. Geol. Survey Water-Supply Paper 424.

Discharge measurements of Winooski River at Montpelier, Vt., during the year ending Sept. 30, 1918.

[Made by M. R. Stackpole.]

Date.	Gage height.	Discharge.	Date.	Gage height.	Discharge.
	Feet.	*Sec.-ft.*		*Feet.*	*Sec.-ft.*
Oct. 31	7.57	3,460	Mar. 1	[a] 6.06	668
Dec. 18	[a] 4.80	389	Mar. 26	[a] 7.23	1,650
Jan. 25	[a] 5.06	275	Apr. 12	5.69	1,510

[a] Stage-discharge relation affected by ice.

Daily discharge, in second-feet, of Winooski River at Montpelier, Vt., for the year ending Sept. 30, 1918.

Day.	Oct.	Nov.	Dec.	Jan.	Feb.	Mar.	Apr.	May.	June.	July.	Aug.	Sept.
1	255	1,500	276	150	155	760	4,200	1,360	960	245	130	200
2	270	1,100	284	155	145	620	5,950	1,300	1,020	360	120	144
3	215	860	320	110	98	560	4,600	1,040	530	270	100	180
4	255	680	345	140	110	470	2,600	850	350	194	60	150
5	500	620	325	130	125	440	2,000	700	310	220	106	136
6	620	560	284	65	125	420	1,700	640	270	150	77	160
7	320	520	268	110	130	400	1,900	640	760	172	92	158
8	280	470	272	75	130	370	2,000	600	670	245	100	130
9	390	440	237	75	115	320	2,450	530	395	215	2,900	152
10	320	390	290	88	130	400	1,960	510	395	200	1,160	120
11	210	500	290	105	150	370	1,580	1,040	330	245	500	124
12	320	370	220	120	155	370	1,440	750	760	260	330	118
13	620	260	260	120	185	400	1,480	1,240	1,120	250	240	154
14	370	195	260	180	250	400	1,780	2,350	620	385	530	225
15	340	240	250	165	310	400	1,900	1,320	435	340	925	154
16	440	300	170	185	310	370	1,900	880	350	240	365	164
17	420	284	250	210	310	400	1,700	720	315	185	275	180
18	280	264	240	185	310	600	1,760	620	295	195	184	325
19	210	312	240	210	310	640	1,360	560	265	200	210	640
20	960	268	240	220	400	1,600	1,180	560	235	165	176	320
21	660	316	240	195	700	1,150	1,180	520	220	106	156	1,080
22	370	345	220	185	580	2,000	1,860	440	225	170	136	690
23	240	345	185	250	480	2,400	1,600	455	430	140	142	395
24	320	312	240	230	380	1,800	1,440	400	405	125	134	890
25	900	231	210	195	360	1,800	1,160	335	340	100	93	780
26	820	219	200	195	910	1,800	1,000	340	260	100	140	2,000
27	720	207	175	170	1,200	1,400	930	660	235	91	128	3,000
28	860	183	185	145	1,050	1,700	960	670	200	74	118	1,420
29	720	185	145	190	......	2,900	1,000	440	156	108	102	900
30	3,700	210	115	140	......	2,300	1,280	840	205	118	114	670
31	3,450	......	125	160	......	2,900	......	660	......	130	102	......

NOTE.—Stage-discharge relation affected by ice Dec. 10 to Mar. 28; discharge for this period computed from gage heights corrected for effect of ice by means of four discharge measurements, observer's notes, and weather records. Discharge estimated Oct. 6-20, Nov. 3-16, 29-30.

Monthly discharge of Winooski River at Montpelier, Vt., for the year ending Sept. 30, 1918.

[Drainage area, 420 square miles.]

Month.	Discharge in second-feet.				Run-off (depth in inches on drainage area).
	Maximum.	Minimum.	Mean.	Per square mile.	
October	3,700	210	657	1.56	1.80
November	1,500	183	423	1.01	1.13
December	845	115	237	.564	.65
January	250	65	157	.374	.43
February	1,200	98	343	.817	.85
March	2,900	320	1,050	2.50	2.88
April	5,950	930	1,930	4.59	5.12
May	2,350	335	773	1.84	2.12
June	1,120	156	435	1.04	1.16
July	385	74	193	.459	.53
August	2,900	60	321	.764	.88
September	3,000	118	525	1.25	1.40
The year	5,950	60	586	1.40	18.96

DOG RIVER AT NORTHFIELD, VT.

LOCATION.—At highway bridge near Norwich University campus in Northfield, Washington County. Union Brook joins Dog River a short distance below station.

DRAINAGE AREA.—47 square miles.

RECORDS AVAILABLE.—May 14, 1909, to September 30, 1918. Records from May 14, 1909, to August 22, 1910, obtained at lower highway bridge; those from August 23, 1910, to date, at present location.

GAGES.—Water-stage recorder on left bank below highway bridge; gage heights referred to gage datum by means of a hook gage inside the well. Inclined staff on left bank read by Florence C. Doyle from August 30 to September 30, 1918.

DISCHARGE MEASUREMENTS.—Made from highway bridge or by wading.

CHANNEL AND CONTROL.—Bed composed of gravel and alluvial deposits; subject to slight shifts.

EXTREMES OF DISCHARGE.—Maximum stage during year, from water-stage recorder. about 5.05 feet on April 2 (discharge, 960 second-feet); minimum stage, from water-stage recorder, 0.85 foot at 11 p. m. August 3 (discharge, 8 second-feet).

1910–1918: Maximum stage recorded at present site, 8.5 feet March 25, 1913 (discharge, 3,400 second-feet); minimum stage recorded, 0.60 foot September 10 and 11, 1913 (discharge, 3.0 second-feet). At the lower gage, 1909–10, flow was practically zero at various times when water was held back by dam above gage.

ICE.—River frozen over during winter; stage-discharge relation affected for short periods.

ACCURACY.—Stage-discharge relation fairly permanent except when affected by ice. Rating curve well defined below 600 second-feet. Operation of water-stage recorder unsatisfactory during a considerable part of the year as shown in footnote to daily discharge table. Daily discharge ascertained by applying to rating table mean daily gage heights determined by inspecting recorder graph, and from observer's readings (staff gage readings to quarter-tenths twice daily). Records fair.

Discharge measurements of Dog River at Northfield, Vt., during the year ending Sept. 30, 1918.

Date.	Made by—	Gage height.	Discharge.	Date.	Made by—	Gage height.	Discharge.
		Feet.	*Sec.-ft.*			*Feet.*	*Sec.-ft.*
Oct. 31	M. R. Stackpole........	3.10	296	Feb. 28	M. R. Stackpole........	a 2.59	162
Nov. 16	.....do..................	1.61	49.5	Apr. 12	.....do..................	2.75	213
Dec. 18	.....do..................	a 1.46	28.6	July 26	H. W. Fear............	.92	9.4
Jan. 24	.....do..................	a 1.29	21.5	Aug. 29	J. W. Moulton..........	1.01	11.8

a Stage-discharge relation affected by ice.

Daily discharge, in second-feet, of Dog River at Northfield, Vt., for the year ending Sept. 30, 1918.

Day.	Oct.	Nov.	Mar.	Apr.	May.	June.	July.	Aug.	Sept.
1	16	198	....	570	157	....	23	10	33
2	12	153	....	760	135	....	28	9.6	16
3	12	127	....	585	119	....	21	8.6	13
4	18	106	....	390	104	....	18	8.4	12
5	30	93	....	302	95	....	16	9.0	13
6	39	89	....	255	90	....	14	8.8	19
7	25	85	....	315	89	....	15	9.0	19
8	19	75	....	315	85	....	14	11	14
9	19	74	....	390	80	....	....	196	14
10	17	73	....	....	81	....	....	66	12
11	16	65	....	....	138	....	....	43	12
12	20	63	....	....	101	....	....	33	14
13	63	54	....	....	158	68	....	22	16
14	34	53	....	....	249	50	....	67	19
15	34	49	....	....	155	40	....	....	14
16	41	51	....	....	124	34	....	....	16
17	32	48	....	....	108	28	....	....	19
18	25	48	....	....	94	27	....	....	32
19	23	50	....	....	83	24	....	....	47
20	44	....	....	....	75	22	....	....	62
21	38	....	237	227	75	20	....	....	107
22	30	....	304	270	67	37	....	....	49
23	27	....	304	229	....	43	....	....	35
24	33	....	281	217	....	37	....	....	54
25	128	....	281	169	....	30	....	....	48
26	61	....	264	146	....	23	9.8	....	268
27	46	....	225	145	....	20	9.8	10	257
28	128	....	235	223	....	19	9.6	11	190
29	79	....	281	186	....	19	10	11	104
30	527	....	315	207	....	32	13	12	83
31	327	....	444	....	....	....	11	12	....

NOTE.—Stage-discharge relation affected by ice from last part of November to about Mar. 20. Water-stage recorder not operating Nov. 20 to Mar. 20, Apr. 10-20, May 5-6, 23-31, June 1-12, July 9-25 and Aug. 15-26.

Monthly discharge of Dog River at Northfield, Vt., for the year ending Sept. 30, 1918.

[Drainage area, 47 square miles.]

Month.	Discharge in second-feet.				Run-off (depth in inches on drainage area).
	Maximum.	Minimum.	Mean.	Per square mile.	
October	527	12	63.3	1.35	1.56
November	198	....	61.0	1.30	1.45
December	....	....	28.0	.596	.69
January	....	....	17.5	.372	.43
February	....	....	42.5	.904	.94
March	444	....	138	2.94	3.39
April	760	145	285	6.06	6.76
May	249	....	99.2	2.11	2.43
June	....	19	39.5	.840	.94
July	....	9.6	16.7	.355	.41
August	196	8.4	25.3	.538	.62
September	268	12	53.7	1.14	1.27
The year	760	....	72.3	1.54	20.89

NOTE.—Discharge estimated by comparison with Winooski River at Montpelier and White River at West Hartford as follows: Nov. 20-30, 25 second-feet; Dec. 1-31, 28 second-feet; Jan. 1-31, 17.5 second-feet; Feb. 1-28, 42.5 second-feet; Mar. 1-20, 55 second-feet; Apr. 10-20, 240 second-feet; May 23-31, 68 second-feet, June 1-12, 51 second-feet; July 9-25, 18 second-feet; Aug. 15-26, 19 second-feet. Use was also made of three discharge measurements obtained during December, January, and February in making estimates of flow during the winter.

LAMOILLE RIVER AT CADYS FALLS, VT.

LOCATION.—About one-fourth mile below power plant of Morrisville Electric Light & Power Co., at what was formerly known as Cadys Falls, 2 miles downstream from Morrisville, Lamoille County.

DRAINAGE AREA.—280 square miles.

RECORDS AVAILABLE.—September 4, 1913, to September 30, 1918. A station was maintained at highway bridge near power plant at Cadys Falls from July 28, 1909, to July 13, 1910.

GAGES.—Friez water-stage recorder on right bank one-fourth mile below highway bridge at Cadys Falls. Gage heights are referred to gage datum by means of a hook gage inside the well; an outside staff gage is used for auxiliary readings.

DISCHARGE MEASUREMENTS.—Made from a cable or by wading.

CHANNEL AND CONTROL.—Channel smooth gravel; well-defined gravel control 500 feet downstream from gage.

EXTREMES OF DISCHARGE.—Maximum open-water stage during year, from water-stage recorder, 10.66 feet at 7.45 p. m. October 30 (discharge, from extension of rating curve, about 7,430 second-feet); minimum stage, from water-stage recorder, 1.85 feet at 1 p. m. August 18 (discharge, 52 second-feet).

1913–1918: Maximum stage recorded October 30, 1917; minimum stage recorded, 1.82 feet August 17, 1914 (discharge, 50 second-feet).

ICE.—River freezes over during extremely cold weather; stage-discharge relation slightly affected by ice. Discharge determined from gage heights with corrections for backwater based on current-meter measurements, observer's notes, and climatic records.

ACCURACY.—Stage-discharge relation practically permanent, except when affected by ice. Rating curve well defined. Operation of water-stage recorder satisfactory throughout year except for periods during the winter when clock would not run on account of extreme cold. Daily discharge ascertained by discharge integrator. Records good.

Discharge measurements of Lamoille River at Cadys Falls, Vt., during the year ending Sept. 30, 1918.

Date.	Made by—	Gage height.	Dis-charge.	Date.	Made by—	Gage height.	Dis-charge.
		Feet.	*Sec.-ft.*			*Feet.*	*Sec.-ft.*
Dec. 15	M. R. Stackpole	[a]2.39	167	Apr. 11	M. R. Stackpole	4.42	1,150
Mar. 2	.....do	[a]3.35	397	11	.....do	4.28	1,080
27	.....do	[a]3.89	804	July 25	H. W. Fear	2.22	147

[a] Stage-discharge relation affected by ice.

Daily discharge, in second-feet, of Lamoille River at Cadys Falls, Vt., for the year ending Sept. 30, 1918.

Day.	Oct.	Nov.	Dec.	Jan.	Feb.	Mar.	Apr.	May.	June.	July.	Aug.	Sept.
1	385	1,500	275	240	190	560	3,150	1,080	590	184	110	196
2	330	1,020	250	200	190	430	4,550	990	630	290	104	140
3	260	820	230	220	200	370	3,900	600	380	300	100	112
4	430	680	240	200	200	370	2,100	620	300	220	90	114
5	740	610	235	240	200	290	1,520	495	265	176	98	112
6	950	550	230	220	220	270	1,260	470	240	170	116	118
7	640	520	230	200	220	250	1,520	495	720	184	110	130
8	480	455	200	200	220	240	1,760	440	550	198	132	100
9	660	420	210	200	240	450	2,100	405	350	184	465	120
10	495	425	220	200	200	450	1,520	385	315	172	330	120
11	400	530	220	190	190	490	1,160	800	280	196	235	112
12	355	445	200	190	170	430	1,040	580	820	164	174	112
13	780	305	200	190	140	350	990	820	1,520	198	164	112
14	530	300	200	190	140	270	1,100	2,250	800	255	152	230
15	485	330	200	170	155	270	1,420	1,080	590	275	162	154
16	720	325	200	170	220	220	1,740	700	435	200	178	136
17	510	270	200	170	240	200	1,520	680	290	158	142	215
18	390	240	200	170	250	270	1,380	410	425	174	114	255
19	350	275	200	155	220	350	990	330	480	158	122	330
20	1,000	260	200	155	290	410	820	325	245	122	120	210
21	700	300	200	155	520	600	820	325	140	95	144	490
22	485	330	200	140	600	1,100	1,520	330	215	87	140	335
23	405	315	220	140	540	1,750	1,460	720	390	116	130	285
24	415	430	200	140	390	1,250	1,240	485	410	124	104	740
25	780	345	190	125	290	970	940	340	345	114	96	770
26	640	260	200	125	440	970	740	250	295	104	110	950
27	510	220	200	140	880	840	570	340	255	99	112	2,500
28	980	220	200	155	780	720	700	380	225	85	85	1,560
29	820	225	200	190	....	900	740	325	220	85	100	540
30	3,800	240	200	170	....	1,500	1,080	710	230	162	104	520
31	4,100	....	220	190	....	1,950	....	550	....	140	110	....

NOTE.—Stage-discharge relation affected by ice from Dec. 10 to Mar. 31; determination of discharge for this period based on gage heights corrected for effect of ice by means of three discharge measurements, observer's notes, and weather records. Discharge estimated Dec. 3, 6–8, and for several short periods during the winter.

Monthly discharge of Lamoille River at Cadys Falls, Vt., for the year ending Sept. 30, 1918.

[Drainage area, 280 square miles.]

Month.	Discharge in second-feet.				Run-off (depth in inches on drainage area).
	Maximum.	Minimum.	Mean.	Per square mile.	
October	4,100	260	791	2.82	3.26
November	1,500	220	440	1.57	1.75
December	275	190	212	.757	.87
January	240	125	179	.639	.74
February	880	140	305	1.09	1.14
March	1,950	200	629	2.25	2.59
April	4,550	570	1,510	5.39	6.01
May	2,250	250	604	2.16	2.49
June	1,520	140	432	1.54	1.72
July	300	85	167	.596	.69
August	465	85	144	.514	.59
September	2,500	100	394	1.41	1.57
The year	4,550	85	483	1.72	23.41

GREEN RIVER AT GARFIELD, VT.

LOCATION.—At site of old dam above highway bridge at Garfield village, town of Hyde Park, Lamoille County. Green River is tributary to Lamoille River about 4 miles east of Morrisville.

DRAINAGE AREA.—20 square miles (roughly approximate).

RECORDS AVAILABLE.—January 3, 1915, to September 30, 1918.

GAGE.—Inclined staff on left bank in pool back of weir; read by P. M. Trescott.

DISCHARGE MEASUREMENTS.—Standard sharp-crested weir of compound section: length of crest at gage height 0.00 is 9.0 feet; at gage height 0.83 foot, length of length of crest is increased 11.17 feet. Current-meter measurements made at footbridge about one-half mile downstream from weir, and at old bridge about one-half mile above weir.

CHANNEL AND CONTROL.—A pool of considerable size is formed in the old mill pond back of the weir; at ordinary stages the velocity of approach to the weir is very small. Some water leaks around the weir in the old tailrace on left bank.

EXTREMES OF DISCHARGE.—Maximum stage recorded during year, 3.03 feet at 9 a. m. October 31 and 5 p. m. April 2 (discharge, from extension of rating curve, about 306 second-feet); minimum stage recorded, 0.29 foot August 28, 30, and 31 (discharge, 4.7 second-feet).

1915–1918: Maximum stage recorded, 3.6 feet at 9 a. m. April 12, 1915 (discharge from extension of rating curve, about 436 second-feet); minimum stage recorded, 0.29 foot August 28, 30, and 31, 1918 (discharge, 4.7 second-feet.)

ICE.—Weir and weir crest kept clear of ice during winter; stage-discharge relation not affected by ice.

REGULATION.—An old timber dam about 2 miles upstream affects flow to some extent. The dam leaks by an amount somewhat greater than the low-water flow. During prolonged low stages the surface of water in pond (103 acres) falls below crest of dam; subsequent increased flow into pond is retained until water again flows over crest, when the increased flow is apparent at gaging station.

ACCURACY.—Stage-discharge relation practically permanent. Rating curve based on weir formula, $Q=3.33\ LH^{3/2}$ with corrections determined from current-meter measurements, and with logarithmic extension above gage height 1.90 feet. Gage read twice daily to hundredths. Daily discharge ascertained by applying to rating table mean daily gage height. Records good below 130 second-feet; at the higher stages the weir is flooded and results are somewhat uncertain.

Cooperation.—Gage-height records furnished by C. T. Middlebrook, consulting engineer, Albany, N. Y.

Discharge measurements of Green River at Garfield, Vt., during the year ending Sept. 30, 1918.

[Made by H. W. Fear.]

Date.	Gage height.	Discharge.
	Feet.	*Sec.-ft.*
July 25 [a]	0.39	6.9
July 25 [b]	.39	7.6

[a] Measurement made at old bridge one-half mile above gage.
[b] Measurement made at footbridge one-half mile below gage.

Daily discharge, in second-feet, of Green River at Garfield, Vt., for the year ending Sept. 30, 1918.

Day.	Oct.	Nov.	Dec.	Jan.	Feb.	Mar.	Apr.	May.	June.	July.	Aug.	Sept.
1	17	126	16	11	9.7	11	163	84	63	20	15	8.4
2	21	81	17	11	9.7	11	271	87	57	20	14	6.3
3	27	62	17	10	9.7	11	286	68	39	19	13	6.0
4	32	47	17	10	9.3	11	207	50	26	18	13	5.7
5	35	43	17	10	8.7	12	163	40	20	17	13	5.7
6	51	38	15	11	8.4	12	138	34	18	16	12	8.7
7	49	34	15	11	8.4	12	149	40	27	19	14	6.6
8	46	32	15	11	8.0	12	170	32	60	18	14	6.3
9	49	29	16	10	8.4	12	172	30	58	19	22	6.0
10	41	28	15	9.7	8.7	13	139	32	32	17	16	6.3
11	37	27	13	9.3	8.7	14	106	62	26	17	14	6.0
12	33	26	13	9.7	9.0	15	91	58	47	17	13	6.0
13	38	25	13	10	9.7	14	79	72	98	17	12	7.1
14	34	24	14	10	10	14	100	210	68	21	12	8.4
15	35	23	14	10	10	14	159	117	51	18	12	7.1
16	49	23	13	10	10	14	197	68	38	15	11	7.1
17	47	22	13	10	9.7	14	181	49	32	13	11	11
18	37	21	13	10	9.7	14	163	39	27	10	10	11
19	31	21	14	10	9.7	15	95	32	25	9.7	10	12
20	46	20	14	9.7	12	17	74	28	23	9.0	9.7	13
21	60	20	14	9.7	12	22	78	30	21	8.4	9.3	25
22	43	21	13	10	11	30	117	28	23	8.0	9.3	21
23	35	22	13	10	11	34	131	33	25	7.7	8.4	22
24	34	21	13	10	10	22	110	32	26	7.4	5.7	39
25	39	22	13	10	10	29	77	28	26	7.4	5.5	64
26	43	18	12	10	13	43	64	26	23	7.1	5.5	68
27	38	17	12	10	12	60	69	35	22	7.1	5.2	188
28	51	17	12	9.7	12	62	70	40	20	6.9	4.9	146
29	56	17	12	9.7	......	65	71	40	25	6.6	5.2	82
30	130	16	11	9.7	......	69	74	70	20	23	4.7	51
31	264	......	11	9.7	......	90	......	64	......	16	4.7	......

Monthly discharge, in second-feet, of Green River at Garfield, Vt., for the year ending Sept. 30, 1918.

Month.	Maximum.	Minimum.	Mean.
October	264	17	49.9
November	126	16	31.4
December	17	11	13.9
January	11	9.3	10.1
February	13	8.0	9.95
March	90	11	25.4
April	286	64	132
May	210	26	53.5
June	98	18	35.5
July	23	6.6	14.0
August	22	4.7	10.6
September	188	5.7	28.7
The year	286	4.7	34.6

MISSISQUOI RIVER NEAR RICHFORD, VT.

LOCATION.—About 3 miles downstream from Richford, Franklin County, 3 miles below mouth of North Branch, and 2 miles above mouth of Trout River.

DRAINAGE AREA.—445 square miles.

RECORDS AVAILABLE.—May 22, 1909, to December 3, 1910, and June 26, 1911, to September 30, 1918.

GAGE.—Gurley water-stage recorder on left bank, about one-fourth mile above highway bridge; chain gage on highway bridge used from June 26, 1911, to July 31, 1915. From May 22, 1909, to December 3, 1910, gage was just below plant of the Sweat-Comings Co. in Richford.

DISCHARGE MEASUREMENTS.—Made from highway bridge or by wading

CHANNEL AND CONTROL.—Channel deep; banks not subject to overflow; stream bed composed of gravel, boulders, and ledge rock. Control is sharply defined by rock outcrop about 100 feet below gage.

EXTREMES OF DISCHARGE.—Maximum stage during year, 17.64 feet on April 1 determined by levels from high-water mark (stage-discharge relation affected by ice); minimum stage, from water-stage recorder, 2.16 feet at 4 p. m. August 30 discharge, 44 second-feet).

1911–1918: Maximum stage recorded April 1, 1918; minimum stage recorded, 4.15 feet by chain gage, July 14, 1911 (discharge, 8 second-feet).

ICE.—Stage-discharge relation usually affected by ice from December to March; discharge determined from gage heights corrected for backwater by means of current-meter measurements, observer's notes, and weather records.

REGULATION.—Considerable daily fluctuation at low stages caused by operation of power plants at Richford.

ACCURACY.—Stage-discharge relation practically permanent except when affected by ice. Rating curve fairly well defined below 6,000 second-feet. Operation of water-stage recorder satisfactory during the year except as indicated in footnote to daily-discharge table. Daily discharge ascertained by applying to rating table mean daily gage height determined by inspecting recorder sheets; determinations for periods for which no record was obtained are based on comparison with records of flow of streams in adjacent drainage basins. Records good for periods when water-stage recorder was in operation, and fair for other periods and during the winter.

Discharge measurements of Missisquoi River near Richford, Vt., during the year ending Sept. 30, 1918.

Date.	Made by—	Gage height.	Discharge.	Date.	Made by—	Gage height.	Discharge.
		Feet.	*Sec.-ft.*			*Feet.*	*Sec.-ft.*
Oct. 11	M. R. Stackpole	4.09	809	Apr. 8	M. R. Stackpole	7.17	3,430
Dec. 12	....do	a 4.26	315	9	....do	7.69	4,090
Jan. 30	....do	a 4.69	160	July 24	H. W. Fear	2.91	234
Mar. 6	....do	a 6.48	760	Aug. 31	J. W. Moulton	2.20	51
Apr. 1	....do	a 13.49	4,730	31	....do	2.35	84
1	....do	a 13.69	4,800				

a Stage-discharge relation affected by ice.

Daily discharge, in second-feet, of Missisquoi River near Richford, Vt., for the year ending Sept. 30, 1918.

Day.	Oct.	Nov.	Dec.	Jan	Feb.	Mar.	Apr.	May.	June.	July.	Aug.	Sept.
1	770	5,280	380	185	82	1,050	5,800	1,720	438	615	324	456
2	1,140	2,590	600	170	160	1,000	9,000	1,880	510	620	258	300
3	890	1,720	420	145	160	960	8,000	1,720	393	446	248	240
4	1,140	1,360	440	130	130	900	6,720	1,480	282	379	186	179
5	1,520	1,100	420	170	94	820	4,270	1,320	248	318	150	168
6	1,680	995	380	170	72	760	3,280	1,240	215	300	1,240	194
7	1,360	890	320	185	120	700	3,170	1,200	482	268	710	272
8	1,060	830	280	160	145	560	4,050	1,170	995	307	575	227
9	1,200	770	300	82	160	500	3,940	890	590	324	698	203
10	960	740	300	120	160	460	3,170	710	395	314	800	200
11	740	680	300	120	170	420	2,340	890	332	290	500	168
12	750	650	320	120	170	360	1,880	1,440	610	258	363	152
13	770	635	320	130	160	380	1,680	2,100	3,060	339	321	203
14	995	565	320	145	130	300	1,880	2,240	2,840	860	282	307
15	1,140	496	320	160	82	280	2,440	1,640	1,480	668	286	395
16	1,920	510	280	220	72	260	2,850	1,140	995	550	237	343
17	1,360	460	300	200	145	260	2,650	830	680	480	200	1,760
18	960	440	300	185	600	300	2,390	710	545	505	170	1,170
19	830	500	300	185	700	340	1,880	570	456	500	179	1,140
20	2,240	500	300	185	900	380	1,680	500	387	400	145	860
21	1,880	575	260	200	960	560	1,700	510	324	282	132	1,600
22	1,200	585	230	185	1,100	1,500	2,500	407	314	234	122	1,840
23	960	860	200	130	1,100	3,200	2,700	325	590	230	140	1,170
24	830	740	230	170	700	2,800	2,440	363	1,280	212	100	1,560
25	1,640	590	300	120	410	2,400	1,970	325	860	170	108	1,640
26	1,880	400	280	130	700	2,200	1,480	310	536	150	125	1,360
27	1,280	350	170	160	1,150	1,550	1,320	318	420	185	125	4,600
28	1,440	320	170	170	1,100	1,050	1,440	367	339	152	100	5,160
29	1,800	320	120	145	....	1,150	1,520	363	339	150	92	3,500
30	5,760	350	120	160	....	1,950	1,600	324	474	209	102	2,200
31	6,720	....	120	130	....	4,000	....	363	....	541	110	....

NOTE.—Stage-discharge relation affected by ice from about Nov. 26 to Apr. 2; determination of discharge for this period based on gage heights corrected for effects of ice by means of five discharge measurements, observer's notes, and weather records. Discharge estimated for following periods for lack of gage-height record: Oct. 12, Nov. 9-10, 18-20, Apr. 3, 16-17, 21-23, May 11-15, and July 18-21.

Monthly discharge of Missisquoi River near Richford, Vt., for the year ending Sept. 30, 1918.

[Drainage area, 445 square miles.]

Month.	Discharge in second-feet.				Run-off (depth in inches on drainage area).
	Maximum.	Minimum.	Mean.	Per square mile.	
October	6,720	740	1,580	3.55	4.09
November	5,280	320	893	2.00	2.23
December	440	120	287	.645	.74
January	220	82	157	.353	.41
February	1,150	72	415	.933	.97
March	4,000	260	1,080	2.43	2.80
April	9,000	1,320	3,060	6.88	7.68
May	2,240	310	947	2.13	2.46
June	3,060	215	713	1.60	1.78
July	860	150	363	.816	.94
August	1,240	92	294	.661	.76
September	5,160	152	1,120	2.52	2.81
The year	9,000	72	906	2.04	27.67

CLYDE RIVER AT WEST DERBY, VT.

Location.—Just below plant of Newport Electric Light Co. at West Derby (Newport), Orleans County, about a mile above mouth of river.

Drainage area.—150 square miles.

Records available.—May 25, 1909, to September 30, 1918.

Gages.—Water-stage recorder on right bank; referred to gage datum by a hook gage inside the well; chain gage fastened to tree is used for auxiliary readings.

Discharge measurements.—Made by wading near gage or from highway bridge one-half mile downstream.

Channel and control.—Stream bed rough and irregular; covered with boulders and ledge rock; fall of river rapid for some distance below gage.

Extremes of discharge.—Maximum stage during year, from water-stage recorder 3.70 feet at 11 p. m. April 3 (discharge, 1,280 second-feet); minimum stage recorded 1.87 feet at 5 a. m. September 1 (discharge, 40 second-feet).

1909–1918: High water of March 25–30, 1913, reached maximum stage of 5.8 feet, as determined by engineers of Geological Survey from high-water marks (discharge about 6,300 second-feet); minimum stage, 1.60 feet at 5.45 p. m. August 25, 1913, 7.30 p. m. July 30, and 4.50 p. m. August 17, 1914 (discharge, 17 second-feet).

Ice.—Ice covers large boulders below gage during greater part of winter and causes some backwater. Winter discharge determined from gage heights, current-meter measurements, observer's notes, and climatic records.

Regulation.—Flow at ordinary stages fully controlled by two dams at West Derby, but power plant is so operated that fluctuations in stage are not great. Distribution of flow affected also by several dams above West Derby. Seymour Lake and several smaller ponds in the basin afforded a large amount of natural storage, but at the present time there is little if any artificial regulation at these ponds.

Accuracy.—Stage-discharge relation practically permanent, except when affected by ice; individual current-meter mesurements occasionally plot erratically, probably because of rough measuring section. Rating curve fairly well defined. Operation of water-stage recorder unsatisfactory during a part of the year, as indicated in footnote to daily-discharge table. Daily discharge ascertained by applying mean daily gage height to rating table, using observer's reading of chain gage when recorder was not in operation. Records fair.

Discharge measurements of Clyde River at West Derby, Vt., during the year ending Sept. 30, 1918.

Date.	Made by—	Gage height (feet).		Discharge (sec.-ft.).	Date.	Made by—	Gage height (feet).		Discharge (sec.-ft.)
		Hook gage.	Chain gage.				Hook gage.	Chain gage.	
Oct. 12	M. R. Stackpole	2.64	2.55	272	Mar. 28	M. R. Stackpole		2.70	357
Dec. 13	do	[a] 2.53	[a] 2.49	138	29	do		2.75	385
Jan. 29	do	[a] 2.15	[a] 2.08	80	July 23	C. H. Pierce	2.32	2.32	157
Mar. 5	do	2.48	2.42	215	Sept. 1	J. W. Moulton	2.15	2.15	98

[a] Stage-discharge relation affected by ice.

Daily discharge, in second-feet, of Clyde River at West Derby, Vt., for the year ending Sept. 30, 1918.

Day.	Oct.	Nov.	Dec.	Jan.	Feb.	Mar.	Apr.	May.	June.	July.	Aug.	Sept.
1	160	950	280	68	74	230	389	655	255	194	204	99
2	230	1,060	260	70	70	250	810	810	288	194	218	96
3	220	1,000	270	80	70	250	1,220	860	278	184	222	102
4	240	850	260	82	70	240	1,120	810	264	198	213	99
5	300	755	210	80	70	217	1,170	702	229	167	187	93
6	360	620	200	80	68	205	1,060	610	209	167	175	102
7	380	500	210	80	76	200	1,010	533	211	164	204	123
8	330	460	175	80	64	195	910	509	213	155	220	99
9	315	411	175	80	52	184	910	478	217	155	245	100
10	330	378	160	80	66	170	810	485	221	146	286	100
11	360	354	120	80	78	160	1,120	471	221	149	292	99
12	310	336	115	82	84	145	1,010	525	304	152	280	99
13	315	310	115	82	100	140	960	493	408	161	259	105
14	290	300	110	82	112	140	702	610	356	264	238	107
15	330	280	90	80	130	140	655	655	304	274	204	113
16	342	264	90	82	167	140	702	702	310	316	182	138
17	354	260	90	82	143	140	810	655	299	304	164	131
18	330	256	84	80	135	140	810	655	274	310	145	152
19	336	248	80	78	138	140	800	610	255	274	128	160
20	397	244	80	76	198	140	860	541	225	245	138	156
21	411	244	80	74	149	150	810	450	200	225	134	200
22	397	244	76	72	140	180	810	415	188	205	105	218
23	384	256	72	70	140	230	702	402	191	191	126	238
24	390	248	68	76	160	275	655	350	209	152	141	286
25	404	236	74	70	177	310	655	288	209	128	191	322
26	378	270	76	68	180	350	610	304	233	119	171	328
27	360	280	76	68	184	363	760	293	217	126	160	422
28	378	290	70	70	205	370	655	264	209	119	145	540
29	372	290	68	80	....	327	610	255	205	107	138	557
30	620	280	66	74	....	389	610	274	209	145	128	565
31	800	....	64	72	....	344	....	269	....	178	76	....

NOTE.—Stage-discharge relation affected by ice Nov. 26 to Dec. 2, and Dec. 7 to Feb. 13; determination of discharge for these periods based on gage heights corrected for effect of ice by means of two discharge measurements, observer's notes, and weather records. Discharge estimated for following periods owing to lack of gage-height records: Oct. 1–8, Nov. 7, Feb. 22–24, 28, Mar. 1–4, 6–8, 10–12, 14–16, 18–19, 21–26, Apr. 19, June 7, 20–21, Aug. 8–9, 31, and Sept. 9–10.

Monthly discharge of Clyde River at West Derby, Vt., for the year ending Sept. 30, 1918.

[Drainage area, 150 square miles.]

Month.	Discharge in second-feet.				Run-off (depth in inches on drainage area).
	Maximum.	Minimum.	Mean.	Per square mile.	
October	800	160	359	2.39	2.76
November	1,060	236	416	2.77	3.09
December	280	64	128	.853	.98
January	82	68	76.7	.512	.59
February	205	52	118	.787	.82
March	389	140	221	1.47	1.70
April	1,220	389	824	5.49	6.12
May	860	255	514	3.43	3.95
June	408	188	247	1.65	1.84
July	316	107	189	1.26	1.45
August	292	76	184	1.23	1.42
September	565	93	198	1.32	1.47
The year	1,220	52	290	1.93	26.19

INDEX.

M.

N.

O.

P.

R.

S.

V.

W.

STREAM-GAGING STATIONS

AND

PUBLICATIONS RELATING TO WATER RESOURCES

PART IV. ST. LAWRENCE RIVER BASIN

STREAM-GAGING STATIONS AND PUBLICATIONS RELATING TO WATER RESOURCES.

INTRODUCTION.

Investigation of water resources by the United States Geological Survey has consisted in large part of measurements of the volume of flow of streams and studies of the conditions affecting that flow, but it has comprised also investigations of such closely allied subjects as irrigation, water storage, water powers, underground waters, and quality of waters. Most of the results of these investigations have been published in the series of water-supply papers, but some have appeared in the bulletins, professional papers, monographs, and annual reports.

The results of stream-flow measurements are now published annually in 12 parts, each part covering an area whose boundaries coincide with natural drainage features, as indicated below:

Part I. North Atlantic slope basins.
II. South Atlantic slope and eastern Gulf of Mexico basins.
III. Ohio River basin.
IV. St. Lawrence River basin.
V. Upper Mississippi River and Hudson Bay basins.
VI. Missouri River basin.
VII. Lower Mississippi River basin.
VIII. Western Gulf of Mexico basins.
IX. Colorado River basin.
X. Great Basin.
XI. Pacific slope basins in California.
XII. North Pacific slope basins, in three volumes:
A, Pacific slope basins in Washington and upper Columbia River basin.
B, Snake River basin.
C, Lower Columbia River basin and Pacific slope basins in Oregon.

HOW GOVERNMENT REPORTS MAY BE OBTAINED OR CONSULTED.

Water-supply papers and other publications of the United States Geological Survey containing data in regard to the water resources of the United States may be obtained or consulted as indicated below:

1. Copies may be obtained free of charge by applying to the Director of the Geological Survey, Washington, D. C. The edition printed for free distribution is, however, small and is soon exhausted.

2. Copies may be purchased at nominal cost from the Superintendent of Documents, Government Printing Office, Washington, D. C., who will on application furnish list giving prices.

3. Sets of the reports may be consulted in the libraries of the principal cities in the United States.

4. Complete sets are available for consultation in the local offices of the water-resources branch of the Geological Survey, as follows:

Boston, Mass., 2500 Customhouse.
Albany, N. Y., 704 Journal Building.
Atlanta, Ga., Post Office Building.
Chicago, Ill., 1404 Kimball Building.
Madison, Wis., care of Railroad Commission of Wisconsin.
Helena, Mont., Montana National Bank Building.
Denver, Colo., 403 New Post Office Building.
Topeka, Kans., Room 23, Federal Building.
Salt Lake City, Utah, 313 Federal Building.
Boise, Idaho, 615 Idaho Building.
Tucson, Ariz., University of Arizona.
Austin, Tex., Capitol Building.
Portland, Oreg., 606 Post Office Building.
Tacoma, Wash., 406 Federal Building.
San Francisco, Calif., 328 Customhouse.
Los Angeles, Calif., 602 Federal Building.
Honolulu, Hawaii, 14 Capitol Building.

A list of the Geological Survey's publications may be obtained by applying to the Director of the United States Geological Survey, Washington, D. C.

STREAM-FLOW REPORTS.

Stream-flow records have been obtained at about 4,500 points in the United States, and the data obtained have been published in the reports tabulated below:

Stream-flow data in reports of the United States Geological Survey.

[A = Annual Report; B = Bulletin; W = Water-Supply Paper.]

Report.	Character of data.	Year.
10th A, pt. 2	Descriptive information only	
11th A, pt. 2	Monthly discharge and discriptive information	1884 to September, 1890.
12th A, pt. 2	do	1884 to June 30, 1891.
13th A, pt. 3	Mean discharge in second-feet	1884 to Dec. 31, 1892.
14th A, pt. 2	Monthly discharge (long-time records, 1871 to 1893)	1888 to Dec. 31, 1893.
B 131	Descriptions, measurements, gage heights, and ratings	1893 and 1894.
16th A, pt. 2	Descriptive information only	1895.
B 140	Descriptions, measurements, gage heights, ratings, and monthly discharge (also many data covering earlier years).	
W 11	Gage heights (also gage heights for earlier years)	1896.
18th A, pt. 4	Descriptions, measurements, ratings, and monthly discharge (also similar data for some earlier years).	1895 and 1896.
W 15	Descriptions, measurements, and gage heights, eastern United States, eastern Mississippi River, and Missouri River above junction with Kansas.	1897.
W 16	Descriptions, measurements, and gage heights, western Mississippi River below junction of Missouri and Platte, and western United States.	1897.
19th A, pt. 4	Descriptions, measurements, ratings, and monthly discharge (also some long-time records).	1897.
W 27	Measurements, ratings, and gage heights, eastern United States eastern Mississippi River, and Missouri River.	1898.
W 28	Measurements, ratings, and gage heights, Arkansas River and western United States.	1898.

Stream-flow data in reports of the United States Geological Survey—Continued.

Report.	Character of data.	Year.
20th A, pt. 4	Monthly discharge (also for many earlier years)	1898.
W 35 to 39	Descriptions, measurements, gage heights, and ratings	1899.
21st A, pt. 4	Monthly discharge	1899.
W 47 to 52	Descriptions, measurements, gage heights, and ratings	1900.
22d A, pt. 4	Monthly discharge	1900.
W 65, 66	Descriptions, measurements, gage heights, and ratings	1901.
W 75	Monthly discharge	1901.
W 82 to 85	Complete data	1902.
W 97 to 100	...do	1903.
W 124 to 135	...do	1904.
W 165 to 178	...do	1905.
W 201 to 214	...do	1906.
W 241 to 252	...do	1907-8.
W 261 to 272	...do	1909.
W 281 to 292	...do	1910.
W 301 to 312	...do	1911.
W 321 to 332	...do	1912.
W 351 to 362	...do	1913.
W 381 to 394	...do	1914.
W 401 to 414	...do	1915.
W 431 to 444	...do	1916.
W 451 to 464	...do	1917.
W 471 to 484	...do	1918.

The records at most of the stations discussed in these reports extend over a series of years, and miscellaneous measurements at many points other than regular gaging stations have been made each year. An index of the reports containing records obtained prior to 1904 has been published in Water-Supply Paper 119.

The following table gives, by years and drainage basins, the numbers of the papers on surface-water supply published from 1899 to 1918. The data for any particular station will, as a rule, be found in the reports covering the years during which the station was maintained. For example, data for Machias River at Whitneyville, Me., 1903 to 1918, are published in Water-Supply Papers 97, 124, 165, 201, 241, 261, 281, 301, 321, 351, 381, 401, 431, 451, and 471, which contains records for the New England streams from 1903 to 1918. Results of miscellaneous measurements are published by drainage basins.

In these papers and in the following lists the stations are arranged in downstream order. The main stem of any river is determined by measuring or estimating its drainage area—that is, the headwater stream having the largest drainage area is considered the continuation of the main stream, and local changes in name and lake surface are disregarded. All stations from the source to the mouth of the main stem of the river are presented first, and the tributaries in regular order from source to mouth follow, the streams in each tributary basin being listed before those of the next basin below.

The exceptions to this rule occur in the records for Mississippi River, which are given in four parts, as indicated on page III, and in the records for the large lakes, where it is simpler to take up the streams in regular order around the rim of the lake than to cross back and forth over the lake surface.

Numbers of water-supply papers containing results of stream measurements, 1899-1918.

Year.	I North Atlantic slope (St. John River to York River).	II South Atlantic slope and eastern Gulf of Mexico (James River to the Mississippi).	III Ohio River basin.	IV St. Lawrence River and Great Lakes basins.	V Hudson Bay and upper Mississippi River basins.	VI Missouri River basin.	VII Lower Mississippi River basin.	VIII Western Gulf of Mexico basins.	IX Colorado River basin.	X Great Basin.	XI Pacific slope basins in California.	XII North Pacific slope basins.		
												Pacific slope basins in Washington and upper Columbia River.	Snake River basin.	Lower Columbia River and Pacific slope basins in Oregon.
1899 [a]	35	[b] 35, 36	36	36	36	[c] 36, 37	37	37	[d] 37, 38	38, [e] 39	38, [f] 39	38	38	38
1900 [g]	47, [h] 48	48	48, [i] 49	49	49	49, [j] 50	50	50	50	51	51	51	51	51
1901	65, 75	65, 75	65, 75	65, 75	[k] 65, 66, 75	66, 75	[k] 65, 66, 75	66, 75	66, 75	66, 75	66, 75	66, 75	66, 75	66, 75
1902	82	[b] 82, 83	83	[l] 82, 83	[k] 83, 85	84	[k] 83, 84	84	85	85	85	85	85	85
1903	97	[b] 97, 98	98	97	[k] 98, 99, [m] 100	99	[k] 98, 99	99	100	100	100	100	100	100
1904	[n] 124, [o] 125, [p] 126	[p] 126, 127	128	129	[k] 128, 130	130, [q] 131	[k] 128, 131	132	133	133, [r] 134	134	135	135	135
1905	[n] 165, [o] 166, [p] 167	[p] 167, 168	169	170	171	172	[k] 179, 173	174	175, [s] 177	176, [r] 177	177	178	178	[t] 177, 178
1906	[n] 201, [o] 202, [p] 203	[p] 203, 204	205	206	207	208	[k] 205, 209	210	211	212, [r] 213	213	214	214	214
1907-8	241	242	243	244	245	246	247	248	249	250, [r] 251	251	252	252	252
1909	261	262	263	264	265	266	267	268	269	270, [r] 271	271	272	272	272
1910	281	282	283	284	285	286	287	288	289	290	291	292	292	292
1911	301	302	303	304	305	306	307	308	309	310	311	312	312	312
1912	321	322	323	324	325	326	327	328	329	330	331	332A	332B	332C
1913	351	352	353	354	355	356	357	358	359	360	361	362A	362B	362C
1914	381	382	383	384	385	386	387	388	389	390	391	392	393	394
1915	401	402	403	404	405	406	407	408	409	410	411	412	413	414
1916	431	432	433	434	435	436	437	438	439	440	441	442	443	444
1917	451	452	453	454	455	456	457	458	459	460	461	462	463	464
1918	471	472	473	474	475	476	477	478	479	480	481	482	483	484

[a] Rating tables and index to Water-Supply Papers 35-39 contained in Water Supply Paper 39. Estimates for 1899 in Twenty-first Annual Report, Part IV.
[b] James River only.
[c] Gallatin River.
[d] Green and Gunnison rivers and Grand River above junction with Gunnison.
[e] Mohave River only.
[f] Kings and Kern rivers and south Pacific coast basins.
[g] Rating tables and index to Water-Supply Papers 47-52 and data on precipitation, wells, and irrigation in California and Utah contained in Water-Supply Paper 52. Estimates for 1900 in Twenty-second Annual Report, Part IV.
[h] Wissahickon and Schuylkill rivers to James River.
[i] Scioto River.
[j] Loup and Platte rivers near Columbus, Nebr., and all tributaries below junction with Platte.
[k] Tributaries of Mississippi from east.
[l] Lake Ontario and tributaries to St. Lawrence River proper.
[m] Hudson Bay only.
[n] New England rivers only.
[o] Hudson River to Delaware River, inclusive.
[p] Susquehanna River to Yadkin River, inclusive.
[q] Platte and Kansas rivers.
[r] Great Basin in California except Truckee and Carson river basins.
[s] Below junction with Gila.
[t] Rogue, Umpqua, and Siletz rivers only.

PRINCIPAL STREAMS.

The St. Lawrence River basin includes streams which drain into the Great Lakes and St. Lawrence River. The principal streams flowing directly or indirectly into Lake Superior from the United States are St. Louis, Ontonagon, Dead, and Carp rivers; streams flowing into Lake Michigan are Escanaba, Menominee, Peshtigo, Oconto, Fox, St. Joseph, and Grand rivers; into Lake Huron flow Thunder Bay, Ausable, Rifle, and Saginaw rivers; into Lake Erie flow Huron, Maumee, Sandusky, Black, and Cuyahoga rivers. Streams flowing into Lake Ontario are Genesee, Oswego, Salmon, and Black rivers. The St. Lawrence receives Oswegatchie and Raquette rivers, Richelieu River (the outlet of Lake Champlain), and St. Francis River, whose principal tributary, Clyde River, reaches it through Lake Memphremagog. The streams of this basin drain wholly or in part the States of Illinois, Indiana, Michigan, Minnesota, New York, Ohio, Pennsylvania, Vermont, and Wisconsin.

In addition to the list of gaging stations and annotated list of publications relating specifically to the section, this part contains a similar list of reports that are of general interest in many sections and cover a wide range of hydrologic subjects, and also brief references to reports published by State and other organizations. (See pp. xvii-xviii.)

GAGING STATIONS.

NOTE.—Dash following a date indicates that station was being maintained September 30, 1918. Period after date indicates discontinuance.

Streams tributary to Lake Superior:
- Brule River at mouth, Minn., 1911.
- Devil Track River at mouth, Minn., 1911.
- Cascade River at mouth, Minn., 1911.
- Poplar River at Lutsen, Minn., 1911-1917.
- Beaver Bay River at Beaver Bay, Minn., 1911-1914.
- St. Louis River near Cloquet, Minn., 1903.
- St. Louis River near Thomson, Minn., 1909-1915.
  - Whiteface River at Meadowlands, Minn., 1909-1912.
  - Whiteface River below Meadowlands, Minn., 1912-1917.
  - Cloquet River at Independence, Minn., 1909-1917.
- Aminicon River near Aminicon Falls, Wis., 1914-1916.
- Brule River near Brule, Wis., 1914-1917.
- Bad River near Odanah, Wis., 1914-
- Montreal River at Ironwood, Mich., 1918-
  - West Branch of Montreal River at Gile, Wis., 1918-
- Ontonagon River near Rockland, Mich., 1903.
- Sturgeon River near Sidnaw, Mich., 1912-1915.
- Perch River near Sidnaw, Mich., 1912-1915.
- Dead River near Negaunee, Mich., 1902-3.
- Dead River at Forestville, Mich., 1898-1902.
- Carp River near Marquette, Mich., 1902-3.

Streams tributary to Lake Michigan:
Escanaba River near Escanaba, Mich., 1903–1915.
Brule River (head of Menominee River) near Florence, Wis., 1914–1916.
Menominee River near Iron Mountain, Mich., 1902–1914.
Menominee River at Lower Quinnesec Falls, Wis., 1898–99.
Menominee River at Koss, Mich., 1902–1909; 1914.
Menominee River below Koss, Mich., 1913–
Iron River near Iron River, Mich., 1900–1905.
Pine River near Florence, Wis., 1914–
Pike River at Amberg., Wis., 1914–
Peshtigo River at High Falls, near Crivitz, Wis., 1912–
Peshtigo River near Crivitz, Wis., 1906–1909.
Peshtigo River at Crivitz, Wis., 1906.
Oconto River near Gillett, Wis., 1906–1909; 1914–
Oconto River at Stiles, Wis., 1906.
Fox River at Berlin, Wis., 1918–
Fox River at Omro, Wis., 1902–3.
Fox River at Oshkosh, Wis., 1902.
Fox River at Wrightstown, Wis., 1902–1904.
Fox River at Rapide Croche dam, Wis., 1896–
Wolf River at Keshena, Wis., 1907–1909; 1911–
Wolf River at White House Bridge, near Shawano, Wis., 1906–7.
Wolf River at Darrows Bridge, near Shawano, Wis., 1906.
Wolf River at New London, Wis., 1913–
Wolf River at Northport, Wis., 1905.
Wolf River at Winneconne, Wis., 1902–3.
West Branch of Wolf River at Neopit, Wis., 1911–1917.
Little Wolf River at Royalton, Wis., 1914–
Little Wolf River near Northport, Wis., 1907–1910.
Waupaca River near Weyauwega, Wis. 1916–17.
Waupaca River near Waupaca, Wis., 1917–
Fond du Lac River, West Branch (head of Fond du Lac River), at Fond du Lac, Wis., 1903.
East Branch of Fond du Lac River at Fond du Lac, Wis., 1903.
Sheboygan River near Sheboygan, Wis., 1916–
Milwaukee River near Milwaukee, Wis., 1914–
Little Calumet River at Harvey, Ill., 1916–
St. Joseph River at Mendon, Mich., 1902–1905.
St. Joseph River near Buchanan, Mich., 1901–1906.
Fawn River at White Pigeon, Mich., 1903–4.
Kalamazoo River near Allegan, Mich., 1901–1907.
Reeds Springs near Albion, Mich., 1904–1906.
Grand River at North Lansing, Mich., 1901–1906.
Grand River at Grand Rapids, Mich., 1901–
Crockery Creek at Slocums Grove, Mich., 1902–3.
Red Cedar River at Agricultural College, Mich., 1902–3.
Muskegon River at Newaygo, Mich., 1901–1906.
Manistee River near Sherman, Mich., 1903–1916.
Boardman River at Traverse City, Mich., 1904.
Streams tributary to Lake Huron:
Thunder Bay River near Alpena, Mich., 1901–1908.
Au Sable River near Lovells, Mich., 1908–1914.
Au Sable River at Bamfield, Mich., 1902–1913.
Rifle River near Sterling, Mich., 1905–1906.

Streams tributary to Lake Huron—Continued.
- Rifle River at Omer, Mich., 1902-3.
- Shiawassee River (head of Saginaw River):
  - Flint River at Flint, Mich., 1903-4.
  - Cass River at Frankenmuth, Mich., 1908-9.
  - Cass River at Bridgeport, Mich., 1908.
  - Tittabawassee River at Freeland, Mich., 1903-1909; 1912-

Streams tributary to Lake Erie:
- Huron River at Dover, Mich., 1904.
- Huron River at Dexter, Mich., 1904-1916.
- Huron River at Barton, Mich., 1914-
- Huron River at Geddes, Mich., 1904-1914.
- Huron River at French Landing, Mich., 1904-5.
- Huron River at Flat Rock, Mich., 1904-
- Maumee River near Sherwood, Ohio, 1903-1906.
- Maumee River near Waterville, Ohio, 1898-1901.
  - St. Marys River at Fort Wayne, Ind., 1905-6.
  - St. Joseph River at Fort Wayne, Ind., 1905-6.
  - Tiffin River near Defiance, Ohio, 1903-1906.
  - Auglaize River near Defiance, Ohio, 1903.
    - Ottawa River at Lima, Ohio, 1902-3.
      - Blanchard River at Ottawa, Ohio, 1902-3.
- Sandusky River near Mexico, Ohio, 1898-1900.
- Sandusky River at Fremont, Ohio, 1898-1901.
- Black River near Elyria, Ohio, 1903-1906.
- Cuyahoga River at Independence, Ohio, 1903-1906.
- Cuyahoga River at Cleveland, Ohio, 1903.
- Cattaraugus Creek at Versailles, N. Y., 1910-

Streams tributary to Lake Ontario:
- Niagara River:
  - Tonawanda Creek:
    - Little Tonawanda Creek near Linden, N. Y., 1912-
- Genesee River at Scio, N. Y., 1916-
- Genesee River at St. Helena, N. Y., 1908-
- Genesee River at Mount Morris, N. Y., 1905-1909.
- Genesee River at Jones Bridge, near Mount Morris, N. Y., 1903-1906; 1908-1913; 1915-
- Genesee River at Rochester, N. Y., 1904-
  - Canaseraga Creek near Dansville, N. Y., 1910-1912; 1915-1917.
  - Canaseraga Creek at Cumminsville, N. Y., 1917-
  - Canaseraga Creek at Groveland Station, N. Y., 1915-
  - Canaseraga Creek at Shakers Crossing, N. Y., 1915-
    - Keshequa Creek at Sonyea, N. Y., 1910-1912; 1917-
    - Keshequa Creek near Sonyea, N. Y., 1915-1917.
  - Hemlock Lake at Hemlock, N. Y., 1894-1902.
    - Canadice Lake outlet near Hemlock, N. Y., 1903-
    - Honeoye Creek at East Rush, N. Y., 1903-1906.
- Seneca River (head of Oswego River) at Baldwinsville, N. Y., 1898-1908.
- Oswego River at Fulton, N. Y., 1900; 1902.
- Oswego River at Battle Island, above Minetto, N. Y., 1900-1906.
- Oswego River at high dam, near Oswego, N. Y., 1897-1901.
  - Seneca Lake at Geneva, N. Y., 1905-6.
  - Cayuga Lake at Ithaca, N. Y., 1905-1908.
    - Fall Creek near Ithaca, N. Y., 1908-9.

Streams tributary to Lake Ontario—Continued.
- Streams tributary to Oswego River—Continued.
    - Owasco Lake outlet near Auburn, N. Y., 1912–
    - Skaneateles Lake at Skaneateles, N. Y., 1890–91.
    - Skaneateles Lake outlet at Willow Glen, N. Y., 1892–1908.
    - Skaneateles Lake outlet at Jordan, N. Y., 1890–1892.
    - Onondaga Lake outlet at Long Branch, N. Y., 1904.
    - West Branch of Onondaga Creek at South Onondaga, N. Y., 1916–
    - Fish Creek, East Branch (through Oneida Lake, head of Oneida River), at Point Rock, N. Y., 1898–99.
    - Oneida River at Brewerton, N. Y., 1899.
    - Oneida River at Oak Orchard, near Euclid, N. Y., 1902–1909.
    - Oneida River at Caughdenoy, N. Y., 1910–1913.
        - Fish Creek:
            - West Branch of Fish Creek at McConnelsville, N. Y., 1898–1901.
            - Oneida Creek at Kenwood, N. Y., 1898–1900.
            - Chittenango Creek at Chittenango, N. Y., 1901–1906.
            - Chittenango Creek at Bridgeport, N. Y., 1898–1901.
- Salmon River at Stillwater Bridge, near Redfield, N. Y., 1911–1913.
- Salmon River near Pulaski, N. Y., 1900–1908; 1910–1914.
    - Orwell Brook near Altmar, N. Y., 1911–1916.
- Black River near Boonville, N. Y., 1911–
- Black River near Felts Mills, N. Y., 1902–1913.
- Black River at Black River, N. Y., 1917–
- Black River at Huntingtonville dam, near Watertown, N. Y., 1897–1901.
    - Forestport feeder near Boonville, N. Y., 1915–
    - Black River canal (flowing south) near Boonville, N. Y., 1915–
    - Moose River at Moose River, N. Y., 1900–
        - Middle Branch of Moose River at Old Forge, N. Y., 1911–
    - Beaver River at State dam near Beaver River, N. Y., 1908–
    - Beaver River at Croghan, N. Y., 1901–1903.

Streams tributary to St. Lawrence River:
- Oswegtachie River, East Branch (head of Oswegatchie River), at Newton Falls, N. Y., 1912–
- Oswegatchie River near Heuvelton, N. Y., 1916–
- Oswegatchie River near Ogdensburg, N. Y., 1903–1916.
    - West Branch of Oswegatchie River near Harrisville, N. Y., 1916–
- Raquette River at Raquette Falls, near Coreys, N. Y., 1908–1912.
- Raquette River at Piercefield, N. Y., 1908–
- Raquette River at South Colton, N. Y., 1904.
- Raquette River at Massena Springs, N. Y., 1903–1916.
    - Bog River near Tupper Lake, N. Y., 1908–1912.
- St. Regis River at Brasher Center, N. Y., 1910–
    - Deer River at Brasher Iron Works (railroad station), Ironton, N. Y., 1912–1916.
- Chateaugay River near Chateaugay, N. Y., 1908.
- Richelieu River at Fort Montgomery, N. Y., 1875–
    - Lake Champlain at Burlington, Vt., 1907–
        - Big Chazy River at Moors, N. Y., 1908.
        - Saranac River at Saranac Lake, N. Y., 1902–3.
    - Saranac River near Plattsburg, N. Y., 1903–
    - Ausable River, West Branch, near Newman, N. Y., 1916–1917.
    - Ausable River at Ausable Forks, N. Y., 1910–
    - Ausable River at Keeseville, N. Y., 1904 and 1908.

Streams tributary to St. Lawrence River—Continued.
Streams tributary to Richelieu River—Continued.
Boquet River at Willsboro, N. Y., 1904 and 1908.
Lake George at Rogers Rock, N. Y., 1913–
Lake George outlet at Ticonderoga, N. Y., 1904–5.
Poultney River at Fairhaven, Vt., 1908.
Mettawee River at Whitehall, N. Y., 1908.
Otter Creek at Middlebury, Vt., 1903–1907; 1910–
East Creek near Rutland, Vt., 1911–1913.
Winooski River above Stevens Branch, near Montpelier, Vt., 1909–1914.
Winooski River at Montpelier, Vt., 1909–
Winooski River at Richmond, Vt., 1903–1907; 1910.
Winooski River near Winooski, Vt., 1903.
Worcester Branch of Winooski River at Montpelier, Vt., 1909–1914.
Dog River at Northfield, Vt., 1909–
Dog River near Montpelier Junction, Vt., 1910.
Mad River at Moretown, Vt., 1910.
Little River near Waterbury, Vt., 1910.
Huntington River at Jonesville, Vt., 1910.
Lamoille River at Morrisville, Vt., 1909–10.
Lamoille River at Cadys Falls, near Morrisville, Vt., 1913–
Lamoille River at Johnson, Vt., 1910–1913.
Lamoille River at West Milton, Vt., 1903.
Green River at Garfield, Vt., 1915–
Missisquoi River at Richford, Vt., 1909–10.
Missisquoi River near Richford, Vt., 1911–
Missisquoi River at Swanton, Vt., 1903.
St. Francis River (by way of Lake Memphremagog and Magog River):
Clyde River at West Derby, Vt., 1909–

REPORTS ON WATER RESOURCES OF THE ST. LAWRENCE RIVER BASIN.[1]

PUBLICATIONS OF THE UNITED STATES GEOLOGICAL SURVEY.

WATER-SUPPLY PAPERS.

Water-supply papers are distributed free by the Geological Survey as long as its stock lasts. An asterisk (*) indicates that this stock has been exhausted. Many of the papers marked in this way may, however, be purchased from the SUPERINTENDENT OF DOCUMENTS, WASHINGTON, D. C. Water-supply papers are of octavo size.

*21. Wells of northern Indiana, by Frank Leverett. 1899. 82 pp., 2 pls. (Continued in No. 26.)

Discusses, by counties, the glacial deposits and the sources of well water; gives many well sections.

*24. Water resources of the State of New York, Part I, by G. W. Rafter. 1899. 99 pp., 13 pls. 15c.

*25. Water resources of the State of New York, Part II, by G. W. Rafter. 1899. 100 pp., 12 pls. 15c.

No. 24 contains descriptions of the principal rivers of New York and their more important tributaries and data on temperature, precipitation, evaporation, and stream flow.

No. 25 contains discussion of water-storage projects on Genesee and Hudson Rivers, power development at Niagara Falls, description and early history of State canals, and a chapter on the use and value of the water powers of the streams and canals; also brief discussion of the water yield of sand areas of Long Island.

*26. Wells of southern Indiana (continuation of No. 21), by Frank Leverett. 1899. 64 pp. 5c.

Discusses, by counties, the glacial deposits and the sources of well water; contains many well sections.

30. Water resources of the Lower Peninsula of Michigan, by A. C. Lane. 1899. 97 pp., 7pls.

Describes lake and river transportation and navigation, water powers and domestic water supplies; discusses climate, topography, geology, and well waters; compares quality and quantity of waters.

*31. Lower Michigan mineral waters, by A. C. Lane. 1899. 97 pp., 4 pls. 10c.

Treats of economic value of mineral waters and discussion and classification of analyses; contains analyses of waters of Lake Superior and of smaller lakes and rivers and of well waters from various geologic formations; also sanitary condition of drinking waters.

*57. Preliminary list of deep borings in the United States, Part I (Alabama-Montana), by N. H. Darten. 1902. 60 pp. (See No. 149.) 5c.

*61. Preliminary list of deep borings in the United States, Part II (Nebraska-Wyoming), by N. H. Darton. 1902. 67 pp. 5c.

Nos. 57 to 61 contain information as to depth, diameter, yield, and head of water in borings more than 400 feet deep; under head "Remarks" give information concerning temperature, quality of water, purposes of boring, etc. The lists are arranged by States, and the States are arranged alphabetically. A second, revised, edition was published in 1906 as Water-Supply Paper 149 (q. v.).

91. The natural features and economic development of the Sandusky, Maumee, Muskingum, and Miami drainage areas in Ohio, by B. H. and M. S. Flynn. 1904. 130 pp. 10c.

Describes the topography, geology, and soils of the areas, and discusses stream flow, dams, water powers, and public water supplies.

---

[1] For stream-measurement reports, see tables on pp. IV, V, VI.

102. Contributions to the hydrology of eastern United States, 1903; M. L. Fuller, geologist in charge. 1904. 522 pp. 30c.

Contains brief reports on wells and springs of Minnesota and of lower Michigan. The report comprises tabulated well records giving information as to location, owner, depth yield, head, etc., supplemented by notes as to elevation above sea, materials penetrated, temperature, use and quality; many miscellaneous analyses.

*103. A review of the laws forbidding pollution of inland waters in the United States, by E. B. Goodell. 1904. 120 pp. Superseded by 152.

Cites statutory restrictions of water pollution.

110. Contributions to the hydrology of Eastern United States, 1904; M. L. Fuller, geologist in charge. 1905. 211 pp., 5 pls. 10c.

Contains:

Water resources of the Watkins Glen quadrangle, New York, by Ralph S. Tarr; pp. 134-140. Discusses the use of the surface and underground waters for municipal supplies and their quality as indicated by examination of Sixmile and Fall creeks, and sanitary analyses of well water at Ithaca.

New artesian water supply at Ithaca, New York, by F. L. Whitney, pp. 55-64.

*114. Underground waters of eastern United States; M. L. Fuller, geologist in charge. 1905. 285 pp., 18 pls. 25c.

Contains brief reports as follows:

Minnesota, by C. W. Hall; Wisconsin district, by Alfred R. Schultz; Lower Michigan; Illinois, by Frank Leverett; Indiana, by Frank Leverett; New York, by F. B. Weels; Ohio, by Frank Leverett.

Each of these reports describes briefly the topography of the area, the relation of the geology to the water supplies, and gives list of pertinent publications; lists also principal mineral springs.

121. Preliminary report on the pollution of Lake Champlain, by M. O. Leighton. 1905. 119 pp., 13 pls. 20c.

Describes the lake and principal inflowing streams and discusses the characteristics of the water and the wastes resulting from the manufacturing processes by which the waters are polluted. Discusses also the effect of mill waste on algæ, bacteria, and fish.

*122. Relation of the law to underground waters, by D. W. Johnson. 1905. 55 pp. 5c.

Cites legislative acts relating to ground waters in Michigan and Wisconsin.

144. The normal distribution of chlorine in the natural waters of New York and New England, by D. D. Jackson. 1905. 31 pp., 5 pls. 10c.

Discusses common salt in coast and inland waters, salt as an index to pollution of streams and wells, the solutions and methods used in chlorine determinations, and the use of the normal chlorine map; gives charts and tables for chlorine in the New England States and New York.

145. Contributions to the hydrology of eastern Unites States, 1905; M. L. Fuller, geologist in charge. 1905. 220 pp., 6 pls. 10c

Contains three brief reports pertaining chiefly to areas in the St. Lawrence River basin:

Two unusual types of artesian flow, by Myron L. Fuller. Describes (1) artesian flows from uniform, unconfined sand on Long Island, N. Y., and in Michigan; and (2) flow from jointed upper portions of limestone and other rocks in southeastern Michigan.

Water resources of the Catatonk area, New York, by E. M. Kindle. Describes topography and geology of areas southeast of Finger Lake region, New York, including part of city of Ithaca; discusses briefly the artesian wells of Ithaca, the quality of the spring water at several small towns, and of the streams used for municipal supplies and for power.

A ground-water problem in southeastern Michigan, by Myron L. Fuller. Discusses causes of failure of wells in certain areas in southeastern Michigan in 1904 and the applications of the conclusions to other regions.

147. Destructive floods in the United States in 1904, by E. C. Murphy and others. 1905. 206 pp., 18 pls. 15c.

Describes flood on Grand River, Mich. (from report of R. E. Horton), discussing streams precipitation, and temperature, discharge, damage, and prevention of future damage.

*149. Preliminary list of deep borings in the United States, second edition, with additions, by N. H. Darton. 1905. 175 pp. 10c.

Gives by States (and within the States by counties) the location, depth, diameter, yield, height of water, and other features of wells 400 feet or more in depth; includes all wells listed in Water-Supply Papers 57 to 61; mentions also principal publications relating to deep borings.

*152. A review of the laws forbidding pollution of inland waters in the United States (second edition), by E. B. Goodell. 1905. 140 pp. 10c.

Cites statutory restrictions of water pollution in Illinois, Indiana, Michigan, Minnesota, New York, Ohio, Pennsylvania, Vermont, and Wisconsin.

*156. Water powers of northern Wisconsin, by L. S. Smith. 1906. 145 pp., 5 pls. 25c.

Describes, by river systems, the drainage, geology, topography, rainfall, and run-off, water powers and dams.

*160. Underground-water papers, 1906; M. L. Fuller, geologist in charge. 1906. 104 pp., 1 pl.

Contains brief report entitled "Flowing well districts in the eastern part of the northern peninsula of Michigan," by Frank Leverett.

*162. Destructive floods in the United States in 1905, with a discussion of flood discharge and frequency and an index to flood literature, by E. C. Murphy and others. 1906. 105 pp., 4 pls. 15c.

Contains accounts of floods on Sixmile Creek and Cayuga Inlet, N. Y. (in 1857, 1901, and 1905) and on Grand River, Mich., and estimate of flood discharge and frequency for Genesee River; gives index to literature on floods in American streams.

*182. Flowing wells and municipal water supplies in the southern portion of the southern peninsula of Michigan, by Frank Leverett and others. 1906. 292 pp., 5 pls. 50c.

*183. Flowing wells and municipal water supplies in the middle and northern portions of the southern peninsula of Michigan, by Frank Leverett and others. 1907. 393 pp., 5 pls. 50c.

Nos. 182 and 183 describe in general the geographic features, water-bearing formations, drainage, quality of water, and subterranean-water temperature, and give details concerning water supplies by counties. The report contains many analyses.

*193. The quality of surface waters in Minnesota, by R. B. Dole and F. F. Wesbrook. 1907. 171 pp., 7 pls. 25c.

Describes by river basins the topography, geology, and soils, the industrial and municipal pollution of the streams, and gives notes on the municipalities; contains many analyses.

*194. Pollution of Illinois and Mississippi rivers by Chicago sewage (a digest of the testimony taken in the case of the State of Missouri *v.* the State of Illinois and the Sanitary District of Chicago), by M. O. Leighton. 1907. 369 pp., 2 pls.

Scope indicated by amplification of title.

236. The quality of surface waters in the United States: Part I, Analyses of waters east of the one hundredth meridian, by R. B. Dole. 1909. 123 pp. 10c.

Describes collection of samples, method of examination, preparation of solutions, accuracy of estimates, and expression of analytical results; gives results of analyses of waters of Lake Superior and Lake Michigan, Kalamazoo and Grand rivers, Lake Huron, Lake Erie, Maumee River and St. Lawrence and Oswegatchie rivers.

239. The quality of the surface waters of Illinois, by W. D. Collins. 1910. 94 pp., 3 pls. 10c.

Discusses the natural and economic features that determine the character of the streams, describes the larger drainage basins and the methods of collecting and analyzing the samples of water, and discusses each river in detail with reference to its source, course, and quality of water includes short chapters on municipal supplies and industrial uses.

254. The underground waters of north-central Indiana, by S. R. Capps, with a chapter on the chemical character of the waters, by R. B. Dole. 1910. 279 pp., 7 pls. 40c.

Describes relief, drainage, vegetation, soils and crops, industrial development, geologic formations; sources, movements, occurence, and volume of ground water; methods of well construction and lifting devices; discusses in detail, for each county, surface features and drainage, geology, and ground water, city, village, and rural supplies, and gives record of wells and analyses of water. Discusses also, under chemical character, methods of analyses and expression of results, mineral constituents, effects of the constituents on waters for domestic, industrial, and medicinal uses, methods of purification and chemical composition; many analyses and field assays.

364. Water analyses from the laboratory of the United States Geological Survey, tabulated by F. W. Clarke, chief chemist. 1914. 40 pp. 5c.

Contains analyses of water from Caledonia Spring, New York, and from the Quincy mine, Mich.

417. Profile surveys of rivers in Wisconsin, prepared under the direction of W. H. Herron, acting chief geographer. 1917. 16 pp., 32 pls. 45c.

Contains brief description of general features of drainage of Wisconsin and of the rivers surveyed, but consists chiefly of maps showing "not only the outlines of the river banks, the islands, the positions of rapids, falls, shoals, and existing dams, and the crossings of all ferries and roads, but the contours of banks to an elevation high enough to indicate the possibility of using the stream."

ANNUAL REPORTS.

Each of the papers contained in the annual reports was also issued in separate form.

Annual reports are distributed free by the Geological Survey as long as its stock lasts. An asterisk (*) indicates that this stock has been exhausted. Many of the papers so marked, however, may be purchased from the SUPERINTENDENT OF DOCUMENTS, WASHINGTON, D. C.

Annual reports 1 to 26 are royal octavo; later reports are octavo.

Fourteenth Annual Report of the United States Geological Survey, 1892–93, J. W. Powell, Director. 1893. (Pt. II, 1894.) 2 parts. *Pt. II. Accompanying papers, xx, 597 pp., 73 pls. $2.10. Contains:

*The potable waters of eastern United States, by W. J. McGee, pp. 1 to 47. Discusses cistern water, stream waters, and ground waters, including mineral springs and artesian wells.

Seventeenth Annual Report of the United States Geological Survey, 1895–96, Charles D. Walcott, Director. 1896. 3 parts in 4 vols. *Pt. II. Economic geology and hydrography, xxv, 864 pp., 113 pls. $2.35. Contains:

*The water resources of Illinois, by Frank Leverett, pp. 695–849, pls. 108–113. Describes the physical features of the State, and the drainage basins, including Illinois, Des Plaines, Kankakee, Fox, Illinois-Vermilion, Spoon, Mackinaw, and Sangamon rivers, Macoupin Creek, Rock River, tributaries of the Mississippi in western Illinois, Kaskaskia, Big Muddy, and tributaries of the Wabash: discusses the rainfall and run-off, navigable waters and water powers, the wells supplying water for rural districts, and artesian wells; contains tabulated artesian well data and water analyses.

Eighteenth Annual Report, United States Geological Survey, 1896–97, Charles D. Walcott, Director. 1897. 5 parts in 6 volumes. *Pt. IV. Hydrography, x, 756 pp., 102 pls. $1.75. Contains:

*The water resources of Indiana and Ohio, by Frank Leverett, pp. 419–560, pls 33–37. Describes Wabash, Whitewater, Great Miami, Little Miami, Scioto, Hocking, Muskingum, and Beaver rivers and lesser tributaries of the Ohio in Indiana and Ohio, the streams discharging into Lake Erie and Lake Michigan, and streams flowing to the Upper Mississippi through the Illinois; discusses shallow and drift wells, the flowing wells from the drift and deeper artesian wells, and gives records of wells at many of the cities; describes the mineral springs and gives analyses of the waters; contains also tabulated lists of cities using surface waters for water works, and of cities and villages using shallow and deep well waters; discusses the source and quality of the city and village supplies, and gives precipitation tables for various points.

Nineteenth Annual Report of the United States Geological Survey, 1897–98, Charles D. Walcott, Director. 1898. (Pts. II, III, and V, 1899.) 6 parts in 7 volumes and separate case for maps with Pt. V. *Pt. IV. Hydrography. $1.85. Contains:

*The rock waters of Ohio, by Edward Orton, pp. 633-717, pls. 71-73. Describes the principal geologic formations of Ohio and the waters from the different strata; discusses the flowing wells at various points and the artesian wells of the deep prelacial channels in Allen, Auglaize, and Mercer counties: discusses city and village supplies; gives analyses of waters from various formations.

MONOGRAPHS.

Monographs are of quarto size. They are not distributed free, but may be obtained from the Geological Survey or from the Superintendent of Documents at the prices given. An asterisk (*) indicates that the Survey's stock of the paper is exhausted. (See Finding lists, pp. 89, 118.)

41. Glacial formations and drainage features cf the Erie and Ohio basins, by Frank Leverett. 1902. 802 pp., 26 pls. $1.75.

Treats of an area extending westward from Genesee Valley in New York across northwestern Pennsylvania and Ohio, central and southern Indiana, and southward from Lakes Ontario and Erie to Allegheny and Ohio rivers.

BULLETINS.

An asterisk (*) indicates that the Geological Survey's stock of paper is exhausted. Many of the papers so marked may be purchased from the SUPERINTENDENT OF DOCUMENTS, WASHINGTON, D. C.

*264. Record of deep-well drilling for 1904, by M. L. Fuller, E. F. Lines, and A. C. Veatch. 1905. 106 pp. 10c.

Discusses the importance of accurate well records to the driller, to owners of oil, gas, and water wells, and to the geologist; describes the general methods of work; gives tabulated records of wells in Illinois, Indiana, Michigan, Minnesota, New York, Ohio, Pennsylvania, and Wisconsin, and detailed record of wells in Onondaga County, N. Y., and Hancock and Wood counties, Ohio. These wells were selected because they gave definite stratigraphic information.

*298. Record of deep-well drilling for 1905, by M. L. Fuller and Samuel Sanford. 1906. 299 pp. 25c.

Gives an account of progress in the collection of well records and samples; contains tabulated records of wells in Illinois, Indiana, Michigan, Minnesota, New York, Ohio, Pennsylvania, Vermont, and Wisconsin, and detailed records of wells in Cook County, Ill.; Erie County, N. Y.; Ottawa, Sandusky, and Summit counties, Ohio; and Manitowoc County, Wis. The wells of which detailed sections are given were selected because they afford valuable stratigraphic information.

GEOLOGIC FOLIOS.

Under the plan adopted for the preparation of a geologic map of the United States the entire area is divided into small quadrangles, bounded by certain meridians and parallels, and these quadrangles, which number several thousand, are separately surveyed and mapped.[1] The unit of survey is also the unit of publication, and the maps and description of each quadrangle are issued in the form of a folio. When all the folios are completed they will constitute the Geologic Atlas of the United States.

A folio is designated by the name of the principal town or of a prominent natural feature within the quadrangle. Each folio includes maps showing the topography, geology, underground structure, and mineral deposits of the area mapped and several pages of descriptive text. The text explains the maps and describes the topographic and geologic features of the country and its mineral products. The topographic map shows roads, railroads, waterways, and, by contour lines, the shapes of the hills and valleys and the height above sea level of all points in the quadrangle. The areal-geology map shows the distribution of the various rocks at the surface. The structural-geology

[1] Index maps showing areas in the St. Lawrence basin covered by topographic maps and by geologic folios will be mailed on receipt of request addressed to the director U. S. Geological Survey, Washington, D. C.

map shows the relations of the rocks to one another underground. The economic-geology map indicates the location of mineral deposits that are commercially valuable. The artesian-water map shows the depth of underground-water horizons. Economic-geology and artesian-water maps are included in folios if the conditions in the areas mapped warrant their publication. The folios are of special interest to students of geography and geology and are valuable as guides in the development and utilization of mineral resources.

Folios 1 to 163, inclusive, are published in only one form (18 by 22 inches), called the library edition. Some of the folios that bear numbers higher than 163 are published also in an octavo edition (6 by 9 inches). Owing to a fire in the Geological Survey building May 18, 1913, the stock of geologic folios was more or less damaged by fire and water, but 80 or 90 per cent of the folios are usable. They will be sold at the uniform price of 5 cents each, with no reduction for wholesale orders. This rate applies to folios in stock from 1 to 184, inclusive (except reprints), also to the library edition of Folio 186. The library edition of Folios 185, 187, and higher numbers sells for 25 cents a copy, except that some folios which contain an unusually large amount of matter sell at higher prices. The octavo edition of Folio 185 and higher numbers sells for 50 cents a copy, except Folio 193, which sells for 75 cents a copy. A discount of 40 per cent is allowed on an order for folios or for folios together with topographic maps amounting to $5 or more at the retail rate.

All the folios contain descriptions of the drainage of the quadrangles. The folios in the following list contain also brief discussions of the underground waters in connection with the economic resources of the areas and more or less information concerning the utilization of the water resources.

An asterisk (*) indicates that the stock of the folio is exhausted.

*81. Chicago, Illinois-Indiana.

Describes an area embracing not only the immediate site of the city but adjacent parts of Cook, Dupage, and Will counties, Ill.; gives an account of the water power, discusses the quality of the waters, and gives analyses of waters from artesian wells; gives also a list of papers relating to the geology and paleontology of the area.

*140. Milwaukee special, Wisconsin, 5c.

Gives analyses of spring waters and of artesian water in Milwaukee; also tabulated data concerning wells.

155. Ann Arbor, Mich. 25c.

Discusses the present lakes, the lakes of the glacial period, and under "Economic geology," the water resources, including the use of the rivers for power and of the underground waters, shallow and artesian, for city and village supplies; discusses the quality of the waters, and gives details by townships.

*169. Watkins Glen-Catatonk, New York.

Includes discussion of water supply at Ithaca.

190. Niagara, N. Y. 50c. either edition.

Gives analyses of mineral water from well at Akron; discusses briefly the municipal supplies of Buffalo, Niagara Falls, Tonawanda, La Salle, and Youngstown, and the use of Niagara River for power development.

205. Detroit, Mich. 50c. either edition.

Discusses surface and ground waters; gives mineral analyses of water from Lake Huron, from rivers near Detroit, and from salt wells.

MISCELLANEOUS REPORTS.

Other Federal bureaus and State and other organizations have from time to time published reports relating to the water resources of the various sections of the country. Notable among those pertaining to the St. Lawrence River basin are the reports of the Chief of Engineers, United States Army, the State Geological Survey of Illinois, the Illinois Water-Supply Commission, the Rivers and Lakes Commission of Illinois, the New York State Conservation Commission and State Water-Supply Commission, and the water-power report of the Tenth Census (vol. 16). The following reports deserve special mention:

The mineral content of Illinois waters, by Edward Bartow, J. A. Udden, S. W. Parr, and George T. Palmer: Illinois State Geol. Survey Bull. 10, 1909.

Chemical and biological survey of waters of Illinois, by Edward Bartow: Univ. Illinois Pubs. 3, 6, 7, 1906-1909.

Chemical survey of the waters of Illinois, report for the years 1897-1902, by A. W. Palmer, with report on geology of Illinois as related to its water supply, by Charles W. Rolfe: Univ. Illinois Pub.

Diversion of the waters of the Great Lakes by way of the Sanitary and Ship canal of Chicago: A brief of the facts and issues, by Lyman E. Cooley, Chicago, 1913.

The State of Missouri *v.* the State of Illinois and the Sanitary district of Chicago, before Frank S. Bright, commissioner of the Supreme Court of the United States, 1904.

The mineral waters of Indiana, their location, origin, and character, by W. S. Blatchley: Indiana Dept. Geology and Nat. Res. Twenty-sixth Ann. Rept., 1901.

Reports of the water resources investigation of Minnesota, by the State Drainage Commission, 1909–1912.

Water powers of Wisconsin, by L. S. Smith: Wisconsin Geol. and Nat. Hist. Survey Bull. 20, 1908.

Report of the Railroad Commission of Wisconsin to the legislature on water powers, 1915.

Hydrology of the State of New York, by George W. Rafter: New York State Mus. Bull. 85, 1905.

Many of these reports can be obtained from the various commissions, and probably all can be consulted in the public libraries of the larger cities.

GEOLOGICAL SURVEY HYDROLOGIC REPORTS OF GENERAL INTEREST.

The following list comprises reports that are not readily classifiable by drainage basins and that cover a wide range of hydrologic investigation:

WATER-SUPPLY PAPERS.

*1. Pumping water for irrigation, by H. M. Wilson. 1896. 57 pp., 9 pls.

Describes pumps and motive powers, windmills, water wheels, and various kinds of engines; also, storage reservoirs to retain pumped water until needed for irrigation.

*3. Sewage irrigation, by G. W. Rafter. 1897. 100 pp., 4 pls. (See Water-Supply Paper 22.) 10c.

Discusses methods of sewage disposal by intermittent filtration and by irrigation; describes utilization of sewage in Germany, England, and France, and sewage purification in the United States.

*8. Windmills for irrigation, by E. C. Murphy. 1897. 49 pp., 8 pls. 10c.

Gives results of experimental tests of windmills during the summer of 1896 in the vicinity of Garden, Kans.; describes instruments and methods and draws conclusions.

*14. New tests of certain pumps and water lifts used in irrigation, by O. P. Hood, 1898. 91 pp., 1 pl.

Discusses efficiency of pumps and water lifts of various types.

*20. Experiments with windmills, by T. O. Perry. 1899. 97 pp., 12 pls. 15c.

Includes tables and descriptions of wind wheels, makes comparisons of wheels of several types, and discusses results.

*22. Sewage irrigation, Part II, by G. W. Rafter. 1899. 100 pp., 7 pls. 15c.

Gives résumé of Water-Supply Paper 3; discusses pollution of certain streams, experiments on purification of factory wastes in Massachusetts, value of commercial fertilizers, and describes American sewage-disposal plants by States; contains bibliography of publications relating to sewage utilization and disposal.

*41. The windmill, its efficiency and economic use, Part I, by E. C. Murphy. 1901. 72 pp., 14 pls.

*42. The windmill, its efficiency and economic use, Part II, by E. C. Murphy. 1901, 75 pp., 2 pls. 10c.

Nos. 41 and 42 give details of results of experimental tests with windmills of various types.

*43. Conveyance of water in irrigation canals, flumes, and pipes, by Samuel Fortier, 1901. 86 pp., 15 pls. 15c.

*56. Methods of stream measurement. 1901. 51 pp., 12 pls. 15c.

Describes the methods used by the Survey in 1901-2. See also Nos. 64, 94, and 95.

*57. Preliminary list of deep borings in the United States, Part I (Alabama-Montana), by N. H. Darton. 1902. 60 pp. (See No. 149.) 5c.

*61. Preliminary list of deep borings in the United States, Part II (Nebraska-Wyoming), by N. H. Darton. 1902. 67 pp. 5c.

Nos. 57 and 61 contain information as to depth, diameter, yield, and head of water in borings more than 400 feet deep; under head "Remarks" gives information concerning temperature, quality of water, purpose of boring, etc. The lists are arranged by States, and the States are arranged alphabetically. A second, revised edition was published in 1905 as Water-Supply Paper 149 (q. v.). 5c.

*64. Accuracy of stream measurements, by E. C. Murphy. 1902. 99 pp., 4 pls. (See No. 95.) 10c.

Describes methods of measuring velocity of water and of measuring and computing stream flow and compares results obtained with the different instruments and methods; describes also experiments and results at the Cornell University hydraulic laboratory. A second, enlarged, edition published as Water-Supply Paper 95.

*67. The motions of underground waters, by C. S. Slichter. 1902. 106 pp., 8 pls. 15c.

Discusses origin, depth, and amount of underground waters; permeability of rocks and porosity of soils; causes, rates, and laws of motion of underground water; surface and deep zones of flow and recovery of waters by open wells and artesian and deep wells; treats of the shape and position of the water table; gives simple methods of measuring yield of flowing well; describes artesian wells at Savannah, Ga.

72. Sewage pollution in the metropolitan area near New York City and its effect on inland-water resources, by M. O. Leighton. 1902. 75 pp., 8 pls. 10c.

Defines "normal" and "polluted" waters and discusses the damage resulting from pollution.

79. Normal and polluted waters in northeastern United States, by M. O. Leighton. 1903. 192 pp. 10c.

Defines essential qualities of water for various uses, the impurities in rain, surface, and underground waters, the meaning and importance of sanitary analyses, and the principal sources of pollution; chiefly, "a review of the more readily available records" of examination of water supplies derived from streams in the Merrimack, Connecticut, Housatonic, Delaware, and Ohio River basins; contains many analyses.

*80. The relation of rainfall to run-off, by G. W. Rafter. 1903. 104 pp. 10c.

Treats of measurements of rainfall and laws and measurements of stream flow; gives rainfall, run-off, and evaporation formulas; discusses effect of forests on rainfall and run-off.

87. Irrigation in India (second edition), by H. M. Wilson. 1903. 238 pp., 27 pls. 25c.

First edition was published in Part II of the Twelfth Annual Report.

93. Proceedings of first conference of engineers of the Reclamation Service, with accompanying papers, complied by F. H. Newell, chief engineer. 1904. 361 pp. 25c.

Contains, in addition to an account of the organization of the hydrographic [water-resources] branch of the United States Geological Survey and the reports of the conference, the following papers of more or less general interest:

Limits of an irrigation project, by D. W. Ross.
Relation of Federal and State laws to irrigation, by Morris Bien.
Electrical transmission of power for pumping, by H. A. Storrs.
Correct design and stability of high masonry dams, by Geo. Y. Wisner.
Irrigation surveys and the use of the plane table, by J. B. Lippincott.
The use of alkaline waters for irrigation, by Thomas A. Means.

*94. Hydrographic manual of the United States Geological Survey, prepared by E. C. Murphy, J. C. Hoyt, and G. B. Hollister. 1904. 76 pp., 3 pls. 10c.

Gives instructions for field and office work relating to measurements of stream flow by current meters. See also No. 95

*95. Accuracy of stream measurements (second, enlarged, edition), by E. C. Murphy. 1904. 169 pp., 6 pls.

Describes methods of measuring and computing stream flow and compares results derived from different instruments and methods. See also No. 94.

*103. A review of the laws forbidding pollution of inland waters in the United States, by E. B. Goodell. 1904. 120 pp. (See No. 152.)

Explains the legal principles under which antipollution statutes become operative, quotes court decisions to show authority for various deductions, and classifies according to scope the statutes enacted in the different States.

110. Contributions to the hydrology of eastern United States, 1904; M. L. Fuller, geologist in charge. 1905. 211 pp., 5 pls. 10c.

Contains the following reports of general interest. The scope of each paper in indicated by its title.

Description of underflow meter used in measuring the velocity and direction of underground water, by Charles S. Slichter.

The California or "stovepipe" method of well construction, by Charles S. Slichter.

Approximate methods of measuring the yield of flowing wells, by Charles S. Slichter.

Corrections necessary in accurate determinations of flow from vertical well casings, from notes furnished by A. N. Talbot.

Experiment relating to problems of well contamination at Quitman, Ga., by S. W. McCallie.

Notes on the hydrology of Cuba, by M. L. Fuller.

113. The disposal of strawboard and oil-well wastes, by R. L. Sackett and Isaiah Bowman. 1905. 52 pp., 4 pls. 5c.

The first paper discusses the pollution of streams by sewage and by trade wastes, describes the manufacture of strawboard, and gives results of various experiments in disposing of the waste. The second paper describes briefly the topography, drainage, and geology of the region about Marion, Ind., the contamination of rock wells and of streams by waste oil and brine.

*114. Underground waters of eastern United States; M. L. Fuller, geologist in charge. 1905. 285 pp., 18 pls. 25c.

Contains report on "Occurrence of underground waters," by M. L. Fuller, discussing sources, amount, and temperature of waters, permeability and storage capacity of rocks, water-bearing formations, recovery of water by springs, wells, and pumps, essential conditions of artesian flows, and general conditions affecting underground waters in eastern United States.

119. Index to the hydrographic progress reports of the United States Geological Survey, 1888 to 1903, by J. C. Hoyt and B. D. Wood. 1905. 253 pp. 15c.

Scope indicated by title.

120. Bibliographic review and index of papers relating to underground waters published by the United States Geological Survey, 1879–1904, by M. L. Fuller. 1905. 128 pp. 10c.

Scope indicated by title.

*122. Relation of the law to underground waters, by D. W. Johnson. 1905. 55 pp. 5c.

Defines and classifies underground waters, gives common-law rules relating to their use, and cites State legislative acts affecting them.

140. Field measurements of the rate of movement of underground waters, by C. S. Slichter. 1905. 122 pp., 15 pls. 15c.

Discusses the capacity of sand to transmit water, describes measurements of underflow in Rio Hondo, San Gabriel, and Mohave River valleys, Calif., and on Long Island, N. Y., gives results of tests of wells and pumping plants, and describes stovepipe method of well construction.

143. Experiments on steel-concrete pipes on a working scale, by J. H. Quinton. 1905. 61 pp., 4 pls. 5c.

Scope indicated by title.

145. Contributions to the hydrology of eastern United States, 1905; M. L. Fuller, geologist in charge. 1905. 220 pp., 6 pls. 10c.

Contains brief reports of general interest as follows:

Drainage of ponds into drilled wells, by Robert E. Horton. Discusses efficiency, cost, and capacity of drainage wells, and gives statistics of such wells in southern Michigan.

Construction of so-called fountain and geyser springs, by Myron L. Fuller.

A convenient gage for determining low artesian heads, by Myron L. Fuller.

146. Proceedings of second conference of engineers of the Reclamation Service, with accompanying papers, compiled by F. H. Newell, chief engineer. 1905. 267 pp. 15c.

Contains brief account of the organization of the hydrographic [water-resources] branch and the Reclamation Service, reports of conferences and committees, circulars of instruction, and many brief reports on subjects closely related to reclamation, and a bibliography of technical papers by members of the service. Of the papers read at the conference those listed below (scope indicated by title) are of more or less general interest:

Proposed State code of water laws, by Morris Bien.
Power engineering applied to irrigation problems, by O. H. Ensign.
Estimates on tunnelling in irrigation projects, by A. L. Fellows.
Collection of stream-gaging data, by N. C. Grover.
Diamond-drill methods, by G. A. Hammond.
Mean-velocity and area curves, by F. W. Hanna.
Importance of general hydrographic data concerning basins of streams gaged, by R. E. Horton.
Effect of aquatic vegetation on stream flow, by R. E. Horton.
Sanitary regulations governing construction camps, by M. O. Leighton.
Necessity of draining irrigated land, by Thos. H. Means.
Alkali soils, by Thos. H. Means.
Cost of stream-gaging work, by E. C. Murphy.
Equipment of a cable gaging station, by E. C. Murphy.
Silting of reservoirs, by W. M. Reed.
Farm-unit classification, by D. W. Ross.
Cost of power for pumping irrigating water, by H. A. Storrs.
Records of flow at current-meter gaging stations during the frozen season, by F. H. Tillinghast.

147. Destructive floods in the United States in 1904, by E. C. Murphy and others. 1905. 206 pp., 18 pls. 15c.

Contains a brief account of "A method of computing cross-section area of waterways," including formulas for maximum discharge and areas of cross section.

*149. Preliminary list of deep borings in the United States, second edition, with additions, by N. H. Darton. 1905. 175 pp. 10c.

Gives by States (and within the States by counties), location, depth, diameter, yield, height of water, and other available information, concerning wells 400 feet or more in depth; includes all wells listed in Water-Supply Papers 57 to 61; mentions also principal publications relating to deep borings.

*150. Weir experiments, coefficients, and formulas, by R. E. Horton. 1906. 189 pp., 38 pls. (See Water-Supply Paper 200.) 15c.

Scope indicated by title.

151. Field assay of water, by M. O. Leighton. 1905. 77 pp., 4 pls. 10c.

Discusses methods, instruments, and reagents used in determining turbidity, color, iron, chlorides, and hardness in connection with the studies of the quality of water in various parts of the United States.

*152. A review of the laws forbidding pollution of inland waters in the United States (second edition), by E. B. Goodell. 1905. 149 pp.

Scope indicated by title.

*160. Underground-water papers, 1906; M. L. Fuller, geologist in charge. 1906. 104 pp., 1 pl.

Gives account of work in 1905; lists of publications relating to underground waters, and contains the following brief reports of general interest:

Significance of the term "artesian," by Myron L. Fuller.
Representation of wells and springs on maps, by Myron L. Fuller.
Total amount of free water in the earth's crust, by Myron L. Fuller.
Use of fluorescein in the study of underground waters, by R. B. Dole.
Problems of water contamination, by Isaiah Bowman.
Instances of improvement of water in wells, by Myron L. Fuller.

*162. Destructive floods in the United States in 1905, with a discussion of flood discharge and frequency and an index to flood literature, by E. C. Murphy and others. 1906. 105 pp., 4 pls. 15c.

*163. Bibliographic review and index of underground-water literature published in the United States in 1905, by M. L. Fuller, F. G. Clapp, and B. L. Johnson. 1906. 130 pp. 15c.

Scope indicated by title.

*179. **Prevention of stream pollution by distillery refuse, based on investigations at Lynchburg, Ohio, by Herman Stabler. 1906. 34 pp., 1 pl. 10c.**

Describes grain distillation, treatment of slop, sources, character, and effects of effluents on streams; discusses filtration, precipitation, fermentation, and evaporation methods of disposal of wastes without pollution.

*180. **Turbine water wheel tests and power tables, by R. E. Horton. 1906. 134 pp. 2 pls. 20c.**

Scope indicated by title.

*185. **Investigations on the purification of Boston sewage, by C.-E. A. Winslow and E. B. Phelps. 1906. 163 pp. 25c.**

Discusses composition, disposal, purification, and treatment of sewages and recent tendencies in sewage-disposal practice in England, Germany, and the United States; describes character of crude sewage at Boston, removal of suspended matter, treatment in septic tanks, and purification in intermittent sand filtration and coarse material; gives bibliography.

*186. **Stream pollution by acid-iron wastes, a report based on investigations made at Shelby, Ohio, by Herman Stabler. 1906. 36 pp., 1 pl.**

Gives history of pollution by acid-iron wastes at Shelby, Ohio, and resulting litigation; discusses effect of acid-iron liquors on sewage purification processes, recovery of copperas from acid iron wastes, and other processes for removal of pickling liquor.

*187. **Determination of stream flow during the frozen season, by H. K. Barrows and R. E. Horton. 1907. 93 pp., 1 pl. 15c.**

Scope indicated by title.

*189. **The prevention of stream pollution by strawboard wastes, by E. B. Phelps. 1906. 29 pp., 2 pls.**

Describes manufacture of strawboard, present and proposed methods of disposal of waste liquors, laboratory investigations of precipitation and sedimentation, and field studies of amount and character of water used, raw material and finished product, and mechanical filtration.

*194. **Pollution of Illinois and Mississippi rivers by Chicago sewage (a digest of the testimony taken in the case of The State of Missouri *v.* The State of Illinois and the Sanitary District of Chicago), by M. O. Leighton. 1907. 369 pp., 2 pls.**

Scope indicated by amplification of title.

*200. **Weir experiments, coefficients, and formulas (revision of paper No. 150), by R. E. Horton. 1907. 195 pp., 38 pls. 35c.**

Scope indicated by title.

*226. **The pollution of streams by sulphite-pulp waste, a study of possible remedies, by E. B. Phelps. 1909. 37 pp., 1 pl. 10c.**

Describes the manufacture of sulphite pulp, the waste liquors, and the experimental work leading to suggestions as to methods of preventing stream pollution.

*229. **The disinfection of sewage and sewage filter effluents, with a chapter on the putrescibility and stability of sewage effluents, by E. B. Phelps. 1909. 91 pp., 1 pl. 15c.**

Scope indicated by title.

*234. **Papers on the conservation of water resources. 1909. 96 pp., 2 pls. 15c.**

Contains the following papers, whose scope is indicated by their titles: Distribution of rainfall, by Henry Gannett; Floods, by M. O. Leighton; Developed water powers, compiled under the direction of W. M. Steuart, with discussion by M. O. Leighton; Undeveloped water powers, by M. O. Leighton; Irrigation, by F. H. Newell; Underground waters, by W. C. Mendenhall; Denudation, by R. B. Dole and Herman Stabler; Control of catchment areas, by H. N. Parker.

*235. **The purification of some textile and other factory wastes, by Herman Stabler and G. H. Pratt. 1909. 76 pp. 10c.**

Discusses waste waters from wool scouring, bleaching and dyeing cotton yarn, bleaching cotton piece goods, and manufacture of oleomargarine, fertilizer, and glue.

236. The quality of surface waters in the United States: Part I, Analyses of waters east of the one hundredth meridian, by R. B. Dole. 1909. 123 pp. 10c.

Describes collection of samples, method of examination, preparation of solutions, accuracy of estimates, and expression of analytical results.

238. The public utility of water powers and their governmental regulation, by René Tavernier and M. O. Leighton. 1910. 161 pp. 15c.

Discusses hydraulic power and irrigation, French, Italian, and Swiss legislation relative to the development of water powers, and laws proposed in the French Parliament; reviews work of bureau of hydraulics and agricultural improvement and the French department of agriculture, and gives résumé of Federal and State water-power legislation in the United States.

*255. Underground waters for farm use, by M. L. Fuller. 1910. 58 pp., 17 pls. 15c.

Discusses rocks as sources of water supply and the relative safety of supplies from different materials; springs and their protection; open or dug and deep wells, their location, yield, relative cost, protection, and safety; advantages and disadvantages of cisterns and combination wells and cisterns.

*257. Well-drilling methods, by Isaiah Bowman. 1911. 139 pp., 4 pls. 15c.

Discusses amount, distribution, and disposal of rainfall, water-bearing rocks, amount of underground water, artesian conditions, and oil and gas bearing formations; gives history of well drilling in Asia, Europe, and the United States; describes in detail the various methods and the machinery used; discusses loss of tools and geologic difficulties; contamination of well waters and methods of prevention; tests of capacity and measurement of depth; and of costs sinking wells.

*258. Underground-water papers, 1910, by M. L. Fuller, F. G. Clapp, G. C. Matson, Samuel Sanford, and H. C. Wolff. 1911. 123 pp., 2 pls. 15c.

Contains the following papers (scope indicated by titles) of general interest:
Drainage of wells, by M. L. Fuller.
Freezing of wells and related phenomena, by M. L. Fuller.
Pollution of underground waters in limestone, by G. C. Matson.
Protection of shallow wells in sandy deposits, by M. L. Fuller.
Magnetic wells, by M. L. Fuller.

259. The underground waters of southwestern Ohio, by M. L. Fuller and F. G. Clapp, with a discussion of the chemical character of the waters, by R. B. Dole. 1912. 228 pp., 9 pls. 35c

Describes the topography, climate, and geology of the region, the water-bearing formations, the source, mode of occurrence, and head of the waters, and municipal supplies; give details by counties; discusses in supplement, under chemical character, method of analysis and expression of results, mineral constitutents, effect of the constitutents on waters for domestic, industrial and medicinal uses, methods of purification, chemical composition; many analyses and field assays. The matter in the supplement was also published in Water-Supply Paper 254 (The underground waters of north-central Indiana).

274. Some stream waters of the western United States, with chapters on sediment carried by the Rio Grande and the industrial application of water analyses, by Herman Stabler. 1911. 188 pp. 15c.

Describes collection of samples, plan of analytical work, and methods of analyses; discusses soap-consuming power of waters, water softening, boiler waters, and water for irrigation; gives results of analyses of waters of the Rio Grande and of Pecos, Gallinas, and Hondo rivers.

*315. The purification of public water supplies, by G. A. Johnson. 1913. 84 pp.; 8 pls. 10c.

Discusses ground, lake, and river waters as public supplies, development of waterworks systems in the United States, water consumption, and typhoid fever; describes methods of filtration and sterilization of water and municipal water softening.

334. The Ohio Valley flood of March–April, 1913 (including comparisons with some earlier floods), by A. H. Horton and H. J. Jackson. 1913. 96 pp., 22 pls 20c.

Although relating specifically to floods in the Ohio Valley, this report discusses also the causes of floods and the prevention of damage by floods.

337. The effects of ice on stream flow, by William Glenn Hoyt. 1913. 77 pp., 7 pls. 15c.

Discusses methods of measuring the winter flow of streams.

*345. Contributions to the hydrology of the United States, 1914. N. C. Grover, chief hydraulic engineer. 1915. 225 pp., 17 pls. 30c.

*(e) A method of determining the daily discharge of rivers of variable slope, by M. R. Hall, W. E. Hall, and C. H. Pierce, pp. 53-65.

364. Water analyses from the laboratory of the United States Geological Survey, tabulated by F. W. Clarke, chief chemist. 1914. 40 pp. 5c.

Contains analyses of waters from rivers, lakes, wells, and springs in various parts of the United States, including analyses of the geyser water of Yellowstone National Park, hot springs in Montana, brines from Death Valley, water from the Gulf of Mexico, and mine waters from Tennessee, Michigan, Missouri and Oklahoma, Montana, Colorado and Utah, Nevada and Arizona, and California.

371. Equipment for current-meter gaging stations, by G. J. Lyon. 1915. 64 pp., 37 pls. 20c.

Describes methods of installing automatic and other gages and of constructing gage wells, shelters, and structures for making discharge measurements and artificial controls.

*375. Contributions to the hydrology of the United States, 1915. N. C. Grover, chief hydraulic engineer. 1916. 181 pp., 9 pls.

(c) The relation of stream gaging to the science of hydraulics, by C. H. Pierce and R. W. Davenport, pp. 77-84.

(e) A method of correcting river discharge for a changing stage, by B. E. Jones, pp. 117-130.

*(f) Conditions requiring the use of automatic gages in obtaining records of stream flow by C. H. Pierce, pp. 131-139.

Three papers presented at the conference of engineers of the water-resources branch in December, 1914.

*400. Contributions to the hydrology of the United States, 1916. N. C. Grover, chief hydraulic engineer.

(a) The people's interest in water-power resources, by G. O. Smith, pp. 1-8.

(c) The measurement of silt-laden streams, by Raymond C. Pierce, pp. 39-51.

(d) Accuracy of stream-flow data, by N. C. Grover and J. C. Hoyt, pp. 53-59.

416. The divining rod, a history of water witching, with a biblography, by Arthur J. Ellis. 1917. 59 pp. 10c.

A brief paper published "merely to furnish a reply to the numerous inquires that are continually being received from all parts of the country" as to the efficacy of the divining rod for locating underground water.

425. Contributions to the hydrology of the United States, 1917; N. C. Grover, chief hydraulic engineer. 1918. Contains:

(c) Hydraulic conversion tables and convenient equivalents, pp. 71-94. 1917.

427. Bibliography and index of the publications of the United States Geological Survey relating to ground water, by O. E. Meinzer. 1918. 169 pp., 1 pl.

Includes publications prepared, in whole or part, by the Geological Survey that treat any phase of the subject of ground water or any subject directly applicable to ground water. Illustrated by map showing reports that cover specific areas more or less thoroughly.

ANNUAL REPORTS.

*Fifth Annual Report of the United States Geological Survey 1883-84, J. W. Powell, Director. 1885. xxxvi, 469 pp., 58 pls. $2.25. Contains:

*The requisite and qualifying conditions of artesian wells, by T. C. Chamberlin, pp. 125-173. Pl. 21. Scope indicated by title.

*Twelfth Annual Report of the United States Geological Survey, 1890-91, J. W. Powell Director. 1891. 2 parts. Pt. II, Irrigation, xviii, 576 pp., 93 pls. $2. Contains:

*Irrigation in India, by H. M. Wilson, pp. 375-561, pls. 107-146. See Water-Supply Paper 87.

Thirteenth Annual Report of the United States Geological Survey, 1891-92, J. W. Powell, Director. 1892. (Pts. II and III, 1893.) 3 parts. *Pt. III, Irrigation, xi, 486 pp., 77 pls. $1.85. Contains:

*American irrigation engineering, by H. M. Wilson, pp. 101-349, pls. 111-145. Discusses the economic aspects of irrigation, alkaline drainage, silt, and sedimentation; gives brief history of legislation; describes perennial canals in Idaho-California, Wyoming, and Arizona; discusses water storage at reservoirs of the California and other projects, subsurface sources of supply, pumping, and subirrigation.

Fourteenth Annual Report of the United States Geological Survey, 1892-93, J. W. Powell, Director. 1893. (Pt. II, 1894.) 2 parts. *Pt. II, Accompanying papers, xx, 597 pp., 73 pls. $2.10. Contains:

*The potable waters of eastern United States, by W. J. McGee, pp. 1-47. Discusses cistern water, stream waters, and ground waters, including mineral springs and artesian wells.

*Natural mineral waters of the United States, by A. C. Peale, pp. 49-88, pls. 3 and 4. Discusses the origin and flow of mineral springs, the source of mineralization, thermal springs, the chemical composition and analysis of spring waters, geographic distribution, and the utilization of mineral waters; gives a list of American mineral spring resorts; contains also some analyses.

Nineteenth Annual Report of the United States Geological Survey, 1897-98, Charles D. Walcott, Director. 1898. (Parts II, III, and V, 1899.) 6 parts in 7 vols. and separate case for maps with Pt. V. *Pt. II, papers chiefly of a theoretic nature, v, 958 pp., 127 pls. $2.65. Contains:

*Principles and conditions of the movements of ground water, by F. H. King, pp. 59-294, pls. 6-16. Discusses the amount of water stored in sandstone, in soil, and in other rocks, the depth to which ground water penetrates; gravitational, thermal, and capillary movements of ground waters, and the configuration of the ground-water surface; gives the results of experimental investigations on the flow of air and water through a rigid, porous media, and through sand, sandstones, and silts; discusses results obtained by other investigators, and summarizes result of observations; discusses also rate of flow of water through sand and rock, the growth of rivers rate of filtration through soil, interference of wells, etc.

*Theoretical investigation of the motion of ground waters, by C. S. Slichter, pp. 295-384, pls. 17. Scope indicated by title.

PROFESSIONAL PAPERS.

*72. Denudation and erosion in the southern Appalachian region and the Monongahela basin, by L. C. Glenn. 1911. 137 pp., 21 pls. 35c.

Describes the topography, geology, drainage, forests, climate and population, and transportation facilities of the region, the relation of agriculture, lumbering, mining, and power development to erosion and denudation, and the nature, effects, and remedies of erosion; gives details of conditions in Holston, Nolichucky, French Broad, Little Tennessee, and Hiwassee river basins, along Tennessee River proper, and in the basins of the Coosa-Alabama system, Chattahoochee, Savannah, Saluda, Broad, Catawaba, Yadkin, New, and Monongahela rivers.

86. The transportation of débris by running water, by G. K. Gilbert, based on experiments made with the assistance of E. C. Murphy. 1914. 263 pp. 3 pls. 70c.

The results of an investigation which was carried on in a specially equipped laboratory at Berkeley, Cal., and was undertaken for the purpose of learning "the laws which control the movement of bed load and especially to determine how the quantity of load is related to the stream slope and discharge and to the degree of comminution of the débris."

A highly technical report.

105. Hydraulic mining débris in the Sierra Nevada, by G. K. Gilbert. 154 pp., 34 pls. 1917.

Presents the results of an investigation undertaken by the United States Geological Survey in response to a memorial from the California Miners' Association asking that a particular study be made of portions of the Sacramento and San Joaquin valleys affected by detritus from torrential streams. The report deals largely with geologic and physiographic aspects of the subject, traces the physical effects, past and future, of the hydraulic mining of earlier decades, the similar effects which certain other industries induce through stimulation of the erosion of the soil, and the influence of the restriction of the area of inundation by the construction of levees. Suggests cooperation by several interests for the control of the streams now carrying heavy loads of débris.

BULLETINS.

*32. Lists and analyses of the mineral springs of the United States (a preliminary study), by A. C. Peale. 1886. 235 pp.

Defines mineral waters, lists the springs by States, and gives tables of analyses so far as as available.

*264. Record of deep-well drilling for 1904, by M. L. Fuller, E. F. Lines, and A. C. Veatch. 1905. 106 pp. 10c.

*298. Record of deep-well drilling for 1905, by M. L. Fuller and Samuel Sanford. 1906. 299 pp. 25c.

Bulletins 264 and 298 discuss the importance of accurate well records to the driller, to owners of oil, gas, and water wells, and to the geologist; describe the general methods of work; give tabulated records of wells by States, and detailed records selected as affording valuable stratigraphic information.

*319. Summary of the controlling factors of artesian flows, by Myron L. Fuller, 1908. 44 pp. 10c.

Describes underground reservoirs, the sources of underground waters, the confining agents, the primary and modifying factors of artesian circulation, the essential and modifying factors of artesian flow, and typical artesian systems.

*479. The geochemical interpretation of water analyses, by Chase Palmer. 1911. 31 pp. 5c.

Discusses the expression of chemical analyses, the chemical character of water and the properties of natural waters; gives a classification of waters based on property values and reacting values, and discusses the character of the waters of certain rivers as interpreted directly from the results of analyses; discusses also the relation of water properties to geologic formations, silica in river water, and the character of the water of the Mississippi and the Great Lakes and St. Lawrence River as indicated by chemical analyses.

616. The data of geochemistry (third edition), by F. W. Clarke. 1916. 821 pp. 45c.

Earlier editions were published as Bulletins 330 and 491. Contains a discussion of the statement and interpretation of water analyses and a chapter on "Mineral wells and springs" (pp. 179-216). Discusses the definition and classification of mineral waters, changes in the composition of water, deposits of calcareous, ocherous, and siliceous materials made by water, vadose and juvenile waters, and thermal springs in relation to volcanism. Describes the different kinds of ground water and gives typical analyses. Includes a brief bibliography of papers containing water analyses.

INDEX BY AREAS AND SUBJECTS.

[A—Annual Report; M—Monograph; B—Bulletin; P—Professional Paper; W—Water-Supply Paper; G F—Geologic folio.]

---

[1] Many of the reports contain brief subject bibliographies. See abstracts.

[2] Many analyses of river, spring, and well waters are scattered through publications, as noted in abstracts.

INDEX OF STREAMS.

O

DEPARTMENT OF THE INTERIOR
ALBERT B. FALL, Secretary

NITED STATES GEOLOGICAL SURVE
GEORGE OTIS SMITH, Director

ACE WATER SUPPLY OF UNITED STATES 1918

. HUDSON BAY AND UPPER MISSIS RIVER BASINS

NATHAN C. GROVER, Chief Hydraulic Engineer
W. G. HOYT, District Engineer

DEPARTMENT OF THE INTERIOR
ALBERT B. FALL, Secretary

UNITED STATES GEOLOGICAL SURVEY
GEORGE OTIS SMITH, Director

Water-Supply Paper 475

SURFACE WATER SUPPLY OF THE UNITED STATES
1918

PART V. HUDSON BAY AND UPPER MISSISSIPPI RIVER BASINS

NATHAN C. GROVER, Chief Hydraulic Engineer
W. G. HOYT, District Engineer

Prepared in cooperation with the States of
MINNESOTA, WISCONSIN, IOWA, and ILLINOIS

WASHINGTON
GOVERNMENT PRINTING OFFICE
1921

CONTENTS.

ILLUSTRATIONS.

SURFACE WATER SUPPLY OF HUDSON BAY AND UPPER MISSISSIPPI RIVER BASINS, 1918.

AUTHORIZATION AND SCOPE OF WORK.

This volume is one of a series of 14 reports presenting records of measurements of flow made on streams in the United States during the year ending September 30, 1918.

The data presented in these reports were collected by the United States Geological Survey under the following authority contained in the organic law (20 Stat. L., p. 394):

> *Provided*, That this officer [the Director] shall have the direction of the Geological Survey and the classification of public lands and examination of the geological structure, mineral resources, and products of the national domain.

The work was begun in 1888 in connection with special studies relating to irrigation in the arid West. Since the fiscal year ending June 30, 1895, successive sundry civil bills passed by Congress have carried the following item and appropriations:

> For gaging the streams and determining the water supply of the United States, and for the investigation of underground currents and artesian wells, and for the preparation of reports upon the best methods of utilizing the water resources.

Annual appropriations for the fiscal years ending June 30, 1895–1919.

1895	$12,500
1896	20,000
1897 to 1900, inclusive	50,000
1901 to 1902, inclusive	100,000
1903 to 1906, inclusive	200,000
1907	150,000
1908 to 1910, inclusive	100,000
1911 to 1917, inclusive	150,000
1918	175,000
1919	148,244.10

In the execution of the work many private and State organizations have cooperated either by furnishing data or by assisting in collecting data. Acknowledgments for cooperation of the first kind are made in connection with the description of each station affected; cooperation of the second kind is acknowledged on page 9.

Measurements of stream flow have been made at about 4,500 points in the United States and also at many points in Alaska and the Hawaiian Islands. In July, 1918, 1,180 gaging stations were being maintained by the Survey and the cooperating organizations. Many miscellaneous discharge measurements are made at other points. In connection with this work data were also collected in regard to precipitation, evaporation, storage reservoirs, river profiles, and water power in many sections of the country and will be made available in water-supply papers from time to time. Information in regard to publications relating to water resources is presented in the appendix to this report.

DEFINITION OF TERMS.

The volume of water flowing in a stream—the "run-off" or "discharge"—is expressed in various terms, each of which has become associated with a certain class of work. These terms may be divided into two groups—(1) those that represent a rate of flow, as second-feet, gallons per minute, miner's inches, and discharge in second-feet per square mile, and (2) those that represent the actual quantity of water, as run-off in depth in inches, acre-feet, and millions of cubic feet. The principal terms used in this series of reports are second-feet, second-feet per square mile, run-off in inches, acre-feet, and millions of cubic feet. They may be defined as follows:

"Second-feet" is an abbreviation for "cubic feet per second." A second-foot is the rate of discharge of water flowing in a channel of rectangular cross section 1 foot wide and 1 foot deep at an average velocity of 1 foot per second. It is generally used as a fundamental unit from which others are computed.

"Second-feet per square mile" is the average number of cubic feet of water flowing per second from each square mile of area drained, on the assumption that the run-off is distributed uniformly both as regards time and area.

"Run-off (depth in inches)" is the depth to which an area would be covered if all the water flowing from it in a given period were uniformly distributed on the surface. It is used for comparing run-off with rainfall, which is usually expressed in depth in inches.

An "acre-foot," equivalent to 43,560 cubic feet, is the quantity required to cover an acre to the depth of 1 foot. The term is commonly used in connection with storage for irrigation.

"Millions of cubic feet" is applied to quantities of water stored in reservoirs, most frequently in connection with studies of flood control.

The following terms not in common use are here defined:

"Stage-discharge relation," an abbreviation for the term "relation of gage height to discharge."

"Control," a term used to designate the section or sections of the stream channel below the gage which determine the stage-discharge relation at the gage. It should be noted that the control may not be the same section or sections at all stages.

The "point of zero flow" for a gaging station is that point on the gage—the gage height—to which the surface of the river falls when the discharge is reduced to zero.

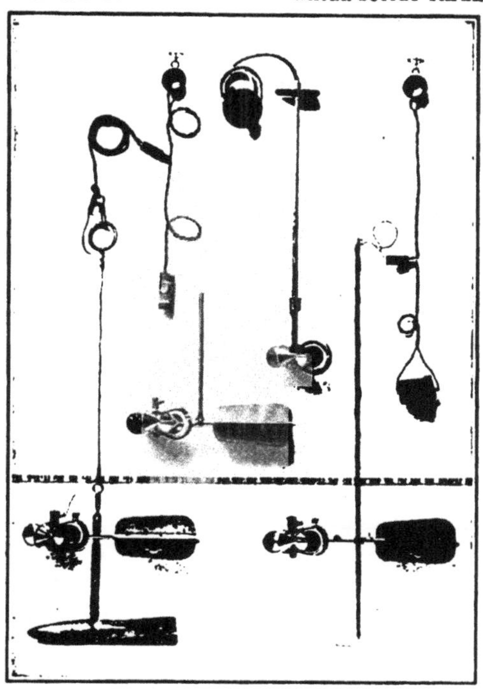

A. PRICE CURRENT METERS.

B. TYPICAL GAGING STATION.

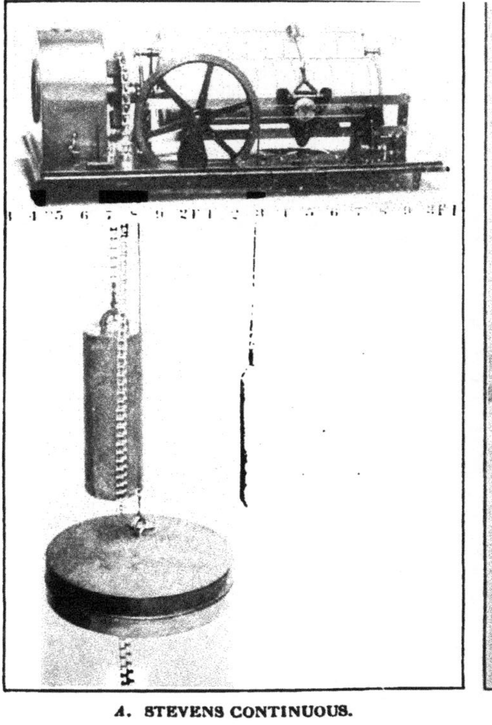

A. STEVENS CONTINUOUS.

B. GURLEY PRINTING.

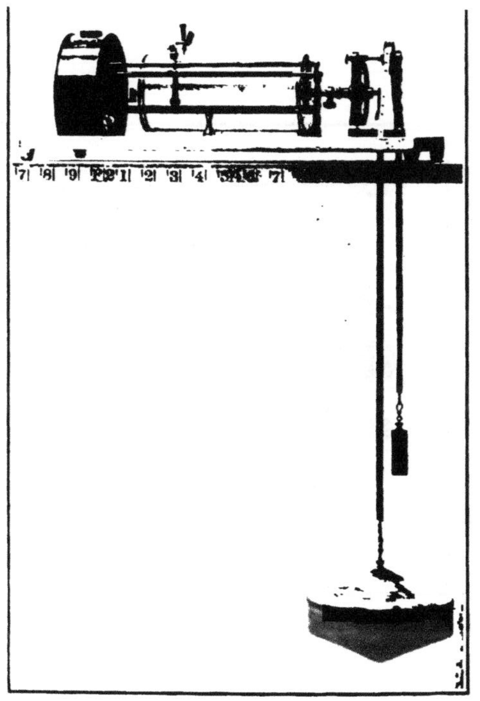

C. FRIEZ.

WATER-STAGE RECORDERS.

EXPLANATION OF DATA.

The data presented in this report cover the year beginning October 1, 1917, and ending September 30, 1918. At the beginning of January in most parts of the United States much of the precipitation in the preceding three months is stored as ground water in the form of snow or ice, or in ponds, lakes, and swamps, and this stored water passes off in the streams during the spring break-up. At the end of September, on the other hand, the only stored water available for run-off is possibly a small quantity in the ground; therefore the run-off for the year beginning October 1 is practically all derived from precipitation within that year.

The base data collected at gaging stations consist of records of stage, measurements of discharge, and general information used to supplement the gage heights and discharge measurements in determining the daily flow. The records of stage are obtained either from direct readings on a staff gage or from a water-stage recorder that gives a continuous record of the fluctuations. Measurements of discharge are made with a current meter. (See Pls. I, II.) The general methods are outlined in standard textbooks on the measurement of river discharge.

From the discharge measurements rating tables are prepared that give the discharge for any stage, and these rating tables, when applied to the gage heights, give the discharge from which the daily, monthly, and yearly means of discharge are determined.

The data presented for each gaging station in the area covered by this report comprise a description of the station, a table giving records of discharge measurements, a table showing the daily discharge of the stream, and a table of monthly and yearly discharge and run-off.

If the base data are insufficient to determine the daily discharge, tables giving daily gage height and records of discharge measurements are published.

The description of the station gives, in addition to statements regarding location and equipment, information in regard to any conditions that may affect the permanence of the stage-discharge relation covering such subjects as the occurrence of ice, the use of the stream for log driving, shifting of control, and the cause and effect of backwater; it gives also information as to diversions that decrease the flow at the gage, artificial regulation, maximum and minimum recorded stages, and the accuracy of the records.

The table of daily discharge gives, in general, the discharge in second-feet corresponding to the mean of the gage heights read each day. At stations on streams subject to sudden or rapid diurnal fluctuation the discharge obtained from the rating table and the mean daily gage height may not be the true mean discharge for the day. If such stations are equipped with water-stage recorders the mean daily discharge may be obtained by averaging discharge at regular intervals during the day, or by using the discharge integrator, an instrument operating on the principle of the planimeter and containing as an essential element the rating curve of the station.

In the table of monthly discharge the column headed "Maximum" gives the mean flow for the day when the mean gage height was highest. As the gage height is the mean for the day it does not indicate correctly the stage when the water surface was at crest height, and the corresponding discharge was consequently larger than given in the maximum column. Likewise, in the column headed "Minimum" the quantity given is the mean flow for the day when the mean gage height was lowest. The column headed "Mean" is the avarage flow in cubic feet per second during the month. On this average flow computations recorded in the remaining columns, which are defined on page 6, are based.

The deficiency table presented for some of the gaging stations shows the number of days in each year on which the mean daily discharge was less than the discharge given in the table. By subtraction the table gives the number of days each year that the mean daily discharge was between the discharges given in the table and, also by subtraction, the number of days that the mean daily discharge was equal to or greater than the discharge given. If one discharge rating table was used throughout the period covered by the deficiency table, gage heights that correspond to the discharges are also given.

ACCURACY OF FIELD DATA AND COMPUTED RECORDS.

The accuracy of stream-flow data depends primarily (1) on the permanence of the stage-discharge relation and (2) on the accuracy of observation of stage, measurements of flow, and interpretation of records.

A paragraph in the description of the station gives information regarding the (1) permanence of the stage-discharge relation, (2) precision with which the discharge rating curve is defined, (3) refinement of gage readings, (4) frequency of gage readings, and (5) methods of applying daily gage heights to the rating table to obtain the daily discharge.[1]

For the rating tables "well defined" indicates, in general, that the rating is probably accurate within 5 per cent; "fairly well defined," within 10 per cent; "poorly defined," within 15 to 25 per cent. These notes are very general and are based on the plotting of the individual measurements with reference to the mean rating curve.

The monthly means for any station may represent with high accuracy the quantity of water flowing past the gage, but the figures

[1] For a more detailed discussion of the accuracy of stream-flow data see Grover, N. C., and Hoyt, J. C. Accuracy of stream-flow data: U. S. Geol. Survey Water-Supply Paper 400, pp. 53–59, 1916.

showing discharge per square mile and depth of run-off in inches may be subject to gross errors caused by the inclusion of large noncontributing districts in the measured drainage area, by lack of information concerning water diverted for irrigation or other use, or by inability to interpret the effect of artificial regulation of the flow of the river above the station. "Second-feet per square mile" and "Run-off (depth in inches)" are therefore not computed if such errors appear probable. The computations are also omitted for stations on streams draining areas in which the annual rainfall is less than 20 inches. All figures representing "Second-feet per square mile" and "Run-off (depth in inches)" previously published by the Survey should be used with caution because of possible inherent sources of error not known to the Survey.

The table of monthly discharge gives only a general idea of the flow at the station and should not be used for other than preliminary estimates; the tables of daily discharge allow more detailed studies of the variation in flow. It should be borne in mind, however, that the observations in each succeeding year may be expected to throw new light on data previously published.

COOPERATION.

In Montana the work was done in cooperation with the United States Reclamation Service. The station on St. Mary River at Kimball, Alberta, was maintained in cooperation with the Canadian Department of Interior.

In Minnesota the work was carried on in cooperation with the State Drainage Commission, E. V. Willard, acting State drainage engineer, under terms of an act of the legislature of 1909 as embodied in joint resolution 19, which reads as follows:

> Whereas the water supplies, water powers, navigation of our rivers, drainage of our lands, and the sanitary condition of our streams and their watersheds generally form one great asset and present one great problem, therefore:
>
> *Be it resolved by the house of representatives, the senate concurring,* That the State drainage commission be, and is hereby, directed to investigate progress in other States toward the solution of said problem in such States, to investigate and determine the nature of said problems in this State.

The International Joint Commission maintained the water-stage recorder and paid the salary of the observer at the station on Kawishiwi River near Winton, and the United States Engineer Corps paid the salaries of the observers at the stations on Minnesota River near Montevideo and Mississippi River at Elk River.

The United States Weather Bureau furnished daily gage readings for the stations on Mississippi River at St. Paul and Minnesota River near Mankato.

In Wisconsin the work was carried on in cooperation with the Railroad Commission of Wisconsin, C. M. Larson, chief engineer, and at certain stations with the Wisconsin-Minnesota Light & Power Co. (Chippewa River at Chippewa Falls, Red Cedar River near Colfax, Red Cedar River at Cedar Falls, Red Cedar River at Menomonie) and Chippewa & Flambeau Improvement Co. (Chippewa River at Bishops Bridge, near Winter).

In Iowa the work was carried on in cooperation with the Iowa Geological Survey, George F. Kay, director; the Mississippi River Power Co., of Keokuk, Iowa, R. H. Bolster, hydraulic engineer; and the Iowa Highway Commission, Thomas H. MacDonald, chief engineer.

In Illinois work was carried on in cooperation with the Division of Waterways of Public Works and Buildings afterward, and at single stations with the United States Army Engineer Corps (Illinois River at Peoria) and the Central Illinois Public Service Co. (South Fork of Sangamon River at power plant near Taylorville).

DIVISION OF WORK.

The data for stations in the Hudson Bay basin, except in Minnesota, were collected and prepared for publication under the direction of W. A. Lamb, district engineer, Helena, Mont., assisted by E. F. Chandler.

The data for stations in the Hudson Bay and Mississippi River basins in Minnesota were collected and prepared for publication under the direction of W. G. Hoyt, district engineer, assisted by S. B. Soulé and E. F. Chandler, assisted by T. G. Bedford, R. B. Kilgore, and H. A. Noble.

For stations in the Mississippi River basin in Wisconsin the data were collected for publication under the direction of W. G. Hoyt, assisted by R. B. Kilgore, T. G. Bedford, J. B. Entringer, L. L. Smith, and F. W. Huels.

For stations in the Mississippi River basin in Iowa the data were collected under the direction of W. G. Hoyt, assisted by R. H. Bolster and R. W. Clyde, assisted by C. Herlofson, A. Davis, P. F. Gregg, and H. C. Hodge.

The data for stations in the Mississippi River basin in Illinois were collected under the direction of W. G. Hoyt, assisted by H. C. Beckman, assisted by A. M. Wohl and H. S. Wohl.

GAGING-STATION RECORDS.

HUDSON BAY DRAINAGE BASIN.

ST. MARY RIVER NEAR BABB, MONT.

[Including diversion from Swiftcurrent Creek.]

LOCATION.—In sec. 27, T. 36 N., R. 14 W., 1,040 feet above headworks of St. Mary canal and 2 miles south of Babb, on Blackfeet Indian Reservation, in Teton County.

DRAINAGE AREA.—278 square miles (including area of Swiftcurrent Creek above point of diversion into St. Mary Lake).

RECORDS AVAILABLE.—April 9, 1902, to September 30, 1918.

GAGE.—Stevens water-stage recorder on left bank; installed June 15, 1918. Prior to that date chain gage on right bank was used; read by Andrew Chevirer from October 1 to August 24 and thereafter by William Olson. During the winter months of 1917 a temporary low-water gage was read, located at site of present automatic gage.

DISCHARGE MEASUREMENTS.—Made from a cable 560 feet below the gage. In September, 1909, the cable was moved from a point about 300 feet downstream. Low-water measurements are made by wading 800 feet below the gage.

CHANNEL AND CONTROL.—Bed of stream composed of gravel and cobblestones. Banks are high and will not be overflowed. The concrete diversion dam for the St. Mary canal, located 1,040 feet below the gage, forms the control. The dam is provided with flashboard sluice gates near the canal head gates. Stage-discharge relation is permanent when the flashboards in the sluice gates remain at the level of the crest of the dam and canal head gates are closed.

EXTREMES OF DISCHARGE.—Maximum stage recorded during year, 6.15 feet June 14 (discharge, 5,200 second-feet); minimum stage 1.02 feet December 10, 11, and 12 (discharge, 66 second-feet).

1902–1918: Maximum stage estimated at 9.4 feet June 5, 1908 (discharge, 7,980 second-feet); minimum stage recorded, 1.0 foot April 3–7, 1904 (discharge, 20 second-feet).

ICE.—Stage-discharge relation affected very little, if any, by ice.

DIVERSIONS.—None.

REGULATION.—Flow is regulated by Sherburne Lake reservoir and natural storage in St. Mary Lakes.

ACCURACY.—Stage-discharge relation affected by placing or removing flashboards on dam and operation of gates. Rating curve used October 1, 1917 to May 31, 1918, and September 8–30 based on measurements made with gates closed and flashboards in place and is well defined between 60 and 5,700 second-feet; curve used June 1 to July 28 based on measurements made with flashboards removed and is well defined between 110 and 5,720 second-feet; indirect method used July 29 to September 8. Gage read daily to half-tenths October 1 to June 15 and to hundredths June 16 to September 30; after June 15 records taken from Stevens continuous water-stage recorder. Daily discharge ascertained by applying daily gage height to rating table. Records good.

The diversion dam below the gaging station was constructed by the United States Reclamation Service for the purpose of diverting water from St. Mary River into St. Mary canal, which carries the water across the divide into North Fork of Milk River. The water then flows in the natural channel of Milk River through Canada, and is finally used for irrigation in the Milk River Valley in Montana. The present capacity of the diversion canal is about 425 second-feet. A storage reservoir is being provided on Swiftcurrent Creek by constructing a dam at the outlet of Sherburne Lake. By means of a diversion channel connecting Swiftcurrent Creek and Lower St. Mary Lake, the run-off from Swiftcurrent Creek is made available for diversion through St. Mary canal.

Discharge measurements of St. Mary River near Babb, Mont., during the year ending Sept. 30, 1918.

Date.	Made by—	Gage height.	Discharge.	Date.	Made by—	Gage height.	Discharge.
		Feet.	*Sec.-ft.*			*Feet.*	*Sec.-ft.*
Nov. 9	Jones and Lamb	1.50	247	June 18	W. A. Lamb	5.40	4,140
Jan. 24	W. A. Lamb	1.56	290	July 7	.....do	2.64	1,140
May 25	R. F. Edwards	2.67	1,110	Aug. 12	.....do	2.11	794
June 15	W. A. Lamb	6.13	5,190	Sept. 6	.....do	1.70	451

Daily discharge, in second-feet, of St. Mary River near Babb, Mont., for the year ending Sept. 30, 1918.

Day.	Oct.	Nov.	Dec.	Jan.	Feb.	Mar.	Apr.	May.	June.	July.	Aug.	Sept.
1	328	222	120	246	120	104	272	758	1,250	1,570	740	572
2	328	246	120	299	104	104	272	798	1,490	1,450	740	541
3	328	246	120	568	88	120	272	958	1,580	1,380	745	524
4	328	246	120	1,120	138	120	272	1,120	1,490	1,320	761	515
5	358	257	120	1,480	199	120	272	2,210	1,440	1,240	778	463
6	358	272	120	1,680	199	120	272	2,880	1,420	1,210	800	448
7	358	257	120	1,630	199	120	272	3,240	1,490	1,150	800	452
8	371	246	120	1,560	191	120	288	3,300	1,880	1,120	788	452
9	371	246	88	1,390	178	120	299	3,120	2,430	1,090	788	458
10	371	246	66	1,210	178	120	328	2,770	3,240	1,080	783	444
11	371	246	66	958	178	120	358	2,320	4,000	1,090	783	390
12	358	246	66	718	165	120	423	1,880	4,700	1,090	805	390
13	358	246	74	643	157	120	568	1,680	4,980	1,070	788	410
14	328	208	88	583	157	127	718	1,600	5,200	1,060	794	423
15	299	208	104	553	157	138	798	1,780	5,120	1,040	788	423
16	299	199	104	458	157	138	798	1,990	4,940	1,030	805	410
17	299	178	104	437	157	199	758	2,100	4,530	1,020	810	390
18	299	178	120	423	157	199	718	2,040	4,210	1,000	783	358
19	299	178	120	390	157	199	643	1,940	3,950	986	772	358
20	299	178	127	358	157	208	643	1,780	3,800	968	761	378
21	288	165	127	346	157	208	643	1,680	3,610	950	735	378
22	272	157	138	328	157	237	703	1,530	3,340	920	725	378
23	272	150	138	311	157	237	758	1,390	3,000	908	720	378
24	246	138	138	288	138	237	838	1,300	2,990	872	695	378
25	222	138	138	272	120	246	838	1,210	2,880	866	690	378
26	222	138	138	257	104	246	838	1,110	2,680	832	665	410
27	199	138	138	237	104	246	838	1,040	2,440	794	680	378
28	199	138	138	208	104	246	758	958	2,220	772	685	371
29	199	127	150	178	....	246	758	958	1,920	735	650	346
30	199	127	157	150	....	257	734	822	1,780	730	625	346
31	199	....	157	120	....	257	....	862	....	735	605	....

Monthly discharge of St. Mary River near Babb, Mont., for the year ending Sept. 30, 1918.

[Drainage area, 278 square miles.]

Month.	Discharge in second-feet.				Run-off.	
	Maximum.	Minimum.	Mean.	Per square mile.	Depth in inches.	Acre-feet.
October	371	199	298	1.07	1.23	18,300
November	272	127	199	.715	.80	11,800
December	157	66	118	.424	.49	7,260
January	1,680	120	626	2.25	2.59	38,500
February	199	88	151	.543	.57	8,390
March	257	104	174	.626	.72	10,700
April	838	272	565	2.03	2.27	33,600
May	3,300	758	1,710	6.15	7.09	105,000
June	5,200	1,250	3,000	10.79	12.04	178,600
July	1,570	730	1,040	3.74	4.31	64,000
August	810	605	745	2.68	3.09	45,800
September	572	346	418	1.50	1.67	24,900
The year	5,200	66	755	2.72	36.87	547,000

ST. MARY RIVER NEAR KIMBALL, ALBERTA.

LOCATION.—In SW. ¼ sec. 25, T. 1 N., R. 25 W. fourth meridian, 1 mile south and 1 mile west from Kimball, Alberta, and 5 miles north of international boundary.

DRAINAGE AREA.—472 square miles (measured on topographic maps).

RECORDS AVAILABLE.—January 1, 1913, to September 30 1918. From September 4, 1902, to December 31, 1912, records were obtained at a point one-quarter of a mile below the boundary line. Records were also obtained by the Irrigation Branch (now the Reclamation Service), Department of the Interior, Canada, at a point half a mile below the present station, from 1905 to 1912. The discharge at the three points is practically the same.

GAGE.—Stevens water-stage recorder with a concrete well and shelter on the right bank used during the open-water season. During the winter months a chain gage, located on the highway bridge 3 miles below the station is used. A staff gage located at cable from which measurements were made was used from October 1, 1917, to November 8, 1917.

DISCHARGE MEASUREMENTS.—Made from a cable 1,200 feet above the gage; low-water measurements made by wading near the gage.

CHANNEL AND CONTROL.—Bed of stream at gage and at control composed of boulders and sandstone ledges. Control is formed by an outcropping ledge of sandstone covered with boulders near left bank.

EXTREMES OF DISCHARGE.—Maximum stage during year from water-stage recorder, 6.35 feet at 11 a. m. June 14 (discharge, 4,970 second-feet); minimum stage, December 13–15 and March 1; flow computed from hydrographic study of winter flow as stage-discharge relation was affected by ice.

1902–1918: Maximum stage recorded, 12.75 feet June 5, 1908 (discharge, 18,000 second-feet, estimated by comparison with record for station near Babb); minimum discharge, 70 second-feet,[1] February 5, 1914.

ICE.—Stage-discharge relation seriously affected by ice December 1 to March 29.

DIVERSIONS.—The St. Mary canal, constructed by the United States Reclamation Service, diverts water from St. Mary River near Babb, Mont., to North Fork of Milk River. During 1918 approximately 58,030 acre-feet were diverted, measurement being made at St. Mary crossing. Seepage from the canal above this point returns directly to the river and is measured at the international boundary. Seepage from the canal between St. Mary crossing and Hudson Bay divide goes into Rolph Creek, which enters St. Mary River below the gaging station at international boundary. The Alberta Railway & Irrigation Co. canal diverts from St. Mary River about 2 miles below the station.

REGULATION.—The flow of Swiftcurrent Creek will be regulated by the Sherburne Lake reservoir, under construction by the United States Reclamation Service.

ACCURACY.—Stage-discharge relation permanent during year except for period affected by ice December 1 to March 29. Rating curve well defined. Daily gage heights obtained from Stevens water-stage recorder records by straight-line method for periods October 1 to December 10, 1917, and March 28 to September 30, 1918. Daily gage heights from December 12 to March 28 from observer's reading to hundredths on chain gage at highway bridge 3 miles below gage. Daily discharge October 1 to November 30 and March 29 to September 30 ascertained by applying mean daily gage height to rating table. Records for this period are good as curve is well defined between 200 and 5,000 second-feet. Daily discharge December 1 to March 28 from winter hydrograph, based upon observer's gage heights and notes on ice, temperature records, and discharge measurements. Records fair.

COOPERATION.—Station maintained jointly with the Reclamation Service, Department of the Interior, Canada.

---

[1] Only estimates of mean monthly flow are available for the winter periods from 1902 to 1912, inclusive, and a lower minimum discharge may have occurred during that time.

Discharge measurements of St. Mary River near Kimball, Alberta, during the year ending Sept. 30, 1918.

Date.	Made by—	Gage height.	Discharge.	Date.	Made by—	Gage height.	Discharge.
		Feet.	*Sec.-ft.*			*Feet.*	*Sec.-ft.*
Oct. 16	A. W. P. Lowrie [a]	2.80	376	June 2	C. H. Ellacott [a]	4.45	1,600
Nov. 6	do	2.65	301	15	do	6.19	4,581
10	B. E. Jones and W. A. Lamb	2.59	276	17	W. A. Lamb	5.92	4,290
19	S. H. Frame [a]	2.33	222	18	V. A. Newhall [a] and D. G. Chadsey [a]	5.83	3,760
Dec. 4	D. G. Chadsey [a]	[b] 3.86	144	July 3	C. H. Ellacott [a]	3.93	1,154
Jan. 1	S. H. Frame	5.46	497	6	B. E. Jones	3.72	859
4	do	5.34	1,341	11	W. A. Lamb	3.55	851
28	do	4.04	224	23	C. H. Ellacott [a]	3.42	740
Feb. 21	A. W. P. Lowrie	4.65	143	Aug. 4	B. E. Jones and R. J. Burley [a]	3.11	[c] 518
Mar. 12	do	4.60	136	7	C. H. Ellacott [a]	3.18	581
29	C. M. O'Neil [a]	2.65	287	10	W. A. Lamb	3.09	512
Apr. 3	do	2.70	306	15	C. H. Ellacott [a]	3.13	541
22	do	3.55	806	31	do	2.90	413
26	do	3.67	928	Sept. 6	B. E. Jones and R. J. Burley [a]	2.85	391
May 10	B. Russell [a]	5.32	2,845	21	C. H. Ellacott [a]	2.94	445
24	W. A. Lamb	4.26	1,360	28	do	2.90	408
28	C. H. Ellacott [a]	3.88	1,110				

[a] Engineer, Department of Interior, Canada.
[b] Stage-discharge relation affected by ice; gage height from staff gage at regular station.
[c] Measurement made below Alberta Railway & Irrigation Co.'s dam, flow of canal included in results.

NOTE.—Stage-discharge relation affected by ice Dec. 12 to Mar. 27. Measurements during this period referred to chain gage on highway bridge 3 miles below gage.

Daily discharge, in second-feet, of St. Mary River near Kimball, Alberta, for the year ending Sept. 30, 1918.

Day.	Oct.	Nov.	Dec.	Jan.	Feb.	Mar.	Apr.	May.	June.	July.	Aug.	Sept.
1	415	284	190	500	200	115	315	846	1,370	1,340	464	508
2	415	315	180	700	200	120	315	972	1,550	1,190	464	495
3	415	315	165	1,000	200	120	315	1,160	1,570	1,120	486	479
4	410	340	145	1,340	200	125	304	1,500	1,510	1,070	520	432
5	410	315	140	1,650	205	125	340	2,280	1,430	995	567	415
6	443	298	140	1,870	205	130	390	2,990	1,460	942	591	396
7	437	294	135	1,850	210	130	375	3,300	1,570	882	573	453
8	448	290	130	1,700	210	130	375	3,380	1,970	860	549	510
9	459	284	130	1,500	215	130	420	3,210	2,720	832	537	567
10	426	277	125	1,300	215	130	476	2,970	3,540	825	520	555
11	400	274	120	1,050	220	130	514	2,520	4,200	825	525	531
12	405	268	120	770	220	135	585	2,120	4,600	818	592	506
13	400	256	115	690	220	135	671	1,890	4,930	811	573	498
14	395	239	115	630	220	140	776	1,820	4,970	797	537	481
15	380	237	115	600	215	145	846	1,980	4,730	790	549	470
16	360	236	120	510	210	150	853	2,160	4,420	790	624	464
17	340	226	125	490	200	160	846	2,340	4,040	762	561	448
18	355	225	130	470	190	170	818	2,200	3,800	762	531	443
19	355	224	140	440	170	180	783	2,080	3,560	755	520	443
20	375	221	150	410	155	190	769	1,990	3,420	727	486	426
21	426	217	155	395	145	205	790	1,780	3,170	727	486	437
22	443	212	160	370	140	220	839	1,710	2,950	713	464	432
23	410	208	170	345	135	230	860	1,570	2,670	727	464	426
24	390	204	180	320	130	240	882	1,450	2,570	713	470	437
25	365	201	190	295	130	255	898	1,370	2,450	664	464	448
26	325	197	200	280	130	270	912	1,260	2,260	650	470	443
27	290	197	215	255	125	275	935	1,170	2,030	592	443	426
28	256	197	230	225	120	280	898	1,100	1,850	508	496	415
29	253	197	250	220	....	285	898	1,070	1,670	459	437	415
30	262	197	320	210	....	290	832	972	1,450	464	415	415
31	268	....	400	205	....	298	....	980	....	464	410	....

Monthly discharge of St. Mary River near Kimball, Alberta, for the year ending Sept. 30, 1918.

Month.	Discharge in second-feet.			Run-off in acre-feet.
	Maximum.	Minimum.	Mean.	
October	459	253	378	23,200
November	340	197	248	14,800
December	400	115	168	10,300
January	1,870	205	729	44,800
February	220	120	183	10,200
March	298	115	182	11,200
April	935	304	661	39,300
May	3,380	846	1,880	116,000
June	4,970	1,370	2,810	167,000
July	1,340	459	793	48,800
August	624	410	509	31,300
September	567	395	460	27,400
The year	4,970	115	739	544,000

Combined daily discharge, in second-feet, of St. Mary River near Kimball, Alberta, and St. Mary canal at St. Mary crossing, near Babb, Mont., for the year ending Sept. 30, 1918.

Day.	Oct.	Nov.	Dec.	Jan.	Feb.	Mar.	Apr.	May.	June.	July.	Aug.	Sept.
1	415	284	190	500	200	115	315	846	1,370	1,340	726	635
2	415	315	180	700	200	120	315	972	1,550	1,190	724	630
3	415	315	165	1,000	200	120	315	1,160	1,570	1,120	745	603
4	410	340	145	1,340	200	125	304	1,500	1,510	1,070	760	567
5	410	315	140	1,650	205	125	340	2,280	1,430	995	806	542
6	443	298	140	1,870	205	130	390	2,990	1,460	1,330	833	516
7	437	294	135	1,850	210	130	375	3,300	1,570	1,270	817	542
8	448	290	130	1,700	210	130	375	3,380	1,970	1,250	814	512
9	459	284	130	1,500	215	130	420	3,210	2,720	1,220	793	567
10	426	277	125	1,300	215	130	476	2,970	3,540	1,220	776	555
11	400	274	120	1,050	220	130	514	2,520	4,200	1,220	764	531
12	405	268	120	770	220	135	585	2,120	4,600	1,210	840	508
13	400	256	115	690	220	135	671	1,890	4,930	1,210	845	498
14	395	239	115	630	220	140	776	1,820	4,970	1,200	830	481
15	380	237	115	600	215	145	846	1,980	4,730	1,190	851	470
16	360	236	120	510	210	150	853	2,160	4,420	1,190	945	464
17	340	226	125	490	200	160	846	2,340	4,040	1,160	897	448
18	355	225	130	470	190	170	818	2,200	3,800	1,160	897	443
19	355	224	140	440	170	180	783	2,080	3,560	1,160	886	443
20	375	221	150	410	155	190	769	1,990	3,420	1,130	841	426
21	426	217	155	395	145	205	790	1,780	3,170	1,130	804	437
22	443	212	160	370	140	220	839	1,710	2,950	1,120	768	432
23	410	208	170	345	135	230	860	1,570	2,670	1,130	725	426
24	390	204	180	320	130	240	882	1,450	2,570	1,120	697	437
25	365	201	190	295	130	255	898	1,370	3,450	1,070	669	448
26	325	197	200	280	130	270	912	1,280	2,260	1,060	661	443
27	290	197	215	255	125	275	935	1,170	2,030	998	618	426
28	256	197	230	225	120	280	898	1,100	1,850	903	624	415
29	253	197	250	220	......	285	898	1,070	1,670	843	569	415
30	262	197	320	210	......	290	832	972	1,450	786	562	415
31	268	......	400	205	......	298	......	980	......	745	551	......

NOTE.—For table of daily discharge of St. Mary canal at St. Mary crossing, see p. 18.

Combined monthly discharge of St. Mary River near Kimball, Alberta, and St. Mary canal at St. Mary crossing, near Babb, Mont., for the year ending Sept. 30, 1918.

Month.	Discharge in second-feet.			Run-off in acre-feet.
	Maximum.	Minimum.	Mean.	
October	459	253	378	23,200
November	340	197	248	14,800
December	400	115	168	10,300
January	1,870	205	729	44,800
February	220	120	183	10,200
March	298	115	182	11,200
April	935	304	661	39,300
May	3,380	846	1,890	116,600
June	4,970	1,370	2,810	167,200
July	1,340	745	1,120	68,900
August	945	551	763	46,900
September	635	415	489	29,100
The year	4,970	115	808	581,700

NOTE.—For table of monthly discharge at St. Mary canal at St. Mary crossing, see p. 18.

ST. MARY CANAL AT INTAKE, NEAR BABB, MONT.

LOCATION.—In SE. ¼ sec. 27, T. 36 N., R. 14 W., 300 feet below headworks of canal and 2 miles south of Babb, on Blackfeet Indian Reservation.

RECORDS AVAILABLE.—June 1 to September 7, 1918.

GAGE.—Staff gage nailed to downstream side of pier of footbridge, 300 feet below headworks of canal. Gage read by United States Reclamation Service employees.

DISCHARGE MEASUREMENTS.—Made from footbridge at gage.

CHANNEL AND CONTROL.—Bed composed of gravel. Repairs to canal may cause slight changes in cross section below gage.

EXTREMES OF DISCHARGE.—Maximum stage recorded during year, 6.3 feet June 27 and 28 (discharge, 626 second-feet).

ICE.—Canal is not operated during winter months.

REGULATION.—Discharge is regulated by the head gates.

ACCURACY.—Stage-discharge relation fairly permanent, but current-meter measurements only fair, due to eddies from bridge piers. Rating curve fairly well defined. Daily discharge ascertained by applying mean daily gage height to rating table. Records fair.

COOPERATION.—Station maintained in cooperation with Reclamation Service, Department of the Interior, Canada.

Discharge measurements of St. Mary canal at intake, near Babb, Mont., during the year ending Sept. 30, 1918.

Date.	Made by—	Gage height.	Dis-charge.	Date.	Made by—	Gage height.	Dis-charge.
		Feet.	*Sec.-ft.*			*Feet.*	*Sec.-ft.*
June 15	W. A. Lamb	5.90	565	Aug. 12	C. H. Ellacott	4.80	319
July 7	do	6.25	613	Sept. 6	W. A. Lamb	3.14	185
Aug. 9	do	4.67	335				

Daily discharge, in second-feet, of St. Mary canal at intake, near Babb, Mont., for the year ending Sept. 30, 1918.

Day.	June.	July.	Aug.	Sept.	Day.	June.	July.	Aug.	Sept
1	208	616	350	187	16	550	615	421	....
2	208	616	350	187	17	550	616	512	....
3	208	616	359	186	18	550	615	522	....
4	238	616	321	188	19	550	615	522	....
5	258	616	321	187	20	550	615	522	....
6	264	616	288	185	21	550	611	421	....
7	270	616	321	69	22	550	611	388	....
8	270	616	357	....	23	550	611	301	....
9	288	616	366	....	24	569	611	294	....
10	342	616	342	....	25	588	615	253	....
11	388	616	314	....	26	588	611	230	....
12	421	616	314	....	27	626	607	230	....
13	457	620	351	....	28	626	607	208	....
14	498	616	388	....	29	623	550	197	....
15	540	613	388	....	30	619	430	197	....
					31	....	380	186	....

NOTE.—Canal gates closed at 9 a. m. Sept. 7.

Monthly discharge of St. Mary canal at intake, near Babb, Mont., for the year ending Sept. 30, 1918.

Month.	Discharge in second-feet.			Run-off in acre-feet.
	Maximum.	Minimum.	Mean.	
June	626	208	450	26,800
July	620	380	599	36,800
August	522	186	340	20,900
September 1-7	188	69	170	2,360
The period	626	69	442	86,860

ST. MARY CANAL AT ST. MARY CROSSING, NEAR BABB, MONT.

LOCATION.—In sec. 19, T. 37 N., R. 13 W., at entrance to flume, 600 feet below outlet of siphon by which canal crosses St. Mary River, 9 miles below headworks, and 6 miles northeast of Babb, on Blackfeet Indian Reservation.

RECORDS AVAILABLE.—July 6 to September 8, 1918.

GAGE.—Stevens water-stage recorder, located on concrete entrance to flume just below outlet to siphon crossing St. Mary River. A staff gage on outside of gage house is also read.

DISCHARGE MEASUREMENTS.—Made from cable 200 feet above gage.

CHANNEL AND CONTROL.—Control is the steel flume several hundred feet long heading at the gage.

EXTREMES OF DISCHARGE.—Maximum stage recorded during year, 6.40 feet July 26 (discharge, 408 second-feet).

ICE.—Canal not operated during winter months.

REGULATION.—Flow is regulated by head gates about 9 miles above.

ACCURACY.—Stage-discharge relation permanent. Rating curve well defined between 180 and 400 second-feet. Daily discharge ascertained by applying mean daily gage height to rating table. Records good.

COOPERATION.—Station maintained in cooperation with Reclamation Service, Department of the Interior, Canada.

Discharge measurements of St. Mary canal at St. Mary crossing, near Babb, Mont., during the year ending Sept. 30, 1918.

Date.	Made by—	Gage height.	Discharge.	Date.	Made by—	Gage height.	Discharge.
		Feet.	*Sec.-ft.*			*Feet.*	*Sec.-ft.*
July 11	W. A. Lamb	6.34	400	Aug. 9	W. A. Lamb	5.06	[illegible]
Aug. 5	Jones and Burley	4.90	242	Sept. 5	Jones and Burley	3.55	126

Daily discharge, in second-feet, of St. Mary canal at St. Mary crossing, near Babb, Mont., for the year ending Sept. 30, 1918.

Day.	July.	Aug.	Sept.	Day.	July.	Aug.	Sept.	Day.	July.	Aug.	Sept.
1		262	132	11	398	239		21	403	318	
2		260	132	12	396	348		22	404	304	
3		259	133	13	400	272		23	403	261	
4		240	135	14	398	293		24	406	227	
5		239	127	15	403	302		25	406	205	
6	391	242	121	16	403	321		26	408	191	
7	392	244	90	17	403	336		27	406	175	
8	394	265	2	18	403	366		28	400	126	
9	390	256		19	403	366		29	384	132	
10	398	256		20	402	355		30	322	147	
								31	281	141	

NOTE.—Discharge for Sept. 7 and 8 computed by hourly method. Canal gates closed Sept. 8.

Monthly discharge of St. Mary canal at St. Mary crossing, near Babb, Mont., for the year ending Sept. 30, 1918.

Month.	Discharge in second-feet.			Run-off in acre-feet.
	Maximum.	Minimum.	Mean.	
July 6–31	408	281	392	20,200
August	366	126	253	15,600
September 1–8	135	2	109	1,730
The period	408	2	291	37,530

ST. MARY CANAL AT HUDSON BAY DIVIDE, NEAR BROWNING, MONT.

LOCATION.—At Douglas bridge on Hudson Bay divide, 3 miles above outlet of canal and 30 miles directly north of Browning, in Blackfeet Indian Reservation.

RECORDS AVAILABLE.—July 3, 1917, to September 30, 1918.

GAGE.—A vertical staff, graduated to tenths, nailed to upstream side of left pier of bridge; read once a day by United States Reclamation Service ditch rider.

DISCHARGE MEASUREMENTS.—Made from upstream side of bridge at gage.

CHANNEL AND CONTROL.—Channel uniform, but slope varies with the stage. Control is a V-shaped concrete drop located 1 mile below gage.

EXTREMES OF DISCHARGE.—Maximum stage recorded during the year, 5.4 feet July 25–29 (discharge, 405 second-feet).

1917–1918: Maximum stage recorded, 5.4 feet July 25–29, 1918 (discharge, 405 second-feet).

REGULATION.—The flow is regulated at the head gates 26 miles above. A small reservoir at Spider Lake eliminates sudden changes at the head gates.

ACCURACY.—Stage-discharge relation practically permanent. Rating curve well defined between 120 and 400 second-feet. Daily discharge ascertained by applying [da]ily gage height to rating table. Records fair.

[COÖPER]ATION.—Station maintained in cooperation with Reclamation Service, Department of the Interior, Canada.

Discharge measurements of St. Mary canal at Hudson Bay divide, near Browning, Mont., during the year ending Sept. 30, 1918.

Date.	Made by—	Gage height.	Discharge.	Date.	Made by—	Gage height.	Discharge.
		Feet.	*Sec.-ft.*			*Feet.*	*Sec.-ft.*
June 18[a]	B. E. Jones	4.96	317	July 9[a]	W. A. Lamb	5.30	403
18[b]	do	4.95	327	Aug. 10[a]	do	4.25	264
20[b]	do	5.16	363	Sept. 7[a]	do	3.02	138

[a] Made at Douglas Bridge. [b] Made at bridge below first drop.

Daily discharge, in second-feet, of St. Mary canal at Hudson Bay divide, near Browning, Mont., for the year ending Sept. 30, 1918.

Day.	June.	July.	Aug.	Sept.	Day.	June.	July.	Aug.	Sept.
1		373	243	138	16	315	394	289	
2		383	254	138	17	329	394	302	
3	114	383	248	138	18	337	394	315	
4	114	383	248	122	19	329	394	345	
5	146	383	233	122	20	329	394	363	
6	183	394	238	122	21	329	394	308	
7	183	394	238	122	22	329	394	289	
8	188	394	238	122	23	329	394	265	
9	193	394	248	72	24	329	394	214	
10	213	394	243		25	337	405	203	
11	218	394	238		26	337	405	183	
12	265	405	238		27	337	405	173	
13	296	383	243		28	354	405	173	
14	302	394	248		29	363	405	114	
15	302	394	265		30	368	363	138	
					31		302	138	

NOTE.—Canal gates closed Sept. 9. Discharge for Sept. 9 computed by hourly method.

Monthly discharge of St. Mary canal at Hudson Bay divide, near Browning, Mont., for the year ending Sept. 30, 1918.

Month.	Discharge in second-feet.			Run-off in acre-feet.
	Maximum.	Minimum.	Mean.	
June 3–30	368	114	277	15,400
July	405	302	390	24,000
August	363	114	241	14,800
September 1–9	138	72	122	2,180
The period	405	72	287	56,400

SWIFTCURRENT CREEK AT MANY GLACIER, MONT.

LOCATION.—In sec. 12, T. 35 N., R. 16 W., at outlet of McDermott Lake at Many Glacier, in Glacier National Park, 14 miles southwest of Babb, in Teton County.

DRAINAGE AREA.—31.4 square miles (measured on topographic map).

RECORDS AVAILABLE.—June 6, 1912, to September 30, 1918.

GAGE.—Stevens water-stage recorder installed June 15, 1918, in shelter built by park officials and Great Northern Railway, and referred to two staff gages, one inside well and one outside. Prior to May 23, 1916, a staff gage on left bank opposite present gage was read. May 23, 1916, to June 15, 1918, a vertical staff at same location as present gage. Gage read by E. Peterson and others twice daily to hundredths.

DISCHARGE MEASUREMENTS.—Made by wading at outlet of lake or below falls. High-water measurements made from highway bridge above power house; measuring section at bridge is very poor.

Channel and control.—Control is a limestone outcrop at outlet of the lake; just below is a fall and a cataract.

Extremes of discharge.—Maximum stage recorded during year, 4.25 feet June 10 (discharge, 1,250 second-feet); minimum stage, 1.48 feet February 25–28 and January 1 and 2 (discharge, 41 second-feet).

1912–1918: Maximum stage recorded, 4.75 feet June 17, 1916 (discharge, 1,550 second-feet); minimum discharge, 10.8 second-feet March 19, 1912, measured by current meter, prior to installation of gage.

Ice.—Stage-discharge affected very little, if any, by ice. Open channel conditions assumed throughout year.

Diversions.—None.

Regulation.—None.

Accuracy.—Stage-discharge relation apparently changed during high water of June, but remained constant during remainder of year. Two rating tables used; one applicable October 1 to June 10, the other June 11 to September 30. The former is well defined between 44 and 825 second-feet; the latter between 60 and 300 second-feet. Gage heights October 1 to June 14 are mean of two readings daily to nearest hundredth; June 15 to September 30 determined by graphic method from Stevens water-stage recorder. Daily discharge ascertained by applying mean daily gage height to rating table. Records good.

The following discharge measurements were made by W. A. Lamb:

July 8, 1918: Gage height, 2.32 feet; discharge, 209 second-feet (subject to some error caused by wave action due to strong wind); September 6, 1918: Gage height, 1.81 feet; discharge, 67 second-feet.

Daily discharge, in second-feet, of Swiftcurrent Creek at Many Glacier, Mont., for the year ending Sept. 30, 1918.

Day.	Oct.	Nov.	Dec.	Jan.	Feb.	Mar.	Apr.	May.	June.	July.	Aug.	Sept.
1	94	56	54	147	48	42	53	440	367	211	192	119
2	94	54	54	455	48	42	54	570	372	242	222	112
3	94	55	54	525	48	44	56	715	367	242	215	89
4	103	56	55	500	46	46	59	740	372	203	188	76
5	128	56	54	460	44	46	59	886	420	188	174	73
6	133	58	53	430	44	48	62	924	590	182	164	71
7	128	59	53	405	46	46	64	935	565	188	155	80
8	118	60	53	353	46	46	65	372	742	207	161	87
9	114	62	53	306	44	46	82	287	1,020	238	158	91
10	94	59	52	278	44	46	144	343	1,250	280	149	91
11	88	59	53	256	44	46	165	324	1,160	293	168	89
12	86	59	53	224	44	46	172	310	1,070	263	242	97
13	86	58	52	88	44	46	189	324	980	226	246	96
14	78	55	53	71	44	46	185	560	860	222	222	95
15	76	52	53	65	44	50	165	666	860	226	199	85
16	74	54	53	65	44	53	155	704	563	226	192	80
17	73	53	53	64	44	53	133	682	563	226	178	75
18	70	54	68	65	44	58	144	732	618	238	168	76
19	67	54	125	65	44	55	162	710	640	246	149	82
20	65	54	125	65	44	53	155	655	618	234	134	85
21	62	54	92	63	44	53	201	555	520	207	129	85
22	60	54	78	60	44	53	224	152	473	182	129	89
23	62	54	63	54	44	53	238	138	484	178	126	89
24	59	54	55	53	44	43	269	130	443	164	129	89
25	59	55	55	53	42	54	287	123	458	137	146	91
26	59	55	55	53	42	58	256	123	389	122	146	84
27	56	55	55	53	42	55	212	121	324	126	143	78
28	63	54	84	51	42	56	204	123	276	134	122	76
29	62	54	136	51	....	65	224	141	234	143	105	75
30	56	54	147	51	....	63	324	287	199	161	101	73
31	56	....	138	48	....	59	....	357	....	178	99	....

Monthly discharge of Swiftcurrent Creek at Many Glacier, Mont., for the year ending Sept. 30, 1918.

[Drainage area, 31.4 square miles.]

Month.	Discharge in second-feet.				Run-off.	
	Maximum.	Minimum.	Mean.	Per square mile.	Depth in inches.	Acre-feet.
October	133	56	81.2	2.59	2.99	4,990
November	62	52	55.7	1.77	1.98	3,310
December	147	52	70.4	2.24	2.58	4,330
January	525	48	177	5.63	6.49	10,900
February	48	42	44.3	1.41	1.47	2,460
March	65	42	50.6	1.61	1.86	3,110
April	324	53	159	5.09	5.68	9,460
May	935	121	456	14.5	16.7	28,000
June	1,250	199	593	18.9	21.1	35,200
July	293	122	204	6.50	7.49	12,500
August	246	99	163	5.19	5.98	10,000
September	119	71	85.9	2.74	3.06	5,110
The year	1,250	42	179	5.70	77.38	129,000

SWIFTCURRENT CREEK AT SHERBURNE, MONT.

Location.—In sec. 35, T. 36 N., R. 15 W., near outlet of Lower Sherburne Lake, in Teton County.

Drainage area.—64 square miles (measured on topographic map).

Records available.—July 1, 1912, to September 30, 1918.

Gage.—Staff gage on left bank about 300 feet below the spillway of Sherburne Lake dam, read by employees of the United States Reclamation Service. From July 1, 1912, to November 9, 1914, a vertical staff gage was maintained on left bank near outlet of lake, and at a different datum from present gage.

Discharge measurements.—Made by wading or from cable 50 feet below gage.

Channel and control.—An outcropping limestone ledge, somewhat broken and irregular, forms control; subject to slight shifts.

Extremes of discharge.—Maximum stage recorded during year, 6.20 feet May 5, June 14–15 (discharge, 1,140 second-feet); minimum stage, gates closed January 11 to March 13; flow only the leakage through gates and small inflow between dam and gage.

1912–1918: Maximum stage recorded, 7.85 feet June 17, 1916 (discharge, 2,280 second-feet); minimum stage, gates closed January 11 to March 13, 1918; flow only the leakage through gates and small inflow between dam and gage.

Ice.—Not seriously affected by ice; gates closed during most of winter season.

Diversions.—None.

Regulation.—The natural flow of the stream was affected by placing and removing flashboards on temporary dam built at outlet of the lake for construction purposes in connection with Sherburne Lake storage dam. See footnote to table of daily discharge. Flow partly regulated by gate operation.

Accuracy.—Stage-discharge relation not permanent during year; affected by changes in control due to landslide. After May 5 control practically permanent. Two rating curves used during year; one from October 1 to January 10 and the other from May 5 to September 30; the former is well defined between 40 and 1,000 second-feet, and the latter between 60 and 1,200 second-feet. Daily gage heights are mean of two readings daily to nearest hundredth. Daily discharge ascertained by applying daily gage heights to rating table, except for period March 14 to April 27, when they were obtained by indirect method for shifting control. Records good.

Discharge measurements of Swiftcurrent Creek at Sherburne, Mont., during the year ending Sept. 30, 1918.

Date.	Made by—	Gage height.	Dis- charge.	Date.	Made by—	Gage height.	Dis- charge.
		Feet.	*Sec.-ft.*			*Feet.*	*Sec.-ft.*
Mar. 16	W. A. Lamb	2.50	148	July 8	W. A. Lamb	3.50	278
May 25	.....do	3.36	245	Aug. 12	.....do	3.00	180
June 17	.....do	6.02	1,070	Sept. 6	.....do	2.48	110

Daily discharge, in second-feet, of Swiftcurrent Creek at Sherburne, Mont., for the year ending Sept. 30, 1918.

Day.	Oct.	Nov.	Dec.	Jan.	Mar.	Apr.	May.	June.	July.	Aug.	Sept.
1	108	246	54	5		151		663	296	137	110
2	113	193	56	4		148		636	278	138	110
3	117	202	56	695		148		458	278	164	110
4	114	180	56	970		148	1,100	348	280	180	110
5	123	142	57	873		146	1,140	348	278	194	110
6	170	176	56	800		145	1,110	350	278	201	110
7	187	180	56	695		145	1,050	496	276	200	110
8	170	148	55	630		145	1,000	816	269	198	110
9	134	134	54	94		153	801	906	252	188	110
10	105	124	55	92		213	448	1,000	238	180	110
11	108	118	55			282	298	1,080	243	180	110
12	104	92	56			415	214	1,110	243	180	110
13	105	82	56			560	234	1,130	245	183	110
14	99	76	56		198	282	324	1,140	245	183	110
15	65	70	32		167	339	673	1,140	245	194	110
16	49	57	10		149	412	649	1,120	247	247	110
17	54	54	42		151	293	554	1,060	247	291	110
18	71	49	93		151	231	422	997	247	278	110
19	60	38	93		148	210	367	967	249	230	109
20	83	46	95		148	202	360	887	260	185	109
21	55	43	98		148	216	302	794	282	164	109
22	59	44			146	282	276	524	287	143	109
23	57	38			153	304	267	502	287	143	108
24	59	40			159	274	256	710	282	143	108
25	58	43			159	276	193	609	265	124	108
26	57	48			158	265	151	513	220	108	108
27	57	51			159	240	148	464	176	108	90
28	58	55			159		148	353	157	108	90
29	55	62	1		159		145	314	144	109	76
30	130	53	1		151		145	317	143	109	60
31	194		3		151		355		137	109	

NOTE.—Entire flow of river held back at Sherburne Lake from Dec. 22-28, Jan. 11 to Mar. 13, and Apr. 28 to May 3.

Monthly discharge of Swiftcurrent Creek at Sherburne, Mont., for the year ending Sept. 30, 1918.

Month.	Discharge in second-feet.			Run-off in acre-feet.
	Maximum.	Minimum.	Mean.	
October	194	49	96.1	5,910
November	246	38	96.1	5,720
December	98	1	40.2	2,470
May	1,140	145	425	26,100
June	1,140	314	725	43,100
July	296	137	244	15,000
August	291	108	171	10,500
September	110	60	106	6,310

NOTE.—Stream partly controlled beginning with 1915, therefore valves for discharge in second-feet per square mile and for run-off, depth in inches, are not computed. June 1-30, 1915, a total of 1,560 acre-feet of water was stored in Sherburne Lake by a temporary construction dam; 134 acre-feet was stored Aug. 25 to Sept. 18, 1915; the latter amount was released Sept. 18-20, 1915.

CANYON CREEK NEAR MANY GLACIER, MONT.

Location.—At the edge of heavy timber area, half a mile above mouth, and 2 miles southeast of Many Glacier, in Teton County.

Drainage area.—7.0 square miles (measured on topographic map).

Records available.—July 12 to September 30, 1918.

Gage.—Stevens water-stage recorder on left bank.

Discharge measurements.—Made from footbridge at gage.

Channel and control.—Bed of stream covered with heavy boulders and cobblestones. Control is riffle about 20 feet below gage; may shift at high stage. Both banks high and can not be overflowed.

Extremes of discharge.—Maximum discharge recorded, 74 second-feet by current-meter measurement, June 16; minimum stage, 0.83 foot September 29 and 30 (discharge, 10 second-feet).

Ice.—Station not operated during winter on account of severe ice effect on stage-discharge relation.

Diversions.—None.

Regulations.—Some natural storage in small lake at head of creek; no artificial regulations.

Accuracy.—Stage-discharge relation practically permanent except for severe ice effect. Rating curve well defined between 15 and 40 second-feet. Daily gage heights obtained from Stevens water-stage recorder graph by the straight-line method, except for period August 4-11 when clock stopped. Daily discharge ascertained by applying mean daily gage height to rating table except for period noted above, for which discharge was interpolated. Records good.

Discharge measurements of Canyon Creek near Many Glacier, Mont., during the year ending Sept. 30, 1918.

Date.	Made by—	Gage height.	Discharge.	Date.	Made by—	Gage height.	Discharge.
		Feet.	*Sec.-ft.*			*Feet.*	*Sec.-ft.*
June 16	W. A. Lamb	(a)	74	Aug. 11	W. A. Lamb	1.12	23.0
27	B. E. Jones	1.35	36	Sept. 6	Jones and Burley [b]	.99	15.0
July 12	W. A. Lamb	1.27	31.0				

[a] Measurement referred to nail in crack in rock.
[b] Engineer, Department of the Interior, Canada.

Daily discharge, in second-feet, of Canyon Creek near Many Glacier, Mont., for the year ending Sept. 30, 1918.

Day.	July.	Aug.	Sept.	Day.	July.	Aug.	Sept.	Day.	July.	Aug.	Sept.
1		26	17	11		24	18	21	24	16	14
2		24	16	12	31	32	17	22	24	15	14
3		21	14	13	29	30	17	23	22	17	14
4		21	15	14	29	27	16	24	20	18	13
5		22	16	15	29	26	15	25	20	18	12
6		22	17	16	30	24	13	26	20	18	11
7		23	17	17	30	23	12	27	20	17	11
8		23	18	18	30	21	13	28	20	14	10
9		23	20	19	29	18	13	29	22	13	10
10		24	19	20	26	17	13	30	23	13	10
								31	24	14	

Monthly discharge of Canyon Creek near Many Glacier, Mont., for the year ending Sept. 30, 1918.

Month.	Discharge in second-feet.			Run-off in acre-feet.
	Maximum.	Minimum.	Mean.	
July 12-31	31	20	25.1	996
August	32	13	20.8	1,280
September	20	10	14.5	863

RED RIVER AT FARGO, N. DAK.

LOCATION.—At dam half a mile above highway bridge connecting Front Street, Fargo, N. Dak., with Moorhead, Minn., 10 miles above mouth of Sheyenne River.

DRAINAGE AREA.—6,020 square miles.

RECORDS AVAILABLE.—May 27, 1901, to September 30, 1918.

GAGE.—Vertical staff attached to tree on left bank about six rods above the dam; vertical staff for use at low stages attached to upper end of fishway at left end of dam; read by F. L. Anders. Prior to September 1, 1914, gage readings were obtained from a vertical staff attached to the breakwater for the center pier of Front Street bridge; this gage is still maintained and used by the Weather Bureau, but can not be read accurately without a field glass and has less permanent control than gage now used. At the same stage, readings on Front Street gage are numerically about 10.4 feet greater than readings on gage now used.

DISCHARGE MEASUREMENTS.—Made from footbridge at gage.

CHANNEL AND CONTROL.—Bed consists of clay and silt, nearly permanent. Control is timber crib dam, rock filled, below gage; has settled slightly during 1918.

EXTREMES OF DISCHARGE.—Maximum stage recorded during year, 3.1 feet March 30 and 31 and May 25 (discharge, 750 second-feet); minimum stage, 1.0 foot February 11 (discharge not computed).

1901–1918: Maximum stage recorded, 19.9 feet April 6, 1916 (stage-discharge relation affected by ice); open channel maximum stage 17.34 feet at 3.30 p. m. July 11, 1916 (discharge, 7,740 second-feet); minimum stage recorded, 1.0 foot February 11, 1918 (discharge not computed).

ICE.—Stage-discharge relation affected by ice December 18 to March 31.

DIVERSIONS.—None.

REGULATION.—No power plants or storage above station within 60 miles; storage not great enough to noticeably affect the discharge at station.

ACCURACY.—Stage-discharge relation affected by settling of dam, and by ice December 18 to March 31. The rating curves used for 1918, one applicable October 1 to December 17 and the other April 1 to September 30; the former is well defined between 150 and 4,000 second-feet, and the latter between 59 and 4,400 second-feet. Gage heights are read to hundredths once daily except during winter when one reading a week is made. Daily discharge ascertained by applying daily gage height to rating tables for days when gage was read; discharges interpolated for intervening days. Open-water records good.

Discharge measurements of Red River at Fargo, N. Dak., during the year ending Sept. 30, 1918.

Date.	Made by—	Gage height.	Discharge.	Date.	Made by—	Gage height.	Discharge.
		Feet.	*Sec.-ft.*			*Feet.*	*Sec.-ft.*
[illegible] 3	E. F. Chandler	1.52	108	July 2	Alf. Hulteng	2.03	[illegible]
[illegible]	Alf. Hulteng	2.33	451	Aug. 27	E. F. Chandler	1.58	134
[illegible]	....do	2.03	357	Sept. 25	....do	1.25	61
[illegible]	E. F. Chandler	2.17	378	25	....do	1.26	75

Daily discharge, in second-feet, of Red River at Fargo, N. Dak., for the year ending Sept. 30, 1918.

Day.	Oct.	Nov.	Dec.	Apr.	May.	June.	July.	Aug.'	Sept.
1	156	98	104	700	414	452	342	218	140
2	142	92	129	550	452	491	324	204	140
3	142	92	156	490	452	530	306	204	140
4	129	131	129	433	452	452	306	208	143
5	116	170	104	378	452	530	306	211	140
6	92	170	88	378	452	490	299	204	140
7	124	142	104	378	452	490	292	198	143
8	156	92	116	378	414	570	248	204	130
9	142	116	116	378	414	570	386	204	116
10	116	142	116	378	378	570	272	238	140
11	70	136	104	378	342	570	245	221	134
12	92	129	97	342	342	550	265	204	116
13	92	185	97	306	342	530	275	172	98
14	86	170	90	342	414	490	277	143	110
15	81	142	83	342	356	471	279	162	124
16	70	129	76	351	299	462	245	166	137
17	81	116	70	360	272	452	258	143	125
18	70	122	....	396	272	414	221	160	116
19	92	129	....	378	289	414	224	178	110
20	92	142	....	360	306	414	218	178	83
21	98	142	....	360	433	396	220	166	105
22	104	185	....	342	610	378	221	265	98
23	104	142	....	342	655	342	162	231	134
24	116	142	....	324	700	324	191	172	110
25	92	142	....	306	750	306	198	172	78
26	116	142	....	299	660	272	158	172	86
27	116	129	....	306	570	324	231	166	87
28	116	70	....	315	490	306	238	169	87
29	116	97	....	324	452	318	245	166	87
30	110	124	....	378	452	330	241	134	88
31	104	....	....	....	452	....	207	140	....

NOTE.—Discharge interpolated for lack of gage readings on following days: Oct. 7, 14, 21, 28, 30, 31; Nov 1, 4, 11, 18, 24, 25, 29; Apr. 7, 16, 28; May 5. 6, 12, 15, 19, 23, 26, 30; June 2, 9, 16, 21, 29, 31; July 4, 14, 21, 28; Aug. 4, 11, 18, 25; Sept. 1, 8, 15, 27–29. Gage read Dec. 29, Jan. 13 and 24; Feb. 11, 22, 28; Mar. 1, 11, 15 to 31; discharge not computed on account of ice.

Monthly discharge of Red River at Fargo, N. Dak., for the year ending ending Sept. 30, 1918.

Month.	Discharge in second-feet.			Run-off in acre-feet.
	Maximum.	Minimum.	Mean.	
October	156	70	108	6,640
November	185	70	132	7,860
December 1-17	156	70	105	3,540
April	700	299	376	22,400
May	750	272	445	27,400
June	570	272	440	26,200
July	342	153	251	15,400
August	265	134	186	11,400
September	143	78	116	6,900

RED RIVER AT GRAND FORKS, N. DAK.

Location.—At Northern Pacific Railway bridge between Grand Forks, N. Dak., and East Grand Forks, Minn., half a mile below mouth of Red Lake River.

Drainage area.—25,000 square miles.

Records available.—May 26, 1901, to September 30, 1918; gage-height records have been kept by the United States Engineer Corps since 1882 and a few discharge measurements were made by them in early years.

Gage.—Chain gage attached to Northern Pacific Railway bridge and vertical staff gages attached to ice breaker below center pier of same bridge; read by H. L. Hayes. The staff gages as used by the United States Engineer Corps and the United States Weather Bureau are on the bridge breakwater at the same place as the staff gage used by the United States Geological Survey and at a datum 5 feet higher.

Discharge measurements.—Made from Great Northern Railway bridge one-quarter of a mile above gage.

Channel and control.—Clay and silt; shifts very slightly.

Extremes of discharge.—Maximum stage recorded during year, 11.3 feet March 28 (discharge 4,480 second-feet); minimum stage, open channel, 3.5 feet September 22–25 and 26–30 (discharge ,440 second-feet); minimum discharge February 21, 186 second-feet (stage discharge relation affected by ice).

1882–1918: Maximum stage recorded; 50.2 feet April 10, 1897 (discharge, 43,000 second-feet); minimum stage, 2.6 feet February 10, 1912 (discharge, 100 second-feeet).

Ice.—Stage-discharge relation affected by ice. The ice cover is usually complete and smooth from late in November until about the beginning of April and the flow steady with few fluctuations; in determining flow during spring break-up, however, corrections amounting to several feet must be applied to gage heights before applying them to open-water rating table, owing to backwater from ice jams.

Diversions.—None.

Regulation.—No power plants above with sufficient storage to cause noticeable variations in the flow.

Accuracy.—Stage-discharge relation affected by ice and by shifting control. Two rating curves used during the year; October 1 to March 26 (open-water season only) well defined between 600 and 16,000 second-feet, and fairly well defined to 26,000 second-feet; March 27 to September 30 well defined between 655 and 16,300 second-feet and fairly well above 16,300. Gage read to quarter-tenths twice daily during open season and three times weekly to tenths during frozen period. Daily discharge ascertained by applying gage height to rating tables, except during ice period when discharge was ascertained by the use of Stout method, temperature records, discharge measurements, and observer's notes Open-water records good; winter records poor.

Discharge measurements of Red River at Grand Forks, N. Dak., during the year ending Sept. 30, 1918.

Date.	Made by—	Gage height.	Discharge.	Date.	Made by—	Gage height.	Discharge.
		Feet.	*Sec.-ft.*			*Feet.*	*Sec.-ft.*
Oct. 16	Chandler and Noble....	3.81	501	May 4	H. A. Noble............	6.71	1,7[illegible]
Dec. 15	.....do..................	4.75	469	June 21	Chandler and Hulteng.	6.58	1,[illegible]
Feb. 23	H. A. Noble............	4.01	186	July 22	E. F. Chandler.........	4.25	702
Mar. 30	Chandler and Noble....	10.48	4,167				

Daily discharge, in second-feet, of Red River at Grand Forks, N. Dak., or the year ending Sept. 30, 1918.

Day.	Oct.	Nov.	Dec.	Jan.	Feb.	Mar.	Apr.	May.	June.	July.	Aug.	Sept.
1	560	622	654	310	200	272	3,520	1,490	2,800	1,160	689	950
2	560	687	654	315	195	297	3,340	1,540	2,740	1,160	689	871
3	560	720	591	320	190	321	3,040	1,590	2,680	1,120	689	833
4	560	687	560	326	186	346	3,160	1,640	2,620	1,070	689	760
5	591	687	530	312	195	371	2,860	1,690	2,560	1,030	689	724
6	622	687	530	298	205	396	2,620	1,740	2,500	990	655	689
7	638	687	530	285	215	420	2,260	1,800	2,440	950	655	655
8	654	754	530	280	211	436	1,910	1,910	2,380	950	622	655
9	622	824	530	275	207	452	1,740	1,910	2,320	950	622	622
10	591	860	530	270	204	468	1,690	1,850	2,260	950	655	622
11	560	897	516	266	200	484	1,640	1,800	2,260	910	655	590
12	560	897	501	260	200	501	1,640	1,690	2,200	871	689	590
13	591	934	492	254	200	516	1,590	1,640	2,140	871	689	558
14	622	934	482	248	200	530	1,590	1,590	2,080	871	724	527
15	591	972	473	248	197	695	1,540	1,540	2,020	833	724	527
16	560	972	420	248	194	860	1,540	1,490	1,970	796	689	497
17	501	934	446	248	192	860	1,490	1,440	1,910	760	655	497
18	530	897	421	262	189	860	1,490	1,440	1,800	760	622	468
19	560	860	396	276	186	934	1,490	1,440	1,760	724	655	468
20	560	824	371	290	186	1,170	1,490	1,490	1,720	724	689	468
21	591	789	360	305	186	1,260	1,390	1,490	1,690	724	724	468
22	560	789	348	286	186	1,720	1,340	1,540	1,590	689	760	440
23	530	789	360	267	186	2,070	1,300	1,590	1,490	689	724	440
24	501	824	371	248	195	2,500	1,250	1,800	1,440	655	724	440
25	560	824	356	240	205	3,120	1,200	2,140	1,340	655	689	440
26	622	789	341	232	125	3,720	1,160	2,380	1,300	689	724	468
27	687	754	326	224	232	4,300	1,160	2,620	1,300	724	760	440
28	720	720	315	215	248	4,480	1,200	2,680	1,250	724	796	440
29	687	654	305	211	....	4,000	1,300	2,740	1,200	724	871	440
30	622	654	305	208	....	4,060	1,390	2,800	1,200	724	1,120	440
31	560	....	305	204	....	3,760	....	2,860	....	689	1,070	....

NOTE.—Discharge interpolated for lack of gage readings Oct. 7; Dec. 11, 13, 14, 18, 19, 21, 23, 25, 26, 28, 30; Jan. 1–3, 5–6, 8–10, 12–13, 15–16, 18–20, 22–23, 25–27, 29–31; Feb. 2, 3, 5, 6, 8–10, 12–13, 15–18, 20, 22, 24, 26, [illegible] Mar. 1–4, 6, 8–11, 13, 15; June 19, 20, 22.

Correction for Stout method used Dec. 3 to Mar. 26 determined from observer's notes, temperature records, and discharge measurements. After applying the Stout correction to gage heights, the discharge was ascertained by applying corrected gage heights to rating table.

Monthly discharge of Red River at Grand Forks, N. Dak., for the year ending Sept. 30, 1918.

[Drainage area, 25,000 square miles.]

Month.	Discharge in second-feet.			Run-off in acre-feet.
	Maximum.	Minimum.	Mean.	
October	720	501	588	36,200
November	972	622	797	47,400
December	654	305	447	27,500
January	326	204	266	16,400
February	248	186	200	11,100
March	4,480	272	1,490	91,600
April	3,520	1,160	1,811	108,000
May	2,860	1,440	1,850	114,000
June	2,800	1,200	1,970	117,000
July	1,160	655	843	51,800
August	1,120	622	723	44,500
September	950	440	568	33,800
The year	4,480	186	966	699,000

DEVILS LAKE NEAR DEVILS LAKE, N. DAK.

LOCATION.—At biologic station of University of North Dakota, near Devils Lake, in Ramsey County, 6 miles southwest of city of Devils Lake.

DRAINAGE AREA.—The theoretical drainage area of the lake is about 3,700 square miles. In years of ordinary rainfall water reaches the lake from only a small part of this area, most of which drains into local depressions and small lakelets, where the water remains until it is lost by evaporation. In 1880 the length of Devils Lake was 35 miles and its area about 120 square miles, but its present area is probably not more than 50 square miles.

RECORDS AVAILABLE.—June 8, 1901, to September 30, 1916 (fragmentary).

GAGE.—Staff gage on pier at the biologic station. Zero of gage, 1416.2 feet above sea level. Previous to 1916 staff gages were placed at convenient points on piers, but it has been necessary to renew them occasionally, sometimes every year, owing to damage caused by ice during the spring break-up. These gages have been reset as near to the correct datum as possible, often by the use of a carpenter's level. Occasionally errors of 0.1 foot in the records have been discovered when accurate checks were made, but no larger errors are likely to occur. The gage is read occasionally by employees of the biologic station.

REGULATION.—The lake has no outlet. The stage of the lake shows the relation between evaporation from the lake surface and the inflow from the surrounding country and gives an indication whether the run-off has been affected by the settlement of the drainage area and cultivation of the land surface.

COOPERATION.—Records furnished by North Dakota Biological Survey.

Gage height of Devils Lake near Devils Lake, N. Dak., during the year ending Sept. 30, 1918.

Date.	Gage height.	Date.	Gage height.	Date.	Gage height.
	Feet.		*Feet.*		*Feet.*
Nov. 10	5.55	May 18	5.45	Oct. 7	4.75
April	5.52	June 8	5.70	Nov. 22[a]	4.70
May 7	5.70	July 30	5.33		

[a] About Nov. 22.

RED LAKE RIVER AT THIEF RIVER FALLS, MINN.

LOCATION.—In Sec. 33, T. 154 N., R. 43 W., one-third mile below dam at Thief River Falls, Pennington County, and 1 mile below mouth of Thief River, which comes in from the right.

DRAINAGE AREA.—3,430 square miles.

RECORDS AVAILABLE.—July 2, 1909, to Sept. 30, 1918.

GAGE.—Inclined staff gage located on left bank; read by Dodrick Knutson.

DISCHARGE MEASUREMENTS.—Made from cable near gage.

CHANNEL AND CONTROL.—Gravel and small boulders; practically permanent.

EXTREMES OF DISCHARGE.—Maximum open-water stage recorded 5.9 feet March 26 (discharge, 995 second-feet); minimum open-water stage about 3.0 feet August 31 (discharge, about 19 second-feet).

1909–1918: Maximum open-water stage recorded 15.0 feet, April 16, 1916 (discharge, 8,000 second-feet); minimum discharge recorded, no flow, July 17 and August 27, 1911.

ICE.—Stage-discharge relation seriously affected by ice.

REGULATION.—A short distance above station is a dam owned by Hansen & Barzen Milling Co. and the city lighting plant. The variation in load on the turbines due to the operation of the lighting plant (at night) and of the mill (chiefly during the day) caused fluctuations in stage at the gage.

ACCURACY.—Stage-discharge relation fairly permanent. Rating curve well defined between 19 and 5,500 second-feet. Gage read to half-tenths once daily. Daily discharge ascertained by applying daily gage heights to rating table, except for periods when stage-discharge relation was affected by ice and when gage was not read, for which it was obtained by comparison with flow of Red Lake River at Crookston and to some extent by weather records. Open-water records good except for extremely low stages, when they are fair; winter records and records for period when gage was not read only roughly approximate.

Discharge measurements of Red Lake River at Thief River Falls, Minn., during the year ending Sept. 30, 1918.

[Made by E. F. Chandler.]

Date.	Gage height.	Discharge.
	Feet.	*Sec.-ft.*
Apr. 13	4.59	413
July 11	4.46	339

Daily discharge, in second-feet, of Red Lake River at Thief River Falls, Minn., for the year ending Sept. 30, 1918.

Day.	Oct.	Nov.	Dec.	Jan.	Feb.	Mar.	Apr.	May.	June.	July.	Aug.	Sept.
1							472	452	582	19	156	183
2							306	538	606	156	183	54
3							266	431	672	242	131	180
4							227	393	650	340	183	306
5						70	306	431	628	306	212	290
6							340	515	650	274	156	183
7							306	560	605	274	212	31
8							375	494	582	306	227	306
9							375	538	582	306	227	290
10							340	472	605	274	227	274
11							357	472	582	274	227	31
12							340	431	560	290	143	19
13							357	412	560	274	119	19
14							375	375	538	306	197	131
15					40		340	340	494	290	169	27
16	215	240	120	80		300	340	375	538	274	131	54
17							375	375	494	242	119	88
18							357	393	494	242	143	19
19							306	340	515	212	156	19
20							340	340	494	212	227	19
21						650	306	417	375	183	227	19
22						650	274	494	375	212	274	88
23						650	340	605	274	156	227	41
24						695	393	605	306	212	227	31
25						840	375	538	393	183	227	41
26						995	375	582	306	212	227	54
27						605	375	560	274	274	227	31
28						605	424	538	306	242	242	70
29					......	616	472	560	242	212	242	88
30					......	628	538	538	202	198	306	131
31		......			......	605	......	560	......	183	19	......

NOTE.—Daily discharges from Oct. 7 to Mar. 20 computed by comparison with other streams, and the mean for the month obtained by averaging those values.

Monthly discharge, in second-feet, of Red Lake River at Thief River Falls, Minn., for the year ending Sept. 30, 1918.

[Drainage area, 3,430 square miles.]

Month.	Maximum.	Minimum.	Mean.
October			215
November			240
December			120
January			80
February			40
March	995		363
April	538	227	358
May	605	340	473
June	672	202	483
July	340	19	238
August	306	19	193
September	306	19	104
The year	995		243

NOTE.—Mean discharge values for the months of October, November, December, January, February, and March obtained from comparison of Red Lake River flow with the flow of adjacent streams.

RED LAKE RIVER AT CROOKSTON, MINN.

LOCATION.—In sec. 31, T. 150 N., R. 46 W., at new Sampson's Addition highway bridge in Crookston, Polk County, a quarter of a mile below dam and power house of Crookston Waterworks Power & Light Co.'s plant. No tributaries enter for several miles.

DRAINAGE AREA.—5,320 square miles.

RECORDS AVAILABLE.—May 19, 1901, to September 30, 1918.

GAGE.—Barret & Lawrence water-stage recorder, on right abutment of bridge; installed in September, 1911, replacing chain gage attached to bridge July 1, 1909. Both gages at same datum. Prior to July 1, 1909, gage was on old Sampson's Addition bridge, about 300 feet farther upstream; this gage read the same as the present one at ordinary stages. Gage attended to by S. V. Holder.

DISCHARGE MEASUREMENTS.—Made from steel highway bridge at gage section.

CHANNEL AND CONTROL.—One channel at all stages. Bed composed of silt, gravel and small boulders; slightly shifting. Control not well defined.

EXTREMES OF DISCHARGE.—Maximum mean daily stage during year from water-stage recorder, 6.2 feet April 2 (discharge, 1,760 second-feet); minimum mean daily stage from water-stage recorder 2.3 feet Sept. 20 (discharge, 50 second-feet).

1901–1918: Maximum mean daily stage recorded during period 21.5 feet April 17, 1916 (discharge, 14,400 second-feet). A minimum discharge of 10 second-feet was recorded by discharge measurement made January 27, 1912. The flow is controlled to such an extent that the minimum recorded discharge has no bearing on the minimum natural flow.

ICE.—Stage-discharge relation seriously affected by ice.

REGULATION.—Considerable diurnal fluctuation at the gage is caused by operation of power plant immediately above station. The plant has little storage, so that the mean monthly flow should represent nearly the natural flow.

ACCURACY.—Stage-discharge relation fairly permanent and changes are small. Two rating curves used during the year; October 1 to March 28 well defined 100 to 10,000 second-feet; March 23 to September 30 well defined 218 to 10,000 second-feet, only fairly well defined below 218 second-feet. Operation of water-stage recorder fairly satisfactory throughout year. Daily discharge ascertained by applying to rating table mean daily gage height obtained by planimeter from the gage-height graph, except during period when stage-discharge relation was affected by ice, for which it was ascertained by applying to the rating table mean daily gage heights corrected for ice effect by means of discharge measurements, observer's notes, and weather records. During open-water periods of the year when gage was not in operation discharge was estimated and interpolated on the basis of flow at Thief River Falls and Grand Forks, N. Dak. Open-water records excellent when gage was in operation, fair for the remainder of period; winter records subject to error.

Discharge measurements of Red Lake River at Crookston, Minn., during the year ending Sept. 30, 1918.

Date.	Made by—	Gage height.	Discharge.	Date.	Made by—	Gage height.	Discharge.
		Feet.	*Sec.-ft.*			*Feet.*	*Sec.-ft.*
Oct. 18	E. F. Chandler	3.20	310	Apr. 13	E. F. Chandler	4.19	673
Nov. 17	H. A. Noble	3.20	381	July 12	....do	3.60	440
Dec. 22[a]	....do	3.43	98	Sept. 26	....do	3.33	283
Feb. 18[a]	....do	3.56	62				

[a] Complete ice cover at control and measuring section.

Daily discharge, in second-feet, of Red Lake River at Crookston, Minn., for the year ending Sept. 30, 1918.

Day.	Oct.	Nov.	Dec.	Jan.	Feb.	Mar.	Apr.	May.	June.	July.	Aug.	Sept.
1	242	275	280	150	100	150	1,760	475	895	397	400	320
2	242	272	280	150	100	200	1,760	495	805	397	436	358
3	245	286	280	150	100	250	1,090	475	895	436	416	320
4	245	275	280	150	100	310	990	675	940	380	358	300
5	249	255	280	150	100	275	940	675	990	320	378	290
6	249	255	250	140	90	210	595	715	1,040	320	397	280
7	249	249	250	140	90	275	358	760	940	284	358	270
8	252	265	250	140	90	346	218	715	940	302	339	260
9	252	249	250	140	90	210	358	675	940	302	339	250
10	255	245	250	140	90	242	440	715	895	284	397	240
11	255	245	200	130	90	383	520	715	805	267	378	220
12	250	252	200	130	90	310	600	715	760	250	416	200
13	250	252	200	130	90	210	675	715	760	300	436	140
14	250	252	200	130	90	242	660	760	715	320	400	68
15	250	328	200	130	90	310	640	715	715	340	360	88
16	250	342	170	120	80	310	630	675	635	340	320	68
17	260	335	170	120	80	500	610	715	675	340	284	99
18	310	324	170	120	80	620	595	715	715	340	267	68
19	310	306	170	120	80	1,100	555	715	675	340	284	88
20	306	310	170	120	80	910	535	715	635	340	284	50
21	303	317	150	120	90	1,260	515	675	575	310	320	68
22	303	314	150	120	90	1,500	495	715	555	310	358	88
23	300	321	150	120	90	1,320	475	760	456	310	397	78
24	296	303	150	120	90	1,320	456	760	535	310	397	140
25	292	328	150	120	90	1,140	475	850	535	310	302	200
26	289	303	140	110	100	1,200	475	805	475	350	320	267
27	289	303	140	110	100	940	475	850	495	350	358	284
28	286	292	140	110	110	1,040	475	895	495	350	339	284
29	282	290	140	110	....	1,140	475	895	456	350	302	284
30	278	290	140	110	....	1,140	475	850	416	350	302	234
31	275	....	140	110	....	1,380	....	895	....	380	320	....

NOTE.—Gage not read Oct. 4, Oct. 7 to Mar. 20, discharge estimated. Gage not read Mar. 22, 23, 29, Apr. 3, 15, 26, 28, May 21, July 30, Aug. 3, 10, 24, discharge interpolated. On July 1, Aug. 31, Sept. 2, 3, 7, 11-16, 18-21, 23-25, water was below gage and discharge has been based on estimate of stage made by observer and other notes regarding flow of river.

Monthly discharge, in second-feet, of Red Lake River at Crookston, Minn., for the year ending Sept. 30, 1918.

[Drainage area, 5,320 square miles.]

Month.	Maximum.	Minimum.	Mean.	Month.	Maximum.	Minimum.	Mean.
October		242	270	May	895	475	726
November	342	245	288	June	1,040	416	712
December			196	July	436		333
January			128	August		267	354
February			91.4	September		50	197
March	1,500	150	669				
April	1,760	218	644	The year	1,760	50	385

MOUSE RIVER AT MINOT, N. DAK.

LOCATION.—At Anne Street footbridge, northeast of Great Northern Railway roundhouse at Minot, in Ward County.

DRAINAGE AREA.—8,400 square miles.

RECORDS AVAILABLE.—May 5, 1903, to September 30, 1918.

GAGE.—Vertical staff attached to pier nearest left end of Anne Street footbridge; read by Ephraim Cox. From 1903 to December, 1909, a vertical staff on old footbridge 20 rods above present site was used. Both gages at 1,534.26 elevation sea level datum.

DISCHARGE MEASUREMENTS.—Made from Anne Street bridge or by wading a few rods below dam at the Soo Railway water tank.

CHANNEL AND CONTROL.—Bed composed of clay and silt; nearly permanent. Dam of the Minneapolis, St. Paul & Sault Ste. Marie Railway Co. forms the low-water control. At higher stages dam is submerged, causing a reversal in rating curve. The crest of dam was slightly changed when repairs were made in spring of 1918.

EXTREMES OF DISCHARGE.—Maximum stage recorded during the year 8.5 feet March 30 (discharge, 790 second-feet); minimum stage, 3.0 feet October 6 (discharge, 0.3 second-foot).

1903–1918: Maximum stage recorded, 21.9 feet April 20, 1904 (discharge, 12,000 second-feet); minimum stage, 1.8 feet February 28, 1913 (discharge, 0.1 second-foot).

ICE.—Stage-discharge relation affected by ice.

REGULATION.—A dam 4 feet high at Minneapolis, St. Paul & Sault Ste. Marie Railway tank, a mile below, raises water at gage about 3 feet at ordinary low stage. The dam being designed merely to give enough depth of water for the intake-pipe suction, has no sluices, but is not absolutely tight. When discharge is less than about 5 second-feet, the water level falls below crest of dam.

ACCURACY.—Stage-discharge relation affected by changes in Soo Railroad dam (low-water control) during the spring break-up and by ice during the winter. Two rating curves used during the year; both fairly well defined below 2,500 second-feet; the first applicable October 1 to March 15, except during ice period; the second March 20 to September 30. Both curves have a decided reversal due to the submergence of Soo Railroad dam above stage of 6.0 feet gage height. Gage read once a week October 1 to March 30, to nearest half-tenth and daily thereafter. Daily discharge ascertained by applying mean daily gage heights to rating table. During period October 1 to March 30, when the gage was read only once a week, the discharge for days of no gage reading was ascertained by interpolation in order to obtain the mean discharge for month. See footnote to table of monthly discharge. Records prior to April 1 poor; thereafter fair.

Discharge measurements of Mouse River at Minot, N. Dak., during the year ending Sept. 30, 1918.

[Made by E. F. Chandler.]

Date.	Gage height.	Discharge.
	Feet.	*Sec.-ft.*
Apr. 7	6.27	278
20	5.67	149
Aug. 9	3.96	1.6

Daily discharge, in second-feet, of Mouse River at Minot, N. Dak., for the year ending Sept. 30, 1918.

Day.	Oct.	Nov.	Dec.	Jan.	Feb.	Mar.	Apr.	May.	June.	July.	Aug.	Sept.
1			40				750	146	36	5.7	4.4	7.0
2					1.8	40	690	164	45	4.4	4.4	11
3		9					606	204	61	4.4	3.2	11
4							469	260	68	5.7	3.2	14
5				4.0			390	284	50	5.7	3.2	14
6	0.3						362	296	45	4.4	2.8	16
7							296	284	45	4.4	3.2	16
8			24				309	137	36	3.8	2.8	16
9					9.0	88	296	120	31	2.8	2.8	14
10		24					194	96	31	2.8	2.4	14
11							184	36	27	1.8	2.4	16
12				6.0			184	20	23	1.8	2.8	14
13	.8						194	9.0	27	2.4	3.2	11
14							164	4.4	27	2.8	3.2	11
15			13				137	5.7	23	3.2	3.8	11
16					13		128	11	23	3.8	4.4	14
17		40					137	27	20	2.8	5.7	11
18							155	50	20	2.4	5.7	11
19				9.0			164	81	14	1.6	5.7	11
20	4.0						174	103	9.0	1.6	7.0	14
21							146	128	7.0	1.8	9.0	14
22			6.0				137	128	7.0	2.4	9.0	16
23					18	390	164	45	5.7	2.8	11	16
24		50					137	40	4.4	3.2	11	14
25							103	45	3.2	4.4	11	14
26				6.0			96	40	3.2	4.4	9.0	11
27	18						103	68	5.7	5.7	9.0	11
28							120	68	7.0	5.7	7.0	11
29			4.0				120	40	5.7	7.0	7.0	9.0
30						790	128	45	4.4	7.0	5.7	9.0
31								40		5.7	7.0	

NOTE.—Gage read once weekly Oct. 1 to Mar. 31. Daily discharge for intervening days ascertained by interpolation, and monthly means computed accordingly.

Monthly discharge of Mouse River at Minot, N. Dak., for the year ending Sept. 30, 1918.

Month.	Discharge in second-feet.			Run-off in acre-feet.
	Maximum.	Minimum.	Mean.	
October			5.24	322
November			31.8	1,890
December			15.3	941
January			5.87	361
February			13.2	733
March	790		258	15,900
April	750	96	241	14,300
May	296	4.4	97.6	6,000
June	68	3.2	23.8	1,420
July	7.0	1.6	3.82	235
August	11.0	2.4	5.55	341
September	16.0	7.0	12.7	756

NOTE.—During winter months, the Stout method of correction for backwater effect used in computations. Record prior to Apr. 1, should be used with caution.

1688°—21—WSP 475——3

EVAPORATION AT UNIVERSITY, N. DAK.[1]

The evaporation gage at University, N. Dak., was established April 17, 1905, on a pool in a ravine called English Coulee, which runs through the campus of the University of North Dakota, immediately west of Grand Forks, N. Dak., and 2 miles west of the Minnesota boundary.

The coulee drains about 60 square miles of very level prairie. Except for brief freshets the flow in the coulee is small, varying from 1 second-foot or less to 20 second-feet. In very dry weather the water lies in pools with scarcely any perceptible flow.

A heavy galvanized-iron tank, 3 feet square and 18 inches deep, is placed in the center of an anchored raft, so that the water in the tank is at the same level as the water surface outside. The tank is filled nearly to the top, to a height precisely marked by the pointed tip of a vertical rod in the center of the tank. Once each day, after the change produced by evaporation or rainfall, the water level is restored to the original height, the precise amount of water transferred being measured with a cup of such size that one cupful of water is equivalent to 0.01 inch depth in the tank.

On the open prairie about 40 rods distant is a standard rain gage. On days of rainfall the difference (which is usually small) between the quantity measured by the rain gage and the surplus in the tank is considered the total evaporation for the day.

Observations were made usually about half an hour before sunset. The temperature of the water recorded is the observation of the water in the tank. As the tank is made of metal, it has been found that at that time of the day there is rarely a perceptible difference in temperature reading between the water within and without the tank. The temperature of the air as recorded is the mean of the readings of the standard self-recording maximum and the self-recording minimum thermometers for the preceding 24 hours.

The following table shows for each 10-day period during the year ending September 30, 1918, the gross evaporation, the total rainfall, and the mean temperatures for the 10 observations of the water and of the air.

Evaporation observations at University, N. Dak., for the year ending Sept. 30, 1918.

Date.	Evaporation.	Rainfall.	Mean temperature (°F.).		Date.	Evaporation.	Rainfall.	Mean temperature (°F.).	
			Water.	Air.				Water.	Air.
1917-18.	*Inches.*	*Inches.*			1917-18.	*Inches.*	*Inches.*		
Oct. 1-10	1.04	0.22	41	43	June 11-20	1.60	0.01	65	65
11-20	.71	.73	33	32	21-30	1.90	.18	67	61
21-31	.17	.20	32	24	July 1 to Aug. 31 [a]	.....	.....	.....	.....
Nov. 1-9	.32	.61	33	38	Sept. 1-10	1.32	.10	55	52
Apr. 9-10	.33	.00	34	47	11-20	.89	.23	49	47
11-20	1.86	.25	42	49	21-30	.92	.13	53	53
21-30	1.78	1.82	41	42	Oct. 1-10	.50	.23	52	51
May 1-10	1.45	.69	52	53	11-20	.75	.19	45	51
11-20	1.32	.44	49	48	21-31	.14	.78	34	37
21-31	.80	2.07	51	54	Nov. 1-10	.20	.60	33	36
June 1-10	1.43	.60	61	59	11-20	.18	.42	32	32

[a] No records available.

KAWISHIWI RIVER NEAR WINTON, MINN.

LOCATION.—In sec. 20, T. 62 N., R. 11 W., in pond above lower dam of St. Croix Lumber Co. at Kawishiwi Falls, 500 feet above Fall Lake, 3,000 feet below Garden Lake, near western line of Lake County, 2½ miles east of Winton, St. Louis County.

DRAINAGE AREA.—1,200 square miles.

RECORDS AVAILABLE.—June 21, 1905, to June 20, 1907; and October 14, 1912, to September 30, 1918.

[1] For complete description of this station and records of evaporation, rainfall, and temperature for 1905 to 1908 see U. S. Geol. Survey Water-Supply Paper 245, pp. 64-67, 1910.

GAGE.—Stevens water-stage recorder installed the last part of September, 1912, by the International Joint Commission in cooperation with the United States Geological Survey, at a point just above right end of dam. Well was attached to timbers, which were bolted to the vertical rock wall of right bank of river. Auxiliary staff gage was also attached to one of these timbers. The gage shelter was supported by timbers, which were bolted to the horizontal portion of the rock wall above all possible high water. On May 27, 1913, the Stevens was replaced by a Friez water-stage recorder. During the high water of June, 1914, the well together with the float and weight were carried away by logs. At this time a concrete well was installed by the International Joint Commission a little below the dam and outside the river channel, and connected with pool above the dam by a pipe through the dam. The gage was repaired and again put in operation about July 1, 1914. Attended to by F. W. Byshe.

DISCHARGE MEASUREMENTS.—Made from cable about 1,000 feet above gage.

CHANNEL AND CONTROL.—At the gage the river flows through a small deep pool formed by a timber dam without openings, which constitutes the control of gage, and is permanent unless dam is destroyed or alterations are made in the crest. About 200 feet above dam is a decided fall. Banks high enough to prevent overflow in vicinity of gage. At measuring section bed of stream is composed of rock and boulders; rather rough; current very swift except at low stages.

EXTREMES OF DISCHARGE.—Maximum stage recorded during year, 5.0 feet June 10 (discharge, 2,890 second-feet). Due to nonoperation of the recording gage, stage of 5.0 feet does not represent the absolute maximum stage; minimum discharge recorded, about 37 second-feet on April 8, 15, and 22.

1905–1907 and 1912-1918: Maximum stage recorded, 7.2 feet April 30 and May 7, 1916 (discharge, 5,370 second-feet); no flow August 24, 25, 30, and 31, September 1, 1915, August 6, 8, 1906, and April 23, 24, and 26, 1907.

ICE.—Discharge relation not seriously affected by ice; open-channel rating curve assumed applicable. The operation of the water-stage recorder is affected by ice, and the flow from December to March, which is very constant during this part of the year, is computed from weekly reading of the staff gage.

REGULATION.—St. Croix Lumber Co. has a dam at the outlet of Garden Lake for controlling the level of water in that lake, and for storing water to be used in driving logs over the stretch of rapids between Garden and Fall lakes. This dam is capable of holding the water in Garden Lake about 7 or 8 feet above its natural level at low water before water will flow over the gates. When the water in Garden Lake is held at a high stage, the elevation of water is considerably higher in Farm Lake, and it is understood that the elevation of the surface of White Iron Lake is somewhat affected by the stage of Garden Lake. During the log-driving season, April to November, the water in Garden Lake is held to the elevation of the top of the gates practically all the time. In November some of the gates are opened so that the lake is drawn down to low-water stage, and remains so until spring. St. Croix Lumber Co. has a dam at the outlet of Birch Lake, which controls its elevation, and is capable of holding the water about 5 feet above low water. This dam is left open during the winter and until the high water of the spring break-up has passed. It is then closed, and the lake held as high as possible during the summer. There are a number of low dams in Stony River used for sluicing logs off rapids, but these have no storage of importance back of them. Large volumes of water are allowed to pass through sluices of dam at the outlet of Garden Lake for a few hours at a time, at irregular intervals, when desired to drive logs from Garden Lake to Fall Lake. At other times these gates are closed so that there is only a slight flow caused by leakage through the dam. At other times some of the gates are partly opened to allow passage of sufficient water to prevent flow over crest of dam.

ACCURACY.—Stage-discharge relation permanent; not usually affected by ice and seldom by logs. Rating curve fairly well defined below 2,890 second-feet. Continuous gage record from recording gage during the open-water period; weekly gage readings during the frozen periods. Daily discharge ascertained as follows: October 1-21, July 28 to September 15 and September 22-30 obtained, by means of discharge integrator, from the recording gage record; October 23 to July 26 based on daily gage reading made by observer. Daily discharge record when recording gage was in operation good. Discharge for periods when water-stage recorder was not in operation not determined except for days when gage was read. Information as to operation of gates in dam at outlet of Garden Lake given in footnote to daily discharge table.

Daily discharge, in second-feet, of Kawishiwi River near Winton, Minn., for the year ending Sept. 30, 1918.

Day.	Oct.	Nov.	Dec.	Jan.	Feb.	Mar.	Apr.	May.	June.	July.	Aug.	Sept.
1	650						80				378	230
2	1,060										380	229
3	1,100		356						2,700			218
4	1,060					180						234
5	920	1,340									1,010	306
6	1,070			314							170	215
7	1,180									866	170	228
8	1,440						57				170	194
9	1,200							590			170	202
10	995	590	163						2,890		204	216
11	1,180				235	163					222	150
12	1,000							356			505	150
13	900			314							648	150
14	805									747	662	150
15	520		163				57				584	150
16	385										342	
17	735										257	
18	400				197	163			1,270		163	
19	815	590						446			538	
20	996			314							459	
21	590									163	584	
22			235				57				163	163
23	163										163	196
24									996		282	176
25					197	163					163	176
26		930								133	330	300
27								2,430			296	189
28										392	230	127
29	1,340		274	274						395	258	127
30							80			430	234	310
31										410	230	

NOTE.—Recording gage not in perfect operation Oct. 22 to July 27 and Sept. 16-21. During this period gage was read once weekly and the following information was obtained regarding operation of gates in dam at outlet of Garden Lake: Oct. 24 to Nov. 30, May 1-26, June 18 to July 25 gates were opened occasionally for purpose of log driving, and mean discharge based on weekly readings may be subject to considerable error. Gates were not operated for log-driving purposes from Dec. 1 to Apr. 30 and from May 27 to June 17; mean discharge based on weekly gage height will give a fair estimate of flow. Gates opened only occasionally Sept. 16-21. Low flow during April due to gates in Garden Lake being closed for the purpose of increasing storage in Garden Lake.

UPPER MISSISSIPPI RIVER BASIN.

MISSISSIPPI RIVER AT ELK RIVER, MINN.

LOCATION.—In sec. 3, T. 121 N., R. 23 W., at highway bridge in town of Elk River, 2,500 feet below mouth of Elk River, in Sherburne County.

DRAINAGE AREA.—14,500 square miles.

RECORDS AVAILABLE.—July 22, 1915, to September 30, 1918.

GAGE.—Chain gage bolted to handrail of bridge, downstream side, near right bank: ---d by W. H. Ebner.

3E MEASUREMENTS.—Made from downstream side of bridge.

, AND CONTROL.—Bed composed of sand and gravel; control not well de- . Banks high and not subject to overflow.

Extremes of discharge.—Maximum stage recorded during year, 6.21 feet at 8.08 a. m. June 1 (discharge, 17,700 second-feet); minimum open-water stage, 2.64 feet at 7.45 p. m. September 28 (discharge, 2,130 second-feet).

1915–1918: Maximum stage recorded under unobstructed channel conditions, 10.8 feet April 7, 1916 (discharge, 27,000 second-feet); minimum open-water stage recorded, 2.64 feet at 7.45 p. m. September 28, 1918 (discharge, about 2,130 second-feet).

Ice.—Stage-discharge relation seriously affected by ice; discharge estimated from records of discharge at Coon Rapids power plant, computed by the Minneapolis General Electric Co., allowance being made for the discharge of Crow and Rum rivers, entering between Coon Rapids and the station. During the greater part of the frozen period 1917–1918 no estimates were made as power plant was not in operation.

Regulation.—Nearest dam above the station on the Mississippi is at St. Cloud, 40 miles upstream. An observed systematic diurnal fluctuation at gage of about 0.1 foot is doubtless due to regulation at St. Cloud; but most of the effect of regulation is eliminated before reaching the station. Flow of the river is controlled by Government dams on the upper river for the purpose of increasing the low-water open-season flow in the interests of navigation.

Accuracy.—Stage-discharge relation permanent except as affected by ice. Rating curve well defined between 4,620 and 12,400, and fairly well defined between 12,700 and 26,300 second-feet. Gage read to quarter-tenths twice daily. Daily discharge ascertained by applying mean daily gage heights to rating table. Open-water records good.

Cooperation.—Gage readings furnished by U. S. Army Engineer Corps.

The following discharge measurement was made by R. B. Kilgore:

October 1, 1917: Gage height, 4.13 feet; discharge, 5,170 second-feet.

Daily discharge, in second-feet, of Mississippi River at Elk River, Minn., for the year ending Sept. 30, 1918.

Day.	Oct.	Nov.	Apr.	May.	June.	July.	Aug.	Sept.
1	4,910	4,620	5,500	3,400	10,200	4,910	3,030	3,030
2	4,910	4,080	4,080	3,400	11,100	4,340	3,210	2,860
3	4,910	3,610	4,910	4,340	10,500	4,910	3,400	2,860
4	5,200	5,200	4,910	5,800	10,500	4,910	3,210	3,030
5	4,910	4,340	4,620	5,500	11,100	5,200	3,030	3,030
6	4,620	4,620	3,840	6,100	10,800	4,620	3,210	2,550
7	4,910	4,910	4,080	4,620	9,840	4,620	3,210	2,700
8	5,200	5,200	4,080	5,200	9,200	4,910	3,210	2,700
9	4,910	5,200	3,840	6,700	10,200	4,080	3,210	2,420
10	4,620	4,620	3,610	7,310	9,840	3,840	3,030	2,550
11	4,910	4,620	3,210	6,700	8,880	5,200	3,210	2,700
12	4,910	4,910	3,210	6,400	8,880	3,840	3,030	2,860
13	4,910	4,620	3,210	6,400	7,620	4,340	3,030	2,550
14	5,800	4,620	3,210	6,700	7,620	4,340	3,210	2,700
15	4,080	4,620	3,030	7,000	7,310	4,080	3,030	2,700
16	4,340	4,080	3,210	7,310	7,000	3,840	2,700	2,420
17	5,500	4,340	3,030	6,100	7,000	3,400	2,860	2,420
18	5,800	4,080	3,210	6,400	6,700	3,840	2,700	2,860
19	5,500	4,080	3,400	6,400	6,100	3,610	2,860	2,550
20	5,500	4,620	3,400	6,400	5,500	3,610	3,210	2,550
21	6,100	3,840	3,210	6,400	5,500	3,610	3,210	2,300
22	5,200	4,080	3,400	7,000	5,500	3,840	3,030	2,420
23	4,620	3,610	3,400	7,000	5,800	3,840	3,030	2,190
24	5,500	3,610	3,210	6,700	5,800	3,610	3,210	2,190
25	5,500	4,080	3,210	7,930	4,910	3,840	3,030	2,420
26	5,500	4,620	3,210	8,880	4,910	3,840	2,860	2,300
27	5,500	3,610	3,400	9,520	4,910	3,610	3,030	2,420
28	5,500	3,400	3,610	8,240	5,200	3,840	2,860	2,300
29	5,500	3,400	3,610	9,520	4,910	3,840	2,860	2,300
30	4,620	2,860	3,210	9,840	4,910	3,610	3,030	2,190
31	4,340	....	....	10,200	....	2,860	3,210	....

Note.—Stage-discharge relation affected by ice Dec. 4 to Mar. 31; discharge not determined.

Monthly discharge of Mississippi River at Elk River, Minn., for the year ending Sept. 30, 1918.

[Drainage area, 14,500 square miles.]

Month.	Discharge in second-feet.				Run-off (depth in inches).
	Maximum.	Minimum.	Mean.	Per square mile.	
October	6,100	4,080	5,100	0.350	0.40
November	5,200	2,860	4,270	.293	.33
April	5,500	3,030	3,640	.250	.28
May	10,200	3,400	6,760	.464	.53
June	11,100	4,910	7,610	.523	.58
July	5,200	2,860	4,090	.281	.32
August	3,400	2,700	3,060	.210	.24
September	3,030	2,190	2,570	.177	.20

MISSISSIPPI RIVER AT ST. PAUL, MINN.

Location.—At Chicago Great Western Railway bridge near foot of Robert Street, St. Paul, 6 miles below mouth of Minnesota River, in Ramsey County.

Drainage area.—35,700 square miles.

Records available.—March 1, 1892, to September 30, 1918. Observation of stage began in 1873 by United States Signal Service and continued by United States Weather Bureau. Many discharge measurements made prior to 1900 by the United States Engineer Corps.

Gage.—Chain gage installed May 9, 1913, on the handrail, downstream side, of Chicago Great Western Railway bridge, near the foot of Robert Street; read by United States Weather Bureau employees. From 1911 to May 9, 1913, the gage was a vertical staff gage attached to a piling on left bank of river about 800 feet upstream from present gage. Prior to 1911 a vertical staff gage on the Diamond Joe Line Wharf, at the foot of Jackson Street, about 400 feet below the chain gage, was used. The datum of all three gages is the same, allowance being made for the slight slope in the river between them.

Discharge measurements.—Up to 1915 made from the Chicago, St. Paul, Minneapolis & Omaha Railway bridge 2 miles above the station; in November, 1915, and April, 1916, measurements were made from the Chicago Great Western Railway bridge to which the gage is attached. Since 1916 measurements have been made from the Wabasha Street highway bridge, about 1,000 feet above station.

Channel and control.—Channel somewhat shifting. Control not well defined. Banks moderately high; have not been overflowed in recent years.

Extremes of discharge.—Maximum stage recorded during year, 7.5 feet March 24 and 25 (discharge not determined); minimum stage recorded, −1.0 foot December 5 (discharge not determined).

1892–1918: Maximum stage recorded, 18.0 feet April 6, 1897 (discharge, 80,800 second-feet); highest known discharge occurred July 22, 1867, and amounted to 117,000 second-feet; minimum stage recorded, −1.0 foot December 5, 1918 (discharge not determined).

Regulation.—During extreme low-water regulation of flow through turbines at the nearest dam in Minneapolis may cause diurnal fluctuation of stage at St. Paul. Flow is regulated by Government reservoirs on the headwaters at Lake Winnebigoshish, Leach Lake, Pokegama Lake, Sandy Lake, Pine River, and Gull Lake to increase the low-water open-season flow in the interests of navigation, but the effect of this regulation is very gradual at St. Paul.

ACCURACY.—Stage-discharge relation changed during the year as indicated by a discharge measurement on November 5, 1918. Change caused by dredging in the vicinity of Daytons Bluff. Sufficient measurements have not been made to develop a rating curve. Gage read once daily to tenths. This perhaps does not represent the mean daily stage accurately on account of artificial regulation at power plants in Minneapolis; occasional additional readings indicate that the error is not large.

COOPERATION.—Gage-height record furnished by United States Weather Bureau.

Daily gage height, in feet, of Mississippi River at St. Paul, Minn., for the year ending Sept. 30, 1918.

Day.	Oct.	Nov.	Dec.	Jan.	Feb.	Mar.	Apr.	May.	June.	July.	Aug.	Sept.
1	2.1	1.8	0.6	1.6	0.7	1.2	4.3	1.6	6.0	2.0	2.3	3.5
2	2.1	1.8	– .2	1.6	.7	1.2	4.1	1.7	6.1	2.0	1.6	3.0
3	2.1	1.5	– .3	1.5	.8	1.3	3.9	1.8	6.4	1.9	1.6	2.7
4	1.9	1.3	– .3	1.3	.9	1.0	3.8	2.1	6.5	2.0	1.3	2.6
5	2.0	1.6	–1.0	1.5	.8	1.8	3.5	2.4	6.2	1.7	1.0	2.5
6	2.0	1.8	– .8	1.3	.9	2.1	3.5	2.4	6.3	2.1	1.1	2.2
7	2.0	1.8	.2	1.1	.9	2.3	3.1	2.6	6.1	1.5	1.1	2.1
8	1.9	1.9	.4	1.0	1.1	2.0	2.9	2.0	5.8	1.1	1.2	1.7
9	2.1	1.8	.1	1.0	1.1	2.2	3.2	2.1	6.0	1.9	1.0	1.1
10	1.9	1.9	.3	.9	.8	1.9	2.8	3.0	5.6	1.7	1.3	1.3
11	2.2	1.9	.3	1.0	.9	2.3	2.6	3.1	5.5	1.3	1.3	1.4
12	2.0	1.6	.9	1.0	1.0	3.9	2.4	3.1	4.8	1.6	1.0	1.3
13	1.9	1.7	.7	.8	.9	3.2	2.3	3.3	4.8	1.4	1.5	1.3
14	1.9	1.7	1.3	.2	1.2	1.8	2.2	3.5	4.3	1.6	1.2	1.1
15	2.0	1.8	1.5	1.8	.2	1.0	1.9	3.8	3.9	1.5	1.1	1.1
16	2.1	1.7	1.4	1.2	1.4	2.2	1.7	3.9	3.8	1.5	1.1	.7
17	1.6	1.6	1.5	1.1	1.0	1.7	1.9	4.0	3.4	1.4	1.1	.9
18	1.8	1.5	1.4	1.0	1.2	2.4	2.1	3.4	3.3	1.1	1.4	.8
19	2.0	1.5	1.5	.9	1.1	3.9	1.8	3.4	3.0	1.4	3.0	.9
20	2.0	1.4	1.4	.8	1.0	4.6	1.8	3.5	3.0	1.4	4.0	1.0
21	2.0	1.3	1.3	1.0	.8	6.7	1.8	3.4	2.7	1.4	4.5	.7
22	2.1	1.2	1.3	.8	1.0	6.9	1.5	3.3	2.6	.9	5.2	.7
23	2.3	1.4	1.2	.8	.9	7.2	1.5	3.5	2.7	1.4	5.2	.1
24	1.8	1.2	.6	.9	1.0	7.5	1.8	3.7	2.3	1.4	5.0	.1
25	2.0	1.2	1.0	.9	.2	7.5	1.9	3.9	2.5	1.1	5.4	.1
26	2.2	1.2	1.3	.9	.8	6.7	1.7	4.6	2.3	1.2	5.4	.0
27	2.4	1.2	1.4	.9	.9	6.1	1.6	4.9	2.0	1.1	5.5	.4
28	2.5	.9	1.8	.9	1.3	5.6	1.6	5.2	2.4	1.1	5.2	.0
29	2.3	.8	1.8	.8	....	5.1	1.4	4.8	2.1	1.3	5.0	.1
30	2.1	.6	1.6	.9	....	4.7	1.3	5.3	2.3	2.3	4.4	–.3
31	2.0	....	1.7	.9	....	4.6	....	5.8	....	2.6	3.9	....

NOTE.—Stage-discharge relation affected by ice from about Dec. 7 to Mar. 19.

MINNESOTA RIVER NEAR MONTEVIDEO, MINN.

LOCATION.—In sec. 17, T. 117 N., R. 40 W., at highway bridge 1 mile south of Montevideo, Chippewa County, 500 feet below mouth of Chippewa River.

DRAINAGE AREA.—6,300 square miles.

RECORDS AVAILABLE.—July 23, 1909, to September 30, 1918.

GAGE.—Chain gage attached to upstream handrail of the bridge, near the left bank; read by Ben O. Brown and Esther Hendricks. Datum of gage lowered 2 feet September 16, 1909, and 1 foot more July 29, 1910, to avoid negative readings. All gage heights referred to latest datum.

DISCHARGE MEASUREMENTS.—Made from upstream side of bridge.

CHANNEL AND CONTROL.—Heavy gravel and sand; fairly permanent. There is a slight rapid just below the gage, but the control section is not well defined. Banks of medium height and will be overflowed at a stage of about 14 feet.

Extremes of discharge.—Maximum stage recorded during year, 8.05 feet March 30 (discharge, about 1,690 second-feet); minimum open-water stage 1.17 feet September 22 (discharge, about 20 second-feet). This is the lowest open-water stage recorded during the period covered by the records.

1909–1918: Maximum stage recorded 15.16 feet at 6 p. m. April 4, 1917 (discharge about 10,200 second-feet); minimum discharge recorded, 6.8 second-feet (measured by current meter February 9, 1912).

Ice.—Stage-discharge relation seriously affected by ice; no measurements made and daily discharge not determined.

Regulation.—No regulation on Minnesota River above station. Regulation on Chippewa River at the plant of the Chippewa Milling Co., in Montevideo, produces a slight fluctuation in the stage of the Minnesota River at gage.

Accuracy.—Stage-discharge relation fairly permanent. Rating curve fairly well defined. Gage read to hundredths twice daily except December 16 to April 19, when it was read at irregular intervals. Daily discharge ascertained by applying mean daily gage height to rating table. Open-water records fair except at extreme low stages for which they are subject to considerable error.

Daily discharge, in second-feet, of Minnesota River near Montevideo, Minn., for the year ending Sept. 30, 1918.

Day.	Oct.	Nov.	Dec.	Mar.	Apr.	May.	June.	July.	Aug.	Sept.
1	123	130	171	....	....	592	708	410	112	130
2	116	162	180	....	....	566	703	386	116	123
3	116	154	171	....	....	566	703	362	106	123
4	114	189	....	....	....	619	731	362	105	116
5	116	216	....	....	....	566	759	338	114	115
6	138	198	....	....	1,270	592	759	316	138	116
7	146	180	....	....	....	592	731	294	105	105
8	114	189	....	....	....	566	731	294	123	104
9	154	162	....	....	....	592	708	294	99	82
10	171	198	....	....	....	619	703	274	82	104
11	180	154	....	....	....	619	703	254	89	94
12	198	207	....	....	....	592	708	244	93	106
13	130	189	....	....	967	675	647	234	105	97
14	146	189	....	....	....	619	647	234	112	97
15	138	171	....	....	....	619	619	234	138	91
16	130	198	....	817	....	566	566	225	189	91
17	138	225	....	....	....	566	566	207	207	79
18	109	198	....	....	....	566	566	207	198	78
19	123	198	....	....	....	514	540	198	198	82
20	130	216	....	....	731	566	566	171	198	80
21	130	189	....	....	675	462	647	154	216	62
22	130	198	....	....	703	436	675	162	216	20
23	116	225	....	1,610	647	462	540	154	225	58
24	130	216	....	....	619	462	566	146	225	73
25	162	116	....	....	566	514	566	154	198	56
26	130	154	....	....	514	566	488	154	198	74
27	130	162	....	....	540	619	514	162	180	67
28	111	154	....	....	566	566	514	154	162	81
29	138	189	....	....	619	566	462	162	171	74
30	171	180	....	1,690	619	619	410	138	146	67
31	146	....	....	....	....	675	....	123	138	....

Note.—Stage-discharge relation affected by ice from about Dec. 4 to Mar. 10. No discharge computations made.

Monthly discharge of Minnesota River near Montevideo, Minn., for the year ending Sept. 30, 1918.

[Drainage area, 6,300 square miles.]

Month.	Discharge in second-feet.				Run-off (depth in inches).
	Maximum.	Minimum.	Mean.	Per square mile.	
October	198	109	136	0.0216	0.02
November	225	116	184	.0292	.03
April 20-30	....	514	618	.0981	.04
May	675	436	572	.0908	.10
June	759	410	624	.0990	.11
July	410	123	232	.0368	.04
August	225	82	239	.0379	.04
September	130	20	88.2	.0140	.02

MINNESOTA RIVER NEAR MANKATO, MINN.

LOCATION.—In sec. 14, T. 108 N., R. 27 W., in Blue Earth County, at Sibley Park, 2 miles above center of Mankato and 1,000 feet below mouth of Blue Earth River.

DRAINAGE AREA.—14,600 square miles.

RECORDS AVAILABLE.—May 20, 1903, to September 30, 1918.

GAGE.—Chain gage on right bank of river, about 1,000 feet below mouth of Blue Earth River; read by Clarence Staley, observer for United States Weather Bureau. The gage support is a substantial cantilever structure, supported by two heavy posts resting in concrete footings, constructed and maintained by the United States Engineer Corps.

DISCHARGE MEASUREMENTS.—Made from new concrete highway bridge in center of Mankato, by wading a short distance below gage, or at extreme high stages, by boat near gage.

CHANNEL AND CONTROL.—Bed composed of sand and light gravel; fairly permanent, except during high stage; banks moderately high and not subject to overflow, except at stages above gage height of 15 feet. Control not well defined.

EXTREMES OF DISCHARGE.—Maximum stage recorded during year, 10.7 feet March 20; minimum stage, 1.2 feet during periods in October and November, December January, and February.

1903–1918: Maximum stage recorded, 21.2 feet, June 26, 1908 (discharge, 43,800 second-feet); minimum stage recorded, 0.5 feet August 31, September 1 and 2, 1911 (discharge, 89 second-feet). The highest known stage occurred in 1881, and is shown in Mankato by a well-marked line, approximately 27 feet above the zero of the present gage (discharge, estimated 65,000 second-feet).

ICE.—Stage-discharge relation seriously affected by ice.

REGULATION.—The nearest dam on the Minnesota River is at Minnesota Falls, 140 miles upstream. A dam on the Blue Earth River at Rapidan, a few miles above the mouth, controls the flow of that river, which is approximately 20 per cent of that at the Mankato station, and produces considerable daily fluctuation at the gage, amounting at times to over 1 foot.

ACCURACY.—Stage-discharge relation not permanent; sufficient measurements have not been made to warrant the publication of daily discharge.

COOPERATION.—Gage-height record furnished by United States Weather Bureau.

Daily gage height, in feet, of Minnesota River near Mankato, Minn., for the year ending Sept. 30, 1918.

Day.	Oct.	Nov.	Dec.	Jan.	Feb.	Mar.	Apr.	May.	June.	July.	Aug.	Sept.
1	1.3	1.2	1.3	1.3	1.2	3.2	4.9	2.2	7.0	2.5	4.7	4.7
2	1.4	1.2	1.3	1.3	1.2	4.9	4.8	2.2	7.0	2.4	4.5	4.7
3	1.4	1.2	1.3	1.3	1.2	5.2	4.8	2.1	7.0	2.4	4.1	4.9
4	1.4	1.3	1.2	1.3	1.2	5.5	4.7	2.3	7.1	2.4	3.9	4.7
5	1.3	1.3	1.2	1.3	1.2	5.7	4.7	2.4	6.8	2.3	3.5	4.5
6	1.3	1.3	1.2	1.3	1.2	5.6	4.7	2.4	6.2	2.3	3.3	4.4
7	1.3	1.3	1.2	1.3	1.2	5.4	4.4	2.5	5.8	2.3	3.4	4.3
8	1.3	1.3	1.2	1.3	1.2	5.3	4.2	2.6	5.5	2.2	3.9	4.1
9	1.3	1.3	1.2	1.3	1.2	5.1	4.0	2.7	5.4	2.2	3.9	4.1
10	1.2	1.3	1.2	1.2	1.3	5.3	3.9	2.7	5.2	2.2	4.1	4.0
11	1.2	1.2	1.3	1.2	1.3	5.5	3.7	2.7	4.9	2.2	4.1	3.9
12	1.3	1.2	1.3	1.2	1.3	5.5	3.6	2.6	4.7	2.2	4.2	3.9
13	1.3	1.2	1.3	1.2	1.3	5.5	3.5	2.6	4.5	2.2	4.2	3.5
14	1.3	1.2	1.3	1.2	1.3	5.6	3.4	3.0	3.9	2.1	4.1	3.5
15	1.2	1.3	1.3	1.2	1.3	5.7	3.3	3.2	3.8	2.1	4.0	3.4
16	1.2	1.3	1.3	1.2	1.3	5.9	3.3	3.3	3.8	2.3	4.4	3.3
17	1.2	1.3	1.3	1.2	1.3	6.8	3.2	3.4	3.1	2.4	6.6	3.2
18	1.5	1.3	1.3	1.2	1.3	7.6	2.7	3.5	3.3	2.5	8.6	3.4
19	1.5	1.3	1.4	1.2	1.3	10.1	2.7	3.5	3.5	2.4	9.1	3.4
20	1.4	1.2	1.4	1.3	1.3	10.7	2.6	3.5	3.5	2.4	9.2	3.3
21	1.3	1.2	1.3	1.3	1.3	10.4	2.6	3.6	3.4	2.5	8.8	3.1
22	1.3	1.2	1.3	1.3	1.3	9.8	2.5	3.8	3.3	2.5	9.8	3.1
23	1.3	1.2	1.3	1.3	1.5	9.1	2.5	3.8	3.4	2.4	10.8	3.0
24	1.3	1.2	1.3	1.3	1.8	7.7	2.4	4.9	3.4	2.4	10.9	3.0
25	1.2	1.3	1.3	1.3	2.0	7.2	2.4	5.3	3.3	2.3	9.8	2.9
26	1.2	1.3	1.3	1.3	2.2	6.8	2.4	5.3	3.3	2.4	9.1	2.9
27	1.3	1.3	1.3	1.3	2.5	6.7	2.3	5.5	3.3	2.5	8.4	2.8
28	1.3	1.3	1.3	1.3	2.6	5.9	2.3	5.7	2.9	3.7	8.1	2.8
29	1.3	1.3	1.3	1.2	......	5.8	2.3	6.0	2.7	4.8	7.5	2.8
30	1.3	1.3	1.3	1.2	......	5.2	2.3	6.3	2.5	5.0	7.1	2.8
31	1.2	......	1.3	1.2	......	5.1	......	6.9	......	4.8	5.2	......

NOTE.—Stage-discharge relation affected by ice about Dec. 6, until the latter part of February or early in March.

ST. CROIX RIVER AT SWISS, WIS.

LOCATION.—In sec. 33, T. 42 N., R. 15 W., at highway bridge near post office of Swiss, Burnett County, 2 miles above point where St Croix River becomes boundary line between Wisconsin and Minnesota and 10 miles northeast of Danbury, Minn., on Minneapolis, St. Paul & Sault Ste. Marie Railway. Namakagon River enters from left 3½ miles above station.

DRAINAGE AREA.—1,550 square miles (measured on map issued by Wisconsin Geological and Natural History Survey, edition of 1911; scale, 1 inch=6 miles).

RECORDS AVAILABLE.—March 20, 1914 to September 30, 1918.

GAGE.—Chain gage attached to downstream side of bridge on May 16, 1918. Prior to that date a cast iron staff gage bolted to concrete pier at left end of bridge was used; gage read by Capt. Richard Goldschmidt.

DISCHARGE MEASUREMENTS.—Made from downstream side of bridge.

CHANNEL AND CONTROL.—Gravel, smooth; aquatic plants during summer months may cause a small amount of backwater at the gage. Right bank high and not subject to overflow; left bank of medium height and may possibly be overflowed during extreme high water.

EXTREMES OF DISCHARGE.—Maximum stage recorded during year, 3.15 feet at 7.30 a. m. June 2 (discharge, 3,000 second-feet); minimum discharge 700 second-feet, February 2.

1914–1918: Maximum stage recorded, 6.73 feet at 6.45 a. m. April 22, 1916 (discharge, 8,480 second-feet); minimum discharge, estimated, 700 second-feet February 2, 1918.

ACCURACY.—Stage-discharge relation practically permanent, except as affected by ice. Two fairly well defined rating curves used during the year. Gage read twice daily, to quarter-tenths. Daily discharge ascertained by applying mean daily gage height to rating table except for period in which stage-discharge relation was affected by ice for which it was ascertained from discharge measurements, observer's notes, and weather records. Open-water records good; winter records fair.

Discharge measurements of St. Croix River at Swiss, Wis., during the year ending Sept. 30, 1918.

[Made by T. G. Bedford.]

Date.	Gage height.	Discharge.	Date.	Gage height.	Discharge.
	Feet.	*Sec.-ft.*		*Feet.*	*Sec.-ft.*
Dec. 18[a]	1.82	792	Feb. 20[a]	2.32	739
Jan. 18[a]	2.02	797	May 16	1.65	1,570

[a] Made through complete ice cover about 200 feet upstream from gage; complete ice cover at control.

Daily discharge, in second-feet, of St. Croix River at Swiss, Wis., for the year ending Sept. 30, 1918.

Day.	Oct.	Nov.	Dec.	Jan.	Feb.	Mar.	Apr.	May.	June.	July.	Aug.	Sept.
1	850	1,130	850	755	725	780	1,220	1,480	2,960	984	942	1,150
2	822	1,130	822	770	700	800	1,220	1,400	2,950	984	924	1,120
3	843	1,130	822	780	705	820	1,220	1,360	2,950	984	906	1,010
4	892	1,100	815	785	710	845	1,220	1,320	2,950	1,150	885	960
5	913	1,060	815	795	705	860	1,180	1,320	2,840	1,150	855	930
6	878	1,060	810	815	700	890	1,070	1,290	2,730	1,150	930	918
7	850	1,020	810	835	710	875	1,220	1,290	2,630	1,120	1,000	895
8	836	976	810	830	715	870	,220	1,320	2,430	,070	1,020	880
9	857	962	810	820	715	860	,180	1,400	2,330	,030	990	870
10	864	955	800	820	720	850	,180	1,600	2,230	,030	972	860
11	892	976	800	820	735	850	1,150	1,640	2,040	996	960	890
12	934	990	800	815	750	850	1,120	1,640	1,860	966	948	936
13	948	990	800	810	760	870	,120	1,600	1,770	948	930	924
14	955	990	800	800	770	890	,080	1,520	1,600	936	906	912
15	934	965	800	795	780	960	,080	1,560	1,440	906	890	918
16	920	962	795	785	785	1,030	1,120	1,600	1,360	900	880	890
17	955	955	795	780	760	1,100	1,180	1,520	1,290	912	875	900
18	1,100	955	795	795	775	1,180	,220	1,480	1,220	924	870	890
19	1,250	955	805	760	755	1,440	,220	1,860	1,180	912	850	880
20	1,250	934	815	760	740	1,690	,220	2,530	1,150	924	840	895
21	1,210	934	830	760	730	1,660	1,180	2,630	1,120	870	855	875
22	1,250	955	850	760	720	1,620	1,150	2,430	1,120	890	948	860
23	1,210	948	825	760	720	1,530	,080	,430	1,070	895	960	850
24	1,170	948	800	760	720	1,450	,080	,230	1,060	906	972	895
25	1,170	934	780	760	730	1,480	,070	,130	1,040	906	960	880
26	1,210	920	755	760	740	1,400	1,070	2,430	1,040	930	948	865
27	1,250	906	745	760	750	1,360	1,060	2,630	1,030	912	936	830
28	1,370	920	730	760	760	1,360	1,150	2,630	1,010	1,030	1,080	800
29	1,330	892	725	760	......	1,290	1,320	2,630	990	1,060	1,290	810
30	1,250	878	720	755	......	1,260	1,480	2,630	1,010	1,010	1,260	820
31	1,130	......	740	750	......	1,220	......	2,630	......	984	1,180	......

NOTE.—Stage-discharge relation affected by ice Dec. 3 to Mar. 25.

Monthly discharge of St. Croix River at Swiss, Wis., for the year ending Sept. 30, 1918.

[Drainage area, 1,550 square miles.]

Month.	Discharge in second-feet.				Run-off (depth in inches).
	Maximum.	Minimum.	Mean.	Per square mile.	
October	1,370	822	1,040	0.671	0.77
November	1,130	878	982	.634	.71
December	850	720	796	.514	.59
January	835	750	783	.505	.58
February	785	700	736	.475	.49
March	1,690	780	1,130	.729	.84
April	1,480	1,060	1,170	.755	.84
May	2,630	1,290	1,880	1.21	1.40
June	2,950	990	1,750	1.13	1.26
July	1,150	870	980	.632	.73
August	1,290	840	960	.619	.71
September	1,150	800	904	.583	.65
The year	2,950	700	1,090	.703	9.57

ST. CROIX RIVER NEAR ST. CROIX FALLS, WIS.

LOCATION.—In sec. 18, T. 34 N., R. 18 W., at power plant of Minneapolis General Electric Co., on Wisconsin side of St. Croix River, near St. Croix Falls, Polk County, Wis., 50 miles above confluence of St. Croix amd Mississippi rivers, near Hastings, Minn. Apple River, draining an area wholly in Wisconsin, enters from left 20 miles below station; Snake River, draining an area in Minnesota, enters from right 35 miles above station.

DRAINAGE AREA.—5,930 square miles.

RECORDS AVAILABLE.—January 10, 1902, to June 30, 1905; January 1, 1910, to September 30, 1918. Data for 1903 published in Water Supply Paper No. 98, pages 176–177, under "St. Croix River near Taylors Falls, Minn."

DISCHARGE.—Determinations of discharge based on kilowatt output of dynamo and exciters, plus flow over dam and spillway, considered as a weir.

EXTREMES OF DISCHARGE.—Maximum daily discharge recorded during year, 10,100 second-feet June 3 and 4; minimum daily discharge recorded, 603 second-feet July 28.

1902–1905, and 1910–1918: Maximum daily discharge recorded, 35,100 second-feet April 23, 1916; minimum daily discharge recorded, 75 second-feet July 17, 1910; the minimum discharge is not natural but caused by regulation.

REGULATION.—Low-water flow controlled by operation of gates of power plant and by storage and release of water at Never's dam several miles upstream.

ACCURACY.—Records have not been checked, nor have discharge measurements been made, by engineers of the United States Geological Survey; probably reliable.

COOPERATION.—Records furnished by Minneapolis General Electric Co.

Daily discharge, in second-feet, of St. Croix River near St. Croix Falls, Wis., for the year ending Sept. 30, 1918.

Day.	Oct.	Nov.	Dec.	Jan.	Feb.	Mar.	Apr.	May.	June.	July.	Aug.	Sept.
1	2,170	3,210	2,370	1,440	1,580	1,880	3,380	3,360	8,920	1,700	2,040	1,030
2	2,330	3,610	1,010	1,370	1,600	1,960	3,980	4,220	8,660	1,960	2,550	1,410
3	2,170	4,010	1,640	1,760	800	1,260	3,080	4,520	10,100	1,890	1,840	2,560
4	2,260	2,310	1,970	1,380	1,540	2,550	3,080	4,240	10,100	1,400	705	2,070
5	2,270	3,500	1,840	1,680	1,930	2,390	2,940	2,020	8,480	2,940	1,540	2,080
6	2,290	3,440	2,150	1,160	1,850	2,000	3,380	3,670	8,690	2,230	1,640	1,900
7	2,000	3,150	2,120	1,500	1,790	2,300	1,080	3,960	7,720	1,340	1,460	1,940
8	2,360	3,140	1,540	1,620	1,700	2,410	2,800	3,050	6,820	1,960	1,540	1,170
9	2,520	3,200	981	1,850	1,720	2,220	3,340	3,560	5,960	2,130	1,600	2,320
10	2,200	3,310	1,740	1,610	1,090	1,390	3,960	4,050	4,920	1,920	1,460	1,940
11	2,690	2,540	1,870	1,680	1,930	2,900	2,880	3,900	4,880	2,020	1,820	1,760
12	3,030	3,070	1,790	1,400	2,140	1,970	3,120	2,050	4,900	1,930	1,660	1,920
13	2,870	3,270	1,470	1,310	2,350	2,180	2,860	4,600	4,820	1,750	2,600	1,930
14	2,170	2,960	1,490	1,910	1,470	1,980	1,230	4,170	4,260	1,060	2,240	1,760
15	2,620	2,710	2,100	1,580	1,510	2,080	2,020	4,040	4,260	1,440	1,730	712
16	3,240	2,930	898	1,750	2,150	1,860	2,560	4,010	2,560	1,820	1,620	1,650
17	3,200	3,170	1,630	1,570	890	1,690	2,430	4,190	3,680	1,800	1,400	1,720
18	3,150	2,870	1,800	1,490	1,980	2,910	2,440	4,060	4,280	1,510	930	1,580
19	3,130	2,820	1,740	1,700	1,980	3,370	2,840	2,140	4,310	1,690	1,220	1,770
20	3,000	2,600	1,880	1,210	1,600	3,810	2,750	6,580	3,900	1,720	1,600	1,570
21	2,200	2,700	1,640	1,610	1,680	3,490	1,330	7,270	4,390	645	1,910	1,620
22	2,970	3,140	2,120	1,540	1,620	3,870	2,170	6,120	2,100	1,740	1,560	660
23	3,190	2,860	617	1,730	2,220	4,250	2,280	7,710	1,390	1,640	2,160	1,550
24	3,090	2,970	1,540	1,630	700	3,100	2,340	7,450	1,710	1,750	1,690	2,000
25	3,230	2,280	467	1,560	1,690	5,960	3,140	6,760	1,630	1,640	1,196	1,390
26	3,800	2,980	1,560	1,710	1,830	4,260	3,300	6,340	2,210	1,590	1,400	1,280
27	4,120	2,770	1,430	808	1,700	4,360	2,620	7,650	2,030	2,000	1,500	1,480
28	3,450	2,610	1,400	1,930	1,860	4,230	1,300	6,730	2,080	603	1,390	1,580
29	4,010	1,740	1,770	1,500	....	4,060	3,180	8,580	1,850	1,590	2,030	1,020
30	3,290	2,170	865	1,640	....	3,870	2,730	7,620	1,240	1,930	2,710	1,770
31	3,210	....	1,830	1,560	....	1,240	....	8,740	....	1,950	2,260	....

Monthly discharge of St. Croix River near St. Croix Falls, Wis., for the year ending Sept. 30, 1918.

[Drainage area, 5,930 square miles.]

Month.	Discharge in second-feet.				Run-off (depth in inches).
	Maximum.	Minimum.	Mean.	Per square mile.	
October	4,120	2,000	2,850	0.481	0.55
November	4,010	1,740	2,930	.494	.55
December	2,370	617	1,590	.268	.31
January	1,930	808	1,550	.261	.30
February	2,350	700	1,680	.283	.29
March	5,960	1,240	2,830	.477	.55
April	3,980	1,080	2,650	.447	.50
May	8,740	2,020	5,080	.857	.99
June	10,100	1,240	4,760	.803	.90
July	2,940	603	1,720	.290	.33
August	2,710	705	1,710	.288	.33
September	2,560	660	1,640	.277	.31
The year	10,100	603	2,590	.437	5.91

NOTE.—Computed by engineers of the U. S. Geological Survey from records of daily discharge furnished by Minneapolis General Electric Co.

NAMAKAGON RIVER AT TREGO, WIS.

LOCATION.—In sec. 35, T. 40 N., R. 12 W., at Chicago & Northwestern Railway bridge at Trego, Washburn County, 20 miles above confluence of Namakagon and Totogatic rivers

DRAINAGE AREA.—420 square miles (measured on map issued by Wisconsin Geological and Natural History Survey, edition of 1911; scale, 1 inch=6 miles).

RECORDS AVAILABLE.—March 11, 1914, to September 30, 1918.

GAGE.—Enameled staff fastened to retaining wall, left bank of river, just above railroad bridge; read by G. E. Krenz.

DISCHARGE MEASUREMENTS.—Made from lower chords of railroad bridge.

CHANNEL AND CONTROL.—Coarse gravel; free from vegetation. Banks medium high and not subject to overflow. Small island downstream with rapids on either side forms the control; channel fairly permanent.

EXTREMES OF DISCHARGE.—Maximum stage recorded during year, 2.6 feet June 6 (discharge, 1,020 second-feet); minimum discharge, 255 second-feet February 23.

1914–1918: Maximum stage recorded, 3.0 feet April 23, 1916 (discharge, 1,330 second-feet); minimum discharge, 235 second-feet December 19, 1916.

ACCURACY.—Stage-discharge relation permanent, except for ice effect. Rating curve well defined between 330 and 1,330 second-feet; below 330 second-feet extended and subject to error. Gage read once daily to half-tenths, except during period December 9 to June 1, when it was read every other day. Daily discharge ascertained by applying daily gage height to rating table except for period in which stage relation was affected by ice, for which it was obtained by applying to rating table daily gage height corrected for ice effect by means of discharge measurements, observer's notes, and weather records. Records good for open-water periods; for winter periods fair.

Dishcarge measurements of Namakagon River at Trego, Wis., during the year ending Sept. 30, 1918.

[Made by T. G. Bedford.]

Date.	Gage height.	Discharge.
	Feet.	*Sec.-ft.*
Dec. 19 [a]	2.41	308
Jan. 19 [a]	2.58	311
Feb. 21 [a]	2.46	261

a Complete ice cover at control and measuring section.

Daily discharge, in second-feet, of Namakagon River at Trego, Wis., for the year ending Sept. 30, 1918.

Day.	Oct.	Nov.	Dec.	Jan.	Feb.	Mar.	Apr.	May.	June.	July.	Aug.	Sept.
1	332	393	332	310	290	290	369	369	908	417	350	369
2	320	332	393	320	290	290	381	369	944	417	332	369
3	332	369	417	330	290	290	393	369	908	417	369	369
4	332	369	369	330	300	290	372	369	944	444	350	369
5	332	369	280	330	300	290	350	369	944	417	332	369
6	320	350	310	350	300	290	372	381	1,020	393	350	369
7	332	369	310	350	300	300	393	393	944	332	369	350
8	332	369	300	350	310	300	405	448	873	350	369	350
9	332	369	300	350	310	310	417	502	803	369	369	320
10	350	350	290	330	310	310	368	517	664	369	369	332
11	350	369	300	330	310	320	320	532	733	369	369	369
12	369	369	310	320	300	330	356	502	698	369	369	350
13	350	350	320	320	300	340	393	472	630	369	369	330
14	350	369	330	320	290	360	362	472	532	369	369	332
15	332	369	330	320	290	370	332	472	502	332	369	332
16	350	369	350	310	290	370	374	458	472	369	369	369
17	332	369	370	310	280	380	417	444	472	369	369	369
18	417	369	390	310	270	390	417	430	472	350	369	369
19	444	369	400	310	270	390	417	417	472	350	320	369
20	417	369	400	310	260	400	417	474	444	350	332	369
21	369	350	400	300	260	410	417	532	444	320	369	369
22	369	369	370	300	260	410	393	532	444	308	393	369
23	369	350	350	300	255	420	369	532	417	332	393	369
24	417	350	330	300	260	440	360	564	417	350	369	369
25	417	350	320	300	270	450	350	597	393	369	332	332
26	417	369	310	290	270	472	335	718	393	369	393	332
27	417	417	310	290	280	472	320	838	369	332	369	320
28	417	369	305	290	290	472	344	820	369	417	472	330
29	417	417	305	290	......	432	369	803	417	417	532	332
30	417	332	300	290	......	393	369	838	417	369	472	332
31	472	......	300	290	......	381	......	873	......	369	417	......

NOTE.—Stage-discharge relation affected by ice Dec. 6 to Mar. 25. Discharge estimated or interpolated every other day Dec. 9 to June 1, as gage was not read.

Monthly discharge of Namakagon River at Trego, Wis., for the year ending Sept. 30, 1918.

[Drainage area, 420 square miles.]

Month.	Discharge in second-feet.				Run-off (depth in inches).
	Maximum.	Minimum.	Mean.	Per square mile.	
October	472	320	372	0.886	1.02
November	417	332	366	.871	.97
December	417	280	336	.800	.92
January	350	290	315	.750	.86
February	310	255	286	.681	.71
March	472	290	367	.874	1.01
April	417	320	375	.893	1.00
May	873	369	529	1.26	1.45
June	1,020	369	615	1.46	1.63
July	444	308	370	.881	1.02
August	532	320	377	.898	1.04
September	369	289	351	.836	.93
The year	1,020	255	389	.926	12.56

APPLE RIVER NEAR SOMERSET, WIS.

LOCATION.—In sec. 21, T. 31 N., R. 19 W., St. Croix County, at power plant of St. Croix Power Co., 3½ miles below Somerset and 2 miles above mouth of river.

DRAINAGE AREA.—550 square miles (measured on map issued by Wisconsin Geological and Natural History Survey, edition of 1911; scale, 1 inch=6 miles).

RECORDS AVAILABLE.—January, 1901, to September 30, 1918.

GAGE.—Vertical staff gage; readings not used in determination of flow.

DISCHARGE.—The discharge of the turbines in second-feet corresponding to the number of kilowatts is determined for each hour during day from a record of the number of wheels in operation and the load; the sum of the discharge divided by 24 gives average discharge through the turbines. To this quantity is added the leakage through the average number of wheels idle each day, the sum giving daily flow through power house. Water is seldom wasted over spillway of dam, but when it is so wasted the quantity is computed from weir formulas and added to the flow through plant. There is a constant leakage through the gate and flashboards amounting to 3 second-feet. This quantity has not been taken into consideration in computing the published records.

EXTREMES OF DISCHARGE.—Maximum daily discharge recorded during the year, 1,160 second-feet, June 3; minimum daily discharge, 63 second-feet, August 1.

1904–1918: Maximum daily discharge, 2,280 second-feet in June, 1905; minimum daily discharge, 38 second-feet May 10, 1910. Due to regulation the minimum discharge has no bearing on the natural minimum flow.

REGULATION.—There are a number of power plants on Apple River above station. The pondage of these plants is small, and though the daily flow may be controlled to some extent the mean monthly flow probably corresponds closely to the natural flow.

ACCURACY.—From 1901 to 1909 the discharge through the plant was determined from tables computed from data collected as tests on one of the turbines made at flume of Holyoke Water-Power Co., Holyoke, Mass. In the summer of 1909 engineers of St. Croix Power Co. made tests on the water flowing through all the wheels as actually installed, by means of a sharp-crested weir 710 inches long located about 60 feet below power house. These tests gave results about 3 per cent larger than the Holyoke tests, and tables based on them have been used in determining the discharge through the plant from 1909 to date. In June, 1914, a series of current meter measurements were made by the Wisconsin Railroad Commission and United States Geological Survey, and a rating curve for the tailrace was developed. Twelve tests were then run with different wheels and loads. It was found that the discharge as determined by the current meter and the discharge as computed by the company agreed very closely, the percentage difference for the twelve tests ranging from − 6.4 per cent to + 1.8 per cent, with an average of − 2.0 per cent; the discharge as determined by the company being 2 per cent less than that determined by the current meter.

COOPERATION.—Records furnished by St. Paul Gas Light Co. of St. Paul, Minn., D. W. Flowers, engineer.

Daily discharge, in second-feet, of Apple River near Somerset, Wis., for the year ending Sept. 30, 1918.

Day.	Oct.	Nov.	Dec.	Jan.	Feb.	Mar.	Apr.	May.	June.	July.	Aug.	Sept.
1	194	251	304	203	184	255	350	244	603	235	63	164
2	202	242	121	199	256	334	292	259	885	225	135	231
3	190	324	233	219	129	282	303	135	1,160	289	223	193
4	210	187	220	213	153	240	383	258	1,020	170	137	206
5	199	239	202	280	196	345	170	151	993	271	155	171
6	281	249	142	138	208	278	342	261	960	268	194	206
7	126	246	196	191	172	280	266	259	869	225	172	219
8	190	255	236	193	207	263	249	234	690	282	159	118
9	214	359	92	191	190	191	276	280	581	235	151	204
10	199	169	113	193	135	194	238	422	686	274	213	228
11	231	211	226	155	157	280	418	462	472	249	132	196
12	247	228	225	272	177	291	141	221	306	95	210	207
13	348	244	250	87	199	258	304	309	336	306	173	219
14	134	249	214	161	132	280	272	378	505	189	168	226
15	296	227	229	199	164	307	257	396	364	235	148	140
16	200	363	131	201	220	430	286	358	230	213	185	202
17	207	251	214	193	131	307	300	391	276	227	213	153
18	219	135	229	328	190	495	290	300	338	161	141	209
19	204	233	202	115	179	642	274	207	343	208	163	181
20	267	268	238	183	164	749	273	310	318	217	159	204
21	219	237	240	204	181	786	214	348	336	111	170	208
22	232	237	262	129	186	618	230	361	349	137	191	153
23	269	243	193	194	244	583	270	318	209	162	227	208
24	240	304	182	196	156	494	249	283	270	176	152	198
25	245	161	159	189	208	208	255	364	284	210	102	151
26	275	248	213	246	229	500	275	591	274	208	162	163
27	309	239	211	164	188	323	265	597	228	211	188	164
28	187	262	171	159	207	328	173	821	238	154	175	191
29	232	149	285	165	....	363	285	705	266	182	251	126
30	261	228	91	238	....	370	309	804	205	348	170	218
31	261	....	189	161	....	239	....	895	....	92	269	....

NOTE.—See note under "Discharge" in station description for method by which these records are obtained.

Monthly discharge of Apple River near Somerset, Wis., for the year ending Sept. 30, 1918.

[Drainage area, 550 square miles.]

Month.	Discharge in second-feet.				Run-off (depth in inches).
	Maximum.	Minimum.	Mean.	Per square mile.	
October	348	126	229	0.416	0.48
November	363	135	241	.438	.49
December	304	91	200	.364	.42
January	328	87	192	.349	.40
February	256	129	184	.335	.35
March	786	191	371	.675	.78
April	418	141	274	.498	.56
May	895	135	385	.700	.81
June	1,160	205	486	.884	.99
July	348	92	212	.385	.44
August	269	63	173	.315	.36
September	231	118	190	.345	.38
The year	1,160	63	262	.476	6.46

KINNIKINNIC RIVER NEAR RIVER FALLS, WIS.

LOCATION.—In sec. 18, T. 27 N., R. 19 W., at Clifton Hollow bridge, a quarter of a mile downstream from dam of Clifton Falls Power Co., 2 miles above mouth of river and 7 miles downstream from River Falls, Pierce County.

DRAINAGE AREA.—170 square miles (measured on map issued by Wisconsin Geological and Natural History Survey, edition of 1911; scale, 1 inch=6 miles).

RECORDS AVAILABLE.—October 23, 1916, to September 30, 1918.

GAGE.—Gurley graph water-stage recorder, in a wooden well fastened to downstream side of right-hand cushing bridge pier.

DISCHARGE MEASUREMENTS.—Made from bridge or by wading.

CHANNEL AND CONTROL.—Channel of rather heavy gravel and sand; control in head of small rapids 150 feet below the gage and is not permanent.

EXTREMES OF DISCHARGE.—Maximum stage recorded during year by recording gage, 6.6 feet at 10 p. m. June 5 (discharge, roughly approximate, 3,080 second-feet). Minimum stage of between 1.7 and 1.8 feet (discharge, approximately 15 second-feet) occurred several times following complete shutdown of power plant. The maximum is about the natural maximum; minimum is caused by regulation at the power house.

ICE.—Stage-discharge relation affected to some extent by ice.

REGULATION.—The daily flow is regulated almost completely by the Clifton power dam just above the station. There are three dams in River Falls which may also have some effect on the daily flow; the storage at these dams is relatively small, and the monthly flow is considered to be nearly the normal flow.

ACCURACY.—Stage-discharge relation not permanent; one rating curve was used throughout the year. Poorly defined between 28 and 470 second-feet. Continuous record obtained by recording gage, except during winter periods and certain other brief periods when gage was not operating properly. Discharge ascertained by fractional day method.

When recording gage was not in operation discharge was based on flow in adjacent drainage basins. Records poor.

Discharge measurements of Kinnikinnic River, near River Falls, Wis., during the year ending Sept. 30, 1918.

Date.	Made by—	Gage height.	Discharge.
		Feet.	*Sec.-ft.*
May 13[a]	T. G. Bedford	2.08	54
13	....do	2.46	160
Aug. 20[a]	S. B. Soulé	2.45	135
20	....do	3.06	336

[a] Made by wading a short distance downstream.

Daily discharge, in second-feet, of Kinnikinnic River near River Falls, Wis., for the year ending Sept. 30, 1918.

Day.	Oct.	Nov.	Dec.	Apr.	May.	June.	July.	Aug.	Sept.
1	76	66	78	80	65	260	75	85	90
2	76	62	90	110	65	110	75	75	90
3	78	72	57	110	65	96	70	75	75
4	68	66	60	110	70	95	70	70	80
5	74	72	60	110	70	680	65	65	80
6	83	95	60	100	80	490	65	75	85
7	64	78	71	100	80	375	60	110	90
8	56	78	....	90	80	260	65	105	90
9	56	80	....	100	100	530	70	75	95
10	66	80	....	100	90	180	75	75	95
11	56	78	....	100	90	115	80	95	95
12	65	76	....	100	40	90	65	80	100
13	58	125	....	100	72	85	70	65	90
14	62	80	....	80	80	70	70	95	90
15	54	80	....	70	80	60	75	95	90
16	53	75	70	56	50	70	75	45	100
17	64	70	70	60	50	75	75	85	95
18	60	104	117	60	50	78	75	55	90
19	52	91	80	60	40	75	100	60	85
20	65	120	80	55	40	75	80	95	85
21	65	104	97	57	50	75	60	70	90
22	52	113	96	60	45	75	52	1100	85
23	56	92	96	60	45	80	65	220	100
24	61	90	....	60	45	75	190	175	100
25	70	88	....	60	65	75	105	125	100
26	61	64	....	60	95	70	60	96	95
27	79	74	....	60	80	75	75	80	90
28	77	69	....	65	95	75	75	85	70
29	75	85	....	65	75	75	55	70	60
30	70	58	....	65	183	75	55	80	80
31	68	....	....	....	400	....	45	85	....

NOTE.—Stage-discharge relation affected by ice and recording gage not in operation from Jan. 1 to Mar. 31; discharge estimated, Jan. 1-31, 60 second-feet; Feb. 1-28, 55 second-feet; Mar. 1-31, 115 second-feet. Recording gage not in operation, discharge estimated Dec. 8 to 15, 24 to 31, 70 second-feet. Recording gage not in perfect operation Nov. 9, 10, 24, Dec. 22, Apr. 6, 7, June 14, Aug. 24, 25.

Monthly discharge of Kinnikinnic River near River Falls, Wis., for the year ending Sept. 30, 1918.

[Drainage area, 170 square miles.]

Month.	Discharge in second-feet.				Run-off (depth in inches on drainage area).
	Maximum.	Minimum.	Mean.	Per square mile.	
October	83	52	65.2	0.384	0.44
November	125	58	82.8	.487	.54
December	....	....	74.3	.437	.50
January	....	....	60	.353	.41
February	....	....	55	.324	.34
March	....	....	115	.677	.78
April	110	55	78.8	.464	.52
May	400	40	81.8	.481	.55
June	680	60	154	.906	1.01
July	190	45	73.9	.435	.50
August	1,100	45	121	.712	.82
September	100	60	88.7	.522	.58
The year	1,100	....	87.7	.516	6.99

CHIPPEWA RIVER AT BISHOP'S BRIDGE, NEAR WINTER, WIS.

Location.—In sec. 23, T. 39 N., R. 6 W., at highway bridge 3 miles downstream from East Fork of Chippewa River (coming in from the left) and 4 miles by road northwest of Winter, Sawyer County.

Drainage area.—775 square miles (measured on map issued by Wisconsin Geological and Natural History Survey, edition of 1911; scale, 1 inch=6 miles).

Records available.—February 23, 1912, to September 30, 1918.

Gage.—Chain gage fastened to highway bridge used since May 23, 1916; read by John Edburg. Gages previously used as follows: February 23, 1912, to January 27, 1914, a wooden staff gage fastened to a wooden pier on right bank just above bridge; datum 3.44 feet above that for chain gage; January 27, 1914, to May 23, 1916, a vertical cast-iron staff gage fastened to same pier; datum same as for chain gage.

Discharge measurement.—Made from downstream side of highway bridge.

Channel and control.—Bed composed of gravel; free from vegetation and not subject to shift. One channel at all stages. Control is head of rapids about 1,000 feet below the gage; practically permanent. Banks not subject to overflow.

Extremes of discharge.—Maximum stage recorded during year 7.24 feet at 4 p. m. June 1 (discharge, 3,040 second-feet); estimated minimum discharge, during January and February, 180 second-feet.

1913–1918: Maximum stage recorded during period, 9.56 feet, April 22, 1916 (discharge, 6,940 second-feet); minimum discharge estimated at 175 second-feet February 17, 1917.

Regulation.—Flow regulated to some extent by operation of storage reservoir in sec. 14, T. 41 N., R. 6 W., about 16 miles above station. This reservoir has a capacity of 550,000,000 cubic feet and is used in connection with reservoirs on upper Flambeau River for the purpose of regulating the flow of Chippewa River.

Accuracy.—Stage-discharge relation permanent except as affected by ice during winter period and by logs during a portion of April and May. Rating curve well defined between 270 and 6,820 second-feet. Gage read to hundredths twice a day. Daily discharge ascertained by applying mean daily gage height to rating table, except for period in which stage-discharge relation was affected by ice, for which it was obtained by applying to the rating table daily gage heights corrected for ice effect by means of discharge measurements, observer's notes, and weather records; discharge for periods of May, when logs were present, interpolated. Excellent records for open-water period except those for May, which are fair; winter records fair.

Discharge measurements of Chippewa River at Bishop's Bridge, near Winter, Wis., during the year ending Sept. 30, 1918.

[Made by T. G. Bedford.]

Date.	Gage height.	Discharge.
	Feet.	*Sec.-ft.*
Dec. 26[a]	5.46	337
Jan. 26[a]	5.34	198
Mar. 1[a]	5.62	216

[a] Made through complete ice cover, 20 feet below gage.

Daily discharge, in second-feet, of Chippewa River at Bishop's Bridge, near Winter, Wis., for the year ending Sept. 30, 1918.

Day.	Oct.	Nov.	Dec.	Jan.	Feb.	Mar.	Apr.	May.	June.	July.	Aug.	Sept.
1	380	710	345	410	180	200	530	790	2,980	340	304	1,050
2	405	675	345	340	180	210	580	830	2,980	322	322	1,000
3	405	640	345	330	180	210	530	790	2,840	340	304	830
4	405	555	345	320	180	220	530	832	2,570	340	304	790
5	405	530	345	320	180	225	505	874	2,570	580	287	675
6	405	530	340	305	180	220	480	916	2,570	455	287	555
7	380	505	340	295	185	210	505	958	2,310	360	340	505
8	380	505	340	285	195	210	480	1,000	2,050	360	390	405
9	360	505	340	280	195	210	480	1,200	1,570	322	405	405
10	380	505	340	270	195	195	480	1,460	1,520	322	390	340
11	405	505	380	255	195	210	455	1,520	1,460	304	380	390
12	405	505	405	240	195	225	430	1,520	1,100	270	430	480
13	405	480	405	240	195	225	430	1,350	1,050	287	380	480
14	430	480	360	240	185	225	430	1,050	915	270	405	480
15	455	480	380	230	180	225	455	1,050	870	304	340	430
16	455	480	405	225	180	225	455	1,050	710	304	304	390
17	480	455	405	225	195	225	505	1,050	640	270	304	430
18	640	455	380	225	195	255	555	1,050	580	287	270	505
19	960	430	360	210	195	270	580	1,050	505	270	287	530
20	960	405	340	210	195	270	580	1,150	505	304	254	430
21	1,000	380	340	210	195	305	610	1,150	480	304	254	430
22	1,150	380	360	210	195	340	610	1,460	405	287	322	430
23	1,150	360	340	210	195	380	555	1,400	430	304	340	405
24	1,100	365	340	210	210	430	640	1,350	380	287	480	455
25	915	360	320	200	225	455	580	1,570	380	304	505	430
26	870	360	340	195	225	480	505	1,980	340	304	480	405
27	915	355	340	190	225	530	455	2,310	322	270	480	430
28	1,000	355	340	180	210	555	455	2,440	322	340	505	380
29	830	350	340	180	......	555	480	2,440	287	322	790	390
30	675	350	410	180	......	530	530	2,570	340	340	790	380
31	750	......	285	180	......	530	......	2,700	......	322	960	......

NOTE.—Stage-discharge relation affected by ice Nov. 24 to Mar. 27. Discharge interpolated because of logs on control, May 4–7.

Monthly discharge of Chippewa River at Bishop's Bridge, near Winter, Wis., for the year ending Sept. 30, 1918.

[Drainage area, 775 square miles.]

Month.	Discharge in second-feet.				Run-off (depth in inches).
	Maximum.	Minimum.	Mean.	Per square mile.	
October	1,150	360	640	0.826	0.95
November	710	350	465	.600	.67
December	410	285	355	.458	.53
January	410	180	245	.316	.36
February	225	180	194	.250	.26
March	555	195	308	.397	.46
April	640	430	512	.661	.74
May	2,700	790	1,380	1.78	2.05
June	2,980	287	1,200	1.55	1.73
July	580	270	322	.415	.48
August	960	254	406	.524	.60
September	1,050	340	506	.653	.73
The year	2,980	180	546	.705	9.56

CHIPPEWA RIVER AT BRUCE, WIS.

Location.—In sec. 4, T. 35 N., R. 7 W., at Minneapolis, St. Paul & Sault Ste. Marie Railway bridge 1 mile east of Bruce, Rusk County. Thornapple River enters from right immediately above station, and Flambeau River from right 21 miles below.

Drainage area.—1,600 square miles (measured on map issued by Wisconsin Geological and Natural History Survey, edition of 1911; scale, 1 inch=6 miles).

Records available—December 31, 1913, to September 30, 1918.

Gage.—Chain gage, attached to downstream side of Minneapolis, St. Paul & Sault Ste. Marie Railroad bridge; read by H. C. Gardner.

Discharge measurements.—Made from downstream side of bridge.

Channel and control.—Bed composed of sand and small gravel; free from vegetation; first and second channels from the west fairly permanent; third channel nearest east bank has a tendency to fill during low stages with sand worked in by Thornapple River. Flow except during extreme high stages is confined within the banks.

Extremes of discharge.—Maximum stage recorded during year, 9.7 feet at 7 a. m. June 2 (discharge, 9,380 second-feet); minimum stage recorded 1.15 feet, morning and afternoon of August 21 (discharge, about 260 second-feet).

1910–1918: Maximum stage recorded during period, 12.3 feet at 5.45 p.m., April 22, 1916 (discharge, 13,400 second-feet); minimum discharge, when river was frozen, approximately 310 second-feet during January and February, 1917; minimum open-water stage recorded 1.15 feet morning and afternoon reading August 21, 1918 (discharge, about 260 second-feet); caused by regulation.

Regulation.—Flow modified to some extent by reservoir on West Fork of Chippewa River, in sec. 14, T. 41 N., R. 6 W. This reservoir has a capacity of 550,000,000 cubic feet, and is used in connection with reservoirs on upper Flambeau River, for the purpose of regulating the flow of Chippewa River. No diurnal fluctuation is observed.

Accuracy.—Stage-discharge relation not permanent; affected by ice during winter periods and changes caused by shifting control during periods of low water. Two rating curves used during the year; the first, which is fairly well defined throughout, is applicable from October 1 to March 28; the second, which is fairly well defined between 390 and 3,100 second-feet, is applicable March 29 to September 30. Gage read twice daily to quarter-tenths. Daily discharge ascertained by applying mean daily gage height to rating table, except for the period in which stage-discharge relation was affected by ice, for which periods it was obtained by applying to rating table mean daily gage heights corrected for ice effect by means of discharge measurements, observer's notes, and weather records. Open-water records fair; winter records subject to error.

Discharge measurements of Chippewa River at Bruce, Wis., during the year ending Sept. 30, 1918.

Date.	Made by—	Gage height.	Discharge.	Date.	Made by—	Gage height.	Discharge.
		Feet.	*Sec.-ft.*			*Feet.*	*Sec.-ft.*
Oct. 24	R. B. Kilgore	3.04	1,630	Feb. 26[a]	T. G. Bedford	3.31	359
Dec. 24[a]	T. G. Bedford	2.82	541	May 5	do	3.77	2,220
Jan. 24[a]	do	2.99	390	Aug. 22	S. B. Soulé	1.78	721

[a] Complete ice cover at control and measuring station.

Daily discharge, in second-feet, of Chippewa River at Bruce, Wis., for the year ending Sept. 30, 1918.

Day.	Oct.	Nov.	Dec.	Jan.	Feb.	Mar.	Apr.	May.	June.	July.	Aug.	Sept.
1	510	1,300	690	455	355	430	935	1,940	8,400	620	620	1,190
2	314	1,100	690	455	340	440	1,270	1,940	9,240	620	374	1,270
3	480	1,100	650	440	330	455	1,430	1,190	7,720	620	417	1,190
4	575	1,020	650	430	330	455	1,270	1,510	6,040	620	480	1,080
5	610	930	540	430	330	455	1,270	1,430	4,970	795	515	1,000
6	610	890	610	430	330	440	1,270	1,430	4,420	900	480	935
7	610	890	610	455	330	430	1,350	2,030	3,870	900	480	830
8	575	850	610	480	330	430	1,350	2,120	3,430	725	690	830
9	540	850	610	455	330	430	1,270	2,700	2,900	690	725	550
10	1,600	810	575	430	340	430	1,190	4,970	2,400	620	655	620
11	1,600	770	610	430	355	430	1,110	5,560	2,210	550	620	480
12	650	770	630	430	355	450	1,110	4,750	1,940	515	620	585
13	610	770	650	420	355	480	1,010	3,870	1,760	515	655	550
14	650	770	630	405	340	480	970	2,909	1,510	480	655	480
15	650	730	610	420	330	480	1,000	2,500	1,430	480	655	515
16	730	690	610	430	320	510	1,080	2,500	1,350	515	320	404
17	930	690	610	415	310	610	1,350	2,300	1,190	515	466	830
18	1,060	690	590	405	330	770	1,510	2,120	1,110	515	473	760
19	1,600	690	575	405	355	1,020	1,510	1,940	1,000	480	445	830
20	1,800	650	540	405	340	1,200	1,350	2,210	900	480	550	473
21	1,800	610	510	405	330	1,500	1,350	2,210	865	480	260	760
22	1,600	540	525	405	330	1,800	1,350	2,300	830	480	585	725
23	1,700	575	540	390	330	2,000	1,270	2,800	795	515	795	725
24	1,600	575	540	380	330	1,900	1,150	2,300	725	515	900	655
25	1,500	575	510	380	340	1,700	1,110	3,210	690	515	830	620
26	1,400	575	510	380	355	1,500	1,110	5,800	690	515	795	655
27	1,600	610	510	380	355	1,200	1,000	7,720	620	480	480	655
28	1,700	690	480	380	360	1,020	935	7,200	620	620	725	620
29	1,700	690	455	370	......	970	970	6,290	585	970	1,040	620
30	1,300	690	450	355	......	970	2,120	5,080	585	760	1,190	585
31	1,060	......	455	350	......	935	......	5,680	......	690	1,190	......

NOTE.—Stage-discharge relation affected by ice Dec. 5 to Mar. 28.

Monthly discharge of Chippewa River at Bruce, Wis., for the year ending Sept. 30, 1918.

[Drainage area, 1,600 square miles.]

Month.	Discharge in second-feet.				Run-off (depth in inches).
	Maximum.	Minimum.	Mean.	Per square mile.	
October	1,800	314	1,090	0.681	0.79
November	1,300	540	770	.481	.54
December	690	450	573	.358	.41
January	480	350	413	.258	.30
February	360	310	338	.211	.22
March	2,000	430	849	.531	.61
April	2,120	935	1,230	.769	.86
May	7,720	1,190	3,310	2.07	2.39
June	9,240	585	2,490	1.56	1.74
July	970	480	603	.377	.43
August	1,190	260	635	.397	.46
September	1,270	404	734	.459	.51
The year	9,240	260	1,090	.681	9.26

CHIPPEWA RIVER AT CHIPPEWA FALLS, WIS.

LOCATION.—In SE. ¼ sec. 6, T. 28 N., R. 8 W., at highway bridge at Chippewa Falls, Chippewa County, 2,500 feet below mouth of Duncan Creek, which comes in from right.

DRAINAGE AREA.—5,600 square miles.

RECORDS AVAILABLE.—June 22, 1888, to September 30, 1918. The gage was originally established by Chippewa Lumber & Boom Co., which has kept a continuous record since 1889. Since 1904 the United States Weather Bureau has obtained gage readings during flood season of each year. On June 1, 1906, the United States Geological Survey began making discharge measurements and maintaining gage readings.

GAGE.—On July 27, 1916, a Gurley graph water-stage recorder replaced a Friez water-stage recorder which was installed in January, 1914, on web between cushing piers supporting first right hand span and about 10 feet upstream from the gage formerly used by the United States Weather Bureau; gage referred to original datum.

DISCHARGE MEASUREMENTS.—Made from downstream side of bridge or by wading.

CHANNEL AND CONTROL.—Heavy gravel; fairly permanent. Both banks high and are rarely overflowed.

EXTREMES OF STAGE.—Maximum stage recorded during year, 12.4 feet at 5 p. m. June 1 (discharge, about 43,700 second-feet); estimated minimum discharge, 175 second-feet January 20; caused by regulation at Wissota dam.

1888–1918: Maximum stage recorded during period, 26.03 feet December 6, 1896. September 10, 1884, a stage of 26.94 feet was reached; discharge not estimated; minimum recorded approximately 40 second-feet February 4, 1917.

ICE.—Stage-discharge relation seriously affected by ice.

REGULATION.—Flow past station controlled to a considerable extent by the operation of the Wissota gates. Large diurnal fluctuation.

ACCURACY.—Stage-discharge relation practically permanent. Rating curve well defined between 530 and 56,200 second-feet; below 530 second-feet poorly defined. Operation of the water-stage recorder was satisfactory throughout the year, except for periods when stage-discharge relation was affected by ice. Daily discharge October 1 to September 30 obtained by discharge integrator. Daily discharge during periods when stage-discharge relation was affected by ice ascertained by applying to rating curve mean daily gage heights corrected for the ice effect by means of discharge measurements, observer's notes, and weather records and to some extent on computations of flow through the Wissota dam. Open-water records good; winter records fair.

Discharge measurements of Chippewa River at Chippewa Falls, Wis., during the year ending Sept. 30, 1918.

Date.	Made by—	Gage height.	Discharge.	Date.	Made by—	Gage height.	Discharge.
		Feet.	*Sec.-ft.*			*Feet.*	*Sec.-ft.*
Dec. 16[a]	Hoyt and Bedford	0.27	1,040	Feb. 18[a]	T. G. Bedford	.50	1,520
Jan. 16[a]	T. G. Bedford	.49	1,320	Aug. 21	S. B. Soulé	.91	2,400

[a] Incomplete ice cover at control; measurement made through complete ice cover.

Daily discharge, in second-feet, of Chippewa River at Chippewa Falls, Wis., for the year ending Sept. 30, 1918.

Day.	Oct.	Nov.	Dec.	Jan.	Feb.	Mar.	Apr.	May.	June.	July.	Aug.	Sept.
1	2,210	4,360	1,820	1,600	1,330	1,440	7,600	7,380	36,300	1,310	3,140	4,090
2	2,280	4,840	1,910	1,750	1,330	1,360	7,320	6,600		1,930	2,880	5,350
3	2,580	3,220	1,750	1,670	986	1,280		5,400		2,400	2,780	5,540
4	2,370	4,030	1,900	1,670	1,340	1,100		7,620		1,400	2,700	3,930
5	2,370	4,010	1,810	1,750	1,170	1,100	6,570	5,280	27 500	1,300	2,130	3,900
6	3,060	3,580	1,660	900	1,280	1,620		5,440		2,120	2,920	3,790
7	1,150	3,700	1,510	1,240	1,190	1,530		6,760		2,120	2,860	3,470
8	2,080	3,610	1,360	1,330	1,190	1,530	6,080	5,680		2,200	2,180	3,450
9	2,260	4,040	1,360	1,330	1,190	2,210	6,500	7,640	10,600	2,150	4,790	4,000
10	2,340	3,850	1,600	1,330	1,040	1,620	5,540	9,060	11,100	2,240	6,370	4,040
11	1,840	2,020	1,440	1,410	1,050	3,010	4,460	14,000	7,120	2,140	5,560	4,120
12	2,080	3,940	1,500	1,580	1,240	3,500	4,920	14,600	7,550	2,250	4,780	4,100
13	2,220	2,970	1,560	1,220	1,260	2,880	4,970	15,100	5,720	2,220	5,010	2,640
14	2,600	2,930	1,380	1,380	1,280	2,620	5,480	12,600	4,710	1,950	5,080	2,740
15	825	2,900	1,280	1,360	1,630	2,550	5,980	9,270	4,300	832	4,550	1,780
16	700	2,740	1,090	1,340	1,480	2,420	5,180	7,500	3,280	1,610	4,050	2,560
17	2,450	3,890	1,340	1,260	1,330	2,300	5,030	8,650	5,870	1,890	3,280	3,440
18	1,660	2,300	1,550	1,160	1,510	3,000	5,670	6,580	4,620	1,770	2,500	3,690
19	4,570	2,860	1,620	815	1,680	3,490	4,720	6,260	3,440	1,900	2,620	3,340
20	6,970	2,840	1,630	175	1,770	3,040	3,960	8,190	3,580	1,720	2,810	3,310
21	8,350	2,840	1,480	210	1,700	4,100	4,200	6,120	3,540	1,420	2,740	3,260
22	6,900	3,340	2,740	1,670	1,620	8,000	5,570	7,440	2,620	828	2,720	2,880
23	6,040	2,850	2,190	1,670	1,440	10,900	5,080	9,460	1,640	1,800	3,200	2,980
24	4,420	3,300	2,210	1,330	1,190	12,000	5,130	9,220	1,530	3,540	7,140	2,970
25	5,330	1,400	1,610	1,160	1,190	13,400	4,400	9,960	1,960	2,740	6,820	3,100
26	5,010	2,720	1,450	1,240	1,440	12,400	4,100	20,300	2,100	2,720	5,200	2,970
27	4,810	2,710	1,560	986	1,440	11,700	4,160	30,600	1,920	2,120	4,800	2,880
28	6,670	2,170	1,670	1,070	1,440	9,780	3,300	34,000	1,920	1,180	4,350	2,640
29	6,350	2,140	1,780	1,240	....	9,110	4,290	33,000	1,730	3,000	4,460	2,120
30	5,960	2,100	1,760	1,330	....	8,440	6,080	28,800	940	3,500	6,300	1,960
31	4,430	....	1,670	1,330	....	7,770	....	30,200	....	3,000	6,560	....

NOTE.—Stage-discharge relation affected by ice Dec. 5 to Mar. 10. Recording gage not in perfect operation Mar. 16, 22–23, 30, Apr. 3–7, 14, June 2–8, July 29–31; discharge partly estimated.

Monthly discharge of Chippewa River at Chippewa Falls, Wis., for the year ending Sept. 30, 1918.

[Drainage area, 5,600 square miles.]

Month.	Discharge in second-feet.				Run-off (depth in inches).
	Maximum.	Minimum.	Mean.	Per square mile.	
October	8,350	700	3,640	0.650	0.75
November	4,840	1,400	3,140	.561	.63
December	2,740	1,090	1,650	.295	.34
January	1,750	175	1,270	.227	.26
February	1,770	986	1,350	.241	.25
March	13,400	1,100	4,880	.871	1.00
April	....	3,300	5,420	.968	1.08
May	34,000	5,280	12,500	2.23	2.57
June	....	940	10,700	1.91	2.13
July	3,540	828	2,040	.364	.42
August	7,140	2,130	4,100	.732	.84
September	5,540	1,780	3,370	.602	.67
The year	....	175	4,520	.807	10.94

FLAMBEAU RIVER NEAR BUTTERNUT, WIS.

LOCATION.—In NW. ¼ SE. ¼ sec. 33, T. 41 N., R. 1 E., Ashland County, 6 miles southeast of Butternut and 7 miles upstream from Park Falls.

DRAINAGE AREA.—660 square miles (measured on map issued by Wisconsin Geological and Natural History Survey, edition of 1911, scale, 1 inch=6 miles).

RECORDS AVAILABLE.—July 30, 1914, to September 30, 1918.

GAGE.—Standard chain gage supported by built-up cantilever, attached to posts set in right bank of river; installed May 26, 1916; read by Miss Mathilda Schulz. Vertical staff gage at same site and datum was used from July 30, 1914, until taken out by ice in spring of 1916.

DISCHARGE MEASUREMENTS.—Made from a cable 1,500 feet downstream from the gage.

CHANNEL AND CONTROL.—Bed at gage composed of mud and rock. Left bank is low and subject to overflow; right bank slopes back gradually to high-water mark. At cable site, 1,500 feet below gage, the bed is rocky and the banks high. Control is at head of Schultz Rapids, about 200 feet below cable and 1,700 feet below gage.

EXTREMES OF DISCHARGE.—Maximum stage recorded during year: 4.5 feet, June 3 (discharge, 1,680 second-feet); minimum discharge estimated at 250 second-feet March 1 to 10.

1914–1918: Maximum stage recorded during period, 9.0 feet, April 22 and 23, 1916 (discharge, 5,430 second-feet); minimum discharge, estimated 250 second-feet, March 1 to 10, 1918.

REGULATION.—Storage reservoirs are maintained by Chippewa & Flambeau Improvement Co. on headwaters of Flambeau River. Of these reservoirs, Rest Lake, in sec. 9, T. 42 N., R. 5 E., with an allowable capacity of approximately 1½ billion cubic feet, is the largest.

ACCURACY.—Stage-discharge relation permanent except as affected by ice. Rating curve well defined between 356 and 3,480 second-feet. Gage read twice daily to quarter-tenths. Daily discharge ascertained by applying mean daily gage height to rating table except for periods in which stage-discharge relation was affected by ice, for which it was obtained by applying to rating table daily gage heights corrected for ice effect by means of discharge measurements, observer's notes, and weather records. Open-water records good; winter records fair.

Discharge measurements of Flambeau River near Butternut, Wis., during the year ending Sept. 30, 1918.

[Made by T. G. Bedford.]

Date.	Gage height.	Discharge.	Date.	Gage height.	Discharge.
	Feet.	*Sec.-ft.*		*Feet.*	*Sec.-ft.*
Dec. 21[a]	2.18	459	Feb. 23[a]	2.44	272
Jan. 22[a]	2.29	322	June 7	3.79	1,240

[a] Complete ice cover at control and measuring section.

Daily discharge, in second-feet, of Flambeau River near Butternut, Wis., for the year ending Sept. 30, 1918.

Day.	Oct.	Nov.	Dec.	Jan.	Feb.	Mar.	Apr.	May.	June.	July.	Aug.	Sept.
1	416	673	500	355	305	250	850	760	1,330	518	592	592
2	400	632	490	330	305	250	850	760	1,500	554	554	592
3	385	592	485	315	305	250	850	716	1,680	592	554	554
4	400	592	475	305	310	250	805	716	1,620	632	554	483
5	400	554	465	290	315	250	716	673	1,560	554	632	466
6	416	554	465	290	320	250	632	716	1,380	554	449	432
7	416	554	465	285	325	250	632	760	1,330	518	432	416
8	400	592	465	285	325	250	632	805	1,280	518	483	385
9	385	632	465	280	330	250	632	805	1,120	483	554	370
10	416	673	465	280	330	250	632	940	1,080	449	554	356
11	518	673	460	280	330	260	592	1,080	985	416	554	385
12	592	673	460	280	330	260	592	1,030	895	416	554	416
13	632	673	460	280	330	270	592	940	805	385	554	449
14	673	632	460	280	320	270	592	850	760	342	518	416
15	673	632	460	280	315	270	592	850	716	356	483	400
16	592	632	460	290	310	270	592	850	716	385	466	385
17	554	632	460	300	305	270	632	805	632	416	449	416
18	805	632	460	300	300	270	673	850	592	416	432	449
19	895	632	460	305	290	280	673	850	554	449	416	483
20	985	592	460	310	280	300	632	940	518	554	416	466
21	985	632	460	315	275	330	632	985	483	554	385	432
22	940	632	450	320	270	340	632	985	466	554	356	432
23	895	632	450	320	270	370	592	1,080	449	554	356	432
24	850	554	450	315	270	400	595	1,080	449	554	329	432
25	805	592	450	315	270	415	592	1,120	400	554	329	416
26	760	554	440	310	270	450	554	1,280	400	554	416	400
27	805	540	430	310	270	480	518	1,500	416	554	416	385
28	760	530	415	310	270	535	518	1,500	385	592	432	356
29	716	520	400	305	....	590	673	1,500	385	632	554	370
30	805	510	390	305	....	670	760	1,380	483	632	632	370
31	673	....	385	305	....	720	....	1,280	....	632	716	....

NOTE.—Stage-discharge relation affected by ice Nov. 27 to Apr. 1.

Monthly discharge of Flambeau River near Butternut, Wis., for the year ending Sept. 30, 1918.

[Drainage area, 660 square miles.]

Month.	Discharge in second-feet.				Run-off (depth in inches).
	Maximum.	Minimum.	Mean.	Per square mile.	
October	985	385	643	0.974	1.12
November	673	510	605	.917	1.02
December	500	385	454	.688	.79
January	355	280	302	.458	.53
February	330	270	302	.458	.48
March	720	250	339	.514	.59
April	850	518	649	.983	1.10
May	1,500	673	977	1.48	1.71
June	1,680	385	846	1.28	1.43
July	632	342	512	.776	.89
August	716	329	488	.739	.85
September	592	356	431	.653	.73
The year	1,680	250	547	.829	11.24

FLAMBEAU RIVER NEAR LADYSMITH, WIS.

Location.—In SE. ¼ sec. 20, T. 35 N., R. 5 W., at H. J. Cornelissen's farm, 6 miles by road northeast of Ladysmith, Rusk County, 21 miles below mouth of South Fork of Flambeau River, which comes in from left, and 28 miles above mouth of river.

Drainage area.—1,940 miles (measured on map issued by Wisconsin Geological and Natural History Survey, edition of 1911; scale, 1 inch=6 miles).

Records available.—January 2, 1914, to September 30, 1918. From February 15, 1903, to December 2, 1906, records were collected at a station in the city of Ladysmith, three-quarters of a mile south of Minneapolis, St. Paul & Sault Ste. Marie Railway station, half a mile below dam of Menasha Pulp Co., and about 6 miles below present station.

Gage.—Chain gage fastened to a cantilever arm, supported by two trees on left bank of river, on the farm of H. J. Cornelissen; read by H. J. Cornelissen.

Discharge measurements.—Made from cable 200 feet below gage.

Channel and control.—Bed composed of gravel and sand; free from vegetation and fairly permanent. At gage section, channel is divided by a small sandy island; at cable section the river flows in one channel. Banks are medium high, wooded, and not subject to overflow. Control not well defined; formed by channel below the gage.

Extremes of discharge.—Maximum open-water stage recorded during year, 7.2 feet June 2 (discharge, 9,520 second-feet); minimum discharge (during frozen period), 540 second-feet in February and March.

1903–1906 and 1914–1918: Maximum discharge recorded during period, 17,400 second-feet April 23, 1916; minimum discharge, 390 second-feet December 4, 1904.

Ice.—Stage-discharge relation seriously affected by large quantities of frazil ice which form on the falls and rapids above the station and fill the channel for a distance of several miles from the gage to pond of the Paper Co.'s dam at Ladysmith.

Regulation.—Chippewa & Flambeau Improvement Co. operates storage reservoirs on Rest Lake and smaller reservoirs on Manitowish and Turtle rivers and Bear Creek. Weekly fluctuations at gage are caused by operation of power plants at Park Falls and storage reservoirs. No daily fluctuation has been observed.

Accuracy.—Stage-discharge relation permanent except as affected by logs and ice. Rating curve well defined between 770 and 17,000 second-feet, approximate above and below these limits. Gage read once daily to quarter-tenths. Daily discharge ascertained by applying daily gage height to rating table, except for periods in which stage-discharge relation was affected by ice and logs, for which discharge was obtained by applying to rating table mean daily gage heights corrected for backwater by means of discharge measurements, observer's notes, and weather records. Open-water records excellent except during July and September, when logs were in river, for which period they are fair; winter records fair.

Discharge measurements of Flambeau River near Ladysmith, Wis., during the year ending Sept. 30, 1918.

[Made by T. G. Bedford.]

Date.	Gage height.	Discharge.	Date.	Gage height.	Discharge.
	Feet.	*Sec.-ft.*		*Feet.*	*Sec.-ft.*
Dec. 22[a]	3.85	646	Feb. 25[a]	4.05	546
Jan. 23[a]	4.00	607	May 18	3.46	2,280

[a] Complete ice cover at control and measuring section.

Daily discharge, in second-feet, of Flambeau River near Ladysmith, Wis., for the year ending Sept. 30, 1918.

Day.	Oct.	Nov.	Dec.	Jan.	Feb.	Mar.	Apr.	May.	June.	July.	Aug.	Sept.
1	1,000	1,620	880	620	580	540	1,340	1,910	8,960	920	1,620	1,340
2	1,000	1,560	870	620	580	540	1,400	1,790	9,520	740	920	1,450
3	960	1,560	860	620	580	550	1,450	1,670	8,400	920	1,000	1,340
4	880	1,560	840	580	580	550	1,560	1,670	7,880	1,090	1,090	1,560
5	1,000	1,560	820	620	580	560	1,670	1,670	8,400	920	1,000	1,240
6	960	1,500	810	620	580	560	1,790	2,030	5,800	1,080	770	1,2[illegible]
7	960	1,500	800	620	580	570	1,910	2,150	4,140	1,240	740	
8	840	1,500	780	620	590	570	2,150	2,390	3,970	920	1,080	
9	840	1,450	770	620	590	580	1,910	2,510	3,640	960	1,160	
10	880	1,340	750	620	600	580	1,620	3,330	3,180	1,040	1,240	
11	920	1,340	740	610	600	590	1,340	3,800	2,640	1,000	1,670	
12	920	1,340	730	610	610	600	2,030	4,480	2,640	920	1,790	
13	1,080	1,400	720	610	610	610	1,790	3,480	2,510	880	1,560	
14	1,240	1,000	710	610	620	620	1,160	3,180	2,510	1,160	1,290	
5	1,240	1,340	700	610	620	620	1,670	4,140	2,150	1,120	1,160	
16	1,340	1,290	690	610	620	640	1,670	2,900	1,670	840	1,080	
17	1,290	1,400	680	610	610	660	1,670	2,900	1,620	1,160	1,000	
18	1,340	1,240	670	600	610	680	1,910	2,510	1,450	1,160	1,000	
19	1,670	1,340	670	600	600	710	1,500	2,390	1,500	1,160	960	870
10	2,900	1,240	660	600	600	740	1,500	1,910	1,080	1,160	880	
22	2,770	1,080	650	600	590	760	1,560	2,510	960	1,050	840	
21	2,510	1,240	640	600	580	770	1,670	2,640	1,080	1,050	960	
23	2,150	1,240	640	600	570	840	1,620	2,510	1,000	1,050	1,290	
24	2,030	1,160	630	600	560	880	1,240	2,770	1,000	1,050	1,340	
25	1,910	1,240	630	600	550	920	1,560	2,770	758	1,160	1,340	
26	1,790	1,040	620	590	550	1,000	1,500	4,140	920	1,160	1,160	
27	2,150	920	620	590	550	1,040	1,560	5,210	920	1,160	1,160	
28	1,910	920	620	590	540	1,080	1,560	5,600	920	1,450	1,120	
29	2,150	920	620	580	......	1,160	1,450	6,000	880	1,670	1,240	
30	2,030	880	620	580	......	1,200	1,790	6,000	920	1,080	1,340	
31	1,670	......	620	580	......	1,240	......	6,220	......	1,120	1,340	......

NOTE.—Stage-discharge relation affected by ice, Nov. 28 to Apr. 6; by logs July 21–25 and Sept. 7 to 30. Gage assumed as reading 1 foot too high Aug. 31 to Sept. 3.

Monthly discharge of Flambeau River near Ladysmith, Wis., for the year ending Sept. 30, 1918.

[Drainage area, 1,940 square miles.]

Month.	Discharge in second-feet.				Run-off (depth in inches).
	Maximum.	Minimum.	Mean.	Per square mile.	
October	2,900	840	1,490	0.768	0.89
November	1,620	880	1,290	.665	.74
December	880	620	712	.367	.42
January	620	580	606	.312	.36
February	620	540	587	.303	.32
March	1,240	540	741	.382	.44
April	2,150	1,160	1,620	.835	.93
May	6,220	1,670	3,200	1.65	1.90
June	9,520	758	3,100	1.60	1.78
July	....	740	1,070	.552	.64
August	1,790	740	1,160	.598	.69
September	....	....	968	.499	.56
The year	....	....	1,380	.711	9.67

JUMP RIVER AT SHELDON, WIS.

LOCATION.—In sec. 26, T. 33 N., R. 5 W., at highway bridge in Sheldon, Rusk County, 11 miles above confluence of Jump and Chippewa rivers.

DRAINAGE AREA.—510 square miles (measured on map issued by Wisconsin Geological and Natural History Survey, edition of 1911; scale, 1 inch = 6 miles).

RECORDS AVAILABLE.—July 22, 1915, to September 30, 1918.

GAGE.—Chain gage bolted to downstream handrail of bridge.

DISCHARGE MEASUREMENTS.—Made from downstream side of bridge.

CHANNEL AND CONTROL.—Bed composed of heavy gravel, clean, and free from vegetation. Right bank high and not subject to overflow; left bank may be overflowed occasionally.

EXTREMES OF DISCHARGE.—Maximum stage recorded during year, 8.95 feet May 27 (discharge, 7,800 second-feet); minimum discharge, estimated 15 second-feet Feb. 3 and 4.

1915–1918: Maximum discharge during period, 8,600 second-feet April 22, 1916; minimum discharge approximately 15 second-feet Feb. 3–4, 1918.

ACCURACY.—Stage-discharge relation permanent except as affected by ice. Rating curve well defined between 45 and 5,930 second-feet. Gage read to quarter-tenths twice daily. Daily discharge ascertained by applying mean daily gage height to rating table except for period in which stage-discharge relation was affected by ice, for which it was obtained by applying to rating table mean daily gage heights corrected for ice effect by means of discharge measurements, observer's notes, and weather records. Open-water records good; winter records fair.

Discharge measurements of Jump River at Sheldon, Wis., during the year ending Sept. 30, 1918.

[Made by T. G. Bedford.]

Date.	Gage height.	Discharge.	Date.	Gage height.	Discharge.
	Feet.	*Sec.-ft.*		*Feet.*	*Sec.-ft.*
Dec. 23 [a]	3.58	42	Feb. 26 [a]	3.90	31
Jan. 24 [a]	3.54	26	May 20	3.80	436

[a] Complete ice cover at control and measuring section.

Daily discharge, in second-feet, of Jump River at Sheldon, Wis., for the year ending Sept. 30, 1918.

Day.	Oct.	Nov.	Dec.	Jan.	Feb.	Mar.	Apr.	May.	June.	July.	Aug.	Sept.
1	148	405	85	35	20	35	1,020	1,020	7,280	60	126	760
2	148	355	85	35	20	35	930	720	6,850	70	122	485
3	133	330	80	35	15	40	840	610	4,540	70	133	355
4	122	305	75	30	15	40	680	540	2,950	70	102	330
5	164	330	70	30	15	45	575	485	1,980	84	88	230
6	235	355	65	30	15	45	485	458	1,400	84	105	200
7	305	330	60	30	15	50	540	458	1,110	70	164	172
8	210	305	55	30	20	50	680	512	840	70	575	148
9	190	330	50	30	20	55	575	575	645	65	1,300	136
10	172	305	45	30	25	60	485	1,620	540	48	1,200	122
11	172	280	40	30	30	65	458	1,860	430	45	800	133
12	185	260	40	30	30	70	405	1,620	355	39	540	156
13	305	250	35	30	30	80	380	1,200	280	38	575	255
14	330	230	30	30	30	90	355	885	240	39	610	235
15	280	230	30	30	30	105	330	720	190	44	485	210
16	270	220	30	30	30	120	355	575	148	68	355	190
17	305	205	30	30	30	130	430	512	133	50	260	180
18	430	200	30	25	25	140	485	458	126	45	176	220
19	1,110	185	40	25	20	180	610	458	108	45	140	610
20	1,110	176	50	25	20	230	575	430	98	42	126	680
21	885	172	60	25	20	540	512	430	88	39	122	575
22	720	172	70	25	25	1,300	485	540	77	36	380	458
23	610	150	70	25	30	2,370	430	760	77	38	2,510	355
24	540	145	60	25	30	2,110	380	645	68	68	2,110	330
25	485	140	50	25	30	1,860	330	1,860	60	74	1,400	355
26	485	130	50	20	30	1,620	305	7,230	58	70	1,020	330
27	645	120	40	20	35	1,510	280	7,800	50	77	512	240
28	720	110	40	20	35	1,300	355	6,660	50	176	540	225
29	645	105	35	20	......	1,200	575	5,750	48	148	1,400	180
30	575	95	35	20	......	1,200	1,020	5,220	50	148	1,510	185
31	458	......	35	20	......	1,110	......	5,220	......	133	1,110	......

NOTE.—Stage-discharge relation affected by ice Nov. 23 to Mar. 28.

Monthly discharge of Jump River at Sheldon, Wis., for the year ending Sept. 30, 1918.

[Drainage area, 510 square miles.]

Month.	Discharge in second-feet.				Run-off (depth in inches).
	Maximum.	Minimum.	Mean.	Per square mile.	
October	1,110	122	422	.827	0.96
November	405	95	231	.453	.51
December	85	30	50.6	.0992	.11
January	35	20	27.3	.0535	.06
February	35	15	24.6	.0482	.05
March	2,370	35	574	1.13	1.30
April	1,020	280	529	1.04	1.16
May	7,800	430	1,870	3.67	4.23
June	7,230	48	1,030	2.02	2.25
July	176	36	69.5	.136	.16
August	2,510	88	664	1.30	1.50
September	760	122	301	.590	.66
The year	7,800	15	486	.953	12.94

EAU CLAIRE RIVER NEAR AUGUSTA, WIS.

LOCATION.—In sec. 12, T. 26 N., R. 6 E., at Trouble Water Bridge, 7 miles northeast of Augusta, Eau Claire County. South Fork of Eau Claire River enters from left 4 miles above station.

DRAINAGE AREA.—500 square miles (measured on map issued by Wisconsin Geological and Natural History Survey, edition of 1911; scale, 1 inch = 6 miles).

RECORDS AVAILABLE.—July 16, 1914, to September 30, 1918.

GAGE.—Chain gage on downstream side of bridge; read by Albert Wagner.

DISCHARGE MEASUREMENTS.—Made from downstream side of bridge or by wading at control about 500 feet downstream from bridge.

CHANNEL AND CONTROL.—Bed at bridge and above is sandy and very shifting. A short distance below the gage the channel narrows and a rock outcrop overlain with large boulders forms the control. Banks are high and not subject to overflow.

EXTREMES OF DISCHARGE.—Maximum open-water stage recorded during year, 9.1 feet at 8 a. m. May 27 (discharge, 5,620 second-feet); minimum discharge, estimated 35 second-feet, from discharge measurements made January 27, 1918.

1914–1918: Maximum open-water stage recorded, 10.6 feet at noon April 1, 1916 (discharge, 7,180 second-feet); minimum open-water stage recorded, 0.10 foot September 2, 1916 (discharge, 40 second-feet); minimum discharge, estimated 35 second-feet, January 27, 1918.

ACCURACY.—Stage-discharge relation practically permanent except as affected by ice Rating curve well defined from 69 to 5,520 second-feet, poorly defined outside these limits. Gage read to quarter-tenths once a day. Daily discharge ascertained by applying daily gage height to rating curve, except for period in which the stage-discharge relation was affected by ice, for which it was obtained by applying to rating table mean daily gage heights corrected for ice effect by means of discharge measurements, observer's notes, and weather records. Open-water records good, except for low stages for which they are fair; winter records fair.

Discharge measurements of Eau Claire River near Augusta, Wis., during the year ending Sept. 30, 1918.

[Made by T. G. Bedford.]

Date.	Gage height.	Discharge.	Date.	Gage height.	Discharge.
	Feet.	*Sec.-ft.*		*Feet.*	*Sec.-ft.*
Dec. 27 [a]	0.95	41	May 10	4.38	1,730
Jan. 27 [a]	2.18	3	Sept. 3 [b]	.28	68

[a] Complete ice cover at control and measuring section.
[b] Made by wading 500 feet downstream from gage.

Daily discharge, in second-feet, of Eau Claire River near Augusta, Wis., for the year ending Sept. 30, 1918.

Day.	Oct.	Nov.	Dec.	Jan.	Feb.	Mar.	Apr.	May.	June.	July.	Aug.	Sept.
1	83	235	87				585	655	4,660	134	87	73
2	69	207	107				550	516	4,370	129	73	69
3	78	201	87			15	466	417	2,430	111	69	62
4	69	201	78				417	323	1,630	118	69	62
5	87	179					353	298	1,120	107	66	66
6	83	179					353	278	930	103	62	66
7	78	174					466	263	1,120	97	66	62
8	78	166				20	655	235	845	87	87	62
9	97	153					533	338	1,810	87	148	62
10	73	141					449	2,220	1,690	87	166	66
11	87	129				25	385	2,290	1,020	83	125	78
12	111	129				30	338	1,570	690	78	118	83
13	129	129				40	323	930	466	73	249	83
14	129	125				45	308	620	417	73	338	78
15	118	118			15	55	293	499	323	78	179	69
16	118	107		20		80	293	466	278	134	141	66
17	107	107	55			85	353	369	229	153	107	66
18	129	107				235	433	323	235	120	118	78
19	207	103				1,130	620	308	221	97	87	87
20	249	97				2,760	550	323	193	87	78	111
21	221	107				2,520	482	323	193	78	78	107
22	201	107				1,960	499	449	166	73	118	97
23	166	125				1,760	449	765	141	73	174	91
24	153	107				1,510	369	620	141	83	158	83
25	174	118				1,460	323	499	134	87	118	78
26	235	129				1,220	293	3,710	129	83	97	69
27	401	97				885	263	5,620	118	78	83	69
28	499	87				805	278	4,750	111	83	87	66
29	369	97				690	620	3,620	107	134	87	66
30	308	87			......	620	845	2,430	125	118	87	66
31	278	......			......	620	......	2,860	......	91	83	......

NOTE.—Stage-discharge relation affected by ice Dec. 5 to Mar. 25.

Monthly discharge of Eau Claire River near Augusta, Wis., for the year ending Sept. 30, 1918.

[Drainage area, 500 square miles.]

Month.	Discharge in second-feet.				Run-off (depth in inches).
	Maximum.	Minimum.	Mean.	Per square mile.	
October	499	69	167	0.334	0.39
November	235	87	135	.270	.30
December	107	....	59.5	.119	.14
January	....	....	20	.040	.05
February	....	....	15	.030	.03
March	2,760	....	604	1.21	1.40
April	845	263	438	.876	.98
May	5,620	235	1,250	2.50	2.88
June	4,660	107	868	1.74	1.94
July	153	73	97.3	.195	.22
August	338	62	116	.232	.27
September	111	62	74.7	.149	.17
The year	5,620	....	323	.646	8.77

RED CEDAR RIVER NEAR COLFAX, WIS.

LOCATION.—In sec. 27, T. 30 N., R. 11 W., at highway bridge 4½ miles north of Colfax, Dunn County. Hay River enters from right 11 miles below station, and Trout Creek, also from right, 3½ miles above.

DRAINAGE AREA.—1,100 square miles (measured on map issued by Wisconsin Geological and Natural History Survey, edition of 1911; scale, 1 inch=6 miles).

RECORDS AVAILABLE.—March 10, 1914, to September 30, 1918.

GAGE.—Chain gage attached to downstream side of bridge; read by Andrew Lundeguam.

DISCHARGE MEASUREMENTS.—Made from downstream side of bridge to which gage is attached.

CHANNEL AND CONTROL.—Bed composed of rock and gravel; small amount of grass growth during summer months. Left bank high and not subject to overflow; right bank medium high and may be overflowed during extremely high water. Control not well defined.

EXTREMES OF DISCHARGE.—Maximum stage recorded during year, 4.05 feet June 1 (discharge, 3,180 second-feet); minimum discharge recorded, 368 second-feet, February 19 (by current-meter measurement).

1914–1918: Maximum stage recorded during period, 6.8 feet at 1 p. m., March 31, 1916 (discharge, 6,990 second-feet); minimum stage recorded 0.80 foot November 19, 1914 (discharge, about 385 second-feet); apparently caused by temporary holding back of the water by ice. Discharge measurement made February 19, 1918, gave a discharge of 368 second-feet.

REGULATION.[4]—The following dams and reservoirs are used to regulate the flow in Red Cedar River. Owing to operation of these reservoirs the flow at station is not natural.

Dam.	Location.	Approximate capacity (millions of cubic feet).
Long Lake	Sec. 24, T. 37 N., R. 11 W	1,008
Cedar Lake	Sec. 21, T. 36 N., R. 10 W	965
Birch Lake	Sec. 25, T. 37 N., R. 10 W	1,174
Bear Lake	Sec. 7, T. 36 N., R. 11 W	280
Chetek Lake	Sec. 20, T. 33 N., R. 10 W	990
		4,417

ACCURACY.—Stage-discharge relation nearly permanent, except as affected by ice, and possibly by grass from June to September. Rating curve well defined between 653 and 4,450 second-feet; curve extended and approximate only outside these limits. Gage read twice daily to quarter-tenths. Daily discharge ascertained by applying mean daily gage height to rating table, except for period in which stage-discharge relation was affected by ice, for which it was ascertained by applying to rating table mean daily gage heights corrected for ice effect by means of discharge measurements, observer's notes, and weather records. Open-water records good; winter records subject to error.

[4] From data on file in Engineering Department of Railroad Commission of Wisconsin.

Discharge measurements of Red Cedar River near Colfax, Wis., during the year ending Sept. 30, 1918.

[Made by T. G. Bedford.]

Date.	Gage height.	Discharge.	Date.	Gage height.	Discharge.
	Feet.	*Sec.-ft.*		*Feet.*	*Sec.-ft.*
Dec. 17[a]	2.17	522	Feb. 19[b]	2.83	368
Jan. 17[b]	3.09	490	May 14	1.45	660

[a] Made from bridge and ice, incomplete ice cover at control section.
[b] Complete ice cover at control and measuring section.

Daily discharge, in second-feet, of Red Cedar River near Colfax, Wis., for the year ending Sept. 30, 1918.

Day.	Oct.	Nov.	Dec.	Jan.	Feb.	Mar.	Apr.	May.	June.	July.	Aug.	Sept.
1	490	820	585				890	635	3,120	450	560	720
2	535	750	635				820	585	2,880	490	585	635
3	512	690	560				785	535	2,880	490	512	662
4	535	690	635				750	490	2,880	535	535	690
5	490	635	350	520	505	770	690	490	1,680	535	490	720
6	490	690					785	560	1,300	690	535	750
7	470	690					820	585	1,040	560	585	720
8	470	690					820	635	1,210	470	690	690
9	535	610					750	855	1,210	490	610	690
10	512	585	455				720	1,040	1,120	535	585	750
11	512	690					690	890	925	512	535	855
12	490	585					690	850	820	490	585	785
13	490	662					662	690	750	490	610	820
14	490	690					662	635	690	450	585	750
15	512	690		510	460		635	635	720	450	585	690
16	635	662				1,430	610	690	635	490	535	635
17	662	690					635	610	610	490	585	690
18	820	635					635	635	635	490	635	690
19	1,040	512					635	635	635	490	585	750
20	820	662					585	585	610	535	560	785
21	820	662					585	585	610	490	585	750
22	750	662				2,200	585	635	585	490	690	750
23	750	635				2,200	662	585	585	535	610	635
24	690	690	540			1,680	610	535	535	960	585	720
25	635	690			440	1,780	635	750	512	690	560	720
26	750	490		535		1,580	635	1,580	512	560	560	750
27	750	690				925	635	3,120	490	512	560	720
28	750	662				750	635	3,120	490	720	610	662
29	610	585			......	820	690	1,980	490	585	635	560
30	690	535			......	820	662	1,480	535	585	720	490
31	785	......			......	785	......	1,780	......	535	720	......

NOTE.—Stage-discharge relation affected by ice Dec. 6 to Mar. 21.

Monthly discharge of Red Cedar River near Colfax, Wis., for the year ending Sept. 30, 1918.

[Drainage area, 1,100 square miles.]

Month.	Discharge in second-feet.				Run-off (depth in inches).
	Maximum.	Minimum.	Mean.	Per square mile.	
October	1,040	470	629	0.572	0.66
November	820	490	654	.595	.66
December	....	....	515	.468	.54
January	....	....	522	.475	.55
February	....	....	470	.427	.44
March	....	....	1,190	1.08	1.24
April	890	585	687	.625	.70
May	3,120	490	949	.863	.99
June	3,120	490	1,060	.964	1.08
July	960	450	542	.493	.57
August	720	490	591	.537	.62
September	855	490	708	.644	.72
The year	3,120	....	711	.646	8.77

RED CEDAR RIVER AT CEDAR FALLS, WIS.

Location.—In sec. 6, T. 28 N., R. 12 W., at highway bridge near Cedar Falls, Dunn County, 4½ miles above crossing of Chicago, St. Paul, Minneapolis & Omaha Railway.

Drainage area.—Not measured.

Records available.—April 1, 1909, to September 30, 1918.

Gage.—Staff gage fastened to bridge pier; read by John G. Wood.

Discharge measurements.—No discharge measurements have been made at this station, which is maintained to determine fluctuation in stage.

Channel and control.—Channel rough and rocky, straight, and free from vegetation. Banks high and not subject to overflow.

Extremes of stage.—Maximum stage recorded during year, 5.15 feet March 19; minimum stage, 1.2 feet, 12 noon October 21.

1909–1918: Maximum stage recorded, 6.1 feet April 1–3, 1916; minimum stage recorded 0.0 foot at 5 p. m. March 11, 1917. Minimum stages are caused by closing gates and wheels in dam above station.

Regulation.—The operation of storage reservoirs in the headwaters of the river (see "Regulation" in station description for Red Cedar River at Colfax, Wis.), together with storage at power plant above gaging station, regulate the flow.

Accuracy.—No measurements have been made, but stage-discharge relation believed permanent. Gage read twice daily to half-tenths. Considerable diurnal fluctuation is observed, so that mean daily gage heights does not represent the average stage.

Cooperation.—Gage-height record furnished by Wisconsin & Minnesota Light & Power Co.

Daily gage height, in feet, of Red Cedar River at Cedar Falls, Wis., for the year ending Sept. 30, 1918.

Day.	Oct.	Nov.	Dec.	Jan.	Feb.	Mar.	Apr.	May.	June.	July.	Aug.	Sept.
1	2.4	2.6	2.7	2.4	2.8	2.5	3.3	2.65	4.65	2.55	2.6	2.35
2	2.55	2.55	1.55	3.1	2.7	2.35	3.1	2.6	4.85	2.6	2.6	1.4
3	2.5	2.65	2.65	3.65	1.4	1.4	3.0	2.6	4.75	2.05	2.35	2.55
4	2.6	1.55	2.65	3.65	2.6	2.45	3.2	2.6	4.5	1.4	1.75	2.6
5	2.5	2.6	2.5	3.35	2.75	3.4	3.15	1.9	4.3	2.45	2.15	2.6
6	2.45	2.55	2.65	1.6	2.8	3.65	3.15	2.5	3.95	2.5	2.55	2.6
7	2.05	2.6	2.45	2.35	2.7	3.7	3.0	2.6	3.85	1.4	2.65	2.65
8	2.6	2.6	2.55	2.3	2.9	3.6	3.1	2.75	3.95	2.55	2.6	2.0
9	2.55	2.55	1.7	2.4	2.7	3.6	3.1	3.2	3.75	2.6	3.35	3.05
10	2.65	2.65	2.45	3.1	1.4	1.9	3.2	3.55	3.8	2.45	2.55	2.55
11	2.65	2.0	2.45	3.5	2.55	2.6	3.25	2.9	3.7	2.55	2.35	2.45
12	2.95	2.6	2.5	2.65	2.9	3.65	3.15	2.5	3.55	2.15	2.65	2.6
13	2.55	2.8	2.65	2.25	2.65	3.5	3.15	2.65	3.55	2.45	2.55	2.45
14	1.8	2.55	2.7	3.05	2.6	3.55	2.9	2.55	3.7	2.25	2.5	2.5
15	2.55	2.6	2.6	2.75	2.7	2.7	3.3	2.5	3.75	2.4	2.6	1.9
16	2.6	3.25	2.25	2.65	2.75	2.65	3.65	2.8	2.0	2.75	2.65	2.7
17	2.5	3.25	2.65	2.55	1.4	1.9	3.0	2.6	2.85	2.45	2.6	2.55
18	2.65	1.85	2.65	3.05	2.7	4.15	2.85	2.65	2.75	2.5	1.4	3.15
19	2.55	2.6	2.8	3.45	2.7	5.15	2.85	1.95	2.7	2.4	2.6	2.6
20	2.4	2.65	2.7	2.2	2.6	5.4	2.7	2.6	2.8	2.45	2.4	2.6
21	1.55	2.55	2.65	2.15	2.6	5.2	1.8	2.6	2.8	1.9	2.65	2.45
22	2.6	2.65	2.85	2.55	2.5	4.9	2.85	2.8	2.55	2.35	2.35	1.4
23	2.6	2.75	1.9	2.9	2.35	4.6	2.75	2.95	2.05	2.45	2.4	2.6
24	2.55	2.65	2.5	2.9	1.4	3.4	2.65	3.25	2.65	2.25	2.45	2.6
25	2.6	2.6	2.0	2.95	2.6	3.55	2.65	2.6	2.55	2.6	1.8	2.35
26	2.45	2.8	3.45	2.8	2.5	3.65	2.75	3.05	2.45	2.55	2.45	2.55
27	2.45	2.65	3.1	1.4	2.6	3.5	2.55	4.35	2.6	2.45	3.7	2.45
28	1.85	2.6	3.2	2.5	2.35	3.3	1.9	4.6	2.3	1.95	2.55	2.55
29	2.6	2.05	3.25	2.85	......	3.25	2.65	4.6	1.85	2.8	2.7	1.4
30	2.55	3.05	1.9	2.7	......	3.05	2.75	4.45	1.4	2.6	2.8	2.6
31	2.65	......	2.6	2.7	......	2.8	......	4.25	......	2.55	2.4	......

RED CEDAR RIVER AT MENOMONIE, WIS.

LOCATION.—In sec. 21, T. 28 N., R. 13 W., 900 feet below power house of Wisconsin & Minnesota Light & Power Co., Menomonie, Dunn County, and 13 miles above confluence of Red Cedar and Chippewa rivers. Wilson Creek discharges from right into service reservoir, just above station.

DRAINAGE AREA.—1,810 square miles (measured on map issued by Wisconsin Geological and Natural History Survey, edition of 1911; scale, 1 inch=6 miles).

RECORDS AVAILABLE.—June 16, 1907, to September 5, 1908; May 9, 1913, to September 30, 1918.

GAGE.—Barrett & Lawrence water-stage recorder installed May 9, 1913, over a wooden well on right bank of river, 1 mile above site of old gage, which was attached to a highway bridge about 200 rods west of Chicago & North Western Railway station west of Menomonie; read from June 16, 1907, to September 5, 1908. No relation between datums of the two gages. Gage inspected by E. Kausrud.

DISCHARGE MEASUREMENTS.—Made from highway bridge, about 1 mile below gage.

CHANNEL AND CONTROL.—Bed at gage composed of heavy gravel; bed at measuring section sandy and liable to shift. Left bank at gage high and not subject to overflow; right bank of medium height and will be overflowed at flood stages; both banks high at measuring section and not subject to overflow.

EXTREMES OF DISCHARGE.—Maximum stage recorded during year, approximately 6.05 feet March 20 (discharge, 7.570 second-feet); minimum stage, 1.65 feet at midnight July 22 (discharge, about 220 second-feet).

1907–8 and 1913–1918: Maximum discharge, 12.700 second-feet March 31 and April 1, 1916; minimum discharge, 100 second-feet November 9, 1907.

REGULATION.—Considerable diurnal fluctuation in stage at gage section is caused by operation of power plants of Wisconsin & Minnesota Light & Power Co. at Menomonie and Cedar Falls. (See "Regulation" in station description for Red Cedar River at Colfax, Wis.)

ICE.—Stage-discharge relation not affected by ice.

ACCURACY.—Stage-discharge relation changed during high water of April, 1916, but has been fairly permanent since with ordinary conditions of flow. Rating curve used well defined between 610 and 1,910 second-feet, and between 3,910 and 9,220 second-feet. Curve extended outside these limits and approximate only. Water-stage recorder gave satisfactory results except for brief periods. Daily discharge records October 1 to September 30, except for brief periods, obtained with Fuller discharge integrator. Records good except for periods when gage was not in operation, for which they are only approximate. Ice does not affect the stage-discharge relation at this station, due to relatively warm water coming from service reservoir.

The following discharge measurement was made by T. G. Bedford·
Gage height, 2.55 feet; discharge, 933 second-feet May 11, 1918.

Daily discharge, in second-feet, of Red Cedar River at Menomonie, Wis., for the year ending Sept. 30, 1918.

Day.	Oct.	Nov.	Dec.	Jan.	Feb.	Mar.	Apr.	May.	June.	July.	Aug.	Sept.
1	695	1,030	1,170	635	1,100	845	1,260	1,110	2,400	460	975	1,100
2	930	1,010	750	1,080	930	890	1,250	640	3,870	690	1,000	770
3	980	1,040	820	1,140	550	900	1,030	870	4,000	550	900	735
4	995	700	1,020	1,210	630	1,000	1,130	835	3,090	470	550	970
5	1,160	770	940	990	1,010	1,390	1,020	490	2,080	535	695	1,120
6	935	935	935	535	1,160	2,150	1,000	670	1,690	575	640	995
7	812	970	900	510	950	1,730	685	820	1,500	480	840	990
8	920	1,090	790	560	1,160	1,760	1,180	820	1,580	565	780	855
9	1,120	1,090	500	620	1,120	1,580	1,160	925	1,240	630	980	585
10	1,020	1,040	640	840	530	1,160	1,040	1,050	1,810	625	1,070	850
11	1,190	615	940	820	775	1,760	1,010	770	1,560	740	760	1,050
12	1,120	715	875	1,220	965	1,020	1,090	640	1,470	735	945	870
13	1,040	1,000	865	690	1,110	1,780	950	1,010	1,220	500	1,030	825
14	670	835	825	870	970	1,730	515	1,160	1,170	455	830	820
15	925	935	915	1,010	1,120	1,370	1,320	880	1,160	600	905	505
16	1,010	1,120	600	1,060	1,160	985	1,330	905	1,110	705	845	850
17	1,040	1,420	730	920	610	900	1,410	885	872	770	710	880
18	975	815	1,070	1,030	800	2,320	1,160	905	1,330	705	440	1,020
19	1,020	720	1,030	1,140	1,090	4,600	650	630	1,260	700	565	1,240
20	805	880	1,170	1,080	1,080	6,970	865	775	1,220	760	620	1,290
21	500	885	1,180	1,020	1,100	5,950	540	1,000	1,080	605	715	1,180
22	775	935	1,090	940	840	4,890	760	1,040	1,060	1,210	775	670
23	905	930	820	870	855	3,120	890	1,120	835	785	925	870
24	1,010	940	935	1,120	470	2,700	880	955	895	730	755	1,030
25	910	915	695	1,120	715	1,720	740	830	1,050	935	620	960
26	1,120	1,090	725	1,120	840	1,560	745	890	820	635	615	1,070
27	985	1,070	1,070	490	1,060	1,570	770	3,100	930	960	750	1,030
28	875	945	1,030	520	875	1,390	450	3,420	830	735	870	1,010
29	975	785	1,110	930	....	1,290	935	2,980	530	825	1,040	785
30	925	1,100	835	1,160	....	1,240	1,160	2,280	425	1,090	1,090	605
31	1,030	....	725	1,160	....	850	....	2,490	....	1,040	1,480	....

Monthly discharge of Red Cedar River at Menomonie, Wis., for the year ending Sept. 30, 1918.

[Drainage area, 1,810 square miles.]

Month.	Discharge in second-feet.				Run-off (depth in inches).
	Maximum.	Minimum.	Mean.	Per square mile.	
October	1,190	500	947	0.523	0.60
November	1,420	615	944	.522	.58
December	1,180	500	893	.493	.57
January	1,220	490	916	.506	.58
February	1,160	470	913	.504	.52
March	6,970	845	2,040	1.13	1.30
April	1,410	450	964	.533	.59
May	3,420	490	1,190	.657	.76
June	4,000	425	1,470	.812	.91
July	1,210	455	703	.388	.45
August	1,480	440	830	.459	.53
September	1,290	505	917	.507	.57
The year	6,970	425	1,060	.586	7.96

TREMPEALEAU RIVER AT DODGE, WIS.

LOCATION.—In sec. 11, T. 19 N., R. 10 W., at highway bridge in Dodge, Trempealeau County, 9 miles above mouth of river.

DRAINAGE AREA.—633 square miles (measured on map issued by Wisconsin Geological and Natural History Survey, edition of 1911; scale, 1 inch=6 miles).

RECORDS AVAILABLE.—December 13, 1913, to September 30, 1918.

GAGE.—Chain gage attached to downstream side of bridge; read by F. E. Shappee and M. W. MacDonald.

DISCHARGE MEASUREMENTS.—Made from downstream side of bridge or by wading.

CHANNEL AND CONTROL.—Bed composed of sand; likely to shift. Banks of medium height and may be overflowed during extreme floods.

EXTREMES OF DISCHARGE.—Maximum stage recorded during year, 8.85 feet at 5 p. m. March 20 (discharge, roughly approximate, 3,360 second-feet); minimum discharge, about 105 second-feet, February 4 and 5.

1914–1917: Maximum stage recorded, 8.35 feet June 9, 1914 (discharge, 3,340 second-feet); minimum discharge, about 105 second-feet, February 4–5, 1918.

ICE.—Stage-discharge relation seriously affected by ice.

REGULATION.—No power plants above station have sufficient capacity to affect natural flow of river.

ACCURACY.—Stage-discharge relation not permanent. A rating curve, fairly well defined between 196 and 3,080 second-feet, was used October 1 to March 10, shifting-channel method used March 11 to September 30. Gage read twice daily to quarter-tenths, except on Sundays, April 14 to September 30. Daily discharge ascertained by applying mean daily gage height to rating table, except during period when stage-discharge relation was affected by ice, for which it was obtained by applying to rating table mean daily gage height corrected for ice effect by means of discharge measurements, observer's notes, and weather records, and except for days when no reading of gage was taken, for which the discharge was interpolated. Records fair.

Discharge measurements of Trempealeau River at Dodge, Wis., during the year ending Sept. 30, 1918.

Date.	Made by—	Gage height.	Discharge.	Date.	Made by—	Gage height.	Discharge.
		Feet.	*Sec.-ft.*			*Feet.*	*Sec.-ft.*
Oct. 8[a]	R. B. Kilgore	1.52	206	Apr. 1	T. G. Bedford	3.03	416
Jan. 14[b]	T. G. Bedford	2.38	146	Sept. 4	W. G. Hoyt	1.37	211
Feb. 15[b]	.....do	3.52	249				

[a] Made by wading 200 feet downstream from gage.
[b] Complete ice cover at control and measuring section.

Daily discharge, in second-feet, of Trempealeau River at Dodge, Wis., for the year ending Sept. 30, 1918.

Day.	Oct.	Nov.	Dec.	Jan.	Feb.	Mar.	Apr.	May.	June.	July.	Aug.	Sept.
1	196	309	257	165	115	980	420	272	747	272	224	197
2	220	309	257	175	110	980	408	248	682	296	236	213
3	220	309	220	170	110	980	333	224	616	320	260	224
4	220	296	196	165	105	1,040	358	202	485	358	248	221
5	220	283	190	170	105	1,090	358	202	433	308	236	213
6	220	283	180	170	110	1,060	333	202	420	272	224	191
7	220	270	175	175	110	875	358	202	396	266	236	181
8	186	257	170	175	115	695	358	136	370	260	248	192
9	220	283	160	175	115	615	358	383	1,050	248	260	202
10	232	283	155	165	120	565	333	433	1,730	236	248	224
11	270	270	145	160	125	537	320	511	2,980	224	254	236
12	296	244	135	155	130	603	296	447	2,400	213	260	248
13	296	270	155	150	140	747	296	383	1,580	202	272	248
14	283	270	170	145	190	890	284	308	864	213	296	213
15	257	257	170	145	250	982	272	308	616	224	272	213
16	270	257	170	140	260	942	296	260	682	272	248	213
17	270	257	170	145	285	903	272	284	747	272	236	225
18	296	244	170	145	310	1,160	296	320	459	248	230	248
19	309	244	170	140	335	2,660	320	296	433	224	224	248
20	309	270	215	135	360	3,280	333	272	408	202	202	236
21	296	257	205	135	385	2,910	308	296	383	208	236	213
22	270	244	205	135	410	2,260	284	320	358	213	272	213
23	309	244	205	135	435	1,520	260	446	352	236	284	213
24	296	244	165	135	460	1,090	260	396	346	202	284	213
25	309	244	165	130	510	773	236	420	320	224	266	202
26	426	220	155	130	615	642	236	800	320	213	248	191
27	426	244	150	130	825	616	213	1,180	296	202	224	181
28	426	244	155	125	1,010	537	236	1,120	272	237	224	181
29	374	220	155	125	......	511	260	1,010	260	272	202	176
30	348	232	150	120	......	459	296	773	266	248	191	171
31	322	......	150	120	......	433	......	642	......	236	181	......

NOTE.—Stage-discharge relation affected by ice Dec. 5 to Mar. 10. Gage not read Apr. 14, 21, 28, May 5, 12, 19, 26, June 2, 9, 16, 23, 30, July 6, 13, 21, 28, Aug. 4, 11, 18, 25, Sept. 1, 8, 15, 22, 29; discharge interpolated.

Monthly discharge of Trempealeau River at Dodge, Wis., for the year ending Sept. 30, 1918.

[Drainage area, 633 square miles.]

Month.	Discharge in second-feet.				Run-off (depth in inches).
	Maximum.	Minimum.	Mean.	Per square mile.	
October	426	186	284	0.449	0.52
November	309	220	262	.414	.46
December	257	135	177	.280	.32
January	175	120	148	.234	.27
February	1,010	105	291	.460	.48
March	3,280	433	1,080	1.71	1.97
April	420	213	306	.483	.54
May	1,180	136	429	.678	.78
June	2,980	260	709	1.12	1.25
July	358	202	246	.389	.45
August	296	181	243	.384	.44
September	248	171	211	.333	.37
The year	3,280	105	366	.578	7.85

BLACK RIVER AT NEILLSVILLE, WIS.

Location.—In sec. 15, T. 24 N., R. 2 W., at lower highway bridge in Neillsville, Clark County. O'Neil Creek enters from left 1 mile above gage and Cunningham Creek, also from left, 1½ miles below.

Drainage area.—774 square miles (measured on map issued by Wisconsin Geological and Natural History Survey, edition of 1911; scale, 1 inch=6 miles).

Records available.—April 7, 1905, to March 31, 1909; December 11, 1913, to September 30, 1918.

Gage.—Chain gage fastened to downstream side of highway bridge; read by A. Bissell.

Discharge measurements.—Made from downstream side of bridge, or by wading in vicinity of bridge.

Channel and control.—Bed composed of heavy gravel and rock; control at head of rapids, a few hundred feet below gage. Banks high and rocky; will not be overflowed at gage section.

Extremes of discharge.—Maximum stage recorded during year, 11.45 feet at 5 p. m. May 26 (discharge, 9,060 second-feet). An estimate of 5 second-feet for minimum discharge may be considerably in error, but discharge must have been low, as shown by flow of 7 second-feet measured January 15, 1918. Station records of Hatfield power station, Wisconsin Railway, Light & Power Co., show that with gates closed and no generation, pond did not raise until February 28.

1905–1909 and 1913–1918: Maximum stage recorded, 19.8 feet June 6, 1905 (discharge, approximately 29,400 second-feet). It is probable that the maximum discharge, which occurred October 6, 1911, exceeded 29,000 second-feet, although data are not available regarding the stage at the gage section during this flood, minimum stage recorded during open-water periods, 2.4 feet October 9, 1905 (discharge, approximately 20 second-feet); an estimated minimum discharge of 5 second-feet during frozen period, February, 1918.

Regulation.—Several dams on Black River and its tributaries upstream from Neillsville are used to create a head for developing power. The operation of these plants causes a diurnal fluctuation at the gage, especially during the winter, when the flow is at a minimum.

Accuracy.—Stage-discharge relation practically permanent except as affected by ice. Rating curve well defined 48 to 14,300 second-feet, fairly well defined below 48 second-feet, and extended above 14,300 second-feet. Gage read twice daily to quarter-tenths. Daily discharge ascertained by applying mean daily gage height to rating table, except for periods in which stage-discharge relation was affected by ice, for which it was obtained by applying to rating table gage heights corrected for ice effect by means of discharge measurements, observer's notes, and weather records. Open-water records good, except at extremely low stages, for which they are fair; winter records fair.

The following discharge measurement was made through a complete ice cover by T. G. Bedford:

January 15, 1918: gage height, 3.66 feet; discharge, 7.4 second-feet.

Daily discharge, in second-feet, of Black River at Neillsville, Wis., for the year ending Sept. 30, 1918.

Day.	Oct.	Nov.	Dec.	Jan.	Feb.	Mar.	Apr.	May.	June.	July.	Aug.	Sept.
1	87	334	116				1,220	1,290	6,280	49	57	97
2	83	296	122				1,080	1,010	5,640	42	57	94
3	88	244	86				950	770	3,680	40	49	84
4	78	244	108				770	585	2,560	44	43	71
5	69	228	118				660	485	1,720	60	88	63
6	86	244	110				560	416	1,800	52	37	76
7	69	260	84				710	374	1,800	47	42	57
8	65	244	48			355	1,080	395	1,570	54	58	40
9	70	241	38				890	890	2,560	49	73	41
10	84	201					710	3,260	1,480	44	69	42
11	90	192					585	2,360	950	43	167	53
12	112	165					1,150	1,720	610	43	201	53
13	116	100					438	1,220	460	41	176	43
14	147	180					374	830	334	43	228	43
15	157	139			5		374	635	260	45	187	56
16	142	147		10			395	485	116	57	130	40
17	144	132					485	416	65	49	98	46
18	170	122					710	374	97	45	73	46
19	296	118					1,290	395	100	43	58	47
20	560	118	25				1,150	460	73	44	57	90
21	485	125				1,930	1,010	460	78	42	53	198
22	416	110					890	1,430	81	42	64	225
23	395	104					770	1,500	76	38	94	144
24	296	102					660	1,150	76	42	87	144
25	257	87					510	1,570	71	40	213	165
26	374	108					416	7,620	48	87	173	100
27	560	110				1,720	374	7,620	45	37	122	92
28	770	94				2,160	560	7,280	42	41	118	84
29	710	116				1,430	1,220	6,120	41	37	90	87
30	510	90			......	1,220	1,570	3,790	132	38	83	76
31	354	......			......	1,150	......	4,290	......	45	83	......

NOTE.—Stage-discharge relation affected by ice Nov. 23-28, Dec. 6 to Apr. 1.

Monthly discharge of Black River at Neillsville, Wis., for the year ending Sept. 30, 1918.

[Drainage area, 774 square miles.]

Month.	Discharge in second-feet.				Run-off (depth in inches).
	Maximum.	Minimum.	Mean.	Per square mile.	
October	770	65	253	0.327	0.38
November	334	87	165	.213	.24
December	122	....	44	.057	.07
January	....	....	10	.013	.01
February	....	....	5	.006	.01
March	2,160	....	1,100	1.42	1.64
April	1,570	374	783	1.01	1.13
May	7,620	374	1,970	2.55	2.94
June	6,280	41	1,090	1.41	1.57
July	60	37	44	.057	.07
August	228	37	99	.128	.15
September	225	40	83	.107	.12
The year	7,620	40	475	.614	8.33

LA CROSSE RIVER NEAR WEST SALEM, WIS.

Location.—In sec. 32, T. 17 N., R. 6 W., La Crosse County, at highway bridge 2 miles west of West Salem and 10 miles above mouth of river. Dutch Creek enters from right 6 miles above station.

Drainage area.—412 square miles (measured on map issued by Wisconsin Geological and Natural History Survey, edition of 1911; scale, 1 inch=6 miles).

Records available.—December 22, 1913, to September 30, 1918.

Discharge measurements.—Made from upstream side of bridge or by wading.

Channel and control.—Bed composed of heavy gravel and rock and free from vegetation. Right bank high and not subject to overflow; left bank above the gage low, and subject to overflow at flood stages. Control for low stages a rocky riffle with a fall of about 6 inches; is apparently drowned out at a stage of about 2.2 feet on gage as shown by a reversal in the rating curve.

Extremes of discharge.—Maximum stage recorded during year, 6.8 feet, at 7 a. m., March 14 (discharge, 2,480 second-feet); minimum discharge about 125 second-feet, December 30.

1913–1918: Maximum stage recorded, 7.4 feet at 5 p. m. March 24, 1917 (discharge, approximately 2,850 second-feet); minimum discharge, about 130 second-feet November 17, 1914, minimum discharge during frozen period, about 125 second-feet, December 30, 1917.

Ice.—Stage-discharge relation seriously affected by ice.

Regulation.—Diurnal fluctuation at gage amounting at low stages to from 0.10 to 0.40 foot, is caused by the operation of power plants, especially the Neshonock dam a few miles above station.

Accuracy.—Stage-discharge relation permanent, except as affected by ice. Rating curve well defined between 181 and 2,300 second-feet. Gage read twice daily to quarter-tenths. Daily discharge ascertained by applying mean daily gage height to rating table except for periods in which stage-discharge relation was affected by ice, for which it was obtained by applying to rating table mean daily gage heights corrected for ice effect by means of discharge measurements, observer's notes, and weather records. Open-water records good, except for low stages, for which they are fair; winter records fair.

Discharge measurements of La Crosse River near West Salem, Wis., during the year ending Sept. 30, 1918.

Date.	Made by—	Gage height.	Discharge.	Date.	Made by—	Gage height.	Discharge.
		Feet.	Sec.-ft.			Feet.	Sec.-ft.
Oct. 7[a]	R. B. Kilgore	1.38	210	Mar. 30	T. G. Bedford	1.72	334
Jan. 13[b]	T. G. Bedford	1.58	152	Sept. 5[c]	W. G. Hoyt	1.37	196
Feb. 14[b]	.....do	3.54	363				

a Made by wading, 1,500 feet downstream from gage.
b Complete ice cover at control and measuring section.
c Made by wading, 500 feet downstream from gage.

Daily discharge, in second-feet, of La Crosse River near West Salem, Wis., for the year ending Sept. 30, 1918.

Day.	Oct.	Nov.	Dec.	Jan.	Feb.	Mar.	Apr.	May.	June.	July.	Aug.	Sept.
1	241	308	248	155	245	1,030	328	308	573	394	268	226
2	230	308	234	165	235	1,060	350	268	506	371	268	230
3	248	308	244	160	155	1,120	350	268	416	371	268	244
4	248	288	248	180	235	1,150	328	248	394	371	241	268
5	234	328	248	175	240	1,060	308	248	350	506	288	230
6	216	328	250	150	245	945	328	248	371	616	268	241
7	209	308	245	215	245	750	350	268	328	484	268	226
8	241	248	240	250	235	640	371	308	308	350	506	248
9	248	196	240	250	205	550	328	308	506	328	638	219
10	244	268	235	215	170	550	308	573	807	308	528	248
11	248	268	230	225	210	528	308	715	1,060	308	328	308
12	288	268	225	225	250	889	308	678	835	288	350	308
13	288	288	220	175	330	1,750	288	416	506	268	328	288
14	244	268	215	245	365	2,240	268	328	416	268	288	248
15	268	268	210	235	415	1,750	288	328	371	308	288	244
16	248	268	160	225	370	889	288	328	350	308	288	230
17	268	268	205	235	330	1,310	308	308	308	308	328	244
18	308	244	190	210	290	2,300	328	394	308	268	288	268
19	268	268	195	185	270	2,060	328	416	328	288	288	288
20	248	268	250	155	250	1,350	308	658	308	288	268	288
21	244	268	240	205	270	1,090	308	551	308	268	248	288
22	248	268	250	210	290	889	328	573	308	268	248	248
23	248	268	185	235	290	715	328	551	288	288	288	226
24	288	268	240	235	460	551	308	438	288	308	248	241
25	328	248	235	240	805	461	288	371	416	328	241	226
26	328	268	150	225	1,000	416	288	328	484	328	248	212
27	371	268	145	140	1,190	394	268	371	350	308	230	244
28	328	268	140	210	1,120	350	268	528	328	288	226	230
29	371	248	150	250	....	350	288	551	350	328	244	216
30	328	248	125	255	....	350	328	461	328	308	248	234
31	308	....	145	245	....	350	....	416	....	288	244	....

NOTE.—Stage-discharge relation affected by ice Dec. 6 to Mar. 10.

Monthly discharge of La Crosse River near West Salem, Wis., for the year ending Sept. 30, 1918.

[Drainage area, 412 square miles.]

Month.	Discharge in second-feet.				Run-off (depth in inches).
	Maximum.	Minimum.	Mean.	Per square mile.	
October	371	209	272	0.660	0.76
November	328	196	273	.663	.74
December	250	125	209	.507	.58
January	255	140	209	.507	.58
February	1,190	155	383	.930	.97
March	2,300	350	962	2.33	2.69
April	371	268	312	.757	.84
May	715	248	411	.998	1.15
June	1,060	288	427	1.04	1.16
July	616	268	332	.806	.93
August	638	226	300	.728	.84
September	308	212	249	.604	.67
The year	2,300	125	362	.879	11.91

WISCONSIN RIVER AT WHIRLPOOL RAPIDS, NEAR RHINELANDER, WIS.

LOCATION.—In sec. 4, T. 35 N., R. 8 E., Lincoln County, at head of Whirlpool Rapids, 1 mile below mouth of outlet of Crescent Lake, which comes in from right. 3 miles downstream from power station of Rhinelander Power Co., and 10 miles southwest of Rhinelander.

DRAINAGE AREA.—1,160 square miles (measured on map issued by Wisconsin Geological and Natural History Survey, edition of 1911; scale, 1 inch=6 miles).

RECORDS AVAILABLE.—September 15, 1915, to September 30, 1918; December 1, 1905, to September 30, 1915, records were collected at a station about 3 miles upstream.

GAGE.—Stevens water-stage recorder, on right bank in wooden shelter, attended by C. W. Jewell.

DISCHARGE MEASUREMENTS.—Made from cable about 150 feet upstream from gage.

CHANNEL AND CONTROL.—Bed of stream composed of heavy gravel and rock. Banks medium high and not subject to overflow. Control is head of rapids, 100 feet downstream from gage; well defined and permanent.

EXTREMES OF DISCHARGE.—Maximum stage recorded during year, 4.2 feet at 11 p. m. June 1 (discharge, 3,030 second-feet); minimum stage recorded 0.65 feet at 8 p. m. July 7 (discharge, 165 second-feet).

1905-1918: Maximum stage recorded, 5.61 feet at 10 p. m. April 22, 1916 (discharge, 5,250 second-feet); minimum discharge recorded, at old station, 0.0 foot during August and September, 1907, and June, 1908. The minimum flows are caused almost entirely by regulation; at the location of new station the discharge will never be zero. Minimum discharge at new location 1915-1918, 0.65 foot 8 p. m. July 7, 1918 (discharge, 165 second-feet).

REGULATION.—Above the station are 14 reservoirs[5] which are operated by the Wisconsin Valley Improvement Co. for the purpose of regulating the flow in Wisconsin River. The aggregate capacity of these reservoirs is 2.8 billion cubic feet during the summer and 3.6 billion cubic feet during the winter. Owing to the operation of these various storage reservoirs and the service reservoirs of three power plants on the river above, the flow at the station is not natural.

ACCURACY.—Stage-discharge relation permanent except as affected by ice. Rating curve well defined between 212 and 5,410 second-feet. Recording gage not in operation December 10 to March 28 and September 10-15. Daily discharge ascertained by use of discharge integrator except during periods when stage-discharge relation was affected by ice or recording gage was not in operation, for which it was obtained from gage readings and discharge measurements at Hat Rapids, weather records, and comparison of flow of Tomahawk River near Bradley and Wisconsin River at Merrill. Open water records excellent, except for periods when recording gage was not in operation, for which they are fair; winter records possibly poor.

Discharge measurements of Wisconsin River at Whirlpool Rapids, near Rhinelander, Wis., during the year ending Sept. 30, 1918.

Date.	Made by—	Gage height.	Discharge.	Date.	Made by—	Gage height.	Discharge.
		Feet.	*Sec.-ft.*			*Feet.*	*Sec.-ft.*
Jan. 14[a]	L. L. Smith	[b]2.76	476	June 10	T. G. Bedford	3.53	2,110
Feb. 18[a]	.....do	[b]3.60	808				

[a] Measurement made at highway bridge below Hat Rapids power plant; nearly complete ice cover.
[b] Chain gage reading at Hat Rapids Bridge.

[5] Information concerning these reservoirs, based on maps and data furnished by A. A. Babcock, manager of the Wisconsin Valley Improvement Co., and data collected by the engineering department of the Railroad Commission of Wisconsin, is contained in Water-Supply Paper 405, p. 127.

Daily discharge, in second-feet, of Wisconsin River at Whirlpool Rapids, near Rhinelander, Wis., for the year ending Sept. 30, 1918.

Day.	Oct.	Nov.	Dec.	Jan.	Feb.	Mar.	Apr.	May.	June.	July.	Aug.	Sept.
1	550	604	690				1,040	961	2,650	652	794	810
2	990	570	488				1,060	915	2,680	832	518	800
3	820	571	649				1,050	916	2,640	770	458	1,030
4	690	766	773				922	923	2,520	292	328	1,140
5	660	601	777				846	706	2,420	376	622	1,090
6	750	740	641				720	758	2,360	916	948	1,240
7	520	749	607				624	962	2,340	530	1,100	489
8	620	669	922				903	912	2,300	876	1,450	426
9	740	760	807				1,080	1,150	1,900	1,160	1,520	724
10	570	680					904	1,200	2,120	1,060	1,820	
11	540	566					800	1,240	1,740	1,020	1,120	
12	590	694					754	887	1,580	977	1,880	1,000
13	884	804					679	1,000	1,350	888	1,740	
14	631	788			720	1,020	426	1,160	1,340	522	1,680	
15	830	738					542	1,080	1,300	628	1,460	420
16	718	759		650			720	1,130	812	878	1,270	782
17	747	842					756	1,110	760	729	1,140	899
18	762	723					822	1,260	990	392	729	1,080
19	914	653					982	785	892	392	765	1,140
20	1,040	691	750				1,070	980	901	588	986	1,080
21	712	712					697	1,280	898	386	828	1,240
22	746	734					985	1,230	852	546	1,120	477
23	1,080	755					1,130	1,280	526	810	1,320	800
24	801	727					924	1,280	590	769	1,540	791
25	812	380					828	1,460	795	796	998	866
26	627	575					770	1,380	814	883	1,220	1,020
27	811	633					747	1,920	810	860	1,340	922
28	611	612					584	2,360	800	450	1,320	940
29	741	510			....	1,420	892	2,300	792	653	1,280	520
30	726	622			....	1,260	1,220	2,310	526	844	1,280	666
31	670	....			....	1,190	....	2,500	....	800	1,220	....

NOTE.—Stage-discharge relation affected by ice Dec. 10 to Mar. 28. Recording gage not in operation Dec. 10 to Mar. 28 and Sept. 10–15; discharge estimated by comparison of flow of Tomahawk River near Bradley, and Wisconsin River at Merrill, and from gage heights at Hat Rapids, and two discharge measurements made at Hat Rapids.

Monthly discharge of Wisconsin River at Whirlpool Rapids, near Rhinelander, Wis. for the year ending Sept. 30, 1918.

[Drainage area, 1,160 square miles.]

Month.	Discharge in second-feet.				Run-off (depth in inches).
	Maximum.	Minimum.	Mean.	Per square mile.	
October	1,080	520	737	0.635	0.73
November	842	380	671	.578	.64
December	....	....	737	.635	.73
January	....	....	650	.560	.65
February	....	....	720	.621	.65
March	....	....	1,050	.905	1.04
April	1,220	426	849	.732	.82
May	2,500	706	1,270	1.09	1.26
June	2,680	526	1,430	1.23	1.37
July	1,160	292	718	.619	.71
August	1,880	328	1,150	.991	1.14
September	1,240	....	870	.750	.84
The year	....	....	906	.781	10.58

WISCONSIN RIVER AT MERRILL, WIS.

Location.—At highway bridge at east end of Merrill, Lincoln County, 1,000 feet below power house of Merrill plant of Wisconsin Valley Lighting Co. and half a mile below mouth of Prairie River, coming in from left.

Drainage area.—2,630 square miles.

Records available.—November 17, 1902, to September 30, 1918.

Gage.—Stevens water-stage recorder installed September 11, 1914; November 17, 1902, to June 17, 1903, staff gage; June 17, 1903, to September 10, 1914, chain gage attached to downstream side of highway bridge; datum same since June 17, 1903. Records prior to June 17, 1903, questionable.

Discharge measurements.—Made from highway bridge a few feet upstream from recording gage.

Channel and control.—Bed composed of heavy gravel and rock; nearly permanent. Small island below gage and small rapids on either side probably constitute control. Both banks fairly high and are rarely overflowed.

Extremes of discharge.—Maximum stage recorded during year, 9.7 feet at 9 a. m. May 28 (discharge, 13,400 second-feet); minimum stage recorded, 3.0 feet at 6 a. m. July 23 (discharge, approximately 450 second-feet).

1912–1918: Maximum stage recorded, approximately 17.5 feet at 5 a. m. July 24, 1912 (discharge, 45,000 second-feet). During the preceding 24 hours 11.25 inches of rain fell in the vicinity of Merrill. According to C. B. Stewart, consulting engineer, Madison, the run-off of the 700 square miles between Merrill and Tomahawk was at the rate of 65 second-feet per square mile. If the estimate is extended to the entire area above Merrill the flow was 17 second-feet per square mile. Minimum stage recorded for the period, 2.7 feet, July 7, 1910 (discharge, approximately 389 second-feet).

Regulation.—Above the gaging station are 17 reservoirs,[b] which are operated by the Wisconsin Valley Improvement Co. for the purpose of regulating the flow in the Wisconsin River. The aggregate capacity of these reservoirs is about 6¼ billion cubic feet. In addition to the above reservoirs there are on Wisconsin and Tomahawk rivers above the station eight dams operated for power.

Accuracy.—Stage-discharge relation practically permanent. Rating curve fairly well defined between 1,600 and 19,400 second-feet. Water-stage recorder gave satisfactory results throughout the year. Daily discharge determined by means of Fuller discharge integrator. Open-water records good; winter records fair.

Discharge measurements of Wisconsin River at Merrill, Wis., during the year ending Sept. 30, 1918.

Date.	Made by—	Gage height.	Discharge.
		Feet.	*Sec.-ft.*
Jan. 11[a]	L. L. Smith	4.76	1,300
Feb. 15[a]	do	5.21	1,470
June 12	T. G. Bedford	5.27	2,280

[a] Made from ice and bridge at bridge section; incomplete ice cover at control.

[b] Information concerning these reservoirs, based on maps and data furnished by the manager of the Wisconsin Valley Improvement Co., and data collected by the engineering department of the Wisconsin Railroad Commission, is contained in Water-Supply Paper 405, p. 127.

Daily discharge, in second-feet, of Wisconsin River at Merrill, Wis., for the year ending Sept. 30, 1918.

Day	Oct.	Nov.	Dec.	Jan.	Feb.	Mar.	Apr.	May.	June.	July.	Aug.	Sept.
1	1,680	2,040	1,360	1,420	1,400	1,630	4,780	3,900	11,200	1,640	1,580	2,740
2	1,450	1,720	1,460	1,300	1,660	1,640	4,610	3,080	11,600	1,710	1,540	2,380
3	1,470	1,800	1,220	1,350	1,490	1,640	3,970	2,960	10,200	1,940	1,810	2,400
4	1,830	1,660	1,150	1,340	1,420	1,730	3,800	2,520	8,120	1,500	1,420	2,010
5	1,540	1,410	1,470	1,280	1,580	1,370	3,400	2,580	6,750	1,920	1,190	2,710
6	1,540	1,970	1,340	1,500	1,680	1,540	3,320	2,260	6,480	2,060	1,500	2,360
7	1,770	1,980	1,470	1,250	1,650	1,420	3,760	2,650	5,880	1,660	2,220	2,480
8	1,580	1,880	1,410	1,340	1,540	1,400	2,900	2,660	5,380	1,540	3,980	2,170
9	1,640	1,910	1,240	1,300	1,560	1,400	3,140	3,660	4,920	1,840	4,520	1,870
10	1,540	2,050	1,040	1,280	1,510	1,420	3,010	4,290	3,830	1,900	4,780	2,070
11	1,620	1,710	1,320	1,320	1,400	1,450	2,740	4,910	4,240	1,900	4,600	2,320
12	1,830	1,320	1,320	1,350	1,360	1,330	2,540	5,540	3,200	1,680	3,170	2,640
13	1,700	1,580	1,170	1,390	1,460	1,300	2,540	3,570	2,860	1,800	3,940	2,390
14	1,750	1,740	1,100	1,280	1,480	1,340	2,170	3,680	2,490	1,600	3,300	2,500
15	1,220	1,960	1,420	1,320	1,480	1,280	1,860	3,000	2,300	1,840	2,780	2,730
16	1,900	1,710	1,520	1,500	1,480	1,290	2,180	2,900	2,360	1,540	2,700	1,580
17	2,000	1,900	1,630	1,360	1,330	1,340	2,600	3,120	1,720	1,570	2,340	2,490
18	2,480	1,720	1,700	1,250	1,290	1,660	3,010	3,370	1,600	1,520	2,200	2,560
19	2,400	1,280	1,810	1,140	1,410	2,130	2,880	3,240	1,980	1,520	1,990	2,610
20	3,100	1,670	1,460	1,200	1,610	3,240	2,890	2,760	1,600	1,430	1,790	2,570
21	2,400	1,740	1,200	1,270	1,520	3,880	2,600	3,030	1,730	1,300	1,790	2,540
22	2,150	1,920	1,200	1,340	1,520	4,830	2,220	3,940	1,660	655	2,530	2,500
23	2,610	1,600	1,090	1,260	1,580	5,080	2,720	3,840	1,730	1,480	3,930	1,820
24	2,080	1,740	1,180	1,220	1,590	5,320	2,420	3,750	1,180	1,360	4,520	1,920
25	2,120	1,540	1,280	1,100	1,640	5,000	2,330	4,510	1,460	1,360	4,460	1,960
26	2,360	1,220	1,540	1,200	1,660	5,620	2,190	7,270	1,590	1,460	3,020	1,980
27	2,300	1,560	1,420	1,260	1,600	5,360	2,160	10,500	1,680	1,450	2,610	1,860
28	2,330	1,550	1,120	1,240	1,610	5,260	2,470	12,900	1,580	1,790	2,780	1,840
29	1,660	1,560	930	1,480	....	5,180	2,640	11,400	1,600	1,820	2,680	1,900
30	1,810	1,450	1,250	1,440	....	4,940	4,020	9,880	1,870	1,640	2,860	1,690
31	1,780	....	1,280	1,580	....	4,910	....	10,100	....	1,800	2,900	....

NOTE.—Stage-discharge relation affected by ice Dec. 9 to Mar. 24. Discharge for May 10, 11, and Sept. 20 and 21 based on gage heights for less than 24-hour period.

Monthly discharge of Wisconsin River at Merrill, Wis., for the year ending Sept. 30, 1918.

[Drainage area, 2,630 square miles.]

Month.	Discharge in second-feet.				Run-off (depth in inches).
	Maximum.	Minimum.	Mean.	Per square mile.	
October	3,100	1,220	1,920	0.730	0.84
November	2,050	1,220	1,690	.643	.72
December	1,810	930	1,330	.506	.58
January	1,580	1,100	1,320	.502	.58
February	1,680	1,290	1,520	.578	.60
March	5,620	1,280	2,800	1.06	1.22
April	4,780	1,860	2,930	1.11	1.24
May	12,900	2,260	4,770	1.81	2.09
June	11,600	1,180	3,830	1.46	1.63
July	2,060	655	1,620	.616	.71
August	4,780	1,190	2,820	1.07	1.23
September	2,740	1,580	2,250	.856	.96
The year	12,900	655	2,400	.913	12.40

WISCONSIN RIVER AT NEKOOSA, WIS.

LOCATION.—In sec. 15, T. 21 N., R. 5 E., 1½ miles below Nekoosa, Wood County. Tenmile Creek enters from left 4 miles below station, and Big Roche a Cri Creek, also from left, 38 miles below.

DRAINAGE AREA.—5,500 square miles (measured on map issued by Wisconsin Geological and Natural History Survey, edition of 1911; scale, 1 inch=6 miles).

RECORDS AVAILABLE.—May 21, 1914, to September 30, 1918.

GAGE.—Stevens water-stage recorder installed July 18, 1916, in wooden shelter on right bank; prior to that date Gurley water-stage recorder at same location. Gage attended by Henry Mans.

DISCHARGE MEASUREMENTS.—Made from cable a short distance above gage house.

CHANNEL AND CONTROL.—Bed composed of gravel; clean; practically permanent. Banks high and will be rarely overflowed.

EXTREMES OF DISCHARGE.—Maximum stage during year, 12.22 feet at 2 a. m. May 30 (discharge, 34,000 second-feet); minimum stage, effective gage height, 0.82 foot 12 noon July 23, (discharge, 1,060 second-feet).

1914–1918: Maximum stage, approximately 15.3 feet during the flood of June 6 to 9, 1914, as determined by levels run to high-water marks after water had receded (discharge, approximately 54,600 second-feet); minimum discharge recorded 0.45 foot at 11 a. m. October 7, 1915 (discharge, 595 second-feet); minimum flow is due to regulation.

ICE.—Stage-discharge relation seriously affected by ice.

REGULATION.—No storage reservoirs discharging into the Wisconsin River between Nekoosa and Merrill. See "Regulation" in station description of Wisconsin River at Merrill (p. 76). Between Nekoosa and Merrill are 12 dams operated for power.

ACCURACY.—Stage-discharge relation practically permanent, except as affected by ice. Rating curve well defined between 1,160 and 52,100 second-feet. Operation of recording gage satisfactory except June 20–22. Daily discharge ascertained by use of discharge integrator. Open-water records excellent; winter records fair.

Discharge measurements of Wisconsin River at Nekoosa, Wis., during the year ending Sept. 30, 1918.

Date.	Made by—	Gage height.	Discharge.
		Feet.	*Sec.-ft.*
Jan. 8[a]	L. L. Smith	2.78	1,410
Feb. 12[a]	do	3.25	1,540
June 17	T. G. Bedford	3.00	4,580

[a] Complete ice cover at gage and measuring section.

Daily discharge, in second-feet, of Wisconsin River at Nekoosa, Wis., for the year ending Sept. 30, 1918.

Day.	Oct.	Nov.	Dec.	Jan.	Feb.	Mar.	Apr.	May.	June.	July.	Aug.	Sept.
1	2,720	3,400	2,320	1,780	2,680	2,270	10,600	9,560	23,800	2,960	3,040	3,240
2	2,110	3,280	2,220	1,960	2,900	2,550	9,710	9,480	23,700	1,880	2,840	3,720
3	1,950	3,400	2,260	2,060	2,410	2,160	9,300	7,080	25,000	2,890	2,780	3,950
4	2,350	4,040	1,960	1,990	2,480	2,120	8,860	6,510	22,600	2,740	2,750	3,700
5	2,560	3,540	2,040	2,180	2,380	2,300	7,620	5,720	17,400	2,800	2,760	3,210
6	2,610	3,380	2,160	1,730	1,980	2,620	6,520	4,820	18,700	2,310	2,060	3,010
7	2,280	2,780	2,460	2,460	1,940	3,300	5,980	5,090	12,300	2,320	2,820	3,090
8	2,790	2,320	4,400	1,630	1,890	3,540	6,000	4,800	12,800	3,390	3,600	4,120
9	2,400	2,440	3,900	1,620	1,840	4,930	7,100	5,200	12,000	2,400	2,770	2,980
10	2,800	3,170	3,390	1,870	1,800	4,770	5,880	9,500	10,600	3,060	5,370	2,260
11	2,380	3,520	2,880	2,210	2,380	5,420	6,070	15,000	8,480	2,480	5,640	3,940
12	2,730	2,820	2,380	2,380	2,160	3,310	5,620	17,900	7,860	3,020	6,520	3,040
13	2,880	3,840	2,260	2,180	1,400	3,860	4,830	14,700	7,110	2,610	6,180	3,130
14	2,760	3,230	2,400	2,810	1,900	4,580	5,180	11,600	5,420	2,300	5,580	[illegible]
15	2,820	3,000	2,460	1,940	2,660	4,080	4,060	7,860	4,920	2,790	4,460	3,920
16	1,980	2,840	2,380	1,890	2,740	4,540	4,860	7,640	4,480	2,730	4,520	3,080
17	2,680	2,640	2,480	2,690	2,410	4,960	4,250	6,280	4,540	2,600	4,420	3,680
18	2,800	3,070	1,720	2,240	2,050	5,320	3,680	6,120	3,560	2,680	3,520	3,380
19	3,080	2,560	1,400	3,020	1,730	6,930	5,320	6,490	3,520	2,530	3,560	2,700
20	4,020	2,380	1,380	2,480	1,440	12,600	6,810	6,910	3,430	2,580	2,620	3,070
21	4,480	2,840	1,420	1,960	1,880	14,300	5,870	7,220	3,340	2,580	3,160	3,630
22	4,560	2,700	1,700	1,670	2,160	14,900	5,970	7,360	3,250	2,000	3,120	3,390
23	4,540	2,450	1,680	2,260	2,500	16,500	6,380	9,480	3,160	1,260	2,800	3,550
24	3,680	3,110	2,150	2,860	2,360	19,600	5,600	10,300	2,080	2,700	3,480	3,340
25	3,710	2,800	1,350	2,250	2,350	20,500	5,580	11,300	1,770	2,340	5,180	3,610
26	3,900	2,360	1,880	2,690	1,790	20,000	4,820	12,700	2,740	2,440	5,760	2,980
27	3,810	1,980	2,400	2,640	1,530	15,600	4,540	20,000	2,940	2,340	4,960	2,730
28	4,740	3,240	2,520	3,220	1,910	12,800	4,290	30,200	2,490	2,080	4,320	2,660
29	4,980	2,420	2,170	2,930	......	11,200	5,180	33,200	2,510	2,540	3,840	2,580
30	5,400	2,400	2,340	1,720	......	10,600	7,600	32,100	2,630	2,600	4,520	2,540
31	4,260	......	2,660	2,160	......	10,400	......	26,800	......	4,070	3,540	......

NOTE.—Stage-discharge relation affected by ice Dec. 9 to Mar. 19. Gage not operating satisfactorily June 20–22; discharge interpolated.

Monthly discharge of Wisconsin River at Nekoosa, Wis., for the year ending Sept. 30, 1918.

[Drainage area, 5,500 square miles.]

Month.	Discharge in second-feet.				Run-off (depth in inches).
	Maximum.	Minimum.	Mean.	Per square mile.	
ctober	5,400	1,960	3,250	0.591	0.68
ovember	4,540	1,960	2,970	.540	.60
ecember	4,400	1,350	2,290	.416	.48
nuary	3,220	1,620	2,240	.407	.47
ebruary	2,900	1,400	2,130	.387	.40
arch	20,500	2,120	8,150	1.48	1.71
pril	10,600	3,680	6,140	1.12	1.25
ay	33,200	4,800	11,900	2.16	2.49
ne	25,000	1,770	8,470	1.54	1.72
ly	4,070	1,260	2,580	.469	.54
ugust	6,520	2,060	3,950	.718	.83
ptember	4,120	2,260	3,230	.587	.65
The year	33,200	1,350	4,790	.871	11.82

WISCONSIN RIVER AT MUSCODA, WIS.

OCATION.—In sec. 1, T. 8 N., R. 1 W., at highway bridge 1 mile north of Muscoda, Grant County. Eagle Mill Creek enters from right half a mile below station and Underwood Creek from left, 4½ miles above.

RAINAGE AREA.—10,300 square miles (measured on map issued by Wisconsin Geological and Natural History Survey, edition of 1911; scale, 1 inch=6 miles).

ECORDS AVAILABLE.—December 21, 1902, to December 31, 1903; December 4, 1913, to September 30, 1918. Gage heights November 1, 1908, to December 31, 1912, published in United States Weather Bureau bulletin, Daily River Stages, parts 9, 10, and 11.

AGE.—Chain gage fastened to hand railing on upstream side of bridge; read by William Hessler. Elevation of zero of present gage approximately 12.62 feet above that of gage maintained December 20, 1902, to December 3, 1913, elevation of gage during period November, 1908, to December 3, 1913, as read and published by United States Weather Bureau was approximately the same as that of present gage, sea-level elevation of which is approximately 666.2 feet.

XTREMES OF DISCHARGE.—Maximum stage recorded during year, 8.04 feet at 5 p. m. June 4 (discharge, about 40,300 second-feet); minimum discharge, estimated 2,000 second-feet, February 11; water apparently held in service reservoir of Prairie du Sac dam.

1903 and 1914–1918: Maximum stage recorded, 22.70 feet September 23, 1903, corresponding to 10.1 feet for present gage datum (discharge, about 60,500 second-feet); minimum open-water stage recorded, 0.7 foot at 5 p. m. December 2, 1914, and July 24, 1915 (discharge, approximately 3,140 second-feet); estimated discharge of 2,000 second-feet, under frozen conditions, February 11, 1918; water apparently held in service reservoir of Prairie du Sac dam.

According to the records of the United States Weather Bureau [7] (see note under "Gage") on June 11, 1881, the river reached a stage of 11.1 feet and during August, 1868, zero on gage; discharge not computed owing to possible changes in channel and datum of gage.

EGULATION.—Nearest power plant above station is at Prairie du Sac, about 40 miles distant; since the latter part of 1915 considerable diurnal fluctuation has been observed at the gage. Owing to regulation by storage in headwaters, the flow at this station is not natural.

[7] Daily river stages, pt. 10, p. 98.

ACCURACY.—Stage-discharge relation not permanent. Two rating curves used during 1918; the first, October 1 to March 23, is fairly well defined between 4,230 and 15,900 second-feet; poorly defined outside these limits; the second, March 24 to September 30, is fairly well defined between 4,500 and 13,700 second-feet; poorly defined outside these limits. Gage read twice a day to quarter-tenths. Daily discharge ascertained by applying mean daily gage height to rating table, except for periods when stage-discharge relation was affected by ice, for which it was obtained by applying to rating table mean daily gage heights corrected for ice effect by means of discharge measurements, observer's notes, and weather records. Open-water records good, except during extreme high and low stages, for which they are fair; winter records roughly approximate.

Discharge measurements of Wisconsin River at Muscoda, Wis., during the year ending Sept. 30, 1918.

Date.	Made by—	Gage height.	Discharge.	Date.	Made by—	Gage height.	Discharge.
		Feet.	*Sec.-ft.*			*Feet.*	*Sec.-ft.*
Jan. 10[a]	Hoyt and Bedford.....	2.72	3,550	Apr 2	T. G. Bedford.........	4.73	17,300
Feb. 12[a]	Bedford and Schwada..	2.80	2,870	Aug. 1	W. G. Hoyt...........	1.67	5,200

[a] Complete ice cover at control and measuring section.

Daily discharge, in second-feet, of Wisconsin River at Muscoda, Wis., for the year ending Sept. 30, 1918.

Day.	Oct.	Nov.	Dec.	Jan.	Feb.	Mar.	Apr.	May.	June.	July.	Aug.	Sept.
1	4,450	7,760	6,380	3,560	3,700	12,200	16,300	9,430	24,200	5,980	5,030	6,310
2	4,690	8,140	6,380	3,530	3,430	11,300	16,300	8,700	34,400	6,640	4,750	5,330
3	4,940	10,100	5,480	3,500	3,580	11,300	17,300	8,350	39,000	6,310	4,750	4,750
4	5,200	7,760	6,380	3,470	2,720	11,300	15,400	8,700	39,900	6,640	4,490	5,980
5	5,200	6,060	6,380	3,450	2,920	12,600	14,600	12,900	39,900	6,640	4,020	5,980
6	4,690	8,900	6,380	3,470	3,220	14,800	14,600	14,600	35,300	6,640	4,750	5,650
7	4,940	8,520	4,940	3,500	3,430	15,300	13,300	11,300	31,800	6,980	5,330	5,330
8	4,450	7,760	4,750	3,520	3,430	14,400	11,700	8,700	30,900	6,310	5,330	5,650
9	4,940	8,140	4,690	3,540	3,160	13,000	12,900	9,060	27,700	6,640	5,330	4,250
10	4,940	7,400	4,630	3,560	3,380	10,900	11,300	9,430	22,400	6,640	5,030	5,030
11	5,480	7,400	4,570	3,380	2,000	10,500	9,800	12,500	19,200	6,980	5,030	5,650
12	5,200	6,060	4,510	3,140	2,850	12,600	9,800	10,200	17,300	6,310	4,750	5,980
13	4,940	7,040	4,450	3,320	3,160	14,800	11,300	11,300	17,300	5,980	5,650	5,330
14	5,200	7,400	4,400	3,280	3,630	18,700	10,900	13,700	17,300	5,980	5,330	5,330
15	4,690	7,400	4,360	3,410	3,820	20,300	9,800	15,900	12,500	5,650	5,980	5,330
16	5,480	7,040	4,300	3,500	3,520	20,300	9,800	20,800	12,900	6,640	8,000	4,020
17	5,760	6,700	4,220	3,450	3,600	19,700	9,430	22,400	13,700	5,980	8,350	5,030
18	6,380	6,700	4,170	3,320	3,120	19,200	9,430	19,200	12,100	6,310	8,350	5,330
19	6,380	5,480	4,150	3,140	2,700	18,700	9,060	18,200	10,200	5,980	6,640	5,030
20	5,480	6,380	4,070	3,000	2,680	17,700	9,060	17,300	9,060	5,980	5,980	5,030
21	5,480	6,700	4,000	2,880	3,010	17,700	9,430	16,800	9,430	5,650	6,310	5,030
22	4,690	7,040	3,950	3,140	3,100	18,700	7,660	17,300	9,060	5,030	5,980	5,980
23	6,060	6,380	3,910	3,160	2,990	19,700	9,060	16,300	8,000	5,650	5,980	4,020
24	6,700	6,060	3,870	3,190	3,080	20,200	11,700	16,800	7,320	5,980	6,310	4,750
25	6,700	5,760	3,820	3,410	4,220	21,900	11,700	16,300	8,000	5,330	5,650	4,490
26	7,400	5,200	3,780	3,000	7,400	24,800	9,430	16,800	7,660	5,980	5,030	4,490
27	7,400	6,060	3,740	3,560	11,300	29,300	9,060	15,400	7,660	5,650	5,650	4,750
28	8,900	6,380	3,710	2,900	13,000	26,200	10,200	17,700	6,310	5,030	5,650	4,490
29	7,400	6,380	3,650	3,380	....	27,700	9,800	20,200	5,980	4,490	5,650	4,750
30	9,300	5,760	3,600	3,410	....	30,100	10,600	20,200	5,650	5,030	6,310	4,490
31	9,700	....	3,560	3,560	....	23,600	....	21,300	....	5,030	6,310	....

NOTE.—Stage discharge relation affected by ice Dec. 8 to Mar. 23.

Monthly discharge of Wisconsin River at Muscoda, Wis., for the year ending Sept. 30, 1918.

[Drainage area, 10,300 square miles.]

Month.	Discharge in second-feet.				Run-off (depth in inches).
	Maximum.	Minimum.	Mean.	Per square mile.	
October	9,700	4,450	5,910	0.574	0.66
November	10,100	5,200	7,000	.680	.76
December	6,380	3,560	4,550	.442	.51
January	3,560	2,880	3,340	.324	.37
February	13,000	2,000	4,010	.389	.41
March	30,100	10,500	18,000	1.75	2.02
April	17,300	9,060	11,400	1.11	1.24
May	22,400	8,350	14,800	1.44	1.66
June	39,900	5,650	18,100	1.76	1.96
July	6,980	4,490	6,000	.583	.67
August	8,350	4,020	5,730	.556	.64
September	6,310	4,020	5,090	.494	.55
The year	39,900	2,000	8,670	.842	11.45

TOMAHAWK RIVER NEAR BRADLEY, WIS.

LOCATION.—In sec. 16, T. 36 N., R. 6 E., 2 miles west of Cassion, 4 miles north of Bradley, Oneida County, 4 miles downstream from mouth of Bearskin Creek, which comes in from right, and 8 miles above mouth of river.

DRAINAGE AREA.—422 square miles.

RECORDS AVAILABLE.—September 18, 1914, to September 30, 1918.

GAGE.—Chain gage fastened to cantilever arm on right bank; read by Frank Sutherland.

DISCHARGE MEASUREMENTS.—Made from cable about half a mile below gage.

CHANNEL AND CONTROL.—Bed at gage and a short distance below sandy and likely to shift; bed at cable section heavy gravel and permanent. Control is formed by rapids about 2,000 feet below the gage. When a head of 15 feet is maintained in Rice Lake storage dam, in secs. 4 and 9, T. 35 N., R. 6 E., backwater will extend halfway up the rapids, which are below gage, and may affect the stage-discharge relation.

EXTREMES OF STAGE.—Maximum stage recorded during year, 4.81 feet, at 7.25 p. m., June 4 (discharge, 1,130 second-feet); minimum stage, 1.45 feet at 6.25 p. m., July 22 (discharge, about 191 second-feet).

1914–1918: Maximum stage recorded, 6.88 feet April 24, 1916 (discharge, 2,120 second-feet); minimum stage, 1.45 feet July 22, 1918 (discharge, about 191 second-feet).

ICE.—Stage-discharge relation seriously affected by ice.

REGULATION.—The following reservoirs are maintained upstream from the station for the purpose of regulating the flow of Wisconsin River:

Dams and reservoirs on Tomahawk River.

Name.	Location of reservoir.	Location of dam.	Area of reservoir.	Drainage area.	Capacity (millions of cubic feet).	
					Summer.	Winter.
			Sq. mi.	*Sq. mi.*		
Squirrel	T. 39 N., R. 5 E.	Sec. 30, T. 39 N., R. 5 E.	3.00	17.07	152	152
Minocqua	Tps. 38–40 N., Rs. 6–7 E.	Sec. 10, T. 39 N., R. 6 E.	11.31	81.60	291	651
Total			14.31	98.67	443	803

ACCURACY.—Stage-discharge relation practically permanent, except as affected by ice and for a few days in April by logs. Rating curve is well defined between 240 and 1,970 second-feet. Gage read twice daily to hundredths. Daily discharge ascertained by applying mean daily gage height to rating table, except for periods in which stage-discharge relation was affected by ice, for which it was ascertained by applying to rating table mean daily gage heights corrected for ice effect by means of discharge measurements, observer's notes, and weather records; and for a few days in April when there was backwater from logs, for which discharge was interpolated. Open-water records good, except at extremely low stages, when they are fair; winter records fair.

Discharge measurements of Tomahawk River near Bradley, Wis., during the year ending Sept. 30, 1918.

Date.	Made by—	Gage height.	Discharge.
		Feet.	*Sec.-ft.*
Jan. 12[a]	L. L. Smith	2.95	305
June 11	T. G. Bedford	3.09	534

[a] Complete ice cover at control and measuring section.

Daily discharge, in second-feet, of Tomahawk River near Bradley, Wis., for the year ending Sept. 30, 1918.

Day.	Oct.	Nov.	Dec.	Jan.	Feb.	Mar.	Apr.	May.	June.	July.	Aug.	Sept.
1	370	460	405	305	270	380	619	434	1,040	258	276	421
2	358	474	400	310	265	385	589	408	1,040	265	300	384
3	354	460	395	310	255	390	604	384	1,080	258	282	354
4	358	434	395	310	245	395	559	370	1,120	258	255	324
5	365	447	395	310	250	395	502	363	1,120	265	237	304
6	384	434	390	310	255	395	488	360	1,040	261	260	289
7	384	434	390	310	260	400	516	372	886	246	261	276
8	367	421	385	305	265	405	516	408	798	234	408	266
9	360	421	385	305	270	415	516	516	697	226	559	258
10	354	408	385	305	275	435	516	604	619	219	589	252
11	356	408	380	305	290	440	530	634	530	212	574	263
12	384	421	380	305	305	460	528	619	460	205	530	338
13	408	447	375	310	310	475	525	604	408	198	460	354
14	421	447	375	310	300	480	523	516	396	201	408	347
15	421	447	370	310	310	490	521	460	384	209	358	332
16	408	460	365	305	310	480	518	434	360	211	324	319
17	408	447	360	305	320	510	516	408	347	227	295	295
18	516	447	350	305	325	540	530	408	328	219	271	354
19	589	447	360	310	330	575	516	408	297	209	255	367
20	619	434	370	310	330	620	512	434	276	204	248	367
21	619	434	385	305	330	650	509	447	265	201	240	354
22	604	434	395	300	330	681	506	460	258	195	328	336
23	574	434	395	295	330	748	502	460	250	202	421	312
24	544	420	395	290	335	815	372	434	242	212	460	297
25	502	415	385	290	345	798	308	460	236	229	434	285
26	502	415	370	285	355	780	297	666	237	240	384	273
27	516	410	360	280	365	850	289	780	236	240	345	265
28	516	410	350	285	375	923	308	850	234	250	332	255
29	516	410	330	290	......	1,000	408	923	232	268	408	249
30	460	410	310	280	......	798	447	961	242	275	460	243
31	460	......	305	275	......	666	......	961	......	271	447	......

NOTE.—Stage-discharge relation affected by ice Nov. 21 to Mar. 21. Stage-discharge relation affected by logs Apr. 9, 12–16, 20–22; discharge interpolated.

Monthly discharge of Tomahawk River near Bradley, Wis., for the year ending Sept. 30, 1918.

[Drainage area, 422 square miles.]

Month.	Discharge in second-feet.				Run-off (depth in inches).
	Maximum.	Minimum.	Mean.	Per square mile.	
October	619	354	452	1.07	1.23
November	474	408	433	1.03	1.15
December	405	305	374	.886	1.02
January	310	275	301	.713	.82
February	375	245	304	.720	.75
March	1,000	330	578	1.36	1.57
April	619	289	486	1.15	1.28
May	961	360	534	1.27	1.46
June	1,120	232	522	1.24	1.38
July	275	195	231	.547	.63
August	589	237	368	.872	1.01
September	421	243	311	.737	.82
The year	1,120	195	408	.967	13.12

PRAIRIE RIVER NEAR MERRILL, WIS.

LOCATION.—On line between secs. 20 and 29, T. 32 N., R. 7 E., at highway bridge 4½ miles northeast of Merrill, Lincoln County and 5½ miles above mouth of river. Haymeadow Creek enters from left 5 miles above station.

DRAINAGE AREA.—164 square miles (measured on map issued by Wisconsin Geological and Natural History Survey, edition of 1911, scale, 1 inch=6 miles).

RECORDS AVAILABLE.—January 18, 1914, to September 30, 1918.

GAGE.—Chain gage attached to upstream side of bridge; read by Mrs. Meta Krause.

DISCHARGE MEASUREMENTS.—From downstream side of bridge to which gage is attached or by wading.

CHANNEL AND CONTROL.—Bed composed of gravel; clean and free from vegetation. Left bank high, not subject to overflow; both banks wooded. Control not well defined.

EXTREMES OF DISCHARGE.—Maximum stage recorded during year, 5.0 feet May 28 (discharge, 1,420 second-feet); minimum discharge, about 75 second-feet, during January and February.

1914–1918: Maximum stage recorded, 6.1 feet April 22, 1916 (discharge, 2,290 second-feet); minimum discharge, 72 second-feet, by discharge measurement made January 4, 1915. Absolute minimum occurred during winter period 1914–1915, and was probably somewhat less than 72 second-feet.

ICE.—Stage-discharge relation seriously affected by ice.

REGULATION.—None.

ACCURACY.—Stage-discharge relation permanent. Rating curve well defined between 103 and 2,200 second-feet. Gage read once a day to half-tenths. Daily discharge ascertained by applying daily gage height to rating table, except for periods in which stage-discharge relation was affected by ice, for which it was obtained by applying to rating table mean daily gage heights corrected for ice effect by means of discharge measurements, observer's notes, and weather records. Open-water records good; winter records fair.

Discharge measurements of Prairie River near Merrill, Wis., during the year ending Sept. 30, 1918.

Date.	Made by—	Gage height.	Discharge.
		Feet.	*Sec.-ft.*
Jan. 11[a]	L. L. Smith	1.84	83
Feb. 16[a]	....do	1.78	80

[a] Incomplete ice cover at control and at measuring section.

Daily discharge, in second-feet, of Prairie River near Merrill, Wis., for the year ending Sept. 30, 1918.

Day.	Oct.	Nov.	Dec.	Jan.	Feb.	Mar.	Apr.	May.	June.	July.	Aug.	Sept.
1	103	137	110	80	80	85	313	421	963	137	110	137
2	101	137	105	80	80	85	313	348	1,010	137	118	137
3	101	133	105	80	80	85	296	348	781	148	137	137
4	110	137	100	80	80	85	278	244	574	148	137	122
5	133	137	100	80	80	85	278	244	458	159	137	118
6	133	137	100	80	80	85	244	212	421	159	137	115
7	122	148	100	80	80	85	244	212	421	133	148	110
8	115	137	100	80	80	90	244	278	366	122	458	110
9	110	137	95	80	80	90	228	313	313	110	535	110
10	115	137	95	80	80	95	212	574	278	106	496	106
11	118	137	95	85	80	95	184	655	228	101	421	118
12	118	137	95	80	75	100	184	614	212	97	402	137
13	115	133	95	80	80	100	159	535	184	91	313	148
14	118	128	95	80	80	105	159	402	159	91	278	148
15	122	128	95	80	80	110	159	313	148	97	228	137
16	122	128	95	80	80	110	172	244	137	103	184	110
17	137	122	95	80	80	115	198	198	128	103	159	118
18	159	122	90	80	80	122	261	212	122	101	137	148
19	159	128	95	80	80	244	244	244	115	97	133	148
20	159	122	90	80	80	348	212	278	106	97	118	137
21	159	122	90	80	80	535	212	244	106	93	110	137
22	159	122	90	80	85	496	184	313	103	91	118	137
23	159	118	85	80	85	458	184	348	103	93	212	137
24	159	118	90	80	85	421	172	296	103	103	458	137
25	159	118	85	75	85	366	159	366	101	118	384	133
26	159	122	85	75	85	348	159	963	103	122	366	128
27	148	110	85	80	85	348	137	1,110	103	115	313	118
28	159	110	90	75	85	348	184	1,420	110	103	244	110
29	172	110	85	75	....	313	366	1,220	118	106	212	106
30	159	115	85	75	....	330	384	1,010	159	148	159	110
31	137	....	85	75	....	330	....	1,110	....	137	137	....

NOTE.—Stage-discharge relation affected by ice Dec. 2 to Mar. 15.

Monthly discharge of Prairie River near Merrill, Wis., for the year ending Sept. 30, 1918.

[Drainage area, 164 square miles.]

Month.	Discharge in second-feet.				Run-off (depth in inches).
	Maximum.	Minimum.	Mean.	Per square mile.	
October	172	101	135	0.823	0.95
November	148	110	128	.781	.87
December	110	85	93.7	.571	.66
January	85	75	79.2	.483	.56
February	85	75	81.1	.495	.52
March	535	85	213	1.30	1.50
April	384	137	224	1.37	1.53
May	1,420	198	493	3.01	3.47
June	1,010	101	274	1.67	1.86
July	159	91	115	.701	.81
August	535	110	242	1.48	1.71
September	148	106	127	.774	.86
The year	1,420	75	185	1.13	15.30

EAU CLAIRE RIVER AT KELLY, WIS.

LOCATION.—In sec. 13, T. 28 N., R. 8 E., at highway bridge three-quarters of a mile below Kelly, Marathon County, 1 mile above mouth of Big Sandy Creek, which enters from right, and 4½ miles above mouth of river.

DRAINAGE AREA.—326 square miles (measured on map issued by Wisconsin Geological and Natural History Survey, edition of 1911; scale, 1 inch=6 miles).

RECORDS AVAILABLE.—January 1, 1914, to September 30, 1918.

GAGE.—Chain gage fastened to downstream side of highway bridge, read by William Woolsey.

DISCHARGE MEASUREMENTS.—Made from downstream side of bridge or by wading below bridge.

Channel and control.—Bed composed of heavy gravel and rock. Gage is in the rapids which form the control. Banks medium high and not subject to overflow.

Extremes of discharge.—Maximum stage recorded during year, 4.4 feet at 4.30 p. m., May 27 (discharge, 2,450 second-feet); minimum discharge estimated 30 second-feet December. 6.

1914–1918: Maximum stage recorded, 5.1 feet April 22 and 23, 1916 (discharge, 3,270 second-feet); minimum open-water stage recorded, 0.45 foot, August 13, 14, 15, October 2 and 3, 1914 (discharge, about 40 second-feet). Discharge December 6, 1917, was estimated as 30 second-feet.

Accuracy.—Stage-discharge relation permanent, except as affected by ice. Rating curve well defined between 71 and 3,150 second-feet. Gage read to quarter-tenths twice daily except Sundays. Daily discharge ascertained by applying mean daily gage height to the rating table, except for periods in which stage-discharge relation was affected by ice, for which it was obtained by applying to rating table mean daily gage heights corrected for ice effect by means of discharge measurements, observer's notes, and weather records; discharge for all Sundays interpolated. Open-water records good; winter records fair.

Discharge measurements of Eau Claire River at Kelly, Wis., during the year ending Sept. 30, 1918.

Date.	Made by—	Gage height.	Discharge.
		Feet.	*Sec.-ft.*
Jan. 10[a]	L. L. Smith	1.20	67
Feb. 14[a]	do	1.29	60
June 13[b]	T. G. Bedford	1.33	218

[a] Complete ice cover at control and measuring section.
[b] Made by wading 80 feet downstream from gage.

Daily discharge, in second-feet, of Eau Claire River at Kelly, Wis., for the year ending Sept. 30, 1918.

Day.	Oct.	Nov.	Dec.	Jan.	Feb.	Mar.	Apr.	May.	June.	July.	Aug.	Sept.
1	85	182	50	55	70	70	499	617	1,130	112	158	126
2	85	188	55	50	70	70	390	471	1,100	139	150	112
3	87	199	55	50	70	70	365	390	1,060	132	134	116
4	89	196	55	50	70	70	320	340	738	104	124	109
5	91	193	40	45	65	70	267	304	557	114	114	104
6	89	182	30	55	55	75	249	267	499	116	106	104
7	91	179	45	55	55	75	332	267	528	120	104	100
8	93	179	50	55	55	75	416	320	471	124	160	98
9	89	177	45	55	55	75	340	443	396	109	188	96
10	93	168	40	65	50	75	300	1,290	320	104	300	104
11	100	164	45	85	45	80	255	1,370	267	104	282	119
12	102	160	45	70	50	80	238	1,020	232	104	264	129
13	100	155	45	70	45	80	227	677	213	104	244	129
14	102	152	50	70	60	80	218	557	196	100	227	129
15	104	145	55	85	60	80	210	416	177	96	188	116
16	100	142	55	70	60	85	218	320	160	104	166	104
17	104	134	55	70	60	85	244	267	142	104	142	106
18	142	134	55	65	65	85	267	232	145	93	128	139
19	252	134	55	70	65	130	340	338	132	87	114	139
20	267	129	55	70	65	300	365	443	116	85	104	139
21	240	134	55	85	65	1,370	342	443	116	85	104	137
22	213	116	55	75	70	1,290	320	557	114	85	129	126
23	182	93	55	85	70	1,130	300	557	109	87	188	116
24	179	79	55	65	70	990	249	443	104	109	284	114
25	185	70	55	60	70	925	216	499	104	160	244	124
26	188	70	55	60	70	862	204	1,470	104	185	204	116
27	238	65	55	65	70	862	199	2,450	109	171	193	109
28	238	60	55	70	70	738	423	2,140	109	184	177	100
29	238	50	55	70	......	708	647	1,460	104	196	166	97
30	227	50	55	70	......	677	677	1,370	108	193	155	94
31	210	......	70	70	......	588	......	1,130	......	177	139	......

Note.—Stage-discharge relation affected by ice Dec. 13 to Apr. 2. Discharge for all Sundays interpolated.

Monthly discharge of Eau Claire River at Kelly, Wis., for the year ending Sept. 30, 1918.

[Drainage area, 326 square miles.]

Month.	Discharge in second-feet.				Run-off (depth in inches).
	Maximum.	Minimum.	Mean.	Per square mile.	
October	267	85	148	0.454	0.52
November	199	50	136	.417	.47
December	70	30	52	.158	.18
January	85	45	65	.201	.23
February	70	45	62	.191	.20
March	1,370	70	385	1.18	1.36
April	677	199	321	.985	1.10
May	2,450	232	738	2.26	2.61
June	1,130	104	322	.988	1.10
July	196	85	122	.374	.43
August	300	104	174	.534	.62
September	139	94	115	.353	.39
The year	2,450	30	221	.678	9.21

BIG EAU PLEINE RIVER NEAR STRATFORD, WIS.

LOCATION.—In sec. 13, T. 27 N., R. 3 E., at highway bridge at Weber Farm, 2 miles north of Stratford, Marathon County, and 1 mile above Chicago & Northwestern Railway bridge. Dill Creek enters from right 5 miles above station.

DRAINAGE AREA.—223 square miles (measured on map issued by Wisconsin Geological and Natural History Survey, edition of 1911; scale, 1 inch=6 miles.)

RECORDS AVAILABLE.—July 24, 1914, to September 30, 1918.

GAGE.—Sloping gage, reading from 1.0 to 15.6 feet, on right bank of the river, and vertical staff gage, reading from 15 to 18 feet, at upper end of sloping gage; read by Christian Weber.

DISCHARGE MEASUREMENTS.—Made by wading about 1,000 feet below gage or from highway bridge.

CHANNEL AND CONTROL.—Bed composed of heavy gravel and rock. Control at head of rapids 400 feet below gage. Both banks at gage are high and will be overflowed only at stage of about 15 feet and above.

EXTREMES OF DISCHARGE.—Maximum open-water stage recorded during year, 8.45 feet at 7.30 p. m. March 19, as ice was leaving river (discharge, about 4,980 second-feet); minimum open-water stage, 1.3 feet at 7 p. m. July 20 (discharge, about 3 second-feet).

1914–1918: Maximum recorded stage 8.85 feet at 6 p. m. April 21 (discharge, 5,540 second-feet); minimum discharge recorded, 3.0 second-feet (by meter measurement) February 5, 1915, and 7 p. m. July 20, 1918. The flood of June, 1914, reached a maximum height of 20.7 feet as determined by levels run to high-water marks.

ACCURACY.—Stage-discharge relation practically permanent, except for ice effect. Rating curve fairly well defined between 150 and 4,000 second-feet; poorly defined outside these limits. Gage read twice daily to quarter-tenths. Daily discharge ascertained by applying daily gage height to rating table, except for periods when discharge relation was affected by ice, December 5 to March 18, for which no daily discharge was estimated. Open-water records for high stages good; for medium and low stages poor.

The following discharge measurement was made by wading a quarter of a mile below gage, May 21, 1918, by T. G. Bedford:

Gage height, 2.28 feet; discharge, 108 second-feet.

Daily discharge, in second-feet, of Big Eau Pleine River near Stratford, Wis., for the year ending Sept. 30, 1918.

Day.	Oct.	Nov.	Dec.	Mar.	Apr.	May.	June.	July.	Aug.	Sept.
1	12	75	18	....	344	280	2,050	16	13	13
2	10	60	18	....	310	208	840	14	10	13
3	12	50	18	....	241	165	447	13	9	12
4	12	60	18	....	182	129	344	14	8	8
5	12	60	18	....	152	107	241	16	6	7
6	12	60	18	....	134	96	1,410	14	5	6
7	12	60	18	....	208	85	642	13	6	5
8	12	55	17	....	182	107	310	10	33	5
9	12	50	....	....	172	178	327	9	43	4
10	15	47	....	....	134	1,730	269	7	27	4
11	17	44	....	....	118	668	182	6	19	6
12	25	40	....	....	103	408	112	5	19	9
13	29	36	....	....	85	233	69	4	112	10
14	27	33	....	....	85	178	53	4	65	10
15	24	31	....	....	85	148	39	5	36	9
16	23	29	....	....	112	118	31	6	25	8
17	24	29	....	....	172	81	27	6	18	6
18	60	29	....	....	295	75	23	5	14	8
19	134	29	....	4,920	382	90	21	5	12	9
20	96	25	....	3,420	269	118	19	4	9	12
21	63	25	....	2,590	228	108	18	3	8	13
22	60	29	....	2,390	241	780	21	3	18	13
23	55	27	....	1,730	190	470	13	6	58	10
24	44	25	....	1,180	141	260	13	12	50	10
25	47	25	....	905	112	424	10	13	27	8
26	101	22	....	694	96	5,190	10	12	19	6
27	295	22	....	494	85	2,790	12	9	13	6
28	220	22	....	424	255	2,130	10	13	13	5
29	158	22	....	344	668	1,110	9	27	13	5
30	112	18	....	327	424	720	12	23	13	5
31	127	....	....	344	....	1,650	....	13	13	....

NOTE.—Stage-discharge relation affected by ice Dec. 8 to Mar. 19; daily discharge not determined.

Monthly discharge of Big Eau Pleine River near Stratford, Wis., for the year ending Sept. 30, 1918.

[Drainage area, 223 square miles.]

Month.	Discharge in second-feet.				Run-off (depth in inches).
	Maximum.	Minimum.	Mean.	Per square mile.	
October	295	10	60.0	0.269	0.31
November	75	18	38.0	.170	.19
December 1-8	18	17	17.9	.080	.02
March 19-31	4,920	327	1,520	6.82	3.30
April	668	85	207	.928	1.04
May	5,190	75	672	3.01	3.47
June	2,050	9	25.3	.113	.13
July	27	3	10.0	.045	.05
August	112	5	23.5	.105	.12
September	13	4	8.2	.037	.41

PLOVER RIVER NEAR STEVENS POINT, WIS.

Location.—In sec. 1, T. 24 N., R. 8 E., Portage County, at Fast Waters highway bridge, 7 miles above mouth of river and 5 miles northeast of Stevens Point.

Drainage area.—136 square miles.

Records available.—January 5, 1914, to September 30, 1918.

Gage.—Metal vertical staff gage bolted to left abutment, downstream side of bridge; read by Ethel Van Order.

Discharge measurements.—Made from downstream side of bridge.

Channel and control.—Bed composed of heavy gravel and small rock; free from vegetation; permanent. At high stages both banks will be overflowed around the bridge. Control not well defined but is probably small rapids below gage.

Extremes of discharge.—Maximum stage recorded during year, 3.3 feet at 5.30 p. m., May 28 (discharge, 670 second-feet); minimum discharge, estimated 55 second-feet, January 1-15 and February 1-15. Observer unable to reach gage May 29 to June 1, so that maximum stage during this period probably was somewhat above the maximum recorded on May 29.

1914-1918: Maximum stage recorded, 4.75 feet, June 5, 1914 (discharge, approximately 1,570 second-feet); minimum discharge estimated 45 second-feet, February 5-7, 1917.

Ice.—Stage-discharge relation seriously affected by ice.

Regulation.—Two dams are used in connection with grist mills above station, but the plants have little pondage, so that the flow at gage, except for brief periods, is nearly natural.

Accuracy.—Stage-discharge relation probably permanent, except as affected by ice. Rating curves well defined between 82 and 410 second-feet; poorly defined outside these limits. Gage read twice daily to quarter-tenths. Daily discharge ascertained by applying mean daily gage height to rating table, except during periods when stage-discharge relation is affected by ice, for which it is ascertained by applying to rating table mean daily gage heights corrected for ice effect by results of discharge measurements, observer's notes, and weather records; daily discharge interpolated October 1-6 and May 29 to June 1, when gage was not read. Open-water records fair, except at extremely low stages, when diurnal fluctuation may cause some error; winter records roughly approximate.

Discharge measurements of Plover River near Stevens Point, Wis., during the year ending Sept. 30, 1918.

Date.	Made by—	Gage height.	Discharge.	Date.	Made by—	Gage height.	Discharge.
		Feet.	*Sec.-ft.*			*Feet.*	*Sec.-ft.*
Oct. 25[a]	R. B. Kilgore..........	1.24	122	Feb. 13[b]	L. L. Smith............	2.76	96
Jan. 9[b]	L. L. Smith............	1.99	63	Mar. 29	T. G. Bedford..........	1.95	264

[a] Made by wading 300 feet upstream from gage.
[b] Complete ice cover at control and measuring section.

Daily discharge, in second-feet, of Plover River near Stevens Point, Wis., for the year ending Sept. 30, 1918.

Day.	Oct.	Nov.	Dec.	Jan.	Feb.	Mar.	Apr.	May.	June.	July.	Aug.	Sept.
1	122	123					234	234	484	212	150	98
2	121	114					234	201	438	132	114	106
3	120	132					201	180	438	132	141	90
4	118	114					201	212	382	132	114	106
5	117	150					190	170	356	170	132	98
6	115	114					170	170	330	150	123	98
7	114	114					190	190	280	114	114	98
8	98	98	60	55	55	215	190	201	330	132	160	82
9	98	114					190	212	280	132	190	98
10	114	114					160	382	256	114	190	98
11	114	98					160	500	256	98	190	98
12	123	98					132	469	245	114	170	114
13	114	123					170	438	223	98	190	98
14	106	98					141	256	223	98	132	114
15	114	106					150	280	212	132	132	132
16	114	98					150	234	150	132	132	132
17	82	98					150	234	132	132	132	114
18	160	98					160	256	150	123	114	123
19	98	106					160	280	114	132	132	132
20	132	98					150	256	150	106	106	132
21	114	114				365	150	256	170	98	114	114
22	106	114					190	330	150	114	114	114
23	98	98			110		190	330	98	132	98	98
24	132	90	65	65			150	301	132	114	114	114
25	132	71					141	280	114	132	123	98
26	123	106				330	150	438	141	141	90	123
27	150	98				268	170	565	132	132	114	114
28	150	114				245	170	670	170	114	114	114
29	150	98				234	234	624	132	150	114	123
30	141	106			......	223	234	577	150	190	114	123
31	114	......			......	234	......	531	......	132	98	......

NOTE.—Stage-discharge relation affected by ice Dec. 1 to Mar. 25. Gage not read Oct. 1-6, and May 26 to June 1; discharge interpolated.

Monthly discharge of Plover River near Stevens Point, Wis., for the year ending Sept. 30, 1918.

[Drainage area, 136 square miles.]

Month.	Discharge in second-feet.				Run-off (depth in inches).
	Maximum.	Minimum.	Mean.	Per square mile.	
October	160	82	119	0.875	1.01
November	150	71	107	.787	.88
December	....	....	63	.460	.53
January	....	....	60	.443	.51
February	....	....	80	.592	.62
March	....	....	271	1.99	2.29
April	234	132	175	1.29	1.44
May	670	170	331	2.43	2.80
June	484	98	227	1.67	1.86
July	212	98	130	.956	1.10
August	190	90	131	.963	1.11
September	132	82	110	.809	.90
The year	....	....	151	1.11	15.05

BARABOO RIVER NEAR BARABOO, WIS.

Location.—In sec. 33, T. 12 N., R. 7 E., at highway bridge 4 miles downstream from Baraboo, Sauk County, 3 miles below creek that rises near Devils Lake and comes in from right, and 15 miles above mouth of river.

Drainage area.—572 square miles (measured on map issued by Wisconsin Geological and Natural History Survey, edition of 1911, scale, 1 inch=6 miles.)

Records available.—December 18, 1913, to September 30, 1918.

Gage.—Chain gage, attached to upstream side of bridge; read by Miss Agnes Schneider.

Discharge measurements.—Made from downstream side of highway bridge to which gage is attached.

Channel and control.—Bed composed of sand and mud. Control not well defined. Water confined to one channel, except at flood stages when right bank is overflowed for a distance of 1,000 feet.

Extremes of discharge.—Maximum stage recorded during year, 15.03 feet at 8 a. m., March 20 (discharge, 3,280 second-feet); minimum stage, 1.15 feet at 4 p. m., December 2 (discharge, about 78 second-feet); caused apparently by temporary holding back of water by ice or otherwise.

1914–1918: Maximum stage recorded, approximately 17.5 feet March 26, 1917 (discharge, 4,200 second-feet); minimum stage, 0.71 foot at 7.30 a. m., July 26, 1916 (discharge, 76 second-feet).

Ice.—Stage-discharge relation seriously affected by ice.

Regulation.—In the vicinity of Baraboo, 4 miles above station, there are four dams and one at Reedsburg, 18 miles above. Smaller plants are also operated on tributaries. Operation of these various plants causes diurnal fluctuation at gage of about 0.3 foot at low-water stages. Estimates of mean monthly discharge probably represent nearly the natural flow.

Accuracy.—Stage-discharge relation changed during high water of March, 1917, and again during May, 1917. Rating curve used October 1 to March 12, 1918, fairly well defined between 150 and 3,270 second-feet; extended and approximate above and below these limits. Curve used March 13 to May 20 poorly defined throughout. Curve used May 21 to September 30 fairly well defined between 167 and 3,270 second-feet; extended and approximate only outside these limits. Gage read to quarter-tenths twice daily. Daily discharge ascertained by applying mean daily gage height to rating table, except for periods when stage-discharge relation was affected by ice, for which it was ascertained by applying to rating table mean daily gage heights corrected for ice effect by means of discharge measurements, observer's notes, and weather records. Open-water records fair; winter records roughly approximate.

Discharge measurements of Baraboo River near Baraboo, Wis., during the year ending Sept. 30, 1918.

Date.	Made by—	Gage height.	Discharge.	Date.	Made by—	Gage height.	Discharge.
		Feet.	*Sec.-ft.*			*Feet.*	*Sec.-ft.*
Nov. 2	R. B. Kilgore	3.15	303	May 9[b]	T. G. Bedford	2.32	230
19	.....do	2.38	203	9[b]	.....do	2.78	276
Jan. 14[a]	W. G. Hoyt	3.29	206	July 13[c]	W. G. Hoyt	2.40	245

[a] Complete ice cover at control and measuring section.
[b] Débris at measuring section.
[c] Tree on downstream side; possibly some backwater.

Daily discharge, in second-feet, of Baraboo River near Baraboo, Wis., for the year ending Sept. 30, 1918.

Day.	Oct.	Nov.	Dec.	Jan.	Feb.	Mar.	Apr.	May.	June.	July.	Aug.	Sept.
1	162	299	198	100	160	890	378	465	1,930	378	226	109
2	150	285	84	150	150	930	364	420	1,330	527	220	158
3	186	285	145	125	120	1,010	350	336	1,090	587	186	164
4	205	299	180	85	180	1,110	336	266	722	557	183	169
5	198	327	180	110	210	1,170	294	266	452	617	166	139
6	180	355	175	80	180	1,240	294	294	407	617	178	154
7	168	355	170	140	180	1,290	308	308	378	662	183	128
8	162	355	160	135	190	1,340	364	246	392	557	188	112
9	154	327	155	145	165	1,390	364	234	322	350	172	174
10	192	299	155	150	165	1,440	350	696	336	287	136	144
11	198	231	150	125	140	1,460	322	1,220	452	322	165	189
12	205	198	150	160	240	1,540	280	1,440	527	294	226	190
13	192	198	150	180	225	1,780	253	1,540	422	239	213	190
14	186	228	145	165	315	2,180	240	1,340	308	206	206	200
15	192	244	130	145	315	2,120	266	728	246	226	193	187
16	205	228	115	175	325	2,050	260	392	226	252	195	213
17	257	218	115	175	315	2,260	260	308	232	246	169	186
18	228	231	145	120	300	2,820	301	1,290	239	239	160	193
19	257	212	145	130	300	3,100	336	1,870	226	226	198	206
20	228	192	145	145	315	3,240	308	2,430	232	183	192	180
21	244	205	145	145	315	3,130	364	2,680	206	177	213	198
22	250	224	160	170	325	3,020	450	2,570	186	206	187	192
23	264	231	185	155	340	2,680	525	2,710	176	206	169	252
24	257	218	190	170	355	2,120	465	2,360	182	206	146	166
25	244	186	230	175	545	1,240	420	1,930	226	194	136	152
26	383	205	250	145	655	760	350	1,230	172	220	171	171
27	470	224	230	100	765	540	308	1,380	512	226	176	154
28	470	231	205	155	910	435	322	1,990	422	239	166	144
29	515	224	145	170	......	406	465	2,290	266	259	163	142
30	425	205	155	190	......	378	495	2,430	294	198	152	162
31	313	......	160	170	......	364	......	2,320	......	220	138	......

NOTE.—Stage-discharge relation affected by ice Dec. 3 to Mar. 12, and Mar. 15.

Monthly discharge of Baraboo River near Baraboo, Wis., for the year ending Sept. 30, 1918.

[Drainage area, 572 square miles.]

Month.	Discharge in second-feet.				Run-off (depth in inches).
	Maximum.	Minimum.	Mean.	Per square mile.	
October	515	150	250	0.437	0.50
November	355	186	251	.439	.49
December	250	84	163	.285	.33
January	190	80	143	.250	.29
February	910	120	311	.544	.57
March	3,240	364	1,590	2.78	3.20
April	525	240	346	.605	.68
May	2,710	234	1,290	2.26	2.61
June	1,930	172	437	.764	.85
July	662	177	320	.559	.64
August	226	136	180	.315	.36
September	252	109	171	.299	.33
The year	3,240	80	457	.799	10.85

KICKAPOO RIVER AT GAYS MILLS, WIS.

LOCATION.—In sec. 28, T. 10 N., R. 4 W., at highway bridge immediately below Norwood Mill, in Gays Mills, Crawford County, 25 miles above mouth of river, and 2 miles below mouth of Tainter Creek, which enters from right.

DRAINAGE AREA.—629 square miles (measured on map issued by Wisconsin Geological and Natural History Survey, edition of 1911, scale, 1 inch=6 miles).

RECORDS AVAILABLE.—December 25, 1913, to September 30, 1918.

GAGE.—Chain gage fastened to downstream side of bridge; read by N. T. Norwood.

DISCHARGE MEASUREMENTS.—Made from downstream side of bridge or by wading a short distance downstream from the gage.

CHANNEL AND CONTROL.—Bed composed of rock covered by a deposit of sand. Banks at gage fairly high and not subject to overflow at ordinary high-water stage. Control is at head of small rapids about 300 feet below gage; not permanent; the plotting of the discharge measurements indicate that at a stage of about 2 feet on the gage the control is charged to some point below, causing a reversal in the curve.

EXTREMES OF DISCHARGE.—Maximum stage recorded during year, 10.15 feet at 5.35 p. m., March 19 (discharge, about 2,900 second-feet); minimum discharge, about 245 second-feet, during January.

1914–1918: Maximum stage recorded, 15.05 feet March 24, 1917 (discharge, approximately 6,300 second-feet); minimum stage for open-water, 0.86 foot at 8 a. m., November 29, 1914 (discharge, 201 second-feet). Absolute minimum was approximately 100 second-feet, and occurred during the later part of January, 1915.

ICE.—Stage-discharge relation seriously affected by ice.

REGULATION.—Mills at Gays Mills immediately above station, Soldiers Grove about 7 miles upstream, and at several points above Soldiers Grove, use comparatively little storage, so that the recorded flow past station represents nearly the natural flow. During low stages a small diurnal fluctuation is observed at the gage

ACCURACY.—Stage-discharge relation not permanent. Shifts occurred during months of March, April, and May. One rating curve used during year; fairly well defined between 285 and 870 second-feet; extended and subject to error outside these limits. Shifting-channel method used March 13 to May 25. Gage read twice daily to nearest quarter-tenth. Daily discharge ascertained by applying mean daily gage height to rating table except for period when stage-discharge relation was affected by ice, for which it was ascertained by applying to the rating table mean daily gage heights corrected for ice effect by discharge measurements, observer's notes, and weather records. Open-water records fair; winter records subject to error.

Discharge measurements of Kickapoo River at Gays Mills, Wis., during the year ending Sept. 30, 1918.

Date.	Made by—	Gage height.	Discharge.	Date.	Made by—	Gage height.	Discharge.
		Feet.	*Sec.-ft.*			*Feet.*	*Sec.-ft.*
Dec. 4	W. G. Hoyt	1.58	349	Apr. 3	T. G. Bedford	1.99	382
Jan. 11[a]	T. G. Bedford	2.18	246	May 31	.....do	3.12	673
Feb. 13[a]	.....do	3.20	405	Aug. 2[b]	W. G. Hoyt	1.40	329

[a] Made through complete ice cover 150 feet downstream from gage.
[b] Made by wading 200 feet downstream from gage.

Daily discharge, in second-feet, of Kickapoo River at Gays Mills, Wis., for the year ending Sept. 30, 1918.

Day.	Oct.	Nov.	Dec.	Jan.	Feb.	Mar.	Apr.	May.	June.	July.	Aug.	Sept.
1	315	445	375	245	300	1,110	420	375	795	565	330	272
2	330	435	360	260	315	945	390	330	700	375	315	272
3	345	445	360	260	285	1,050	390	315	550	420	315	272
4	375	475	375	270	270	1,020	360	300	535	550	315	272
5	360	495	330	270	285	1,230	330	285	535	745	330	285
6	345	515	285	260	285	1,200	375	300	515	720	315	272
7	330	515	285	260	270	1,000	455	315	475	515	300	272
8	315	455	285	270	260	640	455	315	475	405	345	258
9	330	435	285	300	260	515	390	495	475	390	405	272
10	345	435	285	260	260	435	345	2,520	745	405	330	285
11	375	405	285	245	260	500	345	2,100	610	405	405	375
12	420	405	285	245	285	820	315	1,360	495	375	345	390
13	420	405	285	270	405	2,080	315	595	455	345	345	360
14	390	405	300	285	770	2,620	300	535	435	345	345	315
15	375	390	315	285	700	2,420	285	475	420	390	315	300
16	360	375	360	300	625	2,180	315	445	420	435	315	300
17	360	360	390	285	580	1,940	345	625	390	435	345	285
18	405	345	390	260	550	2,500	375	1,290	375	375	315	300
19	405	360	405	285	515	2,740	405	1,560	375	360	345	300
20	405	375	420	245	475	2,740	360	1,710	375	345	285	315
21	375	375	420	245	455	2,380	375	1,710	375	330	285	285
22	390	375	420	270	335	1,550	405	1,500	375	330	285	285
23	405	375	405	260	475	710	435	895	375	315	285	285
24	405	375	375	270	550	588	405	595	375	330	285	272
25	435	345	345	300	720	558	345	565	455	550	285	272
26	515	330	315	300	995	525	315	595	610	455	285	272
27	595	345	285	255	1,110	485	315	1,320	455	435	285	258
28	565	360	270	270	1,140	465	345	1,670	405	390	285	285
29	515	360	270	270	....	445	420	1,290	375	475	285	258
30	495	375	270	285	....	445	405	770	515	375	272	272
31	455	....	260	300	....	420	....	640	....	330	258	....

NOTE.—Stage-discharge relation affected by ice Dec. 8 to Mar. 11.

Monthly discharge of Kickapoo River at Gays Mills, Wis., for the year ending Sept. 30, 1918.

[Drainage area, 629 square miles.]

Month.	Discharge in second-feet.				Run-off (depth in inches).
	Maximum.	Minimum.	Mean.	Per square mile.	
October	595	315	402	0.639	0.74
November	515	330	403	.641	.72
December	420	260	332	.528	.61
January	300	245	270	.429	.49
February	1,140	260	494	.785	.82
March	2,740	420	1,230	1.96	2.26
April	455	285	368	.585	.65
May	2,520	285	897	1.43	1.65
June	795	375	482	.766	.85
July	745	315	426	.677	.78
August	405	258	315	.501	.58
September	390	258	291	.463	.52
The year	2,740	245	494	.785	10.67

MAQUOKETA RIVER BELOW MOUTH OF NORTH FORK OF MAQUOKETA RIVER, NEAR MAQUOKETA, IOWA.

LOCATION.—In southwest corner of NE.¼ sec. 17, T. 84 N., R. 3 E., at Bridgeport Bridge, 3 miles northeast of Maquoketa, Jackson County, 1,200 feet above mouth of Mill Creek, and 2 miles below mouth of North Fork of Maquoketa River.

DRAINAGE AREA.—1,600 square miles (measured on map issued by United States Geological Survey, scale, 1 to 500,000). Drainage area at mouth, 1,960 square miles.

Records available.—September 1, 1913, to September 30, 1918, except October, 1914, to March 20, 1915, when station was temporarily discontinued.

Gage.—Chain gage attached to down stream handrail of bridge 100 feet from right abutment; read by John Strodthoff.

Discharge measurements.—Made from bridge to which gage is attached.

Channel and control.—Bed of stream composed of sand; shifting. Two channels at all stages up to 12 feet, when there is overflow under pile-trestle approach on left side.

Extremes of discharge.—Maximum stage recorded during year, 15.4 feet, February 15, affected by ice; minimum stage recorded 1.75 feet, November 25 and 27 (discharge, 294 second-feet.) Prior to 1918: Maximum stage about 23.5 feet, probably in 1905 (discharge, about 24,300 second-feet).

Diversions.—None.

Regulation.—None.

Accuracy.—Stage-discharge relation not permanent. Two rating curves used during 1918; October 1 to December 4, and June 5 to September 30, well defined between 300 and 20,000 second-feet; February 16 to June 4, well defined between 300 and 20,000 second-feet. Gage read once daily to hundredths. Daily discharge ascertained by applying daily gage heights to rating table, except for days when gage was not read, for which the discharge was interpolated. December 4 to February 15 and February 21 to 23, stage-discharge relation affected by ice; discharge not determined. Open-water records good.

The following discharge measurement was made by Bolster and Gregg:

March 27: Gage height, 2.23 feet; discharge, 505 second-feet.

Daily discharge, in second-feet, of Maquoketa River below mouth of North Fork of Maquoketa River, near Maquoketa, Iowa, for the year ending Sept. 30, 1918.

Day.	Oct.	Nov.	Dec.	Feb.	Mar.	Apr.	May.	June.	July.	Aug.	Sept.
1	324	339	309	....	2,160	439	405	1,270	1,220	469	486
2	294	355	324	....	1,700	439	405	1,270	1,020	452	435
3	324	339	339	....	2,500	422	388	1,060	741	419	419
4	324	339	309	....	3,000	388	372	6,050	655	410	419
5	324	355	....	....	2,750	388	372	6,430	615	402	419
6	309	339	....	....	2,570	372	356	3,160	879	402	402
7	324	339	....	....	1,650	456	456	2,130	879	386	402
8	309	339	....	....	1,220	492	405	1,710	1,280	370	386
9	294	324	....	....	1,110	474	439	1,330	1,380	355	386
10	309	324	....	....	1,220	422	511	1,120	1,020	355	370
11	309	339	....	....	765	405	675	1,020	832	355	435
12	324	339	....	....	860	405	652	879	655	355	577
13	324	324	....	....	959	372	632	786	615	339	577
14	332	324	....	....	3,920	372	530	697	577	355	486
15	339	324	....	....	3,190	356	474	615	540	577	460
16	324	324	....	3,320	1,760	372	456	577		577	435
17	339	324	....	1,820	1,410	405	422	577		2,380	402
18	339	324	....	1,220	1,060	439	2,280	540		4,400	402
19	339	339	....	1,160	1,010	422	3,120	540		1,770	370
20	355	309	....	1,060	959	422	1,430	741	500	1,120	370
21	324	324	....	....	860	422	1,060	927		879	370
22	339	324	....	....	719	439	909	879		741	355
23	324	324	....	....	632	439	3,000	741		655	355
24	324	309	....	2,570	590	439	1,930	655		615	355
25	339	294	....	4,960	570	439	3,510	655	1,000	577	355
26	370	309	....	5,240	531	439	2,450	927	1,330	540	355
27	386	294	....	5,240	492	405	1,590	786	1,280	504	355
28	386	324	....	3,320	474	439	2,750	879	832	469	339
29	386	339	....	....	456	422	2,220	786	615	452	339
30	370	339	....	....	456	422	1,930	786	577	452	339
31	370	....	....	....	422	....	1,430	....	452	522	....

Note.—Discharge interpolated Oct. 14, Nov. 22, Mar. 17 and 26, Apr. 24, May 12, Aug. 4, Sept. 15. Discharge Mar. 3, July 16 to 25, estimated from discharge at Cedar Rapids and Janesville, and from climatologic data. Stage-discharge relation affected by ice Dec. 4 to Feb. 15 and Feb. 21 to 23; discharge not determined.

Monthly discharge of Maquoketa River near Maquoketa, Iowa, for the year ending Sept. 30, 1918.

[Drainage area, 1,600 square miles.]

Month.	Discharge in second-feet.				Run-off (depth in inches).
	Maximum.	Minimum.	Mean.	Per square mile.	
October	386	294	335	0.209	0.24
November	355	294	328	.205	.23
February	5,240	....	2,990	1.87	.69
March	3,920	422	1,350	.844	.97
April	492	356	419	.261	.29
May	3,510	356	1,210	.756	.87
June	6,430	540	1,350	.844	.94
July	1,380	452	758	.474	.55
August	4,400	339	731	.457	.53
September	577	339	405	.253	.28

ROCK RIVER AT AFTON, WIS.

LOCATION.—On line between secs. 22 and 27, T. 2 N., R. 12 E., at highway bridge in Afton, Rock County, 9 miles above Illinois State line. Bass Creek enters from right three quarters of a mile below station.

DRAINAGE AREA.—3,190 square miles (measured on map issued by Wisconsin Geological and Natural History Survey, edition of 1911; scale, 1 inch=6 miles).

RECORDS AVAILABLE.—February 5, 1914, to September 30, 1918.

GAGE.—Chain gage fastened to downstream side of bridge; read by Albert Engelke, and Leslie Seales.

DISCHARGE MEASUREMENTS.—Made from downstream side of bridge, or by wading.

CHANNEL AND CONTROL.—Banks medium high, and will not be overflowed to any extent at flood stages. Bed composed of gravel and clean silt; practically permanent. Control not well defined.

EXTREMES OF DISCHARGE.—Maximum stage recorded during year, 10.51 feet at noon March 26 (discharge, 12,700 second-feet); minimum stage 0.94 feet at 8.30 p. m. August 4 (discharge, 612 second-feet).

1914–1918: Maximum discharge recorded, 10.51 feet at noon March 26, 1918 (discharge, 12,700 second-feet); minimum stage recorded 0.5 foot at 7 a. m., August 16, 1914 (discharge, approximately 459 second-feet).

ICE.—Stage-discharge relation seriously affected by ice.

REGULATION.—Operation of power plants at Janesville and above causes fluctuations at gage during low stages.

ACCURACY.—Stage-discharge relation permanent. Rating curve well defined between 638 and 12,700 second-feet. Gage read twice daily to hundredths. Daily discharge ascertained by applying mean daily gage height to rating table, except for periods when stage discharge relation was affected by ice, for which it was ascertained by applying to rating table mean daily gage heights corrected for ice effect by means of discharge measurements, observer's notes, and weather records; daily discharge interpolated September 28–30 when gage was not read. Open-water records excellent, except at extreme low stages, when they are fair; winter records fair.

Discharge measurements of Rock River at Afton, Wis., during the year ending Sept, 30, 1918.

Date.	Made by—	Gage height.	Discharge.
		Feet.	*Sec.-ft.*
Jan. 3[a]	W. G. Hoyt	3.22	1,020
Feb. 6[a]	T. G. Bedford	2.97	829
Mar. 26	W. G. Hoyt	10.51	12,700

[a] Complete ice cover at measuring section, incomplete at control.

Daily discharge, in second-feet, of Rock River at Afton, Wis., for the year ending Sept. 30, 1918.

Day.	Oct.	Nov.	Dec.	Jan.	Feb.	Mar.	Apr.	May.	June.	July.	Aug.	Sept.
1	1,170	2,850	1,500	1,010	935	3,450	10,700	3,240	2,940	1,040	848	719
2	1,110	3,240	1,460	730	945	3,240	10,700	2,760	2,850	1,000	719	701
3	1,170	2,760	1,500	1,230	735	4,280	10,300	2,850	3,040	1,070	674	724
4	1,080	3,040	1,430	910	750	4,520	10,100	2,670	2,940	1,020	656	710
5	1,140	3,140	1,400	905	845	5,810	8,920	2,490	2,850	1,140	811	719
6	1,100	2,940	1,360	888	780	5,810	8,550	2,940	2,760	1,080	802	710
7	946	2,940	1,350	1,090	850	5,290	8,370	3,040	2,760	1,000	710	719
8	1,100	2,670	1,340	915	840	5,550	8,200	2,760	2,580	1,050	751	701
9	990	2,580	1,330	1,060	865	6,210	7,870	2,850	2,140	1,050	678	714
10	1,110	2,760	1,300	915	835	6,210	7,060	3,240	3,040	1,230	765	728
11	1,060	2,400	1,290	940	855	6,630	6,630	2,850	1,980	1,170	638	728
12	1,070	2,760	1,280	735	855	8,030	6,210	2,670	1,820	1,050	696	737
13	1,040	2,490	1,270	730	875	9,500	5,680	2,760	2,060	1,000	737	728
14	960	2,310	1,280	850	1,400	10,900	5,420	2,670	1,540	995	756	719
15	990	2,310	1,280	880	2,060	8,920	5,160	2,580	1,430	1,030	683	719
16	995	2,140	1,270	905	1,540	8,730	4,900	1,400	1,400	970	714	742
17	1,110	2,140	1,270	875	1,410	9,110	4,640	2,490	936	985	710	733
18	1,110	2,060	1,270	770	1,110	9,900	4,520	2,400	898	769	647	728
19	1,140	2,060	1,280	770	1,060	10,300	4,640	2,490	985	779	825	719
20	1,070	1,980	1,280	820	1,090	10,700	4,400	2,670	1,040	815	674	719
21	1,230	1,980	1,270	905	1,110	11,600	3,920	2,400	1,140	737	710	724
22	1,230	1,980	1,250	765	1,080	12,300	3,920	3,240	995	742	737	728
23	1,400	2,060	1,230	825	1,100	12,500	4,040	5,290	936	706	701	728
24	1,430	1,540	1,200	950	1,140	12,700	4,280	2,940	975	724	701	714
25	1,320	1,460	1,170	865	3,800	12,700	4,040	3,140	1,040	733	710	728
26	1,660	1,540	1,150	800	3,920	12,700	3,560	2,940	1,100	733	710	737
27	1,820	1,580	1,140	775	3,680	12,500	3,340	3,240	1,100	706	719	746
28	1,900	1,500	1,110	860	4,040	12,300	3,450	3,240	1,050	660	719	746
29	1,980	1,540	1,080	880	....	12,500	3,040	3,140	1,080	871	719	746
30	2,490	1,620	1,050	735	....	11,600	3,240	3,040	1,050	1,000	719	746
31	3,140	....	1,010	715	....	11,200	....	3,040	....	917	696	....

NOTE.—Stage-discharge relation affected by ice Dec. 5 to Feb. 26. Gage not read Sept. 28-30; discharge interpolated.

Monthly discharge of Rock River at Afton, Wis., for the year ending Sept. 30, 1918.

[Drainage area, 3,190 square miles.]

Month.	Discharge in second-feet.				Run-off (depth in inches).
	Maximum.	Minimum.	Mean.	Per square mile.	
October	3,140	946	1,320	0.414	0.48
November	3,240	1,460	2,280	.715	.80
December	1,500	1,010	1,270	.398	.46
January	1,230	715	871	.273	.31
February	4,040	735	1,450	.455	.47
March	12,700	3,240	8,960	2.81	3.24
April	10,700	3,040	5,980	1.87	2.09
May	5,290	2,400	2,920	.915	1.05
June	3,040	898	1,750	.549	.61
July	1,230	660	928	.291	.34
August	848	638	720	.226	.26
September	746	701	725	.227	.25
The year	12,700	638	2,440	.765	10.36

ROCK RIVER AT ROCKFORD, ILL.

LOCATION.—In sec. 34, T. 44 N., R. 1 E., at highway bridge at Nelson Avenue, Rockford, Winnebago County, 1 mile below mouth of Kent Creek.

DRAINAGE AREA.—6,520 square miles.

RECORDS AVAILABLE.—July 30, 1914, to September 30, 1918.

GAGE.—Chain gage attached to upstream side of bridge; read by Winston Burrows.

DISCHARGE MEASUREMENTS.—Made from upstream side of bridge.

CHANNEL AND CONTROL.—Coarse gravel and rock; may shift in high stages.

EXTREMES OF DISCHARGE.—Maximum stage recorded during year, 11.3 feet at 8 a. m. March 14 (discharge, 24,600 second-feet); minimum stage, 0.78 foot at 5 p. m. July 28 (discharge, 840 second-feet).

1914-1918: Maximum stage recorded, 15.5 feet February 15, 1915 (discharge not determined because of backwater from ice); maximum open-water stage recorded 13.0 feet March 30 and 31, 1916 (discharge, 32,000 second-feet); minimum discharge recorded, 483 second-feet August 9, 1914.

REGULATION.—Operation of power plant at dam 2 miles upstream in Rockford causes slight fluctuation at gage. During low stages water is stored at night for use in manufacturing plants during day.

ACCURACY.—Stage-discharge relation changed during high water in February; seriously affected by ice during winter. Rating curve used to February 14 fairly well defined; curves used after that date fairly well defined above 1,040 second-feet. Gage read to hundredths twice daily. Daily discharge ascertained by applying mean daily gage heights to rating tables, except for period when stage-discharge relation was affected by ice, for which it was determined from gage heights, observer's notes, weather records, and records of flow of Rock River at Afton, Wis. Records good for medium and high stages during open-water periods; probably somewhat too large for low stages during October, June, and July, on account of gage readings having been taken during day, when flow, due to regulation at dam, was somewhat greater than during night; winter records poor.

Discharge measurements of Rock River at Rockford, Ill., during the year ending Sept. 30, 1918.

[Made by H. C. Beckman.]

Date.	Gage height.	Discharge.	Date.	Gage height.	Discharge.
	Feet.	*Sec-ft.*		*Feet.*	*Sec-ft.*
Nov. 7	4.24	5,020	July 31	2.40	2,060
May 8	4.10	4,470	Aug. 18	1.60	1,300

Daily discharge, in second-feet, of Rock River at Rockford, Ill., for the year ending Sept. 30, 1918.

Day.	Oct.	Nov.	Dec.	Jan.	Feb.	Mar.	Apr.	May.	June.	July.	Aug.	Sept.
1	1,450	4,400	2,500			16,400	9,940	4,210	6,100	1,640	1,640	1,220
2	1,650	4,610	2,360			16,200	10,200	4,030	5,880	1,840	1,540	1,290
3	1,760	4,820	2,360			15,900	10,200	4,210	5,450	1,840	1,450	1,370
4	1,980	4,610	2,640			18,200	10,500	4,030	5,030	1,640	1,160	1,450
5	2,100	4,610	2,640	2,230		19,700	10,500	3,860	4,400	1,540	1,540	1,450
6	2,100	4,610	2,790			18,200	10,700	3,860	4,030	1,450	1,540	1,540
7	1,980	4,610	2,940		2,280	17,900	10,200	3,860	3,690	1,160	1,840	1,540
8	2,230	4,820	3,100			17,600	9,680	4,400	3,530	1,370	1,540	1,040
9	2,500	5,240				17,300	9,430	4,820	3,380	1,640	1,540	1,290
10	2,640	5,240				17,000	8,930	4,610	3,530	1,740	1,450	1,450
11	2,500	5,030				18,200	8,680	4,400	3,380	1,640	1,220	1,450
12	2,500	4,610				19,700	8,430	4,610	3,380	1,640	1,370	1,450
13	2,640	4,200				22,200	8,190	4,610	3,100	1,740	1,370	1,640
14	2,790	4,000				24,200	7.710	4,210	2,680	1,540	1,540	1,840
15	2,940	4,000	2,850		22,600	22,200	7,470	3,860	2,540	1,540	1,640	1,740
16	2,940	3,620		1,920	20,300	20,600	7,230	4,210	2,540	1,540	1,640	1,740
17	2,640	3,440			18,800	20,300	7,000	4,820	2,540	1,540	1,540	1,740
18	2,500	3,270			17,300	19,700	6,540	5,450	2,290	1,740	1,290	1,540
19	2,230	3,270			16,400	19,400	6,540	5,240	2,170	1,840	1,450	1,540
20	1,980	3,270			15,900	19,100	6,320	5,660	1,950	1,740	1,290	1,540
21	2,100	3,270			14,800	18,200	6,100	5,880	1,840	1,290	1,450	1,540
22	2,500	3,270			14,200	17,600	5,880	6,320	1,640	1,450	1,450	1,290
23	2,790	3,100			11,500	17,300	5,660	5,880	1,540	1,640	1,370	1,290
24	2,940	2,940			10,200	16,200	5,450	6,100	1,540	1,740	1,370	1,290
25	3,100	2,790			14,500	15,000	5,340	6,320	1,450	1,450	1,220	1,370
26	3,270	2,940	2,500	1,670	17,300	14,200	5,030	6,100	1,540	1,220	1,220	1,370
27	3,810	2,940			16,400	13,400	4,820	5,240	1,840	1,040	1,220	1,220
28	4,000	2,640			15,900	12,000	4,610	5,450	2,060	880	1,290	1,290
29	4,000	2,640			....	10,700	4,610	5,660	2,170	1,100	1,370	1,100
30	4,200	2,500			....	9,680	4,400	5,880	1,220	1,220	1,290	1,160
31	4,400	....			....	9,940	....	6,320	....	1,450	1,290	....

NOTE.—Discharge Dec. 9 to Feb. 14 estimated, because of ice, from gage heights, observer's notes, weather records, and flow of Rock River at Afton, Wis. Braced figures show mean discharge for periods included.

Monthly discharge of Rock River at Rockford, Ill., for the year ending Sept. 30, 1918.

[Drainage area, 6,520 square miles.]

Month.	Discharge in second-feet.				Run-off (depth in inches).
	Maximum.	Minimum.	Mean.	Per square mile.	
October	4,400	1,450	2,680	0.411	0.47
November	5,240	2,500	3,840	.589	.66
December			2,680	.411	.47
January			1,930	.296	.34
February	22,600		9,220	1.41	1.47
March	24,200	9,680	17,200	2.64	3.04
April	10,700	4,400	7,540	1.16	1.29
May	6,320	3,860	4,970	.762	.88
June	6,100	1,220	2,950	.452	.50
July	1,840	880	1,510	.232	.27
August	1,840	1,160	1,420	.218	.25
September	1,840	1,040	1,430	.219	.24
The year	24,200	880	4,760	.730	9.88

ROCK RIVER AT LYNDON, ILL.

LOCATION.—In sec. 21, T. 20 N., R. 5 E., at highway bridge known as Lyndon Bridge, in eastern part of Lyndon, Whiteside County, 10 miles above Rock Creek and 20 miles below dam at Sterling.

DRAINAGE AREA.—9,010 square miles.

RECORDS AVAILABLE.—November 24, 1914, to September 30, 1918.

GAGE.—Chain gage attached to bridge; read by John Shepard until August 8 and by George Cady thereafter.

DISCHARGE MEASUREMENTS.—Made from downstream side of bridge.

CHANNEL AND CONTROL.—Gravel; may shift.

EXTREMES OF DISCHARGE.—Maximum stage recorded during year, 19.6 feet February 16 (discharge not determined because of backwater from ice); maximum open water stage recorded, 14.4 feet at 6 a. m. March 16 (discharge, 28,600 second-feet); minimum stage recorded, 3.72 feet at 7 a. m. September 27 (discharge, 536 second-feet).

1915–1918: Maximum stage recorded, 19.6 feet February 16, 1918 (discharge not determined because of backwater from ice); maximum open-water stage recorded, 17.0 feet March 28, 1916 (discharge, 39,500 second-feet); minimum stage, 3.72 feet September 27, 1918 (discharge, 536 second-feet).

DIVERSIONS.—Water is diverted at Sterling dam to feed Illinois and Mississippi canal; probably averages about 100 second-feet.

REGULATION.—Flow past gage is regulated by power plants in city of Sterling and above.

ACCURACY.—Stage-discharge relation practically permanent; seriously affected by ice during winter. Rating curve well defined above 1,030 second-feet. Gage read to hundredths twice daily. Diurnal fluctuation at gage rather large during low stages. Daily discharge ascertained by applying mean daily gage height to rating table, except for period when stage-discharge relation was affected by ice, for which it was ascertained from gage heights, observer's notes, weather records, and records of flow of Rock River at Rockford, Ill., and Afton, Wis., discharge interpolated for several days March 1-20. Records good for medium and high stages and fair for low stages, during open-water period; winter records poor.

Discharge measurements of Rock River at Lyndon, Ill., during the year ending Sept. 30, 1918.

[Made by H. C. Beckman.]

Date.	Gage height.	Discharge.	Date.	Gage height.	Discharge.
	Feet.	*Sec.-ft.*		*Feet.*	*Sec.-ft.*
Nov. 9	6.48	4,430	Aug. 9	4.42	1,080
9	6.48	4,490	9	4.83	1,580
May 6	6.63	4,540			

Daily discharge, in second-feet, of Rock River at Lyndon, Ill., for the year ending Sept. 30, 1918.

Day.	Oct.	Nov.	Dec.	Jan.	Feb.	Mar.	Apr.	May.	June.	July.	Aug.	Sept.
1	2,500	4,940	3,030			24,000	13,900	4,740	5,780	2,670	1,740	1,300
2	2,670	4,740	2,670			22,600	13,000	5,570	5,150	2,030	1,480	1,610
3	2,500	4,740	3,210			17,100	12,400	5,780	5,360	2,180	1,880	1,610
4	2,500	4,740	2,850			19,400	11,200	5,150	5,360	2,030	1,880	1,480
5	2,500	5,360	3,030	3,040	2,340	21,800	11,500	4,740	5,150	2,180	1,240	1,360
6	2,500	4,940	3,210			23,600	11,200	4,940	4,540	1,880	1,300	1,540
7	2,340	4,940				25,400	10,500	4,740	4,540	2,340	1,740	1,540
8	1,180	4,740				23,800	10,200	4,150	4,540	2,030	1,300	960
9	2,500	4,740				22,200	10,000	4,540	4,150	3,030	1,480	1,610
10	2,500	4,540				19,200	9,500	6,200	4,540	1,130	1,480	3,030
11	2,500	3,960				19,000	9,250	6,410	4,340	1,740	1,360	1,420
12	2,500	4,540				18,800	8,750	4,540	4,340	2,180	1,610	1,130
13	1,610	2,850	3,950			19,400	8,000	5,150	4,150	2,180	1,540	1,610
14	2,030	5,150				19,900	7,760	4,940	3,960	2,340	1,420	1,360
15	2,500	4,150		2,670	26,000	24,200	7,520	4,940	3,770	1,030	1,740	1,300
16	2,500	3,960				28,600	7,280	4,940	2,850	2,180	1,610	1,740
17	3,030	3,770				25,400	7,050	4,540	3,390	1,300	2,180	1,740
18	5,360	3,390				24,000	6,410	4,540	3,390	2,340	2,340	1,420
19	3,770	3,580				22,600	6,620	4,340	2,500	1,880	3,030	1,480
20	3,770	3,580				22,000	6,410	4,740	2,500	2,030	2,030	2,180
21	3,030	3,030				21,500	5,990	5,150	2,180	2,340	1,880	1,740
22	3,210	3,390				21,000	6,830	4,340	2,500	2,180	2,030	1,540
23	3,210	2,670				19,900	6,620	5,360	2,340	930	2,030	1,740
24	2,850	3,030				18,800	6,410	6,830	2,030	2,340	1,880	1,360
25	2,670	3,210			26,800	17,800	6,200	7,280	1,360	2,340	2,180	1,880
26	2,670	3,210	3,360	2,670		17,100	6,200	7,760	2,340	1,880	1,680	2,030
27	3,770	3,030				15,700	5,990	8,250	2,180	1,480	1,420	610
28	3,770	3,030				16,800	5,990	7,520	3,030	2,670	1,740	765
29	3,960	2,670			....	15,700	5,570	7,520	2,340	1,610	1,240	1,680
30	4,740	3,030			....	15,100	5,360	6,200	2,670	2,030	1,360	885
31	4,540	....			....	14,200	....	6,200	....	1,300	1,480	....

NOTE.—Discharge, Mar. 1, 4, 6, 8, 11, 13, 15, 18, 20 and 21 interpolated, for lack of gage-height record; estimated Dec. 7 to Feb. 28, because of ice, from gage heights, observer's notes, weather records, and records of flow of Rock River at Rockford, Ill., and Afton, Wis. Braced figures show mean daily discharge for period included.

Monthly discharge of Rock River at Lyndon, Ill., for the year ending Sept. 30, 1918.

[Drainage area, 9,010 square miles.]

Month.	Discharge in second-feet.				Run-off (depth in inches).
	Maximum.	Minimum.	Mean.	Per square mile.	
October	5,360	1,180	2,960	0.329	0.38
November	5,360	2,670	3,920	.435	.49
December	....	....	3,560	.395	.46
January	....	....	2,790	.310	.36
February	....	....	17,800	1.98	2.06
March	28,600	14,200	20,500	2.28	2.63
April	13,900	5,360	8,320	.923	1.03
May	8,250	4,150	5,550	.616	.71
June	5,780	1,360	3,580	.397	.44
July	3,030	930	1,990	.221	.25
August	3,030	1,240	1,720	.191	.22
September	3,030	610	1,520	.169	.19
The year	....	610	6,110	.678	9.22

PECATONICA RIVER AT DILL, WIS.

Location.—In sec. 6, T. 1 N., R. 6 E., at Illinois Central Railroad bridge at Dill (Ramona post office), Green County, 1 mile below junction of East and West branches of Pecatonica River and 9 miles above Illinois State line.

Drainage area.—959 square miles (measured on map issued by Wisconsin Geological and Natural History Survey, edition of 1911; scale, 1 inch = 6 miles).

Records available.—February 9, 1914, to September 30, 1918.

Gage.—Chain gage fastened to downstream side of bridge; read by S. A. Frank. Prior to August 2, 1916, vertical staff gage on left abutment.

Discharge measurements.—At low and medium stages made from upstream side of highway bridge about 400 feet above gage; during extremely high water considerable water overflows to left of highway bridge and measurements are made from railroad bridge to which gage is attached.

Channel and control.—Bed composed of sand and mud; undoubtedly shifting. Banks only medium height and will be overflowed at flood stages. Except during extreme flood stages all water passes under railroad bridge to which gage is fastened. There is little fall in river below the gage and no well defined control.

Extremes of discharge.—Maximum stage during year, 13.25 feet at 9 a. m. February 28 (discharge, about 5,850 second-feet); minimum stage, 0.60 foot, at 5 p. m. September 9 (discharge about 176 second-feet).

1914–1918: Maximum stage, 19.1 feet March 27, 1916, determined from flood marks by leveling (discharge, approximately 13,100 second-feet); minimum stage September 9, 1918 (estimated discharge, 176 second-feet).

Ice.—Stage-discharge relation affected by ice.

Regulation.—Operation of dams at Argyle, on East Branch of Pecatonica River, and at Darlington, on West Branch of Pecatonica River, cause little if any diurnal fluctuation at gage.

Accuracy.—Stage-discharge relation apparently permanent, throughout the year. Rating curve fairly well defined between 176 and 1,520 second-feet; poorly defined between 1,520 and 6,000 second-feet. Extension of curve above 6,000 second-feet is based on the flow ot Pecatonica River at Freeport, Ill. Daily discharge ascertained by applying mean daily gage height to rating table, except for period when stage-discharge relation was affected by ice, for which it was ascertained by applying to rating table mean daily gage heights corrected for ice effect by means of discharge measurements, observer's notes, and weather records. Open-water records good; winter records subject to error.

Discharge measurements of Pecatonica River at Dill, Wis., during the year ending Sept. 30, 1918.

Date.	Made by—	Gage height.	Discharge.	Date.	Made by—	Gage height.	Discharge.
		Feet.	*Sec.-ft.*			*Feet*	*Sec.-ft.*
Nov. 9	R. B. Kilgore	1.22	308	May 28	T. G. Bedford	2.64	783
Jan. 4[a]	W. G. Hoyt	1.34	244	Aug. 18	W. G. Hoyt	1.50	360
Feb. 8[a]	T. G. Bedford	1.74	216				

[a] Complete ice cover at control and measuring section.

Daily discharge, in second-feet, of Pecatonica River at Dill, Wis., for the year ending Sept. 30, 1918.

Day.	Oct.	Nov.	Dec.	Jan.	Feb.	Mar.	Apr.	May.	June.	July.	Aug.	Sept.
1	283	328	283	235	230	5,400	340	340	404	352	214	210
2	283	328	272	235	225	4,800	328	316	404	364	230	210
3	283	328	272	240	225	4,580	328	294	390	294	226	212
4	294	340	272	240	225	4,330	316	283	364	283	226	208
5	294	340	272	245	220	4,580	305	294	340	283	222	210
6	283	328	261	245	220	4,380	316	305	328	272	228	210
7	283	328	261	245	215	3,930	352	316	328	294	218	206
8	283	316	261	245	215	2,880	352	316	328	272	205	199
9	283	316	230	245	225	1,800	328	328	316	250	186	182
10	283	305	230	245	235	1,720	316	352	340	283	199	196
11	294	305	230	240	290	1,680	305	340	316	316	352	210
12	305	305	226	240	305	1,920	294	316	294	283	550	226
13	316	305	250	240	920	2,980	294	283	272	250	288	230
14	305	305	250	240	1,720	5,080	294	283	250	250	305	242
15	305	305	250	235	2,330	4,860	283	272	244	250	328	228
16	305	305	250	235	2,980	3,780	294	272	242	272	272	224
17	316	305	250	235	2,880	2,600	305	261	240	294	294	220
18	340	294	250	235	2,510	2,330	340	433	240	283	283	226
19	328	294	250	230	2,150	2,330	364	586	242	272	272	210
20	328	283	260	230	1,520	2,150	364	662	248	261	261	194
21	305	283	325	230	1,160	1,840	364	377	250	250	236	197
22	294	283	390	230	950	1,560	364	3,080	261	244	232	201
23	316	272	400	230	825	990	364	2,330	250	234	220	205
24	328	272	390	230	1,880	780	328	1,040	250	244	212	210
25	352	272	365	225	3,630	586	305	624	261	250	214	208
26	390	272	325	225	4,480	418	283	550	305	272	212	208
27	448	272	290	230	5,680	390	283	586	340	272	210	206
28	480	272	275	230	5,820	364	305	740	294	272	197	210
29	377	283	245	235	....	352	328	586	272	272	190	210
30	352	283	235	230	....	340	352	497	294	261	208	210
31	352	....	235	230	....	340	....	433	....	234	210	....

NOTE.—Stage-discharge relation affected by ice Dec. 13 to Feb. 20.

Monthly discharge of Pecatonica River at Dill, Wis., for the year ending Sept. 30, 1918.

[Drainage area, 959 square miles.]

Month.	Discharge in second-feet.				Run-off (depth in inches).
	Maximum.	Minimum.	Mean.	Per square mile.	
October	480	283	322	0 336	0.39
November	340	272	301	.314	.35
December	400	226	276	.288	.33
January	245	225	236	.246	.28
February	5,820	215	1,580	1.65	1.72
March	5,400	340	2,450	2.55	2.94
April	364	283	323	.337	.38
May	3,080	261	571	.595	.69
June	404	240	297	.310	.35
July	364	234	274	.286	.33
August	550	186	248	.259	.30
September	242	182	211	.220	.25
The year	5,820	182	586	.611	8.31

PECATONICA RIVER AT FREEPORT, ILL.

LOCATION.—In sec. 32, T. 27 N., R. 8 E., at highway bridge at Hancock Avenue, half a mile east of Illinois Central Railroad station at Freeport, Stephenson County, and 2 miles above mouth of Yellow Creek.

DRAINAGE AREA.—1,330 square miles.

RECORDS AVAILABLE.—September 10, 1914, to September 30, 1918.

GAGE.—Chain gage attached to upstream side of bridge; read by W. C. Krueger.

DISCHARGE MEASUREMENTS.—Made from upstream side of bridge.

CHANNEL AND CONTROL.—Bed composed of sand and silt; likely to shift. Left bank of only medium height and is overflowed during high water; at stages above about 16.0 feet part of the flow passes over left bank and through East Freeport.

Extremes of discharge.—Maximum stage recorded during year, 16.4 feet at 4 p. m. February 15 (discharge, 6,880 second-feet); minimum stage, 3.0 feet at 6 p. m. September 7 (discharge, 208 second-feet).

1914–1918: Maximum stage recorded, 19.4 feet March 28, 1916 (discharge, 17,000 second-feet); minimum stage, 3.0 feet September 7, 1918 (discharge, 208 second-feet).

Regulation.—A dam and power plant three-quarters of a mile upstream regulate flow past gage. Only slight diurnal fluctuation is noticeable.

Accuracy.—Stage-discharge relation changed during year; seriously affected by ice during winter. Rating curves well defined between 620 and 6,260 second-feet and fairly well defined beyond these limits. Gage read to hundredths twice daily. Daily discharge ascertained by applying mean daily gage height to rating tables, except for periods when stage-discharge relation was affected by ice, for which it was ascertained by means of occasional gage heights, observer's notes, weather records, and flow of Pecatonica River at Dill, Wis. Open-water records for medium and high stages good; for low stages fair; winter records poor.

Discharge measurements of Pecatonica River at Freeport, Ill., during the year ending Sept. 30, 1918.

[Made by H. C. Beckman.]

Date.	Gage height.	Discharge.
	Feet.	*Sec.-ft.*
Nov. 8	3.97	306
May 7	4.80	537
July 31	4.10	395

Daily discharge, in second-feet, of Pecatonica River at Freeport, Ill., for the year ending Sept. 30, 1918.

Day.	Oct.	Nov.	Dec.	Jan.	Feb.	Mar.	Apr.	May.	June.	July.	Aug.	Sept.
1	324	461	340			5,970	514	572	652	533	332	290
2	324	442	324			6,140	552	533	652	496	332	276
3	308	500	308			5,970	514	514	632	478	276	318
4	324	540	308			5,380	496	478	612	478	290	361
5	324	500	324	335		5,000	496	425	592	392	290	218
6	340	480	278		350	5,120	496	392	572	442	290	218
7	340	442	221			5,380	514	478	552	442	304	256
8	340	406	248			5,000	552	478	533	408	318	256
9	324	372	278			4,000	572	514	514	376	304	243
10	324	372	293			3,750	514	572	442	408	361	330
11	324	372	293			2,670	478	632	693	408	392	345
12	324	372	278			2,520	478	533	735	460	376	315
13	340	372	263		3,060	2,880	442	496	612	425	552	300
14	356	372	210		4,360	5,520	442	460	442	376	496	345
15	340	356	221	320	6,490	5,660	442	442	442	318	425	315
16	340	340			6,310	5,380	442	425	408	262	442	285
17	372	340			5,520	4,770	442	392	425	376	572	300
18	406	372			4,880	4,090	514	408	425	376	514	270
19	406	372			4,270	3,120	533	533	392	361	442	270
20	389	356			3,670	2,220	552	693	376	361	392	270
21	372	340				1,470	514	714	392	346	304	270
22	372	340				1,140	552	1,440	392	332	276	270
23	406	308	400			1,060	592	2,570	392	290	361	270
24	406	308				801	592	2,220	376	376	332	270
25	424	308			3,200	735	552	1,920	376	361	304	256
26	500	293		305		693	496	1,740	361	361	304	256
27	424	293				672	460	1,860	460	361	304	270
28	480	308				632	425	1,650	572	361	304	270
29	600	340			......	592	425	1,340	496	460	290	256
30	620	340			......	572	552	801	392	376	290	276
31	500	......			......	572	......	714	......	376	304	......

Note.—Discharge estimated Dec. 16 to Feb. 12 and Feb. 21-28, because of ice, from gage heights, observer's notes, weather records, and flow of Pecatonica River at Dill, Wis. Braced figures show mean daily discharge for periods indicated.

Monthly discharge of Pecatonica River at Freeport, Ill., for the year ending Sept. 30, 1918.

[Drainage area, 1,330 square miles.]

Month.	Discharge in second-feet.				Run-off (depth in inches).
	Maximum.	Minimum.	Mean.	Per square mile.	
October	620	308	386	0.290	0.33
November	540	293	377	.283	.32
December	....	210	342	.257	.30
January	....		320	.241	.28
February	6,490	....	2,440	1.83	1.91
March	6,140	572	3,210	2.41	2.78
April	592	425	505	.380	.42
May	2,570	392	869	.653	.75
June	735	361	497	.374	.42
July	533	262	393	.295	.34
August	572	276	357	.268	.31
September	361	218	285	.214	.24
The year	6,490	210	823	.619	8.40

SUGAR RIVER NEAR BRODHEAD, WIS.

LOCATION.—In sec. 26, T. 2 N., R. 9 E., at highway bridge 2 miles southwest of Brodhead, Green County, 12 miles above Illinois State line. Jordan Creek enters from right 2 miles below station, and Little Jordan Creek, also from right, 4 miles above.

DRAINAGE AREA.—529 square miles (measured on map issued by Wisconsin Geological and Natural History Survey, edition of 1911; scale, 1 inch = 6 miles).

RECORDS AVAILABLE.—February 7, 1914, to September 30, 1918.

GAGE.—Chain gage attached to downstream side of bridge; read by Arthur Christensen.

DISCHARGE MEASUREMENTS.—Made from upstream side of bridge or by wading.

CHANNEL AND CONTROL.—Bed composed of sand and gravel. Control not well defined. Right bank of medium height; rarely overflowed; left bank at gage overflows at stage of approximately 7 feet on the gage.

EXTREMES OF DISCHARGE.—Maximum stage recorded during year, 7.9 feet March 14 (discharge, 4,350 second-feet); minimum stage recorded, 0.7 foot at 5 p. m. September 8 (discharge, approximately 54 second-feet).

1914–1918: Maximum stage recorded, 11.4 feet September 13, 1915 (discharge, about 13,000 second-feet); minimum stage recorded, 0.7 foot at 5 a. m., September 8, 1918 (water was undoubtedly being held at the dam); discharge determined from extension of rating curve, about 54 second-feet.

ACCURACY.—Stage-discharge relation fairly permanent throughout the year. Control changes somewhat with floods, but not seriously affected during 1918. Rating curve fairly well defined between 108 and 4,500 second-feet. Gage read daily to quarter-tenths. Daily discharge ascertained by applying mean daily gage height to rating table, except for periods when stage-discharge relation is affected by ice, for which it was ascertained by applying to rating table mean daily gage heights corrected for ice effect by means of discharge measurements, observer's notes, and weather records. Open-water records fair; winter records roughly approximate.

Discharge measurements of Sugar River near Brodhead, Wis., during the year ending Sept. 30, 1918.

Date.	Made by—	Gage height.	Discharge.	Date.	Made by—	Gage height.	Discharge.
		Feet.	*Sec.-ft.*			*Feet.*	*Sec.-ft.*
Nov. 9	R. B. Kilgore	1.62	246	May 27	T. G. Bedford	2.08	368
Jan. 4[a]	W. G. Hoyt	2.24	145	Aug. 18[b]	W. G. Hoyt	1.06	121
Feb. 7[a]	T. G. Bedford	2.90	182				

[a] Made through complete ice cover, 600 feet downstream from gage.
[b] Made by wading upstream from bridge.

Daily discharge, in second-feet, of Sugar River near Brodhead, Wis., for the year ending Sept. 30, 1918.

Day.	Oct.	Nov.	Dec.	Jan.	Feb.	Mar.	Apr.	May.	June.	July.	Aug.	Sept.
1	235	306	248	150	140	2,280	322	354	370	306	222	118
2	222	291	197	150	120	2,160	338	354	338	291	197	150
3	235	291	248	145	110	2,100	322	306	322	291	222	173
4	235	262	248	145	150	3,070	276	262	306	210	173	185
5	210	291	248	165	140	3,180	291	235	291	262	235	197
6	197	276	222	140	175	2,490	291	276	291	222	248	150
7	210	276	222	140	185	1,810	370	388	306	197	262	139
8	248	276	222	130	155	1,440	338	388	306	235	235	81
9	248	276	173	125	175	1,190	291	458	276	276	210	150
10	248	276	195	120	130	785	306	458	306	248	197	173
11	262	235	190	130	195	740	291	405	322	235	139	235
12	248	276	190	120	225	965	276	306	338	210	185	197
13	235	276	185	110	195	3,070	262	306	262	222	248	197
14	185	276	185	130	305	4,350	222	291	276	185	235	173
15	235	276	185	120	440	2,880	276	291	248	235	235	150
16	248	262	185	100	660	1,810	262	276	210	235	235	197
17	262	248	190	100	830	1,540	276	262	210	291	235	235
18	322	210	195	160	965	1,290	338	354	248	210	210	162
19	306	248	210	195	1,010	2,160	354	370	248	210	235	185
20	291	248	220	70	965	1,810	262	322	248	210	210	185
21	197	248	235	85	875	1,640	276	291	210	173	210	210
22	276	248	250	85	785	965	354	1,010	248	210	235	162
23	291	248	250	105	660	545	354	875	197	262	210	173
24	306	222	235	95	785	440	306	1,100	276	210	197	197
25	322	173	210	95	1,060	458	276	660	291	197	139	210
26	370	222	195	145	1,340	405	262	475	306	235	222	173
27	528	222	185	130	2,490	388	262	440	291	210	222	185
28	580	222	175	160	2,420	322	262	510	262	128	248	185
29	620	222	175	165	......	354	370	458	235	210	222	150
30	545	248	160	145	......	322	370	405	210	197	185	197
31	338	......	150	130	......	291	......	370	......	197	210	......

NOTE.—Stage-discharge relation affected by ice Dec. 10 to Mar. 2.

Monthly discharge of Sugar River near Brodhead, Wis., for the year ending Sept. 30, 1918.

[Drainage area, 529 square miles.]

Month.	Discharge in second-feet.				Run-off (depth in inches).
	Maximum.	Minimum.	Mean.	Per square mile.	
October	620	185	299	0.565	0.65
November	306	173	255	.482	.54
December	250	150	206	.389	.45
January	195	70	129	.244	.28
February	2,490	110	632	1.19	1.24
March	4,350	291	1,520	2.87	3.31
April	370	222	302	.571	.64
May	1,100	235	428	.809	.93
June	370	197	275	.520	.58
July	306	128	226	.427	.49
August	262	139	215	.406	.47
September	235	81	176	.333	.37
The year	4,350	70	388	.733	9.96

IOWA RIVER AT MARSHALLTOWN, IOWA.

LOCATION.—In sec. 23, T. 84 N., R. 18 W., at Third Avenue highway bridge, 1 mile north of Marshalltown, Marshall County, and about 1 mile below site of old gaging station.

DRAINAGE AREA.—1,380 square miles (measured on map issued by United States Geological Survey; scale, 1 to 500,000).

RECORDS AVAILABLE.—May 21, 1915, to September 30, 1918; February 23, 1903, to August 8, 1903, from old site 1 mile above present station.

Gage.—Chain gage attached to downstream handrail of bridge, 60 feet from right pier; read by B. S. Beehrle.

Discharge measurements.—Made from downstream side of bridge, to which gage is attached.

Channel and control.—Bed of stream sandy and subject to change. Right bank not subject to overflow; left bank will be overflowed at stages about 13 feet.

Extremes of discharge.—Maximum and minimum stages ever recorded occurred during 1918; maximum stage, 17.74 feet June 4 (discharge, 42,000 second-feet); minimum stage recorded, 1.86 feet November 24 (discharge, estimated 2 second-feet).

Ice.—Stage-discharge relation seriously affected by ice December 9 to March 4; observations discontinued during that period.

Accuracy.—Stage-discharge relation not permanent. Three rating curves, none of them very well defined, used during 1918. Gage read once daily to hundredths. Daily discharge ascertained by applying daily gage height to rating table, except for period when stage-discharge relation was affected by ice, for which it was not determined. Open-water records fair.

Discharge measurements of Iowa River at Marshalltown, Iowa, during the year ending Sept. 30, 1918.

Date.	Made by—	Gage height.	Discharge.
		Feet.	*Sec.-ft.*
Mar. 25	Bolster and Gregg	3.35	479
June 7	A. Davis	15.36	18,700

Daily discharge, in second-feet, of Iowa River at Marshalltown, Iowa, for the year ending Sept. 30, 1918.

Day.	Oct.	Nov.	Dec.	Mar.	Apr.	May.	June.	July.	Aug.	Sept.
1	94	97	87	....	300	270	6,820	1,190	427	394
2	81	103	94	....	300	256	6,970	1,060	410	410
3	91	110	100	....	315	256	10,100	977	346	316
4	94	113	68	....	285	241	39,400	852	346	301
5	97	113	65	496	285	241	35,200	1,190	362	272
6	100	113	62	514	270	227	24,600	1,020	362	258
7	12	113	56	532	270	213	6,240	1,280	316	244
8	16	129	5	532	270	227	6,110	1,920	286	230
9	33	113	....	569	256	569	15,100	1,870	258	230
10	44	113	....	496	256	461	9,140	1,620	230	216
11	97	113	....	532	241	362	6,820	1,330	216	230
12	62	129	....	496	241	444	5,270	1,150	202	230
13	72	146	....	645	241	723	3,210	935	202	244
14	44	100	....	684	227	885	2,670	893	202	230
15	86	84	....	803	227	763	2,130	770	202	316
16	36	87	....	763	256	763	1,820	770	176	331
17	47	97	....	885	270	461	1,570	1,020	189	301
18	62	110	....	1,060	315	1,960	1,280	690	189	272
19	69	113	....	1,320	331	1,240	1,240	530	216	244
20	62	113	....	1,010	300	803	2,870	566	286	230
21	97	97	....	927	315	2,030	4,000	530	461	216
22	129	8	....	885	362	2,670	5,060	461	612	202
23	110	33	....	645	346	3,210	1,970	530	690	202
24	100	2	....	569	362	6,380	2,740	495	690	216
25	78	42	....	478	300	6,520	2,080	427	730	189
26	146	65	....	444	285	1,970	1,710	410	770	189
27	129	62	....	394	270	1,710	1,470	410	770	163
28	113	56	....	362	107	1,620	1,470	495	690	150
29	110	110	....	346	256	1,420	1,330	495	566	163
30	103	97	....	315	270	5,860	1,330	530	461	258
31	110	....	....	315	....	8,580	....	495	427	....

Note.—Discharge Nov. 22 and 24, and Dec. 8 affected by storage above Marshalltown. Daily discharge for these dates estimated. Stage-discharge relation affected by ice Dec. 9 to Mar. 4; daily discharge not determined.

Monthly discharge of Iowa River at Marshalltown, Iowa, for the year ending Sept. 30, 1918.

[Drainage area, 1,380 square miles.]

Month.	Discharge in second-feet.				Run-off (depth in inches).
	Maximum.	Minimum.	Mean.	Per square mile.	
October	146	12	79.7	0.058	0.07
November	146	2	92.7	.067	.07
April	362	107	278	.201	.22
May	8,580	213	1,720	1.24	1.43
June	39,400	1,240	7,060	5.11	5.70
July	1,920	410	868	.628	.73
August	770	176	396	.287	.33
September	410	150	248	.179	.21

IOWA RIVER AT IOWA CITY, IOWA.

Location.—In sec. 15, T. 79 N., R. 6 W., at highway bridge 500 feet below Chicago, Rock Island & Pacific Railway main-line bridge; three-quarters of a mile below Iowa State University's power plant, three-quarters of a mile downstream from old gaging station, which was at county highway bridge a short distance above dam.

Drainage area.—3,140 square miles (measured on map issued by United States Geological Survey; scale, 1 to 500,000).

Records available.—October 30, 1913, to September 30, 1918, at present site; June 11, 1903, to July 21, 1906, at old gaging station.

Gage.—Chain gage, attached to upstream handrail of bridge about 40 feet from left-hand end of first span from left bank; read by A. Kostal.

Discharge measurements.—Made from bridge to which gage is attached, or from a boat about 1,000 feet below highway bridge.

Channel and control.—Bed composed of sand; subject to change. Right bank high and will not be overflowed; left bank will be overflowed at high stage under a pile trestle approach to the bridge and beyond left end of the approach at extremely high stage.

Extremes of discharge.—Maximum stage ever recorded occurred this year; gage height 19.45 feet, June 7 (discharge, 36,200 second-feet); minimum stage during this year, 0.15 foot May 10 (discharge, 190 second-feet); minimum discharge of record, 10 second-feet December 26, 1916.

Ice.—Stage-discharge relation affected by ice during winter period; observations discontinued.

Regulation.—Considerable diurnal fluctuation at low stages, owing to operation of power plant above station.

Accuracy.—Stage-discharge relation shifting. Three rating curves used during 1918; the 1917 curve was used to December 5, and is well defined during the period used; curves used March 10 to June 5, and June 6 to September 30, are not well defined. Gage read once daily to half-tenths. Daily discharge ascertained by applying daily gage heights to rating table, except for period when stage-discharge relation was affected by ice, for which the daily discharge was not determined. All records for 1918 at this station are unsatisfactory on account of persistent shifting of the channel both before and after the record-breaking flood of June.

Discharge measurements of Iowa River at Iowa City, Iowa, during the year ending Sept. 30, 1918.

Date.	Made by—	Gage height.	Discharge.
		Feet.	*Sec.-ft.*
Mar. 25	Bolster and Gregg	2.12	1,170
June 6	A. Davis	16.38	36,200

Daily discharge, in second-feet, of Iowa River at Iowa City, Iowa, for the year ending Sept. 30, 1918.

Day.	Oct.	Nov.	Dec.	Mar.	Apr.	May.	June.	July.	Aug.	Sept.
1	218	200	142		710	685	7,840	2,700	775	1,410
2	200	200	262		710	660	4,780	2,440	950	1,080
3	158	207	152		765	398	5,000	2,180	890	890
4	136	218	190		710	535	9,030	1,780	890	775
5	158	236	200		685	442	12,100	1,860	890	365
6	225	225			660	635	24,400	1,860	830	610
7	300	218			635	585	33,300	7,220	830	665
8	262	225			685	442	35,300	5,750	890	775
9	262	200			610	535	30,700	5,490	775	460
10	184	190		1,340	710	190	25,700	6,010	665	410
11	190	184		1,280	610	635	20,900	5,620	775	775
12	174	207		1,340	560	635	16,900	3,550	775	775
13	184	225		1,220	585	635	14,800	3,060	890	560
14	190	243		1,160	310	635	11,600	2,180	775	560
15	174	262		1,100	352	585	9,180	2,350	665	775
16	190	280		1,280	442	610	7,660	2,350	830	460
17	184	225		1,100	610	930	6,800	2,020	1,080	560
18	158	236		1,220	635	3,060	4,760	2,020	950	460
19	152	190		1,220	710	1,880	3,550	1,700	775	460
20	136	168		1,280	685	2,160	3,850	1,630	665	460
21	136	174		1,220	820	2,380	4,400	1,410	775	460
22	152	190		1,280	765	2,300	4,400	1,410	775	460
23	158	200		1,340	765	2,300	4,640	1,270	775	460
24	158	225		1,220	738	6,430	3,960	2,020	775	410
25	152	262		1,160	738	7,120	4,520	1,270	890	410
26	158	236		1,100	738	8,560	4,520	1,200	610	365
27	168	207		1,040	710	14,200	4,640	665	665	365
28	136	207		875	685	13,500	4,290	1,140	560	365
29	152	190		875	738	12,800	3,550	1,080	560	342
30	168	168		765	685	11,900	3,160	1,010	3,850	365
31	184			765		11,300		950	2,790	

NOTE.—Daily discharge at low and medium stages, unsatisfactory; at high stages they are considered reliable; should be used with caution on account of persistent shifting of the channel during the year. Stage-discharge relation affected by ice Dec. 6 to Mar. 9; daily discharge not determined.

Monthly discharge of Iowa River at Iowa City, Iowa, for the year ending Sept. 30, 1918.

[Drainage area, 3,140 square miles.]

Month.	Discharge in second-feet.				Run-off (depth in inches).
	Maximum.	Minimum.	Mean.	Per square mile.	
October	300	136	179	0.057	0.07
November	280	168	213	.068	.08
April	820	310	659	.209	.23
May	14,200	190	3,510	1.13	1.30
June	35,300	3,160	11,000	3.50	3.90
July	7,220	665	2,490	.793	.91
August	3,850	560	955	.304	.35
September	1,410	342	576	.183	.20

IOWA RIVER AT WAPELLO, IOWA.

LOCATION.—In sec. 27, T. 74 N., R. 3 W., at highway bridge half a mile from railroad station at Wapello, Louisa County, and 20 miles from mouth of Iowa River. No large tributaries enter near station.

DRAINAGE AREA.—At gaging station, 12,480 square miles; at mouth, 12,600 square miles (measured on map issued by United States Geological Survey; scale, 1 to 500,000).

RECORDS AVAILABLE.—February 26, 1915, to September 30, 1918.

GAGE.—Chain gage attached near center of first span from right abutment; read by C. W. Warren.

DISCHARGE MEASUREMENTS.—Made from bridge to which gage is attached.

CHANNEL AND CONTROL.—Bed composed of sand and gravel; shifts slightly. Right bank high and will not be overflowed. Levee along left bank broke, causing considerable flooding of cultivated land in June, 1918.

EXTREMES OF DISCHARGE.—Maximum stage recorded during year, 14.94 feet, 6 p. m. June 8 (discharge, 63,100 second-feet); minimum stage recorded, 0 foot December 11 (discharge affected by ice). The flood of June, 1892, was probably much higher than the flood of 1918.

ICE.—Stage-discharge relation seriously affected by ice.

ACCURACY.—Stage-discharge relation nearly permanent. Two rating curves used during 1918; well defined throughout. Gage read once daily to hundredths. Daily discharge ascertained by applying daily gage height to rating table, except for the period February 21–25, when stage-discharge relation was affected by ice, for which it was ascertained from occasional gage readings and temperature records; stage-discharge relation was also affected by ice from December 6 to February 12, but daily discharges were not determined. Open-water records good; winter records fair.

The following discharge measurement was made by Bolster and Gregg:

March 28: Gage height, 3.16 feet; discharge, 7,090 second-feet.

Daily discharge, in second-feet, of Iowa River, at Wapello, Iowa, for the year ending Sept. 30, 1918.

Day.	Oct.	Nov.	Dec.	Feb.	Mar.	Apr.	May.	June.	July.	Aug.	Sept.
1	1,770	1,630	1,560	....	7,790	4,190	2,360	23,400	11,000	7,020	7,880
2	1,770	1,630	1,500	....	7,790	3,990	2,360	22,100	10,700	5,690	6,470
3	1,700	1,770	1,500	....	8,660	3,790	2,280	20,500	11,700	5,440	5,690
4	1,630	1,770	1,440	....	9,260	3,590	2,210	22,100	11,700	4,970	4,530
5	1,630	1,700	1,310	....	8,960	3,590	2,210	28,300	12,100	4,740	4,300
6	1,630	1,700	....	....	7,790	3,400	2,210	37,400	16,100	4,520	4,090
7	1,630	1,700	....	....	7,510	3,400	2,280	55,800	16,900	4,300	3,880
8	1,770	1,630	....	....	6,470	3,400	2,280	59,600	19,200	3,880	3,670
9	1,630	1,630	....	....	5,980	3,210	2,280	60,300	17,700	3,470	3,670
10	1,560	1,630	....	....	5,280	3,030	2,360	58,300	15,700	3,280	3,670
11	1,560	1,630	....	....	5,060	2,850	2,360	53,900	15,400	3,280	3,670
12	1,560	1,700	....	....	4,840	2,680	2,680	46,500	13,900	3,280	3,670
13	1,630	1,630	....	16,900	5,060	2,680	3,210	39,000	11,000	3,280	3,470
14	1,630	1,630	....	22,100	5,060	2,520	4,190	32,200	9,410	3,470	3,470
15	1,630	1,630	....	21,300	5,060	2,520	4,400	24,200	8,480	3,470	3,470
16	1,500	1,630	....	19,200	5,060	2,520	4,620	19,600	8,180	3,470	3,670
17	1,500	1,630	....	13,200	5,280	2,520	4,840	16,500	7,300	3,470	3,670
18	1,560	1,630	....	11,800	5,510	2,520	5,060	14,600	6,740	4,090	3,470
19	1,560	1,630	....	10,200	5,980	2,680	9,570	13,200	6,470	5,940	3,470
20	1,630	1,500	....	7,510	6,470	2,680	7,510	11,700	6,740	6,740	3,280
21	1,560	1,500	....	4,500	7,240	2,680	7,510	11,400	6,470	7,300	3,090
22	1,500	1,500	....	4,000	7,510	2,680	8,660	11,000	6,200	7,590	2,070
23	1,440	1,500	....	5,000	8,660	2,680	8,360	11,000	6,200	8,480	2,730
24	1,440	1,500	....	7,000	8,360	2,680	21,700	11,000	6,470	9,410	2,730
25	1,440	1,500	....	10,000	8,360	2,680	23,800	11,400	6,740	9,410	2,730
26	1,560	1,500	....	9,260	8,360	2,680	26,500	11,700	5,440	9,730	2,560
27	1,630	1,500	....	8,070	7,510	2,680	30,700	11,700	4,740	9,410	2,560
28	1,770	1,560	....	7,790	6,470	2,520	31,200	15,400	4,740	9,100	2,560
29	1,770	1,560	....	....	5,980	2,520	32,700	12,400	5,200	9,100	2,560
30	1,700	1,560	....	....	5,060	2,360	29,200	11,000	5,440	9,100	2,560
31	1,630	....	....	....	4,620	....	26,500	....	5,690	10,000	....

NOTE.—Stage-discharge relation affected by ice Dec. 6 to Feb. 12 and Feb. 21–25; daily discharge for latter period determined from gage heights corrected for ice effect by means of temperature records.

Monthly discharge of Iowa River at Wapello, Iowa, for the year ending Sept. 30, 1918.

[Drainage area, 12,480 square miles.]

Month.	Discharge in second-feet.				Run-off (depth in inches).
	Maximum.	Minimum.	Mean.	Per square mile.	
October	1,770	1,440	1,610	0.129	0.15
November	1,770	1,500	1,610	.129	.14
March	9,260	4,620	6,680	.535	.62
April	4,190	2,360	2,930	.235	.26
May	32,700	2,210	10,300	.825	.95
June	60,300	11,000	25,900	2.07	2.31
July	19,200	4,740	9,670	.775	.89
August	10,000	3,280	6,010	.481	.55
September	7,880	2,070	3,640	.292	.33

CEDAR RIVER AT JANESVILLE, IOWA.

Location.—In sec. 35, T. 91 N., R. 14 W., at Illinois Central Railroad bridge a quarter of a mile below highway bridge and 3 miles above junction with Shellrock River.

Drainage area.—1,660 square miles (measured on map issued by United States Geological Survey; scale, 1 to 500,000).

Records available.—April 26, 1905, to September 30, 1906; May 28, 1915, to September 30, 1918.

Gage.—Chain gage attached to upstream guardrail of bridge about center of left span; read by James Townsend.

Discharge measurements.—Made from upstream side of railroad bridge.

Channel and control.—Bed composed of gravel; shifting. Banks high and not subject to overflow.

Extremes of discharge.—Maximum stage recorded during year, 8.9 feet, March 20 (discharge, 7,220 second-feet); minimum stage recorded, 0.72 foot October 17 (discharge, 165 second-feet).

1905–6 and 1915–1917: Maximum discharge occurred March 28, 1906 (discharge, 22,600 second-feet); minimum stage recorded, 0.72 foot, October 17, 1917 (discharge, 165 second-feet).

Ice.—Stage-discharge relation seriously affected by ice; observations discontinued during winter.

Regulation.—May be slight diurnal fluctuation of water level owing to operation of power plant at Waverly, 9 miles above station.

Accuracy.—Stage-discharge relation nearly permanent. Rating curve used October 1 to July 29, well defined throughout; from July 30 to September 30, a series of transition curves were used to allow for backwater caused by construction of 4 new piers in the gaging section. Gage read once daily to hundredths. Daily discharge ascertained by applying daily gage heights to rating table, except July 30 to September 30. Stage-discharge relation affected by ice December 6 to March 16; daily discharges not determined. Records excellent October to July and fair August and September.

The following discharge measurement was made by Bolster and Gregg:

March 23: Gage height, 4.48 feet; discharge, 2,090 second-feet.

Daily discharge, in second-feet, of Cedar River at Janesville, Iowa, for the year ending Sept. 30, 1918.

Day.	Oct.	Nov.	Dec.	Mar.	Apr.	May.	June.	July.	Aug.	Sept.
1	537	302	211		471	430	2,400	537	720	490
2	410	288	223		583	316	2,320	606	720	510
3	352	371	214		559	334	3,620	537	580	640
4	352	271	220		559	316	4,170	493	580	560
5	410	281	232		493	316	2,480	537	580	490
6	514	267			493	281	2,240	752	580	400
7	371	261			430	274	1,750	802	510	350
8	267	248			352	312	1,390	630	460	370
9	390	410			334	1,570	1,170	559	440	420
10	352	267			430	752	1,120	559	420	380
11	309	271			430	903	1,010	537	380	670
12	242	217			410	1,390	903	430	390	890
13	275	236			410	1,120	703	430	380	670
14	281	255			410	852	703	430	400	640
15	242	232			334	703	679	430	400	540
16	179	267			390	559	654	451	700	680
17	165	239		2,830	430	630	559	703	3,840	680
18	217	255		3,840	410	606	606	728	2,830	560
19	248	236		5,130	390	1,170	728	654	4,060	500
20	205	275		7,220	390	3,210	852	606	4,060	440
21	239	288		5,520	371	3,110	1,010	583	3,020	370
22	242	309		3,110	334	3,940	1,280	537	1,950	310
23	245	232		2,020	390	1,570	955	493	1,690	290
24	226	236		1,750	371	1,750	728	493	1,200	290
25	242	236		1,230	352	1,340	752	537	1,140	310
26	236	248		1,060	352	1,060	752	852	1,020	550
27	261	255		1,010	352	900	679	3,720	850	580
28	255	248		903	352	1,400	703	2,560	620	530
29	223	214		802	371	2,240	654	1,750	850	500
30	371	217		583	390	3,210	703	720	620	450
31	334			630		3,210		780	590	

NOTE.—Discharge May 26 and 27 and Sept. 29 and 30, estimated from Clarksville discharge. Discharge Sept. 17-21 interpolated. Stage-discharge relation affected by ice Dec. 6 to Mar. 16; daily discharge not determined.

Monthly discharge of Cedar River at Janesville, Iowa, for the year ending Sept. 30, 1918.

[Drainage area, 1,660 square miles.]

Month.	Discharge in second-feet.				Run-off (depth in inches).
	Maximum.	Minimum.	Mean.	Per square mile.	
October	537	165	296	0.178	0.20
November	410	214	264	.159	.18
April	583	334	411	.248	.28
May	3,940	274	1,280	.771	.89
June	4,170	559	1,280	.771	.86
July	3,720	430	788	.475	.55
August	4,060	380	1,180	.711	.82
September	890	290	499	.301	.34

CEDAR RIVER AT CEDAR RAPIDS, IOWA.

LOCATION.—In sec. 28, T. 83 N., R. 7 W., in central part of Cedar Rapids, Linn County, half a mile below dam, between electric-railroad bridge and Eighth Avenue bridge.

DRAINAGE AREA.—At gaging station, 6,640 square miles; at junction with Iowa River, 7,930 square miles (measured on map issued by United States Geological Survey; scale, 1 to 500,000).

RECORDS AVAILABLE.—October 26, 1902, to September 30, 1918.

GAGE.—Inclined staff gage fastened to posts driven in right bank of river in rear of plant of Iowa Windmill & Pump Co. plant; read by R. S. Toogood. Elevation of zero of gage from Northwestern Railroad levels, 723.03 feet above sea level.

DISCHARGE MEASUREMENTS.—Made from different bridges in the vicinity of gage, according to the stage.

CHANNEL AND CONTROL.—Bed composed of rock and gravel; free from vegetation; practically permanent.

EXTREMES OF DISCHARGE.—Maximum stage recorded during year, 10.9 feet, June 7 (discharge, 27,800 second-feet); minimum stage recorded during year, 2.65 feet, various dates (discharge, 460 second-feet).

1902–1918: Maximum stage recorded, 17.2 feet April 1, 1912, and March 26, 1917 (discharge, 54,200 second-feet); minimum stage recorded, 2.65 feet, July 24–28, 1911 (discharge, 410 second-feet). Greatest known flood probably occurred in June, 1851, when the maximum stage was about 20 feet, and the discharge about 65,000 second-feet.

ICE.—Stage-discharge relation affected by ice, except in very mild winters, when the swift current and the proximity to power plant keep the measuring section open.

REGULATION.—Power dam above gaging station since 1917 produces marked effect on gage readings. There is no dam below gage which might cause backwater.

ACCURACY.—Stage-discharge relation nearly permanent. Rating curve well defined. Gage read once daily, to tenths. Daily discharge ascertained by applying daily gage height to rating table, except for period when stage-discharge relation was affected by ice, for which discharges were not determined. Open-water records excellent.

COOPERATION.—Gage-height record furnished by United States Weather Bureau.

The following discharge measurement was made by Bolster and Gregg:

March 24: Gage height, 5.86 feet; discharge, 8,300 second-feet.

Daily discharge, in second-feet, of Cedar River at Cedar Rapids, Iowa, for the year ending Sept. 30, 1918.

Day.	Oct.	Nov.	Dec.	Feb.	Mar.	Apr.	May.	June.	July.	Aug.	Sept.
1	1,100	680	805	....	5,180	3,050	1,280	10,000	4,010	5,180	3,320
2	945	945	460	....	5,180	2,550	1,280	14,800	4,010	3,590	2,800
3	945	680	680	....	4,590	2,320	1,280	16,300	3,870	2,550	2,550
4	1,100	805	680	....	4,590	2,080	1,280	17,100	3,590	2,550	2,550
5	1,100	680	680	....	4,010	2,080	1,280	24,200	3,590	2,550	2,800
6	945	680	....	....	3,870	1,860	1,280	23,400	3,870	2,320	2,550
7	945	1,100	....	....	3,870	1,860	1,280	26,200	4,590	2,320	2,320
8	945	680	....	....	3,590	1,660	1,280	26,200	5,790	2,080	2,080
9	945	680	....	....	3,320	1,860	1,460	21,800	7,050	2,080	2,080
10	945	805	....	....	3,320	1,460	2,080	17,900	5,790	2,320	1,860
11	945	805	....	....	3,590	1,280	2,320	14,000	5,790	2,080	2,320
12	805	680	....	....	3,590	1,460	3,870	9,680	4,590	1,860	2,080
13	945	565	....	....	3,870	1,460	5,180	8,670	4,010	1,860	1,860
14	945	680	....	....	3,590	1,460	5,790	6,410	3,870	1,860	2,320
15	805	565	....	....	4,590	1,460	5,790	5,180	3,590	1,860	2,800
16	945	565	....	....	4,010	1,460	5,180	4,590	3,050	2,080	2,550
17	680	680	....	....	4,010	1,460	4,010	4,010	3,050	2,080	2,550
18	945	680	....	....	5,790	1,280	5,180	4,010	3,590	2,080	2,320
19	680	680	....	....	6,410	1,460	5,790	3,870	3,870	4,590	2,320
20	680	945	....	....	7,050	1,660	5,180	3,590	3,590	7,050	2,080
21	680	680	....	....	7,690	1,460	5,790	3,590	3,320	8,340	2,080
22	945	565	....	....	8,340	1,460	7,690	3,870	3,050	8,340	1,860
23	680	680	....	....	10,400	1,460	9,340	4,590	3,050	8,340	1,860
24	680	680	....	....	9,000	1,280	11,400	5,180	2,800	7,690	1,860
25	680	680	....	....	7,050	1,280	12,500	5,180	2,550	6,410	1,860
26	680	680	....	5,790	6,410	1,280	11,100	4,590	2,320	5,790	1,860
27	1,280	565	....	6,410	4,590	1,280	8,340	4,010	2,800	3,590	1,660
28	1,100	680	....	5,790	3,870	1,280	9,680	4,010	3,050	4,010	1,660
29	680	805	....	....	3,590	1,280	7,690	4,590	5,790	4,010	1,660
30	680	805	....	....	3,320	1,280	6,410	4,590	6,410	5,180	1,660
31	805	....	....	....	3,050	....	7,690	....	5,180	3,590	....

NOTE.—Stage-discharge relation affected by ice Dec. 6 to Feb. 25; daily discharge not determined.

Monthly discharge of Cedar River at Cedar Rapids, Iowa, for the year ending Sept. 30, 1918.

[Drainage area, 6,640 square miles.]

Month.	Discharge in second-feet.				Run-off (depth in inches).
	Maximum.	Minimum.	Mean.	Per square mile.	
October	1,260	680	876	0.132	0.15
November	1,100	565	713	.107	.12
March	10,400	3,050	5,010	.755	.87
April	3,050	1,280	1,620	.244	.27
May	12,500	1,280	5,150	.775	.89
June	26,200	3,590	10,200	1.54	1.72
July	7,050	2,320	4,050	.610	.70
August	8,340	1,860	3,880	.584	.67
September	3,320	1,660	2,200	.331	.37

SHELLROCK RIVER NEAR CLARKSVILLE, IOWA.

LOCATION.—In T. 92 N., R. 16 W., at highway bridge 1¼ miles northwest of Clarksville, Butler County, and 25 miles above junction with Cedar River. No large tributaries enter for several miles up and down stream.

DRAINAGE AREA.—1,660 square miles at station and 2,680 square miles at junction with Cedar River (measured on map issued by United States Geological Survey; scale, 1 to 500,000).

RECORDS AVAILABLE.—May 28, 1915, to September 30, 1918.

GAGE.—Chain gage attached to handrail on upstream side of bridge 75 feet from right abutment; read by Mrs. H. H. Sherburne.

DISCHARGE MEASUREMENTS.—Made from downstream side of bridge to which gage is attached.

CHANNEL AND CONTROL.—Bed composed of rock and sand; probably permanent. Right bank high and will not be overflowed; left bank will probably be overflowed during extreme high stage.

EXTREMES OF DISCHARGE.—Maximum stage recorded during year, 10.4 feet, August 17 (discharge, 9,380 second-feet); minimum stage, 1.2 feet November 28 (discharge, 135 second-feet).

1915–1918: Maximum stage recorded, 14.7 feet, March 22, 1917 (probably affected by ice); minimum stage recorded, 1.15 feet October 23, 1916 (discharge, 125 second-feet). In April, 1907, a stage of approximately 16.5 feet was reached (discharge, about 19,000 second-feet).

ICE.—Stage-discharge relation affected by ice and observations discontinued during winter.

ACCURACY.—Stage-discharge relation practically permanent; rating curve well defined between 200 and 10,000 second-feet; not well defined outside these limits. Gage read once daily to hundredths. Daily discharge ascertained by applying daily gage height to rating table, except for the following periods; July 28–29, estimated from Janesville, October 1–8, November 7, 8, 10, 12, 14, and 15, discharge interpolated; December 4 to March 18, discharge not determined because of ice effect. Records excellent, except for extremely low and high stages, which are fair.

The following discharge measurement was made by Bolster and Gregg:

March 24: Gage height, 3.42 feet; discharge, 1,290 second-feet.

Daily discharge, in second-feet, of Shellrock River near Clarksville, Iowa, for the year ending Sept. 30, 1918.

Day.	Oct.	Nov.	Dec.	Mar.	Apr.	May.	June.	July.	Aug.	Sept.
1	203	188	155	....	485	255	2,730	410	460	592
2	197	212	155	....	460	240	2,630	410	410	565
3	191	212	155	....	435	240	2,530	410	365	538
4	185	200	....	....	410	225	3,580	365	345	538
5	179	240	....	....	410	200	2,730	410	325	510
6	173	225	....	....	388	188	2,530	592	308	460
7	167	220	....	....	365	188	2,060	650	308	435
8	161	216	....	....	325	200	1,490	538	290	410
9	155	212	....	....	325	772	1,280	460	290	365
10	155	206	....	....	325	4,040	1,000	435	290	365
11	175	200	....	....	308	3,360	870	365	272	620
12	155	200	....	....	290	1,980	710	345	272	935
13	155	200	....	....	272	1,340	650	325	272	740
14	155	192	....	....	272	1,000	510	325	272	592
15	175	183	....	....	272	805	485	565	272	538
16	165	175	....	....	272	680	435	620	2,830	510
17	155	175	....	....	290	538	410	538	9,090	485
18	188	175	....	....	388	680	388	460	6,570	435
19	188	175	....	3,360	308	935	388	410	4,400	410
20	188	175	....	3,700	308	650	565	365	3,360	410
21	188	175	....	2,340	308	592	435	325	2,440	365
22	188	175	....	1,810	308	592	388	325	2,060	345
23	175	165	....	1,500	290	565	388	290	1,900	345
24	175	165	....	1,280	255	565	388	290	1,730	325
25	175	155	....	1,000	240	510	365	290	1,420	325
26	200	155	....	870	240	460	365	2,240	1,200	308
27	200	145	....	740	225	435	410	2,440	1,000	308
28	240	135	....	650	240	1,980	620	2,000	806	290
29	225	145	....	592	272	3,930	485	1,000	710	272
30	212	165	....	565	255	4,880	388	538	620	272
31	188	....	....	510	....	3,140	....	485	592	....

NOTE.—Discharge July 28 and 29 estimated from Janesville discharge. Discharge Oct. 1 to 8, Nov. 7, 8, 10, 12, 14, and 15 interpolated. Stage-discharge relation affected by ice Dec. 4 to Mar. 19; daily discharge not determined.

Monthly discharge of Shellrock River near Clarksville, Iowa, for the year ending Sept. 30, 1918.

[Drainage area, 1,660 square miles.]

Month.	Discharge in second-feet.				Run-off (depth in inches).
	Maximum.	Minimum.	Mean.	Per square mile.	
October	240	155	182	0.109	0.12
November	240	135	185	.111	.13
April	485	225	318	.192	.22
May	4,880	188	1,170	.704	.81
June	3,580	365	1,070	.645	.74
July	2,440	290	620	.373	.43
August	9,090	272	1,470	.886	1.02
September	935	272	454	.273	.31

SKUNK RIVER AT COPPOCK, IOWA.

LOCATION.—In sec. 36, T. 74 N., R. 8 W., at highway bridge one-eighth of a mile above Chicago, Burlington & Quincy Railroad bridge and a quarter of a mile above junction with Crooked Creek.

DRAINAGE AREA.—2,890 square miles (measured on map issued by United States Geological Survey; scale, 1 to 500,000).

Records available.—October 21, 1913, to September 30, 1918.

Gage.—Chain gage attached to downstream side of bridge; read by J. W. Ricks.

Discharge measurements.—Made from bridge to which gage is attached.

Channel and control.—Bed composed of gravel and sand; shifting.

Extremes of discharge.—Maximum stage recorded during year, 19.7 feet, 7.30 p. m. June 9 (discharge, 19,600 second-feet); minimum discharge recorded, 78 second-feet October 13.

1913–1918: Maximum stage recorded, approximately 24 feet, May, 1903 (discharge, 30,000 second-feet); minimum discharge, 52 second-feet, October 17, 1917.

Ice.—Stage-discharge relation seriously affected by ice; observations discontinued during winter.

Accuracy.—Stage-discharge relation changed during high water of February and again during high water of June, requiring use of two rating curves, one applicable October 1 to December 5 and June 11 to September 30, and the other applicable February 14 to June 10; both are fairly well defined. Gage read once daily to hundredths. Daily discharge ascertained by applying daily gage height to rating table, except for periods when stage-discharge relation was affected by ice, for which daily discharges were not determined. Daily discharge interpolated June 23 and August 15. Open-water records good.

The following discharge measurement was made by Bolster and Gregg:

March 28: Gage height, 3.20 feet; discharge, 348 second-feet.

Daily discharge, in second-feet, of Skunk River at Coppock, Iowa, for the year ending Sept. 30, 1918.

Day.	Oct.	Nov.	Dec.	Feb.	Mar.	Apr.	May.	June.	July.	Aug.	Sept.
1	104	114	104	....	1,160	317	317	4,480	2,960	310	520
2	95	114	104	....	1,280	302	302	3,920	1,760	295	835
3	104	114	104	....	1,780	288	288	3,830	1,300	280	835
4	114	114	104	....	1,780	260	274	3,400	995	265	645
5	104	114	104	....	1,340	260	260	5,700	940	265	405
6	95	104	....	....	1,160	317	260	8,540	885	250	340
7	95	104	....	....	1,000	317	348	10,700	1,420	238	265
8	95	114	....	....	890	317	288	14,000	2,180	226	226
9	95	104	....	....	840	288	288	18,000	1,490	214	214
10	95	104	....	....	645	288	430	18,800	1,680	202	202
11	95	114	....	....	600	274	274	16,200	2,100	202	202
12	86	104	....	....	600	274	246	13,500	1,760	190	190
13	78	104	....	....	560	260	220	11,300	1,300	179	179
14	95	104	....	2,840	560	260	233	9,810	1,060	179	168
15	104	124	....	3,660	520	246	233	8,670	885	179	179
16	95	104	....	3,570	520	317	233	7,100	835	179	168
17	104	104	....	3,230	520	348	220	6,020	735	226	157
18	157	104	....	2,770	520	348	317	4,840	690	340	157
19	124	114	....	1,650	501	332	1,060	3,830	645	310	146
20	104	104	....	....	464	317	690	2,260	600	310	135
21	104	104	....	....	447	348	740	1,680	520	1,620	135
22	95	104	....	....	430	364	1,160	1,560	520	885	124
23	95	104	....	....	430	364	840	1,430	480	600	114
24	86.	104	....	....	430	348	2,120	1,300	690	1,060	114
25	86	104	....	890	396	348	4,200	1,760	560	480	124
26	114	95	....	690	380	348	3,480	1,820	440	440	114
27	208	114	....	740	364	348	3,830	2,330	405	340	114
28	124	104	....	840	364	348	5,170	2,960	388	280	114
29	124	104	....	....	348	348	7,100	3,290	370	250	104
30	114	114	....	....	332	332	7,100	2,480	355	355	114
31	114	....	....	....	332	....	6,620	....	340	370	....

Note.—Daily discharge interpolated June 23 and Aug. 15. Stage-discharge relation affected by ice Dec. 6 to Feb. 13 and Feb. 20–24: daily discharge not determined.

Monthly discharge of Skunk River at Coppock, Iowa, for the year ending Sept. 30, 1918.

[Drainage area, 2,890 square miles.]

Month.	Discharge in second-feet.				Run-off (depth in inches).
	Maximum.	Minimum.	Mean.	Per square mile.	
October	208	78	107	0.037	0.04
November	124	95	108	.037	.04
March	1,780	332	693	.240	.28
April	364	246	314	.109	.12
May	7,100	220	1,580	.547	.63
June	18,800	1,300	6,520	2.25	2.51
July	2,960	340	1,010	.349	.40
August	1,620	179	372	.129	.15
September	835	104	245	.085	.09

SKUNK RIVER AT AUGUSTA, IOWA.

LOCATION.—In sec. 26, T. 69 N., R. 4 W., at highway bridge one-third of a mile from Augusta post office, Des Moines County, and 12.2 miles from mouth of Skunk River, where it empties into pond of Mississippi River Power Co., 32.2 miles above dam at Keokuk, Iowa.

DRAINAGE AREA.—At gaging station, 4,290 square miles; at mouth, 4,350 square miles (measured on map issued by United States Geological Survey; scale, 1 to 500,000).

RECORDS AVAILABLE.—September 30, to November 15, 1913; May 27, 1915, to September 30, 1918.

GAGE.—Chain gage attached to downstream handrail of bridge about 95 feet from left abutment; read once daily by L. E. Williamson. Staff gage attached to downstream left side of middle pier, used by engineers of the Hydraulic Engineering Co. of Maine during 1913. Datum of staff gage approximately 0.73 feet higher than datum of chain gage. Staff gage taken out by ice in spring of 1914.

DISCHARGE MEASUREMENTS.—Made from bridge to which gage is attached or by wading.

CHANNEL AND CONTROL.—Bed of stream sandy and subject to change. Right bank high and will not be overflowed; left bank will only be overflowed at extremely high stage. Remains of old mill dam 600 feet below gage will probably make stage-discharge relation fairly permanent. The riffle at the dam causes a drop of 3 feet at medium low stage. Backwater from the Mississippi may occur once in about 50 years.

EXTREMES OF DISCHARGE.—Prior to 1918: Maximum stage recorded approximately 21 feet about June 1, 1903 (discharge, nearly 45,000 second-feet); minimum discharge recorded, 63 second-feet November 10, 1913; absolute minimum discharge at this station probably 25 second-feet or less.

ICE.—Stage-discharge relation affected by ice December 5 to February 25.

Gage height records withheld from publication until further information can be obtained with which to correct them.

Discharge measurements of Skunk River at Augusta, Iowa, during the year ending Sept. 30, 1918.

Date.	Made by—	Gage height.	Discharge.	Date.	Made by—	Gage height.	Discharge.
		Feet.	*Sec.-ft.*			*Feet.*	*Sec.-ft.*
Mar. 9	Davis and Gregg	3.03	1,020	June 11	A. Davis	16.32	24,400
May 25	.....do	7.02	7,710	13	Bolster and Hodge	13.51	17,100
30	.....do	12.06	14,500	Sept. 25	A. Davis	1.60	105

NOTE.—Gage heights liable to ±0.1 foot error.

DES MOINES RIVER AT KALO, IOWA.

LOCATION.—In sec. 17, T. 88 N., R. 28 W., at highway bridge at Kalo, Webster County, 1½ miles east of Otho, a station on Minneapolis & St. Louis Railroad, and 1½ miles above mouth of Holiday Creek, which enters from left.

DRAINAGE AREA.—4,170 square miles (measured on map issued by United States Geological Survey, scale, 1 to 500,000).

RECORDS AVAILABLE.—October 18, 1913, to September 30, 1918, except October, 1914, to March 21, 1915, when the station was temporarily discontinued.

GAGE.—Chain gage attached to downstream side of bridge in middle of right span; read by S. C. Fuller.

DISCHARGE MEASUREMENTS.—Made from bridge, to which gage is attached, or by wading.

CHANNEL AND CONTROL.—No well-defined control. Bed composed of gravel and is fairly permanent. Point of zero flow estimated to be at gage height −1.0 foot.

EXTREMES OF DISCHARGE.—Maximum stage recorded during year, 9.8 feet June 4 (discharge, 11,400 second-feet); minimum stage recorded, 0.5 foot for various days in October, November, and December (discharge, 128 second-feet).

1913–1918: Maximum stage recorded, 14.0 feet, May 30, 1915 (discharge, 18,500 second-feet); minimum stage, 0.2 foot October 5, 1917 (discharge, 57 second-feet).

ICE.—Stage-discharge relation affected by ice and observations discontinued during winter.

ACCURACY.—Stage-discharge relation permanent throughout year. Rating curve well defined between 200 and 12,000 second-feet; extended below 200 second-feet and only roughly approximate. Gage read once daily to quarter-tenths. Daily discharge ascertained by applying daily gage height to rating table, except for the following periods; June 9, July 4, and September 15, for which discharge was interpolated; December 6 to March 16 when stage-discharge relation was affected by ice for which daily discharges were not determined. Records excellent except below 200 second-feet, which are roughly approximate.

The following discharge measurement was made by Bolster and Gregg:

March 23: Gage height, 4.05; discharge, 2,740 second-feet.

Daily discharge, in second-feet, of Des Moines River at Kalo, Iowa, for the year ending Sept. 30, 1918.

Day.	Oct.	Nov.	Dec.	Mar.	Apr.	May.	June.	July.	Aug.	Sept.
1	236	160	128	....	840	525	3,720	645	677	1,640
2	196	160	128	....	775	370	3,740	615	555	[illegible]
3	178	160	128	....	872	280	2,980	585	498	1,110
4	216	160	128	....	710	525	11,400	664	470	905
5	57	160	128	....	645	420	7,650	742	498	905
6	128	160	....	....	615	370	8,130	525	347	872
7	178	144	....	....	645	302	9,310	775	347	872
8	160	160	....	....	615	280	8,290	1,040	280	775
9	160	178	....	....	585	325	6,470	1,040	325	645
10	160	100	....	....	525	370	4,650	1,180	420	565
11	144	128	....	....	420	585	3,590	1,040	370	840
12	178	160	....	....	302	615	2,860	970	370	535
13	196	128	....	....	470	970	2,300	872	395	585
14	128	144	....	....	470	970	2,000	710	420	710
15	128	128	....	....	420	1,040	1,730	645	280	632
16	128	128	....	....	645	808	1,560	710	555	535
17	160	144	....	1,640	525	585	1,400	445	280	585
18	144	128	....	1,640	585	585	1,180	585	370	710
19	160	114	....	1,730	555	585	1,180	525	710	555
20	160	114	....	2,100	585	615	1,110	710	1,320	565
21	160	114	....	2,520	585	677	1,040	710	1,480	585
22	160	128	....	2,740	645	710	1,040	872	1,730	555
23	144	128	....	2,740	585	710	905	445	1,910	535
24	160	160	....	2,000	585	905	872	555	2,410	470
25	160	100	....	1,560	585	1,320	840	280	2,630	470
26	160	114	....	1,320	555	1,480	808	420	2,860	420
27	144	196	....	1,180	525	2,000	710	470	2,860	256
28	160	196	....	1,110	445	2,100	775	905	2,740	[illegible]
29	160	128	....	1,040	395	3,220	775	970	2,520	[illegible]
30	160	128	....	905	585	3,590	525	872	2,300	296
31	160	....	....	905	....	3,980	....	840	1,910	....

NOTE.—Discharge June 9, July 4, and Sept. 15 interpolated. Stage-discharge relation affected by ice Dec. 6 to Mar. 16; discharge not determined.

Monthly discharge of Des Moines River at Kalo, Iowa, for the year ending Sept. 30, 1918.

[Drainage area, 4,170 square miles.]

Month.	Discharge in second-feet.				Run-off (depth in inches).
	Maximum.	Minimum.	Mean.	Per square mile.	
October	236	57	159	0.038	0.04
November	196	100	142	.034	.04
April	872	302	576	.138	.15
May	3,980	280	1,030	.247	.28
June	11,400	525	3,080	.739	.82
July	1,180	280	721	.173	.20
August	2,860	280	1,120	.268	.31
September	1,640	236	666	.160	.18

DES MOINES RIVER AT DES MOINES, IOWA.

LOCATION.—In T. 78 N., R. 24 W., at Walnut Street Bridge at Des Moines, Polk County, one-third of a mile above mouth of Raccoon River and 205 miles above mouth of Des Moines River.

DRAINAGE AREA.—6,180 square miles. Effective area at high stages, including Raccoon River, 9,770 square miles (measured on map issued by United States Geological Survey; scale, 1 to 500,000).

RECORDS AVAILABLE.—October 2, 1902, to August 3, 1903; October 1, 1914, to September 30, 1918, at Walnut Street Bridge. From May 26, 1905, to July 20, 1906, records were collected at Interurban Bridge near Highland Park, about 5 miles above present station. The United States Weather Bureau has maintained a gage at Locust Street Bridge from July 1, 1897, to January, 1912, and at Walnut Street Bridge from January, 1912, to September 30, 1918.

GAGE.—The original Weather Bureau gage is a staff gage at Locust Street Bridge, one block above Walnut Street Bridge. In January, 1912, a Friez water-stage recorder was installed by the United States Weather Bureau near south end of the second pier from east abutment of Walnut Street Bridge. This gage is set to read the same as Locust Street gage. A copper float in a 9-inch pipe connects with the register at top, which is graduated to record graphically stages from 0 to 33 feet. Gage zero is 774.74 feet above sea level.

DISCHARGE MEASUREMENTS.—Made at any one of several bridges below power dam, according to the stage. Channel satisfactory for accurate measurements.

CHANNEL AND CONTROL.—A sheet-piling dam was constructed about 300 feet above the old mouth of Raccoon River about September, 1913. This dam, called a "beauty dam," is for the purpose of raising low-water stage of river a few feet, thus improving the appearance of the river through the park along the bank. The pooled water from this dam extends past gage to power dam at low water. The dam thus forms a fairly permanent control at low stages. It is drowned out at stages of 8 to 10 feet, depending on the stage in Raccoon River. Dam is now in poor repair, and the stage-discharge relation has been affected thereby.

EXTREMES OF STAGE.—Maximum stage recorded during year, 16.5 feet 1.30 a. m. June 7; minimum stage recorded, 2.6 feet September 29.

1897–1918: Maximum stage recorded, 22.6 feet May 31, 1903; minimum stage recorded, 0.8 foot at various times.

ICE.—The effect of the power dam above station is to improve the conditions of winter flow, but severe winters and occasional ice jams below gage seriously affect stage-discharge relation.

REGULATION.—Edison Power & Light Co.'s dam, about one-quarter of a mile above gage, causes slight diurnal fluctuation of stage. This dam is practically drowned out at a stage of 18 feet, although there is a perceptible ripple with a stage of 21 or 22 feet.

COOPERATION.—The gage-height records are furnished by the United States Weather Bureau. They are the readings shown by the graphic records at 8 a. m. Determinations of discharge withheld until additional data are collected.

The following discharge measurement was made by Bolster and Gregg:

March 21: Gage height, 4.72 feet; discharge, 2,300 second-feet.

Daily gage height, in feet, of Des Moines River at Des Moines, Iowa, for the year ending Sept. 30, 1918.

Day.	Mar.	Apr.	May.	June.	July.	Aug.	Sept.
1	....	3.50	3.20	8.10	3.60	3.40	4.30
2	....	3.40	3.20	7.80	3.50	3.30	4.10
3	....	3.20	3.20	7.50	3.40	3.00	3.90
4	....	3.10	3.10	9.00	3.40	3.00	3.70
5	....	3.10	3.10	10.90	3.50	3.00	3.50
6	....	3.00	3.10	14.90	3.40	3.00	3.40
7	....	3.00	2.90	16.30	3.40	2.90	3.20
8	....	2.90	2.90	16.00	3.50	2.80	3.10
9	....	2.80	2.90	15.40	3.40	2.80	3.00
10	....	2.70	2.80	13.40	3.20	2.90	2.90
11	....	2.70	2.80	10.60	2.80	2.80	2.80
12	....	2.70	2.80	9.00	3.20	2.90	2.80
13	3.80	2.70	3.50	8.20	3.20	2.70	2.80
14	3.80	2.60	3.90	6.60	3.40	2.70	2.80
15	4.00	2.60	4.00	6.10	3.50	2.90	2.80
16	3.80	3.30	3.90	5.70	3.40	3.00	2.90
17	3.70	3.30	3.90	5.40	3.30	2.90	3.00
18	3.70	3.30	3.80	5.00	3.20	2.80	2.90
19	3.90	3.30	3.60	4.80	3.20	2.80	2.90
20	4.20	3.30	3.50	4.50	3.20	2.80	2.80
21	4.60	3.30	3.40	4.30	3.20	2.90	2.80
22	4.90	3.40	3.40	4.50	3.20	3.40	2.80
23	5.10	3.30	3.50	4.50	3.10	3.70	2.80
24	5.20	3.30	4.50	4.40	3.10	3.90	2.80
25	5.00	3.30	5.20	4.00	3.10	4.20	2.80
26	4.60	3.20	4.50	4.00	2.90	4.50	2.70
27	4.30	3.20	4.90	3.90	2.90	5.00	2.70
28	4.00	3.20	5.20	3.80	2.90	4.80	2.70
29	3.80	3.20	5.10	3.70	3.00	4.70	2.60
30	3.70	3.30	5.90	3.60	2.90	4.60	2.70
31	3.60	....	7.60	....	3.10	4.40	....

NOTE.—Water-stage recorder not in operation Oct. 1 to Mar. 13.

DES MOINES RIVER AT OTTUMWA, IOWA.

LOCATION.—At Market Street Bridge, Ottumwa, Wapello County, Iowa. No large tributary within several miles up or down stream.

DRAINAGE AREA.—13,200 square miles (measured from map issued by the United States Geological Survey; scale, 1 to 500,000).

RECORDS AVAILABLE.—Fragmentary high-water observations 1902–1916; daily records March 29, 1917, to September 30, 1918.

GAGE.—Chain gage attached to downstream handrail of bridge. Staff gage painted on northeast face of north pier used prior to August 2, 1917.

DISCHARGE MEASUREMENTS.—Made from Vine Street Bridge about 1,500 feet below gage and by wading.

CHANNEL AND CONTROL.—Channel probably fairly permanent.

EXTREMES OF DISCHARGE.—Maximum stage recorded during the year, 13.9 feet, June 10 (discharge, 41,400 second-feet). Minimum stage recorded 1.3 feet various dates, October and November (discharge, 435 second-feet).

1917–18: Maximum stage recorded, 16.5 feet June 11, 1917 (discharge, 58,700 second-feet; minimum stage, 1.3 feet various days in October and November, 1918. Maximum stage since 1850 and probably in the last century occurred May 31, 1903, and exceeded 100,000 second-feet.

ICE.—Stage-discharge relation seriously affected by ice.

ACCURACY.—Stage-discharge relation probably permanent, except as affected by ice. Rating curve fairly well-defined. Gage read to tenths once daily. Daily discharge ascertained by applying daily gage height to rating table except for periods when stage-discharge relation was affected by ice, for which daily discharges were not determined. Open-water records good except for July, 1917.

COOPERATION.—Gage height record furnished by the United States Weather Bureau.

The following discharge measurement was made by Bolster and Gregg:

March 21: Gage height, 2.62 feet; discharge, 2,210 second-feet.

Daily discharge, in second-feet, of Des Moines River at Ottumwa, Iowa, for the period Mar. 28, 1917, to Sept. 30, 1918.

Day.	Mar.	Apr.	May.	June.	Aug.	Sept.
1917.						
1	....	17,100	14,900	6,600	2,060	845
2	....	13,300	14,500	11,500	2,060	735
3	....	11,500	11,500	19,200	1,900	735
4	....	10,400	14,900	17,500	1,900	735
5	....	10,100	17,100	25,800	1,600	735
6	....	9,540	14,500	30,000	1,600	2,740
7	....	8,180	15,200	38,000	1,600	7,380
8	....	8,180	12,700	40,800	1,460	6,860
9	....	8,180	10,700	41,400	1,460	4,390
10	....	8,180	8,990	55,100	1,460	2,390
11	....	7,910	8,720	58,700	1,460	1,750
12	....	7,910	8,180	52,300	1,330	1,600
13	....	7,640	7,910	56,500	1,330	1,600
14	....	7,640	7,640	52,300	1,330	1,600
15	....	7,640	7,380	45,700	1,330	1,460
16	....	7,640	7,120	39,100	1,330	1,330
17	....	7,640	6,860	33,800	1,330	1,330
18	....	7,380	6,600	17,800	1,200	1,330
19	....	7,120	6,340	17,100	960	1,200
20	....	6,860	6,080	13,300	960	1,200
21	....	6,860	5,830	12,100	960	960
22	....	6,600	6,340	10,900	960	960
23	....	6,600	6,600	9,820	960	735
24	....	7,120	6,080	8,990	960	735
25	....	7,380	6,600	8,180	960	625
26	....	7,120	6,860	7,640	960	625
27	....	7,120	6,860	7,120	845	625
28	22,700	7,120	6,600	8,720	845	625
29	24,200	7,380	6,600	8,180	845	625
30	23,500	9,540	6,600	8,180	845	625
31	20,600	....	6,600	....	845	....

Day.	Oct.	Nov.	Feb.	Mar.	Apr.	May.	June.	July.	Aug.	Sept.
1917-18.										
1	625	625	....	2,220	1,750	1,080	8,990	3,320	1,080	2,740
2	625	625	....	2,390	1,750	960	10,700	2,740	1,080	2,930
3	525	435	....	3,950	1,750	960	11,500	2,220	1,080	4,170
4	525	525	....	3,950	1,460	845	13,300	2,220	1,330	2,930
5	525	525	....	3,520	1,460	845	17,800	2,390	1,200	2,390
6	435	435	....	4,170	1,460	845	21,300	2,390	1,200	1,750
7	525	525	....	4,170	1,460	845	25,400	2,220	1,200	1,600
8	525	525	....	3,120	1,330	845	27,000	2,060	1,200	1,600
9	435	435	....	2,740	1,330	735	33,300	2,060	1,080	1,460
10	525	525	....	2,740	1,460	735	41,400	2,060	1,080	1,460
11	525	525	....	2,740	1,330	735	39,700	2,060	1,080	1,200
12	435	435	5,830	2,220	1,200	735	36,900	2,060	960	1,200
13	435	525	7,910	1,900	1,200	735	27,000	2,390	960	1,200
14	435	525	7,380	2,220	1,080	625	22,000	2,220	845	1,200
15	435	435	6,600	2,220	1,080	625	14,200	2,220	845	1,080
16	435	525	3,730	2,060	1,080	625	12,100	2,060	845	1,080
17	435	525	2,220	2,060	1,080	845	8,720	2,060	960	845
18	435	435	....	2,220	1,080	2,740	7,120	1,750	1,080	845
19	525	525	....	2,220	1,080	2,740	6,340	1,600	1,330	960
20	525	525	....	1,900	1,080	1,750	5,580	1,750	1,750	960
21	435	435	....	2,060	1,080	1,330	4,850	1,750	2,560	960
22	525	525	....	2,060	1,200	2,560	3,950	1,600	1,900	960
23	525	525	....	2,220	1,200	2,220	3,730	1,600	2,560	845
24	435	435	....	2,740	1,200	2,740	15,800	1,330	3,320	845
25	525	525	....	3,120	1,200	3,320	30,400	1,330	2,220	845
26	625	525	2,560	3,320	1,080	7,640	12,100	1,330	2,560	845
27	525	435	1,750	3,320	1,080	6,340	7,120	1,330	2,560	735
28	625	525	1,600	2,560	1,080	6,340	5,330	1,200	2,560	735
29	625	435	....	2,390	1,080	13,900	3,950	1,200	2,740	735
30	525	435	....	2,060	1,080	12,700	3,320	1,200	2,930	735
31	625	....	....	2,060	....	9,540	....	1,080	2,930	....

NOTE.—Regular daily gage readings began Mar. 28, 1917. Daily discharge for July, 1917, doubtful, hence not published. Stage-discharge relation affected by ice Dec. 1 to Feb. 11 and Feb. 18-25, 1918: discharge not determined.

Monthly discharge of Des Moines River at Ottumwa, Iowa, for the period Apr. 1, 1917, to Sept. 30, 1918.

[Drainage area, 13,200 square miles.]

Month.	Discharge in second-feet.				Run-off (depth in inches).
	Maximum.	Minimum.	Mean.	Per square mile.	
1917.					
April	17,100	6,600	8,430	0.638	0.71
May	17,100	5,830	9,010	.683	.79
June	58,700	6,600	25,400	1.92	2.14
July	....	....	....	....	....
August	2,060	845	1,280	.097	.11
September	7,380	625	1,640	.124	.14
1917-18.					
October	625	435	512	0.039	.04
November	625	435	499	.038	.04
March	4,170	1,900	2,670	.202	.23
April	1,750	1,080	1,260	.095	.11
May	13,900	625	2,890	.219	.25
June	41,400	3,320	16,000	1.21	1.35
July	3,320	1,080	1,900	.144	.17
August	3,320	845	1,650	.125	.14
September	4,170	735	1,390	.105	.12

DES MOINES RIVER AT KEOSAUQUA, IOWA.

LOCATION.—In sec. 36, T. 69 N., R. 10 W., at county bridge in Keosauqua, Van Buren County, a quarter of a mile above old dam site and Government locks. No large tributary enters Des Moines River for several miles up or down stream.

DRAINAGE AREA.—At gaging station, 13,900 square miles; at mouth, 14,300 square miles (measured on map issued by United States Geological Survey; scale, 1 to 500,000).

RECORDS AVAILABLE.—May 30, 1903, to July 21, 1906; April 5 to December 31, 1910 (United States Engineer Corps); August 3, 1911, to September 30, 1918.

GAGE.—Chain gage attached to upstream handrail of bridge; read by Frank Schreckengast.

DISCHARGE MEASUREMENTS.—Made from bridge to which gage is attached.

CHANNEL AND CONTROL.—Channel shifts considerably at flood stages. Control is a gravel riffle about a quarter of a mile below gage.

EXTREMES OF DISCHARGE.—Maximum stage recorded during year, 12.95 feet June 25 (discharge, 39,800 second-feet); minimum stage recorded 0.20 foot, several days in November and December (discharge, 760 second-feet).

1903–1918: Maximum stage recorded, 27.9 feet June 1, 1903 (discharge, 97,000 second-feet); minimum stage recorded, zero August 28 to September 6, 1911 (discharge, 160 second-feet). On June 1, 1851, a stage of 24 feet was reached (discharge, 80,000 second-feet).

ICE.—Stage-discharge relation seriously affected by ice. Observations discontinued during winter.

ACCURACY.—Stage-discharge relation fairly permanent for low and medium stages, except as affected by ice. Three fairly well defined rating curves were used. Gage read once daily to half-tenths. Daily discharge ascertained by applying daily gage height to rating tables except for period when stage-discharge relation was affected by ice, for which daily discharge was not determined; daily discharge usually interpolated on Sundays, when no gage reading was taken. Open water records good.

The following discharge measurement was made by Bolster and Gregg:

March 20: Gage height, 1.14 feet; discharge, 2,150 second-feet.

Daily discharge, in second-feet, of Des Moines River at Keosauqua, Iowa, for the year ending Sept. 30, 1918.

Day.	Oct.	Nov.	Dec.	Mar.	Apr.	May.	June.	July.	Aug.	Sept.
1	1,030	890	760	....	1,970	1,150	9,720	3,700	1,140	3,240
2	1,030	825	760	....	1,880	1,080	11,200	2,800	1,140	6,600
3	1,030	890	760	....	1,880	945	12,700	2,600	1,370	5,770
4	1,030	860	760	....	1,700	1,010	14,600	2,300	1,300	3,460
5	960	825	760	....	1,620	1,200	20,200	2,120	1,220	2,400
6	960	890	....	....	1,540	1,380	22,900	2,120	1,220	2,120
7	960	825	....	....	1,500	1,220	25,300	2,080	1,140	1,680
8	960	825	....	....	1,460	1,150	28,400	2,030	1,060	1,520
9	890	825	....	....	1,380	856	33,500	2,030	1,060	1,370
10	890	825	....	....	1,380	856	39,300	1,940	1,060	1,290
11	890	825	....	....	1,300	856	39,300	1,940	925	1,290
12	890	825	....	....	1,220	856	36,700	2,030	790	1,290
13	825	825	....	....	1,220	856	28,800	2,120	995	1,220
14	860	825	....	....	1,190	945	19,800	2,080	790	1,140
15	890	825	....	....	1,150	856	15,600	2,030	790	1,140
16	890	825	....	....	1,220	856	12,400	2,030	790	1,140
17	960	890	....	....	1,220	1,460	9,400	1,860	925	995
18	1,500	890	....	....	1,220	1,790	7,720	1,770	1,060	995
19	1,030	890	....	....	1,220	1,750	6,610	1,680	1,370	995
20	960	825	....	2,050	1,220	1,700	5,800	1,600	1,370	995
21	890	825	....	2,100	1,220	2,990	5,000	1,530	2,400	995
22	825	825	....	2,160	1,220	3,720	4,480	1,450	1,940	960
23	825	825	....	2,360	1,220	2,360	4,200	1,450	1,770	925
24	890	760	....	2,560	1,300	3,720	10,300	2,030	4,190	925
25	825	760	....	2,770	1,220	2,990	40,000	1,450	3,020	858
26	960	760	....	2,770	1,540	4,800	13,800	1,450	2,030	858
27	960	825	....	2,990	1,150	6,610	8,000	1,370	2,300	858
28	960	760	....	2,770	1,190	14,900	4,970	1,500	2,400	858
29	960	760	....	2,460	1,220	25,600	4,450	2,400	2,400	824
30	960	760	....	2,360	1,150	15,900	4,080	1,220	2,600	790
31	890	....	....	2,160	....	10,900	....	1,140	2,800	....

NOTE.—Gage readings usually omitted on Sundays and discharge interpolated, except June 23, July 28 Aug. 18, and Sept. 1, which were estimated on the basis of climatological data. Stage-discharge relation affected by ice Dec. 6 to Mar. 19; daily discharge not determined.

Monthly discharge of Des Moines River at Keosauqua, Iowa, for the year ending Sept. 30, 1918.

[Drainage area, 13,900 square miles.]

Month.	Discharge in second-feet.				Run-off (depth in inches).
	Maximum.	Minimum.	Mean.	Per square mile.	
October	1,500	825	948	0.068	0.08
November	890	760	826	.059	.07
April	1,970	1,150	1,360	.098	.11
May	25,600	856	3,780	.272	.31
June	39,300	4,080	16,600	1.19	1.33
July	3,700	1,140	1,930	.139	.16
August	4,190	790	1,590	.114	.13
September	6,600	790	1,650	.119	.13

RACCOON RIVER AT VAN METER, IOWA.

LOCATION.—In SW. ¼ sec. 22, T. 78 N., R. 27 W., at highway bridge one-third of a mile from railroad station, 1 mile below South Raccoon River, and 30 miles above junction of Raccoon River with Des Moines River.

DRAINAGE AREA.—At gaging station, 3,410 square miles; at mouth, 3,590 square miles (measured on map issued by United States Geological Survey; scale, 1 to 500,000).

RECORDS AVAILABLE.—April 25, 1915, to September 30, 1918.

GAGE.—Chain gage attached to downstream handrail of bridge about 25 feet from right end of bridge; read by Fred Vreeland.

DISCHARGE MEASUREMENTS.—Made from bridge to which gage is attached.

CHANNEL AND CONTROL.—Bed composed of sand and gravel; subject to change. River divided into two channels at low and medium stages by an island with the water surface slightly higher in the left channel than in the right at extreme low water. Right bank high and not subject to overflow; left bank subject to overflow at a stage of about 13 feet; at extreme high stage this overflow will extend for several thousand feet beyond left end of bridge.

EXTREMES OF DISCHARGE.—Maximum stage recorded during year, 13.59 feet June 8 (discharge, 14,600 second-feet); minimum stage, 1.61 feet, September 30 (discharge, 37 second-feet).

1915–1918: Maximum stage recorded, 17.5 feet June 7, 1917 (discharge, 31,800 second-feet); minimum stage recorded, 1.61 feet September 30 (discharge, 37 second-feet).

ICE.—Stage-discharge relation affected by ice December 6 to March 12. Observations discontinued December 13 to February 9.

ACCURACY.—Stage-discharge relation permanent. Rating curve well defined throughout. Gage read once daily to hundredths. Daily discharge ascertained by applying daily gage height to rating table, except for period when stage-discharge relation was affected by ice, for which daily discharges were not determined. Open-water records excellent, except for extremely low stages, for which they are fair.

The following discharge measurement was made by Bolster and Gregg:

March 22: Gage height, 2.86 feet; discharge, 431 second-feet.

Daily discharge, in second-feet, of Raccoon River at Van Meter, Iowa, for the year ending Sept. 30, 1918.

Day.	Oct.	Nov.	Dec.	Mar.	Apr.	May.	June.	July.	Aug.	Sept.
1	188	198	156	....	243	123	3,120	543	204	98
2	179	170	185	....	243	116	3,330	459	194	73
3	210	204	173	....	188	126	3,970	459	201	94
4	188	173	164	....	226	110	7,590	434	164	82
5	185	134	150	....	243	120	7,730	408	150	75
6	194	179	....	....	243	118	14,300	384	123	60
7	150	210	....	....	194	110	13,000	361	116	46
8	108	188	....	....	210	116	14,600	361	108	50
9	123	162	....	....	204	123	13,000	361	86	54
10	118	210	....	....	194	98	11,200	361	75	71
11	123	167	....	....	179	98	9,840	361	91	77
12	116	179	....	....	173	91	10,900	318	98	91
13	134	185	....	633	167	98	8,160	298	91	67
14	136	198	....	697	131	108	4,190	318	86	54
15	93	204	....	665	170	110	3,430	318	89	136
16	110	204	....	602	204	98	2,430	261	98	123
17	120	201	....	514	173	100	2,140	279	108	116
18	159	167	....	514	156	110	1,760	298	3,220	108
19	156	123	....	486	173	123	1,670	261	459	91
20	164	162	....	459	182	194	1,310	243	210	64
21	167	179	....	434	194	486	1,230	210	194	46
22	179	194	....	459	210	434	1,060	210	173	56
23	194	164	....	361	194	834	907	226	159	52
24	170	136	....	340	179	3,640	764	170	150	54
25	150	131	....	361	150	2,330	697	136	142	58
26	145	136	....	318	118	1,670	633	134	136	46
27	150	164	....	279	116	5,860	602	123	150	46
28	134	156	....	279	136	3,750	633	150	145	44
29	164	173	....	279	123	2,430	602	179	139	41
30	188	179	....	261	118	2,720	572	210	136	37
31	243	....	....	279	....	2,920	....	210	145	....

NOTE.—Stage-discharge relation affected by ice Dec. 6 to Mar. 12; daily discharge not determined.

Monthly discharge of Raccoon River at Van Meter, Iowa, for the year ending Sept. 30, 1918.

[Drainage area, 3,410 square miles.]

Month.	Discharge in second-feet.				Run-off (depth in inches).
	Maximum.	Minimum.	Mean.	Per square mile.	
October	243	93	156	0.046	0.05
November	210	123	174	.051	.06
April	243	116	181	.053	.06
May	5,860	91	947	.278	.32
June	14,600	572	4,850	1.42	1.58
July	543	123	292	.086	.10
August	3,220	75	246	.072	.08
September	136	37	69.7	.020	.02

ILLINOIS RIVER AT PEORIA, ILL.

LOCATION.—In sec. 2, T. 8 N., R. 8 E., at foot of Grant Street, Peoria, Peoria County, 3½ miles above station formerly maintained at Peoria & Pekin Union Railroad bridge and 4½ miles above mouth of Kickapoo Creek.

DRAINAGE AREA.—Indeterminate.

RECORDS AVAILABLE.—March 8, 1910, to September 30, 1918; also March 10, 1903, to July 21, 1906, for station at Peoria and Pekin Union Railroad bridge.

GAGE.—Vertical staff gage attached to wooden pile; read by employee of United States Army Engineers.

DISCHARGE MEASUREMENTS.—Made from downstream side of Lower Free bridge, about 2 miles below gage.

CHANNEL AND CONTROL.—Bed of river, which forms control for medium and high stages, composed of mud, and may shift. Dam at Copperas Creek probably forms control for lowest stages; permanent.

EXTREMES OF DISCHARGE.—Maximum stage recorded during year, 19.8 feet February 20 and 21 (discharge, 41,800 second-feet); minimum stage, 10.0 feet September 28 (discharge, 10,000 second-feet).

1910–1918: Maximum stage recorded, 23.2 feet January 25, 1916 (discharge not determined because of backwater from ice); maximum stage recorded during open-water periods, 22.4 feet March 30 to April 2, 1913 (discharge, 55,000 second-feet); minimum stage, 8.0 feet December 14, 1910 (discharge, 7,250 second-feet).

The highest known flood occurred in 1844, when a stage of about 26.6 feet on the present gage was reached.

REGULATION.—The flow at this station includes the water diverted from Lake Michigan through the Chicago Drainage canal. No diurnal fluctuation is noticeable.

ACCURACY.—Stage-discharge relation practically permanent; seriously affected by ice during winter. Rating curve well defined between 11,000 and 40,000 second-feet and fairly well defined beyond these limits. Gage read to half-tenths once daily. Daily discharge ascertained by applying daily gage height to rating table, except for period when stage-discharge relation was affected by ice, for which it was ascertained by applying to rating table daily gage heights corrected for ice effect by means of observer's notes and weather records, and by comparison with flow of adjacent streams. Open-water records good; winter records poor.

COOPERATION.—Gage-height records furnished by the United States Engineer Corps.

Discharge measurements of Illinois River at Peoria, Ill., during the year ending Sept. 30, 1918.

[Made by H. C. Beckman.]

Date.	Gage height.	Discharge.
	Feet.	*Sec.-ft.*
Oct. 16	10.68	10,600
June 12	13.28	16,700
Aug. 23	10.60	10,800

Daily discharge, in second-feet, of Illinois River at Peoria, Ill., for the year ending Sept. 30, 1918.

Day.	Oct.	Nov.	Dec.	Jan.	Feb.	Mar.	Apr.	May.	June.	July.	Aug.	Sept.
1	11,300	12,100	11,900			35,800	20,800	16,900	18,400	13,000	12,500	10,600
2	11,100	12,100	11,900			34,800	20,200	17,200	19,000	13,600	12,300	10,600
3	10,900	11,900	12,300			34,800	20,500	17,500	18,700	13,800	12,300	10,800
4	10,900	12,300	11,900			34,800	19,600	17,500	18,400	14,000	11,900	10,900
5	10,800	12,300	11,900			34,800	18,700	17,500	18,100	14,200	11,600	10,900
6	10,800	12,500	11,900	9,860		34,800	18,400	17,500	17,800	15,000	11,600	10,600
7	10,900	12,700	11,800		11,400	34,800	18,100	17,200	17,800	15,200	11,300	10,600
8	10,900	12,700	11,900			33,900	18,100	17,200	17,200	15,800	11,600	10,400
9	10,900	12,700	11,600			33,900	18,400	16,600	16,400	16,400	11,400	10,600
10	10,900	12,700	11,600			33,400	17,800	17,200	16,400	15,600	11,300	10,800
11	10,800	12,700				31,600	16,900	16,900	16,000	15,600	11,300	10,600
12	10,900	13,000				30,800	16,600	16,900	15,800	15,800	11,300	10,900
13	10,900	12,700				30,800	16,200	16,600	15,400	15,800	10,900	10,800
14	10,900	12,700				29,800	16,000	16,600	15,200	15,600	11,300	10,900
15	11,100	12,500	11,100			29,400	15,800	16,400	15,000	15,200	11,600	10,900
16	11,100	12,700			27,300	29,000	15,400	16,400	14,200	15,000	11,300	10,900
17	11,300	12,700			33,900	29,000	15,400	16,400	14,400	15,000	11,300	10,600
18	11,300	12,700			38,800	28,500	15,200	16,400	14,000	14,800	11,300	10,400
19	11,400	12,700			40,800	28,500	15,400	16,000	13,800	14,600	11,300	10,600
20	11,400	12,300			41,800	28,100	15,400	16,600	14,400	14,200	11,300	10,600
21	11,600	12,300		9,440	41,800	27,700	14,600	16,600	13,400	14,000	10,900	10,600
22	11,400	13,000			41,300	27,700	14,600	16,600	13,400	13,600	10,900	10,300
23	11,600	12,700			39,800	26,900	15,000	16,900	13,000	13,400	10,900	10,300
24	11,600	12,300			38,800	26,100	15,400	17,200	12,500	13,000	10,900	10,300
25	11,600	12,300			38,300	25,300	15,800	17,200	13,000	13,000	10,900	10,400
26	11,100	12,300	10,300		37,800	24,500	16,200	17,200	12,700	13,000	10,800	10,600
27	11,600	12,100			37,300	24,100	16,200	17,800	12,700	13,000	10,800	10,300
28	11,600	11,900			36,800	22,900	16,200	18,400	12,800	12,700	10,400	10,000
29	12,300	12,300			......	22,500	15,400	18,400	13,000	12,800	10,600	10,200
30	11,900	11,900			......	21,700	16,600	18,400	13,000	12,800	10,600	10,200
31	11,900	......			......	20,800	......	18,400	......	12,700	10,800	......

NOTE.—Stage-discharge relation affected by ice Dec. 11 to Feb. 15; daily discharge determined from gage heights corrected for ice effect by means of weather records and comparisons with flow at other stations up stream. Braced figures show mean daily discharge for periods included.

Monthly discharge, in second-feet, of Illinois River at Peoria, Ill., for the year ending Sept. 30, 1918.

Month.	Maximum.	Minimum.	Mean.	Month.	Maximum.	Minimum.	Mean.
October	12,300	10,800	11,200	May	18,400	16,000	17,100
November	13,000	11,900	12,500	June	19,000	12,500	15,200
December	12,300	....	11,100	July	16,400	12,700	14,300
January	....	....	9,580	August	12,500	10,400	11,300
February	41,800	....	23,800	September	10,900	10,000	10,600
March	35,800	20,800	29,400				
April	20,800	14,600	16,800	The year	41,800	....	15,200

KANKAKEE RIVER AT MOMENCE, ILL.

LOCATION.—In sec. 24, T. 31 N., R. 13 E., at highway bridge in Momence, Kankakee County, half a mile below Chicago & Eastern Illinois Railroad bridge and 1½ miles above Tower Creek.

DRAINAGE AREA.—2,340 square miles.

RECORDS AVAILABLE.—February 22, 1905, to July 20, 1906; December 3, 1914, to September 30, 1918.

GAGE.—Chain gage attached to bridge, over left channel; read by Oscar Conrad.

DISCHARGE MEASUREMENTS.—Made from upstream side of bridge across the two channels during medium and high stages, and by wading during low stages.

CHANNEL AND CONTROL.—Bed composed of coarse gravel; may shift. River at gage divided into two channels by an island. Aquatic plants sometimes grow in bed of river during summer.

EXTREMES OF DISCHARGE.—Maximum stage recorded during year, 4.6 feet February 14–18 (discharge not determined because of backwater from ice); maximum stage recorded during open-water period, 4.2 feet at 1 p. m. February 25 (discharge, 6,300 second-feet); minimum stage, 1.44 feet at 11 a. m. August 29 (discharge, 442 second-feet).

1905–6 and 1915–18: Maximum stage recorded, 7.5 feet January 21, 1916 (discharge not determined because of backwater from ice); maximum open-water stage, 6.4 feet January 22, 1916 (discharge estimated from extension of rating curve, 12,600 second-feet); minimum discharge, 360 second-feet, July 13–20, 1906.

ACCURACY.—Stage-discharge relation changed during year; seriously affected by ice during winter. Rating curve used to February 20 well defined; curve used after that date well defined between 550 and 3,100 second-feet, and fairly well defined beyond those limits. Gage read to hundredths once daily. Daily discharge ascertained by appiying daily gage height to rating table, except for period when stage-discharge relation was affected by ice, for which it was obtained by applying to rating daily gage heights corrected for ice effect by means of observer's notes and weather records. Open-water records good; winter records approximate.

Discharge measurements of Kankakee River at Momence, Ill., during the year ending Sept. 30, 1918.

[Made by H. C. Beckman.]

Date.	Gage height.	Discharge.
	Feet.	*Sec.-ft.*
Apr. 20	2.30	1,410
July 23	1.67	621
Aug. 31	1.62	573

Daily discharge, in second-feet, of Kankakee River at Momence, Ill., for the year ending Sept. 30, 1918.

Day.	Oct.	Nov.	Dec.	Jan.	Feb.	Mar.	Apr.	May.	June.	July.	Aug.	Sept.
1	534	1,420	915			5,740	2,570	1,480	1,680	792	592	592
2	534	1,420	945			5,460	2,460	1,480	1,450	792	550	592
3	498	1,420	945			5,460	2,340	1,480	1,390	792	550	592
4	486	1,420	945			5,180	2,220	1,480	1,390	792	550	592
5	486	1,420	945			5,180	2,220	1,480	1,300	792	550	550
6	474	1,420		500	390	4,910	2,110	1,480	1,220	792	550	550
7	474	1,420				4,910	2,000	1,480	1,060	792	512	550
8	462	1,420				4,910	2,000	1,480	1,060	792	512	550
9	462	1,330				4,910	1,890	1,480	980	792	512	550
10	462	1,330				4,640	1,780	1,480	980	735	475	550
11	510	1,330	640			4,640	1,780	1,480	980	735	475	550
12	558	1,330				4,640	1,780	1,480	915	735	475	550
13	570	1,240				4,640	1,680	1,580	915	735	512	550
14	609	1,150				4,910	1,680	1,580	850	792	475	592
15	622	1,150				4,910	1,580	1,580	850	792	475	592
16	622	1,070		390	4,670	4,910	1,580	1,680	792	792	475	592
17	648	1,070				4,640	1,580	1,780	735	792	475	635
18	648	1,070				4,370	1,480	1,780	735	792	550	592
19	674	1,070				4,100	1,480	1,780	685	792	475	592
20	714	1,070				4,100	1,480	1,890	685	792	475	592
21	714	1,070			6,020	4,100	1,480	1,890	685	735	475	550
22	714	990			6,020	3,830	1,480	1,890	685	735	475	550
23	770	990	570		5,740	3,830	1,480	2,110	685	735	475	550
24	826	990			6,020	3,560	1,480	1,890	685	685	475	550
25	826	960			6,300	3,560	1,480	1,890	635	685	470	550
26	900	945		330	6,020	3,300	1,480	1,890	635	685	464	550
27	975	945			6,020	3,300	1,480	1,780	635	685	464	550
28	1,070	945			5,740	3,180	1,480	1,780	635	635	453	550
29	1,150	930			......	2,930	1,480	1,780	685	592	442	350
30	1,240	930			......	2,810	1,480	1,680	792	592	512	550
31	1,330	......			......	2,690	......	1,680	......	592	550	......

NOTE.—Discharge Dec. 6 to Feb. 20 estimated, because of ice, from gage heights, observer's notes, and weather records. Braced figures show mean discharge for periods indicated.

Monthly discharge of Kankakee River at Momence, Ill., for the year ending Sept. 30, 1918.

[Drainage area, 2,340 square miles.]

Month.	Discharge in second-feet.				Run-off (depth in inches).
	Maximum.	Minimum.	Mean.	Per square mile.	
October	1,330	462	696	0.297	0.34
November	1,420	930	1,180	.504	.56
December	945	....	652	.279	.32
January	....	....	404	.173	.20
February	6,300	....	3,520	1.50	1.56
March	5,740	2,690	4,330	1.85	2.13
April	2,570	1,480	1,750	.748	.83
May	2,110	1,480	1,670	.714	.82
June	1,580	635	911	.389	.43
July	792	592	741	.317	.37
August	592	442	499	.213	.25
September	635	550	567	.242	.27
The year	6,300	....	1,400	.598	8.06

KANKAKEE RIVER AT CUSTER PARK, ILL.

LOCATION.—In sec. 19, T. 32 N., R. 10 E., at Wabash Railroad bridge in Custer Park, Will County, half a mile above Horse Creek and 15 miles below dam and power plant at Kankakee.

DRAINAGE AREA.—4,870 square miles.

RECORDS AVAILABLE.—November 6, 1914, to September 30, 1918.

GAGE.—Chain gage attached to bridge; read by J. H. Swords.

DISCHARGE MEASUREMENTS.—Made from downstream side of bridge.

CHANNEL AND CONTROL.—Bed composed of solid rock strewn with boulders and gravel. Right half of channel deep with fissures in bed; left half shallow; may shift slightly.

EXTREMES OF DISCHARGE.—Maximum stage recorded during year, 14.0 feet at 1 and 6 p. m. February 14 (discharge not determined because of backwater from ice); maximum stage recorded during open-water periods, 13.0 feet at 9 a. m., February 16 (discharge, 22,700 second-feet); minimum stage, 4.95 feet October 4 and 5 and August 15 (discharge, 430 second-feet).

1915–1918: Maximum stage recorded, same as for 1918; minimum stage, 4.09 feet November 15, 1914 (discharge not determined; mean discharge for the day, estimated 250 second-feet).

REGULATION.—Operation of power plant at Kankakee causes slight fluctuation at gage.

ACCURACY.—Stage-discharge relation changed slightly during year; seriously affected by ice during winter. Rating curve well defined above and fairly well defined below 1,820 second-feet. Gage read to hundredths twice daily. Daily discharge ascertained by applying mean daily gage height to rating table, except for period when stage-discharge relation was affected by ice, for which it was estimated from occasional gage heights, observer's notes, and weather records. Open-water records good; winter records poor.

Discharge measurements of Kankakee River at Custer Park, Ill., during the year ending Sept. 30, 1918.

[Made by H. C. Beckman.]

Date.	Gage height.	Discharge.
	Feet.	*Sec.-ft.*
Oct. 17	5.07	509
July 16	5.59	1,080
Sept. 7	6.45	2,380

Daily discharge, in second-feet, of Kankakee River at Custer Park, Ill., for the year ending Sept. 30, 1918.

Day.	Oct.	Nov.	Dec.	Jan.	Feb.	Mar.	Apr.	May.	June.	July.	Aug.	Sept.
1	657	2,060	1,000			9,100	2,690	4,630	4,630	5,640	680	940
2	546	2,320	1,070			8,790	2,600	4,390	4,150	4,630	634	758
3	546	2,410	1,000			8,480	2,600	3,680	3,270	3,910	588	784
4	546	2,320	940			7,880	2,690	3,270	2,600	3,070	546	810
5	518	2,150	1,000	680		7,580	2,690	2,880	2,230	2,410	588	875
6	527	2,150			•	7,290	2,880	2,600	1,900	2,060	565	1,440
7	657	2,060				7,000	2,690	2,410	2,060	1,900	536	2,500
8	611	1,980			4,260	6,720	2,600	2,320	1,980	1,900	565	2,500
9	680	1,980				6,440	2,410	2,410	2,410	2,060	565	1,980
10	565	1,900	710			6,170	2,320	3,070	2,320	2,500	600	1,510
11	657	1,740				5,380	2,150	2,880	1,900	2,410	470	1,280
12	565	1,820				5,380	2,150	3,270	1,660	2,060	536	1,210
13	680	1,660				5,130	1,980	3,680	1,360	1,660	498	1,070
14	758	1,580				5,380	1,900	3,910	1,280	1,360	498	1,000
15	634	1,440		600		6,170	1,560	4,150	1,210	1,280	462	940
16	588	1,510			21,300	5,900	1,740	3,680	1,210	1,140	480	940
17	706	1,440			18,400	5,640	1,820	3,270	1,000	1,070	565	940
18	784	1,360			15,000	5,130	1,980	2,690	940	1,000	565	940
19	771	1,360			13,000	4,880	2,150	2,880	810	940	518	940
20	1,000	1,280			13,600	4,390	2,410	2,880	810	940	498	875
21	1,000	1,210			13,600	4,150	3,070	3,070	771	810	480	810
22	1,000	1,360			11,600	3,910	4,150	3,270	693	784	536	784
23	1,070	1,070	680		10,400	3,910	5,130	3,270	668	810	576	758
24	1,070	1,210			9,410	3,910	4,630	3,270	693	668	556	706
25	1,070	1,210		470	8,790	3,680	3,910	4,150	657	680	470	745
26	1,210	1,210			8,180	3,680	3,470	4,150	622	810	611	758
27	1,140	1,070			8,480	3,470	3,270	3,910	646	940	565	732
28	1,210	1,070			8,480	3,270	3,270	3,470	1,000	758	527	634
29	1,360	1,070			....	3,070	4,150	2,880	1,740	940	518	680
30	1,440	1,070			....	2,880	4,630	3,070	3,680	940	745	634
31	1,740	......			....	2,690	......	4,150	......	745	680	......

NOTE.—Discharge Dec. 6 to Feb. 15 estimated, because of ice, from gage heights, observer's notes, and weather records. Braced figures show mean discharge for periods indicated.

Monthly discharge of Kankakee River at Custer Park, Ill., for the year ending Sept. 30, 1918.

[Drainage area, 4,870 square miles.]

Month.	Discharge in second-feet.				Run-off (depth in inches).
	Maximum.	Minimum.	Mean.	Per square mile.	
October	1,740	518	849	0.174	0.20
November	2,410	1,000	1,600	.329	.37
December	1,070	....	742	.152	.18
January	....	....	580	.119	.14
February	....	....	8,000	1.64	1.71
March	9,100	2,690	5,400	1.11	1.28
April	5,130	1,580	2,860	.587	.65
May	4,630	2,320	3,340	.686	.79
June	4,630	622	1,700	.349	.39
July	5,640	668	1,700	.349	.40
August	745	462	556	.114	.13
September	2,500	634	1,050	.216	.24
The year	....	462	2,320	.476	6.48

DES PLAINES RIVER AT LEMONT, ILL.

LOCATION.—In sec. 20, T. 37 N., R. 11 E., at concrete highway bridge at Stephens Street, a quarter of a mile north of main section of Lemont, Cook County; 8 miles above junction of Des Plaines River and Chicago Drainage canal.

DRAINAGE AREA.—705 square miles.

RECORDS AVAILABLE.—November 4, 1914, to September 30, 1918.

Gage.—Enamel staff gage attached to bridge; read by William Weck, jr.

Discharge measurements.—Made from downstream side of bridge or by wading below dam.

Channel and control.—A concrete dam, forming a new control and changing the former stage-discharge relation, was built across the channel about 500 feet below the gage August 20, 1916; permanent.

Extremes of discharge.—Maximum stage recorded during year, 6.6 feet at 4 p. m. February 16 (discharge not determined because of backwater from ice); maximum stage recorded during open-water period, 5.4 feet March 2 (discharge, 2,700 second-feet); minimum stage, 2.44 feet August 12 and 28 (discharge, 6 second-feet).

1915–1918: Maximum stage recorded, 6.6 feet February 16, 1918 (discharge not determined because of backwater from ice); maximum stage recorded during open-water periods, 5.9 feet June 10, 1916 (discharge, 3,380 second-feet); minimum discharge, 3.9 second-feet (measured by current meter), November 26 1914.

Accuracy.—Stage-discharge relation permanent; affected by ice February 14 to 28. Rating curve well defined between 120 and 2,220 second-feet. Gage read to hundredths once daily. Daily discharge ascertained by applying daily gage heights to rating table, except for periods noted in footnote to daily-discharge table. Open-water records good for medium and high stages, fair for low stages; winter records fair.

The following discharge measurement was made by H. C. Beckman while river was frozen across but crest of dam was clear of ice:

January 29, 1918: Gage height, 2.54 feet; discharge, 21 second-feet.

Daily discharge, in second-feet, of Des Plaines River at Lemont, Ill., for the year ending Sept. 30, 1918.

Day.	Oct.	Nov.	Dec.	Jan.	Feb.	Mar.	Apr.	May.	June.	July.	Aug.	Sept.
1	22	180	63	22	10	2,580	400	852	445	40	28	22
2	10	150	63	20	10	2,700	357	760	400	33	31	22
3	9	150	52	17	10	2,580	301	715	280	33	28	17
4	22	150	52	17	10	2,460	245	625	232	33	28	22
5	17	138	70	17	14	2,460	245	492	212	33	15	28
6	22	138	63	17	17	2,460	180	476	174	44	9	33
7	22	120	63	17	17	2,340	150	415	180	40	9	22
8	22	110	63	17	17	1,980	193	385	156	31	7	10
9	17	110	22	17	17	1,860	238	357	132	24	9	9
10	22	95	22	17	17	1,740	206	385	132	28	9	10
11	28	95	22	17	17	1,570	180	422	120	28	7	9
12	33	85	22	17	1,050	1,410	198	385	100	31	6	22
13	28	85	22	17	1,740	1,460	168	415	80	19	7	33
14	22	70	20	17		1,860	144	422	70	15	24	28
15	22	95	17	17		2,460	120	408	110	17	15	22
16	22	95	20	17		2,580	80	329	70	48	15	17
17	33	95	22	17		2,340	132	245	63	55	48	17
18	95	70	25	17		2,220	174	232	48	33	31	10
19	85	52	28	17		1,860	212	174	44	28	15	22
20	70	52	40	14		1,740	219	174	31	28	10	22
21	63	44	52	10	2,300	1,460	301	193	22	19	19	6
22	52	70	58	10		1,250	430	212	31	22	19	6
23	52	63	63	10		1,100	625	371	24	22	19	9
24	63	52	66	10		900	670	805	24	19	15	17
25	52	52	70	10		805	540	805	40	31	9	22
26	52	52	61	10		670	500	715	31	40	9	10
27	95	52	52	10		540	492	625	24	66	9	6
28	95	44	42	10		524	445	540	33	110	6	9
29	120	63	33	10	......	492	625	500	40	66	7	10
30	120	52	28	10	......	460	805	524	28	40	9	9
31	138	......	22	10	......	460	......	476	......	40	28	......

Note.—No gage reading, every other day Nov. 10 to Jan. 24, Jan. 27, 29, 31, and Feb. 2, 3, 5, 7, 9, and 10; daily discharge interpolated. Mean daily discharge estimated Feb. 14–28, because of backwater from ice, from gage heights, observer's notes, and weather records.

Monthly discharge of Des Plaines River at Lemont, Ill., for the year ending Sept. 30, 1918.

[Drainage area, 705 square miles.]

Month.	Discharge in second-feet.				Run-off (depth in inches).
	Maximum.	Minimum.	Mean.	Per square mile.	
October	138	9	49.2	0.070	0.08
November	180	44	89.3	.127	.14
December	70	17	42.5	.060	.07
January	22	10	14.7	.021	.02
February	....	10	1,340	1.90	1.98
March	2,700	460	1,660	2.35	2.71
April	805	80	319	.452	.50
May	852	174	466	.661	.76
June	445	22	113	.160	.18
July	110	15	36.0	.051	.06
August	48	6	16.1	.023	.03
September	33	6	16.7	.024	.03
The year	....	6	340	.482	6.56

DES PLAINES RIVER AT JOLIET, ILL.

LOCATION.—In NE. ¼ sec. 9, T. 35 N., R. 10 E., at Jackson Street Bridge, Joliet, Will County, 1,200 feet upstream from Cass Street Bridge.

DRAINAGE AREA.—Not measured.

RECORDS AVAILABLE.—December 3, 1914, to September 30, 1918; on original chain gage September 5 to December 19, 1914.

GAGE.—Gurley seven-day water-stage recorder, installed December 3, 1914. Chain gage attached to upstream side of bridge at Cass Street read from September 5 to December 19, 1914.

DISCHARGE MEASUREMENTS.—Made from upstream side of Cass Street Bridge.

CHANNEL AND CONTROL.—Channel excavated in solid rock, with a concrete wall on either side; permanent.

EXTREMES OF DISCHARGE.—Maximum mean daily discharge during days of record for the year, 12,500 second-feet, February 15; minimum mean daily discharge, 6,960 second-feet, February 3.

1914–1918: Maximum mean daily discharge during days of record, 13,200 second-feet, June 10, 1916; minimum mean daily discharge, 5,420 second-feet, April 25, 1915.

DIVERSIONS.—Water is diverted to the Illinois & Michigan canal at dam No. 1, about 100 feet above the gage.

REGULATION.—Flow past the gage is largely regulated by the operation of the power plant of the Chicago sanitary district at Lockport, which utilizes the flow of the Chicago Drainage canal and, to a lesser extent, by the operation of Economy Light & Power Co.'s plant, about 100 feet above gage.

ACCURACY.—Stage-discharge relation permanent; not affected by ice during winter. Rating curve well defined. Operation of the water-stage recorder satisfactory except as noted in the table of daily discharge. Daily discharge ascertained by use of discharge integrator. Records excellent.

Discharge measurements of Des Plaines River at Joliet, Ill., during the year ending Sept. 30, 1918.

[Made by H. C. Beckman.]

Date.	Gage height.	Discharge.	Date.	Gage height.	Discharge.
	Feet.	*Sec.-ft.*		*Feet.*	*Sec.-ft.*
Mar. 19 [a]	5.02	9,180	July 23 [b]	....	348
19 [b]	....	379	Sept. 14 [b]	....	348
Nov. 23 [b]	....	526			

a Made in Des Plaines River. b Made in Illinois & Michigan canal.

Daily discharge, in second-feet, of Des Plaines River at Joliet, Ill., for the year ending Sept. 30, 1918.

Day.	Oct.	Nov.	Dec.	Jan.	Feb.	Mar.	Apr.	May.	June.	July.	Aug.	Sept.
1	8,800	8,740	8,730	7,080	7,730	10,300	8,230	9,060	9,470	9,160	9,270	7,110
2	8,500	8,760	8,280	7,880	7,450	9,960	8,480	8,680	8,900	8,540	8,720	7,470
3	8,370	8,540	8,680	7,890	6,960	10,700	8,680	a8,850	8,980	9,780	8,580	8,270
4	a8,700	8,040	8,560	7,600	7,370	11,100	8,300	a9,090	8,960	8,620	8,640	(b)
5	9,070	8,270	8,800	7,480	7,840	11,100	8,170	a8,620	9,460	9,510	8,340	(b)
6	9,250	8,450	8,750	8,130	7,740	11,700	8,120	8,520	9,820	a8,710	8,940	(b)
7	7,810	8,670	8,700	7,860	7,830	11,200	7,650	8,520	9,920	8,860	9,300	(b)
8	9,070	8,540	9,100	8,060	8,420	10,200	8,090	8,540	9,010	9,720	9,360	7,390
9	8,950	8,400	7,280	7,900	8,320	9,840	8,450	8,400	8,590	9,600	8,920	7,330
10	8,960	8,630	8,470	8,140	7,800	10,100	8,610	8,740	8,980	9,710	8,680	7,660
11	8,860	7,880	8,280	8,230	7,730	9,340	8,400	8,450	8,490	9,820	8,700	7,500
12	8,780	8,620	8,470	6,560	10,200	9,180	7,960	8,440	a8,370	9,840	9,780	7,470
13	9,070	8,570	7,910	6,670	9,900	9,570	7,920	8,620	a8,320	9,500	8,580	8,300
14	7,530	8,950	8,060	6,480	10,700	10,700	7,830	8,380	8,740	9,680	8,800	a8,560
15	8,710	8,660	7,880	(b)	12,500	11,200	7,920	8,470	9,100	9,390	9,390	a8,200
16	8,500	8,820	7,030	(b)	11,600	10,400	8,030	8,460	8,840	9,360	8,610	a8,300
17	8,700	(b)	8,060	7,160	11,000	10,400	8,190	8,200	9,510	9,230	8,860	a8,150
18	8,700	(b)	8,090	a7,060	10,300	10,300	8,400	9,440	9,870	8,280	8,800	a8,150
19	8,740	(b)	8,000	a7,660	10,600	10,100	8,350	8,690	8,910	8,180	8,620	a8,240
20	8,530	(b)	8,070	6,970	10,400	9,950	8,650	8,370	8,780	8,000	8,520	a8,430
21	8,100	(b)	8,220	7,510	10,200	9,480	8,370	8,560	8,700	8,140	8,300	a8,530
22	8,190	(b)	8,110	7,190	10,200	9,680	8,050	8,510	9,110	9,420	8,860	a7,580
23	8,700	(b)	6,970	7,790	9,900	9,120	a8,520	8,580	8,790	9,540	8,970	a7,830
24	9,160	(b)	7,740	7,490	9,640	9,400	a8,700	9,000	9,550	8,840	8,180	a8,170
25	8,900	8,180	7,260	7,860	9,960	8,810	a8,900	9,000	9,620	8,590	8,1 0	a8,220
26	8,880	9,000	8,050	7,610	9,870	8,940	9,060	8,670	9,920	8,440	8,470	a8,200
27	9,210	8,980	7,870	7,630	9,870	8,720	8,660	9,120	9,380	8,700	8,630	(b)
28	7,830	9,250	7,820	7,610	10,200	8,540	a9,000	9,520	9,600	8,820	8,200	8,180
29	8,640	7,650	8,080	7,610	....	8,340	a8,720	9,440	9,220	9,540	8,180	8,020
30	8,840	8,980	7,030	7,640	....	8,040	a9,300	9,210	8,590	9,450	8,080	8,360
31	8,910	....	8,170	7,760	....	7,540	....	10,300	....	8,780	7,980	....

a Discharge partly estimated because of incomplete gage record. b No record.

NOTE.—Daily discharge in the above table does not include the flow in the Illinois & Michigan canal. (See "Diversions" in the station description.)

Monthly discharge, in second-feet, of Des Plaines River at Joliet, Ill., for the year ending Sept. 30, 1918.

Month.	Maximum.	Minimum.	Mean.
October	9,250	7,530	8,680
December	9,100	6,970	8,080
February	12,500	6,960	9,370
March	11,700	7,540	9,800
April	9,300	7,650	8,390
May	10,300	8,200	8,790
June	9,920	8,320	9,130
July	9,840	8,000	9,090
August	9,780	7,980	8,690

NOTE.—Discharge in the above table does not include flow of the Illinois & Michigan canal, which diverts water around the gage. See "Diversions" in station description and measurements of flow in the canal.

FOX RIVER AT ALGONQUIN, ILL.

LOCATION.—In NW. ¼ sec. 34, T. 43 N., R. 8 E. third principal meridian, at Chicago, Street Bridge in Algonquin, McHenry County, 100 feet above Public Service Co.'s dam and 500 feet above Crystal Lake outlet.

RECORDS AVAILABLE.—October 1, 1915, to September 30, 1918.

DRAINAGE AREA.—1,340 square miles (measured on map of United States Geological Survey; scale, 1 to 500,000).

GAGE.—Enamel staff gage attached to concrete abutment of bridge; read by Edward Pederson.

CHANNEL AND CONTROL.—Control is a concrete dam about 100 feet below gage; appears to be cracking, and may settle.

DISCHARGE MEASUREMENTS.—Made from upstream side of bridge or by wading below dam.

Extremes of discharge.—Maximum stage recorded during year, 4.4 feet at 7 a. m. and 6 p. m. March 14 (discharge, 5,600 second-feet); minimum stage, 0.59 foot at 7 a. m. and 6 p. m. August 31 (discharge, 67 second-feet).

1916–1918: Maximum stage recorded, 5.3 feet March 31, 1916 (discharge, 7,120 second-feet); minimum stage, 0.59 foot August 31, 1918 (discharge, 67 second-feet).

Diversions.—Water is diverted to operate grist mill at dam, which runs on average of about 4 hours a day, except Sundays, during September to March, inclusive, and one day a week during remainder of year. If total used for each day were uniformly distributed, it would probably average less than 5 second-feet and never exceed 8 second-feet.

Accuracy.—Stage-discharge relation changed during year; not affected by ice during winter. Rating curve used to March 5 fairly well defined; curve used after that date well defined above and fairly well defined below 750 second-feet. Gage read to hundredths twice daily. Storage pond is large, so the small amount of water used by grist mill does not noticeably affect the gage heights. Daily discharge ascertained by applying mean daily gage height to rating tables. Records good.

Discharge measurements of Fox River at Algonquin, Ill., during the year ending Sept. 30, 1918.

[Made by H. C. Beckman.]

Date.	Gage height.	Discharge.	Date.	Gage height.	Discharge.	Date.	Gage height.	Discharge.
	Feet.	*Sec.-ft.*		*Feet.*	*Sec.-ft.*		*Feet.*	*Sec.-ft.*
Mar. 15	4.03	4,790	Apr. 8	2.08	1,440	July 5	0.95	268
15	4.08	5,010	15	1.68	970	5	.94	259

Daily discharge, in second-feet, of Fox River at Algonquin, Ill., for the year ending Sept. 30, 1918.

Day.	Oct.	Nov.	Dec.	Jan.	Feb.	Mar.	Apr.	May.	June.	July.	Aug.	Sept.
1	423	1,200	664	312	255	1,760	2,560	1,080	702	288	185	72
2	423	1,200	620	312	255	2,090	2,390	1,080	702	280	178	72
3	423	1,260	567	305	255	2,600	2,230	1,020	702	265	172	77
4	430	1,260	525	305	255	3,300	2,070	960	702	250	162	82
5	430	1,260	461	298	255	4,250	1,840	960	653	272	151	82
6	423	1,200	401	292	250	5,400	1,610	905	625	272	141	86
7	415	1,200	344	292	250	5,200	1,540	905	588	265	130	91
8	415	1,200	292	286	250	5,000	1,470	905	551	265	120	96
9	415	1,200	255	286	250	5,000	1,400	905	525	265	120	101
10	415	1,200	220	286	255	4,800	1,330	905	507	250	110	106
11	415	1,200	188	279	267	4,600	1,260	905	490	250	110	110
12	423	1,140	188	279	279	4,600	1,200	905	472	250	106	120
13	423	1,140	194	279	292	4,010	1,080	905	455	242	101	130
14	423	1,080	199	279	305	5,600	1,020	850	439	235	101	141
15	430	1,020	204	279	318	5,000	905	800	422	235	106	151
16	430	1,020	209	273	331	5,000	850	750	406	229	110	162
17	446	967	215	273	344	4,800	800	702	389	222	110	172
18	461	914	220	273	358	4,800	850	653	373	222	110	185
19	477	860	226	273	372	4,800	850	634	357	216	106	191
20	500	810	244	273	387	4,800	905	625	242	210	106	197
21	534	810	267	273	401	4,600	905	625	326	204	101	197
22	567	759	292	273	415	4,600	960	702	310	197	101	204
23	620	712	318	267	430	4,400	960	960	310	197	91	210
24	712	712	344	267	509	4,200	905	905	310	197	91	210
25	759	712	358	267	664	4,010	905	850	302	191	82	210
26	914	712	365	267	810	3,820	850	800	302	191	82	204
27	1,020	712	358	261	967	3,630	850	750	295	185	77	197
28	1,080	712	351	261	1,400	3,450	905	750	295	185	77	197
29	1,140	664	344	261	......	3,270	960	750	295	185	72	191
30	1,140	664	331	261	......	2,910	1,080	750	288	185	72	185
31	1,200	......	318	261	......	2,730	......	750	......	191	67	......

Note.—The above table does not include small amount of water used to operate grist mill. (See "Diversions" in station description.)

Monthly discharge of Fox River at Algonquin, Ill., for the year ending Sept. 30, 1918.

[Drainage area, 1,340 square miles.]

Month.	Discharge in second-feet.				Run-off (depth in inches).
	Maximum.	Minimum.	Mean.	Per square mile.	
October	1,200	415	591	0.441	0.51
November	1,260	664	983	.734	.82
December	664	188	325	.243	.28
January	312	261	279	.208	.24
February	1,400	250	406	.303	.32
March	5,600	1,760	4,160	3.10	3.57
April	2,560	800	1,250	.933	1.04
May	1,080	625	837	.625	.72
June	702	288	448	.334	.37
July	288	185	229	.171	.20
August	185	67	111	.083	.10
September	210	72	148	.110	.12
The year	5,600	67	818	.610	8.29

FOX RIVER AT WEDRON, ILL.

Location.—In sec. 9, T. 34 N., R. 4 E., at highway bridge at Wedron, LaSalle County, 1,000 feet above Buck Creek.

Drainage area.—2,500 square miles.

Records available.—November 5, 1914, to September 30, 1918.

Gage.—Chain gage attached to bridge; read by Nels Mathias to January 31 and by T. W. Server after that date.

Discharge measurements.—Made from upstream side of bridge.

Channel and control.—Bed or river at measuring section is soft and probably shifts. Control about 1,000 feet downstream composed of coarse gravel and large boulders; seldom shifts.

Extremes of discharge.—Maximum stage recorded during year, 13.4 feet at 8 a. m. February 15 (discharge, 15,500 second-feet); minimum stage, 5.40 feet at 6 a. m. and 6 p. m. September 4 (discharge, 145 second-feet).

1915–1918: Maximum stage recorded, 15.4 feet February 3, 1916 (discharge not determined because of backwater from ice); maximum open-water stage recorded, 13.8 feet March 28, 1916 (discharge, 16,700 second-feet); minimum discharge recorded, 105 second-feet November 20, 1914 (measured by current meter).

Regulation.—Slight diurnal fluctuation is caused by operation of power plants at and above Montgomery.

Accuracy.—Stage-discharge relation changed during high water in February; seriously affected by ice during winter. Rating curve used to February 12 well defined above and fairly well defined below 1,130 second-feet; curve used after that date well defined between 275 and 11,300 second-feet. Gage read to hundredths twice daily. Diurnal fluctuation only slight. Daily discharge ascertained by applying mean daily gage height to rating tables, except for period when stage-discharge relation was affected by ice, for which it was estimated from occasional gage heights, observer's notes, and weather records. Open-water records good for medium and high stages, and fair for low stages; winter record poor.

Discharge measurements of Fox River at Wedron, Ill., during the year ending Sept. 30, 1918.

[Made by H. C. Beckman.]

Date.	Gage height.	Discharge.	Date.	Gage height.	Discharge.
	Feet.	*Sec.-ft.*		*Feet.*	*Sec.-ft.*
Oct. 18	6.76	853	July 8	5.86	371
Nov. 16	7.01	1,170	Aug. 21	5.78	318
Nov. 16	7.02	1,150			

Daily discharge, in second-feet, of Fox River at Wedron, Ill., for the year ending Sept. 30, 1918.

Day.	Oct.	Nov.	Dec.	Jan.	Feb.	Mar.	Apr.	May.	June.	July.	Aug.	Sept.
1	384	1,590	940			3,680	2,870	1,610	1,310	428	301	185
2	580	1,670	852			4,660	2,720	1,710	997	317	349	228
3	652	1,670	852			5,750	2,570	1,660	830	405	285	194
4	544	1,610	852			8,270	2,570	1,560	997	405	296	145
5	510	1,610	810			7,230	2,170	1,460	871	388	306	194
6	580	1,670		400	680	7,230	3,500	1,460	922	285	228	185
7	510	1,790				6,710	1,930	1,410	997	440	502	296
8	372	1,670				5,980	1,820	1,260	997	376	247	296
9	510	1,610				5,980	1,710	1,310	790	394	301	280
10	615	1,670				6,220	1,710	1,360	712	405	296	224
11	580	1,440				4,660	1,660	1,310	871	417	247	275
12	544	1,330				5,520	1,560	1,220	922	440	206	376
13	544	1,380	550		11,300	5,520	1,460	1,220	712	371	194	360
14	510	1,330			13,100	7,750	1,310	1,360	535	382	202	322
15	372	1,230			15,200	7,230	1,260	1,180	502	285	202	371
16	580	1,180		370	8,010	6,220	1,310	1,080	751	280	285	338
17	896	1,130			4,660	5,980	1,220	997	471	382	411	233
18	940	985			3,860	5,750	1,310	954	568	388	638	228
19	940	940			5,080	5,520	1,260	922	535	388	417	354
20	852	1,180			5,980	5,750	1,310	871	471	388	301	317
21	690	1,080			3,860	5,750	1,610	997	471	382	285	256
22	652	1,030			3,500	5,300	1,820	1,260	423	311	285	296
23	1,030	1,030			3,500	5,520	1,930	1,410	376	228	285	306
24	1,030	1,030			4,050	5,080	1,660	1,610	275	266	266	206
25	1,130	896			4,660	4,870	1,510	1,610	266	354	270	266
26	1,230	769	450	330	4,660	4,450	1,560	1,560	405	502	252	327
27	1,330	940			4,450	4,250	1,560	1,360	535	502	177	399
28	1,330	896			4,050	4,050	1,510	1,360	922	394	198	332
29	1,440	769			....	3,680	1,660	1,360	603	327	185	322
30	1,550	730			....	3,500	1,820	1,360	471	237	252	285
31	1,550	....			....	3,170	....	1,310	....	311	228	....

NOTE.—Discharge Dec. 6 to Feb. 12 estimated, because of ice, from gage heights, observer's notes, and weather records. Braced figures show mean daily discharge for periods included.

Monthly discharge of Fox River at Wedron, Ill., for the year ending Sept. 30, 1918.

[Drainage area, 2,500 square miles.]

Month.	Discharge in second-feet.				Run-off (depth in inches).
	Maximum.	Minimum.	Mean.	Per square mile.	
October	1,550	372	806	0.322	0.37
November	1,790	730	1,260	.504	.56
December	940	....	565	.226	.26
January	....	....	365	.146	.17
February	15,200	....	3,860	1.54	1.60
March	8,270	3,170	5,520	2.21	2.55
April	3,500	1,220	1,800	.720	.80
May	1,710	871	1,330	.532	.61
June	1,310	266	684	.274	.31
July	502	228	367	.147	.17
August	638	177	287	.115	.13
September	399	145	280	.112	.12
The year	15,200	145	1,410	.564	7.66

VERMILION RIVER NEAR STREATOR, ILL.

LOCATION.—In sec. 1, T. 30 N., R. 3 E. third principal meridian, at highway bridge known as Bridge No. 3, 1½ miles south of Streator, La Salle County, and 100 feet below Santa Fe Railway bridge.

DRAINAGE AREA.—1,080 square miles.

RECORDS AVAILABLE.—July 27, 1914, to September 30, 1918.

GAGE.—Chain gage attached to highway bridge; read by Mathew Reid until March 31, and by Floyd Leslie after that date.

Discharge measurements.—Made from downstream side of bridge or by wading.

Channel and control.—Gravel and rocks; probably permanent.

Extremes of discharge.—Maximum stage recorded during year, 11.0 feet at 4 p. m. February 15 (discharge, 5,080 second-feet); minimum stage, 0.53 foot at 4 p. m. June 27 (discharge, 1.4 second-feet).

1914–1918: Maximum stage recorded, 22.4 feet January 21, 1916 (discharge estimated from extension of rating curve, 16,000 second-feet); minimum stage 0.45 foot August 16 and 17, 1914 (discharge, 0.7 second-foot).

Accuracy.—Stage-discharge relation permanent; seriously affected by ice during winter. Rating curve well defined between 300 and 2,500 second-feet, and fairly well defined between 10 and 300 second-feet and above 2,500 second-feet. Gage read to hundredths once daily. Daily discharge ascertained by applying daily gage heights to rating table, except for period when stage-discharge relation was affected by ice, for which it was estimated from occasional gage heights, observer's notes, and weather records. Records good, except for periods of extreme low stages and period of ice effect, for which they are poor.

Discharge measurements of Vermilion River near Streator, Ill., during the year ending Sept. 30, 1918.

[Made by H. C. Beckman.]

Date.	Gage height.	Discharge.	Date.	Gage height.	Discharge.
	Feet.	*Sec.-ft.*		*Feet.*	*Sec.-ft.*
July 9	6.23	1,790	Aug. 21	0.80	8.5
Aug. 21	.80	8.2			

Daily discharge, in second-feet, of Vermilion River near Streator, Ill., for the year ending Sept. 30, 1918.

Day.	Oct.	Nov.	Dec.	Jan.	Feb.	Mar.	Apr.	May.	June.	July.	Aug.	Sept.
1	19	15	12			684	146	1,500	783	970	39	75
2	18	31	12			495	127	1,050	588	716	34	68
3	18	39	17			557	291	818	464	620	15	39
4	3.8	34	12			652	557	716	360	495	21	34
5	3.8	30	9			652	620	620	291	404	30	83
6	5.2	24			2	652	652	557	228	252	26	39
7	3.8	28				557	557	526	216	652	15	39
8	3.0	23				495	526	464	346	1,750	12	26
9	2.4	23				495	464	557	526	1,700	15	127
10	1.8	20				375	419	588	419	1,700	9.4	119
11	1.8	17				346	375	684	318	1,450	12	113
12	1.9	17			4,200	346	332	749	265	818	81	44
13	1.8	16	3		4,680	332	304	588	228	854	21	78
14	1.8	21			5,000	404	240	684	156	818	18	59
15	3.3	18			5,080	360	216	684	127	495	3.8	109
16	2.4	15		2	3,200	346	204	620	53	434	15	167
17	19	12			2,150	332	228	557	80	291	14	49
18	22	20			1,700	291	304	557	59	216	14	74
19	30	13			1,600	265	464	495	48	169	13	34
20	28	9.4			1,400	265	620	2,450	47	146	6.0	30
21	24	9.4			1,170	216	1,350	1,050	42	131	6.9	26
22	16	10			930	216	1,800	818	74	91	10	21
23	14	8.6			818	216	1,650	684	30	39	10	23
24	9.4	9.4			818	240	1,650	1,350	33	15	8.6	18
25	9.4	12			620	193	1,010	1,700	14	49	6.0	24
26	17	12	12		557	204	930	1,300	2.2	9.4	5.2	18
27	15	9.4			495	193	930	818	1.4	434	9.4	21
28	13	13			526	204	1,090	783	131	216	6.9	21
29	13	12			......	193	818	652	652	193	8.6	21
30	15	9.4			......	165	1,250	588	970	150	193	21
31	12	......			......	156	......	818	......	51	167	......

Note.—Discharge for Dec. 6 to Feb. 11 estimated, because of ice, from gage heights, observer's notes, and weather records. Braced figures show mean discharge for periods indicated.

Monthly discharge of Vermilion River near Streator, Ill., for the year ending Sept. 30, 1918.

[Drainage area, 1,080 square miles.]

Month.	Discharge in second-feet.				Run-off (depth in inches).
	Maximum.	Minimum.	Mean.	Per square mile.	
October	30	1.8	11.2	0.010	0.01
November	39	8.6	17.7	.016	.02
December			7.71	.0071	.008
January			2.00	.0019	.002
February	5,080		1,250	1.16	1.21
March	684	156	358	.331	.38
April	1,800	127	671	.621	.69
May	2,450	464	840	.778	.90
June	970	1.4	252	.233	.26
July	1,750	9.4	527	.488	.56
August	193	3.6	27.1	.025	.03
September	167	18	54.0	.050	.06
The year	5,080		328	.304	4.13

SPOON RIVER AT SEVILLE, ILL.

Location.—In sec. 24, T. 6 N., R. 1 E. fourth principal meridian, at Toledo, Peoria & Western Railway bridge, a quarter of a mile east of railway station at Seville, Fulton County.

Drainage area.—1,600 square miles.

Records available.—July 24, 1914, to September 30, 1918.

Gage.—Chain gage attached to bridge; read by C. D. Bartlett until July 1 and by R. M. Boales after that date.

Discharge measurements.—Made from downstream side of bridge; low-water measurements are made by wading below dam at railroad station.

Channel and control.—Control is a loose rock dam, about 2 miles downstream from gage, used to create a reservoir for the pumping station of Toledo, Peoria & Western Railway.

Extremes of stage.—Maximum stage recorded during year, 15.3 feet at 9 a. m. February 16; minimum stage, 2.55 feet at 7 a. m. January 25.

1914–1918: Maximum stage recorded, 26.0 feet January 23, 1916; minimum stage, 1.35 feet July 31 and August 28 and 29, 1914.

Ice.—Stage-discharge relation affected by ice during winter.

Data inadequate for determination of discharge.

Discharge measurements of Spoon River at Seville, Ill., during the year ending Sept. 30, 1918.

[Made by H. C. Beckman.]

Date.	Gage height.	Discharge.	Date.	Gage height.	Discharge.
	Feet.	*Sec.-ft.*		*Feet.*	*Sec.-ft.*
Oct. 15	2.73	74	July 10	15.02	6,270
15	2.73	76	Aug. 22	4.27	492
June 11	3.68	317			

Daily gage height, in feet, of Spoon River at Seville, Ill., for the year ending Sept. 30, 1918.

Day.	Oct.	Nov.	Dec.	Jan.	Feb.	Mar.	Apr.	May.	June.	July.	Aug.	Sept.
1	3.1	5.7	2.95	2.7	......	6.8	3.6	5.9	6.0	6.4	4.4	7.5
2	3.0	5.7	2.95	......	......	6.8	3.6	5.8	5.7	......	3.8	5.8
3	2.9	5.6	2.95	2.7	......	......	6.6	5.8	5.5	......	3.7	4.5
4	2.9	5.4	2.95	......	......	8.8	3.6	5.8	5.5	......	3.5	5.4
5	3.4	5.4	......	2.7	......	6.7	3.6	4.1	5.5	......	3.5	5.1
6	3.3	5.3	2.85	......	......	6.7	3.6	4.1	6.9	......	3.4	4.8
7	3.3	5.3	2.95	2.6	......	5.5	4.8	3.8	6.7	......	3.3	4.6
8	4.4	5.1	......	......	3.3	5.4	5.0	3.8	5.9	......	3.2	4.2
9	4.1	4.2	......	2.6	3.8	5.4	5.2	3.8	5.9	......	3.2	4.0
10	3.6	4.0	......	......	......	5.4	4.8	3.8	......	14.7	3.2	3.9
11	3.3	3.8	2.7	2.6	7.3	4.3	4.8	3.8	3.7	10.6	3.6	3.8
12	3.1	3.8	......	......	13.4	4.3	4.8	3.8	4.0	7.7	3.4	4.0
13	2.7	3.7	2.7	......	14.0	4.2	4.6	3.8	4.2	6.4	3.3	4.4
14	2.8	3.6	......	2.6	......	4.2	5.1	3.8	3.9	5.8	3.1	4.2
15	2.7	3.6	2.7	......	14.8	4.2	5.6	3.8	2.4	5.4	4.0	4.0
16	7.2	3.5	......	2.6	15.3	4.2	6.1	4.2	3.1	5.2	6.8	3.8
17	8.4	3.4	2.7	......	5.5	4.3	6.1	4.4	3.3	5.4	5.6	3.7
18	7.2	3.4	......	2.6	5.5	4.3	6.1	4.5	2.2	4.8	6.4	3.6
19	7.2	3.3	2.7	......	5.4	4.3	6.1	4.6	3.2	4.6	9.7	3.5
20	7.2	3.3	......	......	5.3	4.2	6.1	4.8	3.2	4.4	8.0	3.5
21	7.0	3.2	......	2.6	5.3	4.2	6.0	5.4	3.0	4.2	5.1	3.4
22	6.9	3.1	3.1	......	5.1	4.2	5.9	6.8	3.0	4.1	4.4	3.4
23	6.9	3.1	......	2.6	5.1	4.2	5.9	7.2	3.0	4.0	4.0	3.3
24	6.7	2.9	......	......	5.1	3.7	5.9	8.1	3.3	3.9	3.8	3.3
25	6.5	2.8	4.3	2.55	4.8	3.8	5.8	10.4	5.7	3.8	3.7	3.3
26	6.5	2.8	......	......	4.8	3.6	5.8	9.8	6.8	3.8	3.6	3.3
27	6.4	2.8	......	......	4.6	3.6	5.8	9.1	10.1	3.8	3.6	3.2
28	6.4	2.8	......	......	6.8	3.6	6.0	8.1	12.4	3.8	3.4	3.2
29	6.3	2.8	3.6	......	......	3.6	5.2	8.2	12.7	3.8	3.3	3.1
30	6.0	2.9	......	......	......	3.6	5.9	7.6	11.3	6.6	3.5	3.1
31	5.9	......	3.6	......	......	3.6	......	8.8	......	6.2	6.3	......

NOTE.—Stage-discharge relation probably affected by ice about Dec. 5 to Feb. 25. Sudden drop in stage Feb. 16 probably caused by breaking of ice jam.

SANGAMON RIVER AT MONTICELLO, ILL.

LOCATION.—In sec. 12, T. 18 N., R. 5 E. third principal meridian, at Illinois Central Railroad bridge half a mile west of Monticello, Piatt County.

DRAINAGE AREA.—550 square miles.

RECORDS AVAILABLE.—February 4, 1908, to December 31, 1912; June 23, 1914, to September 30, 1918.

GAGE.—Chain gage attached to downstream side of bridge; read by David Coay.

DISCHARGE MEASUREMENTS.—Made from downstream side of bridge and wooden trestle approach during medium and high stages, and by wading during low stages.

CHANNEL AND CONTROL.—Measuring section is at a pool. Control consists of fine gravel; likely to shift.

EXTREMES OF DISCHARGE.—Maximum stage recorded during year, 14.4 feet at 8 a. m., February 14 (discharge, 6,180 second-feet); minimum stage, 1.85 feet October 10–12 and December 12 and 14 (discharge, 11 second-feet).

1908–1912 and 1914–1918: Maximum stage recorded 15.2 feet May 14, 1908 (discharge, 9,280 second-feet); maximum stage during flood of March to April, 1913, 17.7 feet March 25 (discharge not known); minimum stage recorded, 1.5 feet July 31, August 1 and 3, 1914 (discharge, 1 second-foot).

ACCURACY.—Stage-discharge relation changed slightly several times during year; seriously affected by ice during winter. Rating curve fairly well defined below 4,000 second-feet. Gage read to quarter-tenths once daily. Daily discharge ascertained by applying daily gage height to rating table, except for period when stage-discharge relation was affected by ice for which it was estimated from occasional gage heights, observer's reports, and weather records and except for days noted in table of daily discharge. Open-water records good for low and medium stages, fair for very high stages; winter records poor.

Discharge measurements of Sangamon River at Monticello, Ill., during the year ending Sept. 30, 1918.

[Made by H. C. Beckman.]

Date.	Gage height.	Discharge.	Date.	Gage height.	Discharge.	Date.	Gage height.	Discharge.
	Feet.	*Sec.-ft.*		*Feet.*	*Sec.-ft.*		*Feet.*	*Sec.-ft.*
Oct. 10	1.86	11.1	June 25	10.56	1,910	Aug. 26	3.03	108
10	1.86	10.7	July 15	5.16	324	30	2.42	41
Feb. 16	10.95	2,090						

Daily discharge, in second-feet, of Sangamon River at Monticello, Ill., for the year ending Sept. 30, 1918.

Day.	Oct.	Nov.	Dec.	Jan.	Feb.	Mar.	Apr.	May.	June.	July.	Aug.	Sept.
1	12	59	19			226	114	1,320	508	675	42	a 72
2	12	59	a 19			a 211	114	1,100	a 375	637	36	a 86
3	12	56	19			a 196	258	891	242	473	34	100
4	12	a 44	19			181	490	758	194	a 382	a 32	354
5	a 12	32	19	15	300	181	354	a 660	159	290	29	490
6	12	29				170	290	562	159	226	25	675
7	a 12	27				170	a 274	a 628	128	a 708	21	862
8	12	25				159	258	695	148	1,190	21	a 778
9	12	25				210	226	675	a 171	1,440	17	695
10	11	25				a 218	194	675	194	1,440	17	599
11	11	a 24			1,810	226	181	675	148	1,320	a 19	338
12	11	23			3,100	194	170	a 716	114	920	21	290
13	a 11	23	12		4,270	181	159	758	100	618	17	226
14	a 12	23			6,180	170	a 141	1,040	87	a 494	17	148
15	12	21		8	4,270	148	123	1,040	71	371	21	a 138
16	12	21			3,100	148	114	862	a 64	322	21	128
17	12	21			a 2,070	a 138	170	715	56	258	21	128
18	12	a 20			1,040	128	258	580	48	226	a 28	114
19	14	19			715	109	422	a 571	45	194	34	100
20	16	19			618	100	695	562	45	170	36	100
21	a 16	19			562	96	a 1,090	599	45	a 144	32	100
22	16	19			526	96	1,480	526	45	118	21	a 92
23	16	19			490	96	1,610	456	a 40	104	21	83
24	16	19			a 430	a 102	1,480	388	34	96	25	71
25	15	a 19			371	109	862	338	1,360	87	a 35	71
26	16	19	30	5	322	118	1,190	a 298	2,270	75	45	67
27	17	19			274	138	1,440	258	1,260	75	71	56
28	a 22	19			226	128	a 1,500	226	1,040	a 71	45	48
29	27	19			......	114	1,560	194	1,010	67	a 40	a 46
30	32	19			......	100	1,440	a 266	a 842	48	36	45
31	59	......			......	a 107	......	338	......	42	59	......

a Discharge interpolated because of no gage-height record.

NOTE.—Discharge estimated for Dec. 6 to Feb. 10, because of ice, from gage heights, observer's notes, and weather records. Braced figures show mean daily discharge for periods included.

Monthly discharge of Sangamon River at Monticello, Ill., for the year ending Sept. 30, 1918.

[Drainage area, 550 square miles.]

Month.	Discharge in second-feet.				Run-off (depth in inches).
	Maximum.	Minimum.	Mean.	Per square mile.	
October	59	11	15.9	0.029	0.03
November	59	19	26.2	.048	.05
December	....	....	19.5	.035	.04
January	....	....	9.19	.017	.02
February	6,180	....	1,190	2.16	2.25
March	226	96	151	.275	.32
April	1,610	114	622	1.13	1.26
May	1,320	194	625	1.14	1.31
June	2,270	34	367	.667	.74
July	1,440	42	428	.778	.90
August	71	17	30.3	.055	.06
September	862	45	237	.431	.48
The year	6,180	....	303	.551	7.46

SANGAMON RIVER AT RIVERTON, ILL.

Location.—In southeast corner of SW. ¼ sec. 9, T. 16 N., R. 4 W. third principal meridan, at Wabash Railroad bridge a quarter of a mile west of Riverton, Sangamon County, and 2½ miles below mouth of South Fork.

Drainage area.—2,560 square miles.

Records available.—February 13, 1908, to December 31, 1912; August 7, 1914, to September 30, 1918.

Gage.—Chain gage attached to bridge; read by J. J. Washburn.

Discharge measurements.—Made from downstream side of bridge or by wading.

Channel and control.—Measuring section is at a pool. Control consists of fine gravel; shifts slightly.

Extremes of discharge.—Maximum stage recorded during year, 22.8 feet at 4 p. m. May 11 (discharge, 9,980 second-feet); minimum stage, 7.19 feet February 2, 3, and 6 (discharge estimated, 16 second-feet).

1908–1912 and 1914–1918: Maximum stage recorded, 27.8 feet February 3, 1916 (discharge, 20,800 second-feet;) high water of 1883 reached a height of approximately 32 feet on present gage, and that of 1875 is said to have been one-half foot lower (discharge not estimated;) minimum stage recorded, 6.9 feet October 3–15, 1915 (discharge, 3 second-feet).

Accuracy.—Stage-discharge relation changed slightly during year; affected by ice during winter. Rating curve well defined between 94 and 4,350 second-feet, and fairly well defined beyond these limits. Gage read to hundredths once daily. Daily discharge ascertained by applying daily gage heights to rating table, except for period when stage-discharge relation was affected by ice, for which it was estimated from occasional gage heights, observer's notes, and weather records. Open-water records good; winter records poor.

Discharge measurements of Sangamon River at Riverton, Ill., during the year ending Sept. 30, 1918.

[Made by H. C. Beckman.]

Date.	Gage height.	Discharge.	Date.	Gage height.	Discharge.
	Feet.	*Sec.-ft.*		*Feet.*	*Sec.-ft.*
Oct. 11	7.56	36	June 14	9.26	343
11	7.56	36	Aug. 24	8.46	190

Daily discharge, in second-feet, of Sangamon River at Riverton, Ill., ,or the year ending Sept. 30, 1918.

Day.	Oct.	Nov.	Dec.	Jan.	Feb.	Mar.	Apr.	May.	June.	July.	Aug.	Sept.
1	64	102	75			862	368	4,630	1,100	1,570	262	227
2	55	106	69			736	416	4,910	736	1,530	205	250
3	55	115	62			706	465	4,000	736	1,530	184	284
4	52	129	45			706	416	3,300	736	1,370	168	1,100
5	51	136	48	50	800	706	862	2,880	736	1,170	162	1,490
6	43	127				706	1,210	2,280	798	862	147	1,450
7	51	122				619	1,370	1,930	798	1,210	140	1,330
8	43	112				566	1,100	1,650	556	3,240	113	1,250
9	41	102				515	1,060	4,910	440	3,790	92	1,170
10	39	82				515	1,030	9,280	416	3,860	85	1,170
11	37	86			5,700	465	767	9,980	416	3,720	84	862
12	40	85			7,320	440	647	9,100	392	3,600	84	767
13	41	90	30		9,800	392	566	9,440	344	3,420	80	592
14	43	88			9,620	392	566	8,300	320	3,180	79	490
15	43	82		25	8,160	416	540	7,580	296	2,830	79	404
16	50	80			6,120	416	515	8,160	284	1,610	79	416
17	52	76			6,560	416	676	6,930	238	1,330	76	592
18	53	75			7,860	416	566	5,240	227	927	75	619
19	51	71			8,010	404	995	4,490	216	894	380	676
20	50	73			7,320	368	1,250	4,280	164	1,060	392	619
21	174	69			4,070	356	1,780	3,240	154	619	182	706
22	320	66			2,430	344	2,330	2,940	151	566	90	465
23	490	68			1,830	344	2,880	1,980	151	465	80	416
24	184	65			1,410	332	3,000	1,370	145	440	178	368
25	122	57			1,290	490	3,540	1,210	296	404	154	320
26	102	51	85	15	1,250	490	3,930	1,490	440	368	140	296
27	86	50			960	465	4,700	1,250	1,100	356	113	238
28	84	52			960	392	4,770	1,060	862	273	105	227
29	75	58			......	344	4,910	995	1,450	174	94	216
30	96	76			......	344	4,910	894	1,780	205	113	184
31	113	......			......	320	......	767	......	250	238	......

NOTE.—Discharge estimated for Dec. 6 to Feb. 10, because of ice, from gage heights, observer's notes, and weather records. Braced figures show mean daily discharge for period indicated.

Monthly discharge of Sangamon River at Riverton, Ill., for the year ending Sept. 30, 1918.

[Drainage area, 2,560 square miles.]

Month.	Discharge in second-feet.				Run-off (depth in inches).
	Maximum.	Minimum.	Mean.	Per square mile.	
October	490	37	90.3	0.035	0.04
November	136	50	85.0	.033	.04
December	....	....	54.3	.021	.02
January	....	....	29.5	.012	.01
February	9,800	....	3,520	1.38	1.44
March	862	320	483	.189	.22
April	4,910	368	1,740	.680	.76
May	9,980	767	4,210	1.64	1.89
June	1,780	145	549	.214	.24
July	3,860	174	1,510	.590	.68
August	392	75	144	.056	.06
September	1,490	184	640	.250	.28
The year	9,980	....	1,070	.418	5.68

SANGAMON RIVER NEAR OAKFORD, ILL.

LOCATION.—In sec. 6, T 19 N., R. 7 W. third principal meridian, at highway bridge 3 miles northeast of Oakford, Menard County, 2½ miles above Chicago, Peoria & St. Louis Railroad bridge, and 1¼ miles above mouth of Crane Creek.

DRAINAGE AREA.—5,000 square miles.

RECORDS AVAILABLE.—October 26, 1909, to June 30, 1911; December 10, 1911, to March 31, 1912; and August 25, 1914, to September 30, 1918.

GAGE.—Chain gage attached to bridge; read by R. W. Schnell from October 1 to December 31, by Henry Chesser from January 1 to June 30, and by Frank Dick from July 1 to September 30.

DISCHARGE MEASUREMENTS.—Made from downstream side of bridge.

CHANNEL AND CONTROL.—Bed composed of sand and fine gravel; shifting. The river for some distance above and below station has been dredged and straightened, thus increasing the slope considerably and disturbing the regimen of flow.

EXTREMES OF DISCHARGE.—Maximum stage recorded during year, 12.2 feet February 15 and 17 (discharge, 10,500 second-feet); minimum stage, 1.28 feet October 19-22 (discharge, 183 second-feet).

1914-1918: Maximum stage recorded, 19.9 feet June 8 and 9, 1917 (discharge determined from extension of rating curve, 33,300 second-feet); minimum stage recorded, 0.65 foot September 27, 1916 (discharge, 128 second-feet). Minimum discharge recorded, 85 second-feet August 30-31. November 27, and December 2, 1914. Maximum and minimum discharges recorded during periods of record, same as above.

ACCURACY.—Stage-discharge relation practically permanent; seriously affected by ice during winter. Rating curve fairly-well defined. Gage read to quarter-tenths once daily. Daily discharge ascertained by applying daily gage heights to rating table, except for period when stage-discharge relation was affected by ice, for which it was estimated from occasional gage heights, observer's notes, and weather records; discharge interpolated, because of no gage-height record July 7, August 10-23 and 25, and September 8. Open-water records good; winter records poor.

Discharge measurements of Sangamon River near Oakford, Ill., during the year ending Sept. 30, 1918.

[Made by H. C. Beckman.]

Date.	Gage height.	Discharge.	Date.	Gage height.	Discharge.
	Feet.	*Sec.-ft.*		*Feet.*	*Sec.-ft.*
Oct. 13	1.32	185	June 13	3.12	1,040
13	1.32	190	Aug. 24	2.31	644
Nov. 17	1.42	248			

Daily discharge, in second-feet, of Sangamon River near Oakford, Ill., for the year ending Sept. 30, 1918.

Day.	Oct.	Nov.	Dec.	Jan.	Feb.	Mar.	Apr.	May.	June.	July.	Aug.	Sept.
1	294	315	214			1,810	851	6,400	2,980	2,100	696	452
2	294	315	210			1,810	746	6,040	1,810	1,810	647	524
3	272	315	210			1,600	851	5,060	1,670	1,950	598	549
4	259	294	210			1,530	851	4,940	1,670	2,100	574	696
5	243	294	206	130	1,000	1,530	1,200	3,770	1,670	2,260	524	2,260
6	230	315				1,390	1,530	3,770	1,600	2,420	500	2,740
7	230	315				1,390	1,950	3,460	1,810	4,120	452	2,580
8	222	315				1,390	1,670	3,590	1,600	5,820	428	2,420
9	210	294				1,390	1,670	6,040	1,390	6,880	405	2,260
10	202	294	155			1,320	1,600	6,040	1,320	7,000		1,810
11	190	272			6,280	1,390	1,460	7,390	1,260	7,000		1,670
12	190	272			6,400	1,200	1,320	7,650	1,140	6,640		1,600
13	198	264			9,080	1,140	1,320	9,920	1,200	5,710		1,390
14	202	255			10,100	1,140	1,200	10,300	1,020	4,940		1,200
15	198	251		65	10,500	1,140	1,080	9,780	906	4,230		1,020
16	190	251			10,300	1,200	1,020	8,820	906	3,590	350	906
17	190	243			10,500	906	1,080	8,300	798	3,060		906
18	194	238			9,640	1,020	1,140	7,390	906	2,340		906
19	183	230			8,820	962	1,260	10,200	851	2,020		906
20	183	230			8,300	962	1,460	8,560	1,020	1,810		962
21	183	230			7,390	906	2,260	5,820	962	1,600		1,020
22	183	226			7,000	906	2,980	5,380	549	1,390		906
23	264	222	230		5,050	962	3,500	3,950	574	1,260		851
24	405	214			3,060	906	3,770	3,460	746	1,200	1,080	798
25	405	210		45	2,980	851	3,680	2,180	549	1,080	814	696
26	359	210			2,260	851	4,230	2,020	2,260	1,020	549	647
27	315	214			2,100	906	4,630	2,180	2,580	962	549	598
28	272	218			1,810	851	5,930	2,340	3,060	906	476	549
29	272	218			....	851	6,520	2,740	3,500	851	428	500
30	405	214			....	906	6,520	3,950	3,950	798	405	476
31	315	....			....	851	....	3,460	....	696	428	....

NOTE.—Discharge interpolated for July 7, Aug. 10–23 and 25, and Sept. 8, because of no gage-height record; estimated for Dec. 6 to Feb. 10, because of ice, from gage heights, observer's notes, and weather records.

Monthly discharge of Sangamon River near Oakford, Ill., for the year ending Sept. 30, 1918.

[Drainage area, 5,000 square miles.]

Month.	Discharge in second-feet.				Run-off (depth in inches).
	Maximum.	Minimum.	Mean.	Per square mile.	
October	405	183	250	0.050	0.06
November	315	210	258	.052	.06
December	....	....	203	.041	.05
January	....	....	78.9	.016	.02
February	10,500	....	4,700	.940	.98
March	1,810	851	1,160	.232	.27
April	6,520	746	2,310	.462	.52
May	10,300	2,020	5,640	1.13	1.30
June	3,950	549	1,540	.308	.34
July	7,000	696	2,890	.578	.67
August	....	....	466	.093	.11
September	2,740	452	1,160	.232	.26
The year	10,500	....	1,700	.340	4.64

SOUTH FORK OF SANGAMON RIVER AT POWER PLANT, NEAR TAYLORVILLE, ILL.

LOCATION.—In sec. 14, T. 13 N., R. 3 W., at Chicago & Illinois Midland Railroad bridge, 6 miles northwest of Taylorville, Christian County, 500 feet east of power plant of Central Illinois Public Service Co., 5 miles below mouth of Bear Creek and 8 miles below station formerly maintained at Wabash Railroad bridge.

DRAINAGE AREA.—510 square miles. (Measured on map issued by the United States Geological Survey; scale, 1: 500,000.)

RECORDS AVAILABLE.—May 18, 1917, to September 30, 1918.

GAGE.—Chain gage attached to bridge; read by H. Hendricks.

Discharge measurements.—Made from upstream side of bridge or by wading.

Channel and control.—Soft mud; likely to shift.

Extremes of discharge.—Maximum stage recorded during year, 18.0 feet at 8 a. m., May 11 (discharge, 2,960 second-feet); minimum discharge, 3.1 second-feet, October 9.

1917–18: Maximum stage recorded, 26.6 feet June 6, 1917 (discharge, 10,400 second-feet); minimum discharge, 3.1 second-feet, October 9. A stage of about 27.3 feet on the present gage is said to have been reached January 31, 1916 (discharge, 11,300 second-feet).

Diversions.—An average of about half a second-foot is used for boiler feed and other purposes at the power plant.

Accuracy.—Stage-discharge relation changed slightly during high water in February; seriously affected by ice during winter. Rating curves fairly well defined above 16 second-feet. Gage read to hundredths once daily. Daily discharge ascertained by applying daily gage heights to rating tables, except for periods noted in footnote to daily-discharge table. Open-water records good for medium and high stages, fair for low stages; winter records poor.

Discharge measurements of South Fork of Sangamon River at power plant near Taylorville, Ill., during the year ending Sept. 30, 1918.

[Made by H. C. Beckman.]

Date.	Gage height.	Discharge.	Date.	Gage height.	Discharge.
	Feet.	*Sec.-ft.*		*Feet.*	*Sec.-ft.*
Oct. 12	3.72	4.0	June 15	4.33	24.5
12	3.72	4.1	Aug. 26	4.28	25.9

Daily discharge, in second-feet, of South Fork of Sangamon River at power plant near Taylorville, Ill., for the year ending Sept. 30, 1918.

Day.	Oct.	Nov.	Dec.	Jan.	Feb.	Mar.	Apr.	May.	June.	July.	Aug.	Sept.
1	9.0	112	19			125	62	1,180	93	150	14	46
2	8.2	96	17			125	117	829	85	109	12	85
3	7.4	84	18			125	125	478	73	66	11	177
4	6.9	84	17			125	1,000	387	66	38	9.6	522
5	6.6	80	14		2	125	1,000	317	58	29	8.6	733
6	6.2	23		3		125	1,020	267	55	25	7.6	646
7	6.6	21				125	868	258	49	73	6.6	663
8	6.8	19			267	117	327	267	49	327	4.6	522
9	3.1	11			1,440	117	297	934	46	511	3.8	437
10	3.6	9			2,110	109	240	2,710	36	489	3.8	387
11	4.4	30			2,430	93	186	2,960	40	557	3.8	281
12	4.2	26			2,860	85	159	2,860	38	646	3.8	175
13	4.0	25	7		2,860	77	159	2,590	31	570	3.3	69
14	4.4	24			2,860	73	141	2,430	29	377	3.3	58
15	5.0	21			2,430	85	125	2,190	27	168	3.3	58
16	6.0	19		2	2,430	109	109	1,870	27	109	3.8	73
17	5.0	17			2,110	93	277	1,560	27	89	4.6	267
18	6.5	17			1,800	81	599	1,120	25	77	5.6	317
19	250	17			1,480	77	697	769	24	66	4.6	229
20	340	17			1,160	73	630	437	22	55	6.6	140
21	230	20			846	66	806	347	21	46	9.6	53
22	76	11			529	73	956	277	19	40	18	141
23	59	19			213	85	978	249	19	36	15	109
24	52	16			200	81	956	222	18	52	14	89
25	38	16			186	77	806	196	18	52	12	80
26	28	18	25	1	177	69	1,120	177	16	36	11	71
27	40	14			168	66	1,530	159	16	29	29	61
28	32	17			141	55	1,540	141	14	27	17	52
29	46	19			......	49	1,560	126	73	24	10	43
30	52	21			......	46	1,470	109	126	20	10	38
31	47	......			......	43	......	97	......	16	133	......

Note.—Discharge interpolated, because of no gage-height record, for Feb. 17–22 and 24, Apr. 28, May 2, and Sept. 11, 12, 19, 20, and 25–28; estimated for Dec. 6 to Feb. 7, because of ice, from gage heights, observer's notes, and weather records.

Monthly discharge of South Fork of Sangamon River at power plant near Taylorville, Ill., for the year ending Sept. 30, 1918.

[Drainage area, 510 square miles.]

Month.	Discharge in second-feet.				Run-off (depth in inches).
	Maximum.	Minimum.	Mean.	Per square mile.	
October	340	3.1	45.0	0.088	0.10
November	112	9	30.8	.060	.07
December	....	....	15.0	.029	.03
January	....	....	1.97	.0039	.004
February	2,860	....	1,020	2.00	2.08
March	125	43	89.5	.175	.20
April	1,560	62	662	1.30	1.45
May	2,960	97	920	1.80	2.08
June	125	14	41.3	.081	.09
July	646	16	158	.310	.36
August	133	3.3	13.0	.025	.03
September	733	38	221	.433	.48
The year	2,960	....	263	.516	6.97

KASKASKIA RIVER AT VANDALIA, ILL.

Location.—In sec. 16, T. 6 N., R. 1 E. third principal meridian, at highway bridge at east end of Main Street, Vandalia, Fayette County, 3½ miles above Hickory Creek.

Drainage area.—1,980 square miles.

Records available.—February 26, 1908, to December 31, 1912; August 11, 1914, to September 30, 1918.

Gage.—Chain gage attached to bridge; read by Wilson Haley.

Discharge measurements.—Made from downstream side of bridge or by wading.

Channel and control.—Measuring section is at a pool; likely to shift.

Extremes of discharge.—Maximum stage recorded during year, 18.5 feet May 11 and 12 (discharge, 8,460 second-feet); minimum stage recorded, 0.91 foot at 1 p. m. October 17 (discharge, 38 second-feet).

1908–1912 and 1914–1918: Maximum stage recorded, 23.0 feet June 5, 1917 (discharge, 16,400 second-feet); minimum stage, 0.38 foot August 12, 1914 (discharge, 13 second-feet).

Accuracy.—Stage-discharge relation changed during high water in February; seriously affected by ice during winter. Rating curves well defined above and fairly well defined below 368 second-feet. Gage read to hundredths once daily. Daily discharge ascertained by applying daily gage heights to rating tables, except for period when stage-discharge relation was affected by ice, for which it was estimated from occasional gage heights, observer's notes, and weather records.

Discharge measurements of Kaskaskia River at Vandalia, Ill., during the year ending Sept. 30, 1918.

[Made by H. C. Beckman.]

Date.	Gage height.	Discharge.	Date.	Gage height.	Discharge.	Date.	Gage height.	Discharge.
	Feet.	*Sec.-ft.*		*Feet.*	*Sec.-ft.*		*Feet.*	*Sec.-ft.*
Oct. 9	1.09	50	Feb. 16	17.28	7,280	July 15	9.02	2,240
Nov. 19	2.10	162	June 17	3.10	386	Aug. 27	2.41	267

Daily discharge, in second-feet, of Kaskaskia River at Vandalia, Ill., for the year ending Sept. 30, 1918.

Day.	Oct.	Nov.	Dec.	Jan.	Feb.	Mar.	Apr.	May.	June.	July.	Aug.	Sept.
1	67	770	120			991	332	5,490	933	2,280	297	609
2	67	685	114			1,050	332	4,710	818	2,200	263	585
3	67	657	108			904	314	4,090	790	2,080	246	1,020
4	62	552	102			818	407	3,390	710	1,820	229	1,500
5	62	458	102	65		818	736	2,920	684	1,820	213	4,140
6	58	413			1,320	790	585	2,360	634	1,750	213	5,140
7	54	369				736	515	1,820	585	1,540	198	5,350
8	46	307				684	585	1,540	818	1,890	184	4,890
9	43	307				634	538	1,880	585	3,740	177	2,640
10	42	268				538	538	5,070	492	4,530	164	1,920
11	42	249				492	515	8,460	470	4,710	157	1,680
12	41	231			6,270	448	492	8,460	427	4,650	144	1,470
13	41	222	55		7,320	515	448	7,800	387	3,740	138	1,230
14	40	204			8,060	609	427	7,210	368	2,520	138	991
15	39	196			7,800	585	427	4,770	332	1,750	150	846
16	38	180			6,770	561	407	4,830	332	1,750	132	710
17	38	172			6,010	515	2,320	3,740	368	1,440	184	962
18	38	165			5,350	492	3,840	3,120	314	1,170	126	1,680
19	46	151			4,290	448	2,720	2,800	280	962	121	1,890
20	222	151			3,240	427	2,480	3,080	263	818	116	1,960
21	481	151		45	2,440	407	4,830	2,980	246	736	110	1,640
22	391	144			1,780	368	5,140	2,580	229	634	157	1,360
23	327	138			1,400	350	4,040	2,160	198	470	121	1,140
24	222	138			1,330	350	3,290	1,750	213	538	121	991
25	204	132			1,200	368	3,440	1,500	246	470	157	875
26	196	132	150		1,110	368	5,420	1,330	448	448	184	763
27	180	132			1,050	368	7,440	1,200	2,000	407	263	684
28	172	126			962	350	7,100	1,080	2,040	368	164	634
29	222	126			......	350	6,460	962	1,780	350	157	585
30	657	120			......	332	6,270	933	2,120	332	138	538
31	713	......			......	332	......	1,020	......	297	634	......

NOTE.—Discharge estimated for Dec. 6 to Feb. 11, because of ice, from gage heights, observer's notes, and weather records. Braced figures show mean discharge for periods indicated.

Monthly discharge of Kaskaskia River at Vandalia, Ill., for the year ending Sept. 30, 1918.

[Drainage area, 1,980 square miles.]

Month.	Discharge in second-feet.				Run-off (depth in inches).
	Maximum.	Minimum.	Mean.	Per square mile.	
October	713	38	159	0.080	0.09
November	770	120	268	.135	.15
December	....	....	97.5	.049	.06
January	....	....	51.5	.026	.03
February	8,060	....	2,890	1.46	1.52
March	1,050	332	548	.277	.32
April	7,440	314	2,410	1.22	1.36
May	8,460	933	3,390	1.71	1.97
June	2,120	198	670	.338	.38
July	4,710	297	1,680	.848	.98
August	634	110	187	.094	.11
September	5,350	538	1,680	.848	.95
The year	8,460	38	1,150	.581	7.92

KASKASKIA RIVER AT NEW ATHENS, ILL.

LOCATION.—In W. ½ NE. ¼ sec. 28, T. 2 S., R. 7 W. third principal meridian, at Illinois Central Railroad bridge 600 feet north of railroad station at New Athens, St. Clair County, 1 mile below mouth of Silver Creek and 3 miles above mouth of Lively Creek.

DRAINAGE AREA.—5,220 square miles.

RECORDS AVAILABLE.—January 23, 1907, to December 31, 1912; June 22, 1914, to September 30, 1918. Gage height of river was taken on Wednesday and Thursday mornings from January 23, 1907, to October 28, 1909, by C. J. von Roth Roffy for the New Athens Journal, and by whom they were published. Record authentic. Gage heights have been reduced to the present datum; maximum error probably not more than 0.4 foot, decreasing with increase of stage.

GAGE.—Chain gage attached to bridge; read by Henry Hoffman.

DISCHARGE MEASUREMENTS.—Made from downstream side of bridge to which gage is attached, or from highway bridge about 500 feet downstream.

CHANNEL AND CONTROL.—Sand and gravel; may shift.

EXTREMES OF DISCHARGE.—Maximum stage recorded during year, 24.5 feet at noon April 30 (discharge, 20,300 second-feet); minimum stage recorded, 2.47 feet at noon October 18 (discharge, 107 second-feet).

1907–1912 and 1914; 1918: Maximum stage recorded, 35.7 feet August 26, 1915 (discharge, 63,100 second-feet); minimum stage, 2.08 feet August 10, 1914 (discharge, 102 second-feet).

ACCURACY.—Stage-discharge relation changed during high water in February; seriously affected by ice during winter; also affected by backwater from Mississippi River about April 4–8 and June 1–20. Rating curves used during periods of no backwater from Mississippi River fairly well defined. Gage read to hundredths once daily. Daily discharge ascertained by applying daily gage height to rating tables, except for periods noted in footnote to table of daily discharge. Open-water records fair; winter records and records during period of backwater poor.

Published estimates of discharge for the following periods may be considerably too large, the excess depending on the amount of backwater produced at New Athens: January 21–28, June 14–18, July 19 to August 3, 1907; May 17 to July 23, 1908; March 14, April 21 to May 1, May 11–17, June 12 to July 27, 1909; May 10–13, June 12–15, 1910; March 22 to May 11, June 19–22, 1912.

Discharge measurements of Kaskaskia River at New Athens, Ill., during the year ending Sept. 30, 1918.

[Made by H. C. Beckman.]

Date.	Gage height.	Discharge.	Date.	Gage height.	Discharge.
	Feet.	*Sec.-ft.*		*Feet.*	*Sec.-ft.*
Oct. 8	2.71	145	June 18 a	5.98	557
Feb. 15	22.49	13,300	Aug. 28	4.52	606

a Made during backwater from Mississippi River.

Daily discharge, in second-feet, of Kaskaskia River at New Athens, Ill., for the year ending Sept. 30, 1918.

Day.	Oct.	Nov.	Dec.	Jan.	Feb.	Mar.	Apr.	May.	June.	July.	Aug.	Sept.
1	187	1,060	295			2,190	556	18,900	1,300	2,240	669	458
2	178	910	316			2,190	556	17,000	1,230	2,440	556	748
3	178	880	316			2,420	582	15,500	1 240	2,490	505	1,450
4	170	850	274			2,640	348	14,100	1,150	2,490	480	1,410
5	170	790	244	180		2,440	444	13,200	1,060	2,440	456	2,340
6	161	730	224		8,000	2,090	464	12,200	906	2,290	432	4,670
7	153	700	215			1,990	500	11,500	1,160	2,140	409	7,100
8	145	612				1,840	742	10,800	2,860	1,990	365	6,780
9	145	530				1,610	1,130	10,100	1,920	1,940	365	6,780
10	137	478				1,410	950	8,910	1,360	1,840	344	6,620
11	137	429				1,250	834	6,700	1,060	2,390	323	5,720
12	129	405				1,130	834	9,820	952	3,300	323	5,020
13	129	360			12,600	1,040	776	13,000	911	3,740	323	4,190
14	129	338	160		13,700	1,010	748	13,900	794	3,960	323	3,970
15	129	316			13,900	950	692	14,100	693	4,070	283	2,400
16	122	316			13,900	892	664	14,100	649	4,070	283	1,840
17	114	295			13,700	863	2,640	13,900	596	3 630	304	1,650
18	107	295			12,700	863	7,340	13,300	561	2,860	324	1,370
19	122	295			11,800	834	8,810	13,000	460	2,240	344	2,490
20	114	295		125	10,900	834	10,300	13,000	422	1,790	530	3,080
21	122	274			10,100	776	11,000	12,800	664	1,490	556	4,530
22	129	234			9,340	748	11,800	12,600	609	1,220	530	5,030
23	137	234			8,910	720	11,800	12,700	556	1,100	409	4,790
24	145	234			8,610	664	12,000	12,700	505	1,010	323	3,580
25	234	224	350		8,310	664	13,200	12,700	480	892	387	2,390
26	360	224			7,260	636	15,200	11,800	406	834	530	1,740
27	360	234			4,250	636	17,000	10,700	456	805	582	1,410
28	338	234			2,640	609	18,100	7,600	480	748	505	1,250
29	382	254			....	582	19,800	3,630	556	692	892	1,100
30	880	274			....	582	20,300	2,440	1,330	636	748	1,030
31	1,400	....			....	556	....	1,940	....	609	556	....

NOTE.—Discharge interpolated for Oct. 21, Mar. 3, Apr. 21, Aug. 17-18, and Sept. 15, for lack of gage-height record; estimated for Dec. 7 to Feb. 12, because of ice, from gage heights, observer's notes, and weather records; determined from daily gage heights at Chester and New Athens, by slope method described in Water Supply Paper 345, p. 35, for Apr. 4-8 and June 1-20, because of backwater from Mississippi River.

Monthly discharge of Kaskaskia River at New Athens, Ill., for the year ending Sept. 30, 1918.

[Drainage area, 5,220 square miles.]

Month.	Discharge in second-feet.				Run-off (depth in inches).
	Maximum.	Minimum.	Mean.	Per square mile.	
	1,400	107	237	0.045	0.05
November	1,060	224	443	.085	.09
December	....	....	252	.048	.06
January	....	....	143	.027	.03
February	13,900	....	7,090	1.36	1.42
March	2,640	556	1,210	.232	.27
April	20,300	348	6,340	1.21	1.35
May	18,900	1,940	11,600	2.22	2.56
June	2,860	422	913	.175	.20
July	4,070	609	2,080	.398	.46
August	892	283	448	.086	.10
September	7,100	456	3,190	.611	.68
The year	20,300	107	2,790	.534	7.27

BIG MUDDY RIVER AT PLUMFIELD, ILL.

LOCATION.—In W. ½ sec. 20, T. 7 S., R. 2 E., at highway bridge at Plumfield, Franklin County, 6 miles west of West Frankfort, 1½ miles below mouth of Middle Fork, and 2 miles below station formerly maintained at Chicago, Burlington & Quincy Railroad bridge.

DRAINAGE AREA.—753 square miles.

RECORDS AVAILABLE.—August 18, 1914, to September 30, 1918; June 16, 1908 to September 30, 1912, and November 1, to December 31, 1912, maintained at Chicago, Burlington & Quincy Railroad bridge.

GAGE.—Chain gage attached to bridge; read by Louis Robertson.

DISCHARGE MEASUREMENTS.—Made from downstream side of bridge or by wading.

CHANNEL AND CONTROL.—Probably permanent. Control is about a quarter of a mile below gage. Point of zero flow is at a stage of about 0.6 foot.

EXTREMES OF DISCHARGE.—Maximum stage recorded during year, 24.3 feet at 6 a. m. and 6 p. m. May 15 (discharge, 9,330 second-feet); minimum stage, 0.84 foot at 6 a. m. October 12 (discharge, 2.4 second-feet).

1914–1918: Maximum stage recorded, 30.2 feet February 1, 1916 (discharge, 16,300 second-feet); minimum stage August 18 to 26, 1914, when there was no flow past the gage.

ACCURACY.—Stage-discharge relation practically permanent; seriously affected by ice during winter. Rating curve fairly well defined above 43 second-feet. Gage read to hundredths once daily. Daily discharge ascertained by applying daily gage heights to rating table, except for periods noted in footnote to daily-discharge table. Open-water records good except for low stages, for which they are fair; winter records poor.

Discharge measurements of Big Muddy River at Plumfield, Ill., during the year ending Sept. 30, 1918.

[Made by H. C. Beckman.]

Date.	Gage height.	Discharge.
	Feet.	*Sec.-ft.*
June 19	1.26	9.6
19	1.26	8.4

Daily discharge, in second-feet, of Big Muddy River at Plumfield, Ill., for the year ending Sept. 30, 1918.

Day.	Oct.	Nov.	Dec.	Jan.	Feb.	Mar.	Apr.	May.	June.	July.	Aug.	Sept.
1	4.0	386	438			113	24	7,930	62	452	300	108
2	3.6	348	360			312	9	7,190	42	252	174	264
3	3.3	207	143			760	196	6,020	35	133	90	240
4	3.5	103	114			790	123	4,600	26	62	58	373
5	3.2	62	85	150		494	67	3,370	26	39	24	565
6	2.9	42	62		830	466	43	2,180	133	24	16	700
7	2.8	32	43			480	33	1,820	288	17	10	522
8	2.8	26				324	26	790	399	14	8.2	264
9	2.7	20				218	21	412	286	11	6.0	123
10	2.7	18				153	18	185	174	9.1	5.0	67
11	2.6	15				113	16	730	128	6.8	4.7	42
12	2.4	13				94	14	3,070	80	6.2	4.2	28
13	2.8	11			5,600	72	13	6,380	43	5.2	3.9	20
14	2.6	9.1	15		6,020	58	12	9,030	29	5.0	3.8	16
15	2.6	8.4		45	5,930	46	12	9,330	22	4.7	3.8	13
16	2.5	7.6			5,680	42	14	8,930	16	3.9	3.6	26
17	2.5	6.6			5,040	39	153	8,030	13	4.4	3.2	67
18	3.1	6.0			4,110	37	536	6,920	10	4.0	3.8	336
19	3.1	5.2			3,170	32	965	4,880	7.9	4.0	3.5	494
20	3.0	4.8			2,150	30	1,280	3,910	7.0	3.9	3.1	312
21	8.8	4.7			1,300	26	1,540	2,680	6.8	4.1	3.0	163
22	7.9	4.6			685	24	1,730	1,940	6.0	4.0	30	252
23	6.6	4.4			264	26	1,880	1,480	5.0	3.8	34	185
24	4.5	4.2			153	30	2,090	1,340	5.0	3.9	31	98
25	4.4	4.3	140	10	133	46	2,640	1,300	5.6	4.2	33	58
26	4.5	4.2			118	57	3,220	1,280	5.6	4.0	58	37
27	3.8	7.0			118	54	4,460	1,060	6.6	3.8	300	24
28	3.2	12			113	50	6,470	480	6.4	3.9	264	19
29	4.2	46			......	38	7,460	240	12	3.7	229	14
30	15	324			......	32	8,030	153	580	3.6	196	11
31	62	......			......	28	......	94	......	3.0	143	......

NOTE.—Discharge interpolated for Nov. 13 and 15, Dec. 4, June 9, and Aug. 8; estimated for Dec. 8 to Feb. 12, because of ice, from gage heights, observer's notes, and weather records.

Monthly discharge of Big Muddy River at Plumfield, Ill., for the year ending Sept. 30, 1918.

[Drainage area, 753 square miles.]

Month.	Discharge in second-feet.				Run-off (depth in inches).
	Maximum.	Minimum.	Mean.	Per square mile.	
October	62	2.4	5.92	0.008	0.009
November	386	4.2	58.2	.077	.09
December	438		96.1	.128	.15
January			66.5	.088	.10
February	6,020		1,800	2.39	2.49
March	790	24	164	.218	.25
April	8,030	9	1,440	1.91	2.13
May	9,330	94	3,460	4.59	5.29
June	580	5.0	82.2	.109	.12
July	452	3.0	35.6	.047	.05
August	300	3.0	66.1	.088	.10
September	700	11	181	.240	.27
The year	9,330	2.4	614	.815	11.06

BIG MUDDY RIVER AT MURPHYSBORO, ILL.

LOCATION.—In SW. ¼ sec. 8, T. 9 S., R. 2 W., at lower highway bridge on South Twentieth Street, a quarter of a mile below mouth of Louis Creek at Mobile & Ohio Railway bridge.

RECORDS AVAILABLE.—December 6, 1916, to September 30, 1918.

DRAINAGE AREA.—2,170 square miles (measured on map issued by United States Geological Survey; scale, 1 to 500,000).

GAGE.—Chain gage attached to bridge; read by E. W. Jacobs.

CHANNEL AND CONTROL.—Bed composed of heavy clay; likely to shift.

DISCHARGE MEASUREMENTS.—Made from downstream side of bridge or by wading.

EXTREMES OF DISCHARGE.—Maximum stage recorded during year, 33.9 feet at 8 p. m. May 16 (discharge not determined because of backwater from Mississippi River); maximum stage recorded during periods not affected by backwater, 27.7 feet at 5 p. m. February 16 (discharge, 10,000 second-feet); minimum stage recorded during year, 1.64 feet at 1 p. m. October 11 (discharge, 3.9 second-feet).

1917–1918: Maximum discharge, estimated 15,600 second-feet January 10, 1917; minimum discharge, 3.9 second-feet, October 11, 1917. About February 2, 1916, the river reached a height of 39.6 feet—the highest known stage—on the present gage (discharge ascertained from extension of rating curve, 28,000 second-feet).

ACCURACY.—Stage-discharge relation changed during year; seriously affected by ice during winter; also affected by backwater from Mississippi River whenever height on gage of United States Weather Bureau at Chester, Ill., is above about 10.0 feet. Rating curve used until March 4 fairly well defined between 45 and 9,000 second-feet; curve used after that date fairly well defined above 68 second-feet. Gage read to hundredths once daily. Daily discharge ascertained by applying daily gage heights to rating tables, except for periods noted in footnote to table of daily discharge. Open-water records good for medium stages, fair for very low and high stages; winter records poor.

Discharge measurements of Big Muddy River at Murphysboro, Ill., during the year ending Sept. 30, 1918.

[Made by H. C. Beckman.]

Date.	Gage height.	Discharge.
	Feet.	*Sec.-ft.*
Oct. 8	1.87	7.3
8	1.87	7.4
Aug. 2[a]	9.40	1,770

Daily gage height, in feet, of Big Muddy River at Murphysboro, Ill., for the year ending Sept. 30, 1918.

Day.	Oct.	Nov.	Dec.	Jan.	Feb.	Mar.	Apr.	May.	June.	July.	Aug.	Sept.
1	......	5.3	5.1	......	......	3.8	5.2	30.9	7.9	......	1.85	......
2	......	5.8	......	3.35	4.4	......	6.8	31.2	......	8.6	2.22	4.8
3	2.15	5.8	5.1	......	......	......	7.4	30.6	10.0	8.4	2.75	6.1
4	2.04	4.0	4.8	3.2	......	3.2	8.6	28.5	9.6	8.0	......	8.8
5	1.96	3.6	4.2	3.55	4.9	......	8.6	......	9.1	7.6	2.85	9.2
6	1.94	3.5	3.6	3.55	......	7.4	8.5	26.2	8.8	7.0	2.9	9.6
7	......	3.0	3.15	......	5.6	......	......	24.9	8.5	......	2.85	9.9
8	1.84	2.90	2.92	6.2	......	6.4	8.4	22.2	8.2	6.0	2.5	8.4
9	1.70	2.80	......	6.2	19.0	6.1	7.5	19.4	......	5.2	1.96	7.1
10	1.66	2.72	2.80	6.4	21.0	4.6	6.7	16.8	8.0	5.3	2.04	5.7
11	1.64	......	2.70	......	......	4.4	6.1	17.1	8.2	5.4	2.12	5.1
12	1.72	2.96	2.66	6.6	24.4	4.1	5.3	27.3	9.1	5.5	1.98	4.4
13	1.70	3.7	2.62	......	......	3.8	4.6	30.5	10.9	5.2	1.85	3.0
14	......	3.05	2.59	......	27.0	3.6	......	32.2	10.6	......	1.78	2.96
15	2.10	2.72	2.57	5.0	27.0	3.4	3.6	33.5	10.4	5.0	1.74	......
16	2.05	2.54	......	......	27.7	3.25	4.4	33.9	......	4.0	1.88	4.2
17	2.00	2.48	2.55	4.1	......	......	7.6	33.5	9.9	3.2	2.12	5.4
18	2.30	......	2.52	......	26.2	3.4	8.6	31.9	9.2	2.7	......	5.9
19	2.29	2.40	2.65	3.7	24.6	3.5	10.6	......	8.9	2.2	3.7	6.4
20	2.20	2.32	3.1	......	......	3.55	12.8	29.9	8.6	2.12	4.6	6.0
21	......	2.22	3.4	......	21.0	3.5	17.3	28.4	8.0	......	4.0	5.2
22	2.60	2.14	3.55	3.45	......	......	17.8	26.5	7.7	2.06	3.4	4.6
23	2.70	2.08	......	......	14.9	3.1	18.2	24.4	......	2.00	3.4	4.2
24	2.64	2.04	4.8	3.4	......	......	......	22.3	6.4	1.92	4.0	3.8
25	2.60	......	......	......	7.2	3.6	21.6	20.2	5.5	1.85	4.2	3.5
26	2.70	2.02	4.4	4.4	......	3.45	24.5	......	6.1	1.88	6.9	3.35
27	2.60	2.22	4.3	......	4.3	3.5	26.4	18.2	7.1	1.92	9.2	3.05
28	......	2.70	4.2	......	......	......	29.2	15.1	7.7	1.98	......	2.85
29	2.60	4.2	4.2	3.45	......	3.8	30.2	11.5	8.1	1.92	......	......
30	2.80	5.0	......	......	......	3.9	......	8.1	......	1.90	4.8	2.6
31	3.20	......	3.7	4.0	......	......	......	7.8	......	1.88	4.2	......

NOTE.—Stage-discharge relation affected by ice Dec. 11 to Feb. 8 and by backwater from Mississippi River Mar. 5 to July 21 and Sept. 5–10.

Daily discharge, in second-feet, of Big Muddy River at Murphysboro, Ill., for the year ending Sept. 30, 1918.

Day.	Oct.	Nov.	Dec.	Jan.	Feb.	Mar.	July.	Aug.	Sept.
1	a 26	495	455			226	......	18	a 404
2	a 22	598	a 455			......	......	43	469
3	18	598	455			......	......	94	797
4	13	255	395			1,170	......	a 100	1,590
5	10	195	287	325	250	......	......	106	
6	9.2	180	195			......	......	112	
7	a 8.0	105	128			......	......	106	1,000
8	6.8	92	95			......	......	68	
9	4.5	79	a 87		5,650	......	......	25	
10	4.1	69	79		6,730	......	......	30	
11	3.9	a 84			7,680	......	......	35	539
12	4.8	100			8,640	......	......	26	381
13	4.5	210			a 9,420	......	......	18	126
14	a 10	112			10,200	......	......	14	118
15	15	69	40	310	a 10,200	......	......	12	a 229
16	13	49			10,700	......	......	20	340
17	11	43			10,200	......	......	35	612
18	27	a 39			9,720	......	......	a 140	743
19	20	35			8,760	......	......	244	880
20	20	29			a 7,740	......	......	424	770
21	a 38	21			6,730	......	......	300	563
22	56	17			a 5,200	......	31	190	424
23	67	14			3,660	......	27	190	340
24	60	13			a 2,290	......	22	300	262
25	56	a 12	260	30	920	......	18	340	208
26	67	12			a 612	......	20	1,020	182
27	56	21			304	......	22	1,710	132
28	a 56	67			a 264	......	26	a 1,400	106
29	56	287			......	......	22	a 882	a 92
30	79	435			......	......	21	469	78
31	135	......			......	......	20	340	......

a Discharge interpolated.

NOTE.—Discharge estimated for Dec. 11 to Feb. 8 because of ice and for Sept. 5–10 because of backwater from Mississippi River. Discharge March 5 to July 21 not determined owing to backwater. Braced figures show mean discharge for periods indicated.

Monthly discharge of Big Muddy River at Murphysboro, Ill., for the year ending Sept. 30, 1918.

[Drainage area, 2,170 square miles.]

Month.	Discharge in second-feet.				Run-off (depth in inches).
	Maximum.	Minimum.	Mean.	Per square mile.	
October	135	3.9	31.5	0.015	0.02
November	598	12	144	.066	.07
December	455		190	.088	.10
January			215	.099	.11
February	10,700		4,560	2.10	2.19
August	1,710	12	284	.131	.15
September		78	546	.252	.28

MISCELLANEOUS MEASUREMENTS.

Miscellaneous discharge measurements in Hudson Bay drainage basin during the year ending Sept. 30, 1918.

Date.	Stream.	Tributary to—	Locality.	Gage height.	Discharge.
				Feet.	*Sec.-ft.*
June 16	Allen Creek	Swift Current Creek	Trail crossing on Many Glacier-Canyon Creek trail.		17.6
July 14	do	do	do	0.29	[a] 3.0
Aug 1	do	do	do	.06	2.6
Sept. 6	do	do	do	.03	1.2

[a] Temporary gage set under foot log at trail crossing.

Miscellaneous discharge measurements in Upper Mississippi River drainage basin during the year ending Sept. 30, 1918.

Date.	Stream.	Tributary to—	Locality.	Gage height.	Discharge.
				Feet.	*Sec.-ft.*
June 5	Iowa River	Mississippi River	Belle Plain, Iowa		28,000

INDEX.

V.

W.

Z.

STREAM-GAGING STATIONS

AND

PUBLICATIONS RELATING TO WATER RESOURCES

PART V. HUDSON BAY AND UPPER MISSISSIPPI RIVER DRAINAGE BASINS

STREAM-GAGING STATIONS AND PUBLICATIONS RELATING TO WATER RESOURCES.

INTRODUCTION.

Investigation of water resources by the United States Geological Survey has consisted in large part of measurements of the volume of flow of streams and studies of the conditions affecting that flow, but it has comprised also investigation of such closely allied subjects as irrigation, water storage, water powers, underground waters, and quality of waters. Most of the results of these investigations have been published in the series of water-supply papers, but some have appeared in the bulletins, monographs, professional papers, and annual reports.

The results of stream-flow measurements are now published annually in 12 parts, each part covering an area whose boundaries coincide with natural drainage features as indicated below:

Part I. North Atlantic slope basins.
II. South Atlantic slope and eastern Gulf of Mexico basins.
III. Ohio River basin.
IV. St. Lawrence River basin.
V. Upper Mississippi River and Hudson Bay basins.
VI. Missouri River basin.
VII. Lower Mississippi River basin.
VIII. Western Gulf of Mexico basins.
IX. Colorado River basin.
X. Great basin.
XI. Pacific Slope basins in California.
XII. North Pacific slope basins, published in three volumes:
A, Pacific slope basins in Washington and upper Columbia River basin.
B, Snake River basin.
C, Lower Columbia River basin and Pacific slope basins in Oregon.

HOW GOVERNMENT REPORTS MAY BE OBTAINED OR CONSULTED.

Water-supply papers and other publications of the United States Geological Survey containing data in regard to the water resources of the United States may be obtained or consulted as indicated below:

1. Copies may be obtained free of charge by applying to the Director of the Geological Survey, Washington, D. C. The edition printed for free distribution is, however, small and is soon exhausted.

2. Copies may be purchased at nominal cost from the Superintendent of Documents, Government Printing Office, Washington, D. C., who will on application furnish lists giving prices.

3. Sets of the reports may be consulted in the libraries of the principal cities in the United States.

4. Complete sets are available for consultation in the local offices of the water-resources branch of the Geological Survey, as follows:

Boston, Mass., 2500 Customhouse.
Albany, N. Y., 704 Journal Building.
Harrisburg, Pa., Care of Water Supply Commission.
Asheville, N. C., 32-35 Broadway.
Chattanooga, Tenn., Temple Court Building.
Madison, Wis., care of Railroad Commission of Wisconsin.
Chicago, Ill., 1404 Kimball Building.
Ames, Iowa, care of State Highway Commission.
Helena, Mont., Montana National Bank Building.
Topeka, Kans., 23 Federal Building.
Austin, Tex., Capitol Building.
Denver, Colo., 403 New Post Office Building.
Salt Lake City, Utah, 313 Federal Building.
Boise, Idaho, 615 Idaho Building.
Idaho Falls, Idaho, 228 Federal Building.
Portland, Oreg., 606 Post Office Building.
Tacoma, Wash., 406 Federal Building.
San Francisco, Calif., 328 Customhouse.
Los Angeles, Calif., 619 Federal Building.
Honolulu, Hawaii, 14 Capitol Building.

A list of the Geological Survey's publications may be obtained by applying to the Director of the United States Geological Survey, Washington, D. C.

STREAM-FLOW REPORTS.

Stream-flow records have been obtained at more than 4,510 points in the United States, and the data obtained have been published in the reports tabulated below:

Stream-flow data in reports of the United States Geological Survey.

[A=Annual Report; B=Bulletin; W=Water-Supply Paper.]

Report.	Character of data.	Year.
10th A, pt. 2	Descriptive information only	
11th A, pt. 2	Monthly discharge and descriptive information	1884 to Sept., 1890.
12th A, pt. 2	do	1884 to June 30, 1891.
13th A, pt. 3	Mean discharge in second-feet	1884 to Dec. 31, 1892.
14th A, pt. 2	Monthly discharge (long-time records, 1871 to 1893)	1888 to Dec. 31, 1893.
B 131	Descriptions, measurements, gage heights, and ratings	1893 and 1894.
16th A, pt. 2	Descriptive information only	
B 140	Descriptions, measurements, gage heights, ratings, and monthly discharge (also many data covering earlier years).	1895.
W 11	Gage heights (also gage heights for earlier years)	1896.
18th A, pt. 4	Descriptions, measurements, ratings, and monthly discharge (also similar data for some earlier years).	1895 and 1896.
W 15	Descriptions, measurements, and gage heights, eastern United States, eastern Mississippi River, and Missouri River above junction with Kansas.	1897.
	Descriptions, measurements, and gage heights, western Mississippi River below junction of Missouri and Platte, and western United States.	1897.
t. 4.	Descriptions, measurements, ratings, and monthly discharge (also some long-time records).	1897.
	Measurements, ratings, and gage heights, eastern United States, eastern Mississippi River, and Missouri River.	1898.
	Measurements, ratings, and gage heights, Arkansas River and western United States.	1898.

Stream-flow data in reports of the United States Geological Survey—Continued.

Report.	Character of data.	Year.
20th A, pt. 4	Monthly discharge (also for many earlier years)	1898.
W 35 to 39	Descriptions, measurements, gage heights, and ratings	1899.
21st A, pt. 4	Monthly discharge	1899.
W 47 to 52	Descriptions, measurements, gage heights, and ratings	1900.
22d A, pt. 4	Monthly discharge	1900.
W 65, 66	Descriptions, measurements, gage heights, and ratings	1901.
W 75	Monthly discharge	1901.
W 82 to 85	Complete data	1902.
W 97 to 100	....do	1903.
W 124 to 135	....do	1904.
W 165 to 178	....do	1905.
W 201 to 214	....do	1906.
W 241 to 252	....do	1907-8.
W 261 to 272	....do	1909.
W 281 to 292	....do	1910.
W 301 to 312	....do	1911.
W 321 to 332	....do	1912.
W 351 to 362	....do	1913.
W 381 to 394	....do	1914.
W 401 to 414	....do	1915.
W 431 to 444	....do	1916.
W 451 to 464	....do	1917.
W 471 to 484	....do	1918.

NOTE.—No data regarding stream flow are given in the 15th and 17th annual reports.

The records at most of the stations discussed in these reports extend over a series of years, and miscellaneous measurements at many points other than regular gaging stations have been made each year. An index of the reports containing records obtained prior to 1904 has been published in Water-Supply Paper 119.

The following table gives by years and drainage basins the numbers of the papers on surface-water supply published from 1899 to 1918. The data for any particular station will in general be found in the reports covering the years during which the station was maintained. For example, data for Machias River at Whitneyville, Me., 1903 to 1917, are published in Water-Supply Papers 97, 124, 165, 201, 241, 261, 281, 301, 321, 351, 381, 401, 431, 451, and 471, which contain records for the New England streams from 1903 to 1918. Results of miscellaneous measurements are published by drainage basins.

In these papers and in the following lists the stations are arranged in downstream order. The main stem of any river is determined by measuring or estimating its drainage area—that is, the headwater stream having the largest drainage area is considered the continuation of the main stream, and local changes in name and lake surface are disregarded. All stations from the source to the mouth of the main stem of the river are presented first, and the tributaries in regular order from source to mouth follow, the streams in each tributary basin being listed before those of the next basin below.

In exception to this rule the records for Mississippi River are given in four parts, as indicated on page III, and the records for large lakes are taken up in order of streams around the rim of the lake.

Numbers of water-supply papers containing results of stream measurements, 1899–1918.

Year.	I North Atlantic slope basins (St. John River to York River).	II South Atlantic slope and eastern Gulf of Mexico basins (James River to the Mississippi).	III Ohio River basin.	IV St. Lawrence River and Great Lakes basins.	V Hudson Bay and upper Mississippi River basins.	VI Missouri River basin.	VII Lower Mississippi River basin.	VIII Western Gulf of Mexico basins.	IX Colorado River basin.	Great Basin.	XI Pacific slope basins in California.	XII North Pacific drainage basins: Pacific slope basins in Washington and upper Columbia River basin.	XII Snake River basin.	XII Lower Columbia River basin and Pacific slope basins in Oregon.
1899 a	35	b 35, 36	36	36	36	c 36, 37	37	37	d 37, 38	38, e 39	38, f 39	38	38	38
1900 g	47, h 48	48	48, i 49	49	49	49, j 50	50	50	50	51	51	51	51	51
1901	65, 75	65, 75	65, 75	65, 75	k 65, 66, 75	66, 75	k 65, 66, 75	66, 75	66, 75	66, 75	66, 75	66, 75	66, 75	66, 75
1902	82	b 82, 83	83	l 82, 83	k 83, 85	84	k 83, 84	84	85	85	85	85	85	85
1903	97	b 97, 98	98	97	k 98, 99, m 100	99	k 98, 99	99	100	100	100	100	100	100
1904	n 124, o 125, p 126	p 126, 127	128	129	k 128, 130	130, q 131	k 128, 131	132	133	133, r 134	134	135	135	135
1905	n 165, o 166, p 167	p 167, 168	169	170	171	172	k 169, 173	174	175, s 177	176, r 177	177	178	178	t 177, 178
1906	n 201, o 202, p 203	p 203, 204	205	206	207	208	k 205, 209	210	211	212, r 213	213	214	214	214
1907–8	241	242	243	244	245	246	247	248	249	250, r 251	251	252	252	252
1909	261	262	263	264	265	266	267	268	269	270, r 271	271	272	272	272
1910	281	282	283	284	285	286	287	288	289	290	291	292	292	292
1911	301	302	303	304	305	306	307	308	309	310	311	312	312	312
1912	321	322	323	324	325	326	327	328	329	330	331	332-A	332-B	332-C
1913	351	352	353	354	355	356	357	358	359	360	361	362-A	362-B	362-C
1914	381	382	383	384	385	386	387	388	389	390	391	392	393	394
1915	401	402	403	404	405	406	407	408	409	410	411	412	413	414
1916	431	432	433	434	435	436	437	438	439	440	441	442	443	444
1917	451	452	453	454	455	456	457	458	459	460	461	462	463	464
1918	471	472	473	474	475	476	477	478	479	480	481	482	483	484

a Rating tables and index to Water-Supply Papers 35–39 contained in Water-Supply Paper 39. Tables of monthly discharge for 1899 in Twenty-first Annual Report, Part IV.
b James River only.
c Gallatin River.
d Green and Gunnison rivers and Grand River above junction with Gunnison.
e Mohave River only.
f Kings and Kern Rivers and south Pacific slope drainage basins.
g Rating tables and index to Water-Supply Papers 47–52 and data on precipitation, wells, and irrigation in California and Utah contained in Water-Supply Paper 52. Tables of monthly discharge for 1900 in Twenty-second Annual Report, Part IV.
h Wissahickon and Schuylkill Rivers to James River.
i Scioto River.
j Loup and Platte Rivers near Columbus, Nebr., and all tributaries below junction with Platte.
k Tributaries of Mississippi from east.
l Lake Ontario and tributaries to St. Lawrence River proper.
m Hudson Bay only.
n New England rivers only.
o Hudson River to Delaware River, inclusive.
p Susquehanna River to Yadkin River, inclusive.
q Platte and Kansas Rivers.
r Great Basin in California, except Truckee and Carson River basins.
s Below junction with Gila.
t Rogue, Umpqua, and Siletz Rivers only.

PART V.—HUDSON BAY AND UPPER MISSISSIPPI RIVER DRAINAGE BASINS.

PRINCIPAL STREAMS.

The Hudson Bay and upper Mississippi River basins include streams whose waters reach Hudson Bay and the Mississippi above its junction with the Ohio (except the Missouri). The principal streams flowing into Hudson Bay from the United States are St. Mary River, Red River, and Rainy River. The principal tributaries of the upper Mississippi are Crow Wing, Sauk, Crow, Rum, Minnesota, St. Croix, Chippewa, Zumbro, Black, Root, Wisconsin, Wapsipinicon, Rock, Iowa, Des Moines, Illinois, and Kaskaskia rivers. These streams drain wholly or in part the States of Illinois, Indiana, Iowa, Minnesota, Missouri, Montana, North Dakota, South Dakota, and Wisconsin.

In addition to the list of gaging stations and the annotated list of publications relating specifically to the section, these pages contain a similar list of reports that are of general interest in many sections and cover a wide range of hydrologic subjects, and also brief references to reports published by State and other organizations. (See p. XVII.)

GAGING STATIONS.

NOTE.—Dash after a date indicates that station was being maintained September 30, 1918. Period after a date indicates discontinuance.

HUDSON BAY DRAINAGE BASIN.

St. Mary River near Babb (formerly dam site), Mont., 1902–
St. Mary River below Swiftcurrent Creek, at Babb, Mont., 1901–2; 1910–1915.
St. Mary River near Kimball, Alberta, 1902–
    U. S. Reclamation Service, St. Mary canal at intake, near Babb, Mont., 1918–
    U. S. Reclamation Service, St. Mary canal at St. Mary crossing, near Babb, Mont., 1918–
    U. S. Reclamation Service, St. Mary canal at Hudson Bay Divide, near Browning, Mont., 1917–
    Swiftcurrent Creek at Many Glacier, Mont., 1912–
    Swiftcurrent Creek at Sherburne, Mont., 1912–
    Swiftcurrent Creek near Babb (formerly Wetzel), Mont., 1902–1910.
    Canyon Creek near Many Glacier, Mont., 1918–
    Kennedy Creek near Babb (formerly Wetzel), Mont., 1903–1907.
Ottertail River at German Church, near Fergus Falls, Minn., 1913–1917.
Ottertail River near Fergus Falls, Minn., 1904–1913.
Red River near Fergus Falls, Minn., 1909–10.
Red River at Fargo, N. Dak., 1901–
Red River at Grand Forks, N. Dak., 1901–
Red River at Pembina, N. Dak., 1901.

Red River at Emerson, Manitoba, 1900–1902.
Mustinka River near Wheaton, Minn., 1916; 1917.
Pelican River near Fergus Falls, Minn., 1909–1912.
Sheyenne River at Haggart, N. Dak., 1902–1907.
Wild Rice River at Twin Valley, Minn., 1909–1917.
Devils Lake near Devils Lake, N. Dak., 1901–
Red Lake River at Thief River Falls, Minn., 1909–
Red Lake River at Crookston, Minn., 1901–
Thief River near Thief River Falls, Minn., 1909–1917.
Clearwater River at Red Lake Falls, Minn., 1909–1917.
South Branch of Two Rivers at Hallock, Minn., 1911–1914.
Pembina River at Neche, N. Dak., 1903–1915.
Roseau River at Dominion City, Canada, 1912.
Roseau River near Caribou, Minn., 1917.
West Branch of Roseau River near Malung, Minn., 1911–1914.
Mouse River near Foxholm, N. Dak., 1904–1906.
Mouse River at Minot, N. Dak., 1903–
Des Lacs River at Foxholm, N. Dak., 1904–1906.
Rainy Lake at Rainier, Minn., 1910–1917.
Rainy River at International Falls, Minn., 1907–1917.
Kawishiwi River near Winton, Minn., 1905–1907; 1912–
Vermilion River below Lake Vermilion, near Tower, Minn., 1911–1917.
Little Fork at Little Fork, Minn., 1909–1917.
Big Fork at Big Falls, Minn., 1909–1912.
Big Fork at Laurel, Minn., 1909.
Black River near Loman, Minn., 1909.

UPPER MISSISSIPPI RIVER BASIN.

Mississippi River above Sandy River, Minn., 1895–1915.
Mississippi River near Fort Ripley, Minn., 1909–10.
Mississippi River near Sauk Rapids, Minn., 1903–1906.
Mississippi River at Elk River, Minn., 1915–
Mississippi River at Anoka, Minn., 1905–1914.
Mississippi River at St. Paul, Minn., 1873–
Sandy River below Sandy Lake reservoir, Minn., 1893–1916.
Pine River below Pine River reservoir, Minn., 1886–1916.
Prairie River near Grand Rapids, Minn., 1909.
Crow Wing River at Nimrod, Minn., 1910–1914.
Crow Wing River at Motley, Minn., 1909; 1913–1917.
Crow Wing River at Pillager, Minn., 1903; 1909–1913.
Long Prairie River near Motley, Minn., 1909–1917.
Sauk River near St. Cloud, Minn., 1909–1913.
Elk River near Big Lake, Minn., 1911–1917.
Crow River at Rockford, Minn., 1909–1917.
Crow River near Dayton, Minn., 1906.
North Fork of Crow River near Rockford, Minn., 1909–10.
South Fork of Crow River near Rockford, Minn., 1909–1912.
Rum River at Onamia, Minn., 1909–1912.
Rum River at Cambridge, Minn., 1909–1914.
Rum River at St. Francis, Minn., 1903.
Rum River near Anoka, Minn., 1905–6; 1909.
Minnesota River near Odessa, Minn., 1909–1913.
Minnesota River near Montevideo, Minn., 1909–

Mississippi River tributaries—Continued.
- Minnesota River near Mankato, Minn., 1903–
  - Whetstone River near Big Stone, S. Dak., 1910–1912.
  - Lac qui Parle River at Lac qui Parle, Minn., 1910–1914.
  - Chippewa River near Watson, Minn., 1909–1917.
  - Redwood River near Redwood Falls, Minn., 1909–1914.
  - Cottonwood River near New Ulm, Minn., 1909–1913.
  - Blue Earth River, at Rapidan Mills, Minn., 1909–10.
- St. Croix River at Swiss, Wis., 1914–
- St. Croix River near St. Croix Falls, Wis., 1902–1905; 1910–
  - Namakagon River at Trego, Wis., 1914–
  - Yellow River at Webster, Wis., 1914.
  - Kettle River near Sandstone, Minn., 1908–1917.
  - Snake River at Mora, Minn., 1909–1913.
  - Snake River near Pine City, Minn., 1913–1917.
  - Apple River near Somerset, Wis., 1901–
  - Kinnikinnic River near River Falls, Wis., 1916–
- Cannon River at Welch, Minn., 1909–1914.
- Chippewa River at Bishops Bridge, near Winter, Wis., 1912–
- Chippewa River near Bruce, Wis., 1913–
- Chippewa River at Chippewa Falls, Wis., 1888–
- Chippewa River near Eau Claire, Wis., 1902–1909.
  - West Fork of Chippewa River near Winter, Wis., 1911–1916.
  - Flambeau River near Butternut, Wis., 1914–
  - Flambeau River near Ladysmith, Wis., 1914–
  - Flambeau River at Ladysmith, Wis., 1903–1906.
  - Jump River at Sheldon, Wis., 1915–
  - Eau Claire River near Augusta, Wis., 1914–
  - Eau Claire River near Eau Claire, Wis., 1913–14.
  - Red Cedar River near Colfax, Wis., 1914–
  - Red Cedar River at Cedar Falls, Wis., 1909–
  - Red Cedar River at Menominee, Wis., 1907–8; 1913–
- Zumbro River at Zumbro Falls, Minn., 1909–1917.
  - South Branch of Zumbro River near Zumbro Falls, Minn., 1911–1917.
- Trempealeau River at Dodge, Wis., 1913–
- Black River at Neillsville, Wis., 1905–1909; 1913–
- Black River at Melrose, Wis., 1902–3.
- La Crosse River near West Salem, Wis., 1913–
- Root River near Houston, Minn., 1909–1917.
  - North Branch of Root River near Lanesboro, Minn., 1910–1917.
- Upper Iowa River near Decorah, Iowa, 1913–14.
- Wisconsin River near Rhinelander, Wis., 1905–1915.
- Wisconsin River at Whirlpool Rapids, near Rhinelander, Wis., 1915–
- Wisconsin River at Merrill, Wis., 1902–
- Wisconsin River near Nekoosa, Wis., 1914–
- Wisconsin River near Necedah, Wis., 1902–1914.
- Wisconsin River at Muscoda, Wis., 1902–3; 1913–
  - Tomahawk River near Bradley, Wis., 1914–
  - Prairie River near Merrill, Wis., 1914–
  - Little Rib River near Wausau, Wis., 1914–1916.
  - Eau Claire River at Kelley, Wis., 1914–
  - Big Eau Pleine River near Stratford, Wis., 1914–
  - Plover River near Stevens Point, Wis., 1914–
  - Baraboo River near Baraboo, Wis., 1913–
  - Kickapoo River at Gays Mills, Wis., 1913–

Mississippi River tributaries—Continued.
- Turkey River at Garber, Iowa, 1913–1916.
- Maquoketa River above mouth of North Fork, near Maquoketa, Iowa, 1913–14.
- Maquoketa River at Manchester, Iowa, 1903.
- Maquoketa River below mouth of North Fork, near Maquoketa, Iowa, 1913–
- Wapsipinicon River at Stone City, Iowa, 1903–1914.
- Rock River at Watertown, Wis., 1914.
- Rock River at Afton, Wis., 1914–
- Rock River above mouth of Pecatonica River, at Rockton, Ill., 1903.
- Rock River below mouth of Pecatonica River, at Rockton, Ill., 1903–1909.
- Rock River at Rockford, Ill., 1914–
- Rock River near Nelson, Ill., 1906.
- Rock River at Sterling, Ill., 1905–6.
- Rock River at Lyndon, Ill., 1914–
  - Catfish River at Madison, Wis., 1902–3.
  - Lake Mendota at Madison, Wis., 1902–3.
  - Yahara River near Edgerton, Wis., 1916–17.
  - Pecatonica River at Dill, Wis., 1914–
  - Pecatonica River at Freeport, Ill., 1914–
    - Sugar River near Brodhead, Wis., 1914–
- Iowa River near Iowa Falls, Iowa, 1911–1914.
- Iowa River at Marshalltown, Iowa, 1903; 1915–
- Iowa River at Iowa City, Iowa, 1903–1906; 1913–
- Iowa River at Wapello, Iowa, 1915–
  - Cedar River near Austin, Minn., 1909–1914.
  - Cedar River at Janesville, Iowa, 1905–6; 1915–
  - Cedar River at Cedar Rapids, Iowa, 1902–
    - Shellrock River near Clarksville, Iowa, 1915–
- Skunk River at Coppock, Iowa, 1913–
- Skunk River at Augusta, Iowa, 1913; 1915–
- Des Moines River at Jackson, Minn., 1909–1913.
- Des Moines River at Fort Dodge, Iowa, 1905–6; 1911–1913.
- Des Moines River at Kalo, Iowa, 1913–
- Des Moines River at Des Moines, Iowa, 1902–3; 1905–6; 1914–
- Des Moines River at Ottumwa, Iowa, 1917–
- Des Moines River at Keosauqua, Iowa, 1903–1906; 1911–
  - Raccoon River near Des Moines, Iowa, 1902–3.
  - Raccoon River at Van Meter, Iowa, 1915–
- Illinois River near Minooka, Ill., 1902–1904.
- Illinois River near Seneca, Ill., 1902–3.
- Illinois River near Ottawa, Ill., 1902–1904.
- Illinois River near La Salle, Ill., 1902–3.
- Illinois River at Peoria, Ill., 1910–
- Illinois River near Peoria, Ill., 1903–1906.
  - Kankakee River at Davis, Ind., 1905–6.
  - Kankakee River at Momence, Ill., 1905–6; 1914–
  - Kankakee River at Custer Park, Ill., 1914–
    - Yellow River at Knox, Ind., 1905–6.
  - Des Plaines River at Riverside, Ill., 1896–1898.
  - Des Plaines River above mouth of Jackson Creek, near Channahon, Ill., 1903–1906.
  - Des Plaines River above Kankakee River, near Channahon, Ill., 1902–3.
  - Des Plaines River at Lemont, Ill., 1914–

Mississippi River tributaries—Continued.
- Illinois River tributaries—Continued.
  - Des Plaines River at Romeo, Ill., 1914.
  - Des Plaines River at Joliet, Ill., 1914–
  - Fox River at Algonquin, Ill., 1915–
  - Fox River at South Elgin, Ill., 1914–15.
  - Fox River at Aurora, Ill., 1914.
  - Fox River at Sheridan, Ill., 1905–6.
  - Fox River at Wedron, Ill., 1914–
  - Fox River at Ottawa, Ill., 1903.
  - Vermilion River near Streator, Ill., 1914–
  - Spoon River at Seville, Ill., 1914–
  - Sangamon River at Monticello, Ill., 1908–1912; 1914–
  - Sangamon River at Decatur, Ill., 1905.
  - Sangamon River at Riverton, Ill., 1908–1912; 1914–
  - Sangamon River at Springfield, Ill., 1903.
  - Sangamon River near Oakford, Ill., 1909–1912; 1914–
  - Sangamon River near Chandlerville, Ill., 1908–9.
    - South Fork of Sangamon River near Taylorville, Ill., 1908–1912; 1914–1917.
    - South Fork Sangamon River at power plant, near Taylorville, Ill., 1917–
    - Salt Creek near Kenny, Ill., 1908–1912.
- Cahokia Creek at Poag, Ill., 1909–1912.
- Kaskaskia River near Arcola, Ill., 1908–1912.
- Kaskaskia River at Shelbyville, Ill., 1908–1912; 1914.
- Kaskaskia River at Vandalia, Ill., 1908–1912; 1914–
- Kaskaskia River at Carlyle, Ill., 1908–1912; 1914–15.
- Kaskaskia River at New Athens, Ill., 1907–1912; 1914–
  - Shoal Creek near Breese, Ill., 1909–1912; 1914.
  - Silver Creek near Lebanon, Ill., 1908–1912; 1914.
- Big Muddy River near Cambon, Ill., 1908–1912.
- Big Muddy River at Plumfield, Ill., 1914–
- Big Muddy River at Murphysboro, Ill., 1917–
  - Beaucoup Creek near Pinckneyville, Ill., 1908–1912; 1914.

REPORTS ON WATER RESOURCES OF THE HUDSON BAY AND UPPER MISSISSIPPI RIVER BASINS.

PUBLICATIONS OF THE UNITED STATES GEOLOGICAL SURVEY.

WATER-SUPPLY PAPERS.

Water-supply papers are distributed free by the Geological Survey as long as its stock lasts. An asterisk (*) indicates that this stock has been exhausted. Many of the papers marked in this way may, however, be purchased (at prices quoted) from the SUPERINTENDENT OF DOCUMENTS, Washington, D. C. Omission of the price indicates that the report s not obtainable from Government sources. Water-supply papers are of octavo size.

*21. Wells of northern Indiana, by Frank Leverett. 1899. 82 pp., 2 pls.
Discusses, by counties, glacial deposits and sources of well waters; many well sections.

*44. Profiles of rivers in the United States, by Henry Gannett. 1901. 100 pp., 11 pls. 15c.
Gives elevations and distances along Red River (of the North), and Minnesota, Skunk, Iowa, Des Moines, Illinois, and Rock rivers; also brief descriptions.

*57. Preliminary list of deep borings in the United States, Part I (Alabama-Montana), by N. H. Darton. 1902. 60 pp. 5c.

*61. Preliminary list of deep borings in the United States, Part II (Nebraska-Wyoming), by N. H. Darton. 1902. 67 pp. 5c.
A revised edition of Nos. 57 and 61, was published in 1905 as Water-Supply Paper 149 (q. v.)

96. Destructive floods in the United States in 1903, by E. C. Murphy. 1904. 81 pp., 13 pls. 15c.
Contains notes on early floods in Mississippi Valley.

102. Contributions to the hydrology of eastern United States, 1903; M. L. Fuller, geologist in charge. 1904. 522 pp. 30c.
Contains brief reports on wells and springs of Minnesota and Missouri.
The reports comprise tabulated well records giving information as to location, owner, depth, yield, head, etc., supplemented by notes as to elevation above sea, material penetrated, temperature, use, and quality; many miscellaneous analyses.

*103. A review of the laws forbidding pollution of inland waters in the United States, by E. B. Goodell. 1904. 120 pp.
Cites statutory restrictions of water pollution in Iowa, Illinois, North Dakota, South Dakota, and Wisconsin. Superseded by 152.

*114. Underground waters of eastern United States; M. L. Fuller, geologist in charge. 1905. 285 pp., 18 pls. 25c.
Contains brief reports as follows: Missouri, by E. M. Shepard; Iowa, by W. H. Norton; Minnesota, by C. W. Hall; Wisconsin district, by Alfred R. Schultz; Illinois, by Frank Leverett; Indiana, by Frank Leverett; each of these reports describes briefly the topography of the area, the relation of the geology to the water supplies, and gives list of pertinent publications; lists also principal mineral springs.

117. The lignite of North Dakota and its relation to irrigation, by F. A. Wilder. 1905. 59 pp., 8 pls. 10c.
Describes the thickness, extent, variations, and fuel value of the lignite and its use for pumping water, the area, soils, and lignite of the river flats, and the status of irrigation in the State.

*122. Relation of the law to underground waters, by D. W. Johnson. 1905. 55 pp. 5c.
Cites legislative acts affecting underground waters in South Dakota and Wisconsin.

145. Contributions to the hydrology of eastern United States, 1905; M. L. Fuller, geologist in charge. 1905. 220 pp., 6 pls. 10c.

Contains two reports relating to areas draining to Hudson Bay or upper Mississippi River.

Water resources of Mineral Point quadrangle, Wisconsin, by U. S. Grant. Describes springs, streams, and shallow and deep wells.

Water supplies at Waterloo, Iowa, by W. H. Norton. Summarizes results of investigations to determine availability of artesian water to replace the surface supply from Cedar River; discusses necessity of test wells, supplementary supplies, artesian head, and permanency of flow.

*149. Preliminary list of deep borings in the United States, second edition, with additions, by N. H. Darton. 1905. 175 pp. 10c.

Gives by States (and within the States by counties), the location, depth, diameter, yield, height of water, and other features of wells 400 feet or more in depth; includes all wells listed in Water-Supply Papers 57 and 61; mentions also principal publications relating to deep borings.

*152. A review of the laws forbidding pollution of the inland waters in the United States (second edition), by E. B. Goodell. 1905. 149 pp. 10c.

Cites statutory restrictions of water pollution in Iowa, Illinois, North Dakota, South Dakota, and Wisconsin.

*156. Water powers of northern Wisconsin, by L. S. Smith. 1906. 145 pp., 5 pls. 25c

Describes by river systems the drainage, geology, topography, rainfall and run-off, water powers, and dams.

*162. Destructive floods in the United States in 1905, with a discussion of flood discharge and frequency and an index of flood literature, by E. C. Murphy and others. 1906. 105 pp., 4 pls. 15c.

Contains accounts of floods in southeastern Minnesota, on Devils Creek, Iowa, and in Des Moines County, Iowa; gives estimates of flood discharge and frequency on Illinois River and on Mississippi River at St. Paul.

*193. The quality of surface waters in Minnesota, by R. B. Dole and F. F. Westbrook. 1907. 171 pp., 7 pls. 25c.

Describes by river basins the topography, geology, and soils, the individual and municipal pollution of the streams, and gives notes on the municipalities; contains many analyses.

*194. Pollution of Illinois and Mississippi Rivers by Chicago sewage (a digest of the testimony taken in the case of the State of Missouri *v.* the State of Illinois and the Sanitary District of Chicago), by M. O. Leighton. 1907. 369 pp., 2 pls.

Scope indicated by amplification of title.

*195. Underground waters of Missouri, their geology and utilization, by E. M. Shepard, 1907. 224 pp., 6 pls. 30c.

Describes the topography and geology of the State, the waters of the various formations, and discusses the water supplies by districts and counties, gives statistics of city water supplies, analyses of waters, and many well records.

*227. Geology and underground waters of South Dakota, by N. H. Darton. 1909. 156 pp., 15 pls. 40c.

Describes physical features, geologic formations, water horizons, and, by counties, deep wells and well prospects; gives notes on construction and management of artesian wells.

*236. The quality of surface waters in the United States: Part I, Analyses of waters east of the one hundredth meridian, by R. B. Dole. 1909. 123 pp. 10c.

Describes collection of samples, methods of examination, preparation of solutions, accuracy of estimates and expression of analytical results; gives results of analyses of waters of Mississippi, Minnesota, Chippewa, Wisconsin, Rock, Iowa, Cedar, Des Moines, Illinois, Kankakee, Fox, Sangamon, Kaskakia, and Big Muddy rivers.

239. The quality of the surface waters of Illinois, by W. D. Collins. 1910. 94 pp., 3 pls. 10c.

Discusses the natural and economic features that determine the character of the streams, describes the larger drainage basins, and the methods of collecting and analyzing the samples of water, and discusses each river in detail with reference to its source and course and the quality of water; includes short chapters on municipal supplies and industrial uses.

254. The underground waters of north-central Indiana, by S. R. Capps, with a chapter on the chemical character of the waters, by R. B. Dole. 1910. 279 pp., 7 pls. 40c.

Describes relief, drainage, vegetation, soils, and crops, industrial development, geologic formations; sources, movements, occurence, and volume of ground water; methods of well construction and lifting devices; discusses, in detail for each county, surface features and drainage, geology and ground water, city, village, and rural supplies, and gives records of wells and analyses of waters. Discusses also, under chemical character, methods of analyses and expression of results, mineral constituents, effect of the constituents on waters for domestic, industrial, and medicinal uses, methods of purification, chemical composition; many analyses and field assays.

256. Geology and underground waters of southern Minnesota, by C. W. Hall, O. E. Meinzer, and M. L. Fuller. 1911. 406 pp., 18 pls. 60c.

Discusses the physiography of the State, geologic formations and their water-bearing capacity, artesian conditions, the mineral quality of the underground waters, types of wells, finishing wells in sand, drilling in quartzite, fluctuation in yield and head, "blowing" and "breathing" wells, freezing of wells, drainage by wells, hydraulic rams, and scientific prospecting for water, municipal supplies, power, storage and distribution, consumption of water, prices, sanitation. Gives by counties details concerning surface features, rocks, yield, head, and quality of water, and summaries and analyses.

293. Underground water resources of Iowa, by W. H. Norton, W. S. Hendrixson, H. E. Simpson, O. E. Meinzer, and others. 1912. 994 pp., 18 pls. 70c.

Describes the relief, drainage, temperature, and precipitation of the State and the geologic formations; discusses the geologic occurrence of ground waters, artesian phenomena and yield of artesian wells, the chemical composition of ground waters, municipal, domestic, and industrial water supplies, and mineral waters; gives details concerning topography, geology, ground waters, and city and village supplies by districts and counties.

*345. Contributions to the hydrology of the United States, 1914. N. C. Grover, chief hydraulic engineer. 1915. 225 pp., 17 pls. 30c. Contains:

(f) Gazetteer of surface waters of Iowa, by W. G. Hoyt and H. J. Ryan, pp. 169-221.

364. Water analyses from the laboratory of the United States Geological Survey, tabulated by F. W. Clarke, chief chemist. 1914. 40 pp. 5c.

Contains analyses of spring and well waters from Nashville and Macomb, Ill., and Story City, Iowa.

417. Profile surveys of rivers in Wisconsin, prepared under the direction of W. H. Herron, acting chief geographer. 1917. 16 pp., 32 pls. 45c.

Contains brief description of general features of drainage of Wisconsin and of the rivers surveyed, but consists chiefly of maps showing "not only the outlines of the river banks, the islands, the position of rapids, falls, shoals, and existing dams, and the crossings of all ferries and roads but the contours of banks to an elevation high enough to indicate the possibilities of using the stream" for the development of power by low or medium heads.

ANNUAL REPORTS.

Each of the papers contained in the annual reports was also issued in separate form.

Annual reports are distributed free by the Geological Survey as long as its stocks lasts. An asterisk (*) indicates that this stock has been exhausted. Many of the papers so marked, however, may be purchased, from the SUPERINTENDENT OF DOCUMENTS, WASHINGTON, D. C.

Sixteenth Annual Report of the United States Geological Survey, 1894-95. 4 parts.
*Pt. II. Papers of an economic character, xix, 598 pp., 43 pls. $1.25. Contains:

The public lands and their water supply, by F. H. Newell, pp. 457-533, pls. 35 to 39. Describes general character of the public lands, the lands disposed of (railroad, grant, and swamp lands and private miscellaneous entries), lands reserved (Indian, forest, and military reservations) the vacant lands, and the rate of disposal of vacant lands; discusses the streams, wells, and reservoirs as sources of water supply; gives details for each State.

Seventeenth Annual Report of the United States Geological Survey, 1895–96, Charles D. Walcott, Director, 1896; 3 parts in 4 vols. *Pt. II. Economic geology and hydrography, xxv, 864 pp., 113 pls. $2.35. Contains:

Preliminary report on artesian waters of a portion of the Dakotas, by N. H. Darton, pp. 603–694, pls. 69 to 107. Gives an outline of the geologic relations; describes the water horizons and the extent of the artesian water, and gives details concerning wells and prospects by counties; discusses the origin, amount, pressure, head, and composition of the artesian waters, the use of artesian water for power, and gives details concerning artesian irrigation by counties; contains also remarks on the construction and management of artesian wells.

*The water resources of Illinois, by Frank Leverett, pp. 695–849, pls. 108 to 113. Describes the physical features of the State, and the drainage basins, including Illinois, Des Plaines, Kankakee, Fox, Illinois-Vermilion, Spoon, Mackinaw, and Sangamon rivers, Macoupin Creek, Rock River, tributaries of the Mississippi in western Illinois, Kaskaskia, Big Muddy, and tributaries of the Wabash; discusses the rainfall and run-off, navigable waters and water powers, the wells supplying waters for rural districts, and artesian wells; contains tabulated artesian well data and water analyses.

Eighteenth Annual Report of the United States Geological Survey, 1896–97, 5 parts in 6 vols. *Pt. IV, Hydrography, x, 756 pp., 102 pls. $1.75. Contains:

*The water resources of Indiana and Ohio, by Frank Leverett, pp. 419–560, pls. 33 to 37. Describes the Wabash, Whitewater, Great Miami, Little Miami, Scioto, Hocking, Muskingum, and Beavers rivers, and lesser tributaries of the Ohio in Indiana and Ohio, the streams discharging into Lake Erie and Lake Michigan, and streams flowing to the upper Mississippi through the Illinois; discusses shallow and drift wells, the flowing wells, from the drift and deeper artesian wells, and gives records of wells at many of the cities; describes the mineral springs, and gives analyses of the waters; contains also tabulated lists of cities using surface waters for water works, and of cities and villages using shallow and deep-well waters; discusses the source and quality of the city and village supplies, and gives precipitation tables for various points.

MONOGRAPHS.

Monographs of quarto size. They are not distributed free, but may be obtained from the Geological Survey or from the SUPERINTENDENT OF DOCUMENTS, WASHINGTON, D. C., at the prices indicated. An asterisk (*) indicates that the Survey's stock of the paper is exhausted.

25. The glacial Lake Agassiz, by Warren Upham. 1896. 658 pp., 38 pls. $1.70.

Contains a chapter (pp. 523–582) on "Artesian and common wells of the Red River Valley," which discusses the sources of artesian water, the fresh waters in the drift sheets, the saline and alkaline waters in the Dakota sandstone, and the use of artesian water for irrigation; contains analyses of waters from wells, streams, and lakes in Red River Valley and the adjoining region; and gives notes on wells in Clay, Kittson, Marshall, Norman, Polk, Traverse, and Wilkin counties, in Minnesota; in Cass, Grand Forks, Pembina, Richland, Traill, and Walsh counties, in North Dakota; and in a part of the area covered by Lake Agassiz, in Manitoba. The monograph includes numerous maps relating to the Pleistocene geology of the region and a map (Pl. XXXVII) showing the distribution and depths of artesian wells in glacial drift and bedrock.

38. The Illinois glacial lobe, by Frank Leverett. 1899. 817 pp., 24 pls. $1.60.

Includes a chapter (pp. 550–788) on "Wells of Illinois," which contains a general discussion of artesian and other wells, a table of municipal water supplies derived from underground sources, and a detailed description of wells and ground-water conditions in practically every county in the State. The monograph includes maps showing the geology, the distribution of wells, the intake areas of "Potsdam" and St. Peter sandstones, and the relation of glacial drift to groundwater supplies.

PROFESSIONAL PAPERS.

Professional papers are distributed free by the Geological Survey as long as its stock lasts. An asterisk (*) indicates that this stock has been exhausted. Many of the papers marked with an asterisk may, however, be purchased from the SUPERINTENDENT OF DOCUMENTS, WASHINGTON, D. C. Professional papers are of quarto size.

*32. Preliminary report on the geology and underground-water resources of the central Great Plains, by N. H. Darton. 1905. 433 pp., 72 pls. $1.80.

Covers South Dakota, Nebraska, central and western Kansas, eastern Colorado, and eastern Wyoming. Describes the geography, geology, and water horizons; gives deep-well data and well prospects by counties; also describes other mineral resources. Includes maps showing the geology, location of deep wells, structure of the Dakota sandstone, depths to this sandstone head of artesian water, and areas of artesian flow.

BULLETINS.

An asterisk (*) indicates that the Geological Survey's stock of the paper is exhausted. Many of the papers so marked may be purchased from the SUPERINTENDENT OF DOCUMENTS, WASHINGTON, D. C.

*264. Record of deep-well drilling for 1904, by M. L. Fuller, E. F. Lines, and A. C. Veatch. 1905. 106 pp. 10c.

Discusses the imporance of accurate well records to the driller, to owners of oil, gas, and water wells, and to the geologist; describes the general methods of work; gives tabulated records of wells in Illinois and Iowa, and detailed records of wells in Boone, Dupage, Henry, and La Salle counties, Ill., and Des Moines and Scott counties, Iowa. These wells were selected because they give definite stratigraphic information.

*298. Record of deep-well drilling for 1905, by M. L. Fuller and Samuel Sanford. 1906. 299 pp. 25c.

Gives an account of progress in the collection of well records and samples; contains tabulated records of wells in Illinois, Indiana, Iowa, Minnesota, Missouri, North Dakota, South Dakota, and Wisconsin; and detailed records of wells in Brown, Hancock, La Salle, Pike, and Schuyler counties, Ill.; Blackhawk, Floyd, Louisa, Mahaska, Scott, and Wapello counties, Iowa; and Hennepin, Ottertail, and Pine counties, Minn. The wells of which detailed sections are given were selected because they afford valuable stratigraphic information.

GEOLOGIC FOLIOS.

Under the plan adopted for the preparation of a geologic map of the United States the entire area is divided into small quadrangles bounded by certain meridians and parallels, and these quadrangles, which number several thousand, are separately surveyed and mapped.[1] The unit of survey is also the unit of publication, and the maps and description of each quadrangle are issued in the form of a folio. When all the folios are completed they will constitute the Geologic Atlas of the United States.

A folio is designated by the name of the principal town or of a prominent natural feature within the quadrangle. Each folio includes maps showing the topography, geology, underground structure, and mineral deposits of the area mapped and several pages of descriptive text. The text explains the maps and describes the topographic and geologic features of the country and its mineral products. The topographic map shows roads, railroads, waterways, and, by contour lines, the shapes of hills and valleys and the height above sea level of all points in the quadrangle. The areal-geology map shows the distribution of the various rocks at the surface. The structural-geology map shows relations of the rocks to one another underground. The economic-geology map indicates the location of mineral deposits that are commercially valuable. The artesian-water map shows the depth to underground-water horizons. Economic-geology and artesian-water maps are included in folios if the conditions in the areas mapped warrant their publication. The folios are of special interest to students of geography and geology and are valuable as guides in the development and utilization of mineral resources.

The folios numbered from 1 to 163, inclusive, are published in only one form (18 by 22 inches), called the library edition. Some of the folios that bear numbers higher than 163 are published also in an octavo edition (6 by 9 inches). Owing to a fire in the Geological Survey building May 18, 1913, the stock of geologic folios was more or less damaged by fire and water, but 80 or 90 per cent of the folios are usable. They will be sold at the uniform price of 5 cents each, with no reduction for wholesale orders. This rate applies to folios in stock from 1 to 184, inclusive (except reprints), also to the library edition of folio 186. The library edition of folios 185, 187, and higher numbers sells for 25 cents a copy, except that some folios which contain an unusually large amount of matter sell at higher prices. The octavo edition of folio 185 and higher numbers sells for 50 cents a copy. A discount of 40 per cent is allowed on an order for folios or for folios together with topographic maps amounting to $5 at the retail rate.

[1] Index maps showing areas in the Hudson Bay and upper Mississippi River basins covered by topographic maps and by geologic folios will be mailed on receipt of request addressed to the Director, U. S. Geological Survey, Washington, D. C.

All the folios contain descriptions of the drainage of the quadrangles. The folios in the following list contain also a brief discussion of the underground waters in connection with the economic resources of the areas and more or less information concerning the utilization of the water resources.

An asterisk (*) indicates that the stock of the folio is exhausted.

117. Casselton-Fargo, North Dakota-Minnesota. 5c.

Gives a somewhat detailed account of the water supply, including descriptions and logs of principal wells and tabulated well records, contains artesian-water maps showing areas which will probably yield flowing wells.

*145. Lancaster-Mineral Point, Wisconsin-Iowa-Illinois.

Discusses the springs, shallow and deep wells, streams and water power; gives analyses of artesian water from well at Dubuque, Iowa.

168. Jamestown-Tower (Jamestown, Eckelson, and Tower quadrangles), North Dakota. 5c.

Discusses shallow, deep and artesian wells; head, pressure, power, volume, and character of the water, and gives a tabulated list of representative wells, contains an artesian-water map showing areas in which flowing wells may probably be obtained.

185. Murphysboro-Herrin, Illinois.[a] Library edition, 25c., octavo edition, 50c.

188. Tallula-Springfield, Illinois.[a] Library edition, 25c., octavo edition, 50c.

Discusses wells and the wholesomeness of the water; gives analyses of water from wells in the city of Springfield.

195. Belleville-Breese, Illinois. 25c.

Discusses wells and gives analyses of water from springs and wells.

200. Galena-Elizabeth, Illinois-Iowa. 25c.

201. Minneapolis-St. Paul, Minnesota.[a] Library edition, 25c., octavo edition, 50c.

MISCELLANEOUS REPORTS.

Other Federal bureaus and the State and other organizations have from time to time published reports relating to the water resources of the various sections of the country. Notable among those pertaining to the Hudson Bay and upper Mississippi River basins are the reports of the State surveys of Illinois and North Dakota, the Wisconsin Geological and Natural History Survey and the Railroad Commission of Wisconsin, the Illinois Water-Supply Commission, and the Rivers and Lakes Commission of Illinois, and the water-power report of the Tenth Census (vol. 17). The following reports deserve special mention:

Contributions to the physical geography of the United States, Part I. On the physical geography of the Mississippi Valley, with suggestions for the improvement of navigation of the Ohio and other rivers, by Charles Ellet, jr.: Smithsonian Pub. 13, Washington, 1850.

The Mississippi and Ohio rivers, by Charles H. Ellet. 1853.

Report upon the physics and hydraulics of the Mississippi River, by A. A. Humphreys and H. L. Abbott.

The mineral content of Illinois waters, by Edward Barstow, J. A. Udden, S. W. Parr, and George T. Palmer: Illinois State Geol. Survey Bull. 10, 1909.

Water resources of the East St. Louis district, by Isiah Bowman: Illinois State Geol. Survey Bull. 5, 1907.

Chemical and biological survey of waters of Illinois, by Edward Barstow: Univ. Illinois Pub. 3, 6, 7, 1906–1909.

[a] Issued in two editions; specify which edition is wanted.

Chemical survey of the waters of Illinois, report for the years 1897–1902, by A. W. Palmer, with report on geology of Illinois as related to its water supply, by Charles W. Rolfe: Univ. Illinois Pub.

Report and plans for the reclamation of lands subject to overflow in the Kaskaskia River Valley, Illinois; begun under the direction of the Internal Improvement Commission; completed and published under the direction of the Rivers and Lakes Commission of Illinois, by Jacob A. Harman. 1912.

Diversion of the waters of the Great Lakes by way of the sanitary and ship canal of Chicago: A brief of the facts and issues, by Lyman E. Cooley, Chicago. 1913.

The State of Missouri *vs.* the State of Illinois and the Sanitary district of Chicago. before Frank S. Bright, Commissioner of the Supreme Court of the United States. 1904.

The mineral waters of Indiana, their location, origin, and character, by W. S. Blatchley: Indiana Dept. Geology and Nat. Res. Twenty-sixth Ann. Rept., 1901.

Report of the water-resources investigation of Minnesota by the State drainage commission, 1910.

Report of the commission on conservation [Montana] on bills relating to the public lands, water rights, and the protection and preservation of the forests, 1911.

Governor's message relating to conservation [in Montana] on bills relating to public lands, water rights, and the protection and preservation of the forests.

Water resources of the Devils Lake region, North Dakota, by E. J. Babcock: North Dakota Geol. Survey, Second Bienn. Rept., 1903.

The water powers of Wisconsin, by Leonard S. Smith: Wisconsin Geol. and Nat. Hist. Survey Bull. 20. Madison, Wis., 1908.

Report of the Railroad Commission of Wisconsin to the legislature on water powers. Madison, Wis., 1915.

Many of these reports can be obtained by applying to the several organizations, and most of them can be consulted in the public libraries of the larger cities.

GEOLOGICAL SURVEY HYDROLOGIC REPORTS OF GENERAL INTEREST.

The following list comprises reports not readily classifiable by drainage basins and covering a wide range of hydrologic investigations:

WATER-SUPPLY PAPERS.

*1. Pumping water for irrigation, by H. M. Wilson. 1896. 57 pp., 9 pls.

Describes pumps and motive powers, windmills, water wheels, and various kinds of engines; also storage reservoirs to retain pumped water until needed for irrigation.

*3. Sewage irrigation, by G. W. Rafter. 1897. 100 pp., 4 pls. (See Water-Supply Paper 22.) 10c.

Discusses methods of sewage disposal by intermittent filtration and by irrigation; describes utilization of sewage in Germany, England, and France, and sewage purification in the United States.

*8. Windmills for irrigation, by E. C. Murphy. 1897. 49 pp., 8 pls. 10c.

Gives results of experimental tests of windmills during the summer of 1896 in the vicinity of Garden, Kansas; describes instruments and methods and draws conclusions.

*14. New tests of certain pumps and water lifts used in irrigation, by O. P. Hood. 1898. 91 pp., 1 pl.

Discusses efficiency of pumps and water lifts of various types.

*20. Experiments with windmills, by T. O. Perry. 1899. 97 pp., 12 pls. 15c.

Includes tables and descriptions of wind wheels, compares wheels of several types, and discusses results.

*22. Sewage irrigation, Part II, by G. W. Rafter. 1899. 100 pp., 7 pls. 15c.

Gives résumé of Water-Supply Paper 3; discusses pollution of certain streams, experiments on purification of factory wastes in Massachusetts, value of commercial fertilizers, and describes American sewage-disposal plants by States; contains bibliography of publications relating to sewage utilization and disposal.

*41. The windmill; its efficiency and economic use, Part I, by E. C. Murphy. 1901. 72 pp., 14 pls. 5c.

*42. The windmill; its efficiency and economic use, Part II, by E. C. Murphy. 1901. 75 pp. (73–147), 2 pls. (15–16). 10c.

Nos. 41 and 42 give details of results of experimental tests with windmills of various types.

*43. Conveyance of water in irrigation canals, flumes, and pipes, by Samuel Fortier. 1901. 86 pp., 15 pls. 15c.

*56. Methods of stream measurement. 1901. 51 pp., 12 pls. 15c.

Describes the methods used by the Survey in 1901–2. (See also Nos. 64, 94, and 95.)

*64. Accuracy of stream measurements, by E. C. Murphy. 1902. 99 pp., 4 pls. (See No. 95.) 10c.

Describes methods of measuring velocity of water and of measuring and computing stream flow, and compares results obtained with the different instruments and methods; describes also experiments and results at the Cornell University hydraulic laboratory. A second, enlarged, edition published as Water-Supply Paper 95.

*67. The motions of underground waters, by C. S. Slichter. 1902. 106 pp., 8 pls. 15c.

Discusses origin, depth, and amount of ground waters; permeability of rocks and porosity of soils; causes, rates, and laws of motions of ground waters; surface and deep zones of flow, and recovery of waters by open wells and artesian and deep wells; treats of the shape and position of the water table; gives simple methods of measuring yields of flowing wells; describes artesian wells at Savannah, Ga.

72. Sewage pollution in the metropolitan area near New York City and its effect on inland water resources, by M. O. Leighton. 1902. 75 pp., 8 pls. 10c.

Defines "normal" and "polluted" waters and discusses the damage resulting from pollution.

*80. The relation of rainfall to run-off, by G. W. Rafter. 1903. 104 pp. 10c.

Treats of measurements of rainfall and laws and measurements of streams flow; gives formulas for rainfall, run-off, and evaporation; discusses effects of forests on rainfall and run-off.

87. Irrigation in India (second edition), by H. M. Wilson. 1903. 238 pp., pls. 25c.

First edition was published in Part II of the Twelfth Annual Report.

93. Proceeding of first conference of engineers of the Reclamation Service, with accompanying papers, compiled by F. H. Newell, Chief Engineer. 1904. 361 pp. 25c. [Requests for this report should be addressed to the U. S. Reclamation Service.]

Contains the following papers of more or less general interest:
Limits of an irrigation project, by D. W. Ross.
Relation of Federal and State laws to irrigation, by Morris Bien.
Electical transmission of power for pumping, by H. A. Storrs.
Correct design and stability of high masonry dams, by Geo. Y. Wisner.
Irrigation surveys and use of the planetable, by J. V. Lippincott.
The use of alkaline waters for irrigation, by Thomas H. Means.

*94. Hydrographic manual of the United States Geological Survey, prepared by E. C. Murphy, J. C. Hoyt, and G. B. Hollister. 1904. 76 pp., 3 pls. 10c.

Gives instruction for field and office work relating to measurements of stream flow by current meters. (See also No. 95.)

*95. Accuracy of stream measurements (second, enlarged edition), by E. C. Murphy. 1904. 169 pp., 6 pls.

Describes methods of measuring and computing stream flow and compares results derived from different instruments and methods. (See also No. 94.)

*103. A review of the laws forbidding pollution of inland water in the United States, by E. B. Goodell. 1904. 120 pp. (See No. 152.)

Explains the legal principles under which antipollution statutes become operative, quotes court decisions to show authority for various deductions, and classifies according to scope the statutes enacted in the different States.

*110. Contributions to the hydrology of Eastern United States; 1904, M. L. Fuller, geologist in charge. 1905. 211 pp., 5 pls. 10c.

Contains the following reports of general interest. The scope of each paper is indicated by its title.
Description of under flow meter used in measuring the velocity and direction of underground water, by Charles S. Slichter.
The California or "stovepipe" method of well construction, by Charles S. Slichter.
Approximate methods of measuring the yield of flowing wells, by Charles S. Slichter.
Corrections necessary in accurate determinations of flow from verticals well casings, from notes furnished by A. N. Talbot.

113. The disposal of strawboard and oil-well wastes, by R. L. Sackett and Isaiah Bowman. 1905. 52 pp., 4 pls. 5c.

The first paper discusses the pollution of stream by sewage and by trade wastes, describes the manufacture of strawboard, and gives results of various experiments in disposing of the waste. The second paper describes briefly the topography, drainage, and geology of the region about Marion, Ind., and the contamination of rock wells and of streams by waste oil and brine

*114. Underground waters of eastern United States; M. L. Fuller, geologist in charge. 1905. 285 pp., 18 pls. 25c.

Contains reports on "Occurrence of underground waters," by M. L. Fuller, discussing sources, amount, and temperature of waters, permeability and storage capacity of rocks, water bearing formations, recovery of water by springs, wells, and pumps, essential conditions of artesian flows, and general conditions affecting ground waters in eastern United States.

119. Index to the hydrographic progress reports of the United States Geological Survey, 1888 to 1903, by J. C. Hoyt and B. D. Wood. 1905. 253 pp. 15c.

120. Bibliographic review and index of papers relating to underground waters published by the United States Geological Survey, 1879–1904, by M. L. Fuller. 1905. 128 pp. 10c.

*122. Relation of the law to underground waters, by D. W. Johnson. 1905. 55 pp. 5c.

Defines and classifies underground waters, gives common-law rules relating to their use, and cites State legislative acts affecting them.

140. Field measurements of the rate of movement of underground waters, by C. S. Slichter. 1905. 122 pp., 15 pls. 15c.

Discusses the capacity of sand to transmit water, describes measurements of underflow in Rio Hondo, San Gabriel, and Mohave River Valleys, Cal., and on Long Island, N. Y., gives results of tests of wells and pumping plants, and describes stovepipe method of well construction.

143. Experiments on steel-concrete pipes on a working scale, by J. H. Quinton. 1905. 61 pp., 4 pls. 5c.

Scope indicated by title.

145. Contributions to the hydrology of eastern United States, 1905; M. L. Fuller, geologist in charge. 1905. 220 pp., 6 pls. 10c.

Contains brief reports of general interest, as follows:

Drainage of ponds into drilled wells, by Robert E. Horton. Discusses efficiency, cost, and capacity of drainage wells, and gives statistics of such well in Southern Michigan.

Construction of so-called fountain and geyser springs, by Myron L. Fuller

A convenient gage for determining low artesian heads, by Myron L. Fuller.

146. Proceedings of second conference of engineers of the Reclamation Service, with accompanying papers, compiled by F. H. Newell, Chief Engineer. 1905. 267 pp. 15c. [Inquiries concerning this report should be addressed to the Reclamation Service.]

Contains brief account of the organization of the hydrographic [water-resources] branch and the Reclamation Service, reports of conferences and committees, circulars of instruction, and many brief reports on subjects closely related to reclamation, and a bibliography of technical papers by members of the service. Of the papers read at the conference those listed below (scope indicated by title) are of more or less general interest.

Proposed State code of water laws, by Morris Bien.
Power engineering applied to irrigation problems, by O. H. Ensign.
Estimates on tunneling in irrigation projects, by A. L. Fellows.
Collection of steam-gaging data, by N. C. Grover.
Diamond-drill methods, G. A. Hammond
Mean-velocity and area curves, by F. W. Hanna.
Importance of general hydrogrophic data concerning basins of streams gaged, by R. E. Horton.
Effect of aquatic vegetation on stream flow, by R. E. Horton
Sanitary regulations governing construction camps, by M. O. Leighton.
Necessity of draining irrigated land, by Thos. H. Means.
Alkali soils, by Thos. H. Means.
Cost of stream gaging work, by E. C. Murphy.
Equipment of a cable gaging station, by E. C. Murphy.
Silting of reservoirs, by W. M. Reed
Farm-unit classification, by D. W. Ross.
Cost of power for pumping irrigated water, by H. A. Storrs.
Records of flow at current-meter gaging stations during the frozen season, by F. H. Tillinghast.

147. Destructive floods in United States in 1904, by E. C. Murphy and others. 1905. 206 pp., 18 pls. 15c.

Contains a brief account of "A method of computing cross-section area of waterways," including formulas for maximum discharge and areas of cross section.

*150. Weir experiments, coefficients, and formulas, by R. E. Horton. 1906. 189 pp., 38 pls. (See Water-Supply Paper 200.) 15c.

Scope indicated by title.

151. Field assay of water, by M. O. Leighton. 1905. 77 pp., 4 pls.

Discusses methods, instruments, and reagents used in determining turbidity, color, iron, chlorides, and hardness in connection with the studies of the quality of water in various parts of the United States

*152. A review of the laws forbidding pollution of inland waters in the United States, second edition, by E. B. Goodell. 1905. 149 pp. 10c.

Scope indicated by title.

*155. Fluctuations of the water level in wells, with special reference to Long Island. N. Y., A. C. Veatch. 1906. 83 pp., 9 pls. 25c.

Includes general discussion of fluctuations due to rainfall and evaporation, barometric changes, temperature changes, changes in rivers, changes in lake level, tidal changes, effects of settlement, irrigation, dams, underground-water developments, and to indeterminate causes.

*160. Underground water papers. 1906; M. L. Fuller, geologist in charge. 1906. 104 pp., 1 pl.

Gives account of work in 1905; lists publications relating to underground waters, and contains the following brief reports of general interest:
Significance of the term "artesian," by Myron L. Fuller.
Representation of wells and springs on maps, by Myron L. Fuller.
Total amount of free water in the earth's crust, by Myron L. Fuller.
Use of fluorescein in the study of underground waters, by R. B. Dole.
Problems of water contamination, by Isaiah Bowman.
Instances of improvement of water in wells, by Myron L. Fuller.

*162. Destructive floods in the United States in 1905, with a discussion of flood discharge and frequency and an index to flood literature, by E. C. Murphy and others. 1906. 105 pp., 4 pls. 15c.

*163. Bibliographic review and index of underground-water literature published in the United States in 1905, by M. L. Fuller, F. G. Clapp, and B. L. Johnson. 1906. 130 pp. 15c.

Scope indicated by title.

*179. Prevention of stream pollution by distillery refuse, based on investigations at Lynchburg, Ohio, by Herman Stabler. 1906. 34 pp., 1 pl. 10c.

Describes grain distillation, treatment of slop, sources, character, and effects of effluents on streams; discusses filtration, precipitation, fermentation, and evaporation methods of disposal of wastes without pollution.

*180. Turbine water-wheel tests and power tables, by R. E. Horton. 1906. 134 pp., 2 pls. 20c.

Scope indicated by title.

*185. Investigations on the purification of Boston sewage, * * * with a history of the sewage-disposal problem, by C.-E. A. Winslow and E. B. Phelps. 1906. 163 pp. 25c.

Discusses composition, disposal, purification, and treatment of sewages and tendencies in sewage-disposal practice in England, Germany, and the United States; describes character of crude sewage at Boston, removal of suspended matter, treatment in septic tanks, and purification by intermittent sand filtration and in beds of coarse material; gives bibliography.

*186. Stream pollution by acid-iron wastes, a report based on investigations made at Shelby, Ohio, by Herman Stabler. 1906. 36 pp., 1 pl.

Gives history of pollution by acid-iron wastes at Shelby, Ohio, and of resulting litigation; discusses effect of acid-iron liquors of sewage-purification processes, recovery of copperas from acid-iron wastes, and other processes for removal of pickling liquor.

*187. Determination of stream flow during the frozen season, by H. K. Barrows and R. E. Horton. 1907. 93 pp., 1 pl. 15c.

Scope indicated by title.

*189. The prevention of stream pollution by strawboard waste, by E. B. Phelps. 1906. 20 pp., 2 pls.

Describes manufacture of strawboard, present and proposed methods of disposal of waste liquors, laboratory investigations of precipitation and sedimentation, and field studies of amounts and character of water used, raw material and finished product, and mechanical filtration.

*194. Pollution of Illinois and Mississippi rivers by Chicago sewage (a digest of the testimony taken in the case of the State of Missouri *v.* The State of Illinois and the Sanitary District of Chicago), by M. O. Leighton. 1907. 369 pp., 2 pls.

Scope indicated by amplification of title.

*200. Weir experiments, coefficients, and formulas (revision of paper No. 150), by R. E. Horton. 1907. 195 pp., 1 pl. 35c.

Scope indicated by title.

*226. The pollution of streams by sulphite-pulp waste, a study of possible remedies by E. B. Phelps. 1909. 37 pp., 1 pl. 10c.

Describes manufacture of sulphite pulp, the waste liquors, and the experimental work leading to suggestions as to methods of preventing stream pollution.

*229. The disinfection of sewage and sewage filter effluents, with a chapter on the putrescibility and stability of sewage effluents, by E. B. Phelps. 1909. 91 pp., 1 pl. 15c.

Scope indicated by title.

*234. Papers on the conversion of water resources. 1909. 96 pp., 2 pls. 15c.

Contains the following papers, whose scope is indicated by their titles: Distribution of fall, by Henry Gannett; Floods, by M. O. Leighton; Developed water powers, compiled under the direction of W. M. Steuart, with discussion by M. O. Leighton; Undeveloped water powers, by M. O. Leighton; Irrigation, by F. H. Newell; Underground waters, by W. C. Mendenhall; Denudation, by R. B. Dole and Herman Stabler; Control of catchment areas, by H. N. Parker.

*235. The purification of some textile and other factory wastes, by Herman Stabler, and G. H. Pratt. 1909. 76 pp. 10c.

Discusses waste waters from wool-scouring, bleaching, and dyeing cotton yarn, bleaching cotton piece goods, and manufacture of oleomargarine, fertilizer, and glue.

*236. The quality of surface waters in the United States: Part I, Analyses of waters east of the one hundredth meridian, by R. B. Dole. 1909. 123 pp. 10c.

Describes collection of samples, methods of examination, preparation of solutions, accuracy of estimates, and expression of analytical results.

238. The public utility of water powers and their governmental regulation, by René Tavernier and M. O. Leighton. 1910. 161 pp. 15c.

Discusses hydraulic power and irrigation, French, Italian, and Swiss legislation relative to the development of water powers, and laws proposed in the French Parliament; reviews work of bureau of hydraulics and agricultural improvement of the French department of agriculture and gives résumé of Federal and State water-power legislation in the United States.

*255. Underground waters for farm use, by M. L. Fuller. 1910. 58 pp., 17 pls. 15c.

Discusses rocks as sources of water supply and the relative saftey of supplies from different materials; springs, and their protection; open or dug and deep wells, their location, yields, relative cost, protection, and safety; advantages and disadvantages of cisterns and combination wells and cisterns.

*257. Well-drilling methods, by Isaiah Bowman. 1911. 139 pp., 4 pls. 1oc.

Discusses amount, distribution, and disposal of rainfall, water-bearing rocks, amount of ground water, artesian conditions, and oil and gas bearing formations; gives history of well drilling in Asia, Europe, and the United States; describes in detail the various methods and the machinery used; discusses loss of tools and geologic difficulties; contamination of well waters and methods of prevention; tests of capacity and measurement of depth; and costs of sinking wells.

*258. Underground water papers, 1910, by M. L. Fuller, F. G. Clapp, G. C. Matson, Samuel Sanford, and H. C. Wolff. 1911. 123 pp., 2 pls. 15c.

Contains the following papers (scope indicated by titles) of general interest:
Drainage by wells, by M. L. Fuller.
Freezing of wells and related phenomena, by M. L. Fuller.
Pollution of underground waters in limestone, by G. C. Matson.
Protection of shallow wells in sandy deposits, by M. L. Fuller.
Magnetic wells, by M. L. Fuller.

274. Some stream waters of the western United States, with chapters on sediment carried by the Rio Grande and the industrial application of water analyses, by Herman Stabler. 1911. 188 pp. 15c.

Describes collection of samples, plan of analytical work, and methods of analyses; discusses soap-consuming power of waters, water softening, boiler waters, and water for irrigation.

*315. The purification of public water supplies, by G. A. Johnson. 1913. 84 pp., 8 pls. 10c.

Discusses ground, lake, and river waters as public supplies, development of waterworks systems in the United States, water consumption, and typhoid fever; describes methods of filtration and sterilization of water, and municipal water softening.

334. The Ohio Valley flood of March–April, 1913 (including comparisons with some earlier floods), by A. H. Horton and H. J. Jackson. 1913. 96 pp., 22 pls. 20c.

Although relating specifically to floods in the Ohio Valley, this report discusses also the causes of floods and the prevention of damage by floods.

*337. The effects of ice on stream flow, by William Glenn Hoyt. 1913. 77 pp., 7 pls. 15c.

Discusses methods of measuring the winter flow of streams.

*345. Contributions to the hydrology of the United States, 1914; N. C. Grover, chief hydraulic engineer. 1915. 225 pp., 17 pls. 30c. Contains:

* (e) A method of determining the daily discharge of rivers of variable slope, by M. R. Hall, W. E. Hall, and C. H. Pierce, pp. 53-65.

*364. Water analyses from the laboratory of the United States Geological Survey, tabulated by F. W. Clarke, chief chemist. 1914. 40 pp. 5c.

Contains analyses of waters from rivers, lakes, wells, and springs in various parts of the United States, including analyses of the geyser water of Yellowstone National Park, hot springs in Montana, brines from Death Valley, water from the Gulf of Mexico, and mine waters from Tennessee, Michigan, Missouri and Oklahoma, Montana, Colorado, and Utah, Nevada and Arizona, and California.

371. Equipment for current-meter gaging stations, by G. J. Lyon. 1915. 64 pp., 37 pls. 20c.

Describes methods of installing automatic and other gages and of constructing gage wells, shelters, and structures for making discharge measurements and artificial controls.

*375. Contributions to the hydrology of the United States, 1915; N. C. Grover, chief hydraulic engineer. 1916. 181 pp., 9 pls. 15c.

Contains three papers presented at the conference of engineers of the water-resources branch in December, 1914.

* (c) The relation of stream gaging to the science of hydraulics, by C. H. Pierce and R. W. Davenport, pp. 77-84.

(e) A method for correcting river discharge for changing stage, by B. E. Jones, pp. 117-130.

(f) Conditions requiring the use of automatic gages in obtaining records of stream flow, by C. H. Pierce, pp. 131-139.

*400. Contributions to the hydrology of the United States, 1916; N. C. Grover, chief hydraulic engineer. 1917. 108 pp., 7 pls. Contains

(a) The people's interest in water-power resources, by G. O. Smith, pp. 1-8.

* (c) The measurement of silt-laden streams, by R. C. Pierce, pp. 39-51.

(d) Accuracy of stream-flow data, by N. C. Grover and J. C. Hoyt, pp. 53-59.

416. The divining rod, a history of water witching, with a bibliography, by Arthur J. Ellis. 1917. 59 pp. 10c.

A brief paper published "merely to furnish a reply to the numerous inquiries that are continually being received from all parts of the country" as to the efficacy of the divining rod for locating underground water.

*425. Contributions to the hydrology of the United States, 1917; N. C. Grover, chief hydraulic engineer. 1918. Contains:

* (c) Hydraulic conversion tables and convenient equivalents, pp. 71–94. 1917.

427. Bibliography and index of the publications of the United States Geological Survey relating to ground water, by O. E. Meinzer. 1918. 169 pp., 1 pl.

Includes publications prepared, in whole or part, by the Geological Survey that treat any phase of the subject of ground water or any subject directly applicable to ground water. Illustrated by map showing reports that cover specific areas more or less thoroughly.

ANNUAL REPORTS.

*Fifth Annual Report of the United States Geological Survey, 1883–84, J. W. Powell, Director. 1885. xxxvi, 469 pp., 58 pls. $2.25. Contains:

*The requisite and qualifying conditions of artesian wells, by T. C. Chamberlin, pp. 125–173, pl. 21. Scope indicated by title.

*Twelfth Annual Report of the United States Geological Survey, 1890–91, J. W. Powell, Director. 1891. 2 parts. *Pt. II, Irrigation, xviii, 576 pp., 93 pls. $2. Contains:

*Irrigation in India, by H. M. Wilson, pp. 363–561, pls. 107 to 146. See Water-Supply Paper 87.

Thirteenth Annual Report of the United States Geological Survey, 1891–92, J. W. Powell, Director. 1892. (Pts. II and III, 1893.) 3 parts. *Pt. III, Irrigation, xi, 486 pp., 77 pls. $1.85. Contains:

*American irrigation engineering, by H. M. Wilson, pp. 101–349, pls. 111 to 146. Discusses the economic aspects of irrigation, alkaline drainage, silt and sedimentation; gives brief history of legislation; describes perennial canals in Idaho, California, Wyoming, and Arizona; discusses water storage at reservoirs of the California and other projects, subsurface sources of supply, pumping, and subirrigation.

Fourteenth Annual Report of the United States Geological Survey, 1892–93, J. W. Powell, Director. 1893. (Pt. II, 1894.) 2 parts. *Pt. II, Accompanying papers, xx, 597 pp., 73 pls. $2.10. Contains:

*Potable waters of the eastern United States, by W J McGee, pp. 1 to 47. Discusses cistern water, stream waters, and ground waters, including mineral springs and artesian wells.

*Natural mineral waters of the United States, by A. C. Peale, pp. 49–88, pls. 3 and 4. Discusses the origin and flow of mineral springs, the source of mineralization, thermal springs, the chemical composition and analysis of spring waters, geographic distribution, and the utilization of mineral waters; gives a list of American mineral spring resorts; contains also some analyses.

Nineteenth Annual Report of the United States Geological Survey, 1897–98, Charles D. Walcott, Director. 1898. (Parts II, III, and V, 1899.) 6 parts in 7 vols. and separate case for maps with Pt. V. *Pt. II.—Papers chiefly of a theoretic nature, v, 958 pp., 172 pls. $2.65. Contains:

*Principles and conditions of the movements of ground water, by F. H. King, pp. 59–294, pls. 6 to 16. Discusses the amount of water stored in sandstone, in soil, and in other rocks, the depth to which ground water penetrates; gravitational, thermal, and capillary movements of ground waters, and the configuration of the ground-water surface; gives the results of experimental investigations on the flow of air and water through a rigid, porous medium and through sands, sandstones, and silts; discusses results obtained by other investigators, and summarizes results of observations; discusses also rate of flow of water through sand and rock, the growth of rivers, rate of filtration through soil, interference of wells, etc.

*Theoretical investigation of the motion of ground waters, by C. S. Slichter, pp. 295–384, pl. 17. Scope indicated by title.

INDEX OF STREAMS.

O

Lightning Source UK Ltd.
Milton Keynes UK
UKHW010617110219
337000UK00006B/223/P